T0249497

Desk Encyclopedia of
GENERAL VIROLOGY

DESK ENCYCLOPEDIA OF GENERAL VIROLOGY

EDITOR-IN-CHIEF

Dr MARC H V VAN REGENMORTEL

AMSTERDAM • BOSTON • HEIDELBERG • LONDON • NEW YORK • OXFORD
PARIS • SAN DIEGO • SAN FRANCISCO • SINGAPORE • SYDNEY • TOKYO
Academic Press is an imprint of Elsevier

Academic Press is an imprint of Elsevier
Linacre House, Jordan Hill, Oxford, OX2 8DP, UK
525 B Street, Suite 1900, San Diego, CA 92101-4495, USA

British Library Cataloguing in Publication Data
A catalogue record for this book is available from the British Library

Library of Congress Cataloguing in Publication Data
A catalogue record for this book is available from the Library of Congress

ISBN: 978-0-12-375146-1

For information on all Elsevier publications
visit our website at books.elsevier.com

EDITOR-IN-CHIEF

Marc H V van Regenmortel PhD
Emeritus Director at the CNRS,
French National Center for Scientific Research,
Biotechnology School of the University of Strasbourg,
Illkirch, France

ASSOCIATE EDITORS

PREFACE

This volume, the *Desk Encyclopedia of General Virology*, is one of a series of four volumes that contain many of the entries in the third edition of the *Encyclopedia of Virology* edited by Brian W J Mahy and Marc H V van Regenmortel, published by Academic Press/Elsevier in 2008.

This first volume consists of 78 chapters that give an overview of the development of virology during the last 10 years. Fifteen chapters discuss general topics such as the nature, origin, phylogeny, and evolution of viruses, and describe the history of virology. The current official taxonomy and classification of viruses, based on the eighth report of the International Committee on Taxonomy of Viruses published in 2005, is also included.

The second section of 10 chapters summarizes the considerable progress in our understanding of the structure and assembly of virus particles and describes how this knowledge was obtained using techniques such as X-ray crystallography, cryoelectron microscopy, and image reconstruction. The next section is devoted to the genetic material of viruses and discusses topics such as the exchange of DNA and RNA fragments among parental viral genomes through recombination, the occurrence of retrotransposable DNA elements and defective-interfering RNAs and the newly discovered antiviral role of RNA interference triggered by viral double-stranded RNA.

The fourth section of 11 chapters describes the different mechanisms used by viruses to infect, and replicate in, their host cells. Viral pathogenesis and current advances in the development of antiviral compounds are described, as well as methods used in disease surveillance and in the diagnosis of viral infections.

The fifth section of 13 chapters is devoted to various aspects of the immune response to viruses and describes innate immunity, antigen presentation, the role of cytokines and chemokines, and the phenomenon of antigenic variation in viruses. Four chapters discuss viral vaccines and the experimental strategies used for developing them, as well as the mechanisms by which antiviral antibodies neutralize viral infectivity.

The final section of 23 chapters presents an overview of some major groups of viruses, with particular attention being given to our current knowledge of their molecular biology.

By bringing all these topics under one cover, it is hoped that this volume will succeed in presenting an up-to-date introduction to recent advances in virology.

As the chapters initially appeared in an Encyclopedia, the style of presentation is fairly didactic and does not require prior specialized knowledge, which might make this volume suitable for an introductory course in virology.

Three other volumes prepared from material that appeared in the third edition of the *Encyclopedia of Virology* are also ready. Students and scientists interested in human viruses may find it useful to consult the second volume, which deals with the viral diseases of man. Those interested in viruses infecting animals and bacteria may want to consult the third volume, while those interested in plant and fungal viruses may find that reading the fourth volume, which is devoted to those viruses, with this volume, gives them a background on virology along with information on plant viral diseases.

Marc H V van Regenmortel

CONTRIBUTORS

H-W Ackermann
Laval University, Quebec, QC, Canada

G M Air
University of Oklahoma Health Sciences Center, Oklahoma City, OK, USA

J W Almond
sanofi pasteur, Lyon, France

H Attoui
Université de la Méditerranée, Marseille, France

L A Babiuk
University of Alberta, Edmonton, AB, Canada

S Babiuk
National Centre for Foreign Animal Disease, Winnipeg, MB, Canada

S C Baker
Loyola University of Chicago, Maywood, IL, USA

T S Baker
University of California, San Diego, La Jolla, CA, USA

Y Bao
National Institutes of Health, Bethesda, MD, USA

C F Basler
Mount Sinai School of Medicine, New York, NY, USA

K I Berns
University of Florida College of Medicine, Gainesville, FL, USA

C D Blair
Colorado State University, Fort Collins, CO, USA

G W Blissard
Boyce Thompson Institute at Cornell University, Ithaca, NY, USA

J J Bujarski
Northern Illinois University, DeKalb, IL, USA and Polish Academy of Sciences, Poznan, Poland

A J Cann
University of Leicester, Leicester, UK

S Casjens
University of Utah School of Medicine, Salt Lake City, UT, USA

J T Chang
Baylor College of Medicine, Houston, TX, USA

V G Chinchar
University of Mississippi Medical Center, Jackson, MS, USA

W Chiu
Baylor College of Medicine, Houston, TX, USA

J-M Claverie
Université de la Méditerranée, Marseille, France

J R Clayton
Johns Hopkins University Schools of Public Health and Medicine, Baltimore, MD, USA

C J Clements
The Macfarlane Burnet Institute for Medical Research and Public Health Ltd., Melbourne, VIC, Australia

J Collinge
University College London, London, UK

A Collins
University of Wisconsin School of Medicine and Public Health, Madison, WI, USA

K M Coombs
University of Manitoba, Winnipeg, MB, Canada

J A Cowley
CSIRO Livestock Industries, Brisbane, QLD, Australia

J DeRisi
University of California, San Francisco, San Francisco, CA, USA

A Domanska
University of Helsinki, Helsinki, Finland

A J Easton
University of Warwick, Coventry, UK

L Enjuanes
CNB, CSIC, Madrid, Spain

K J Ertel
University of California, Irvine, CA, USA

J L van Etten
University of Nebraska–Lincoln, Lincoln, NE, USA

D J Evans
University of Warwick, Coventry, UK

C M Fauquet
Danforth Plant Science Center, St. Louis, MO, USA

F Fenner
Australian National University, Canberra, ACT, Australia

H J Field
University of Cambridge, Cambridge, UK

K Fischer
University of California, San Francisco, San Francisco, CA, USA

T R Flotte
University of Florida College of Medicine, Gainesville, FL, USA

P Forterre
Institut Pasteur, Paris, France

R S Fujinami
University of Utah School of Medicine, Salt Lake City, UT, USA

F García-Arenal
Universidad Politécnica de Madrid, Madrid, Spain

E Gellermann
Hannover Medical School, Hannover, Germany

S A Ghabrial
University of Kentucky, Lexington, KY, USA

W Gibson
Johns Hopkins University School of Medicine, Baltimore, MD, USA

A E Gorbalenya
Leiden University Medical Center, Leiden, The Netherlands

A Grakoui
Emory University School of Medicine, Atlanta, GA, USA

M V Graves
University of Massachusetts–Lowell, Lowell, MA, USA

D E Griffin
Johns Hopkins Bloomberg School of Public Health, Baltimore, MD, USA

R J de Groot
Utrecht University, Utrecht, The Netherlands

J M Hardwick
Johns Hopkins University Schools of Public Health and Medicine, Baltimore, MD, USA

L E Harrington
University of Alabama at Birmingham, Birmingham, AL, USA

T J Harrison
University College London, London, UK

S Hilton
University of Warwick, Warwick, UK

A Hinz
UMR 5233 UJF-EMBL-CNRS, Grenoble, France

R Hull
John Innes Centre, Norwich, UK

A D Hyatt
Australian Animal Health Laboratory, Geelong, VIC, Australia

P Jardine
University of Minnesota, Minneapolis, MN, USA

J E Johnson
The Scripps Research Institute, La Jolla, CA, USA

Y Kapustin
National Institutes of Health, Bethesda, MD, USA

K M Keene
Colorado State University, Fort Collins, CO, USA

P J Klasse
Cornell University, New York, NY, USA

W B Klimstra
Louisiana State University Health Sciences Center at Shreveport, Shreveport, LA, USA

R J Kuhn
Purdue University, West Lafayette, IN, USA

I Kusters
sanofi pasteur, Lyon, France

P F Lambert
University of Wisconsin School of Medicine and Public Health, Madison, WI, USA

G Lawrence
The Children's Hospital at Westmead, Westmead, NSW, Australia and University of Sydney, Westmead, NSW, Australia

E J Lefkowitz
University of Alabama at Birmingham, Birmingham, AL, USA

P Leinikki
National Public Health Institute, Helsinki, Finland

J Lenard
University of Medicine and Dentistry of New Jersey (UMDNJ), Piscataway, NJ, USA

K N Leppard
University of Warwick, Coventry, UK

A Lescoute
Université Louis Pasteur, Strasbourg, France

R Ling
University of Warwick, Coventry, UK

M Luo
University of Alabama at Birmingham, Birmingham, AL, USA

D B McGavern
The Scripps Research Institute, La Jolla, CA, USA

P P C Mertens
Pirbright Laboratory, Woking, UK

R F Meyer
Centers for Disease Control and Prevention, Atlanta, GA, USA

E S Mocarski
Emory University School of Medicine, Atlanta, GA, USA

S A Morse
Centers for Disease Control and Prevention, Atlanta, GA, USA

N Nathanson
University of Pennsylvania, Philadelphia, PA, USA

C K Navaratnarajah
Purdue University, West Lafayette, IN, USA

A R Neurath
Virotech, New York, NY, USA

M L Nibert
Harvard Medical School, Boston, MA, USA

N Noah
London School of Hygiene and Tropical Medicine, London, UK

W F Ochoa
University of California, San Diego, La Jolla, CA, USA

M R Odom
University of Alabama at Birmingham, Birmingham, AL, USA

M B A Oldstone
The Scripps Research Institute, La Jolla, CA, USA

K E Olson
Colorado State University, Fort Collins, CO, USA

P Palese
Mount Sinai School of Medicine, New York, NY, USA

P Palukaitis
Scottish Crop Research Institute, Invergowrie, UK

A E Peaston
The Jackson Laboratory, Bar Harbor, ME, USA

M M Poranen
University of Helsinki, Helsinki, Finland

M H V van Regenmortel
CNRS, Illkirch, France

A Rezaian
University of Adelaide, Adelaide, SA, Australia

F J Rixon
MRC Virology Unit, Glasgow, UK

L Roux
University of Geneva Medical School, Geneva, Switzerland

R W H Ruigrok
CNRS, Grenoble, France

R J Russell
University of St. Andrews, St. Andrews, UK

K D Ryman
Louisiana State University Health Sciences Center at Shreveport, Shreveport, LA, USA

R M Sandri-Goldin
University of California, Irvine, Irvine, CA, USA

R Sanjuán
Instituto de Biología Molecular y Cellular de Plantas, CSIC-UPV, Valencia, Spain

A Schneemann
The Scripps Research Institute, La Jolla, CA, USA

G Schoehn
CNRS, Grenoble, France

U Schubert
Klinikum der Universität Erlangen-Nürnberg, Erlangen, Germany

T F Schulz
Hannover Medical School, Hannover, Germany

B L Semler
University of California, Irvine, CA, USA

M L Shaw
Mount Sinai School of Medicine, New York, NY, USA

E J Snijder
Leiden University Medical Center, Leiden, The Netherlands

J A Speir
The Scripps Research Institute, La Jolla, CA, USA

T Tatusova
National Institutes of Health, Bethesda, MD, USA

D A Theilmann
Agriculture and Agri-Food Canada, Summerland, BC, Canada

C Upton
University of Victoria, Victoria, BC, Canada

A Urisman
University of California, San Francisco, San Francisco, CA, USA

R Vainionpää
University of Turku, Turku, Finland

P A Venter
The Scripps Research Institute, La Jolla, CA, USA

R A Vere Hodge
Vere Hodge Antivirals Ltd., Reigate, UK

L P Villarreal
University of California, Irvine, Irvine, CA, USA

J Votteler
Klinikum der Universität Erlangen-Nürnberg, Erlangen, Germany

J D F Wadsworth
University College London, London, UK

E K Wagner
University of California, Irvine, Irvine, CA, USA

R Warrier
Purdue University, West Lafayette, IN, USA

S C Weaver
University of Texas Medical Branch, Galveston, TX, USA

F Weber
University of Freiburg, Freiburg, Germany

W Weissenhorn
UMR 5233 UJF-EMBL-CNRS, Grenoble, France

J T West
University of Oklahoma Health Sciences Center, Oklahoma City, OK, USA

E Westhof
Université Louis Pasteur, Strasbourg, France

A J Zajac
University of Alabama at Birmingham, Birmingham, AL, USA

J Ziebuhr
The Queen's University of Belfast, Belfast, UK

E I Zuniga
The Scripps Research Institute, La Jolla, CA, USA

CONTENTS

GENERAL TOPICS

History of Virology: Bacteriophages

H-W Ackermann, Laval University, Quebec, QC, Canada

Glossary

Bacteriophage 'Eater of bacteria' (Greek, phagein).
Eclipse Invisible phase during phage replication.
Lysogenic (1) Bacterium harboring a prophage. (2) Bacteriophage produced by such a bacterium.
Lysogeny Carriage of a prophage.
Metagenomics Sequencing of total genomic material from an environmental sample.
One-step growth Single cycle of phage multiplication.
Phage typing Testing bacteria with a set of phages of different host ranges.
Prophage Latent bacteriophage genome in a lysogenic bacterium.
T-even Phage of the T-series with an even number (T2, T4, T6).

Introduction

Phages or bacteriophages, also referred to as bacterial viruses, include eubacterial and archaeal viruses and are now more appropriately called 'viruses of prokaryotes'. Phages are ubiquitous in nature and may be the most numerous viruses on this planet. They have a long evolutionary history, are considerably diverse, and include some highly evolved and complex viruses. Approximately 5500 phages have been examined in the electron microscope and they constitute the largest of all virus groups. The present system of phage taxonomy comprises 13 officially recognized families.

Following the discovery of phages, their study entered a symbiosis with the nascent science of molecular biology and then expanded to cover a multitude of subjects. The knowledge on phages grew by accretion and represents the efforts of thousands of scientists. For example, it took over 50 years and hundreds of publications to establish the genome map and sequence of coliphage T4. Phage literature is voluminous and comprises over 11 400 publications for the years 1915–65 and, to the author's reckoning, about 45 000 for 1965 to today. The volume of phage research is evident in the fact that there are two books each for single viruses, λ and T4, and that, by mid-2006, the journal *Virology* has published over 2000 phage articles since its start in 1955. The history of phages may be divided into three, partly overlapping periods. Because of the volume of phage research and the many scientists involved, the development of phage research is largely presented in tabular form. Any mention of individual scientists and facts is selective.

The Early Period: 1915–40

Discovery

Several early microbiologists had observed bacterial lysis. The British bacteriologist Ernest H. Hankin reported in 1896 that the waters of the Jumna and Ganges rivers in India killed *Vibrio cholerae* bacteria. The responsible agent passed bacteriological filters and was destroyed by boiling. In 1898, the Russian microbiologist Nikolai Gamaleya reported the lysis of *Bacillus anthracis* bacteria by a transmissible 'ferment'. It is probable that these scientists would have discovered bacterial viruses, if suitable techniques had been available.

Frederick William Twort (1877–1950), a British pathologist and superintendent of the Brown Institution in London, tried to propagate vaccinia virus on bacteriological media. He observed that 'micrococcus' colonies, in fact staphylococci, often became glassy and died. The agent of this transformation was infectious, passed porcelain filters, and was destroyed by heating. Twort proposed several explanations: an amoeba, an ultramicroscopic virus, a living protoplasm, or an enzyme with the power of growth. Subsequently, Twort joined the army and participated in the British Salonika campaign. Back at the Brown Institution, he published five more articles on phages and spent the rest of his career trying to propagate animal viruses on inert media.

Félix Hubert d'Herelle (1873–1949), born in Montreal, went to France as a boy and returned to Canada, styled himself a 'chemist', studied medicine without obtaining a degree, and finally went to Guatemala as a bacteriologist with minimal knowledge of this science. Later in Yucatan, he isolated the agent of a locust disease, the *Coccobacillus acridiorum* (now *Enterobacter aerogenes)*, was sent to the Pasteur Institute of Paris, returned to Mexico and France, and went to fight locusts in Argentine, Turkey, and Tunisia. He noticed the appearance of holes ('plaques') in coccobacillus cultures. In 1915, again in Paris, he investigated an epidemic of bacillary dysentery in military recruits. He noted that his *Shigella* cultures were destroyed by a filterable, plaque-forming agent, whose appearance coincided with convalescence of the recruits. His discovery was published in 1917. D'Herelle understood immediately that he had found a new category of viruses and he used phages for the treatment of bacterial infections. D'Herelle then went to Vietnam, Holland, Egypt, India (Assam), the USA, the Soviet Union (Tbilisi), and finally to Paris. He coined the term 'bacteriophage', devised the plaque test, and stated that there was only one kind of phage, with many races, the *Bacteriophagum intestinale.* He promoted and practiced phage therapy in many countries and founded several phage institutes. The Tbilisi Institute, which he co-founded with G. Eliava in 1934, still exists.

Twort and D'Herelle could not have been more different personalities. Twort was a medical doctor and a self-effacing person; he traveled little, and showed minimal interest in his discovery. He published only five more articles on phages. D'Herelle was self-educated, peripathetic, worldly, and combative. He devoted the rest of his career to phages and wrote five books and nearly 110 articles on them. The almost simultaneous discovery of bacteriophages by two different scientists is a remarkable coincidence.

Phage Research: General

The years 1920–40 were characterized by research on the nature of phages and lysogeny, phage therapy, and the detection of phages in a wide variety of bacteria and a wide range of habitats (**Table 1**). D'Herelle and his followers asserted that bacteriophages were viruses. A major argument was that they formed plaques. Others, observing that some bacteria produced lytic agents that could be transmitted from one culture to another, held that 'bacteriophages' were generated spontaneously and were endogenous in nature. The famous Belgian immunologist, J. Bordet, proposed that phages were enzymes produced by 'lysogenic' bacteria. He stated bluntly: "The invisible virus of d'Herelle does not exist." When Bordet discovered Twort's paper from 1915, a nasty controversy on the nature of phages and the priority of phage discovery ensued.

Table 1 Phage discovery: phages of selected host organisms

Year	*Investigators*	*Host species or group*
1915	Twort	*Staphylococcus* sp.
1917	D'Herelle	*Shigella dysenteriae*
1918	D'Herelle	*Salmonella typhi*
1921	D'Herelle	*Bacillus subtilis, Corynebacterium diphtheriae, Escherichia coli, Pasteurella multocida, Vibrio cholerae, Yersinia pestis*
1934	Cowles	Clostridia
1937	Whitehead and Hunter	Lactic acid bacteria
1947	Reilly and coll.	Streptomycetes
1963	Safferman and Morris	Cyanobacteria
1970	Gourlay	Mycoplasmas
1974	Torsvik and Dundas	Halophilic archaea
1983	Janekovic and coll.	Hyperthermophilic archaea

Phages Become Viruses

The viral nature of phages was demonstrated by physicochemical studies. W. J. Elford and C. H. Andrewes (1932) showed that phages could differ in size. M. F. Burnet and colleagues (1933) demonstrated that phages differed in antigenicity and resistance to inactivation by heat, urea, and citrate. M. Schlesinger (1934) found that phages were made of DNA and protein. The publication of phage electron micrographs (1940) settled the matter of phage nature and made the enzyme theory untenable. It was now clear that phages were particulate and morphologically diverse. In retrospect, however, both d'Herelle and his opponents were right: phages are viruses with a dual nature.

Phage Therapy

Phage therapy was practiced by D'Herelle himself (1919) soon after phage discovery. He treated human dysenteria, chicken cholera, bovine hemorrhagic fever, bubonic plague, cholera, and a variety of staphylococcal and streptococcal infections. His phage laboratory in Paris was devoted to phage therapy. Many people embarked on phage therapy. Spectacular successes and dismal failures ensued. The failures had many reasons: use of crude, uncontrolled, inactivated, or endotoxin-containing lysates, narrow host specificity of some phages; absence of bacteriological controls; and plain charlatanism. The literature on phage therapy peaked in 1930 and totaled 560 publications for 1920–40. At this time, antibiotics were introduced and interest in phage therapy quickly waned.

Phage Typing and Ecology

Host range tests had shown that bacteria, even of the same species, usually differed in phage sensitivity. Some strains were lysed and some were not. By testing bacteria with a battery of phages, they could be subdivided into 'phage types', which greatly facilitated the tracing of bacterial infections to their sources and gave epidemiology an enormous boost. The first phage typing scheme, introduced in 1938, was for *Salmonella typhi* bacteria, which were subdivided by means of a series of adapted derivatives of phage ViII. For the first time, viruses were found to have a practical use. Phage typing developed rapidly in the following decades.

New phages were sought and found everywhere. It appeared that most bacterial pathogens had phages (**Table 1**) and that these viruses were omnipresent in water and soil, on plants, in food, on the skin of humans and animals, and in their body cavities and excreta. It also was discovered that bacteriophages caused faulty fermentations in the dairy industry. A few bacteriophages

Table 2 Phage discovery: individual phages and phage groups

Year	Investigators	Phage	Family[a]	Host genus[b]
1915	Twort	Twort?	M	*Staphylococcus*
1926	Clark and Clark	C1	P	*Streptococcus*
1933	Burnet and McKie	C16	M	Enterics
		S13	*Microviridae*	Enterics
1935	Sertic and Boulgakov	φX174	*Microviridae*	Enterics
1945	Demerec and Fano	T1 to T7	M, P, S	Enterics
1951	E. Lederberg	λ	S	Enterics
	G. Bertani	P1 and P2	M	Enterics
	Zinder and J. Lederberg	P22	P	Enterics
1960	Loeb	f1	*Inoviridae*	Enterics
1961	Loeb and Zinder	f2	*Leviviridae*	Enterics
	Takahashi	PBS1	M	Bacillus
1963	Taylor	Mu	M	Enterics
1964	Okubo	SPO1	M	Bacillus
1965	Reilly	φ29	P	Bacillus
1967	Schito	N4	P	Enterics
1968	Espejo and Canelo	PM2	*Corticoviridae*	*Pseudoalteromonas*
	Riva and coll.	SPP1	S	*Bacillus*
1970	Gourlay	L1 (MVL1)	*Plectrovirus*	*Acholeplasma*
1971	Gourlay	L2 (MVL2)	*Plasmaviridae*	*Acholeplasma*
1973	Six and Klug	P4	M	Enterics
	Vidaver and coll.	φ6	*Cystoviridae*	*Pseudomonas*
1974	Olsen and coll.	PRD1	*Tectiviridae*	*Enterics*
1982	Martin and coll.	SSV1	*Fuselloviridae*	*Sulfolobus*
1983	Janekovic and coll.	TTV1	*Lipothrixviridae*	*Thermoproteus*
1994	Zillig and coll.	SIRV1	*Rudiviridae*	*Sulfolobus*
2002	Rachel and coll.	ATV	*Bicaudaviridae*[c]	*Acidianus*
2004	Häring and coll.	PSV	*Globuloviridae*[c]	*Pyrobaculum*
2005	Häring and coll.	ABV	*Ampullaviridae*[c]	*Acidianus*

[a]Tailed phages: M, *Myoviridae*; P, *Podoviridae*; S, *Siphoviridae*.
[b]Enterobacteria are considered as a single host 'genus' because of frequent cross-reactions.
[c]Archaeal viruses awaiting formal classification.

isolated during this early period are still available today: a streptococcal and a T4-like phage, two microviruses, and, possibly, the original phage isolated by Twort himself (**Table 2**, **Figure 1**).

The early period ended with the rise of dictatorships and World War II, first in the Soviet Union and later in Europe. Scientists were silenced, persecuted, forced into exile, or outright killed (G. Eliava was shot in 1937 by Stalin's secret police and Eugène and Elisabeth Wollman disappeared in 1943 in a Nazi concentration camp). Phages now had their martyrs. D'Herelle himself, a Canadian citizen, was interned by the Vichy Government and prevented from pursuing phage therapy. Phage research was obliterated in much of Europe.

The Intermediate Period: 1939–62

Phages as Tools and Products of Molecular Biology

Phage research survived and attained new heights in the USA. It helped to start molecular biology and provided crucial insights into vertebrate and plant virology. In turn,

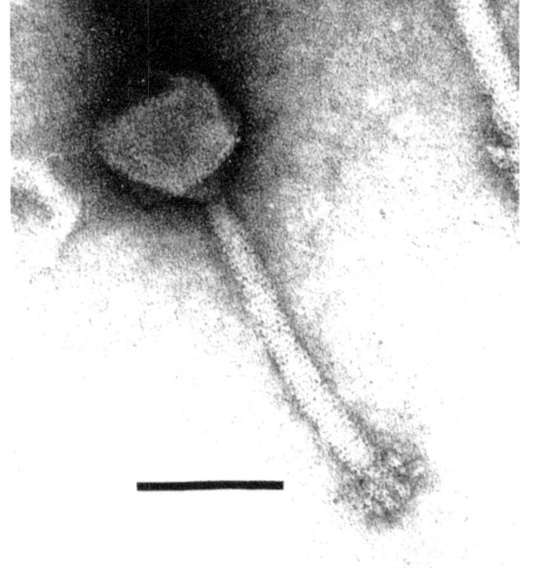

Figure 1 Phage Twort, uranyl acetate. Scale = 100 nm. The phage was reportedly deposited in 1948 by Twort himself in the collection of the Pasteur Institute of Paris.

Table 3 Basic phage research

Year	*Investigators*	*Phages*	*Event or observation*
(a) The rise of molecular biology			
1939	Ellis and Delbrück	Coliphage	One-step growth experiment
1940	Pfankuch and Kausche; Ruska	Coliphage	Phages seen in the electron microscope
1945	Demerec and Fano	T1–T7	Differentiation by host range
1942	Luria and Anderson	T1, T2	Two different morphologies
1942–53	Anderson and coll.	T1–T7	Morphological studies
1946	Cohen and Anderson	T2	Contains 37% of DNA
1947	Luria	T1–T7	Multiplicity reactivation
1948	Putnam and Cohen	T6	Transfer of DNA phosphorus from parent to progeny phage
1949	Hershey and Rothman	T2	Genetic recombination between phage mutants
1950	Anderson	T4	DNA is located in the head and separated from protein by osmotic shock
	Lwoff and Gutmann		Clarification of the nature of lysogeny
1951	Cohen	T-even	Phage synthesis in stages: protein first, DNA later
	Herriott	T2	Empty phages consist of protein
1951	Doermann	T3, T-even	Eclipse period in phage multiplication
1952	Anderson	T2, T4, T5	Phages adsorb tail first
	Cohen	T-even	DNA contains 5-hydroxymethylcytosine
	Hershey and Chase	T2	DNA carries genetic information
	Zinder and J. Lederberg	P22	General transduction
1953	Lwoff		Review paper on lysogeny
1955	Lennox	P1	General transduction
1959	Adams		Summary of phage research
	Brenner and Horne		Introduction of negative staining
1960	Doty and coll.		DNA–DNA hybridization
1962	Campbell	λ	Model of prophage insertion
1963	Taylor	Mu	Acts as a transposon
(b) After 1965			
1965, 1967	Bradley	General	Phage classification
1967	Eisenstark	General	First phage count
	Wood and Edgar	T4	Assembly pathway completed
	Ikeda and Tomizawa	P1	Prophage is a plasmid
	Ptashne	λ	Isolation of repressor
1968	Signer	λ	Prophage may be a plasmid
1969	Six and Klug	P2	Depends on helper phage P2
1973	Arber	P1	Restriction endonucleases identified
1992	Young	General	Clarification of phage lysis
2000	Wikoff and coll.	HK97	Capsid structure
2005	Fokine and coll., Morais and coll.	T4, HK97, ϕ29	Common ancestry of tailed phages
(c) Phage-coded bacterial toxins			
1951	Freeman	β	Diphtheria toxin
1964	Zabriskie	A25	Erythrogenic toxin of streprococci
1970	Inoue and Iida	CEβ	Botulinus toxin C
1971	Uchida	β	Diphtheria toxin gene is part of phage
1984	O'Brien and coll.	H-19J, 933W	Shiga-like toxins of *E. coli*
1996	Waldor and Mekalanos	CTXϕ	Cholera toxin

phage research benefited enormously from both disciplines. The next two decades became a legendary period of phage research, as a small number of researchers produced results of fundamental scientific importance (**Tables 3** and **4**).

In 1939, E. L. Ellis devised the 'one-step growth' experiment to measure the kinetics of phage infection and to establish the viral nature of bacteriophages. He was joined by Max Delbrück (1906–81), a German immigrant and physicist. Delbrück proposed to concentrate phage research on the seven coliphages T1–T7, collected and studied by F. Demerec and U. Fano in 1945 and all easily propagated. Fortunately, these phages were both diverse and partially related, thus allowing useful comparisons. Electron microscopy showed that the T phages (T for type) belonged to four morphotypes: T1, T3–T7, T2–T4–T6, and T5 (**Figure 2**). Morphotypes correlated with growth parameters and antigenic properties. The T phages became

Table 4 Basic research: Milestones in genome sequencing

Year	*Investigators*	*Phages*	*Family*[a]	*Host*
1976	Fiers and coll.	MS2	*Leviviridae*	Enterics
1978	Sanger and coll.	ϕX174	*Microviridae*	Enterics
1980	Van Wezenbeek and coll.	M13	*Inoviridae*	Enterics
1981	Mekler	Qβ	*Leviviridae*	Enterics
1982	Sanger and coll.	λ	S	Enterics
1983	Dunn and Studier	T7	P	Enterics
1986	Vlcek and Paces	ϕ29[b]	P	*Bacillus*
1988	Mindich	ϕ6[b]	Cystoviridae	*Pseudomonas*
1994	Kutter and coll.	T4[b]	M	Enterics
1996	Grimaud	Mu	M	Enterics
1998	Christie and coll.	P2[b]	M	Enterics
1999	Mannisto and coll.	PM2	Corticoviridae	*Pseudoalteromonas*
2000	Vanderbyl and Kropinski	P22[b]	P	Enterics
2002	Mesyanzhinov and coll.	ϕKZ	M	*Pseudomonas*
2004	Lobocka and coll.	P1[b]	M	Enterics

[a]Tailed phages: M, *Myoviridae*; P, *Podoviridae*; S, *Siphoviridae*.
[b]Sequence completed.

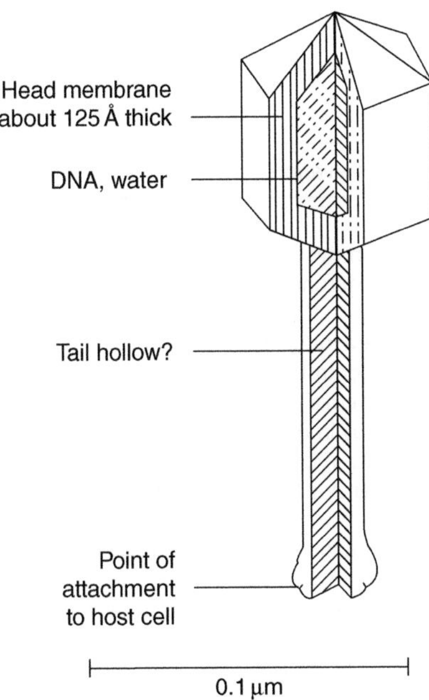

Figure 2 Drawing of phage T2 by T. F. Anderson (1952). Reproduced from Anderson TF, Rappaport C, and Muscatine NA (1953) On the structure and osmotic properties of phage particles. *Annales de l'Institut Pasteur* 84: 5–15, with permission from Elsevier.

the raw material for fundamental studies which, although limited to tailed phages of the T series, provided a general understanding of the phage life cycle. It became clear that:

1. Phages have different morphologies.
2. Phages contain DNA and transmit it from parent to progeny. The DNA is the carrier of heredity and genetic information. Phage coats consist of proteins.
3. Phages adsorb to bacteria by their tails.
4. Phage DNA enters the bacteria while the protein coat remains outside.
5. Phage synthesis occurs in stages, one of which is the 'eclipse'. Neither phages nor particulate precursors can be detected during the eclipse period. Novel phages are liberated ready-made in a single event ('burst').

Lysogeny

The problem of lysogeny, a leftover from the previous period, was solved by the French bacteriologist André Lwoff (1902–94) from the Pasteur Institute of Paris. The Wollmans had already postulated the dual nature of some phages: they were infectious outside the bacterium and noninfectious inside. Starting in 1949, Lwoff isolated single cells of a phage-carrying strain of *Bacillus megaterium* with a micromanipulator and transferred them into microdrops of fresh medium. After 19 successive generations, there was still no lysis. When lysis occurred, it was sudden and phages were plentiful in the microdrops. Lysis and phage production were inducible, especially by ultraviolet (UV) light. It was concluded that lysogenic bacteria perpetuate latent phages as 'prophages' in a repressed state. They do not secrete phages. Spontaneously or after induction, the prophage enters a vegetative phase, new phages are produced, and the bacterium dies. A major review article published in 1953 clarified the nature of lysogeny. The prophage became the origin of the 'provirus' concept in vertebrate virology.

In 1951, Esther Lederberg isolated the lysogenic coliphage λ. Small and easily manipulated, it was to have an extraordinary future and became perhaps the best studied of all viruses. As the prototypic lysogenic phage, it was the subject of countless publications on the nature of lysogeny, genetic regulation, and transduction. It was shown that defective prophages could carry specific bacterial genes and insert them into bacteria (specific transduction). Certain phages such as P22 could transfer any bacterial genes and thus mediate 'general transduction'. This led to an understanding of phage conversion. It appeared that the converting phages were lysogenic and that their prophages converted bacteria by coding for novel properties, such as the production of antigens. As early as 1951, V. J. Freeman reported that the production of diphtheria toxin depended on the presence of prophages. The study of lysogeny culminated in 1962 when A. Campbell proposed his model of prophage insertion by a crossover process.

Varied and Interesting

In 1959, Brenner and Horne revolutionized electron microscopy by introducing negative staining with phosphotungstic acid. This produced pictures of unprecedented quality. They showed that T4 tails were contractile and that tail contraction was part of the T4 infective process. The same year, single-stranded (ss) DNA was discovered in the minute phage ϕX174 *(Microviridae)*. Filamentous phages *(Inoviridae)* and phages with ssRNA *(Leviviridae)* were isolated in 1960 and 1961, respectively. True marine phages and phages of cyanobacteria, then called 'blue-green algae', were also found. This offered some hope to control 'algal' blooms.

While phage therapy disappeared, phage typing became a major epidemiological tool (**Table 5**). In addition to the *S. typhi* typing scheme, international typing sets for *S. paratyphi B* and *Staphylococcus aureus* were devised and are still in use. Typing schemes for many medically important bacteria (e.g., corynebacteria, various *E. coli* and *Salmonella* serotypes, shigellae, *Pseudomonas aeruginosa*, and *Vibrio cholerae*) appeared. A literature survey by H Raettig lists no less than 2000 articles on phage typing, mostly in the years 1940–65. Phage research was considerably promoted by M H Adams' book *Bacteriophages*, which summarized phage research until 1957 and introduced the agar-double-layer method of phage propagation.

Table 5 Applied and environmental phage research

Year	*Investigators*	*Phages*	*Event or observation*
(a) Phage therapy			
1919	D'Herelle		Treats human dysentery and chicken cholera
1921	Bruynoghe and Maisin		Treat human furunculosis
1925	D'Herelle		Treats plague in Egypt
1927–28	D'Herelle		Treats cholera in Assam
	Moore		Phages for control of plant disease
1938	Fisk		Treats septicemia in mice
1981–87	Slopek and coll.		Phage therapy in Poland
1982–87	Smith and Huggins		Treat diarrhea in calves, lambs, piglets
2001	Nelson and coll.		Phage lytic enzymes for elimination of streptococci in mice
(b) Diagnostics, industry, and biotechnology			
1936	Whitehead and Cox		Faulty fermentations in the dairy industry
1938	Craigie and Yen	ViII	Phage typing of *Salmonella typhi*
1942	Fisk		Phage typing of *Staphylococcus aureus*
1954	Cherry and coll.	O1	Identification of salmonellae
1955	Brown and Cherry	γ	Identification of *Bacillus anthracis*
1973	Lobban and Kaiser	T4	Ligase for cloning
	Studier	T7	Promoter for cloning
1977	Blattner and coll.	λ	Charon vectors for cloning
1977–83	Messing and coll., others	M13, fd	Cloning and sequencing vectors
1985	Smith	M13	Phage display
1987	Ulitzur and Kuhn	λ	Luciferase reporter phages
1990	Sternberg	P1	Cloning of large DNA fragments
(c) Phages and the environment			
1920	Dumas		Phages in soil
1923	Gerretsen and coll.		Phages in rhizobia
	Van der Hoeden		Phages in meat
1924	Zdansky		Phages in sewage
1955	Spencer		Indigenous marine phage
1966	Coetzee		Phages as pollution indicators
1974	Martin and Thomas		Phages as groundwater tracers
1987	Goyal and coll.		Book on phage ecology
1989	Bergh and coll.		Large numbers of phages in seawater
2002	Brüssow and Hendrix		Global phage population estimated at 10^{31}

Diversification and the Ongoing History of Bacteriophages: 1965–Today

Phage research now expanded explosively and divided into many, partly overlapping fields. No particular time periods are discernible; all fields developed gradually and are still expanding. As judged from literature listings, the present number of phage publications is 800–900 articles or monographs per year. The following is a selection of research activities.

Basic Phage Research

1. The present period of phage research started in 1965 (**Table 3**) after D. E. Bradley published a seminal, still cited paper on phage morphology (republished in 1967). Bradley classified phages into six basic types defined by morphology and nature of nucleic acid. Phages were divided into tailed phages (with contractile, long and noncontractile, or short tails), isometric phages with ssDNA or ssRNA, and filamentous phages. This became the basis of present-day phage classification.
2. Novel phages were isolated at a rate of about 100 per year. A first phage survey, published in 1967, listed 111 phages examined by electron microscopy, 102 of which were tailed and 9 isometric or filamentous. Most infected enterobacteria. Presently, the number of phages with known morphology exceeds 5500 and the number of host genera has grown to 155. It includes mycoplasmas, cyanobacteria, and archaea.
3. Replication and assembly of most phage families were studied. A highlight was the description of the assembly pathway of phage T4 (1967), obtained by a combination of electron microscopy and genetics.
4. Many phage genomes were sequenced. The first was that of ssRNA phage MS2 (*Leviviridae*), the first viral genome ever to be sequenced. This was followed by the sequences of various small viruses of the families *Microviridae*, *Inoviridae*, and *Leviviridae*. A landmark was the sequencing of the complete λ genome (1982), followed by that of coliphage T7. The complete sequence of the T4 genome was established in 1994. This was the beginning of large-scale phage genome sequencing (**Table 3(b)**) and of the novel science of 'genomics', devoted to computer-assisted comparisons of phage and other genomes. By 2006, the genomes of representatives of all phage families and of many individual tailed phages had been sequenced.
5. Bacteriophages appeared as major factors of bacterial virulence, coding for bacterial exotoxins (e.g., botulinus and Shiga toxins), antigens, and other virulence factors. A momentous discovery made in 1996 was that filamentous phages in their double-stranded replication form were able to convert nontoxigenic cholera strains to toxigenic forms. Virulence genes were found to be present in converting phages and integrated prophages. Sequencing of bacterial genomes showed that prophages with virulence factors were generally organized into 'pathogenicity islands'.
6. Cryoelectron microcopy and three-dimensional image reconstruction revealed the capsid structure of various isometric phages (*Microviridae*, *Leviviridae*, *Tectiviridae*) and of the tail and base plate structure of phage T4. It was also shown that the capsids of adenoviruses and tectiviruses and of representatives of all three tailed phage families (T4, HK97, ϕ29) were structurally related. This indicated phylogenetic relationships between adeno- and tectiviruses and provided the long-sought evidence that tailed phages are a monophyletic evolutionary group.

Applied Research

See **Table 5** for a listing of research activities, arranged chronologically.

1. In the dairy industry, phages of lactococci and *Streptococcus thermophilus* were shown to cause major economic losses by destroying starter cultures and causing faulty fermentations. The phages involved were studied in detail by electron microscopy and genome sequencing.
2. Phages were investigated for use as indicators of fecal pollution. Phages with ssRNA (*Leviviridae*), somatic coliphages, and tailed phages of *Bacteroides fragilis* were proposed as indicators, but their usefulness was not conclusively proven. On the other hand, phages of many types proved useful as tracers of groundwater movements.
3. Phages were harnessed for the detection of bacteria, particularly by the construction of 'reporter' phages carrying luminescent (luciferase) genes, which are expressed when these phages infect a bacterium.
4. In biotechnology, phages are used as cloning and sequencing vectors, notably the filamentous phage M13 *(Inoviridae)*, phages λ and P1, and hybrids of phages and plasmids (named 'cosmids' or 'phagemids'). Phage products such as T4 ligase and T7 promoters are used in cloning reactions and phage lysozymes are used to free useful intracellular bacterial enzymes.
5. The technique of 'phage display', introduced in 1985, mainly uses filamentous phages (M13, fd). Foreign peptide genes are fused to phage coat proteins. This powerful technique led to the development of antibody technology and has many medical and pharmaceutical applications.
6. Phage therapy reappeared. In France, it had always been practiced on a small scale. In the mid-1980s, a team in Wroclaw, Poland, conducted phage treatments of various human diseases (mainly pyogenic

infections). After the fall of communism, the West learned with surprise that phage therapy had been practiced in the Soviet Union since the 1940s and that the institute founded in Tbilisi par d'Herelle and Eliava produced phages for the Soviet Union with up to 1200 employees engaged in production. Phage therapy is still practiced in Georgia.

Phage Ecology

1. Marine phages were investigated worldwide, including Arctic and Antarctic waters, largely by electron microscopy and the novel epifluorescence technique. Marine phages were found in vast numbers and appeared as part of a food web of phages, bacteria, and flagellates. Many marine phages were cyanophages and, all those investigated so far have tailed particles. Sequencing of uncultured virus DNA ('metagenomics') detected known entities such as T7-like phages and showed that only 25% of sequences matched known genes. Seawater appears as an immense phage reservoir (**Table 5**).
2. Phages were sought and found in extreme habitats such as volcanic hot springs, alkaline lakes, or hypersaline lagoons. This led to the discovery of several families of archaeal viruses (**Table 2**).

See also: Assembly of Non-Enveloped Virions; History of Virology: Vertebrate Viruses.

Further Reading

Ackermann H-W (2006) 5500 bacteriophages examined in the electron microscope. *Archives of Virology* 152: 227–243.
Ackermann H-W and DuBow MS (1987) *Viruses of Prokaryotes, Vol. I. General Properties of Bacteriophages.* Boca Raton, FL: CRC Press.
Adams MH (1959) *Bacteriophages.* New York: Interscience Publishers.
Anderson TF, Rappaport C, and Muscatine NA (1953) On the structure and Ormotic properties of phage particles. *Annales de l'Institut Pasteur* 84: 5–15
Cairns J, Stent GS, and Watson JD (eds.) (1992) *Phage and the Origins of Molecular Biology.* (expanded edn.) Cold Spring Harbor, NY: Cold Spring Harbor Laboratory Press.
Calendar R (2006) *The Bacteriophages,* 2nd edn., New York: Oxford University Press.
Goyal SM, Gerba CP, and Bitton G (eds.) (1987) *Phage Ecology.* New York: Wiley.
Kutter E and Sulakvelidze A (eds.) (2004) *Bacteriophages: Biology and Applications.* Boca Raton, FL: CRC Press.
Marks T and Sharp R (2000) Bacteriophages and biotechnology: A review. *Journal of Chemical Technology and Biotechnology* 75: 6–17.
Raettig H (1958) *Bakteriophagie 1917 Bis 1956. Vol. I. Einführung, Sachregister, Stichwortverzeichnis. Vol. II. Autorenregister.* Stuttgart: Gustav Fischer.
Raettig H (1967) *Bakteriophagie 1957–1965. Vol. I. Einführung, Sachregister, Stichwort-Index. Introduction, Index of Subjects, Reference Word Index. Vol. II. Bibliography.* Stuttgart: Gustav Fischer.
Summers WC (1999) *Felix d'Herelle and the Origins of Molecular Biology.* New Haven, CT: Yale University Press.
Tidona CA, Darai G, and Büchen-Osmond C (2001) *The Springer Index of Viruses.* Berlin: Springer.
Twort A (1993) *In Focus, Out of Step: A Biography of Frederick William Twort F.R.S. 1877–1950.* Stroud, UK: A. Sutton.

History of Virology: Plant Viruses

R Hull, John Innes Centre, Norwich, UK

Introduction

History shows that the study of virus infections of plants has led the overall subject of virology in the development of several major concepts including that of the entity of viruses themselves. To obtain a historical perspective of plant virology, five major (overlapping) ages can be recognized.

Prehistory

The earliest known written record describing what was almost certainly a plant virus disease is a poem in Japanese written by the Empress Koken in AD 752 and translated by T. Inouye as follows:

In this village
It looks as if frosting continuously
For, the plant I saw
In the field of summer
The color of the leaves were yellowing

The plant, identified as *Eupatorium lindleyanum,* has been found to be susceptible to tomato yellow leafcurl virus, which causes a yellowing disease.

In Western Europe in the period from about 1600 to 1660, many paintings and drawings were made of tulips that demonstrate flower symptoms of virus disease. These are recorded in the Herbals of the time and in the still-life paintings of artists such as Johannes Bosschaert in 1610. During this period, blooms featuring such striped patterns were prized as special varieties leading to the phenomenon

Table 1 Tulipomania: the goods exchanged for one bulb of Viceroy tulip

4 t of wheat	4 barrels of beer
8 t of rye	2 barrels of butter
4 fat oxen	1000 lb cheese
8 fat pigs	1 bed with accessories
12 fat sheep	1 full dress suit
2 hogsheads of wine	1 silver goblet

of 'tulipomania'. The trade in infected tulip bulbs resulted in hyperinflation with bulbs exchanging hands for large amounts of money or goods (**Table 1**).

In describing an experiment to demonstrate that sap flows in plants, Lawrence reported in 1714 the unintentional transmission of a virus disease of jasmine by grafting. The following quotation from Blair in 1719 describes the procedure and demonstrates that even in this protoscientific stage, experimenters were already indulging in arguments about priorities of discovery.

> The inoculating of a strip'd Bud into a plain stock and the consequence that the Stripe or Variegation shall be seen in a few years after, all over the shrub above and below the graft, is a full demonstration of this Circulation of the Sap. This was first observed by Mr. Wats at Kensington, about 18 years ago: Mr. Fairchild performed it 9 years ago; Mr. Bradly says he observ'd it several years since; though Mr. Lawrence would insinuate as if he had first discovered it.

Recognition of Viral Entity

In the latter part of the nineteenth century, the idea that infectious disease was caused by microbes was well established, and filters were available that would not allow the passage of known bacterial pathogens. Mayer in 1886 showed that a disease of tobacco (*Mosaikkrankheit*; now known to be caused by tobacco mosaic virus; TMV) could be transmitted to healthy plants by inoculation with extracts from diseased plants; Iwanowski demonstrated in 1892 that sap from such tobacco plants was still infective after it had been passed through a bacteria-proof filter candle. This work did not attract much attention until it was repeated by Beijerinck who in 1898 described the infectious agent as *contagium vivum fluidum* (Latin for contagious living fluid) to distinguish it from contagious corpuscular agents. Beijerinck's discovery is considered to be the birth of virology. In 1904 Baur showed that an infectious variegation of *Abutilon* could be transmitted by grafting, but not by mechanical inoculation. Beijerinck and Baur used the term 'virus' in describing the causative agents of these diseases, to contrast them with bacteria; this term had been used as more or less synonymous with bacteria by earlier workers. As more diseases of this sort were discovered, the unknown causative agents came to be called 'filterable viruses'.

Between 1900 and 1935, many plant diseases thought to be caused by filterable viruses were described, but considerable confusion arose because adequate methods for distinguishing one virus from another had not yet been developed.

The original criterion of a virus was an infectious entity that could pass through a filter with a pore size small enough to hold back all known cellular agents of disease. However, diseases were soon found that had virus-like symptoms not associated with any pathogen visible in the light microscope, but that could not be transmitted by mechanical inoculation. With such diseases, the criterion of filterability could not be applied. Their infectious nature was established by graft transmission and sometimes by insect vectors. Thus, certain diseases of the yellows and witches'-broom type, such as aster yellows, came to be attributed to viruses on quite inadequate grounds. Many such diseases are now known to be caused by phytoplasma or spiroplasma, and a few by bacteria or rickettsia.

The Biological Age

During most of the period between 1900 and 1935, attention was focused on the description of diseases using the macroscopic symptoms and cytological abnormalities as revealed by light microscopy, host ranges, and methods of transmission which were the only techniques available. The influence of various physical and chemical agents on virus infectivity was investigated, but methods for the assay of infective material were primitive. Holmes showed that the local lesions produced in some hosts following mechanical inoculation could be used for the rapid quantitative assay of infective virus. This technique enabled properties of viruses to be studied much more readily and paved the way for the isolation and purification of viruses a few years later. Until about 1930, there was serious confusion by most workers regarding the diseases produced by viruses and the viruses themselves. This was not surprising, since virtually nothing was known about the viruses except that they were very small. In 1931 Smith made an important contribution that helped to clarify this situation. Working with virus diseases in potato, he realized the necessity of using plant indicators, plant species other than potato, which would react differently to different viruses present in potatoes. Using several different and novel biological methods to separate the viruses, he was able to show that many potato virus diseases were caused by a combination of two viruses with different properties, which he named virus X (potato virus X, PVX) and virus Y (potato virus Y, PVY). As PVX was not transmitted by the aphid *Myzus persicae*, whereas PVY was, PVY could be separated from PVX. He obtained PVX free of PVY by needle inoculation of the mixture to *Datura stramonium* which does not support PVY. Furthermore, Smith observed that PVX from different sources fluctuated markedly in the severity of symptoms it produced in various hosts leading to the concept of strains.

An important practical step forward was the recognition that some viruses could be transmitted from plant to plant by insects. Fukushi recorded the fact that in 1883 a Japanese rice grower transmitted what is now known to be rice dwarf virus, RDV) by the leafhopper *Recelia dorsalis*. However, this work was not published in any available form and so had little influence. In 1922 Kunkel first reported the transmission of a virus by a planthopper; within a decade, many insects were reported to be virus vectors leading to the recognition of specific virus–vector interactions.

Since Fukushi first showed in 1940 that RDV could be passed through the egg of a leafhopper vector for many generations, there has been great interest in the possibility that some viruses may be able to replicate in both plants and insects. It is now well established that plant viruses in the families *Rhabdoviridae* and *Reoviridae* and the genera *Tenuivirus*, *Tospovirus*, and *Marafivirus* multiply in insects as well as in plants.

The Biochemical/Biophysical Age

Beale's recognition in 1928 that plants infected with TMV contained a specific antigen opened the age in which the biochemical nature of viruses was elucidated. In the 1930s Gratia showed that plants infected with different viruses contained different specific antigens and Chester demonstrated that different strains of TMV and PVX could be distinguished serologically.

The high concentration at which certain viruses occur in infected plants and their relative stability turned out to be of crucial importance in the first isolation and chemical characterization of viruses, because methods for extracting and purifying proteins were not highly developed. In the early 1930s, various attempts were made to isolate and purify plant viruses using methods similar to those that had just been developed for purifying enzymes. Following detailed chemical studies suggesting that the infectious agent of TMV might be a protein, Stanley announced in 1935 the isolation of this virus in an apparently crystalline state. At first Stanley considered that the virus was a globulin containing no phosphorus but in 1936 Bawden *et al.* described the isolation from TMV-infected plants of a liquid crystalline nucleoprotein containing nucleic acid of the pentose type. They showed that the particles were rod-shaped, thus confirming the earlier suggestion of Takahashi and Rawlins based on the observation that solutions containing TMV showed anisotropy of flow.

Electron microscopy and X-ray crystallography were the major techniques used in early work to explore virus structure, and the importance of these methods has continued to the present day. Bernal and Fankuchen applying X-ray analysis to purified preparations of TMV obtained accurate estimates of the width of the rods. The isolation of other rod-shaped viruses, and spherical viruses that formed crystals, soon followed. All were shown to consist of protein and pentose nucleic acid.

Early electron micrographs confirmed that TMV was rod-shaped and provided approximate dimensions, but they were not particularly revealing because of the lack of contrast between the virus particles and the supporting membrane. The application of shadow-casting with heavy metals greatly increased the usefulness of the method for determining the overall size and shape of virus particles but not structural detail. With the development of high-resolution microscopes and of negative staining in the 1950s electron microscopy became an important tool for studying virus substructure. From a comparative study of the physicochemical properties of the virus nucleoprotein and the empty viral protein shell found in TYMV preparations, Markham concluded in 1951 that the RNA of the virus must be held inside a shell of protein, a view that has since been amply confirmed for this and other viruses by X-ray crystallography. Crick and Watson suggested that the protein coats of small viruses are made up of numerous identical subunits arrayed either as helical rods or as a spherical shell with cubic symmetry. Subsequent X-ray crystallographic and chemical work has confirmed this view. Caspar and Klug formulated a general theory that delimited the possible numbers and arrangements of the protein subunits forming the shells of the smaller isodiametric viruses.

Until about 1948, most attention was focused on the protein part of the viruses. Quantitatively, the protein made up the larger part of virus preparations. Enzymes that carried out important functions in cells were known to be proteins, and knowledge of pentose nucleic acids was rudimentary. No function was known for them in cells, and they generally were thought to be small molecules primarily because it was not recognized that RNA is very susceptible to hydrolysis by acid, by alkali, and by enzymes that commonly contaminate virus preparations. In 1949 Markham and Smith isolated turnip yellow mosaic virus (TYMV) and showed that purified preparations contained two classes of particles, one an infectious nucleoprotein with about 35% of RNA, and the other an apparently identical protein particle that contained no RNA and that was not infectious. This result clearly indicated that the RNA of the virus was important for biological activity. Analytical studies showed that the RNAs of different viruses have characteristically different base compositions while those of related viruses are similar. About this time, it came to be realized that viral

RNAs might be considerably larger than had been thought. A synthetic analog of the normal base guanine, 8-azaguanine, when supplied to infected plants was incorporated into the RNA of TMV and TYMV, replacing some of the guanine. The fact that virus preparations containing the analog were less infectious than normal virus gave further experimental support to the idea that viral RNAs were important for infectivity. However, it was the classic experiments in the mid-1950s of Gierer and Schramm, and Fraenkel-Conrat and Williams that demonstrated the infectivity of naked TMV RNA and the protective role of the protein coat. These discoveries ushered in the era of modern plant virology.

In the early 1950s Brakke developed density gradient centrifugation as a method for purifying viruses. Together with a better understanding of the chemical factors affecting the stability of viruses in extracts, this procedure has allowed the isolation and characterization of many viruses. The use of sucrose density gradient fractionation enabled Lister to discover the bipartite nature of the tobacco rattle virus genome. Since that time, density gradient and polyacrylamide gel fractionation techniques have allowed many viruses with multipartite genomes to be characterized. Their discovery, in turn, opened up the possibility of carrying out genetic reassortment experiments with plant viruses leading to the allocation of functions to many of the viral genes. Density gradient fractionation of purified preparations of some other viruses revealed noninfectious nucleoprotein particles containing subgenomic RNAs. Other viruses have been found to have associated with them satellite viruses or satellite RNAs that depend on the 'helper' virus for some function required during replication.

Further developments in the 1970s included improved techniques related to X-ray crystallographic analysis and a growing knowledge of the amino acid sequences of the coat proteins allowed the three-dimensional structure of the protein shells of several plant viruses to be determined in molecular detail.

For some decades, the study of plant virus replication had lagged far behind that of bacterial and vertebrate viruses mainly because there was no plant system in which all the cells could be infected simultaneously to provide the basis for synchronous 'one-step growth' experiments. However, following the initial experiments of Cocking in 1966, Takebe and colleagues developed protoplast systems for the study of plant virus replication. Although these systems had significant limitations, they greatly increased our understanding of the processes involved in plant virus replication.

Another important technical development has been the use of *in vitro* protein-synthesizing systems such as that from wheat germ, in which many plant viral RNAs act as efficient messengers. Their use allowed the mapping of plant viral genomes by biochemical means to begin.

The Molecular Biology Age

The molecular age opened in 1960 with the determination of the full sequence of 158 amino acids in the coat protein of TMV. The sequence of many naturally occurring strains and artificially induced mutants was also determined at about the same time. This work made an important contribution to establishing the universal nature of the genetic code and to our comprehension of the chemical basis of mutation.

Our understanding of the genome organization and functioning of viruses has come from the development of procedures whereby the complete nucleotide sequence of viruses with RNA genomes can be determined. In 1982 the genomes of both the first plant RNA virus (TMV) and DNA virus cauliflower mosaic virus (CaMV) were sequenced. Since then, the genomes of representatives of all the plant virus genera have been sequenced and there are many sequences of virus species.

The late 1980s and 1990s was a period when molecular biological techniques were applied to a wide range of aspects of plant virology. These included the ability to prepare *in vitro* infectious transcripts of RNA viruses derived from cloned viral cDNA allowing techniques such as site-directed mutagenesis to be applied to the study of genome function and reverse genetics being used to elucidate the functions of viral genes and control sequences. These approaches, together with others such as yeast systems for identifying interacting molecules, the expression of viral genes in transgenic plants, and labeling viral genomes in such a manner that their sites of function within the cell are known, were revealing the complexities of the interactions between viruses and their hosts. Nucleotide sequence information has had, and continues to have, a profound effect on our understanding of many aspects of plant virology, including (1) the location, number, and size of the genes in a viral genome; (2) the amino acid sequence of the known or putative gene products; (3) the molecular mechanisms whereby the gene products are transcribed; (4) the putative functions of a gene product, which can frequently be inferred from amino-acid-sequence similarities to products of known function encoded by other viruses; (5) the control and recognition sequences in the genome that modulate expression of viral genes and genome replication; (6) the understanding of the structure and replication of viroids and of the satellite RNAs found associated with some viruses; (7) the molecular basis for variability and evolution in viruses, including the recognition that recombination is a widespread phenomenon among RNA viruses and that viruses can acquire host nucleotide sequences as well as genes from other viruses; and (8) the beginning of a taxonomy for viruses that is based on evolutionary relationships. On the host side, advances in plant genome sequencing are identifying plant genes that confer resistance to viruses.

During the 1980s, major advances were made on improved methods of diagnosis for virus diseases, centering on serological procedures and on methods based on nucleic acid hybridization. Since the work of Clark and Adams reported in 1977, the enzyme-linked immunosorbent assay (ELISA) technique has been developed with many variants for the sensitive assay and detection of plant viruses. Monoclonal antibodies against TMV lead to a very rapid growth in their use for many kinds of plant virus research and for diagnostic purposes. The late 1970s and the 1980s also saw the start of application of the powerful portfolio of molecular biological techniques to developing other approaches to virus diagnosis, to a great increase in our understanding of the organization and strategy of viral genomes, and to the development of techniques that promise novel methods for the control of some viral diseases. The use of nucleic acid hybridization procedures for sensitive assays of large numbers of samples and the polymerase chain reaction, also dependent on detailed knowledge of genome sequences, are being increasingly used in virus diagnosis. Most recently DNA chips are being developed for both virus diagnostics and studying virus infection.

In the early 1980s, it seemed possible that some plant viruses, when suitably modified by the techniques of gene manipulation, might make useful vectors for the introduction of foreign genes into plants. Some plant viruses have been found to contain regulatory sequences that can be very useful in other gene vector systems, notably the widely used CaMV 35S promoter. Another practical application of molecular techniques to plant viruses has been the modification of viral genomes so that products of interest to industry and medicine can be produced in plants. This is being done either by the introduction of genes into the viral genome or by modification of the coat protein sequence to enable epitopes to be presented on the virus.

Early attempts (early to mid-1900s) to control virus diseases in the field were often ineffective. They were mainly limited to attempts at general crop hygiene, roguing of obviously infected plants, and searching for genetically resistant lines. Developments since this period have improved the possibilities for control of some virus diseases. Heat treatments and meristem tip culture methods have been applied to an increasing range of vegetatively propagated plants to provide a nucleus of virus-free material that then can be multiplied under conditions that minimize reinfection. Such developments frequently have involved the introduction of certification schemes. Systemic insecticides, sometimes applied in pelleted form at the time of planting, provide significant protection against some viruses transmitted in a persistent manner by aphid vectors. It has become increasingly apparent that effective control of virus disease in a particular crop in a given area usually requires an integrated and continuing program involving more than one kind of control measure.

However, such integrated programs are not yet in widespread use. Cross-protection (or mild-strain protection) is a phenomenon in which infection of a plant with a mild strain of a virus prevents or delays infection with a severe strain. The phenomenon has been used with varying success for the control of certain virus diseases, but the method has various difficulties and dangers. In 1986 Powell-Abel and co-workers considered that some of these problems might be overcome by the application of the concept of pathogen-derived resistance of Sandford and Johnston. Using recombinant DNA technology, they showed that transgenic tobacco plants expressing the TMV coat-protein gene either escaped infection following inoculation or showed a substantial delay in the development of systemic disease. These transgenic plants expressed TMV coat-protein mRNA as a nuclear event. Seedlings from self-fertilized transformed plants that expressed the coat protein showed delayed symptom development when inoculated with TMV. Thus, a new approach to the control of virus diseases emerged. However, this approach revealed some unexpected results which led to the recognition that plants have a defense system against 'foreign' RNA. This defense system, initially termed post-translational gene silencing and now called RNA silencing or RNA interference (RNAi) was first recognized in plants and had, and is still having, a great impact on molecular approaches as diverse as disease control and understanding gene functions. Among the tools that have arisen from understanding this new phenomenon is virus-induced gene silencing which is being used to determine the functions of genes in plants and animals by turning them off.

However, the RNAi defense system in plants which targets double-stranded RNA, an intermediate RNA virus replication, raised the question of how RNA viruses replicated in plants. Studies on gene functions revealed that many plant viruses contain so-called 'virulence' genes. Many of these have been shown to suppress host RNA silencing, thus overcoming the defense system. Suppression of gene silencing is widespread among plant viruses and examples are being found in animal viruses. Thus, as noted at the beginning of this article, plant virology is still providing insights onto phenomena applicable to virology in general.

See also: Nature of Viruses.

Further Reading

Baulcombe DC (1999) Viruses and gene silencing in plants. *Archives of Virology* 15(supplement): 189–201.

Caspar DLD and Klug A (1962) Physical principles in the construction of regular viruses. *Cold Spring Harbor Symposium on Quantitative Biology* 27: 1–24.

Clark MF and Adams AN (1977) Characteristics of the microplate method of enzyme-linked immunosorbent assay for the detection of plant viruses. *Journal of General Virology* 34: 475–483.
Fischer R and Emans N (2000) Molecular farming of pharmaceutical proteins. *Transgenic Research* 9: 279–299.
Hull R (2001) *Matthews' Plant Virology*. San Diego: Academic Press.
Li F and Ding SW (2006) Virus counterdefense: Diverse strategies for evading the RNA silencing mechanism. *Annual Review of Microbiology* 60: 507–531.
Lindbo JA and Dougherty WG (2005) Plant pathology and RNAi: A brief history. *Annual Review of Phytopathology* 43: 191–204.
Pavord A (1999) *The Tulip*. London: Bloomsbury.
Van der Want JPH and Dijkstra J (2006) A history of plant virology. *Archives of Virology* 151: 1467–1498.

History of Virology: Vertebrate Viruses

F Fenner, Australian National University, Canberra, ACT, Australia

Introduction

As in other branches of experimental science, the development of our knowledge of viruses has depended on the techniques available. Initially, viruses were studied by pathologists interested in the causes of the infectious diseases of man and his domesticated animals and plants and these concerns remain the main force advancing the subject. The idea that viruses might be used to probe fundamental problems of biology arose in the early 1940s, with the development of the knowledge of bacterial viruses. These studies helped establish the new field of molecular virology in the period 1950–70, and this has revolutionized the study of viruses and led to an explosion of knowledge about them.

The Word Virus

Since antiquity the term virus had been synonymous with poison, but during the late nineteenth century it became a synonym for microbe (Pasteur's word for an infectious agent). It did not acquire its present connotation until the 1890s, after the bacterial or fungal causes of many infectious diseases had been discovered, using the agar plate, effective staining methods, and efficient microscopes. It then became apparent that there were a number of infectious diseases of animals and plants from which no bacterium or fungus could be isolated or visualized with the microscope. After the introduction in 1884 of the Chamberland filter, which held back bacteria, Loeffler and Frosch demonstrated that the cause of foot-and-mouth disease was a filterable (or ultramicroscopic) virus. The first compendium of all then known viruses was edited by T. M. Rivers of the Rockefeller Institute and published in 1928. Entitled *Filterable Viruses*, this emphasized that viruses required living cells for their multiplication. In the 1930s chemical studies of the particles of tobacco mosaic virus and of bacteriophages showed that they differed from all cells in that at their simplest they consisted of protein and nucleic acid, which was either DNA or RNA. Gradually, the adjectives filterable and ultramicroscopic were dropped and the word viruses developed its present connotation.

Early Investigations

Foot-and-Mouth Disease Virus

In 1898 F. J. Loeffler and P. Frosch described the filterability of an animal virus for the first time, noting that "the filtered material contained a dissolved poison of extraordinary power or that the as yet undiscovered agents of an infectious disease were so small that they were able to pass through the pores of a filter definitely capable of retaining the smallest known bacteria." Although the causative agent of foot-and-mouth disease passed through a Chamberland-type filter, it did not go through a Kitasato filter which had a finer grain. This led to the conclusion that the causative virus, which was multiplying in the host, was a corpuscular particle. Loeffler and Frosch gave filtration a new emphasis by focussing attention on what passed-through the filter rather than what was retained and established an experimental methodology which was widely adopted in the early twentieth century in research on viral diseases.

Yellow Fever Virus

Following the acceptance of the notion of filterable infectious agents, pathologists investigated diseases from which no bacteria could be isolated and several were soon shown to be caused by viruses. One of the most fruitful investigations, in terms of new concepts, was the work of the United States Army Yellow Fever Commission headed by Walter Reed in 1900–01. Using human volunteers, they demonstrated that yellow fever was caused by a filterable virus which was transmitted by mosquitoes and that the principal vector was *Aedes aegypti*. They also showed that infected persons were infectious for mosquitoes only during the first few days of the disease and that mosquitoes were not infectious until 7–10 days after imbibing infectious blood, thus defining the extrinsic incubation period and establishing essentially all of the basic principles of the epidemiology of what came to be called arboviruses (arthropod-borne viruses).

Physical Studies of Viruses

Further advances in understanding the nature of viruses depended on physical and chemical studies. As early as 1907 H. Bechhold in Germany developed filters of graded permeabilities and a method of determining their pore size. Subsequently, W. J. Elford, in London, used such membranes for determining the size of animal virus particles with remarkable accuracy.

In Germany, on the eve of World War II, H. Ruska and his colleagues had produced electron microscopic photographs of the particles of tobacco mosaic virus, bacteriophages, and poxviruses. Technical improvements in instrumentation after the war, and the introduction of negative staining for studying the structure of viruses by Cambridge scientists H. E. Huxley in 1957 and S. Brenner and R. W. Horne in 1959 resulted in photographs of the particles of virtually every known kind of virus. These demonstrated the variety of their size, shape, and structure, and the presence of common features such as the icosahedral symmetry of many viruses of animals.

Following a perceptive paper on the structure of small virus particles by F. H. C. Crick and J. D. Watson in 1956, in 1962 D. L. D Caspar of Boston and A. Klug of Cambridge, England, produced a general theory of the structure of regular virus particles of helical and icosahedral symmetry. Structure of the virus particle became one of the three major criteria in the system of classification of viruses that was introduced in 1966.

The Chemical Composition of Viruses

If the ultramicroscopic particles found in virus-infected hosts were the pathogens, what did these particles consist of? Following observations on plant and bacterial viruses in 1935, in 1940 C. L. Hoagland and his colleagues at the Rockefeller Institute found that vaccinia virus particles contained DNA but no RNA. Thus evidence was accumulating that viruses differed from bacteria not only in size and their inability to grow in lifeless media, but in that they contained only one kind of nucleic acid, which could be either DNA or RNA.

The development of restriction-endonuclease digestion of DNA, based on studies of phage restriction by W. Arber (Nobel Prize, 1978) of Basel in the 1960s, and then elaborated by biochemist H. O. Smith of Baltimore in 1970, has simplified the mapping of the genomes of DNA viruses, a study initiated by D. Nathans (Nobel Prize, 1978), also of Baltimore, using Simian virus 40. The development of the polymerase chain reaction in 1985 by K. B. Mullis (Nobel Prize, 1993) revolutionized sequencing methods for both DNA and RNA, leading to the availability of the complete genomic sequences of most viruses and the ability to make diagnoses using minute amounts of material.

Investigations of bacterial viruses were motivated by scientific curiosity, but animal virology was developed by pathologists studying the large number of diseases of humans and livestock caused by viruses, and it has retained this practical bias. Over the first two decades of the twentieth century, testing of filtrates of material from a number of infected humans and animals confirmed that they were caused by viruses. Among the most important was the demonstration in 1911 by P. Rous (Nobel Prize, 1966), at the Rockefeller Institute, that a sarcoma of fowls could be transmitted by a bacteria-free filtrate.

The Cultivation of Animal Viruses

The first systematic use of small animals for virus research was Pasteur's use of rabbits, which he inoculated intracerebrally with rabies virus in 1881. It was not until 1930 that mice were used for virus research, with intracerebral inoculations of rabies and yellow fever viruses. By using graded dilutions and large numbers of mice, quantitative animal virology had begun. The next important step, initiated by E. W. Goodpasture in 1931–32, was the use of chick embryos for growing poxviruses. This was followed a few years later by the demonstration by F. M. Burnet that many viruses could be titrated by counting the pocks that they produced on the chorioallantoic membrane, whereas others grew well in the allantoic and/or amniotic cavities of developing chick embryos.

Tissue cultures had first been used for cultivating vaccinia virus in 1928, but it was the discovery by J. F. Enders, T. H. Weller, and F. C. Robbins of Harvard University (Nobel Prize, 1954) in 1949 that poliovirus would grow in non-neural cells that gave a tremendous stimulus to the use of cultured cells in virology. Over the next few years their use led to the cultivation of medically important viruses such as those causing measles (by J. F. Enders and T. C. Peebles in 1954) and rubella (by T. H. Weller and F. C. Neva in 1962). Even more dramatic was the isolation of a wide variety of new viruses, belonging to many different families. Also the different cytopathic effects produced by different viruses in monolayer cell cultures were found to be diagnostic.

The next great advance, which greatly increased the accuracy of quantitative animal virology, occurred in 1952, when the plaque assay method for counting phages was adapted to animal virology by R. Dulbecco (Nobel Prize, 1975), using a monolayer of chick embryo cells in a petridish. In 1958 H. Temin and H. Rubin applied Dulbecco's method to Rous sarcoma virus, initiating quantitative studies of tumor viruses. Biochemical studies of animal virus replication were simplified by using continuous cell lines and by growing the cells in suspension.

Biochemistry

During the 1950s virus particles were thought to be 'inert' packages of nucleic acid and proteins, although in 1942 G. K. Hirst of the Rockefeller Institute had shown that particles of influenza virus contained an enzyme, later identified by A. Gottschalk of Melbourne as a neuraminidase.

Further advances in viral biochemistry depended on methods of purification of virus particles, especially the technique of density gradient centrifugation. In 1967 J. Kates and B. R. McAuslan demonstrated the presence of a DNA-dependent RNA polymerase in purified vaccinia virus virions, a discovery that was followed the next year by the demonstration of a double-stranded (ds) RNA-dependent RNA polymerase in reovirus virions and then of single-stranded (ss) RNA-dependent RNA polymerases in virions of paramyxoviruses and rhabdoviruses. In 1970 came the revolutionary discovery by D. Baltimore of Boston (Nobel Prize, 1975) and H. Temin of Wisconsin (Nobel Prize, 1975), independently, of the RNA-dependent DNA polymerase, or reverse transcriptase, of Rous sarcoma virus. Many other kinds of enzymes were later identified in the larger viruses; for example, no less than 16 enzymes have now been identified in vaccinia virus virions.

Just as investigations with bacterial viruses were of critical importance in the development of molecular biology, studies with animal viruses have led to the discovery of several processes that have proved to be important in the molecular biology of eukaryotic cells, although many of these discoveries are too recent for serious historical appraisal. Thus, in 1968 M. Jacobson and D. Baltimore of Boston showed that the genomic RNA of poliovirus was translated as a very large protein, which was then cleaved by a protease that was subsequently found in the viral replicase gene. RNA splicing was discovered in 1977, independently by P. A. Sharp (Nobel Prize, 1993) and L. T. Chow and their respective colleagues during studies of adenovirus replication. Work with the same virus by Rekosh and colleagues in London in 1977 resulted in the definition of viral and cell factors for initiation of new DNA strands using a novel protein priming mechanism. Capping of mRNA by m^7G and its role in translation was discovered in 1974, during work with reoviruses by A. J. Shatkin. Other processes first observed with animal viruses and now known to be important in eukaryotic cells were the existence of 3′ poly A tracts on mRNAs, the pathway for synthesis of cell surface proteins, and the role of enhancer elements in transcription.

Analysis of viral nucleic acids progressed in parallel with that of the viral proteins. Animal viruses were found with genomes of ssRNA, either as a single molecule, two identical molecules (diploid retroviruses), or segmented; dsRNA; ssDNA, dsDNA, and partially dsDNA. The so-called unconventional viruses or prions of scrapie, kuru, and Creutzfeld–Jacob disease, which have been extensively investigated since 1957 by D. C. Gajdusek of the US National Institutes of Health (Nobel Prize, 1976) and S. B. Prusiner of the University of California at San Francisco (Nobel Prize, 1997), appear to be infectious proteins. Soon after the discovery that the isolated nucleic acid of tobacco mosaic virus was infectious (see below), it was shown that the genomic RNAs of viruses belonging to several families of animal viruses were infectious; namely those with single, positive-sense RNA molecules. Then in 1973 came the discovery of recombinant DNA by P. Berg of Stanford University and his colleagues, using the animal virus SV40 and bacteriophage l.

Structure of the Virion

Using more sophisticated methods of electron microscopy, many animal virus virions have been shown to be isometric icosahedral structures, or roughly spherical protein complexes surrounded by a lipid-containing shell, the envelope, which contains a number of virus-coded glycoprotein spikes. X-ray crystallography of crystals of purified isometric viruses has revealed the molecular structure of their icosahedra; similar revealing detail has been obtained for the neuraminidase and hemagglutinin spikes of influenza virus.

Tumor Virology

After the discovery of Rous sarcoma virus in 1911, there was a long interval before the second virus to cause tumors, rabbit papilloma virus, was discovered by R. E. Shope in 1933. However, after that viruses were found to cause various neoplasms in mice, and in 1962 J. J. Trentin of Yale University showed that a human adenovirus would produce malignant tumors in hamsters. Since then, direct proof has been obtained that several DNA viruses (but not adenoviruses) and certain retroviruses can cause tumors in humans. Molecular biological studies have shown that oncogenicity is largely caused by proteins they produce that are encoded by viral oncogenes, a concept introduced by R. J. Huebner and G. Todaro of the US National Institutes of Health in 1969 and corrected, refined, and greatly expanded since the mid-1970s by J. M. Bishop, H. Varmus (Nobel Prize, 1989), and R. A. Weinberg of Boston.

Only one group of RNA viruses, the retroviruses, which replicate through an integrated DNA provirus, cause neoplasms, whereas viruses of five groups of DNA viruses are tumorigenic. Study of these oncogenic viruses has shed a great deal of light on the mechanisms of carcinogenesis. Oncogenic DNA viruses contain oncogenes as an essential part of their genome, which when integrated into the host cell DNA may promote cell transformation, whereas the oncogenes of retroviruses are derived from proto-oncogenes of the cell.

Impact on Immunology

Immunology arose as a branch of microbiology and several discoveries with animal viruses were important in the development of important concepts in immunology. It was the discovery in the 1930s by E. Traub in Tübingen of persistent infection of mice with lymphocytic choriomeningitis virus that led to the development by F. M. Burnet (Nobel Prize, 1960) of the concept of immunological tolerance in 1949. It was work with the same virus that led to the discovery of MHC restriction by P. C. Doherty and R. Zinkernagel in Canberra (Nobel Prize, 1996) in 1974.

Vaccines and Disease Control

From the time of Pasteur's use of rabies vaccine in 1885, a major concern of medical and veterinary virologists has been the development of vaccines. Highlights in this process have been the development of the 17D strain of yellow fever virus by M. Theiler (Nobel Prize, 1951) of the Rockefeller Foundation laboratories in 1937, the introduction of influenza vaccine in 1942, based on Burnet's 1941 discovery that the virus would grow in the allantois, the licensing in 1954 of inactivated poliovaccine developed by J. Salk and in 1961 of the live poliovaccine introduced by A. B. Sabin, both based on the cultivation of the virus in 1949 by Enders, and development of the first genetically engineered human vaccine with the licensing of yeast-grown hepatitis B vaccine in 1986. Finally, 1977 saw the last case of natural smallpox in the world and thus the first example of the global eradication of a major human infectious disease, the result of a 10-year campaign conducted by the World Health Organization with a vaccine directly derived from Jenner's vaccine, first used in 1796.

Recognition of HIV-AIDS

In 1981 a new disease was described in the male homosexual populations of New York, San Francisco, and Los Angeles. It destroyed the immune system, and the disease was called the acquired immune deficiency syndrome (AIDS) and the causal virus, human immunodeficiency virus (HIV), a retrovirus, was first isolated by L. Montagnier, of the Pasteur Institute in Paris, in 1983. A sexually transmitted disease, it has now spread all over the world, with about 40 million cases worldwide, and without very expensive chemotherapy is almost always fatal. It is now clear that it arose from an inapparent but persistent infection of chimpanzees in Africa. In spite of enormous efforts over the past 20 years, it has so far been impossible to devise an effective vaccine.

Arthropod-Borne Viruses of Vertebrates

Arthropods, mainly insects and ticks, were early shown to be important as vectors of virus diseases of vertebrates (the arboviruses) and of plants. Sometimes, as in myxomatosis, carriage was found to be mechanical, but in most cases the virus was found to multiply in the vector as well as the vertebrate concerned. The development of methods of growing insect cells in culture by T. D. C. Grace in Canberra in 1962 opened the way for the molecular biological investigation of the replication of insect viruses and arboviruses in invertebrate cells.

Taxonomy and Nomenclature

After earlier tentative efforts with particular groups of viruses, viral nomenclature took off in 1948, with the production of a system of latinized nomenclature for all viruses by the plant virologist F. O. Holmes. This stimulated others, in particular C. H. Andrewes of London and A. Lwoff of Paris, to actions which resulted in the setting up of an International Committee on Nomenclature (later Taxonomy) of Viruses in 1966. Adopting the kind of viral nucleic acid, the strategy of replication, and the morphology of the virion as its primary criteria, this committee has now achieved acceptance of its decisions by the great majority of virologists. One interesting feature of its activities is that its rules avoid the controversies about priorities that plague taxonomists working with fungi, plants, and animals.

The Future

The future will see an explosive expansion of understanding and knowedge of viruses of vertebrates, with the application of techniques of genetic engineering, nucleic acid sequencing, the polymerase chain reaction, and the use of monoclonal antibodies. All these discoveries are addressed in the appropriate sections of this encyclopedia. Molecular biology, which was initially conceived during studies of bacterial viruses, is now being used to study all the viruses of vertebrates and the pathogenesis and epidemiology of animal viral diseases. The expansion in knowledge of these subjects in the next decade can be expected to exceed that of the previous century.

See also: History of Virology: Bacteriophages; History of Virology: Plant Viruses; Nature of Viruses.

Further Reading

Bos L (1995) The embryonic beginning of virology: Unbiased thinking and dogmatic stagnation. *Archives of Virology* 140: 613.

Cairns J, Stent GS, and Watson JD (eds.) (1962) *Phage and the Origins of Molecular Biology.* Cold Spring Harbor, NY: Cold Spring Harbor Laboratory of Quantitative Biology.

Fenner F and Gibbs AJ (eds.) (1988) *Portraits of Viruses. A History of Virology.* Basel: Karger.

Horzinek MC (1995) The beginnings of animal virology in Germany. *Archives of Virology* 140: 1157.

Hughes SS (1977) *The Virus. A History of the Concept.* London: Heinemann Educational Books.

Joklik WK (1996) Famous institutions in virology. The Department of Microbiology, Australian National University and the Laboratory of Cell Biology, National Institute of Allergy and Infectious Diseases. *Archives of Virology* 141: 969.

Matthews REF (ed.) (1983) *A Critical Appraisal of Viral Taxonomy.* Boca Raton, FL: CRC Press.

Porterfield JS (1995) Famous institutions in virology. The National Institute of Medical Research, Mill Hill. *Archives of Virology* 141: 969.

van Helvoort T (1994) History of virus research in the twentieth century: The problem of conceptual continuity. *History of Science* 32: 185.

Waterson AP and Wilkinson L (1978) *An Introduction to the History of Virology.* Cambridge: Cambridge University Press.

Nature of Viruses

M H V Van Regenmortel, CNRS, Illkirch, France

The Discovery of Viruses

By the end of the nineteenth century, it had been established that many infectious diseases of animals and plants were caused by small microorganisms which could be visualized in the light microscope and could be cultivated as pure cultures on synthetic nutritive media. However, in the case of several infectious diseases, it had not been possible to identify the causative agent in spite of many carefully executed experiments. These results suggested that some other type of infectious agent existed which was not identifiable by light microscopy and could not be cultured on conventional bacteriological media.

The first evidence that such infectious agents, later called viruses, were different from pathogenic bacteria was obtained by filtration experiments with porcelain Chamberland filters used for sterilization. These bacteria-retaining filters, which had been used by Louis Pasteur to remove pathogenic microorganisms from water, were used by Dmitri Ivanovsky in St. Petersburg in 1892 in his study of the tobacco mosaic disease. Ivanovky showed that when sap from a diseased tobacco plant was passed through the filter, the filtrate remained infectious and could be used to infect other tobacco plants. Although Ivanovsky was the first person to show that the agent causing the tobacco mosaic disease passed through a sterilizing filter, all his publications show that he did not grasp the significance of his observation and that he remained convinced that he was dealing with a small bacterium rather than with a new type of infectious agent. More than 10 years after his initial experiment he still believed that the filter he used might have had fine cracks which allowed small spores of a microorganism to pass through it.

The same filtration experiment was repeated on 1898 by Martinus Beijerinck in Delft, Holland, who, unaware of Ivanovsky's papers, again showed that the filtered tobacco sap from diseased plants was infectious. However, Beijerinck went further and demonstrated that the infectious agent was able to diffuse through several millimeters of an agar gel. From this, he concluded that the infection was not caused by a microbe but by what he called a *contagium vivum fluidum* or contagious living liquid. He established that the agent could reproduce itself in a tobacco plant and called it a virus. In the same year, a similar type of filtration experiment was done by Friedrich Loeffler and Paul Frosch who were investigating the important foot-and-mouth disease of cattle. These German investigators reported that although the causative agent of the disease passed through a Chamberland-type filter, it did not go through a Kitasato filter which had a finer grain than the Chamberland filter. From this result, they concluded that the causative virus, which was multiplying within the host, was a corpuscular particle and not a soluble agent as claimed by Beijerinck. Within a few years, it became generally accepted that filterable viruses that remained invisible in the light microscope represented a new class of pathogenic agent different from bacteria.

Although all historical accounts of the beginnings of virology mention the work of Ivanovsky, Beijerinck, and Loeffler, there is disagreement among various authors about who should be credited with the discovery that viruses were a new type of infectious agent. This is an interesting debate since it concerns the nature of what is a scientific discovery. It is indeed not sufficient to make a novel observation such as the filterability of an infectious agent but it is also necessary to interpret the observation correctly and to grasp the significance of an unexpected experimental finding. Ivanovsky was the first to observe the filterability of a virus but he did not recognize that he was dealing with a new type of infectious agent. Beijerinck realized that he was dealing with something different from a microbe but he thought that the virus was an infectious liquid rather than a small corpuscular particle. Only Loeffler correctly concluded that the virus causing foot-and-mouth disease was a small particle stopped by a fine-grain Kitasato filter and he therefore came closest to the modern concept of a virus. It was only the work of William Elford with graded collodion membranes, done more than 30 years later, which established that different viruses had particle diameters in the range of 20–200 nm. Only with the advent of the electron microscope was it finally possible to determine the actual morphology of virus particles.

Viruses as Chemical Objects

The perception of what viruses are changed dramatically in 1935 when Wendel Stanley, working at the Rockefeller Institute in Princeton, showed that tobacco mosaic virus (TMV), the agent studied by Ivanovsky and Beijerinck, could be crystallized in the form of two-dimensional paracrystals. This led to the view that viruses were actually chemical objects rather than organisms and it stimulated an intense interest in viruses which now seemed to be entities at the borderline between chemistry and biology. Many scientists became fascinated with viruses and viewed them as 'living molecules' since they seemed to be able to reproduce themselves. The needle-shaped crystals of TMV, visualized by Stanley, suggested that a dead protein could be a living infectious agent and it was believed by some that viruses might hold the key to the origin of life. Stanley had initially reported that TMV was a pure protein, but in 1936 Frederick Bawden and Norman Pirie in Britain showed that TMV also contained phosphorus and carbohydrate and was in fact an RNA-containing nucleoprotein. However, it took nearly another 20 years before it was established that it was the RNA in the virus that was the infectious entity.

Whether or not viruses should be regarded as living organisms has been regarded by some to be only a matter of taste. A definite answer to this question requires that one has a clear understanding of what is meant by 'life'.

What Is Life?

Life is not a material entity, nor a force, nor a property, but a conceptual object made up of the collection of all living systems, past, present, and future. All living systems possess the property of 'being alive' and the concept 'life' corresponds to the abstract, mental representation of this property. Philosophers say that 'life' is the extension of the predicate 'is alive', the extension of a concept being the objects that the concept refers to. Instead of analyzing the concept 'life', it is thus more relevant to ask which characteristics of biological organisms give them the property of being alive. One needs to ascertain to what objects the concept of 'life' refers to and thus to provide an answer to the question: what is a living organism?

What Is an Organism?

A simple answer to this question would be to say that organisms are living agents but this is unsatisfactory for several reasons, one of them being that it would rule out dead organisms. A better answer would be to say that organisms are living agents at some point of their existence. The related claim that all living agents are organisms is equally problematic since organs like hearts or kidneys are clearly not organisms although they are considered to be alive. Equating organisms with living agents is thus not satisfactory and these two concepts need to be differentiated.

A useful approach is to consider the class of living agents as a cluster concept. Such a concept is defined by a cluster of properties, the majority of which have to be present in all members of the class although some properties can be absent in individual members. Many of the properties tend to be present simultaneously because of the existence of underlying relationships between them: this means that if an individual member possesses any one of these properties, it increases the probability that it also possesses some of the others. Such a cluster concept, known as a polythetic class, is useful to group biological entities that show inherent variation and it has also been used to define virus species.

Living agents can be defined by a cluster of properties that include:

1. compositional and structural properties such as the presence of nucleic acids and proteins, and of heterogenous and specialized parts;
2. functional properties such as the capacity to grow and develop, to reproduce, and to repair themselves; and
3. properties such as metabolism and environmental adaptation that arise from the interaction of a living agent with its environment.

Biosystems interact selectively with the environment through a membrane boundary which restricts the types of exchanges that can occur between components of the system and items in its environment. The environment of a system consists of those things that can be influenced by the system or that may act upon the system. The notion of environment is thus limited to the immediate environment of the system and it exists only relative to a given system.

Any agent in order to be living must necessarily possess a sufficient subset of this cluster of properties, although it need not possess them all. For instance, sterile organisms that do not reproduce or plant seeds with a completely dormant metabolism can be included in the class of living agents.

In addition to having many of the properties listed above, living agents in order to qualify as organisms must also belong to a reproductive lineage characterized by a life cycle. Organs, therefore, are not organisms since they do not reproduce themselves as members of a lineage. Organs are replicated when the entire organism reproduces but they lack a life cycle. The various entities that comprise the life cycle of an organism correspond to developmental stages in the life of that organism.

Finally, in order to qualify as organisms, living agents must also possess a functional autonomy that allows them to exercise control over themselves and to be at least partly independent from other organisms and environmental influences. Organs, tissues, or the leaves of a plant, for instance, are living things but they are not organisms since they are not functionally autonomous, their life being dependent on that of the organisms they belong to.

It should be clear from the preceding discussion that things such as DNA molecules and other biochemical constituents of cells, as well as organelles such as ribosomes are neither living agents nor organisms. For the same reasons, viruses, although they are biological systems, are neither living agents nor *a fortiori* microorganisms.

Viruses Should Not Be Confused with Virus Particles or Virions

A virus has both intrinsic properties such as the size of the virus particle and relational properties such as having a host or vector, the second type of property existing only by virtue of a relation with other objects. Relational properties are also called emergent properties because they are possessed only by the viral system as a whole and are not present in its constituent parts. During its multiplication cycle, a virus takes on various forms, for instance, as a replicating nucleic acid in the host cell. One stage of what is metaphorically called the 'life cycle' of a virus corresponds to the virus particle or virion which can be characterized by intrinsic, structural, and chemical properties such as size, mass, chemical composition, and sequence of both coat proteins and nucleic acids.

Compared to a virion, a virus also possesses a number of relational properties that become actualized during transmission and infection, for instance, when the virus becomes integrated into the host cell during the viral replication cycle. A virus cannot be thus reduced to the physical constituents and chemical composition of a virion and it is necessary to include in its description the various biotic interactions and functional activities that make the virus a biological system. Confusing 'virus' with 'virion' is somewhat similar to confusing the entity 'insect', which includes several different life stages, with a single one of these stages such as a pupa, a caterpillar, or a butterfly.

When Eckhard Wimmer of Stony Brook University in New York gives the chemical formula of poliovirus as $C_{332,652}$ $H_{492,288}$ $N_{98,245}$ $O_{131,196}$ $P_{7,501}$ $S_{2,340}$, he provides only the chemical composition of one poliovirus particle. However, this reductionist chemical description of a virion does not amount to a characterization of the entity poliovirus.

The chemical composition or even the sequence of the coat protein of the virion does not give information on the tertiary and quaternary structural elements in the outer capsid that form the three-dimensional assembly of atoms that will be recognized by host cell receptors, allowing the virus to infect its host. The receptor-binding site of the virion is actually a relational, emergent biological entity defined by its ability to be recognized by the complementary receptor molecules present in certain host cells. The viral receptor-binding site is thus not an intrinsic feature of the virion that could be defined independently of an external cellular receptor since its identity as a site depends on the existence of a specific relationship with a host cell. This relationship arose during the process of biological evolution that culminated in the capacity of the virus to infect certain host cells. Such a specific functional relationship with a host is the essential feature that gives viruses their unique status as molecular genetic parasites capable of using cellular systems for their own replication.

The Nature of Viruses

Although viruses are not living organisms, they are considered to be biological entities because they possess some of the properties of living systems such as having a genome and being able to adapt to particular hosts and biotic habitats. However, viruses do not possess many of the essential attributes of living organisms such as the ability to capture and store free energy and they lack the characteristic autonomy that arises from the presence of integrated, metabolic activities. Viruses do not replicate or self-replicate themselves, but are 'being replicated' (i.e., passively rather than actively) through the metabolic activities of the cells they have infected. The replication of viruses involves a process of copying done by certain constituents of host cells. In contrast, cells do not replicate but reproduce themselves by a process of fission.

A virus becomes part of a living system only after its genome has been integrated in an infected host cell and it is not more alive than other cellular constituents such as genes, macromolecules, and organelles. The consensus among biologists is that the simplest system that can be said to be alive is a cell. Viruses are thus nonliving, infectious agents which can be said, at best, to lead a kind of borrowed life. Duncan McGeoch has described viruses metaphorically as "mistletoes on the Tree of Life."

The difference between viruses and various types of organisms becomes obvious when the functional roles of the proteins found in viruses and in organisms are compared. When proteins are divided into three broad functional categories corresponding to energy utilization, information carriers, and communication mediators, the proportion of each protein class found in viruses is markedly different from that found in living organisms.

Viruses have the highest proportion of proteins involved in information processes related to the control and expression of genetic information but have very few proteins of the energy and communication classes (**Figure 1**). This distribution is due to the fact that viruses utilize the metabolic machinery of the host cell and rely entirely on the energy supply systems of the host they infect. In contrast, bacteria have the highest proportion of proteins of the energy class involved in small-molecule transformation, whereas animals have a high proportion of proteins involved in intra- and intercellular communication.

Some authors have argued that viruses are living microorganisms because they share with certain parasitic organisms the property of being obligate parasites. However, the way viruses depend on their cellular hosts for replication is a type of molecular parasitism that is totally different from the metabolic dependency shown, for instance, by rickettsia or by bacteria that colonize the gut of certain animals. Obligate parasitism on its own is clearly not a sufficient criterion for establishing that an entity is a living organism.

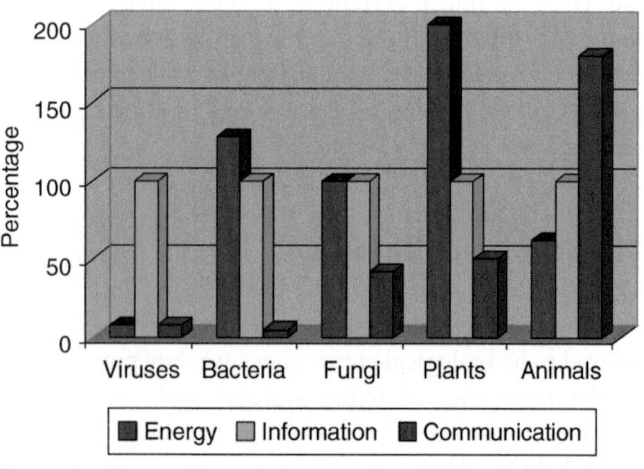

Figure 1 The proteins of viruses, microorganisms, plants, and animals have different functional roles. The vertical bars represent the proportion of proteins in the categories of energy utilization and carrier of information, relative to those in the category of mediator of information. Viruses have the highest proportion of proteins involved in information processes related to the control and expression of genetic information. Reproduced from Patthy L (1999) *Protein Evolution*. Oxford: Blackwell Science, with permission from Blackwell Publishing.

Viruses are subcellular infectious agents which at one stage of their replication cycle in the infected cell are reduced to their nucleic acid component. The role of the virion is to allow transmission to new hosts while it also protects the viral genome from degradation by nucleases and from other environmental attack.

It is customary to distinguish acute viral infection which is associated with active replication and production of virions from the asymptomatic type of specific virus–host relationship characterized by latency, persistence, and absence of disease. In the latter case the viral genome is maintained in the host but no virions are produced for long periods of time and no antiviral host immune response is elicited. Small amounts of virions are produced episodically which is sufficient for transmission of the virus to new hosts. If the virus happens to switch to a new type of host, this may involve a changeover from no disease in the latent host to the appearance of disease following reactivation in the new host. This host-switching phenomenon is sometimes responsible for the appearance of new emerging viral diseases.

Fitness and Viral Evolution

Fitness is usually taken to be the property of an organism that ensures its survival and reproductive success in a variable and unpredictable environment. Since viruses are not living organisms, the fitness of a virus is usually measured in terms of virus progeny, that is, the number of virions produced during an acute virus infection. In the case of a latent, persistent viral infection, viral fitness has no equivalent quantitative definition. Whereas for acute virus infections, fitness is measured at the expense of the host (since infected cells die), in the case of latent infections, fitness of the virus merges with fitness of the host and the concept loses its usefulness for describing a differential capacity of the virus.

Even in the case of living organisms, the concept of fitness is of limited value for explaining Darwinian evolution. Natural selection has been defined as the process whereby organisms of a given variety outnumber those of other varieties and prevail in the long run, mainly due to their greater fertility and adaptness to changing environments. In fact, selection amounts only to differential survival and since fitness is defined as anything that promotes the chances of survival, survival of the fittest and natural selection amounts to no more than survival of the survivors.

Because of the small number of phenotypic characters that can be studied with viruses, virus evolution has always been difficult to study. Most studies have concentrated on phylogenetic analyses based on the comparison of genome sequences, with the expectation that genes will diverge in sequence as they evolve from a common ancestor. Like other genomes, viral genomes evolve through such mechanisms as mutation, recombination, and gene reassortment. Molecular analysis of viral evolution has concentrated on measurements of variation in the genotype, without much clarity about what this variation means in terms of survival value for the virus. In many cases, it seems that viruses may in fact have co-evolved with their hosts or vectors, due to the maintenance of host- or vector-restricted molecular constraints on otherwise much higher rates of evolution of the virus itself.

Viral Isolates, Strains, and Serotypes

The term virus isolate refers to any particular virus culture that is being studied and it is thus simply an instance of a given virus.

A viral strain is a biological variant of a virus that is recognizable because it possesses some unique phenotypic properties that remain stable under natural conditions. Characteristics that allow strains to be recognized include (1) biological properties such as a particular disease symptom or a particular host, (2) chemical or antigenic properties, and (3) the genome sequence when it is known to be correlated with a unique phenotypic character. If the only difference between a 'wild type' virus taken as reference and a particular variant is a small difference in genome sequence, such a variant or mutant is not given the status of a separate strain in the absence of a distinct phenotypic characteristic.

Strains that possess unique, stable antigenic properties are called serotypes. Serotypes necessarily also possess unique structural, chemical, and genome sequence properties that are related to the differences in antigenicity. Serotypes constitute stable replicating lineages which allow them to remain distinct over time. The infectivity of individual serotypes of animal viruses can be neutralized only by their own specific antibodies and not by antibodies directed to other serotypes. This inablility of serotype-specific antibodies to cross-neutralize other serotypes is important in the case of animal viruses that are submitted to the immunological pressure of their hosts.

See also: Origin of Viruses; Principles of Virion Structure; Virus Species.

Further Reading

Grafe A (1991) *A History of Experimental Virolgy.* Heidelberg: Springer.
Mahner M and Bunge M (1997) *Foundations of Biophilosophy.* Berlin, Springer.
Patthy L (1999) *Protein Evolution.* Oxford: Blackwell Science.
Van Regenmortel MHV (2003) Viruses are real, virus species are man-made, taxonomic constructions. *Archives of Virology* 148: 2481–2488.
Van Regenmortel MHV (2007) The rational design of biological complexity. A deceptive metaphor. *Proteomics* 7: 965–975.
Villarreal LP (2005) *Viruses and the Evolution of Life.* Washington, DC: ASM Press.
Waterson AP and Wilkinson L (1978) *An Introduction to the History of Virology.* Cambridge: Cambridge University Press.
Wilson RA (2005) *Genes and the Agents of Life.* Cambridge: Cambridge University Press.
Witz J (1998) A reappraisal of the contribution of Friedrich Loeffler to the development of the modern concept of virus. *Archives of Virology* 143: 2261–2263.

Origin of Viruses

P Forterre, Institut Pasteur, Paris, France

Glossary

Archaea A domain of prokaryotic microorganisms whose informational mechanisms (DNA replication, transcription, and translation) are closely related to those of eukaryotes.

Homology Two biological structures are homologs if they originated from a common ancestral structure. The homology cannot be quantified (a protein is or is not homologous to another one). Many biologists without an evolutionary background still confuse homology and similarity (only the latter can be quantified).

LUCA (Last Universal Common Ancestor) The most recent common ancestor shared by all modern cellular organisms. The modern genetic code was already established in LUCA. Other features of LUCA (cellular vs. acellular, RNA vs. DNA genome) are still highly controversial.

RNA world The period of life evolution before the appearance of DNA. Depending on the authors, the RNA world is viewed as either cellular (a world of RNA cells) or acellular (a world of free-living macromolecules).

Universal tree of life The tree based on 16S/18S rRNA comparison, in which the cellular living world is divided into three domains: Archaea, Bacteria, and Eukarya. The evolutionary relationships between these domains are controversial.

The Classical View of Virus Origin and Its Consequences

The origin of viruses and their evolutionary relationships with cellular organisms are still enigmatic, but recent advances from comparative genomics and structural biology have produced a new framework to discuss these issues on firmer grounds. Historically, three hypotheses have been proposed to explain the origin of viruses: (1) they originated in a precellular world ('the virus-first hypothesis'); (2) they originated by reductive evolution from parasitic cells ('the reduction hypothesis'); and (3) they originated from fragments of cellular genetic material that escaped from cell control ('the escape hypothesis'). All these hypotheses had specific drawbacks. The virus-first hypothesis was usually rejected firsthand, since all known viruses require a cellular host. The reduction hypothesis was difficult to reconcile with the observation that the most reduced cellular parasites in the three domains of life, such as Mycoplasma in Bacteria, Microsporidia in Eukarya, or Nanoarchaea in Archaea, do not look like intermediate forms between viruses and cells. Finally, the escape hypothesis failed to explain how such elaborate structures as complex capsids and nucleic acid injection mechanisms evolved from cellular structures, since we do not know any cellular homologs of these crucial viral components.

Because of these drawbacks, the problem of virus origin was for a long time considered untractable and not worth of serious consideration (similarly, the study of bacterial evolution was considered a hopeless and futile task prior to the pioneering work of Carl Woese). However, since the problem of the origin is so entrenched in the human mind, it was never completely ignored. Much like the concept of prokaryotes became the paradigm on how to think about bacterial evolution, the escape hypothesis became the paradigm favored by most virologists to solve the problem of virus origin. This scenario was chosen mainly because it was apparently supported by the observation that modern viruses can pick up genes from their hosts. In its classical version, the escape theory suggested that bacteriophages originated from bacterial genomes and eukaryotic viruses from eukaryotic genomes (**Figure 1(a)**). This led to a damaging division of the virologist community into those studying bacteriophages and those studying eukaryotic viruses, 'phages' and viruses being somehow considered to be completely different entities. The artificial division of the viral world between 'viruses' and bacteriophages also led to much confusion on the nature of archaeal viruses. Indeed, although most of them are completely

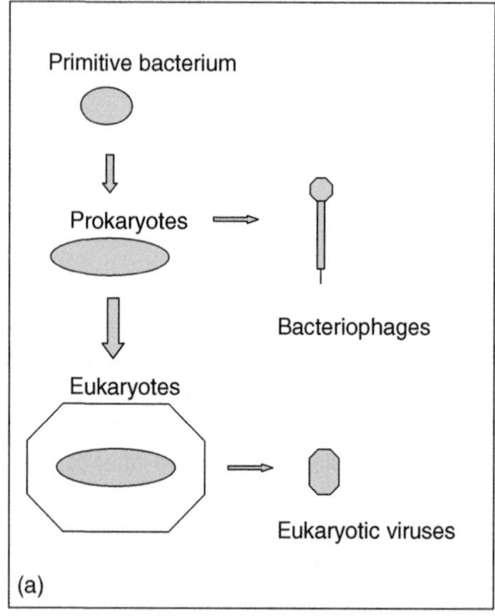

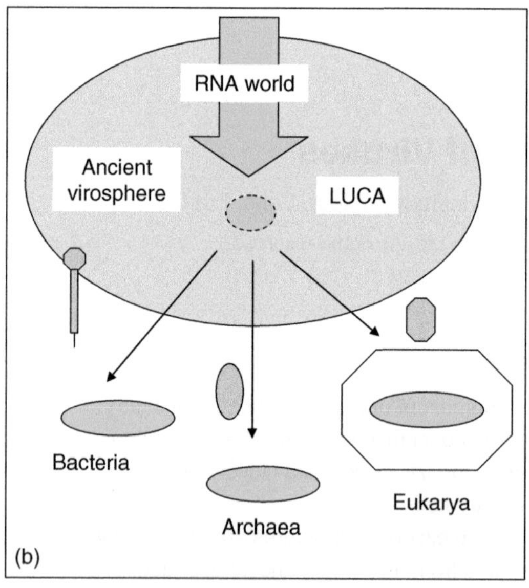

Figure 1 Two conflicting views of virus origin and evolution.

unrelated to bacterial viruses, they are often called 'bacteriophages', since archaea (formerly archaebacteria) are still considered by some biologists as 'strange bacteria'. For instance, archaeal viruses are grouped with bacteriophages in the drawing that illustrates viral diversity in the last edition of the *Virus Taxonomy Handbook.* Hopefully, these outdated visions will finally succumb to the accumulating evidence from molecular analyses.

Viruses Are Not Derived from Modern Cells

Abundant data are now already available to discredit the escape hypothesis in its classical adaptation of the prokaryote/eukaryote paradigm. This hypothesis indeed predicts that proteins encoded by bacterial viruses (avoiding the term bacteriophage here) should be evolutionarily related to bacterial proteins, whereas proteins encoded by viruses infecting eukaryotes should be related to eukaryotic proteins. This turned out to be wrong since, with a few exceptions (that can be identified as recent transfers from their hosts), most viral encoded proteins have either no homologs in any cell or only distantly related homologs. In the latter cases, the most closely related cellular homolog is rarely from the host and can even be from cells of a domain different from the host. More and more biologists are thus now fully aware that viruses form a world of their own, and that it is futile to speculate on their origin in the framework of the old prokaryote/eukaryote dichotomy. As for all other aspects of microbiology, the problem of the nature and origin of viruses made a great leap forward when this dichotomy was successfully challenged by the trinity concept introduced by Carl Woese, that is, the division of the living cellular world into three domains, Archaea, Bacteria, and Eukarya.

The building of a universal tree of life based on rRNA sequence comparisons, and the idea that all living organisms did not diverge from a 'primitive bacterium', as previously assumed in most texbooks, but from a less-defined last universal common ancestor (LUCA) opened the way to think about the origin and evolution of viruses in a new context (**Figure 1(b)**). Indeed, the main problem with the initial formulation of the three major hypotheses for the origin of viruses was that all of them were based on our knowledge of modern cells (themselves viewed as either prokaryotes or eukaryotes). Hence, modern viruses need modern cells to replicate, modern cells cannot regress to viral forms, and free RNA or DNA does not recruit today's proteins from modern cells to form capsids and other elaborated viral structures. The major innovation introduced by the work of Carl Woese was to open a window on the possibility of ancient worlds populated by cells most likely very different from modern ones. As early as 1980, Woese and co-workers coined the term urkaryote to name the cellular lineage that gave rise to modern eukaryotic cells (prior to the mitochondrial endosymbiosis and possibly before the origin of the nucleus itself). Later on, Woese discussed in detail the fact that the organisms that populated the basal branches of the universal tree were probably not yet modern cells, and suggested that some proteins with puzzling phylogenetic patterns could have originated in ancestral cell lineages that have now disappeared. These theoretical considerations laid the ground to the interpretation of observations that could not have been understood in the classical linear and dichotomic view of cellular evolution, such as the discovery that viruses infecting either prokaryotic or eukaryotic hosts could share homologous features (see below). In the new framework introduced by Woese's universal tree of life, these observations led to the idea that viruses might be relics of lost domains, or else that viruses predated LUCA. As a result,viruses now appear in the literature as additional branches in the universal tree or the universal tree itself is immersed in a 'viral ocean'.

The metaphor of the viral ocean, proposed by Dennis Bamford, illustrates both the concept of virus antiquity and their predominance in the modern biosphere. Another major breakthrough in recent viral research was indeed the realization that viruses are much more abundant than cells and much more diverse than previously suspected. It is thus currently assumed that viral genomes represent the major throve of genetic diversity on Earth. All these trends make it irrelevant to ask whether viruses are alive. As recently pointed out by Jean-Michel Claverie, the question of the nature of viruses has been for a long time obscured by the confusion between virus and virus particle, whereas the major components of the virus life cycle correspond to the intracellular viral factory. All this concurs to put again the question of virus origin on the agenda. Thus a brief summary of the main data that presently point to an ancient origin of viruses and the discussion of how the three major hypotheses explaining the origin of viruses have been rejuvenated in this new framework are presented here.

Viruses Are Ancient

For a long time, virologists thought that the various virus families were evolutionary unrelated, indicating a polyphyletic origin. In recent years, this view has progressively changed with the identification of more and more relationships (sometimes totally unexpected) between different viral lineages, making it possible to define a limited number of large viral groups whose hosts encompass the three cellular domains. For example, it has been clearly established that some double-stranded RNA viruses infecting bacteria are homologous to those infecting eukarya, suggesting that they predated the divergence between bacteria and eukaryotes. Since the RNA replicases/transcriptases of these double-stranded RNA viruses are homologous to those of single-stranded RNA viruses, all RNA viruses presently known seem to be evolutionary related (at least in term of their replication apparatus). In the case of DNA viruses, structural analyses of their capsid proteins have revealed an unexpected close evolutionary relationship between some bacterial spherical viruses, eukaryotic viruses of the nucleocytoplasmic-large-DNA viruses (NCLDV) superfamily, and an archaeal virus isolated at Yellowstone, USA. Structural comparative analyses also indicate that head and tailed bacterial viruses (*Caudavirales*) share homologous features with herpes viruses. These data clearly indicate that some viral specific proteins originated before the divergence between the three domains (hence before LUCA). It is tempting to suggest that a similar ancient origin also explains why most viral proteins either have no cellular homologs (except for plasmid versions or viral remnants in cellular genomes) or are only very distantly related to their cellular homologs. All these considerations thus push back the origin of viruses before the emergence of modern cells. We will now discuss how the three classical hypotheses for the origin of viruses have been revisited in this new context.

The Virus-First Hypothesis

The virus-first hypothesis was for a long time politically incorrect. It clashes with the cellular theory of life and the traditional assumption that viruses are nonliving entities. This hypothesis was first revived in the 1980s by Wolfram Zillig, who suggested that viruses originated in a prebiotic word, using the 'primitive soup' as a host. Such hypothesis has gained strength in recent years, in parallel with the suggestion that cellular organisms originated only at a late stage of life evolution. The idea that 'life' first evolved in an acellular context can be traced back to the first version of the RNA world theory. More recently, it was boosted by the discovery that archaeal lipids are dramatically different from the bacterial ones (with an opposite stereochemistry of the glycerol backbone linkages and a different type of carbon chains). To explain this dichotomy, several authors have proposed that LUCA was not a cellular entity and that cellular membranes originated independently after the divergence of Archaea and Bacteria. A more elaborate version of this scenario has been proposed by William Martin and Eugene Koonin, who suggested that life originated and evolved in the cell-like mineral compartments of a warm hydrothermal chimney. In that model, viruses emerged from the assemblage of self-replicating elements using these inorganic compartments as the first hosts. The formation of true cells occurred twice independently only at the end of the process (and at the top of the chimney), producing the first archaea and bacteria. The latter escaped from the same chimney system as already fully elaborated modern cells. In the model, viruses first co-evolved with acellular machineries producing nucleotide precursors and proteins (**Figure 2(a)**). This acellular 'life' evolved by competition between different machineries and associated viruses to infect more and more compartments of the hydrothermal system.

Cellular versus Acellular Evolution of Early Life

The acellular model of early life evolution (up to LUCA) raises several problems. First, comparative genomics analyses indicate that some membrane proteins (ATP synthetases, signal recognition particle receptors) are homologous and ubiquitous in the three domains of life, hence were probably already present in LUCA. Some authors have further stressed that the emergence of the RNA world involves at least the existence of complex mechanisms to produce ATP, RNA, and proteins. This means an elaborated metabolism to produce ribonucleotide triphosphate (rNTP) and amino acids, RNA polymerases and ribosomes, as well as an ATP-generating system. If such a complex metabolism was present, it appears unlikely that it was unable to produce lipid precursors, hence membranes. If this is correct, then 'modern' viruses did not predate cells, but originated in a world populated by primitive cells. The proponents of this scenario consider that it fits better with the contention that Darwinian selection requires competition between well-defined individual entities. It has been often assumed that RNA viruses are relics of the RNA world. In that case, viruses might have originated in a world of primitive cells with RNA genomes. In that context, it is even possible that cellularization occurred before the emergence of the modern protein synthesizing machinery and that RNA cells existed that contained no proteins (at least no proteins produced by an RNA machinery related to modern ribosomes). Modern viroids may be relics of this stage, whereas true viruses might have only appeared after the establishment of the ribosome-based mechanism for protein synthesis. In such a cellular scenario, now one has to explain how RNA viruses originated from RNA cells. Interestingly, this has led to a revival of the reduction and escape hypothesis, but in a new context.

The Reduction Hypothesis

The reduction hypothesis revisited in the context of pre-LUCA cells posits that RNA viruses originated by reduction from parasitic RNA cells, by losing progressively their own machinery for protein synthesis and for energy

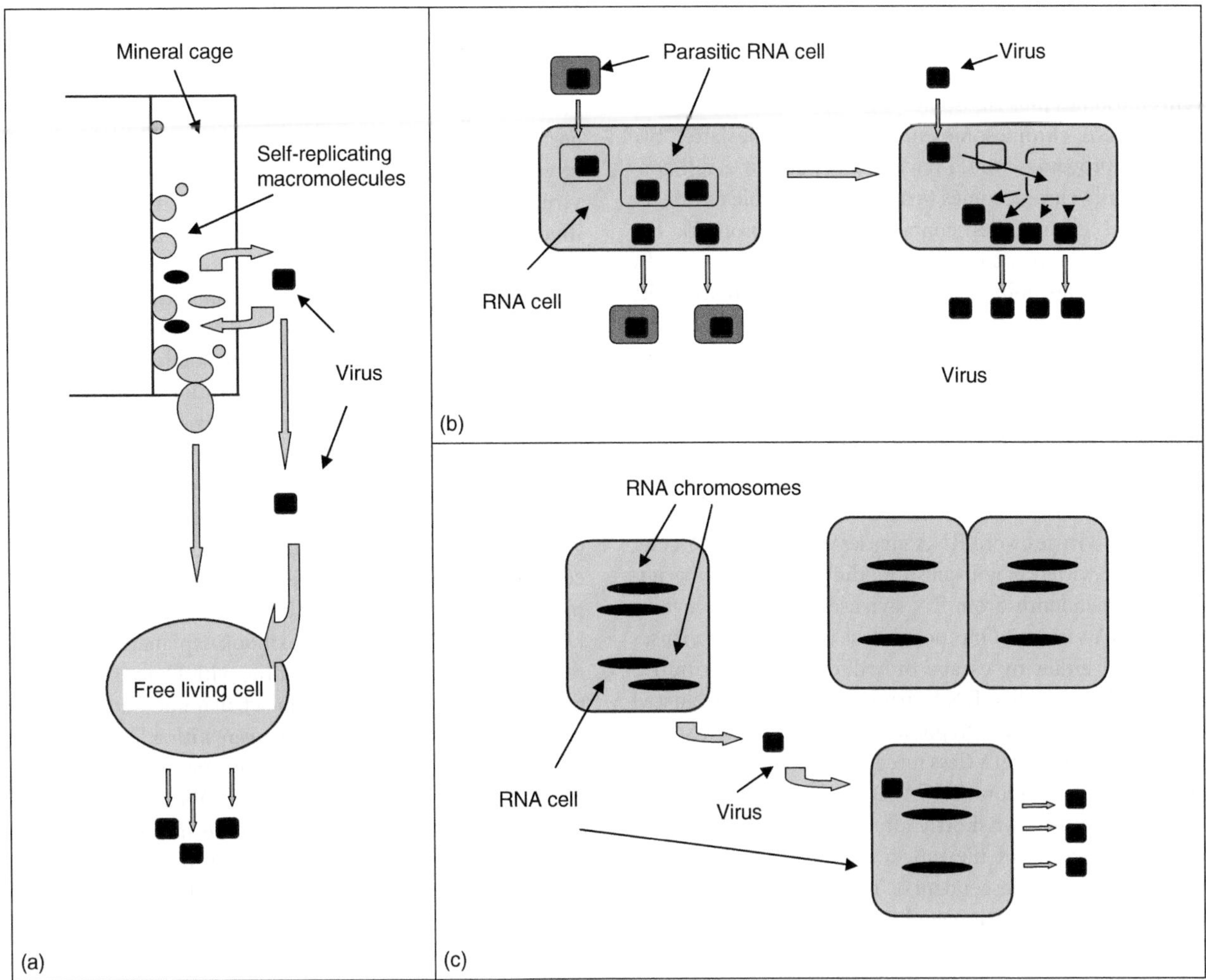

Figure 2 The three revisited hypotheses for virus evolution: (a) the virus-first hypothesis; (b) the reduction hypothesis; (c) the escape hypothesis.

production (**Figure 2(b)**). An analogy for the possible mechanism of reduction can be seen in the reductive evolution that led to modern *Chlamydia.* Indeed, an interesting parallel can be drawn between the viral particles (the virion) and the infectious form of this bacterium (the elementary body) that is small and metabolically inactive. The main difference between viruses and *Chlamydia* is that the intracellular form of the latter, called the 'reticulate body', is a fully developed intracellular bacterium that uses its own ribosomes for protein synthesis, whereas the intracellular viral factories (although often physically separated from the host cytoplasm) have somehow a direct access to the host ribosomes. A first step in the evolution of a parasitic RNA cell toward a viral state might thus have been the division of its cell cycle between an inert extracellular stage (the protovirion) and an intracellular stage. The second step would have been the dissolution of the membrane of the intracellular parasitic cell to gain access to the protein machinery of the host. One can argue that the transformation of an intracellular parasitic cell into a viral-type factory became impossible with modern cells (such as *Chlamydia*) because, as stated by Carl Woese, the latter are too complex and too integrated to be 'deconstructed' into free-living subentities. In contrast, small parasitic RNA cells reproducing inside larger RNA cells were probably much simpler and could have been easily reduced into a viral factory by loss of their own membrane and translation machinery.

The Escape Hypothesis

The escape hypothesis is also easier to defend in the context of an ancestral world of RNA cells. It has been often argued that the genomes of ancestral RNA cells may have been fragmented and composed of semiautonomous chromosomes that were replicated independently and transferred randomly from cell to cell. The coupling between the segregation of the cellular genome and of the cellular machinery for protein synthesis (the genotype and the phenotype) was probably not so efficient in these

RNA cells than it is in modern cells. The reproduction of such primitive RNA cells could have produced a mixture of progenies, some of them containing both systems (chromosomes plus ribosomes) but others containing only either RNA chromosomes or ribosomes. The latter two types of progenies would have died, except if a cell containing only chromosomes turned out to be able to infect a complete cell (or a cell containing only ribosomes). The RNA chromosomes carrying genes facilitating specifically their infectious ability and/or protecting their integrity during their resting stages would have been selected in such a situation (**Figure 2(c)**).

Relationships between RNA and DNA Viruses

If the first viruses were RNA viruses infecting RNA cells, one is left with a major question; did DNA viruses originate independently from RNA viruses or did they evolve from RNA viruses? One possibility is that DNA viruses originated either by escape or reduction from primitive DNA cells, much like RNA viruses could have originated from RNA cells. Such hypothesis supposes the existence of primitive DNA cells (less integrated than modern ones) that lived either before LUCA (if the latter was already a cellular organism) or shortly after LUCA (corresponding to early branches of the universal tree which have disappeared without descendants). The possibility that some large DNA viruses originated by reduction from extinct DNA cells evolutionarily related to early eukaryotic cells was recently boosted by the discovery of the giant mimivirus whose genome size (1.2 Mb) is three times larger than the smallest genomes of parasitic archaea or bacteria. This virus encodes a few components of the translation system that could be relics of ancient cellular lineages now extinct. On the other hand, the huge diversity of DNA viruses suggests that different lineages of DNA viruses could have originated at different periods and by different mechanisms. Indeed, there are arguments to suggest that at least some DNA viruses originated from RNA viruses. In particular, the RNA replicases/transcriptases of RNA viruses are homologous to the reverse transcriptase of retroviruses and to DNA polymerases of the A family encoded by many DNA viruses. Similarly, RNA and DNA viruses encode homologous RNA/DNA helicases. The hypothesis of an evolutionary transition from RNA to DNA viruses could explain the existence of intermediate forms such as retroviruses (with an RNA genome and an RNA–DNA–RNA cycle) and hepadnaviruses (with a DNA genome and a DNA–RNA–DNA cycle). Interestingly, retroviruses and hepadnaviruses are evolutionarily related, suggesting that the transition from RNA to DNA occurred in the virosphere.

Viruses and the Origin of DNA

Considering the possibility that at least some DNA viruses originated from RNA viruses, it has been suggested that DNA itself could have appeared in the course of virus evolution (in the context of competition between viruses and their cellular hosts). Indeed, DNA is a modified form of RNA, and both viruses and cells often chemically modify their genomes to protect themselves from nucleases produced by their competitor. It is usually considered that DNA replaced RNA in the course of evolution simply because it is more stable (thanks to the removal of the reactive oxygen in position 2′ of the ribose) and because cytosine deamination (producing uracil) can be corrected in DNA (where uracil is recognized as an alien base) but not in RNA. The replacement of RNA by DNA as cellular genetic material would have thus allowed genome size to increase, with a concomitant increase in cellular complexity (and efficiency) leading to the complete elimination of RNA cells by the ancestors of modern DNA cells. This traditional textbook explanation has been recently criticized as incompatible with Darwinian evolution, since it does not explain what immediate selective advantage allowed the first organism with a DNA genome to predominate over former organisms with RNA genomes. Indeed, the newly emerging DNA cell could not have immediately enlarged its genome and could not have benefited straight away from a DNA repair mechanism to remove uracil from DNA. Instead, if the replacement of RNA by DNA occurred in the framework of the competition between cells and viruses, either in an RNA virus or in an RNA cell, modification of the RNA genome into a DNA genome would have immediately produced a benefit for the virus or the cell. It has been argued that the transformation of RNA genomes into DNA genomes occurred preferentially in viruses because it was simpler to change in one step the chemical composition of the viral genome than that of the cellular genomes (the latter interacting with many more proteins). Furthermore, modern viruses exhibit very different types of genomes (RNA, DNA, single-stranded, double-stranded), including highly modified DNA, whereas all modern cellular organisms have double-stranded DNA genomes. This suggests a higher degree of plasticity for viral genomes compared to cellular ones. The idea that DNA originated first in viruses could also explain why many DNA viruses encode their own enzymes for deoxynucleotide triphosphate (dNTP) production, ribonucleotide reductases (the enzymes that produce deoxyribonucleotides from ribonucleotides), and thymidylate synthases (the enzymes that produce deoxythymidine monophosphate (dTMP) from deoxyuridine monophosphate (dUMP). Because, in modern cells, dTMP is produced from dUMP, the transition from RNA to DNA occurred likely in two steps, first with the appearance of ribonucleotide reductase and production of U-DNA (DNA containing uracil), followed by the appearance of thymidylate synthases and formation of T-DNA (DNA containing thymine). The existence of a few bacterial viruses with U-DNA genomes has ben taken as evidence that they could be relics of this period of evolution.

If DNA first appeared in the ancestral virosphere, one has also to explain how it was later on transferred to cells. One scenario posits the co-existence for some time of an RNA cellular chromosome and a DNA viral genome (episome) in the same cell, with the progressive transfer of the information originally carried by the RNA chromosome to the DNA 'plasmid' via retro-transposition.

New Hypotheses about the Role of Viruses in the Origin of Modern Cells

The idea that viruses 'invented' DNA implies that they have been major players in the origin of modern cells. Indeed, several provocative hypotheses have been proposed in recent years that put viruses as central players of various evolutionary scenarios In the context of an ancient DNA virosphere, it has been argued that different lineages of DNA viruses would have 'invented' different enzymatic activities to replicate, repair, and recombine their DNA, explaining why the proteins dealing with DNA are now so diverse, often belonging to several nonhomologous protein families. It has thus been proposed that many (possibly all) cellular enzymes involved today in cellular DNA replication, repair, and/or recombination first originated in viruses before being transferred to cells. More specifically, several authors suggested that either the bacterial DNA replication mechanism, the eukaryotic/archaeal ones, or both, are of viral origin, in order to explain why the major proteins of the DNA replication machineries in eukaryotes and archaea (DNA polymerase, helicase, and primase) are not homologous to their functional analogs in bacteria (suggesting that LUCA had still an RNA genome). In order to explain why the archaeal and eukaryotic DNA replication machineries also exhibit some crucial differences (besides a core of homologous proteins), it was even suggested that the three cellular domains (Archaea, Bacteria, and Eukarya) originated from the independent fusions of three RNA cells and three large DNA viruses. In the latter scenario, the replacement of an RNA genome by a DNA genome at the onset of domain formation would have produced a drastic reduction in the rate of protein and rRNA evolution, explaining why proteins evolved apparently much less rapidly after the formation of the three domains than during the period between LUCA and the last common ancestor of each domain. The formation of each domain from three different types of RNA cells could have also selected three groups of RNA and DNA viruses specific for each domain, those that were able to infect these three ancestral RNA cells and their immediate descendants. This could explain a paradox in the modern biosphere, that each domain is characterized by its own set of viruses (for instance, NCLDVs are specific for eukaryotes) despite the fact that these viruses probably originated from a virosphere that predated LUCA. It must be remembered that NCLDVs share homologous capsid proteins with some bacterial and archaeal viruses.

Another area in which evolutionists have now recruited viruses for help is the problem of eukaryote origin. The 'viral eukaryogenesis' hypothesis posits that the eukaryotic nucleus originated from a large DNA virus, possibly related to NCLDV. This hypothesis was inspired by the analogies between the life cycle of the nucleus and those of poxviruses. In particular, both the eukaryotic nucleus and poxviruses build their membrane from the endoplasmic reticulum. Again, the discovery of the mimivirus (a member of the NCLDV family) led credence to such hypothesis. The relationships between the eukaryotic nucleus and giant viruses might have been even more complex. Hence, it was also recently proposed that a bidirectional evolutionary pathway was operating early on, with both large DNA viruses producing nuclei by infecting ancestral proto-eukaryotic cells, and also infectious nuclei producing new large DNA viruses. Finally, it was suggested that several different viruses might have been involved in eukaryogenesis to explain the presence of multiple RNA and DNA polymerases in eukaryotic cells.

Although most hypotheses previously discussed will probably always lack definitive proof, comparative genomics analyses have recently revealed a clear case of viral intervention in the formation of modern eukaryotic cells, that is, the viral origin of the DNA transcription and replication apparatus of mitochondria. This was inferred from the discovery that the RNA polymerase, DNA polymerase, and DNA helicase operating in mitochondria are of viral origin. These enzymes probably originated from a provirus that was integrated into the genome of the α-proteobacterium at the origin of mitochondria, since proviruses encoding homologs of these enzymes have been detected in the genome of several proteobacteria.

Conclusions

The idea that modern viruses are not simple extensions of prokaryotic or eukaryotic cells but derived from an ancient virosphere whose evolution encompassed the RNA world and the period of the RNA-to-DNA transition has far-reaching consequences. One of the most important in terms of practical consequences for all biologists is that modern viruses (and plasmids, which most likely originated from them) would have inherited from this ancient virosphere many molecular mechanisms that have disappeared from modern DNA cells. This would explain why the molecular biology of the viral world for transcription, replication repair, and recombination is more diverse than that of the cellular world (despite the fact that we have only explored a tiny fraction of the modern virosphere). If this view is correct, many still unknown molecular mechanisms (and their associated proteins) remain to be discovered in viruses. The exploration of viral diversity will be for sure one of the major challenges of biology in this new century.

See also: Evolution of Viruses; Nature of Viruses.

Further Reading

Bamford DH, Grimes JM, and Stuart DI (2006) What does structure tell us about virus evolution? *Current Opinion in Structural Biology* 15: 655–663.

Bell PJL (2001) Viral eukaryogenesis: Was the ancestor of the nucleus a complex DNA virus? *Journal of Molecular Evolution* 53: 251–256.

Claverie JM (2006) Virus takes center stage in cellular evolution. *Genome Biology* 7: 1–10.

Fauquet CM, Mayo MA, Maniloff J, Desselberger U, and Ball LA (eds.) (2005) *Virus Taxonomy: Eighth Report of the International Committee on Taxonomy of Viruses.* San Diego, CA: Elsevier Academic Press.

Filée J and Forterre P (2005) Viral proteins functioning in organelles: A cryptic origin? *Trends in Microbiology* 13: 510–513.

Forterre P (2005) The two ages of the RNA world, and the transition to the DNA world, a story of viruses and cells. *Biochimie* 87: 793–803.

Forterre P (2006) Three RNA cells for ribosomal lineages and three DNA viruses to replicate their genomes: A hypothesis for the origin of cellular domain. *Proceedings of the National Academy of Sciences, USA* 103: 3669–3674.

Forterre P and Krish H (2003) Viruses: Origin, evolution and biodiversity. *Research in Microbiology (Special Issue)* 154: 223–311.

Hamilton G (2006) Virology: The gene weavers. *Nature* 441: 683–685.

Koonin EV and Martin W (2005) On the origin of genomes and cells within inorganic compartments. *Trends in Genetics* 21: 647–654.

Ortmann AC, Wiedenheft B, Douglas T, and Young M (2006) Hot crenarchaeal viruses reveal deep evolutionary connections. *Nature Reviews Microbiology* 4: 520–528.

Prangishvili D, Forterre P, and Garrett RA (2006) Viruses of the Archaea: A unifying view. *Nature Reviews Microbiology* 4: 837–848.

Raoult D, Audic S, Robert C, *et al.* (2004) The 1.2-megabase genome sequence of mimivirus. *Science* 306: 1344–1350.

Villarreal LP (2005) *Viruses and the Evolution of Life.* Washington: ASM Press.

Woese CR, Kandler O, and Wheelis ML (1990) Towards a natural system of organisms: Proposal for the domains Archaea, Bacteria, and Eucarya. *Proceedings of the National Academy of Sciences, USA* 12: 4576–4579.

Ribozymes

E Westhof and A Lescoute, Université Louis Pasteur, Strasbourg, France

Glossary

Nucleolytic Qualifies ribozymes which undergo cleavage and ligation following a nucleophilic attack of the 2′ hydroxyl group on the adjacent 3′ phosphate group.

Transesterification Transfer of a phosphoryl group.

Transpeptidation Formation of a peptide bond.

Concatemer Multiple copies of a DNA sequence arranged end to end in tandem.

Introduction

The discovery in the 1980s of RNA molecules with catalytic activity has revolutionized molecular biology. Ribozymes, molecules able to catalyze various chemical reactions, are widespread in biology. Large- and small-size ribozymes can be distinguished. The large ribozymes comprise the group I introns, the group II introns, and ribonuclease P (RNaseP). A step of maturation of pre-mRNAs in some bacteria or in nuclear eukaryotic rRNAs implicates the self-splicing of group I introns which use a guanosine molecule as a cofactor to catalyze a two-step transesterification reaction. The crystal structures of group I introns show that the catalytic core is conserved and stabilized by long-range interactions that involve different peripheral elements. At the same time the structures of the specificity domain of two RNaseP RNAs, responsible for the maturation of pre-tRNA, have been solved and show that the global architecture of the molecule is conserved partly due to the long-range interactions. Most importantly, it was established that the rRNA 23S is responsible for catalyzing the transpeptidation reaction during protein synthesis. Among the simplest catalytic RNA molecules are the small nucleolytic RNA species. The small ribozymes include the Varkud satellite ribozyme (VS ribozyme), the hairpin ribozyme, the hepatitis delta virus (HDV), and hammerhead ribozymes. They carry out reversible cleavage and ligation reactions in the presence of physiological concentrations of magnesium ions. Ribozyme activity is necessary during the replication of some plant pathogen RNA genomes.

The Chemical Reactions

The reaction catalyzed by small nucleolytic ribozymes takes place at a specific phosphodiester bond by a transesterification reaction involving the 2′-hydroxyl group for cleavage and the 5′-hydroxyl group for ligation. The cleavage reaction proceeds by activation of the 2′-hydroxyl adjacent to the scissile phosphate. The 2′-hydroxyl group performs a nucleophilic attack on the adjacent phosphorus atom, which leads to a pentacoordinate transition state. After capturing a proton, the 5′-oxygen atom leaves leading to the formation of a cyclic 2′,3′-phosphate group (**Figure 1**). Ligation is the reverse reaction of cleavage and involves: (1) activation of the 5′-hydroxyl group by a base; (2) attack of the 2′,3′-cyclic phosphate; and (3) liberation and protonation of the 2′-hydroxyl group. Magnesium ions are generally necessary for the native folding of RNA molecules. Whether magnesium ions participate actively or indirectly in the catalysis reaction is still a matter of debate. During recent years, it has become clear, however, that the nucleic acid bases themselves can play an active role in acid–base catalysis. Nowadays, the crucial roles for efficient catalysis of tertiary contacts between peripheral elements, even in small ribozymes, have been firmly established.

The Ribozymes of Plant Viroids and Virusoids

Replication by the Rolling-Circle Mechanism

Two of the autocatalytic RNAs, the hammerhead and hairpin ribozymes, were discovered first in some viroids and virusoids responsible for several economically important infectious diseases of plants. These pathogens are small circular single-stranded RNAs, without a capsid, possess no open reading frame (ORF) and the RNAs are classified as noncoding RNAs (**Figure 2**). Their replication takes place (1) in the cytoplasm with a helper virus in the case of virusoids; (2) in an autonomous way (without helper virus) in chloroplasts, or in the nucleus in the case of viroids. The replication of these circular genomes is made according to a 'rolling-circle' mechanism, which involves the copying of the circular genome positive (+) strand by an RNA polymerase of the host or helper virus to give a multimeric negative-strand RNA (−) (concatemer) (**Figure 3**). In the case of the viroid family *Pospiviroidae*, the synthesized linear concatemers of the (−) strands are used as matrix for the synthesis of (+) concatemers; the circularization of those which form unit genomes seems to require an RNA ligase as indicated by the presence in some instances of a 2′ phosphomonoester and a 3′,5′ phosphodiester bond.

In the family *Avsunviroidae*, the (−) concatemer undergoes an autocleavage reaction involving a hammerhead ribozyme. The mechanism of circularization of the (−) strand to be used as a matrix is still controversial. Indeed, two hypotheses have been proposed: either the intervention of an RNA ligase or the ribozyme catalyzes the opposite reaction and ligates the RNA. In the end, the (−) strand is used as matrix for the synthesis of (+) concatemers by the

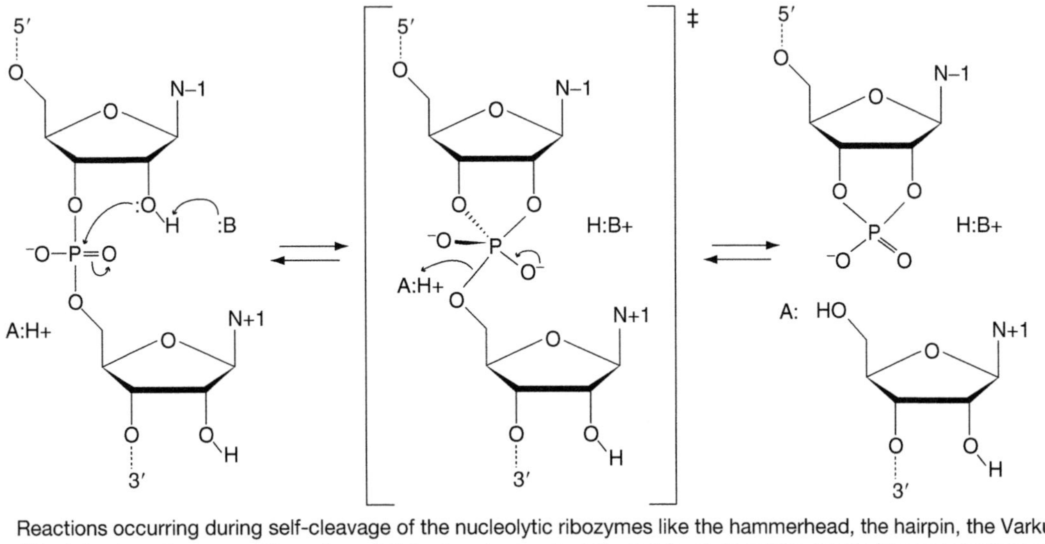

Figure 1 Reactions occurring during self-cleavage of the nucleolytic ribozymes like the hammerhead, the hairpin, the Varkud, or the Hepatitis delta virus ribozymes. The adjacent 2′-hydroxyl group is activated for nucleophilic attack. In a concomitant way, a proton is given to the leaving 5′-oxygen group.

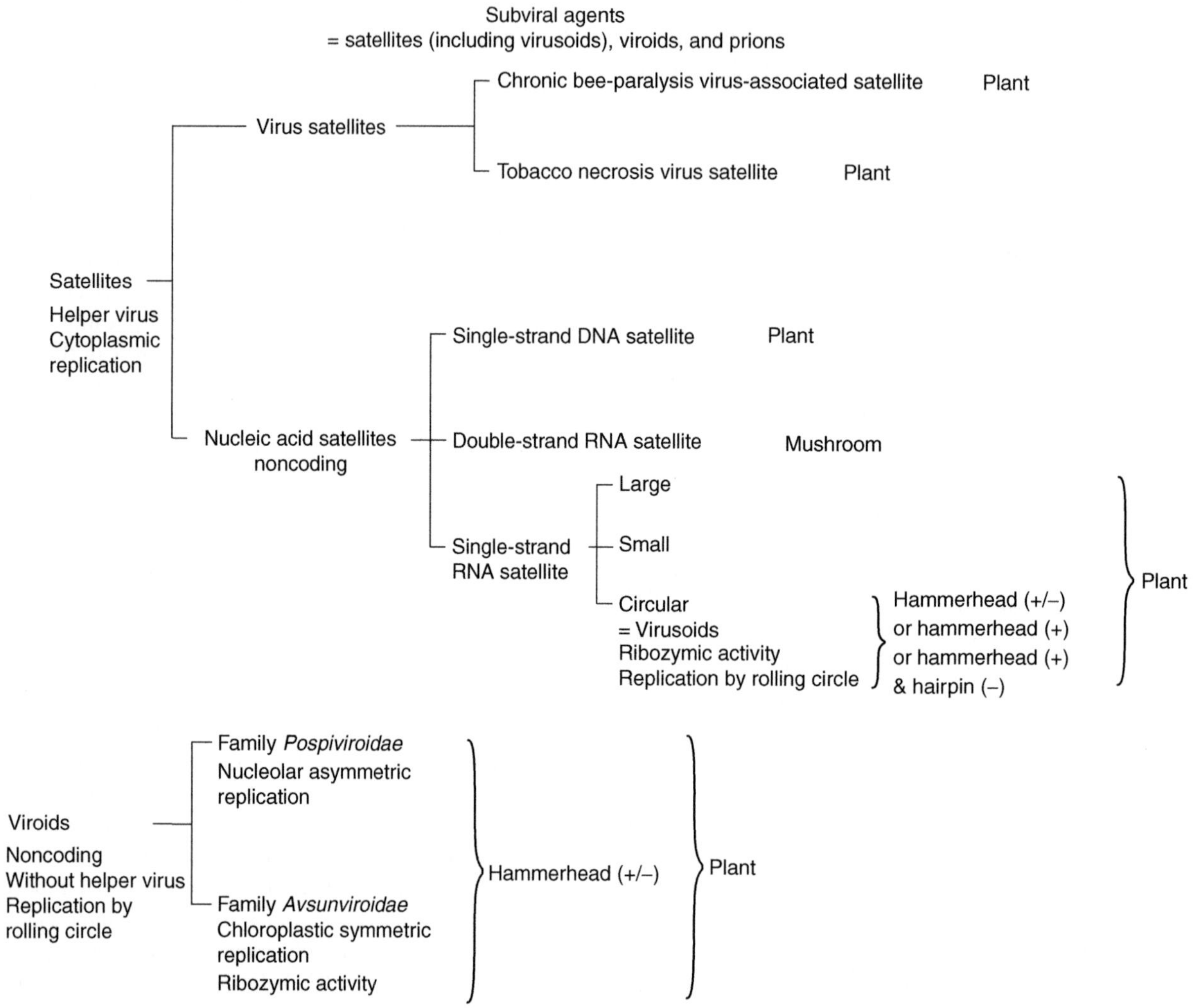

Subviral agents
= satellites (including virusoids), viroids, and prions

- Satellites — Helper virus, Cytoplasmic replication
  - Virus satellites
    - Chronic bee-paralysis virus-associated satellite — Plant
    - Tobacco necrosis virus satellite — Plant
  - Nucleic acid satellites noncoding
    - Single-strand DNA satellite — Plant
    - Double-strand RNA satellite — Mushroom
    - Single-strand RNA satellite
      - Large
      - Small
      - Circular = Virusoids, Ribozymic activity, Replication by rolling circle — Hammerhead (+/–) or hammerhead (+) or hammerhead (+) & hairpin (–)
    - (Large, Small, Circular) — Plant
- Viroids — Noncoding, Without helper virus, Replication by rolling circle
  - Family *Pospiviroidae* — Nucleolar asymmetric replication
  - Family *Avsunviroidae* — Chloroplastic symmetric replication, Ribozymic activity
  - (both families) — Hammerhead (+/–) — Plant

Figure 2 Classification of the satellites and viroids. The viroids proliferate and are copied in an autonomous way (without helper virus). Their circular RNA genomes do not have an open reading frame. The virusoids are satellite RNAs with circular single-stranded genomes. They do not code for protein, contrary to the satellite genomes of viruses which code for capsid proteins.

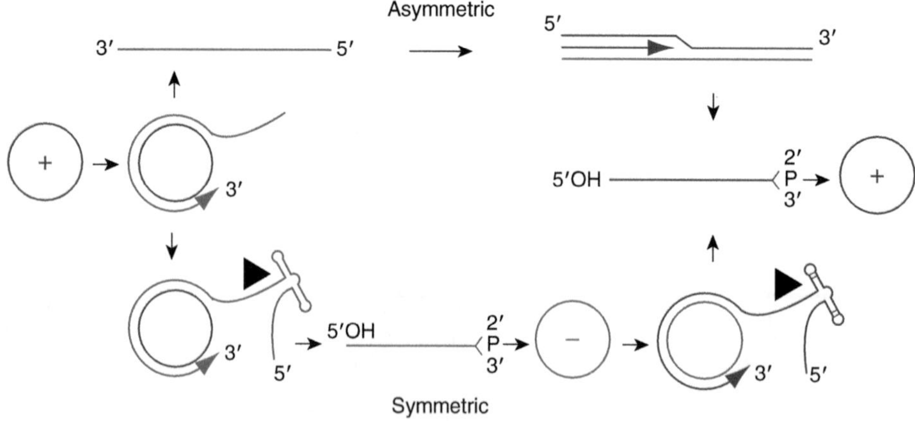

Figure 3 Mechanism of replication using the rolling-circle mechanism.There are two main ways for replication: an asymmetrical way (top) which requires only one rolling circle as seen in the family *Pospiviroidae* and a symmetrical way which requires two rolling circles specific to the family *Avsunviroidae*. The (+) strands are drawn in red and the (–) strands in blue. The hammerhead ribozymes in the cleavage of the concatemers utilize only the symmetrical way.

mechanism of rolling circle. After cleavage by the hammerhead ribozyme, the (+) strands are circularized. Thus, only the RNA intermediaries are implicated in this mechanism of replication since the viroid genome does not code for proteins. Thus, the various biological properties of the viroids, such as the ability to identify the host, depend exclusively on the sequence and the structure of their RNAs.

The hammerhead ribozyme plays a fundamental role in the replication of these pathogenic RNA genomes. Yet, its role is not limited to this function as it has been identified in genomes of the triton (*Notophtalamus viridescens*), in parasitic schistosomes, and in cave crickets (*Dolichopoda* sp.). Two hammerhead ribozymes have been found also in the *Arabidopsis thaliana* genome.

The Hairpin Ribozyme

The hairpin ribozyme is involved in the processing of RNA molecules generated in the replication cycles of various species. The tobacco ringspot virus satellite RNA (sTRSV), for example, carries a hammerhead ribozyme on the (+) strand and a hairpin ribozyme on the (−) strand. The crystal structure of the hairpin ribozyme shows that it is comprised of two important structural elements: a four-way helical junction which is necessary for the optimal function of the ribozyme and two internal loops within the interacting helices (**Figure 4**). The minimal ribozyme (i.e., without the other two arms) catalyzes the cleavage reaction but requires 3 orders of magnitude higher concentration of magnesium cations (Mg^{2+}) than the natural form. Single-molecule studies show that the natural ribozyme exists in three states: an undocked form (where helices do not contact each other), an intermediate state (with helices docked but without contact between the internal loops), and a docked form (with intimate contacts between the loops). Although the junction is not essential to the hairpin ribozyme activity, it greatly enhances the rate of folding of the ribozyme in an active form at physiological concentration of Mg^{2+} ions. These auxiliary elements play the same role as in the hammerhead ribozyme and appear to act as 'folding enhancers'. However, there is no evidence for the direct implication of metal ions in the chemistry of the hairpin ribozyme.

The Hammerhead Ribozyme

Classification

The natural hammerhead ribozymes can be classified into two main types according to the helix numbering which carries the RNA 3′ and 5′ ends. In type 1, the 5′ end is carried by helix I which always presents an internal loop. Helix II presents in some instances an internal loop before the terminal loop. Helix III does not present a terminal loop but possibly an internal loop. In type 3, the 5′ end is carried by helix III. Helices I and II present sometimes an internal loop preceding the terminal loop. Ribozymes

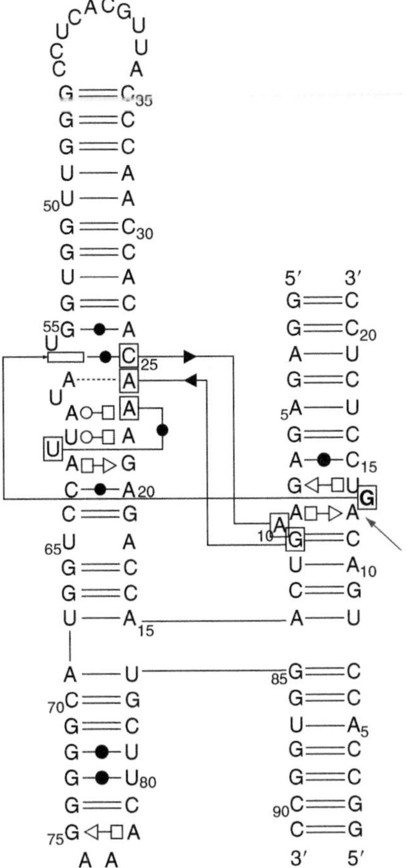

Figure 4 Schematic diagram of the interactions present in the three-dimensional crystal structure of the hairpin ribozyme (PDB ID 1HP6). The red arrow indicates the cleavage site.

where the 5′ end is carried by helix II have, so far, never been observed in any genomes.

Structure and folding

The structure of the catalytic core of the hammerhead ribozyme was determined by crystallography using an RNA/DNA hybrid as well as in the 'all RNA' form with a methyl group on the 2′-hydroxyl of the site of cleavage which prevents the cleavage reaction. The two practically identical structures show a catalytic core composed of 11 nucleotides essential for catalysis, organized in a three-way junction from which the helices I, II, and III leave. Helix II is stacked on helix III, while helix I is parallel to helix II. Helix II, composed of Watson–Crick base pairs, is stacked on three non-Watson–Crick base pairs.

Many structural studies of the ribozyme in solution have shown the importance of Mg^{2+} ions for folding. Indeed, in absence of divalent ions, the core of the ribozyme is not structured and the helices are equidistant from each other, revealing a disorganization of stacking, whereas in the presence of divalent ions the ribozyme folds in two stages to an active structure. At a low concentration, helices II and III are stacked but the angle between helices I and II is very large and the catalytic core is not structured; the ribozyme is inactive. At higher concentration, helix I reorients to a position near helix II and the ribozyme is then in an active conformation. However, the majority of the studies on hammerhead ribozyme folding were carried out on minimal ribozymes presenting the conserved and presumably essential elements, that is, 13 nucleotides of the catalytic core of which 11 cannot be mutated without causing loss of activity, and three helices I, II, and III with variable length and termination. These minimal ribozymes are active in an optimal way at high concentrations of Mg^{2+} ions (>10 mM) but almost inactive at physiological concentration of Mg^{2+} (0.1–0.3 mM).

However, recent studies showed that the tertiary contacts between the terminal loops of helices I and II, peripheral to the catalytic core and nonessential for catalysis, are essential for the activity of hammerhead ribozymes in physiological conditions (low concentration of Mg^{2+} ions). The role of the peripheral structures and their interactions in the catalytic activity of the hammerhead ribozymes is expected from previous observations on longer ribozymes such as group I introns. The progressive deletion of the peripheral area of group I introns results in increasing requirements of salts and ultimately in loss of activity. The various categories of group I intron all have the same catalytic core, although the outlying areas and the tertiary interactions differ and they all contribute to the stability of the catalytic core. This suggests that during evolution, several folding solutions were found which ensured the optimal catalytic activity of the ribozyme. In the same manner, the peripheral area of the natural hammerhead ribozymes evolved so as to optimize the folding and the catalytic activity for biological needs under natural conditions of Mg^{2+} ion levels. The tertiary interactions between these regions reduce the conformational space accessible to the molecule and facilitate the folding of the ribozyme, thereby increasing its catalytic activity. It is clear that the catalytic core of the ribozyme had to undergo structural changes, which explains the increase in its catalytic efficiency although the precise nature of the changes is unknown. The structure of the full-length *Schistosoma mansoni* hammerhead ribozyme reveals the structural basis for the effectiveness of natural ribozymes and explains most of the apparent inconsistencies between the structure of the minimalist ribozymes and the biochemical data.

Role of Peripheral Elements in the Mechanism of Ligation

Like any enzyme, the hammerhead and hairpin ribozymes are able to carry out the reaction opposite to cleavage, that is, ligation. However, in majority of conditions, cleavage is favored at least 100-fold compared to ligation. The various stages for ligation are (1) activation of the 5′-hydroxyl by a base, (2) attack of the 2′,3′-cyclic phosphate, and (3) release and protonation of 2′-hydroxyl, restoring the 3′,5′-phosphodiester bond.

The Hairpin Ribozyme

The studies of the dynamics of the hairpin ribozyme made it possible to determine the conformational kinetic properties of the reactions of ligation and cleavage. They show that after cleavage the four-way junction of the ribozyme moves into an undocked state. The reaction of cleavage is definitely slower than the reaction of ligation, but ligation is slower than helical unstacking. The product is thus normally released before ligation takes place. However, if the product of cleavage remains bound to the enzyme, the ribozyme alternates between docked and undocked state before ligation takes place. These experiments also show that the nature of the product is a factor influencing the speed of ribozyme unfolding. Indeed, the presence of a 2′,3′-cyclic phosphate increases the unfolding of the ribozyme compared to a product having either a 3′-phosphate or 3′-hydroxyl. Therefore, the product has important effects on structural dynamics of the ribozyme, and the loss of covalent continuity in the backbone is not the only factor increasing the undocking of the ribozyme. The undocking promoted by the presence of cyclic phosphate, involves detachment of the product. This is in agreement with the biology of the ribozyme in its natural environment where it cleaves the RNA concatemer (produced by rolling circle) and releases a monomeric 2′,3′-cyclic phosphate product. In the case of the hairpin ribozyme (−) strand of the satellite RNA 'sTRSV', the helix formed by the product and the enzyme, with a length of six base pairs, has a weak dissociation constant. Single-molecule spectroscopy shows that docking–undocking events can take place several times before the ligation does take place. The equilibrium constant of cleavage and ligation shows a significant bias toward ligation with an internal equilibrium constant of $K_{int} = k_L/K_C = 34$. This preference for ligation ensures that during virus replication a certain number of circular RNAs will be maintained which will be used as matrix for the synthesis of the concatemer. On the other hand, the rate of undocking is greater than the rate of ligation which allows cleavage of concatemers because of the rapid undocking that follows cleavage.

The Hammerhead Ribozyme

For the hammerhead ribozyme, it has been shown that the kinetic constant of ligation ($k_{-2} = 0.008\ min^{-1}$) is 100 times lower than the constant of cleavage ($k_2 = 1\ min^{-1}$); the

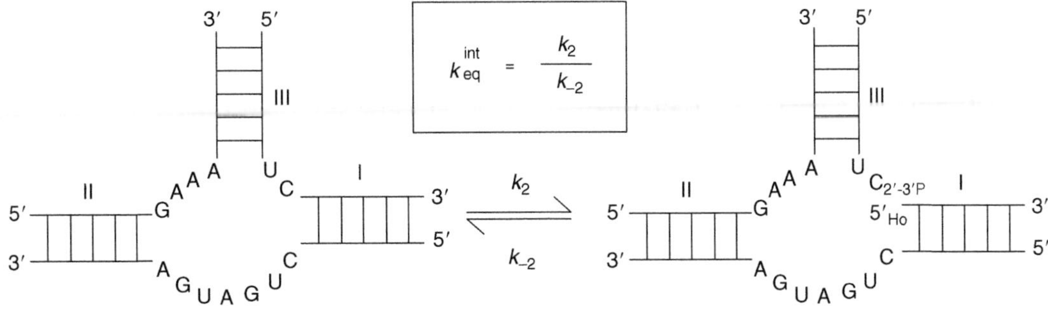

Figure 5 Secondary structure and reaction of the catalytic core of a minimal hammerhead ribozyme. The reaction of cleavage is defined by the constant of cleavage k_2 and the reaction of ligation is defined by the constant of ligation k_{-2}. The definition of the equilibrium constant is framed.

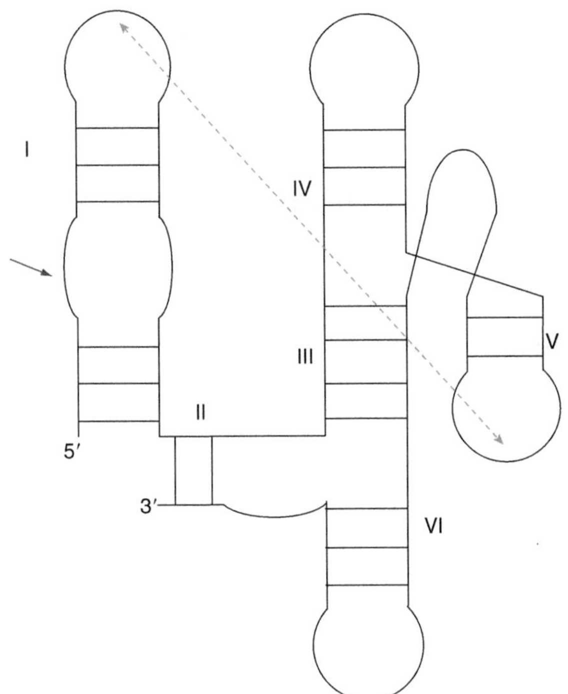

Figure 6 Schematic representation of the secondary structure of the VS ribozyme. The pseudoknot is indicated by the green arrow. The red arrow indicates the cleavage site.

equilibrium constant $K_{eq} = k_2/k_{-2}$ thus has a value of 125 (**Figure 5**). This is explained by the more favorable thermodynamic state of the enzyme/product complex than of the complex enzyme/substrate which favors the formation of the latter by driving the reaction toward cleavage. The tertiary interactions, by limiting the accessible conformational space (i.e., by stabilizing the folding of the complex enzyme/product), increase the ligation activity of the ribozyme. The same phenomena are observed for the hairpin ribozyme, since the presence of the native four-way junction with four helices moves the internal equilibrium toward binding and increases the rate of folding.

The Mechanism of Monomer Circularization after Replication

The peripheral elements together with the catalytic core are not only involved in the reaction of cleavage but also in the reaction of ligation. Each monomeric copy of the genome, product of the replication by rolling circle and cleavage by the hammerhead ribozyme, must be circularized to be used in turn as a replication matrix. Little is known about the *in vivo* circularization of the monomer. The viroid RNAs belonging to the family *Pospiviroidae* are probably ligated by an RNA ligase as indicated by the presence of a 2′-phosphomonoester and a 3′,5′-phosphodiester bond at the site of ligation. On the other hand, information on the ligation of viroid RNAs in the family *Avsunviroidae* is limited. In the case of the circularization of the genome of peach latent mosaic viroid (PLMVd), a 2′,5′-phosphodiester bond has been observed *in vitro* and *in vivo*. Such a bond would prevent cleavage of the ribozyme and would ensure the maintenance of an essential circular RNA matrix by the rolling-circle mechanism.

Yeast and Human Satellite Viruses

The *Neurospora crassa* versus Ribozyme

The VS RNA is transcribed from the mitochondrial DNA of *Neurospora crassa*. The VS RNA contains a ribozyme of around 150 nt in length acting in the processing of replication intermediates. There is only limited information available on its structure. The VS ribozyme comprises five helices organized in two three-way junctions (2–3–6 and 3–4–5 junctions) to form a Y shape. A pseudoknot is formed between the terminal loop of the substrate helix and the helix V of the ribozyme (**Figure 6**). The scissile phosphate is within an internal loop of the substrate stem–loop. The catalyzed reaction requires divalent Mg^{2+}cations.

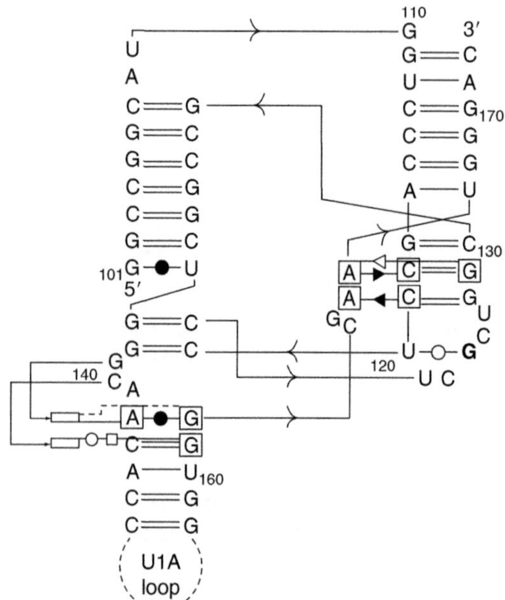

Figure 7 Schematic diagram of the interactions present in the three-dimensional crystal structure of HDV ribozyme after self-cleavage (PDB ID 1DRZ).

The Human HDV

The HDV is a satellite virus of hepatitis B virus (HBV) on which it is dependent for its replication cycle. The genome of HDV is a 1700-nt-long circular single-stranded RNA. Its replication is RNA directed without DNA intermediates. Like certain pathogenic subviral RNAs that infect plants, HDV RNA replicates by a rolling-circle mechanism. Both genomic (+) and complementary (–) strands of RNA contain a single ribozyme of about 85 nt. The HDV ribozyme folds into a compact structure comprising a double pseudoknot and its disruption results in a marked loss of activity (**Figure 7**). The X-ray structures of the two states of the ribozyme (pre- and post-cleaved states) reveal a significant conformational change in the RNA after cleavage and point to the role of a divalent metal ion in the cleavage mechanism.

Conclusions

The hammerhead and hairpin ribozymes were initially studied in their minimalist form, which resulted in weak activities due to the presence of numerous conformations in dynamic exchange, most of them inactive, which required high concentrations of divalent ions for activity. However, the peripheral elements are essential for optimal activity in native conditions. These peripheral regions, by interacting with each other, facilitate and stabilize folding into a single active structure. They are, therefore, necessary in physiological conditions although they are not directly implied in the catalysis. These results put into question the conclusions drawn previously from data obtained with molecular systems that had been simplified and reduced to the extreme. Full understanding of the chemical and biological actions of ribozymes is certainly not yet achieved, however, and surprises are still lurking. Recently, a ribozyme structurally derived from group I introns was shown to catalyze the formation of a 2′,5′-linkage leading to the formation of a lariat structure mimicking the cap of mRNA. Our present knowledge underlines the fundamental links existing between the folding pathway, the selection and stabilization of a single native state by tertiary interactions between peripheral elements and the catalytic activity of ribozymes.

See also: Interfering RNAs; Satellite Nucleic Acids and Viruses.

Further Reading

Blount KF and Uhlenbeck OC (2005) The structure–function dilemma of the hammerhead ribozyme. *Annual Review of Biophysics and Biomolecular Structure* 34: 415–440.

Daros JA, Elena SF, and Flores R (2006) Viroids: An Ariadne's thread into the RNA labyrinth. *EMBO Reports* 7: 593–598.

Doudna JA and Cech TR (2002) The chemical repertoire of natural ribozymes. *Nature* 418: 222–228.

Gesteland RF, Cech TR, and Atkins JF (eds.) (2006) *The RNA World.* New York: Cold Spring Harbor Laboratory Press.

Khvorova A, Lescoute A, Westhof E, and Jayasena SD (2003) Sequence elements outside the hammerhead ribozyme catalytic core enable intracellular activity. *Nature Structural and Molecular Biology* 10: 708–712.

Lafontaine DA, Norman DG, and Lilley DM (2002) The global structure of the VS ribozyme. *EMBO Journal* 21: 2461–2471.

Leontis NB and Westhof E (2001) Geometric nomenclature and classification of RNA base pairs. *RNA* 7: 499–512.

Lilley DM (2004) The Varkud satellite ribozyme. *RNA* 10: 151–158.

Lilley DM (2005) Structure, folding and mechanisms of ribozymes. *Current Opinion in Structural Biology* 15: 313–323.

Martick M and Scott WG (2006) Tertiary contacts distant from the active site prime a ribozyme for catalysis. *Cell* 126: 309–320.

McKay DB (1996) Structure and function of the hammerhead ribozyme: An unfinished story. *RNA* 2: 395–403.

Westhof E (2007) A tale in molecular recognition: The hammerhead ribozyme. *Journal of Molecular Recognition* 20: 1–3.

Virus Species

M H V Van Regenmortel, CNRS, Illkirch, France

Glossary

Ecological niche The ecological niche of a virus refers to certain biological properties of the virus, such as host range, tissue tropism in the host, and type of vector, rather than a geographical space or a particular environment. In the absence of the virus, its ecological niche property is also absent and the notion of a vacant niche is thus meaningless.

Polythetic class A polythetic class consists of members that exhibit overall similarity and have a large number of common properties. However, the members of a polythetic class do not all share a common character that constitutes a single defining property of the class. In contrast, the members of a monothetic class, such as those of the order *Mononegavirales*, possess such a single defining property, that is, the presence of a negative-sense, single-stranded RNA (ssRNA) genome.

Introduction

The question of what is a virus species is related to the general problem of how the world of viruses should be partitioned and ordered for achieving a coherent scheme of distinct and easily recognizable viral entities. In view of the variability of viruses arising from the error-prone process of genomic replication, it is often difficult to decide whether a newly encountered virus is the same as one seen previously, for this requires that we give an answer to the vexing question: "How different must two virus isolates be in order to be considered different viruses, rather than the same virus?" Virologists usually have no difficulty in distinguishing pathogenic variants of a virus, while still recognizing these variants as the same kind of virus. Even without realizing it, they pass judgment on the significance of the observed differences between individual virus isolates and, if the extent of difference appears small enough, they will consider the variant to be the same virus. In taxonomic parlance, they will say that the variant belongs to the same virus species.

'Species' is the universally accepted term for the lowest taxonomic clustering of living organisms. Although viruses are not organisms, they are considered to be biological entities because they possess some of the properties of biological agents, such as having a genome and being able to adapt to particular hosts and changing environments. The classification system of viruses therefore uses the same hierarchical ranks of order, family, genus, and species used in all biological classifications.

Although the species category is the most fundamental one in any biological classification, it took many years before virologists started to assign viruses to separate species. Virus genera and families were readily accepted but there was considerable opposition to the idea that the species concept could be applied to viruses. One reason for this reluctance was the lack of agreement among biologists about what a species actually is. It is in fact remarkable that a century and a half after the appearance of Charles Darwin's *The Origin of Species*, there is still no general agreement about what constitutes a plant, animal, or microbial species. In different branches of biology, no less than 22 different species concepts have been applied.

Distinguishing between Abstract Classes and Real Objects

A basic requirement for clear thinking is the ability to distinguish between real, tangible objects, such as viruses, and the abstract classes used in taxonomy, such as families and species, which are mental constructs that exist only in the minds of people who think about them. Although a taxonomic class is defined by properties possessed by concrete objects, it is only an abstract thought that has no real existence outside of human minds. For this reason, viral species, genera, and families cannot cause diseases and cannot be purified, sequenced, or seen through electron microscopes. Only the viruses themselves can be studied in this manner. Classes used in biological classification have a hierarchical structure. A class, such as a particular species, can belong only to one higher rank immediately above it such as a particular genus and that genus, in turn, can only belong to one family. The relationship known as 'class membership' is the relationship between an abstract class or thought and the members of that class which are real, concrete objects located in space and time. Class membership is thus able to bridge two different logical categories, the abstract and the concrete. It is important to be able to establish a certain link between these two logical categories, as it is impossible for an object such as a virus to be 'part' of an abstraction, such as a species. Similarly, an abstract thought cannot be 'part of' a concrete object. The correct statement is to say that the viruses that are handled by virologists are members of certain abstract constructs, such as species or genera. These constructs are invented by taxonomists for the purpose of introducing some order to the bewildering variety of viruses.

A universal class, also known as an Aristotelian class, is defined by properties that are constant and immutable. This allows members of such a class to be recognized with absolute certainty, because one or more property is necessarily present in every member of the class. Virus families, for instance, are universal classes because they consist of members, all of which share a number of defining properties that are both necessary and sufficient for class membership. Allocating a virus to a family is thus an easy task, because a few structural or chemical attributes will suffice to allocate the virus to a particular family. For instance, all the members of the family *Adenoviridae* are nonenveloped viruses that have an icosahedral particle and double-stranded DNA, with projecting fibers at the vertices of the protein shell.

Unfortunately, not all properties of members of classes correspond to unambiguous and stable properties such as the presence or absence of a DNA genome or of a particular particle morphology. Many qualitative properties of concrete objects are inherently vague and do not possess precise borderlines. For instance, in the description of viruses, properties used for recognizing members of species, such as the degree of genome sequence similarity between virus isolates or the nature of the symptoms induced by a virus in its host, whether mild or severe, tend to be inherently imprecise and fuzzy. As a result the species classes that can be conceptualized on the basis of such properties will themselves be fuzzy and membership in the class will then be a matter of convention or stipulation. This explains some of the difficulties one encounters when dealing with the species level in any biological classification.

When fuzziness is accepted as an unavoidable characteristic of species, it becomes possible to describe species in terms of continuums with hazy boundaries. In a similar way, colors can be distinguished conceptually in spite of the continuous nature of the spectrum of electromagnetic waves, and mountain peaks are given names in spite of the absence of sharp boundaries in geological rock formations.

Species in Biology

The traditional view of species is that they correspond to groups of similar organisms that can breed among themselves and produce fertile offspring. The classical definition of so-called 'biological species' states that "species are groups of interbreeding natural populations which are reproductively isolated from other such groups." The reproductive isolation often simply reflects a geographic isolation or a behavioral incompatibility. This definition is only applicable to organisms that reproduce sexually and it has limited value in the plant kingdom because of the high frequency of hybridization between different species of plants. Even in animals, the criterion of reproductive isolation does not always hold, as shown by the ability of dogs, wolves, and coyotes to interbreed although they are members of different species of the same genus.

In order to make the definition of biological species applicable to asexual organisms, it was later modified as follows: "a species is a reproductive community of populations, reproductively isolated from others, that occupies a specific niche in nature." Some authors reject the view that asexual organisms can form biological species but most biologists disagree with that view, as this would render the species concept inapplicable to a large portion of the biological realm.

Another species concept is that of evolutionary species, which has been defined as "a single lineage of ancestor–descendant populations which maintains its identity from other such lineages and which has its own evolutionary tendencies and historical fate." Such a concept does not provide any guidance on how far back in time a species can be traced. Life on Earth is a biological and historical continuum and it is as difficult to demarcate boundaries in time that would separate individual evolutionary species as it is to define clear-cut breeding discontinuities that would separate different biological species. These difficulties led to the view that species are similar to fuzzy sets with unclear boundaries and that it is impossible to draw sharp boundaries between them, as is done with universal classes such as genera and families. In practice this means that it is not possible to rely on a single defining property for differentiating between two species. This led to the proposal that species are so-called 'polythetic classes' defined by a combination of properties, each of which might occur outside any given class yet could be absent in a member of the particular class (**Figure 1**).

The members of a polythetic class exhibit overall similarity and have a large number of characters in common. However, in contrast to the members of a monothetic or universal class, they need not all share a common character that could be used as a defining property of the class. A polythetic class is actually a cluster concept based on the concept of family resemblance introduced in philosophy by Ludwig Wittgenstein.

What Is a Virus Species?

In the past, many virologists were opposed to the introduction of species in virus classification because they assumed that the only legitimate species concept was that of biological species defined by sexual reproduction, gene pools, and reproductive isolation. Such a concept is

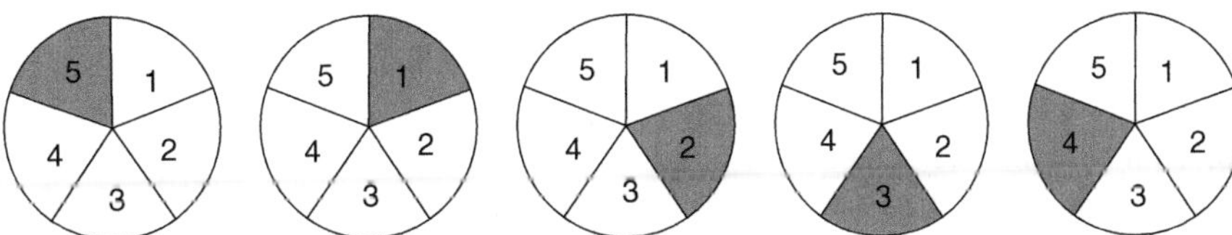

Figure 1 Schematic representation of the distribution of properties (1 to 5) in five members of a polythetic class. Each member possesses several of these properties (4 out of 5) but no single property is present in all the members of the class. The missing property in each case is represented by the gray section.

obviously not applicable to viruses that are replicated as clones. Another reason for the reluctance to adopt the species category in virus classification was the absence of a satisfactory definition of a virus species. Various definitions of virus species had been proposed over the years but none gained general acceptance. One definition, for instance, stated that "virus species are strains whose properties are so similar that there seems little value in giving them separate names." Such a definition is not very helpful as it simply replaces undefined species by undefined strains and suggests that attributing names to viruses is the same activity as constructing a taxonomy.

In 1991, the International Committee on Taxonomy of Viruses (ICTV), which is the body established by the International Union of Microbiological Societies to make decisions on matters of virus classification and nomenclature, endorsed the following definition which was proposed by the author: "a virus species is a polythetic class of viruses that constitute a replicating lineage and occupy a particular ecological niche." This definition is not only based on phenetic similarities among members of species but also stresses the internal cohesion present in biological lineages that share a common biotic niche. Another important aspect of the definition is that a virus species is defined as a polythetic class rather than as a traditional universal class (see **Figure 1**). The ICTV has been using the concept of polythetic class for the creation of separate species taxa.

The total DNA or RNA sequence found in a virion corresponds to the viral genome and is part of the viral phenotype because it corresponds to a portion of the virion's chemical structure. The phenome of the virus corresponds to its observable physical properties, including the morphology and molecular constitution of the virion, as well as the biochemical activities and relational properties of the virus. The phenotype has a temporal dimension as it changes over time, for instance during replication.

A classification based on genome sequences is actually a phenotypic classification with the characters being molecular, rather than morphological or relational. Although putative phylogenetic relationships can be inferred from sequences, they remain tentative and depend on many tacit assumptions. There is little justification for the common assumption that when species are demarcated on the basis of a proposed phylogeny, this produces a classification that is necessarily more correct or useful than a classification based on other phenotypic characters.

Defining Species as a Taxonomic Class Is Not the Same As Demarcating Individual Viral Species

As a definition is the explanation of the meaning of a word or concept, it is possible to give a definition of the concept species. In contrast, it is not possible to give a definition of an object such as a particular virus, as objects can only be described by enumerating their properties and cannot be 'defined'. When viruses or any other object are given proper names, their names do not have a meaning that can be captured in a definition. When we define the name of an abstract species class by means of defining properties, we state the conditions for that name to apply to any given member of the species.

Two types of species definitions must be distinguished. One definition concerns the taxonomic species category corresponding to the lowest level in a hierarchical system of classification. The definition of virus species accepted by the ICTV is a definition of that type and it was introduced because many virologists thought that the species concept was not applicable to viruses. However, such a definition of the category species should not be confused with the definitions or demarcations of each of the 1950 species classes listed in the Eighth ICTV Report published in 2005. Confusing the two meanings of species, that is, as a particular class which has certain viruses as members and as a taxonomic category, respectively, would amount to confuse the concept of chemical element (the class of all elements) with the class corresponding to a single element, such as gold defined by its atomic number 79.

It must be emphasized that once the taxonomic category of virus species had been accepted by virologists, this did not do away with the difficult task of demarcating and defining the hundreds of individual species classes that had to be created on the basis of different combinations of properties for each polythetic species class. In order to demarcate individual species, it was necessary to rely on properties that were not present in all the members of the genus to which the species belongs, because such properties obviously would not permit individual species to be differentiated. Characteristics such as virion morphology, genome organization, method of replication, and number and size of viral proteins are shared by all the members of a genus and are thus not useful for distinguishing individual species in a genus. The properties that are useful for discriminating between individual species within a genus are the natural host range, cell and tissue tropism, pathogenicity, mode of transmission, certain physicochemical and antigenic properties of virions, and small differences in genome sequence. Unfortunately, these properties can be altered by only one or a few mutations and they may therefore vary in different members of the same species. This is one of the reasons species demarcation is not an easy task and often requires drawing boundaries across a continuous range of genomic and phenotypic variability. Taxonomic decisions at the level of species are very much a matter of opinion and adjudication rather than of logical necessity. There is indeed no precise degree of genome difference that could be used as a cutoff point to differentiate between two species nor is there a simple quantitative relationship between the extent of genome similarity and the similarity in phenotypic and biological characteristics of a virus.

There is also a strong subjective element in virus classification, as the practical need to distinguish between individual viruses is not the same in all areas of virology. From a human perspective, not all infected hosts are equally relevant. Human pathogens or pathogens that infect animals and plants of economic importance tend to be studied more intensively than the very many viruses that infect insects or marine organisms. As a result fine distinctions based on minor differences in host range or pathogenicity may be emphasized in the case of viruses of particular interest to humans and the criteria used for differentiating individual species may thus depend on the type of host that is infected. There is, of course, no reason the criteria used for distinguishing between species should be the same in all the virus genera and families, because the need to make certain distinctions are not the same in all fields of virology.

Virus Identification

In order to identify a virus as a member of a species, use is made of so-called diagnostic properties that allow the virus to be recognized as being similar to other viruses. This, of course, means that the species had to be created and defined by taxonomists beforehand. Only when the properties of many members of an established species are compared, is it possible to discover which diagnostic property or set of properties will discriminate between the members of that species and other species, thereby allowing viruses to be identified. A single diagnostic character may sometimes suffice for virus identification but this character should not be confused with the collective set of combined properties which has been used initially to demarcate and define the species as a polythetic class. Species are not defined by means of one or more diagnostic properties used for virus identification but are defined by taxonomists, who stipulate which covariant sets of shared properties have to be present in most members of the species. It is the frequent, combined occurrence of these properties in individual members of a species that allows one to predict many of the properties of a newly discovered virus once it has been identified as a member of a particular species. If a species could be defined monothetically instead of polythetically, that is, by a single diagnostic character and nothing else, the identification of a virus as a member of a species would not be very informative. A single diagnostic character, such as a nucleotide motif or the reactivity with a monoclonal antibody, may suffice to identify a virus but such a property should not be mistaken for a single defining characteristic of the species.

This apparent contradiction between the need for many characters to define and delineate a species and the fact that a single property may suffice to identify a virus disappears when it is realized that defining an abstraction, such as a species, is a task different from identifying a concrete object, such as a virus. Providing a definition of an abstract taxonomic class and identifying concrete objects are not equivalent tasks and they abide by different logical rules.

Species Names and How to Write Them

After the category species was accepted as the lowest taxonomic class to be used in viral taxonomy, the ICTV decided in 1998 that the existing English common names of viruses were to become the official species names and that, to denote the difference, species names should be written in italics with a capital initial letter. This typography, which is similar to that used for other taxa such as virus genera and families, makes it possible to distinguish virus species officially recognized by the ICTV and written in italics from other viral entities such as tentative species and virus strains written in Roman characters.

When ICTV introduced italicized virus species names, it did not intend that the existing names of viruses written in Roman characters should be abandoned. In their writings, virologists need to refer continuously to the viruses they study and because these viruses are not taxonomic abstractions, their names should be written in Roman characters.

In most cases it is necessary only once in a scientific paper to draw attention to the taxonomic position of the virus and this can be done by mentioning that the virus under study, for example, measles virus, is a member of the species *Measles virus*, genus *Morbillivirus*, and family

Paramyxoviridae. In publications written in languages other than English, the common names of the virus remain those used in that language. It will suffice, for instance, to say once: "le virus de la rougeole est un membre de l'espèce *Measles virus*."

It is correct to say that a virus, a strain, or an isolate of the species *Cucumber mosaic virus* has been sequenced but incorrect to say the species itself has been sequenced. Unfortunately, some authors confuse viruses with species and write that the species *Cucumber mosaic virus* (instead of the virus cucumber mosaic virus) has been isolated from a tobacco plant, is transmitted by an aphid vector, is the causal agent of a mosaic disease, and has been sequenced. This is of course incorrect because virus species are man-made taxonomic classes and do not have hosts, vectors, or sequences. For the same reason, the genus *Cucumovirus* or the family *Paramyxoviridae* cannot be isolated, visualized in an electron microscope, or sequenced.

At a time when English is more widely understood than Latin, the use of italicized English instead of italicized Latin for the names of virus species is more practical and is in line with the emergence of English as the modern language of international scientific communication. Most virologists are opposed to the Latinization of species names and welcomed the introduction of well-known English species names. They certainly preferred this to the creation of an entirely new set of Latin names for the 1950 different virus species that are recognized at present.

Binomial Names for Virus Species

A problem with the current way of naming virus species is that the name of the species (for instance *Measles virus*) differs only in typography from the name of the virus (measles virus). One way to facilitate the distinction between virus names and species names would be to change the current names of virus species into nonlatinized binomial names. Such a system, which has been advocated by plant virologists for many years, consists in replacing the word '*virus*' appearing at the end of the existing species name by the genus name which also ends in '*-virus*'. *Measles virus* then becomes *Measles morbillivirus*, *Hepatitis A virus* becomes *Hepatitis A hepatovirus*, and *Tobacco mosaic virus* becomes *Tobacco mosaic tobamovirus*. The advantage of such a system, which could be implemented without problems for about 98% of all virus species names, is that inclusion of the genus name in the species name provides additional information about the properties of the virus. A changeover to binomial species names would not affect the common names of viruses in English or other languages since names such as measles virus or 'virus de la rougeole' would remain the same. The ICTV is currently debating the possibility of introducing binomial species names and some decision is likely to be made in the near future.

Given that the common names of viruses are used repeatedly in scientific texts there is a need for abbreviating them and the ICTV has published several lists of recommended acronyms for virus names. Since the names of virus species are used only very seldom in publications, there is no need to abbreviate them. If binomial names of virus species were introduced in the future, the abbreviations of common names of viruses will of course not be affected.

See also: Nature of Viruses; Phylogeny of Viruses; Quasispecies; Taxonomy, Classification and Nomenclature of Viruses.

Further Reading

Calisher CH and Mahy BWJ (2003) Taxonomy: Get it right or leave it alone. *American Journal of Tropical Medicine and Hygiene* 68: 505–506.

Claridge MF, Duwah HA and Wilson MR (eds.) (1997) *Species: The Units of Biodiversity*. London: Chapman and Hall.

Fauquet CM, Mayo MA, Maniloff J, Desselberger U and Ball LA (eds.) (2005) *Virus Taxonomy, Classification and Nomenclature of Viruses: Eighth Report of the International Committee on Taxonomy of Viruses*, San Diego, CA: Elsevier Academic Press.

Lewontin RC (1992) Genotype and phenotype. In: Keller EF and Lloyd EA (eds.) *Keywords in Evolutionary Biology*, pp. 137–144. Cambridge, MA: Harvard University Press.

Pigliucci M (2003) Species as family resemblance concepts: The (dis-)solution of the species problem? *Bioessays* 25: 596–602.

Van Regenmortel MHV (1990) Virus species, a much overlooked but essential concept in virus classification. *Intervirology* 31: 241–254.

Van Regenmortel MHV (2003) Viruses are real, virus species are man-made taxonomic constructions. *Archives of Virology* 148: 2481–2488.

Van Regenmortal MHV (2006) Virologists, taxonomy and the demands of logic. *Archives of Virology* 151: 1251–1255.

Van Regenmortel MHV (2007) Virus species and virus identification: Past and current controversies. *Infection, Genetics and Evolution* 7: 133–144.

Van Regenmortel MHV and Mahy BWJ (2004) Emerging issues in virus taxonomy. *Emerging Infectious Diseases* 10: 8–13.

Quasispecies

R Sanjuán, Instituto de Biología Molecular y Cellular de Plantas, CSIC-UPV, Valencia, Spain

Glossary

Adaptive landscape A graphical representation of fitness as a function of the genotype, first used by the population geneticist S. Wright.
Antagonistic epistasis Genetic interaction that makes mutations have larger effects alone than in combination.
Antagonistic pleiotropy The situation in which a particular mutation that is beneficial in one environment becomes harmful in another environment.
Back-mutation Mutation that regenerates the ancestral genotype.
Biological fitness The number of descendants.
Degenerate quasispecies Quasispecies with more than a single fittest type.
Effective population size The number of individuals of a population that actually contribute to the next generation.
Epistasis Interaction between genetic loci, implying that mutations have nonindependent effects.
Error catastrophe A situation in which the mutation rate is too high to be effectively counteracted by selection and the population becomes a pool of randomly drifting sequences.
Error threshold The critical mutation rate that marks the transition from mutation–selection balance to error catastrophe.
Genetic drift Changes in the genetic composition of finite populations due to random sampling of genotypes between generations.
Master sequence The fittest and most abundant sequence in a nondegenerate quasispecies.
Muller's ratchet Accumulation of deleterious mutations in small populations due to genetic drift that becomes irreversible in the absence of recombination or back-mutations.
Multiplicative landscape An adaptive landscape in which there is no epistasis, that is, mutational effects are independent.
Mutation–selection balance The dynamic equilibrium between the generation of variability through deleterious mutation and its elimination through natural selection.
Mutational robustness Ability to tolerate mutations.
Phase transition A sudden change in the properties of a physical system in response to a change in one or more parameters.
Quasispecies A population at the mutation–selection balance composed by a master sequence and an ensemble of deleterious mutants.
Selection coefficient Denoted s, the relative fitness difference between a genotype and the wild type.
Sequence space A discrete space with as many dimensions as the genome length and four values on each dimension corresponding to the four possible nucleotides.
Wild type An arbitrary reference genotype, typically the most abundant one.

The Quasispecies Theory

Historical Context

The word quasispecies, borrowed from chemistry, refers to a population of quasi-identical molecules, by opposition to a molecular species, which is made of fully identical molecules. The term is probably unfortunate in the biological context because first, it could be erroneously thought of as if it made reference to the biological concept of species and second, biological populations are generally variable, making its usage redundant. For these reasons, some virologists have discarded it, while most keep making use of it to refer to a highly variable viral population. However, this is mainly a matter of semantics. Quasispecies constitute indeed a formal evolutionary theory first introduced in the early 1970s by Manfred Eigen and intended to capture the population dynamics of hypothetical replicons at the origin of life. These replicons were small and displayed a high rate of self-copying error due to the lack of proofreading mechanisms. Pioneer experiments with RNA bacteriophages showed high levels of genetic variability, suggesting error-prone replication and, later on, RNA viruses were confirmed to show high per-base mutation rates and short genomes. The quasispecies theory offered thus a relevant conceptual framework for RNA virus evolution and became a virological paradigm.

Scientific Context

Quasispecies theory is a population genetics theory. Population genetics, which took its first steps in the beginning of the twentieth century, addresses how changes in gene frequencies depend on measurable parameters such as selection coefficients, population sizes, mutation rates, or recombination rates, and provides a well-grounded theory for the study of evolution. Nevertheless, evolution is a very complex process and its mathematical treatment requires the use of simplifying assumptions. Choosing the appropriate assumptions is critical to build accurate models of evolution. This choice, which depends on the specific questions to be addressed, usually makes the difference between evolutionary models.

One central question in population genetics is how much variability can be stably maintained from the interplay between mutation and natural selection. In all populations, the continuous production of deleterious mutations through error-prone replication is counterbalanced by the elimination of these mutations through selection. When a population is sufficiently adapted to its environment, the observed variability is mainly determined by the interplay between the above two opposing factors which, in the long term, reach the so-called mutation–selection balance. Quasispecies theory mainly addresses this issue, though it had already been addressed decades before by the founders of population genetics. Since quasispecies come from the field of physical chemistry, their terminology is somewhat different, but most concepts are fully equivalent to those of population genetics (**Table 1**).

The Fundamental Model

The basic quasispecies model describes the population dynamics of macromolecular sequences maintained by error-prone replication. Mutations occurring during the copying process can modify the replication rate of the newly arisen sequences, allowing Darwinian selection to operate. The system can be described by a value matrix **W** in which each diagonal element W_{kk} equals the number of nonerroneous copies resulting from exact replication of sequence k and each off-diagonal element W_{kj} equals the number of copies of sequence k resulting from erroneous replication of sequence j. The stationary sequence distribution is called the quasispecies and consists of a master sequence along with a distribution of mutants. Although the slower replicators cannot sustain their abundance level by themselves, they are constantly recreated from other sequences that mutate into them. At the stationary state, the average biological fitness of the population equals the largest eigenvalue of **W** and the abundance of each sequence is given by the associated eigenvector.

In practice, given the huge dimension of **W** (equal to the genome length), exact solutions of the quasispecies equations cannot be found. If all sequences in a given error class (i.e., all one-error mutants, all two-error mutants, etc.) are assumed to have identical fitness, it becomes possible to estimate the frequency of each sequence as a function of the per-base mutation rate, its number of mutations, its relative fitness, and a recursively defined function that accounts for the fitness of all possible mutant intermediates. However, the fitness of all intermediates is not known in practice. The problem can be solved by assuming that mutations have independent fitness effects, or using population genetics terms, that there is no epistasis. This implies that the fitness of each mutant is simply the product of fitness values associated to each single mutant and thus, that the mean frequency of mutants carrying d mutations relative to that of the master (q_d/q_o), can be simply obtained by summing over all possible combinations of d mutations in a genome of length L. Using the binomial distribution, this can be written as $q_d/q_o = (1-\mu)^{-LBi(d|\mu,L)s^{-d}}$, where μ is the per-base mutation rate.

Assuming that L is large enough to be treated as infinite can be useful to illustrate the equivalence between quasispecies and population genetics models. The binomial distribution can then be substituted by a Poisson distribution of parameter $U = \mu L$, where U is the genomic mutation rate, and $(1-\mu)^L$ can be replaced by e^{-U}. Hence, $q_d/q_o = (U/s)^{d/d!}$ and, setting $q_o = e^{-U/s}$, it follows that q_d is Poisson-distributed with parameter $\lambda = U/s$. Hence, the average number of mutations per genome is U/s and the mean relative fitness of the population is e^{-U}, which are classical population genetics results.

Table 1 Conceptually equivalent terminology in quasispecies and population genetics theories

Quasispecies theory	*Population genetics theory*
Quasispecies	Population at the mutation–selection balance
Stationary state	Mutation–selection balance
Master sequence	Wild type
Superiority of the master	Average selection coefficient
Phase transition	Selective sweep
Excess production	Progeny number
Quality factor	Replication fidelity

Common Assumptions

Although the quasispecies theory is very general in its basic formulation, some simplifications have to be done for the sake of tractability, going to the detriment of generality. Common assumptions of the original quasispecies models are that: (1) there is a single fittest sequence; (2) mutation rates are constant over time and genotypes; (3) mutants show invariant fitness regardless of the number of mutations they carry; (4) molecules are degraded at a constant rate; (5) the environment is constant; (6) the population size is infinite; and (7) there is no genetic recombination. All of these assumptions were made in most of the analyses originally published by Eigen and co-workers. The situation in which there is

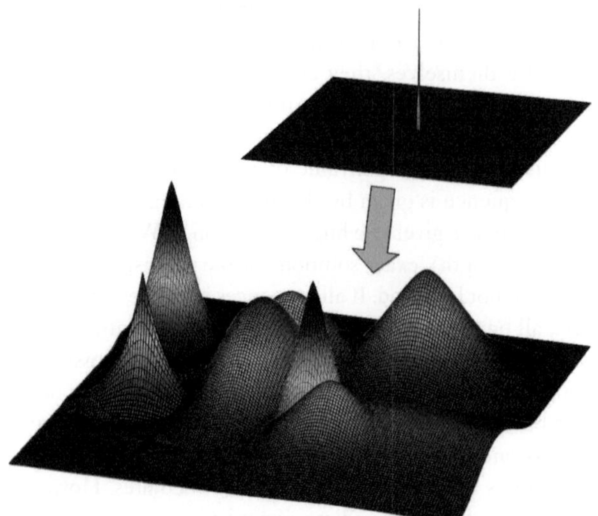

Figure 1 Two hypothetical adaptive landscapes. Sequence spaces are highly dimensional and need to be reduced to only two dimensions for visualization. The third dimension represents fitness, such that peaks in the landscape are local or global fitness maxima, whereas valleys are low-fitness regions. Above, the two-class single-peak landscape originally used by Eigen, in which there is a single fittest sequence and all other types show a lower, constant fitness value. A more complex landscape, with several local peaks, is represented below. Ruggedness appears as consequence of frequent epistasis of variable sign and intensity. The lower example provides a more realistic picture than the above two-class single-peak case, though it might still be much simpler than real landscapes.

a single fittest sequence and all other genotypes have constant fitness can be represented by a two-class single-peak landscape, although real landscapes can be much more complex, with several local adaptive peaks separated by low-fitness regions (**Figure 1**).

Later on, some of the above assumptions were relaxed. For example, Peter Schuster and co-workers studied the case in which there is more than one fittest sequence and called it a degenerate quasispecies. The two master sequences can be located at distant regions of the sequence space separated by low-fitness regions, hence being effectively unconnected. Different generalizations of the two-class single-peak landscape have been studied, including multidimensional landscapes. A simple though useful case is the multiplicative landscape, in which epistasis is assumed to be null and hence fitness decays exponentially with mutation number. Others have also investigated quasispecies in variable environments or with finite population sizes.

Relevant Predictions

Although population genetics and quasispecies are fundamentally equivalent, the assumptions that are typically done in quasispecies models often differ from those made in classical population genetics. Importantly, different assumptions can lead to different predictions. Below, we focus on two relevant examples, concerning the frequency of mutations and the topology of adaptive landscapes.

The evolution of mutational robustness

The Haldane–Muller principle states that the average fitness of the population at the mutation–selection balance does not depend on the fitness effects of mutations or, more specifically, on selection coefficients. The reason is that the more deleterious a mutation is, the more it reduces the average population fitness but, on the other hand, the more its frequency is lowered by selection. These two factors cancel out and the average relative fitness of the population simply writes e^{-U} (see above). This result is based on neglecting back-mutations, which is justified for large genomes replicating at low per-base error rates. However, quasispecies were intended to capture the population dynamics of small replicons with high per-base mutation rates, implying that secondary and back-mutations could not be systematically neglected. The reason is that the probability of occurrence of back-mutations relative to forward mutations increases in short genomes and, at high genomic mutation rates, the evolutionary fate of any given mutant genotype can depend not only on its fitness effect, but also on secondary mutations, or on the influx of mutations regenerating the genotype from neighbors in sequence space.

When secondary and back-mutations are taken into account, the average fitness of the population does not anymore strictly depend on the deleterious mutation rate, but also on the average selection coefficient. In other words, the fitness of a genotype is not only determined by its own replication rate, but also by those of its mutational neighbors. Everything else being equal, the lower the selection coefficient the higher the average fitness. A consequence is that, at high mutation rates, a slower replicator can potentially outgrow a faster competitor by virtue of its higher mutational robustness, a phenomenon called the 'survival of the flattest' (**Figure 2**). This effect rapidly vanishes as the mutation rate decreases and the genome length increases, but it can be magnified in small populations, in which genetic drift favors the accumulation of mutations. It should be noted that the statement that the average fitness is not a property of the master but also of the neighboring sequences does not mean that the genotypes constituting the quasispecies establish any kind of cooperation, complementation, or mutualism. Experiments showing cooperative interactions between genotypes should therefore not be considered as supportive of the quasispecies predictions.

The error catastrophe

The existence of an error catastrophe is probably the most widely known prediction of the quasispecies theory. An error threshold is the critical error rate beyond which

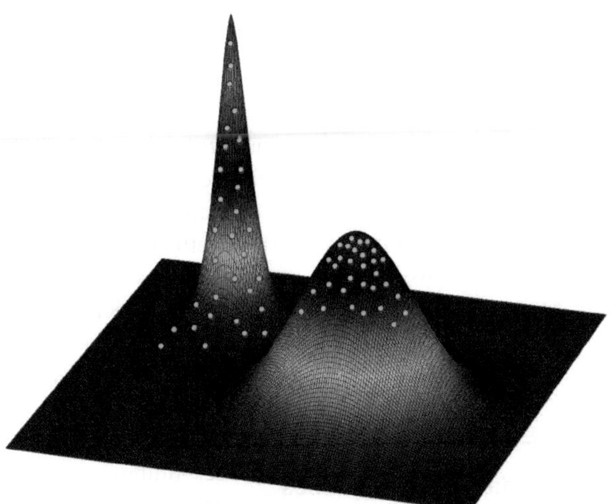

Figure 2 Representation of survival of the flattest in two-peak landscape. One population is located at a higher peak, which should confer it a selective advantage. However, the other population is located at a flatter peak, implying that neighbors in the sequence space have similar fitness and, thus, that the population is more robust to mutation. A prediction of the quasispecies theory is that, at high mutation rates, increased robustness can provide a selective advantage to the latter population.

Darwinian selection cannot further maintain the genetic integrity of the quasispecies. Below the threshold, the quasispecies is stably localized around one or more sequences, but if the mutation rate exceeds the threshold value, the variety of mutants that are in amounts comparable to the fittest types becomes very large and the population is said to enter into error catastrophe. According to the original definition given by Eigen, the threshold satisfies $W_{00} = \bar{w}_{k\neq0}$, where W_{00} is the nonmutated progeny of the master and $\bar{w}_{k\neq0}$ is the average fitness (mutated and nonmutated progeny) of all other sequences present in the population. If $W_{00} < \bar{w}_{k\neq0}$, the master is overgrown by its own mutational cloud and, in the absence of back-mutation, goes extinct through error catastrophe (although assumptions about back-mutation do not have a bearing on the existence or magnitude of error catastrophes).

The existence of an error threshold, however, is not an obligate prediction of the quasispecies theory. For example, in a multiplicative landscape, the frequency of the master is given by $q_0 = e^{-U/s}$, a quantity that decreases as the mutation rate increases but never reaches zero, hence there is no error threshold. The original quasispecies model produces an error threshold because all mutants are assumed to have a constant fitness regardless of their mutational distance to the master. In this case, $\bar{w}_{k\neq0}$ is a constant, $\bar{w}_{k\neq0} = 1 - s$, and, therefore, an error threshold deterministically takes place when $L = \log\sigma_0/\mu$, where σ_0 is the superiority of the master over all mutants. It must be noted that, even in this model, though there is an error threshold, no sharp transition in average fitness occurs. Average fitness smoothly decreases with increasing mutation rate until it reaches a plateau value equal to $\bar{w}_{k\neq0}$. In general, whether there is an error threshold critically depends on the assumed adaptive landscape. Some generalizations of the two-class single-peak landscape predict error catastrophes, but, in general, the outcome mainly depends on the sign of epistasis. Error catastrophes have a greater probability to exist if epistasis is antagonistic, such that mutations tend to have progressively less fitness effects as they accumulate (notice that the two-class single-peak landscape represents an extreme form of antagonistic epistasis). However, recent work has shown that with null epistasis, there can be an error catastrophe if lethal mutations are taken into account. This is a complex theoretical issue which still remains to be fully resolved.

Experimental Evolution of RNA Viruses

RNA viruses are useful tools for experimentally addressing fundamental evolutionary issues while they still allow us to pay attention to the molecular aspects. Based on scientific but also historical reasons, the quasispecies theory is the *de facto* standard for interpreting evolutionary experiments with RNA viruses. RNA viruses are characterized by their error-prone replication and small genomes, which makes the quasispecies an *a priori* relevant theoretical approach. However, even if some predictions are specific to the quasispecies theory, most are equivalent to those made by previously proposed and simpler population genetics models. Furthermore, quasispecies do not provide a theoretical underpinning to some experimental aspects of viral evolution which are better explained by nondeterministic population genetics models. After three decades of experimental work, some general evolutionary properties of RNA virus populations have been elucidated, most of them being compatible with both quasispecies and classical population genetics. A few data, though, support the need to use the more complex quasispecies formulation.

Key Evolutionary Parameters of RNA Viruses

High per-base mutation rates

In pioneer evolutionary experiments performed in the late 1970s, RNA bacteriophage Qβ populations serially passaged in laboratory conditions were found to be highly variable at the nucleotide level, each viable phage genome differing on average in one to two positions relative to the reference sequence. It was suggested that these populations were in a dynamic equilibrium, with viable mutants arising at a high rate and being strongly selected against, an equilibrium known as the mutation–selection balance.

From then on, many evolutionary experiments have been done with different model RNA viruses, including the bacteriophage φ6, the vesicular stomatitis virus (VSV), poliovirus-1, or foot-and-mouth disease virus (FMDV). It has been estimated that the mutation rate of RNA viruses is within the range 10^{-6}–10^{-4} substitutions per replication event, which is orders of magnitude higher than that of DNA-based organisms. Biochemical experiments with the avian myeloblastosis virus, VSV, and human immunodeficiency virus type 1 (HIV-1) have established that their polymerases lack 3′ exonuclease activity, providing a basis for error-prone replication. Recent experiments have demonstrated that poliovirus-1 is capable of counteracting the mutagenic effect of base analogs such as ribavirin by evolving specific genotypic changes that confer increased replication fidelity. This suggests that the high mutation rate of RNA viruses is not only a consequence of fundamental biochemical restrictions, but also the product of evolutionary optimization.

Compacted genomes

Another seemingly general property of RNA viruses is their small genome size (e.g., 3569 nt for MS2 phage to 11 162 nt for VSV), with many examples of overlapping reading frames and multifunctional genes. In contrast to multicellular eukaryotes, noncoding regions occupy a small fraction of the genome, hence making silent mutations relatively infrequent. A consequence of genome compactness is that RNA viruses are extremely sensitive to deleterious mutations. For VSV, it has been estimated that single random nucleotide substitutions have average selection coefficients of 0.5 (fitness is decreased by 50%), at least an order of magnitude higher than in DNA-based organisms, and it has been estimated that up to 40% of random mutations inactivate the virus. A related consequence of genome compactness is that epistasis is antagonistic on average, because in such compact genomes, one or few changes are enough to disrupt most of the encoded functions, and hence additional mutations can only produce a comparatively smaller effect.

Rapid growth

RNA viruses critically rely on fast replication for survival. Their rapid infection cycles allow them to reach high titers before the onset of host defense mechanisms. In cellular cultures, phages complete an infection cycle in less than an hour, whereas mammalian viruses do it in a few hours. Each infected cell typically releases hundreds to few thousands of infectious particles and viruses rapidly reach population sizes of several billion particles per milliliter. Intuitively, high particle counts indicate that the number of available genotypic variants is high, which should allow Darwinian selection to operate more efficiently. However, it must be noted that although population sizes are huge, the relevant parameter in evolutionary terms is the effective population size, since it indicates the actual number of particles that contribute to the next generation. Demographic fluctuations, originated by transmission bottlenecks, host defense mechanisms, or antiviral treatments make the effective population considerably smaller than particle counts. For example, in typical passage regimes with mammalian lytic viruses, if 10^4 particles were inoculated into 10^5 cells, effective population sizes would be in the range of 10^4–10^5 particles, whereas particle counts would typically be in the order of 10^9.

Main Evolutionary Properties of RNA Viruses

Rapid adaptation dynamics

Similar to what is found with other microorganisms, a frequent observation in long-term evolution experiments is that RNA viruses show an initially fast adaptation that tends to decelerate over time. Such dynamics indicates that, after being placed in a new environment, populations are evolving from a region of low fitness toward an adaptive peak or plateau. High population sizes make it more likely that beneficial mutations are created and, together with fast replication rates, this explains the observed rapid adaptation. Also, elevated selection coefficients increase the probability that any given beneficial mutation becomes fixed in the population. Finally, high mutation rates produce high levels of genetic variability on which selection can operate, though it must be noted that, since most mutations are deleterious, increasing the mutation rate does not necessarily imply increasing the rate of adaptation. Maximal adaptation rates should indeed be reached at intermediate mutation rates, the optimal mutation value being mainly determined by the selection coefficient against deleterious mutations and the rate of recombination.

Evolutionary parallelisms and convergences

In several serial passage experiments with bacteriophages, plant viruses, and mammalian viruses, sequencing of lineages independently adapted to the same environment has revealed a large amount of evolutionary parallelisms and convergences at the genetic level, both synonymous and nonsynonymous. This pattern indicates that, upon facing identical selective pressures, viruses often find the same adaptive pathways. It is possible that in simple and compacted genomes, such as those of RNA viruses, the number of alternative evolutionary solutions is limited, or that, due to their elevated adaptive potential, RNA viruses are able of systematically finding the highest peak in the adaptive landscape. Both scenarios would indicate that viral evolution at high population sizes has an important deterministic component, as proposed by quasispecies models. However, it is also possible that the common ancestor was located near a local peak surrounded by low-fitness regions and that, since Darwinian selection always pushes populations uphill, these populations would have remained trapped in their local peak.

Fitness tradeoffs

Despite their remarkable adaptability, RNA viruses face some evolutionary constraints. For example, shifts between vertebrate and arthropod hosts impose drastic environmental fluctuations to arboviruses. Several studies have confirmed that adaptation to a novel host often decreases competitive ability in the former host. These tradeoffs can arise by antagonistic pleiotropy or antagonistic epistasis among beneficial mutations. Pleiotropy, epistasis, and hence fitness tradeoffs are expected to be common in compacted genomes with many overlapping functions. A particularly important example of tradeoff is that exerted on mutation rate. Most mutations are deleterious, and hence there is a short-term selective pressure for reducing mutation rates toward whatever limit is imposed by biochemical restrictions, but, on the other hand, mechanisms of replication fidelity should come at a kinetic cost and thus be selected against. Since RNA viruses critically rely on fast replication, the cost of proofreading functions in terms of replication rate might be particularly strong.

Transmission bottlenecks

Considering the fact that RNA viruses show large selection coefficients and high population sizes, it could be concluded that selection is the main factor driving their evolution. However, in nature, viral populations experience strong bottlenecks upon transmission and therefore, effective population sizes are well below particle counts. Low effective population sizes increase the strength of genetic drift, hence favoring the accumulation of deleterious mutations and potentially jeopardizing the survival of the population. This stochastic process is known as Muller's ratchet, named after the population geneticist who first anticipated it in the 1930s. The expected rate of fitness loss equals the product of the mutation rate and the average selection coefficient for deleterious mutations, thus making RNA viruses especially sensitive to Muller's ratchet. This prediction has been experimentally validated by performing serial plaque-to-plaque passages in a variety of RNA viruses, including bacteriophage $\phi 6$, VSV, FMDV, and HIV-1.

Is the Quasispecies Theory Relevant to the Evolution of RNA Viruses?

There has been some controversy as to whether quasispecies are relevant to RNA virus evolution. Detractors have sometimes argued that the theory overlaps with previous population genetics theory, whereas supporters have claimed that it goes beyond population genetics. Many virologists believe that the elevated genetic variability of RNA viruses proves their quasispecies nature, but the mere observation of genetic variability is a trivial consequence of mutation and hence can hardly be considered supportive of any biological theory. Most of the empirically inferred properties of RNA viruses are already well explained by classical population genetics, as for example the expected high adaptability of large populations, the most likely deleterious effects of mutation, the complex relation between mutation and adaptation rates, the existence and evolutionary implications of fitness tradeoffs and epistasis, or the genetic contamination of small populations throughout Muller's ratchet. To justify the necessity of the extra complexity introduced by the quasispecies theory, evidence for its specific predictions should be provided.

The evolution of robustness

RNA viruses are highly sensitive to mutation compared to more complex microorganisms. Genetic hypersensitivity and large population sizes make selection very efficient at purging deleterious mutations and hence promote the preservation of the wild type. Although this seems to be the predominant survival strategy among RNA viruses, quasispecies theory predicts that at high mutation rates and low population sizes, mutational robustness can be selectively advantageous. The first experimental evidence in favor of this prediction came from work with the bacteriophage $\phi 6$ showing that the evolution of a genotype can be influenced by the topology of the neighboring regions in the adaptive landscape and that robustness can be a selectable character. A more direct approach was undertaken with viroids, plant pathogens consisting of small, noncoding RNA molecules. Two different viroid species were competed in common host plants; one was characterized by fast population growth and genetic homogeneity, whereas the other showed slow population growth and a high degree of variation. When the mutation rate was artificially increased using UVC radiation, the fitness of the slower replicator increased relative to that of the faster competitor, probably due to its higher mutational robustness, which might support the survival of the flattest hypothesis. Mutagens are especially suited for observing this quasispecies prediction, since they simultaneously increase mutation rates and decrease viral titers. Recent experiments with VSV populations indicate that base analogs such as 5-fluorouracil can provide a selective advantage to robust genotypes.

Error catastrophe and lethal mutagenesis

The consequences of artificially increasing error rates have been explored in cell culture experiments

Table 2 Differences between lethal mutagenesis and error catastrophe processes

	Lethal mutagenesis	*Error catastrophe*
Nature of the process	Population extinction through mutation accumulation	Delocalization of the quasispecies due to mutation
Name of the threshold	Extinction threshold	Error threshold
Key parameters	Mutation rate, fitness of the wild type, selection coefficient	Mutation rate and selection coefficient
Demography	Population size declines	No changes in population size are specified
Fate of the wild type	Not necessarily extinguished until the end of the process	Extinguished while other variants survive
Dependence on mutation rate	Extinction is more likely and occurs faster at higher mutation rates	Beyond the error threshold, further increases in mutation rate have no effect
Typically assumed effect of mutations on fitness	The fitness of mutant genotypes decays with mutation number	The fitness of mutant genotypes does not decay with mutation number
Mutational signature	No specific changes are required in the consensus sequence	The consensus sequence randomly drifts through time

with a variety of RNA viruses, including VSV, HIV-1, poliovirus-1, FMDV, and lymphocytic choriomeningitis virus (LCMV). All of these studies have shown that mutagens are detrimental to viral fitness and, in some cases, lead to extinction. For example, in HIV-1, the addition of the base analog 5-hydroxydeoxycytidine can result in a loss of infectivity after 10–20 serial passages. Virologists, assuming that error catastrophe is a form of lethal mutagenesis, have often interpreted these experimental observations in terms of error catastrophe. However, the two concepts are not equivalent, because lethal mutagenesis implies extinction, whereas the error catastrophe is a shift in sequence space without a bearing on the survival of the population (**Table 2**). The results from lethal mutagenesis experiments are easily explained by classical population genetics, provided fitness is expressed in absolute rather than relative terms. A simple criterion for extinction is $e^{-U}R$, where R is the progeny number per infectious particle and e^{-U} equals the mean equilibrium fitness of the population (this expression simply states that if the absolute mean fitness is below one, population size will deterministically decline). Therefore, current experimental evidence does not support the error catastrophe prediction. More specific observations would be required (**Table 2**), although it must be noted that the error catastrophe is not a necessary prediction of the quasispecies theory because it critically depends on the assumed adaptive landscape.

See also: Antigenic Variation; Antiviral Agents; Evolution of Viruses; Nature of Viruses; Virus Species.

Further Reading

Bull JJ, Sanjuan R, and Wilke CO (2007) Theory of lethal mutagenesis for viruses. *Journal of Virology* 81: 2930–2939.

Domingo E, Biebricher CK, Eigen M, and Holland JJ (2001) *Quasispecies and RNA Virus Evolution: Principles and Consequences.* Austin, TX: Landes Bioscience.

Domingo E, Sabo D, Taniguchi T, and Weissmann C (1978) Nucleotide sequence heterogeneity of an RNA phage population. *Cell* 13: 735–744.

Eigen M, McCaskill J, and Schuster P (1988) Molecular quasi-species. *Journal of Physical Chemistry* 92: 6881–6891.

Elena SF and Lenski RE (2003) Evolution experiments with microorganisms: The dynamics and genetic bases of adaptation. *Nature Reviews in Genetics* 4: 457–469.

Hartl DL and Clark AG (2006) *Principles of Population Genetics,* 4th edn., Sunderland, MA: Sinauer.

Holmes EC and Moya A (2002) Is the quasispecies concept relevant to RNA viruses? *Journal of Virology* 76: 460–465.

Lenski RE (2000) Evolution, theory and experiments. In: Alexander M, Bloom BR, Hopwood DA, *et al.* (eds.) *Encyclopedia of Microbiology,* 2nd edn., pp. 283–298. New York: Academic Press.

Moya A, Holmes EC, and Gonzalez-Candelas F (2004) The population genetics and evolutionary epidemiology of RNA viruses. *Nature Reviews in Microbiology* 2: 279–288.

Sanjuan R, Cuevas JM, Furio V, Holmes EC, and Moya A (2007) Selection for robustness in mutagenized. *PLoS Genetics* 3(6): e93.

Wagner A (2005) *Robustness and Evolvability in Living Systems.* Princeton, NJ: Princeton University Press.

Wilke CO (2005) Quasispecies theory in the context of population genetics. *BMC Evolutionary Biology* 5: 44.

Wolf JB, Brodie ED III, and Wade MJ (2000) *Epistasis and the Evolutionary Process.* Oxford: Oxford University Press.

Virus Databases

E J Lefkowitz and M R Odom, University of Alabama at Birmingham, Birmingham, AL, USA
C Upton, University of Victoria, Victoria, BC, Canada

Glossary

Annotation The process of adding descriptive information to a data set and the individual data elements that comprise that data set. May include both computational analyses and manual curation.
Curation The process of manually adding descriptive information or explanatory material to a data set usually based on a review of the existing literature.
Database A collection of information, usually stored electronically on a computer.
Database, hierarchical (XML) Extensible markup language. A formally defined hierarchical organization of data.
Database management system (DBMS) A database software package designed to operate on a particular computer platform and operating system.
Database schema A map of the database structure, including all of the fields, tables, and relationships that comprise the framework for storage of information within that database.
Database, relational A database structure that provides the ability to define formal relationships between all data elements stored within the database.
Ontology A controlled vocabulary used to formally describe (for our purposes) a biological process or entity.
SQL Structured Query Language. A computer language used to query relational databases for information.

Introduction

In 1955, Niu and Frankel-Conrat published the C-terminal amino acid sequence of tobacco mosaic virus capsid protein. The complete 158-amino-acid sequence of this protein was published in 1960. The first completely sequenced viral genome published was that of bacteriophage MS2 in 1976 (GenBank accession number V00642). Sanger used DNA from bacteriophage phiX174 (J02482) in developing the dideoxy sequencing method, while the first animal viral genome, SV40 (J02400), was sequenced using the Maxam and Gilbert method and published in 1978. Viruses therefore played a pivotal role in the development of modern-day sequencing methods, and viral sequence information (both protein and nucleotide) formed a substantial subset of the earliest available biological databases. In 1965, Margaret O. Dayhoff published the first publicly available database of biological sequence information. This *Atlas of Protein Sequence and Structure* was available only in printed form and contained the sequences of approximately 50 proteins. Establishment of a database of nucleic acid sequences began in 1979 through the efforts of Walter Goad at the US Department of Energy's Los Alamos National Laboratory (LANL) and separately at the European Molecular Biology Laboratories (EMBL) in the early 1980s. In 1982, the LANL database received funding from the National Institutes of Health (NIH) and was christened GenBank. In December of 1981, the Los Alamos Sequence Library contained 263 sequences of which 50 were from eukaryotic viruses and 12 were from bacteriophages. By its tenth release in 1983, GenBank contained 1865 sequences (1 827 214 nucleotides) of which 449 (457 721 nucleotides) were viral. In August of 2006, GenBank (release 154) contained approximately 59 000 000 records, including 367 000 viral sequences.

The number of available sequences has increased exponentially as sequencing technology has improved. In addition, other high-throughput technologies have been developed in recent years, such as those for gene expression and proteomic studies. All of these technologies generate enormous new data sets at ever-increasing rates. The challenge, therefore, has been to provide computational systems that support the storage, retrieval, analysis, and display of this information so that the research scientist can take advantage of this wealth of resources to ask and answer questions relevant to his or her work. Every article in this encyclopedia contains knowledge that has been derived in part from the analysis of large data sets. The ability to effectively and efficiently utilize these data sets is directly dependent on the databases that have been developed to support storage of this information. Fortunately, the continual development and improvement of data-intensive biological technologies has been matched by the development and improvement of computational technologies. This work, which includes both the development and utilization of databases as well as tools for storage and analysis of biological information, forms a very important part of the bioinformatics field. This article provides an overview of database structure and how that structure supports the storage of biological information. The different types of data associated with the analysis of viruses are discussed, followed by a review of some of the various online databases that store general biological information as well as virus-specific information.

Databases

Definition

A database is simply a collection of information, including the means to store, manipulate, retrieve, and share that information. For many of us, lab notebook fulfilled our initial need for a 'database'. However, this information storage vehicle did not prove to be an ideal place to archive our data. Backups were difficult, and retrieval more so. The advent of computers – especially the desktop computer – provided a new solution to the problem of data storage. Though initially this innovation took the form of spreadsheets and electronic notebooks, the subsequent development of both personal and large-scale database systems provided a much more robust solution to the problems of data storage, retrieval, and manipulation. The computer program supplying this functionality is called a 'database management system' (DBMS). Such systems provide at least four things: (1) the necessary computer code to guide a user through the process of database design; (2) a computer language that can be used to insert, manipulate, and query the data; (3) tools that allow the data to be exported in a variety of formats for sharing and distribution; and (4) the administrative functions necessary to ensure data integrity, security, and backup. However, regardless of the sophistication and diverse functions available in a typical modern DBMS, it is still up to the user to provide the proper context for data storage. The database must be properly designed to ensure that it supports the structure of the data being stored and also supports the types of queries and manipulations necessary to fully understand and efficiently analyze the properties of the data.

Data

The development of a database begins with a description of the data to be stored, all of the parameters associated with the data, and frequently a diagram of the format that will be used. The format used to store the data is called the database schema. The schema provides a detailed picture of the internal format of the database that includes specific containers to store each individual piece of data. While databases can store data in any number of different formats, the design of the particular schema used for a project is dependent on the data and the needs and expertise of the individuals creating, maintaining, and using the database. As an example, we will explore some of the possible formats for storing viral sequence data and provide examples of the database schema that could be used for such a project.

Figure 1(a) provides an example of a GenBank sequence record that is familiar to most biologists. These records are provided in a 'flat file' format in which all of the information associated with this particular sequence is provided in a human-readable form and in which all of the information is connected in some manner to the original sequence. In this format, the relationships between each piece of information and every other piece of information are only implicitly defined, that is, each line starts with a label that describes the information in the rest of the line, but it is up to the investigator reading the record to make all of the proper connections between each of the data fields (lines). The proper connections are not explicitly defined in this record. As trained scientists, we are able to read the record in **Figure 1(a)** and discern that this particular amino acid sequence is derived from a strain of Ebola virus that was studied by a group in Germany, and that this sequence codes for a protein that functions as the virus RNA polymerase. The format of this record was carefully designed to allow us, or a computer, to pull out each individual type of information. However as trained scientists, we already understand the proper connections between the different information fields in this file. The computer does not. Therefore, to analyze the data using a computer, a custom software program must be written to provide access to the data.

Extensible markup language (XML) is another widely used format for storing database information. **Figure 1(b)** shows an example of part of the XML record for the Ebola virus polymerase protein. In this format, each data field can be many lines long; the start and end of a data record contained within a particular field are indicated by tags made of a label between two brackets ('<label>...</label>'). Unlike the lines in the GenBank record in **Figure 1(a)**, a field in an XML record can be placed inside of another, defining a structure and a relationship between them. For example, the *TSeq_orgname* is placed inside of the *TSeq* record to show that this organism name applies only to that sequence record. If the file contained multiple sequences, each *TSeq* field would have its own *TSeq_orgname* subfield, and the relationship between them would be very clear. This self-describing hierarchical structure makes XML very powerful for expressing many types of data that are hard to express in a single table, such as that used in a spreadsheet. However, in order to find any piece of information in the XML file, a user (with an appropriate search program) needs to traverse the whole file in order to pull out the particular items of data that are of interest. Therefore, while an XML file may be an excellent format for defining and exchanging data, it is often not the best vehicle for efficiently storing and querying that data. That is still the realm of the relational database.

GenBank record

```
LOCUS       NP_066251               2212 aa            linear   VRL 30-MAR-2006
DEFINITION  polymerase [Zaire ebolavirus].
ACCESSION   NP_066251
VERSION     NP_066251.1  GI:10313999
DBSOURCE    REFSEQ: accession NC_002549.1
KEYWORDS    .
SOURCE      Zaire ebolavirus (ZEBOV)
  ORGANISM  Zaire ebolavirus
            Viruses; ssRNA negative-strand viruses; Mononegavirales;
            Filoviridae; Ebola-like viruses.
REFERENCE   1  (residues 1 to 2212)
  AUTHORS   Volchkov,V.E., Volchkova,V.A., Chepurnov,A.A., Blinov,V.M.,
            Dolnik,O., Netesov,S.V. and Feldmann,H.
  TITLE     Characterization of the L gene and 5' trailer region of Ebola virus
  JOURNAL   J. Gen. Virol. 80 (Pt 2), 355-362 (1999)
   PUBMED   10073695
...
REFERENCE   8  (residues 1 to 2212)
  AUTHORS   Volchkov,V.E.
  TITLE     Direct Submission
  JOURNAL   Submitted (20-AUG-1998) Institute of Virology, Philipps-University
            Marburg, Robert-Koch-Str. 17, Marburg 35037, Germany
COMMENT     PROVISIONAL REFSEQ: This record has not yet been subject to final
            NCBI review. The reference sequence was derived from AAD14589.
            Method: conceptual translation.
FEATURES             Location/Qualifiers
     source          1..2212
                     /organism="Zaire ebolavirus"
                     /strain="Mayinga"
                     /db_xref="taxon:186538"
     Protein         1..2212
                     /product="polymerase"
                     /function="synthesis of viral RNAs; transcriptional RNA
                     editing"
                     /note="L"
                     /calculated_mol_wt=252595
     CDS             1..2212
                     /gene="L"
                     /locus_tag="ZEBOVgp7"
                     /coded_by="NC_002549.1:11581..18219"
                     /citation=[1]
                     /db_xref="GeneID:911824"
ORIGIN
        1 matqhtqypd arlsspivld qcdlvtracg lyssyslnpq lrncklpkhi yrlkydvtvt
       61 kflsdvpvat lpidfivpvl lkalsgngfc pveprcqqfl deiikytmqd alflkyylkn
...
     2101 ndynqqrqsr tqtyhfirta kgritklvnd ylkfflivqa lkhngtwqae fkklpelisv
     2161 cnrfyhirdc nceerflvqt lylhrmqdse vklierltgl lslfpdglyr fd
//
```

(a)

XML record

```
<?xml version="1.0"?>
<!DOCTYPE TSeqSet PUBLIC "NCBI_TSeq.dtd">
<TSeqSet>
<TSeq>
  <TSeq_seqtype value="protein"/>
  <TSeq_gi>10313999</TSeq_gi>
  <TSeq_accver>NP_066251.1</TSeq_accver>
  <TSeq_taxid>186538</TSeq_taxid>
  <TSeq_orgname>Zaire ebolavirus</TSeq_orgname>
  <TSeq_defline>polymerase [Zaire ebolavirus]</TSeq_defline>
  <TSeq_length>2212</TSeq_length>
  <TSeq_sequence>
  MATQHTQYPDARLSSPIVLDQCDLVTRACGLYSSYSLNPQLRNCKLPKHIYRLKYDVTVTKFLSDVPVAT
  LPIDFIVPVLLKALSGNGFCPVEPRCQQFLDEIIKYTMQDALFLKYYLKNVGAQEDCVDEHFQEKILSSI
  ...
  QLQIQRSPYWLSHLTQYADCELHLSYIRLGFPSLEKVLYHRYNLVDSKRGPLVSITQHLAHLRAEIRELT
  NDYNQQRQSRTQTYHFIRTAKGRITKLVNDYLKFFLIVQALKHNGTWQAEFKKLPELISVCNRFYHIRDC
  NCEERFLVQTLYLHRMQDSEVKLIERLTGLLSLFPDGLYRFD
  </TSeq_sequence>
</TSeq>
</TSeqSet>
```

(b)

Figure 1 Data formats. Examples of two different formats for organizing sequence data are shown. (a) An example of a GenBank flat file sequence record. (b) Part of the same record using a hierarchical XML format.

'Relational database management systems' (RDBMSs) are designed to do two things extremely well: (1) store and update structured data with high integrity, and (2) provide powerful tools to search, summarize, and analyze the data. The format used for storing the data is to divide it into several tables, each of which is equivalent to a single spreadsheet. The relationships between the data in the tables are then defined, and the RDBMS ensures that all data follow the rules laid out by this design. This set of tables and relationships is called the schema. An example diagram of a relational database schema is provided in **Figure 2**. This Viral Genome Database (VGD) schema is an idealized version of a database used to store viral genome sequences, their associated gene sequences, and associated descriptive and analytical information. Each box in **Figure 2** represents a single object or concept, such as a genome, gene, or virus, about which we want to store data and is contained in a single table in the RDBMS. The names listed in the box are the columns of that table, which hold the various types of data about the object. The 'gene' table therefore contains columns holding data such as the name of the gene, its coding strand, and a description of its function. The RDBMS is able to enforce a series of rules for tables that are linked by defining relationships that ensure data integrity and accuracy. These relationships are defined by a foreign key in one table that links to corresponding data in another table defined by a primary key. In this example, the RDMS can check that every gene in the 'gene' table refers to an existing genome in the 'genome' table, by ensuring that each of these tables contains a matching 'genome_id'. Since any one genome can code for many genes, many genes may contain the same 'genome_id'. This defines what is called a one-to-many relationship between the 'genome' and 'gene' tables. All of these relationships are identified in **Figure 2** by arrows connecting the tables.

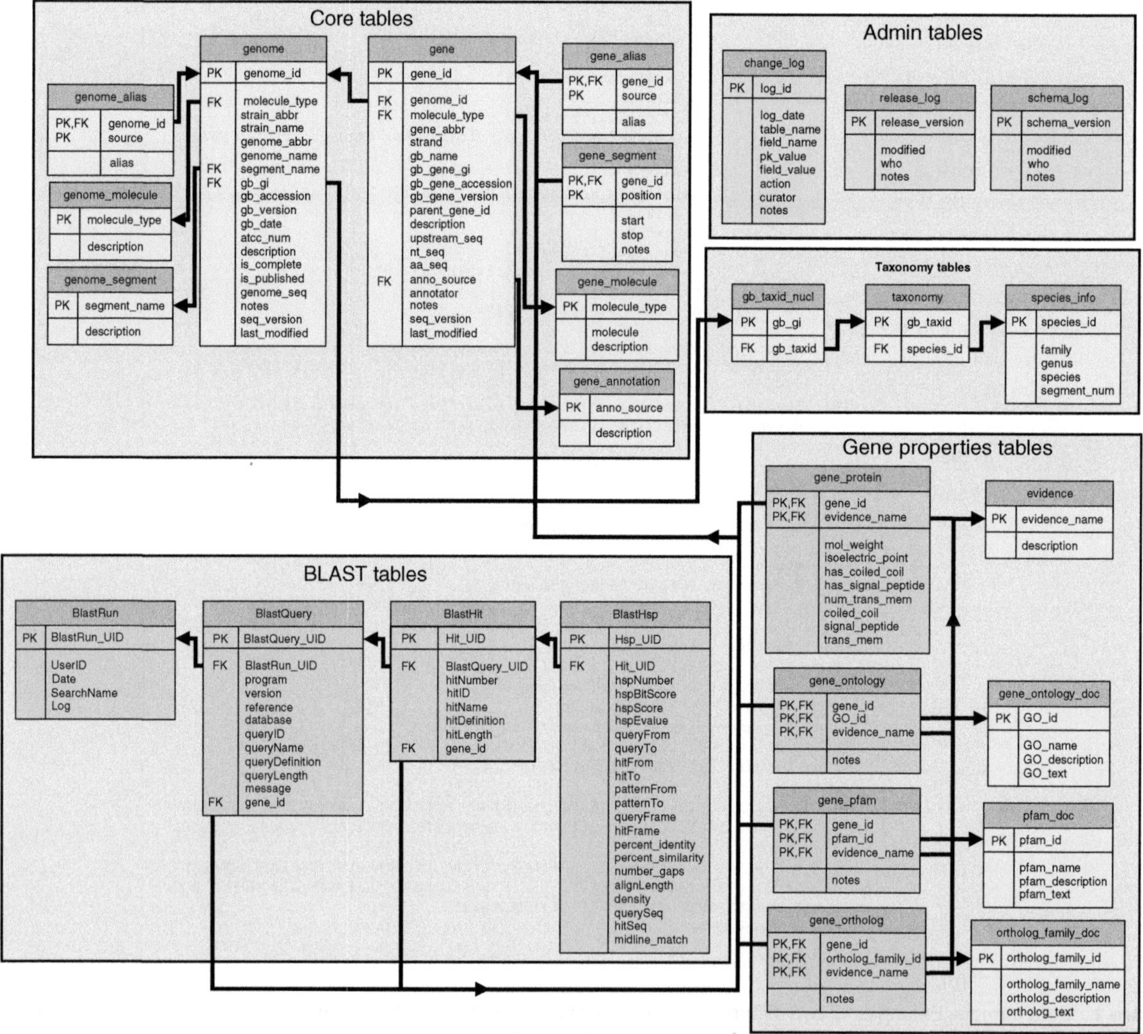

Figure 2 VGD Database schema. Underlying structure of a Viral Genome Sequence Database (VGD). Data tables are grouped according to the type of information contained in each set of tables. Each table contains a set of fields that hold a particular type of data. Lines and arrows display the relationships between fields as defined by the foreign key (FK) and primary key (PK) that connect two tables. (Each arrow points to the table containing the primary key.) Tables are color-coded according to the source of the information they contain: yellow, data obtained from the original GenBank sequence record and the ICTV Eighth Report; pink, data obtained from automated annotation or manual curation; blue, controlled vocabularies to ensure data consistency; green, administrative data.

Because viruses have evolved a variety of alternative coding strategies such as splicing and RNA editing; it is necessary to design the database so that these processes can be formally described. The 'gene segment' table specifies the genomic location of the nucleotides that code for each gene. If a gene is coded in the traditional manner – one ORF, one protein – then that gene would have one record in the 'gene_segment' table. However, as described above, if a gene is translated from a spliced transcript, it would be represented in the 'gene_segment' table by two or more records, each of which specifies the location of a single exon. If an RNA transcript is edited by stuttering of the polymerase at a particular run of nucleotides, resulting in the addition of one or more nontemplated nucleotides, then that gene will also have at least two records in the 'gene_segment' table. In this case, the second 'gene_segment' record may overlap the last base of the first record for that gene. In this manner, an extra, nontemplated base becomes part of the final gene transcript. Other more complex coding schemes can also be identified using this, or similar, database structures.

The tables in **Figure 2** are grouped according to the type of information they contain. Though the database itself does not formally group tables in this manner, database schema diagrams are created to benefit database designers and users by enhancing their ability to understand the structure of the database. These diagrams make it easier to both populate the database with data and query the database for information. The core tables hold basic biological information about each viral strain and its genomic sequence (or sequences if the virus contains segmented genomes) as well as the genes coded for by each genome. The taxonomy tables provide the taxonomic classification of each virus. Taxonomic designations are taken directly from the *Eighth Report of the International Committee on Taxonomy of Viruses* (ICTV). The 'gene properties' tables provide information related to the properties of each gene in the database. Gene properties may be generated from computational analyses such as calculations of molecular weight and isoelectric point (pI) that are derived from the amino acid sequence. Gene properties may also be derived from a manual curation process in which an investigator might identify, for example, functional attributes of a sequence based on evidence provided from a literature search. Assignment of 'gene ontology' terms (see below) is another example of information provided during manual curation. The BLAST tables store the results of similarity searches of every gene and genome in the VGD searched against a variety of sequence databases using the National Center for Biotechnology Information (NCBI) BLAST program. Examples of search databases might include the complete GenBank nonredundant protein database and/or a database comprised of all the protein sequences in the VGD itself. While most of us store our BLAST search results as files on our desktop computers, it is useful to store this information within the database to provide rapid access to similarity results for comparative purposes; to use these results to assign genes to orthologous families of related sequences; and to use these results in applications that analyze data in the database and, for example, display the results of an analysis between two or more types of viruses showing shared sets of common genes. Finally, the 'admin' tables provide information on each new data release, an archive of old data records that have been subsequently updated, and a log detailing updates to the database schema itself.

It is useful for database designers, managers, and data submitters to understand the types of information that each table contains and the source of that information. Therefore, the database schema provided in **Figure 2** is color-coded according to the type and source of information each table provides. Yellow tables contain basic biological data obtained either directly from the GenBank record or from other sources such as the ICTV. Pink tables contain data obtained as the result of either computational analyses (BLAST searches, calculations of molecular weight, functional motif similarities, etc.) or from manual curation. Blue tables provide a controlled vocabulary that is used to populate fields in other tables. This ensures that a descriptive term used to describe some property of a virus has been approved for use by a human curator, is spelled correctly, and when multiple terms or aliases exist for the same descriptor, the same one is always chosen.

While the use of a controlled vocabulary may appear trivial, in fact, misuse of terms, or even misspellings, can result in severe problems in computer-based databases. The computer does not know that the terms 'negative-sense RNA virus' and 'negative-strand RNA virus' may both be referring to the same type of virus. The provision and use of a controlled vocabulary increases the likelihood that these terms will be used properly, and ensures that the fields containing these terms will be easily comparable. For example, the 'genome_molecule' table contains the following permissible values for 'molecule_type': 'ambisense ssRNA', 'dsRNA', 'negative-sense ssRNA', 'positive-sense ssRNA', 'ssDNA', and 'dsDNA'. A particular viral genome must then have one of these values entered into the 'molecule_type' field of the 'genome' table, since this field is a foreign key to the 'molecule_type' primary key of the 'genome_molecule' table. Entering 'double-stranded DNA' would not be permissible.

Annotation

Raw data obtained directly from high-throughput analytical techniques such as automated sequencing, protein interaction, or microarray experiments contain little-to-no information as to the content or meaning. The process of adding value to the raw data to increase the knowledge content is known as annotation and curation. As an example, the results of a microarray experiment may provide an indication that individual genes are up- or down-regulated under certain experimental conditions. By annotating the properties of those genes, we are able to see that certain sets of genes showing coordinated regulation are a part of common biological pathways. An important pattern then emerges that was not discernable solely by inspection of the original data. The annotation process consists of a semiautomated analysis of the information content of the data and provides a variety of descriptive features that aid the process of assigning meaning to the data. The investigator is then able to use this analytical information to more closely inspect the data during a manual curation process that might support the reconstruction of gene expression or protein interaction pathways, or allow for the inference of functional attributes of each identified gene. All of this curated information can then be stored back in the database and associated with each particular gene.

For each piece of information associated with a gene (or other biological entity) during the process of annotation and curation, it is always important to provide the evidence used to support each assignment. This evidence may be described in a Standard Operating Procedure (SOP) document which, much like an experimental protocol, details the annotation process and includes a description of the computer algorithms, programs, and analysis pipelines that were used to compile that information. Each piece of information annotated by the use of this pipeline might then be coded, for example, 'IEA: Inferred from Electronic Annotation'. For information obtained from the literature during manual curation, the literature reference from which the information was obtained should always be provided along with a code that describes the source of the information. Some of the possible evidence codes include 'IDA: Inferred from Direct Assay', 'IGI: Inferred from Genetic Interaction', 'IMP: Inferred from Mutant Phenotype', or 'ISS: Inferred from Sequence or Structural Similarity'. These evidence codes are taken from a list provided by the Gene Ontology (GO) Consortium (see below) and as such represent a controlled vocabulary that any data curator can use and that will be understood by anyone familiar with the GO database. This controlled evidence vocabulary is stored in the 'evidence' table, and each record in every one of the gene properties tables is assigned an evidence code noting the source of the annotation/curation data.

As indicated above, the use of controlled vocabularies (ontologies) to describe the attributes of biological data is extremely important. It is only through the use of these controlled vocabularies that a consistent, documented approach can be taken during the annotation/curation process. And while there may be instances where creating your own ontology may be necessary, the use of already available, community-developed ontologies ensures that the ontological descriptions assigned to your database will be understood by anyone familiar with the public ontology. Use of these public ontologies also ensures that they support comparative analyses with other available databases that also make use of the same ontological descriptions. The GO Consortium provides one of the most extensive and widely used controlled vocabularies available for biological systems. GO describes biological systems in terms of their biological processes, cellular components, and molecular functions. The GO effort is community-driven, and any scientist can participate in the development and refinement of the GO vocabulary. Currently, GO contains a number of terms specific to viral processes, but these tend to be oriented toward particular viral families, and may not necessarily be the same terms used by investigators in other areas of virology. Therefore it is important that work continues in the virus community to expand the availability and use of GO terms relevant to all viruses. GO is not intended to cover all things biological. Therefore, other ontologies exist and are actively being developed to support the description of many other biological processes and entities. For example, GO does not describe disease-related processes or mutants; it does not cover protein structure or protein interactions; and it does not cover evolutionary processes. A complementary effort is under way to better organize existing ontologies, and to provide tools and mechanisms to develop and catalog new ontologies. This work is being undertaken by the National Center for Biomedical Ontologies, located at Stanford University, with participants worldwide.

Access (Searching for Information)

The most comprehensive, well-designed database is useless if no method has been provided to access that database, or if access is difficult due to a poorly designed application. Therefore, providing a search interface that meets the needs of intended users is critical to fully realizing the potential of any effort at developing a comprehensive database. Access can be provided using a number of different methods ranging from direct query of the database using the relatively standardized 'structured query language' (SQL), to customized applications designed to provide the ability to ask sophisticated questions regarding the data contained in the database and mine the data for meaningful patterns. Web pages may be designed to provide simple-to-use forms to access and query data stored in an RDBMS.

Using the VGD schema as a data source, one example of an SQL query might be to find the gene_id and name of all the proteins in the database that have a molecular weight between 20 000 and 30 000, and also have at least one transmembrane region.

Many database providers also provide users with the ability to download copies of the database so that these users may analyze the data using their own set of analytical tools.

Output (Utilizing Information)

When a user queries a database using any of the available access methods, the results of that query are generally provided in the form of a table where columns represent fields in the database and the rows represent the data from individual database records. Tabular output can be easily imported into spreadsheet applications, sorted, manipulated, and reformatted for use in other applications. But while extremely flexible, tabular output is not always the best format to use to fully understand the underlying data and the biological implications. Therefore, many applications that connect to databases provide a variety of visualization tools that display the data graphically, showing patterns in the data that may be difficult to discern using text-based output. An example of one such visual display is provided in **Figure 3** and shows conservation of synteny between the genes of two different poxvirus species. The information used to generate this figure comes directly from the data provided in the VGD. Every gene in the two viruses (in this case crocodilepox virus and molluscum contagiosum virus) has been compared to every other gene using the BLAST search program. The results of this search are stored in the BLAST tables of the VGD. In addition, the location of each gene within its respective genomic sequence is stored in the 'gene_segment' table. This information, once extracted from the database server, is initially text but it is then submitted to a program running on the server that reformats the data and creates a graph. In this manner, it is much easier to visualize the series of points formed along a diagonal when there are a series of similar genes with similar genomic locations present in each of the two viruses. These data sets may contain gene synteny patterns that display deletion, insertion, or recombination events during the course of viral evolution. These patterns can be difficult to detect with text-based tables, but are easy to discern using visual displays of the data.

Information provided to a user as the result of a database query may contain data derived from a combination of sources, and displayed using both visual and textual feedback. **Figure 4** shows the web-based output of a query designed to display information related to a particular virus gene. The top of this web page displays the location of the gene on the genome visually showing surrounding genes on a partial map of the viral genome. Basic gene information such as genome coordinates, gene name, and the nucleotide and amino acid sequence are also provided. This information was originally obtained from the original GenBank record and then stored in the VGD database. Data added as the result of an automated annotation

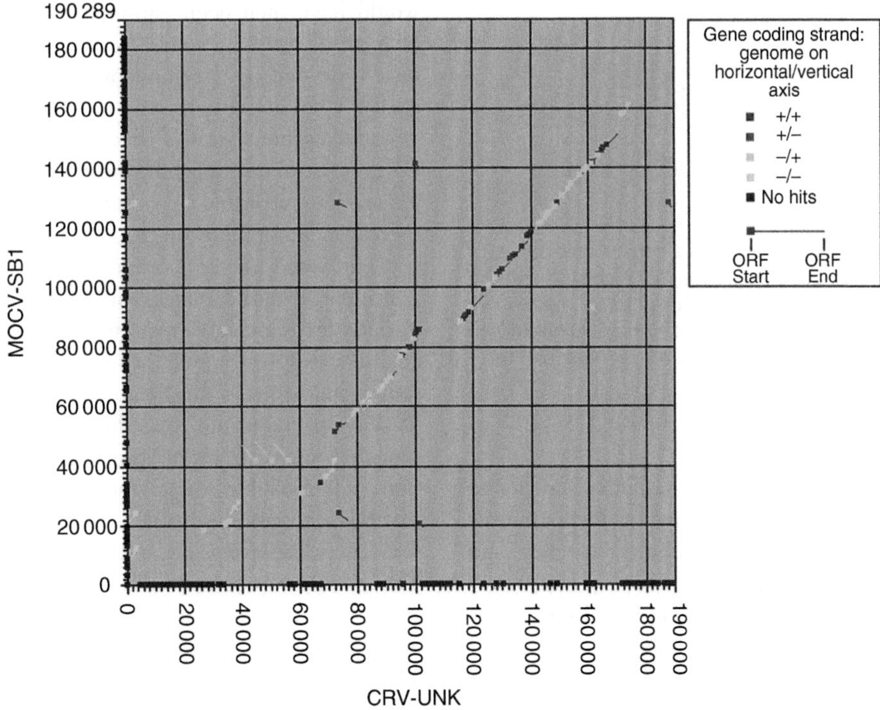

Figure 3 Gene synteny plot. A comparison of gene sequence and genomic position conservation between crocodilepox virus (horizontal axis) and molluscum contagiosum virus (vertical axis). All predicted proteins encoded by each virus were compared to each other using BLASTP. Each pair of proteins showing some measure of similarity, as determined by a BLAST expect (*E*) value <0.00001, was plotted according to the location of each gene on each respective genome. The color of the points reflects the identity of the coding strand of each gene. Black points along either of the two axes represent proteins unique to that genome.

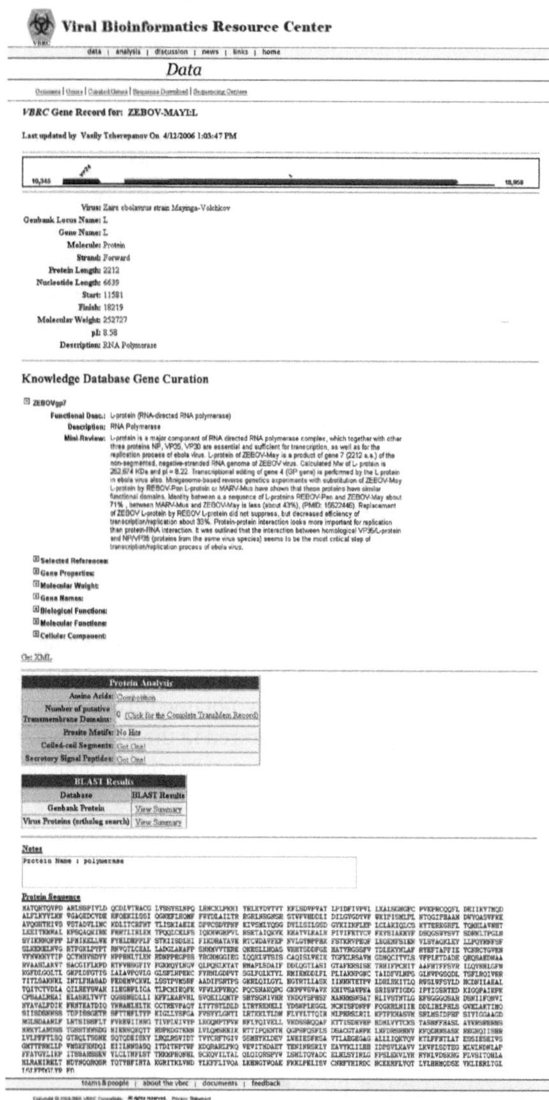

Figure 4 Gene record. An example of a gene record derived from data in the Viral Genome Database. The record contains a map showing the gene's location; basic descriptive information; analytical data; manually curated descriptive information; links to protein sequence analyses and BLAST similarity data; and the gene sequence itself.

pipeline are also displayed. This includes calculated values for molecular weight and pI; amino acid composition; functional motifs; BLAST similarity searches; and predicted protein structural properties such as transmembrane domains, coiled-coil regions, and signal sequences. Finally, information obtained from a manual curation of the gene through an extensive literature search is also displayed. Curated information includes a mini review of gene function; experimentally determined gene properties such as molecular weight, pI, and protein structure; alternative names and aliases used in the literature; assignment of ontological terms describing gene function; the availability of reagents such as antibodies and clones; and also, as available, information on the functional effects of mutations. All of the information to construct the web page for this gene is directly provided as the result of a single database query. (The tables storing the manually curated gene information are not shown in **Figure 2**.) Obviously, compiling the data and entering it into the database required a substantial amount of effort, both computationally and manually; however, the information is now much more available and useful to the research scientist.

Errors

No discussion of databases would be complete without considering errors. As in any other scientific endeavor, the data we generate, the knowledge we derive from the data, and the inferences we make as a result of the analysis of the data are all subject to error. These errors can be introduced at many points in the analytical chain. The original data may be faulty: using sequence data as one example, nucleotides in a DNA sequence may have been misread or miscalled, or someone may even have mistyped the sequence. The database may have been poorly designed; a field in a table designed to hold sequence information may have been set to hold only 2000 characters, whereas the sequences imported into that field may be longer than 2000 nucleotides. The sequences would have then been automatically truncated to 2000 characters, resulting in the loss of data. The curator may have mistyped an Enzyme Commission (EC) number for an RNA polymerase, or may have incorrectly assigned a genomic sequence to the wrong taxonomic classification. Or even more insidious, the curator may have been using annotations provided by other groups that had justified their own annotations on the basis of matches to annotations provided by yet another group. Such chains of evidence may extend far back, and the chance of propagating an early error increases with time. Such error propagation can be widespread indeed, affecting the work of multiple sequencing centers and database creators and providers. This is especially true given the dependencies of genomic sequence annotations on previously published annotations. The possible sources of errors are numerous, and it is the responsibility of both the database provider and the user to be aware of, and on the lookout for, errors. The database provider can, with careful database and application design, apply error-checking routines to many aspects of the data storage and analysis pipeline. The code can check for truncated sequences, interrupted open reading frames, and nonsense data, as well as data annotations that do not match a provided controlled vocabulary. But the user should always approach any database or the output of any application with a little healthy skepticism. The user is the final arbiter of the accuracy of the information, and it is their responsibility to look out for inconsistent or erroneous results that may indicate either a random or systemic error at some point in the process of data collection and analysis.

Virus Databases

It is not feasible to provide a comprehensive and current list of all available databases that contain virus-related information or information of use to virus researchers. New databases appear on a regular basis; existing databases either disappear or become stagnant and outdated; or databases may change focus and domains of interest. Any resource published in book format attempting to provide an up-to-date list would be out-of-date on the day of publication. Even web-based lists of database resources quickly become out-of-date due to the rapidity with which available resources change, and the difficulty and extensive effort required to keep an online list current and inclusive. Therefore, our approach in this article is to provide an overview of the types of data that are obtainable from available biological databases, and to list some of the more important database resources that have been available for extended periods of time and, importantly, remain current through a process of continual updating and refinement. We should also emphasize that the use of web-based search tools such as Google, various web logs (Blogs), and news groups, can provide some of the best means of locating existing and newly available web-based information sources. Information contained in databases can be used to address a wide variety of problems. A sampling of the areas of research facilitated by virus databases includes

- taxonomy and classification;
- host range, distribution, and ecology;
- evolutionary biology;
- pathogenesis;
- host–pathogen interaction;
- epidemiology;
- disease surveillance;
- detection;
- prevention;
- prophylaxis;
- diagnosis; and
- treatment.

Addressing these problems involves mining the data in an appropriate database in order to detect patterns that allow certain associations, generalizations, cause–effect relationships, or structure–function relationships to be discerned. **Table 1** provides a list of some of the more useful and stable database resources of possible interest to virus researchers. Below, we expand on some of this information and provide a brief discussion concerning the sources and intended uses of these data sets.

Data

Major repositories of biological information

The two major, overarching collections of biological databases are at the NCBI, supported by the National Library of Medicine at the NIH, and the EMBL, part of the European Bioinformatics Institute. These large data repositories try to be all-inclusive, acting as the primary source of publicly available molecular biological data for the scientific community. In fact, most journals require that, prior to publication, investigators submit their original sequence data to one of these repositories. In addition to sequence data, NCBI and EMBL (along with many other data repositories) include a large variety of other data types, such as that obtained from gene-expression experiments and studies investigating biological structures. Journals are also extending the requirement for data deposition to some of these other data types. Note that while much of the data available from these repositories is raw data obtained directly as the result of experimental investigation in the laboratory, a variety of 'value-added' secondary databases are also available that take primary data records and manipulate or annotate them in some fashion in order to derive additional useful information.

When an investigator is unsure about the existence or source of some biological data, the NCBI and EMBL websites should serve as the starting point for locating such information. The NCBI Entrez Search Engine provides a powerful interface to access all information contained in the various NCBI databases, including all available sequence records. A search engine such as Google might also be used if NCBI and EMBL fail to locate the desired information. Of course PubMed, the repository of literature citations maintained at NCBI, also represents a major reference site for locating biological information. Finally, the journal *Nucleic Acids Research* (NAR) publishes an annual 'database' issue and an annual 'web server' issue that are excellent references for finding new biological databases and websites. And while the most recent NAR database or web server issue may contain articles on a variety of new and interesting databases and websites, be sure to also look at issues from previous years. Older issues contain articles on many existing sites that may not necessarily be represented in the latest journal publication, but are nevertheless still available and current.

There are several websites that serve to provide general virus-specific information and links of use to virus researchers. One of these is the NCBI Viral Genomes Project, which provides an overview of all virus-related NCBI resources including taxonomy, sequence, and reference information. Links to other sources of viral data are provided, as well as a number of analytical tools that have been developed to support viral taxonomic

Table 1 Virus databases and other related information

Information resource name	*Sponsor*	*URL*	*Description*
Indices of databases			
All the Virology on the WWW	David M. Sander; Tulane University	http://www.virology.net/	Collection of virology information on the Internet
Nucleic Acids Research 2006 Database Issue	Journal: Nucleic Acids Research	http://nar.oxfordjournals.org/content/vol34/suppl_1/index.dtl	Issue describing molecular biology databases
Nucleic Acids Research 2006 Web Server Issue	Journal: Nucleic Acids Research	http://nar.oxfordjournals.org/content/vol34/suppl_2/index.dtl	Issue describing molecular biology websites
General sites			
NCBI	National Center for Biotechnology Information (NCBI), National Library of Medicine, National Institutes of Health (NIH)	http://www.ncbi.nlm.nih.gov/	Resource for molecular biology information
PubMed	National Library of Medicine, NIH	http://www.pubmed.gov/	Biomedical literature database
GenBank Sequence Database	NCBI	http://www.ncbi.nih.gov/Genbank/	Annotated collection of publicly available nucleotide sequences
NCBI Trace Archive	NCBI	http://www.ncbi.nlm.nih.gov/Traces/	NCBI repository of sequence trace files
NCBI Assembly Archive	NCBI	http://www.ncbi.nlm.nih.gov/Traces/assembly/	NCBI repository of sequence assembly files
EMBL Sequence Database	European Molecular Biology Laboratory (EMBL)	http://www.ebi.ac.uk/embl/	Annotated collection of publicly available nucleotide sequences
DDBJ Sequence Database	DNA Data Bank of Japan	http://www.ddbj.nig.ac.jp/	Annotated collection of publicly available nucleotide sequences
UniProt (Universal Protein Resource)	EMBL; Protein Information Resource, Georgetown University; Swiss Institute of Bioinformatics	http://www.ebi.ac.uk/uniprot/	Protein sequence repository
Taxonomy			
ICTVnet	International Committee on Taxonomy of Viruses (ICTV)	http://www.danforthcenter.org/iltab/ictvnet/	ICTV official website
ICTVdb	ICTV	http://www.ncbi.nlm.nih.gov/ICTVdb	ICTV database of virus taxonomy
NCBI Taxonomy Browser	NCBI	http://www.ncbi.nlm.nih.gov/Taxonomy/	Taxonomy-based retrieval of sequence data
Virus-oriented			
NCBI Viral Genomes Resource	NCBI	http://www.ncbi.nlm.nih.gov/genomes/VIRUSES/viruses.html	NCBI resources for the study of viruses
ASV	American Society for Virology (ASV)	http://www.asv.org/	Society to promote the exchange of information and stimulate discussion and collaboration among scientists active in all aspects of virology
IUMS-Virology	International Union of Microbiological Societies	http://www.iums.org/divisions/divisions-virology.html	Society to promote the study of microbiological sciences internationally
Viruses: From Structure to Biology	ASV; Washington University School of Medicine	http://medicine.wustl.edu/~virology/	Historical overview of virus research

BRCs (Bioinformatics Resource Centers for Biodefense and Emerging or Re-Emerging Infectious Diseases)	National Institute of Allergy and Infectious Diseases (NIAID), NIH	http://www.brc-central.org/	NIAID/NIH-funded Bioinformatics Resource Centers providing web-based genomics resources to the scientific community on viruses in category A, B, and C priority pathogens
VBRC (Viral Bioinformatics Resource Center)	University of Alabama at Birmingham; NIAID	http://www.vbrc.org/	Bioinformatics resources directed at *Arenaviridae, Bunyaviridae, Flaviviridae, Filoviridae, Paramyxoviridae, Poxviridae*, and *Togaviridae*
Viral Bioinformatics-Canada	University of Victoria, Canada	http://www.virology.ca/	Viral genomics resources
PATRIC (PathoSystems Resource Integration Center)	Virginia Bioinformatics Institute; NIAID	http://patric.vbi.vt.edu/	Bioinformatics resources directed at *Caliciviridae, Coronaviridae*, hepatitis A virus, hepatitis E virus, *Lyssaviridae*
VirOligo	Oklahoma State University	http://viroligo.okstate.edu/	Virus-specific oligonucleotides for PCR and hybridization
VirGen	Bioinformatics Centre, University of Pune, India	http://bioinfo.ernet.in/virgen/	Annotated and curated database for complete viral genome sequences
VIDE (Virus Identification Data Exchange)	Australian Centre for International Agricultural Research; University of Idaho	http://image.fs.uidaho.edu/vide/	Online descriptions and lists of plant viruses
RNAs and Proteins of dsRNA Viruses	Institute for Animal Health, UK Biotechnology and Biological Sciences Research Council	http://www.iah.bbsrc.ac.uk/dsRNA_virus_proteins/	Databases for the study of *Reoviridae, Cystoviridae, Birnaviridae, Totiviridae*
euHCVdb (European HCV Database)	Institute for the Biology and Chemistry of Proteins, CNRS and Lyon University	http://euhcvdb.ibcp.fr/	European hepatitis C virus database
HCV Database Project	Los Alamos National Laboratories; NIAID	http://hcv.lanl.gov/	HCV sequence and immunology database
Hepatitis Virus Database	National Institute of Genetics (Japan)	http://s2as02.genes.nig.ac.jp/	Database for the study of hepatitis B, C, and E
HERVd (Human Endogenous Retrovirus Database)	Institute of Molecular Genetics, Academy of Sciences of The Czech Republic	http://herv.img.cas.cz/	Human endogenous retrovirus database
HIV Drug Resistance Database	Stanford University	http://hivdb.stanford.edu/	HIV drug resistance database
HIV Sequence Database	Los Alamos National Laboratories; NIAID	http://www.hiv.lanl.gov/	HIV sequence, resistance, immunology, and vaccine trials databases
HIV-1, Human Protein Interaction Database	Southern Research Institute, Birmingham, Alabama; NCBI	http://www.ncbi.nlm.nih.gov/RefSeq/HIVInteractions/	Interactions of HIV-1 proteins with those of the host cell
BioHealthBase (Influenza)	University of Texas Southwestern Medical Center; NIAID	http://www.biohealthbase.org/	Bioinformatics resources directed at influenza
Influenza Sequence Database	Los Alamos National Laboratories	http://www.flu.lanl.gov/	Influenza sequence database

Continued

Table 1 Continued

Information resource name	*Sponsor*	*URL*	*Description*
Influenza Virus Resource	NCBI	http://www.ncbi.nlm.nih.gov/genomes/FLU/	Influenza sequence database
IVDB (Influenza Virus Database)	Beijing Genomics Institute	http://influenza.genomics.org.cn/	Integrated information resource and analysis platform for influenza virus
Influenza Genome Sequencing Project	NIAID	http://www.niaid.nih.gov/dmid/genomes/mscs/influenza.htm	NIAID influenza sequencing project
Poxvirus Bioinformatics Resource Center	University of Alabama at Birmingham; NIAID	http://www.poxvirus.org/	Poxvirus genomic sequences, gene annotations, and analysis
SARS-CoV RNA SSS Database	Center for Modern Biology, School of Life Sciences, Yunnan University, Kunming, China	http://www.liuweibo.com/sarsdb/	SARS Secondary Structural Sequence database
ACLAME (A Classification of Genetic Mobile Elements)	Department of Macromolecular Conformation and Bioinformatics, Free University of Brussels, Belgium	http://aclame.ulb.ac.be/	Database of mobile genetic elements
Subviral RNA Database	University of Ottawa	http://subviral.med.uottawa.ca/	Viroids and viroid-like RNAs
Genomes of the T4-like phages	Department of Biochemistry, Tulane University	http://phage.bioc.tulane.edu/	Sequences of T4-like bacteriophages
DPVweb (Descriptions of plant viruses)	Association of Applied Biologists (UK)	http://www.dpvweb.net/	Plant virus database
ILTAB (International Laboratory for Tropical Agricultural Biotechnology)	Danforth Plant Sciences Center, St. Louis, Missouri	http://www.danforthcenter.org/iltab/	*Caulimoviridae*, *Geminiviridae*, and *Potyviridae* database
Functional motifs			
InterPro	European Bioinformatics Institute (EBI)	http://www.ebi.ac.uk/interpro/	Database of protein families, domains, and functional sites
CDD (Conserved Domain Database)	NCBI	http://www.ncbi.nlm.nih.gov/Structure/cdd/cdd.shtml	Conserved Domain Database and search service
PFAM (Protein families)	Washington University in St. Louis	http://pfam.wustl.edu/	Database of protein families and Hidden Markov models
PRINTS (Protein fingerprints)	University of Manchester, England	http://www.bioinf.man.ac.uk/dbbrowser/PRINTS/	Conserved motifs used to characterize a protein family
PROSITE	Swiss Institute of Bioinformatics	http://ca.expasy.org/prosite/	Database of protein families and domains
VIDA (Virus database of homologous protein families)	Wohl Virion Centre, University College London and Genome Informatics Research Lab, Universitat Pompeu Fabra, Barcelona	http://www.biochem.ucl.ac.uk/bsm/virus_database/	Homologous protein families from complete and partial virus genomes
VOCs (Viral Orthologous Clusters)	University of Victoria, Canada	http://athena.bioc.uvic.ca/workbench.php?tool=vocs	Viral orthologous clusters database
VOGs (Viral Clusters of Orthologous Groups)	NCBI	http://www.ncbi.nlm.nih.gov/genomes/VIRUSES/vog.html	Viral clusters of orthologous groups database

Structural			
PDB (Protein Data Bank)	Research Collaboratory for Structural Bioinformatics	http://www.rcsb.org/	Three-dimensional structure database
Macromolecular Structure Database	EBI	http://www.ebi.ac.uk/msd/	Macromolecular structure database
Dali	Institute of Biotechnology, University of Helsinki, Finland	http://ekhidna.biocenter.helsinki.fi/dali/	Database of 3D structure comparisons
The Big Picture Book of Viruses	David M. Sander; Tulane University	http://www.virology.net/Big_Virology/	Catalog of virus pictures on the Internet
VIPERdb (Virus particle explorer)	Scripps Research Institute Structural Biology	http://viperdb.scripps.edu/	Icosahedral virus capsid structures
Virus World	Institute for Molecular Virology, University of Wisconsin-Madison	http://virology.wisc.edu/virusworld/	Images of virus capsids from X-ray crystallography and cryo-EM data
Pathway			
Reactome	Cold Spring Harbor Laboratory, EBI, Gene Ontology Consortium	http://www.reactome.org/	Curated knowledgebase of biological pathways
BioCyc	SRI International, Menlo Park, California	http://www.biocyc.org/	Collection of databases describing genome and metabolic pathways
KEGG (Kyoto Encyclopedia of Genes and Genomes)	Kyoto University and the Human Genome Center at University of Tokyo	http://www.genome.jp/kegg/	Database of biological pathways and genes
Other			
GO (Gene Ontology Database)	Gene Ontology Consortium	http://www.geneontology.org/	Development of controlled vocabularies (ontologies) that describe gene products in terms of their associated biological processes, cellular components, and molecular functions
National Center for Biomedical Ontology	Stanford University	http://bioontology.org/	Development of innovative technologies and methods that allow scientists to create, disseminate, and manage biomedical ontologies
GEO (Gene Expression Omnibus)	NCBI	http://www.ncbi.nlm.nih.gov/geo/	A gene expression/molecular abundance data repository
IEDB (Immune Epitope Database and Analysis Resource)	La Jolla Institute for Allergy and Immunology	http://www.immuneepitope.org/	Dissemination of immune epitope information

Continued

Table 1 Continued

Information resource name	*Sponsor*	*URL*	*Description*
IntAct (Protein interactions)	EBI	http://www.ebi.ac.uk/intact/	Database and analysis system for protein interactions
Analytical tools			
ExPASy	Swiss Institute of Bioinformatics (SIB)	http://ca.expasy.org/	Tools for the analysis of protein sequences, structures, etc.
PASC (Pairwise Sequence Comparison)	NCBI	http://www.ncbi.nlm.nih.gov/sutils/pasc/	Analysis of pairwise identity distribution within viral families
NCBI Viral Genotyping Tool	NCBI	http://www.ncbi.nlm.nih.gov/projects/genotyping/	Identification of viral sequence genotypes
Medically oriented			
AIDSinfo	US Department of Health and Human Services	http://aidsinfo.nih.gov/	Information on HIV/AIDS treatment, prevention, and research
AMA Infectious diseases	American Medical Association	http://www.ama-assn.org/ama/pub/category/1797.html	Online information resources
CDC NCID (National Center for Infectious Diseases)	US Centers for Disease Control and Prevention (CDC)	http://www.cdc.gov/ncidod/diseases/	Centers for Disease Control and Prevention (USA) disease index
CIPHI Fact Sheets	Canadian Institute of Public Health Inspectors	http://action.web.ca/home/ciphiont/readingroom.shtml	Information pages for public health, including viral diseases
CSEI (Center for the Study of Emerging Infections)	Institute for BioSecurity, St Louis University, School of Public Health	http://www.emerginginfections.slu.edu/	Information on emerging infections
Diseases & Conditions	New York State Department of Health	http://www.health.state.ny.us/diseases/	General public information
ENIVD (European Network for Diagnostics of "Imported" Viral Diseases)	European Commission	http://www.enivd.org/	European Network for Diagnostics of 'Imported' Viral Diseases
EPR (Epidemic and Pandemic Alert and Response)	World Health Organization (WHO)	http://www.who.int/csr/disease/en/	WHO integrated global alert and response system for epidemics and other public health emergencies
European Centre for Disease Prevention and Control	The European Union	http://www.ecdc.eu.int/	European Centre for Disease Prevention and Control
MedScape	WebMD	http://www.medscape.com/	Information and educational tools for specialists
Merck Manual of Diagnosis and Therapy, Infectious Diseases	Merck & Co., Inc.	http://www.merck.com/mrkshared/mmanual/section13/sec13.jsp	Online reference of infectious diseases
MMWR (Morbidity and Mortality Weekly Report)	CDC	http://www.cdc.gov/mmwr/	CDC Morbidity and Mortality Weekly Report
NIH Vaccine Research Center	Dale and Betty Bumpers' Vaccine Research Center, NIAID, NIH	http://www.vrc.nih.gov/	NIH vaccine research

classification and sequence clustering. Another useful site is the All the Virology on the WWW website. This site provides numerous links to other virus-specific websites, databases, information, news, and analytical resources. It is updated on a regular basis and is therefore as current as any site of this scope can be.

Taxonomy and classification

One of the strengths of storing information within a database is that information derived from different sources or different data sets can be compared so that important common and distinguishing features can be recognized. Such comparative analyses are greatly aided by having a rigorous classification scheme for the information being studied. The International Union of Microbiological Societies has designated the International Committee on Taxonomy of Viruses (ICTV) as the official body that determines taxonomic classifications for viruses. Through a series of subcommittees and associated study groups, scientists with expertise on each viral species participate in the establishment of new taxonomic groups, assignment of new isolates to existing or newly established taxonomic groups, and reassessment of existing assignments as additional research data become available. The ICTV uses more than 2600 individual characteristics for classification, though sequence homology has gained increasing importance over the years as one of the major classifiers of taxonomic position. Currently, as described in its Eighth Report, the ICTV recognizes 3 orders, 73 families, 287 genera, and 1950 species of viruses. The ICTV officially classifies viral isolates only to the species level. Divisions within species, such as clades, subgroups, strains, isolates, types, etc., are left to others. The ICTV classifications are available in book form as well as from an online database. This database, the ICTVdb, contains the complete taxonomic hierarchy, and assigns each known viral isolate to its appropriate place in that hierarchy. Descriptive information on each viral species is also available. The NCBI also provides a web-based taxonomy browser for access to taxonomically specified sets of sequence records. NCBI's viral taxonomy is not completely congruent with that of ICTV, but efforts have been under way to ensure congruency with the official ICTV classification.

Nucleotide sequence data

The primary repositories of existing sequence information come from the three organizations that comprise the International Nucleotide Sequence Database Collaboration. These three sites are GenBank (maintained at NCBI), EMBL, and the DNA Data Bank of Japan (DDBJ). Because all sequence information submitted to any one of these entities is shared with the others, a researcher need query only one of these sites to get the most up-to-date set of available sequences. GenBank stores all publicly available nucleotide sequences for all organisms, as well as viruses. This includes whole-genome sequences as well as partial-genome and individual coding sequences. Sequences are also available from large-scale sequencing projects, such as those from shotgun sequencing of environmental samples (including viruses), and high-throughput low- and high-coverage genomic sequencing projects. NCBI provides separate database divisions for access to these sequence datasets. The sequence provided in each GenBank record is the distillation of the raw data generated by (in most cases these days) automated sequencing machines. The trace files and base calls provided by the sequencers are then assembled into a collection of contiguous sequences (contigs) until the final sequence has been assembled. In recognition of the fact that there is useful information contained in these trace files and sequence assemblies (especially if one would like to look for possible sequencing errors or polymorphisms), NCBI now provides separate Trace File and Assembly Archives for GenBank sequences when the laboratory responsible for generating the sequence submits these files. Currently, the only viruses represented in these archives are influenza A, chlorella, and a few bacteriophages.

An important caveat in using data obtained from GenBank or other sources is that no sequence data can be considered to be 100% accurate. Furthermore, the annotation associated with the sequence, as provided in the GenBank record, may also contain inaccuracies or be out-of-date. GenBank records are provided and maintained by the group originally submitting the sequence to GenBank. GenBank may review these records for obvious errors and formatting mistakes (such as the lack of an open reading frame where one is indicated), but given the large numbers of sequences being submitted, it is impossible to verify all of the information in these records. In addition, the submitter of a sequence essentially 'owns' that sequence record and is thus responsible for all updates and corrections. NCBI generally will not change any of the information in the GenBank record unless the sequence submitter provides the changes. In some cases, sequence annotations will be updated and expanded, but many, if not most, records never change following their initial submission. (These facts emphasize the responsibility that submitters of sequence data have to ensure the accuracy of their original submission and to update their sequence data and annotations as necessary.) Therefore, the user of the information has the responsibility to ensure, to the extent possible, its accuracy is sufficient to support any conclusions derived from that information. In recognition of these problems, NCBI established the Reference Sequence (RefSeq) database project, which attempts to provide reference sequences for genomes, genes, mRNAs, proteins, and RNA sequences that can be used, in NCBI's words, as "a stable reference for gene characterization, mutation analysis, expression studies, and polymorphism discovery".

RefSeq records are manually curated by NCBI staff, and therefore should provide more current (and hopefully more accurate) sequence annotations to support the needs of the research community. For viruses, RefSeq provides a complete genomic sequence and annotation for one representative isolate of each viral species. NCBI solicits members of the research community to participate as advisors for each viral family represented in RefSeq, in an effort to ensure the accuracy of the RefSeq effort.

Protein sequence data

In addition to the nucleotide sequence databases mentioned above, UniProt provides a general, all-inclusive protein sequence database that adds value through annotation and analysis of all the available protein sequences. UniProt represents a collaborative effort of three groups that previously maintained separate protein databases (PIR, SwissProt, and TrEMBL). These groups, the National Biomedical Research Foundation at Georgetown University, the Swiss Institute of Bioinformatics, and the European Bioinformatics Institute, formed a consortium in 2002 to merge each of their individual databases into one comprehensive database, UniProt. UniProt data can be queried by searching for similarity to a query sequence, or by identifying useful records based on the text annotations. Sequences are also grouped into clusters based on sequence similarity. Similarity of a query sequence to a particular cluster may be useful in assigning functional characteristics to sequences of unknown function. NCBI also provides a protein sequence database (with corresponding RefSeq records) consisting of all protein-coding sequences that have been annotated within all GenBank nucleotide sequence records.

Virus-specific sequence databases

The above-mentioned sequence databases are not limited to viral data, but rather store sequence information for all biological organisms. In many cases, access to nonviral sequences is necessary for comparative purposes, or to study virus–host interactions. But it is frequently easier to use virus-specific databases when they exist, to provide a more focused view of the data that may simplify many of the analyses of interest. **Table 1** lists many of these virus-specific sites. Sites of note include the NIH-supported Bioinformatics Resource Centers for Biodefense and Emerging and Reemerging Infectious Diseases (BRCs). The BRCs concentrate on providing databases, annotations, and analytical resources on NIH priority pathogens, a list that includes many viruses. In addition, the LANL has developed a variety of viral databases and analytical resources including databases focusing on HIV and influenza. For plant virologists, the Descriptions of Plant Viruses (DPV) website contains a comprehensive database of sequence and other information on plant viruses.

Structural information

The three-dimensional structures for quite a few viral proteins and virion particles have been determined. These structures are available in the primary database for experimentally determined structures, the Protein Data Bank (PDB). The PDB currently contains the structures for more than 650 viral proteins and viral protein complexes out of 38 000 total structures. Several virus-specific structure databases also exist. These include the VIPERdb database of icosahedral viral capsid structures, which provides analytical and visualization tools for the study of viral capsid structures; Virus World at the Institute for Molecular Virology at the University of Wisconsin, which contains a variety of structural images of viruses; and the Big Picture Book of Viruses, which provides a catalog of images of viruses, along with descriptive information.

Functional motifs and orthologous clusters

Ultimately, the biology of viruses is determined by genomic sequence (with a little help from the host and the environment). Nucleotide sequences may be structural, functional, regulatory, or protein coding. Protein sequences may be structural, functional, and/or regulatory, as well. Patterns specified in nucleotide or amino acid sequences can be identified and associated with many of these biological roles. Both general and virus-specific databases exist that map these roles to specific sequence motifs. Most also provide tools that allow investigators to search their own sequences for the presence of particular patterns or motifs characteristic of function. General databases include the NCBI Conserved Domain Database; the Pfam (protein family) database of multiple sequence alignments and hidden Markov models; and the PROSITE database of protein families and domains. Each of these databases and associated search algorithms differ in how they detect a particular search motif or define a particular protein family. It can therefore be useful to employ multiple databases and search methods when analyzing a new sequence (though in many cases they will each detect a similar set of putative functional motifs). InterPro is a database of protein families, domains, and functional sites that combines many other existing motif databases. InterPro provides a search tool, InterProScan, which is able to utilize several different search algorithms dependent on the database to be searched. It allows users to choose which of the available databases and search tools to use when analyzing their own sequences of interest. A comprehensive report is provided that not only summarizes the results of the search, but also provides a comprehensive annotation derived from similarities to known functional domains. All of the above databases define functional attributes based on similarities in amino acid sequence. These amino acid similarities can be used to classify proteins into functional families. Placing proteins into common functional families is also frequently performed by grouping the proteins into orthologous families based on the overall similarity of their amino acid sequence as determined by pairwise BLAST comparisons. Two virus-specific databases of orthologous gene families are the Viral Clusters of Orthologous Groups database (VOGs) at NCBI, and the Viral Orthologous Clusters database (VOCs) at the Viral Bioinformatics Resource Center and Viral Bioinformatics, Canada.

Other information

Many other types of useful information, both general and virus-specific, have been collected into databases that are available to researchers. These include databases of gene-expression experiments (NCBI Gene Expression Omnibus – GEO); protein–protein interaction databases, such as the NCBI HIV Protein-Interaction Database; The Immune Epitope Database and Analysis Resource (IEDB) at the La Jolla Institute for Allergy and Immunology; and databases and resources for defining and visualizing biological pathways, such as metabolic, regulatory, and signaling pathways. These pathway databases include Reactome at the Cold Spring Harbor Laboratory, New York; BioCyc at SRI International, Menlo Park, California; and the Kyoto Encyclopedia of Genes and Genomes (KEGG) at Kyoto University in Japan.

Analytical Tools

As indicated above, the information contained in a database is useless unless there is some way to retrieve that information from the database. In addition, having access to all of the information in every existing database would be meaningless unless tools are available that allow one to process and understand the data contained within those databases. Therefore, a discussion of virus databases would not be complete without at least a passing reference to the tools that are available for analysis. To populate a database such as the VGD with sequence and analytical information, and to utilize this information for subsequent analyses, requires a variety of analytical tools including programs for

- sequence record reformatting,
- database import and export,
- sequence similarity comparison,
- gene prediction and identification,
- detection of functional motifs,
- comparative analysis,
- multiple sequence alignment,
- phylogenetic inference,
- structural prediction, and
- visualization.

Sources for some of these tools have already been mentioned, and many other tools are available from the same websites that provide many of the databases listed in **Table 1**. The goal of all of these sites that make available data and analytical tools is to provide – or enable the discovery of – knowledge, rather than simply providing access to data. Only in this manner can the ultimate goal of biological understanding be fully realized.

See also: Evolution of Viruses; Phylogeny of Viruses; Taxonomy, Classification and Nomenclature of Viruses; Virus Classification by Pairwise Sequence Comparison (PASC).

Further Reading

Ashburner M, Ball CA, Blake JA, *et al.* (2000) Gene ontology: Tool for the unification of biology. The Gene Ontology Consortium. *Nature Genetics* 25(1): 25–29.

Bao Y, Federhen S, Leipe D, *et al.* (2004) National Center for Biotechnology Information Viral Genomes Project. *Journal of Virology* 78(14): 7291–7298.

Fauquet CM, Mayo MA, Maniloff J, Desselberger U, and Ball LA (2005) *Virus Taxonomy: Classification and Nomenclature of Viruses. Eighth Report of the International Committee on Taxonomy of Viruses.* San Diego, CA: Elsevier Academic Press.

Galperin MY (2006) The Molecular Biology Database Collection: 2006 update. *Nucleic Acids Research* 34 (database issue): D3–D5.

Joshi-Tope G, Gillespie M, Vastrik I, *et al.* (2005) Reactome: A knowledgebase of biological pathways. *Nucleic Acids Research* 33(database issue): D428–D432.

Kellam P and Alba MM (2002) Virus bioinformatics: Databases and recent applications. *Applied Bioinformatics* 1(1): 37–42.

Korber B, LaBute M, and Yusim K (2006) Immunoinformatics comes of age. *PLoS Computational Biology* 2(6): e71.

Kuiken C, Korber B, and Shafer RW (2003) HIV sequence databases. *AIDS Reviews* 5(1): 52–61.

Kuiken C, Mizokami M, Deleage G, *et al.* (2006) Hepatitis C databases, principles and utility to researchers. *Hepatology* 43(5): 1157–1165.

Lefkowitz EJ, Upton C, Changayil SS, *et al.* (2005) Poxvirus bioinformatics resource center: A comprehensive *Poxviridae* informational and analytical resource. *Nucleic Acids Research* 33(database issue): D311–D336.

Lesk AM (2005) *Database Annotation in Molecular Biology.* Chichester, UK: Wiley.

Lindler LE, Lebeda FJ, and Korch G (2005) *Biological Weapons Defense: Infectious Diseases and Counterbioterrorism.* Totowa, NJ: Humana Press.

Natarajan P, Lander GC, Shepherd CM, *et al.* (2005) Exploring icosahedral virus structures with VIPER. *Nature Reviews Microbiology* 3(10): 809–817.

Rubin DL, Lewis SE, Mungall CJ, *et al.* (2006) National Center for Biomedical Ontology: Advancing biomedicine through structured organization of scientific knowledge. *Omics* 10(2): 185–198.

Yusim K, Richardson R, Tao N, *et al.* (2005) Los Alamos Hepatitis C Immunology Database. *Applied Bioinformatics* 4(4): 217–225.

Phylogeny of Viruses

A E Gorbalenya, Leiden University Medical Center, Leiden, The Netherlands

Introduction: Evolution, Phylogeny, and Viruses

Biological species, including viruses, change through generations and over time in the process known as evolution. These changes are first fixed in the genome of successful individuals that give rise to genetic lineages. Due to either limited fidelity of the replication apparatus copying the genome or physico-chemical activity of the environment, nucleotides may be changed, inserted, or deleted. Genomes of other origin may also be a source of innovation for a genome through the use of specially evolved mechanisms of genetic exchange (recombination). Accepted changes, known as mutations, may be neutral, advantageous, or deleterious, and depending on the population size and environment, the mutant lineage may proliferate or go extinct. Overall, advantageous mutations and large population size increase the chances for a lineage to succeed. The lineage fit is constantly reassessed in the ever-changing environment and lineages that, due to mutation, became a success in the past could be unfit in the new environment. Due to the growing number of mutations accumulating in the genomes, lineages diverge over time, although occasionally, due to stochastic reasons or under similar selection pressure, they may converge.

The relationship between biological lineages related by common descent is called phylogeny; the same term also embodies the methodology of reconstructing these relationships. Phylogeny deals with past events and, therefore, it is reconstructed by quantification of differences accumulated between lineages. Due to the lack of fossils and (relatively) high mutation rate, viruses were not considered to provide a recoverable part of phylogeny until the advent of molecular data proved otherwise. Comparison of nucleotide and amino acid sequences, and, occasionally, other quantitative characteristics such as distances between three-dimensional structures of biopolymers, have been used to reconstruct virus phylogeny. Results of phylogenetic analysis are commonly depicted in the form of a tree that may be used as a synonym for phylogeny. For instance, all-inclusive phylogeny of cellular species is depicted as the Tree of Life (ToL).

With few exceptions, virus phylogeny follows the theory and practice developed for phylogeny of cellular life forms. For inferring phylogeny, differences between the sequences of species members, assumed to be of a discernable common origin, are analyzed. If species in all lineages evolve at a uniform constant rate, like clock ticks, their evolution conforms to a molecular clock model. The utility of this model in relation to viruses may be very limited. Rather, related virus lineages may evolve at different and fluctuating rates and some sites may mutate repeatedly with each new mutation erasing a record about the prior change. As a result, the accumulation of inter-species differences may progress nonlinearly with the time elapsed. At present, our understanding of these parameters of virus evolution is poor and this limits our ability to assess the fit between a reconstructed phylogeny and the true phylogeny, with the latter practically remaining unknown for most virus isolates. This gap in our knowledge does not eliminate the conceptual strength of phylogentic analysis for reconstructing the relationships between biological species.

The ultimate goal of virus phylogeny is reconstructing the relationships between 'all' virus isolates and species. In contrast to cellular species, which form three compact domains (kingdoms) and whose origin is traced back to a common ancestor in the ToL, major virus classes may combine species that have originated from different ancestors. Thus, reconstructing the comprehensive virus phylogeny requires comparisons that involve genomes of virus and cellular origins. This formidable task remains largely 'work in progress'. In fact, most efforts in virus phylogeny are invested in reconstructing the relationships at the micro, rather than grand, scale and they focus on well-sampled lineages that have practical (e.g., medical) relevance. Phylogeny itself or in combination with other data may provide a deep insight into virus evolution and diverse aspects of virus life cycles, including virus interactions with their hosts.

Our knowledge about contemporary virus diversity has been steadily advancing with new viruses being constantly described by systematic efforts as well as occasional discoveries. These developments indicate that only a small part of virus diversity has so far been unraveled and has become available for phylogenetic studies. It is also likely that many more lineages existed in the past; some of these lineages are likely to have ancestral relationships with contemporary lineages.

Tree Definitions

Species share similarity that varies depending on the rate of evolution and time of divergence. The entire process of generating contemporary species diversity from a common ancestor is believed to proceed through a chain of intermediate ancestors specific for different subsets of the analyzed species. The relationship between the common ancestor, intermediate ancestors, and contemporary species may be likened to the relationship between, respectively, root, internal nodes, and terminal nodes (leaves) of a tree, an abstraction that is widely used for the visualization of this relationship. Trees are also part of graph theory, a branch of mathematics, whose apparatus is used in phylogeny. Formally and due to a strong link between phylogeny and taxonomy, leaves may be called operational taxonomy units (OTUs) and internal nodes and roots, since they have not been directly observed, are known as hypothetical taxonomy units (HTUs). Nodes are connected by branches or edges.

The tree may be characterized by topology, length of branches, shape, and the position of the root. The topology is determined by relative position of internal and terminal nodes; it defines branching events leading to contemporary species diversity. If two or more trees obtained for different data sets feature a common topology, these trees are called congruent. The branch length of a tree may define either the amount of change fixed or the time passed between two nodes connected in a tree, and is known as 'additive' or 'ultrametric', respectively. The tree shape may be linked to particulars of evolutionary process and reflect changes in the population size and diversity due to genetic drift and natural selection. The position of the root at the tree defines the direction of evolution. Species that descend from an internal node in a rooted tree form a cluster and the node is called most recent common ancestor (MRCA) of the cluster that thus has a monophyletic origin. The branch lengths and the root position may be left undefined for a tree that is then called 'cladogram' and 'unrooted tree', respectively.

Phylogenetic Analysis

Multiple alignment of polynucleotide or amino acid sequences representing analyzed species and maximized for similarity is traditionally used as input for phylogenetic analysis. The quality of alignment is among the most significant factors affecting the quality of phylogenetic inference. Due to the redundancy of the genetic code, changes in polynucleotide sequences are accumulated at a higher rate than those in amino acid sequences. In viruses, including RNA viruses, this difference is not counterbalanced by other constraints linked to dinucleotide frequency and RNA secondary (tertiary) structure. Because of these differences, phylogeny of closely related species is commonly inferred using polynucleotide sequences, while protein sequences, preserving better phylogenetic signal, may be used to infer phylogeny of distantly related species.

Differences between species, as calculated from alignment, may be quantified as either pairwise distances forming a distance matrix or position-specific substitution columns (discrete characters of states of alignment), the latter preserving the knowledge about location of differences. The respective methods dealing with these quantitative characteristics are known as distance and discrete (character state). The distance methods are praised for their speed and are considered a technique of choice for analysis of large data sets. They are often designed to converge on a unique phylogeny, with none others being even considered. The unweighted pair group method with arithmetic means (UPGMA) in which a constantly recalculated distance matrix is used to define the hierarchy of similarities through systematic and stepwise merging of most similar pairs at a time was the first technique introduced for clustering. The neighbor-joining (NJ) method uses a more sophisticated algorithm of clustering that minimizes branch lengths, and is the most popular among distance methods. Although different trees may be compared in how they fit a distance matrix, it is distance character-based methods that are routinely used to assess numerous alternative phylogenies in search for the best one in a computationally very intensive process. Due to the calculation time involved, assessing all possible phylogenies is found to be impractical for data sets including more than 10 sequences; for larger data sets different heuristic approximations are used that may not guarantee a recovered phylogeny to be the best overall. There are two major criteria for selecting the best phylogeny using character-state based information through either maximum parsimony (MP) or maximum likelihood (ML). In MP analysis, a phylogeny with a minimal number of substitutions separating analyzed species is sought. The ML analysis offers a statistical framework for comparing the likelihood of fitting different trees into the data in search for one with the best fit. The latter approach is mathematically robust and its statistical power may also be used in combination with other techniques of tree generation. Most recently, a Bayesian variant of the ML approach has gained popularity. It utilizes prior knowledge about the evolutionary process in combination with repeated sampling from subsequently derived hypotheses.

After a tree is chosen, it is common to assign support for internal nodes through assessing nodes' persistence in trees related to the chosen tree. One particular technique, called bootstrap analysis, in which trees are generated for numerous randomly modified derivatives of the original data set, is most frequently used. Each internal node in the original tree is characterized by a so-called bootstrap value that is equal to the number of nodes appearing in all tested trees. Although the relationship between bootstrap and statistical values is not linear, nodes with very high bootstrap values are considered to be reliable. If species evolve according to a molecular clock model, the root position in a tree could directly be calculated from the observed inter-species differences as a midpoint of cumulative inter-species differences. Alternatively, the root position may be assigned to a tree from knowledge about analyzed species that was gained independently from phylogenetic analysis. Commonly, this knowledge comes in the form of a single or more species which are assumed (or known) to have emerged before the 'birth' of the analyzed cluster. These early diverged species are collectively defined as 'outgroup', while the analyzed species may be called 'in-group'. Also, a tree may be generated unrooted, a common practice in phylogenetic analysis of viruses for which the applicability of the molecular clock model remains largely untested and reliable outgroups may not be routinely available. In the unrooted tree, grouping of species in separate clusters may be apparent, although these clusters may not be treated as monophyletic as long as the direction of evolution has not been defined. These challenges are addressed by the development of new approaches that infer rooted trees without artificially restricting species evolution to a constant rate (known as relaxed molecular clock models).

Virus phylogeny can be inferred using genomes or distinct genes and each of these approaches, standard in phylogenomics, may be considered as complementary. Under the first approach, genome-wide alignments are used for analysis. Due to complexities of the evolutionary process that may be region specific, reliable genome-wide alignments can routinely be built only for relatively closely related viruses whose analysis, however, may be further complicated by recombination events (see below). Using the second approach, genes with no evidence for recombination may be merged (concatenated) in a single data set that may be used to produce a superior phylogenetic signal compared to those generated for distinct genes or entire genomes. For viruses with small genomes or for a diverse set of viruses, it is common practice to use a single gene to infer virus phylogeny. Although the results produced may be the best models describing evolutionary history of a group of viruses, the validity of this gene-based approach for the genome-wide extrapolation remains a point of debate.

When the tree is reconstructed for (part of) genomes, an underlying assumption is that the analyzed data set has a uniform phylogeny. This condition may be violated due to homologous recombination between (closely) related viruses. In phylogenetic terms, recombination may be revealed through incongruency of trees built for a genome region, where recombination occurs, and other regions. Trees may also become incongruent due to various technical reasons related to the size and diversity of a virus data set and deviations of the evolutionary process among lineages. These characteristics complicate interpretation of the congruency test, which is widely used in different programs to identify recombination in viruses.

Applications of Phylogeny in Virology

Phylogenetic analysis is used in a wide range of studies to address both applied and fundamental issues of virus research, including epidemiology, diagnostics, forensic studies, phylogeography, origin, evolution, and taxonomy of viruses. First question to be answered during an outbreak of a virus epidemic concern the virus identity and origin. Answers to these questions form the basis for implementing immediate practical measures and prospective planning enabling specific and rapid virus detection and epidemic containment, which may include the use and development of antiviral drugs and vaccines. Among different analyses performed for virus identification at the early stage of a virus epidemic, the phylogenetic characterization is used for determining the relationship of a newly identified virus with all other previously characterized and sequenced viruses.

Results of this analysis may be sufficient to provide answers to the questions posed, as regularly happens with closely monitored viruses that include most human viruses of high social impact, for example, influenza, human immunodeficiency virus (HIV), hepatitis C virus (HCV), poliovirus, and others. For these viruses, there exist large databases of previously characterized isolates and strains that comprehensively cover the natural diversity. Should a newly identified virus belong to one of these species, chances are that it has evolved from a previously characterized isolate or a close variant and this immediately becomes evident in the clustering of these viruses in the phylogenetic tree. Combining the results of gene-specific and genome-wide phylogenetic analysis allows one to determine whether recombination contributed to the isolate origin. For instance, recombination was found to be extremely uncommon in the evolution of HCV, but not for poliovirus lineages that recombine promiscuously, also with closely related human coxsackie A viruses, both of which belong to human enteroviruses.

When an emerging infection is caused by a new never-before-detected virus, the phylogenetic analysis is instrumental for classification of this virus and in the case of a zoonotic infection, for determining the dynamic of virus introduction into the (human) population and initiating the search for the natural virus reservoir. This was the case with many emerging infections including those caused by most recently introduced Nipah virus, a paramyxovirus, and SARS coronavirus (SARS-CoV). With the latter virus, poor sampling of the coronavirus diversity in the SARS-CoV lineage at the time, some uncertainty over the relationship between phylogeny and taxonomy of coronaviruses, and the complexity of phylogenetic analysis of a virus data set including isolated distant lineages led to considerable controversy over the exact evolutionary position of SARS-CoV among coronaviruses. Since then, the matter has largely been resolved but this experience illustrates some challenges in inferring virus phylogeny.

The search for a zoonotic reservoir of an emerging virus may involve a significant and time-consuming effort that requires numerous phylogenetic analyses of ever-expanding sampling of the virus diversity generated in pursuit of the goal. In this quest, phylogenetic analysis canalizes the effort and provides crucial information for reconstructing parameters of major evolutionary events that promoted the virus origin and spread. For instance, intertwining HIV and simian immunodeficiency virus (SIV) lineages in the primate lentivirus tree led to postulation that the existing diversity of HIV in the human population originated from several ancestral viruses independently introduced from primates over a number of years. Similar phylogenetic reasoning was used to trace the origin of a local HIV outbreak to a common source of HIV introduction through dental practice (known as 'HIV dentist' case). These are typical examples illustrating the utility of phylogenetic analysis for epidemiological and forensic studies.

Geographic distribution of places of virus isolation is another important characteristic relative to which virus phylogeny may be evaluated. This field of study belongs to phylogeography. The evolution of human JC polyomavirus provides an example of confinement of circulation of virus clusters to geographically isolated areas, represented by three continents. Recent identification of West Nile virus in the USA illustrates geographical expansion of an Old World virus into the New World. Analysis of phylogenies of field isolates of rabies virus of the family *Rhabdoviridae* sampled from different animals over Europe led to the recognition that interspecies virus expansion is occuring faster when compared to geographical expansion.

Phylogenies also reveal information about the relative strength of the virus–host association over time. In some virus families (e.g., the *Coronaviridae*) host-jumping events may be relatively frequent, including the emergence of at least two human viruses, dead-end SARS-CoV and successfully circulating human coronavirus OC43 (HCoV-OC43). At the other end of the spectrum one finds the family *Herpesviridae*. Extensive phylogenetic analysis of herpesviruses and their hosts showed a remarkable congruency of topologies of trees indicating that this virus family may have emerged some 400 million years ago and that herpesviruses cospeciate with their hosts.

Phylogenetic analysis becomes increasingly important in virus classification (taxonomy) and relies on complex multicharacter rules applied to separate virus families by respective 'study groups'. For viruses united in high-rank taxa above the genus level, phylogenetic clustering for most conserved replicative genes is commonly observed and used in the decision making process. For instance, human hepatitis E virus, originally classified as a calicivirus using largely virion properties, was eventually expelled from the family due to poor fit of genome characteristics, including results of phylogenetic analysis. Phylogenetic considerations also played an important role in forming recently established families, for example, the *Marnaviridae* and *Dicistroviridae*. In contrast, phylogenetic analysis has been of relatively little use in the taxonomy of large DNA phages which has been developed in such a way that existing families may unite phages with different gene layouts and phylogenies. The relationship between phylogeny and taxonomy is evolving and in future one might hope for important advancements that improve cross-family consistency in relation to phylogeny.

See also: Evolution of Viruses; Origin of Viruses; Taxonomy, Classification and Nomenclature of Viruses; Virus Classification by Pairwise Sequence Comparison (PASC); Virus Databases; Virus Species.

Further Reading

Dolja VV and Koonin EV (eds.) (2006) Comparative genomics and evolution of complex viruses. *Virus Research* 117: 1–184.
Domingo E (2007) Virus evolution. In: Knipe DM, Howley PM, Griffin DE, *et al.* (eds.) *Fields Virology*, 5th edn., pp. 389–421. Philadelphia, PA: Wolters Kluwer, Lippincott Williams and Wilkins.
Domingo E, Webster RG, and Holland JJ (eds.) (1999) *Origin and Evolution of Viruses.* San Diego: Academic Press.
Fauquet CM, Mayo MA, Maniloff J, Desselberger U, and Ball LA (2005) *Virus Taxonomy: Eighth Report of the International Committee on Taxonomy of Viruses.* San Diego, CA: Elsevier Academic Press.
Felsenstein J (2004) *Inferring Phylogenies.* Sunderland, MA: Sinauer Associates, Inc.
Gibbs AJ, Calisher CH, and Garcia-Arenal F (eds.) (1995) *Molecular Basis of Virus Evolution.* Cambridge: Cambridge University Press.
Moya A, Holmes EC, and Gonzalez-Candelas F (2004) The population genetics and evolutionary epidemiology of RNA viruses. *Nature Reviews Microbiology* 2: 279–288.
Page RD and Holmes EC (1998) *Molecular Evolution. A Phylogenetic Approach.* Boston: Blackwell Publishing.
Salemi M and Vandamme AM (eds.) (2003) *The Phylogenetic Handbook. A Practical Approach to DNA and Protein Phylogeny.* Cambridge: Cambridge University Press.
Villarreal LP (2005) *Viruses and Evolution of Life.* Washington, DC: ASM Press.

Evolution of Viruses

L P Villarreal, University of California, Irvine, Irvine, CA, USA

Glossary

Dendrogram A schematic line drawing, often tree-like, that represents evolutionary relationships between species.

Error catastrophe A threshold at which a high error rate of genome replication can no longer maintain the integrity of essential genetic information.

ERV An endogenous retroviral-like genetic element, containing a long terminal repeat that is found in the genomes of many organisms.

Fitness landscape A hypothetical representation of a three-dimensional surface that represents relative fitness of genetic variants.

Muller's ratchet A decrease in fitness resulting from a repetition of a genetically restricted founder population.

Quasispecies A population of viral genomes that are the product of error-prone replication usually envisioned as a swarm or cloud of related genomes.

Red Queen hypothesis An evolutionary concept in which an organism must evolve at high rates in order to maintain its competitive advantage.

Reticulated evolution A pattern of evolution in which elements are derived from different ancestors represented as a tree with cross-connections between distinct branches.

Sequence space A multidimensional representation of all possible sequences for a given genome.

Introduction

The initial study of virus evolution sought to explain how virus variation affects viral and host survival and to understand viral disease. However, we now realize that virus evolution is a basic issue, impacting all life in some way. In general, the principles of virus evolution are very much the same Darwinian principles of evolution for all life, involving genetic variation, natural selection, and survival of favorable types. However, virus evolution also entails features such as high error rates, quasispecies populations, and genetic exchange across vast reticulated gene pools that extend the traditional concepts of evolution. Evolution simply means a noncyclic change in the genetic characteristics of a virus; and viruses are the most rapidly evolving genetic agents for all biological entities. Principles of virus evolution provide an integrating framework for understanding the diversity of viruses and the relationships with host as well as providing explanations for the emergence of new viral disease. Most emerging viral diseases are due to species jumps from persistently infected hosts that have long-term virus–host evolutionary histories (called natural reservoirs). Since early human populations (i.e., small bands of hunter gatherers) could not have supported the great human viral plagues of civilization (e.g., smallpox virus/variola and measles), these viruses must also have originated from species jumps that adapted to humans. In recent history, the emergence of HIV demonstrates that virus evolution continues to impact human populations. Early observation established that viruses often show significant variation in virulence. Such variation was used as an early, yet risky, form of vaccination (i.e., variolation against smallpox). The variation that occurred with passage into alternate tissue and host was also used to make various vaccines (rabies in 1880s) or attenuate viral virulence, such as live yellow fever vaccine. But variation also allows some viruses, such as influenza A, to escape neutralization by vaccination. However, variation in viral disease or virulence does not provide a quantitative basis to study evolution.

The study of virus variation and evolution is an applied science that allows the observation of evolutionary change in real time. For example, human individuals (or populations) infected with either human immunodeficiency virus 1 (HIV-1) or hepatitis C virus (HCV) show progressive or geographical evolutionary adaptation associated with the emergence of specific viral clades that affect disease therapy and progression (such as resistance to antiviral drugs). **Figure 1** shows HIV variation in an individual human patient whereas **Figure 2** shows HCV variation in the human population. Virus evolution is also important for the commercial growth of various organisms, such as the dairy industry (lactose fermenting bacteria), the brewing industry, agriculture, aquaculture, and farming. In all these applied cases major losses can result from virus adaptation to the cultivated species, often from viruses of wild species. Some organisms appear much less prone to viral adaptations (e.g., nematodes, ferns, sharks). Virus evolution can also be applied to technological innovation, as in phage display. This is a process in which a terminal surface protein of some filamentous bacterial virus can be genetically engineered for novel surface protein expression. By generating

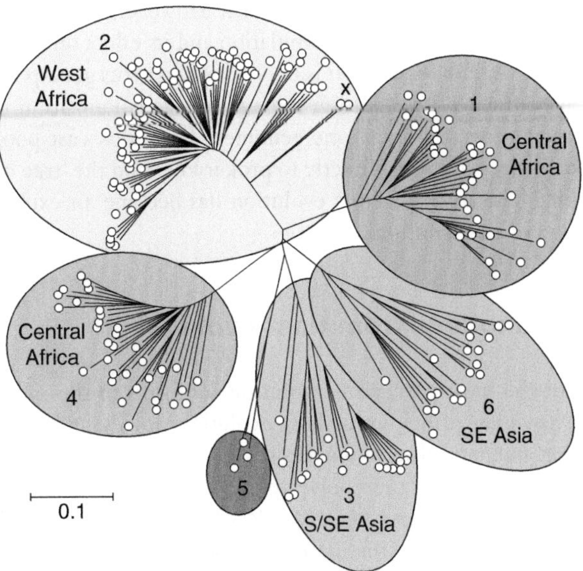

Figure 2 Unrooted phylogenetic analysis of HCV nucleotide sequences from globally distributed human isolates. Adapted from Simmonds P (2001) Reconstructing the origins of human hepatitis viruses. *Philosophical Transactions of the Royal Society of London* 356: 1013–1026, with permission from Royal Society Publishing.

diversity *in vitro* (with up to 10^{15} types), and applying the principles of evolution (random variation) to biochemical selection (such as binding to a chemical substrate), a reiterative amplification can find solutions to problems in biochemistry, such as surface interactions or catalytic activity.

Virus Evolution as a Basic Science

Virus variation is a global issue. In the last decade it has been established that viruses are the most numerous biological entities on the planet. The oceans and soil harbor vast numbers of viral-like particles (VLPs), mostly resembling the tailed DNA viruses of bacteria. In addition, some of these environmental viruses are unexpectedly large and complex, such as the phycodnaviruses of algae or mimivirus of amoeba, a 1.2 Mb DNA virus that can encode nearly 1000 genes. Thus viruses represent a vast

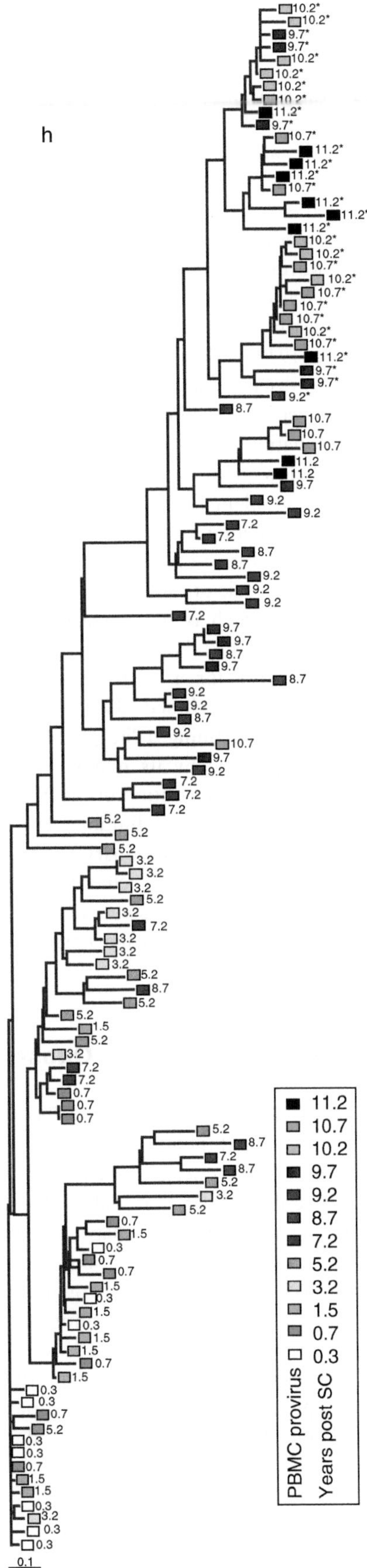

Figure 1 HIV population analysis from an infected individual. Shown is a neighbor-joining phylogram derived from maximum likelihood distance between all sequences. Sequences are represented by a square for PBMC sequences or a triangle for plasma sequences. The arbitrary color gradient corresponds to the time of sampling. Adapted from Shankarappa R, Margolick JB, Gange SJ, *et al.* (1999) Consistent viral evolutionary changes associated with the progression of human immunodeficiency virus Type 1 infection. *Journal of Virology* 73: 10489–10502, with permission from the American Society for Microbiology.

and diverse source of novel genes. However, the evolutionary dynamics of this population and its effect on hosts is not well understood. It is likely that this virus gene pool also affects host evolution since prophage colonization is known for all prokaryotic genomes. Thus, this vast pool of viruses connects directly to prokaryotes and the 'tree of life'. The study of virus evolution has become an extension of all evolution.

Distinctions from Host Evolution

For the most part, virus evolution conforms to the same Darwinian principles as host evolution, involving variation and natural selection. However, viruses have multiple origins, and are thus polyphyletic. There are six major categories of viruses (+RNA, –RNA, dsRNA, retro, small DNA, large DNA) that have no common genes and hence have no common ancestor. However, these categories all have conserved hallmark genes (i.e., capsid proteins, Rd RNA pol, RT, primase, helicase, and RCR initiation proteins) which are all monophyletic and which may trace their origins to primordial host/viral gene pools. Many DNA viruses, for example, have replication strategies and polymerases that are clearly distinct from that of their host, which depends on such hallmark genes. Thus, virus evolution appears ancient but inextricably linked to its host. Also, in contrast to host evolution, viral quasispecies show a population-based adaptability that extends the selection of the fittest to include populations of otherwise unfit genomes (described below). Viruses can clearly cross the usual host-species barriers so that viral evolution can be reticulated in vast genes pools. For example, bacterial DNA viruses and +RNA viruses can show high rates of recombination across viruses that infect numerous host species. Accordingly, tailed DNA phage of bacteria appear to represent one single vast gene pool. Viruses can also violate concepts of death and extinction, reassembling genomes from parts and/or repairing lethal damage by multiplicity reactivation. In addition, damaged (defective) viruses can also affect virus and host evolution. Such defectives are found in many types of viruses and can also be found in many host genomes (as defective prophage or defective endogenous retroviruses). Such defective viruses can clearly affect host survival. In all these characteristics, viruses extend the Darwinian principles of host evolution.

A History of Virus Evolution

The coherent study of virus evolution awaited the development of sequence technology to measure mutations and genetic variation in viral populations. Concepts of natural selection, fitness, and propagation of favorable variation had long been established in the evolutionary biology literature prior to the growth of virology. Thus, mathematical models, such as Fisher population genetics, concerned with gene frequency in a (sexually exchanging) species population, had been well developed and seemed directly applicable to virus evolution, which resemble that of host genes. Here, viral fitness was typically expressed as relative replication rates (replicative fitness) but sometimes host virulence and disease were also used. However, as presented below, a comprehensive definition of viral fitness remains problematic. The first quantitative measurements of the rate of virus mutation was done in the 1940s with bacterial phage. The mutation rates were expressed as a set of ordinary differential equations that were subsequently used to develop the quasispecies equations as applied to error-prone RNA genomes (see below). However, the species definition of a virus poses a problem for evolutionary thinking and challenges how we define kinship in viruses. Unlike the sex-defined host, a virus species is currently defined as a polythetic class: a mosaic of related parts of which not all elements are shared (such as host range, genome relatedness, antigenic properties). No specific defining characteristic or gene exists for a virus species and sexual exchange need not be included. This is an inherently fuzzy definition, like defining a 'heap', which although clear, cannot be specified by its number of parts. The ensuing molecular characterization of many virus populations supports this species definition. The challenge then is to understand viral evolutionary patterns working with such fuzzy definitions. Yet, conserved patterns of virus evolution are still seen, some of which suggest viruses are indeed an ancient lineage, possibly extending into the primordial RNA world.

Error-Prone Replication and Quasispecies

In the 1970s, Manfred Eigen and also Peter Shuster developed a fundamental theoretical model of virus evolution. A set of ordinary differential equations was published that described what was called 'quasispecies'. Starting from measurements of phage mutation rates, they considered the consequences of high error rates as expected from RNA replication (an error-prone noncorrecting replication process). The resulting population shared many properties and was called quasispecies, a society (or community) of individuals that are the error products of replication. The name 'quasispecies' thus describes a chemically diverse set of molecules and was not intended to refer to a biological species (i.e., genetic exchanging). However, as discussed above, the fuzzy definition of virus species and quasispecies overlaps somewhat, which has been a source of confusion. Several premises were used to develop this theory: (1) the individual products ignore one another and interact only as individuals; (2) the system is not at equilibrium and resources are not limiting. Based on relative replication, the growth of favorable types is described which provides a mathematical definition of replicative fitness. The original equations represent an idealized generalized system of infinite population size and are not directly applicable to the real world, although they provide valuable insights into real world systems. The equations do not address variable mortality (longevity), interference, exclusion, competition, complementation, and persistence, or how such issues affect nonreplicative fitness definitions. The issue of mortality and fitness is interesting from the perspective of viruses. For example, an interfering defective virus can be considered dead, but can clearly interfere with and drive the extinction wild-type template replication in quasispecies. In some cases, the quasispecies equations appear to be mathematically equivalent to classical Wright and Fisher population genetics equations as applied by Kimura and Maroyama to asexual haploid populations at the mutation-selection balance. However, these two approaches begin with distinct perspectives, and it was the assumption of high error in the quasispecies equations that had a major impact on experimentation and our current understanding of virus evolution. This has also led to some counter intuitive conclusions, such as the concept of selection of 'the fittest' compared to the consensus character of the master template. Quasispecies from a virus with high error rates (such as HIV-1) might be composed of all mutant progeny RNAs such that the consensus template (the mean, the fittest, or the master template) may not actually exist. With classical population genetics, an asexual clonal population should fix the clonal sequence. With quasispecies, this is not observed. The first laboratory measurements of viral quasispecies were made using Qβ RNA polymerase *in vitro*. Error estimates ranged from 10^{-3} to 10^{-4} substitutions per site per year (an error rate applicable to most RNA viruses). With Qβ, the replication of many nonviable mutants generated a genetic spectra that had a characteristic makeup. For example, separate DNA clones of Qβ were initially distinct from each other but quickly generated the same RNA quasispecies as before cloning. Additional lab measurements have shown that quasispecies can have significant adaptive fitness (above the cloned master template) and display memory; that is, they can retain information of prior selections in a minority of the population. Complementation, interference, competition suppression, and extinction have all been measured in various quasispecies, thus indicating a collective form of evolution and violating an original premise of genome independence in these equations. In addition, the sequence diversity in a quasispecies is now seen as a source of adaptive potential, not simply error (see below). Despite these results, the concept of quasispecies has still been highly useful, and not simply as a theoretical development. For example, the live poliovirus vaccine is clearly a heterogeneous quasispecies. Within this population exists a minority of neurovirulent variants that are suppressed by the majority avirulent virus. A main point of the quasispecies concept is that it provides an understanding of high adaptability from a population that has many, even lethal, mutations. It is interesting that the proposed early RNA world would have also been a collective quasispecies world.

Error Catastrophe, Sequence Space

Quasispecies theory also predicts a situation known as error catastrophe, defined as an error rate threshold at which information is lost and the system decays. If error rates are too high, or the information content (genome length) too extensive, the system will be unable to maintain its information integrity. This predicts a basic limit on the size of RNA genomes, consistent with the observation that the largest RNA genomes are only about 27–32 kbp (coronaviruses). There is a possible therapeutic use of error catastrophe: drugs (possibly Ribavarin and 5-fluorouracil) that increase the error rates of RNA polymerase can potentially push a virus beyond its error threshold and induce a catastrophe. Quasispecies is an inherently fuzzy and dynamic population that has no sharp boundaries or specific members and has been metaphorically referred to as swarms and clouds. Here, cloud is a metaphor for the population landscape that exists in high-dimensional hyperspace and cannot be readily visualized. The concept of sequence space has been used to represent topography of the distribution of all mutants. Kinship relationships between mutants can be measured by Hamming distance; the minimal steps needed to specify the difference between two mutants. In spite of high error and adaptation rates (and sometimes high recombination rates), RNA viruses are not able to explore all potential sequence space. Selection significantly limits the quasispecies, since the potential sequence space is hyperastronomical even for a moderately sized virus. For example, an RNA virus with 10 000 nt would correspond to 10^{6000} possible sequences, well beyond what could be explored by even the potentially vast number of viruses over the lifetime of the world. In addition, there are clearly mechanistic constraints that prevent many possible sequences, such as necessary domains of ± strand RNA folds, physical association with ribonucleotide proteins, virion packaging and assembly – all in addition to usual selection for gene fitness (function) that all severely limit possible adaptations. This creates a multipeak 'fitness' landscape in hyperspace (see **Figure 3**). Assuming fitness itself can have a single definition (i.e., replicative fitness, not subjected to variable and stochastic competition), we can visualize this space as many steep valleys and ledges

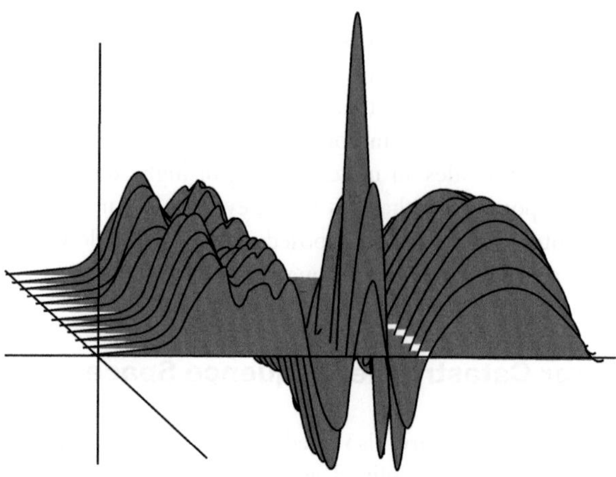

Figure 3 Hypothetical fitness landscape for an RNA virus. Assuming one definition for a nonrelativistic fitness (such as replicative fitness), the coordinates indicate relative fitness. Those below the *y*-axis are interfering or lethal variants.

(in this case with 10 000 dimensions). Normally we think that adaptation by natural selection is the force to explore and move through fitness landscape. But as the deep valleys are often lethal, they cannot be explored via natural selection. Here we see the major adaptive power of the quasispecies collective. Since random, even lethal, errors and drift are inherent in a quasispecies, lethal valleys can be readily crossed by such variable genomes, allowing the master genome to adapt by natural selection to a new fitness peak. Thus, error-prone replication and the generation of mutant clouds allows for much better exploration of sequence space and eventual adaptability.

When viruses are transmitted to new hosts, they can experience a genetic bottleneck since a relatively small number of viral genomes could be involved (aka low multiplicity passage). If this process is serially repeated, a phenomenon known as 'Muller's Ratchet' can result in lost competitiveness as the essentially clonal RNA virus accumulates deleterious mutations (sometimes measured as pfu/plaque). However, in lab studies, virus extinction from serial passage does not occur, presumably due to plaque selection for a restored phenotype. Even a single plaque is in reality a small population (due to nonideal particle/pfu ratios and ID_{50}). However, lost competitiveness with other viruses is seen with clonal laboratory passage. However, if a quasispecies population is passed, this generally results in increased competitive fitness. Such passage can produce a seemingly never-ending better version of the virus that outcompetes all prior versions of the same virus (although virion yields and absolute replication are not necessarily improved). This has been likened to the Red Queen hypothesis in that the viruses are evolving at high rates, simply to maintain their competitive position, so as not to be displaced as the dominant viral type. Virus–virus competition is thus a crucial selection.

Never-Ending Adaptation

A real world example of the potentially never-ending virus adaptation is shown in **Figure 4**. The HA and NA genes of human influenza A virus have been monitored for several decades. As shown, the prevalent master template of the virus circulating in the human population has been continually changing, due to immune selection and stochastic viral immigration, necessitating yearly vaccine changes (also shown). Although such a population dynamic has been stably maintained in the human population, all prior versions have essentially become extinct, as they do not reappear.

Not all RNA virus populations show this dynamic of a continual change or even the diversity expected from quasispecies. Even in influenza virus A, avian isolates from natural host (waterfowl) can be genetically stable. Some RNA (and retro) viruses with high error rates can nevertheless maintain stable populations in specific hosts. For example, measles virus shows much less antigenic drift in human infections compared to influenza virus A. Hepatitis G virus (a human prevalent and distant relative of HCV) shows little variation in even isolated human populations. The filoviruses (Ebola virus and Marburg virus) have shown no genetic variation in Zaire isolates from 20 years apart. Hendra virus, isolated from Australian fruit bats, and Nipah virus from Malaysian fruit bats also show little genetic diversity. Arenaviruses and hantavirus are also genetically stable in their natural rodent host. The reasons for such population stability have not been well evaluated. In some cases (measles), purifying selection would seem likely. In other cases, persistence and low replication rates seem to apply. For example, simian foamy virus (SFV) and human T-lymphotropic virus II (HTLV-II) generate only about 10^{-8} substitutions per site per year, probably due to low replication rates.

Virus–Host Congruence and RNA Stability

There are now many examples of species-specific RNA virus/host coevolution, indicating very slow rates of virus evolution. Since error rates must be similar, this appears to be at odds with the quasispecies theory. For example, Hantavirus (genus *Bunyavirus*) coevolution with its rodent host, suggests a 20 million year association. Arenaviruses (ssRNA bisegmented ambisense) also coevolve in Old/New World rodents. These viruses are of special interest with regard to emergence as they represent the source of five hemorrhagic human fevers (such as Lassa virus). In all

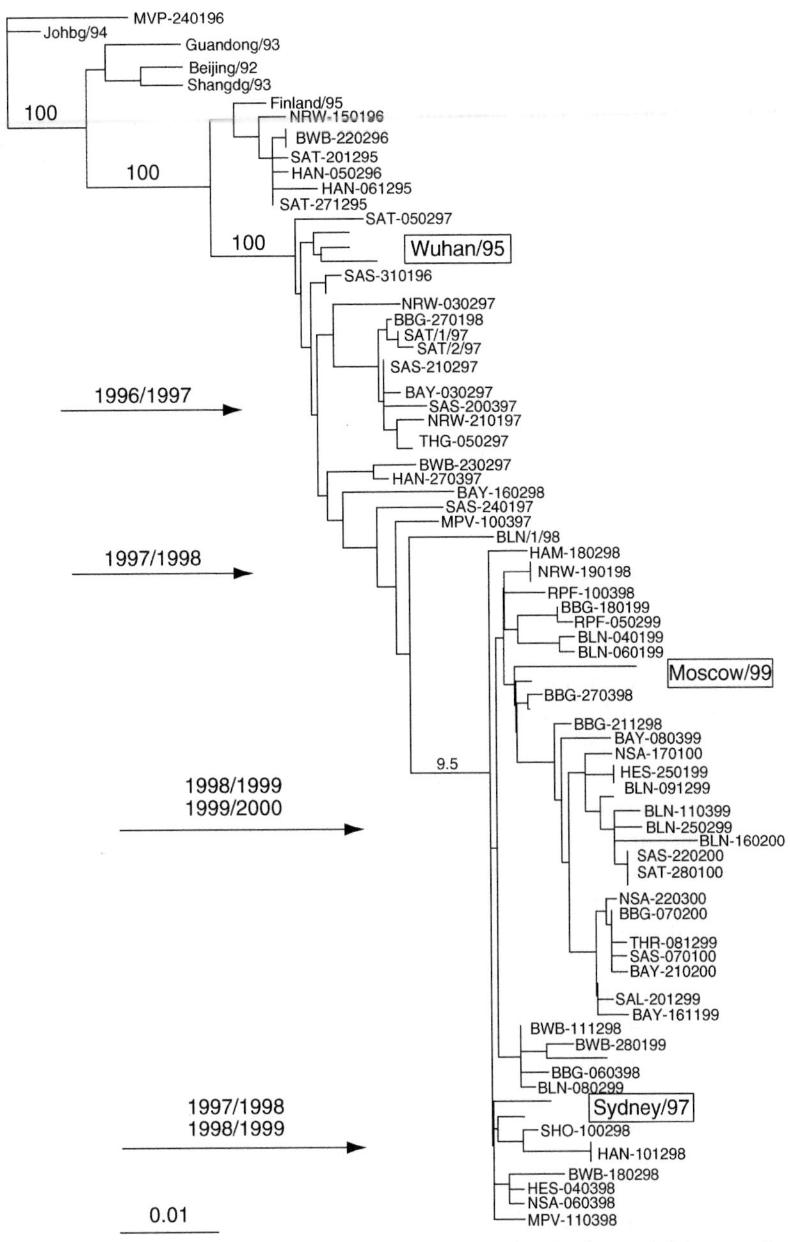

Figure 4 Phylogenetic tree of yearly influenza A/H3N2 viruses variants based on the hemaglutinin gene. Locations of specific vaccine strains are boxed. Reproduced from Schweiger B, Zadow I, and Heckler R (2002) Antigenic drift and variability of influenza viruses. *Medical Microbiology and Immunology* 191: 133–138, with permission from Springer-Verlag.

these examples, however, it appears that the virus causes a persistent unapparent infection in its natural host and that human disease is due to species jump.

Tools

Although viral genomes were the first to be sequenced, the initial focus was simply to identify similarity between viral genes, not to evaluate distant evolutionary relationships. The most popular tool for finding similarity is BLAST (Basic Local Alignment Search Tool) from the National Centers for Biotechnology Information (NCBI), which calculates similarity between query sequences and infers a probability based on a matrix database. Various versions of BLAST are the most used tool in bioinformatics to trace evolution. Although BLAST will identify similar genes, it is also necessary to compile and evaluate the similarities in sets of the related sequences. Multiple sequence alignment software, such as ClustalW, is used for this purpose. Phylogenetic relationships are then inferred from tree-building software. This software includes maximum parsimony, neighbor-joining, and maximum likelihood methods. The statistical significance of the tree (relative to all possible trees) can then be evaluated by algorithms such as bootstrap. More recently, Bayesian analyses, such as Bali-Phy, which implements a Markov Chain Monte Carlo (MCMC) method and calculates joint posterior probabilities of phylogeny and alignment have become popular. This software has the added potential of using sliding windows and evaluating multiple trees, such as in virus–host coevolution. These methods, however, use the single master consensus sequence as its query and do not evaluate quasispecies, collective-based populations. Also, the clearly reticulated or hybrid character of some virus evolution is problematic. Also, distant evolutionary relationships are no longer preserved in the same sequences. Here, conservation of structural motifs, assembly patterns, gene order, and replication strategy are used to identify distant kinship.

Patterns of RNA Virus Evolution

The sometimes extreme variation of RNA virus sequence has led some to propose that most family lineages appear to be only about 10 000 years old, which is clearly at odds with much older estimates. The +RNA viruses in particular show a remarkable diversity of genomes and replicator mechanisms. These families also show much evidence of recombination and a tendency to cross host barriers. About 38 families of +RNA viruses with up to four segments are known. There are four distinct classes of replicase recognized in viruses that also share a common genetic plan. These RNA viruses have three helicase superfamilies, two protease superfamilies, and two jelly roll capsid domains. For the most part, capsid and RdRpol sequences are congruent except for members of the families *Luteoviridae* and *Tetraviridae*, which appear to have undergone recombination between these two gene lineages. The smallest +RNA virus is a member of the bacterial virus family *Leviviridae*, which has only four genes. This simple virus appears to represent the ancestral +RNA virus. Curiously, no RNA virus has yet been found to infect archaebacteria. The largest +RNA viruses belong to the genus *Coronavirus* (27–32 kbp). The most recently described +RNA viruses are members of the family *Marnaviridae* infecting bats and marine organisms sea life, which appear to be basal to evolution of picornaviruses. However, natural populations of some +RNA viruses can be stable. For example, dengue virus (*Flaviviridae*) shows low rates of amino acid substitution (e.g., nonsynonymous to synonymous ratios). Since it is an acute arbovirus infection with high error rates, strong selective constraints likely account for this stability involving multiple (systemic) tissues and vector transmission.

Negative-strand RNA viruses have distinct patterns of evolution, which is traced via their polymerase genes. Gene order tends to be highly conserved. The unsegmented viruses, such as rhabdoviruses, lyssaviruses, and paramyxoviruses, do not undergo significant recombination so their variation tends to be by point mutations and deletions. Although high error rates, variation, and quasispecies generation can be seen in laboratory settings, natural isolates, such as lyssaviruses and measles virus, tend to be relatively homogeneous. For example, lyssaviruses show a slow rate of evolution (5×10^{-5}/site/year). Lyssavirus persistent infections in natural host might contribute to this stability. However, measles virus is a strictly human-specific acute infection so its stability is likely due to purifying selection.

Patterns of DNA Virus Evolution: Tailed Phage

Large DNA viruses of bacteria, archaea, and eukaryotes appear to be evolutionarily linked. Although little sequence conservation can be identified between the T4 phage of bacteria, the halophage of archaea, the members of the family *Phycodnaviridae* infecting algae, and the herpes viruses of vertebrate eukaryotes, all show similarities in their gene programs, DNA polymerase types, capsid structures, and capsid assembly, consistent with a common ancestor. For example, both the Enterobacteria phage T4 (T4) and herpes simplex virus 1 (HSV-1) have $T = 1$ capsid symmetry with 60 copies of capsid protein. The bacterial DNA viruses would appear to represent the ancestor of all these viruses, but the origins of these phages now appear lost in the primordial gene pool. These DNA viruses can have large genomes that could not be sustained by error-prone replication. Thus, many DNA viruses do have error-correcting DNA replication, with error rates that approach or equal those of their host cells (10^{-8}). Giant bacterial phage genomes (*Bacillus megaterium* phage G, of about 600 genes), and algal phycodnaviruses have now been characterized. Even larger DNA viruses of amoeba (acanthamoeba polyphaga mimivirus) coding for more than 1000 genes are known to be abundant in some water habitats.

The tailed phage of bacteria have been called the dark matter of genetics, due to their numerical dominance ($\sim 10^{31}$ *en toto*). This corresponds to about 10^{24} productive infections per second on a global scale. Most host-restricted phage lineages clearly conserve sets of core proteins (especially capsid genes), but others (the broader T-even phages) do not conserve any hallmark genes. Hallmark genes, when present, are usually recognized by conserved domains within proteins, such as replication and structural proteins. Replicator strategy and gene order are also frequently conserved. Phage also tends to conserve genes that are active against other phage (i.e., DNA modification, lambda RexA, T4 rII). With the sequencing of numerous phage genomes, however, a large number of

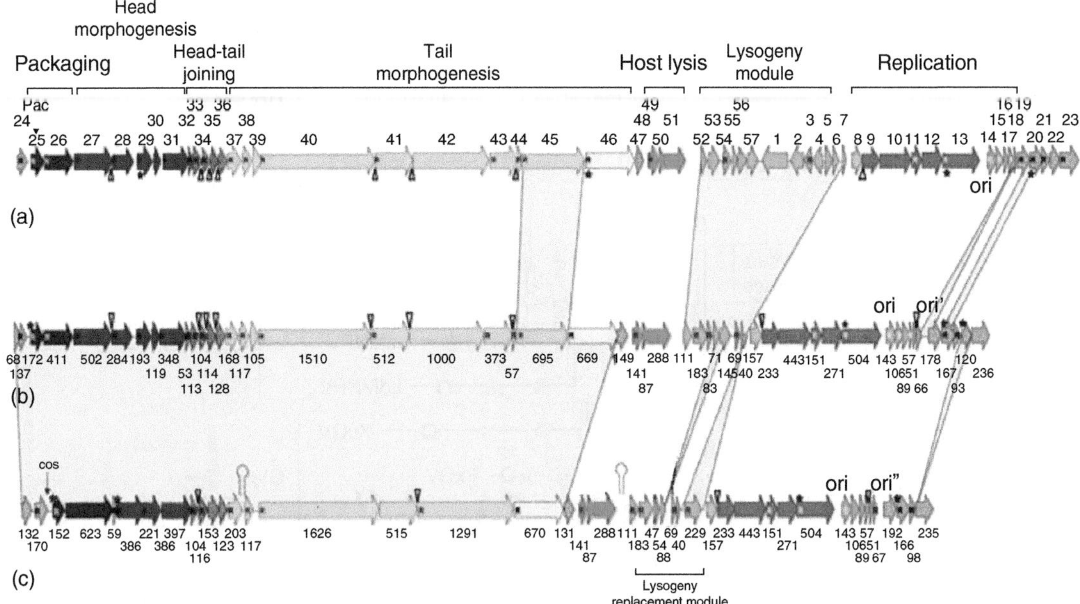

Figure 5 Genome comparison of temperate *S. thermophilus* phage 01205, virulent *S. thermophilus* Sfi11, and the virulent *S. thermophilus* Sfi19. Probable gene functions are indicated and genomes have been divided into functional units. Genes belonging to the same module are indicated with the same color. Areas of shading indicate regions of major difference. From Desiere F, Lucchini S, Canchaya C, Ventura M, and Brüssow H (2002) Comparative genomics of phages and prophages in lactic acid bacteria. *Antonie van Leeuwenhoek* 82: 73–91.

novel genes have been identified. Currently, 350 full genomes of tailed phage and 400 prophage from bacterial genomes have been sequenced. In general, large DNA viruses are tenfold overrepresented in small single-domain genes (~100 aa). Comparative genomics, especially of lactobacterial phages, suggest that most phage genomes evolve as mosaics, with sharp boundaries between genes as well as at protein domains within genes (see **Figure 5**). Recombination between lytic, temperate, and cryptic prophages appears to account for this gene and subgene domain variation. Some specific phages have mechanisms to generate specific gene diversity (such as bordetella phage using RT for surface receptor diversity), but most diversity is the product of recombination. Two broad patterns of phage variation have been observed corresponding to host-unassociated lytic and host-associated (congruent) temperate phage. In most bacterial genomes (ECOR *Escherichia coli* collection, cyanobacteria, *B. subtilis*), patterns of prophage colonization account for significant genetic distinctions between closely related host strains. The general picture for tailed phage of bacteria is that they are not the products of reduction of host genomes.

Large Eukaryotic DNA Viruses

As noted, evolutionary links between tailed phage and large DNA viruses of eukaryotes are apparent. The phycodnaviruses of unicellular green algae clearly have many phage-like characteristics, including the presence of restriction/modification enzymes, homing endonucleases, the injection of viral DNA, and the external localization of the viral capsid. They also have many characteristics of eukaryotic DNA viruses, such as a clearly herpes virus-related DNA polymerase, PCNA proteins, nonintegrating DNA, and numerous signal-transduction proteins. Thus, phycodnaviruses show hybrid characteristics of prokaryotic and eukaryotic viruses.

The evolutionary pattern of the large DNA viruses of eukaryotes is generally best traced by comparing their respective DNA-dependent DNA polymerases (DdDp). These exist in distinct classes that are typically specific for each viral lineage and are usually the most highly conserved of the set of core genes within a viral lineage. However, some viruses, such as the white spot syndrome virus (WSSV) infecting shrimps, have almost no genes in common with other DNA viruses. Generally, the specific set of core genes is clade specific. The first fully sequenced viral genome tree was that of the baculoviruses (see **Figure 6**). The overall pattern of evolution shows the conservation of the core set in which most clades can be differentiated from one another mainly by acquisition of several novel viral genes (although some lineage-specific gene loss is also apparent). In another example, coccolithoviruses differ from related phycodnaviruses by the acquisition of 100 kbp gene set, including six subunits of DdDp core genes. Similar patterns of divergence can be seen with the herpesvirus family members. In addition,

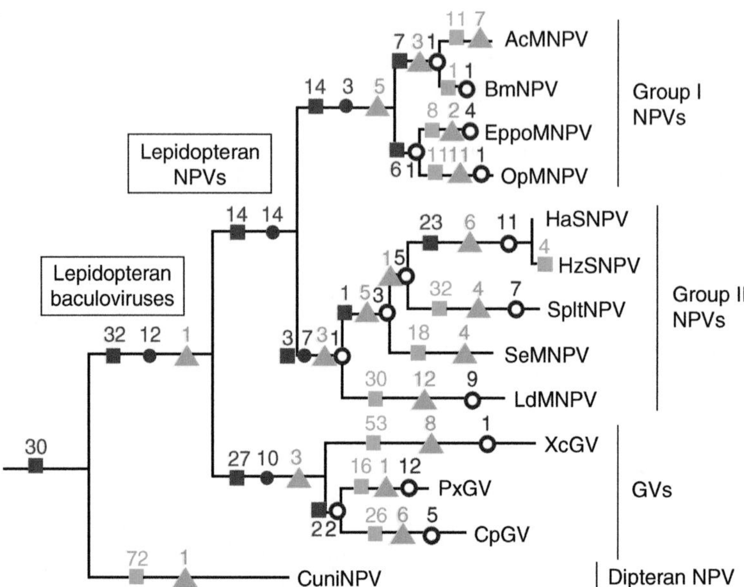

Figure 6 Gene content map of 13 complete sequences of baculoviruses, including the genus *Granulovirus*. The tree shows the most parsimonious hypothesis of changes in gene content during baculovirus evolution. Colors and shapes indicate gene conservation, acquisition, and loss. Reproduced from Herniou EA, Olszewski JA, Cory JS, O'Reilly DR (2003) The genome sequence and evolution of baculoviruses. *Annual Review of Entomology* 48: 211–234, with permission from Annual Reviews.

most herpesvirus clades also show coevolution with their host. However, the poxviruses (orthopoxviruses), show a different overall evolutionary pattern and are not congruent with host. The more ancestral orthopoxvirus members, such as cowpox virus and mousepox virus, have greater gene numbers that appear to have been lost in the human-specific and virulent smallpox virus. Avipoxviruses have even greater gene diversity but the entomopoxviruses are the most complex and diverse of all. The complexity and brick shape of the poxviruses originally inspired the view that these viruses might evolve from bacterial cells following the reduction of complexity. However, DNA sequencing makes it clear that viral core genes have no bacterial analogs. In some instances, viral lineages have clearly fused with other viral and host lineages. For example, the baculovirus Autographa californica MNP virus (AcMNPV) has acquired a gypsy-like retrovirus (e.g., TED), an endogenous retrovirus associated with moth development. The polydnaviruses (circular DNA viruses) are fused into their host genomes (as endogenous DNA viruses) of some parasitoid wasps, essential for survival of the wasp larvae.

Small DNA Viruses

The small, double-stranded, circular DNA viruses (*Papillomaviridae* and *Polyomaviridae*) show evolutionary patterns that are highly host linked. Virus and host evolution are mostly congruent, and virus evolution tends to be slow. For example, approximately 100 human papillomaviruses show congruent evolution with human (and primate) host. This seems to be due to both a highly species- and tissue-specific virus replication, as well as a tendency to establish persistent infections. However, the rolling circular replicon (RCR) viruses, such as parvoviruses, can have distinct evolution patterns. Both mouse minute virus (MMV) and canine parvoviruses can have quasispecies-like populations, which can show evolutionary rates at 10^{-4} substitutions per site per year. Such rates are at the lower end of those seen with RNA viruses. In bacteria, RCR viruses and RCR plasmids appear to represent a common gene pool. Other poorly characterized small eukaryotic DNA viruses, such as human torque teno virus, are asymptomatic but show high variation during persistence for unknown reasons.

Endogenous and Autonomous Retroviruses

Retroviruses present a special problem in understanding patterns of eukaryotic virus evolution. Like prophage of bacteria, retroviruses both stably colonize their host as endogenous or genomic retroviruses (ERVs) that are often defective, but may also sometimes emerge from their host (especially rodents) to produce autonomous virus. In addition, retroviruses are polyphyletic and prone to generating quasispecies due to high error rates as well as high rates of recombination. The most common conserved retrovirus genome elements are domains within the long terminal repeats (LTRs), RT, integrase, protease, gag protein, and env protein. Of these, env are the most often altered or deleted in host genomes. In addition, tRNA primer sites (such a lys [K] tRNA) are also often conserved and used for classification (i.e., human endogenous retrovirus, HERV-K family). However, each of these retroviral elements can potentially have distinct patterns of evolution and conservation, generating distinct dendrograms. Vertebrates, especially mammals, seem to host many retroviral elements within their genomes. Their autonomous retroviruses have a tendency to infect cells of the immune system. Murine leukemia virus (MLV) is the best-studied simple autonomous retrovirus, but many endogenous MLV relatives also exist. Retroviruses are present in genomes of early eukaryotes but significantly expanded in vertebrates. Gypsy-like retroviruses (aka chromoviruses, defined via RT and gag similarity) are often found conserved as full-length elements including env genes in most lower eukaryote genomes (e.g., *Caenorhabditis elegans*), but were mostly lost from tetrapods. Some lower eukaryotes clearly prevent colonization by ERVs, such as *Neurospora* fungi (via the RIP exclusion system). Many endogenous retroviruses are congruent with host evolution, whereas other ERVs are recently acquired and highly host specific. In terms of gene diversity, the retroviral *env* are the most diverse. There are five RT-based families recognized such as *Retroviridae, Hepadnaviridae, Caulimovoridae, Pseudoviridae,* and *Metaviridae*, the latter three being especially prevalent as genomic elements in flowering plants (especially Gypsy). Yet not all retroviruses seem able to colonize host germ line. For example, lentiviruses (such as simian immunodeficiency virus (SIV) and HIV) show no examples of endogenization compared to the simpler MLV-related viruses that can be both autonomous (i.e., MLV, Gibbon ape leukemia virus) and endogenous (i.e., *Mus dunni* ERV, koala ERV). The converse can also be true, since no autonomous versions of HERV K, for example, are known.

Early views proposed that retroviruses evolved from nonviral retroposons (LTR RT elements, non-LTR LINE-like elements). These non-LTR elements, have distinct nonretroviral mechanistic features and core protein domains, but retain some virus-like domains of RT; thus, they appeared to predate retroviruses. However, we now know that gypsy-like retroviruses were present in the earliest eukaryotes. In addition, some LTR-containing elements, such as Gypsy, had initially been considered ancestral to retroviruses because all copies seemed to be defective. However, it is now established that complete gypsy retroviruses are conserved as ERVs in some yeast and *Drosophila* strains. Thus, although endogenous and exogenous retroviruses appear to evolve from each other, there is no evidence that exogenous retroviruses have emerged from non-LTR LINE-like elements.

The congruence between ERVs and host eukaryote evolution is sometimes striking. For example, all mammals have acquired their own peculiar versions of ERVs (and LINES). Recently, it has become clear that the placental mammals have conserved several families of ERV-derived *env* genes that provide an essential function for placental tissue (ERV W-syncytin 1, ERV FRD syncytin 2, enJSRV). Clearly, retrovirus evolution is highly intertwined with that of their hosts.

Never-Ending Emergence

A remaining concern of virus evolution is to understand the emergence of new viral pathogens. The unpredictable and stochastic nature of such virulent adaptations makes predictions difficult, as the link between virulence and evolution is vague. For example, the genetic changes that made the SARS virus (persisting in bats) into an acute human pathogen are still not predictable. Viral fitness and selection, and how they change from persistent states with acute species jumps, are not yet defined. However, some variables contribute to the likelihood of viral emergence, such as virus ecology. The population density and dynamics of the new host and the ecological interactions-between new and stable viral host are often crucial. In addition, virus–virus interactions can be important, allowing for recombination and/or reassortments or lowering immunological selective barriers via immunosuppression. The emergence of HIV-1 from different, persistent SIVs of African monkeys through chimpanzees into a new human disease, for example, includes the same issues. Also, the potential emergence of pandemic human influenza from avian (Anatiformes) sources, such as H5N1, remains a great concern. Thus, virus evolution will continue to interest us as we seek to predict, control, or eradicate viral agents of disease.

See also: Antigenic Variation; Coronaviruses: Molecular Biology; Origin of Viruses; Phylogeny of Viruses; Picornaviruses: Molecular Biology; Quasispecies; Retrotransposons of Vertebrates; Virus Databases; Virus Species.

Further Reading

Desiere F, Lucchini S, Canchaya C, Ventura M, and Brussow H (2002) Comparative genomics of phages and prophages in lactic acid bacteria. *Antonie Van Leeuwenhoek* 82(1–4): 73–91.

Domingo E (2006) *Current Topics in Microbiology and Immunology, Vol. 299: Quasispecies: Concept and Implications for Virology.* Berlin: Springer.

Herniou EA, Olszewski JA, Cory JS, and O'Reilly DR (2003) The genome sequence and evolution of baculoviruses. *Annual Review of Entomology* 48: 211–234.

Hurst CJ (2000) *Viral Ecology.* San Diego: Academic Press.

Schweiger B, Zadow I, and Heckler R (2002) Antigenic drift and variability of influenza viruses. *Medical Microbiology and Immunology* 191: 133–138.

Shankarappa R, Margolick JB, Gange SJ, *et al.* (1999) Consistent viral evolutionary changes associated with the progression of human immunodeficiency virus type 1 infection. *Journal of Virology* 73: 10489–10502.
Simmonds P (2001) Reconstructing the origins of human hepatitis viruses. *Philosophical Transactions of the Royal Society of London* 356: 1013–1026.
Roossinck MJ (2005) Symbiosis versus competition in plant virus evolution. *Nature Reviews Microbiology* 3(12): 917–924.
Villarreal LP (2005) *Viruses and the Evolution of Life.* Washington, DC: ASM Press.
Zanotto PM, Gibbs MJ, Gould EA, and Holmes EC (1996) A reevaluation of the higher taxonomy of viruses based on RNA polymerases. *Journal of Virology* 70(9): 6083–6096.

Taxonomy, Classification and Nomenclature of Viruses

C M Fauquet, Danforth Plant Science Center, St. Louis, MO, USA

Glossary

Classification The categorization of organisms into defined groups on the basis of identified characteristics.
Family A category in the virus taxonomic classification of related organisms, comprising one or more genera.
Genus A category in the virus taxonomic classification of related organisms, comprising one or more species.
Nomenclature The assigning of names to organisms in a scientific classification system.
Order A category in the virus taxonomic classification of related organisms, comprising one or more families.
Species A virus species is a polythetic class of viruses that constitutes a replicating lineage and occupies a particular ecological niche.
Taxonomy The science of classifying plants, animals, and microorganisms into increasingly broader categories based on shared features. The practice or principles of classification.

Introduction

Virus taxonomy is a very important but controversial field of science. It was ranked as the first constraint for the modern development of virus databases, and the exponential increase in virus sequencing is worsening the situation. However, substantial progress has been made particularly in the last 20 years, both on the conceptual framework and practical implication of virus taxonomy. The International Committee on Taxonomy of Viruses (ICTV) is the only committee of the Virology Division, of the International Union of Microbiological Societies (IUMS), in charge of that task since 1966, for the international virology community. Virus Taxonomy Reports have been published regularly by ICTV and they became the reference in virus taxonomy and nomenclature. This article aims at providing some historical information about the establishment and changes in virus taxonomy and describes the current status of virus classification, nomenclature, and orthography.

There is no such thing as a 'natural' or a 'biological' classification; by essence any classification is an arbitrary human invention and viruses are no exception. The question really is: "How can we classify viruses in such way that it makes sense and is useful to many scientists?" The need for virus classification is not only supported by the common human need for organization, but also as a scientific tool to compare viruses and extrapolate useful information from one virus to another and from one family to another. Crucial biological information can be extrapolated directly from human viruses like picornaviruses to plant comoviruses and vice versa, when the classification indicates many structural and genomic common characteristics. When a newly discovered virus is assigned to a particular taxon, this virus can immediately be granted a number of *a priori* properties that only need confirmation and that have an immediate impact on specific virological studies. Furthermore, although it is not clearly stated by ICTV that the current classification is thought to reflect virus evolution, it is accepted that virus taxonomy is aiming at this objective and could become a tool by itself to study and evaluate virus evolution.

Virus nomenclature cannot be dissociated from classification. There must be a coherent system for naming viruses accompanied with a system for classifying viruses. Furthermore, using correct orthography and typography of virus taxa is not simply an excercise meant to complicate the task of virologists but is based on rules that help scientists to extract useful information from what is written down. It is therefore important to establish and follow guidelines for orthography, nomenclature, and classification of viruses. For all these reasons, virus classification and nomenclature have always been very controversial and have led to passionate discussions over the past four decades.

Historical Background of Virus Taxonomy

The first evidence for the existence of viruses was shown by Beijerinck in 1898, but it was not until the 1920s that virologists began to classify viruses. The first system referred to pathogenic properties of animal and human viruses and to symptoms for plant viruses. For example, viruses sharing the pathogenic properties causing hepatitis (e.g., hepatitis A virus, hepatitis B virus, yellow fever virus, Rift Valley fever virus) were grouped together as 'the hepatitis viruses'. In 1939, Holmes published a classification of plant viruses dependent on host reactions and differential host species using a binomial–trinomial nomenclature based on the name of the infected plant. It was only in the 1950s, with the utilization of the electron microscope, that the first real virus classification was established. Naturally the shape and size of virus particles became major criteria for virus classification. Because of that powerful and rapid technology the number of newly discovered viruses increased rapidly and several hundreds of new viruses were listed in a short period of time.

In 1966 in Moscow, at the International Congress for Microbiology, 43 virologists created the International Committee on Nomenclature of Viruses (ICNV) with the aim of developing a worldwide recognized taxonomy and nomenclature system for all viruses. The name of the ICNV was changed in 1974 to the more appropriate ICTV. The ICTV, which is the only committee of the Virology Division of the International Union of Microbiological Societies, is now recognized as the official international body that decides on all matters related to taxonomy and nomenclature of viruses.

Since 1966, virologists have agreed that all viruses isolated from different organisms should be classified together in a unique system, separate from that of microorganisms such as fungi, bacteria, and mycoplasma. However, there has been much controversy on how to achieve this aim. Lwoff, Horne, and Tournier in 1962 proposed the adoption of a system classifying viruses into subphyla, classes, orders, suborders, and families. It was also proposed that the hierarchical classification would be based on the type of nucleic acid (DNA or RNA), the strandedness (single = ss or double = ds), the presence or absence of an envelope, the capsid symmetry, the type of replication cycle (with or without an RNA intermediate for DNA viruses), and the number of genome segments. This hierarchical classification system has never been recognized by the ICTV, but most of the criteria used to demarcate the major classes of viruses formed the basis of the universal taxonomy system now in place, and all published ICTV reports have used this scheme with only minor changes.

It is only in the last 15 years that a hierarchical classification level higher than the family was proposed and accepted. A first order, *Mononegavirales*, was accepted in 1990, and the orders *Caudovirales* and *Nidovirales* were adopted in 1996. In 2005, ICTV considered introducing four new orders namely *Picornavirales, Herpesvirales, Reovirales,* and *Retrovirales*, and these may become accepted in the near future.

It is important to note that the species category for viruses was only adopted in 1991. From then onwards viruses were assigned to species or tentative species. In addition, a list of species demarcation criteria has been established for each family. It is anticipated that by 2010 the ICTV will have introduced species criteria for all viruses and that some level of homogeneity will have been reached, although it is perfectly acceptable to have different sets of criteria for different families of viruses.

Since the establishment of the ICTV, a total of eight virus taxonomic reports have been published. At the first meeting of the Committee in Mexico City in 1970, two families with corresponding two genera and 24 floating genera were adopted to begin grouping the vertebrate, invertebrate, and bacterial viruses together, and in addition, 16 plant virus 'groups' were introduced. Although virologists working with vertebrate viruses had assigned viruses to genera and families for many years, plant virologists until 1993 used the term 'group' to designate viruses with similar properties. It was only in 1995 that the ICTV adopted a uniform system for all viruses, encompassing 2644 assigned viruses. The *Eighth ICTV Report on Virus Taxonomy*, published in 2005 describes a Universal Virus Classification that comprises 3 orders, 73 families, 9 subfamilies, 287 genera, and 5450 viruses belonging to 1950 species (**Table 1**).

Over the past four decades the number of classified viruses, as well as the number of each type of taxa, has increased exponentially and continues to grow (**Figure 1**). Because DNA sequencing has become a routine technique, it seems likely that the number of recognizes viruses and viral taxa will continue to grow exponentially. Furthermore, virus genome sequences provide qualitative and quantitative criteria for defining the molecular variability of viruses that are useful for classification purposes. Sequencing will also permit identification and classification of many viruses that are difficult to isolate and characterize by other methods.

Organization and Structure of ICTV

The ICTV is the only Committee of the Virology Division of the International Union of Microbiological Societies. It is a non-profit organization composed of volunteered

Table 1 The order of presentation of viruses in the eighth ICTV report

Order	Family	Subfamily	Genus	Type species	Host
The DNA viruses					
The dsDNA viruses					
Caudovirales					
	Myoviridae		"T4-like viruses"	*Enterobacteria phage T4*	B
			"P-like viruses"	*Enterobacteria phage P*	B
			"P-like viruses"	*Enterobacteria phage P*	B
			"Mu-like viruses"	*Enterobacteria phage Mu*	B
			"SP1-like viruses"	*Bacillus phage SP1*	B
			"ϕH-like viruses"	*Halobacterium phage ϕH*	Ar
	Siphoviridae		"λ-like viruses"	*Enterobacteria phageλ*	B
			"T1-like viruses"	*Enterobacteria phage T1*	B
			"T5-like viruses"	*Enterobacteria phage T5*	B
			"L5-like viruses"	*Mycobacterium phage L5*	B
			"c2-like viruses"	*Lactococcus phage c2*	B
			"ψM1-like viruses"	*Methanobacterium phage ψM1*	Ar
			"ϕC31-like viruses"	*Streptomyces phage ϕC31*	B
			"N15-like viruses"	*Enterobacteria phage N15*	B
	Podoviridae		"T7-like viruses"	*Enterobacteria phage T7*	B
			"P2-like viruses"	*Enterobacteria phage P2*	B
			"ϕ29-like viruses"	*Bacillus phage ϕ29*	B
			"N4-like viruses"	*Enterobacteria phage N4*	B
	Tectiviridae		*Tectivirus*	*Enterobacteria phage PRD1*	B
	Corticoviridae		*Corticovirus*	*Pseudoalteromonas phage PM2*	B
	Plasmaviridae		*Plasmavirus*	*Acholeplasma phage L2*	B
	Lipothrixviridae		*Alphalipothrixvirus*	*Thermoproteus tenax virus 1*	Ar
			Betalipothrixvirus	*Sulfolobus islandicus filamentous virus*	Ar
			Gammalipothrixvirus	*Acidianus filamentous virus 1*	Ar
	Rudiviridae		*Rudivirus*	*Sulfolobus islandicus rod-shaped virus 2*	Ar
	Fuselloviridae		*Fusellovirus*	*Sulfolobus spindle-shaped virus 1*	Ar
			Salterprovirus	*His1 virus*	Ar
	Guttaviridae		*Guttavirus*	*Sulfolobus newzealandicus droplet-shaped virus*	Ar
	Poxviridae	*Chordopoxvirinae*			
			Orthopoxvirus	*Vaccinia virus*	V
			Parapoxvirus	*Orf virus*	V
			Avipoxvirus	*Fowlpox virus*	V
			Capripoxvirus	*Sheeppox virus*	V
			Leporipoxvirus	*Myxoma virus*	V
			Suipoxvirus	*Swinepox virus*	V
			Molluscipoxvirus	*Molluscum contagiosum virus*	V
			Yatapoxvirus	*Yaba monkey tumor virus*	V
		Entomopoxvirinae			

		Alphaentomopoxvirus	*Melolontha melolontha entomopoxvirus*	I
		Betaentomopoxvirus	*Amsacta moorei entomopoxvirus 'L'*	I
		Gammaentomopoxvirus	*Chironomus luridus entomopoxvirus*	I
Asfarviridae		*Asfivirus*	*African swine fever virus*	V, I
Iridoviridae		*Iridovirus*	*Invertebrate iridescent virus 6*	I
		Chloriridovirus	*Invertebrate iridescent virus 3*	I
		Ranavirus	*Frog virus 3*	V
		Lymphocystivirus	*Lymphocystis disease virus 1*	V
		Megalocytivirus	*Infectious spleen and kidney necrosis virus*	V
Phycodnaviridae		*Chlorovirus*	*Paramecium bursaria Chlorella virus 1*	Al
		Coccolithovirus	*Emiliania huxleyi virus* 86	Al
		Prasinovirus	*Micromonas pusilla virus SP*	Al
		Prymnesiovirus	*Chrysochromulina brevifilum virus PW1*	Al
		Phaeovirus	*Ectocarpus siliculosus virus 1*	Al
		Raphidovirus	*Heterosigma akashiwo virus 01*	Al
Baculoviridae		*Nucleopolyhedrovirus*	*Autographa californica multiple nucleopolyhedrovirus*	I
		Granulovirus	*Cydia pomonella granulovirus*	I
Nimaviridae		*Whispovirus*	*White spot syndrome virus 1*	I
Herpesviridae	*Alphaherpesvirinae*			
		Simplexvirus	*Human herpesvirus 1*	V
		Varicellovirus	*Human herpesvirus 3*	V
		Mardivirus	*Gallid herpesvirus 2*	V
		Iltovirus	*Gallid herpesvirus 1*	V
	Betaherpesvirinae			
		Cytomegalovirus	*Human herpesvirus 5*	V
		Muromegalovirus	*Murid herpesvirus 1*	V
		Roseolovirus	*Human herpesvirus 6*	V
	Gammaherpesvirinae			
		Lymphocryptovirus	*Human herpesvirus 4*	V
		Rhadinovirus	*Saimiriine herpesvirus 2*	V
		Ictalurivirus	*Ictalurid herpesvirus 1*	V
Adenoviridae		*Mastadenovirus*	*Human adenovirus C*	V
		Aviadenovirus	*Fowl adenovirus A*	V
		Atadenovirus	*Ovine adenovirus D*	V
		Siadenovirus	*Frog adenovirus*	V
		Rhizidiovirus	*Rhizidiomyces virus*	F
Polyomaviridae		*Polyomavirus*	*Simian virus 40*	V
Papillomaviridae		*Alphapapillomavirus*	*Human papillomavirus 32*	V
		Betapapillomavirus	*Human papillomavirus 5*	V
		Gammapapillomavirus	*Human papillomavirus 4*	V
		Deltapapillomavirus	*European elk papillomavirus*	V
		Epsilonpapillomavirus	*Bovine papillomavirus 5*	V
		Zetapapillomavirus	*Equine papillomavirus 1*	V
		Etapapillomavirus	*Fringilla coelebs papillomavirus*	V
		Thetapapillomavirus	*Psittacus erithacus timneh papillomavirus*	V

Continued

Table 1 Continued

Order	Family	Subfamily	Genus	Type species	Host
			Iotapapillomavirus	*Mastomys natalensis papillomavirus*	V
			Kappapapillomavirus	*Cottontail rabbit papillomavirus*	V
			Lambdapapillomavirus	*Canine oral papillomavirus*	V
			Mupapillomavirus	*Human papillomavirus 1*	V
			Nupapillomavirus	*Human papillomavirus 41*	V
			Xipapillomavirus	*Bovine papillomavirus 3*	V
			Omikronpapillomavirus	*Phocoena spinipinnis papillomavirus*	V
			Pipapillomavirus	*Hamster oral papillomavirus*	V
	Polydnaviridae		*Bracovirus*	*Cotesia melanoscela bracovirus*	I
			Ichnovirus	*Campoletis sonorensis ichnovirus*	I
	Ascoviridae		*Ascovirus*	*Spodoptera frugiperda ascovirus 1a*	I
	Unassigned		*Mimivirus*	*Acanthamoeba polyphaga mimivirus*	Pr, V
The ssDNA viruses					
	Inoviridae		*Inovirus*	*Enterobacteria phage M13*	B
			Plectrovirus	*Acholeplasma phage L51*	B
	Microviridae		*Microvirus*	*Enterobacteria phage* ϕ*X174*	B
			Chlamydiamicrovirus	*Chlamydia phage 1*	B
			Bdellomicrovirus	*Bdellovibrio phage MAC1*	B
			Spiromicrovirus	*Spiroplasma phage 4*	B
	Geminiviridae		*Mastrevirus*	*Maize streak virus*	P
			Curtovirus	*Beet curly top virus*	P
			Topocuvirus	*Tomato pseudo-curly top virus*	P
			Begomovirus	*Bean golden yellow mosaic virus*	P
	Circoviridae		*Circovirus*	*Porcine circovirus-1*	V
			Gyrovirus	*Chicken anemia virus*	V
	Unassigned		*Anellovirus*	*Torque teno virus*	V
	Nanoviridae		*Nanovirus*	*Subterranean clover stunt virus*	P
			Babuvirus	*Banana bunchy top virus*	P
	Parvoviridae	*Parvovirinae*			
			Parvovirus	*Minute virus of mice*	V
			Erythrovirus	*Human parvovirus B9*	V
			Dependovirus	*Adeno-associated virus 2*	V
			Amdovirus	*Aleutian mink disease virus*	V
			Bocavirus	*Bovine parvovirus*	V
		Densovirinae			
			Densovirus	*Junonia coenia densovirus*	I
			Iteravirus	*Bombyx mori densovirus*	I
			Brevidensovirus	*Aedes aegypti densovirus*	I
			Pefudensovirus	*Periplaneta fuliginosa densovirus*	I
The DNA and RNA reverse transcribing viruses					
	Hepadnaviridae		*Orthohepadnavirus*	*Hepatitis B virus*	V
			Avihepadnavirus	*Duck hepatitis B virus*	V
	Caulimoviridae		*Caulimovirus*	*Cauliflower mosaic virus*	P

			Petuvirus	*Petunia vein clearing virus*	P
			Soymovirus	*Soybean chlorotic mottle virus*	P
			Cavemovirus	*Cassava vein mosaic virus*	P
			Badnavirus	*Commelina yellow mottle virus*	P
			Tungrovirus	*Rice tungro bacilliform virus*	P
	Pseudoviridae		*Pseudovirus*	*Saccharomyces cerevisiae Ty1 virus*	F, P
			Hemivirus	*Drosophila melanogaster copia virus*	F, I
			Sirevirus	*Glycine max SIRE1 virus*	P
	Metaviridae		*Metavirus*	*Saccharomyces cerevisiae Ty3 virus*	F, P, I
			Errantivirus	*Drosophila melanogaster Gypsy virus*	I
			Semotivirus	*Ascaris lumbricoides Tas virus*	I
	Retroviridae	*Orthoretrovirinae*			
			Alpharetrovirus	*Avian leukosis virus*	V
			Betaretrovirus	*Mouse mammary tumor virus*	V
			Gammaretrovirus	*Murine leukemia virus*	V
			Deltaretrovirus	*Bovine leukemia virus*	V
			Epsilonretrovirus	*Walleye dermal sarcoma virus*	V
			Lentivirus	*Human immunodeficiency virus 1*	V
		Spumaretrovirinae			
			Spumavirus	*Simian foamy virus*	V
The RNA viruses					
The dsRNA viruses					
	Cystoviridae		*Cystovirus*	*Pseudomonas phage* ϕ*6*	B
	Reoviridae		*Orthoreovirus*	*Mammalian orthoreovirus*	V
			Orbivirus	*Bluetongue virus*	V, I
			Rotavirus	*Rotavirus A*	V
			Coltivirus	*Colorado tick fever virus*	V, I
			Seadornavirus	*Banna virus*	V
			Aquareovirus	*Aquareovirus A*	V
			Idnoreovirus	*Idnoreovirus 1*	I
			Cypovirus	*Cypovirus 1*	I
			Fijivirus	*Fiji disease virus*	P, I
			Phytoreovirus	*Wound tumor virus*	P, I
			Oryzavirus	*Rice ragged stunt virus*	P, I
			Mycoreovirus	*Mycoreovirus 1*	F
	Birnaviridae		*Aquabirnavirus*	*Infectious pancreatic necrosis virus*	V
			Avibirnavirus	*Infectious bursal disease virus*	V
			Entomobirnavirus	*Drosophila X virus*	I
	Totiviridae		*Totivirus*	*Saccharomyces cerevisiae virus L-A*	F
			Giardiavirus	*Giardia lamblia virus*	Pr
			Leishmaniavirus	*Leishmania RNA virus 1–1*	Pr
	Partitiviridae		*Partitivirus*	*Atkinsonella hypoxylon virus*	F
			Alphacryptovirus	*White clover cryptic virus 1*	P
			Betacryptovirus	*White clover cryptic virus 2*	P

Continued

Table 1 Continued

Order	*Family*	*Subfamily*	*Genus*	*Type species*	*Host*
	Chrysoviridae		*Chrysovirus*	*Penicillium chrysogenum virus*	F
	Hypoviridae		*Hypovirus*	*Cryphonectria hypovirus 1*	F
	Unassigned		*Endornavirus*	*Vicia faba endornavirus*	P
The negative-stranded ssRNA viruses					
Mononegavirales					
	Bornaviridae		*Bornavirus*	*Borna disease virus*	V
	Rhabdoviridae		*Vesiculovirus*	*Vesicular stomatitis Indiana virus*	V, I
			Lyssavirus	*Rabies virus*	V
			Ephemerovirus	*Bovine ephemeral fever virus*	V, I
			Novirhabdovirus	*Infectious hematopoietic necrosis virus*	V
			Cytorhabdovirus	*Lettuce necrotic yellows virus*	P, I
			Nucleorhabdovirus	*Potato yellow dwarf virus*	P, I
	Filoviridae		*Marburgvirus*	*Lake Victoria marburgvirus*	V
			Ebolavirus	*Zaire ebolavirus*	V
	Paramyxoviridae	*Paramyxovirinae*			
			Rubulavirus	*Mumps virus*	V
			Avulavirus	*Newcastle disease virus*	V
			Respirovirus	*Sendai virus*	V
			Henipavirus	*Hendra virus*	V
			Morbillivirus	*Measles virus*	V
		Pneumovirinae			
			Pneumovirus	*Human respiratory syncytial virus*	V
			Metapneumovirus	*Avian metapneumovirus*	V
	Unassigned		*Varicosavirus*	*Lettuce big-vein associated virus*	P
			Ophiovirus	*Citrus psorosis virus*	P
	Orthomyxoviridae		*Influenzavirus A*	*Influenza A virus*	V
			Influenzavirus B	*Influenza B virus*	V
			Influenzavirus C	*Influenza C virus*	V
			Thogotovirus	*Thogoto virus*	V, I
			Isavirus	*Infectious salmon anemia virus*	V
	Bunyaviridae		*Orthobunyavirus*	*Bunyamwera virus*	V, I
			Hantavirus	*Hantaan virus*	V
			Nairovirus	*Dugbe virus*	V, I
			Phlebovirus	*Rift Valley fever virus*	V, I
			Tospovirus	*Tomato spotted wilt virus*	P, I
	Unassigned		*Tenuivirus*	*Rice stripe virus*	P, I
	Arenaviridae		*Arenavirus*	*Lymphocytic choriomeningitis virus*	V
			Deltavirus	*Hepatitis delta virus*	V
The positive-stranded ssRNA viruses					
	Leviviridae		*Levivirus*	*Enterobacteria phage MS2*	B
			Allolevivirus	*Enterobacteria phage Qβ*	B
	Narnaviridae		*Narnavirus*	*Saccharomyces 20S narnavirus*	F

	Mitovirus	*Cryphonectria mitovirus 1*	F
Picornaviridae	*Enterovirus*	*Poliovirus*	V
	Rhinovirus	*Human rhinovirus A*	V
	Cardiovirus	*Encephalomyocarditis virus*	V
	Aphthovirus	*Foot-and-mouth disease virus*	V
	Hepatovirus	*Hepatitis A virus*	V
	Parechovirus	*Human parechovirus*	V
	Erbovirus	*Equine rhinitis B virus*	V
	Kobuvirus	*Aichi virus*	V
	Teschovirus	*Porcine teschovirus*	V
Unassigned	*Iflavirus*	*Infectious flacherie virus*	I
Dicistroviridae	*Cripavirus*	*Cricket paralysis virus*	I
Marnaviridae	*Marnavirus*	*Heterosigma akashiwo RNA virus*	F
Sequiviridae	*Sequivirus*	*Parsnip yellow fleck virus*	P
	Waikavirus	*Rice tungro spherical virus*	P
Unassigned	*Sadwavirus*	*Satsuma dwarf virus*	P
	Cheravirus	*Cherry rasp leaf virus*	P
Comoviridae	*Comovirus*	*Cowpea mosaic virus*	P
	Fabavirus	*Broad bean wilt virus 1*	P
	Nepovirus	*Tobacco ringspot virus*	P
Potyviridae	*Potyvirus*	*Potato virus Y*	P
	Ipomovirus	*Sweet potato mild mottle virus*	P
	Macluravirus	*Maclura mosaic virus*	P
	Rymovirus	*Ryegrass mosaic virus*	P
	Tritimovirus	*Wheat streak mosaic virus*	P
	Bymovirus	*Barley yellow mosaic virus*	P
Caliciviridae	*Lagovirus*	*Rabbit hemorrhagic disease virus*	V
	Norovirus	*Norwalk virus*	V
	Sapovirus	*Sapporo virus*	V
	Vesivirus	*Vesicular exanthema of swine virus*	V
Unassigned	*Hepevirus*	*Hepatitis E virus*	V
Astroviridae	*Avastrovirus*	*Turkey astrovirus*	V
	Mamastrovirus	*Human astrovirus*	V
Nodaviridae	*Alphanodavirus*	*Nodamura virus*	I
	Betanodavirus	*Striped jack nervous necrosis virus*	V
Tetraviridae	*Betatetravirus*	*Nudaurelia capensis* β *virus*	I
	Omegatetravirus	*Nudaurelia capensis* ω *virus*	I
Unassigned	*Sobemovirus*	*Southern bean mosaic virus*	P
Luteoviridae	*Luteovirus*	*Barley yellow dwarf virus-PAV*	P
	Polerovirus	*Potato leafroll virus*	P
	Enamovirus	*Pea enation mosaic virus-1*	P
Unassigned	*Umbravirus*	*Carrot mottle virus*	P
Tombusviridae	*Dianthovirus*	*Carnation ringspot virus*	P
	Tombusvirus	*Tomato bushy stunt virus*	P

Continued

Table 1 Continued

Order	Family	Subfamily	Genus	Type species	Host
			Aureusvirus	*Pothos latent virus*	P
			Avenavirus	*Oat chlorotic stunt virus*	P
			Carmovirus	*Carnation mottle virus*	P
			Necrovirus	*Tobacco necrosis virus A*	P
			Panicovirus	*Panicum mosaic virus*	P
			Machlomovirus	*Maize chlorotic mottle virus*	P
Nidovirales					
	Coronaviridae		*Coronavirus*	*Infectious bronchitis virus*	V
			Torovirus	*Equine torovirus*	V
	Arteriviridae		*Arterivirus*	*Equine arteritis virus*	V
	Roniviridae		*Okavirus*	*Gill-associated virus*	I
	Flaviviridae		*Flavivirus*	*Yellow fever virus*	V, I
			Pestivirus	*Bovine viral diarrhea virus 1*	V
			Hepacivirus	*Hepatitis C virus*	V
	Togaviridae		*Alphavirus*	*Sindbis virus*	V, I
			Rubivirus	*Rubella virus*	V
	Unassigned		*Tobamovirus*	*Tobacco mosaic virus*	P
			Tobravirus	*Tobacco rattle virus*	P
			Hordeivirus	*Barley stripe mosaic virus*	P
			Furovirus	*Soil-borne wheat mosaic virus*	P
			Pomovirus	*Potato mop-top virus*	P
			Pecluvirus	*Peanut clump virus*	P
			Benyvirus	*Beet necrotic yellow vein virus*	P
	Bromoviridae		*Alfamovirus*	*Alfalfa mosaic virus*	P
			Bromovirus	*Brome mosaic virus*	P
			Cucumovirus	*Cucumber mosaic virus*	P
			Ilarvirus	*Tobacco streak virus*	P
			Oleavirus	*Olive latent virus 2*	P
	Unassigned		*Ourmiavirus*	*Ourmia melon virus*	P
			Idaeovirus	*Rasberry bushy dwarf virus*	P

	Tymoviridae	*Tymovirus*	*Turnip yellow mosaic virus*	P
		Marafivirus	*Maize rayado fino virus*	P, I
		Maculavirus	*Grapevine fleck virus*	P
	Closteroviridae	*Closterovirus*	*Beet yellows virus*	P
		Ampelovirus	*Grapevine leafroll-associated virus 3*	P
		Crinivirus	*Lettuce infectious yellows virus*	P
	Flexiviridae	*Potexvirus*	*Potato virus X*	P
		Mandarivirus	*Indian citrus ringspot virus*	P
		Allexivirus	*Shallot virus X*	P
		Carlavirus	*Carnation latent virus*	P
		Foveavirus	*Apple stem pitting virus*	P
		Capillovirus	*Apple stem grooving virus*	P
		Vitivirus	*Grapevine virus A*	P
		Trichovirus	*Apple chlorotic leaf spot virus*	P
	Barnaviridae	*Barnavirus*	*Mushroom bacilliform virus*	F
Unassigned viruses				
Unassigned Vertebrate Viruses				V
Unassigned Invertebrate Viruses				I
Unassigned Prokaryote Viruses				B
Unassigned Fungus Viruses				F
Unassigned Plant Viruses				P
The subviral agents: Viroids, satellites and agents of spongiform encephalopathies (prions)				
Viroids				
	Pospiviroidae	*Pospiviroid*	*Potato spindle tuber viroid*	P
		Hostuviroid	*Hop stunt viroid*	P
		Cocadviroid	*Coconut cadang-cadang viroid*	P
		Apscaviroid	*Apple scar skin viroid*	P
		Coleviroid	*Coleus blumei viroid 1*	P
	Avsunviroidae			5
		Avsunviroid	*Avocado sunblotch viroid*	P
		Pelamoviroid	*Peach latent mosaic virus*	P
Satellites				
Vertebrate Prions				V
Fungi prions				F

Virus hosts: Al, Algae; Ar, Archaea; B, Bacteria; F, Fungi; I, Invertebrates; P, Plants; Pr, Protozoa; V, Vertebrates.

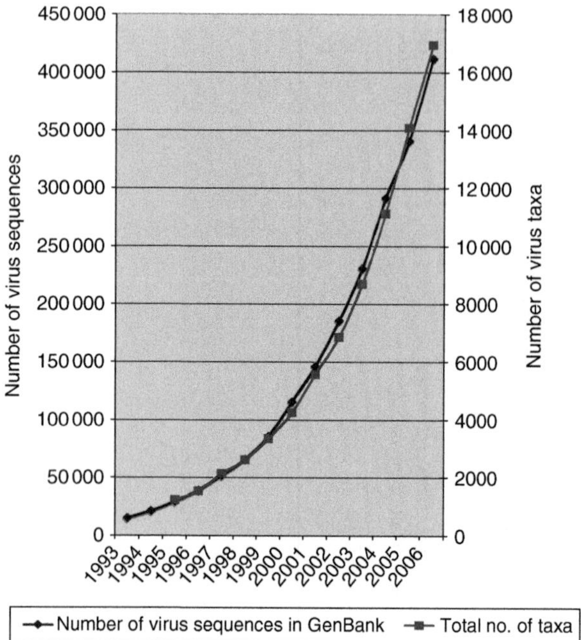

Figure 1 Number of virus taxa (including isolates) and virus sequences stored at GenBank since 1993.

virologists from many countries who make decisions on virus names and taxa through a democratic process. The ICTV operates through subcommittees and study groups consisting of more than 500 virologists with expertise in human, animal, insect, protozoa, archaea, bacteria, mycoplasma, fungi, algae, and plant viruses.

Taxonomic proposals are initiated and formulated by study groups or by single individuals. The proposals are examined, offered to public scrutiny, accepted by the relevant subcommittee and presented for approval by the Executive Committee of the ICTV. All decisions are ratified by postal vote, where all members of the ICTV and more than 50 national microbiological societies are represented. Presently, there are 75 study groups working in concert with six subcommittees: one each for the vertebrate, invertebrate, plant, bacterial, and fungal viruses and one for the virus ICTV DataBase (ICTVdB). The ICTV does not impose taxa but ensures that all propositions are compatible with the International Code for Virus Classification and Nomenclature for accuracy, homogeneity, and consistency. The ICTV regularly publishes reports describing all existing virus taxa and containing a complete list of classified viruses with their abbreviations. The ICTV published its Eighth Report in 2005. An internet website is also maintained where all new taxonomic proposals are loaded and where the most important information relative to virus taxonomy is made available and updated regularly. The increasing number of virus species and virus strains being identified, along with the explosion of data on many descriptive aspects of viruses and viral diseases, particularly sequence data, has led the ICTV to launch an international virus database project (ICTVdB) and a Taxonomic Proposal Management System specifically to handle taxonomic proposals.

Polythetic Classification and Demarcation Criteria

There are currently two systems in use for classifying organisms: the Linnean and the Adansonian systems. The Linnean system is the monothetic hierarchical classification applied by Linnaeus to plants and animals, while the Adansonian is a polythetic hierarchical system. Although convenient to use, the Linnean system has shortcomings when applied to the classification of viruses because there is no obvious reason to privilege one criterion over another. The Adansonian system considers all available criteria at once and makes several classifications, taking the criteria successively into consideration. Criteria leading to the same classifications are considered correlated and are therefore not discriminatory. Subsequently, a subset of criteria is considered, and the process is repeated until all criteria can be ranked to provide the best discrimination of the species. Furthermore, qualitative and quantitative data can be simultaneously considered when building such a classification. In the case of viruses, the method is not used on a systematic basis, although it has been shown that at least 60 characters are needed for a complete virus description (**Table 2**).

The increasing number of reported viral nucleic acid sequences allows the construction of phylogenetic trees based on a single gene or a group of genes. Sequence comparisons by themselves have not satisfactorily provided a clear classification of all viruses together but are widely used at the order, family, and genus levels. Recently the National Center for Biotechnology Information (NCBI) in Washington developed a system of pairwise sequence comparisons (the so-called PASC system) between viral sequences which allows a new virus to be assigned to known taxa. It seems probable that, in future, virus classification will make increasing use of sequence data.

For more than 40 years, the ICTV has been classifying viruses essentially at the family and genus levels using a nonsystematic polythetic approach. Viruses are first clustered in genera and then in families. A subset of characters including physicochemical, structural, genomic, and biological criteria is then used to compare and group viruses. This subset of characters may change from one family to another according to the availability of the data

Table 2 Virus family descriptors used in virus taxonomy

I. Virion properties
  A. Morphology properties of virions
    1. Virion size
    2. Virion shape
    3. Presence or absence of an envelope and peplomers
    4. Capsomeric symmetry and structure
  B. Physical properties of virons
    1. Molecular mass of virions
    2. Buoyant density of virions
    3. Sedimentation coefficient
    4. pH stability
    5. Thermal stability
    6. Cation (Mg^{++}, Mn^{++}) stability
    7. Solvent stability
    8. Detergent stability
    9. Radiation stability
  C. Properties of genome
    1. Type of nucleic acid – DNA or RNA
    2. Strandedness – single stranded or double stranded
    3. Linear or circular
    4. Sense – positive, negative, or ambisense
    5. Number of segments
    6. Size of genome or genome segments
    7. Presence or absence and type of 5′-terminal cap
    8. Presence or absence of 5′-terminal covalently linked polypeptide
    9. Presence or absence of 3′-terminal poly(A) tract (or other specific tract)
    10. Nucleotide sequence comparisons
  D. Properties of proteins
    1. Number of proteins
    2. Size of proteins
    3. Functional activities of proteins (especially virion transcriptase, virion reverse transcriptase, virion hemagglutinin, virion neuraminidase, virion fusion protein)
    4. Amino-acid-sequence comparisons
  E. Lipids
    1. Presence or absence of lipids
    2. Nature of lipids
  F. Carbohydrates
    1. Presence or absence of carbohydrates
    2. Nature of carbohydrates
II. Genome organization and replication
    1. Genome organization
    2. Strategy of replication of nucleic acid
    3. Characteristics of transcription
    4. Characteristics of translation and post-translational processing
    5. Site of accumulation of virion proteins, site of assembly, site of maturation and release
    6. Cytopathology, inclusion body formation
III. Antigenic properties
    1. Serological relationships
    2. Mapping epitopes
IV. Biological properties
    1. Host range, natural and experimental
    2. Pathogenicity, association with disease
    3. Tissue tropisms, pathology, histopathology
    4. Mode of transmission in nature
    5. Vector relationships
    6. Geographic distribution

and depending on the importance of a particular character for a particular family. Obviously, there is no homogeneity in this respect in the current virus classification system, and virologists weigh the criteria in a subjective process. Nevertheless, over time, there has been a great stability of the current classification at the genus and family levels. It is also clear that hierarchical classification above the family level will encounter conflicts between phenotypic and genotypic criteria and that virologists may have to reconsider the entire classification process in order to progress at this level.

Virus Taxa Descriptions

Virus classification continues to evolve with the technologies available for describing viruses. The first wave of descriptions, those before 1940, took into account mostly the visual symptoms of viral diseases along with modes of viral transmission. A second wave, between 1940 and 1970, brought together an enormous amount of information from studies of virion morphology (electron microscopy, structural data), biology (serology and virus properties), and physicochemical properties of viruses (nature and size of the genome, number and size of viral proteins). The impact of descriptions on virus classification has been particularly influenced by electron microscopy and the negative-staining technique for virions in the 1960s and 1970s. With this technique, viruses could be identified from poorly purified preparations of all tissue types and information about size, shape, structure, and symmetry could be quickly provided. As a result, virology progressed simultaneously for all viruses infecting animals, insects, plants, and bacteria. Since 1970, the virus descriptors list has included genome and replication information (sequence of genes, sequence of proteins), as well as molecular relationships with virus hosts.

The most recent wave of information used to classify viruses is virus genome sequences. Genome sequence comparisons are becoming more and more prevalent in virus taxonomy as exemplified by the presence of a significant number of phylogenetic trees in the *Eighth ICTV Report*. Some scientists promote the concept of quantitative taxonomy, aimed at demonstrating that virus genome sequences contain all the coding information required for all the biological properties of the viruses. This is in complete agreement with the polythetic concept of virus species definition if one considers that the unique sequence of a genome contains in fact all the information of the virus to perform all the steps of its replication cycle with structural and nonstructural genes and all of its biological functions. A good example of quantitative taxonomy is the re-classification of flaviviruses from the genus *Flavivirus* in the family *Togaviridae* into the new family *Flaviviridae* based upon sequencing of the yellow fever virus genome and comparisons with the gene sequence arrangement of members of the genus *Alphavirus* in the family *Togaviridae.* Another recent example is the merging of the genera *Rhinovirus* and *Enterovirus* in the family *Picornaviridae,* based on the fact that phylogenetic trees and pairwise comparisons did not support the continued distinction between the two genera.

There is a correlative modification of the list of virus descriptors, and **Table 2** lists the family and genus descriptors which are used in the current ICTV report. **Table 2** lists 45 different types of properties where each property (e.g., morphology) can take on different individual states (e.g., filamentous, icosahedral, etc.). A universal lists of virus descriptors has been established which is used by the ICTVdB. It contains a common set of descriptors for all viruses and subsets for specific viruses in relation to their specific hosts (human, animal, insect, plant, and bacterial).

The Order of Presentation of the Virus Classification

Currently, and for practical reasons only, virus classification is structured according to the 'Order of Presentation of Viruses' indicated in **Table 1**. The presentation of virus orders, families, and genera in this particular order reflects convenience rather than any hierarchical or phylogenetic consideration. The Order of Presentation of Viruses follows four criteria: (1) the nature of the viral nucleic acid, (2) the strandedness of the nucleic acid (single stranded (ss) or double stranded (ds)), (3) the use of a reverse transcription process (DNA or RNA), and (4) the sense of gene coding on the encapsidated genome (positive, negative, or ambisense). These four criteria give rise to six clusters comprising the 86 families and unassigned genera (genera without a designated family). Within each cluster, families and unassigned genera have been listed according to their possible affinities. For example, the families *Picornaviridae, Dicistroviridae, Sequiviridae, Comoviridae,* and *Potyviridae* are listed one after another because they share a number of similarities in their genome organization and sequence relatedness and they may form the basis for a proposed order in the future.

A New Virus Taxon: The Virus Species

For many years, virologists debated the existence of virus species which was a very controversial issue and a series of definitions surfaced at regular intervals but none was adopted. However, in 1991, the ICTV Executive Committee accepted the species concept and the adopted definition is "A virus species is a polythetic class of viruses that constitutes a replicating lineage and occupies a particular ecological niche." This simple definition has already and will continue to have a profound effect on virus classification. In the *Eighth Report of the ICTV,* the 'List of Species' and the 'List of Tentative Species' are accompanied by a 'List of Species Demarcating Criteria' provided for each genus. Naturally, this list of criteria should follow the polythetic nature of the species definition, and more than one criterion should be used to determine a new species. It is obvious that most criteria are shared among the different genera, within and across families. These shared criteria include host range, serological relationships, vector transmission type, tissue tropism, genome rearrangement, and sequence homology (**Table 3**). However, while the nature of the criteria is similar, the levels of demarcation clearly differ from one family to another. This may reflect differences in appreciation from one family to another, but most likely reflects the differential ranking of a particular criterion in different families. The huge differences in sequence homologies (up to 30%) among lentivirus nucleoprotein sequences may not have the same biological significance as small differences for potyvirus capsid protein sequences (0–10%), and therefore universal levels of sequence identity for similar genes may not exist for viruses. However,

Table 3 List of criteria demarcating different virus taxa

I. Order
- Common properties between several families including:
  - Biochemical composition
  - Virus replication strategy
  - Particle structure (to some extent)
  - General genome organization

II. Family
- Common properties between several genera including:
  - Biochemical composition
  - Virus replication strategy
  - Nature of the particle structure
  - Genome organization

III. Genus
- Common properties within a genus including:
  - Virus replication strategy
  - Genome size, organization, and/or number of segments
  - Sequence homologies (hybridization properties)
  - Vector transmission

IV. Species
- Common properties within a species including:
  - Genome arrangement
  - Sequence homologies (hybridization properties)
  - Serological relationships
  - Vector transmission
  - Host range
  - Pathogenicity
  - Tissue tropism
  - Geographical distribution

it is important to note that the nature of the demarcating criteria at the genus level will probably not change since they have passed the test of years. Despite the fact that they were mostly established using biochemical and structural criteria, most of them have remained valid when correlated with genome organization and sequence data.

A Uniform Nomenclature for All Virus Taxa

Nomenclature is tightly associated with classification, in the sense that the taxonomic names indicate, to some extent, the nature of the taxa. Similarly for viruses, the ICTV has set rules for virus nomenclature and the orthography of taxonomic names that are regularly revised and improved. The international virus species names end in 'virus', international genus names in '...virus', international subfamily names in '...virinae', international family names in '...viridae', and international order names in '...virales'. In formal taxonomic usage, the virus order, family, subfamily, genus, and species names are printed in italics (or underlined) and the first letter is capitalized. For all taxa except species, new names are created following ICTV guidelines. Because of the difficulty in creating new official international names for virus species, it has been decided in 1998 by the ICTV to use the existing English vernacular virus names. However, to differentiate virus species names from virus names it has also been decided that their typography would be different, that is, the species names would be italicized, and the first letter of the name capitalized while the virus names would not. In addition ICTV had created an additional category called 'Tentative Species Names' to accommodate viruses that seemed to belong to a new species, but did not have enough data to support this decision; it was also a way to 'reserve' a name already used in literature. In 2005, ICTV decided to replace this category by 'Unassigned Viruses' in the genus.

Latinized binomials for virus names have been supported by animal and human virologists of the ICTV for many years but have never been implemented. Their recommendation was in fact withdrawn from ICTV nomenclature rules in 1990, and consequently, such names as *Herpesvirus varicella* or *Polyomavirus hominis* should not be used. For several years, plant virologists adopted a different nomenclature, using the vernacular name of a virus but replacing the word 'virus' by the genus name: for example, *Cucumber mosaic cucumovirus* and *Tobacco mosaic tobamovirus*. This system is called 'the non-latinized binomial system', although the binomial order is the opposite of the typical latinized binomial system where the genus name ends with the virus name. Though this usage is favored by many scientists, and examples of such a practice can be found for human, animal, and insect viruses (e.g., *Human rhinovirus*, *Canine calicivirus*, and *Acheta densovirus*), it has not yet been adopted as a universal system by the ICTV; however, it has been decided that each study group would decide what is best for the viruses they deal with and the new names would have to be ratified through a formal taxonomic proposal by the ICTV.

In formal usage, the name of the taxon precedes the name of the taxonomic unit: for example, "the family *Picornaviridae*" or "the genus *Rhinovirus*". In informal vernacular usage, virus order, family, subfamily, genus, and species names are written in lower case roman script; they are not capitalized or italicized (or underlined) – for example 'animal reoviruses'. To avoid ambiguous identifications, it has been recommended to journal editors that published virological papers follow ICTV guidelines for proper virus identification and nomenclature and that viruses should be cited with their full taxonomic terminology when they are first mentioned in an article, for example, order *Mononegavirales*, family *Paramyxoviridae*, subfamily *Pneumovirinae*, genus *Pneumovirus*, species *Human respiratory syncytial virus*.

A Universal Classification System

The present universal system of virus taxonomy is set arbitrarily at the hierarchical levels of order, family, subfamily, genus, and species. Lower hierarchical levels, such as suborder, subgenera, and subspecies, may be considered in the future if need arises. Hierarchical levels under the species level such as strains, serotypes, variants, and pathotypes are established by international specialty groups and/or by culture collections, but not by the ICTV.

Species

The species taxon is always regarded as the most important taxonomic level in classification but has proved difficult to apply to viruses. In 1991, the ICTV accepted the definition of species, stated above, proposed by Marc van Regenmortel. The major advantage of this definition is that it can accommodate the inherent variability of viruses and is not dependent on the existence of a unique set of characteristics. Members of a polythetic class are defined by more than one property and no single property is absolutely indispensable. Thus, in each family, it might be possible to determine the set of properties of the class 'species' and thus to verify if the family members are representatives of the class 'species' or if they belong to a different taxonomic level.

Many practical matters are related to the definition of a virus species. These include (1) homogeneity of the different taxa across the classification, (2) diagnostic-related matters, (3) virus collections, (4) evolution studies, (5) biotechnology, (6) sequence database projects, (7) virus database projects, (8) publication matters, and also (9) intellectual property rights.

Genera

There is no formal ICTV definition for a genus, but it is commonly considered as "a population of virus species that share common characteristics and are different from other populations of species." Although this definition is somewhat elusive, this level of classification seems stable and useful. Some genera have been moved from one family to another (or from one family to an unassigned genus status such as the genus *Hepevirus*) over the years, but the composition and description of these genera has remained very stable. The characters defining a genus are different from one family to another. The use of subgenera has been abandoned in current virus classification.

Families

Genera are usually clustered in families, and most of the time, when a new genus, obviously not belonging to any existing family, is created, virologists also create a new family. Even after the creation of the ICTV, plant virologists have continued to classify plant viruses in 'groups', refusing to place them in genera and families. This position was mostly caused by a refusal to accept a binomial nomenclature. However, because of obvious similarities, plant reoviruses and rhabdoviruses had been integrated into the families *Reoviridae* and *Rhabdoviridae* (**Table 1**). Plant virologists subsequently accepted in 1995 the placing of plant viruses into species, genera, and families. The number of unassigned genera is regularly decreasing with time; the most recent clustering is the creation of the family *Flexiviridae* with the genera *Potexvirus*, *Carlavirus*, *Mandarivirus*, *Foveavirus*, *Capillovirus*, *Allexivirus*, *Vitivirus*, and *Trichovirus*. However, there are still 22 unassigned genera that do not belong to any family. Their presence originates mostly from the preference of plant virologists for accumulating data on virus species and genera before clustering genera in families. The unassigned genus status is now being used by animal virologists as a convenient temporary classification status. Examples are the unassigned genera *Anellovirus*, that is close to the family *Circoviridae* but different enough to be separated, the previously unassigned genus *Cripavirus* that has been upgraded to full family status (*Dicistrovirida*), and another unassigned genus *Iflavirus* that has been created to accommodate new picorna-like viruses that are not typical picornaviruses.

Orders

As mentioned above, the higher hierarchical levels for virus classification are extremely difficult to establish. To date only three orders have been accepted: *Caudovirales*, *Mononegavirales*, and *Nidovirales*. The first order, *Mononegavirales*, was established in 1990 and comprises the nonsegmented ssRNA negative-sense viruses, namely, the families *Bornaviridae*, *Filoviridae*, *Paramyxoviridae*, and *Rhabdoviridae*. This order was formed because of the great similarity between these families over many criteria, including their replication strategies. A second order, *Caudovirales*, contains all families of dsDNA phages possessing a tail, including the *Myoviridae*, *Podoviridae*, and *Siphoviridae*. The order *Nidovirales* comprises the families *Coronaviridae*, *Arteriviridae*, and *Roniviridae* and was created because it was clear that the viruses belonging to these families share many properties and yet are so different that they cannot be placed together in the same family. Many members of the ICTV advocate the creation of many more orders, and as a matter of fact four new orders encompassing the families *Herpesviridae*, *Picornaviridae*, *Reoviridae*, *and Retroviridae* have been proposed and provisionally named but it has been decided to proceed cautiously in this area so as to avoid creation of short-lived orders. The creation of formal taxa higher than the orders – for example kingdoms, classes, and subclasses – has not been considered by the ICTV.

See also: Nature of Viruses; Phylogeny of Viruses; Virus Classification by Pairwise Sequence Comparison (PASC); Virus Databases; Virus Species.

Further Reading

Adams MJ, Antoniw JF, Bar-Joseph M, *et al.* (2004) The new plant virus family *Flexiviridae* and assessment of molecular criteria for species demarcation. *Archives of Virology* 149: 1045–1060.

Dolja VV, Boyko VP, Agranovsky AA, and Koonin EV (1991) Phylogeny of capsid proteins of rod-shaped and filamentous RNA plant viruses: Two families with distinct patterns of sequence and probably structure conservation. *Virology* 184: 79–86.

Dolja VV and Koonin EV (1991) Phylogeny of capsid proteins of small icosahedral RNA plant viruses. *Journal of General Virology* 72: 1481–1486.

Fauquet CM, Mayo MA, Maniloff J, Desselberger U, and Ball LA (eds.) (2004) *Virus Taxonomy: Eigth Report of the International Committee on Taxonomy of Viruses,* p.1258. San Diego, CA: Elsevier Academic Press.

Francki RIB, Milne RG, and Hatta T (1985) *Atlas of Plant Viruses.* Boca Raton, FL: CRC Press.

Harrison BD, Finch JT, Gibbs AJ, *et al.* (1971) Sixteen groups of plant viruses. *Virology* 45: 356–363.

Koonin EV (1991) The phylogeny of RNA-dependent RNA polymerases of positive-strand RNA viruses. *Journal of General Virology* 72: 2197–2206.

Mayo MA and Horzinek MC (1998) A revised version of the International Code of Virus Classification and Nomenclature. *Archives in Virology* 143: 1645–1654.

Van Regenmortel MHV (1990) Virus species, a much overlooked but essential concept in virus classification. *Intervirology* 31: 241–254.
Van Regenmortel MHV, Bishop DHL, Fauquet CM, Mayo MA, Maniloff J, and Calisher CH (1997) Guidelines to the demarcation of virus species. *Archives of Virology* 142: 1505–1518.

Relevant Website

http://www.ictv.ird.fr – ICTV; Taxonomic Proposal Management System.

Virus Classification by Pairwise Sequence Comparison (PASC)

Y Bao, Y Kapustin, and T Tatusova, National Institutes of Health, Bethesda, MD, USA

Glossary

Demarcation A mapping of ranges of pairwise distances into taxonomic categories.

Introduction

Virus classification is very important for virus research. It is also an extremely difficult task for many virus families. Traditionally, virus classification relied on properties such as virion morphology, genome organization, replication mechanism, serology, natural host range, mode of transmission, and pathogenicity. Yet viruses sharing the above properties can reveal tremendous differences at the genome level. For example, classification of many phages is currently based on presence, structure, and length of a tail, and this approach has been shown not to correlate with genomic information, leading to a very difficult situation and hundreds of unclassified phages.

Molecular virus classification based on virus sequences has been used increasingly in recent years, thanks to the growing number of viral sequences available in the public sequence databases. The most commonly used sequence comparison methods include multiple sequence alignment and phylogenetic analysis. Another molecular classification method that has drawn more and more attention from virologists is pairwise sequence comparison (PASC). In this article, we briefly describe various sequence comparison methods, introduce the PASC tool, and compare it with other methods.

Sequence Comparison Methods

A universal approach to compare biological sequences, in a sense of producing meaningful results at various levels of divergence, is in the realm of sequence alignment. An alignment is an arrangement of residues of two or more sequences in a way that reveals their possible relatedness, with space characters inserted into the sequences to indicate single-residue insertions and deletions. A variety of algorithms and programs are available to suit a wide range of problems requiring sequence alignments as parts of their solutions. Depending on the specifics of a problem, different types of algorithms or their combinations may work best. Most alignment algorithms can be broadly categorized by the scope of their application on sequences (local vs. global), or by the number of sequences involved (pairwise vs. multiple).

Each pairwise alignment can be viewed as an array of per-residue operations transforming one sequence to the other. These operations are substitutions (called matches and mismatches in nucleotide alignments), insertions, and deletions. A generalization of this concept to multiple alignments is possible. Alignments are scored using a scoring scheme appropriate for a biological context. The widely used affine scheme assigns substitution scores to substitutions and a penalty to each space, and an additional penalty to each gap defined as a maximal consecutive run of spaces. Given a set of sequences, an alignment is called optimal if it has the maximal score over all possible alignments. Optimal alignments are not necessarily unique; two or more alignments can be tied with the same score.

Local algorithms are capable of detecting similarities between arbitrary parts of sequences. Applications involving local alignments are numerous, including search for orthologous genes or conserved protein domains. Alignments produced with local algorithms are tractable to mathematical analysis, which allowed the building of tools that evaluate statistical significance of the alignments. The algorithm for computing optimal local alignments is known as Smith–Waterman and it runs in time proportional to the product of the sequences' lengths. Since this is too slow for large-scale searches, many heuristic methods (with BLAST being the most popular) have been developed, allowing matching typical queries against gigabase-sized archives of sequence in a matter of seconds.

Fast as they are, algorithms like BLAST are not suitable for all applications. Although they are capable of picking out segments of high similarity, no segment of input sequences is guaranteed to be a part of the resulting alignments, and some segments may belong to more than one individual alignment. Additional post-processing steps are often required in order to produce consistent sets of local alignments. This complicates the use of local alignments in applications involving uniform computing of identities.

When sequences in the set are expected to align end-to-end, global alignment algorithms are applicable. The strict algorithm for computing an optimal global alignment is known as Needleman–Wunsch. With the running time estimate being the same as in the Smith–Waterman, global alignments allow straightforward evaluation of identities as they provide unambiguous mapping for every residue. An important end-space free variant is used when one of the sequences is expected to align in the interior of another. Global algorithms are normally not suitable for applications aiming to capture rearrangement events.

Multiple sequence alignments can technically be viewed as a generalization of the pairwise case. However, they often serve different goals, such as a detection of weak and/or dispersed similarities over a set of sequences known to share a common function or structure. Computing an optimal multiple sequence alignment is a computationally costly task, and most implementations use various heuristics to approximate the alignment in a reasonable time. Note that even when the optimal multiple alignment is available, pairwise alignments inferred from it are not guaranteed to be optimal. There are many multiple sequence alignment tools available: CLUSTALW, DIALIGN, MAFFT, MUSCLE, PROBCONS, ProDA, and T-COFFEE, etc.

Phylogenetic analysis is probably the most frequently used molecular virus classification tool. A phylogeny or evolutionary tree is a mathematical structure which is used to model the historical relationships between groups of organisms or sequences. Main types of methods used to construct phylogenetic trees include distance-based methods (such as neighbor-joining), parsimony, maximum likelihood, and other probabilistic inference techniques. The most common distance-based methods utilize multiple sequence alignments to estimate the evolutionary distance between each pair of sequences and reconstruct the tree from the distances. Either protein sequences or DNA sequences can be used.

Phylogenetic analysis was used in the vast majority of virus families described in the Eighth Report of the International Committee on the Taxonomy of Viruses (ICTV) to support their classification. It has also been applied to the classification of a large group of distantly related viruses. For example, phages consist of many different families. Therefore, the conventional phylogenetic analysis that uses genomic sequences or individual protein sequences would not work for the classification of phages as a whole group. A 'phage proteomic tree' was developed to classify phages by the overall similarities of all protein sequences present in the phage genomes.

Although phylogenetic analysis is well established as a tool for virus classification, it is usually computationally intensive, and requires expertise to perform the analysis and to interpret the results. A more robust method is preferred so that researchers without an advanced computer system and advanced knowledge about the phylogenetic analysis can also use it. Also, as discussed below, despite the fact that some of the sequence alignment methods (such as BLAST) are very fast and easy to carry out, their results will not reveal the taxonomic relationships between the two viruses. A method that can place a new virus in the appropriated taxonomic position is desired. PASC is a good combination of the two methods.

The Principle of PASC

In the PASC system, pairwise global alignment is performed on complete genomes or particular protein sequences for each viral family, and their percentage of identity is calculated. The number of virus pairs at each percentage is plotted. The distribution of the identities is not evenly spread, but rather clustered into groups of peaks for viruses at strain, species, genus, and subfamily levels. The percentage range of each peak serves as a good reference for taxonomic classification based on sequence similarities. This method has been applied to polioviruses using the protein and nucleotide sequences of the VP1 gene, as well as the whole genome sequence, coronaviruses using the protein sequences of the polymerase and helicase, potyviruses using the protein and nucleotide sequences of the complete ORF and the coat protein gene, geminiviruses using the complete sequences of DNA-A, flexiviruses using the protein and nucleotide sequences of the three major viral gene products (replication protein, triple gene block and coat protein), papillomaviruses using the nucleotide sequences of the L1 gene, and poxviruses using the protein sequences of the DNA polymerase. There is an increasing interest to expand PASC to other virus families.

In order to apply PASC to a larger number of virus families and be used by a wide range of virologists, the following should be considered:

1. The same algorithm and sequence dataset should be used to determine the demarcations and to place new viruses in the right taxonomic position.
2. The sequence dataset used for demarcation determination and the virus taxonomy database should be updated frequently to reflect the most recent status.

3. The algorithm should be robust and fast enough for large sequence sets.
4. The system should be readily accessible to researchers worldwide and easy to use.

PASC Implementation at NCBI

The National Center for Biotechnology Information (NCBI) has developed a web-based PASC system that meets all of the criteria mentioned above. In this implementation, complete viral genomes are organized into groups corresponding to broad taxonomic entities such as family or floating genus. Within each group, alignments are pre-computed and stored in a database for each pair of the genomes. The alignments are used to evaluate identities, defined as the ratio of matching residues over the total alignment length. The alignments are computed using the pairwise global algorithm with the affine scoring scheme assigning one to matches and minus one to mismatches and nonterminal spaces and gaps. Since the genomes vary in lengths, terminal spaces are not penalized during the alignment computing but taken into account when computing the identity.

PASC interface is built around a histogram of pairwise identities. The primary feature of the interface is the comparison of an external sequence such as a newly sequenced viral genome, with genomes in a user-selected group. After the sequence is submitted, PASC will start computing the alignments, or extracting them from the database if the query is a member of the group. At the end of the process, a user is presented with a list of closest matches. Matches can be selected to visualize their positions on the identity distribution chart.

The PASC system at NCBI not only reproduced results for virus families for which PASC had been applied to, but also generated data with well-separated identity distributions that can be used as taxonomy demarcations for other virus families such as *Caliciviridae*, *Flaviviridae*, and *Togaviridae* among others.

Applications of PASC

PASC can be used to define taxonomy demarcations for many viral families. Two examples are shown here. In the first example, 45 complete genomes from the family *Luteoviridae* were used to construct the distribution of the pairwise identities among the genomes (**Figure 1**). Pairs located at 88% and up are all from different strains in the same species; pairs between 53% and 85% are mostly from different species in the same genus; and pairs located at 52% and lower are all from different genera. These percentages can therefore serve as demarcations

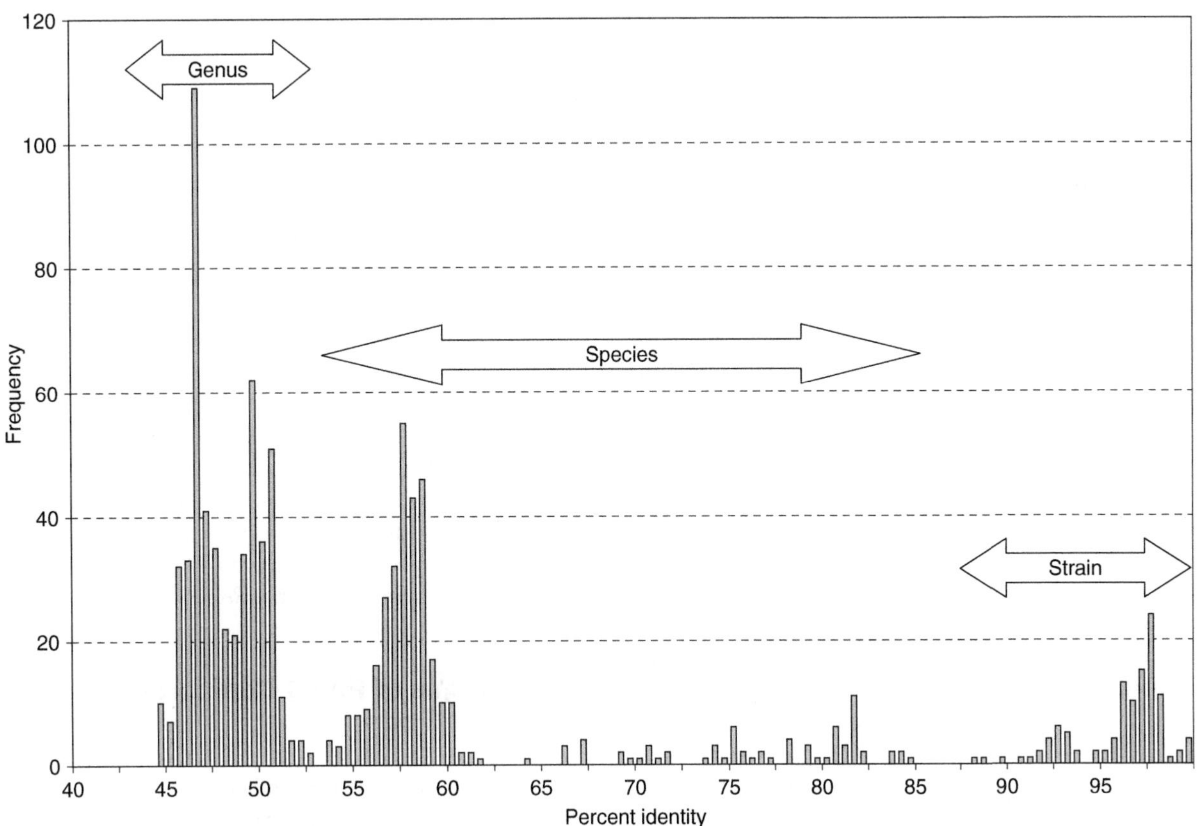

Figure 1 Frequency distribution of pairwise identities from the complete nucleotide sequence comparison of 45 luteoviruses.

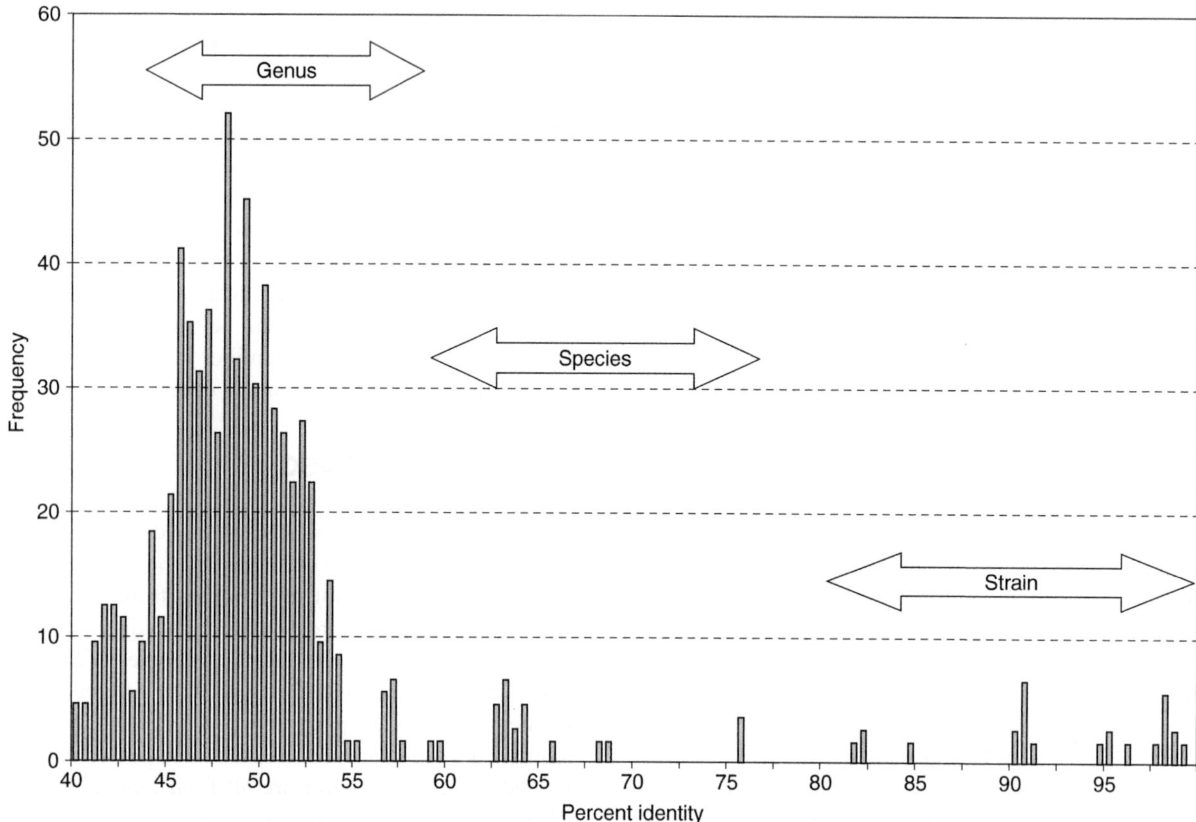

Figure 2 Frequency distribution of pairwise identities from the nucleotide sequence comparison of the RdRp segments of 38 reoviruses.

for classification of luteoviruses. In the second example, the pairwise identities of 38 RNA-dependent RNA polymerase (RdRp) gene segment of viruses in the family *Reoviridae* were calculated and plotted (**Figure 2**). Similar to the above mentioned example, 81% and up, between 59% and 76%, and 57% and below can be used as boundaries for strains, species, and genera of reoviruses based on the identities between their polymerase genes. It should be noted that the determination of demarcation using PASC is not always straightforward. For some virus families, phenotypic characteristics of the viruses are required to be taken into account.

PASC can place newly sequenced viruses into the correct taxonomy group. For example, a genomic sequence of a luteovirus (GenBank accession number AY956384) appeared recently in the international sequence databases with the name chickpea chlorotic stunt virus, which is not an official ICTV species name. When this sequence was tested with the luteoviruses in the PASC system, it was found that the virus with the highest sequence similarity to it is cucurbit aphid-borne yellows virus in the genus *Polerovirus*. The similarity is 61.7%, which is in the demarcation of different species in the same genus. It can thus be suggested that this virus is a member of a new species in the genus *Polerovirus*.

Finally, PASC can identify possible questionable classifications in the existing groups when the peaks on the graph are very well separated. **Table 1** lists the pairs of the RdRp segment of reoviruses whose identities are between 54% and 54.5% in **Figure 2**. From the description above, this region represents pairs of viruses from different genera. This is true for most of the pairs in the table. However, the group also includes a pair containing St. Croix river virus and palyam virus, which both currently belong to the genus *Orbivirus*. Further investigation using PASC revealed that palyam virus has identities higher than 59% with many other viruses in the family, and therefore is indeed an orbivirus. The highest identity of St. Croix river virus with other viruses is only about 54%, which is lower than the demarcation for species in the same genus. This suggests that the species *St. Croix river virus* should probably be placed in a new genus in the *Reoviridae* family.

Advantages of PASC Compared to Other Methods

Unlike other classification methods based on the phenotypic properties of viruses, PASC is a quantitative tool. For those virus families that are suitable for PASC analyses,

Table 1 Pairs of the RdRp segment of the reoviruses with identities between 54% and 54.5%

Identity	*Same genus?*	*Same species?*	*Genome 1*[a]	*Genome 2*
0.544895	No	No	20279540 Coltivirus|Eyach virus	37514915 Mycoreovirus|Mycoreovirus 3
0.542353	No	No	8574569 Seadornavirus|Banna virus	24286507 Cypovirus|Cypovirus 1|Dendrolimus punctatus cypovirus 1
0.541812	No	No	8574569 Seadornavirus|Banna virus	32470626 Orthoreovirus|Mammalian orthoreovirus|Mammalian orthoreovirus 3
0.541744	No	No	8574569 Seadornavirus|Banna virus	25808995 Orbivirus|Bluetongue virus|Bluetongue virus 2|Corsican bluetongue virus
0.541257	Yes	No	50253405 Orbivirus|St. Croix River virus	50261332 Orbivirus|Palyam virus
0.541096	No	No	22960700 Coltivirus|Colorado tick fever virus	32349409 Mycoreovirus|Mycoreovirus 1|Cryphonectria parasitica mycoreovirus-1 (9B21)
0.540999	No	No	8574569 Seadornavirus|Banna virus	14993633 Cypovirus|Cypovirus 1
0.54089	No	No	14993610 Cypovirus|Cypovirus 14	20177438 Fijivirus|Nilaparvata lugens reovirus

[a]The numbers correspond to sequences in GenBank. The taxonomy lineages from the genus level of the viruses are also shown.

demarcations can be easily determined and new viruses can be clearly placed into the correct taxonomy. However, there are times when PASC alone cannot give a definite classification, and other viral properties have to be considered.

Compared with another quantitative approach, phylogenetic analysis, PASC is less computationally intensive and can be easily updated with new sequence data. In addition, PASC results are relatively easier to interpret, which can be potentially done by a computer program without human intervention. It would therefore be possible to set up an automatic system for high throughput classification.

Many researchers use BLAST to search sequence databases to find best matches for viral sequences of interest. Although BLAST is readily available and easy to run, it is not the best tool for virus classification. This is because BLAST is a local alignment program, and, as discussed above, it may not take highly variable regions into account when calculating identities. Even when an output from BLAST covers the whole sequence as a single alignment, information about the taxonomy relationship between the query virus and the virus closest to it is not immediately available. For example, an identity of 75% could be within the range of the same species in one viral family, but in the range of different species in another family. On the contrary, PASC suggests explicitly whether the query virus is in the same species as some existing viruses, or if it should be assigned within a new species or genus in the family.

The total number of virus sequences in the GenBank/EMBL/DDBJ databases is more than 4 times now than 5 years ago (from about 109 000 to about 446 000 in October 2006). It is possible to test the PASC system on many virus families now. New sequencing technology makes it possible to generate large amounts of virus sequences from environmental samples without the need to isolate and purify virus particles. In such cases, a molecular based method is the only way to classify the viral sequences, and PASC can be very useful for this purpose.

Limitations of PASC

Although PASC has been applied successfully to several virus families, the approach has some limitations.

First of all, PASC is not suitable for virus families whose current classification is largely based on virus morphologies, such as phages in the families *Siphoviridae* and *Podoviridae*. The whole genome PASC may not work well for virus families with highly diverse sequences. This includes viruses with low overall sequence similarities or large differences in genome sizes and organization. For example, in the family *Herpesviridae*, the percentage of identity between different species in the genus *Varicellovirus* ranges from 39% (between cercopithecine herpesvirus 9 and suid herpesvirus 1) to 83% (between bovine herpesvirus 1 and bovine herpesvirus 5), while the percentage of identity between a virus in the genus *Simplexvirus* and one in the genus *Varicellovirus* could be as high as 54% (between cercopithecine herpesvirus 2 and bovine herpesvirus 5). The overlap of such identities makes it impossible to determine the species and genus demarcations for herpesviruses. In the family *Poxviridae*, the largest genome (canarypox virus) is almost 3 times as big as the smallest one (bovine papular stomatitis virus). The huge differences in the genomes sizes will introduce large artifacts when calculating the identities. In such cases, single gene or a cluster of genes needs to be used in PASC instead of whole genomes. The polymerase protein sequences of poxviruses have been used to perform PASC and a good result was obtained. However, not every single gene can be used for PASC. The genes must be present in all viruses of a family and be very conserved. They need to be tested extensively and accepted by the research community as a useful taxonomic criterion before being applied to the PASC system.

Second, for those families whose PASC were constructed with whole genomes, the query sequences to be tested on PASC to determine their taxonomic positions have to be complete genome sequences as well in order to get an accurate prediction. Although a percentage of identity can be obtained when a partial genome is used, the value will be smaller than what it really should be if a complete genome were used because of the way percentage of identity is calculated in this system. In addition, this value obtained with a partial sequence may not reflect the real taxonomic position of this virus, if, for example, recombination is frequent and important for this virus. This will reduce the number of sequences that can be tested by PASC.

Last, it is almost impossible to get identical PASC results when different methods are used. As mentioned above, there are many pairwise sequence alignment programs available. Even within a single program, variation of parameters can affect alignments. After an alignment is obtained, there can be various ways to calculate the distance. For demarcations computed using different types of distances, it may be difficult to choose a rational reason to privilege one demarcation threshold over another. Moreover, once a demarcation is adopted and a researcher uses a different definition to measure the distance between a new virus and an existing one, the comparison with the histogram will sometimes be misleading. These issues can be overcome by using a centralized PASC system where the alignment identities of new viruses are computed using the same algorithm and the same parameter set that were used to compute identities within the family and to create the demarcation.

Conclusion

PASC is a molecular classification tool for many virus families. It calculates the pairwise identities of virus sequences within a virus family and displays their distributions, and can help determine the demarcations at strains, species, genera, and subfamilies level. PASC has many advantages over conventional virus classification methods. The tool has been successfully applied to many virus families, although it may not work well for virus families with highly diverse sequences. The PASC tool at NCBI established distributions of identity for a number of virus families. A new virus sequence can be tested with this system within a few minutes to suggest the taxonomic position of the virus in these families. This system eliminates the potential discrepancies in the results caused by different algorithms and/or different data used by the virology community. Data in the system can be updated automatically to reflect changes in virus taxonomy and additions of new virus sequences to the public database. The web interface of the tool makes it easy to navigate and perform analyses.

See also: Taxonomy, Classification and Nomenclature of Viruses; Virus Databases; Virus Species.

Further Reading

Edgar RC and Batzoglou S (2006) Multiple sequence alignment. *Current Opinion in Structural Biology* 16: 368–373.

Fauquet CM, Mayo MA, Maniloff J, Desselberger U and Ball LA (eds.) (2005) *Virus Taxonomy, Classification and Nomenclature of Viruses: Eighth Report of the International Committee on the Taxonomy of Viruses.* London: Academic Press.

Felsenstein J (2004) *Inferring Phylogeny.* Sunderland, MA: Sinauer Associates.

Gusfield D (1997) *Algorithms on Strings, Trees, and Sequences: Computer Science and Computational Biology.* Cambridge: Cambridge University Press.

Page RD and Holmes EC (1998) *Molecular Evolution: A Phylogenetic Approach.* Oxford: Blackwell Science.

van Regenmortel MHV (2007) Virus species and virus identification: Past and Current Controversies. *Infection, Genetics and Evolution* 7: 133–144.

van Regenmortel MHV, Bishop DH, Fauquet CM, *et al.* (1997) Guidelines to the demarcation of virus species. *Archives of Virology* 142: 1505–1518.

Relevant Website

http://www.ncbi.nlm.nih.gov – NCBI, PASC.

Prions of Vertebrates

J D F Wadsworth and J Collinge, University College London, London, UK

Glossary

Codon 129 polymorphism There are two common forms of *PRNP* encoding either methionine or valine at codon 129; a major determinant of genetic susceptibility to and phenotypic expression of prion disease.
Conformational selection model A hypothetical model which explains transmission barriers on the basis of overlap of permissible conformations of PrP^{Sc} (prion strains) between mammalian species.
Molecular strain typing A means of rapidly differentiating prion strains by biochemical differences in PrP^{Sc}.
Prion The infectious agent causing prion diseases.
Prion incubation period The interval between exposure to prions and the development of neurological signs of prion disease; typically months even in rodent models and years to decades in humans.
Prion protein (PrP) A glycoprotein encoded by the host genome and expressed in many tissues but especially on the surface of neurons.
Prion strain Distinct isolates of prions originally identified and defined by biological characteristics which breed true in inbred mouse lines.
PRNP The human prion protein gene; mouse gene is designated *Prnp.*
Protein-only hypothesis The prions that lack a nucleic acid genome, are composed principally or solely of abnormal isomers of PrP (PrP^{Sc}), and replicate by recruitment of host PrP^{C}.
PrP Prion protein.
PrP^{C} The normal cellular isoform of PrP rich in α-helical structure.
PrP^{Sc} The '*scrapie*' or disease-associated isoform of PrP which differs from PrP^{C} in its conformation and is generally found as insoluble aggregated material rich in β-sheet structure.
Subclinical infection A state where host prion propagation is occurring but which does not produce clinical disease during normal lifespan; essentially a carrier state of prion infection.
Transmission barrier This describes the observation that transmission of prions from one species to another is generally inefficient when compared to subsequent passage in the same host species

Introduction

The prion diseases are a closely related group of neurodegenerative conditions which affect both humans and animals. They have previously been described as the subacute spongiform encephalopathies, slow virus diseases, and transmissible dementias, and include scrapie in sheep, bovine spongiform encephalopathy (BSE) in cattle, and the human prion diseases, Creutzfeldt–Jakob disease (CJD), variant CJD (vCJD), Gerstmann–Sträussler–Scheinker disease (GSS), fatal familial insomnia (FFI), and kuru. While rare in humans, prion diseases are an area of intense research interest. This is first because of their unique biology, in that the transmissible agent appears to be devoid of nucleic acid and to consist of a post-translationally modified host protein. Secondly, because of the ability of these and related animal diseases to cross from one species to another, sometimes by dietary exposure, there has been widespread concern that the exposure to the epidemic of BSE poses a distinct and conceivably a severe threat to public health in the United Kingdom and other countries. The extremely prolonged and variable incubation periods of these diseases, particularly when crossing a transmission barrier, means that it will be some years before the parameters of any human epidemic can be predicted with confidence. In the meantime, we are faced with the possibility that significant numbers in the population may be incubating this disease and that they might pass it on to others via blood transfusion, blood products, tissue and organ transplantation, and other iatrogenic routes.

Abberant Prion Protein Metabolism Is the Central Feature of Prion Disease

The nature of the transmissible agent in prion disease has been a subject of heated debate for many years. The understandable initial assumption that the causative agent of 'transmissible dementias' must be some form of virus was challenged by the failure to directly demonstrate such a virus (or indeed any immunological response to it) and by the remarkable resistance of the transmissible agent to treatments that inactivate nucleic acids. These findings led to suggestions that the transmissible agent may be devoid of nucleic acid and might be a protein. Subsequently in 1982, Prusiner and co-workers isolated a protease-resistant sialoglycoprotein, designated the prion protein (PrP), that was the major constituent of infective fractions and was found to accumulate in affected brain.

The term prion (from *pro*teinaceous *in*fectious particle) was proposed by Prusiner to distinguish the infectious pathogen from viruses or viroids and was defined as "small proteinaceous infectious particles that resist inactivation by procedures which modify nucleic acids."

Initially, PrP was assumed to be encoded by a gene within the putative slow virus thought to be responsible for these diseases; however, amino acid sequencing of part of PrP and the subsequent recovery of cognate cDNA clones using an isocoding mixture of oligonucleotides led to the realization that PrP was encoded by a single-copy chromosomal gene rather than by a putative viral nucleic acid.

Following these seminal discoveries, a wealth of data has now firmly established that the central and unifying hallmark of the prion diseases is the aberrant metabolism of PrP, which exists in at least two conformational states with different physicochemical properties. The normal cellular form of the protein, referred to as PrP^C, is a highly conserved cell surface glycosylphosphatidylinositol (GPI)-anchored sialoglycoprotein that is sensitive to protease treatment and soluble in detergents. In contrast, the disease-associated scrapie isoform, designated as PrP^{Sc}, is found only in prion-infected tissue as aggregated material, partially resistant to protease treatment, and insoluble in detergents. Due to its physicochemical properties, the precise atomic structure of the infectious particle or prion is still undetermined but considerable evidence argues that prions are composed largely, if not entirely, of an abnormal isoform of PrP. The essential role of host PrP for prion propagation and pathogenesis is demonstrated by the fact that knockout mice lacking the PrP gene ($Prnp^{o/o}$ mice) are entirely resistant to prion infection, and that reintroduction of PrP transgenes restores susceptibility to prion infection in a species-specific manner.

Human Prion Diseases Are Biologically Unique

Human prion diseases are biologically unique and can be divided etiologically into inherited, sporadic, and acquired forms. Approximately 85% of cases of human prion disease occur sporadically as Creutzfeldt–Jakob disease (sporadic CJD) at a rate of roughly 1 case per million population per year across the world, with an equal incidence in men and women. The etiology of sporadic CJD is unknown, although hypotheses include somatic *PRNP* mutation, or the spontaneous conversion of PrP^C into PrP^{Sc} as a rare stochastic event. Polymorphism at residue 129 of human PrP (encoding either methionine (M) or valine (V)) powerfully affects genetic susceptibility to human prion diseases. About 38% of Europeans are homozygous for the more frequent methionine allele, 51% are heterozygous, and 11% homozygous for valine. Homozygosity at *PRNP* codon 129 predisposes to the development of sporadic and acquired CJD. Most sporadic CJD occurs in individuals homozygous for this polymorphism. This susceptibility factor is also relevant in the acquired forms of CJD, most strikingly in vCJD where all clinical cases studied so far have been homozygous for codon 129 methionine of the PrP gene *PRNP*. Additionally, a *PRNP* susceptibility haplotype has been identified indicating additional genetic susceptibility to sporadic CJD at or near the *PRNP* locus.

Approximately 15% of human prion diseases are associated with autosomal dominant pathogenic mutations in *PRNP*. How pathogenic mutations in *PRNP* cause prion disease is yet to be resolved, however, in most cases the mutation is thought to lead to an increased tendency of PrP^C to form a pathogenic PrP isoform. However experimentally manipulated mutations of the prion gene can lead to spontaneous neurodegeneration without the formation of detectable protease resistant PrP. These findings raise the question of whether all inherited forms of human prion disease invoke disease through the same mechanism, and in this regard it is currently unknown whether all are transmissible by inoculation.

Although the human prion diseases are transmissible diseases, acquired forms have, until recently, been confined to rare and unusual situations. The two most frequent causes of iatrogenic CJD occurring through medical procedures have arisen as a result of implantation of dura mater grafts and treatment with human growth hormone derived from the pituitary glands of human cadavers. Less frequent incidences of human prion disease have resulted from iatrogenic transmission of CJD during corneal transplantation, contaminated electroencephalographic (EEG) electrode implantation, and surgical operations using contaminated instruments or apparatus. The most well-known incidences of acquired prion disease in humans resulting from a dietary origin have been kuru that was caused by cannibalism among the Fore linguistic group of the Eastern Highlands in Papua New Guinea, and more recently the occurrence of variant CJD in the United Kingdom and some other countries that is causally related to human exposure to BSE in cattle. Incubation periods of acquired prion diseases in humans can be extremely prolonged, and it remains to be seen if a substantial epidemic of vCJD will occur within the UK and elsewhere.

PRNP codon 129 genotype has shown a pronounced effect on kuru incubation periods and susceptibility, and most elderly survivors of the kuru epidemic are heterozygotes. The clear survival advantage for codon 129 heterozygotes provides a powerful basis for selection pressure in the Fore. However, an analysis of worldwide haplotype diversity and allele frequency of coding and noncoding polymorphisms of *PRNP* suggests that balancing selection at this locus (in which there is more variation than expected because of heterozygote advantage) is much older and more geographically widespread. Evidence for balancing selection has been shown in only a few human genes. With biochemical and physical evidence of cannibalism on five continents, one explanation is that cannibalism resulted in several prion disease epidemics in human prehistory, thus imposing balancing selection on *PRNP*.

The Protein-Only Hypothesis of Prion Propagation

Despite extensive investigation, no evidence for a specific prion-associated nucleic acid has been found. Instead, a wide body of data now supports the idea that infectious prions consist principally or entirely of an abnormal isoform of PrP. PrP^{Sc} is derived from PrP^{C} by a post-translational mechanism and neither amino acid sequencing nor systematic study of known covalent post-translational modifications have shown any consistent differences between PrP^{C} and PrP^{Sc}. The protein-only hypothesis, in its current form, argues that prion propagation occurs through PrP^{Sc} acting to replicate itself with high fidelity by recruiting endogenous PrP^{C} and that this conversion involves only conformational change. However, the underlying molecular events during infection that lead to the conversion of PrP^{C} to PrP^{Sc} and how PrP^{Sc} accumulation leads to neurodegeneration remain poorly defined.

The most coherent and general model thus far proposed is that the protein, PrP, fluctuates between a dominant native state, PrP^{C}, and a series of minor conformations, one or a set of which can self-associate in an ordered manner to produce a stable supramolecular structure, PrP^{Sc}, composed of misfolded PrP monomers. Once a stable 'seed' structure is formed, PrP can then be recruited leading to an explosive, autocatalytic formation of PrP^{Sc}. Such a mechanism could underlie prion propagation and account for the transmitted, sporadic, and inherited etiologies of prion disease. Initiation of a pathogenic self-propagating conversion reaction, with accumulation of aggregated PrP, may be induced by exposure to a 'seed' of aggregated PrP following prion inoculation, or as a rare stochastic conformational change, or as an inevitable consequence of expression of a pathogenic PrP^{C} mutant which is predisposed to form misfolded PrP. It is now clear that a full understanding of prion propagation will require knowledge of both the structure of PrP^{C} and PrP^{Sc} and the mechanism of conversion between them.

Structure and Putative Function of PrP^{C}

PrP is highly conserved among mammals, has been identified in marsupials, birds, amphibians, and fish, and may be present in all vertebrates. It is expressed during early embryogenesis and is found in most tissues in the adult with the highest levels of expression in the central nervous system, in particular in association with synaptic membranes. PrP is also widely expressed in cells of the immune system. As a GPI-anchored cell surface glycoprotein, it has been speculated that PrP may have a role in cell adhesion or signaling processes, but its precise cellular function has remained obscure.

Mice lacking PrP as a result of gene knockout ($Prnp^{o/o}$ mice) show no gross phenotype; however, these mice are completely resistant to prion disease following inoculation and do not replicate prions. $Prnp^{o/o}$ mice do however show subtle abnormalities in synaptic physiology and in circadian rhythms and sleep. While the relative normality of $Prnp^{o/o}$ mice was thought to result from effective adaptive changes during development, data from *Prnp* conditional knockout mice suggest this is not the case. These mice undergo ablation of neuronal PrP expression at 9 weeks of age and remain healthy without evidence of neurodegeneration or an overt clinical phenotype. Thus, acute loss of neuronal PrP in adulthood is tolerated and the pathophysiology of prion diseases appears to be unrelated to loss of normal PrP function in neurons.

Nuclear magnetic resonance (NMR) measurements and crystallographic determination of PrP from numerous mammalian species, including human PrP, show that they have essentially the same conformation. Following cleavage of an N-terminal signal peptide, and removal of a C-terminal peptide on addition of a GPI anchor, the mature PrP^{C} species consists of an N-terminal region of about 100 amino acids which is unstructured in the isolated molecule in solution and a C-terminal segment, also around 100 amino acids in length. The C-terminal domain is folded into a largely α-helical conformation (three α-helices and a short antiparallel β-sheet) and stabilized by a single disulfide bond linking helices 2 and 3. There are two asparagine-linked glycosylation sites. The N-terminal region contains a segment of five repeats of an eight-amino-acid sequence (the octapeptide-repeat region), expansion of which by insertional mutation leads to inherited prion disease. While unstructured in the isolated molecule, it seems likely that the N-terminal region of PrP may acquire coordinated structure *in vivo* through coordination of either Cu^{2+} or Zn^{2+} ions.

Structural Properties of PrP^{Sc}

PrP^{Sc} is extracted from affected brains as highly aggregated, detergent insoluble material that is not amenable to high-resolution structural techniques. However, Fourier transform infrared spectroscopic methods show that PrP^{Sc}, in sharp contrast to PrP^{C}, has a high β-sheet content. PrP^{Sc} is covalently indistinguishable from PrP^{C} but can be distinguished from PrP^{C} by its partial resistance to proteolysis and its marked insolubility in detergents. Under conditions in which PrP^{C} exists as a detergent-soluble monomer and is completely degraded

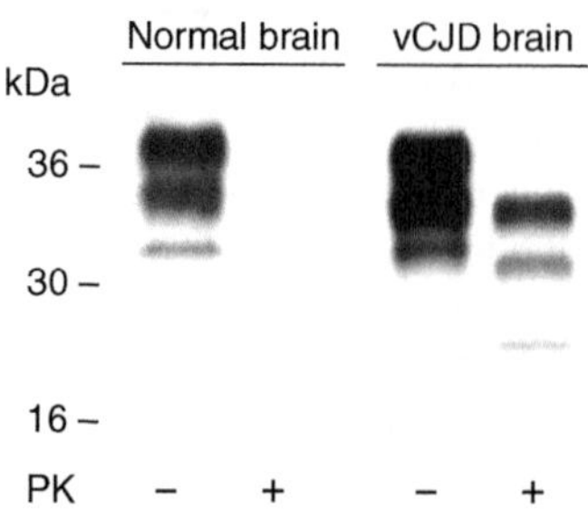

Figure 1 PrP analysis by immunoblotting. Immunoblot analysis of normal human brain and vCJD brain homogenate before and after treatment with proteinase K (PK). PrP^C in both normal and vCJD brain is completely degraded by proteinase K, whereas PrP^{Sc} present in vCJD brain shows resistance to proteolytic degradation leading to the generation of N-terminally truncated fragments of di-, mono-, and nonglycosylated PrP.

by the nonspecific protease, proteinase K, PrP^{Sc} exists in an aggregated form with the C-terminal two-thirds of the protein showing marked resistance to proteolytic degradation leading to the generation of N-terminally truncated fragments of di-, mono-, and nonglycosylated PrP (**Figure 1**). While there is no evidence for a specific prion-associated nucleic acid or other protein components, purified prion rods do however contain an inert polysaccharide scaffold.

Defining the precise molecular events that occur during the conversion of PrP^C to the infectious isoform of PrP is of paramount importance as this process is a prime target for therapeutic intervention. Direct *in vitro* mixing experiments have been performed in an attempt to produce PrP^{Sc}. In such experiments, PrP^{Sc} is used in excess as a seed to convert PrP^C to a protease-resistant form, designated PrP^{Res}. While there are now many examples in the literature of conditions that generate PrP^{Res}, historically such reactions have not been able to demonstrate *de novo* production of prion infectivity. Recently, however, a protein misfolding cyclic amplification system has demonstrated substantial amplification of PrP^{Res} and prion infectivity and may now provide the means to systematically investigate whether the generation of an infectious PrP isoform requires additional, as yet unknown, cofactors for the acquisition of infectivity.

The difficulty in performing structural studies on native PrP^{Sc} has led to attempts to produce soluble β-sheet-rich forms of recombinant PrP which may be amenable to NMR or crystallographic structure determination. Conditions have now been identified in which the PrP polypeptide can be converted between alternative folded conformations representative of PrP^C and PrP^{Sc}. At neutral or basic pH PrP adopts an α-helical fold representative of PrP^C and this conformation is locked by the presence of the native disulfide bond. Upon reduction of the disulfide bond, PrP^C rearranges to a predominantly β-sheet structure. This alternative conformation, designated β-PrP, is only populated at acidic pH with the PrP^C conformation predominating at neutral pH. Importantly, β-PrP shares overlapping properties with PrP^{Sc}, including partial resistance to proteolysis and a propensity to aggregate into fibrils. Success in producing disease in experimental animals that can be serially propagated following inoculation with PrP^{Sc}-like forms derived from recombinant PrP would not only prove the protein-only hypothesis, but would also provide an essential model by which the mechanism of prion propagation can be understood in molecular detail. In this regard, it has been recently reported that intracerebral injection of a β-sheet-rich fibrillar preparation of N-terminally truncated recombinant mouse PrP (comprising residues 89–230) into mice overexpressing PrP with the same deletion caused neurological disease after about 520 days.

Prion Disease Pathogenesis

Although the pathological consequences of prion infection occur in the central nervous system and experimental transmission of these diseases is most efficiently accomplished by intracerebral inoculation, most natural infections do not occur by these means. Indeed, administration to sites other than the central nervous system is known to be associated with much longer incubation periods, which in humans may extend to 50 years or more. Experimental evidence suggests that this latent period is associated with clinically silent prion replication in lymphoreticular tissue, whereas neuroinvasion takes place later. The M-cells in the intestinal epithelium appear to mediate prion entry from the gastrointestinal lumen into the body, and follicular dendritic cells (FDCs) are thought to be essential for prion replication and for accumulation of disease-associated PrP^{Sc} within secondary lymphoid organs. B-cell-deficient mice are resistant to intraperitoneal inoculation with prions probably because of their involvement with FDC maturation and maintenance. However, neuroinvasion is possible without FDCs, indicating that other peripheral cell types can replicate prions. The interface between FDCs and sympathetic nerves represents a critical site for the transfer of lymphoid prions into the nervous system; however, the mechanism by which this is achieved remains unknown. Distinct forms of prion disease show differences in lymphoreticular involvement that may be related to the etiology of the disease or to divergent properties of distinct prion strains. For example, the tissue distribution of PrP^{Sc} in vCJD differs strikingly from that in classical CJD, with uniform and prominent involvement of lymphoreticular tissues, with the highest amounts (up to 10% of brain concentrations) in tonsil. In contrast, in sporadic CJD, PrP^{Sc} has only been irregularly detected by immunoblotting in noncentral nervous system tissues at very much lower levels. Tonsil biopsy is used for diagnosis of vCJD and to date has shown 100% sensitivity and specificity for diagnosis of vCJD at an early clinical stage; tonsil is the tissue of choice for prospective studies investigating the prevalence of vCJD prion infection within the UK and other populations. The demonstration of extensive lymphoreticular involvement in the peripheral pathogenesis of vCJD raises concerns that iatrogenic transmission of vCJD prions through medical procedures may be a major public health issue. Prions resist many conventional sterilization procedures and surgical stainless steel-bound prions transmit disease with remarkable efficiency when implanted into mice. Disturbingly, cases of transfusion-associated vCJD prion infection have also now been reported. In contrast, there is no epidemiological evidence of transmission of classical CJD via blood transfusion or blood products.

Microscopic examination of the central nervous system of humans or animals with prion disease reveals typical characteristic histopathologic changes, consisting of neuronal vacuolation and degeneration, which gives the cerebral gray matter a microvacuolated or 'spongiform' appearance, and a reactive proliferation of astroglial cells (**Figure 2**). Demonstration of abnormal PrP immunoreactivity, or more specifically biochemical detection of PrP^{Sc} in brain material by immunoblotting techniques, is diagnostic of prion disease (**Figures 1** and **2**) and some forms of prion disease are characterized by deposition of amyloid plaques composed of insoluble aggregates of PrP. The histopathological features of vCJD are remarkably consistent and distinguish it from other human prion diseases with large numbers of PrP-positive amyloid plaques that differ in morphology from the plaques seen in kuru and GSS in that the surrounding tissue takes on a microvacuolated appearance, giving the plaques a florid appearance (**Figure 2**).

Prion Strains

A major problem for the 'protein-only' hypothesis of prion propagation has been to explain the existence of multiple isolates, or strains, of prions. Prion strains are distinguished by their biological properties: they produce distinct incubation periods and patterns of neuropathological targeting (so-called lesion profiles) in defined inbred mouse lines. As they can be serially propagated in inbred mice with the same *Prnp* genotype, they cannot be encoded by differences in PrP primary structure. Usually, distinct strains of conventional pathogen are explained by differences in their nucleic acid genome. However, in the absence of such a scrapie genome, alternative possibilities must be considered.

Support for the contention that prion strain specificity may be encoded by PrP itself was provided by study of two distinct strains of transmissible mink encephalopathy prions which can be serially propagated in hamsters, designated 'hyper' and 'drowsy'. These strains can be distinguished by differing physicochemical properties of the accumulated PrP^{Sc} in the brains of affected hamsters. Following limited proteolysis, strain-specific migration patterns of PrP^{Sc} on Western blots are seen which relate to different N-terminal ends of PrP^{Sc} following protease treatment implying differing conformations of PrP^{Sc}. Distinct PrP^{Sc} conformations are now recognized to be associated with other prion strains and, similarly, different human PrP^{Sc} isoforms have been found to propagate in the brain of patients with phenotypically distinct forms of CJD.

The different fragment sizes seen on Western blots, following treatment with proteinase K, suggests that there are several different human PrP^{Sc} conformations, referred to as molecular strain types. These types can be further classified by the ratio of the three PrP bands seen after protease digestion, corresponding to N-terminally

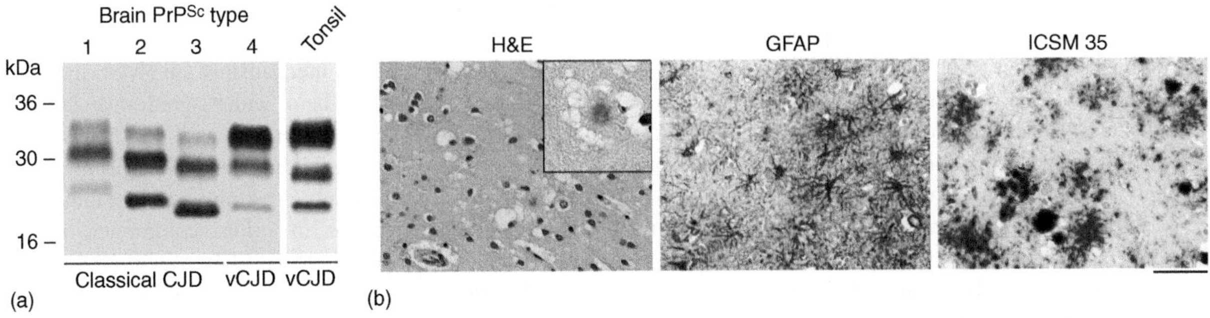

Figure 2 Characterization of disease-related prion protein in human prion disease. (a) Immunoblots of proteinase K-digested tissue homogenate with monoclonal antibody 3F4 showing PrP^{Sc} types 1–4 in human brain and type PrP^{Sc} type 4t in vCJD tonsil. Types 1–3 PrP^{Sc} are seen in the brain of classical forms of CJD (either sporadic or iatrogenic CJD), while type 4 PrP^{Sc} and type 4t PrP^{Sc} are uniquely seen vCJD brain or tonsil, respectively. (b) Brain from a patient with vCJD showing spongiform neurodegeneration following hematoxylin- and eosin staining (H&E), reactive proliferation of astroglial cells following staining with a monoclonal antibody recognizing glial-fibrillary acidic protein (GFAP), and abnormal PrP immunoreactivity following immunohistochemistry using anti-PrP monoclonal antibody ICSM 35 (ICSM 35). Scale (main panels) = 100 μm. Inset, high-power magnification of florid PrP plaques. Courtesy of Professor Sebastian Brandner.

truncated cleavage products generated from di-, mono-, or nonglycosylated PrP^{Sc}. Four types of human PrP^{Sc} have now been commonly identified using molecular strain typing (**Figure 2**), although much greater heterogeneity seems likely. Efforts to produce an unified international classification and nomenclature of human PrP^{Sc} types has been complicated by the fact that the N-terminal conformation of some PrP^{Sc} subtypes seen in sporadic CJD can be altered *in vitro* via changes in metal-ion occupancy or solvent pH. Although agreement is yet to be reached on methodological differences, nomenclature and the biological importance of relatively subtle biochemical differences in PrP^{Sc}, there is strong agreement between laboratories that phenotypic diversity in human prion disease relates to the propagation of disease-related PrP isoforms with distinct physicochemical properties. Polymorphism at *PRNP* residue 129 appears to dictate the propagation of distinct PrP^{Sc} types in humans and it has now become clear that prion strain selection and the propagation of distinct PrP^{Sc} types may also be crucially influenced by other genetic loci of the host genome.

The hypothesis that alternative conformations or assembly states of PrP^{Sc} provide the molecular substrate for clinicopathological heterogeneity seen in human prion diseases (and that this relates to the existence of distinct human prion strains) has been strongly supported by transmission experiments to conventional and transgenic mice. Transgenic mice expressing only human PrP with either valine or methionine at residue 129 have shown that this polymorphism constrains both the propagation of distinct human PrP^{Sc} conformers and the occurrence of associated patterns of neuropathology. Biophysical measurements suggest that this powerful effect of residue 129 on prion strain selection is likely to be mediated via its effect on the conformation of PrP^{Sc} or its precursors or on the kinetics of their formation, as it has no measurable effect on the folding, dynamics, or stability of PrP^{C}. These data are consistent with a conformational selection model of prion transmission barriers and strongly support the 'protein only' hypothesis of infectivity by suggesting that prion strain variation is encoded by a combination of PrP conformation and glycosylation. These findings also provide a molecular basis for *PRNP* codon 129 as a major locus influencing both prion disease susceptibility and phenotype in humans.

The identification of strain-specific PrP^{Sc} structural properties now allows an etiology-based classification of human prion disease by typing of the infectious agent itself. Stratification of all human prion disease cases by PrP^{Sc} type will enable rapid recognition of any change in relative frequencies of particular PrP^{Sc} subtypes in relation to either BSE exposure patterns or iatrogenic sources of vCJD prions. This technique may also be applicable in determining whether BSE has been transmitted to other species thereby posing a threat to human health.

Neuronal Cell Death in Prion Disease

Although various mechanisms have been proposed to explain neuronal death in prion disease, the precise structure of the infectious agent and the cause of neuronal cell death in prion disease remains unclear. While PrP^{C} is absolutely required for prion propagation and neurotoxicity, knockout of PrP^{C} in adult brain and in embryonic models has no overt phenotypic effect, effectively excluding loss of of PrP^{C} function in neurons as a significant mechanism in prion neurodegeneration. Notably, however, there is also considerable evidence that argues that PrP^{Sc} and indeed prions (whether or not they are identical) may not themselves be highly neurotoxic. Consequently, it is now hypothesized that the neurotoxic prion molecule may not be PrP^{Sc} itself, but a toxic intermediate that is produced in the process of conversion of PrP^{C} to PrP^{Sc}, with PrP^{Sc}, present as highly aggregated material, being a relatively inert end product. The steady-state level of such a toxic monomeric or oligomeric PrP intermediate, designated PrP^{L} (for lethal), could determine the rate of neurodegeneration. Subclinical prion infection states may generate the toxic intermediate at extremely low levels below the threshold required for neurotoxicity. Recently, direct support for this hypothesis has been demonstrated by depleting endogenous neuronal PrP^{C} in mice with established neuroinvasive prion infection. This depletion of PrP^{C} reverses early spongiform change and prevents neuronal loss and progression to clinical prion disease despite the accumulation of extraneuronal PrP^{Sc} to levels seen in terminally ill wild-type mice. These data establish that propagation of non-neuronal PrP^{Sc} is not pathogenic, but arresting the continued conversion of PrP^{C} to PrP^{Sc} within neurons during scrapie infection prevents prion neurotoxicity. Importantly, this model also validates PrP^{C} as a key therapeutic target in prion disease.

Future Perspective

The novel pathogenic mechanisms involved in prion propagation are likely to be of wider significance and may be relevant to other neurological and non-neurological illnesses. Indeed, advances in understanding prion neurodegeneration are already casting considerable light on related mechanisms in other, commoner, neurodegenerative diseases such as Alzheimer's, Parkinson's, and Huntington's disease. While the protein-only hypothesis of prion propagation is supported by compelling experimental data and now appears also able to encompass the phenomenon of prion strain diversity, the goal of systematically producing prions *in vitro* remains. Success in producing disease in experimental animals that can be serially propagated following inoculation with PrP^{Sc}-like forms derived from recombinant PrP would provide an essential model by which the mechanism of prion propagation can be understood in molecular detail.

Further Reading

Caughey B and Baron GS (2006) Prions and their partners in crime. *Nature* 443: 803–810.
Collinge J (1999) Variant Creutzfeldt–Jakob disease. *Lancet* 354: 317–323.
Collinge J (2001) Prion diseases of humans and animals: Their causes and molecular basis. *Annual Review of Neuroscience* 24: 519–550.
Collinge J (2005) Molecular neurology of prion disease. *Journal of Neurology, Neurosurgery, and Psychiatry* 76: 906–919.
Collinge J, Sidle KCL, Meads J, Ironside J, and Hill AF (1996) Molecular analysis of prion strain variation and the aetiology of 'new variant' CJD. *Nature* 383: 685–690.
Griffith JS (1967) Self replication and scrapie. *Nature* 215: 1043–1044.
Hill AF and Collinge J (2003) Subclinical prion infection. *Trends in Microbiology* 11: 578–584.
Mabbott NA and MacPherson GG (2006) Prions and their lethal journey to the brain. *Nature Reviews Microbiology* 4: 201–211.
Mallucci G and Collinge J (2005) Rational targeting for prion therapeutics. *Nature Reviews Neuroscience* 6: 23–34.
Prusiner SB (1982) Novel proteinaceous infectious particles cause scrapie. *Science* 216: 136–144.
Prusiner SB (1998) Prions. *Proceedings of the National Academy of Sciences, USA* 95: 13363–13383.
Soto C (2004) Diagnosing prion diseases: Needs, challenges and hopes. *Nature Reviews Microbiology* 2: 809–819.
Weissmann C (2004) The state of the prion. *Nature Reviews Microbiology* 2: 861–871.
Wuthrich K and Riek R (2001) Three-dimensional structures of prion proteins. *Advances in Protein Chemistry* 57: 55–82.

Viruses and Bioterrorism

R F Meyer and S A Morse, Centers for Disease Control and Prevention, Atlanta, GA, USA

Published by Elsevier Ltd.

Introduction

Man has known that biological organisms and toxins were useful as weapons of war long before the germ theory of disease was understood. However, as the twentieth century came to a close, the perceived difficulties in production, weaponization, and deployment of these biological weapons as well as a belief that moral restraints would preclude the use of these weapons gave many a false sense of security. Recently, a number of events have served to focus attention on the threat of terrorism and the potential for the use of biological, chemical, or nuclear weapons against the military, civilian populations, or agriculture for the purpose of causing illness, death, or economic loss. This possibility became a reality in October 2001 when someone sent spores of *Bacillus anthracis* to media companies in New York City and Boca Raton, Florida, resulting in five deaths, considerable panic throughout the United States and other countries, and raised the awareness of our vulnerability.

There are more than 1400 species of infectious organisms that are known to be pathogenic for humans; many additional organisms are capable of causing disease in animals or plants. Realistically, only a few of these infectious agents pose serious problems or are capable of affecting human, animal, or plant health on a large scale. Even fewer of these agents are viruses. Viruses that could be used as weapons against humans, animals, or plants generally possess traits including ease of production and dissemination, transmissibility, environmental stability, and high morbidity and mortality rates.

Definitions

The use of biological agents is often characterized by the manner in which they are used. For the purposes of this article 'biological warfare' is defined as a special type of warfare conducted by a government against a target; 'bioterrorism' is defined as the threat or use of a biological agent (or toxin) against humans, animals, or plants by individuals or groups motivated by political, religious, ecological, or other ideological objectives. Furthermore, terrorists can be distinguished from other types of criminals by their motivation and objective; however, criminals may also be driven by psychological pathologies and may use biological agents. When criminals use biological agents for murder, extortion, or revenge, it is called a 'biocrime'.

Historical Perspective

The use of viral agents for biological warfare has a long history, which predates their recognition and isolation by culture. Their early use is consistent with what, at the time, was known about infectious diseases, particularly smallpox. In the sixteenth century, the Spanish explorer, Francisco Pizarro, presented the indigenous peoples of South America with variola-contaminated clothing, which resulted in widespread epidemics of smallpox. During the French and Indian War (1745–67), Sir Jeffrey Amherst, commander of the British forces in North America, suggested the deliberate use of smallpox to 'reduce' Native

American tribes hostile to the British. Captain Ecuyer (one of Amherst's subordinates), fearing an attack on Ft. Pitt from Native Americans, acquired two variola-contaminated blankets and a handkerchief from a smallpox hospital and, in a gesture of good will, distributed them to the Native Americans. As a result, several outbreaks of smallpox occurred in various tribes in the Ohio River valley. In 1775, during the Revolutionary War, the British attempted to spread smallpox among the Continental forces by inoculating (variolation) civilians fleeing Boston. In the South, there is evidence that the British were going to distribute slaves who had escaped during hostilities, and were sick with smallpox, back to the rebel plantations in order to spread the disease.

The use of viruses other than Variola major is a more recent phenomenon and reflects an increased knowledge of how to grow and stabilize viruses for delivery purposes. Allegations have been made by the government of Cuba that the CIA was responsible for the massive outbreaks of swine fever in 1971 and dengue fever in 1980 that ravaged the country. However, subsequent investigations have failed to find substantive proof of CIA involvement in these outbreaks. The Aum Shinrikyo, a religious cult responsible for the 1995 release of sarin gas in the Tokyo subway system, was also involved in biological warfare activity and sent a team of 40 people to Zaire to acquire Ebola virus. Fortunately, they were unsuccessful in this endeavor. In 1997, unknown farmers in New Zealand deliberately and illegally introduced rabbit hemorrhagic disease virus (a calicivirus) onto the south island as an animal control tool to kill feral rabbits.

Over the past two decades, the human immunodeficiency virus (HIV) has been involved in a number of biocrimes. This most likely reflects the availability of HIV-contaminated blood as a source of this virus. For example, in 1990, Graham Farlow, an asymptomatic HIV-positive inmate at a prison in New South Wales, Australia, injected a guard with HIV-contaminated blood. The guard became infected with HIV; Farlow subsequently died of AIDS. In 1992, Brian T. Stewart, a phlebotomist at a St. Louis, MO hospital, injected his 11-month-old son with HIV-contaminated blood during a fight over payment of child support. In 1993, Iwan E. injected his former girlfriend with 2.5 ml of HIV-contaminated blood after she broke up with him. In 1994, Dr. Richard J. Schmidt, a married Louisiana gastroenterologist, injected a former lover with HIV-contaminated blood. Molecular typing of the HIV strains demonstrated that she contracted the same strain of HIV as found in one of Dr. Schmidt's patients. In perhaps the most famous case, Dr. David Acer, a Florida dentist infected with HIV, transmitted the disease to six of his patients between 1987 and 1990. The intentional infection of these patients is a possibility although there is no direct evidence. In spite of these incidents, HIV has not been included on lists of threat agents for public health bioterrorism preparedness. However, some contend that HIV has great weapon potential if the goal is to destabilize a society.

Viruses have also been involved in suspected incidents or hoaxes. In 1999, an article appeared suggesting that the CIA was investigating whether Iraq was responsible for causing the outbreak of West Nile fever in the New York City area. The story relied heavily on a previous story written by an Iraqi defector, claiming that Saddam Hussein planned to use West Nile virus strain SV 1417 to mount an attack. The investigation indicated that there was no known evidence of bioterrorism involved in the spread of West Nile virus. A fictional 'virus' was also involved in one of the largest bioterrorism hoaxes in 2000. According to e-mail messages widely circulated on the Internet, an organization known as the Klingerman Foundation was mailing blue envelopes containing sponges contaminated with a fictional pathogen called the 'Klingerman virus'. According to the e-mail alert, 23 people had been infected with the virus, including 7 who died.

Viruses as Bioweapons

Advances in viral culture and virus stabilization made during the second half of the twentieth century facilitated large-scale production of viral agents for aerosol dissemination. A report for the United Nations on chemical and biological weapons and the effects of their possible use gave estimates on the numbers of casualties produced by a hypothetical biological attack (**Table 1**). Three viruses (Rift Valley fever virus, Tick-borne encephalitis virus, and Venezuelan Equine Encephalomyelitis (VEE) virus) were evaluated in a scenario in which 50 kg of the agent was released by aircraft along a 2 km line upwind of a population center of 500 000. The viral agents produced fewer casualties and impacted a smaller area when compared with the bacterial agents used in this hypothetical model. Of note,

Table 1 Estimates of casualties produced by hypothetical biological attack

Agent	*Downwind Reach (km)*	*Dead*	*Incapacitated*
Rift Valley fever	1	400	35 000
Tick-borne encephalitis	1	9500	35 000
VEE	1	200	19 800
Francisella tularensis	>20	30 000	125 000
Bacillus anthracis	>20	95 000	125 000

Note. These estimates are based on the following scenario: release of 50 kg of agent by aircraft along a 2 km line upwind of a population center of 500 000.

smallpox was apparently not evaluated because it had not yet been eradicated and level of vaccine-induced immunity in the population was high.

Viral agents were part of the biological weapons arsenal of both the Soviet Union and the United States (**Table 2**). VEE virus was stockpiled by both countries as an incapacitating agent; Variola major and Marburg viruses were stockpiled as lethal agents by the Soviet Union. The Soviet Union reportedly conducted a live field test of Variola major virus on Vozrozhdeniye Island in the Aral Sea in the 1970s, in which 400 g of the virus was released into the atmosphere by explosion. Unfortunately, a laboratory technician who was collecting plankton samples from an oceanographic research vessel 15 km from the island became infected. It was reported that after returning home to Aralsk, she transmitted the infection to several people including children. All those infected died. A number of other viruses that infect humans (e.g., Ebola virus, Lassa fever virus, enterovirus 70) or livestock (e.g., foot and mouth disease virus, rinderpest, Newcastle disease virus) have also been studied for their offensive capabilities or for the development of medical and veterinary countermeasures.

Today, with the increased level of concern, a number of viruses have been cited as possible weapons for use against humans or animals (**Table 2**). The requirements for an ideal biological warfare agent include availability,

Table 2 Classification of viral agents that are considered to be of concern for bioterrorism and biowarfare and those that have been weaponized or studied for offensive or defensive purposes as part of former or current national biological weapons programs

Nucleic acid	*Family*	*Genus*	*Species*
Negative-sense single-stranded RNA	Arenaviridae	Arenaviruses	Lassa fever[a,b]
			Junin[a,b]
			Machupo[a,b]
			Sabia
			Guanarito
	Bunyaviridae	Phlebovirus	Rift Valley fever[b]
		Nairovirus	Crimean-Congo HF
		Hantavirus	Hantaan and related viruses[b]
			Sin Nombre
	Orthomyxoviridae	Influenzaviruses	Influenza A[b]
	Filoviridae	Filovirus	Ebola[a]
			Marburg[c]
	Paramyxoviridae	Henipavirus	Nipah virus
		Morbillivirus	Rinderpest[a,b,d,e,f]
		Avulavirus	Newcastle disease virus[b]
Positive-sense single-stranded RNA	Flaviviridae	Flavivirus	Yellow fever[a,b,d]
			Dengue[b]
			Tick-borne encephalitis virus[a]
			Japanese encephalitis virus[a]
			Omsk hemorrhagic fever virus
	Togaviridae	Alphavirus	Venezuelan equine encephalomyelitis virus[c,g]
			Eastern equine encephalomyelitis virus[b]
			Western equine encephalomyelitis virus[b]
			Chikungunya virus[b]
	Picornaviridae	Enterovirus	Enterovirus 70[h]
		Hepatovirus	Hepatitis A virus
		Apthovirus	Foot and mouth disease virus[f,i]
Double-stranded DNA	Poxviridae	Orthopoxvirus	Variola major[c,b,j]
			Camelpox[h]
	Asfarviridae	Asfivirus	African swine fever virus[a]

[a]Studied by the Soviet Union BW program.
[b]Studied by the U.S. BW program.
[c]Weaponized by the Soviet Union BW program.
[d]Studied by the Canada BW program.
[e]Studied by the France BW program.
[f]Studied by the Germany BW program.
[g]Weaponized by the U.S. BW program.
[h]Studied by the Iraq BW program.
[i]Studied by the Iran BW program.
[j]Studied by the North Korea BW program.

ease of production, stability after production, a susceptible population, absence of specific treatment, ability to incapacitate or kill the host, appropriate particle size in aerosol so that the virus can be carried long distances by prevailing winds and inhaled deeply into the lungs of unsuspecting victims, ability to be disseminated via food or water, and the availability of a vaccine to protect certain groups. Other factors such as the economic and psychological impact of an attack on animal agriculture with a viral agent must also be considered.

Variola major is considered to be the major viral threat agent for humans. Thus, considerable effort has been expended toward preparing the public health and medical communities for the possibility that this agent will be employed by a terrorist. Variola major is considered to be an ideal terrorist weapon because it is highly transmissible by the aerosol route from infected to susceptible persons; the civilian populations of most countries contain a high proportion of susceptible persons; the disease is associated with a high morbidity and about 30% mortality; initially, the diagnosis of a disease that has not been seen for almost 30 years would be difficult; and, other than the vaccine, which may be effective in the first few days post infection, there is no proven treatment available.

Alphaviruses (**Table 2**) are also of concern because they can be produced in large amounts in inexpensive and unsophisticated systems; they are relatively stable and highly infectious for humans as aerosols, and strains are available that produce incapacitating (e.g., VEE) or lethal infections (EEE case fatality rates range from 50–75%). Furthermore, the existence of multiple serotypes of VEE and EEE viruses, as well as the inherent difficulties of inducing efficient mucosal immunity, make defensive vaccine development difficult.

The filoviruses and arenaviruses that cause hemorrhagic fever have also been considered as agents that might be used by terrorists because of their high virulence and capacity for causing fear and anxiety. The filoviruses, Ebola and Marburg, can also be highly infectious by the airborne route. Humans are generally susceptible to infection with these viruses with fatality rates greater than 80%, and infection can be transmitted between humans through direct contact with virus-containing body fluids. There are five species of arenaviruses (Lassa fever, Junin, Machupo, Guanarito, and Sabia) that can cause viral hemorrhagic fevers with a case fatality rate of about 20%. Large quantities of these viruses can be produced by propagation in cell culture. Infection occurs via the respiratory pathway suggesting that dissemination via aerosol might be used by a terrorist. Human to human transmission has also been reported with aerosol transmission the most likely route for at least some of the secondary cases. The filoviruses and arenaviruses discussed above are BSL-4 agents and diagnostic capacities for infections caused by these viruses are limited.

Impact of Biotechnology

Because the nucleic acid of many viruses, including some that are currently not threats, can be manipulated in the laboratory, the potential for genetic engineering remains a serious threat. Biotechnology, which has had a tremendous impact on the development of medicines, vaccines, and in the technologies needed to counter the threat of naturally occurring disease, can also be used to modify viruses with unintended consequences or even for the development of novel biological agents. Several examples involving viruses are presented below.

Mousepox Virus

An Australian research group was investigating virally vectored immunocontraceptive vaccines based on ectromelia virus, the causative agent of the disease termed mousepox. They created a recombinant virus, which expressed the mouse cytokine IL-4 in order to enhance the antibody-mediated response to other recombinant antigens carried on the virus vector. Instead, the ectromelia virus vector expressing IL-4 altered the host's immune response to this virus resulting in lethal infections in normally genetically resistant mice (e.g., C57BL/6). Additionally, this virus also caused lethal infections in mice previously immunized against infection with ectromelia virus. The creation of this 'supermousepox' virus led to speculation that similar genetic engineering could be performed on Variola major leading to a biological weapon that would be effective against an immunized population.

Pandemic Influenza

The influenza pandemic of 1918–19, which followed World War I, was uniquely severe, causing an estimated 20–40 million deaths globally. This pandemic happened before the advent of viral culture and very little was known about the virus until the discovery of the polymerase chain reaction (PCR). Recently, the complete coding sequences of all eight viral RNA segments has been determined by using reverse transcription-PCR (RT-PCR) to amplify the viral RNA sequences from formalin-fixed and frozen tissue samples from individuals who died during this pandemic in an effort to shed light on both the reasons for its extraordinary virulence and evolutionary origin. More recently, researchers reconstructed the 1918 Spanish influenza pandemic virus using reverse genetics and observed that this reconstructed virus exhibited exceptional virulence in the model systems examined and that the 1918 hemaglutinin and polymerase genes were essential for optimal virulence.

Synthetic Genomes

A full-length poliovirus complementary DNA (cDNA) (*c.* 7500 bp) has been synthesized in the laboratory by assembling oligonucleotides of plus- and minus-strand polarity. The synthetic poliovirus cDNA was transcribed by RNA polymerase into viral RNA, which translated and replicated in a cytoplasmic extract of uninfected HeLa S3 cells, resulting in the *de novo* synthesis of infectious poliovirus. The publication of this research raised concerns that more complicated viruses (e.g., Variola major or Ebola) could be synthesized from scratch based on publicly available sequences, or that viruses could be created that do not exist in the wild.

Recognition, Response, and Deterrence

An effective defense requires a comprehensive approach that includes: prevention of access to viral stocks; improved means of detecting deliberately induced disease outbreaks; rapid medical recognition of specific syndromes (e.g., hemorrhagic fever syndrome); rapid laboratory identification of viruses in patient specimens; prevention of person–person transmission; reliable decontamination procedures; development of effective vaccines; and development of effective antiviral therapy.

Rapid and accurate detection of biological threat agents is the basis of an effective public health response to bioterrorism. In order to address this issue, CDC in collaboration with other partners established a national network of laboratories called the Laboratory Response Network (LRN), which was provided with the tools to accomplish this mission. Rapid assays utilizing advanced molecular and immunological technologies for detection of agents such as variola virus, as well as emerging public health threats such as SARS coronavirus and H5N1 influenza virus, were distributed to member laboratories. Equipment, training, and proficiency testing are elements of the LRN and contribute to a uniform operational plan. The importance of high-quality standardized testing for detection of these agents is exemplified by the rapid need for medical countermeasures to protect or treat civilian populations. Accurate laboratory analysis is a major element in the decision process for deployment of the Federal Government's Strategic National Stockpile (SNS) of medical countermeasures.

As part of the effort to deter biological terrorism and strengthen the law enforcement response to such an act, the US recently established a microbial forensic laboratory known as the National Bioforensics Analysis Center that operates in partnership with the Federal Bureau of Investigation. Scientists are already developing methods for the forensic investigation of incidents involving viruses.

Summary

For the terrorist, the use of a viral agent would pose a challenge due to problems associated with acquisition, cultivation, and dissemination. The target for an attack with a viral agent can range from humans to animals and plants. Therefore, agricultural targets are also a major concern. Nature has provided many challenges to combating viral diseases. Viral agents are much more prone to genetic variation and mutation, and can be manipulated or created in the laboratory to take on desired characteristics. Differentiating between natural and intentional viral disease outbreaks can be challenging. Unlike bacterial diseases, many of which are treatable, there are fewer medical countermeasures to employ when dealing with viral infections. Laboratory diagnostic methods and reagents must continuously be refined to account for genetic changes and variants. Thus, the challenge of developing bioterrorism countermeasures is significant. Fortunately, this effort contributes to combating natural disease events more effectively, which has global benefits.

Further Reading

Bray M (2003) Defense against filoviruses used as biological weapons. *Antiviral Research* 57: 53–60.

Breeze RG, Budowle B and Schutzer SE (eds.) (2005) *Microbial Forensics.* Burlington, MA: Academic Press.

Carus WS (2002) *Bioterrorism and Biocrimes. The Illicit Use of Biological Agents since 1900.* Amsterdam, The Netherlands: Fredonia Books.

Cello J, Paul AV, and Wimmer E (2002) Chemical synthesis of poliovirus cDNA: Generation of infectious virus in the absence of natural template. *Science* 297: 1016–1018.

Charrel RN and de Lamballerie X (2003) Arena viruses other than Lassa virus. *Antiviral Research* 57: 89–100.

Enserink M (2002) Did bioweapons test cause a deadly smallpox outbreak? *Science* 296: 2116–2117.

Henderson DA, Inglesby TV, Gartlett JG, *et al.* (1999) Smallpox as a biological weapon. *JAMA* 281: 2127–2137.

Hopkins DR (1983) *Princes and Peasants. Smallpox in History.* Chicago, IL: University of Chicago Press.

Jackson RJ, Ramsay AJ, Christensen SD, Beaton S, Hall DF, and Ramshaw IA (2001) Expression of mouse interleukin-4 by a recombinant ectromelia virus suppresses cytolytic lymphocyte responses and overcomes genetic resistance to mousepox. *Journal of Virology* 75: 1205–1210.

Lanciotti RS, Roehrig JT, Duebel V, *et al.* (1999) Origin of the West Nile virus responsible for an outbreak of encephalitis in the northeastern United States. *Science* 286: 2333–2337.

Metzker ML, Mindell DP, Liu X, Ptak RG, Gibbs RA, and Hillis DM (2002) Molecular evidence of HIV-1 transmission in a criminal case. *Proceedings of the National Academy of Science, USA* 99: 14292–14297.

Rotz LD, Khan AS, Lillibridge SR, Ostroff SM, and Hughes JM (2002) Public health assessment of potential biological terrorism agents. *Emerging Infectious Diseases* 8: 225–229.

Sidwell RW and Smee DF (2003) Viruses of the Bunya- and Togaviridae families: Potential as bioterrorism agents and means of control. *Antiviral Research* 57: 101–111.

Tumpey TM, Basler CF, Aguilar PV, *et al.* (2005) Characterization of the reconstructed 1918 Spanish influenza pandemic virus. *Science* 310: 77–80.

World Health Organization(1970) *Health Aspects of Chemical and Biological Weapons.* Geneva: WHO.

VIRIONS

Principles of Virion Structure

J E Johnson and J A Speir, The Scripps Research Institute, La Jolla, CA, USA

Introduction

The virion is a nucleoprotein particle designed to move the viral genome between susceptible cells of a host and between susceptible hosts. An important limitation on the size of the viral genome is its container, the protein capsid. The virion has a variety of functions during the virus life cycle (**Table 1**); however, the principles dictating its architecture result from the need to provide a container of maximum size derived from a minimum amount of genetic information. The universal strategy evolved for the packaging of viral nucleic acid employs multiple copies of one or more protein subunit types arranged about the genome. In most cases the subunits are arranged with well-defined symmetry, but there are examples where assembly intermediates and sometimes mature particles are not globally symmetric. The assembly of subunits into nucleoprotein particles is, in many cases, a spontaneous process that results in a minimum free energy structure under intracellular conditions. The two broad classes of symmetric virions are helical rods and spherical particles.

Helical Symmetry

The nucleoprotein helix can, in principle, package a genome of any size. Extensive studies of tobacco mosaic virus (TMV) show that protein subunits will continue to add to the extending rod as long as there is exposed RNA. Protein transitions required to form the TMV helix from various aggregates of subunits are now understood at the atomic level. It is clear that subunits forming the helix display significant polymorphism in the course of assembly; however, excluding the two ends of the rod, all subunits are in identical environments in the mature helical virion. This is the ideal protein context for a minimum free energy structure. In spite of these packaging and structural attributes the helical virion must be deficient in functional requirements that are common for animal viruses because they are found only among plant, bacterial, and archeal viruses. Even among plant viruses, only 7 of the 25 recognized groups are helical. The large majority of all viruses are roughly spherical in shape.

Icosahedral Symmetry

The architectural principles for constructing a 'spherical' virus were first articulated by Crick and Watson in 1956. They suggested that identical subunits were probably distributed with the symmetry of Platonic polyhedra (the tetrahedron, 12 equivalent positions; the octahedron, 24 equivalent positions; or the icosahedron, 60 equivalent positions). Subunits distributed with the symmetry of the icosahedron (**Figure 1(a)**) provide the maximum sized particle, for a given sized subunit, in which all copies of a subunit lie in identical positions. The repeated interaction of chemically complementary surfaces at the subunit interfaces leads naturally to such a symmetric particle. The 'instructions' required for assembly are contained in the tertiary structure of the subunit (**Figure 1(b)**). The actual assembly of the protein capsids is a remarkably accurate process. The use of subunits for the construction of organized complexes places strict control on the process and will naturally eliminate defective units. The reversible formation of noncovalent bonds between properly folded subunits leads naturally to error-free assembly and a minimum free energy structure.

Virus Structures

Crystallographic studies of nearly 90 unique, icosahedral viruses have demonstrated that there are a limited number of folds utilized in forming viral capsids. By far the most common fold is the eight-stranded antiparallel β-sandwich shown schematically in **Figure 1(b)**. The details of all the folds observed in nonenveloped virus capsids and their distribution are discussed elsewhere in this encyclopedia.

Early ideas explaining spherical virus architecture were extended on the basis of physical studies of small spherical RNA plant viruses. The large yields and ease of preparation made them ideal subjects for investigations requiring substantial quantities of material. Protein subunits forming virus capsids of this type are usually 20–40 kDa. An example of a virus consistent with the Crick and Watson hypothesis is satellite tobacco necrosis virus (STNV) which is formed from 60 identical 25 kDa subunits. The particle's outer radius is 80 Å and the radius of the internal cavity is 60 Å providing a volume of 9×10^5 Å^3 for packaging RNA. A single hydrated ribonucleotide in a virion will occupy on average roughly 600–700 Å^3. Thus, the STNV volume is adequate to package a genome of only 1200–1300 nt. STNV is a satellite virus and the packaged genome codes for only the coat protein. The 'helper virus', tobacco necrosis virus, supplies proteins required for RNA replication. Most simple ribovirus genomes contain coding capacity for at least two proteins; roughly, 1200 nt for the capsid protein and 2500 nt for an RNA-directed RNA polymerase. The inner radius required to package such a minimal genome is 90 Å. Consistent with this requirement were experimental studies showing that the vast majority of simple spherical viruses had outer radii of at least 125 Å, which corresponds to inner radii of roughly 100 Å for a typical shell thickness for a 400 residue subunit. Such particles had to be formed from more than 60 subunits, yet X-ray diffraction patterns of crystalline tomato bushy stunt virus (TBSV) and turnip yellow mosaic virus (TYMV) were consistent with icosahedral symmetry.

Table 1 Functions of the virus capsid in simple RNA viruses

Assembly	Subunits must assemble to form a protective shell for the RNA
Package	Subunits specifically package the viral RNA
Binding to receptors	The capsid may actively participate in virus infection processes binding to receptors and mediating cell entry (animal) and disassembly
Transport	Virion transport within the host (plant)
Mutation	Capsid protein mutation to avoid the immune system
RNA replication	Some capsid proteins function as a primer for viral RNA replication

Quasi-Equivalent Virus Capsids

Although a number of investigators developed hypotheses explaining the apparent inconsistent observations, Caspar and Klug, in 1962, derived a general method for the construction of icosahedral capsids that contained multiples of 60 subunits. The method systematically enumerates all possible quasi-equivalent structures and is similar to that derived by Buckmeister Fuller to construct geodesic domes. The quasi-equivalent theory of Caspar and Klug explained the distribution of morphological units (features identifiable at low resolution by electron microscopy often

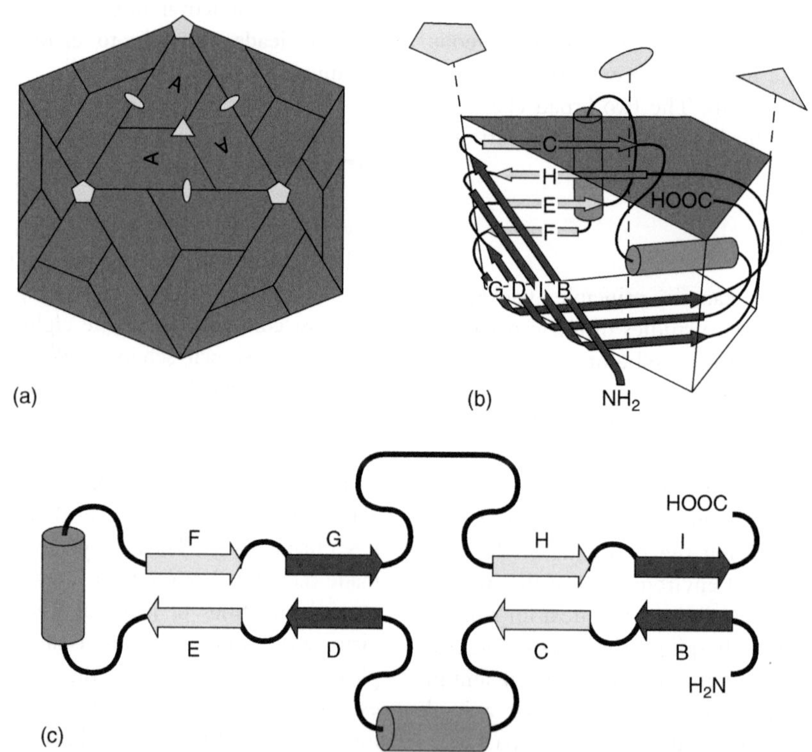

Figure 1 (a) The icosahedral capsid contains 60 identical copies of the protein subunit (blue) labeled A. These are related by fivefold (yellow pentagons at vertices), threefold (yellow triangles in faces), and twofold (yellow ellipses at edges) symmetry elements. For a given sized subunit this point group symmetry generates the largest possible assembly (60 subunits) in which every protein lies in an identical environment. (b) A schematic representation of the subunit building block found in many RNA and some DNA viral structures. Such subunits have complementary interfacial surfaces which, when they repeatedly interact, lead to the symmetry of the icosahedron. The tertiary structure of the subunit is an eight-stranded β-barrel with the topology of the jellyroll (see c-, β-strand and helix coloring is identical to b). Subunit sizes generally range between 20 and 40 kDa with variation among different viruses occurring at the N- and C-termini and in the size of insertions between strands of the β-sheet. These insertions generally do not occur at the narrow end of the wedge (B–C, H–I, D–E, and F–G turns). (c) The topology of viral β-barrel showing the connections between strands of the sheets (represented by yellow or red arrows) and positions of the insertions between strands. The green cylinders represent helices that are usually conserved. The C–D, E–F, and G–H loops often contain large insertions.

corresponding to hexamer, pentamer, trimer, or dimer aggregates of the subunits) on all structures observed to date, but the results from high-resolution crystallographic studies have shown some remarkable inconsistencies with the microscopic principles upon which the theory is based.

Quasi-equivalence is best visualized graphically. Formally, subunits forming quasi-equivalent structures must be capable of assembling into both hexamers (which are conceptually viewed as planar) and pentamers (which are convex because one subunit has been removed from the planar hexamer and yet similar (quasi-equivalent) contacts are maintained). Caspar and Klug proposed that, logically, atomic interactions of fivefold-related and sixfold-related subunits could be closely similar, as only the dihedral angle between subunits would change. They argued that this difference could be accommodated by allowed variations in noncovalent bond angles between residues stabilizing the oligomers. This prediction was correct for some virus capsids, as illustrated below for cowpea chlorotic mottle virus (CCMV), but in many viruses there is an explicit molecular switch that changes the oligomeric state. If subunits assembled as all hexamers the result would be a sheet of hexamers and a closed shell could not form (**Figure 2(a)**). The rules of quasi-equivalence described a systematic procedure for inserting pentamers into the hexagonal net in such a way as to form a closed shell with exact icosahedral symmetry. **Figure 2** illustrates this principle and the selection rules for inserting pentamers. **Figure 3** illustrates how the morphogenesis of such an assembly may occur using the crystallographic structure of CCMV as inspiration. CCMV was the first virus structure determined that agreed in detail with the predictions of Caspar and Klug.

The quasi-equivalence theory has been universally successful in describing surface morphology of spherical viruses observed in the electron microscope and, prior to the first high-resolution crystallographic structure of a virus, it was assumed that the underlying assumptions of Caspar and Klug were essentially correct. The structure of TBSV determined at 2.9 Å resolution revealed an unexpected variation from the concept of quasi-equivalence which was defined as "any small, non random, variation in a regular bonding pattern that leads to a more stable structure than does strictly equivalent bonding." Unlike CCMV, the structure of TBSV showed that differences occurring between pentamer interactions and hexamer interactions were not small variations in bonding patterns, but almost totally different bonding patterns. **Figure 4** shows diagrammatically the subunit interactions in the shell of TBSV, SBMV, BBV, and TCV. These high-resolution structures revealed that the mathematical concept of quasi-equivalence predicted surface lattices accurately, but not for the reasons expected. Bonding contacts between quasi-threefold-related subunits are maintained with little deviation from exact symmetry while quasi-twofold contacts and icosahedral twofold contacts (which are predicted to be very similar) are quite different. The hexamer quasi-symmetry is better described as a trimer of dimers in the TBSV and related structures. Unlike the conceptual model and the CCMV capsid, the particle curvature in TBSV results from both pentamers and hexamers.

The high-resolution $T=3$ structures showed that the overall features of the quasi-equivalent theory were correct, but the underlying concepts of quasi-equivalent bonding had to be revised. The first low-resolution structure (22.5 Å) of a $T=7$ virus required an even greater conceptual adjustment to the underlying principles of quasi-equivalence. Rayment, Baker, and Caspar reported, in 1982, that the polyomavirus capsid contained 72 capsomers as previously reported from electron microscopy studies, but that all the capsomers were pentamers of protein subunits even when they were located at hexavalent lattice points. The $T=7$ surface lattice predicts 12 pentamers and 60 hexamers; thus, the prediction of the number and position of the morphological units is correct, but the fine structure of the morphological units is incorrect. Although the result was highly controversial when first reported, additional electron microscopy studies and the 3.5 Å resolution X-ray structure of polyomavirus have fully confirmed the all-pentamer structure. This result clearly shows the limits of theory in predicting virus structure and indicates that further understanding of capsid structure will come only from experimental studies. The structure of the polyomavirus and its relatives illustrates an important concept when considering surface lattice formation. The feature of greatest importance is not the symmetry of the morphological unit positioned on the hexamer sites, but only its ability to accommodate six neighbors (it is a hexavalent position). Normally morphological units with sixfold symmetry accomplish this, but here, rather acrobatic molecular switching has permitted a pentamer of subunits to accommodate six neighbors. Although the all pentamer capsid has been observed for the $T=7$ structure of papilloma viruses, cauliflower mosaic virus appears to have the hexamer/pentamer distribution predicted by quasi-equivalent theory as do the $T=7$ capsids of the lambda-like bacteriophage, HK97. A substantial number of complex virus structures have been determined by cryo-electron microscopy and the surface lattices agree well with the predictions of quasi-equivalence. Thus there is considerable confidence in the lattice assignments, but the capsomer structure and therefore number of subunits must be carefully confirmed.

Picorna-Like Virus Capsids

A number of viral capsids are constructed with pseudo $T=3$ symmetry. These structures contain β-barrel

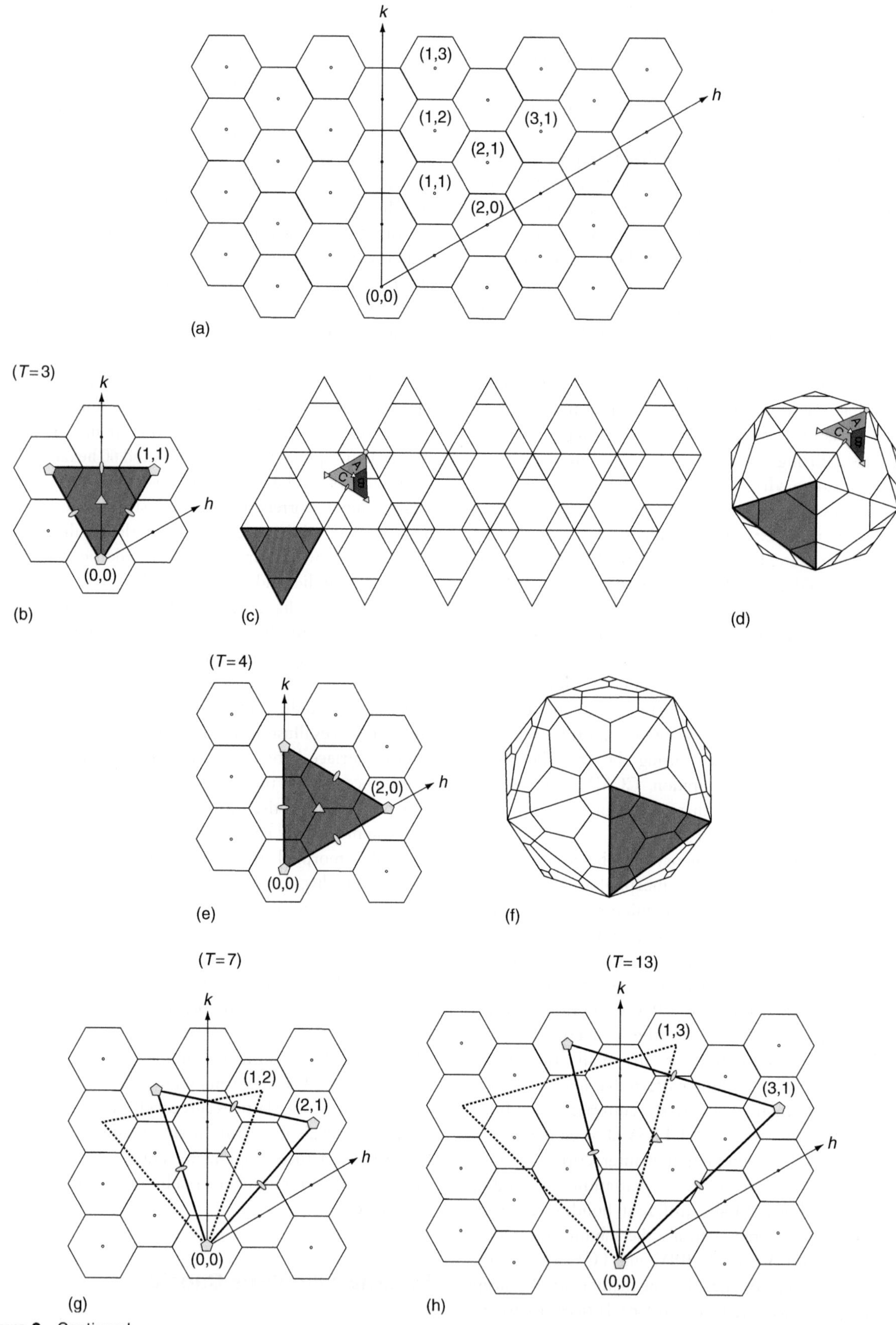

Figure 2 Continued

subunits (**Figure 1(b)**) in the quasi-equivalent environments formed in $T=3$ structures, but each of the three β-barrels in the asymmetric unit has a unique amino acid sequence. Rather than 180 identical subunits, the $P=3$ particles contain 60 copies each of three different subunits (**Figure 5**). These structures do not require quasi-equivalent bonding because each unique interface will have different amino acids interacting, rather than the same subunits forming different contacts. The animal picornaviruses have capsids of this type. Animal virus capsids undergo rapid mutation which avoids recognition by the circulating immune system. Capsids composed of three subunit types could mutate in one subunit without affecting the other two. This would be less likely to affect assembly or other functions of the particle in $P=3$ shells than it would in $T=3$ shells. At least one plant virus group displays $P=3$ shells, the comoviruses. An interesting variation occurs in these capsids when compared with the picornaviruses. Two of the domains forming the shell are contained in a single polypeptide chain. This phenomenon is readily understood in the context of the synthesis of the subunits in picorna and comoviruses. In both cases the proteins are synthesized as a polyprotein that is subsequently cleaved into functional proteins by a virally encoded protease. Clearly one of the cleavage sites in picornaviruses is missing in the comoviruses, resulting in these two domains being still a 'polyprotein'. Tobacco ringspot virus (TRV) is an example where all three subunits are linked in a single polypeptide chain like 'beads on a string'. It is interesting to note that the X-ray structure of TRV showed that the linkage of subunits in the capsid is the same as that predicted for the precursor polyprotein of picornaviruses based on the their X-ray structures, supporting the divergent evolution of the $P=3$ capsids from plant viruses to animal viruses.

Virus Particle Maturation

While $T=3$ and $P=3$ animal viruses undergo maturation processes that confer infectivity, there is little change in their particle morphology during this process. In contrast,

Figure 2 Geometric principles for generating icosahedral quasi-equivalent surface lattices. These four constructions show the relation between icosahedral symmetry axes and quasi-equivalent symmetry axes. The latter are symmetry elements that hold only in a local environment. (a) It is assumed in quasi-equivalence theory that hexamers and pentamers (a hexamer contains six units and a pentamer contains five units) can be interchanged at a particular position in the surface lattice. Hexamers are initially considered planar (an array of hexamers forms a flat sheet as shown) and pentamers are considered convex, introducing curvature in the sheet of hexamers when they are inserted. Inserting 12 pentamers at appropriate positions in the hexamer net generates the closed icosahedral shell, composed of hexamers and pentamers. The positions at which hexamers are replaced by pentamers are defined by the indices *h* and *k* measured along the labeled axes. The values of (h,k) used in the following examples are labeled. To construct a model of a particular quasi-equivalent lattice, one face of an icosahedron (equilateral triangles colored orange in (b–f)) is generated in the hexagonal net. The origin (0,0) is replaced with a pentamer, and the (h,k) hexamer is replaced by a pentamer. The third replaced hexamer is identified by threefold symmetry (i.e., complete the equilateral triangle). Each quasi-equivalent lattice is identified by a number $T=h^2+hk+k^2$ where *h* and *k* are the indices used above. *T* indicates the number of quasi-equivalent units in the icosahedral asymmetric unit. For the purpose of these constructions it is convenient to choose the icosahedral asymmetric unit as 1/3 of an icosahedral face defined by the triangle connecting a threefold axis to two adjacent fivefold axes. Other asymmetric units can be chosen such as the triangle connecting two adjacent threefold axes and an adjacent fivefold axis (see (c) and **Figure 4**). The total number of units in the particle is $60T$, given the symmetry of the icosahedron. The number of pentamers must be 12 and the number of hexamers is $(60T-60)/6=10(T-1)$. (b) One face of the icosahedron for a $T=3$ surface lattice is identified by the orange triangle with the bold outline. The yellow symmetry labels are the same as those defined in **Figure 1**. The hexamer replaced has coordinates $h=1, k=1$. The icosahedral asymmetric unit is 1/3 of this face and it contains three quasi-equivalent units (two units from the hexamer coincident with the threefold axis and one unit from the pentamer). (c) Arranging 20 identical faces of the icosahedron as shown can generate the three-dimensional model of the quasi-equivalent lattice. Three quasi-equivalent units labeled A (blue), B (red), and C (green) are shown. These correspond to the three quasi-equivalent units defined in **Figures 3** and **4** rather than the alternative definition used in (a) and (b). (d) The folded icosahedron is shown with hexamers and pentamers outlined. The orange face represents the triangle originally generated from the hexagonal net. The $T=3$ surface lattice represented in this construction has the appearance of a soccer ball. The trapezoids labeled A, B, and C identify quasi-equivalent units in one icosahedral asymmetric unit of the rhombic triacontahedron discussed in **Figure 4**. (e) An example of a $T=4$ icosahedral face $(h=2, k=0)$. In this case the hexamers are coincident with icosahedral twofold axes. (f) A folded $T=4$ icosahedron with the orange face corresponding to the face outlined in the hexagonal net. Note that folding the $T=4$ lattice has required that the hexamers have the curvature of the icosahedral edges. (g) A single icosahedral face generated from the hexagonal net for a $T=7$ lattice. Note that there are two different $T=7$ lattices ($h=2, k=1$ in bold outline; and $h=1, k=2$ in dashed outline). These lattices are the mirror images of each other. To fully define such a lattice, the arrangement of hexamers and pentamers must be established as well as the enantiomorph of the lattice. (h) A single icosahedral face for a $T=13$ lattice is shown. The two enantiomorphs of the quasi-equivalent lattice ($h=3, k=1$ – bold; and $h=1, k=3$ – dashed) are outlined. The procedure for generating quasi-equivalent models described here does not exactly correspond to the one described by Caspar and Klug in 1962. Caspar and Klug distinguish between different icosadeltahedra by a number $P=h^2+hk+k^2$ where *h* and *k* are integers that contain no common factors but 1. The deltahedra are triangulated to different degrees described by an integer *f* that can take on any value. In their definition $T=Pf^2$. The description in this figure has no restrictions on common factors between *h* and *k*; thus, $T=h^2+hk+k^2$ for all positive integers. The final models are identical to those described by Caspar and Klug.

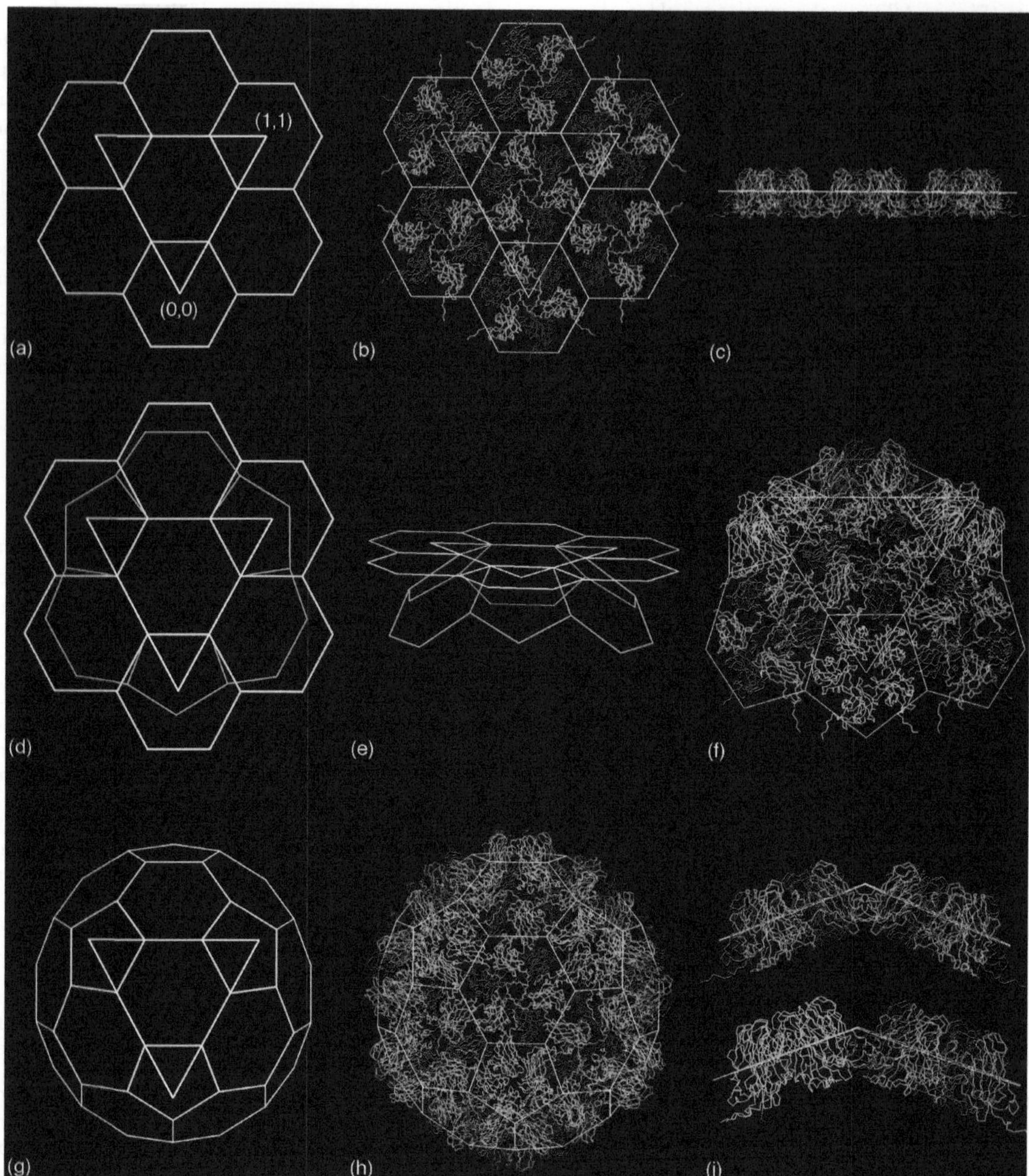

Figure 3 Molecular graphics construction of a $T = 3$ quasi-equivalent icosahedron. (a) Hexagonal sheet overlaid with the triangular coordinates (white) for a theoretical $T = 3$ quasi-equivalent icosahedron ($h = 1$, $k = 1$, see **Figure 2(b)**). The sheet has true sixfold rotational symmetry about axes passing through the hexamer centers, which are normal to the sheet. (b) Copies of the hexamer coordinates from the CCMV X-ray structure (colored by asymmetric unit position, see **Figures 2(c)–2(d)** and **4**) can be positioned in the sheet by simple translations. (c) A side view of the modeled sheet demonstrates its planarity. (d) Hexamers at the corners of the white ($h = 1$, $k = 1$) triangle become pentamers. The planar sheet (yellow model) takes on curvature to maintain contacts between the polygons (green model). (e) The magnitude of the pentamer-induced curvature is displayed in the side view of the partial polyhedron. (f) Coordinates of the CCMV X-ray structure fit this construction without any manipulation. (g) A completed $T = 3$ icosahedral model. The 12 pentamers generate curvature that closes the structure. This cage (a truncated icosahedron) accurately describes the geometric morphology of CCMV (h) which is composed of modular, planar, pentamers (12) and hexamers (20). Angular pentamer–hexamer and hexamer–hexamer interfaces (i) stabilize curvature in the absence of convex pentamers used to construct the soccer ball of **Figure 3(d)** (see also **Figure 4**).

bacteriophages, such as HK97, display dramatic subunit reorganization during maturation. (The head (capsid) of the HK97 bacteriophage first assembles as a metastable prohead (prohead I and II), and then undergoes a complex maturation process that involves large-scale macromolecular transitions that expand the protein shell. There are three well-defined expansion intermediates (EI-I/II, EI-III, EI-IV) that end with the EI-IV state, which is the

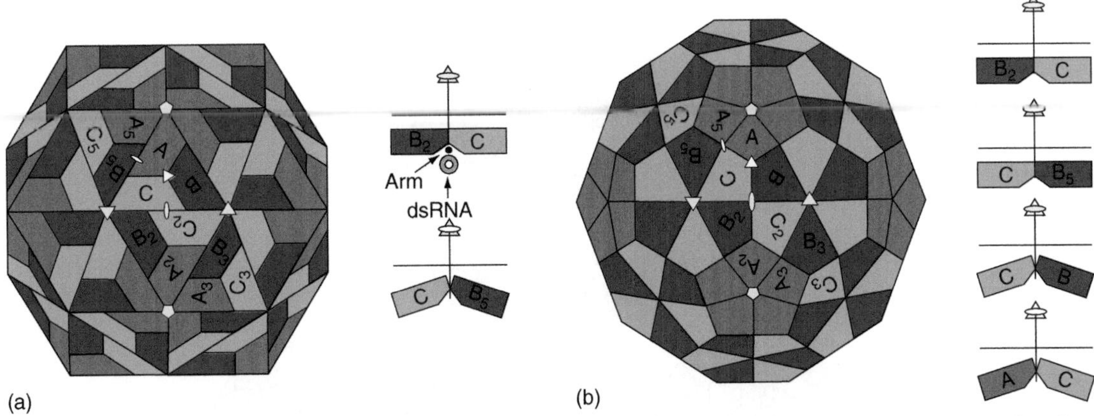

Figure 4 Although quasi-equivalence theory can predict, on geometrical principles, the organization of hexamers and pentamers in a viral capsid, the detailed arrangement of subunits can only be established empirically. High-resolution X-ray structures of $T=3$ plant and insect viruses show that the particles are organized like the icosahedral rhombic triacontahedron or truncated icosahedron. A convenient definition of the icosahedral asymmetric unit for both geometrical shapes is the wedge defined by icosahedral threefold axes left and right of the particle center and an icosahedral fivefold axis at the top. The icosahedral asymmetric unit contains three subunits labeled A (blue), B (red), and C (green) (see **Figures 2(c)–2(d)**). The asymmetric unit polygons represent chemically identical protein subunits that occupy slightly different geometrical (chemical) environments as indicated by differences in their coloring. Polygons with subscripts are related to A, B, and C by icosahedral symmetry (i.e., A to A_5 by fivefold rotation). The shapes of the $T=3$ soccer ball model in **Figure 2(d)**, truncated icosahedron in **Figure 3**, and rhombic triacontahedron are all different; however, the quasi-symmetric axes are in the same positions relative to the icosahedral symmetry axes for all three models. Quasi-threefold and quasi-twofold axes are represented by the white symbols. The quasi-sixfold axes are coincident with the icosahedral threefold axes in $T=3$ particles as shown in **Figures 2(b)–2(d)** and **3**. (a) The rhombic triacontahedron is constructed by placing rhombic faces perpendicular to icosahedral twofold symmetry axes (yellow ellipse). Thus, the A, B, and C polygons are co-planar within each asymmetric unit. The shape of the subunit in $T=3$ plant and insect viruses is nearly identical to the shape of the subunit in the $T=1$ virus and they pack in a very similar fashion. The $T=1$ subunits in one face (**Figure 1(a)**) are related by an icosahedral threefold axis while the $T=3$ subunits in one face are related by a quasi-threefold axis. The dihedral angle between subunits C and B_5 (juxtaposed across quasi-twofold axes) is 144° and is referred to as a bent contact (bottom right image), while the dihedral angle between subunits C and B_2 (juxtaposed across icosahedral twofold axes) is 180° and is referred to as a flat contact (top-right image). Two dramatically different contacts between subunits with identical amino acid sequences are generated by the insertion of an extra polypeptide from the N-terminal portion of the A or C subunits into the groove formed at the flat contact. This polypeptide is called an 'arm'. The flat contact can also be upheld by insertion of nucleic acid structure into the same groove. Neither N-terminal arms nor nucleic acid structure has been observed in the groove across the quasi-twofold axis; thus, C and B_5 are in direct contact as in, for example, the X-ray structure of flock house virus (FHV). (b) A truncated icosahedron achieves curvature at different interfaces compared to the rhombic triacontahedron. Interactions between B_2–C and between C–B_5 polygons are both defined by 180° dihedral angles (side view at top right), whereas bends similar in magnitude occur within the asymmetric unit at the B–C and C–A polygon interfaces (138° and 142°, respectively; side view at bottom right). This creates the planar pentamer and hexamer morphological units characteristic of the truncated icosahedron and the CCMV X-ray structure (**Figure 3(h)**).

final step before head maturation is complete. The crosslinks between the 420 subunits are formed during these steps, except for the last 60 situated around the 12 pentamers. Formation of the final 60 crosslinks is the final step that creates head II, the mature capsid.) The X-ray structure of mature HK97 showed that each subunit in the $T=7l$ capsid makes contact with nine other subunits, displaying molecular promiscuity at an unprecedented level. These interactions, together with covalent crosslinks between subunits create an exceptionally stable particle. It is clear, however, that such a structure cannot be achieved in a single step. It is too complicated. The solution is to assemble in stages. The initial assembly product is called a procapsid. This is formed by three gene products: 415 copies of the capsid protein, 12 copies of the portal subunit, and about 60 copies of a virally encoded protease that is inside the 500 Å diameter particle. This is an equilibrium assembly that is similar to the assembly of simpler viruses. Subunits are confined to discrete volumes and contact only immediate neighbors, again, as observed with simple viruses. Completion of assembly signals protease activation, and 103 residues are cut from each capsid protein and the protease auto-digests, leaving an empty particle that is a substrate for DNA packaging enzymes. When the terminase complex, also virally encoded, attaches to the double-stranded DNA viral genome and the portal, DNA packaging ensues and after ~10% of the DNA is packaged, the particle undergoes an extraordinary reorganization involving 40° rotations of subunits and 40 Å translations, totally remodeling the contacts resulting in the highly interlocked structure in the mature virion. Maturation of this type is entirely programmed into the original tertiary and quaternary structures and is an excellent example of a molecular machine.

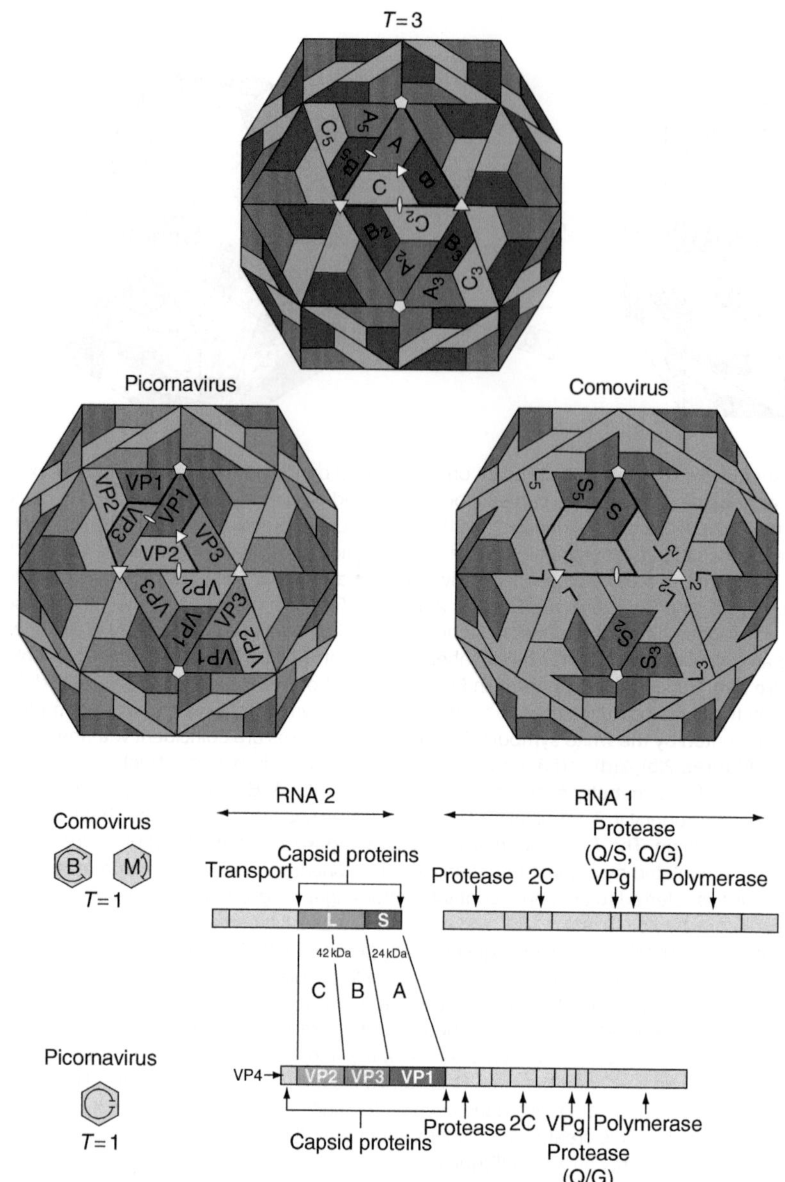

Figure 5 A comparison of $T=3$, picornavirus, and comovirus capsids. In each case, one trapezoid represents a β-barrel and the icosahedral asymmetric units are outlined in bold. The icosahedral asymmetric unit of the $T=3$ shell contains three identical subunits labeled A, B, and C (see **Figure 4**). The asymmetric unit of the picornavirus capsid contains three β-barrels, but each has a characteristic amino acid sequence labeled VP1, VP2, and VP3. The comovirus capsid is similar to the picornavirus capsid except that two of the β-barrels (corresponding to the green VP2 and VP3 units) are covalently linked to form a single polypeptide, the large protein subunit (L), while the small protein subunit (S) corresponds to VP1 (note the similar color shading). The individual subunits of the comovirus and picornavirus capsids are in identical geometrical (chemical) environments (e.g., VP1 and S are always pentamers) technically making these $T=1$ capsids, but they are often referred to as pseudo $T=3$, or $P=3$ capsids. Comoviruses and picornaviruses have a similar gene order, and the nonstructural 2C and polymerase genes display significant sequence homology. The relationship between the capsid subunit positions in these viruses and their location in the genes is indicated by color-coding and the labels A, B, and C in the gene diagram.

Concluding Remarks

This overview of virus structure and assembly illustrates the remarkable level of sophistication associated with one or a few gene products acting in concert. The requirement for maximum utility from minimal genetic information has led to the evolution of proteins with exceptionally high functional density that are studied to develop better understanding of virus-associated disease and as accessible paradigms for cellular function.

See also: Assembly of Bacterial Viruses; Structure of Non-Enveloped Virions.

Further Reading

Caspar DLD (1980) Movement and self-control in protein assemblies. Quasi-equivalence revisited. *Biophysics Journal* 32: 103–138.
Caspar DLD and Klug A (1962) Physical principles in the construction of regular viruses. *Cold Spring Harbor Symposia on Quantitative Biology* 27: 1–24.
Crick FHC and Watson JD (1956) Structure of small viruses. *Nature* 177: 473–475.
Johnson JE and Speir JA (1997) Quasi-equivalent viruses: A paradigm for protein assemblies. *Journal of Molecular Biology* 269: 665–675.
Natarajan P, Lander GC, Shepherd CM, Reddy VS, Brooks CL, III, and Johnson JE (2005) Exploring icosahedral virus structure with VIPER. *Nature Reviews Microbiology* 3: 809–817.
Speir JA, Bothner B, Qu C, Willits DA, Young MJ, and Johnson JE (2006) Enhanced local symmetry interactions globally stabilize a mutant virus capsid that maintains infectivity and capsid dynamics. *Journal of Virology* 80: 3582–3591.
Tang L, Johnson KN, Ball LA, Lin T, Yeager M, and Johnson JE (2001) The structure of pariacoto virus reveals a dodecahedral cage of duplex RNA. *Nature Structural Biology* 8: 77–83.
Zandi R, Reguera D, Bruinsma RF, Gelbart WM, Rudnick J, and Reiss H (2001) Origin of icosahedral symmetry in viruses. *Proceedings of the National Academy of Sciences, USA* 101: 15556–15560.
Zlotnick A (2005) Theoretical aspects of virus capsid assembly. *Journal of Molecular Recognition* 18: 479–490.

Structure of Non-Enveloped Virions

J A Speir and J E Johnson, The Scripps Research Institute, La Jolla, CA, USA

Introduction

Nonenveloped viruses provide model systems for high-resolution structural study of whole virus capsids (and their components) and the principles of large-scale nucleoprotein tertiary and quaternary interactions. In many cases, they are relatively small, stable, and highly symmetric particles that can be purified as homogeneous samples in large quantities. These properties make them suitable for the techniques of X-ray crystallography, X-ray fiber diffraction, and cryoelectron microscopy (cryo-EM) combined with image reconstruction. Enveloped virus particles generally have greater flexibility and asymmetry, two features that inhibit application of high-resolution techniques.

The first crystal structures of intact virus capsids published in the late 1970s and early 1980s are examples of the simplest forms of replicating organisms. They were small, spherical, positive-sense single-stranded (ss) RNA plant viruses about 170–350 Å in diameter with $T=1$ or $T=3$ symmetry, which were easy to purify in gram quantities. By the mid- to late 1980s, the structures of small ssRNA animal viruses (poliovirus and the common cold virus, rhinovirus) and the intact, 3000 Å-long ssRNA helical tobacco mosaic virus (TMV) were also completed. The range of virus types, sizes, and complexity addressed by X-ray crystallography has grown significantly with advances in X-ray technology and computing. Detailed polypeptide and nucleic acid models have now been fitted into data between 4–1.8 Å resolution. Crystal structures of 800 Å diameter virus capsids with $T=25$ icosahedral symmetry have now been determined and show quaternary polypeptide interactions as intricate and varied as their biology.

Spherical particles larger than 800 Å in dimension have been studied by cryo-EM providing detailed images of some of the most complex structures ever visualized at high resolution. This technique is also readily applied to smaller viruses like those described above. Typical cryo-EM studies produce 15–30 Å resolution image reconstructions, but under ideal conditions sub-nanometer resolution can be obtained. The upper diameter limit for applying the cryo-EM technique to viruses is yet to be hit; however, with increased sample thickness (e.g., particle diameter), comes an increase in data collection and analysis restrictions. Currently, the two largest spherical virus structures determined with cryo-EM are the 2200 Å diameter PpV01 algae virus, that has an icosahedral protein shell with $T=219$ quasi-symmetry, and the 5000 Å diameter mimivirus, found in amoebae, that has an icosahedral protein shell with estimated $T=1179$ quasi-symmetry. While the PpV01 structure was determined at 30 Å resolution, the thickness of ice needed to embed mimivirus (7500 Å for shell + attached fibers) limited the image reconstruction to 75 Å resolution, which is the reason for the estimated T number. Thus, cryo-EM can provide detailed images of symmetrical virus particles of almost any size and is an increasingly valuable tool in virus structural studies.

In cases where capsid size or other factors prohibit obtaining near-atomic resolution data, structures of the individual proteins or their assembly products can be determined by nuclear magnetic resonance (NMR) and/or X-ray crystallography. Capsid subunit folds, nucleic acid-binding sites, and some details of quaternary structure can be gathered in this way. Exciting leaps in understanding virus capsid assembly, interactions, and dynamics have come from combining individual structures with cryo-EM particle images to build complete pseudo-atomic models for the larger, more-complex virus capsids. The combined approach generates discoveries that could not have been obtained from either method alone. This article highlights some of the important features of the more than 200 high-resolution nonenveloped virus structures determined to date.

The Building Blocks

The β-Barrel Fold

The earliest spherical virus capsid structures, both of animal and plant viruses, revealed a strikingly common fold for the structural subunits that continues to be found in new virus structures: the antiparallel β-barrel (**Figure 1**). It has been found mainly in icosahedral particles, but is also known to form the bacilliform particles of alfalfa mosaic virus (AMV). The barrel is formed by two back-to-back antiparallel β-sheets with a jelly-roll topology defining a protein shell approximately 30 Å thick. The β-barrels position in the capsid shells in a variety of orientations with the insertions between β-strands having a great range of sizes, giving each virus the ability to evolve a distinct structure, function, and antigenic identity. Indeed, the structure of nudaurelia capensis omega virus (NωV), an insect tetravirus, revealed an entire 133-residue immunoglobin-like domain inserted between strands E and F that forms a major portion of the particle surface. Just as prevalent, the very N- and C-termini of the capsid proteins are often extended polypeptides that do not form part of the barrel. Those portions of the extended termini close to the barrel can have ordered structure to various degrees depending on the local (quasi-)symmetry environment (see **Figure 1**), and tend to engage in varied types of interactions, such as binding one or more nearby subunits and/or nucleic acid, that play critical roles in particle assembly, stability, and nucleic acid packaging.

Other Folds

While the β-barrel remains a common element in a wide range of virus structures, other capsid subunit folds have been discovered in both enveloped (not shown) and non-enveloped particles (**Figure 2**). The nonenveloped subunits also form icosahedral shells as well as helical rods. The structure determination of the $T=3$, ssRNA bacteriophage MS2 in 1990 revealed the first new fold for an icosahedral virus capsid subunit. It consists of a five-stranded β-sheet that faces the inside of the particle, with a hairpin loop and two helices facing the exterior. The only structural similarity to the particles formed by β-barrels was their assembly into a $T=3$ quasi-icosahedral shell. Just 10 years later, the head (capsid) structure of the $T=7$, dsDNA tailed bacteriophage HK97 showed another new capsid subunit fold, and the same fold has also been found in bacteriophage T4 capsids. Other than the high probability that a capsid subunit with a β-barrel is going to form an icosahedral particle, no other general themes can be derived from the variety of subunit folds to predict particle size, quasi-symmetry, or host specificity.

Completely Symmetric Capsids

$T=1$ Icosahedral Particles

The simplest of the spherical particles, the $T=1$ capsids, have only one protein subunit in each of the 60 positions related by icosahedral symmetry. This has been the symmetry found in three of the four known encapsulated satellite viruses, one of which is the 180 Å diameter, ssRNA satellite tobacco mosaic virus (STMV) particle (**Figures 3(a)** and **3(b)**). The small, compact STMV capsid subunits (159 a.a.) have the β-barrel fold with the β-strands running roughly tangential to the particle curvature with the B–C, D–E, F–G, and H–I turns pointing toward the fivefold axes. This is a common subunit orientation and will be simply called 'tangential' going forward. The turns between strands are all short (no insertions), with only an extended N-terminus jutting 60 Å from the barrel that helps form both the threefold and twofold symmetry related interactions (trimers and dimers) via hydrogen bonds and extension of antiparallel β-sheets. The strongest inter-subunit contacts occur at the capsid subunit dimers, where a large number of both hydrophobic and hydrophilic interactions form across a large buried surface between the subunits. The interior surface of the STMV capsid dimers binds a large section of ordered double-helical RNA. Two additional symmetry-related subunits interact with the RNA via their extended N-termini. The protein–RNA interactions are split between direct side-chain bonds to the ribose sugars, and water-mediated hydrogen bonds by residues of both the local subunit dimer β-barrels and N-termini. Altogether the ordered RNA accounts for about 59% of the packaged nucleic acid. Although STMV is one of the simplest icosahedral viruses, its features demonstrate many of the basic protein–protein and protein–RNA bonding schemes that have been observed in larger and more complex particles.

While an advantage of the $T=1$ capsid is its equivalent subunit environments, its major limitation is the small particle volume. To create larger shells, the subunits either have to be larger or there have to be more of them. The former is seen in the crystal structure of the ssDNA bacteriophage ΦX174 capsid (**Figures 3(c)** and **3(d)**). Its 426 a.a. F proteins form a 340 Å diameter $T=1$ icosahedral particle, which has nearly fourfold greater volume compared to the satellite virus capsids. Each subunit has the β-barrel fold, but it accounts for only one-third of the F protein. The remainder of the capsid subunit is composed of long loops inserted between the

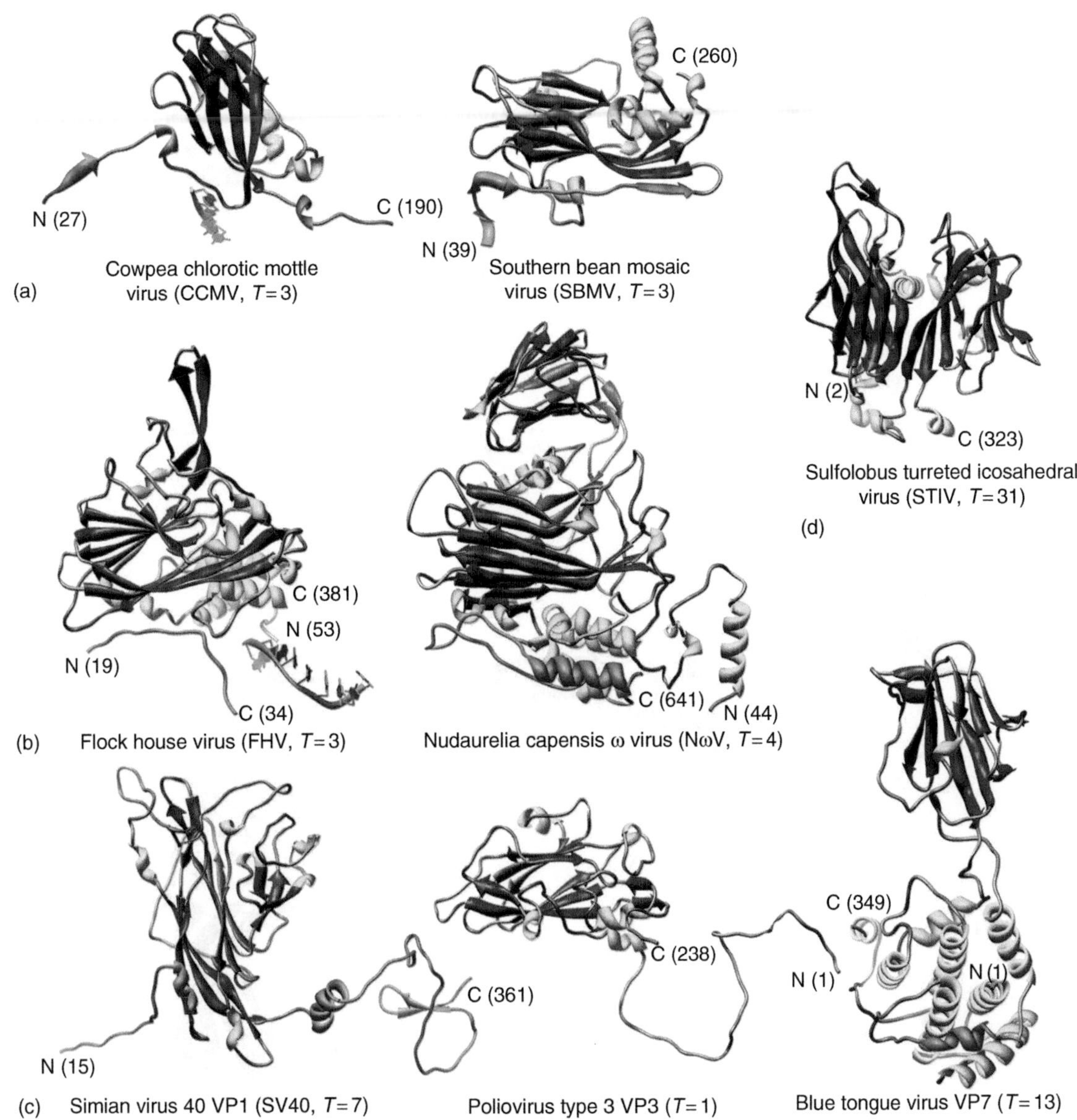

Figure 1 Structure of (a) plant, (b) insect, (c) vertebrate, and (d) archaea virus protein subunits that assemble into icosahedral shells. The name of the virus appears below the corresponding protein subunit along with the capsid triangulation number. The N- and C-termini are labeled with the residue numbers in brackets. The N-termini (and sometimes the C-termini) usually consist of highly charged residues associated with the interior of the virus particle and/or the packaged nucleic acid. As such, they do not follow icosahedral symmetry and are regularly disordered (invisible) in the final structures. Many virus subunit structures determined to near atomic resolution have the β-barrel fold and/or insertions with nearly all β-structure (colored magenta). Multiple copies (from 180 to 900) of the single subunit shown for each virus, except for that of poliovirus, form the entire icosahedral protein shell. Assembly of icosahedral virus particles with more than 60 subunits requires quasi-symmetric interactions often involving subtle to extensive differences in structure at the subunit N- and C-termini. The subunit regions involved in quasi-symmetric interactions critical to virion structure and assembly are colored green (only a single variation is shown for each virus; not yet known for STIV). The 'switch' in structure between identical subunits is a response to differences in the local chemical environment, defined by the number of subunits forming the icosahedral shell, in order to maintain similar bonding between neighboring subunits. The structural variations include the presence or absence of highly ordered RNA structure (green stick models) in FHV and CCMV. Poliovirus utilizes multiple copies of two additional subunits highly similar to VP3 to form a complete virion. Thus, there is no quasi-symmetry in poliovirus (note the absence of any green highlights) since neighboring subunits are different proteins.

β-strands that make up the majority of inter-subunit interactions and capsid surface. ΦX174 has additional capsid proteins, but the contiguous protein shell is made up of only the F protein. While a larger subunit can create a viable amount of enclosed volume, native $T=1$ particles are uncommon. This may be due to inefficiencies in dealing with the larger amount of genome dedicated to the major capsid protein and the more complex structural requirements of a single large subunit to build a stable and functional particle.

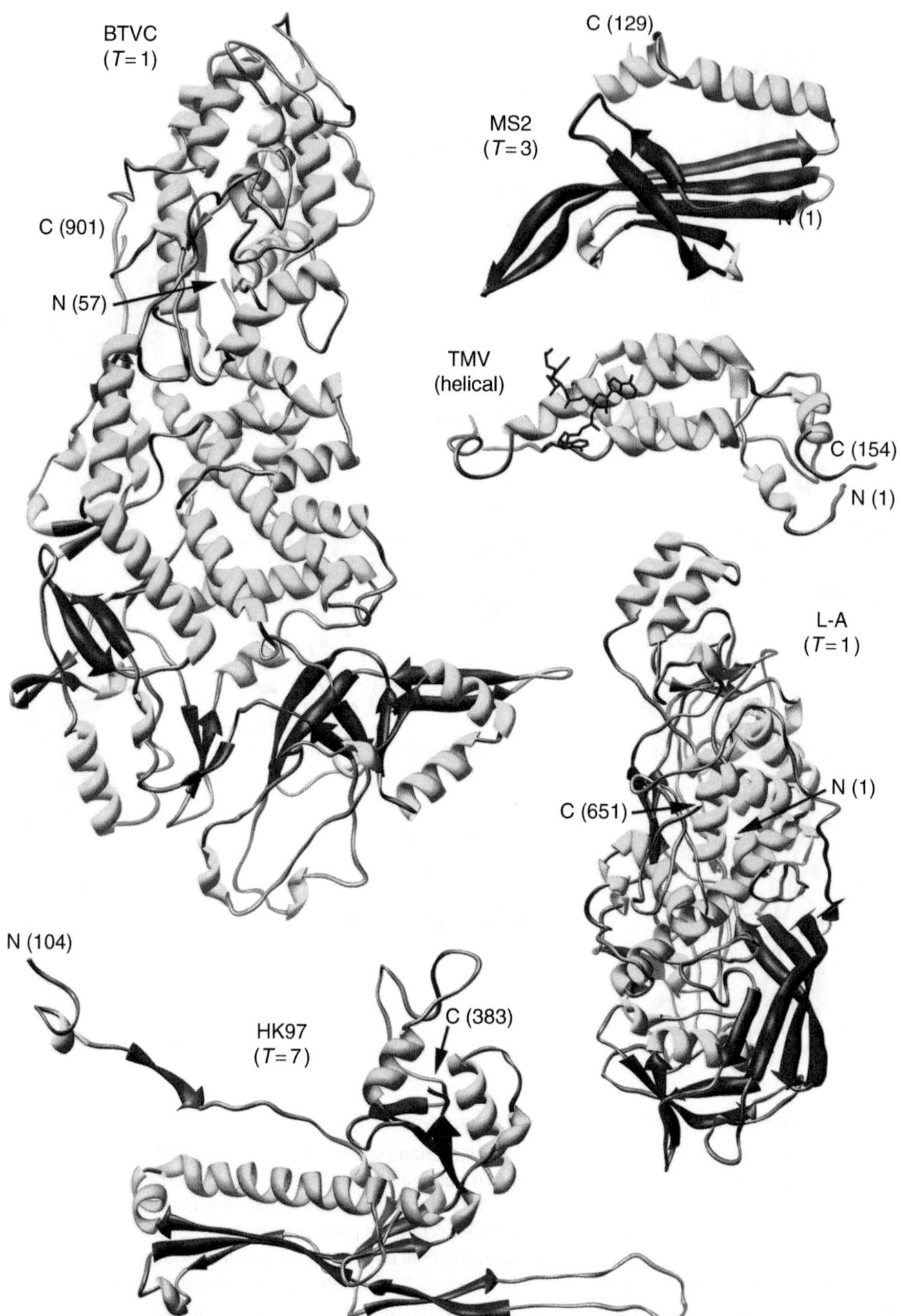

Figure 2 Structure of capsid subunits with folds other than the β-barrel. The name of the virus appears next to the corresponding protein subunit along with the capsid triangulation number (if applicable). The N- and C-termini are labeled with the residue numbers in brackets. With the exception of MS2, helical structure is dominant, with some β-sheet formations mixed into the folds. The subunit of the helical TMV rod is mostly composed of a four antiparallel α-helix bundle that forms the core of the rods, but has a small β-sheet away from the axis that forms the rod surface. The three bound nucleotides per TMV subunit are shown as a blue stick model (see the section titled 'Helical particles'). Both the BTVC (inner shell) and L-A subunits form $T=1$ particles with two copies of the subunit in the icosahedral asymmetric unit (discussed in text), but do not share any detectable structural homology. The BTVC inner shell subunit is the largest capsid subunit determined to near-atomic resolution.

Helical Particles

The only high-resolution structures of helical viruses are from three members of the genus *Tobamovirus*, whose type member is TMV. The tobamoviruses form rod-shaped particles 3000 Å long and 180 Å in diameter, with a central hole about 40 Å across. The particles are built from over 2100 identical capsid protein subunits that are 158 a.a. in size, and arranged in a right-handed helix

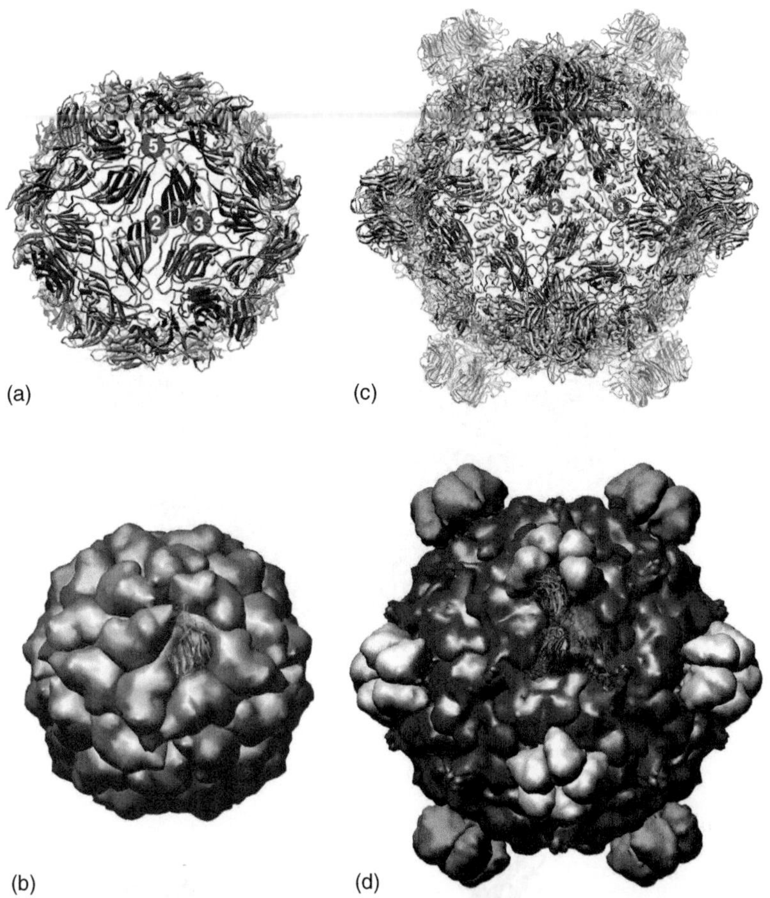

Figure 3 Structures of two $T=1$ particles viewed down an icosahedral twofold rotation axis (approximately to scale). STMV is displayed in (a) and (b), and ΦX174 in (c) and (d). The top images show all 60 capsid subunits rendered as ribbon drawings (β-strands – magenta or pink; helices – yellow), with the icosahedral asymmetric unit (IAU) shown all in blue. The fivefold, threefold, and twofold icosahedral axes are labeled with semitransparent gray circles containing the appropriate rotation number in white. The images are depth-queued such that features further away fade into the white background. The back of the particles are purposely faded (invisible) to make a clearer representation. The bottom images are surface representations of the same structures above, with the IAU shown as a transparent surface to show the ribbon drawing of the capsid subunits. Note the more compact look of STMV due to the close approach of complementary surfaces on the β-barrels vs. the more spread-out look of ΦX174 due to its use of extended loops to create a contiguous capsid. The exterior of the ΦX174 shell includes the G-protein (pink subunits), which surrounds hydrophilic channels at the fivefold axes, and has a β-barrel fold that is oriented nearly parallel to the fivefold axes such that they form 12 broad spikes on the particle surface. This will be called the 'axial' orientation. This increases the particle dimensions at these axes by 60 Å but does not contribute to the enclosed volume as they sit atop an essentially complete protein shell. (b, d) Courtesy of the Viperdb web site.

with 49 subunits in every three turns and a pitch of 23 Å. The subunit has a narrow, elongated structure composed of a four-helix bundle on one end, and a small β-sheet on the other (**Figure 2**). The subunits sit perpendicular to the particle axis, with the helix end of the subunit nearest the center and the β-sheet end forming the outer surface of the rod. Neighboring subunits stack upon one another in the particle helix such that two helices of each protein interact with two others from the neighboring subunit, forming another four-helix bundle to build the rod.

The structure of the TMV particle determined in the late 1980s revealed the first details of virus capsid protein interactions with nucleic acid. A long loop inserted between two of the subunit helices forms the interior channel of the particle that includes a continuous RNA binding site at the edge of the helices. The entire RNA genome is well ordered in the TMV structure by conforming to the helical symmetry of the coat protein stacking. There are three nucleotides bound to each protein subunit (**Figure 2**) over the entire length of the rod, which adds subunits during assembly until the entire genome is encapsidated. The RNA is bound at its phosphate groups by direct, calcium-mediated, and probably water-mediated interactions with nearby arginine and aspartic acid residues, and at its sugars and bases by hydrophobic complementary, polar interactions, and a few hydrogen bonds from local arginine

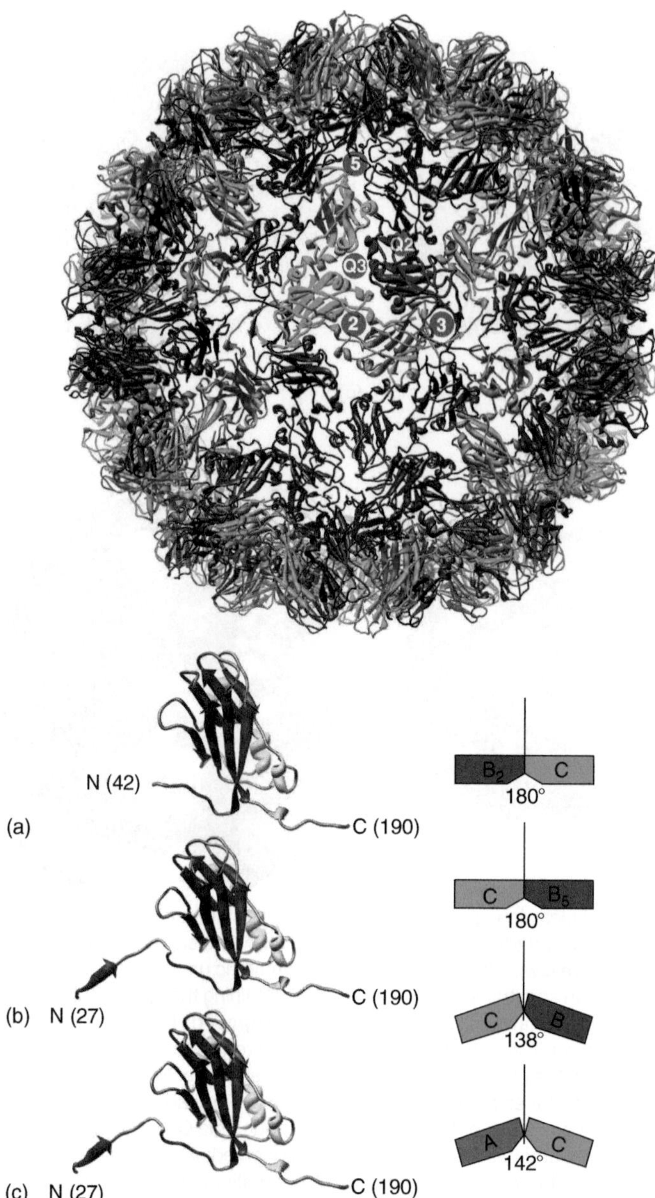

Figure 4 Structure of the $T=3$ CCMV particle viewed down an icosahedral twofold axis. Icosahedral and quasi-symmetry (Q) axes are labeled as in **Figure 3**. The quasi-sixfold axes are not labeled as they coincide with the icosahedral threefold axes. (top) The entire CCMV particle with A-subunits colored blue, B colored red, and C colored green. An IAU is shown above the twofold axis in brighter colors and larger ribbons. The calcium sites between quasi-threefold related subunits are strongly implicated in CCMV particle expansion, which occurs in the absence of divalent metal ions at high pH. The CCMV crystal structure shows that under these conditions the charged residues of the calcium-binding site would cause a repulsion between quasi-threefold related subunits, and a cryo-EM structure of the swollen form of CCMV confirmed the opening of large holes at the quasi-threefolds. Indeed, swollen CCMV breaks apart into subunit dimers and RNA in high salt concentrations, and under various solution conditions, can reassembe into a number of structures including $T=1$, $T=3$, and $T=7$ particles, long tubes, multi-shelled particles and tubes, and laminar plates. It is one of the most pleomorphic viral subunits ever studied, but amazingly only makes $T=3$ particles in plant infections. (Bottom left) Individual structures of the proteins in the A, B, and C subunit environments colored as in **Figure 2**. Note they are closely identical except for the lack of extended N-terminal structure in A, and slightly varying directions taken by the C-termini away from the β-barrels (not readily visible in these views). (Bottom right) Diagrammatic representation of the flat and bent subunit/ capsomer interfaces. Individual blocks represent the subunits as viewed tangentially to the particle curvature. Colors and letters designate the subunit type with the dihedral angle between them given underneath.

and aspartic acid residues. This mode of nucleic acid binding showed that, since the viral genome sequences do not share the symmetry of their enclosing particles, interactions with the coat protein are complementary to the nucleic acid, but non-base-specific to ensure that the entire genome is encapsidated. The same general principles continue to be found in the ordered protein–nucleic acid interactions observed in icosahedral particle structures.

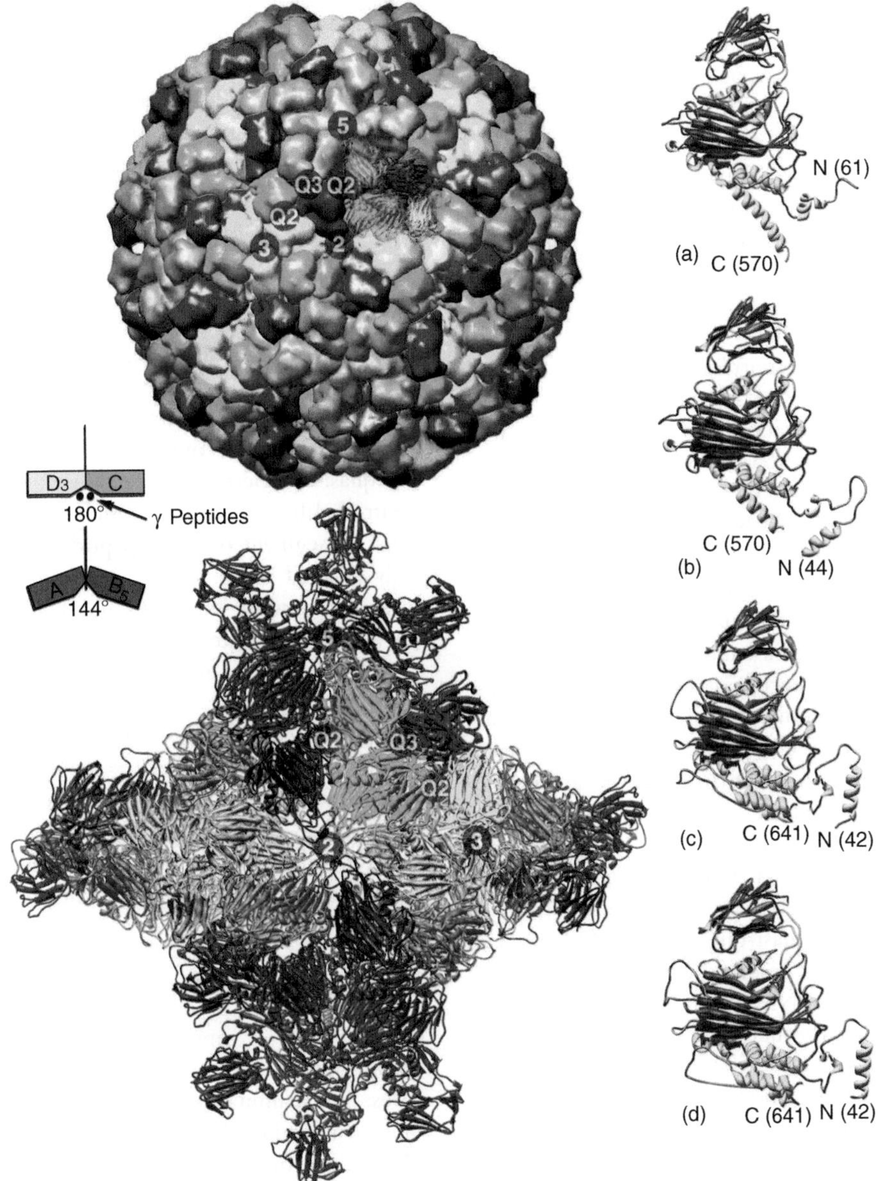

Figure 5 Structure of the $T=4$ NωV particle viewed down an icosahedral twofold axis. Icosahedral and quasi-symmetry (Q) axes are labeled as in **Figure 3**. The low-fidelity quasi-sixfold axes are not labeled as they coincide with the icosahedral twofold axes. (Right) Individual structures of the proteins in the A, B, C, and D subunit environments colored as in **Figure 2**. While the basic folds and domains are similar, they differ considerably at both the N- and C-termini, in loop structure at one end of the barrel, and slightly in position and orientation of the Ig domains. In particular, the C- and D-subunits have additional ordered polypeptide in the helical domain that dictates switching between bent and flat contacts in the capsid architecture (see below and text). (Left) Surface representation of the entire $T=4$ particle (as in **Figure 1**) and a more detailed ribbon representation of the area between the four fivefold axes surrounding the center twofold axis. A diagrammatic representation of the subunit/capsomer interfaces is also shown as in **Figure 3**. Complete pentamers are only shown on the top and bottom of the ribbon image. The A-, B-, and C-subunits are colored as in **Figure 3**, and the D-subunits are colored yellow. An asymmetric unit is shown above and to the right of the twofold axis in brighter colors and larger ribbons (also shown in surface representation above as ribbons). The $T=4$ arrangement can be thought of as two sets of triangles: one set is formed by the quasi-threefold related A-, B-, and C-subunits, and the other is formed by the icoshedral threefold related D-subunits. The dihedral angles of the interfaces between two ABC units (144°), and between ABC and DDD units (180°; both are quasi-twofold related contacts) creates flat triangles with fivefold axes at the corners, and bending between them along the lines from fivefold to twofold axes. Surface image courtesy of the Viperdb web site.

Quasi-Symmetric Capsids

T = 3 Icosahedral Particles

Icosahedral particles built from multiples of 60 identical capsid subunits exist in great numbers, and are arranged with many variations on the quasi-symmetry first predicted by Caspar and Klug in 1962. In one of the closest matches to their predictions, 90 coat protein dimers (180 total coat proteins) of the ssRNA plant cowpea chlorotic mottle virus (CCMV) assemble with nearly ideal $T=3$ quasi-symmetry into the shape of a 280 Å diameter truncated icosahedron, which is defined by planar pentamer (formed by A subunits) and hexamer (formed by B and C subunits) morphological units (capsomers) at the fivefold and threefold (quasi-sixfold) axes, respectively (**Figure 4**, top). The 190 a.a. capsid subunit folds into a small β-barrel with short loops between strands, but has both extended N- and C-termini (**Figure 4**, bottom left). All 180 subunits have the axial orientation with the N-termini extending toward the center of the pentamers and hexamers, and the C-termini extending outward into neighboring capsomers by reaching across the twofold axes. The very ends of the C-termini are clamped in between the β-barrel and N-terminal extension of the twofold related subunit, forming the largest number of interactions between capsid subunits and tying the capsomers together. This occurs at both the twofold and quasi-twofold axes with high fidelity. The dimer interactions also define the dihedral angles between the planar capsomers; the angles between hexamer–hexamer (twofold) and pentamer–hexamer (quasi-twofold) interfaces only differ by 4° (**Figure 4**, bottom right), producing a smooth particle curvature. A triangular unit of A, B, and C subunits defines the icosahedral asymmetric unit (IAU), and also a high-fidelity quasi-threefold axis. Calcium and RNA binding sites are preserved at all three interfaces (calcium site exterior, RNA site interior, and under the calcium site) with nearly equivalent positions of the coordinating side chains.

The pentamer and hexamer units show more differentiation. Besides subtle contact changes between the β-barrels, the N-termini of the hexamer subunits form a nearly prefect sixfold symmetric parallel β-tube about the threefold axes in an impressive display of quasi-sixfold interactions. The N-termini of the pentamer subunits extend a bit toward the fivefold axes but form no structure due to disorder after residue 42. Interestingly, deletion of the β-hexamer in one study did not prevent hexamers from forming, but did destabilize the particle. Other than the differences seen at the capsomer centers, the remainder of the interactions between the A, B, and C subunits are subtly changed to accommodate the differences in chemical environments, rather than switched from one form to another, and may explain the highly pleomorphic properties of the CCMV capsid subunit.

T = 4 Icosahedral Particles

The ssRNA insect tetraviruses remain the only known viruses that have a nonenveloped, $T=4$ capsid. The crystal structure of the 400 Å diameter NωV capsid in 1996 showed the 644 a.a. capsid subunit has three domains: an exterior Ig-like fold, a central β-barrel in tangential orientation, and an interior helix bundle (**Figure 5**, right side). The helix bundle is created by the N- and C-termini of the polypeptide before and after the β-barrel fold, and the Ig-like domain is an insert between the E and F β-strands. All 240 of the NωV capsid subunits undergo autoproteolysis at residue 571 after assembly, leaving the cleaved portion of the C-terminus, called the γ-peptide, associated with the capsid. The fold and tangential orientation of the β-barrel together with the autoproteolysis revealed a strong relationship between tetraviruses and the $T=3$ insect nodaviruses. Indeed, the folds and catalytic sites superimpose with little variation, and a γ-peptide also exists in the nodaviruses. Unlike in CCMV, the quasi-symmetry of these insect particles is obviously controlled by a molecular switch (**Figure 5**, left side). In NωV, a segment of the γ-peptide (residues 608–641) is only ordered in the C and D subunits, and functions as a wedge between the ABC and DDD units to prevent curvature and create a flat contact. The interface between two ABC units is bent partly due to the lack of ordered γ-peptide. A similar situation exists in the nodaviruses, except that the ordered polypeptide between twofold related C-subunits in the $T=3$ particle comes from the A-subunit N-terminus, and ordered duplex RNA also fills the interface. The structural information from the tetravirus group has defined important relationships and some common themes in the biology of the insect viruses.

T = 7 Icosahedral Particles

All three $T=7$ capsid structures determined at near atomic resolution are dsDNA viruses with unique features: two mammalian polyomaviruses, SV40 and polyoma virus, and the head of the λ-like tailed bacteriophage HK97 from the family *Siphoviridae.* The 500 Å diameter SV40 capsid is constructed of 72 pentamers arranged with $T=7d$ quasi-symmetry, not 12 pentamers and 60 hexamers as predicted by quasi-equivalence theory. Thus, 60 of the SV40 pentamers are actually in hexavalent environments as they are located in positions that are occupied by hexamers in an ideal $T=7$ lattice. As the valency of a true $T=7$ lattice is present, it was a surprise to discover it could be achieved without the expected number of identical subunits or domains (i.e., five subunits instead of six). The pentamers contain five copies of the 364 a.a. VP1 protein, which have the axial β-barrel orientation and extended C-terminal arms. Interlocking secondary structures of the G-strands and DE loops with clockwise neighbors form the intimately associated pentamers. The C-arms of all five subunits in each pentamer reach outward to one of three specific types of neighboring pentamers depending on the local environment (threefold and two types of twofold clusters), and form another β-strand between the N-terminus and BIDG sheet of the neighboring subunit (a similar clamp to that of CCMV). This ties the neighboring pentamers together in a general way determined by the position and valency of each pentamer. Remarkably, the VP1 proteins and clamp structures in the pentamer cores are structurally identical, and only the direction and ordered structure of the C-terminal arms are changing as they switch contacts between the nonequivalent pentamers.

By comparison, the 650 Å diameter HK97 capsid is a true $T=7l$ icosahedron, composed of 420 subunits arranged in 12 concave pentamers and 60 planar hexamers. Strikingly, the 385 a.a. protein subunit has a new virus fold (**Figure 6**). Each subunit forms two covalent isopeptide bonds (cross-links) with neighboring subunits on opposite sides of the protein to create topologically linked protein rings arranged with icosahedral symmetry, also a new find in protein structure studies. Notably, the cross-links are formed between subunits that encircle the pentamers and hexamers (one from each neighboring capsomer), not between the subunits forming them. This defines how their complex interweaving occurs at the particle threefold and quasi-threefold axes (**Figure 6**). The fact that this is the final mature state of the HK97 head particle, and that the cross-links are not yet formed in particles at the early to mid-stages of the capsid maturation process, has led to informative studies on capsid dynamics using HK97 as a model system. The final step of HK97 maturation involves formation of the final 60 cross-links in the transition from expansion intermediate IV (EI-IV) to the mature capsid (head II). The pentons appear to be dynamic, oscillating up and down relative to the contiguous capsid shell, until cross-linked. See the article 'Principles of Virus Structure' for animations of the entire transition from the prohead state through head II.

$T=13$ Icosahedral Particles

The 700 Å diameter dsRNA blue tongue virus core (BTVC) particle remains the largest molecular structure ever determined by X-ray crystallography to near-atomic resolution. The core particle has two shells: an inner $T=1$ particle, composed of 120 large (901 a.a.) VP3 proteins (two in each icosahedral asymmetric unit), which packages a few RNA polymerases and 10 strands of dsRNA, and an outer $T=13l$ particle composed of 780 VP7 proteins (349 a.a.), for a total of 900 capsid proteins and a molecular weight of over 54 000 000 Da. The VP7 subunits form a trimer in solution. The trimer structure was determined by X-ray crystallography and combined with cryo-EM images of BTVC to determine the entire BTVC crystal structure. The VP7 subunit has two domains, an outer β-barrel and an inner helix bundle (**Figure 1**), both of which have extensive contacts in the trimer. In an interesting twist, the β-barrel is actually an insert between two helices in the linear VP7 sequence, and the helix bundle is responsible for the majority of interactions between trimers in the assembled shell. The VP7 trimers are postulated to 'crystallize' or assemble on the inner shell's surface into the 12 pentamers and 120 hexamers, which follow quasi-symmetry with high fidelity having very little differences in their intercapsomer interactions. In contrast, the inner $T=1$ shell breaks from quasi-equivalence theory by using large-scale flexing of the two huge VP3 proteins to generate similar shapes of interactions in the absence of similar contacts, and build a nearly featureless hydrophobic surface for interaction with VP7. Together, these shells present a dazzling array of protein interactions that go from near perfect conformity with quasi-equivalent theory to defining a new category of viral subunit interactions in the $T=1$ particle.

$T=25$ and Larger Icosahedral Particles: The Double-Barrel Subunit

Virus capsids having to assemble with $T=25$ quasi-symmetry or higher have avoided the enormous challenge of accurately placing 1500 or more individual proteins in a shell by reducing the problem to just a few related hexon locations. The crystal structure of the hexon from the dsDNA, $T=25$ adenovirus capsid revealed it is a trimer of three identical 967 a.a. proteins. Each protein has two axially oriented β-barrel domains. In the trimer, the barrels are almost precisely related by a pseudo-sixfold axis such that each forms a corner of the hexon base. The hexon is the building block for the particle and is extremely stable due to interpenetrating polypeptides in the trimer. Thus, a stable trimer satisfies the hexavalent lattice positions in $T=25$ adenovirus as the stable pentamers do in $T=7$ SV40. Combined with cryo-EM images of the capsid, these structures revealed that there are only four different hexon-binding sites in adenovirus (pentons have a different subunit), and that much larger hexon facets can be constructed in this manner. In a recent series of groundbreaking discoveries that surprised virus researchers, a combination of both cryo-EM and crystallography methods revealed that the hexon structures of large dsDNA viruses in all domains of life have similar 'double-barrel trimers' forming pseudo-hexagonal shapes. These include adenovirus (eukarya), $T=25$ PRD1 (bacteria), $T=169$ paramecium bursaria chlorella 1 virus (PBCV-1, green algae), and, most recently, $T=31$ sulfolobus turreted icosahedral virus (STIV, archaea) (**Figure 7**). When the double-barrel subunits from all four viruses are superimposed, over 46% of their polypeptides

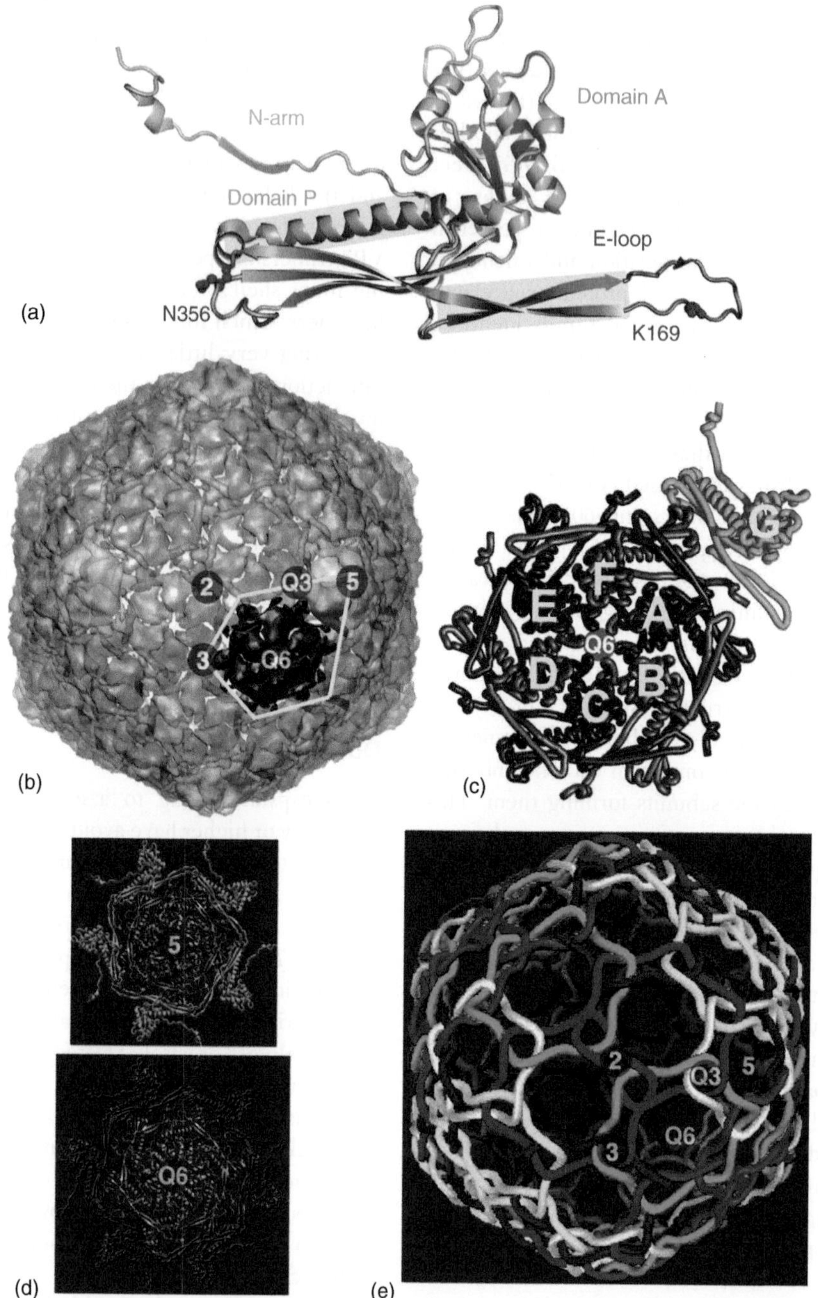

Figure 6 Structure of the $T=7$ HK97 particle. Icosahedral and quasi-symmetry (Q) axes are labeled as in **Figure 3**. (a) HK97 subunit fold. The small HK97 subunits have mixed α/β structure, forming two domains in the shape of an L that have a continuous hydrophobic core, and placing the two residues involved in cross-linking other subunits (red) on opposite sides of the protein. (b) Surface representation of the entire HK97 capsid viewed down an icosahedral twofold axis, with pentamers colored separately from hexamers. The IAU is outlined in yellow. Unlike the $T=3$ and $T=4$ structures, none of the hexamers sit on icosahedral axes. (c) Ribbon diagram of the seven subunits in the IAU showing how they align such that the P-domains and E-loops overlap to define the sides of the local capsomers, but place their tips (with the cross-linking residues) outside the capsomer to be accessible to neighboring subunits. (d) The two types of cross-linked subunit 'circles' or 'rings' in HK97. The surrounding cross-linked subunits (one from each neighboring capsomer) are shown in green (5-circle) or blue (Q6-circle), with the cross-link shown in red. (e) Diagrammatic representation of the interlocking cross-links in HK97. The 5- or Q6-circles, now with subunit and cross-links just represented by tubes, are colored separately showing how they form molecular chainmail by interweaving at the icosahedral and quasi-threefold axes. This molecular chainmail stabilizes the unusually thin 18-Å-wide protein shell (only half the width usually formed by capsids composed of β-barrel subunits), which can be described as a protein balloon. The capsids are also stabilized by other extensive inter-subunit interactions with up to nine other coat proteins. The N-arm, which is not involved in the cross-links, extends 67 Å away from the subunit to contact subunits in neighboring capsomers. (a–d) Courtesy of Dr. Lu Gan. (e) Courtesy of Gabe Lander.

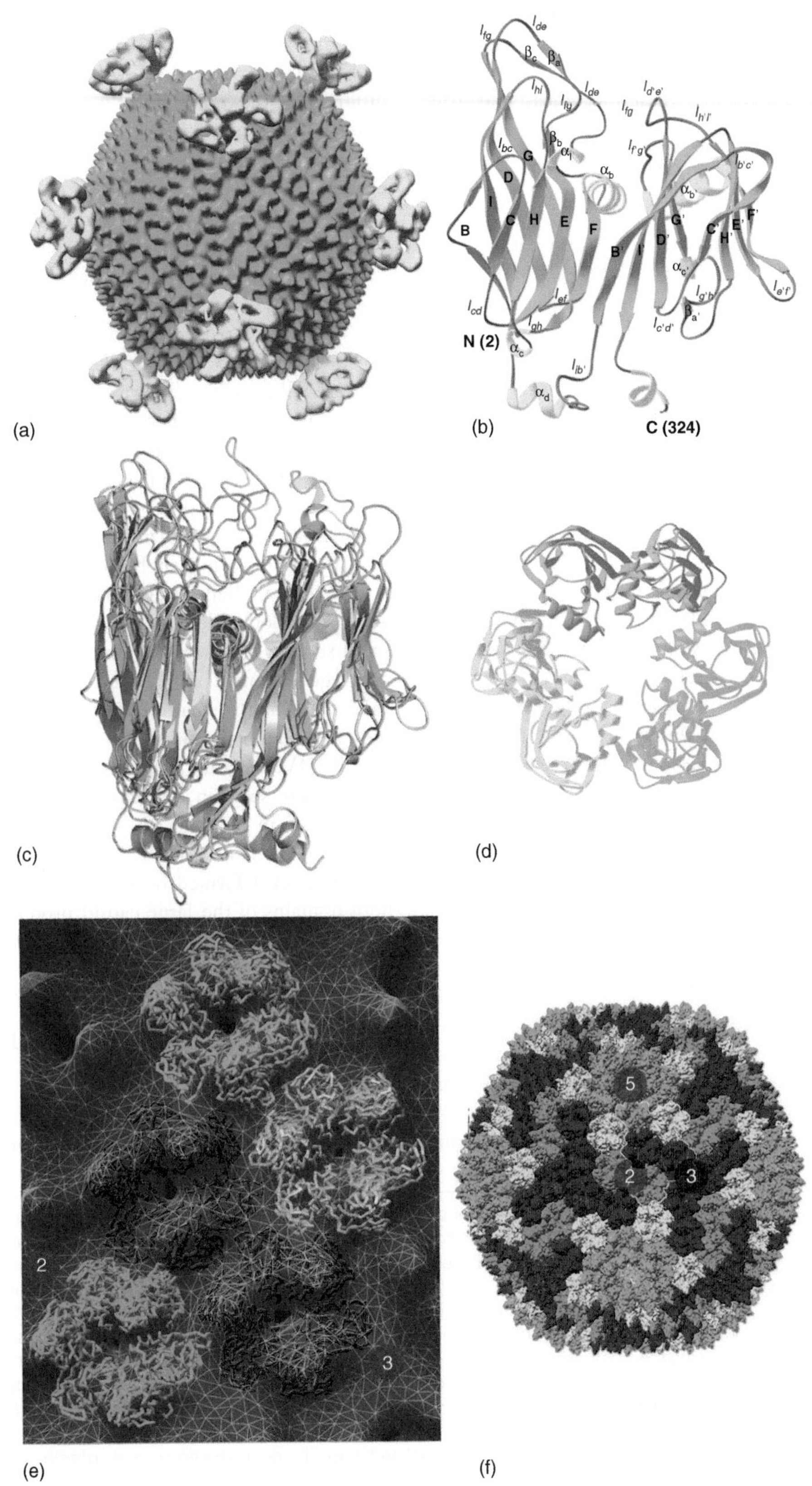

Figure 7 Continued

are within 2.2 Å r.m.s.d. of each other. For the first time, these apparently complicated and unrelated viruses infecting quite different hosts have been solidly linked by their structures, supporting the hypothesis that at least one virus lineage originated before life separated into three domains.

Bacilliform Particles

The ssRNA plant alfalfa mosaic virus (AMV), a member of the family *Bromoviridae* that includes CCMV, forms four types of bacilliform particles composed of one of the four genomic RNAs and a single 220 a.a. capsid subunit. The particles are 180 Å in diameter and range in length from 300 to 570 Å. The spherical ends have $T=1$ symmetry, while the cylindrical bodies have a repeating sixfold symmetry with large holes at the hexamer centers. After a mild trypsin treatment and removal of RNA, the capsid subunit readily forms $T=1$ particles that crystallize and their structure has been determined. The $T=1$ particle is 220 Å in diameter with large holes 35 Å across at the fivefold axes. The subunit is a simple β-barrel in the axial orientation with extended termini at the base. The C-termini reach across the twofold axes to hook between the N-terminus and β-barrel of the symmetry-related subunit in a strikingly similar fashion to the structure of the CCMV particle. Thus, this structure of the AMV endcaps helps demonstrate how the highly pleomorphic capsid proteins of bromoviruses can assemble into several particle forms via strong but flexible dimer contacts depending on solution conditions.

Psuedo-Symmetric Capsids

$P=3$ Icosahedral Particles

Instead of having three copies of a single protein in slightly different chemical environments, a number of viruses have evolved with two or three different proteins in each environment that have unique primary structures but similar folds. The coat protein arrangements are similar to that proposed for $T=3$ lattices such that they are given the $P=3$ (pseudo $T=3$) designation. Unlike the single protein in T capsid arrangements, the different capsid proteins in P arrangements are free to mutate, function, and form interactions independent of one another but more of the genome must be used to encode multiple capsid proteins. Two well-studied families with $P=3$ capsids are the ssRNA animal picornaviruses and plant comoviruses, which have a common genome organization. The structures of several human rhinovirus (HRV) serotypes (common cold virus) are 320 Å diameter particles with the VP1, VP2, and VP3 capsid subunits in a $T=3$-like IAU (**Figure 8**). These subunits all have the β-barrel fold oriented tangentially and similar overall structure except for some loops and extended protein termini. The exterior of the particle has a deep cleft circling the fivefold axes (viral canyon) that binds the ICAM receptor on the surface of cells and which is also bound by neutralizing antibodies.

The capsid of the plant bean-pod mottle virus (BPMV) is also 320 Å in diameter, but has only two subunits, small and large. Both fold into β-barrel domains: one in the small subunit and two that remain covalently attached in the large subunit. The subunits assemble into a $T=3$-like shell such that the small subunit is equivalent to VP1, and the large subunit to VP2 and VP3 in the HRV capsids. The BPMV structure had the first ordered protein–RNA interactions seen in an icoshedral virus. Seven ribonucleotides in a single RNA strand were visible inside the capsid shell near the icosahedral threefold axis in a shallow pocket formed by nine different segments from both domains of the large capsid protein. A few specific interactions and a larger number of nonbonded contacts bind the ordered RNA segment.

Other Icosahedral Particles with Uncommon Architectures

While shell arrangements other than T lattices are possible with multiple single or unique capsid proteins in

Figure 7 Structure of the $T=31$ STIV capsid and major coat protein (MCP). (a) The cryo-EM image reconstruction of the entire STIV capsid viewed down an icosahedral twofold axis. The shell formed by 900 copies (300 trimers) of the MCP is shown in blue, and the 200 Å tall turrets that compose the fivefold axes are shown in yellow. (b) Crystal structure of the MCP, with β-strands colored cyan and helices yellow. Like other large dsDNA viruses, the protein folds into two covalently linked β-barrels, one shown on the left and the other on the right. Interestingly, the large capsid subunit of the plant comoviruses is the only one outside the large dsDNA virus families with the double-barrel fold. (c) Superposition of the major coat proteins from STIV (yellow), PRD1 (cyan), and PBCV-1 (green) in approximately the same orientation as (b), showing the closely similar double-barrel folds. (d) Model of the STIV MCP trimer, based on that seen in the other dsDNA viruses. Each protein is colored separately, and each barrel makes the corner of one of the five unique hexon units seen in the assembled $T=31$ virus. (e) Top view of the STIV asymmetric unit from the cryo-EM structure (shown as a blue lattice) with Cα traces of the MCP trimer fitted in the hexon positions. Icosahedral axes are labeled as in previous figures, with the blue hexon just below the fivefold axis. The highly correlated fit between the atomic model and cryo-EM reconstruction reveals possible inter-subunit interactions, quasi-symmetry, and how the main protein shell is constructed, shown in (f). (f) Pseudo-atomic model of the complete MCP shell using the same color scheme as in (e), with the asymmetric unit outlined in white. Note the modularity of the hexons, which theoretically can be exploited to form capsids of extremely large size such as the 5000 Å diameter mimivirus. Courtesy of Dr. Reza Khayat *et al.*, as published in Khayat R, Tang L, Larson ET, *et al.* (2005) Structure of an archaeal virus capsid protein reveals a common ancestry to eukaryotic and bacterial viruses. *Proceedings of the National Academy of Sciences, USA* 102: 18944–18949.

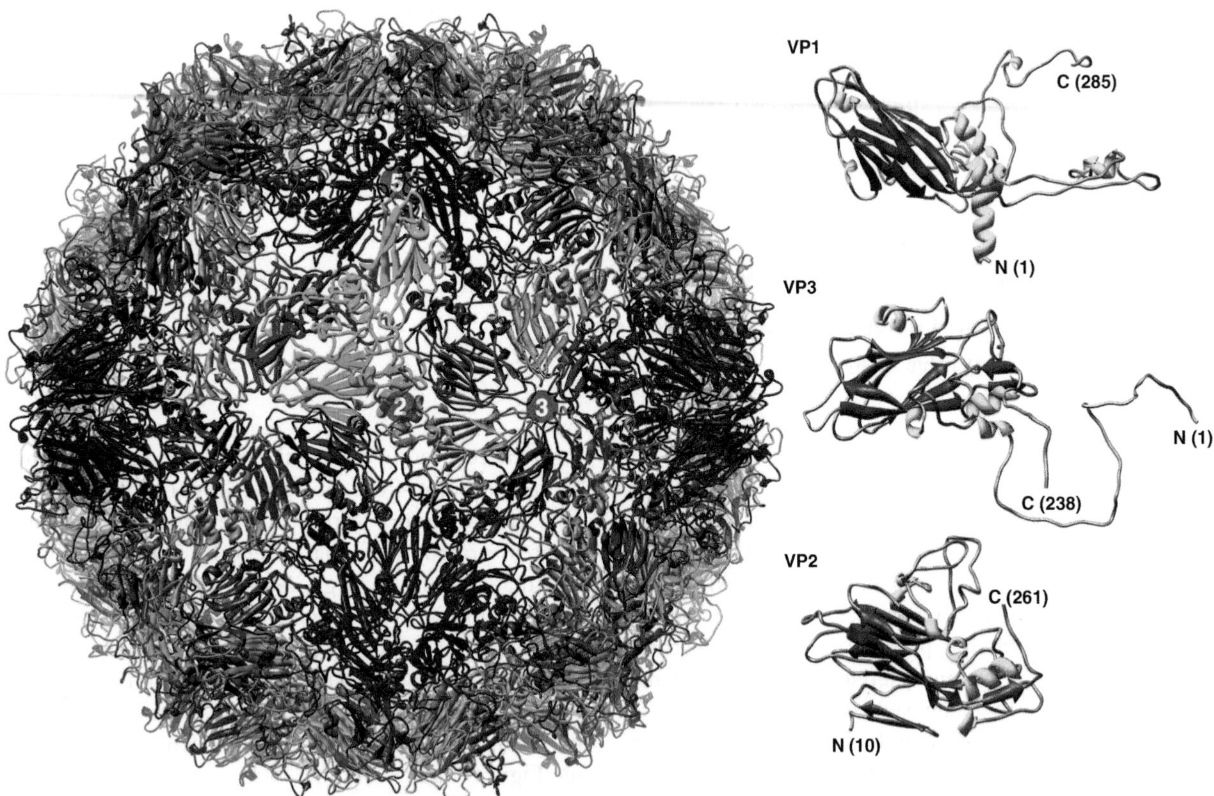

Figure 8 Structure of the $P=3$ HRV-16 capsid viewed down an icosahedral twofold axis. Icosahedral axes are labeled as in **Figure 3**, but note there are is no quasi-symmetry since the proteins have unique primary structures. The VP1 proteins are colored blue, VP2 green, and VP3 red. An asymmetric unit is shown above and to the left of the twofold axis in brighter colors and larger ribbons (note the color and asymmetric unit definitions are a bit different than in $T=3$ particles). Structures of the three different proteins are shown to the right and colored as in **Figure 2**. Note the slight tilt of the VP1 subunits that helps to create the canyon surrounding the pentamers.

the IAU, these are so far rare. The inner $T=1$ shell of the BTVC is described above in the $T=13$ virus section. A very similar $T=1$ shell has been described for the 440 Å diameter dsRNA yeast L–A virus capsid structure, which also has two copies of the single 680 a.a. in the IAU. This arrangement is emerging as a common structure among the inner protein shells of dsRNA viruses that transcribe mRNA inside the particle and extrude the nascent strands through holes in the shell.

Concluding Remarks

A great deal of our early knowledge of virus structure came from the nonenveloped viruses, and they continue to provide valuable data in understanding the virus life cycle. The important role of viruses in disease makes it imperative we understand the chemistry and biology of their capsids in ever greater detail. While some overall themes emerge, such as the general nature of capsid protein–protein interactions, protein–RNA interactions, and protein shell organizations, their remarkable variation and complexity defies attempts to form a comprehensive understanding of how they assemble and function. Recent structural evidence pointing to the existence of a primordial virus may lead to important new insights as virus structural studies address even more complex particles at an accelerated rate.

See also: Assembly of Bacterial Viruses; Cryo-Electron Microscopy; Electron Microscopy of Viruses; Principles of Virion Structure.

Further Reading

Burnett RM (2006) More barrels from the viral tree of life. *Proceedings of the National Academy of Sciences, USA* 103: 3–4.

Chiu W and Johnson JE (eds.) (2003) *Advances in Protein Chemistry, Vol. 64: Virus Structure.* San Diego: Academic Press.

Freddolino PL, Arkhipov AS, Larson SB, McPherson A, and Schulten K (2006) Molecular dynamics simulations of the complete satellite tobacco mosaic virus. *Structure* 14: 437–449.

Khayat R, Tang L, Larson ET, *et al.* (2005) Structure of an archaeal virus capsid protein reveals a common ancestry to eukaryotic and bacterial viruses. *Proceedings of the National Academy of Sciences, USA* 102: 18944–18949.

Rossmann MG, Morais MC, Leiman PG, and Zhang W (2005) Combining X-ray crystallography and electron microscopy. *Structure* 13: 355–362.

Roy P, Maramorosch K and Shatkin AJ (eds.) (2005) *Advances in Virus Research, Vol. 64: Virus Structure and Assembly.* San Diego: Academic Press.

Relevant Website

http://viperdb.scripps.edu/ – Viperdb Web Site (A Database of Virus Structures with Many Visualization and Analysis Tools).

Electron Microscopy of Viruses

G Schoehn and R W H Ruigrok, CNRS, Grenoble, France

Introduction

Most viruses are too small to be observed with light microscopy and many are too large or too irregular to crystallize. Therefore, electron microscopy (EM) is the method of choice for the direct visualization of viruses and viral proteins or subviral particles. However, electrons do not travel far in air and therefore the inside of the microscope has to be under very high vacuum (10^{-5} to 10^{-8} torr). This implies that biological samples have to be dehydrated and/or fixed. It is important that the biological sample is stabilized (or fixed) so that its ultrastructure remains as close as possible to that in the biologically active material when exposed to the vacuum.

The penetrating power of electrons in biological material is also very limited, which means that the specimens must either be very thin or must be sliced into thin sections (50–100 nm) to allow electrons to pass through. This is not a real problem for the imaging of viruses because the sizes of viruses are usually in this range but if we want to study viruses interacting with cells during infection or replication, sectioning is necessary.

Contrast in the transmission electron microscope basically depends on the atomic number of the atoms in the specimen; the higher the atomic number, the more electrons are scattered and the larger the contrast. Biological molecules are composed of atoms of very low atomic number (carbon, hydrogen, nitrogen, phosphorus, and sulfur). Staining methods using heavy metal salts or image analysis have to be used to enhance the contrast in electron micrographs.

In order to make this article easier to read we will illustrate the various EM-preparation techniques with images of intact adenovirus particles, individual capsid proteins or protein complexes from adenoviruses. Adenoviruses are double-stranded DNA viruses with an icosahedral capsid with a pseudo T number of 25 and a total protein mass of around 125 MDa. The major capsid components are the hexon, a trimeric protein with a basal hexagonal shape, and the penton that is a noncovalent complex between the pentameric penton base and the trimeric fiber protein. Two hundred and forty hexons form the 20 facets of the icosahedron, whereas the pentons form and project from the 12 vertices. The fiber binds to host cell receptors with its C-terminal knob domain whereas protein loops that extend from the penton base and that contain an Arg-Gly-Asp (RGD) sequence are required to bind to a secondary cell receptor to trigger endocytosis of the virus. The capsid also contains minor components, proteins IIIa, VI, VIII and IX that glue the hexons and pentons together at specific positions in the capsid. For a schematic drawing of the virus, see **Figure 1(a)**.

Techniques for Single Particle Imaging

Before any sample preparation takes place, one should be informed about the nature of the sample: its molecular weight, some information on its physical size, and the chemical nature of the components (proteins, nucleic acids, lipid membranes). Further, it is important to verify by sodium dodecyl sulfate polyacrylamide gel electrophoresis (SDS PAGE) whether the protein(s) in the sample is (are) intact and not totally or partially degraded. Information on possible contamination of protein samples with nucleic acids can be obtained by measuring the optical density spectrum at 220–350 nm. Such a spectrum can also give information on whether a protein sample contains aggregated protein since a significant absorption at wavelengths above 300 nm is then observed. Such controls are important because the imaging techniques that are described here will always lead to a picture. However, one wants to know how the picture relates to the active biological sample.

Negative Staining

For negative staining the sample is placed on a support film, bathed in a solution of a heavy metal salt that is subsequently blotted, and the sample is then left to dry in air (**Figure 2**). The stain will dry around the molecules, and the area where the molecules happen to be will

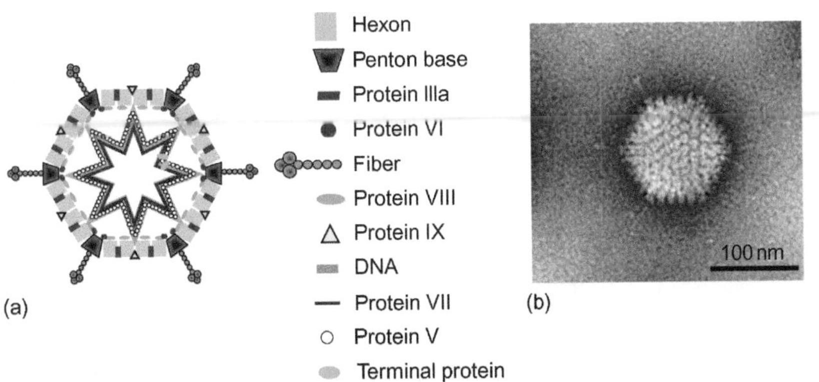

Figure 1 The adenovirus particle. (a) Schematic view of an adenovirus particle. A schematic view through a sliced adenovirus showing the different structural proteins of the virus. The adenovirus particle is an icosahedral particle with on each vertex a noncovalent complex between a trimeric fibre and a pentameric base. This complex is called the penton. (b) Electron micrograph of an adenovirus particle negatively stained with a 1% solution of ammonium molybdate pH 7.5. The particle is seen in the same orientation as in (a). The approximate molecular weight of an adenovirus is 150 MDa and the size of the particle is 1000 Å in diameter.

appear light and the space around it dark, that is, opaque for electrons. As support films one can use plastic films, commercially available or home-made, but these films tend to become hydrophobic after some time and need to be cleaned first with a plasma cleaner. Plastic films are mechanically resistant but are often rather thick and give a high background noise in the images. Thin carbon films give better results. These films are evaporated onto the surface of freshly cleaved mica by heating carbon at high temperatures under vacuum.

For sample preparation it is important to know the concentration of the sample. For small molecules like isolated capsomer proteins, a concentration between 0.05 and 0.1 mg ml^{-1} often gives the best results. When the concentration is too low the stain will not be spread out evenly on the support film and if the concentration is too high the molecules will be one on top of the other and the picture will be difficult to interpret. For intact virus particles such as adenoviruses and influenza viruses, a concentration of 1 mg ml^{-1} should be used. Using a drawn-out glass Pasteur pipette (the sample is taken up into the pipette by capillary force) 1–2 μl of the sample is touched at the carbon–mica interface of a small piece of precut carbon on mica (**Figure 2(a)**). The surface of the carbon on the outside is hydrophobic but at the interface the carbon is clean and hydrophilic and the sample is sucked in between the carbon and mica layers. Then the carbon film with the adsorbed sample is floated off in a pool of negative-stain (usually uranyl acetate (UA), ammonium molybdate, or sodium silicotungstate, SST) and a copper EM grid is placed on the top of the floating film. The grid plus carbon film is then picked up by placing a small piece of paper on the grid and, as the paper wets, it is lifted up taking grid and carbon film up from the pool of stain. In this way the edges of the carbon film that stick out over the rim of the copper grid do not fold back onto the bottom of the grid, creating carbon layers at both sides of the grid with an inclusion of stain. The paper-grid-carbon-sample sandwich is placed on filter paper to blot and air-dry for at least 5 min before insertion into the column of an electron microscope (**Figure 2(a)**).

Different heavy atom salts that are available commercially can be used for negative staining. However, chemical reactions between the salt and the specimen are possible. When the salt binds specifically to the sample molecules one may obtain positive staining rather than negative staining. With positive staining one does not obtain an image of a clear sample molecule on a dark background but a dark molecule on a clear background. It should also be taken into consideration that some salts for staining might have non-neutral pH values. Finally, stain molecules have defined sizes and the resolution of the image may be limited by the grain size of the stain molecules.

Uranyl acetate has a low pH (pH 4.4) and is known to cross-link and stabilize fragile macromolecular assemblies. This stain is especially useful for enveloped viruses since the binding of uranyl ions to the negatively charged lipid heads stabilizes the membranes. This stain does not stay close to the support film but will also stain the top of virus particles so that one obtains a superimposed image of the surface features that touch the support film plus the features on the opposite side of the virus particle. It is not such a good stain for thin objects that lie flat on the support film because the contrast is reduced by the stain lying on top of the object. In **Figure 2** the molecules in plate C (right) were stained with uranyl acetate. Note that uranyl acetate, because of its ability to chemically react with proteins, lipids, and nucleic acids, is known to cause positive staining as well as negative staining. One serious problem with this salt is that it is no longer available in some countries since even nonenriched uranium may be

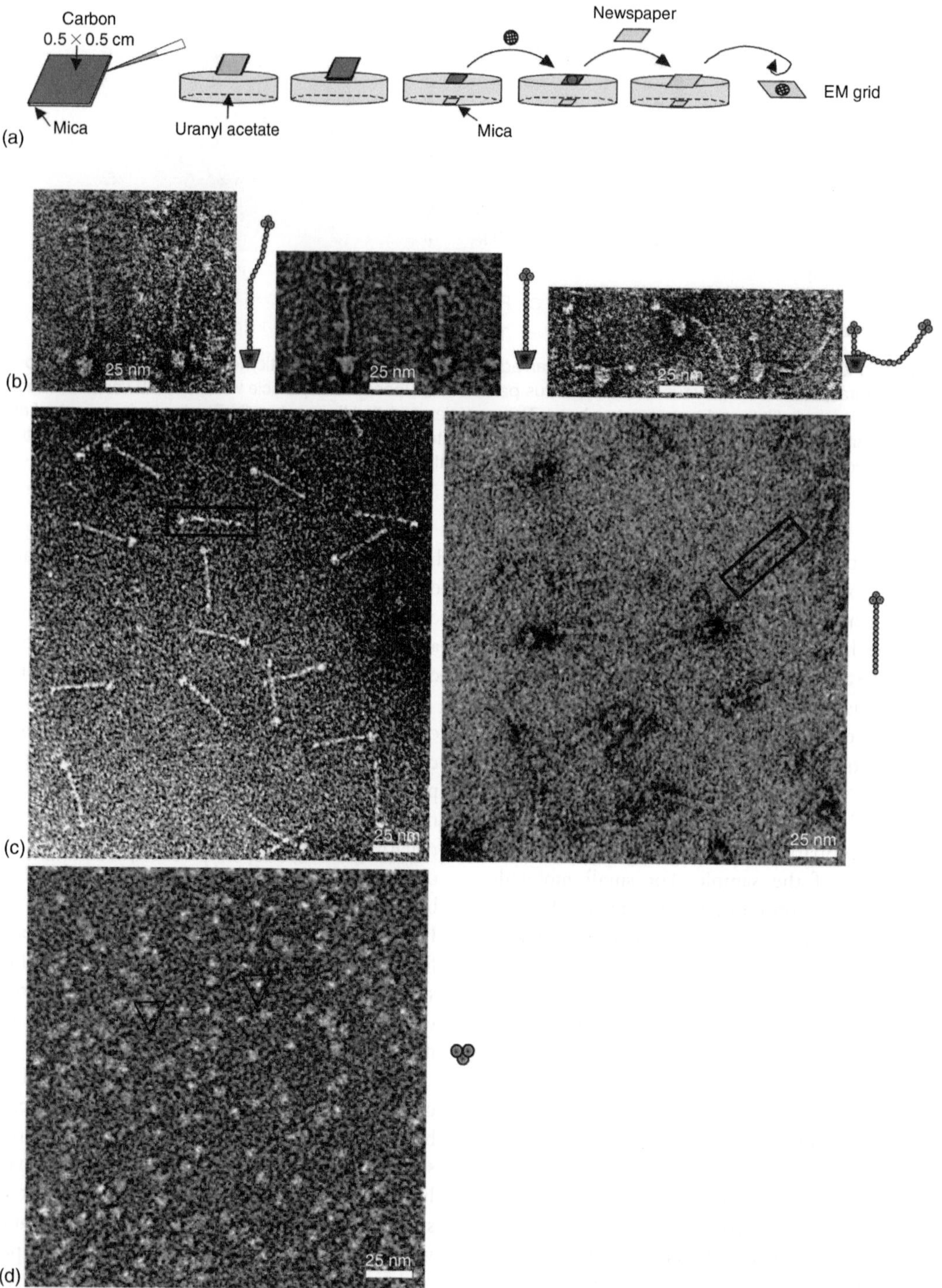

Figure 2 Negative staining technique and results. (a) The different steps of the negative staining method: the sample is adsorbed onto the carbon support film by touching a pipette with the sample to the mica and carbon interface. Then the sample/carbon is floated off the mica onto the surface of a solution of a heavy atom salt. An EM grid is placed on the surface of the carbon and then a piece of paper is used to pick up the grid with carbon film. The paper with grid, carbon, and sample is blotted on filter paper and the grid is then air-dried before observation in the microscope. (b) Electron micrographs of different adenovirus pentons negatively stained with sodium silicotungstate. From left to right one can see a bovine adenovirus 3 penton base (very long and bent fiber), a human adenovirus 41 penton with the long fiber, and an avian adenovirus penton (one base with two different fibers sticking out). The penton base is around 300 kDa is size, whereas the fiber is between 120 and 350 kDa. (c) Electron micrograph of an isolated adenovirus 2 fiber stained with sodium silicotungstate (left) and uranyl acetate (right). In each case, a single fiber is highlighted by a black rectangle. The molecular weight of one fiber is 190 kDa. (d) Electron micrograph of a recombinant canine adenovirus fiber-head expressed in *Escherichia coli*, stained with sodium silicotungstate. The molecular weight of the trimeric particle is around 60 kDa and it is sometimes possible to recognize the triangular shape of the particle (black triangles).

considered to be of military importance. Another problem is that it is slightly radioactive. Sodium silicotungstate is a chemically inert stain with a neutral pH. The stain stays very close to the support film and outlines only those features of a virus or protein that are in contact with the film and it is therefore often used to stain small proteins. The size of the stain molecule is rather large (i.e., 10 Å in diameter) and this feature may limit the resolution of image reconstructions. The stain is often sold as silicotungstic acid. The acid needs to be solubilized in water, neutralized with NaOH and the salt is precipitated with cold ethanol. In **Figure 2** plates (b), (c) (left), and (d) were stained with SST. This stain allows visualization (and sometimes determination of their oligomeric state) of proteins as small as 60 kDa (**Figure 2(d)**).

Phosphotungstic acid (PTA) can be used at pH values between 5 and 7. The staining characteristics are similar to those of SST although PTA often does not spread out well during air-drying and can remain in thick puddles. The stain in these puddles is so thick that protein structures cannot be visualized. This stain can be rather destabilizing for lipid membranes and enveloped viruses often show membrane blebs that are squeezed out of the regular lipid envelope during air-drying.

Ammonium molybdate has a neutral pH. During drying, the stain stays quite high on the support film. It is therefore especially useful for the staining of large particles (**Figure 1(b)**). This stain can be used for cryo-negative staining when very high stain concentrations (16%) are used. The preparation of the grids is as described below for regular cryo-microscopy. The advantage over regular cryo-microscopy is the much higher contrast. The disadvantage is that of all negative staining techniques: only the outside of the particles is visualized by negative staining without any information on the structures inside the virus particles.

Advantages and limitations

With the negative staining technique it is possible to visualize small proteins and sometimes determine their oligomeric state in a very rapid manner. The main limitation of this technique results from the air-drying step. If the sample is fragile, which is the case with enveloped viruses, some flattening can occur during drying and lipid blebs can be squeezed out. Finally, the molecular forces that act upon proteins, complexes, or viruses during the adsorption phase onto the support film can also induce conformational changes.

Cryo-Electron Microscopy

In order to overcome the adsorption step, stain artifacts and, most importantly, the drying step, Dr. Jacques Dubochet and collaborators developed a new sample preparation technique: the frozen-hydrated sample imaging or cryo-EM (**Figure 3**). For this technique the sample concentration must be much higher than that for negative staining because there is no direct support film that has a concentrating effect. The support film for the sample in solution is a carbon film with holes, either home-made (rather irregular, **Figure 3(b)**, left film) or commercially obtained (these films have very regular, evenly spaced holes, **Figure 3(b)**, right film). An aliquot of 2–4 μl of purified sample at a concentration of 0.5–1 mg ml^{-1} is pipetted onto a copper grid covered by the carbon film. Excess liquid is removed by brief blotting with a piece of blotting paper and the grid is then plunged into liquid ethane cooled by a liquid nitrogen bath (**Figure 3(a)**). The rate at which the temperature of the buffer in the holes is lowered is so high that there is no time for the formation of ice crystals; the water forms vitreous ice (also called amorphous ice). Because of the absence of ice crystals, the sample is not deformed but stays in its native state and the vitreous ice is translucent for electrons. The frozen grids have to be transferred into a cryo-holder and inserted into the microscope. The temperature of the sample has to stay very low (i.e., under −160 °C) in order to prevent the vitreous ice to transform into hexagonal or cubic ice. If this irreversible conversion occurs the structure of the sample will be destroyed. All cryo-EM work has to be done at low temperature and the microscope has to be equipped with cryo-EM tools (cryo-holder, anti contaminator blades around the specimen in the microscope in order to avoid the deposition of water and other vapors onto the sample). The ranges of salt concentration and pH that can be used are quite large and include physiological conditions.

Advantages and limitations

Because no heavy atoms are added to the sample, the contrast in the images is very low (**Figure 3(c)**) (the contrast is partly generated by the difference in density between the frozen water and the protein, i.e., 1 g/ml vs. 1.3 g ml^{-1}). This low contrast makes the use of image analysis indispensable. The combination of low contrast and noise also limits the minimum size of the protein/macromolecular assemblies that can be studied by cryo-EM to about 200 kDa. Radiation damage is also a big problem in cryo-EM. As soon as the sample is exposed to the electron beam its starts to be destroyed (**Figure 3(c)**), right). However, minimal dose systems have been developed to limit the electron dose that hits the sample by exposing it only at the nominal magnification during the photographic or charged-coupled device (CCD) camera exposure time. Basically, the electron dose has to be less than 10 $e^{-}Å^{-2}$ in order to preserve the structure of the sample.

There have been a number of recent technological advances that have made the use of this technique easier. The blotting time of the grid is one of the important steps in sample freezing, but is also one of the less reproducible

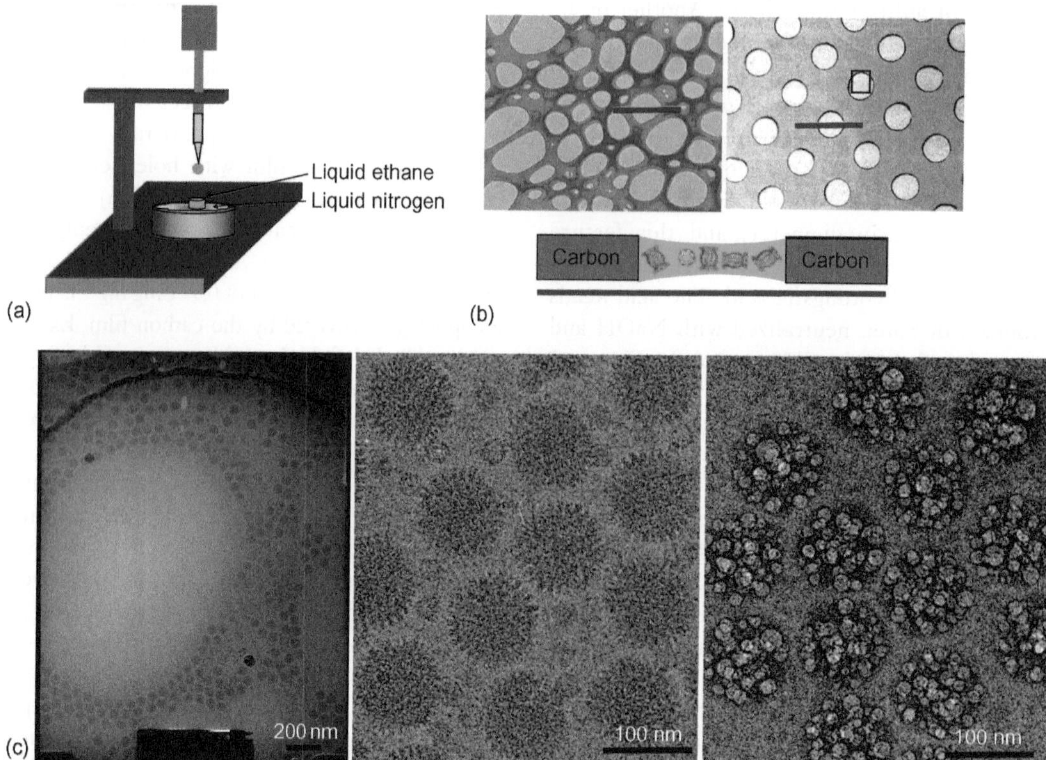

Figure 3 Cryo-electron microscopy (EM) technique and results. (a) Plunger used for rapid freezing of the sample. A grid covered with a holey carbon film is loaded with the sample just before blotting and plunging it very rapidly into liquid ethane cooled down by liquid nitrogen. (b) Examples of home-made holey carbon (left) and commercial quantifoil grid (right). The red bar represents a slice through a hole showing the sample in the frozen vitrified water (bottom). The rectangle on the quantifoil grid represents the portion of field indicated by the red bar above. (c) Cryo-EM images of adenovirus particles. The sample was prepared on a quantifoil film. The picture on the left part is an entire micrograph showing the repartition of the adenovirus particles according to the ice thickness (one part of the hole is totally empty due to the fact that the ice is too thin. The middle part shows a higher magnification of the cryo-EM micrograph. Details are visible on the virus particle; the smaller particles are adenovirus-associated viruses. On the right, the same adenovirus sample is shown after an 1 min exposure to the electron beam. Irradiation damage (bubbles on the picture) is clearly visible.

ones. The thickness of the ice in the hole has to be as close as possible to the diameter of the sample. Automated procedures for producing reproducible cryo-EM grids have been developed and commercial devices for obtaining reproducible ice thickness are now available. The introduction of support films with regularly spaced holes in parallel with automated freezing will make automated data acquisition possible.

Cryo-Electron Tomography

Electron tomography allows the study of the 3D organization of thin individual cell organelles and bacterial cells at nanometer resolutions without slicing them. This technique is also able to reconstruct the 3D structure of nonregular viruses. In contrast to all other EM reconstruction techniques, the samples are unique objects. For tomography, a tilt-series is taken from a single object (from −70° to +70°). CCD cameras are used for the recording of many images of the same sample without serious radiation damage. The second step is to align all the images coming from the tilt series in order to generate a 3D model of the object. Tomography holds the potential to routinely enable reconstructions that will allow the fitting of intermediate-resolution EM-structures or high-resolution X-ray-structures into the features of the tomography reconstruction.

Metal Shadowing

The angular shadowing of protein samples consists of evaporation under vacuum of a thin film of metal at an angle over a dried sample on support film (**Figure 4(a)**). The sample is mixed with glycerol and then deposited on the surface of a mica sheet. The mica is then introduced into an evaporator and dried. The sample is then tilted for evaporation. Different kinds of metal can be evaporated onto the surface of the sample (titanium, gold, platinum-paladium alloy). Subsequently a carbon film is evaporated onto the mica, which is then floated off on a water bath, much like the methods used for negative staining above, and the metal comes off the mica with the carbon. Different metals will lead to different grain sizes. Differences in

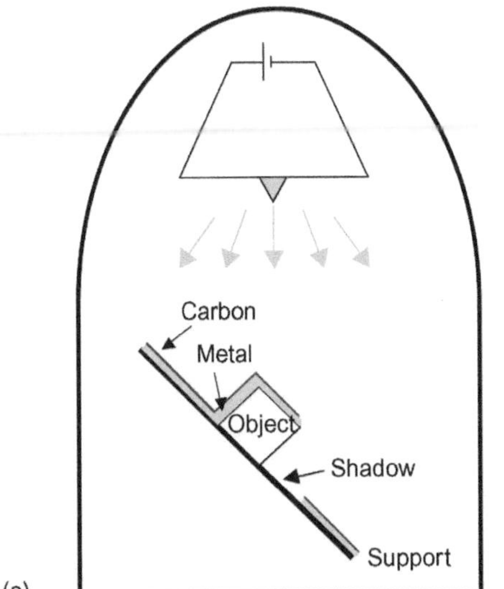

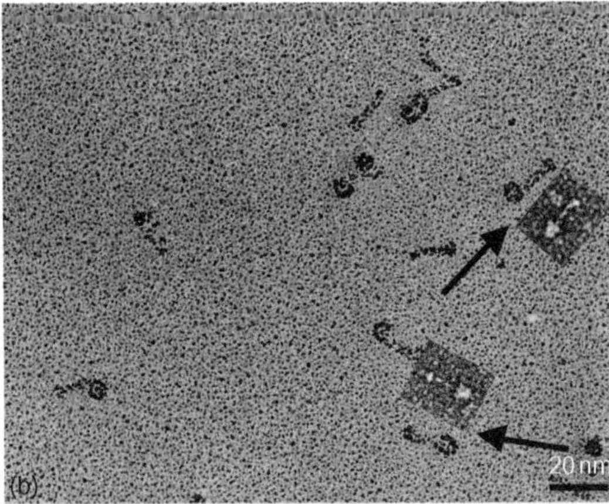

Figure 4 Principle of rotary shadowing. (a) Principle of shadowing. The shadowing process occurs under high vacuum. The specimen is in a tilted position. At the top, the metal to be evaporated is shown in yellow. It is heated until evaporation by a high-intensity current and will be deposited onto the sample. For rotating shadowing, the specimen is not fixed in a tilted position but rotated while keeping the same tilt angle. (b) Human adenovirus 3 penton base was rotatively shadowed using platinum. For comparison the same object negatively stained using sodium silicotungstate has been placed on the same figure (insets are arrowed).

sample height will lead to differences in metal deposit and, hence in contrast (see **Figure 4(b)**). This technique is the only technique that can determine the handedness of a helix and distinguish, for example, between icosahedral 7*d* or 7*l* surface lattices. Rotational shadowing is based on the same principle as angular shadowing but the sample is rotated during evaporation (a fixed angle of evaporation is kept). This technique is especially helpful for the visualization of small proteins or nucleic acids (**Figure 4(b)**). Note that the dimensions of shadowed proteins are difficult to interpret because the metal shadow will add significantly to the thickness of the protein.

Preparation of Cells or Very Large Particles for EM Observation

One of the most interesting aspects of virus studies is the analysis of virus infection, trafficking, and particle formation in cells. Unfortunately, cells and tissues are generally too thick to be directly observed by EM and they first have to be sectioned.

Thin and Ultrathin Sectioning

Classical thin sectioning

The cells or tissues are first stabilized by chemical fixation (usually with aldehydes such as formaldehyde or gluteraldehyde) in order to immobilize the protein and nucleic acid components of the cell (**Figure 5**). Subsequently, fixation with osmium tetroxide will immobilize and stain the membranes of the cell. Note that all these fixatives are very toxic. Then the material is dehydrated in ethanol or acetone and embedded in plastic. Ultrathin sections (60 nm) cut with diamond knives using an ultramicrotome are floated on water, transferred to specimen support grids and examined in the transmission electron microscopy (TEM). Before observation, the sections are further contrasted with uranyl acetate and lead citrate. There are many different types of plastic resins. The hardest resins (epoxy) will best preserve the ultrastructure of the cells during sectioning but will not allow antibody labeling of sections because the antibodies have no or little access to their antigens. There exist soft or spongy resins (Lowicryl) that do permit antibody labeling but these resins are not hard enough to guarantee ultrastructure conservation.

Application: Antibody labeling

Successful immunocytochemistry is only possible when antibodies (or other affinity markers) are able to bind to their target ligands. For extracellular antigens this is not a problem, but for molecules inside cells the markers must be able to access the antigens. For EM it is important to use preparation methods that open the cells and allow introduction of antibodies without destroying the normal cellular organization. The antigens must also be preserved during the different sectioning steps.

The cryo-section technique is now one of the two most important techniques for subcellular immunocytochemistry. Specialized resin formulas have been developed specifically for use in immunocytochemistry. These include the Lowicryl resins, LR White and LR

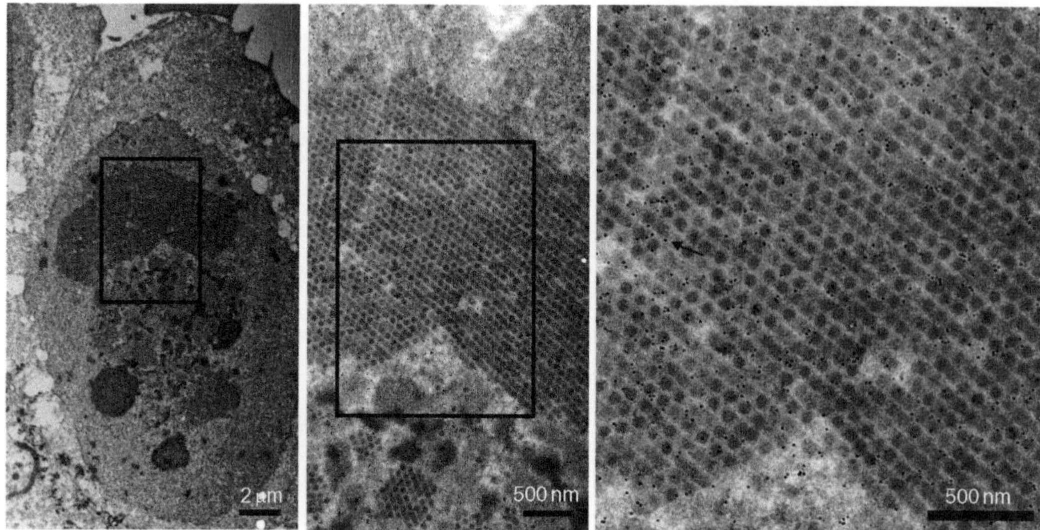

Figure 5 Thin sectioned, adenovirus infected cells. Going from left to right one can see an increase of magnification going from a general view of the infected nucleus to clearly recognizable particles. On the right, the infected cells have been labeled with anti-adenovirus-hexon (major capsid component) antibody coupled with colloidal gold (arrow). Cells were fixed with 2.5% glutaraldehyde in 100 mM HEPES buffer pH 7.4, post-fixed with 1% osmium tetraoxide and dehydrated with ethanol and then embedded in Epon prior to slicing. Grids were stained with saturated uranyl acetate in 50% ethanol and then with 1 M lead citrate.

Gold, Unicryl, and MonoStep. They all have low viscosity and can be polymerized, at low temperatures, with ultraviolet light.

In practice, this method is used with electron opaque markers, which bind to the antibodies and show their location within the cell. In this way, subcellular antigens recognized by antibodies can be localized directly with TEM. Colloidal gold coupled to protein A (a protein from bacterial cell walls that binds to the Fc portion of antibodies) has been used extensively in recent years to localize antibodies on resin and frozen sections of biological materials (**Figure 5**). The ability to produce homogeneous populations of colloidal gold with different particle sizes enabled the use of these probes to co-localize different structures on the same section.

Cryo-sectioning

As for single particles, it is possible to freeze biological material fast enough to vitrify the water present inside the cells. Vitrification of water occurs when the freezing is so fast that ice crystals have no time to form. Vitrified biological material can be sectioned at low temperatures. Because no drying occurs and no salt has been added, cryo-sections better represent the native state of cells although the contrast in the sample is low. Presently, there are seven main rapid freezing methods possible:

1. Immersion freezing – the specimen is plunged into the cryogen.
2. Slam (or metal mirror) freezing – the specimen is impacted onto a polished metal (usually copper) surface cooled with liquid nitrogen or helium.
3. Cold block freezing – two cold, polished metal blocks attached to the jaws of a pair of pliers squeeze-freeze the specimen.
4. Spray freezing – a fine spray of sample in liquid suspension is shot into the cryogen (usually liquid propane).
5. Jet freezing – a jet of liquid cryogen is sprayed onto the specimen.
6. Excision freezing – a cold needle is plunged into the specimen, simultaneously freezing and dissecting the sample.
7. High-pressure freezing – freezing the specimen at very high pressure avoids the expansion of water that occurs during freezing.

CEMOVIS: Cryo-EM of vitreous sections

Dr. Jacques Dubochet in Lausanne has developed the cryo-EM of vitreous sections (CEMOVIS) technique. This technique which consists of cutting very thin sections of a vitrified sample requires good vitrification of the sample and good cutting. Progress in these two domains has been of crucial importance. The first one is high-pressure freezing, which made it possible to increase the thickness of a typical biological sample that can be vitrified from 10 to 100 µm. The second progress came from a better understanding of the cutting process which allowed optimization of the cutting. It is now possible to obtain reproducibly ribbons of homogeneous, 50 nm thick sections in which biological structures are remarkably preserved. It is also possible to observe for the first time, at a resolution approaching 2 nm, the ultrastructure of cells and tissues in their native state.

The next step using CEMOVIS is to combine it with electron tomography. Recording a large number of tilted images (typically 100) should allow the calculation of a 3D model of the specimen to a resolution approaching 4 nm. This is a real challenge for studying virus replication.

Image Analysis

As mentioned above, in tomography, about a hundred images of the same particle are combined in a computer in order to calculate a 3D reconstruction of a single particle. The resolution obtained is presently around 30 Å. However, computer analysis of images is also needed for cryo-EM images because the contrast in cryo EM is too low. The images of thousands of particles can be combined to obtain a high-contrast model of the particle provided that all particles have the same composition and conformation. The resolution of models derived with this method can be as good as 7 Å for large protein complexes or intact virus particles. A large number of mathematical approaches are available to calculate such 3D

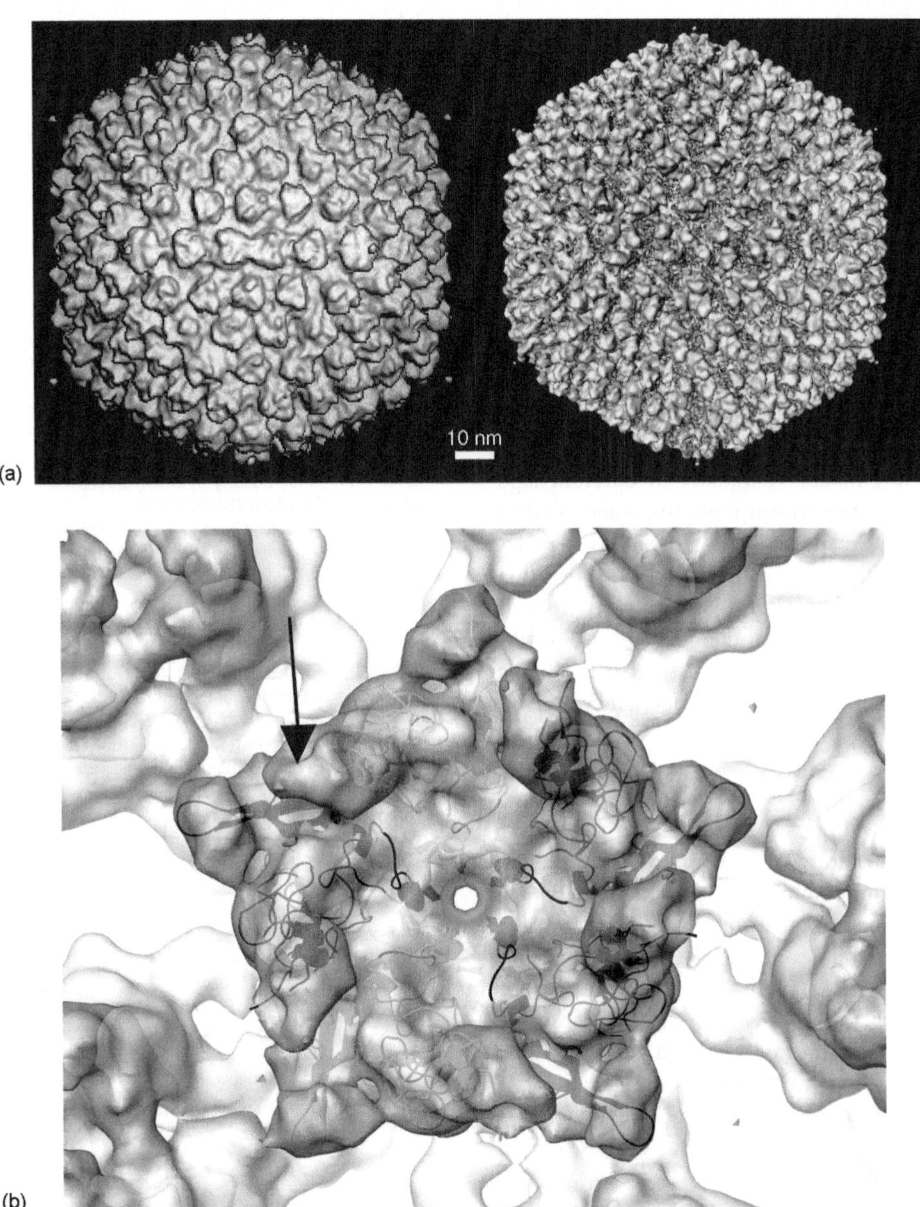

Figure 6 Image analysis and combination with other techniques (a) 3D reconstruction of adenovirus 5 particle after imaging in negative staining (left) and cryo-EM (right). Respectively 150 and 3000 individual negatively stained or cryo-EM images were combined to obtain 3D reconstructions at 35 Å (negative staining) and 10 Å (cryo-EM) resolution. (b) Combination of EM with X-ray crystallography. Fitting of the X-ray structure of the adenovirus penton base into the EM density. The arrow highlights a mobile part of the protein that is resolved in electron microscopy but not in X-ray crystallography.

reconstructions. Different approaches exist to calculate models of nonsymmetrical complexes, for helical structures, and for icosahedral virus capsid structures. These techniques requires classical EM photographs to be digitized by scanning the negatives or the images have to be directly recorded in the microscope using a CCD camera.

Figure 6 shows 3D reconstructions of adenovirus type 5 particles from negatively stained particles and from cryo-EM images. The two models are identical but the cryo-model has a higher resolution (around 10 Å compared to 35 Å for the negative stain model). Because the contrast in the negatively stained images is high, such a model from negatively stained particles can be used as starting model for the cryo-reconstruction. The major difference between the two models is that in the model derived from the stained particles only the outside of the virus capsid is visualized, whereas for the cryo-model all proteins inside and outside of the capsid are visible. The cryo-EM model of the capsid was combined with the atomic resolution data derived from the crystal structure of the two main adenovirus capsomers, the trimeric hexon and the pentameric penton base. By fitting the crystal structures into the EM density, a quasi-atomic model can be derived that describes the exact position of all capsomers and the interacting surfaces and protein loops between the capsomers. **Figure 6** shows details of the fit of the crystal structure of the penton base into a penton base from the EM model. A number of different algorithms are available to perform these fits, some of which take into account the possibility of flexibility between protein domains. The fit in **Figure 6** (arrow) shows density in the EM model that is not filled by density from the crystal structure. Some loops of the molecule are flexible and therefore invisible in the crystal structure, whereas the lower resolution of the EM model does allow imaging.

The Future

In our opinion, the future of EM in virology lies in an extended use of the combination of structural data as shown above. Fitting crystal structures of intact proteins or domains into protein complexes (viral complexes or complexes between viral and host proteins) will show how proteins interact in the infected cell. Although the EM models have a resolution of only around 10 Å, the information on the interface of the proteins in the complex is of near-atomic resolution. These techniques can be applied to regular viruses. For irregular viruses, such as influenza and paramyxoviruses, tomography will allow the construction of models with a resolution of around 40 Å that will show how the proteins in the virus particle interact. In particular, it will be interesting to see the packaging of the nucleocapsids in these viruses. At present we do not know how the DNA is packaged in adenovirus particles and it is possible that the icosahedral averaging techniques result in a loss of this information if the packing symmetry of the DNA is in fact not icosahedral. The same problem exists for capsid features that are not icosahedral such as unique capsid vertices that may be used for the packaging of DNA in adenoviruses and herpesviruses. Such problems may be overcome by using reconstruction techniques that are not based on icosahedral symmetry.

See also: Structure of Non-Enveloped Virions; Principles of Virion Structure.

Further Reading

Al-Amoudi A, Dubochet J, and Studer D (2002) Amorphous solid water produced by cryosectioning of crystalline ice at 113 K. *Journal of Microscopy* 207: 146–153.

Al-Amoudi A, Chang JJ, Leforestier A, *et al.* (2004) Cryo-electron microscopy of vitreous sections. *EMBO Journal* 23: 3583–3588.

Chiu W, Baker ML, Jiang W, Dougherty M, and Schmid MF (2005) Electron cryomicroscopy of biological machines at subnanometer resolution. *Structure* 13: 363–372.

Dubochet J, Adrian M, Chang JJ, *et al.* (1988) Cryo-electron microscopy of vitrified specimens. *Quarterly Reviews of Biophysics* 21: 129–228.

Fabry CM, Rosa-Calatrava M, Conway JF, *et al.* (2005) A quasi-atomic model of human adenovirus type 5 capsid. *EMBO Journal* 24: 1645–1654.

Griffiths G, McDowall A, Back R, and Dubochet J (1984) On the preparation of cryosections for immunocytochemistry. *Journal of Ultrastructure Research* 89: 65–78.

Grunewald K and Cyrklaff M (2006) Structure of complex viruses and virus-infected cells by electron cryo tomography. *Current Opinion in Microbiology* 9(4): 437–442.

Harris JR (1997) *Royal Microscopy Society Handbook 35: Negative Staining and Cryoelectron Microscopy.* Oxford: Bios Scientific Publishers.

Sartori N, Richter K, and Dubochet J (1993) Vitrification depth can be increased more than 10-fold by high pressure freezing. *Journal of Microscopy* 172: 55–61.

Tokuyasu KT (1986) Application of cryoultramicrotomy to immunocytochemistry. *Journal of Microscopy* 143: 139–149.

Tokuyasu KT and Singer SJ (1976) Improved procedures for immunoferritin labeling of ultrathin frozen sections. *Journal of Cell Biology* 71(3): 894–906.

Wepf R, Amrein M, Burkli U, and Gross H (1991) Platinum/iridium/carbon: A high-resolution shadowing material for TEM, STM and SEM of biological macromolecular structures. *Journal of Microscopy* 163: 51–64.

Cryo-Electron Microscopy

W Chiu and J T Chang, Baylor College of Medicine, Houston, TX, USA
F J Rixon, MRC Virology Unit, Glasgow, UK

Introduction

X-ray crystallography has yielded atomic models of icosahedral viruses of up to 700 Å diameter (e.g., bluetongue virus, reovirus, and bacteriophage PRD1). However, crystallography has not been successful with larger viruses, is not yet able to yield high-resolution details of nonicosahedral components in intact virus particles, and is of limited use in examining partially purified or low-concentration specimens or samples with mixtures of different conformational states. Many of these limitations can be overcome by using cryoelectron microscopy (cryoEM) and image reconstruction approaches. Furthermore, cryoEM has been useful not only to virologists for understanding virus assembly and virus–antibody and virus–receptor interactions, but also to electron microscopists for driving developments in electron imaging technology.

In the early 1970s, DeRosier, Crowther, and Klug introduced the method for icosahedral particle reconstruction, which marked the beginning of three-dimensional (3-D) electron microscopy for spherical viruses. Initially, these studies used stained and dried specimens. However, in the early 1980s, Dubochet and co-workers developed vitrification methods that use rapid freezing of solutions to preserve the intact structures of biological macromolecules for electron imaging. They demonstrated the feasibility of reconstructing the 3-D structure of a virus particle in its native conformation without using stain or fixative.

Both the quality of the electron images and the resolving power of the single-particle reconstruction algorithms have been continuously improved in the last 20 years to the extent that it is now possible to routinely obtain virus structures at subnanometer resolutions (6–10 Å). Although not yet equal to X-ray crystallographic resolution, this does allow identification of secondary structure elements such as long α-helices and large β-sheets in protein subunits of virus particles.

All the subnanometer resolution cryoEM structures of viruses solved so far, have made use of the icosahedral properties of the particle. However, the assembly and infection processes in many viruses depend on structures which break the icosahedral symmetry of the particle. Details of the organization of these nonicosahedral components are absent from almost all crystal structures of virions due to the icosahedral averaging used in the data processing. Similarly, cryoEM reconstructions from single-particle images that are dependent on icosahedral averaging do not show the nonicosahedral components.

Recently, image reconstruction techniques have been introduced to compute a density map without imposing icosahedral symmetry. This approach has been successfully used to visualize all the structural components of bacteriophages T7, Epsilon15, and P22 at moderate resolution (17–20 Å).

Another imaging method called electron cryotomography (cryoET) has been used to reconstruct large enveloped viruses such as human immunodeficiency virus (HIV), vaccinia virus, and herpes simplex virus (HSV) at 100–60 Å resolution. Subsequent single-particle averaging of certain computationally extracted components has yielded their 3-D structures to 50–30 Å resolution. This approach is ideal for obtaining low-resolution information from virus particles containing both conformationally constant and variable components.

CryoEM of Single Virus Particles

CryoEM of single particles consists of four basic steps: cryo-specimen preparation, imaging in an electron cryomicrosope, image reconstruction, and structural interpretation. Because there are excellent reviews that describe the detailed procedures involved, in this article the focus is on the fundamental concepts of this methodology, including some practical considerations where appropriate.

Cryo-Specimen Preparation

Specimens in the electron microscope have to be examined in a high vacuum (10^{-6} torr) environment, which would dehydrate and thus destroy the native structure of any normal biological specimen. Fortunately, the rapid freezing methods developed by Dubochet and co-workers can preserve specimens embedded in a thin layer of vitreous ice. Using this procedure, the virus particles are maintained in their original solution state. Vitrification is the process by which water solidifies without undergoing crystallization and it requires extremely rapid cooling. This can be achieved by applying the sample to a microscope grid that has been overlaid with a thin film of holey carbon, blotting away excess sample, and plunging the very thin aqueous film into liquid ethane that has been cooled to liquid nitrogen temperature. An ideal situation is for the ice to be slightly thicker than the virus particles. In practice, there is always variation in ice thickness even across a single microscope grid. For any new specimen, a process of systematic trial and error will eventually identify freezing conditions that yield optimal ice thickness in at least some parts of a grid.

An important practical consideration in cryo-specimen preparation is the concentration of the sample, which ideally should be in the order of 10^{10} particles ml^{-1}. A subnanometer-resolution icosahedral reconstruction generally needs images of several thousands of particles in different orientations. Therefore, a high particle concentration will reduce the number of micrographs that have to be collected, which will, in turn, reduce the number of data sets to be processed. An added benefit is that having a large number of particles per micrograph facilitates the determination of the defocus value of each micrograph, which is necessary for subsequent data processing. However, such concentrations can be difficult to obtain with some virus samples, in which case the necessary data collection and processing can become very time consuming.

Low-Dose Single-Particle Imaging

Following sample preparation, the grids containing the ice-embedded virus particles can be stored in liquid nitrogen until placed into the electron microscope for imaging. Throughout all stages of the process, the frozen-hydrated specimens must be handled extremely carefully to avoid any ice contamination from moisture in the atmosphere. The specimen is kept at or below liquid nitrogen temperature during the electron microscopy session. Frozen-hydrated biological specimens are extremely sensitive to radiation and must be scrutinized using a very low electron dose. They are examined initially at low magnification (100–1000×) to find a suitable area on the grid and then images are recorded at a higher magnification (40 000–80 000×) from the chosen area (**Figure 1**).

The total electron dose that can be tolerated by a specimen depends only slightly on the specimen temperature and the accelerating electron voltage. For most practical purposes, a typical dose is about 15–25 electrons $Å^{-2}$ for single-particle imaging aimed at a resolution of better than 20 Å. A major variable in recording any electron micrograph is the defocus setting for the objective lens, which will vary according to the targeted resolution of the experiment and the type of microscope used. Higher defocus increases image contrast while smaller defocus yields higher-resolution information. Typically, the data for reconstruction are collected with defocus values between 1 and 3 μm in order to produce visible

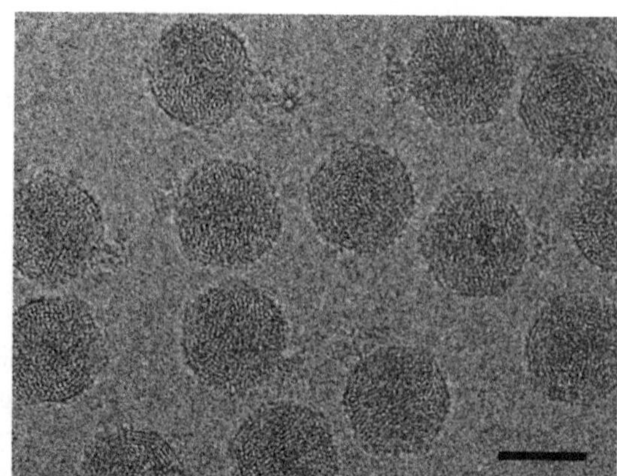

Figure 1 CryoEM imaging. Electron image of frozen-hydrated Epsilon15 phage taken at 300 kV displays the capsids as dark circles. The viral DNA inside is observed as a fingerprint pattern. Scale = 500 Å. Courtesy of Joanita Jakana, Baylor College of Medicine.

image contrast in the raw micrograph while preserving the higher-resolution features retrievable through image processing.

There are two ways to record an electron micrograph: on photographic film or with a charge-coupled device (CCD) camera. The advances in CCD camera technology now enable most medium-resolution data (down to 6–8 Å) to be recorded directly in digital format ready for subsequent steps of image processing. Given the rapid growth in CCD technology, it is likely that photographic film will eventually be replaced completely.

Icosahedral Reconstruction

Many viruses are composed of spherical protein shells arranged icosahedrally. **Figure 1** shows a representative cryoEM image of this type of particle. It is important that the particles are oriented randomly in the matrix of embedding ice because the task of image reconstruction involves combining particles from different orientations into a 3-D density map. The separate steps in image reconstruction are outlined in **Figure 2** and described briefly below. To simplify the technical aspects of the procedure, numerous image-processing software packages that can be used for icosahedral reconstruction have been developed.

Initially, each micrograph captured from a CCD camera or digitized from photographic film is evaluated for its image quality based on the power spectrum derived from the average of multiple Fourier transforms of computationally isolated particle images. This power spectrum contains information about the structure of the particles, the imaging conditions used, and the image quality. The ring pattern in the power spectrum (**Figure 3(a)**) reflects the imaging condition known as the contrast transfer

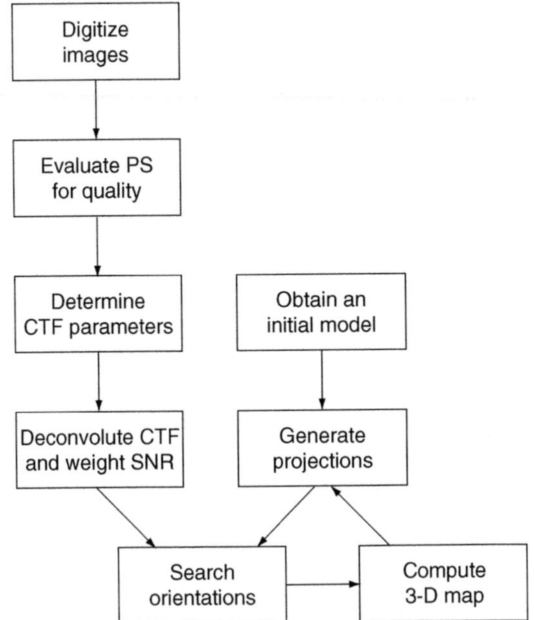

Figure 2 A 3-D image reconstruction scheme for icosahedral particles. The quality of digitized images is evaluated by analyzing their power spectrum (PS). Next, the contrast transfer function (CTF) parameters are determined for each image and used to deconvolute and weigh them according to their signal-to-noise ratio (SNR). The orientations of the deconvoluted images are determined by comparing them against projections from a 3-D map generated from a starting model. The oriented images are used to reconstruct a better 3-D map, which is subsequently used to generate better projections. The new projections are used to improve the accuracy of the search for the image orientations. The 3-D map improves after each iteration until it converges and the processing can stop.

function (CTF). It can vary in appearance according to factors such as defocus, astigmatism, specimen drift, and signal decay as a function of resolution, all of which are attributable to various instrumental factors. The signal-to-noise ratio (SNR) in each micrograph can be computed from the circularly averaged power spectrum. The SNR plot (**Figure 3(b)**) indicates the potential signal strength at different resolutions and gives an indication of the ultimately achievable resolution in the reconstruction.

In order to obtain a correct structure, each raw particle image has to be deconvoluted by its CTF and be compensated for the signal decay at high resolution (**Figure 2**). The image deconvolution and signal-to-noise weighing are usually carried out for each raw particle image before the orientation of the particles is determined. Following orientation determination, the corrected particle images are merged to synthesize the 3-D map. Alternatively, the deconvolution can be carried out on the preliminary 3-D maps computed from each micrograph, with the deconvoluted maps being merged subsequently to produce the final map.

The most computationally demanding aspect of image processing is the estimation and refinement of the five spatial parameters (three for rotations around *x*-, *y*-, and *z*-axes, and two for center in *x*- and *y*-directions) for each particle image (**Figure 2**). There are a number of different algorithms and software packages that can be used to perform this critical task. Most of these programs require an initial model of the particle, from which they generate multiple projection images showing the particle model in different orientations. These projection images are then compared with each of the raw particle images. The

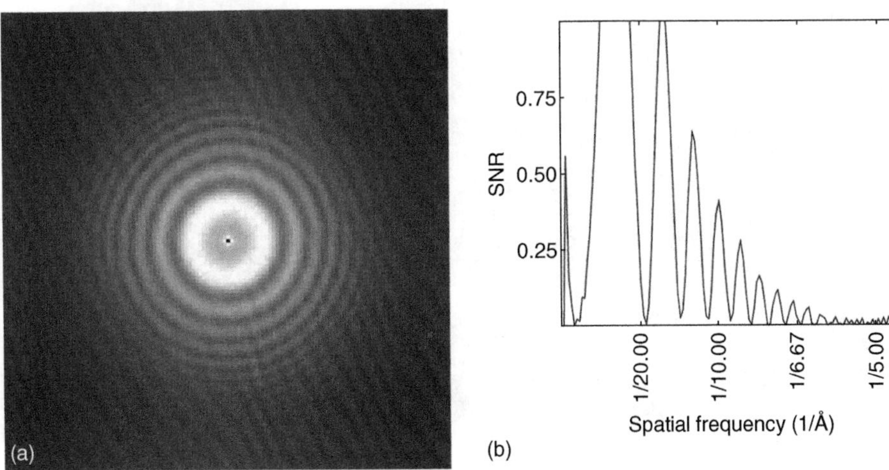

Figure 3 Evaluating image quality. (a) The power spectrum derived from averaging the Fourier transforms of particles boxed out from electron images such as those in **Figure 1**. (b) SNR plot of the boxed particle images as a function of spatial frequency (resolution) computed from the circularly averaged power spectrum. For a conservative measure of 10% SNR, the usable data extends beyond 8 Å resolution. Courtesy of Donghua Chen, Baylor College of Medicine.

projection image that gives the closest match is assumed to have the same orientation as the virus particle that produced the raw image. Each algorithm uses different criteria for comparing the reference and raw particle images to derive the spatial parameters. Throughout a series of iterative refinement steps, the model is continuously updated and its resolution is gradually increased (**Figure 2**).

During the steps of orientation and center determination, and 3-D reconstruction, the algorithms assume that the particle has icosahedral symmetry. This means that the particle is made up of 60 repeated (asymmetric) units, which are related to each other by five-, three-, and twofold symmetry operations. If the particle does not have the expected symmetry, the algorithm will not yield a consistent map and will not be able to derive its structure. On the other hand, it is possible that in some particles the icosahedral symmetry is preserved only up to a certain resolution, which would limit the ultimate resolution of the reconstruction. This is analogous to a protein crystal only diffracting up to some resolution limit beyond which the crystal structure cannot be resolved.

Interpreting CryoEM Single-Particle Density Maps

The challenge in interpreting a cryoEM density map varies according to its resolution. Maps of different resolutions reveal different structural features of the icosahedral shell proteins (**Figure 4**). A low-resolution map (down to 30 Å) can yield the size, shape, and organization (triangulation number) of the icosahedral particle. In fact, cryoEM is probably the most reliable technique for providing such information. In a moderate resolution map (20–10 Å), the subunit boundaries may begin to be resolved. The segmentation of the map into subunits can be done qualitatively by visual inspection of the structure in front of a graphics terminal or quantitatively by a software program. The accuracy of the segmentation is ultimately dictated by the resolution of the map.

When a cryoEM density map reaches subnanometer resolution, long α-helices and large β-sheets become visually recognizable and can be assessed quantitatively (**Figure 5**). In intermediate resolution (7–9 Å) density

Figure 4 CryoEM maps of Epsilon15 phage at different resolutions. The surface-rendered images of the whole capsid (top row) and a close-up of the asymmetric unit (bottom) show that (a) at 40 Å resolution, the overall morphology and triangulation number can be distinguished; (b) at 20 Å resolution, it is possible to visualize the subunit separation in the hexons and pentons; and (c) at 9.5 Å resolution, α-helices are discernible as rod-like densities (EMD-1176). The display was colored radially.

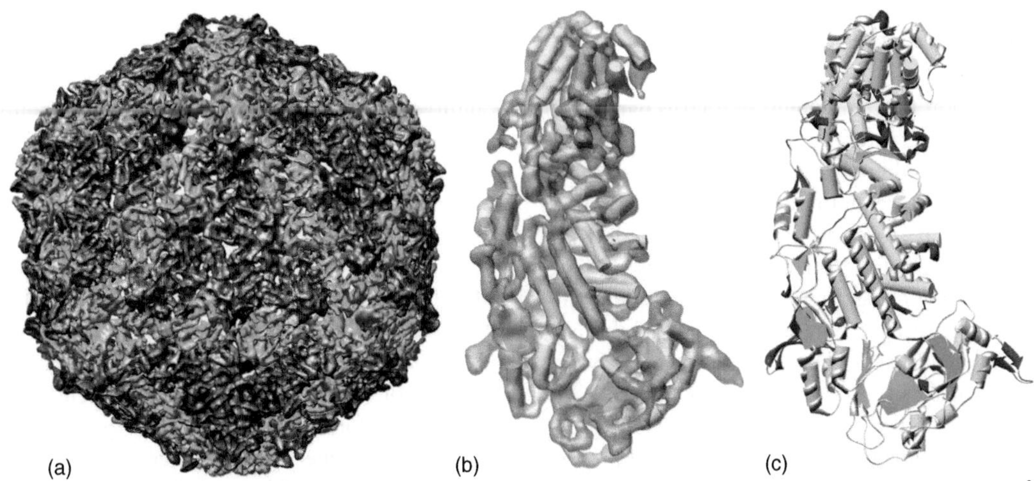

Figure 5 CryoEM map of rice dwarf virus matches the crystal structure. (a) The inner capsid map segmented from a 6.8 Å resolution cryoEM reconstruction of the entire virus, with the two different conformations of the capsid protein P3A and P3B colored green and violet, respectively (EMD-1060). (b) Segmented inner capsid shell protein P3A with predicted secondary structure elements annotated as cylinders (α-helices) and surfaces (β-sheets). (c) Overlaying the crystal structure (ribbons, 1UF2) onto the secondary structure prediction shows a good match. (a) Reproduced from Zhou ZH, Baker ML, Jiang W, *et al.* (2001) Electron cryomicroscopy and bioinformatics suggest protein fold models for rice dwarf virus. *Nature Structural Biology* 8: 868–873. (c) Reproduced from Baker ML, Ju T, and Chiu W (2007) Identification of secondary structure elements in intermediate-resolution density maps. *Structure* 15: 7–19, with permission from Elsevier.

maps, it has been shown that it is possible to correctly identify and annotate nearly all of the helices with $\geq$2.5 turns and sheets with $\geq$2 strands. For example, the match (**Figure 5(c)**) between the secondary structural elements determined from the 6.8 Å cryoEM map of the rice dwarf virus and the crystal structure that was obtained later demonstrated the reliability of the cryoEM reconstruction and of the structural feature identification.

Asymmetric Reconstruction of an Icosahedral Particle with a Unique Vertex

Though icosahedral reconstruction methods have been very successful in providing structural information on capsid shell proteins, they reveal nothing about any non-icosahedral components. Recently, however, several image-processing procedures have emerged for reconstructing spherical virus particles with tail structures without imposing symmetry. Roughly speaking, the reconstruction procedure uses an initial icosahedral reconstruction of the particle as a starting model for the asymmetric reconstruction. Because only one vertex of the virus particle has the additional mass of the tail, a crude model can be generated by computationally adding a cylinder of density at one of the fivefold vertices of the icosahedral reconstruction. This tailed model is projected in each of the 60 icosahedrally equivalent views and compared with the raw particle image to identify the view that produces the best match. Once the orientations of all the images have been determined, they are used to reconstruct the map without imposing any symmetry. This process of projection matching is iterated using a constantly updated reconstruction, until the cylindrical symmetry of the tail is broken, revealing its true structure. **Figure 6** shows the asymmetric reconstruction of Epsilon15 phage where, in addition to the capsid shell, multiple nonicosahedrally ordered molecular components and the internal DNA can be visualized.

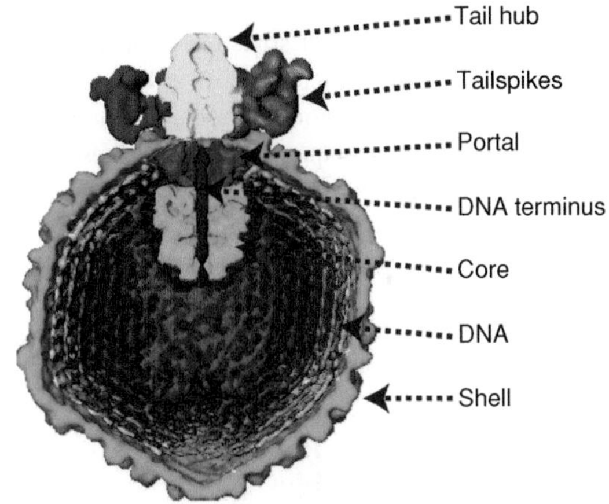

Figure 6 Asymmetric reconstruction of Epsilon15 phage. The map reveals all the structural components of the capsid shell, DNA, core, portal, tailspikes, and tail hub (EMD-1175). Reproduced from Jiang W, Chang J, Jakana J, Weigele P, King J, and Chiu W (2006) Structure of Epsilon15 phage reveals organization of genome and DNA packaging/injection apparatus. *Nature* 439(7076): 612–616.

Modeling Virus Particles with CryoEM Density Map Constraints

It is not uncommon for components of a virus particle to be crystallized and structurally determined while only a low- or medium-resolution cryoEM map of the entire virus particle is available. In these cases, a pseudo-atomic model of the virus particle can often be built by docking the crystal structures of the solved components into the cryoEM map. The docking can be done visually or quantitatively using a fitting program. Examples of this are provided by adenovirus and herpesvirus. In neither case has the intact capsid been crystallized, but inserting the crystal structures of the adenovirus hexon and penton base proteins, or a domain of HSV-1 major capsid protein into the respective cryoEM structures, has provided us with pseudo-atomic models that reveal important features of the capsids. An outstanding example of how cryoEM and crystallographic data can be merged to establish a pseudo-atomic model is provided by the T4 bacteriophage baseplate (**Figure 7**). Here, the crystal structures of six separate molecular components have been fitted into the reconstruction of the baseplate, leading to a proposal for the mechanism of DNA injection into the host cell.

Where there is no crystal structure available, it may still be possible to construct a set of structural models using either comparative modeling or *ab initio* modeling based on the amino acid sequence alone. In these cases, a cryoEM density map can be used as a template to select the most native-like model. A domain of the VP26 capsid protein of HSV, which has a molecular mass of ~10 kDa, has been modeled *ab initio* using the cryoEM structure of the HSV capsid as a selection constraint. Although the validity of this approach has not yet been confirmed by crystal data, it is encouraging that the model fitting the cryoEM data was consistent with published genetic and biochemical data for VP26.

Low-Dose Cryoelectron Tomographic Imaging of Virus Particles

Particles of some viruses have components that assume different conformations in different particles, making it impossible to generate a 3-D density map by averaging these particles. For this type of particle, the only option available is tomography. While the reconstruction techniques described above work by combining views of multiple particles in different orientations, tomographic reconstruction combines multiple views of the same particle as its orientation is changed by rotating it between consecutive exposures using a tilting microscope holder. Therefore, this technique allows a low-resolution map to be generated by collecting a tilt series from any cryoEM specimen, and it has been used successfully with a range of viruses including HSV virions and capsids, vaccinia virus, and HIV. Because of the constraints on the number of images that can be acquired due to the effects of radiation damage, cryotomography yields only low-resolution data. Therefore, the strategy is to record multiple images of the specimens at low magnification and under low-dose

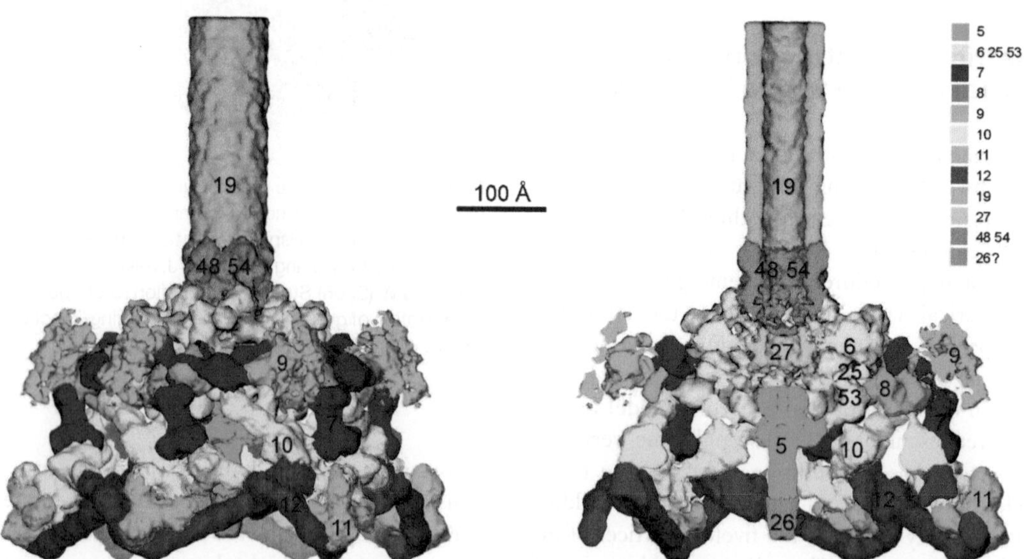

Figure 7 Pseudo-atomic models. Side and cut-away view of the T4 phage tube–baseplate complex. Six crystallized components of the T4 baseplate were docked into the cryoEM map, which permitted identification of the baseplate components (color-coded gene products with legend on right) (EMD-1086). Courtesy of M. G. Rossmann at Purdue University and reproduced with permission from Rossmann MG, Mesyanzhinov VV, Arisaka F, and Leiman PG (2004) The bacteriophage T4 DNA injection machine. *Current Opinion in Structural Biology* 14(2): 171–180.

condition. Typically, a magnification of 10 000–25 000× and a dose of 1–2 electrons $Å^{-2}$ per micrograph are used (**Figure 8**).

Once captured, the tilt series images are aligned and merged to generate a 3-D tomogram. Because the grid supporting the specimen obscures the electron beam, the range of tilts that can be used is limited to about ±70°. As a result, there is a missing wedge of data, which means that each tomogram is subject to distortion and must be interpreted cautiously.

Although tomographic reconstructions are generally of low resolution (100–60 Å), post-tomographic averaging of components with constant conformations can yield higher resolutions. For example, post-tomographic data sorting, mutual alignment, and averaging have been used to generate a model of the herpesvirus capsid with an averaged view of the single portal vertex (**Figure 9(a)**). Similarly, the spike of the simian immunodeficiency virus (SIV) particle has been aligned and averaged to produce a map that permitted docking of the crystal structures of

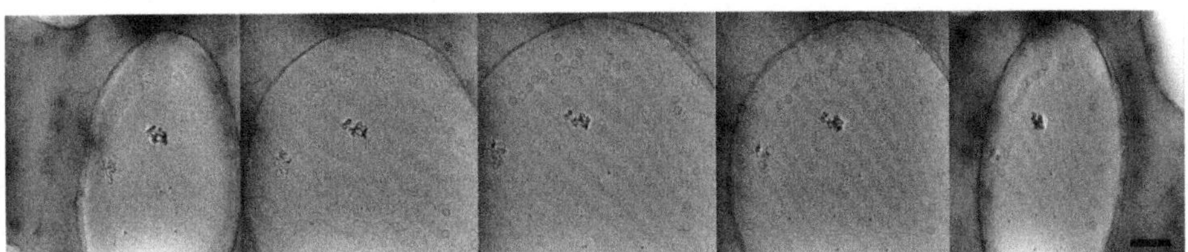

Figure 8 Tomography images. Selected CCD frames recorded at −60°, −30°, 0°, +30°, and +60° tilts (left to right) show herpesvirus capsids forming an arc in the top half of each image. Scale = 5000 Å.

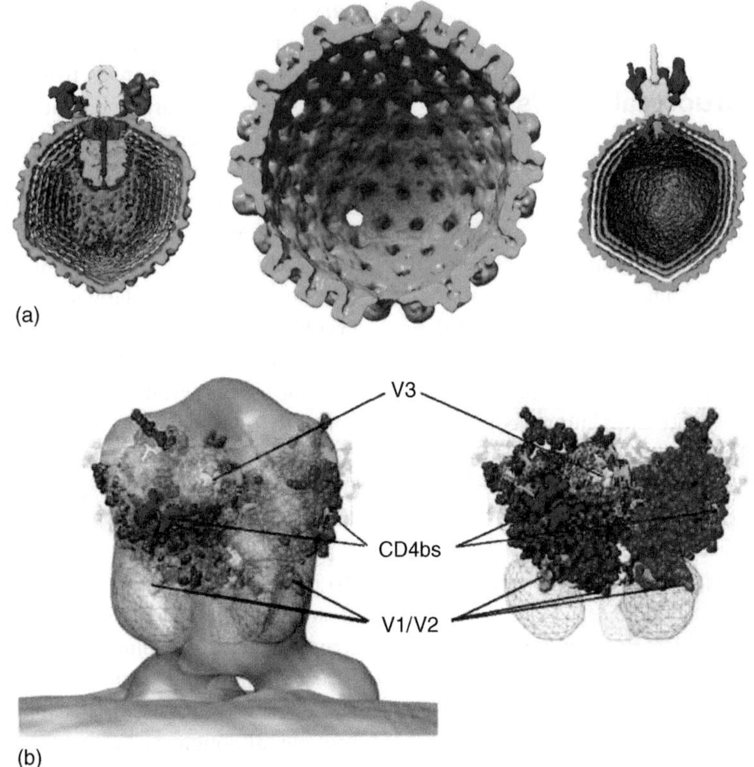

Figure 9 Increasing resolution by averaging subvolumes from tomographic reconstructions. (a) Structure of the herpes simplex virus-1 capsid (center) shown after removal of the pentons to reveal the unique portal vertex (purple). This structure was derived by averaging 13 subvolumes in different tomographic reconstructions (EMD-1308). The arrangement of the portal in the capsid floor is similar to that seen in the tailed double-stranded DNA (dsDNA) phages Epsilon15 (left, EMD-1175) and P22 (right, EMD-1222). (b) Averaging membrane spikes from tomographic reconstructions of SIV particles and then imposing threefold symmetry improves the resulting map, thereby allowing the crystal structure of the spike protein to be fitted. The V1/V2 (orange) and V3 (yellow) loops and CD4 binding sites (red) are labeled in this model. (a) Reproduced from Chang JT, Schmid MF, Rixon FJ, and Chiu W (2007) Electron cryotomography reveals the portal in the herpesvirus capsid. *Journal of Virology* 81(4): 2065–2068, with permission from American Society for Microbiology. (b) Courtesy of K. Taylor and K. H. Roux at Florida State University and reproduced with permission from Zhu P, Liu J, Bess J, *et al.* (2006) Distribution and three-dimensional structure of AIDS virus envelope spikes. *Nature* 441(7095): 847–852.

the spike proteins (**Figure 9(b)**). In this type of approach, the final resolution is limited by a number of factors, including the accuracy of the alignment, the number of independent views of the components included in the averaging, the symmetry of the components, the proper image deconvolution, the correct compensation for the missing wedge data, and the conformational uniformity of the components.

Biological Insights from CryoEM Maps

The extent of biological information that can be obtained from cryoEM reconstruction varies according to the resolution of the map as well as the design of the experiment. As shown in **Figures 4** and **5**, different structural features can be derived from the cryoEM maps at varying resolutions, ranging from shape, size, triangulation number, capsomere morphology and number, quaternary, tertiary, and secondary structural features of individual protein subunits. Here, a small and not exhaustive set of examples is provided to demonstrate the usefulness of cryoEM maps in the context of virology.

Conservation of Structural Motifs among Viruses

Questions regarding the origin of viruses and their distant evolutionary relationships are handicapped by the low sequence conservation generally found in these small and rapidly evolving entities. Therefore, structural comparison has been used as an alternative indicator of their relatedness. The best-known example of structural analysis revealing a link between supposedly unrelated viruses is that of bacteriophage PRD1 and human adenovirus. Although early cryoEM analysis had shown that the PRD1 capsid structure was very similar to that of adenovirus, it was not until the crystal structures of their respective major capsid proteins were compared that the evolutionary relationship between these viruses was firmly established.

However, in cases where crystal structures are not available, cryoEM maps determined at subnanometer resolution can provide sufficient information on the pattern of secondary structure elements and thus of the protein folds. **Figure 10** shows a structural comparison of the capsid proteins of the eukaryote-infecting herpesvirus (*Herpesviridae*) and the prokaryote-infecting tailed DNA bacteriophages (*Caudovirales*). Although these proteins have no detectable sequence homology, it is clear from the similar spatial distributions of secondary structural elements that they share a characteristic fold. CryoEM-derived structural data have also revealed conserved features in the unique portal vertices of the two virus types, including a characteristic subunit arrangement and a similar location within the capsid shell (**Figure 9(a)**). These observations have led to the conclusion that these very distinct, extant viruses must have arisen from the same primordial progenitors.

Structural Polymorphism during Virus Maturation

Many virus particles undergo structural changes during their morphogenesis. For example, tailed bacteriophages typically form their capsid shell around an internal scaffold of proteins. This initial assembly product (the procapsid) then undergoes a maturation process during which the DNA is packaged, the internal scaffold is lost, and the capsid shell undergoes extensive structural reconfiguration. **Figure 11** demonstrates the structural changes in P22 bacteriophage as revealed from reconstructions of the procapsid and of the mature virion. Not only can changes in the overall size and shape of the capsid be observed but also correlative changes in the quaternary and tertiary structures of individual subunits can be noted. For instance, the organization of subunits in the hexon capsomere changes from a skewed shape to a six-fold symmetrical arrangement. In addition to the overall domain movements in the hexon subunit that bring about this change, the cryoEM structures also reveal more detailed changes such as the refolding of an N-terminal α-helix. Large-scale structural changes have also been visualized during maturation of the HSV capsid. In addition to comparing relatively stable forms of a particle, cryoEM can be used to follow rapid structural changes (e.g., by changing the pH of the buffer immediately before freezing) as observed in Semliki forest virus, Sindbis virus, and La Crosse virus.

Viral Genome Organization

To date, most information on how viral nucleic acids are arranged within particles has come from cryoEM reconstructions. Low-resolution cryoEM reconstruction has revealed that small sections of the RNA genomes in some RNA viruses are bound icosahedrally to the capsid shell proteins, although the bulk of the genome was not resolved even in these cases. However, in general, viral genomes are not packaged icosahedrally within particles and so cannot be visualized by icosahedral reconstruction techniques.

Those double-stranded DNA viruses that package their genomes into preformed capsids pose a particularly interesting dilemma because this must be accomplished without organizing proteins (such as histones) to counter the electrostatic charge of the DNA and the organization must allow for efficient entry and ejection without

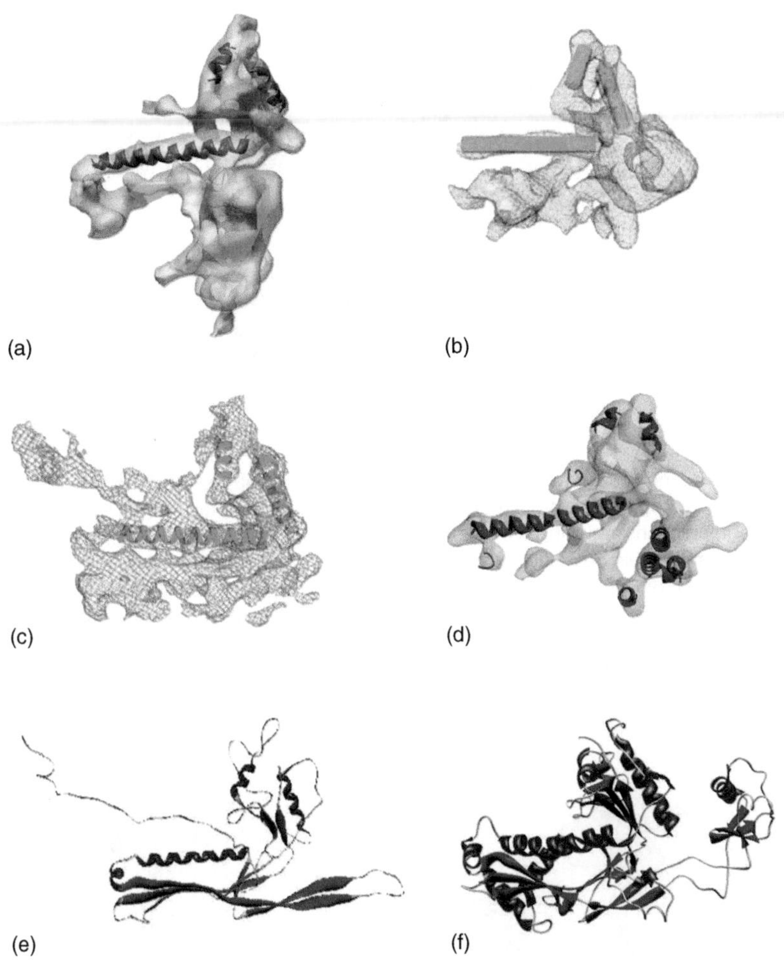

Figure 10 Evolutionary conservation of a structural motif in dsDNA virus particles. CryoEM structure of (a) Epsilon15 phage shell protein, with annotated helices and sheets (EMD-1176), (b) P22 phage coat protein gp5 (EMD-1101), (c) Phi29 phage head protein gp8 (EMD-1120), (d) HSV-1 major capsid protein VP5 floor domain, and crystal structures of (e) HK97 phage head protein gp5 (1OGH) and (f) T4 phage head protein gp24 (1YUE) all have similar structural signatures, which suggest a common ancestry even though they have little sequence similarity. Reproduced from Jiang W, Chang J, Jakana J, Weigele P, King J, and Chiu W (2006) Structure of Epsilon15 phage reveals organization of genome and DNA packaging/injection apparatus. *Nature* 439(7076): 612–616.

forming knots or tangles. The arrangement of the nucleic acids in such capsids has been extensively studied and debated. Proposed models include the co-axial spool, spiral fold, liquid crystal, and folded toroid. CryoEM has led the way in shedding light on this question. Analysis of a bacteriophage T7 mutant that was found to orient preferentially in vitreous ice showed the capsids having either concentric rings (axial view) of DNA spaced about 25 Å apart or punctate arrays (side view) suggesting that the DNA was organized in concentric layers. Subsequently, it became possible to examine the DNA in more detail using asymmetric reconstruction algorithms. In Epsilon15 phage, for example, DNA strands are packed as concentric layers in the capsid cavity suggesting a co-axial spool arrangement (**Figure 12(a)**), and a slice normal to the DNA strands and parallel to the tail axis shows hexagonal features (**Figure 12(b)**). A DNA strand connecting each layer has not yet been observed, but this could be the result of errors in the alignment of the particle images during reconstruction, or due to heterogeneity in where the DNA winding commences. The goal of such studies is to trace the path of the entire nucleic acid molecule within the virion. Asymmetric reconstruction methods at higher resolution are well poised for these determinations.

Virus–Antibody and Cell Receptor Complexes

There is increasing interest in using cryoEM to analyze aspects of virus biology involving interactions with other molecules such as receptors or antibodies. A good example of the value of this approach is provided by work on virus–receptor interactions in the small icosahedral picornaviruses, notably poliovirus and the common cold virus (rhinovirus). In order to recognize and enter their

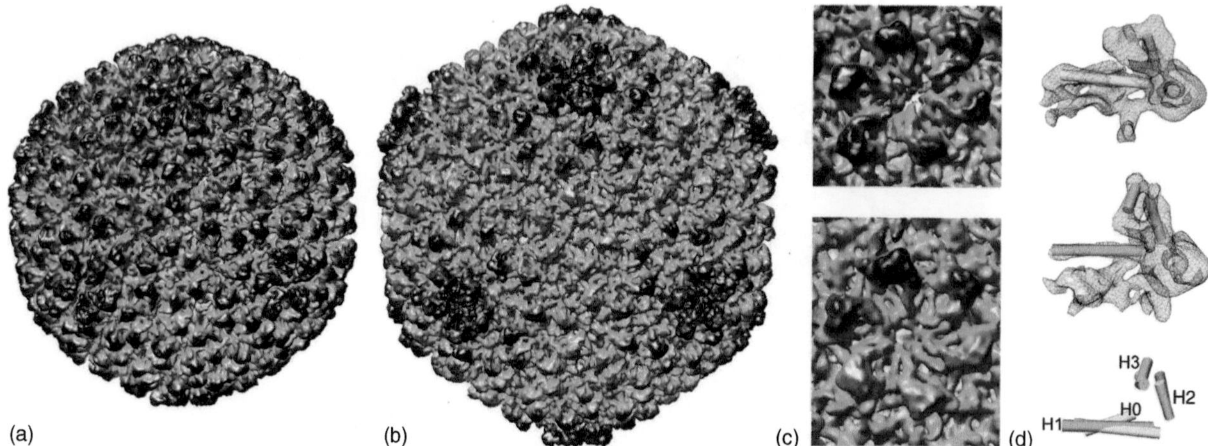

Figure 11 Structural changes during virus maturation. (a) The spherical procapsid of P22 is ~600 Å in diameter, and its hexons are elongated, with a hole in the center. (b) The angular mature capsid is ~700 Å in diameter, and its hexon subunits are arranged hexagonally with more lateral interactions, resulting in a smaller hole (EMD-1101). (c) A close-up view of the procapsid (top) and mature (bottom) hexon. (d) The P22 capsid protein gp5 rearranges when maturing from the procapsid (top, annotated secondary structures in yellow) to the capsid (middle, annotated secondary structures in green), resulting in changes to secondary, tertiary, and quaternary structure, as shown by overlapping the annotated secondary structures (bottom). (d) Reproduced from Jiang W, Li Z, Zhang Z, Baker ML, Prevelige PE, and Chiu W (2003) Coat protein fold and maturation transition of bacteriophage p22 seen at subnanometer resolutions. *Nature Structural Biology* 10:131–135.

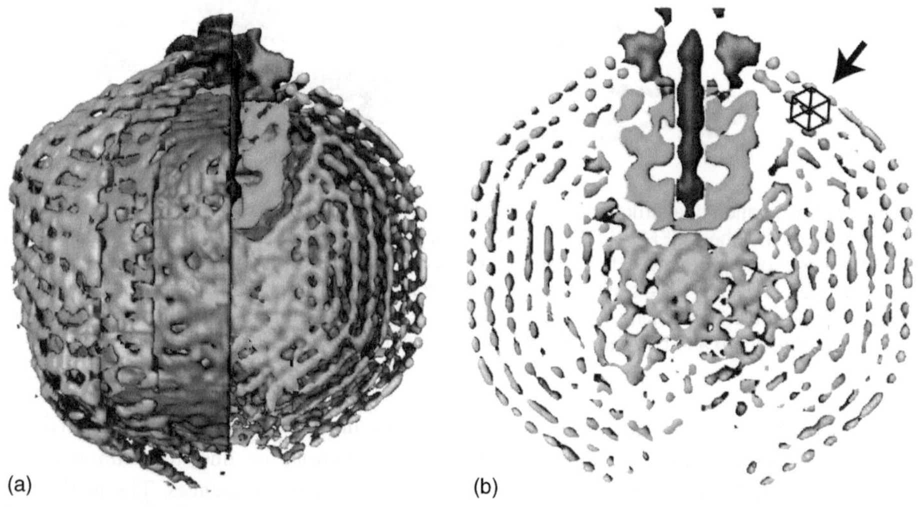

Figure 12 Viral genome organization in tailed dsDNA phages. (a) In Epsilon15 phage the dsDNA (blue) is arranged in concentric layers surrounding a protein core (green) and portal (purple) (EMD-1175). (b) A slice parallel to the tail axis and normal to the dsDNA strands shows hexagonal features (arrow over polygon).

target cells, viruses must usually bind to specific receptor molecules on the cell surface. When the crystal structure of the rhinovirus particle was first determined, it revealed the existence of a depression or 'canyon' around each fivefold axis of the particle, which was proposed as the binding site for the receptor. However, direct confirmation of this proposal was not obtained until icosahedral reconstruction was carried out on cryoEM images of virus particles bound to their receptors. Although virus receptors are components of the cell membrane, these initial binding studies were carried out using soluble fragments of receptor molecules.

Typical eukaryote cells are too large to freeze rapidly by normal methods making it difficult to study virus binding *in situ*. However, by embedding receptor molecules in artificial liposomes that were small enough to vitrify (**Figure 13(a)**), the interaction between the poliovirus particle and its receptor could be visualized in the context of a membrane. Comparison of the micrographs with projection images of an existing icosahedral reconstruction

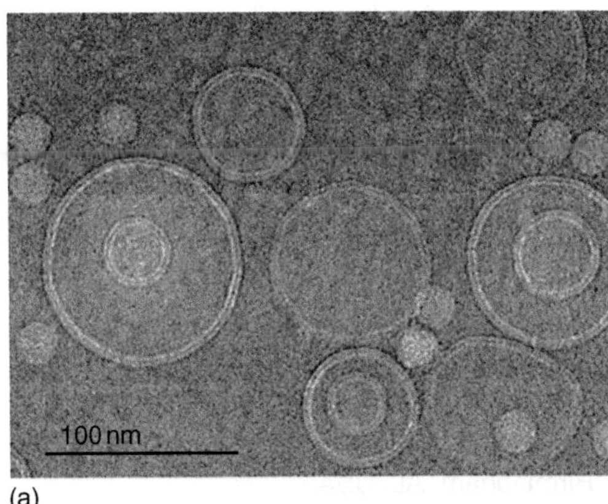

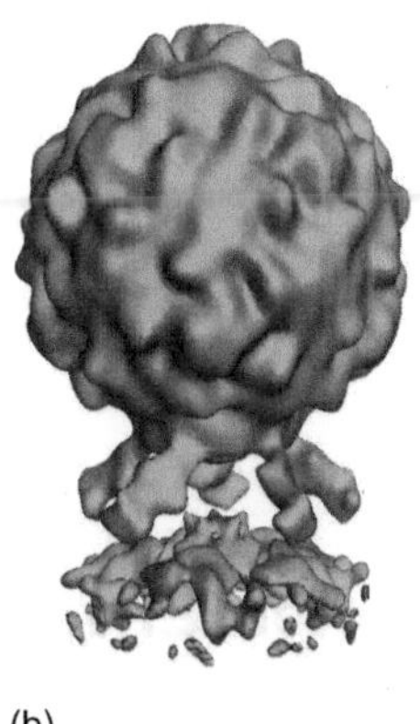

(a) (b)

Figure 13 Virus–receptor interactions. (a) CryoEM image of poliovirus attached to liposomes with embedded receptors. (b) The reconstruction shows the icosahedral capsid binding to five receptors on the membrane, which orients the capsid so that a fivefold vertex is directed toward the membrane. Courtesy of J. M. Hogle at Harvard Medical School and reproduced with permission from Bubeck D, Filman DJ, and Hogle JM (2005) Cryo-electron microscopy reconstruction of a poliovirus–receptor-membrane complex. *Nature Structural and Molecular Biology* 12: 615–618.

allowed the orientation of the virus particles bound to the liposome to be determined. Then, in a process roughly analogous to that used in the asymmetric reconstruction of tailed bacteriophage particles (described earlier), an artificial density was introduced at the point on the membrane closest to the virus particle and the reconstruction was repeated without imposing icosahedral symmetry. The structure that resulted from this procedure showed the virus particle, receptor molecules, and adjacent parts of the liposome membrane (**Figure 13(b)**), thereby demonstrating that the particle adopts a consistent alignment with respect to the lipid membrane. This not only provided information on how the receptor molecules interact with the virus particle in the context of a membrane, but also revealed the presence of a perturbation to the membrane induced by virus binding. This perturbation might represent an early stage in the formation of the pore, which is necessary for release of the viral genome into the cell.

Concluding Remarks

CryoEM is an important technique that has been used to determine the structures of many viruses. In many cases, achieving subnanometer resolutions has revealed new insights into the structural organization and biological properties of viruses. CryoEM results have been complemented by docking crystal structures or by molecular modeling to provide additional information. Furthermore, conformational changes, structural transformations, and virus–receptor interactions have been observed. Even in cases where the sample is pleomorphic, electron cryotomography has revealed important information. Overall, the study of viruses has been enormously enriched by the use of cryoEM.

Acknowledgment

Research support has been provided by National Institutes of Health, National Science Foundation, and the Robert Welch Foundation.

See also: Electron Microscopy of Viruses; Principles of Virion Structure; Structure of Non-Enveloped Virions.

Further Reading

Adrian M, Dubochet J, Lepault J, and McDowall AW (1984) Cryo-electron microscopy of viruses. *Nature* 308(5954): 32–36.

Baker ML, Ju T, and Chiu W (2007) Identification of secondary structure elements in intermediate-resolution density maps. *Structure* 15(1): 7–19.

Baker TS, Olson NH, and Fuller SD (1999) Adding the third dimension to virus life cycles: Three-dimensional reconstruction of icosahedral viruses from cryo-electron micrographs. *Microbiology and Molecular Biology Reviews* 63(4): 862–922.

Chang JT, Schmid MF, Rixon FJ, and Chiu W (2007) Electron cryotomography reveals the portal in the herpesvirus capsid. *Journal of Virology* 81(4): 2065–2068.

Crowther RA, Amos LA, Finch JT, DeRosier DJ, and Klug A (1970) Three dimensional reconstructions of spherical viruses by Fourier synthesis from electron micrographs. *Nature* 226(5244): 421–425.

Dubochet J, Adrian M, Chang JJ, *et al.* (1988) Cryo-electron microscopy of vitrified specimens. *Quarterly Reviews of Biophysics* 21(2): 129–228.

Jiang W, Chang J, Jakana J, Weigele P, King J, and Chiu W (2006) Structure of Epsilon15 phage reveals organization of genome and DNA packaging/injection apparatus. *Nature* 439(7076): 612–616.

Jiang W and Chiu W (2006) Electron cryomicroscopy of icosahedral virus particle. In: Kuo J (ed.) *Methods in Molecular Biology*, pp. 345–363. Totowa, NJ: The Humana Press.

Johnson JE and Rueckert R (1997) Packaging and release of the viral genome. In: Chiu W, Burnett RM, and Garcea RL (eds.) *Structural Biology of Viruses*, pp. 269–287. New York: Oxford University Press.

Lucic V, Forster F, and Baumeister W (2005) Structural studies by electron tomography: From cells to molecules. *Annual Review of Biochemistry* 74: 833–865.

Rossmann MG, He Y, and Kuhn RJ (2002) Picornavirus–receptor interactions. *Trends in Microbiology* 10(7): 324–331.
Rossmann MG, Mesyanzhinov VV, Arisaka F, and Leiman PG (2004) The bacteriophage T4 DNA injection machine. *Current Opinion in Structural Biology* 14(2): 171–180.
Smith TJ (2003) Structural studies on antibody–virus complexes. *Advances in Protein Chemistry* 64: 409–454.
Zhou ZH, Baker ML, Jiang W, *et al.* (2001) Electron cryomicroscopy and bioinformatics suggest protein fold models for rice dwarf virus. *Nature Structural Biology* 8: 868–873.
Zhou ZH, Dougherty M, Jakana J, He J, Rixon FJ, and Chiu W (2000) Seeing the herpesvirus capsid at 8.5 Å. *Science* 288(5467): 877–880.

Assembly of Non-Enveloped Virions

M Luo, University of Alabama at Birmingham, Birmingham, AL, USA

Architecture of Viruses

Nonenveloped viruses have two essential components: protein and nucleic acid. The protein forms a coat called 'capsid' that packages the nucleic acid, which may be DNA or RNA. This complex constitutes a virion or a virus particle. The nucleic acid is the viral genome that encodes all the virus-specific genes required for viral replication. The protein capsid packages the viral genome during replication, and transmits it for the next round of infection. When the virion reaches the host cell, the capsid usually recognizes a specific receptor that helps the virion to enter. Once inside the host cell, the capsid has to release the viral genome so replication can begin. The size of a viral genome is usually limited, so only a few genes can be encoded. It is more efficient if only one or a small number of genes encode for capsid proteins that can self-assemble into a complete shell by use of many copies of the same proteins. The assembly of the capsid proteins follows a specific type of symmetry that allows a small protein unit to assemble into a large particle. The protein capsid can have a helical (filamentous virus) or icosahedral (spherical virus) symmetry. Helical symmetry is described by the diameter, d, the pitch, P, and the number of subunits per turn. There are as many capsid proteins as necessary for completely covering the nucleic acid genome. Icosahedral symmetry is defined by 12 fivefold axes, 20 threefold axes, and 30 twofold axes. A number, T, called the 'triangulation number', indicates how many quasi-symmetrical subunit interactions are within one asymmetrical region of the icosahedron. There are a total of 60 T copies of the capsid proteins in one icosahedral capsid.

For example, a picornavirus (pico-(small)-RNA virus) has a single positive-stranded RNA genome of about 8000 nucleotides. The RNA genome encodes a long polyprotein that is processed into individual viral proteins after translation directly from the positive RNA genome. Three of the 10 proteins are capsid proteins: viral protein 1 (VP1), viral protein 3 (VP3), and viral protein 0 (VP0). VP0 is processed to VP2 and VP4 after virus assembly. VP1, VP2, and VP3 are the three capsid proteins that form the body of the coat while VP4 is entirely inside the coat. There are 60 copies of each capsid protein in the coat. A picornavirus is, therefore, a $T=1$ particle. However, the structure of VP1, VP2, and VP3 is highly homologous. If the small differences in each major capsid protein are ignored, the three proteins can be considered as the same building block. A picornavirus is thus called a pseudo-$T=3$ particle (**Figure 1**). The interactions between the capsid proteins in the coat are similar despite their different symmetry locations. For instance, VP1 (blue) interacts around the fivefold symmetry axis, whereas VP2 (green) and VP3 (red) interact with each other around the threefold symmetry axis. Since VP2

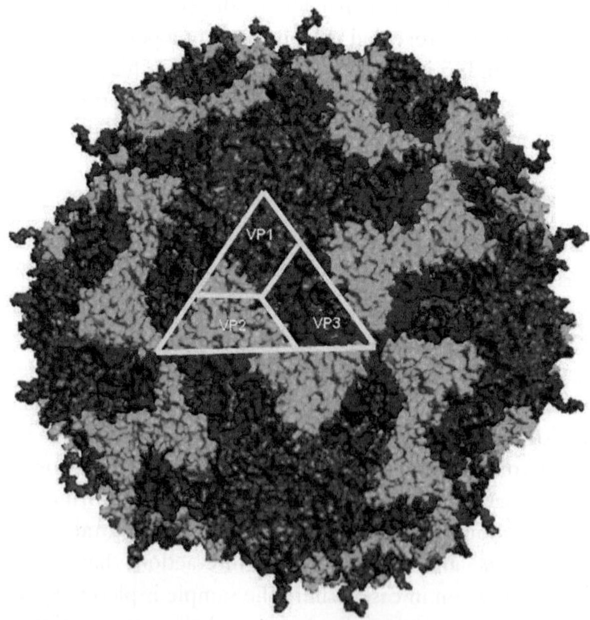

Figure 1 The capsid of Theiler's murine encephalomyelitis virus (TMEV), a picornavirus. The triangle outlines one set of 60 copies of each capsid protein.

and VP3 are similar, the symmetry around the threefold axis is like a sixfold symmetry axis. This symmetry axis is therefore called a quasi-sixfold axis. The interactions between the neighboring VP1 proteins are considered quasi-equivalent (means more or less similar) to the interactions between VP2 and VP3 even though VP1 subunits are around a fivefold axis, and VP2/VP3 subunits are around a quasi-sixfold axis. This quasi-equivalence allows the closed capsid to be assembled using the same building block, the capsid proteins. It should be emphasized, however, that the true differences that exist in the capsid proteins are critically correlated with functions such as virus entry and release or packaging the viral genome.

Methods of Structure Determination

X-ray diffraction is the common technique used for studying the atomic structure of proteins and nucleic acids. When X-rays strike on the electrons of the atoms in a stationary specimen, a diffraction pattern of spots with different intensities is generated and recorded. By analysis of the diffraction pattern and the spot intensities, a three-dimensional electron density map (EDM) can be calculated by Fourier transformation. A three-dimensional chemical structure could be built based on the interpretation of the EDM. Two types of X-ray diffraction experiments are useful for virus structure studies: fiber diffraction (for filamentous viruses) and crystallography (for spherical viruses and globular viral proteins).

Another common technique is electron microscopy. Recent advances in electron microscopy allow researchers to determine relatively high-resolution three-dimensional structures of viral particles by use of image reconstruction of cryo-electron micrographs and electron tomography. This technique is particularly suitable for large viral particles that are difficult to crystallize. Electron microscopy and X-ray crystallography are therefore complementary to each other.

Atomic Structure of Helical Viruses

The disk of the tobacco mosaic virus (TMV) coat protein has been crystallized and its atomic structure resolved by X-ray crystallography. The intact TMV structure containing its nucleic acid could only be determined by X-ray fiber diffraction experiments, as also was that of Pf2 phage. The coat proteins of TMV and Pf2 contain mainly α-helices and the nucleic acid interacts with the coat protein by one base (Pf2) or three bases (TMV) per protein unit. The protein subunit is arranged in a super-helical structure coincident with a super-helical structure of the nucleic acid. The protein subunits form the outer layer of the helix to protect the nucleic acid from the environment. The nucleic acid is embedded between the protein layers that are stacked up in the spiral structure. The axis of the coat protein helix coincides with that of the nucleic acid. Since the symmetry of the super-helix puts no limit on the length of the helix, the virus particle can grow as long as the length of the nucleic acid genome; in this way all bases in the genome are covered by the coat protein. The coat proteins of TMV have many aggregation forms, depending on pH or ionic strength. In natural conditions, the TMV coat protein forms a disk with two layers of 17 protein subunits in each. The subunit of the TMV capsid protein contains four antiparallel α-helices with connecting long loops. Each protein subunit takes the shape of a shoe. When the disks of the capsid proteins are stacked, the RNA fits into the grooves formed by the protein subunits. The backbone of the RNA has charged interactions with the side chains of the amino acids while the bases of the RNA are accommodated by the hydrophobic space. The TMV RNA genome is inserted into the center of the coat protein disk to begin virus assembly. More and more disks are added to the top of the growing virus particle to pull the RNA through the center of the super-helix. The assembly is completed when the complete RNA is pulled into the virus particle. The coat protein of Pf2 has a different shape that contains an extended α-helix. The coat protein is added one by one to the DNA helix emerging from the bacterial membrane. The length of the α-helix is in the direction of the virus particle.

Atomic Structure of Spherical Viruses

Nonenveloped spherical viruses form large single crystals under proper conditions. Their atomic structure can be determined by X-ray crystallography with aid of fast computers and synchrotron X-ray sources. Since 1978, there are numerous atomic structures of viruses reported. Today, atomic structures have been determined for every major family of nonenveloped viruses, especially human pathogens.

Most capsid proteins of these viruses contain an antiparallel, eight-stranded β-barrel folding motif (**Figure 2**). The motif has a wedge-shaped block with four β-strands (BIDG) on one side and another four (CHEF) on the other. There are also two conserved α-helices (A and B): one is between βC and βD, and the other between βE and βF. In animal viruses, there are large loops inserted between the β-strands. These loops form the surface features of individual viruses. The common presence of the β-barrel motif in viral capsid proteins is the result of structural requirements for capsid assembly. It also points to a common ancestor of different virus families.

A virus capsid may contain multiple copies of the β-barrel fold with the same amino acid sequence (such

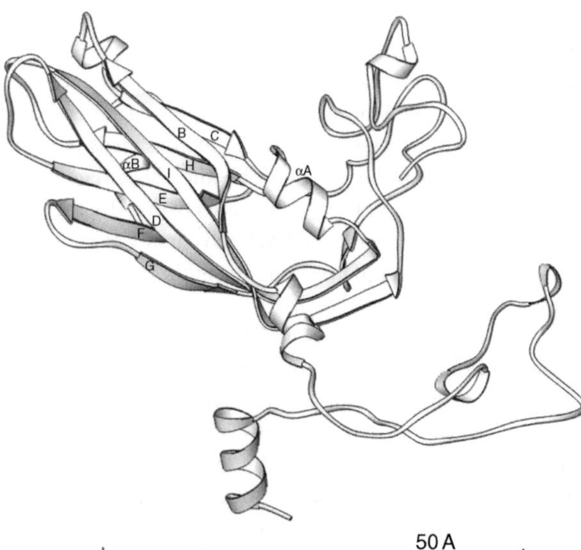

Figure 2 The β-barrel fold found in spherical viruses. The fold contains eight essential β-strands: BIDG on one side and CHEF on the other side. The figure was generated with rhinovirus 16 VP1 (PDB code 1AVM).

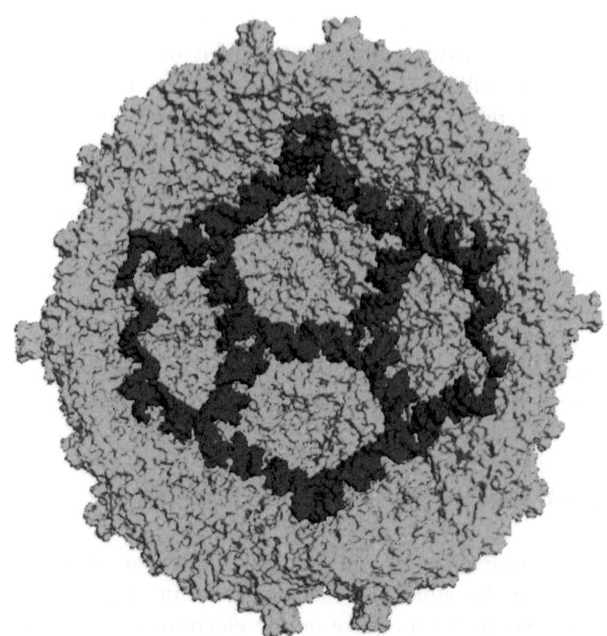

Figure 3 The interior of the pariacoto virus capsid (green). The structured RNA genome is shown in red. The figure was generated with coordinates of pariacoto virus (PDB code 1F8V).

as $T=3$ calicivirus or $T=1$ canine parvovirus) or with different amino acid sequences (such as pseudo $T=3$ picornavirus). In some cases, there are two β-barrel folds in a single polypeptide (such as adenovirus hexon). Capsid proteins of spherical viruses can have other motifs such as α-helices in reovirus and hepadnavirus.

Nucleic Acid–Protein Interaction

The viral nucleic acid genome is always packaged inside the protein capsid. Positively charged patches formed by the side chains of arginines and lysines are found on the interior surface of the protein capsid. These positively charged areas are the preferred binding sites for the nucleotides. Usually the structure of the nucleic acid cannot be observed in a single-crystal X-ray diffraction experiment because of the random orientation of the icosahedral particles in the crystal. However, in rare cases, the nucleic acid might assume icosahedral symmetry by interacting with the protein capsid. Fragments of the complete genome assume the same conformation, although with different nucleotide sequences, at locations related by icosahedral symmetry. Such structures have been seen in bean pod mottle virus, tobacco mosaic virus, pariacoto virus and flock house virus (RNA viruses), and canine parvovirus (DNA virus). The bases are stacked either as an A-type RNA helix or form a coiled conformation to fit the interactions with the protein capsid. In the case of pariacoto virus, the RNA genome forms a cage that reassembles the icosahedral symmetry of the capsid (**Figure 3**). These viruses readily form empty virus particles and have a hydrophobic pocket on the interior surface of the capsid. The nucleic acid generally interacts nonspecifically with the capsid protein.

Evolution

The highly conserved β barrel motif of the viral capsid protein indicates that many viruses must have evolved from a single origin. The unique three-dimensional structure of this motif is required for capsid assembly and it is generally conserved over a longer period of time than the amino acid sequence. The superposition of the capsid proteins from different viruses can be used to estimate the branch point in the evolutionary tree for each virus group. The structure alignment not only relates plant viruses to animal viruses, RNA viruses to DNA viruses, but also viruses to other proteins such as concanavalin A that has a similar fold and competes with poliovirus for its cellular receptor. The evolutionary relationship of these viruses is supported by amino acid sequence alignment of more conserved viral proteins such as the viral RNA polymerase. The structural similarities of the capsid proteins support the notion of a common evolutionary origin among nonenveloped viruses.

Assembly

The icosahedral capsid is assembled from smaller units made of several protein subunits. For picornaviruses, a protomeric unit is first formed with one copy of each polypeptide after translation. The termini of the subunits are intertwined with each other to hold the subunits together in the protomer. The protomers are then associated as pentamers which in turn form the complete icosahedral virion while encapsidating the viral RNA. In $T=3$ or $T=1$ plant RNA viruses, the pentamers are formed by dimers of the capsid proteins. In adenovirus and SV40, the capsid proteins form hexon units (three polypeptides, each has two β-barrels) or pentamers before they assemble into an icosahedral shell. The nucleic acid could be packaged at different stages of assembly.

Host Receptor Recognition Site

Animal viruses have to recognize a specific host cellular receptor for entry during infection. Host receptor binding is the initial step of virus life cycle and could be an effective target for preventing virus infection. Based on the atomic structure of animal viruses, it was found that the receptor recognition site is located in an area surrounded by hyper-variable regions of the antigenic sites. Usually, the area is in a depression (called the 'canyon') on the viral surface that may be protected from recognition by host antibodies. This structural feature is, for instance, present in human rhinovirus (also known as the common cold virus), and the active site of influenza virus hemagglutinin (HA). The receptor-binding site on influenza virus HA does not have a deep depression, but it is surrounded by antigenic sites. The receptor-binding area on the surface of the viral capsid is conserved for recognition by the receptor, whereas the sites recognized by antibodies are distinct from the receptor-binding site and keep changing from strain to strain. By this mechanism, the virus can escape the host immune system by mutating the antibody epitopes, and at the same time maintain a constant receptor-binding site to continue its infection of the host cells. Evidence supports that this is a general mechanism that viruses use to evade the host immune defense.

Antigenic Sites

Antibodies are the first line of defense by the immune system against a viral infection. The epitopes combined with the neutralizing antibodies are mapped on a few isolated locations on the surface of viral proteins. The structure of human rhinovirus complexed with Fab fragments showed that the antibody makes contact with an area about 6 nm^2 and that the epitope spans different discontinuous polypeptides. Therefore, an effective vaccine usually needs to include a complete viral protein or a large fragment. The binding of the antibodies does not significantly change the structure of the antigen. The exact mechanism by which antibodies neutralize antigens is dependent upon the binding site and processes of the virus replication.

Antiviral Agents

Viral infectious diseases can be cured if an agent can be administered to stop viral infection. Such agents have been synthesized and shown to bind to the capsid of rhinovirus in the crystal structure. The compounds were inserted into the hydrophobic pocket within the β-barrel of the major capsid protein VP1. Binding of the compounds stops uncoating of the virion and the receptor binding, which resulted in the failure of releasing the viral RNA into the cytoplasma. These compounds inhibit infections of several other RNA viruses and may be effective against other viruses after modification since the β-barrel structure exists in many viruses.

The most successful antiviral drugs are the HIV protease inhibitors which are developed based on the atomic structure of the protease. Through iterative cycles of computer modeling, chemical synthesis and structural studies of the protein-inhibitor complexes, a panel of clinical effective drugs has been brought to the market and has shown great benefits to patients. Inhibitors of influenza virus neuraminidase have also been developed by the same method and marketed as antiviral drugs.

See also: Assembly of Enveloped Virions.

Further Reading

Knipe DM, Howley PM, Griffin DE, *et al.* (eds.) (2002) *Fields Virology*, 4th edn. Philadephia, PA: Lippincott Williams and Wilkins.

Rossmann MG and Johnson JE (1989) Icosahedral RNA virus structure. *Annual Review of Biochemistry* 58: 533–573.

Schneemann A (2006) The structural and functional role of RNA in icosahedral virus assembly. *Annual Review of Microbiology* 60: 51–67.

Assembly of Enveloped Virions

C K Navaratnarajah, R Warrier, and R J Kuhn, Purdue University, West Lafayette, IN, USA

Glossary

Core-like particle (CLP) Subviral particles assembled from recombinantly expressed capsid protein are referred to as core-like particles. They are morphologically similar to authentic cores isolated from viruses or infected cells.

Cryo-EM Cryo-electron microscopy and image reconstruction techniques are used to elucidate the structures of viruses and other macromolecular structures.

Immature virion Viruses usually produce noninfectious particles which require a maturation step in order to form the infectious, mature virus. The maturation step often involves the proteolytic cleavage of a precursor protein.

Nucleocapsid core (NC) Capsid protein packages the nucleic acid genome to form a stable protein–nucleic acid complex which is then enveloped with a lipid bilayer.

Introduction

Viruses have long been distinguished by their physical features, usually visualized by electron microscopy or analyzed biochemically. One feature that has been frequently used to categorize viruses is the presence or absence of a lipid bilayer. Many animal viruses are surrounded by a lipid bilayer that is acquired when the nucleocapsid buds through cell membranes, usually at a late stage of virus assembly. While the protein coat of nonenveloped viruses plays a crucial role in protecting the genome from the environment, for enveloped viruses the lipid membrane partially fulfills this role. The lipid membranes are decorated with virus-encoded envelope proteins that are important for the subsequent infectivity of the virus, although some viruses also incorporate cellular proteins in their membrane. Virus envelopment can take place after the assembly of an intact nucleocapsid structure (betaretroviruses) or capsid assembly and envelopment can occur concomitantly (orthomyxovirus). Specific or nonspecific interactions between the viral envelope glycoproteins and the proteins that make up the nucleocapsid mediate the envelopment of the core or nucleoprotein–nucleic acid complex. Enveloped viruses acquire their lipid bilayer from a variety of locations within the cell, but a given virus will usually bud from one specific cellular membrane (**Table 1**). Enveloped viruses can take advantage of the cellular secretory pathway in order to assemble and bud out of the cell. In contrast, nonenveloped viruses usually exit infected cells by disrupting the plasma membrane. Thus, budding provides enveloped viruses with a nonlytic method of exiting infected cells, and they must do so while the cell is still alive.

Viral Envelope

The main component of the viral envelope is the host-derived lipid bilayer. The precise composition of this lipid membrane varies, as different viruses acquire their envelopes from different cellular membranes. The choice of membrane from which the virus buds is often determined by the specific targeting and accumulation of

Table 1 List of enveloped virus families and the origin of the envelope

Virus family[a]	*Membrane*
Arenaviridae	Plasma membrane
Arterivirus	Endoplasmic reticulum
Asfarviridae	Endoplasmic reticulum and plasma membrane
Baculoviridae	Plasma membrane
Bunyaviridae[b]	Golgi complex
Coronaviridae	ER Golgi intermediate compartment
Cystoviridae[c]	Plasma membrane
Deltavirus	Endoplasmic reticulum
Filoviridae	Plasma membrane
Flaviviridae	Endoplasmic reticulum
Fuselloviridae[d]	Plasma membrane
Hepadnaviridae	Endoplasmic reticulum
Herpesviridae	Nuclear envelope
Hypoviridae	Plasma membrane
Iridoviridae	Plasma membrane
Lipothrixviridae[d]	Plasma membrane
Orthomyxoviridae	Plasma membrane
Paramyxoviridae	Plasma membrane
Plasmaviridae[c]	Plasma membrane
Polydnaviridae	Plasma membrane
Poxviridae	ER Golgi intermediate compartment
Retroviridae	Plasma membrane
Rhabdoviridae[b]	Plasma membrane
Togaviridae	Plasma membrane

[a]Animal viruses unless otherwise noted.
[b]This group also includes plant viruses.
[c]Bacteriophage.
[d]Archaea.

the envelope proteins at a particular site in the secretory pathway (**Table 1**). There are examples of viruses that bud from the plasma membrane (togaviruses, rhabdoviruses, paramyxoviruses, orthmyxoviruses, and retroviruses), endoplasmic reticulum (ER) (coronaviruses and flaviviruses) and the Golgi complex (bunyaviruses). There are also examples of viruses that undergo transient envelopment and reenvelopment (herpesvirus).

Viral proteins are found embedded in the lipid membrane. The majority of these proteins are transmembrane glycoproteins. The viral envelope glycoproteins mediate the interaction of the virus with cell receptors and promote the fusion of the viral and cellular membranes during infection of susceptible cells. Viral glycoproteins are also crucial for the assembly of the virion. They can make important lateral contacts with each other, thus driving oligomerization and also capturing other viral components such as the capsid or matrix protein. The majority of enveloped viruses contain one or more glycoproteins that are usually found as oligomers embedded within the lipid membrane. High-resolution structural information is available for many glycoproteins such as the hemagglutinin and neuraminidase proteins of influenza A virus, the gp120 of HIV, and the E protein of dengue virus. Based on these structural and biochemical studies, it has been shown that most glycoproteins are primed for the conformational changes that are required in order to gain entry to the host cell during an infection.

Some enveloped viruses contain integral membrane proteins that have multiple membrane-spanning regions that oligomerize to form channels in the membrane. The influenza A virus M2 protein forms an ion channel and plays an important role in the assembly and entry of the virus particle. Some viruses such as the retroviruses also incorporate cellular membrane proteins into the viral envelope. In a majority of the cases the host proteins that are present at the sites of assembly or budding are incorporated in a passive, nonselective manner. However, there are examples where the virus actively recruits specific host proteins that may help in evading the defenses of the immune system or enhance infectivity.

Icosahedral Enveloped Viruses: Alphaviruses and Flaviviruses

Alphaviruses, and more recently, flaviviruses have served as model systems to study the assembly and budding of simple enveloped viruses. These positive-strand RNA viruses consist of a single RNA genome that is encapsidated by multiple copies of a capsid protein to form the nucleocapsid core (NC). The envelopment of the NC is mediated by the interaction between the envelope glycoproteins and this core. The assembly and budding of these two simple enveloped viruses will be described in detail in order to present common themes in the assembly of icosahedral enveloped viruses.

Alphavirus Assembly

Alphavirus life cycle

Alphaviruses are members of the family *Togaviridae*, which also includes the genus *Rubivirus*. Alphaviruses enter the host cell by receptor-mediated endocytosis via the clathrin-coated endocytic pathway. Following fusion at low pH with the endosomal membrane, the NC is released into the cytoplasm. The NC has been proposed to uncoat by transfer of capsid proteins (CPs) to ribosomes. This releases the genome RNA into the cytoplasm which is translated to produce the nonstructural proteins. The nonstructural proteins transcribe a negative-sense copy of the genome RNA. This RNA serves as template for genomic and subgenomic RNA. The subgenomic RNA, which is synthesized in greater amounts than the genomic RNA, codes for the structural proteins of the virus. CP is found at the N-terminus of the structural polyprotein, followed by proteins PE2 (E3+E2), 6K, and E1. Two hundred and forty copies of CP, E2, and E1 assemble to form the alphavirus virion (**Figure 1(a)**). The transmembrane E1 glycoprotein functions during entry to mediate the fusion of the viral membrane with that of the endosomal membrane, while the transmembrane E2 glycoprotein is responsible for cell receptor binding. E1 and PE2, a precursor of E2 and E3, are processed together as a heterodimer in the ER and Golgi, and are transported to the cell surface in the form of spikes that are each composed of three heterodimers of E1/E2. E3 serves as a chaperone to promote the correct folding of E2, as well as to prevent the premature fusion of E1 in the acidic environment of the late Golgi. The maturation cleavage of PE2 to generate E3 and E2 by a furin-like protease in a late Golgi compartment primes the glycoprotein spike complex for subsequent fusion during virus entry. The function of 6K is unclear but it does promote infectivity of the particle. A single copy of the genome RNA is packaged by 240 copies of the CP to form an icosahedral NC in the cytoplasm of infected cells. The NC interaction with the E1/E2 trimeric spikes at the plasma membrane results in the budding of the mature virus from the cell membrane.

Alphavirus virion structure

Cryo-electron microscopy (cryo-EM) and image reconstruction techniques have revolutionized the understanding of the molecular architecture of alphaviruses (**Figure 1(a)**). Studies with Ross River, Semliki Forest, Venezuelan equine encephalitis, Aura, and Sindbis have shown that these viruses consist of an outer protein layer made up of the glycoproteins E1 and E2 (**Figures 1(a)** and **1(c)**). The membrane spanning regions of these glycoproteins traverse

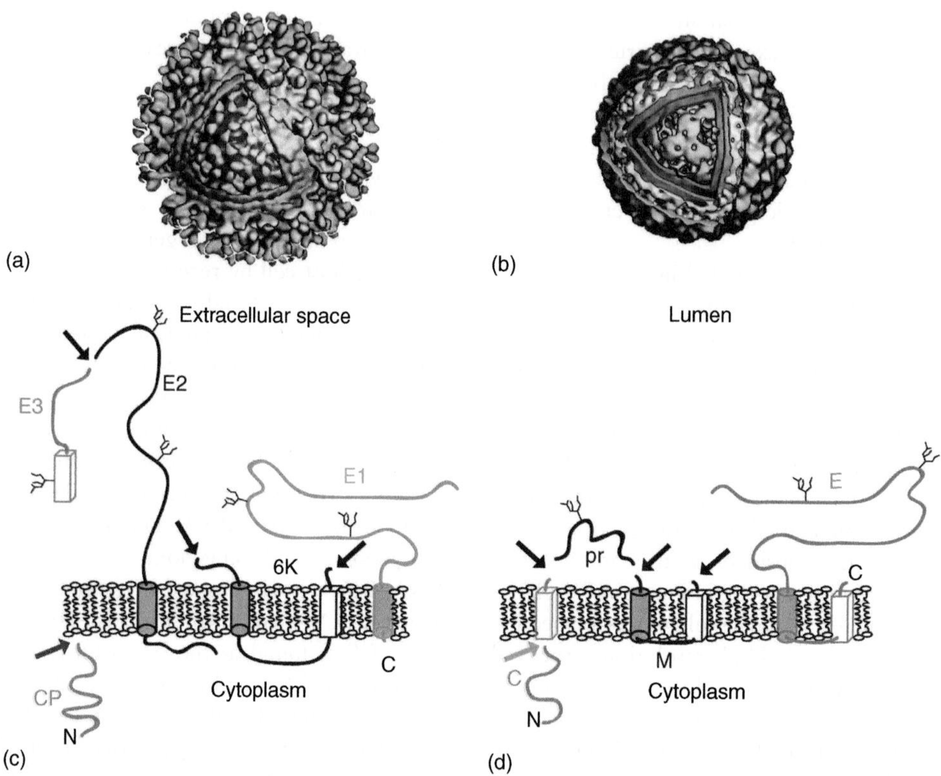

Figure 1 Cryo-EM reconstructions and glycoprotein topology of alphaviruses and flaviviruses. (a) Surface shaded view of a Cryo-EM reconstruction of Ross River virus (alphavirus). The glycoprotein envelope shown in blue has been cut away to reveal the lipid bilayer (green) and the nucleocapsid core (orange). (b) Cryo-EM reconstruction of dengue virus (flavivirus). The glycoprotein envelope shown in blue has been cut away to reveal the lipid bilayer (green) and the nucleocapsid core (orange/purple). (c) The topology of the alphavirus glycoproteins is shown. The rectangular cubes represent signal sequences while the cylinders represent stop transfer sequences. The glycosylation sites are represented by branched structures. The arrows indicate cleavage sites. The black arrows indicate signalase cleavage sites, the purple arrow indicates the furin cleavage site, and the red arrow indicates the CP cleavage site. (d) The topology of the flavivirus structural proteins is shown. The rectangular cubes represent signal sequences while the cylinders represent stop transfer sequences. The glycosylation sites are represented by branched structures. The arrows indicate cleavage sites. The black arrows indicate signalase cleavage sites, the purple arrow indicates the furin cleavage site, and the green arrow indicates the NS2B/NS3 (viral protease) cleavage site.

a host-derived lipid bilayer that surrounds the NC of the virus. The CP and glycoprotein layers interact with one another and are arranged symmetrically in a $T=4$ icosahedral configuration. Fitting of the atomic coordinates of the crystal structures of the ectodomain of E1 and amino acids 106–264 of the CP into the cryo-EM density of Sindbis virus allowed a pseudo-atomic model of the virus to be generated. The fitting of the E1 structure into the cryo-EM density reveals that E1 forms an icosahedral scaffold on the surface of the viral membrane. E1 is positioned almost tangential to the lipid bilayer, whereas E2 has a more radial arrangement. The bulk of E2 lies on top of E1 and caps the fusion peptide, thereby preventing premature fusion with cell membranes. This arrangement of the glycoproteins is in agreement with the function of each protein, where the surface-exposed E2 interacts with cellular receptors and protects E1 until it is required for fusion. The fusion peptide is only exposed when the E1–E2 heterodimer dissociates in the presence of low pH in the endosome. Fitting of amino acids 106–264 of the CP into the cryo-EM density of Sindbis virus showed that each subunit of the projecting pentamers and hexamers (known as capsomeres) observed in the NC layer is made up of the CP protease domain consisting of amino acids 114–264. There is very little interaction between amino acids 114–264 of the CP either within the capsomere or in between capsomeres. Thus, the major contributors to the stability of the NC in the absence of glycoproteins are CP–RNA and RNA–RNA interactions that take place in the RNA–protein layer below the projecting capsomeres.

Alphavirus assembly and budding

Alphavirus virions always contain an NC and it is likely that this promotes and is required for budding through direct interactions with the glycoproteins. Thus, the first step in assembly is for the alphavirus CP to specifically recognize and encapsidate the genome RNA to form NCs in the cytoplasm of infected cells (**Figure 2**). The N-terminus of the CP (amino acids 1–80, SINV numbering) is largely basic and thought to be involved in charge

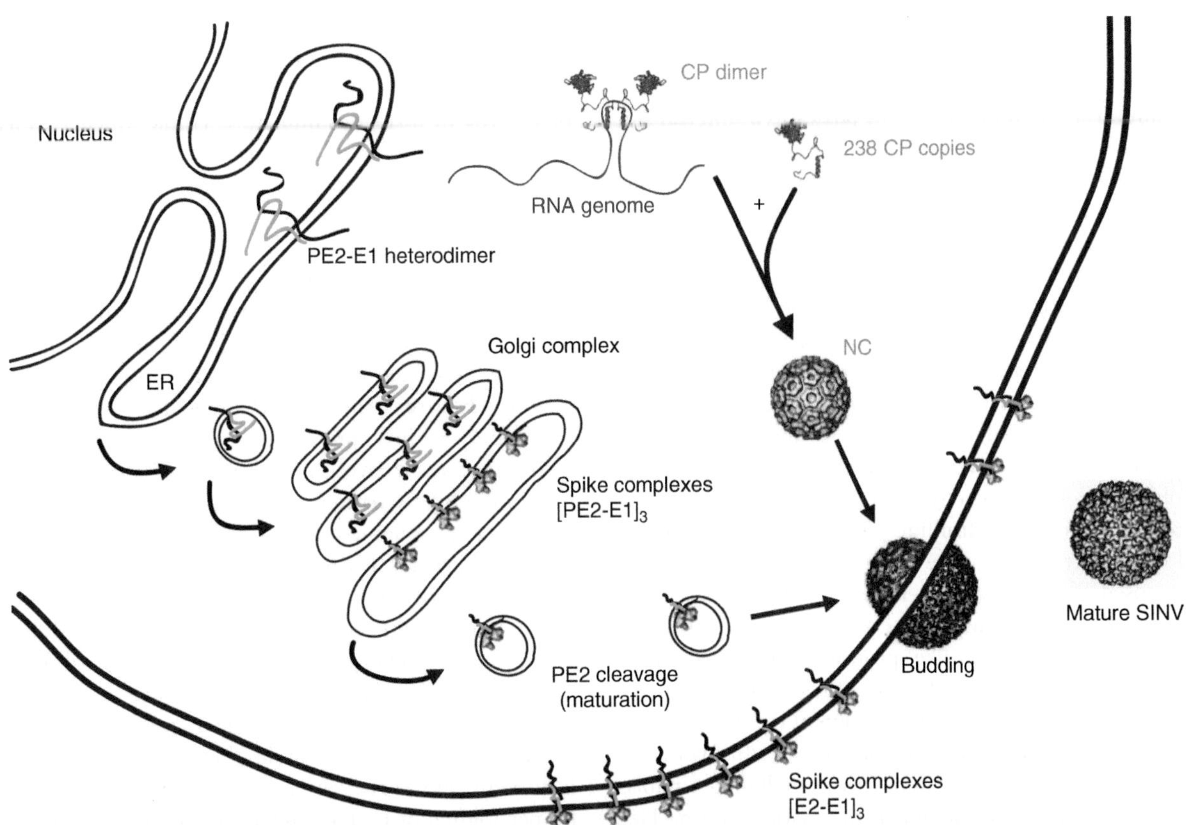

Figure 2 Alphavirus assembly. Two capsid proteins (CPs) bind the encapsidation signal of the genome RNA to form a CP dimer-RNA complex in the cytoplasm. The CP dimer is stabilized by coiled–coil helix I interactions. The subsequent steps of nucleocapsid core assembly have not been elucidated but cores form and accumulate in the cytoplasm. The glycoproteins PE2 and E1 form heterodimers in the endoplasmic reticulum (ER). The glycoproteins are folded, glycosylated, and palmitoylated as they are transported through the ER and Golgi. The PE2–E1 heterodimers form spike complexes $[PE2–E1]_3$ in the Golgi. E3 is cleaved from the spikes by a furin-like protease before they are transported to the plasma membrane. The NC interacts with the cytoplasmic domain of E2 in the spike complexes, driving the budding of the mature virus from the plasma membrane.

neutralization of the genome RNA. Amino acids 38–55 are conserved uncharged residues that form a coiled-coil alpha-helix (helix I) important in dimerization of the CP during the assembly process. While the process of virus assembly is difficult to study in the complex cellular environment, the development of an *in vitro* assembly system based on bacterial expression of CP has led to advances in understanding NC assembly. These studies suggest that the initial event of NC assembly is the binding of CP amino acids 81–112 to the encapsidation signal on the genome RNA corresponding to nucleotides 945–1076 (SINV numbering) (**Figure 2**). This interaction exposes a second site on the encapsidation signal where another molecule of CP binds and forms a dimer with the first CP molecule. Amino acids 114–264 constitute the previously mentioned chymotrypsin-like serine protease that autoproteolytically cleaves CP from the nascent structural polyprotein. This region is involved in binding residues from the cytoplasmic domain of E2, thus linking the outer icosahedral glycoprotein shell with the NC across the lipid bilayer (**Figure 1(a)**).

Other lines of evidence support the dimer model of NC assembly. Helix I of CP, which is required for core accumulation in infected cells, may be functionally substituted by a GCN4 helix that forms dimeric coiled–coil interactions but not by a GCN4 helix that has a propensity to form trimeric coiled–coil interactions. In addition, helix I acts as a checkpoint in NC assembly whereby incompatible helices prevent the formation of core-like particles (CLPs) *in vitro*. Furthermore, a portion of CPs in either NCs or CLPs may be cross-linked into dimers by DMS, a lysine specific cross-linker with a 12 Å cross-linking distance. Cross-linking enabled an assembly deficient helix mutation of CP to assemble into NCs, suggesting that the cross-link can functionally replace the helix interaction.

While assembly of the CP into NCs proceeds in the cytoplasm, the processing and assembly of the glycoproteins occur in the ER and Golgi (**Figures 1(c)** and **2**). The autocatalytic cleavage of the CP reveals a signal sequence on the N-terminus of the newly cleaved structural polyprotein that directs it to the ER (**Figure 1(c)**). PE2 is translocated into the ER until it reaches a

26-amino-acid stop transfer signal which anchors PE2 in the membrane. The C-terminal 33 residues of E2 then act as a second signal sequence to direct the next protein, 6K (55 amino acids), into the ER. 6K possesses a stop transfer sequence which anchors it in the membrane, and the C-terminus of 6K acts as signal sequence for the translocation of E1. E1 is anchored in the ER membrane by a final stop transfer sequence close to its C-terminus. The release of PE2 and E1 by cellular signalase cleavage allows the formation of PE2-E1 heterodimers in the ER (**Figure 2**). PE2 and E1 are each glycosylated in all alphaviruses, but the number and location of the modifications vary. In addition to glycosylation, the glycoproteins are palmitoylated in the Golgi apparatus. As the heterodimers are processed and transported through the ER and Golgi, they undergo a series of folding intermediates that are mediated by disulfide exchange and chaperones. Ultimately, they associate to form spikes which are composed of trimers of PE2-E1 dimers (**Figure 2**).

The final maturation event is the cleavage of PE2 into E3 and E2 by a furin-like protease (**Figure 2**). This cleavage occurs in a late Golgi or post-Golgi compartment and results in the destabilization of the heterodimer enabling the mature virus to fuse more readily with the target membrane. In most alphaviruses including Sindbis, E3 is released and not found in the mature virion. The final destination for the spike complexes is the plasma membrane, where the cytoplasmic domain of E2 (cdE2) recruits NCs assembled in the cytoplasm (**Figure 2**). Structural studies show that cdE2 residues Tyr400 and Leu402 bind into a hydrophobic pocket of the CP. Mutation of Tyr400 negatively impacted virus budding, while protein translation and core accumulation were at wild-type levels. This interaction of cdE2 with the hydrophobic pocket in the CP is thought to drive the budding of the mature virus at the cell membrane.

Flavivirus Assembly

Flavivirus life cycle

Flaviviruses belong to the family *Flaviridae* of positive-strand RNA viruses which also consist of the pestiviruses and the hepaciviruses. The flaviviruses comprise more than 70 members including important human pathogens such as yellow fever virus, dengue virus, and West Nile virus. Flaviviruses enter cells via receptor-mediated endocytosis. The low pH environment of the endosomal membrane triggers the conformational change of the envelope glycoprotein which results in the fusion of the viral and endosomal membranes releasing the genome into the cytoplasm. The viral proteins are translated from the RNA genome as a single polyprotein. Signal sequences and stop transfer sequences result in the translocation of the nascent polyprotein to the ER membrane. The polyprotein is processed by a combination of cellular and viral proteases to produce the mature structural and nonstructural proteins. Genome replication and virion assembly occur in ER membrane-bound vesicles. The structural proteins and the genome bud into the lumen of the ER to form the immature virion which is transported through the secretory pathway. Prior to secretion of the virion, a furin cleavage converts the immature virus into the mature, infectious form of the virus.

Flavivirus virion structure

The flavivirus virion is made up of three structural proteins: capsid (C), pre-membrane (prM), and envelope (E) that are translated from the 5′ one-third of the RNA genome (**Figures 1(b)** and **1(d)**). Signal sequences at the C-terminus of the C protein and prM serve to translocate prM and E respectively into the ER (**Figure 1(d)**). The role of the highly basic C protein (12 kDa) is to encapsidate the viral genome during virion assembly. In contrast to the alphavirus CP which exhibits no membrane association, the flavivirus C protein is anchored to the membrane, at least transiently (**Figure 1(d)**). However, Rubella virus, the sole member of the genus *Rubivirus* within the family *Togaviridae*, has a membrane anchored CP, perhaps indicating a common origin for the *Togaviridae* and *Flaviviridae*. prM is a glycoprotein that associates with the E protein and serves as a chaperone to facilitate the proper folding of E. The immature virions that bud into the ER consist of prM-E heterodimers. prM prevents premature fusion from occurring in the acidic environment of the ER and Golgi. Thus, prM has several functions analogous to the E3 glycoprotein of the alphaviruses. Cleavage of prM into pr and M by a furin-like protease triggers the rearrangement of the prM-E heterodimers into E–E homodimers, resulting in a radical change in size and shape required for the formation of the mature virus particle. The E glycoprotein is responsible for host cell receptor binding and for fusion of the viral and cellular membranes. The E glycoprotein is also critical for the assembly of the virion. High-resolution structures of the ectodomains of several flavivirus E proteins are available. The ectodomain is divided into three domains. Domain II constitutes the dimerization domain as it contains most of the intradimeric contacts between E–E homodimers. Domain II also contains the fusion peptide, a glycine-rich hydrophobic sequence that initiates fusion by insertion into the target cell membrane. Domain III comprises the immunoglobulin-like domain responsible for receptor binding. In addition to the dramatic conformational and translational changes that the E protein undergoes during the virion maturation process, it also changes conformation during membrane fusion. The low pH of the endosome during infection triggers a conformational change which results in the formation of E homotrimers. In this arrangement, the fusion peptides are exposed and available to insert into cellular membranes. Interestingly, the structure of the E protein was found to be very similar to the structure of the

Semliki Forest virus E1 protein, the fusion protein of the alphaviruses.

The structures of two flaviviruses, dengue and West Nile virus, have been solved by cryo-EM and image reconstruction techniques and have been shown to be similar (**Figure 1(b)**). The mature virion is ~50 nm in diameter and exhibits a smooth outer surface in contrast to the alphaviruses which have distinctive spike structures (cf. **Figures 1(a)** and **1(b)**). The E proteins are arranged parallel to the surface of the virus, with 90 E dimers arranged in groups of three to form a 'herringbone' pattern on the viral surface. This arrangement of the E proteins completely covers the surface of the virus, thus rendering the lipid bilayer inaccessible. Domain III of E protrudes slightly from the viral surface, allowing interaction with cell receptors. The membrane-spanning regions of E and M proteins form antiparallel helices while the stem regions are arranged parallel to the membrane.

The immature virus particle exhibits a dramatically different glycoprotein organization compared to the mature virion. Cryo-EM and image reconstruction of dengue and yellow fever virus immature virions have revealed that these particles are larger (~60 nm) and have spikes that protrude from the surface of the virus. These spikes are composed of trimers of prM-E heterodimers. The pr peptide covers the fusion peptide of E in this arrangement, similar to E2 covering E1 in alphaviruses, thus protecting it from premature fusion as the immature particle is transported through the acidic environment of the secretory pathway.

The NC is found below the viral envelope and is composed of a single copy of the genome RNA and multiple copies of the C protein. Cryo-EM reconstructions of the virion have shown that in contrast to the alphaviruses, there is no apparent organization to the flavivirus NC. This may be because there is no direct interaction between the C proteins in the core and the glycoproteins in the viral envelope since the E and M proteins do not penetrate below the inner leaflet of the membrane. Furthermore, no NCs have been observed in the cytoplasm of infected cells and attempts to establish an *in vitro* assembly system analogous to the alphavirus *in vitro* assembly system have failed. The lack of coordination between the C protein and the viral envelope proteins suggests that the assembly of virions is driven by the lateral interactions of the E and M proteins in the viral envelope and not by the C protein. This is supported by the observation that flavivirus infections result in the production of noninfectious subviral particles which are composed of just the viral envelope (E and M) and the lipid bilayer. Thus, the flavivirus glycoproteins are sufficient to induce particle budding.

Flavivirus assembly and budding

Virus-induced membrane structures called vesicle packets, which are continuous with the ER membrane, are the sites of flavivirus replication and assembly (**Figure 3**). Within these structures the structural proteins are in intimate contact with the genome RNA. The C protein associates with the genome RNA via interactions between the positive charges distributed throughout the protein and the negatively charged phosphate backbone of the RNA. It is not yet clear how the C protein specifically recognizes the genome RNA; unlike for alphaviruses, a packaging signal has not been conclusively identified for flaviviruses. Coupling between genome replication and assembly within the vesicle packets has been proposed as a mechanism to ensure the specific encapsidation of the genome RNA. It has been shown that one or more non-structural proteins (NS2A and NS3) are involved in genome packaging and NC assembly. The NC lacks a defined icosahedral structure as described above. Therefore, core formation is probably concomitant with the association of the C protein and RNA genome with the viral glycoproteins and budding into the ER lumen, thus giving rise to the immature particle (**Figure 3**). The immature virion is transported from the ER to the Golgi where the viral glycoproteins are post-translationally modified. The cleavage of the prM protein in the trans-Golgi network triggers the dramatic reorganization of the viral glycoproteins that results in the formation of the mature virion (**Figure 3**). The mature virion is then released from the host cell by exocytosis.

Conclusion

Following from the discussion of alphavirus and flavivirus assembly, it is apparent that the assembly of even the simplest enveloped viruses requires the complex interaction of viral and host factors in order to produce a virus particle which is at once stable and at the same time primed for disassembly. The whole range of the cell's machinery including the translation apparatus, polymerases, chaperones, and post-translational modification enzymes are co-opted by viruses in order to replicate the viral components necessary for assembly. Enveloped viruses have evolved to utilize different cellular membranes and cellular compartments for assembly and they take advantage of the secretory pathway to produce their viral glycoproteins. A majority of viruses bud from the plasma membrane (**Table 1**). This is the case with alphaviruses, where NC assembly occurs in the cytoplasm and the final assembly of the mature virion occurs at the plasma membrane. The high concentration of viral proteins, often concentrated at specific sites allows for the efficient interaction and assembly of virions. In contrast to alphaviruses, NC assembly and glycoprotein assembly is coupled in the flaviviruses and occurs in vesicle packets associated with the ER. Thus, the whole flavivirus virion is transported through the ER and Golgi while in the case of the alphaviruses only the glycoproteins are transported through the secretory pathway. These exit strategies are not unique and

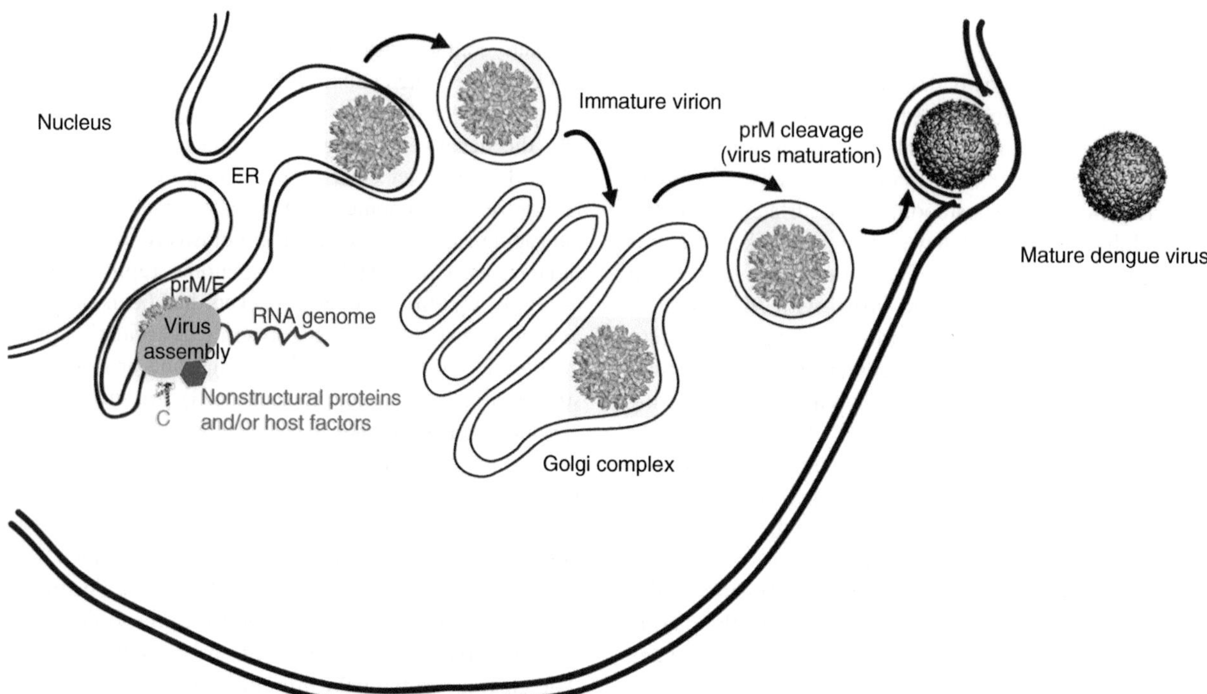

Figure 3 Flavivirus assembly. Flavivirus assembly occurs on ER-associated membranes known as vesicle packets. Assembly and genome replication are coupled and the sites of assembly consist of the capsid proteins (C), the glycoproteins prM and E, the RNA genome, and one or more nonstructural proteins and/or host factors. The immature virion buds into the ER and is transported to the Golgi and trans-Golgi network. The glycoproteins are post-translationally modified as the immature virus is transported through the secretory pathway. Furin cleavage of prM results in the formation of the mature virus which then exits the cell by exocytosis.

thus serve as model systems to study enveloped virus assembly and release.

Proteolytic cleavage of glycoproteins in order to convert them from stable oligomeric structures to metastable structures primed for fusion are common themes in enveloped virus structure and assembly. Cleavage of PE2 into E3 and E2 by a furin-like protease primes the alphavirus spike complex for fusion. A similar cleavage of prM triggers a dramatic conformational change of the flavivirus glycoproteins resulting in the formation of the mature virion which is now infectious.

Alphavirus budding requires the specific interaction of the NC with the E1–E2 spike complexes at the plasma membrane, thus ensuring that all virions have a genome packaged into them. However, the flaviviruses only require the interaction of the envelope proteins for budding, giving rise to subviral particles devoid of the C protein and genome RNA. Thus, the flavivirus envelope proteins alone are sufficient to drive budding of virus particles and the close coupling of genome replication and the C protein (perhaps mediated by replication proteins and host factors) is required to package the genome into virus particles. A third strategy for budding is exhibited by the retroviruses where capsid assembly has been shown to be sufficient to drive budding of the virus. In this case, targeting of the envelope proteins to these sites of CP assembly is essential to ensure the incorporation of the glycoproteins into the virion.

Although much has already been discovered about enveloped virus assembly, there are still many processes yet to be described. There is an increasing interest in the assembly pathway of viruses partly fueled by the potential to develop successful therapeutic agents targeting virus specific assembly processes. Advances in the field of structural biology will further help attempts to understand the assembly pathway of this important class of viruses.

See also: Assembly of Non-Enveloped Virions.

Further Reading

Garoff H, Hewson R, and Opstelten DJ (1998) Virus maturation by budding. *Microbiology and Molecular Biology Reviews* 62(4): 1171–1190.

Harrison SC (2006) Principles of virus structure. In: Knipe DM, Howley PM, Griffin DE, *et al.* (eds.) *Fields Virology,* 5th edn., pp. 53–98. Philadelphia, PA: Lippincott Williams and Wilkins.

Hunter E (2006) Virus assembly. In: Knipe DM, Howley PM, Griffin DE, *et al.* (eds.) *Fields Virology,* 5th edn., pp. 141–168. Philadelphia, PA: Lippincott Williams and Wilkins.

Mukhopadhyay S, Kuhn RJ, and Rossmann MG (2005) A structural perspective of the flavivirus life cycle. *Nature Reviews Microbiology* 3(1): 13–22.

Assembly of Bacterial Viruses

S Casjens, University of Utah School of Medicine, Salt Lake City, UT, USA

Glossary

Coat protein The protein which is the structural building block for the nucleic acid container part of the virion.

DNA encapsidation or DNA packaging The process of enclosing a nucleic acid inside the protein coat of a virion.

Head The icosahedrally symmetric portion of *Caudovirales* virions that contains the nucleic acid.

Portal protein The *Caudovirales* protein through which DNA enters and exits the coat protein shell; it is present at only one of the 12 otherwise identical icosahedral vertices.

Procapsid or prohead The virion precursor particle that contains no nucleic acid; it is the preformed container into which the nucleic acid is packaged.

Scaffolding protein A protein that directs the proper assembly of the coat protein, but is not present in the completed virion.

Tail The portion of *Caudovirales* virions that binds to cells bearing the correct surface receptor and through which DNA traverses during delivery from the virion into the cell. The tail is attached to the portal protein of the head in the complete virion.

Triangulation (*T*) number Formally, *T* is the number of smaller equilateral triangles (facets) into which each of the 20 icosahedral faces is subdivided in an icosadeltahedron. In viruses, this is directly related to the number of coat protein subunits (*S*) per virion ($S = 60T$), since there are typically three subunits per facet and 20 faces in icosahedral viruses.

Virion The stable virus particle that is released from infected cells and is capable of binding to and infecting other sensitive cells.

History and Overview

The structure of bacterial viruses (bacteriophages or phages) remained a mystery for over a quarter of a century after their original discovery in about 1910, and even then early electron micrographs of phage T4 showed only a vague outline of its complex features. The invention in the late 1950s of negative staining with heavy metal salts and high-resolution metal evaporation techniques allowed electron microscopy to be used to visualize virions with a resolution of a few nanometer – a big step forward, but most individual protein components could still not be resolved. Meanwhile, viruses were shown to be made up primarily of protein on the outside and nucleic acid inside. It is worth noting that phage T2 was a major participant in the proof that DNA is in fact the genetic material, when Hershey and Chase used it to show that during an infection the virion's DNA entered the cell and the bulk of the protein remained outside. After their discovery that DNA is the genetic material, in 1953, Crick and Watson also recognized that virus chromosomes could not encode a single protein molecule large enough to enclose their nucleic acid molecules, and they correctly postulated that virions had to be constructed with repeating arrays of protein molecules. They also noted that there are only two ways of building virions with symmetric arrays of coat proteins that enclose a space suitable for a virion's nucleic acid, and those are arrays built with cubic or helical symmetry (the former includes icosahedral, octahedral, and tetrahedral arrays which would be built from 60, 24, and 12 identical protein subunits, respectively). In 1962, Caspar and Klug deduced that many viruses must be built from icosahedral shells that contain more than 60 subunits, and developed a geometric theory for such arrangements if the subunits were given some flexibility in bonding to their neighbors (referred to as 'quasi-equivalence', since the flexibility gives rise to subunits that are not quite identical in their conformation and/or bonding properties). Some viruses, especially the tailed phages (*Caudovirales*), which are in large part the subject of this article, were found not to be this 'simple' and their repeating arrays of proteins are embellished with parts that are not icosahedrally symmetric. These asymmetric parts can be very complex; for example, phage T4 virions contain over 40 different protein species, of which a substantial majority are not icosahedrally arranged like the coat protein. Recent determination of the complete atomic structure of virions, procapids, or parts of them by X-ray diffraction (usually 2–4 Å resolution) and three-dimensional (3-D) reconstruction by superposition of electron micrographs of many particles and averaging of the resulting structure about the icosahedral rotational symmetry axes (current best is 6–8 Å resolution) has given rise to much detailed information about the structure of protein arrays in viruses in general and phages in particular, and very recent advances have allowed structural determination of virions' asymmetric parts to below 20 Å in advantageous cases (**Figure 1**).

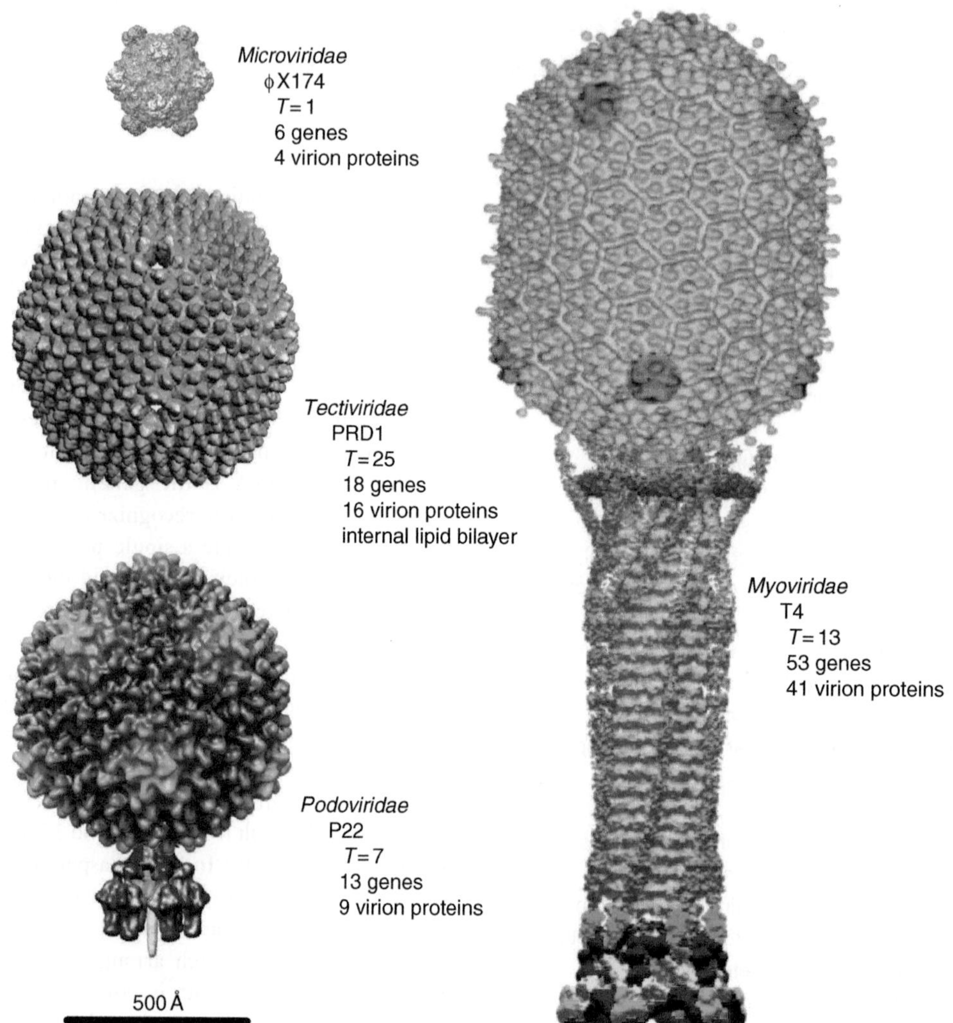

Figure 1 Bacterial virus structures. The structures of representatives from four bacteriophage families are shown at the same scale. The images are shown at 20 ± 5 Å resolution. The 'number of genes' indicated in the figure is the number of genes currently known to be involved in virion assembly and the 'virion proteins' is the number of these present in the completed virion. In φX174, DNA packaging into the virion is dependent upon DNA replication but the genes that encode the replication machinery are not included. In each case, different structural proteins visible from the outside are shown as different colors. The PRD1 structure shown is not the actual virion but only the arrangement of coat proteins. The T4 virion is shown in its 'tail fiber up' conformation, in which the tail fibers (dark green) are folded up and the whiskers are folded down, and both are lying along the body of the tail. The T4 image was provided by V. Kostyuchenko, A. Fokine, and M. Rossmann, the P22 image was provided by G. Lander and J. Johnson, and the other two images are from the Virus Particle Explorer website (http://viperdb.scripps.edu/index.php).

As structural information about virus particles and other macromolecular assemblies accumulated, scientists naturally wondered how such structures were built. Animal and plant viruses were difficult to study inside infected cells, and so the assembly of the bacterial viruses in the easier-to-study bacterial cells became an important model for understanding the mechanisms utilized in the assembly of many macromolecular structures in addition to the viruses themselves. Critical in these studies was the determination of 'parts lists' for the bacteriophages under study, and these were determined both biochemically by delineation of the different protein components and genetically by identification of the genes that encode the protein components; both approaches were crucial in understanding the true nature of phage virions and how they assemble. Although some eukaryotic virions do include host encoded proteins, this has not been found to be the case for phages; all of their protein components are encoded by their own genomes. Although small organic molecules (e.g., putrescine) have been found in phage virions, none have been shown to be essential structural components.

Phages are known that contain any one of the four types of nucleic acid, single-stranded DNA (ssDNA), double-stranded DNA (dsDNA), single-stranded RNA

Table 1 Types of bacteriophages

Phage type	*Best-studied phages*	*Major virion symmetry*	*Virion nucleic acid*	*Comments*
Microviridae	ϕX174	Icosahedral; protein fold A[a]	Circular ssDNA; 4–6 kbp	$T=1$
Inoviridae	f1, M13	Helical	Circular ssDNA; 5–9 kbp	Special proteins at both ends
Caudovirales Myoviridae	T4, P2	Icosahedral head; protein fold B[a]; long helical contractile tail	Linear dsDNA; 30–500 kbp[b]	*Caudovirales* heads range from $T=3$ to $T=52$ and can be isometric or elongated
Caudovirales Siphoviridae	λ, HK97, SPP1	Icosahedral head; protein fold B[a]; long noncontractile tail	Linear dsDNA; 22–140 kbp[b]	Heads can be isometric or elongated
Caudovirales Podoviridae	P22, T7, ϕ29	Icosahedral head; protein fold B[a]; short tail	Linear dsDNA; 18–75 kbp[b]	Heads can be isometric or elongated
Tectiviridae	PRD1	Icosahedral; protein fold A[a]	Linear dsDNA; 13–15 kbp	$T=25$; contains internal lipid bilayer
Corticoviridae	PM2	Icosahedral	Circular dsDNA; 10 kbp	$T=21$; contains internal lipid bilayer
Plasmaviridae	L2	Pleomorphic	Circular dsDNA; 2 kbp	Contains lipid bilayer
Leviviridae	MS2, R17	Icosahedral; protein fold C[a]	Linear ssRNA; 3.5–4.3 kbp	$T=3$
Cystoviridae	ϕ6	Icosahedral; protein fold D[a]	Linear dsRNA; 13–14 kbp	$T=13$; three different dsRNA molecules in virion; contains external lipid bilayer

[a]The protein folds (named arbitrarily in the table) of typical *Microviridae*, *Caudovirales* (*Siphoviridae*), *Leviviridae*, and *Tectiviridae* coat proteins have been determined by X-ray diffraction; that of *Cystoviridae* is suspected from its arrangement in virions which is similar to other dsRNA viruses.
[b]Approximate size ranges – *Caudovirales* phages are extremely diverse and the limits of genome sizes are not fully known.

(ssRNA), and double-stranded RNA (dsRNA), and they are built with icosahedral symmetry, helical symmetry, and a combination of the two. **Table 1** lists the eight formally named phage orders which appear to be largely unrelated to one another, and one of these, the *Caudovirales*, has been divided into three families with different tail types. Note that there are poorly studied phages that infect the bacteria class Mollicutes that have not yet been formally classified, and viruses that infect the Archeas are not discussed here. Given this great diversity of phage structures and assembly mechanisms, it is not possible to describe them all in detail here, and the reader is urged to consult other articles in this volume for more specific information on the various bacteriophage systems. In this article, we explore phage virion structure and the strategies and mechanisms by which virions assemble. Focus is largely upon the *Caudovirales*, since their assembly is the best understood, but other phage phyla will be mentioned as they are relevant.

Assembly Pathways

A major theme that emanates from the early study of phage assembly is the idea of an 'assembly pathway'. The complex phages (*Caudovirales*), which can have as many as 50 different proteins in their virions, typically do not synthesize these different proteins in the temporal order of their assembly. Instead, the different proteins are all synthesized simultaneously, and the inherent properties of the proteins dictate their order of assembly. Thus, virion assembly proceeds via a specific 'pathway' in which the components assemble in an orderly manner. How is such a program of assembly achieved? The generally accepted mechanism for this type of control, which is now known to occur in the construction of many macromolecular structures, not just phages, is that the component proteins change conformation upon binding to the growing structure. Thus, once an initiator complex (of say protein A) is generated, protein B binds to it and changes its conformation so that protein C can bind, protein C in turn changes its conformation so protein D can bind, etc. In some cases, the precise juxtaposition of proteins B and C on the growing structure may also contribute to the creation of a binding site for D. In such a scheme, the unassembled proteins B, C, and D have no affinity for each other and so do not assemble in alternate ways. Thus, an ordered 'assembly pathway' of protein A first creating an initiation complex, then B, then C and then D binding in this defined order is determined solely by the properties of A, B, C, and D proteins. Because of these properties, when one component is missing, in most cases the assembly intermediate to which the missing component would have bound accumulates, and downstream components remain unassembled. This has allowed the examination of many such intermediates in the characterization of phage assembly pathways. The isolation in relevant genes of conditionally lethal nonsense mutations, which are particularly tractable in phage systems, made this approach extremely productive. Several issues regarding the assembly pathway strategy remain poorly understood, and these include such questions as: What controls initiation of the assembly process? Control could simply be the synthesis and proper folding of a limiting amount of protein A in the example above, but in some situations such as procapsid assembly (below) it appears to be more complex. And what is the exact nature of the conformational changes that occur upon binding? These remain fertile ground for research.

Assembly Strategies and Mechanisms

Branched Pathways, Subassemblies, and Quality Control

Phages have apparently evolved ways of maintaining quality-control systems in virion assembly. One such system is assembly though 'subassemblies' or discrete assembly intermediates that are unable to join with other subassemblies unless they are properly assembled. Thus, if an incorrectly built subassembly (say missing one of 10 protein molecules) is defective and unusable, it is much less costly than if a complete virion (containing hundreds to thousands of protein molecules) were rendered defective due to one missing or defective protein molecule. Thus in the *Caudovirales*, for example, head precursors called procapsids are built first and most likely only package DNA if they are correctly assembled, tails are built independently (though a series of smaller subassemblies), and DNA-filled heads and tails only join when both are successfully completed. A second quality-control system may be that phage proteins have evolved to minimize assembly problems by ensuring that the participating proteins do not assemble unless their C-terminus is intact. Such a strategy ensures that prematurely terminated polypeptides (arguably the most probable translation error) are unable to participate, since they are missing their assembly site. Although this is an evolutionary argument, the fact that among the many phage assembly genes and proteins that have been studied, very few of the N-terminal fragments have 'dominant negative' genetic characteristics or are incorporated into the assembling virion.

Assembly of Symmetrical Protein Arrays

Assembly and maturation of icosahedral arrays

The use of an icosahedral protein shell to enclose phages' virion chromosomes appears to have evolved independently at least four times. Phage coat proteins are sufficiently ancient that even those with common ancestors are often not recognizably similar in amino acid sequence. We are thus left with comparison of their folded (secondary and tertiary) structures to deduce common or separate ancestries, and such structural information is not known for most phages. Nonetheless, high-resolution atomic structures are known by X-ray diffraction for a few members of several groups and those from the *Caudovirales*, *Tectiviridae*, *Microviridae*, and *Leviviridae* form three very different coat protein folding groups that are certainly not related. But do coat proteins from all members within each of these groups have a common ancestor? Our current limited knowledge suggests that this may well be true. High-resolution 3-D reconstructions of electron micrographs of several members of the *Podoviridae* and *Myoviridae* strongly suggest that all *Caudovirales* have the same unusual protein fold that has been determined for phage HK97. Finally, the *Cystoviridae* are thought to have a fourth type of subunit that is related to dsRNA viruses that infect eukaryotes.

Although much is understood about the actual structures of the phage (and other virus) icosahedral arrays that encase the nucleic acid in so many virions, in different viruses the size and shape of these arrays can vary greatly and control of these variations is poorly understood. No octahedral or tetrahedral viruses have been discovered. The geometry of icosahedral arrays is well understood, and if strict conformational and bonding identity is maintained among the participating chemically identical coat proteins, only exactly 60 subunits can be accommodated in the shell. However, if some flexibility (quasi-equivalence, above) is allowed, then the number of subunits (S) allowed in an isometric icosahedral shell is equal to $60T$ where $T = h^2 \times hk \times k^2$, where h and k are any integer including zero. This gives rise to the infinite series $T = 1, 3, 4, 7, 9, 12, \ldots$ and $S = 60, 180, 240, 420, 540, 720, \ldots$, respectively. Such shells can be thought of as having pentamers of the coat protein at each of the 12 icosahedral vertices and different numbers of hexamers making up the 20 triangular faces in different-sized shells. Phage heads are known with T numbers ranging from 1 to 52. Little is understood about exactly how the assembly of such shells is initiated or how their size is controlled. In many cases, the coat protein by itself will assemble into several sizes of shell as well as long tubes and irregular structures. In some cases (notably phages T4 and ϕ29), the icosahedral shell is elongated along the tail axis, and evidence from the systems under study suggests that both portal protein and scaffolding protein (see below) can be involved in this control, so factors beyond the innate coat protein properties clearly participate in the control of the spatial aspect of shell assembly.

Coat protein shells of the dsDNA phages assemble into a metastable 'immature' procapsid shell which undergoes a maturation that consists of a conformational transition to its final more stable state at about the time of DNA packaging. This maturation transition is typically accompanied by a significant expansion of the shell, the release of the scaffolding protein, and any proteolytic cleavage of the head proteins that might occur. This transition is surprisingly complex, and in the case of phage HK97, where it has been studied at the highest resolution, it proceeds through at least five separable coat protein conformational/organizational stages, in which there is little or no refolding of the main structural domain of coat protein, but there are significant changes in the orientations of the coat protein and the conformations of surface parts of the protein that are not within the central domain (**Figure 2**). The trigger(s) for this transition is not fully understood, but it clearly requires the successful completion of procapsid assembly. DNA entry and/or head protein proteolysis may be the natural signal for the transition to occur. This transition is universal in the assembly of tailed phages, and it has been suggested that topological difficulties in the construction of a closed shell with building units that have uneven surfaces might demand building a loosely bonded shell first, which subsequently rearranges into a more tightly bonded structure. The transition could also be part of a quality-control mechanism to ensure the structure is correctly built, and the concomitant expansion increases the available internal space for occupation by the DNA.

Symmetry disruptions and symmetry mismatches

The tails of the tailed phages (*Caudovirales*) and the portal of the *Tectiviridae* disrupt the icosahedral symmetry of coat protein shells of these virions, by being present at only one of the 12 otherwise identical vertices. How does this occur? Where it is known (phages P22, T4, T7, ϕ29), this 'disruption' replaces exactly one of the coat pentamers and it is present in the procapsid before the tail joins to the head. In procapsids, a dodecameric (12-mer) ring of portal protein occupies the position of the coat protein pentamer and tails will eventually bind to this portal protein. A single unique vertex is most simply explained

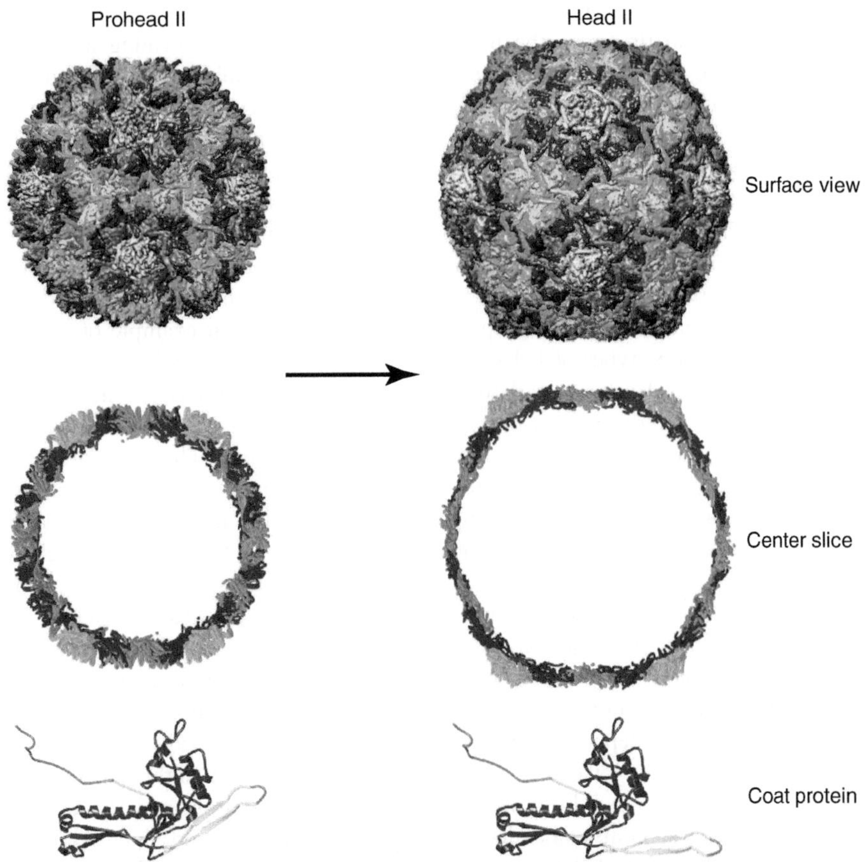

Figure 2 The maturation transition of the bacteriophage HK97 head. HK97 coat protein assembles into procapsids, its N-terminal 102 residues are removed by proteolytic cleavage, and the resulting structure is called prohead II. This coat protein shell undergoes a several-stage transition to become head II, which has expanded in diameter and has a thinner shell. In the surface view, the seven different 'quasi-equivalent' coat subunit locations are given different colors. In the slice view, the penton subunits are green and hexon subunits red or blue. In addition to subunit reorientation during the transition, the loop (called the E loop) shown at the lower right of the coat protein ribbon diagrams (the bottom tier) undergoes a major change in conformation during the transition. The images were provided by Lu Gan, Jack Johnson, and Roger Hendrix.

if the portal protein forms the site that initiates coat shell assembly, and, in agreement with this idea, it has been shown that the absence of portal protein affects the spatial aspects of coat protein assembly in several phages (e.g., T4, λ, ϕ29). However, in some other phages, such as P22, coat assembles quite normally in the absence of portal protein. In all these systems, other factors, especially scaffolding protein (below), clearly play an as yet imprecisely defined role in initiation of coat protein assembly and in creating the unique portal vertex. The 12-fold symmetry of the portal ring and the fivefold symmetry of the capsid at the portal vertex means that there is a 'symmetry clash' between these two parts of the virion, and thus there cannot be identical interactions between the individual coat and portal subunits; it has been suggested that this could allow rotation of the portal with respect to the coat shell, but this has not yet been shown experimentally to be the case.

Helical arrays

Cylindrically shaped helical and stacked-ring protein arrays are present in the long tails of *Myoviridae* and *Siphoviridae*. In both types of viruses, these arrays assemble (in the cases studied) from the tail tip up to the head proximal end to give rise to the tail subassembly. Both types of tails have a complex baseplate (or tail tip) structure at their head-distal ends. Baseplates assemble first, and when they are complete, they form the site at which the helical tail shaft begins to assemble. But helical arrays can be indefinitely long; so how are the discrete tail lengths programmed? Phage tails use a protein template (discussed in more detail below) to determine tail length, and when this length is attained, other proteins bind to the growing tip to block further growth. There are three types of helix (or stacked ring) building proteins, the *Myoviridae* tail tube (inner) and sheath (outer) proteins and the *Siphoviridae* tail shaft protein. Of these three, the sheath proteins are most highly conserved, and in many (but not all) tailed phages they are recognizably related in amino acid sequence. The other two are extremely variable, and with no structural information it is impossible to know whether or not they form discrete groups of distantly related proteins. Thus, there is no convincing evidence that the *Myoviridae* tail core and sheath and the *Siphovirdiae* tail shaft subunits have a common ancestor, and their different functionalities and roles suggests that they may have independent origins.

The *Inoviridae* or filamentous phages, unlike most other bacterial viruses, do not escape the host cell by lysing it. Their helical coat is assembled from subunits embedded in the cell membrane as the DNA is extruded through a complex assembly machine built into the membranes of the infected cell. Since they assemble in a such a different manner from the other phages under discussion here, and the detailed mechanism of their assembly is not understood in detail, they will not be considered in depth here.

Catalysis and Assembly

Scaffolds, templates, and jigs

Scaffolding proteins are proteins that assemble in fairly large numbers into procapsids but are not present in the completed virion. They act transiently to help in the proper assembly of coat protein shells and in the tailed-phage cases in the determination of the one unique vertex (above). The precise mechanisms by which scaffolding proteins perform these functions is unknown, but typically the interior space of *Caudovirales* procapsids contains several hundred molecules of scaffolding protein that are essential for proper procapsid assembly. They are then either proteolytically destroyed or released intact before DNA is packaged. Among the latter type, the scaffold of phage P22 has been shown to participate in an average of five rounds of procapsid assembly and so clearly acts catalytically in the assembly process. Stable oligomeric complexes between coat and scaffold molecules have not been found, yet scaffold 'nucleates' coat assembly (in the absence of all other proteins) at concentrations at which coat fails to assemble by itself. Scaffold remains as small oligomers in the absence of coat protein. It is not clear whether scaffold is forming a template whose surface guides coat assembly or is transiently binding and modifying the coat conformation so that it assembles correctly. *Microviridae*, in spite of their 'simple' $T=1$ coat protein shell, have both an internal and external scaffolding protein. In addition to roles in both temporal and spatial control of coat protein assembly, scaffolds may also function to fill the interior and thus exclude cellular macromolecules from the procapsid interior.

Perhaps the best example of an assembly template is the assembly of the *Caudovirales* tail shafts. *Siphoviridae* tail shaft subunits require two things to assemble, an initiation site on the completed baseplate and a template along which to assemble. This template has been best studied in phage λ, but even there tail shaft assembly is complex and not yet fully understood. It has been shown that the phage λ gene *H* (or 'tape measure') protein serves as a template for the assembly of the shaft subunit outward from the distal tip structure, and it is the length of the extended tape measure protein that determines the length of the tail shaft; shortening the tape measure protein causes the formation of a commensurately shorter tail. Interestingly, in these virions, the tape measure is thought to occupy the lumen of the tail, through which DNA must pass during injection, and the tape measure protein is ejected from the virion along with the DNA during the infection process. A second type of tape measure has been suggested for assembly of the phage PRD1 $T=25$ coat protein shell. The protein product of PRD1 gene *30* is stretched along the face edges of the icosahedral structure and may determine the length of the face and thus the size of the icosahedron.

A third type of spatial control of assembly is the 'jig', of which the phage T4 whisker is the prime example. This fiber protrudes from the head–tail junction, and its distal end binds to the elbow of the bent tail fiber, holding it in the correct position for joining of the tail fiber to the baseplate. The use of such a jig here may reflect some inherent difficulty in successfully attaching the tail fibers (see below).

Assembly enzymes and chaperonins

Aside from the scaffolding proteins, there are other well-studied examples of proteins that act catalytically during phage virion assembly. Perhaps the most curious of these is the gene *63* product of phage T4. This protein is required for the addition of tail fibers to the otherwise completed virion, but it is not present in virions. Its exact mechanism of action is not known, but it has been suggested that it and the jig described above are required to allow the creation of a flexible ball-and-socket joint between the fiber and the rest of the virion, which in turn might allow the fiber to 'search' more efficiently for its binding site on the surface of bacterial cells. In addition, T4 gene *38* protein appears to bind to and catalyze efficient trimerization of the subunit that forms the outer half of the bent tail fiber, but then it is released and not found in virions. Chaperonins are required in the folding of many proteins and so can be critical to phage assembly in that the proteins must be properly folded to participate in the assembly process. It is interesting to note that the chaperonin protein-folding function was discovered during the genetic analysis of the host's role in phage λ head assembly, and that some phages (e.g., T4) encode replacement subunits for the GroE chaperonin to ensure that their proteins are folded correctly.

Other catalytic actions during phage assembly are the ATP cleavage-dependent action of DNA translocases that move the DNA into the procapsid through the portal ring and the cleavage of overlength replicated DNA to virion length by the phage-encoded enzyme called 'terminase' in the *Caudovirales* (the nucleic acid packaging process is less well understood in other phages). Both of these enzymatic functions are essential for tailed-phage assembly, and the reader is referred to the article in this volume on nucleic acid encapsidation for a more detailed discussion. In addition, enzymatic proteolytic modification of virion proteins is often essential during assembly (below).

Nucleic Acid Encapsidation

The packaging of nucleic acid within virions is covered in more depth in other articles of this volume, but it is noted here that different phages utilize very different strategies to build a virion with nucleic acid on the inside and protein (and lipid in some cases) surrounding that nucleic acid. The packaging of an ssRNA within the virion molecule by members of the *Leviviridae* family occurs by concomitant assembly of coat protein molecules into a $T=3$ icosahedral structure and condensation of the RNA (which is predicted to be rather compact due to extensive secondary structure) by a process that is not understood in detail. Here, as apparently in all phages, recognition of a particular sequence in the viral nucleic acid is required to initiate packaging, thereby ensuring that only virus-specific nucleic acid is encapsidated. The other phage types either assemble a procapsid into which the nucleic acid is inserted by an energy-consuming nucleic acid translocase (*Caudovirales*, *Tectiviridae*, *Cystoviridae*) or assemble the virion around the ssDNA as it is extruded from the cell (*Inoviridae*).

Covalent Protein Modifications during Assembly

Protein cleavage

A common, but not universal, feature of phage assembly is the controlled proteolytic cleavage of some protein participants. This is best studied in the *Caudovirales*, and so they will be discussed here. It was discovered early that an N-terminal portion of all of phage T4's coat protein molecules are removed by proteolytic cleavage during assembly, and that this cleavage is dependent upon proper assembly of the coat protein shell. It has been suggested that the role of such coat protein trimming, which is common, is to make assembly irreversible, allow coat protein to find a more stable folded state, to alter coat protein function between virion assembly and DNA delivery, to help create room inside for the DNA, to be a kind of quality control, and/or to simply to remove part of the virion that is no longer needed after successful assembly. These of course are not mutually exclusive roles. Assembly-dependent cleavage of other virion assembly proteins such as portal proteins, scaffolding proteins and tape measure proteins, as well as the phage-encoded protease itself, have also been observed. Where it has been studied, a phage-encoded protease co-assembles into the procapsid and subsequently proteolyzes the head proteins, including itself. Tail protein cleavage is less well understood, and the protease that trims, for example, the C-terminus of phage λ tape measure protein has not been identified, but a host protease is responsible for the essential but non-assembly-dependent removal of the N-terminal region of a phage P22 virion protein that is injected with the DNA.

Protein cross-linking

Escherichia coli phage HK97 typifies a subset of *Siphoviridae* that are known to covalently cross-link all of their coat proteins as the final step in head shell maturation. Other covalent cross-links have been described between a few of the coat protein molecules of phage λ and a fragment of its putative head protease, but their role in assembly is not known. In HK97, each coat subunit is joined to its neighbors through lysine–asparagine side-chain isopeptide bonds. The topology of these cross-links is such that they form covalent rings of five and six subunits that are interlocked, thus making a sort of molecular 'chain mail'. This cross-linking is catalyzed by the coat protein itself after the shell has expanded, and the cross-links contribute to the stability of the head.

Lipid Membrane Acquisition

Several bacterial virus families have lipid bilayers in their virions (**Table 1**). These have been studied most in the *Tectiviridae* phage PRD1, which has a lipid bilayer between the icosahedral capsid shell and the internal dsDNA. This layer may aid in the protection of the intravirion nucleic acid, but also participates in DNA delivery into sensitive cells. Members of the *Tectiviridae*, *Cystoviridae*, and *Corticoviridae* all appear to acquire their lipid membrane in the cytoplasm from the host's membrane by as yet poorly understood mechanisms, after virus-encoded proteins have been inserted into the host membrane. Information is available on some of the virion proteins that interact with the lipid bilayers, but little is known about the detailed mechanism of the assembly of the membrane into these phages' virions. The *Plasmaviridae* phage L2, which infects a member of the bacterial *Mycoplasma* family, appears to bud from the host cell membrane. This kind of release may only be possible in the mycoplasmas, since they are the only bacteria that have no cell wall, but the morphogenesis of L2 has not been studied in detail.

Summary

Clearly, there is no single 'mechanism of bacteriophage capsid assembly'. Bacterial viruses are varied and complex, and they utilize many different mechanisms on the road to assembling completed virions. Two major take home lessons regarding phage assembly are (1) the ubiquitous use of obligate pathways in assembly processes and (2) the use of proteins that are essential for assembly but which are not present in the completed virion. The latter means that assembly cannot be understood by simply examining the properties of the components of the completed virion. In addition, the protein and lipid parts of virions are not simple containers designed solely for the protection of the nucleic acid inside. They are also sophisticated molecular devices that are designed to deliver their nucleic acid payloads into sensitive cells by mechanisms that are as varied as the ways in which they are assembled. No doubt, this requirement that virions be metastable 'spring-loaded' structures, that can spontaneously release their nucleic acid when the right external signal is received, is largely responsible for the complexity of phage virion structure and assembly. Because of this diversity and complexity, the study of phage assembly has shed considerable light on the mechanisms of many other macromolecular assembly processes.

Acknowledgments

The author thanks Lu Gan, Roger Hendrix, Jack Johnson, Victor Kostyuchenko, and Michael Rossmann for material for the figures.

See also: Assembly of Non-Enveloped Virions; Genome Packaging in Bacterial Viruses; Principles of Virion Structure; Structure of Non-Enveloped Virions.

Further Reading

Abrescia N, Cockburn J, Grimes J, *et al.* (2004) Insights into assembly from structural analysis of bacteriophage PRD1. *Nature* 432: 68–74.
Calendar R (ed.) (005) *The Bacteriophages*, 2. New York: Oxford University Press.
Casjens S (1985) An introduction to virus structure and assembly. In: Casjens S (ed.) *Virus Structure and Assembly*, pp. 1–28. Boston: Jones & Bartlett.
Casjens S and Hendrix R (1988) Control mechanisms in dsDNA bacteriophage assembly. In: Calendar R (ed.) *The Bacteriophages*, vol. 1, pp. 15–91. New York: Plenum.
Caspar D and Klug A (1962) Physical principles in the construction of regular viruses. *Cold Spring Harbor Symposium on Quantitative Biology* 27: 1–24.
Chiu W, Burnett R, and Garcea R (eds.) (1997) *Strucutral Biology of Viruses*. New York: Oxford University Press.
Fane B and Prevelige P , Jr. (2003) Mechanism of scaffolding-assisted viral assembly. *Advances in Protein Chemistry* 64: 259–299.
Fokine A, Chipman P, Leiman P, Mesyanzhinov V, Rao V, and Rossmann M (2004) Molecular architecture of the prolate head of bacteriophage T4. *Proceedings of the National Academy of Sciences, USA* 101: 6003–6008.
Hendrix R (2005) Bacteriophage HK97: Assembly of the capsid and evolutionary connections. *Advances in Virus Research* 64: 1–14.
Horn W, Tars K, Grahn E, *et al.* (2006) Structural basis of RNA binding discrimination between bacteriophages Qβ and MS2. *Structure* 14: 487–495.
Kostyuchenko V, Chipman P, Leiman P, Arisaka F, Mesyanzhinov V, and Rossmann M (2005) The tail structure of bacteriophage T4 and its mechanism of contraction. *Nature Structural and Molecular Biology* 12: 810–813.
Lander G, Tang L, Casjens S, *et al.* (2006) A protein sensor for headful viral chromosome packaging is activated by spooled dsDNA. *Science* 312: 1791–1795.
Poranen M, Tuma R, and Bamford D (2005) Assembly of double-stranded RNA bacteriophages. *Advances in Virus Research* 64: 15–43.
Steven A, Heymann J, Cheng N, Trus B, and Conway J (2005) Virus maturation: Dynamics and mechanism of a stabilizing structural transition that leads to infectivity. *Current Opinion in Structural Biology* 15: 227–236.
Wikoff W, Conway J, Tang J, *et al.* (2006) Time-resolved molecular dynamics of bacteriophage HK97 capsid maturation interpreted by electron cryo-microscopy and X-ray crystallography. *Journal of Structural Biology* 153: 300–306.

Genome Packaging in Bacterial Viruses

P Jardine, University of Minnesota, Minneapolis, MN, USA

Glossary

Chromosome Encapsidated nucleic acid polymer that is the genomic component of the phage virion.
Concatamer A long nucleic acid polymer made up of tandemly linked genomes.
Packaging ATPase Enzyme complex that binds to, or is part of, the procapsid and translocates the chromosome from the outside to the inside of the capsid.
Procapsid The preformed, precursor capsid into which the genome is packaged.
Prohead See procapsid.
Virion The mature, infectious phage particle.

Introduction

The parasitic phase of the virus life cycle begins with the delivery of a viral genome into the host cell. Therefore, the bringing together of the virus chromosome with the viral capsid can be considered the culmination of the process of virus assembly that yields the infectious particles that will continue the next round of host-dependent replication. Bacteriophages have evolved complex mechanisms by which they ensure that this process is both effective and efficient.

As in other virus types, the encapsidation of the bacteriophage chromosome occurs by one of two distinct pathways. The first is the co-assembly of the components of the proteinaceous capsid with the chromosome at the terminus of viral replication. By this mechanism, capsid components assemble around the phage chromosome via the principles of self-directed assembly, with large classes of single-stranded DNA (ssDNA) and single-stranded RNA (ssRNA) phages being general examples. The ssDNA filamentous phages of the family *Inoviridae*, such as M13, achieve co-assembly of chromosome and capsid via an extrusion process by which ssDNA chromosome, coated with binding protein, is translocated through the cell membrane. The chromosome picks up external protein components of the capsid at membrane associated assembly sites. This extrusion/co-assembly is ATP dependent. In contrast, ssRNA phages of the family *Leviviridae*, such as MS2, present a co-assembly of an icosahedral capsid around a highly structured ssRNA chromosome.

In the second general pathway of genome encapsidation, the focus of this article, capsid assembly and genome replication remain separated until they converge in an event during which the phage chromosome is packaged into a preformed capsid using a process distinct from those that drive co-assembly. Unlike co-assembly, genome packaging is often mediated by mechanisms that energetically drive the chromosome from the outside to the inside of the capsid. In many phage systems, this event requires the assembly of specialized molecular machinery and the input of energy to push the nucleic acid chromosome into a confined space against a concentration gradient that result in the compaction of the nucleic acid polymer by several orders of magnitude.

There are three types of genome packaging that represent three large groups of phages and are distinguished by the type of nucleic acid substrate that is packaged. These consist of (1) the double-stranded DNA (dsDNA) phages of the order *Caudovirales*, which includes the families *Podo-*, *Sipho-*, and *Myoviridae*, and others groups including the family *Tectiviridae*; (2) the ssDNA phages of the family *Microviridae*; and (3) the double-stranded RNA (dsRNA) phages of the family *Cystoviridae*. All three of these groups of phages have evolved complex and highly efficient strategies to select and translocate viral chromosomes from the cytosol of the host cell into a preformed capsid.

General Considerations

One of the most crucial aspects of genome packaging in bacteriophage, as in other viruses, is the appropriate selection of virus chromosomal nucleic acid polymer from the milieu of nucleic acid inside the host cell. This selectivity is achieved by the binding and recognition of precursor viral chromosomes by components of the genome packaging apparatus and targeting them to the preassembled precursor capsid. This targeting event can be coupled to other events such as genome replication, as in the family *Microviridae*, be part of the chromosomal maturation process, as in the dsDNA phages, or precede the final replicative stages of the chromosome as found in the family *Cystoviridae*. Nomenclature can be diverse throughout the literature, with precursor capsids being referred to as procapsids, proheads, or preheads, depending on the system. This article refers to the precursor capsid targeted for packaging as the prohead or procapsid as is the convention for each system.

Additionally, distinction must be made between the bacteriophage genome and the packaged chromosome that becomes part of the infectious progeny virion. In some cases, these two terms are synonymous, as in the family *Microviridae* and the dsDNA phages lambda, ø29, and PRD1, since the chromosome is limited to a single copy of the genome. Exceptions arise for two reasons: (1) the genome is segmented into multiple chromosomes, as in the family *Cystoviridae*; and (2) the chromosome is terminally redundant as a result of a headful packaging mechanism using a concatamer as a substrate, as in caudoviruses such as the T-phages, P22, SPP1, etc. A singular exception is the Mu-like bacteriophage, in which the chromosome is comprised of a complete copy of the phage genome flanked by host-derived DNA sequences, a result of excision of the phage DNA from the host chromosome.

Finally, the inclusion of the packaging motor complex into the mature virion is variable, depending on the phage. In general, the enzyme complex responsible for chromosome translocation detaches from the packaged capsid in the dsDNA phages and is replaced by tail organelles. In the family *Microviridae*, the packaging complex is part of the DNA replication complex that is similarly not retained on the capsid surface. In contrast, the packaging complex of the family *Cystoviridae* remains attached to the capsid and is part of the mature virion. A summary of the key components and processes involved in packaging is presented in **Table 1**.

dsDNA Packaging

Since dsDNA phages are the most prevalent biological pathogens in the biosphere, the dominant type of genome packaging is the dsDNA packaging seen in the families *Podo-*, *Sipho-*, *Tecti-*, and *Myoviridae* (**Figure 1**). As with other packaging motifs, dsDNA phages pre-assemble an empty, immature capsid that docks with the phage chromosome for packaging. Universally, DNA packaging in dsDNA phages involves three viral components: (1) the preassembled prohead or procapsid, (2) the matured DNA chromosome, and (3) the ATP-dependent DNA-translocating enzymes.

The dsDNA phage chromosome is packaged into an immature, icosahedral prohead. The architecture of the capsid in these dsDNA phages is different from other icosahedral phages. At a unique fivefold vertex of the prohead, a portal structure is embedded, replacing one of the 12 pentamers of the icosahedron. This portal is a 12-fold homo-oligomer of the phage-encoded portal protein forming a channel through the capsid wall. In phages with prolate heads, the portal vertex is always at one end of the long axis of the prohead. In most cases, it is believed that the portal is responsible for initiation of capsid polymerization in conjunction with an associated scaffold structure. The portal provides the binding site for the rest of the DNA packaging machinery and is the channel through which the DNA passes into, and ultimately out of, the capsid. After packaging, it is the axis around which the neck and tail components assemble, hence the designation connector for this structure in some systems.

The dsDNA phage prohead lattice is in an immature form prior to genome packaging. The immature prohead lattice is often physically less stable than in the mature virion. In some cases, accessory domains of the major capsid protein required for lattice assembly are proteolytically cleaved prior to packaging. In most phages, the conformational switch of the capsid lattice to the mature form occurs during packaging. This rearrangement of the capsid lattice is accompanied by an increase in the sixfold symmetry of the capsid hexons, and an overall increase,

Table 1 Packaging facts

Phage family	*Genome type*	*Point of entry into procapsid*	*Enzymes involved*	*Nucleic acid polymer translocated*	*Completion events*
Sipho-, Podo-, Myo-, Tectiviridae	dsDNA	Unique fivefold vertex via portal	Large and small packaging enzyme complex	Concatameric dsDNA (lambda, P22, SPP1, T-phages)	Cleavage of concatamer, motor detachment, and tailing
				Unit-length genome inserted into host chromosome (Mu)	Cleavage of chromosome from host genome, motor detachment, and tailing
				Unit-length chromosome with terminal proteins (phi29, PRD1)	Motor detachment and tailing
Microviridae	ssDNA	Twofold axis	Replication complex, J protein	ssDNA	
Cystoviridae	dsRNA	Fivefold vertex	Translocating NTPase complex	ssRNA	Replication of ssRNA to dsRNA

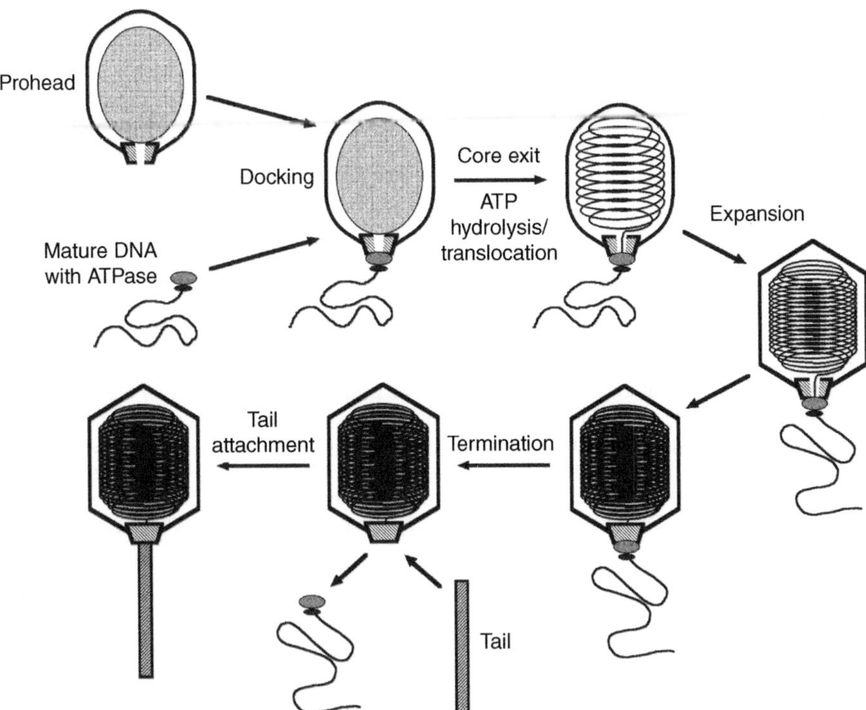

Figure 1 Schematic of generalized dsDNA phage assembly. A prohead interacts with the packaging ATPase holoenzyme–DNA complex via its head–tail connector. ATP hydrolysis powers translocation of the mature DNA, and at some point the scaffold core is ejected, either whole or following proteolysis. After an amount of DNA enters the head the shell capsomeres rearrange, making the head more angular and, in most phages, increasing the head volume. DNA translocation continues until a full complement of DNA enters the head, determined by either the unit length of the DNA, sequence recognition of the DNA length, or a headful mechanism. The ATPase–DNA complex detaches from the connector and is replaced by neck, tail, and/or tail fiber components, yielding a mature, infectious virion. Reproduced from Calendar R (ed.) (2006) *The Bacteriophages*. New York: Oxford University Press, with permission from Oxford University Press.

or expansion, of the head shell volume and consequent thinning of the shell thickness. Expansion ranges from the unperceivable, as in ø29, to the dramatic, as in HK97 where the capsid volume doubles.

Considerable diversity in the structural organization and maturation pathway of the DNA chromosome exists in these phages. DNA maturation converges at the point of chromosome translocation in that all dsDNA phages studied employ a common mechanism for genome translocation. The differences in DNA chromosome structure in the dsDNA phages arise as a result of differing strategies that have evolved to replicate phage dsDNA inside the cell. The central challenge, not faced by RNA phages, is the inability to initiate DNA-dependent DNA replication from a linear dsDNA without loss of information from the 5′ ends of the replicated DNA polymer. This 'riddle-of-the-ends' is solved by dsDNA phages by using one of four strategies: (1) replication of a circular form of the genome (P1); (2) producing multiple, tandemly linked copies of the genome, termed concatamers, in a single replicating molecule such that complete chromosomes can be produced by cleavage (P22, T-phages, SPP1); (3) replication of the genome as part of integrated segment imbedded in the host cell DNA genome (Mu); and (4) initiate replication of linear DNA using a priming mechanism dependent on a covalently linked terminal protein that does not lead to degeneracy of the chromosome (ø29, PRD1). The first three replication strategies require a DNA cleavage reaction as part of the packaging initiation process, since only linear dsDNA of a certain size can be packaged. Bacteriophage lambda combines the first two replication strategies in that early replication is by theta replication of its circular genome that switches to the lambda rolling circle mechanism to produce long, multigenome concatamers which serve as the packaging substrate. The fourth replication mechanism above does not require a cleavage event since the DNA substrate is present in the cell in unit length form.

The DNA substrate is recruited from the cytosol by the enzymatic components of the packaging apparatus. In general, there are two proteins involved, one large and one small, that form a holoenzyme complex. Their stochiometry, relative to each other and to the procapsid, is still unclear, with reports from various systems ranging from four complexes per motor up to six. The 'small' subunit is a DNA binding and recognition protein, while the 'large'

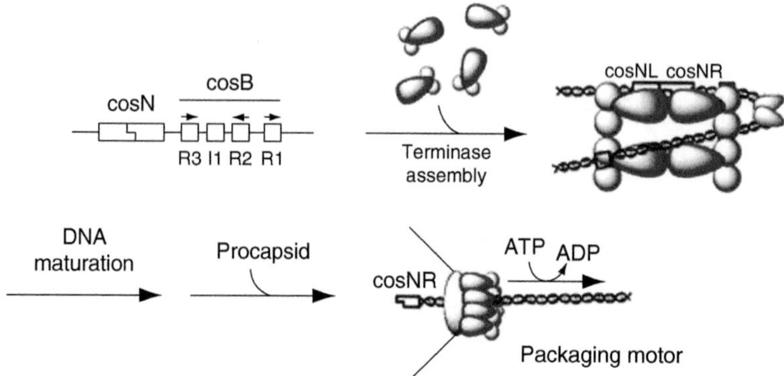

Figure 2 Model for terminase assembly at the cos site. The terminase protomer is a heterotrimer composed of one gpA subunit (red lobes) tightly associated with a gpNu1 dimer (blue spheres). Four protomers assemble at cos, resulting in bending of the duplex into a 'packasome' complex. Duplex bending by a gpNu1 dimer (purple lobes) bound to the R3 and R2 elements is central to the assembly of the packaging machinery. DNA maturation includes duplex nicking and separation of the DNA strands. Reproduced from Ortega M and Catalano CE (2006) Bacteriophage lamda gpNu1 and *E. coli* proteins cooperatively bind and bend viral DNA: Implications for the assembly of a genome packaging motor. *Biochemistry* 45: 5180–5189, with permission from ACS Publications.

component is an ATPase. These proteins assemble on the DNA substrate, with the small subunit recognizing a specific sequence in the DNA termed the pac site. This process has been particularly well described in phage lambda (**Figure 2**). It is believed that initial DNA-binding step is mediated by other factors of cellular origin (integration host factor (IHF) in the case of lambda). (In phages with terminal proteins at the DNA ends, the terminal protein is analogous to the small subunit.) The specificity of this recognition ensures that only phage chromosomes will be packaged into the receptive phage prohead. In the case of phages with either circular- or concatameric-replicated DNA, the large subunit of these enzymes cut the DNA to produce a free end. This endonculease activity is also responsible for production of a second endonuclease cut of the DNA at the end of packaging in phages that package segments of concatameric DNA, thus the designation 'terminase' for this enzyme ensemble in many phages (P22, lambda, T-even, T-odd, SPP1, Mu, etc). The designation terminase is not relevant to phages with unit length chromosomes such as ø29 and PRD1 since no cleavage is required; thus, these enzymes are more accurately termed packaging ATPases.

Once formed, the packaging ATPase–DNA complex docks with the prohead via the portal complex. The initial events that mediate the insertion of the DNA into the portal are unclear. Once initiated, DNA translocation occurs via an ATP hydrolysis-mediated process. The energy requirement for this process is clear when one considers the forces against which this molecular machine is working. First, entropy dictates that the concentration of DNA outside and inside the prohead should be equal. Thus, entropy works against dsDNA packaging throughout the process. Second, dsDNA is a relatively stiff polymer, with a persistence length (the minimum curvature of relaxed dsDNA) on the same scale as the capsid into which the DNA is being packaged in most phages. Therefore, energy is required to bend the DNA within the confines of the prohead. Finally, DNA has a net negative surface charge owing to the phosphate backbone of the polymer. As the DNA becomes compacted into the prohead, DNA strands are forced into close contact and electrostatic repulsion builds. Once again, energy must be invested into the DNA to achieve this end state. The end result is that the compacted DNA is pressurized inside the capsid to several atmospheres, requiring that the packaging motor complex must exert a translocating force on the order of 100 pN, making these molecular motors some of the strongest described in any biological systems. It is not clear how the DNA is organized to accommodate the degree of packing density, which approaches that of crystalline DNA. It is clear that the DNA must be pressed together in some form of hexagonal array that follows the contour of the capsid (**Figure 3**), which is the most energy-efficient conformation. Several models, including solenoids, liquid crystals, and folded toroids, have been suggested, but to date no structural study has been able to clearly resolve the structure of the packaged chromosome.

Packaging is terminated either by simply reaching the end of the unit-length chromosome (ø29, PRD1), or by a second DNA cleavage event necessitated by the translocation of a concatameric packaging substrate. In the latter, this cleavage event relies on the signaling from the packaged DNA density inside the capsid to the translocating complex on the outside of the capsid. In some cases, the near filling of the capsid permits the recognition and cleavage of the next approaching pac sequence, as with the lambda cos sequence, such that all progeny have identical genomes. In other instances, the second cleavage event is strictly 'headful', meaning the restriction of the DNA being packaged is sequence independent and relies on the detection of packaged DNA density inside the head. In this

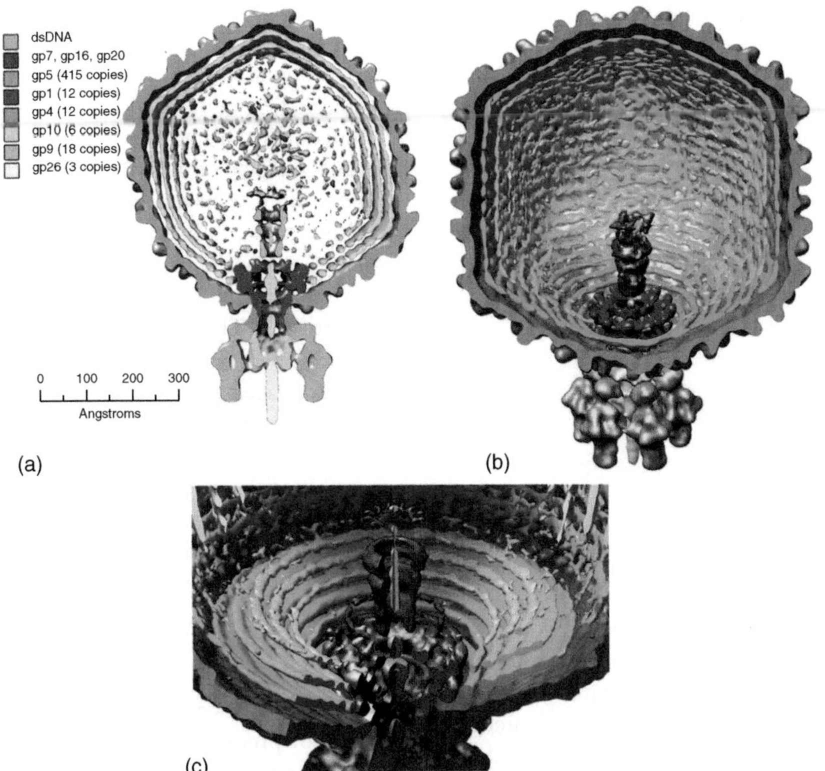

Figure 3 The interior features of the P22 virion. (a) The locations, deduced from many previous molecular biological studies, of the assembled gene products within a cutaway view of the reconstructed density of the P22 virion. Gene products 1, 4, 9, 10, and 26 make up the tail machine. Layers of dsDNA (green) are clearly visible as concentric shells within the capsid; they break into distinct rings of density near the portal vertex. Density (green) in the center of the channel formed by the ejection proteins (purple) could be the end of the P22 chromosome; however, density on this axis within the portal protein ring (red) does not appear to be consistent with DNA. (b) A cutaway view of the internal portion of the asymmetrically reconstructed particle contoured at 3, showing the 12-fold symmetry of the portal (red), the putative ejection proteins (purple), and individual strands of dsDNA (green). (c) Close-up view of the packaged interior upon 12-fold averaging along the tail tube axis. Although the E-proteins (purple) themselves in reality may or may not exhibit 12-fold symmetry, this view demonstrates the channel-like nature of the structure they form in the virion, as well as the dsDNA (green) that may be seated within their channel. Three concentric shells of spooled DNA are clearly visible. Reproduced from Lander GC, Tang L, Casjens SR, *et al.* (2006) The structure of an infections P22 vision shows the signal for headful DNA packaging. *Science* 312(5781): 1791–1795, with permission from American Association for the Advancement of Science.

case, the capsid volume has evolved to exceed the capacity required to package a chromosome the length of the genome; thus the packaged DNAs are terminally redundant. This terminal redundancy is crucial to the progeny in that it will buffer the genome against loss during the next replicative infection. How this signal is transmitted from the portal to the packaging enzymes is unclear.

As described above, prohead assembly often involves the assembly of a precursor capsid that is structurally distinct from the mature capsid. In many cases, the prohead is yet to undergo ejection of the core-scaffold components that mediate head shell polymerization into the size and shape required by the virus. Similarly, final conformational rearrangements that stabilize the capsid shell allow the binding of accessory stabilizing proteins to the head surface, or release of the packaging motor and assembly of neck and tail components required to make the completed virion infectious. The end result of the separation of these final events in assembly permits the specific targeting of immature proheads by the DNA packaging machinery and completion of assembly (i.e., packaging motor detachment and neck/tail assembly) only after packaging has been completed. Upon completion of packaging, the ATPase complex detaches from the packaged head and neck and tail components assemble. This generally involves the assembly of several neck proteins, followed by the assembly of the tail in completed form (e.g., T4, lambda) or by the stepwise addition of tail components to the packaged head (P22, ø29).

The mechanism of DNA translocations in these motors remains unresolved, owing mostly to the complex nature of the portal motor complex. The role of the ATPases is clear in that they bind and hydrolyze ATP, transferring the energy of this reaction to the physical movement of the DNA substrate. It is widely held that the ATPases therefore have an active role in translocation, and function in a manner similar to the helicases, as is clear in the dsRNA phage packaging complex. What is unclear is whether the portal protein actively participates in this process. Regardless of a direct role in translocation, the portal protein ring does appear to play a role in signaling the end of translocation, as illustrated by mutants in the portal proteins of SPP1 and P22 phages which package altered chromosome lengths. This suggests the head-full sensor for these phages lies in the portal protein, and that the signal for the cleavage of the DNA upon completion of packaging is transmitted to the ATPase by the portal ring.

ssDNA Packaging

The archetype ssDNA phage assembly system is øX174 due to its extensive genetic and structural characterization over recent decades. Unlike the dsDNA phages, genome packaging in ssDNA phages is coupled to genome replication. The substrate genome for replication is a closed, circular dsDNA that is copied using a complex of host cell enzymes. This replicating complex switches to rolling circle replication upon interaction with viral proteins A and C, a host helicase (the *rep* protein), and the procapsid. As the positive polarity DNA is extruded from the rolling circle, it is transferred into the procapsid through a depression in the twofold axis of the preformed icosahedral shell. Unlike dsDNA phages, this point of entry is not structurally unique like a portal, but the replication/packaging complex binding confers exclusion to the other twofold axis on the capsid surface. Packaging in øX174 is dependent on an ssDNA binding protein J, which enters the capsid during packaging, and binds the DNA via charge interaction. J protein also mediates interaction of the packaged DNA with the inner surface of the capsid. Changes in capsid size and stability also occur during packaging, as with the dsDNA phages. In contrast to the dsDNA phages, phages of the family *Microviridae* do not require the high chromosome density inside the capsid that requires extensive input of energy for packaging. Thus, the energy of translocation harnessed from coupled DNA replication and DNA binding is sufficient in generating the required packaging force to efficiently produce infectious progeny.

ssRNA Packaging

A large class of dsRNA phages, the family *Cystoviridae*, translocate their genome into preformed capsids. These phages are analogous in their replication and structure to the eukaryotic reoviruses, and are best characterized in the ø6 group. Unlike the dsDNA phages, the cystoviruses translocate a single-stranded nucleic acid polymer into the procapsid. This circumstance is the product of the replication strategy of the virus, during which positive-strand RNA synthesized during infection serves as both template for translation and substrate for genome packaging. Conversion of the ssRNA to dsRNA occurs inside the capsid after translocation at the end of the assembly pathway. Also, unlike the dsDNA or ssDNA phages, the dsRNA genome is segmented into three chromosomes of different lengths (S, M, and L for small, medium, and large, respectively). This requires the translocation apparatus to not only select chromosomal nucleic acid polymer as in the dsDNA phages, but to select and process multiple ssRNAs in sequence (designated s, m, and l in the ssRNA form prior to conversion to S, M, and L) to allow for fidelity in the assembly of the mature virion.

The translocating NTPase of this class of viruses, the P4 enzyme, has been extensively characterized by biochemical and structural analysis. (Unlike the ATPases in the dsDNA phages, the packaging enzymes in the family *Cystoviridae* are not limited to ATP, but can use other NTPs depending on the phage.) A hexameric complex of this enzyme is assembled at each fivefold vertex of the precursor procapsid. The mechanism by which only one of the 12 vertices is selected to translocate a bound ssRNA substrate is unclear. The complex is believed to function in a manner analogous to helicases, translocating ssRNA through the hexamer and into the capsid. The P4 prohead complex recognizes stem loop structures at the 5′ end of the ssRNAs to be packaged, presumably through binding sites on the capsid surface adjacent to the ATPase. The order of binding, and thus packaging, is controlled by the alteration of the capsid binding site that occurs as RNA is packaged. Thus, a naive procapsid binds the s fragment specifically, packages it, and in doing so undergoes a conformational change that alters the procapsid RNA binding domain such that it now binds the m fragment, which is then packaged, followed by the l fragment (**Figure 4**). The determinate of these conformational changes in the capsid to allow recognition of the variant 5′ ends of the chromosomes is the length of packaged RNA, rather than sequence.

Concluding Remarks

The phenomenon of genome packaging remains one of the most intensively studied processes of virion assembly, bringing together diverse but complementary fields of biochemistry, molecular genetics, structural biology, and biophysics. Since many of the phage studied have evolved mechanism analogous to eukaryotic viruses, efforts to resolve the mechanisms involved in genome packaging in phage promise to be reflected in their eukaryotic counterparts. Since genome translocation has no direct

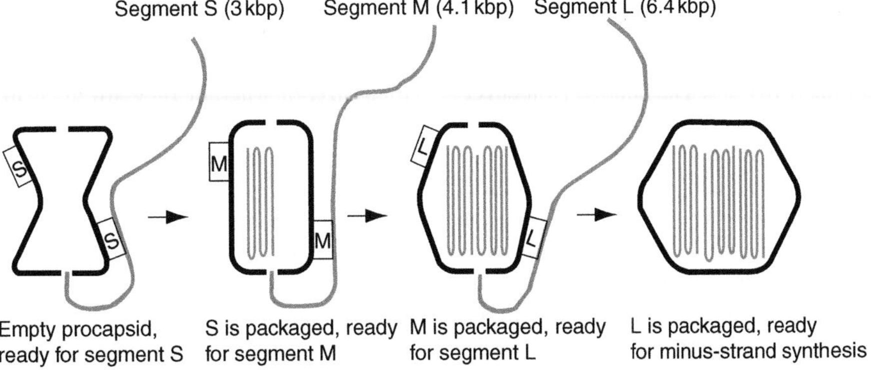

Figure 4 The packaging model of ø6. The empty procapsid shows only binding sites for S. After a full-size S is packaged, the S sites disappear and M sites appear. After a full-size M is packaged, the M sites disappear and L sites appear. After a full-size L is packaged, minus-strand synthesis commences. Reproduced from Calendar R (ed.) (2006) *The Bacteriophages*. New York: Oxford University Press, with permission from Oxford University Press.

link to most cellular processes, inhibition of genome packaging has been long held to serve as a future target for antiviral therapy once the mechanisms of chromosome and capsid maturation and chromosome translocation are resolved. Regardless of this practical consideration, the efficiency, complexity, and power of this process will remain a target of investigation for its own sake.

See also: Structure of Non-Enveloped Virions.

Further Reading

Calendar R (ed.) (2006) *The Bacteriophages*. New York: Oxford University Press.

Lander GC, Tang L, Casjens SR, *et al.* (2006) The structure of an infections P22 virion shows the signal for headful DNA packaging. *Science* 312(5781): 1791–1795.

Ortega M and and Catalano CE (2006) Bacteriophage lamda gpNu1 and *E. coli* IHF proteins cooperatively bind and bend viral DNA: Implications for the assembly of a genome packaging motor. *Biochemistry* 45: 5180–5189.

Viral Membranes

J Lenard, University of Medicine and Dentistry of New Jersey (UMDNJ), Piscataway, NJ, USA

Introduction

Viruses of many kinds possess lipids as integral components of their structure. Lipid-containing, or enveloped, viruses include: *Baculo-*, *Bunya-*, *Corona-*, *Filo-*, *Herpes-*, *Lenti-*, *Orthomyxo-*, *Paramyxo-*, *Pox-*, *Retro-*, *Rhabdo-*, and *Togaviridae*. Despite the great diversity of these viruses in regard to structure, replicative strategy, host range and pathogenicity, the function of the lipids in each case is the same – to form a membrane surrounding the encapsidated viral RNA or DNA genome. The lipids of the viral envelope form a continuous bilayer that functions as a permeability barrier protecting the viral nucleocapsid from the external milieu. Embedded in this bilayer are numerous copies of a limited number of virally encoded transmembrane proteins (often just one or two) that are required for virus entry into a potential host cell. These proteins mediate two essential functions: attachment of the virion to the cell surface; and fusion of the viral envelope with a cell membrane, resulting in accession of the viral nucleocapsid containing the genome to the cellular cytoplasm.

The membrane is acquired during viral assembly within an infected cell. This generally occurs by the budding of the previously assembled viral nucleocapsid through a cell membrane, incorporating host cell lipids and viral membrane proteins, but excluding virtually all cell membrane proteins. The particular cell membrane through which viral budding occurs is characteristic for each virus. Many viruses bud through the plasma membrane, but bunyaviruses bud through the Golgi apparatus, coronaviruses chiefly through the endoplasmic reticulum, and herpesviruses through the nuclear membrane. Poxviruses, which are among the largest and most complex animal viruses, are unique in acquiring two discrete membranes through a series of interactions with different cellular organelles, in a process that is not well understood.

Viral membranes have been extensively studied for many years, and for many reasons. The budded viral envelopes themselves represent isolated, readily purifiable subdomains of discrete cellular membrane bilayers. The envelope proteins that mediate entry of viruses into infectible cells are regarded as prototypes of membrane fusion machines. The initial binding of virions to potential host cells is mediated by prototypical ligand–receptor interactions in which viral proteins (ligands) bind to cell surface proteins (receptors). These properties of viral envelopes represent potential targets for antiviral therapy; they also provide models for important cellular phenomena.

Viral Bilayers

For many non-pathogenic viruses that can be readily purified in sufficient quantities in the laboratory, the bilayer arrangement of lipids in the viral envelope has been directly demonstrated using physical methods. It is assumed that the lipids in all viral envelopes are similarly arranged in a bilayer, based on those observations, and on the fact that the host cell membranes from which the viruses bud all contain bilayers, and also because it is the only physically reasonable arrangement of lipids that could supply the required protection from environmental stresses. Intact virions are impermeant to proteases and other enzymes. Indeed, virions can swell and shrink in response to changes in osmolarity, showing that viral envelopes are impermeant to small molecules and ions as well as large proteins, and thus must consist of intact bilayers that completely surround the encapsidated viral genome.

The lipid composition of various viruses grown under different conditions, and in different cell types, has been studied in detail. These have shown that wide variations in lipid composition are tolerated in many viruses; they have provided no evidence that any substantial fraction of envelope lipids is bound to viral envelope proteins specifically, or exists in a nonbilayer conformation.

Recent studies have shown that many viruses bud from specialized regions of the plasma membrane known as 'rafts'. Rafts are regions of the plasma membrane characterized by high concentrations of sphingolipids (sphingomyelin and sphingoglycolipids) and cholesterol. These components participate in the formation of separate, partially miscible phases, distinguishable from the more fluid phase(s) which are relatively enriched in unsaturated phospholipids. Raft and non-raft phases co-exist as contiguous bilayers, but diffusion is relatively restricted within raft phases, and exchange of molecules between phases occurs more slowly than diffusion within a single phase. Although the existence of local heterogeneity in lipid and protein structure has been demonstrated in cell membranes using a variety of physical and detergent extraction techniques, a precise definition of rafts has not been achieved. There is little agreement regarding their size, or whether they are nucleated by lipids or by proteins. Despite these uncertainties, certain membrane constituents are clearly concentrated in specialized regions rich in sphingolipids and cholesterol, and these have been identified as the sites of important cellular signaling and transport functions, and of assembly and budding for many enveloped viruses. Detailed lipid analyses of purified budded virions have confirmed their origin in rafts, and have in turn helped to define the lipid composition of virus-associated rafts.

Viral Membrane Proteins

The proteins of viral membranes, like those of other membranes, may be classified as either integral or peripheral. Integral proteins are those that cross the membrane bilayer at least once, and thus cannot be solubilized without disrupting the bilayer, that is, without using detergents. Peripheral proteins are also membrane associated, but they do not cross the membrane and they can be removed from it by treatment with aqueous salts, high pH or chaotropic agents, which leave intact the hydrophobic interactions that stabilize the bilayer.

Most integral membrane proteins of enveloped viruses span the bilayer only once, although exceptions exist. Each transmembrane (anchoring) domain is a sequence of 18–27 predominantly hydrophobic amino acid residues. Because transmembrane sequences are inherently insoluble in water, integral membrane proteins require detergents for extraction from the bilayer and solubilization. When detergents and lipids are both removed from purified viral proteins, they tend to aggregate into rosettes, forming a kind of protein micelle, with the transmembrane sequences clustered together at their centers in order to maximize hydrophobic interactions and minimize contact with water. Purified viral membrane proteins can be reinserted into lipid bilayers of defined composition by mixing the detergent-solubilized protein with lipids, then removing the detergent by dialysis. These reconstituted viral membranes ('virosomes') often possess the native receptor binding and fusion activities.

The functions of the major integral viral membrane proteins are: first, to attach the virus to the uninfected host cell; and second, to effect penetration of the genome into the host cell cytoplasm through membrane fusion of the viral envelope with a host cell membrane. As much as 90% of each viral receptor binding/fusion protein is external to the viral membrane and thus accessible to removal and/or degradation by added proteases. Viral membrane proteins are often morphologically identifiable in electron micrographs as 'spikes' on the outer surface of membrane particles. Under favorable conditions, nearly the entire external domain may be rendered soluble and recovered intact and correctly folded after proteolytic removal from the virion, facilitating crystallization and structural analysis. These domains possess oligosaccharide side chains identical to those of cellular integral proteins in structure and attachment sites, and often possess disulfide bonds as well, reflecting the viral proteins' normal procession through the cell's endoplasmic reticulum–Golgi system (**Figure 3**).

While these proteins constitute the major fraction (>95%) of viral integral membrane proteins, there is an additional class of small integral proteins that oligomerize within the bilayer to form channels that facilitate the transport of ions or small molecules. These proteins have been called 'viroporins'. They include the M2 protein of influenza, the 6K protein of alphaviruses, and Vpu of HIV-1. They are thought to function in various ways to facilitate the assembly and release of new viral particles from the infected cell.

Peripheral membrane proteins are attached to the viral membrane by a combination of hydrophobic and electrostatic interactions. They may penetrate the bilayer to some extent, but they do not cross it as integral proteins do. Viral peripheral membrane proteins include M1 of influenza, M of paramyxoviruses, and MA of retroviruses. All enveloped viruses except the togaviruses encode an M-like peripheral protein that functions to bring together the envelope and nucleocapsid components during viral assembly.

Cellular Virus Receptors: Virus Membranes as Ligands

The first step in infection, attachment of the virus to the outer surface of the host cell, is performed by specific membrane proteins of enveloped viruses. One or more unique cellular 'receptors' are recognized by each species or strain of virus. The presence of a specific cellular receptor is often the major factor determining the susceptibility of a particular species to infection; it also determines the infectibility of different tissues or cells within infected individuals.

Different viruses may bind to any of a large number of different cell surface proteins, carbohydrates, or lipids. Binding serves several purposes. Most generally, it attaches the virus to the uninfected cell, maintaining proximity, and increasing effective viral concentration on the cell surface. More specifically, interaction of viral spikes with specific cell surface proteins may initiate conformational changes that activate the viral proteins for fusion. Binding to certain cell surface proteins may also promote endocytosis of the virus, by any of several cellular pathways. Endocytosis introduces the viral envelope into the lower pH of the endosome, which is required for activation of some viral fusion proteins. Activation of certain cell surface receptors by virus binding may also initiate specific signaling cascades within the cell, which may be useful to the virus during subsequent steps of infection.

Each enveloped virus exhibits unique binding specificities of its membrane proteins with particular cell surface features, resulting in a unique combination of these effects. For example, the HIV-1 recognition protein gp120 exhibits a near total specificity for binding to the CD4 receptor on immune cells. Further activation of the virus' fusion protein gp40 occurs by interaction with a co-receptor, either the chemokine receptor CCR5 or CXCR4. In contrast, orthomyxo- and many paramyxoviruses have much broader specificity. Their recognition proteins (HA or HN, respectively) bind to sialic acid residues attached to various cell surface proteins or lipids. Different strains show preference for sialic acid in different covalent linkages. Rhabdoviruses such as vesicular stomatitis or rabies virus are still less specific, binding indiscriminately to negative charge clusters, whether created by lipids, proteins, oligosaccharides, or surface-bound polyanions. This nonspecific binding property helps to account for the broad host range of these viruses, although some have also been reported to bind specifically to acetylcholine receptors, which may explain their neurotropism.

Viral Fusion

Fusion of the viral envelope with a cell membrane is facilitated by integral viral membrane proteins. The best studied viruses (HIV-1, orthomyxo-, paramyxo-, retro-, toga-, and rhabdoviruses) each possess a single fusogenic glycoprotein, but herpes- and poxviruses may possess several that work together. Virosomes consisting only of a purified viral fusion protein reconstituted into a lipid bilayer vesicle fuse readily with protein-free lipid bilayers, suggesting that the fusion proteins can act on host cell lipids and do not require participation by host cell proteins.

Because of their ready availability and ease of purification, viral fusion proteins have served as prototypes for understanding biological fusion reactions. Several general principles have emerged, which have been found to apply to at least one major class of cellular fusion reactions (those mediated by proteins called SNAREs) as well as all characterized viral fusions.

First, both viral and cellular fusion proteins act directly on the lipid bilayer to facilitate rearrangements identical to those that occur during protein-free lipid bilayer fusion. Fusion is thought to occur through a series of steps, constituting the so-called 'stalk–pore' pathway (**Figure 1**). The two closely apposed bilayers (**Figure 1(a)**) dimple towards each other to form the 'stalk' (**Figure 1(b)**). This thins to form a 'hemifusion diaphragm' (**Figure 1(c)**),

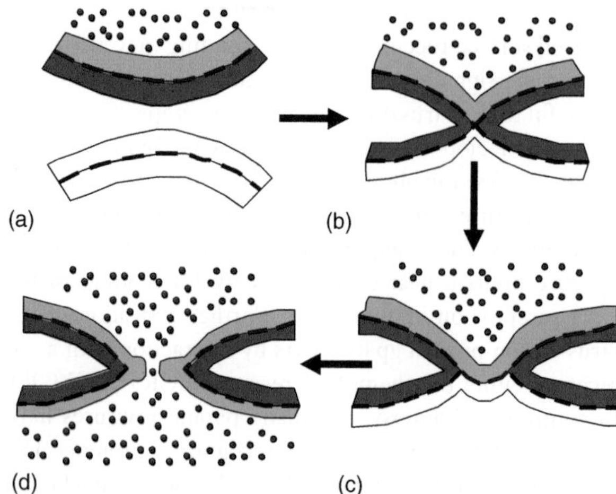

Figure 1 The stalk–pore mechanism of membrane bilayer fusion. (a) Initial pre-fusion state. The two leaflets of the cell membrane bilayer are colored green and red. The cytoplasm is indicated by blue dots. The viral membrane is colorless. (b) The stalk structure. In this state, continuity is established between the outer leaflets of the two membrane bilayers, allowing lipid mixing. (c) The fusion diaphragm constitutes a single bilayer separating the two aqueous compartments, comprising the inner leaflets of the cellular and viral membrane bilayers. The outer leaflets have already fused, and their lipids have mixed. (d) The fusion pore arises from rearrangement of the limited-area fusion diaphragm, perhaps facilitated by fusion proteins. Expansion of the fusion pore allows complete mixing of aqueous compartments and completes fusion. Fusion can occur even between protein-free lipid bilayers under certain conditions, but fusion proteins increase efficiency, probably by acting at each step in this pathway. Reproduced from Chernomordik LVand Kozlov MM (2005) Membrane hemifusion: Crossing a charm in two leaps. *Cell* 123: 375–382.

which now separates the two aqueous compartments by a single membrane bilayer in place of the two that separated them in the pre-fusion state. The hemifusion bilayer consists of the inner leaflets of the two original bilayers, the outer leaflets having already fused with consequent mixing of their lipid components. The hemifusion diaphragm can then rearrange to form a 'fusion pore' (**Figure 1(d)**), which must stabilize and widen to allow aqueous mixing, and thus complete the fusion reaction.

Precisely how viral membrane proteins promote fusion via the stalk–pore mechanism remains a subject of active research. Several properties of viral fusion proteins are known to be essential for activity, however. All viral and cellular fusion proteins are oligomeric, usually trimeric for virus fusion proteins. In most cases, several (probably 5–7) trimers must act cooperatively in order to complete the fusion reaction. One attractive idea is that the several fusion protein trimers encircle a limited area of bilayer, into which the fusion proteins can then transfer the energy released by their ensuing conformational transitions (**Figure 2**) in order to effect the lipid rearrangements required for fusion. Viral fusion proteins might potentiate any or all of the fusion steps: initial stalk formation, hemifusion diaphragm formation from the stalk, fusion pore formation from the hemifusion diaphragm, expansion of the initial fusion pore to complete fusion.

Within these general principles two distinct classes of viral fusion proteins have been recognized, possessing radically different architecture. Type I viruses include orthomyxo-, paramyxo-, retro-, and coronaviruses. Type II viruses include toga- and flaviviruses. Despite their different structures, these two classes of proteins facilitate the same lipid modifications during fusion.

All virus fusion proteins must remain inactive during biosynthesis and assembly so as to prevent premature, indiscriminate, counterproductive fusion within the infected cell. Both type I and type II fusion proteins are activated by a two-step process potentiated by interactions with a host cell. The inactive type I precursor protein is first cleaved at a specific site by limited proteolysis during assembly, generating a metastable, active form (**Figure 2(a)**). This then undergoes a conformational change mediated either by interaction with a specific cellular co-receptor (HIV-1), or by interaction with viral recognition proteins (paramyxoviruses) or by the low pH inside an endosome (orthomyxoviruses). Type II viral fusion proteins acquire their active conformation by an incompletely understood rearrangement of viral envelope proteins to form fusion protein trimers followed by interactions with specific lipids, notably cholesterol.

Activated fusion proteins of either type possess the following three structural features, which are required for complete fusion (**Figure 2**):

1. *Transmembrane domain.* This is the helical hydrophobic sequence (around 20 residues) that defines the fusion protein as an integral protein, and fixes it irreversibly in the viral bilayer. The transmembrane domain is inserted into the membrane bilayer during synthesis on membrane-bound ribosomes and appears not to rearrange during subsequent processing, activation or fusion. Type I and type II fusion proteins possess similar transmembrane domains. Not all transmembrane domains can participate in fusion reactions; a certain amount of conformational flexibility is required.

The transmembrane domain is required to complete the fusion reaction. Constructs in which the external, fusogenic domain of the influenza HA protein was attached only to the outer leaflet of the viral bilayer by a covalent bond with a lipid were capable of inducing hemifusion only, but were unable to complete the reaction.

2. *Fusion peptide.* This is a second relatively hydrophobic sequence that inserts into the cell membrane bilayer, and thus serves to bind the virus and cell membranes together. Exposure of the fusion peptide, enabling it to penetrate the target cell membrane, is an essential aspect of fusion protein activation, and requires a conformational change from a precursor form (**Figure 2(b)**). In type I proteins, this results

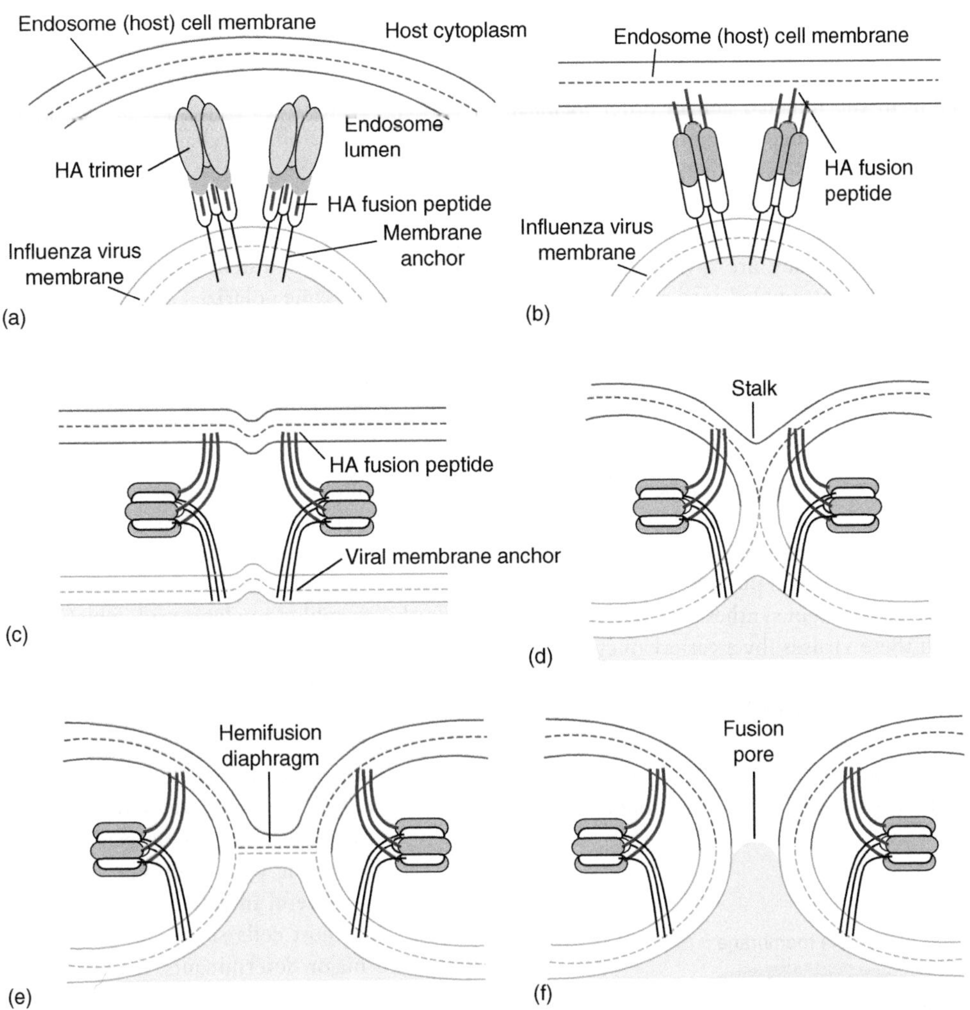

Figure 2 Proposed stalk–pore mechanism for membrane fusion mediated by a class I virus fusion protein, influenza hemagglutinin, HA. (a) The influenza virus has been internalized into the host cell endosome by receptor-mediated endocytosis (not shown). (b) Acidification of the endosome causes a conformational change in HA, which causes the fusion peptide (red) to become exposed; it subsequently inserts into the endosomal membrane. (c) Further refolding and clustering of HA trimers leads to bending of the two membranes toward each other. The resulting stalk (d) rearranges into the hemifusion diaphragm (e), where the virus interior is separated from the cytoplasm by a single bilayer, composed of the internal leaflets of the original membranes. This eventually rearranges, resulting in the formation of the fusion pore (f), which initially flickers and then dilates to complete the fusion reaction (not shown). Intra-leaflet lipid mixing, illustrated in **Figure 1**, is not shown. From Cross KJ, Burleigh LM, and Steinhauer DA (2001) Mechanism of cell entry by influenza virus. *Expert Reviews in Molecular Medicine* 6: 1–18, Cambridge University Press.

from proteolytic activation; the influenza fusion peptides, for example, comprise the newly created N termini. Type II fusion peptides are located in a protruding loop of the protein structure, which is exposed by a poorly understood interaction with lipids, notably cholesterol. As with the transmembrane domain, a certain amount of conformational flexibility is required in the fusion peptide.

3. *A rigid, oligomeric rod-like structure connecting the fusion peptide with the transmembrane domain.* This consists of a helical coiled-coil in type I proteins (**Figure 2(b)**), and an arrangement of β-sheet domains in type II proteins. Once the fusion peptide has inserted into the target membrane, fusion is completed by the rearrangement of this metastable structure to its lowest free energy form (**Figures 2(d)–2(e)**). In order to assume this form, the rigid oligomer folds back upon itself, forming a 'hairpin' (**Figures 2(d)–2(e)**), thus dragging the cell membrane, tethered by the fusion peptide, toward the viral membrane, tethered by the transmembrane domain. The free energy released by this rearrangement is transferred to the lipid bilayers, providing the energy necessary to complete the fusion reaction. In the final fused product, the fusion peptide and the transmembrane domain are adjacent to each other in the same membrane, held in proximity by the fully stable, rigid hairpin (**Figure 2(f)**).

Membrane Synthesis and Viral Assembly

Viruses generally use the housekeeping mechanisms already operating in the infected cell in order to make maximal use of their limited genomes. Hence, viral membrane protein synthesis is carried out on host cell membrane-bound ribosomes, which inserts them into the endoplasmic reticulum membrane in the correct orientation (**Figure 3**). There they are glycosylated by the host cell machinery and assembled into appropriate oligomers, as directed by their own primary amino acid sequence. Most viral glycoproteins are then passed on to the Golgi by cellular mechanisms, where they are further glycosylated by host cell enzymes. For this reason, the envelope proteins of vesicular stomatitis virus (VSV), influenza, and a few other enveloped viruses have provided valuable tools to study the glycosylation and transport of membrane proteins through the cellular endoplasmic reticulum–Golgi–plasma membrane system. Because host cell protein synthesis is often inhibited by infection with these viruses (by a variety of cytopathic mechanisms), large amounts of a single viral membrane protein are produced and correctly processed in infected cells, without competition by cellular proteins.

Likewise, viral proteins are targeted to specific cellular locations by cellular processes. The viral proteins display the same amino acid sequence 'addresses' as host cell proteins, which are recognized by the host cell glycosylation and transport machinery. For example, the single VSV glycoprotein, named G, is glycosylated in the endoplasmic reticulum, the oligosaccharide is modified in the Golgi, and the mature protein is targeted to the basolateral plasma membrane of polarized cells after entry into the late Golgi. The influenza HA protein, on the other hand, is glycosylated and delivered to the apical plasma membrane of the same polarized cells after passage through the intracellular membrane system. The retention of coronavirus glycoproteins by the endoplasmic reticulum, and of bunyavirus glycoproteins by the Golgi, reflects the operation of the same cellular mechanisms that retain resident cellular proteins in these organelles. The localization of viral membrane proteins in turn, determines the location of viral assembly and budding.

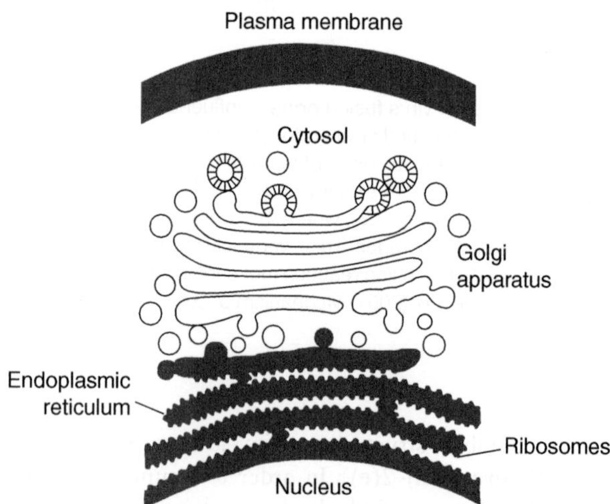

Figure 3 The endoplasmic reticulum–Golgi–plasma membrane system of a cell. All viral and cellular integral membrane proteins are synthesized by ribosomes bound to the endoplasmic reticulum membrane. Proteins destined for the plasma membrane are transported first to the proximal region of the Golgi (the *cis* face), then sequentially through the Golgi cisternae, to the *trans* face and out to the plasma membrane. In polarized cells, targeting to the apical or basolateral surface of the cell occurs from the *trans* face of the Golgi. Assembly and budding of different enveloped viruses occurs at characteristic points within this membrane system.

The budding process consists of the wrapping of a viral glycoprotein-enriched piece of membrane around the previously assembled nucleocapsid, which contains the viral genome (**Figure 4**). Remarkably, the completed viral envelope contains viral proteins and host cell lipids, with host cell membrane proteins being almost completely excluded. There is much less discrimination, however, between different viral proteins than between viral and cellular proteins, since viral envelope proteins of one kind can assemble with nucleocapsids of another, resulting in the formation of pseudotype virions. Pseudotypes have proven useful in redirecting specific viral genomes to alternate host cells since membrane attachment proteins are major determinants of host cell specificity (see above). It has been suggested that the discrimination between viral and cellular membrane proteins may arise from the exclusion of cellular proteins from virus-associated raft-like lipid phases.

Different viruses have been described, that bud at every stage in the endoplasmic reticulum–Golgi–plasma membrane pathway (**Figure 3**). While paramyxo-, orthomyxo-, rhabdo-, and togaviruses (and many others) generally bud from the plasma membrane, they have also been shown to bud intracellularly under certain conditions. Some retroviruses assemble at the plasma membrane, while others do not; this has provided a classical basis for distinguishing between different types of retroviruses. Other viruses normally bud intracellularly, from the endoplasmic reticulum or Golgi apparatus, for example, coronaviruses and bunyaviruses respectively, but these have also been observed to bud further down the pathway. In these cases, the nucleocapsid assembles on the cytoplasmic face of the membrane, and then buds into the intracellular organelle. The newly formed virion may then be secreted out of the cell through the normal secretory pathway, although this does not always occur efficiently.

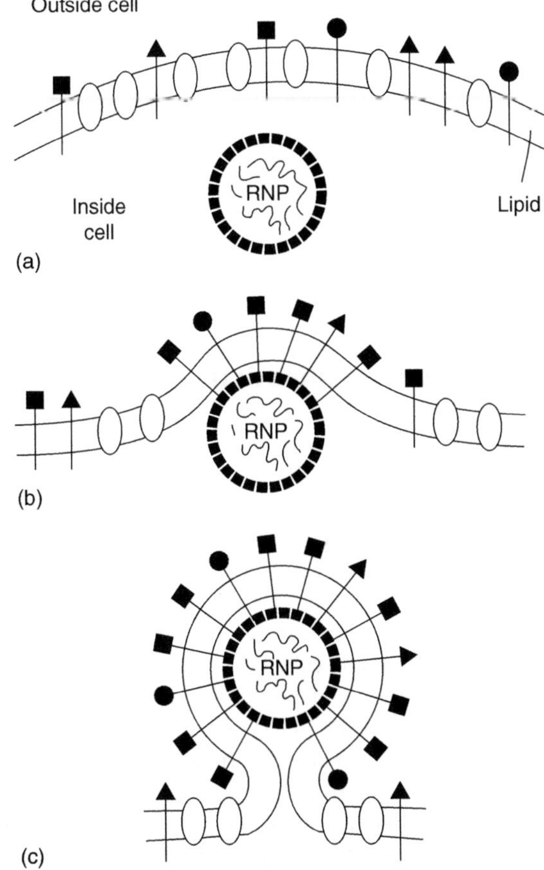

Figure 4 One kind of virus budding. Viral glycoproteins, inserted into the cellular membrane at the endoplasmic reticulum and processed through the Golgi to the plasma membrane (see **Figure 3**), associate with the assembled viral nucleocapsid. The direct association pictured here is characteristic of togaviruses. For other viruses, possessing helical nucleocapsids, the association is mediated by a peripheral membrane protein. Cellular membrane proteins are excluded from the envelope of the mature virion. This may occur during assembly, as pictures, or by prior formation of a viral membrane patch (or raft), before the nucleocapsid arrives at the membrane.

In all enveloped viruses except togaviruses, budding is mediated by a peripheral membrane protein, usually called M or MA, which links the glycoprotein-containing patch of lipids with the viral nucleocapsid, containing the viral genome. The M proteins interact specifically with nucleocapsids of their own viral species, but they do not always interact specifically with the corresponding viral glycoproteins. Instead, they may concentrate on the cytoplasmic side of the raft-like lipid phases that accumulate various viral glycoproteins. This could provide the structural basis for the formation of pseudotype virions, and might explain why many viruses contain widely varying ratios of glycoproteins to M or nucleocapsid proteins.

In contrast, togaviruses, which lack any M protein, possess an icosahedral nucleocapsid, which interacts directly with the cytoplasmic domain of the viral membrane protein. Completed virions contain an equal number of nucleocapsid and membrane protein molecules. Both are in a similar geometric arrangement, mediated by specific protein–protein interactions between them.

As described above, the lipids of the viral membrane are taken from the host cell membrane during budding. No new lipids are specifically synthesized in response to viral infection. Alterations in cellular lipid metabolism have been reported to result from some viral infections in cultured cells, but these are probably secondary to other cytopathic effects; there is no indication that they play an important role in the progress of infection.

See also: Baculoviruses: Molecular Biology of Nucleopolyhedroviruses; Orthomyxoviruses: Molecular Biology.

Further Reading

Alberts B, Johnson A, Lewis J, Raff M, Roberts K, and Walter P (2002) *Molecular Biology of the Cell,* 4th edn., chs. 10–13. New York: Garland Science (Taylor and Francis Group).

Briggs JAG, Wilk T, and Fuller SD (2003) Do lipid rafts mediate virus assembly and pseudotyping? *Journal of General Virology* 84: 757–768.

Chernomordik LV and Kozlov MM (2005) Membrane hemifusion: Crossing a chasm in two leaps. *Cell* 123: 375–382.

Cross KJ, Burleigh LM, and Steinhauer DA (2001) Mechanism of cell entry by influenza virus. *Expert Reviews in Molecular Medicine* 6: 1–18, Cambridge University Press.

Gonzalez ME and Carrasco L (2003) Viroporins. *FEBS Letters* 552: 28–34.

Harrison SC (2006) Principles of virus structure. In: Knipe DM, Howley PM, Griffin DE, *et al.* (eds.) *Fields Virology,* 5th edn, pp. 59–98. Philadelphia: Lippincott Williams and Wilkins.

Jardetzky TS and Lamb RA (2004) A class act. *Nature* 427: 307–308.

Kielian M and Rey FA (2006) Virus membrane fusion proteins: More than one way to make a hairpin. *Nature Reviews Microbiology* 4: 67–76.

Simons K and Vaz WLC (2004) Model systems, lipid rafts and cell membranes. *Annual Review of Biophysics and Biomolecular Structure* 33: 269–295.

Smith AE and Helenius A (2004) How viruses enter animal cells. *Science* 304: 237–242.

Mimivirus

J-M Claverie, Université de la Méditerranée, Marseille, France

Glossary

16S rDNA The gene encoding the small ribosomal RNA molecule, a universal component of all cellular prokaryotic organisms, the sequence of which is used for identification and classification purposes.

Aminoacyl-tRNA synthetases The highly specific enzymes responsible for the loading of a given amino acid onto its cognate tRNA(s). These enzymes are at the center of the use of the genetic code, together with the specific tRNA/mRNA recognition mediated by the anticodon/codon pairing on the ribosome.

COG Cluster of orthologous groups. Families of evolutionarily conserved genes, usually associated with well-defined protein functions and 3-D structures.

NCLDV Nucleocytoplasmic large DNA viruses, a group of large dsDNA viruses, the replication of which involves a stage the host cell cytoplasm and the host cell nucleus. Although presently very diverse in size, host range, and morphology, these are suspected to have originated from a common ancestor, contemporary to the emergence of the eukaryote lineage.

Paralogs Copies of similar genes (originated by duplication) found in the same genome.

Proteomics New methodogical approaches allowing the whole protein complement of an organism (or a virus) to be identified.

Introduction

The recent discovery in 2003 of acanthamoeba polyphaga mimivirus and the analysis of its complete genome sequence sent a shock wave through the community of virologists and evolutionists. Its particle size (750 nm), genome length (1.2 million bp), and large gene repertoire (911 protein coding genes) blur the established boundaries between viruses and parasitic cellular organisms. In addition, the analysis of its genome sequence identified many types of genes never before encountered in a virus, including aminoacyl-tRNA synthetases and other central components of the translation machinery previously thought to be the signature of cellular organisms. The information available on this giant double-stranded DNA (dsDNA) virus mostly consists of electron microscopy (EM) images, its genome sequence, and proteomic data. Very little is yet known about its pathogenicity.

History and Classification

In 1992, following a pneumonia outbreak in the West Yorkshire mill town of Bradford (England), a routine investigation for legionella (a pneumonia causing intracellular parasitic bacteria) within the amoeba colonizing the water of a cooling tower led Timothy Rowbotham from the Britain's Public Health Laboratory Service to discover a microorganism resembling a small Gram-positive coccus (initially called Bradford coccus). Cultivation attempts failed, and no amplification product was obtained with universal 16S rDNA bacterial primers at that time (**Figure 1**). The mysterious sample was stored in a freezer for about 10 years, until it reached the laboratory of Prof. Didier Raoult, at the school of medicine in Marseilles, France. There, EM of infected *Acanthamoeba polyphaga* cells provided the first hints that Bradford coccus was in fact a giant virus, with mature icosahedral particules *c.* 0.7 μm in diameter, a size comparable to that of mycoplasma cells (**Figure 2**). The viral nature of

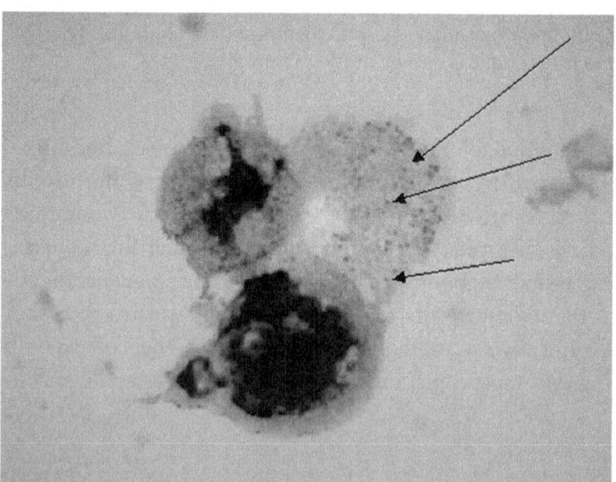

Figure 1 Coccus-like appearance of mimivirus particles (indicated by arrows) under the light microscope after Gram coloration in infected amoeba. Reproduced from La Scola B, Audic S, Robert C, *et al.* (2003) A giant virus in amoebae. *Science* 299: 2033, with permission from American Association for the Advancement of Science.

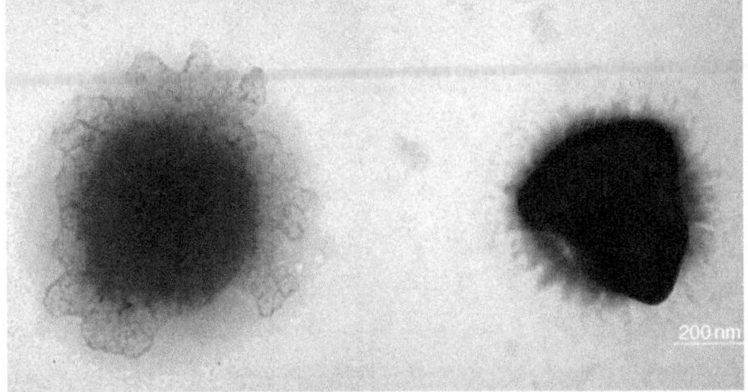

Figure 2 Comparable size of a mimivirus particle and a mycoplasma cell. EM picture of a co-culture. Reproduced from La Scola B, Audic S, Robert C, *et al.* (2003) A giant virus in amoebae. *Science* 299: 2033, with permission from American Association for the Advancement of Science.

the agent was definitively established by the demonstration of an eclipse phase during its replication, and the analysis of several gene sequences exhibiting a clear phylogenetic affinity with nucleocytoplasmic large DNA viruses (NCLDVs), a group of viruses including members of the *Poxviridae*, the *Iridoviridae*, the *Phycodnaviridae*, and the *Asfarviridae*. This new virus was named mimivirus (for mimicking microbe) and is now classified by ICTV as the first member of the *Mimiviridae*, a new family within the NCLDV. The size of its particles makes mimivirus the largest virus ever described.

Host Range and Pathogenicity

Among a large number of primary or established cell lines from vertebrates or invertebrates that were tested for their ability to support mimivirus infection and replication, only cells from the species *Acanthamoeba polyphaga*, *A. castellanii*, and *A. mauritaniensis* could be productively infected by a cell-free viral suspension. This narrow range of target cell specificity, restricted to protozoans, apparently conflicts with other reports suggesting that mimivirus might be an amoebal pathogen-like legionella, causing pneumonia in human. Numerous seroconversions to mimivirus have been documented in patients with both community- and hospital-acquired pneumonia. In addition, mimivirus DNA was found in respiratory samples of a patient with hospital-acquired pneumonia and a laboratory infection of a technician by mimivirus has also been reported. The patient's serum sample was found to strongly react with many mimivirus proteins, most of them unique to the virus, making cross-reactions with other pathogens unlikely. Isolation of the mimivirus from an infected patient is yet to be accomplished in order to formally establish that it is indeed pathogenic to human. In the meantime, as a precautionary measure, mimivirus should be considered a potential pneumonia agent and manipulated as a class 2 pathogen. In this context, it is worth noticing that mimivirus particles have been shown to remain infectious for at least 1 year at temperature of 4–32 °C in a neutral buffer.

Replication Cycle

Upon infection of *A. polyphaga* cells, mimivirus exhibits a typical viral replication cycle with an eclipse phase of 4 h post infection (p.i.), followed by the appearance of newly synthesized virions at 8 h p.i. in the cytoplasm, leading to the clustered accumulation of viral particles filling up most of the intracellular space, until infected amoebae are lysed at 24 h p.i. Little more is known about the details of the various stages of the replication cycle. Combining various hints from genomic and proteomic analysis as well as from EM suggests the following infection scenario (**Figure 3**):

1. free virus particles mimicking bacteria (by their micron size and perhaps the lipopolysaccharide (LPS)-like layer surrounding the capsid) are taken up as putative food by the amoeba;
2. the LPS-like layer is digested within the amoeba endocytic vacuole, making protein at the surface of the capsid accessible for a specific interaction at the vacuole membrane;
3. the content of the capsid is then injected into the amoeba cytoplasm, eventually using a specialized apparatus visible as a 600 Å 'vertex' (see below) leaving the empty capsid in the endocytic vacuole (documented by EM); and
4. early transcription events occur in the cytoplasm, as suggested by the proteomic content of the capsid.

It is not clear if, how, and when the virus genome gets into the amoeba nucleus (for its replication) or if the synthesis of particle components and capsid assembly only proceeds from 'virus factories' located in the cytoplasm.

Genome Organization

The mimivirus genome (**Figure 4**) consists of a single dsDNA, 1 181 404 bp long, within which 911 protein-coding genes are predicted. Two inverted repeats of about 900 nt are found near both extremities of the sequence, suggesting that the genome might adopt a circular topology as result of their annealing. The genome nucleotide composition is 72% A + T and exhibits some level of local strand asymmetry revealed by plotting the A + C strand excess as well as the gene excess (number of genes expressed from one strand minus the number of genes expressed from the other strand). Both graphs exhibit a slope reversal near

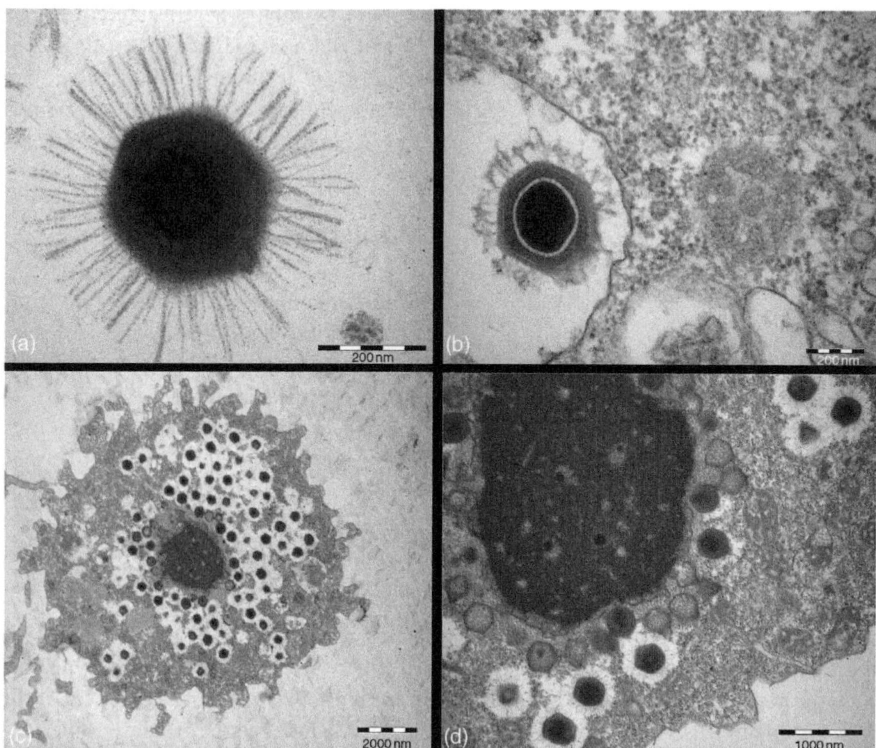

Figure 3 EM pictures of mimivirus particles at various maturation stages. (a) Free particle (see **Figure 6** for a different aspect of the fibers recovering the capsid). (b) Mimivirus particle in an amoeba vacuole, exhibiting a damaged (digested) fiber layer. (c) Clustered particles in the cytoplasm of an amoeba, 8 h p.i. (d) Close-up of a virus factory surrounded by particles at various stages of maturation: empty particles, filled particle (with a dark central core) without their translucent capsular material, mature particles (central dark core, white halo of cross-linked fibers). Reproduced from Suzan-Monti M, La Scola B, and Raoult D (2006) Genomic and evolutionary aspects of mimivirus. *Virus Research* 117: 145–155, with permission from Elsevier.

position 380 000 of the genome, as found in bacterial genomes, and associated with the location of the origin of replication. Mimivirus genes are preferentially transcribed in the orientation going away from this putative origin of replication, thus defining a 'leading strand' (with 578 genes) and a 'lagging strand' (with 333 genes). Such a strong asymmetrical distribution of genes is unique among known nucleocytoplasmic large DNA viruses. Despite this local asymmetry, the total number of genes on both strands are very similar (450 'R' vs. 461 'L' genes). The overall amino acid composition of the predicted proteome is strongly biased in favor of residues encoded by codons rich in A+T such as isoleucine, asparagine, and tyrosine. The relative usage of synonymous codons for any given amino acid is also biased by the high A+T percentage. This has made the production of recombinant mimivirus proteins difficult in traditional *Escherichia coli* expression systems. Paradoxically, the preferred codon usage in mimivirus genes is almost the opposite to the one exhibited by its host *Acanthamoeba castellanii* or *A. polyphaga* (the most frequently used codons being the rarest in the amoebal genes). It is also quite different from the more even distribution observed in human and vertebrate genes.

Mimivirus genome contains multiple traces of gene duplications, including a few ancient en bloc duplication events, many dispersed individual gene duplications, and recurrent tandem duplication events leading to large families of co-localized paralogs. Overall, one-third of mimivirus genes have at least one paralog. The analysis of gene collinearity indicates that genome segments [1–110 000] and [120 000–200 000] have been duplicated at positions approximately symmetrical with respect to the chromosome center. The largest paralogous families include an ankyrin-repeat-containing protein (with 66 members), BTB/POZ-domain-containing proteins (26 members), and a 14-paralog family of proteins of unknown function, of which 12 are arranged in a perfect tandem repeat. In contrast to parasitic/endosymbiotic bacteria with genome of comparable size (such as *Rickettsia*), the mimivirus genome exhibits no pseudogene, and thus no sign of an ongoing genome reduction/degradation process. Mimivirus does not appear to be under any evolutionary pressure to decrease the size of its genome. The segmental duplication and massive individual gene duplication explain the origin of a large part of the mimivirus genome, without the need to invoke an exceptional propensity for horizontal gene transfers.

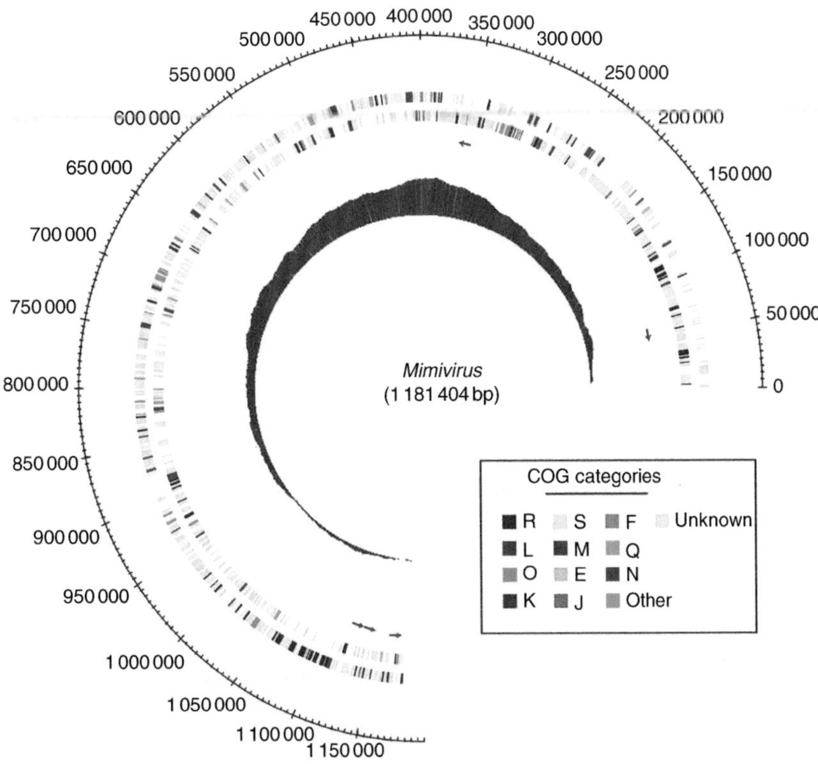

Figure 4 Map of the mimivirus chromosome. The predicted protein-coding sequences are shown on both strands and colored according to the function category of their matching COG. Genes with no COG match are shown in gray. Abbreviations for the COG functional categories are as follows: E, amino acid transport and metabolism; F, nucleotide transport and metabolism; J, translation; K, transcription; L, replication, recombination, and repair; M, cell wall/membrane biogenesis; N, cell motility; O, post-translational modification, protein turnover, and chaperones; Q, secondary metabolites biosynthesis, transport, and catabolism; R, general function prediction only; S, function unknown. Small red arrows indicate the location and orientation of tRNAs. The A + C excess profile is shown on the innermost circle, exhibiting a peak around position 380 000. The leading strand is defined as the one directed outward, from this point. Reproduced from Raoult D, Audic S, Robert C, *et al.* (2004) The 1.2-megabase genome sequence of mimivirus, *Science* 306: 1344–1350, with permission from American Association for the Advancement of Science.

Promoter Structure

A unique feature of the mimivirus genome is the presence of the motif AAAATTGA in the 150 nt upstream region of nearly 50% of the protein-coding genes. From its strong preferential occurrence in the 5′ upstream region (mostly at positions ranging from −50 to −75 from the initiator ATG), it is assumed that this motif corresponds to a specific promoter signal. The presence of the AAAATTGA motif significantly correlates with genes transcribed from the leading strand (313/578 = 54% vs. 133/313 = 40%). It also correlates with predicted gene functions normally associated with the early and late/early phase of the virus replication (nucleotide metabolism, transcription, translation). As the AAAATTGA motif is not prevalent in the genome sequence of amoebal organisms, it is hypothesized that this motif is a core promoter element specifically recognized by the transcription preinitiation complex encoded by the mimivirus genome (mainly the large and small RNA polymerase subunits and a remote homolog of the TFIID 'TATA box-binding' initiation factor). The reasons behind the lack of variability of this mimivirus promoter element, usually quite degenerate among eukaryotes and eukaryotic viruses, are unknown.

Gene Content

As is usual for viruses, only a third of the 911 protein-coding genes of mimivirus have been associated with functional attributes. About 200 of these genes exhibit significant matches to 108 distinct COG families. The genome also encodes six tRNAs (three $tRNA_{leu}$, one $tRNA_{trp}$, one $tRNA_{cys}$, and one $tRNA_{his}$).

Typical NCLDV Core Genes in Mimivirus

The comparative genomic study of NCLDV suggested the monophyletic origin of four viral families: the *Asfarviridae*, the *Iridoviridae*, the *Phycodnaviridae*, and the *Poxviridae.* These studies identified sets of core genes (shared across these viral families) with various levels of conservation: class I core genes are found in all species from all families, class II core genes are absent in some species from a given family, while class III ones are shared by three families.

A strong argument for classifying mimivirus as a *bona fide* member of the NCLDV is the presence of 9/9 class I core gene homologs in its genome, 6/8 of the class II core genes, and 11/14 of the class III. Remarkably, the two class II core genes missing from mimivirus genomes encode two enzymes central to DNA synthesis: thymidylate kinase (catalyzing the formation of dTDP from dTMP) and the quasi-universal detoxifying enzyme dUTPase (dUTP pyrophosphatase, transforming dUTP into dUMP, to avoid the misincorporation of uracyl in DNA).

Phylogenetic analyses using class I core gene sequences let to mimivirus being assigned to a new family, the *Mimiviridae*, distinct from the four previously defined NCLDV families (**Figure 5**).

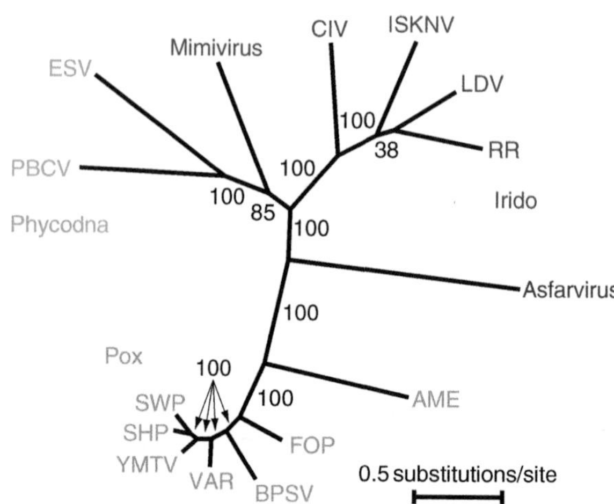

Figure 5 Phylogenetic position of mimivirus among established NCLDV families. Viral species representing the diverse families of NCLDV are included as follows: *Acanthamoeba polyphaga mimivirus, Phycodnaviridae* (*Paramecium bursaria chlorella virus* (PBCV) and *Ectocarpus siliculosus virus* (ESV)), *Iridoviridae* (*Chilo iridescent virus* (CIV), *Regina ranavirus* (RR), *Lymphocystis disease virus type 1* (LDV), and *Infectious spleen and kidney necrosis virus* (ISKNV)), *Asfarviridae* (*African swine fever virus*), and *Poxviridae* (*Amsacta moorel entomopoxvirus* (AME), *Variola virus* (VAR), *Fowlpox virus* (FOP), *Bovine papular stomatitis virus* (BPSV), *Yaba monkey tumor virus* (YMTV), *Sheeppox virus* (SHP), and *Swinepox virus* (SWP)). This tree was built with the use of maximum likelihood and based on the concatenated sequences of eight conserved proteins (NCLDV class I genes): vaccina virus (VV) D5-type ATPase, DNA polymerase family B, VV A32 virion packaging ATPase, capsid protein, thiol oxidoreductase, VV D6R helicase, serine/threonine protein kinase, and A1L transcription factor. One of the class I genes (VV A18 helicase) is absent in LDV and was not included. The alignment contains 1660 sites without insertions and deletions. A neighbor-joining tree and a maximum-parsimony tree exhibited similar topologies. Bootstrap percentages are shown along the branches. Reproduced from Raoult D, Audic S, Robert C, *et al.* (2004) The 1.2-megabase genome sequence of mimivirus, *Science* 306: 1344–1350, with permission from American Association for the Advancement of Science.

Remarkable Genes Found in Mimivirus

Translation-related genes

The mimivirus genome exhibits numerous genes encoding central components of the protein translation system. This was most unexpected, as viruses are traditionally distinguished from cellular organisms by their inability to perform protein synthesis independently from their host. This incomplete virally encoded translation subsystem includes four amynoacyl-tRNA synthetases (ArgRS, TyrRS, CysRS, and MetRS). These enzymes, responsible for the accurate loading of a given amino acid on its cognate tRNAs, are the true enforcers of the genetic code. The enzymatic activity of the TyrRS has been experimentally validated, as well as its strict specificity for tyrosine and its eukaryotic $tRNA_{Tyr}$.

In addition, mimivirus genome encodes two translation initiation factors: a GTP-binding elongation factor (EF-Tu) and a peptide chain release factor. Finally, mimivirus exhibits the first virus-encoded tRNA (Uracil-5)-methyltransferase. This enzyme is responsible for the U to T modification characterizing the T-loop in all tRNAs, a region involved in ribosome recognition, as well as aminoacylation and the binding of EF-Tu. Besides these unique translation-related enzymes, mimivirus encodes six tRNAs, a common feature in some NCLDV, such as phycodnaviruses.

DNA repair-related genes

The mimivirus genome encodes a comprehensive set of DNA repair enzymes, covering all types of DNA damage: alkylating agents, ultraviolet (UV) light, or ionizing radiations. In particular, mimivirus is the first virus to exhibit a formamidopyrimidine-DNA glycosylase (used to excise oxidized purines), a UV-damage endonuclease, a 6-*O*-methylguanine-DNA methyltransferase (used to get rid of O^6-alkylguanine), and a homolog to the DNA mismatch repair enzyme MutS. Mimivirus also possesses DNA topoisomerases, the enzymes required for solving entanglement problems associated with DNA replication, transcription, recombination, and chromatin remodeling. There are three mimivirus-encoded topoisomerases: first the type IA topoisomerase identified in a virus, the usual type IIA topoisomerase (such as found in most NCLDVs), and a type IB topoisomerase as found in the family *Poxviridae*.

DNA synthesis

The mimivirus genome exhibits the only known virally encoded nucleoside diphosphate kinase (NDK) protein. This enzyme catalyzes the synthesis of nucleoside triphosphate (other than ATP) from ATP. Its activity has been demonstrated experimentally, and exhibits a specific affinity for deoxypyrimidine nucleotides. This enzyme may help alleviate limited supplies in dTTP and dCTP for DNA synthesis. The genome also encodes a deoxynucleoside kinase (DNK), a thymidylate synthase (dUMP to dTMP), as well as the salvage enzyme thymidine kinase (TMP synthesis from thymidine).

Host signaling interfering pathways

The mimivirus genome encodes a large number of putative proteins exhibiting ankyrin repeats (66), a BTB/POZ domain (26), an F-box domain (10), and a kinase domain (9). Genes harboring these features are frequently involved in protein–protein interactions (ankyrin repeat), ubiquitin-mediated interactions (BTB/POZ, F-Box), and intracellular signaling (kinases). Remarkably, mimivirus exhibits three proteins exhibiting both a cyclin and a cyclin-dependent kinase (CDK) domain. It is likely that these proteins, by interfering with the cell signaling pathways, play a central role in turning the amoebal host into an efficient virus factory. On the other hand, identifying the cellular targets of these viral proteins, and their precise mode of action, might reveal original and valuable new directions for the therapeutic manipulation of eukaryotic cell metabolism.

Miscellaneous metabolic pathways

Previous analyses of NCLDV genomes, in particular from phycodnaviruses, revealed that these viruses possess biosynthetic abilities going well beyond the minimal requirements of viral DNA replication, transcription, and virion-packaging systems.

Mimivirus builds on this established trend by exhibiting a wealth of amino acid-, lipid-, and sugar-modifying enzymes, albeit all of them in the form of apparently incomplete (virus-encoded) pathways. For instance, mimivirus encodes three enzymes related to the metabolism of glutamine: an asparagine synthase (glutamine hydrolyzing), a glutamine synthase, and a glutamine-hydrolyzing guanosine 5′-monophosphate synthase. The reasons behind the special affinity of the virus for glutamine are unknown. Mimivirus also encodes three lipid-modifying enzymes: a cholinesterase, a lanosterol 14α-demethylase, and a 7-dehydrocholesterol reductase. Although these enzymes might be involved in the disruption of the host cell membrane, none of them have been described as participating in the infection process in other viruses. Finally, mimivirus exhibits an impressive array of enzymes normally involved in the synthesis of complex polysaccharides, such as perosamine, a high molecular weight capsular material found in bacteria. Mimivirus particles (see below) retain the Gram stain, a unique phenomenon that might be related to the presence of a densely reticulated (lipo)polysaccharide layer at their surface. In this context, the mimivirus-encoded procollagen-lysine, 2-oxoglutarate 5-dioxygenase, an enzyme that hydroxylates lysine residues in collagen-like peptides, might be central both to the attachment of carbohydrate moieties and to the formation of intermolecular cross-links. It is tempting to speculate that the unique hairy-like appearance of the virion, together with its resistance to chemical and mechanical disruption, is directly linked to the presence of a layer of heavily cross-linked lipo-proteo-polysaccharide material covering the 'regular' capsid.

Intein and introns

Inteins are protein-splicing domains encoded by mobile intervening sequences. They catalyze their own excision from the host protein. Although found in all domains of life (Eukarya, Archaea, and Eubacteria), their distribution is highly sporadic. Mimivirus is one of the few dsDNA viruses exhibiting an intein, inserted within its DNA polymerase B gene. Mimivirus intein is closely related to the one found in the DNA polymerase of heterosigma akashiwo virus (HaV), a large dsDNA virus infecting the single-cell bloom-forming raphidophyte (golden brown alga) *H. akashiwo*. Both appear monophyletic to the archaeal inteins.

Type I introns are self-splicing intervening sequences that are excised at the mRNA level. One type IB intron has been identified in several chlorella viruses species, but they remain rare in eukaryotic viruses. Mimivirus exhibits four self-excising (predicted) introns: one in the largest RNA polymerase subunit gene, the other three in the second-largest subunit. Given that introns are mostly detected when they interrupt the coding sequence of known proteins, some additional instances located within anonymous open reading frames (ORFs) might have escaped detection.

Particle Structure

Morphology

The discovery of the characteristic (i.e., icosahedral) viral morphology exhibited by mimivirus particles (initially mistaken as small Gram-positive intracellular bacteria) under the electron microscope was the turning point in its correct identification as a *bona fide* virus. Despite its unprecedented size, the icosahedral symmetry of the particle is good enough to allow a computer-generated three-dimensional (3-D) reconstruction at a resolution of about 75 Å, from series of cryoelectron microscopy (cryo-EM) images. Mimivirus has a capsid with a diameter of *c.* 0.5 μm, covered by 0.125-μm-long, closely packed fibers (**Figure 6**). The total diameter of a free particle is thus about 0.75 μm, consistent with its visibility in the light microscope following Gram staining (**Figure 1**). The pseudo-triangulation for mimivirus particles is estimated at $T \approx 1180$, predicting that the whole capsid is constituted

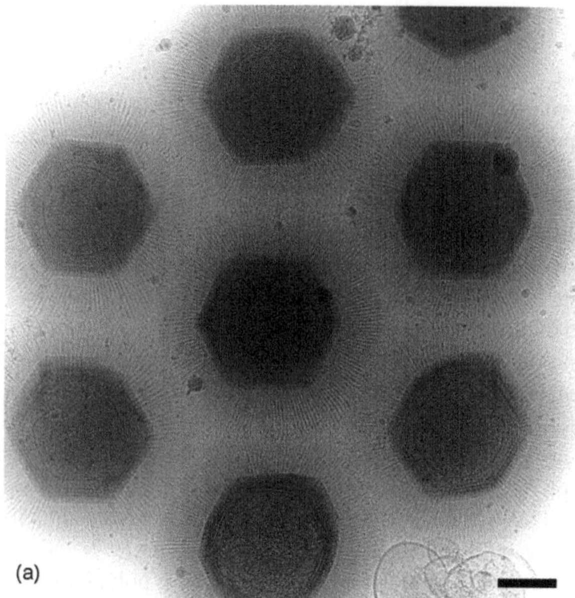

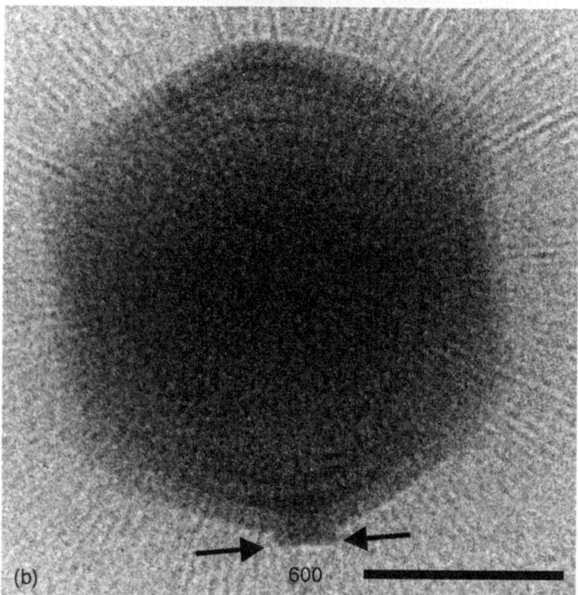

Figure 6 Cryo-EM high-quality images of mimivirus particles. (a) Cluster of mature particles, exhibiting a solid and compact fiber layer. (b) Close-up of one particle (0.75 μm across) exhibiting a densely packed layer of cross-linked fibers, and a unique vertex. Reproduced from Xiao C, Chipman PR, Battisti AJ, *et al.* (2005) Cryo-electron microscopy of the giant mimivirus. *Journal of Molecular Biology* 353: 493–496, with permission from Elsevier.

of *c.* 70 000 individual molecules of the L425 ORF-encoded major capsid protein. Inside a 70-Å-thick protein shell, cryo-EM images suggest the presence of two 40-Å-thick lipid membranes, a structure also found in other NCLDVs such as the African swine fever virus and some poxviruses.

Virion Proteomics

A detailed study of purified isolated virions using various electrophoresis separation methods followed by mass spectrometry analysis indicated that 114 different mimivirus genes are present in the viral particle. Many of these genes correspond to multiple products, exhibiting a variety of post-translational modifications including glycosylation (such as observed for the major capsid protein L425), proteolysis, and, most likely, phosphorylation.

Besides the expected *bona fide* 'structural' proteins (i.e., capsid proteins, major core protein, A16L-like virion associated protein, lipocalin-like lipoprotein, and a 'spore coat' assembly factor), a large number of proteins with enzymatic functions have been identified. For instance, mimivirus particles appear to possess a complete transcriptional apparatus including all virus-encoded DNA-directed RNA polymerase subunits (five), as well as four transcription factors, its mRNA capping enzyme, and two helicases. Nine gene products all related to oxidative pathways constitute the next largest functional group of particle-associated enzymes. These enzymes are probably important to cope with the oxidative stress generated by the host defense. Five proteins associated to DNA topology and damage repair are also present in the particle. Finally, seven enzymes associated to lipid (lipase and phosphoesterase) or protein modification (kinases and phosphatase) complete the pool of particle enzymes, together with the two tRNA methyltransferases. Yet, the majority of particle-associated proteins are the products of genes (65) of unknown function, of which 45 exhibit no convincing similarity in databases. It is worth noticing that only a small fraction (13/114) of the products associated with the particle are encoded by genes harboring the AAAATTGA promoter signal, confirming its correlation with early and late/early viral functions.

The functions predicted for the particle-associated gene products suggest a possible scenario for the early stage of the virus replicative cycle. First, it is likely that the lipolytic and proteolytic enzymes are part of the process by which the virus penetrates the amoeba cytoplasm from its vacuolar location. The two topoisomerases (IA and IB) might then be involved in the unpacking of the DNA from the particle, and its injection into the cytoplasm. The DNA repair enzymes might then get into action prior to starting transcription. The complete transcription machinery found in the viral particle is very similar to that in poxviruses. This suggests that some transcription could take place in the host cytoplasm immediately upon mimivirus infection.

Particle-Associated mRNAs

In a way reminiscent of members of the family *Herpesviridae*, a number of virus-encoded mRNAs are found associated with the virus particle. They include the messengers for DNA polymerase (R322), the major capsid protein (L425), the TFII-like transcription factor (R339), the tyrosyl-, cysteinyl-, and arginyl-tRNA synthetase (L124, L164, and R663), and four proteins of unknown function. These immediate early gene transcripts may be needed to initiate the first step of the replicative cycle, prior to any viral gene expression.

Evolution: The Position of Mimivirus in the Tree of Life

Of the 63 homologous genes common to all known unicellular organisms from the three domains of life (Eukarya, Eubacteria, and Archaea), seven have been identified in the mimivirus genome: three aminoacyl-tRNA synthetases, the two largest RNA polymerase subunits, the sliding clamp subunit of the DNA polymerase (PCNA), and a 5′-3′ exonuclease. The unrooted phylogenetic tree built from the concatenated sequences of the corresponding proteins indicates that mimivirus branches out near the origin of the Eukaryota domain (**Figure 7**). This position is consistent with several competing hypotheses proposing that large dsDNA viruses either predated, participated in, or closely followed, the emergence of the eukaryotic cell.

Other Possible Members of the Family *Mimiviridae*

Mimivirus was serendipitously discovered within *A. polyphaga*, a free-living ubiquitous amoeba, prevalent in aquatic environments. Many of the mimivirus core genes exhibit a phylogenetic affinity with members of the family *Phycodnaviridae* (algal and phytoplankton viruses). It can be expected therefore that additional viruses belonging to the family *Mimiviridae* will be found in aquatic/marine environments, for instance in marine protists. The bioinformatic analysis of environmental (metagenomics) DNA sequences strongly suggests that large viruses evolutionarily closer to mimivirus than any known virus species may be found among randomly collected bacteria-sized (passing through 3 μm but retained on 0.3 μm filters) marine microorganisms. Close ecological encounters of mimivirus ancestors with gorgonian octocorals (genus *Leptogorgia*) are also suggested by clear evidence of a lateral transfer of a mismatch repair (MutS) gene. Since very little is known about viruses infecting coral-associated microbial populations, it seems likely that new species of *Mimiviridae* might be hiding within this complex ecological niche.

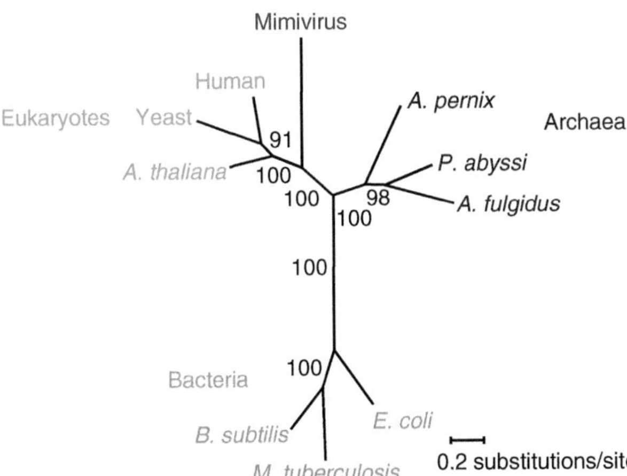

Figure 7 A phylogenetic tree of species from the three domains of life (Eukaryota, Eubacteria, and Archaea) and mimivirus. The tree was inferred with the use of a maximum-likelihood method based on the concatenated sequences of seven universally conserved protein sequences: arginyl-tRNA synthetase, methionylt-RNA synthetase, tyrosyl-tRNA synthetase, RNA polymerase II largest subunit, RNA polymerase II second largest subunit, PCNA, and 5′-3′ exonuclease. Bootstrap percentages are shown along the branches. Reproduced from Raoult D, Audic S, Robert C, *et al.* (2004) The 1.2-megabase genome sequence of mimivirus. *Science* 306: 1344–1350, with permission from American Association for the Advancement of Science.

See also: Nature of Viruses; Origin of Viruses.

Further Reading

Abergel C, Chenivesse S, Byrne D, Suhre K, Arondel V, and Claverie JM (2005) Mimivirus TyrRS: Preliminary structural and functional characterization of the first amino-acyl tRNA synthetase found in a virus. *Acta Crystallographica Section F* 61: 212–215.

Benarroch D, Claverie J-M, Raoult D, and Shuman S (2006) Characterization of mimivirus DNA topoisomerase IB suggests horizontal gene transfer between eukaryal viruses and bacteria. *Journal of Virology* 80: 314–321.

Berger P, Papazian L, Drancourt M, La Scola B, Auffray JP, and Raoult D (2006) Amoeba-associated microorganisms and diagnosis of nosocomial pneumonia. *Emerging Infectious Diseases* 12: 248–255.

Claverie JM, Ogata H, Audic S, Abergel C, Suhre K, and Fournier P-E (2006) Mimivirus and the emerging concept of 'giant' virus. *Virus Research* 117: 133–144.

Ghedin E and Claverié JM (2005) Mimivirus relatives in the Sargasso Sea. *Virology Journal* 2: 62.

La Scola B, Audic S, Robert C, *et al.* (2003) A giant virus in amoebae. *Science* 299: 2033.

La Scola B, Marrie TJ, Auffray JP, and Raoult D (2005) Mimivirus in pneumonia patients. *Emerging Infectious Diseases* 11: 449–452.

Ogata H, Raoult D, and Claverie JM (2005) A new example of viral intein in mimivirus. *Virology Journal* 2: 8.

Raoult D, Audic S, Robert C, *et al.* (2004) The 1.2-megabase genome sequence of mimivirus. *Science* 306: 1344–1350.

Raoult D, Renesto P, and Brouqui P (2006) Laboratory infection of a technician by mimivirus. *Annals of Internal Medicine* 144: 702–703.

Renesto P, Abergel C, Decloquement P, *et al.* (2006) Mimivirus giant particles incorporate a large fraction of anonymous and unique gene products. *Journal of Virology* 80: 11678–11685.

Suhre K (2005) Gene and genome duplication in Acanthamoeba polyphaga mimivirus. *Journal of Virology* 79: 14095–14101.

Suhre K, Audic S, and Claverie JM (2005) Mimivirus gene promoters exhibit an unprecedented conservation among all eukaryotes. *Proceedings of the National Academy of Sciences, USA* 102: 14689–14693.

Suzan-Monti M, La Scola B, and Raoult D (2006) Genomic and evolutionary aspects of mimivirus. *Virus Research* 117: 145–155.

Xiao C, Chipman PR, Battisti AJ, *et al.* (2005) Cryo-electron microscopy of the giant mimivirus. *Journal of Molecular Biology* 353: 493–496.

NUCLEIC ACIDS

Recombination

J J Bujarski, Northern Illinois University, DeKalb, IL, USA and Polish Academy of Sciences, Poznan, Poland

Introduction

Genetic recombination of viruses could be defined as the exchange of fragments of genetic material (DNA or RNA) among parental viral genomes. The result of recombination is a novel genetic entity that carries genetic information in nonparental combinations. Biochemically, recombination is a process of combining or substituting portions of nucleic acid molecules. Recombination has been recognized as an important process leading to genetic diversity of viral genomes upon which natural selection can function. Depending on the category of viruses, recombination can occur at the RNA or DNA levels. Since these processes are different for DNA and RNA viruses, they are described separately.

Recombination in DNA Viruses

In many DNA viruses, genetic recombination is believed to occur by means of cellular DNA recombination machinery. Cellular DNA recombination events are of either homologous (general recombination) or nonhomologous types. The nonhomologous recombination events occur relatively rarely and are promoted by special proteins that interact with special DNA signal sequences. In general, homologous recombination events occur much more often and they are most commonly known as genetic crossing-over that happens in every DNA-based organism during meiosis.

The biochemical pathways responsible for DNA crossing-over are well established. General elements involved in general recombination include DNA sequence identity, complementary base-pairing between double-stranded DNA molecules, heteroduplex formation between the two recombining DNA strands, and specialized recombination enzymes. The best-studied recombination system of *Escherichia coli* involves proteins such as recA, and RecBCD, and it has led to a large amount of literature. Interestingly, related DNA recombination proteins have been characterized in eukaryotes, including yeast, insects, mammals, and plants.

Yet certain DNA virus species encode their own recombination proteins, and some of these viruses serve as model system with those to study the recombination processes. One of the best-known systems is of certain bacteriophages that recombine independently from the host mechanisms. These independent pathways are used for repairing damaged phage DNA and for exchanging DNA to increase diversity among the related phages. In Enterobacteria phage M13, high recombination frequency was observed within the origin of phage DNA replication in the *E. coli* host. There, the crossovers have occurred at the nucleotide adjacent to the nick at the replication origin, because of joining to a nucleotide at a remote site in the genome. These results implicated a breakage-and-religation mechanism of such apparently illegitimate cross-overs.

Importantly enough, many of these phage recombination mechanisms are analogous to the pathways operating in the host bacteria. For instance, Rec proteins of phages T4 and T7 are analogous to bacterial RecA, RecG, RuvC, or RecBCD proteins, while RecE pathway in the rac prophage of *E. coli* K-12 or the phage 1 red system influenced the studies of bacterial systems. A correlation of different stages of DNA recombination with transcription and DNA replication during Enterobacteria phage T4 growth cycle is shown in **Figure 1**.

Phage lambda (λ) has a recombination system that can substitute for the RecF pathway components in *E. coli.* The Enterobacteria phage λ moves its viral genome into and out of the bacterial chromosome using site-specific recombination. Based on crystal structures of the reaction intermediates, it is clear how the Enterobacteria phage λ integrase interacts with both core and regulatory DNA elements (**Figure 2**).

Recombination between viral DNA and host genes can lead to acquisition of cellular genes by DNA viruses. For instance, tRNA genes are present in Enterobacteria phage T4. Interestingly, these tRNA sequences contain introns suggesting that Enterobacteria phage T4 must have passed through a eukaryotic host during evolution. Similar viral–host recombination events were observed for retroviruses in eukaryotic cells.

Genetic recombination in DNA viruses is often studied using functional marker mutations. In single-component DNA viruses recombination occurs by exchanging DNA fragments, whereas in segmented DNA viruses, additional events rely on reassortment of the entire genome segments. This complicates the recombination behavior observed among mutants. One method of recombination analysis utilizes so-called conditional-lethal types, where the cells are infected with two variants and the recombinants are selected after application of nonpermissive conditions (two-factor crosses). This allows the mutants to be organized into complementation groups with the relative positions of mutations being placed on a linear map. Another method is called three-factor crosses. Here three

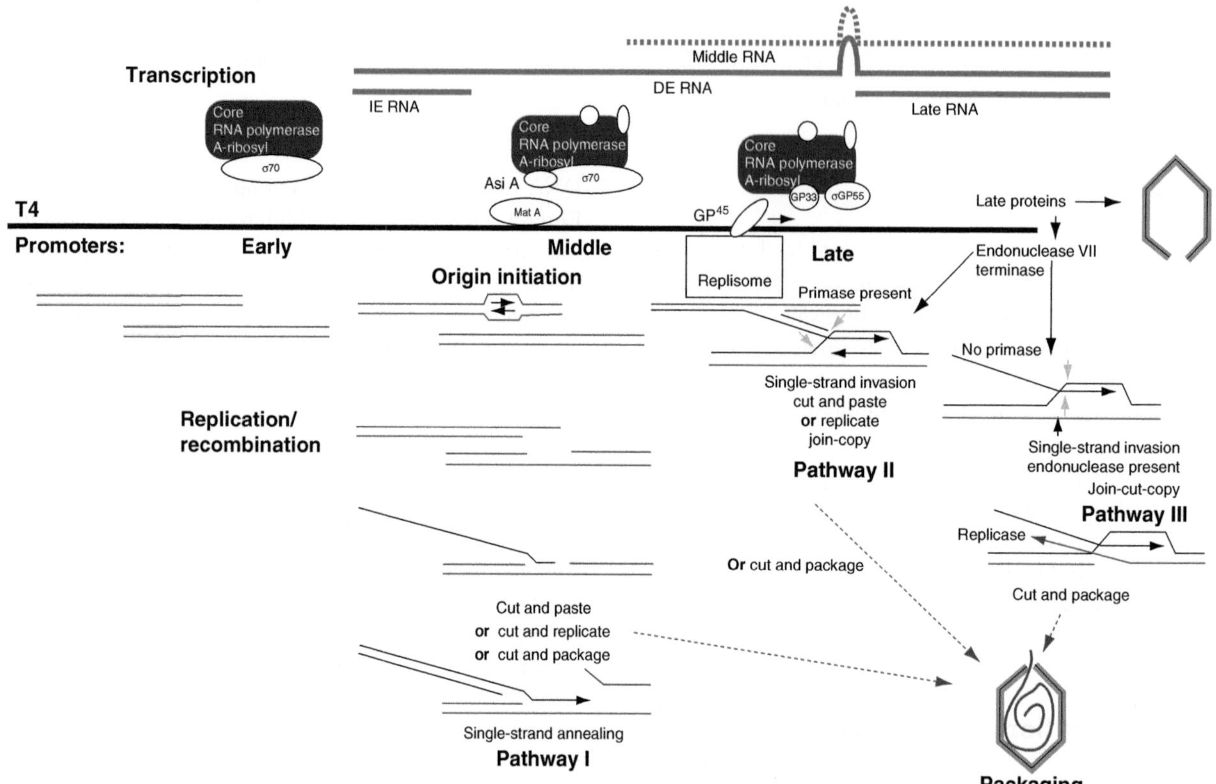

Figure 1 Diagram of the relationship between the Enterobacteria phage T4 transcriptional pattern and the different mechanisms of DNA replication and recombination. (a) Shows the transcripts initiated from early, middle, and late promoters by sequentially modified host RNA polymerase. Hairpins in several early and middle transcripts inhibit the translation of the late genes present on these mRNAs. (b) Depicts the pathways of DNA replication and recombination. Hatched lines represent strands of homologous regions of DNA, and arrows point to positions of endonuclease cuts. Reproduced from Mosig G (1998) Recombination and recombination-dependent DNA replication in enterobacteria phage T4. *Annual Review of Genetics* 32: 379–413, with permission from Annual Reviews.

mutations are employed, with crossing-over occurring between two mutations while the third mutation is not selected. This allows for determination of linkage relationships among mutants and of the order of marker mutations. Due to reassortment, both the two-factor and the three-factor crosses are of less use in segmented DNA viruses.

DNA viruses of eukaryotes also recombine their genomic material. For instance, herpes simplex virus (HSV) was found to support recombination while using pairs of temperature-sensitive mutants (two-factor crossings). In fact, a recombination-dependent mechanism of HSV-1 DNA replication has been described. The recombination frequency was proportional to the distance between mutations which suggested the lack of specific signal sequences responsible for the crossing-over. By using three-factor crossing, the HSV system involved two ts mutants and a syncytial plaque morphology as an unselectable marker. Similarly, in case of adenoviruses, the host range determined by the helper function of two mutations has been used as a third marker between ts mutants. Here, intertypic crosses between ts mutants have been identified based on segregation patterns and the restriction enzyme polymorphism.

Epstein–Barr virus (EBV) is a member of the family *Herpesviridae*, and it carries a long double-stranded genomic DNA, that shows a high-degree variation among strains. These variations include single base changes, restriction site polymorphism, insertions, or deletions. Based on tracking these mutations, it was found that some EBV strains arose due to DNA recombination.

Poxviruses represent the largest DNA viruses known (except those of algae and the mimivius). Homologous recombination was detected in the genome of vaccinia virus (VV), based on the high frequency of intertypic crossovers, the marker rescue, and the sequencing of recombinants. These processes could be both intra- and intermolecular, and they depend on the size of the DNA target. It has been suggested that either viral DNA replication itself or the activity of the viral DNA polymerase might participate in VV DNA recombination. Indeed, some VV proteins with DNA strand transfer activity have been identified.

The DNA genome of Simian virus 40 (SV40, *Papovaviridae*) was found to recombine in somatic cells. The artificially constructed recombinant circular oligomers were used to find high general recombination frequency

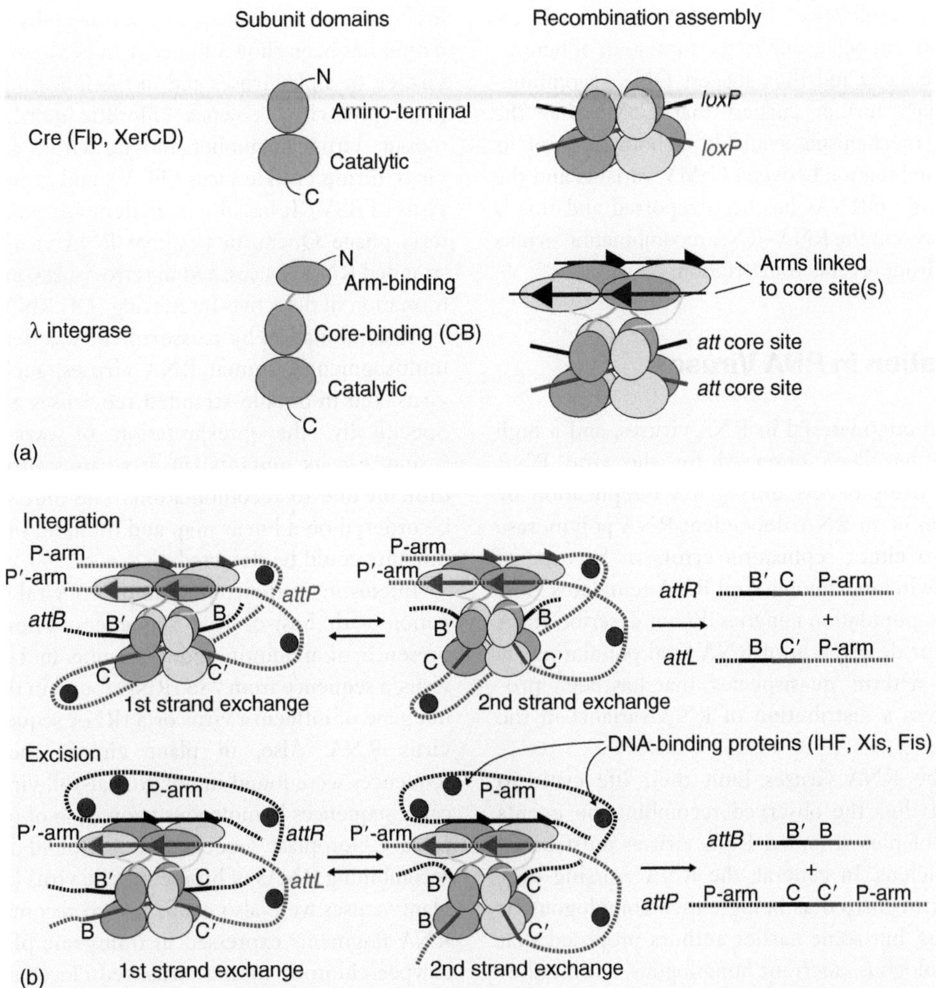

Figure 2 (a) Enterobacteria phage λ integrase compared to the simpler recombinases. Tyrosine recombinases such as Cre have two domains that bind the core recombination sites and carry out recombination on their own. Enterobacteria phage λ integrase has a third, amino-terminal 'arm binding' domain that binds to the arm region of the attachment site. The DNA complex cartoon for Enterobacteria phage λ integrase (lower right) represents the new crystal structures. (b) Integration and excision by Enterobacteria phage λ integrase. The first and second strand exchange cartoons represent the first and second halves of the recombination reaction, respectively. In the first half of integration, for example, Enterobacteria phage λ integrase brings *attP* and *attB* sites together and exchanges the first pair of strands to generate a Holliday junction intermediate. In the second half of the reaction, the Holliday intermediate has isomerized to form a distinct quaternary structure and exchange of the second pair of strands generates recombinant *attL* and *attR* products. Reproduced from Van Duyne GD (2005) Enterobacteria phage λ integrase: Armed for recombination. *Current Biology* 15: R658–R660, with permission from Elsevier.

of SV40 DNA. However, homologous recombination events were rare.

Among plant DNA viruses, genetic recombination was studied in case of geminiviruses and caulimoviruses. The geminiviruses carry a single-stranded DNA genome, composed of either one or two circular DNA molecules. Frequent intermolecular crossing-over events were observed by using mutant combinations. Homologous crossovers were detected to occur intramolecularly between tandem repeats of a geminivirus DNA using agro-infected tobacco plants. The mechanism may involve either homologous crossing-over events or copy-choice processes that rely on template switching by DNA replicase. Moreover, deletions, insertions, and more profound rearrangements have been detected in the geminivirus DNA. These are the illegitimate recombination processes that may involve aberrant breakage-and-religation events or errors in DNA replication, that could occur either inter- or intramolecularly.

Cauliflower mosaic virus (CaMV) belongs to a family of plant double-stranded (ds) DNA pararetroviruses that replicate via reverse transcription. A high recombination rate was observed during CaMV infection *in planta*. These crossovers could occur at the DNA level (thus in the nucleus) or at the RNA level (thus more likely during reverse transcription in cytoplasm). However, features such as recombinational hot spots and mismatch repair might indicate replicative (i.e., RNA) step, whereas mismatch repair can occur due to the formation of heteroduplex intermediates and thus suggest DNA recombination. These data further suggest that CaMV has the recombination mechanisms available at both steps of its life cycle. Recombination between CaMV variants and the CaMV transgenic mRNAs has been reported and this is believed to represent the RNA–RNA recombination events that happen during reverse transcription.

Recombination in RNA Viruses

RNA is the genetic material in RNA viruses, and a high mutation rate has been observed for the viral RNA genome. This likely occurs during RNA replication by means of action of an RNA-dependent RNA polymerase enzyme due to either replication errors or because of the replicase switching among viral RNA templates. The terms of classic population genetics do not describe RNA viruses. A better description of RNA viral populations is provided with a term 'quasispecies' that has been proposed to address a distribution of RNA variants in the infected tissue.

Many of the RNA viruses limit their life cycle to cytoplasm and thus the observed recombination events among RNAs of plus-stranded RNA viruses must occur outside the nucleus. In general, the RNA crossing-over processes are categorized as being either homologous or nonhomologous, but some earlier authors proposed that there are homologous, aberrant homologous, and nonhomologous RNA recombination types. Aberrant homologous recombination involves crossovers between related RNAs, but the crosses occur at not-corresponding sites leading to sequences insertions or deletions. More recently, mechanistic models were utilized to define the following RNA–RNA recombination classes: (1) The 'similarity-essential' recombination, where substantial sequence similarity between the parental RNAs is required as the major RNA determinant; (2) The 'similarity-nonessential' recombination does not require sequence similarity between the parental RNAs, although such regions may be present; and (3) There is the 'similarity-assisted' recombination where sequence similarity can influence the frequency or the recombination sites but additional RNA determinants are also critical.

Genetic RNA recombination has been described in many RNA virus groups. In particular, sequence data reveal RNA rearrangements reflecting RNA–RNA crossover events during RNA virus evolution. For instance, RNA rearrangements were demonstrated in the genomes of dengue virus-type I, flock house virus, hepatitis D virus, bovine viral diarrhea virus, and equine arthritis virus RNA. For plant RNA viruses, this has been demonstrated in potyviruses such as yam mosaic virus, sugarcane yellow leaf virus, and luteoviruses. Experimentally, RNA recombination has been shown to occur in picornaviruses, coronaviruses, or alphaviruses and in the following plant viruses: plum pox virus, cowpea chlorotic mottle virus, alfalfa mosaic virus, cucumber mosaic virus, tobacco mosaic virus, turnip crinkle virus (TCV), and tomato bushy stunt virus (TBSV). It has also been demonstrated in enterobacteria phage Qbeta, in negative RNA viruses, in double-stranded RNA viruses, and in retroviruses, as well as during formation of defective-interfering (DI) RNAs.

Recombination by reassortment was demonstrated for multisegmental animal RNA viruses, such as influenza virus, and in double-stranded reoviruses and orbiviruses. Specifically, the interpretation of two-factor crosses (using, e.g., ts mutants) in reoviruses turned out to be difficult due to recombination. The mutant sites cannot be ordered on a linear map and often no linkage between mutants could be detected.

Interestingly, there are examples of viral RNA recombination with host-derived sequences. These include the presence of uniquitin-coding region in bovine diarrhea virus, a sequence from 28S rRNA found in the hemagglutinin gene of influenza virus or a tRNA sequence in Sindbis virus RNA. Also, in plant viruses the host-derived sequences were found in potato leaf-roll virus isolates that carry sequences homologous to an exon of tobacco chloroplast. Chloroplast sequences were found in the actively recombining RNAs of brome mosaic virus (BMV). Several plant viruses were also confirmed to recombine with viral RNA fragments expressed in transgenic plants, including cowpea chlorotic mottle virus, red clover necrotic mottle virus, potato virus Y virus, and plum pox virus.

The existence of several RNA virus recombination systems has made possible the studies of the molecular mechanisms of RNA recombination. The majority of RNA recombination models predict copy-choice mechanisms, either due to primer extension (in flaviviruses, carmoviruses), at the subgenomic promoter regions (BMV, poliovirus), or by strand translocation (in nidoviruses). In retroviruses, there are three copy-choice mechanisms: (1) forced (strong stop) strand transfer, (2) pause-driven strand transfer, and (3) pause-independent (RNA structure-driven) strand transfer. However, in enterobacteria phage Qbeta a breakage-and-religation mechanism has been described. Details of some of these systems are discussed below.

The molecular mechanisms of the formation of both nonhomologous and homologous RNA recombinants have been studied using an efficient system of BMV. In order to increase recombination frequency, the BMV RNA3-based constructs were generated where the 3′ noncoding region was extended, while carrying partial deletions. This debilitated the replication of RNA such that the sequence got repaired by recombination with the sequences of other

BMV RNA segments. It appeared that short base-paired regions between the two parental BMV RNA molecules could target efficient nonhomologous crossovers. A proposed model predicted that the formation of local RNA–RNA heteroduplexes could function because they brought together the RNA substrates and because they slowed down the approaching replicase enzyme complexes.

These early studies also analyzed the molecular requirements of homologous recombination by inserting the BMV RNA2-derived sequences into the recombination vector. This revealed the accumulation of both precise and imprecise RNA2–RNA3 recombinants and that the recombination frequencies depended upon the composition of nucleotide sequences within the region of recombination. The crossovers tended to happen at stretches of GC-rich regions alternating with AU-rich sequences suggesting the RNA replicase switching between RNA templates. Elements capable of forming strand-specific, stem–loop structures were inserted at the modified 3′ noncoding regions of BMV RNA3 and RNA2 in either positive or negative orientations, and various combinations of parental RNAs were tested for patterns of the accumulating recombinant RNA3 components. This provided experimental evidence that homologous recombination between BMV RNAs more likely occurred during positive-rather than negative-strand synthesis.

True homologous recombination crossing-over has been observed among the RNA molecules of the same segments during BMV infections. By using nonselective marker mutations at several positions, it was demonstrated that RNA1 and RNA2 segments crossed-over at 5–10% frequency, whereas the intercistronic region in RNA3 supported an unusually high recombination frequency of 70%. The subsequent use of various deletion constructs has revealed that the high-frequency crossing-over mapped to the subgenomic promoter (sgp) region, and in particular to its internal polyA tract. Further studies have shown that sgp-mediated crossing-over has occurred at the minus-strand level (i.e., during plus-strand synthesis), most likely by discontinuous process, where the replicase complex detached from one strand and reinitiated on another strand. This process is most likely primed by a sg RNA3a intermediate, which prematurely terminates on the polyU (in minus strand) tract and re-anneals to this region on another minus RNA3 template (**Figure 3**). Also, it turned out that the frequency of crossing-over and the process of initiation of transcription of sg RNA4 were reversely linked, suggesting a competition between these two reactions.

The role of replicase proteins in BMV RNA recombination has been studied by using well-characterized 1a and 2a protein mutants. A ts mutation in protein 1a 5′ shifted the crossover sites indicating the participation of helicase domain of 1a. Likewise, mutations at several regions of 2a affected the frequency of nonhomologous recombination. The relationship between replication and recombination was studied by using BMV variants that

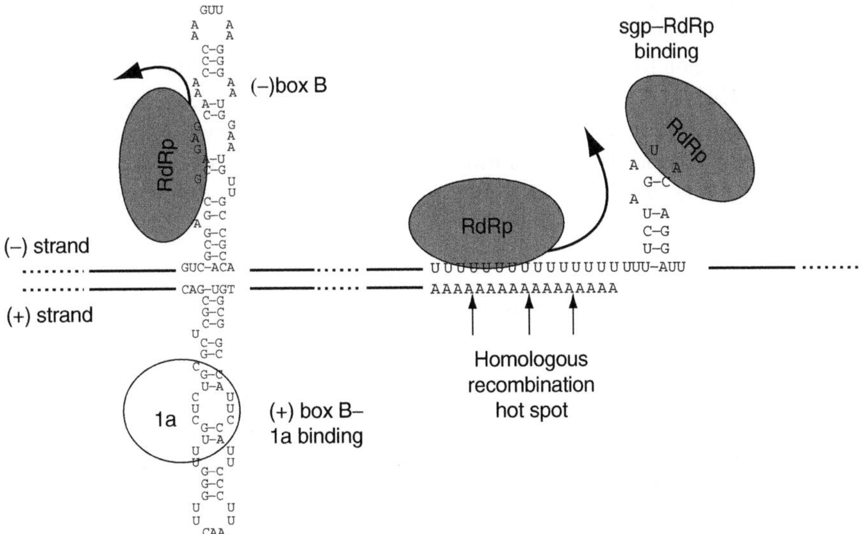

Figure 3 Model illustrating the synthesis of sg RNA3a in view of multiple functions of the intergenic region in (–) RNA3. The BMV RdRp enzyme complex (represented by gray ovals) migrates alongside the (–)-strand RNA template and pauses (represented by curved arrows) at the secondary structure or, most notably, at the oligoU tract, leading to the formation of subgenomic sgRNA3a. Yet another molecule of the RdRp enzyme binds to the sgp and initiates the *de novo* synthesis of sgRNA4. Also, the rehybridization of the sgRNA3a oligoA tail to the RNA3 (–) template can resume full-length copying which primes the observed RNA3–RNA3 recombination (5, 69). The (+) and (–) RNA strands are represented by thick lines and both the oligoU tract in the (–) strand template and the oligoA 3′ termini are exposed. The stem-and-loop structures adopted by the (+) and the (–) strands upstream to their oligoU (A) tracts are shown. The binding region to protein 1a via the box B of the stem–loop structure in (+) strands is shown. Reproduced from Wierzchoslawski R, Urbanowicz A, Dzianott A, Figlerowicz M, and Bujarski JJ (2006) Characterization of a novel 5′ subgenomic RNA3a derived from RNA3 of brome mosaic bromovirus. *Journal of Virology* 80: 12357–12366, with permission from American Society for Microbiology.

carried mutations in 1a and 2a genes. This revealed that the 1a helicase and the 2a N-terminal or core domains were functionally linked during both processes *in vivo* and *in vitro*. Also, it was shown that the characteristics of homologous and nonhomologous crossovers could be modified separately by mutations at different protein sites. All these studies confirmed the involvement of replicase proteins in recombination and supported the template-switching model.

More recently, the role of host factors in BMV recombination was addressed by using both yeast and Arabidopsis systems. In yeast, transient co-expression of two derivatives of BMV genomic RNA3 supported intermolecular homologous recombination at the RNA level but only when parental RNAs carried the *cis*-acting replication signals. The results implied that recombination occurred during RNA replication. In *Arabidopsis*, the use of gene-knock-out mutations in the RNA interference pathway revealed that BMV can recombine according to both the copy-choice template-switching and to the breakage-and-religation mechanisms.

The role of replicase proteins in RNA recombination has also been studied in other RNA viruses. For TCV, a small single-stranded RNA virus, a high-frequency recombination was observed between satellite RNA D and a chimeric subviral RNA C. The crossing-over most likely relied on viral replicase enzyme switching templates during plus-strand synthesis of RNA D which reinitiated RNA elongation on the acceptor minus-strand RNA C template. The participation of replicase proteins was demonstrated *in vitro*, where a chimeric RNA template containing the *in vivo* hot-spot region from RNA D joined to the hot-spot region from RNA C. Structural elements such as a priming stem in RNA C and the replicase binding hairpin, also from RNA C, turned to play key roles during recombination, probably reflecting late steps of RNA recombination such as strand transfer and primer elongation. The host factors related to the host-mediated viral RNA turnover have been found to participate in tombusvirus RNA recombination. The screening of essential yeast genes mutants identified host genes that affected the accumulation of TBSV recombinants, including genes for RNA transcription/metabolism, and for protein metabolism/transport. Suppression of TBSV RNA recombination was observed by the yeast Xrn1p $5'$–$3'$ exoribonuclease, likely due to rapid removal of the $5'$ truncated RNAs, the substrates of recombination. These $5'$ truncated viral RNAs are generated by host endoribonucleases, such as the Ngl2p endoribonuclease.

Coronavirus RNAs were found to recombine between the genomic and DI RNA molecules. It was postulated that recombination has occurred due to the nonprocessive nature of the coronavirus RNA polymerase enzyme (**Figure 4**) and an efficient protocol for targeted recombination has been developed.

Similarly, in nodaviruses, the two-partite RNA viruses, recombination processes were found to occur between RNA segments at a site that potentially could secure base pairing between the nascent strand and the acceptor template. The recombination sites might have been chosen based on factors such as the similarity to the origin of replication or special secondary structures. A postulated model implies the polymerase to interact directly with the acceptor nodavirus RNA template.

A double-stranded RNA Pseudomonas phage Phi6 was hypothesized to recombine its RNA based on a copy-choice template switching mechanism, where the crossovers would have occurred inside the virus capsid structures at regions with almost no sequence similarity. Interestingly, the frequency of recombination was enhanced by conditions that prevented the minus-strand synthesis. Experiments were designed to reveal the effects of drift on existing genetic

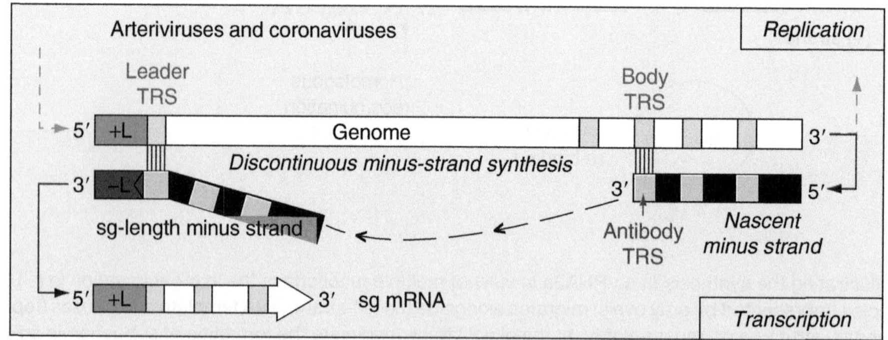

Figure 4 Models for discontinuous transcription from minus-strand sg-length templates in arteriviruses and coronaviruses. These viruses have a common 59 leader sequence on all viral mRNAs. Discontinuous extension of minus-strand RNA synthesis has been proposed as the mechanism to produce sg-length minus-strand templates for transcription. The replicase/transcriptase can attenuate at one of the body TRSs in the 39-proximal part of the genome, after which the nascent minus strand extends with the anti-leader ('L') sequence. Next, the completed sg-length minus strands serve as templates for transcription. Reproduced from Pasternak AO, Spaan WJM, and Snijder EJ (2006) Nidovirus transcription: How to make sense...? *Journal of General Virology* 87: 1403–1421, with permission from Society for General Microbiology.

variation by minimizing the influence of variation on beneficial mutation rate. The segmented genome of the pseudomonas phage Phi6 has allowed to present the first empirical evidence that the advantage of sex during adaptation increases with the intensity of drift.

The enterobacteria phage Qbeta, a small single-stranded RNA virus, could recombine both *in vivo* and *in vitro*. Here, the mechanism of recombination was not based on a template-switching by the replicase, but rather via a replicase-mediated splicing-type religation of RNA fragments. The system produced nonhomologous recombinants, whereas the frequency of homologous crossovers was low. These data suggested an RNA trans-esterification reaction catalyzed by a conformation acquired by enterobacteria phage Qbeta replicase during RNA synthesis. In summary, the results on various plus-strand RNA virus systems demonstrate the availability of a variety of template-switch mechanisms, the mutual-primer-extension on two overlapping RNA strands, the primer-extension on one full-length RNA strand, as well as both replicative and nonreplicative trans-esterification mechanisms where a piece of another RNA is added to the 3′ terminus of an RNA either by viral RdRp or by other enzymes (e.g., RNA ligase), respectively.

The recombination events in retroviruses contribute significantly to genetic variability of these viruses. The crossovers do occur by reverse transcriptase jumpings between the two genomic RNA molecules inside virion capsids. Apparently, the virally encoded reverse transcriptase enzymes have been evolutionarily selected to prone the jumpings between templates during reverse transcription. It turns out that the recombinant jumpings between RNA templates are responsible for both inter- and intramolecular template switchings and also for the formation of defective retroviral genomes. It has been found that the most stable interactions between two copies of retrovirus RNAs were within the 5′ nucleotides 1–754. There is experimental evidence demonstrating that the template 'kissing' interactions effectively promote recombination within the HIV-I 5′ untranslated region. The possibilities of recombination in retroviruses at the DNA level (of the integrated provirus sequences) were discussed earlier in this article.

Defective-Interfering RNAs

There is a variety of subviral RNA molecules that are linked to viral infections. Those derived from the viral genomic RNAs and interfering with the helper virus accumulation or symptom formation are called as DI RNAs. First reports (in 1954) about DI RNAs coexisting with viral infection was provided by Paul von Magnus with influenza virus. Thereafter, numerous both animal and plant viruses were found to generate DI RNAs. Naturally occurring DI RNAs have been identified in coronavirus infections. These molecules appear to arise by a polymerase strand-switching mechanism. The leader sequence of the DI RNAs was found to switch to the helper-virus derived leader sequence, indicating that helper virus-derived leader was efficiently utilized during DI RNA synthesis. Also, the leader switching likely occurred during positive-strand DI RNA synthesis, and the helper-virus positive-strand RNA synthesis tended to recognize double-stranded RNA structures to produce positive-strand DI RNAs. The parts of the coronavirus RNA required for replication and packaging of the defective RNAs were investigated, with both the 5′- and the 3′-terminal sequences being necessary and sufficient. The coronavirus DI RNAs have been utilized to study the mechanism of site-specific RNA recombination. This process relies on the acquisition of a 5′ leader that is normally used for production of numerous coronavirus sg RNAs. Also, these DI RNAs have been used as vehicles for the generation of designed recombinants from the parental coronavirus genome.

In case of plant viruses, tombusviruses and carmoviruses were found to accumulate DI RNAs, which maintain a consistent pattern of rearranged genomic sequences flanked by the 5′ and 3′ unchanged replication signals. In some cases, the base pairing between a partial nascent strand and the acceptor template can lead to the appearance of the rearranged regions in DI RNAs.

In addition to rearranged DI RNAs, some RNA viruses accumulate defective RNAs due to a single internal deletion in the genomic RNA of the helper virus. Such examples include beet necrotic wheat mosaic furovirus, peanut clump furovirus, clover yellow mosaic potexvirus, sonchus yellow net rhabdovirus, and tomato spotted wilt tospovirus. Features such as the ability to translate or the magnitude of the defective RNA seem to affect the selection of the best-fit sizes of DI RNAs during infection.

Another type of single-deletion DI RNAs are produced during broad bean mottle bromovirus (BBMV) infections from the RNA2 segment. A model has been proposed where local complementary regions bring together the remote parts of RNA2 which then facilitates the crossover events. Similar RNA2-derived DI RNAs have been reported to accumulate during the cucumovirus infections.

The closteroviruses, the largest known plant RNA viruses, form multiple species of defective RNAs, including the citrus tristeza virus defective RNAs that arise from the recombination of a subgenomic RNA with distant 5′ portion of the virus genomic RNA (**Figure 5**). Apparently, closteroviruses can utilize sg RNAs for the rearrangement of their genomes.

Negative-strand RNA viruses also form DI RNAs. For instance, in vesicular stomatitis virus (VSV), a rhabdovirus, the *cis*-acting RNA replication terminal elements participate in the formation of the 5′-copy-back DI RNAs, reflecting likely communication between distant portions of the VSV genome.

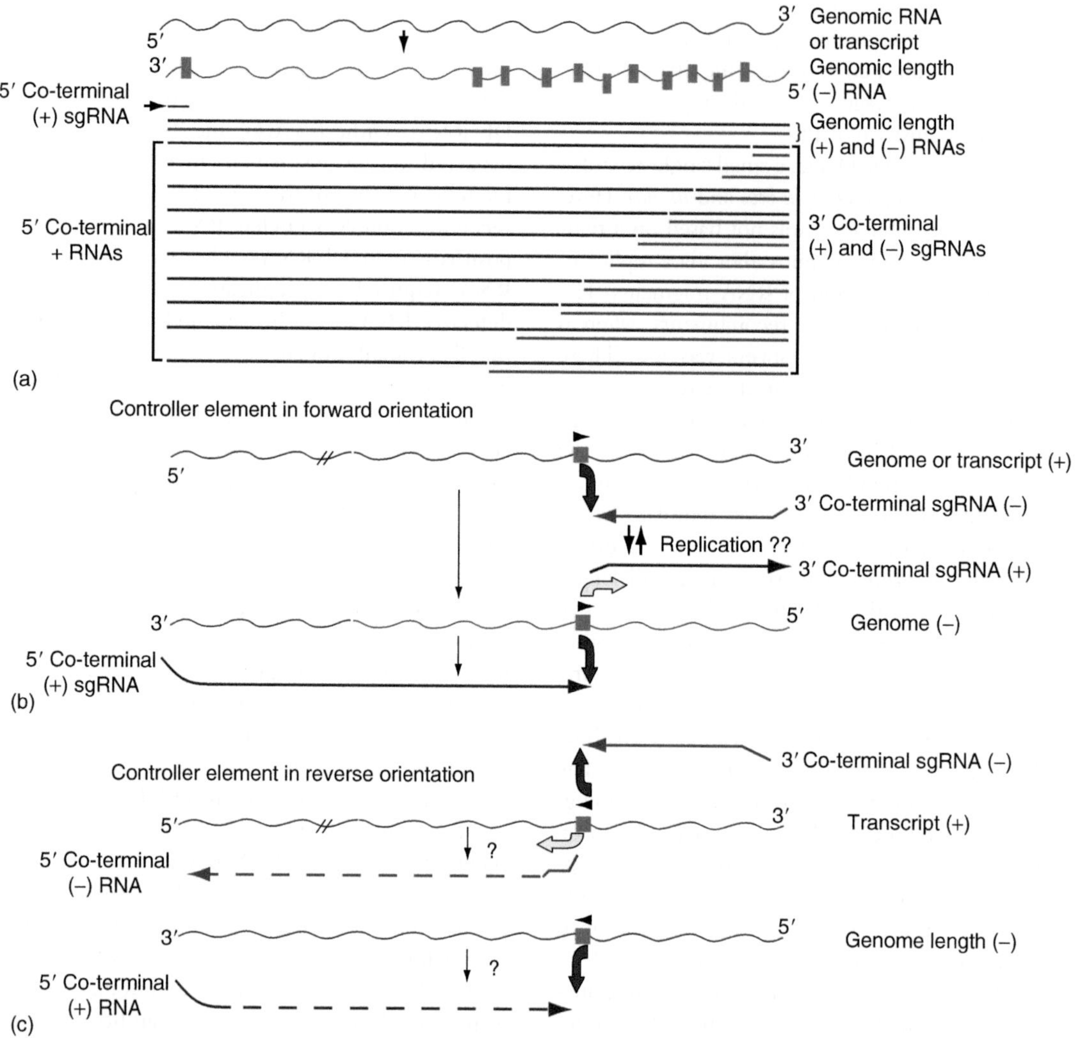

Figure 5 (a) The outline of different species of genomic RNA and 59 and 39 terminal sgRNAs potentially produced in CTV-infected cells. The positive-sense RNAs are shown in blue and the negative-sense RNAs are shown in red. The wavy line represents the genomic RNA or the plus-sense transcript (blue) and the genomic length minus-sense RNA (red) produced from the plus-sense RNA. The solid green boxes on the genomic negative-stranded RNA represent the sgRNA controller elements. The solid lines represent the full array of plus- and minus-stranded genomic and 39- and 59-terminal sgRNAs potentially produced during replication of CTV. (b, c) Models predicting the generation of 59- and 39-terminal positive- and negative-sense sgRNAs with the controller element present in normal and reverse orientation. One control region is shown for clarity. The wavy blue line represents the transcript (blue) containing the control region (green box) in normal and reverse orientation (the direction of the arrowheads above the controller element indicates the orientation of the controller element). The thick curved arrows represent the transcription termination (vertical direction, red) or promotion (horizontal direction, yellow). The solid blue lines with arrowheads represent the positive-sense 39- and 59-terminal sgRNAs and the solid red line with arrowhead indicates the 39-terminal negative-sense sgRNA. The dashed lines with arrowheads indicate the potential 59 terminal positive (blue)- and negative (red)-sense sgRNAs. Reproduced from Gowda S, Satyanarayana T, Ayllon MA, *et al.* (2001) Characterization of the *cis*-acting elements controlling subgenomic mRNAs of *Citrus tristeza virus:* Production of positive- and negative-stranded 39-terminal and positive-stranded 59 terminal RNAs. *Virology* 286: 134–151, with permission from Elsevier.

Summary and Conclusions

Genetic recombination is a common phenomenon among both DNA and RNA viruses. The recombination events have been observed based on natural rearrangements of the sequenced viral genomes. Also, experimental systems demonstrate the occurrence of recombination events that play important roles in securing the genetic diversity during viral infection. Different molecular mechanisms are involved in DNA versus RNA viruses. Many DNA viruses utilize host cellular machinery of general homologous recombination (such as meiotic crossing-over), whereas some encode their own proteins that are responsible for recombination. In addition, certain groups of DNA viruses support site-specific (nonhomologous) recombination events. In general, the virus DNA recombination mechanisms seem to involve post-DNA replication molecular events.

For RNA viruses the majority of known homologous and nonhomologous RNA recombination events appear to be integrally linked to RNA replication machinery. Various types of copy-choice (template switching) mechanisms were proposed to describe the easy formation of RNA recombinants in numerous RNA virus systems. The roles of both special RNA signal sequences and viral proteins have been elucidated, reflecting the variety of the recombination strategies used by RNA virus groups. The involvement of host cell genes in RNA virus recombination has begun to get elucidated in several RNA viruses. Besides replicational copy-choice mechanisms, some RNA viruses use the breakage-and-religation mechanism where viral RNA gets regenerated by religation from RNA fragments, as shown experimentally for Enterobacteria phage Qbeta. New venues of RNA recombination research just emerge including our better understanding of the involvement of RNA *cis*-acting signals, the role of RNA replication, and the importance of cellular host genes such as RNA ribonucleases or RNA interference.

See also: Evolution of Viruses.

Further Reading

Agol VI (2006) Molecular mechanisms of poliovirus variation and evolution. *Current Topics in Microbiology and Immunology* 299: 211–259.

Briddon RW and Stanley J (2006) Subviral agents associated with plant single-stranded DNA viruses. *Virology* 344: 198–210.

Chetverin AB (2004) Replicable and recombinogenic RNAs. *FEBS Letters* 567: 35–41.

Cromie GA, Connelly JC, and Leach DR (2001) Recombination at double-strand breaks and DNA ends: Conserved mechanisms from phage to humans. *Molecular Cell* 8: 1163–1174.

Figlerowicz M and Bujarski JJ (1998) RNA recombination in brome mosaic virus, a model plus strand RNA virus. *Acta Biochimica Polonica* 45(4): 847–868.

Galetto R and Negroni M (2005) Mechanistic features of recombination in HIV. *AIDS Reviews* 7(2): 92–102.

Gowda S, Satyanarayana T, Ayllon MA, *et al.* (2001) Characterization of the *cis*-acting elements controlling subgenomic mRNAs of *Citrus tristeza virus*: Production of positive- and negative-stranded 39-terminal and positive-stranded 59 terminal RNAs. *Virology* 286: 134–151.

Koonin EV, Senkevich TG, and Dolja VV (2006) The ancient Virus World and evolution of cells. *Biology Direct* 1(29): 1–27.

Masters PS and Rottier PJ (2005) Coronavirus reverse genetics by targeted RNA recombination. *Current Topics in Microbiology and Immunology* 287: 133–159.

Miller ES, Kutter E, Mosig G, Arisaka F, Kunisawa T, and Ruger W (2003) Bacteriophage T4 genome. *Microbiology and Molecular Biology Reviews* 67: 86–156.

Miller WA and Koev G (1998) Getting a handle on RNA virus recombination. *Trends in Microbiology* 6: 421–423.

Mosig G (1998) Recombination and recombination-dependent DNA replication in Enterobacteria phage T4. *Annual Review of Genetics* 32: 379–413.

Nagy PD and Simon AE (1997) New insights into the mechanisms of RNA recombination. *Virology* 235: 1–9.

Noueiry AO and Ahlquist P (2003) Brome mosaic virus RNA replication: Revealing the role of the host in RNA virus replication. *Annual Review of Phytopathology* 41: 77–98.

Pasternak AO, Spaan WJM, and Snijder EJ (2006) Nidovirus transcription: How to make sense...? *Journal of General Virology* 87: 1403–1421.

Poon A and Chao L (2004) Drift increases the advantage of sex in RNA bacteriophage Phi6. *Genetics* 166: 19–24.

Van Duyne GD (2005) Lambda integrase: Armed for recombination. *Current Biology* 15: R658–R660.

Weigel C and Seitz H (2006) Bacteriophage replication modules. *FEMS Microbiology Reviews* 30: 321–381.

White KA and Nagy PD (2004) Advances in the molecular biology of tombusviruses: Gene expression, genome replication, and recombination. *Progress in Nucleic Acid Research and Molecular Biology* 78: 187–226.

Wierzchoslawski R, Urbanowicz A, Dzianott A, Figlerowicz M, and Bujarski JJ (2006) Characterization of a novel 5′ subgenomic RNA3a derived from RNA3 of Brome Mosaic Bromovirus. *Journal of Virology* 80: 12357–12366.

Wilkinson DE and Weller SK (2003) The role of DNA recombination in herpes simplex virus DNA replication. *IUBMB Life* 55(8): 451–458.

Worobey M and Holmes EC (1999) Evolutionary aspects of recombination in RNA viruses. *Journal of General Virology* 80: 2535–2543.

Satellite Nucleic Acids and Viruses

P Palukaitis, Scottish Crop Research Institute, Invergowrie, UK
A Rezaian, University of Adelaide, Adelaide, SA, Australia
F García-Arenal, Universidad Politécnica de Madrid, Madrid, Spain

Glossary

Satellite-like RNA A subviral genome dependent upon another virus for replication and encapsidation, but is required for vector transmission of the helper virus.

Satellite RNA or satellite DNA A subviral genome dependent on a helper virus for both replication and encapsidation.

Satellite virus A subviral genome dependent on a helper virus for its replication, but encoding its own capsid protein.

Virus-associated nucleic acid A subviral genome that depends on another virus for encapsidation and transmission, but not for its replication.

Introduction

Satellites of viruses constitute a heterogeneous collection of subviral agents. Satellites can be differentiated from other subviral nucleic acids, such as subgenomic (sg) RNAs, defective (D) or defective interfering (DI) RNAs, and viroids, by their molecular, biological, and genetic nature. Unlike sg RNAs, D RNAs, and DI RNAs, satellites and viroids have little or no sequence similarity to any known virus. Whereas viroids are replicated by host polymerases, RNA satellites are replicated by the polymerase of a virus, referred to as the helper virus (HV). Satellites have been found associated with DNA and RNA viruses, and in the latter case, with both ssRNA and dsRNA viral genomes, the satellites being of the same nucleic acid type as the HV. While, in general, the HV can exist independent of the satellite, there are exceptions where a satellite contributes to the transmission of the HV and thus is referred to as being satellite-like. In addition, some viral RNAs are found in association with other viral genomes on which they depend for their encapsidation and transmission, but not their replication. These viral RNAs are referred to here as virus-associated RNAs. Satellites are divided basically into two main groups; those that encode their own capsid protein (CP) are called satellite viruses, while those that require their HV for both replication and encapsidation are referred to as satellite RNAs or DNAs. There are also satellites of satellites, in which some satellite RNAs are replicated by the HV, but are encapsidated by the CP of a satellite virus. In addition, in the case of the carmovirus turnip crinkle virus (TCV), which replicates both satellite RNAs and DI RNAs, it also produces a chimeric RNA (referred to as satC) that is in part a DI RNA and in part a satellite RNA (satD). While the vast majority of satellites are found in association with plant viruses, a few have been found in association with animal or fungal viral genomes. Most plant viruses do not contain satellites associated with them, but as some satellites can significantly affect the disease induced by the HV, the presence of satellites has important consequences for viral-induced diseases.

History

The use of the term satellite as a subviral agent was first conceptualized in 1962 by Kassanis, to describe the relationship of a 17-nm diameter viral particle found in association with some isolates of the 26-nm diameter necrovirus tobacco necrosis virus (TNV). The smaller particle was dependent on TNV for its accumulation, but was serologically unrelated to TNV. It became known as the satellite virus of TNV (now known as tobacco necrosis satellite virus, TNSV). A few other satellite viruses have been described since then, but, in general, they are rare. By contrast, satellite RNAs are more common. The first satellite RNA was described by Schneider in 1969, in association with the nepovirus tobacco ringspot virus (TRSV), the satellite RNA being encapsidated by the CP of the HV. The symptoms induced by TRSV were attenuated dramatically by the presence of the satellite RNA. This is not always the case, with some satellite RNAs having no effect on either the HV accumulation or disease induced by the HV, and a few satellite RNAs exacerbating the HV-induced disease. Most of these satellite RNAs contain ssRNA genomes, while a few contain genomes of dsRNA. Some of the satellite RNAs with ssRNA genomes are translated to produce proteins, which may or may not be required for their replication, depending on the particular satellite RNA. Satellite DNAs were first described in 1997 by Dry and colleagues, in association with the geminivirus tomato leaf curl virus (TLCV). It was isolated from field-infected tomato plants in northern Australia. Satellite DNAs also are encapsidated by the CP of their HV. Since then, satellite DNAs referred to as satellite DNA β have been described in association with many geminiviruses, principally from Asia.

Geographical Distribution

Since satellite RNAs are only found associated with their specific HVs, they are limited to the distribution range of the HV and the vectors of the HV. On the other hand, not all isolates of the HV have satellites associated with them, further delimiting the distribution of satellites. Nevertheless, as many of the HVs are distributed worldwide, so are many of the satellites. The distribution of satellite DNA β has so far been linked to the geminiviruses of the Old World, an exception being DNA β in honeysuckle that appears to have been distributed through vegetative plant material.

Classification

There is no correlation between the taxonomy of their HV and the presence of satellites. Satellites are associated with viruses belonging to at least 17 genera, and with only a limited number of species within those genera. Satellites do not constitute a homogeneous group in terms of their nucleic acid type, size, sequence, structure, or translatability. Most satellites are composed of RNA, but some consist of DNA. Most satellites are linear molecules, but some RNA satellites and all of the DNA satellites are circular in structure. Some satellites may encode proteins, while many do not. Thus, satellites are classified primarily into categories based on the features of the above properties, vis-à-vis their nucleic acid form and their genetic capacity. All satellites are grouped into the following categories, first differentiated on the basis of whether they are satellite viruses (**Table 1**) or satellite nucleic acids (**Table 2**), but having in common the fact that they are not required for the replication of the HV.

1. Satellite viruses:
   - subgroup 1 – chronic bee-paralysis virus-associated satellite virus
   - subgroup 2 – satellites that resemble tobacco necrosis virus
2. Satellite nucleic acids:
   - ssDNA satellites
   - dsRNA satellites
   - ssRNA satellites: (1) subgroup 1 – large, ssRNA satellites; (2) subgroup 2 – small, linear, ssRNA satellites; (3) subgroup 3 – circular, ssRNA satellites.

Some RNAs previously described as satellite RNAs contribute to the natural means of transmission of the HV and thus are considered satellite-like RNAs, rather than true satellites (**Table 3**). In other cases, there are nucleic acids that are dependent on the HV for encapsidation, but not replication. These subviral agents are considered here as virus-associated nucleic acids (**Table 3**) and will be described and differentiated below.

General Properties and Effects of Satellites

Satellite Viruses

Best characterized of the satellite viruses (**Table 1**) are the satellites that resemble TNSV, all of them depending, for their replication, on plant viruses with ssRNA genomes. All satellite viruses in this subgroup contain an ssRNA of 800–1200 nt. Particles are isometric, 17 nm in diameter, with a $T=1$ symmetry, built of 60 protein subunits. The particle structure has been determined at high resolution by X-ray diffraction for TNSV, tobacco mosaic satellite virus (TMSV), and panicum mosaic satellite virus (PMSV), with ~80% of the encapsidated RNA in stem–loop structures. Particle structure differs from that of their HV. In spite of the different structure of CP and virus particle, the particles of both the satellite and HV may share important properties; for example, the particles of TNSV and TNV are able to bind specifically to the zoospores of the vector fungus *Olpidium brassicae*. The 1000–1200 nt RNA of TNSV does not have a methylated cap structure or a genome-linked protein (VPg) in its 5′ end. Unlike most plant virus RNAs it has a phosphorylated 5′ end. The 3′ termini may share a structure with that of the HV genomic RNA (gRNA); for instance, in both tobacco mosaic virus and TMSV, the 3′ termini form tRNA-like structures and are aminoacylable with histidine. *Cis*-acting sequences necessary for RNA amplification have been mapped at the 5′ and 3′ nontranslated regions (NTRs) of PMSV RNA as well as within the CP open reading frame (ORF), and are conserved in a D RNA, which is maintained by PMSV. A translational

Table 1 Satellite viruses

Helper virus/satellite virus	*Particle size (nm)*	*CP[a]*	*Satellite RNA size (nt)*	*Accession no.*
Subgroup 1				
Chronic bee-paralysis virus (CBPV)/ CBPV-associated satellite virus (CBPVA)	17	NR[b]	~1100 (three species)	
Subgroup 2				
Necrovirus				
Tobacco necrosis virus (TNV)/satellite TNV (STNV)	17	~21 600	1239	J02399
Sobemovirus				
Panicum mosaic virus (PMV)/satellite PMV (SPMV)	16	~17 500	824–826	M17182
Tobamovirus				
Tobacco mosaic virus (TMV)/satellite TMV (STMV)	17	~17 500	1059	M24782
Nodavirus				
Macrobranchium rosenbergii nodavirus (MrNV)/extra small virus (XSV)	15	~17 000	796	AY247793
Unassigned				
Maize white line mosaic virus (MWLMV)/ satellite MWLMV (SMWLMV)	17	23 961	1168	M55012

[a]CP, capsid protein MW.
[b]NR, not reported.

Table 2 Satellite nucleic acids

Helper virus/satellite	*Satellite size*	*Encoded protein (aa/kDa)*	*Accession no.*
ssDNA satellites (circular)			
Ageratum yellow vein virus (AYVV)/AYVV satellite DNA β	1347	118/13.7	AJ252072
Bhendi yellow vein mosaic virus (BYVMV)/BYVMV satellite DNA β	1353	140/	AJ308425
Cotton leaf curl Bangalore virus (CLCuBV)/CLCuBV satellite DNA β	1355	118/	AY705381
Cotton leaf curl Gezira virus (CLCGV)/CLCGV satellite DNA β	1348	117/13.6	AY077797
Cotton leaf curl Multan virus (CLCMV)/CLCMV satellite DNA β	1349	118/13.7	AJ298903
Eupatorium yellow vein virus (EuYVV)/EuYVV satellite DNAβ	1356	116/13.5	AJ438938
Honeysuckle yellow vein virus (HYVV)/HYYV satellite DNA β	1344	116/13.5	AJ316040
Malvastrum yellow vein virus (MYVV)/MYVV satellite DNA β	1350	118/	AJ786712
Tobacco curly shoot virus (TCSV)/TCSV satellite DNA β	1354	118/	AJ421484
Tomato yellow leaf curl China virus (TYLCCV) /TYLCCV satellite DNA β	1336	118/	AJ421621
Tomato leaf curl virus (TLCV)/TLCV satellite DNA	682	None	U74627
dsRNA satellites (linear)			
L-A ds RNA virus of *Saccharomyces cerevisiae*/M satellite RNAs	1801 bp	NR	U78817
	0.5–1.8 kb	NR	
Trichomonas vaginalis T1 virus (TVTV)/TVTV satellite RNA	497 bp	NR	U15991
	~700 bp	NR	NR
	~1700 bp	NR	NR
Ophiostoma novo-ulmi mitovirus 3a (OnuMV3a)/OnuMV3a S-dsRNA	738–767 bp	4 small ORFs	AY486119
Large ssRNAs satellites			
Benyvirus			
Beet necrotic yellow vein virus (BNYVV)/BNYVV RNA 5	1342-1447 nt	~26 kDa	U78292
Nepovirus			
Arabis mosaic virus (ArMV)/ArMV large satellite RNA	1104 nt	~39 kDa	D00664
Chicory yellow mottle virus (CYMV)/CYMV large satellite RNA	1145 nt	~39 kDa	D00686
Grapevine Bulgarian latent virus (GBLV)/GBLV satellite RNA	~1500 nt	NR	NR
Grapevine fanleaf virus (GFLV)/GFLV satellite RNA	1114 nt	~37 kDa	D00442
Myrobalan latent ringspot virus (MLRV)/MLRV satellite RNA	~1400 nt	~45 kDa	NR
Strawberry latent ringspot virus (SLRV)/SLRV satellite RNA	1118 nt	~36 kDa	X69826
Tomato black ring virus (TBRV)/TBRV satellite RNA	1372-1375 nt	~48 kDa	X05689
Potexvirus			
Bamboo mosaic virus (BaMV)/BaMV satellite RNA	836 nt	~20 kDa	L22762
Small, linear ssRNAs satellites			
Carmovirus			
Turnip crinkle virus (TCV)/TCV satellite RNA	194 nt, 230 nt, 356 nt		X12749
Cucumovirus			
Cucumber mosaic virus (CMV)/CMV satellite RNA	333–405 nt		M18872
Peanut stunt virus (PSV)/PSV satellite RNA	393 nt		Z98198
Necrovirus			
Beet black scorch virus (BBSV)/BBSV satellite RNA	615 nt		AY394497
Tobacco necrosis virus (TNV)/TNV small satellite RNA	620 nt		NR
Nepovirus			
Chicory yellow mottle virus (CYMV)/CYMV small satellite RNA	457 nt		NC006453
Sobemovirus			
Panicum mosaic virus (PMV)/PMV satellite RNA	350 nt		NR
Tombusvirus			
Artichoke mottled crinkle virus (AMCV)/AMCV satellite RNA	~700 nt		NR
Cymbidium ringspot virus (CymRSV)/CymRSV satellite RNA	621 nt		D00720
Carnation Italian ringspot virus (CIRV)/CIRV satellite RNA	~700 nt		NR
Pelargonium leaf curl virus (PLCV)/PLCV satellite RNA	~700 nt		NR
Petunia asteroid mosaic virus (PAMV)/PAMV satellite RNA	~700 nt		NR
Tomato bushy stunt virus (TBSV)/TBSV satellite RNA	612 nt, 822 nt		AF022788
Umbravirus			
Pea enation mosaic virus (PEMV)/PEMV satellite RNA	717 nt		U03564
Circular ssRNAs satellites			
Polerovirus			

Continued

Table 2 Continued

Helper virus/satellite	*Satellite size*	*Encoded protein (aa/kDa)*	*Accession no.*
Cereal yellow dwarf virus (CYDV-RPV)/CYDV-RPV satellite RNA	322 nt		M63666
Nepovirus			
Arabis mosaic virus (ArMV)/ArMV satellite RNA	300 nt		NC001546
Tobacco ringspot virus (TRSV)/TRSV satellite RNA	359 nt		M14879
Sobemovirus			
Lucerne transient streak virus (LTSV)/LTSV satellite RNA	324 nt		X01984
Rice yellow mottle virus (RYMV)/RYMV satellite RNA	220 nt		NC003380
Solanum nodiflorum mottle virus (SNMV)/SNMV satellite RNA	377 nt		J02386
Subterranean clover mottle virus (SCMV)/SCMV satellite RNA	332 nt, 388 nt		M33000
Velvet tobacco mottle virus (VTMoV)/VTMoV satellite RNA	365–366 nt		J02439

Table 3 Satellite-like and virus-associated nucleic acids

Helper virus/satellite-like/virus-associated nucleic acid	*Size of nucleic acid*	*Encoded protein*
Satellite-like ssRNAs[a]		
Benyvirus		
Beet necrotic yellow vein virus (BNYVV)/BNYYV RNA 3	1754 nt	~25 kDa
BNYVV/BNYVV RNA 4	1467 nt	~31 kDa
Umbravirus		
Groundnut rosette virus (GRV)/GRV satellite RNA	895–903 nt	
Virus-associated nucleic acids[b]		
Hepadnavirus		
Hepatitis B virus (HBV)/hepatitis delta virus (HDV)	1979 nt	~22 kDa (δAg-S) ~24 kDa (δAg-L)
Luteovirus		
Beet western yellows virus (BWYV) /	2843 nt	~85 kDa, plus others
BWYV-associated RNA (BWYVaRNA)		
Carrot red leaf virus (CRLV) /		
CRLV-associated RNA (CLRVaRNA)	2835 nt	~85 kDa, plus others
Begomovirus		
Cotton leaf curl virus (CLCuV)/		
CLCuV-associated DNA 1 (CLCuVaDNA 1)	1376 nt	~33 kDa
Nanovirus		
Banana bunchy top virus (BBTV)/BBTV-associated DNAs S1, S2, and S3	~1100	
Faba bean necrotic yellows virus (FBNYV)/FBNYV-associated DNAs C1, C7, C9, and C11	~1000	~33 kDa
Milk vetch dwarf virus (MDV)/MDV-associated DNAs C1, C2, C3, and C10	~1000	~33 kDa
Subterranean clover stunt virus (SCSV)/SCSV-associated DNAs C2 and C6	~1000	~33 kDa

[a]Dependent on a helper virus for replication and encapsidation, but is essential for vector transmission of the helper virus.
[b]Dependent on a helper virus for encapsidation, but not for replication.

enhancer domain has been mapped in the 3′ NTR of TNSV. In addition to the ORF encoding the CP, some satellite viruses contain other ORFs; it remains unclear if the encoded products have any role *in vivo*.

Interference with the accumulation of the HV has been described for TNSV. Satellite viruses can also modify the symptoms induced by the HV, as is the case with PMSV, which in co-infection with panicum mosaic virus (PMV) enhances the mild symptoms of PMV to cause a severe mosaic and chlorosis. Symptom induction is due to the PMSV CP, and a chlorosis-inducing domain has been mapped. In addition to its structural and symptom-inducing functions, the CP of PMSV is involved in systemic movement, binds PMV particles, and counters the effects of post-transcriptional gene silencing suppressors.

The particles of TNSV may contain a noncoding satellite RNA of about 620 nt, which depends on TNV for replication and on TNSV for encapsidation. A satellite RNA with similar dependence relationships has been described for PMV and PMSV. These systems are good examples of the complexity of dependence relationships in satellitism.

The other subgroup of satellite viruses contains satellites found associated with chronic bee-paralysis virus (CBPV). The satellites consist of three RNA species of about 1.1 kbp, which can be encapsidated either in 17-nm isometric particles by the satellite-encoded CP, or by the CP of the HV, CBPV. The satellites interfere with CBPV replication.

Satellite Nucleic Acids

ssDNA Satellites

All the DNA satellites reported to date are associated with a single virus genus, *Begomovirus*, in the family *Geminiviridae* (**Table 2**). This group of over 100 viruses is transmitted by whiteflies and their geminate particles encapsidate either one or two ssDNA species, each of about 2700 nt, as well as satellite DNAs that may occur. The first satellite DNA was found in association with TLCV and is a single-stranded (ss), circular DNA molecule of 682 nt, which can be supported for replication by a number of begomovirus species. It lacks any significant ORF or identifiable promoter element and does not contribute to the infection of TLCV. More recently, a large group of ss, satellite DNAs termed DNA β has been isolated. They are related to TLCV satellite DNA but are about twice the size of the TLCV satellite DNA and encode a protein known as βC1. The C1 protein is expressed from a single complementary-sense transcript with conserved regulatory elements.

The search for the causal agents of Ageratum yellow vein disease led to the discovery of the first DNA β satellite because it was required for disease symptom expression. Subsequently, a similar role was established for cotton leaf curl virus (CLCuV) DNA β, which is part of a disease complex causing major crop losses in Pakistan. Some geminivirus satellite DNAs reported as DNA β are of similar size to TLCV satellite DNA and lack a βC1 gene. Geminivirus satellite DNAs therefore can be considered a single group in which some species are defective for encoding a protein.

Geminivirus satellite DNAs can exert a drastic effect on the symptoms produced by their HV. The pathogenesis is mediated by the βC1 protein. The mechanism of βC1-mediated pathogenesis is not clear; however, the protein causes a drastic disease-like phenotype when expressed transgenically. This severe effect on host plants suggests changes to growth pattern and is accompanied by vein thickening, enations, and development of leaf-like structures. The protein has been demonstrated to be a suppressor of gene silencing, raising the possibility that it may control host functions by microRNA regulation.

Geminivirus satellite DNAs do not show a strict affinity for HV. TLCV satellite DNA is supported for replication by viruses as diverse as tomato yellow leaf curl virus, African cassava mosaic virus, and even beet curly top virus, which belongs to a different genus. In another case, the DNA β associated with Ageratum yellow vein disease was found to be maintained in experimental plants infected with Sri Lankan cassava mosaic virus (SLCMV), which has a bipartite genome. This interaction altered the host range of SLCMV to include Ageratum. Interestingly, the satellite could substitute for DNA B component implying a functional similarity. Other evidence supporting the role of geminivirus satellite DNAs in movement has also been obtained. The DNA A component of tomato leaf curl New Delhi virus (TLCNDV) is not capable of systemic infection in the absence of DNA B and remains confined to the sites of inoculation where it is capable of replication; however, CLCuV DNA β can substitute DNA B of TLCNDV to restore systemic infection in tomato. Surprisingly, systemic infection of DNA A component alone, accompanied by symptoms, could also be mediated by transient expression of the βC1 protein.

dsRNA Satellites

A number of examples of dsRNA satellites have been described, but only two are recognized officially at this time (**Table 2**): one group is associated with the yeast *Saccharomyces cerevisiae* and is designated the M satellites of L-A dsRNA virus; the other group of three dsRNAs is associated with the Trichomonas vaginalis T1 virus (TVTV) of the eponymous protozoan. Both HVs are in the family *Totiviridae*. In the case of the three TVTV satellite RNAs of 497, ~700, and ~1700 bp, the ds satellite RNAs are maintained only by some isolates of TVTV, which both replicate and encapsidate these dsRNAs. The L-A dsRNA virus M satellites consist of several dsRNAs varying from 1.0 to 1.8 kbp in size. These dsRNA satellites are dependent on genes of the HV for both replication and encapsidation. The M1 dsRNA satellite encodes a toxin which kills other yeasts not harboring this dsRNA. In its prototoxin form, this protein provides immunity to the yeast secreting the toxin. Other L-A dsRNA viruses have different M satellites associated with them, all of them also encoding toxin and immunity systems.

All of the putative dsRNA satellites were found with fungi, in association with dsRNA viruses in the families *Hypoviridae*, *Narnaviridae*, *Partitiviridae*, and *Totiviridae*. One of these dsRNA satellites, from Ophiostoma novo-ulmi mitovirus 3a (OnuMV3a) in the genus *Mitovirus*, of the family *Narnaviridae* consists of a group of dsRNAs of 738–767 bp, designated OnuMV3a S-dsRNA (**Table 2**). These dsRNA satellites did not affect the hypovirulence associated with their HV in the fungus *Sclerotinia homeocarpa*. Their nucleotide sequences indicate that these OnuMV3a S-dsRNAs did not have the coding capacity for their own replicase, and thus presumably depend on the replicase of the HV.

ssRNA Satellites

Subgroup 1: Large ssRNA satellites

Subgroup I satellites have messenger RNA properties. These satellite RNAs are between 0.8 and 1.5 kb in size and encode nonstructural proteins that are expressed *in vivo* (**Table 2**). Most satellites in this subgroup are associated with nepoviruses. Large ssRNA satellites of nepoviruses share the 5′ and 3′ structural features of the gRNAs of the HV: a 3′-terminal poly(A) sequence and a 5′-terminal VPg that, in the analyzed instances, is indistinguishable from that on the HV genome and thus, is encoded by it. The encoded nonstructural protein of different satellite RNAs are basic proteins of 36–48 kDa. For the satellite RNAs of tomato black ring virus (TBRV), grapevine fanleaf virus (GFLV), and arabis mosaic virus (ArMV), it has been shown that the encoded proteins are needed for the replication of the satellite RNA. Most large satellite RNAs of nepoviruses do not seem to have an effect on the accumulation or pathogenicity of the HV. However, this may depend on the experimental system, as the large satellite RNA of ArMV was shown to modulate the symptoms of the HV depending on the species of host plant.

Large satellite RNAs (1342–1347 nt) have been found to be associated with many isolates of the benyvirus beet necrotic yellow vein virus (BNYVV). These satellite RNAs, designated BNYVV RNA 5, encode a protein of 26 kDa, which is responsible for intensification of the rhizomania disease induced by infection of BNYVV RNAs 1 and 2, plus its two satellite-like RNAs 3 and 4.

Another large satellite RNA has been found to be associated with the potexvirus bamboo mosaic virus (BaMV), the only satellite RNA encapsidated into rod-shaped particles. BaMV satellite RNA encodes a 20-kDa protein that is expressed *in vivo* but is not needed for satellite RNA replication. Interestingly, this protein shares significant sequence similarity with the structural protein of the PMSV, and binds cooperatively to RNA, with a preference for the satellite RNA. The presence of BaMV satellite RNA significantly reduces the accumulation of BaMV RNA. *Cis*-acting sequences required for BaMV satellite RNA replication have been mapped at the 5′ NTR, and comprise a stem–loop structure that is conserved among BaMV satellite RNA variants.

Subgroup 2: Small, linear ssRNA satellites

This subgroup contains ss satellite RNAs associated with seven genera of plant viruses (**Table 2**), although many of the members are associated with nepoviruses. There are no circular forms of these satellite RNAs present in infected cells, although dimer or higher multimer forms may be detected in virions and/or in infected cells. These satellite RNAs vary in size from ~200 to ~800 nt and do not appear to encode proteins, although isolates of some satellite RNAs contain nonconserved ORFs. These RNAs appear to have highly ordered secondary structures, which is probably responsible for the high stability and infectivity of these RNAs, as well as their biological properties. Most of the small ss satellite RNAs either have no effect on or reduce the accumulation of their HV. This reduction in HV accumulation often, but not always, results in a decrease in the disease symptoms induced by the HV. Moreover, in a few cases, the presence of the satellite RNA reduced or had no effect on the HV accumulation, but intensified the disease induced by the HV. One of the latter, a tomato necrosis-inducing satellite RNA of the cucumovirus cucumber mosaic virus (CMV), could even induce necrosis in tomato independent of the HV, if expressed in the complementary-sense orientation within the genome of virus vector but not when expressed as a transgene; however, in the case of other satellite RNAs of CMV, different strains of CMV and the related HV tomato aspermy virus (TAV) have been used to show that the HV makes a contribution to the disease syndrome associated with the satellite RNA. The basis for satellite RNA-mediated attenuation of disease symptoms is not clear, but may not be due simply to reduction of the HV titer, since CMV satellite RNAs supported by TAV show an attenuation of symptoms, but no effect on the accumulation of the HV. Similarly, the reduction of HV level may not be due simply to competition between the HV and the satellite RNA for the same replicase, since data from TCV and CMV satellite RNAs indicate that satellite RNAs may affect the suppression of RNA silencing function of their associated HV, leading to a reduction in the titer of the virus due to RNA silencing. A general model for satellite- and viroid-mediated intensification of disease symptoms has been proposed, caused by silencing inducing RNAs generated from such subviral RNAs acting on specific host mRNAs.

Subgroup 3: Small, circular ssRNA satellites

This subgroup of satellite RNAs contains members that vary in size from 220 to 388 nt, and are associated with three genera of plant viruses (**Table 2**). The satellite RNAs exist in circular as well as linear forms in infected plant cells, but only the circular satellite RNA form is encapsidated by the helper sobemovirus, while only the linear satellite RNA form is encapsidated by the helper polerovirus and nepoviruses. Multimeric, linear forms of these satellites are produced in infected cells and, in some cases, these are also packaged into virus particles. The multimeric forms are generated via a rolling-replication mechanism and the unit-length linear as well as circular forms are generated via ribozymes. There is no mRNA activity associated with these various satellite RNAs, and they all have a high degree of secondary structure. The satellite RNAs either have no effect on (sobemovirus satellites) or reduce (nepovirus and polerovirus satellites) the accumulation of their HV. In addition, the presence of satellite RNAs either reduces (nepovirus and polerovirus satellites) or exacerbates (sobemovirus satellites) the symptoms induced by the helper virus. The molecular basis for symptom attenuation or intensification is unknown.

Satellite-Related Nucleic Acids

A number of subviral nucleic acids have been described that have in some instances been referred to as satellite RNAs or DNAs. However, as these nucleic acids either contribute to the infection cycle of the HV, or are not dependent upon an HV for their replication, they are better described as satellite-like or virus-associated nucleic acids, respectively (**Table 3**).

Satellite-like ssRNAs

This group of subviral RNAs includes the satellite RNAs of groundnut rosette virus (GRV) and RNAs 3 and 4 of BNYVV. In the case of satellite RNAs of GRV, they are required for transmission of GRV by the luteovirus groundnut assistor virus. How these satellite RNAs of 895–903 nt affect transmission is unknown, since they do not code for any functional proteins. Some of the satellite RNAs of GRV also affect accumulation of the HV and the symptoms induced by the HV, causing either attenuation due to severely reduced replication of the HV, or exacerbation, producing unique symptoms. In the case of BNYVV, where only RNAs 1 and 2 are essential for replication, movement within leaves and particle assembly, RNA 3 (1774 nt) is required for spread of the virus into roots and RNA 4 (1467 nt) is essential for virus transmission by the soil-fungus vector of BNYYV. Thus, BNYYV RNAs 3 and 4 are considered satellite-like, in that they contribute to the natural infection cycle of the virus. By contrast, BNYYV RNA 5 is a true satellite, in that it is not required for any phase of the infection cycle of BNYYV, although its encoded 26 kDa protein in combination with the 25 kDa protein encoded on RNA 3 intensifies symptom expression of the disease rhizomania.

Virus-associated nucleic acids

This group of subviral agents does not fit the above definition of satellites, but as they are peripheral to but dependent on an HV for their encapsidation, they have sometimes been referred to as satellite-like RNAs or DNAs. However, as a functional distinction exists between these subviral nucleic acids and those of the satellite-like agents described above, which are dependent on an HV for their replication, these subviral agents are collectively referred to as (virus-) associated nucleic acids.

Hepatitis delta virus

Hepatitis delta virus (HDV) is a subviral agent of the human pathogen, hepatitis B virus (HBV). HDV is replicated in the nucleus by the host DNA-dependent RNA polymerase II, but is dependent upon the envelope proteins of HBV for its encapsidation and transmission, and is thus not a satellite. The HDV (1679 nt) RNA folds into an unbranched rod-like structure similar to that of viroids, and in common with the circular ssRNA satellites, contains ribozymes that are involved in the processing and maturation of this RNA during replication. The HDV RNA encodes two forms of a protein designated the delta antigen (δAg). The smaller, 22 kDa form (δAg-S), which is a nuclear phosphoprotein, is required for replication of HDV RNA, while the 24 kDa larger form (δAg-L), which contains an additional 19 aa at the C-terminus and is produced late in infection, is required along with δAg-S for assembly of the HDV RNA into HBV envelope particles. δAg-L also acts as an inhibitor of HDV replication. HDV intensifies the severity of liver disease caused by HBV.

Luteovirus-associated RNAs

Some isolates of two luteoviruses, beet western yellows virus and carrot red leaf virus (CRLV) were found to have associated with them additional RNAs of size 2.8 kb. These associated RNAs encoded their own viral replicases, but required the respective luteoviruses for their movement, encapsidation, and transmission. These RNAs, designated BWYVaRNA (2843 nt) and CRLVaRNA (2835 nt), have similar genome organizations, are about 50% similar in sequence, and both intensify the disease caused by their HV. CLRVaRNA together with CLRV and the umbravirus carrot mottle virus induced carrot motley dwarf disease.

Begomovirus- and nanovirus-associated DNAs

Circular ssDNA species of about 1300 nt, unrelated to geminivirus genomes, have been isolated from plants infected with some Old World begomoviruses. These molecules, named DNA 1, resemble a DNA component of nanoviruses. CLCuV-associated DNA1 is an example (**Table 3**). They encode their own replicase-associated protein (Rep) and replicate independent of the HV. In the absence of a nanovirus CP which allows aphid transmission, the virus-associated DNAs are encapsidated in helper geminiviruses and are transmitted by their whitefly vector. Nanoviruses themselves may also contain virus-associated DNAs. These are about 1.0 or 1.1 kbp in size and capable of autonomous replication but require the respective nanovirus for their movement, encapsidation, and transmission by aphids.

Replication and Structure

Satellite DNAs

The replication of geminivirus satellite DNAs mimics that of the HV. Unlike RNA viruses, geminiviruses do not encode a polymerase and depend on host DNA polymerases for replication. The only geminivirus-encoded protein essential for replication is Rep. Geminivirus satellite DNAs utilize the HV Rep and contain the conserved sequence for DNA nicking by the Rep. Replication takes place by a combination of rolling circle and recombination dependent mechanisms and the DNA is encapsidated in the HV virion. The replication is independent of βC1 expression.

Sequence comparisons at the nucleotide and amino acid levels reveal a number of conserved structural features, despite a significant level of sequence divergence. These include an A-rich region, a sequence of about 80 nt referred to as the satellite conserved region, and putative stem–loop structure containing a mononucleotide motif, also present in geminivirus genomes.

Satellite RNAs

A general property of satellite RNAs is that their replication depends on the replication machinery of the HV, which involves virus- and host-encoded factors. Hence, replication of satellite RNAs depends on interactions with both the HV and the host plant. Satellite RNAs may or may not share structural features at their 5′ and 3′ termini in common with the HV RNA. For instance, the large satellite RNAs of the nepoviruses have an HV-encoded VPg at the 5′ end and a poly(A) tail at the 3′ end, as do their HV RNAs, but the satellite RNAs of CMV have a 5′ cap structure, as do the CMV RNAs, but unlike the HV, the satellite RNAs do not have a tRNA-like structure that can be valylated at the 3′ end. Structural differences and, often, differences in the replication process between the satellite RNA and the HV RNA, indicate that the replication machinery of the HV may need to be adapted to replicate the satellite RNA, in ways which are not completely understood. The HV replication complex could be modified by satellite-encoded factors, as has been hypothesized for the large satellite RNAs of nepoviruses, or by unidentified host factors. It should be pointed out that in the best characterized systems, the efficiency of replication depends on the host plant. For CMV satellite RNAs, replication efficiency also depends on the strain of HV. Similarly, while the expression of the HV RNA-dependent RNA polymerase was enough for the replication of cereal yellow dwarf virus (CYDV) satellite RNA in the homologous host, this was not the case for cymbidium mosaic virus satellite RNA in a heterologous host system, in which, however, a DI RNA was amplified.

Replication has been studied best in the small noncoding satellite RNAs. For the small linear satellite RNAs of TCV and CMV, multimeric forms of both positive (arbitrarily defined as the encapsidated sense) and negative sense are found in infected tissues. The junction between monomers can be perfect or have deletions. Circular forms are not found, and replication does not proceed through a rolling-circle mechanism. Regulatory sequences for RNA replication have been mapped in detail in the satellite RNAs of TCV. In the hybrid satC molecule, a hairpin structure is a replication enhancer and has a role in depressing the accumulation of the HV. Also, structural motifs favoring recombination have been analyzed extensively in satD and satC of TCV.

The replication of small, circular satellite RNAs is by a rolling-circle mechanism; upon, infection, linear, multimeric forms of the negative strand are synthesized. The multimeric plus strand is then synthesized on this template or on negative strand circular monomers, depending on the satellite RNA. The positive strand multimer is cleaved autocatalytically and ligation occurs for those satellite RNAs that are encapsidated as a circular form. Hammerhead and hairpin ribozyme structures are found in these satellite RNAs monomers or dimers, which catalyze hydrolysis or hydrolysis and ligation, respectively. Other structures required for replication and encapsidation have been mapped in CYDV satellite RNA.

For one circular satellite RNA, a DNA counterpart has been shown to occur in the host genome, as a retroid element. HDV, which depends on HBV for encapsidation, shares structural features and replication mechanisms with the small, circular satellite RNAs.

Structural analyses have been done mostly with small satellite RNAs. Models for the *in vitro* secondary structure have been proposed for several satellite RNAs (e.g., CMV satellite RNA, TCV satellite RNAs, TRSV satellite RNA, and CYDV satellite RNA) based on nuclease sensitivity and chemical modification of bases data. For CMV satellite RNA, *in vivo* models have also been proposed. All analyzed satellite RNAs have a high degree of secondary structure, with more than 50% and up to 70% of bases involved in pairing. The high degree of secondary structure could explain the high stability of these molecules as well as the very high infectivity reported for some of them, such as CMV satellite RNA and TCV satellite RNA. The high degree of secondary structure may also be related to their biological activity, which for these noncoding molecules must depend on structural features. As has been detailed in other sections, sequences, structural domains, and tertiary interactions involved in pathogenicity and replication have been well characterized for some small satellite RNAs.

Sequence Variation and Evolution

Sequence variation and evolution has been analyzed most for CMV satellite RNA. Early experiments under controlled conditions showed a high variability and genetic plasticity. Populations started from cDNA clones rapidly evolved to a swarm of sequences whose master sequence changed upon passage on different hosts. Analysis of genetic variants from natural populations showed again a high diversity for CMV satellite RNA. In natural populations, satellite RNA diversification and evolution proceeded by mutation accumulation and by recombination. Constraints to genetic variation were analyzed, and were related to the maintenance of base pairing in secondary structure elements. Population diversity of CMV satellite RNA was higher than for the HV, and the population structures of CMV and the satellite RNA were not correlated. These analyses also showed that CMV satellite RNA behaved in nature as a molecular hyperparasite spreading epidemically on the CMV populations. The incidence of CMV satellite RNA in CMV populations has been shown to be low, except during episodes of epidemics of tomato necrosis. High diversity of natural populations has also been shown for PMSV RNA and for BaMV satellite RNA. For BaMV satellite RNA, the diversity was highest in the NTRs. Conversely, the population of rice yellow mottle virus satellite RNAs over sub-Saharan Africa showed little sequence diversity. The incidence of these three satellite RNAs in the analyzed populations of the respective HV was, in all cases, high.

Phylogenetic analyses divide geminivirus satellites broadly into two categories, those isolated from host plants of the family Malvaceae and the rest isolated from other plants, mostly in the Solanaceae. Compared to satellite RNAs, satellite DNAs do not exhibit a wide diversity. They are structurally similar, and lack strict specificity for HV species. Geminivirus satellite DNA sequences analyzed exhibit a minimum overall similarity of about 47% and 37%, at the nucleotide level and amino acid level, respectively. To date, over 120 DNA β sequences have been reported in data banks; however, many of these are members of the same species because their overall sequence similarity is well above 90%. Some examples of distinct geminivirus satellites that have been adequately characterized are listed in **Table 2**.

There has been much speculation on the origin of satellite RNAs. For CMV satellite RNA and TCV satellite RNA, it has been suggested that they could have been generated out of small sequences synthesized upon the HV RNA as a template by the HV polymerase. This hypothesis is no longer sustained by those that proposed it. Whatever the mechanism of generation of the satellite RNAs, it seems that the satellitism as a phenomenon, that is, the dependence on a certain virus for replication, has evolved independently several times. This is suggested by the lack of correlation between satellite and HV taxonomy, and by the existence of subviral nucleic acids with different degrees of dependence on an HV. For instance, an evolutionary line from nondependent viruses to satellites such as BaMV satellite RNA through satellite viruses could have proceeded by size and information content reduction. A phylogenetic relationship between viroids and small circular satellite RNAs has also been proposed.

Expression of Foreign Sequences from Satellite Vectors

The satellite RNAs of BaMV and BNYVV, as well as the satellite DNA of tomato yellow leaf curl China virus isolate Y10 (TYLCCNV-Y10), all of which express proteins not required for the replication or spread of the satellite or HV, have been used as expression vectors of foreign sequences. In the case of the satellite RNA of BaMV, the expression of such sequences and the accumulation of the satellite RNA were reduced considerably in systemically infected leaves. Nevertheless, this satellite expression system has been useful for studying *cis*- and *trans*-acting replication signals. The 26 kDa protein encoded by RNA 5 of BNYYV has been replaced with the sequence encoding the green fluorescent protein, which was expressed in both the inoculated leaves and systemically infected leaves. The small size of the satellite DNA of TYLCCNV-Y10 precludes use of this system for expression of most genes, but the system has been used to express plant gene segments inducing RNA silencing in several plants species. The TMSV system also has been used to express plant gene sequences, in place of its CP, resulting in RNA silencing of those plant genes, but not of the satellite virus or its HV. The modified TMSV expression vector was able to spread through the plant in the absence of its CP.

Satellite-Mediated Control of Viruses

The ability of some satellite RNAs to attenuate the symptoms induced by their HV has been utilized to develop strategies for disease control. In one strategy, the satellite RNA of CMV has been used in combination with mild strains of CMV as a 'vaccine', to pre-inoculate tomato plants and ameliorate symptoms induced by the replicating challenge virus (through the attenuating ability of the vaccine satellite RNA), in fields and greenhouses. Moreover, the vaccine could also prevent or reduce the frequency of infection of both the HV (through cross-protection by the HV of the vaccine strain) and of satellite RNAs that exacerbate the disease caused by the HV (through cross-protection by the vaccine satellite RNA). In a second strategy, the satellite RNAs of CMV, TRSV, and GRV have been expressed in transgenic plants, and provided either resistance to infection by the respective HV, or tolerance to the disease induced by the HV and/or related pathogenic satellites.

Further Reading

Briddon RW and Stanley J (2006) Subviral agents associated with plant single-stranded DNA viruses. *Virology* 344: 198–210.

Dry IB, Krake LR, Rigden J, and Rezaian MA (1997) A novel subviral agent associated with a geminivirus: The first report of a DNA

satellite. *Proceedings of the National Academy of Sciences, USA* 94: 7088–7093.
Gossele V, Fache I, Meulewaeter F, Cornelissen M, and Metzlaff M (2002) SVISS – A novel transient gene silencing system for gene function discovery and validation in tobacco plants. *Plant Journal* 32: 859–566.
Palukaitis P and García-Arenal F (2003) Cucumoviruses. *Advances in Virus Research* 62: 241–323.
Simon AE, Roossinck MJ, and Havelda Z (2004) Plant virus satellite and defective interfering RNAs: New paradigms. *Annual Reviews in Phytopathology* 42: 415–437.
Tao X and Zhou X (2004) A modified satellite that suppresses gene expression in plants. *Plant Journal* 38: 850–860.
Vogt PK and Jackson AO (eds.) (1999) *Satellites and Defective Viral RNAs*. Heidelberg: Springer.

Retrotransposons of Vertebrates

A E Peaston, The Jackson Laboratory, Bar Harbor, ME, USA

Glossary

Ancestral retrotransposon A retrotransposon present in the genome of a common ancestor of two or more host groups. Also referred to as ancestral repeat.

Apurinic/apyrimidinic Endonuclease enzyme that catalyzes the cleavage of a phosphodiester bond in a DNA molecule.

Autonomous retrotransposon A retrotransposon encoding proteins required for its reverse transcription and transposition.

Clade A group of organisms consisting of a single common ancestor and all its descendents.

DDE transposases A class of transposase enzymes containing a highly conserved amino acid motif, aspartate–aspartate–glutamate (DDE), required for metal ion coordination in catalyzing integration of retrotransposon cDNA into the host DNA, thus also known as integrase.

Exaptation In broad terms, a feature conferring evolutionary fitness on an organism but which was originally nonfunctional or designed for some other function in the organism. Thus, the genes of a retrotransposon newly inserted in the host genome maybe, over evolutionary time, co-opted for function in the host; in this new role they are called exaptations.

Homoplasy A structure arising in two or more species as the result of a convergent evolution, and not as a result of common descent which indicates that the feature in one species is homologous to that in the other.

Lineage-specific retrotransposon
A retrotransposon introduced into the genome of one but not another host grouping after their evolutionary divergence.

Nucleotide substitution rate The rate at which single nucleotide mutations occur within regions of a genome not subject to selection. One way to estimate this is by comparison of the consensus sequences of ancestral repeats with their remnant sequences in different host genomes.

Y-transposase, also known as tyrosine recombinase A class of transposase enzymes which can use a conserved tyrosine to cut and rejoin its DNA substrates by a 3′ phosphotyrosine linkage; used by some retrotransposons to integrate their cDNA into the host DNA, and thus sometimes called integrase.

History

Transposable elements are defined segments of DNA which replicate and move to other loci within the genome by a variety of mechanisms. The vast majority of these mobile elements in vertebrates are retrotransposons, which replicate by means of an RNA intermediate that is reverse-transcribed into DNA and inserted in a new location within the genome. Retrotransposons are found in all vertebrate genomes, but have been intensively studied in relatively few.

Retrotransposons were initially detected as discrete fragments of genomic DNA that rapidly reannealed after denaturation, and were recognized to be repetitive sequence elements interspersed within genomic DNA. The advent of DNA analysis by restriction enzyme fragmentation led to the identification of discrete families of repetitive elements whose members shared particular sets of internal restriction sites. The most highly repeated long and short sequences were named long interspersed repetitive elements (LINEs) and short interspersed repetitive elements (SINEs), respectively.

The LTR elements, a third general type of retrotransposon resembling the integrated form of proviruses, was similarly discovered. These endogenous retroviruses and retrovirus-like elements are generally restricted to an intracellular life cycle and vertical transmission through the germline and, unlike 'true' retroviruses, are not infectious. However, notable exceptions blur this distinction.

The advent of whole genome sequencing and sequence analysis is rapidly providing detailed pictures of vertebrate retrotransposon landscapes, such as those of a pufferfish (*Takifugu rubripes*), the chicken (*Gallus gallus*), the laboratory mouse (*Mus musculus*), human (*Homo sapiens*), dog (*Canis lupus familiarus*), and others. New lineage-specific retrotransposons of vertebrates such as primate-specific SVA (SINE-R, VNTR, Alu), are coming to light, as well as the discovery within selected vertebrate lineages of ancient retrotransposons, such as the Penelope-like elements (PLEs) that are present in the genomes of metazoans, fungi, and amoebozoans. Analysis of retroelement phylogenies can assist in disentangling evolutionary relationships of their hosts.

Nomenclature

Different structural and functional classification schemes for retrotransposable elements have been proposed. No scheme has been generally accepted by both the virology and the retroelement communities, and at the level of individual elements, nomenclature is frequently based on historical tradition and can be confusing. A simple classification scheme based primarily on genetic structure is used here, with acknowledgment that, with the exception of reverse transcriptase, coding domains are not uniformly retained across different groups. Detailed discussion of classification and individual elements is outside the scope of this article, but the subject can be pursued through references listed in Further Reading.

Retrotransposon Structural and Functional Features

Retrotransposons are usually subdivided into two structural superfamilies based on the presence or absence of terminal directly repeated sequences of several hundred nucleotides, the LTRs. The LTR-retrotransposon superfamily includes elements resembling integrated or endogenous proviruses, the non-LTR superfamily includes LINEs, SINEs, and SVA elements. In each superfamily, there are autonomous elements, which encode reverse transcriptase (RT) as well as a variable set of other proteins necessary for replication and transposition of the element. The superfamilies also include nonautonomous elements whose open reading frames (ORFs) usually encode no proteins, or putative proteins lacking homology to other transposable element proteins and lacking known function. Nonautonomous elements always have an autonomous partner to provide the necessary proteins *in trans*. A third group of retrotransposable elements, the PLEs, were originally described in invertebrates but have more recently also been reported in vertebrates. PLEs have distinct genomic and transcript structural features which preclude a neat fit in either the LTR or non-LTR superfamilies; consequently, this group will be treated independently in this article.

Fundamental structural features of non-LTR and LTR retrotransposons are illustrated schematically in **Figure 1**. A useful repository of consensus sequences of retrotransposons from different animal species, and their classification is curated by the Genetic Information Research Institute.

LTR Superfamily

Four major groups of LTR retrotransposons have been identified in vertebrates to date: Ty3/*gypsy*; BEL; the vertebrate retrovirus group; and Ty1/*copia*. The hepadnaviruses, a fifth group encoding RT similar to LTR retrotransposons but which are circular and lack LTRs, are discussed elsewhere in this encyclopedia. The LTR retrotransposon groups are distinguished from one another by amino acid sequence comparison of their homologous enzymatic domains, primarily RT, and by features specific to one or two groups. Nonautonomous retrotransposon families are distinguished from one another by DNA sequence comparisons of their internal domain and of their LTRs.

The DIRS1 (Dictyostelium repeat sequence 1) subclass LTRs are either inverted repeats, or split direct repeats. In addition, these elements lack the typical integrase and protease coding domains, containing instead a domain encoding a bacteriophage lambda recombinase-like protein also known as tyrosine-recombinase or Y-transposase. These features of DIRS1 make its inclusion in the LTR superfamily problematic, although its RT has homology to the more typical elements of the superfamily.

Structure

LTR elements are typically constructed as two direct LTRs flanking an internal sequence of variable coding content (**Figure 1**). The LTRs are necessary and sufficient for promoter activity and transcription of the retroelement. Functional features common to LTRs include: 5′ TG and 3′ CA dinucleotides necessary for integration into the host genome; Pol II promoter elements and a transcription start site; and a polyadenylation and cleavage signal.

In typical autonomous LTR retrotransposons, the internal sequence contains a variable but small number of ORFs. Most elements contain an ORF including Gag

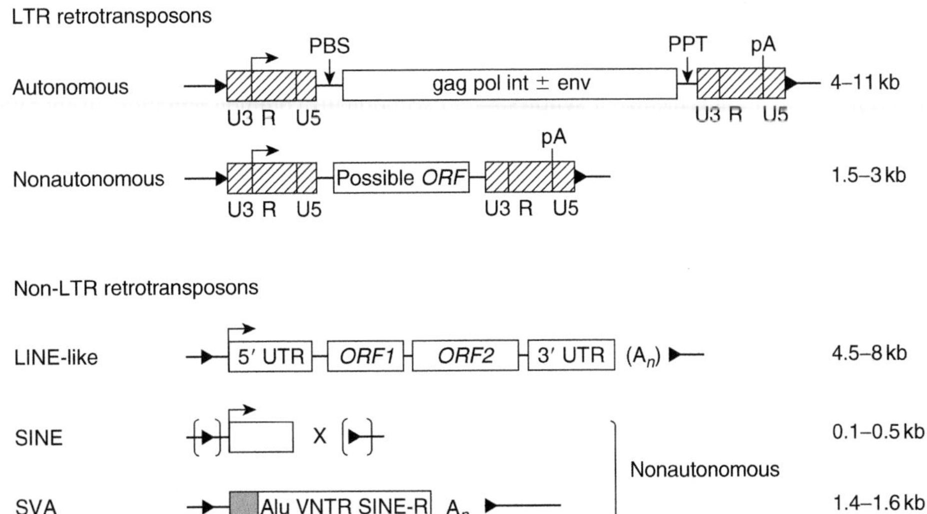

Figure 1 General structure of selected retrotransposons. A typical LTR retrotransposon is flanked by target site duplications (black arrowheads) and its LTR domains (hatched) contain a pol II promoter, transcription start site and a polyadenylation and cleavage signal. The internal domain of autonomous elements contains ORFs encoding homologs of retroviral gag and pol genes, and occasionally an additional env gene homolog. The short ORF of nonautonomous elements does not encode these proteins, if any. LINE-like elements terminate in family-specific 3′ UTRs with or without poly (A) tails, the poly (A) tail (A_n) indicated here is typical of L1. Flanking target site duplications typical of the L1 clade may or may not be present in other LINE-like clades. The autonomous LINE-like elements contain an internal pol II promoter and ORFs encoding structural and reverse transcriptase (RT) genes required for retrotransposition. SINEs contain an internal pol III promoter but lack an ORF, and may or may not be flanked by target site duplications (in parentheses), depending on their specific LINE-like partner. SVA consists of a 5′ variable number hexameric repeat (gray box), followed by an Alu-like element, a VNTR domain, the SINE-R element derived from the human endogenous retrovirus HERVK10, and a poly (A) tail. SVA elements are thought to be mobilized *in trans* by L1 proteins, and are flanked by target site duplications typical of L1. Right angled arrow, transcription start site; pA, polyadenylation and cleavage signal; PBS, primer binding site for initiating minus strand cDNA synthesis; PPT, polypurine tract used for initiation of plus strand cDNA synthesis. Element sizes are approximate ranges. The diagrams are not to scale.

(group-specific antigen) and Pol (polymerase) domains, and some additionally contain an ORF for Env (envelope). Gag encodes a structural polyprotein integral to the formation of cytoplasmic virus-like particles in which reverse transcription takes place. Pol encodes several enzymatic activities: (1) protease required to cleave the translated products of the gag–pol transcript into their functional forms; (2) RT for transcribing the retrotransposon RNA into double-stranded cDNA; (3) ribonuclease H (RNase H) for processing the RNA template prior to plus strand cDNA synthesis; and (4) integrase incorporating an aspartate (D), aspartate, glutamate (E) (DDE)-type transposase activity. The integrase sequence of the chromovirus genus of the Ty3/*gypsy* group additionally encodes a chromodomain, a domain found in chromatin complex structural proteins or chromatin remodeling proteins. Further information regarding retroviral genes and structure can be found elsewhere in the encyclopedia. The Pol domain order is usually protease, RT, RNase H, integrase, except for the Ty1/Copia group and the Gmr1-like elements of the Ty3/Gypsy group in which integrase is upstream of RT.

In addition to coding sequence, the internal sequence contains a 5′ primer binding site (PBS) for first-strand DNA synthesis and a 3′ polypurine tract which serves as the primer binding site for second-strand DNA synthesis. These two features are conserved in the internal sequence region of nonautonomous LTR retrotransposons.

Only a few LTR retrotransposons in any genome are intact, a majority of them being inactivated by different mutations. These may be due to intrinsically error-prone, reverse transcription by RT, to mutation associated with the nucleotide substitution rate of the specific genome, to mutation associated with methylation of CpG dinucleotides of the element, to insertional mutation by other transposable elements, or to recombination and deletion. Solitary LTRs usually far outnumber full-length elements in the genome and are understood to be the result of homologous recombination between the 5′ and 3′ LTRs of a full-length element, with excision and loss of the intervening sequence. Over time, older transpositionally active elements of a retrotransposon lineage are inactivated by mutation, becoming functionally extinct. Younger functionally intact members of the lineage continue the retrotransposition activity, or if none exist then that lineage becomes extinct. Transposition-incompetent elements in the genome can be viewed as molecular fossils of past transposition activity.

Transposition mechanism

Most LTR-retrotransposons are believed to replicate and transpose using a complex process very similar to that of infectious retroviruses, as covered elsewhere in this encyclopedia. In the DIRS1 group, the presence of a Y-transposase and lack of target site duplications suggests a different mechanism, involving RT-mediated synthesis of a closed circular cDNA, and insertion into the target using Y-mediated recombination.

Non-LTR Superfamily

Retrotransposons included in this superfamily are the autonomous LINEs and nonautonomous SINEs and SVA elements. One clade of LINEs is the LINE-1 clade (L1), and to minimize confusion, the general category of 'LINE' retrotransposons will be referred to hereafter as LINE-like.

LINE-like retrotranposon structure

Multiple clades of LINE-like retrotransposons have so far been recognized on the basis of RT and other protein sequence comparisons. At least nine clades are represented in vertebrate genomes. Common structural features of LINE-like retrotransposons include a 5′ untranslated region (UTR) containing a (G+C)-rich promoter region, followed by two nonoverlapping open reading frames (ORF1 and ORF2), and a 3′ UTR.

However, different clades are distinguished by many structural differences. A brief description of the 3′ UTR and ORF differences between the first three identified vertebrate LINE-like clades (LINE-1 (L1), CR1, and RTE) is given below to convey a sense of the structural features distinguishing LINE-like retrotransposons.

Variable-sized target-site duplications flank full-length elements of the L1 and RTE lineages, but not the CR1 lineage. The 3′ UTR of L1 characteristically contains an AATAAA polyadenylation signal and terminates in a 3′ poly (A) tail. In contrast, the 3′ UTR of CRI elements contains no A or AT-rich regions and terminates in a [(CATTCTRT) $(GATTCTRT)_{1-3}$] motif, while the very short 3′ UTRs of the RTE clade are dominated by A/T-rich trimer, tetramer, and/or pentamer repeats.

In the L1 clade, ORF1 encodes a single-stranded, nucleic acid-binding protein with nucleic acid-chaperone activity. A similar activity is proposed for the ORF1 product of CR1, while the 43-amino-acid ORF1 of RTE is thought to be too short to encode these components. In most LINE-like clades, ORF2 encodes a large protein with an N-terminal endonuclease and a C-terminal RT domain. As in most LINE-like clades, the endonuclease is of the apurinic/apyrimidinic type (AP) in L1, CR1, and RTE elements. A site-specific endonuclease typifies a minority of clades. A third domain in the ORF2-encoded protein, a COOH-terminal cysteine-rich domain, is conserved in mammalian L1 elements.

As with LTR retrotransposons, most LINE-like retrotransposons in a genome are transpositionally disabled. For example, of the half million or so L1 copies in the human genome, less than 1% are intact full-length elements. In all LINE-like lineages, 5′ truncation is an exceptionally common mutation, likely reflecting the rarity of successful full-length element reverse transcription and insertion. An interesting speculation is that, by providing opportunity for L1 to acquire novel 5′ ends, 5′ truncation may be an evolutionary strategy for L1 to evade suppression of transcription by the host.

LINE-like retrotransposon transposition

LINE-like retrotransposons are thought to all use a target-primed mechanism of transposition, reverse transcribing their RNA directly on the new chromosomal integration site, although the details likely vary between the different clades and are not well understood for all. The human L1 is perhaps the best-studied vertebrate LINE-like retrotransposon and is used here to outline the process. Genetic evidence and recent cell culture experiments suggest the following model (**Figure 2**). Following transcription, nuclear export, and translation of the L1 mRNA in the cytoplasm, L1 proteins preferentially associate with their encoding transcript ('*cis* preference') forming a cytoplasmic ribonucleoprotein particle. Upon access of the particle to the nucleus, L1 endonuclease nicks the host DNA at a loose consensus sequence (3′-AA/TTTT-5′) in the minus strand. The L1 transcript poly (A) tail base pairs with the nicked target and the target thus primes first strand synthesis using L1 RT with L1 mRNA as template. A nick in the target plus strand, offset from the minus-strand nick, is made by an unknown enzyme, possibly L1 endonuclease. Second strand synthesis proceeds using the first strand as template, perhaps primed by microhomology-mediated priming between the first strand and the 3′ end of the plus strand. Host enzymes are thought to be involved in completion of synthesis of both new strands and DNA repair linking them to the host DNA, creating the flanking, perfect target-site duplications typical of L1. L1 uses host ribonuclease H to degrade L1 RNA template, but other clades of LINE-like retrotransposons encode their own RNase H.

SINE structure

SINEs are small nonautonomous retrotransposons, usually derived from different cellular structural RNAs, such as 7SLRNA or 5S rRNA; however, most eukaryotic SINE families are derived from tRNAs. A SINE generally consists of a 5′ structural RNA-like region containing an internal polymerase III promoter, a region unrelated to structural RNA, and a 3′ end derived from a LINE-like retrotransposon in the same genome. An exception is Alu elements, whose 3′ ends are not shared with L1, their autonomous partner, unless one counts the poly (A) tail. SINEs are usually flanked by target site duplications typical of the element from which their 3′ end is derived.

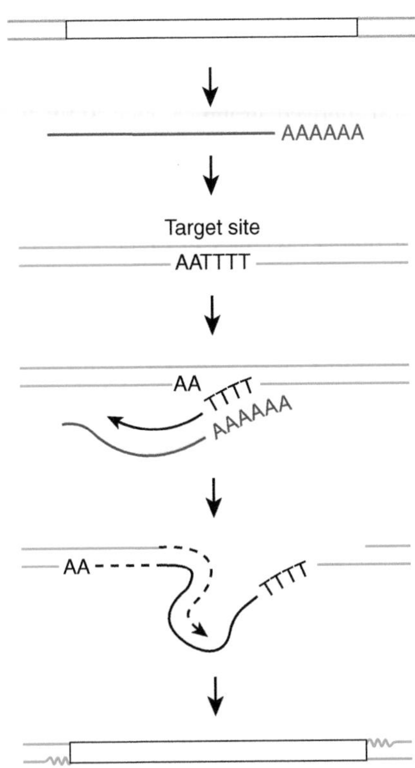

Figure 2 Retrotransposition mechanism of L1 elements. Polyadenylated L1 transcripts (red line) are transported to the cytoplasm, packaged in their own translation products, and then transported back to the nucleus. In the nucleus, L1 endonuclease nicks the minus strand of the new insertion site (blue line) at the target consensus sequence 5′ TTTT/AA 3′. The T-rich 3′-OH minus strand primes reverse transcription and first strand synthesis (black line). A nick in the plus strand, staggered in position to the minus strand nick, exposes 3′ plus strand target DNA which is thought to prime L1 second strand transcription (black broken line) using the first strand as template, after RNA degradation by RNase H and strand switching by RT. Finally, host enzymes are thought to participate in repairing the gaps in the host DNA, creating target-site duplications flanking the L1 (zig-zag lines indicate duplicated host DNA).

SVA structure

SVA elements, exclusively found in hominoid primates, are chimeric elements named for their principal components, SINE-R, VNTR, and Alu. A 5′ (CCCTCT) hexamer repeat is followed by an antisense Alu sequence and then, multiple copies of a variable number tandem repeat (VNTR). The 3′ portion consists of the primate-specific SINE-R, derived from an LTR-retrotransposon thought to be human endogenous retrovirus K-10 (HERVK-10), and finally a polyadenylation and cleavage signal and poly (A) tail.

Transposition of SINE and SVA

Experimental evidence from several vertebrate species indicates that, as long suspected, the 3′ end of SINE transcripts is recognized *in trans* by the cognate LINE-like machinery for transposition. For example, transcripts of Alu elements, and the mouse SINEs, B1, B2, and ID, are recognized and transposed by L1 machinery, whereas UnaSINE1, an eel SINE, is retrotransposed by UnaL2. Similarly, the characteristics of SVA insertions strongly suggest they use the L1 machinery for retrotransposition.

Penelope-Like Elements

Structure

In a few species of vertebrates, a limited number of elements with intact ORFs resembling the Penelope element of *Drosophila virilis* have been described. The elements are flanked by short target site duplications, and usually consist of LTRs flanking anINT. The LTR sequences do not resemble those of LTR retrotransposons, and are thought to represent tandem arrangement of two copies of the element with variable 5′ truncation of the upstream copy. The upstream LTR may be preceded by an inverted LTR fragment. The single ORF includes an N-terminal domain containing a conserved DKG amino acid motif, followed by the RT domain, a variable length linker sequence thought to contain a nuclear localization signal, and an endonuclease domain. The endonuclease is of the GIY-YIG type, otherwise unreported in eukaryotes.

Transposition

It is not clear how PLEs are transposed. The presence of introns in genomic copies of some PLEs found in invertebrates, and their absence from cDNA of the element, argues against an L1-like '*cis* preference' action of PLE proteins. Other possibilities include unconventional transposition of full-length unspliced mRNA, or use of a DNA template for transposition.

Evolutionary Features

Origin

Phylogenetic analyses of autonomous retrotransposable elements have historically relied on amino acid sequence alignments of RT, the only protein coding domain common to all elements. The evidence indicates that LINE-like retrotransposons are divided into 17 or more clades, many of which have wide distribution in eukaryotes. These analyses, and abundant representation of LINE-like retrotransposons in basal eukaryote genomes, support the origin of ancestral LINE-like retrotransposons close to the emergence of the eukaryotic crown group in the Proterozoic eon. Combined analyses of RT, RNaseH, and endonuclease domains suggest that ancestral LINE-like retrotransposons evolved from group II introns in genomes of eubacteria, and fungal and plant organelles, and originally possessed a single ORF for RT and a site-specific endonuclease. Early branching lineages such as L1 acquired a second ORF (ORF1), and the site-specific endonuclease was replaced by a relatively non-site-specific AP endonuclease probably acquired from the host DNA repair machinery. Later branching lineages also acquired an RNaseH domain, most likely from their eukaryotic host, but elements within these lineages have not all retained this domain. Most evidence suggests that LINE-like retrotransposon transmission is strictly vertical through the germline. Evidence supporting horizontal transmission of RTE from snakes to ruminants has been reported, although the mechanism is unknown, and the topic remains controversial.

Phylogenetic studies of LTR elements, together with their absence from some basal eukaryotic genomes, suggest they arose more recently than LINE-like retrotransposons, although this is not a settled topic. It has been speculated that LTR retrotransposons arose as a chimera between a non-LTR retrotransposon carrying RT and RNaseH and a DNA transposon carrying an integrase domain. The acquisition of additional ORFs distinguishes many clades. Evidence suggests the vertebrate retrovirus group acquired a second RNaseH domain, and the primary sequence of the first degenerated, maintaining some structural information but not the catalytic site. The acquisition of Env, enabling infectious transfer is a striking feature of some lineages within the vertebrate retroviral group, although the origin, potentially ancient, of Env in this group is obscure. Strong phylogenetic evidence from invertebrate genomes indicates that other LTR element groups independently acquired different Env-like genes from infectious viruses, but vertebrate representatives of these are yet to be described. The evolutionary success of the Chromovirus clade of the Ty3/*gypsy* group, found in plants and fungi as well as vertebrates, has been attributed to its acquisition of the chromodomain. As with LINE-like retrotransposons, LTR elements have had varying success in colonizing vertebrate genomes, although all major LTR groups are represented by an active element in at least one species of vertebrates (**Table 1**).

New autonomous endogenous LTR elements are acquired through horizontal transfer of exogenous elements, which can then invade the genome and be transmitted through the germline, blurring the distinction between exogenous and endogenous elements. An example of current interest is the ongoing infectious epidemic and endogenization of the Koala retrovirus. However, some genera of the vertebrate retrovirus group, such as the lentiviruses, appear incapable of generating endogenous elements. This appears to be the result of Env mutations that disable their ability to infect germ cells. New nonautonomous elements arise through recombination, and all are transmitted in the germline. In general, LTR elements seem to be active within a genome over short evolutionary scales relative to LINE-like lineages.

As previously mentioned, SINEs appear to have arisen from structural RNA sequences. The primate Alu, mouse B1, and related families were derived from 7SL RNA. However, almost all other SINEs are derived from tRNA, and are placed as an evolutionarily older family than the 7SL-derived group. A new SINE-like family of diverse low-copy-number species- or lineage-specific retrotransposons derived from small nucleolar RNA was recently described in vertebrates.

Construction of a RT-based PLE polygeny is problematic since PLE RT differs from all of other retroelement RTs and more closely resembles telomerase. This, together with their distinct structure, has led to general agreement that PLEs form a separate group from the LTR and non-LTR retrotransposons. A relatively early origin for PLEs is supported by grouping of PLEs found in many eukaryotic genomes. Degenerate, and full-length PLEs have been reported from fungi and a wide range of invertebrates. In vertebrates, they have so far been found in teleost fish, sharks, and amphibia, not always as degenerate molecular fossils. It is possible that some are still active in the fish *Tetraodon nigroviridis* and *Danio rerio*.

Retrotransposons as Phylogenetic Markers

At any point in time, relatively few copies of a retrotransposon are capable of replication and retrotransposition. As these copies accumulate mutations, the mutations are inherited by subsequent members of the retroelement lineage in the host genome. Once the host lineage splits, new insertions will occur independently in each descendant lineage. Since stably integrated elements are identical by descent, and the probability of parallel independent insertions into a genome is low, retrotransposons can be considered to be homoplasy-free characters, offering unique utility as markers to study evolution of host species. Retrotransposons have been used in several species of fish, mammals, birds, and reptiles, to clarify phylogenetic relationships. As suggested above, these analyses rest on the assumption that, in general, retrotransposons integrate randomly into genomes, with an exceptionally low probability that an element would independently insert in orthologous positions in two species. Another assumption is that, unless removed by segmental deletion, an insertion remains in its locus once fixed in the genome, eventually becoming unrecognizable as a result of accumulated mutations. Evidence that these assumptions do not always hold indicates that, as with analyses based on other genome elements, care is required in the conduct and interpretation of phylogenetic analyses based on retroelements.

Retrotransposon Effects on the Genome

Transposable elements have profoundly affected the structure and function of vertebrate genomes in many different ways.

Table 1 Retrotransposon clades identified in some vertebrate genomes

Type of retrotransposon	*Cartilaginous fish*[a,b]	*Teleost fish*[c] *(Tn, Tr, Dr)*	*Mammals*[d] *(Hs, Mm, Cf)*	*Birds (chicken)*[e]
Non-LTR retrotransposons[f]	2.15	0.78, 1.32, 0.39	20.42, 19.2, 16.49	6.5
Restriction-enzyme like				
NeSL	n.a.	0.08, 0.01, <0.01	n.d., n.d., n.d.	n.d.
R2	n.a.	n.d., n.d., <0.01	n.d., n.d., n.d.	n.d.
R4	n.a.	Fossils,[e] 0.09, n.d.	n.d., n.d., n.d.	n.d.
Apurinic/apyrimidinic				
R1	*	n.d., n.d., n.d.	n.d., n.d., n.d.	n.d.
L1/TX1	*	0.03, 0.06, 0.02	16.9, 18.8, 14.5	n.d.
RTE/Rex3[g]	n.a.	0.18, 0.39, 0.2	n.d., n.d., n.d.	n.d.
L2/Maui	***	0.04, 0.53, 0.11	3.22, 0.38, 1.84	0.1
L3/CR1[h]	***	n.d., n.d., n.d.	0.31, 0.05, 0.15	6.4
Rex1/Babar	***	0.45, 0.25, 0.05	n.d., n.d., n.d.	n.d.
I/Bgr	n.a.	0.01, fossils, 0.01	n.d., 0.01, n.d.	n.d.
Jockey	*	n.d., n.d., n.d.	n.d., n.d., n.d.	n.d.
LOA	*	n.d., n.d., n.d.	n.d., n.d., n.d.	n.d.
LTR retrotransposons	0.1	0.12, 0.30, 0.40	8.29, 9.87, 3.25	
Vertebrate retroviruses	*	0.03, 0.09, <0.01	8.29, 9.87, 3.25	1.3
TY1/Copia	n.a.	0.02, 0.01, <0.01	n.d., n.d., n.d.	n.d.
TY3/Gypsy[i]	*	0.06, 0.17, 0.13	n.d., n.d., n.d.	n.d.
BEL	*	Fossils, 0.02, 0.01	n.d., n.d., n.d.	n.a.
DIRS1[h]	n.a.	0.02, 0.01, 0.25	n.d., n.d., n.d.	n.a.
Penelope-like elements	n.a.	0.06, 0.09, <0.01	n.d., n.d., n.d.	n.a.
SINEs[h]	1.36		13.14, 8.22, 9.12	Fossils

[a]The estimated percentage of the genome occupied by the respective elements is shown. Where numeric estimates are unavailable, *indicates detected at low frequency, ***indicates detected at high frequency. There are some differences in calculation of the estimates as indicated below.
[b]*Callorhincus milii.* The estimates are based on analysis of 18 Mb of random sequence from this fish; copy number is uncertain and additional retrotransposon clades may be identified when the complete sequence is available. From Venkatesh B, Tay A, Dandona N *et al.* (2005) A compact cartilaginous fish model genome. *Current Biology* 15: R82–R83.
[c]Teleost fish considered here include *T. nigroviridis*, *T. rubripes*, and *D. rerio* (Tn, Tr, Dr respectively); the percentage is the percentage of RT gene-containing sequence from whole genome shotgun sequences. From Volff JN, Bouneau L, Ozouf-Costas C, and Fischer C (2003). Diversity of retrotransposable elements in compact pufferfish genomes. *Trends in Genetics* 19: 674–678.
[d]Mammals here include human, mouse and domestic dog (Hs, Mm, Cf, respectively) From Waterston RH, Lindblad-Toh K, Birney E, *et al.* (2002) Initial sequencing and comparative analysis of the mouse genome. *Nature* 420: 520–562; Lander ES, Linton LM, Birren B, *et al.* (2001) Initial sequencing and analysis of the human genome. *Nature* 409: 860–921; Kirkness EF, Bafna V, Halpern AL, *et al.* (2003) The dog genome: Survey sequencing and comparative analysis. *Science* 301: 1898–1903.
[e]From Hillier LW, Miller W, Birney E, *et al.* (2004) Sequence and comparative analysis of chicken genome provide unique perspectives on vertebrate evolution. *Nature* 432: 695–716.
[f]Non-LTR retrotransposons are subdivided into a phylogenetically older group encoding a restriction enzyme-like endonuclease, and a younger group with an apurinic/apyrimidinic endonuclease.
[g]Members of this clade detected in reptiles.
[h]Members of this clade detected in reptiles and amphibians.
[i]Fossils of this clade detected in reptiles, potentially active elements detected in amphibians.
n.a. indicates data not known. n.d. indicates the element was not detected after searching. Fossils are elements deemed to be extinct, having lost the means of retrotransposition or lacking evidence of recent retrotransposition.

Retrotransposon Content

Vertebrate genomes differ markedly in the total quantity and diversity of retrotransposons they contain, and the evolutionary trajectory of different elements. Genome size has been correlated to the quantity of transposable elements in the genome. Retrotransposons occupy approximately 35–50% or more of mouse, human, and domestic dog genomes. The chicken genome is approximately 39% of the size of mouse and human genomes, but only about 8% of the genome is recognizable retrotransposable elements. In the very compact genomes of smooth pufferfish *T. nigroviridis* and *Takifugu rubripes*, roughly 12% the size of mouse and human genomes, retrotransposable elements occupy less than 5% of the DNA. The enormous retrotransposon copy number accumulation in species with large genomes suggests that these species lack some constraint on retrotransposon activity that is present in animals with small genomes. It has been suggested that retrotransposons physically organize the genome through higher order chromatin structuring, provision of dispersed regulatory units, and other means. Thus, differences in genome retrotransposon content could significantly affect the operation of different genomes.

The pufferfish genomes, considering their small size, contain a remarkable diversity of retrotransposons in comparison with mammals. In mammalian genomes, 4 LINE-like clades have been identified (L1, L2, RTE, and CR1) with L1 the major currently active element, while in pufferfish genomes seven clades have been identified (NeSL, R4, L1, RTE, I, L2, Rex1) most of which have been recently active. Within the L1 clade alone, a single lineage has dominated L1 activity in mouse and human genomes since the mammalian radiation and comprises about 20% of the genome, whereas multiple L1 lineages predating the mammalian L1 emergence are active in several fish species, although present in very low copy numbers. All the major groups of LTR retrotransposons are represented in pufferfish genomes, as are PLEs. In contrast, only the vertebrate retrovirus group (endogenous retroviruses) and a few molecular fossils of the Ty3/Gypsy and DIRS1 groups are present in the mouse and human genomes, and there is no evidence for the presence of PLEs. The zebrafish, *D. rerio*, with a larger genome than the pufferfish, also hosts a great variety of retrotransposons. Whether these obvious differences in retrotransposon evolutionary biology between the three fish genomes and the mouse/human genomes represent the general case between teleosts and mammals is as yet unknown. Birds, as exemplified by the chicken, are different again. A single LINE-like retrotransposon, CR1, comprises about 90% of all identified chicken retrotransposons, L2/MIRs and endogenous retroviruses equally comprise the remainder. Curiously, the chicken genome lacks SINEs although it contains faint remnants of ancient SINEs pre-dating the bird-mammal split. A variety of retrotransposons have been reported from reptile genomes, some revealing lineage-specific retrotransposons such as the Sauria SINE derived from a LINE-like element of the RTE clade.

Mammalian genomes *per se* can differ markedly from one another in their retrotransposon content and activity, reflecting the evolutionary trajectory of different elements. For example, LINE, SINE, and LTR elements occupy similar percentages of mouse and human genomes. However, endogenous retroviruses are almost extinct in humans, while multiple families of endogenous retroviruses are active in rodents. The L1 lineage appears to be still active in most mammalian genomes, but recent evidence indicates its extinction in several tribes of sigmodontine rodents at or after their divergence from the earliest extant genus. As would be predicted, L1 extinction was linked to extinction of B1, a rodent SINE thought to be transposed by L1. Unexpectedly, vigorous expansion of an endogenous retrovirus, MysTR, to very high copy numbers unprecedented in any other endogenous retrovirus group, was also linked to L1 extinction. Whether there is any relationship between L1 activity and endogenous retrovirus activity is unknown.

Genomic Distribution of Retrotransposons

Diverse patterns of retrotransposon distribution in genomes are dependent on the type of element and the host genome. In the mouse and human, retrotransposons are generally dispersed widely through the genome. In contrast, retroelements strongly cluster in heterochromatic gene-poor regions in *T. nigroviridis*, similar to the distribution in *Drosophila*. Whether this extremely uneven distribution is specific to small genomes, or teleosts, is unknown, and how and why it might arise is the subject for some speculation.

At a smaller scale, much variation is evident within genomes. The density of retrotransposons varies among different chromosomes in individual vertebrate species, and in mammals is usually highest on the X and Y chromosomes. A higher density of L1 on X chromosomes than autosomes is also observed in the Ryuku spiny rat, *Tokudaia osimensis*, in which both males and females have an XO karyotype, arguing against the hypothesis that evolutionary selection of a high density of L1 on X is due to involvement of L1 in X-inactivation. Although LTR elements are distributed more or less uniformly in mouse and human genomes, L1 occurs at much higher density in gene-poor, AT-rich regions and the human SINE, Alu, occurs at high density in GC-rich regions. Closer analysis demonstrates young Alus preferentially occur in AT-rich regions, whereas in older Alus a stronger bias toward GC-rich regions emerges. One of several possible explanations for this skewed distribution of LINEs and Alus is selective targeting of L1 and Alu to AT-rich regions, and subsequent positive selection for Alus in gene-rich GC-rich regions. A significant antisense bias observed for many older intact LTR elements and solitary LTRs located in mammalian introns, is thought to arise from negative selection of sense-oriented elements bearing strong splice acceptor or donor sites, or strong transcriptional regulatory function.

Effect on Genes

Through their specific exaptation for use by the host, or through incorporation within genes, retrotransposons significantly contribute to gene evolution, and some examples follow. Sequences from LTR elements alone occupy about 1.5% of mouse and 0.8% of human genes, and genes containing these elements tend to be newly evolved genes. Indeed, LTRs drive developmentally regulated expression of cellular genes in early mouse embryos, perhaps an evolving example fitting with the hypothesis that randomly distributed retrotransposons provide a means to set up, over evolutionary time, co-ordinated transcriptional regulatory circuits. An interesting example of exaptation is the independent selection by sheep, primate, and rodent lineages of Env expression from lineage-specific endogenous retroviruses for function in placental syncytiotrophoblast morphogenesis. At least one ultra-conserved sequence in mammalian genomes has proved to be an ancient SINE whose multiple insertions, predating the divergence of amniotes and amphibians, have been exapted for use in transcriptional regulation or as a conserved exon in multiple unrelated genes.

New insertions can also disrupt normal gene function through interference with transcriptional regulation, through physical alteration of transcripts by aberrant splicing or premature termination, and through exon shuffling by inadvertent transduction of non-retrotransposon sequences. While L1 generates processed pseudogenes in the human genome, the paucity of processed pseudogenes in the chicken genome indicates that CR1 does not, suggesting that its retrotransposition machinery does not recognize mRNA. Thus, different vertebrate LINE-like retrotransposons may differ in their effects on different genomes.

Retrotransposons also act as substrates for recombination, fostering genomic instability, and are involved in both duplications and deletions within genomes, and other structural rearrangements. For example, in the human genome, SVA elements are reported to be associated with about 53 kbp of genomic duplications in the human genome, including duplication of entire genes and the creation of new genes. Recently, recombination hot spots in mouse and human genomes were linked with a sequence, CCTCCCT, found in an ancient nonautonomous LTR element (THE1) in humans. The LTR element itself is not recognizable in mouse, likely having been mutated beyond recognition by the higher nucleotide substitution rate in this species. Finally, illegitimate recombination between Alu elements in primate genomes has been linked with occasional genomic deletions. However, the effect of different retrotransposons on genome integrity may vary according to the retrotransposon type and host species. For example, high karyotypic variation in sigmodontine rodent species with extinct L1 compared with those maintaining active L1 supports the notion that L1 proteins may be important for DNA break–repair in these animals. Interestingly, a single L1 element, L1_MM, was enriched in regions of low recombination activity in the mouse, and L1 is underrepresented in human recombination hot-spots.

Host Responses to Retrotransposons

The inherent tendency of retrotransposons to amplify their copy number creates mutagenic insertions potentially harmful to the host, leading to the view of retrotransposons as genomic parasites. An adaptive response by the host would selectively encourage the evolution of repressive mechanisms directed against retroelements, and this in turn would exert selective pressure on retroelements to resist repression. It has been suggested that evolution of the APOBEC3 family of cytidine deaminases, cellular inhibitors of retrotransposition of LINE-like, SINE and LTR retrotransposons, may have been driven in part as a genome response to invasion of mammalian genomes by retrotransposons. In addition to random mutation of active elements within the genome, other strategies to inhibit retrotransposon proliferation include transcriptional silencing through epigenetic modifications to chromatin and DNA, and post-transcriptional silencing through RNA interference. From time to time, retroelements escape repression and undergo expansion in the affected genome. Thus, continued activity of a retroelement in the genome, such as of endogenous retroviruses in the mouse, may indicate that a host is lagging in the evolutionary race to control the genome invader. Some retrotransposons have been astoundingly successful in certain genomes, for example, there are over 1×10^6 Alu copies in the human genome, consisting of one family with about 20 subfamilies. This, together with the low frequency of pathogenic effects for many retroelements, and the adoption by hosts of many elements for their own biology, suggest the controversial idea that many retrotransposons have formed or are forming a symbiotic rather than parasitic relationship with their hosts.

Acknowledgments

This work was supported in part by VSPHS NIH (R01HD037102).

Further Reading

Aparicio S, Chapman J, Stupka E, *et al.* (2002) Whole-genome shotgun assembly and analysis of the genome of *Fugu rubripes*. *Science* 297: 1301–1310.

Craig NL, Craigie R, Gellert M, and Lambowitz AM (eds.) (2002) *Mobile DNA II*. Herndon, VA: ASM Press.

Curcio MJ and Derbyshire KM (2003) The outs and ins of transposition: From mu to kangaroo. *Nature Reviews Molecular Cell Biology* 4: 865–877.

Eickbush TH and Furano AV (2002) Fruit flies and humans respond differently to retrotransposons. *Current Opinion in Genetics and Development* 12: 669–674.

Furano AV, Duvernell DD, and Boissinot S (2004) L1 (LINE-1) retrotransposon diversity differs dramatically between mammals and fish. *Trends in Genetics* 20: 9–14.

Goodwin TJ and Poulter RT (2001) The DIRS1 group of retrotransposons. *Molecular Biology and Evolution* 18: 2067–2082.

Hedges DJ and Deininger PL (2006) Inviting instability: Transposable elements, double-strand breaks, and the maintenance of genome integrity. *Mutation Research: Fundamental Mechanisms of Mutagenesis* doi:10.1016/j.mrfmmm.2006.11.021.

Hillier LW, Miller W, Birney E, *et al.* (2004) Sequence and Comparative analysis of chicken genome provide unique perspectives on vertebrate evolution. *Nature* 432: 695–716.

Kazazian HH, Jr. (2004) Mobile elements: Drivers of genome evolution. *Science* 303: 1626–1632.

Kordis D (2005) A genomic perspective on the chromodomain-containing retrotransposons: Chromoviruses. *Gene* 347: 161–173.

Kirkness EF, Bafna V, Halpern AL, *et al.* (2003) The dog genome: Survey sequencing and comparative analysis. *Science* 301: 1898–1903.

Lander ES, Linton LM, Birren B, *et al.* (2001) Initial sequencing and analysis of the human genome. *Nature* 409: 860–921.
Malik HS and Eickbush TH (2001) Phylogenetic analysis of ribonuclease H domains suggests a late, chimeric origin of LTR retrotransposable elements and retroviruses. *Genome Research* 11: 1187–1197.
Ostertag EM and Kazazian HH, Jr. (2001) Biology of mammalian L1 retrotransposons. *Annual Review of Genetics* 35: 501–538.
Piskurek O, Austin CC, and Okada N (2006) Sauria SINEs: Novel short interspersed retroposable elements that are widespread in reptile genomes. *Journal of Molecular Evolution* 62: 630–644.
Roy-Engel AM, Carroll ML, El-Sawy M, *et al.* (2002) Non-traditional Alu evolution and primate genomic diversity. *Journal of Molecular Biology* 316: 1033–1040.
van de Lagemaat LN, Medstrand P, and Mager DL (2006) Multiple effects govern endogenous retrovirus survival patterns in human gene introns. *Genome Biology* 7: R86.
Venkatesh B, Tay A, Dandona N, *et al.* (2005) A compact cartilaginous fish model genome. *Current Biology* 15: R82–R83.
Volff JN (ed.) (2005) *Cytogenetic and Genome Research, Vol. 110: Retrotransposable Elements and Genome Evolution.* Basel: S Karger AG.
Voff JN, Bouneau L, Ozouf-Costas C, and Fischer C (2003) Diversity of retrotransposable elements in compact Pufferfish genomes. *Trends in Genetics* 19: 674–678.
Waterston RH, Lindblad-Toh K, Birney E, *et al.* (2002) Initial sequencing and comparative analysis of the mouse genome. *Nature* 420: 520–562.
Weber MJ (2006) Mammalian small nucleolar RNAs are mobile genetic elements. *Public Library of Science Genetics* 2: e205doi:10.1371/journal.pgen.0020205.

Relevant Website

http://www.girinst.org – Genetic Information Research Institute, Mountain View.

Defective-Interfering Viruses

L Roux, University of Geneva Medical School, Geneva, Switzerland

This article is reproduced from the previous edition, volume 1, pp 371–375, © 1999, Elsevier Ltd.

History

In 1943, Henle and Henle reported the decreased infectivity for mice of influenza virus stocks obtained after a series of undiluted passages in embryonated chicken eggs. In the early 1950s, von Magnus showed that such undiluted passages generate incomplete virus particles capable of limiting the growth of infectious virus (hence exhibiting interference). This first characterization was soon followed by similar reports by Mims, on the one hand, and Cooper and Bellet, on the other, dealing respectively with Rift Valley fever virus and vesicular stomatitis virus (VSV). In the late 1950s, Cooper and Bellet went so far as to assign interference to sedimentable particles, but failed to identify them as antigenically related to VSV. From the mid-1960s on, the characterization of other positive- and negative-stranded RNA virus defective particles continued. In 1970, a review by A. Huang and D. Baltimore set the basic definition of defective interfering (DI) particles and emphasized their widespread occurrence. Since then, DI particles have been described for almost all the known DNA and RNA viruses, including plant and even fungal viruses.

Structure

DI particles have the same protein composition as their homologous nondefective 'parents', often called St. However, they differ from the St particles in the primary structure of their genome. As emphasized later, DI genomes lack part of the genetic information. They may or may not serve as coding sequences. However, they always conserve the *cis*-sequences needed for replication initiation (origins of replication), sometimes present in more than one copy, and sequences involved in encapsidation. Foreign sequences can also be inserted. DI particles can sometimes be separated from St particles on the basis of size, when the size of the particle closely corresponds to the size of the genome (for instance rhabdovirus), or on the basis of particle density differences (changes in nucleic acid to protein ratios). Often, however, only viral stocks enriched in DI particles are available owing to the size heterogeneity of the virus particles.

Generation of DI Genomes

DI DNAs very likely arise from various recombinational events not necessarily linked to genome replication, and which result in deletion, tandem duplication, insertion of host DNA, and polymerization of small monomer sequences. DI RNAs have been proposed to arise almost exclusively during genome replication by a mechanism of 'leaping polymerase' consisting of polymerase stop/fall-off or slippage/reinitiation events (**Figure 1**). In this

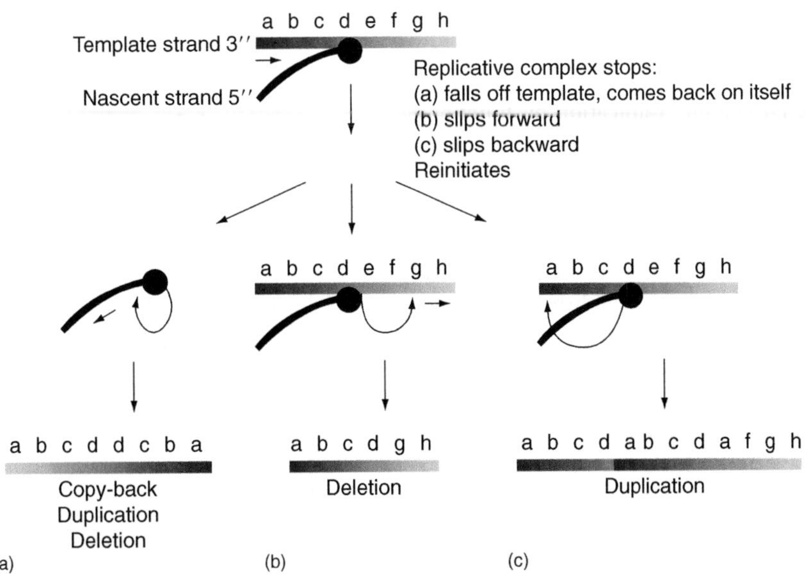

Figure 1 Defective RNA genome generation.

model the replicase complex moves with the nascent RNA still attached to it. Depending on where reinitiation takes place, and on the number and the direction of the leaps, the resulting molecules can be of the copy-back type, with more or less intramolecular inverted complementary sequences (a), of the internal deletion type (b), and of the duplication type (c). Multisteps (b) or (c) and combinations of steps (b) and (c) can, moreover, lead to various mosaic types. Insertion of host RNA is also observed, especially in plant DI RNA. The frequency at which the polymerase leaps and resumes its synthesis is unknown. The probability for this exercise to be successful in producing a viable DI genome has been estimated for VSV to range in the order of 10^{-7}–10^{-8} per genome replication.

Defectiveness

The DI genomes contain interrupted or rearranged open reading frames. They partly or completely lack the full coding capacity of the viral genome. They are therefore defective, and depend for their replication and for their propagation (formation of virus particles) on the functions provided by the homologous standard virus (helper). Co-infection of cells with DI and standard particles is therefore essential for DI particle multiplication. Consequently, low-multiplicity infections, and particularly plaque purification, represent conditions which decrease, and potentially eliminate, DI particles from a viral preparation.

Interference

As stated earlier, the generation of a defective genome is likely to represent a rare event. This event would never be seen unless it was successfully amplified. During this amplification step the defective genome is preferentially replicated over the nondefective genome. This ability to replicate efficiently at the expense of the nondefective genome is called interference. The mechanisms of interference are not completely understood. They obviously change depending on the specificity of the viruses, and appear to be also affected by the host cell types. In general, interference involves an early step in genome replication, and can be pictured as a competition for limiting replication 'factors' (viral replicase, encapsidation proteins, host cell factors). Reiterated origins of replication or encapsidation sites on DI DNAs, presence of higher affinity sites for the replicase or for the encapsidation on both positive and negative polarity DI RNAs, and shorter length of the replicating units, higher availability for replication of molecules not involved in transcription, have been shown or postulated as taking part in the interference mechanism.

Defective Interfering versus Defective Viruses

Based on the outcome of experimental co-infections of defective with nondefective viruses, a distinction has been made between defective interfering or defective noninterfering particles, according to the ability of the defective viruses to selectively restrict nondefective virus replication. This distinction may not apply during the first events following the generation of a defective genome. As this is bound to be a rare event, an interference mechanism has to be invoked any time this defective genome is amplified to the point it can be detected, or become predominant.

Cyclic Variations of Defective Interference

The dependence of DI genome replication on functions provided by the nondefective genome on the one hand, and the interference exhibited by DI genome on the other hand, result in out-of-phase cyclic variations of both DI and St genome replication. As illustrated in **Figure 2**, efficient St genome replication must precede extensive DI genome replication. This in turn establishes conditions of high interference which results in inhibition of St genome replication. Decrease of helper function availability leads to DI genome replication dampening, and therefore to release from interference, allowing efficient St genome replication to resume. These cyclic variations have been observed in serial passages of St and DI viruses in cell culture, as well as in persistent infections. The periodicity of a complete cycle is generally a matter of days or of a few serial passages.

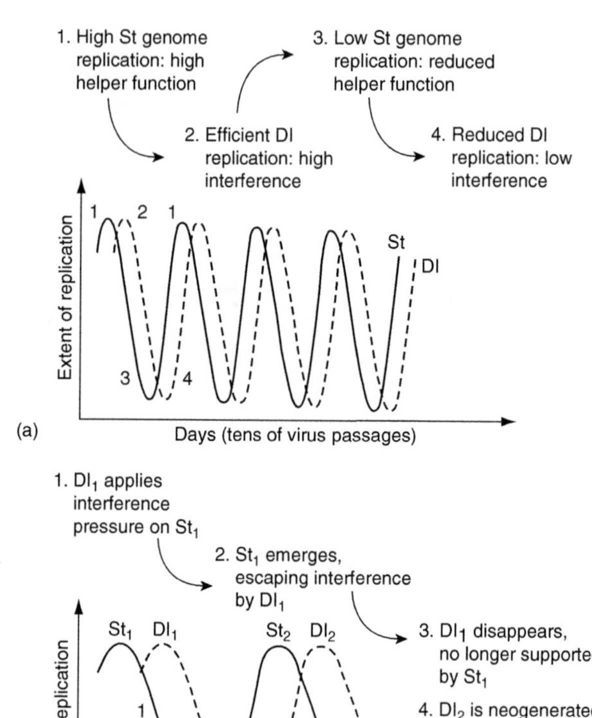

Figure 2 Cyclic variation of DI replication in (a) days, and (b) months.

Assay for DI Particles

DI particles can be detected by physical separation on velocity or density gradients when applicable (see the section titled 'Structure'). The presence of subgenomic nucleic acids in viral stocks or in infected cells (distinct from viral messengers) can also be diagnostic. The ability to decrease the infectivity of a viral stock or to protect infected cells from the lytic infectious virus (see the section titled, 'Biological effects') are used in various biological assays to estimate quantitatively and qualitatively the DI particle composition of a viral stock. These assays, although appropriate to characterize DI particle preparations, are generally not sensitive enough to exclude, when negative, the presence of DI particles in a viral preparation. The test that still remains the most dependable to assess presence or absence of DI particles consists of multiple independent serial undiluted passages. It is based on the observation that a viral stock contaminated with an undetectable amount of DI particles will, on subsequent independent serial passages, promote in each series amplification of the same contaminating DI particles. A DI particle-free stock, on the other hand, will yield in each series different DI particles or different sets of DI particles.

Biological Effects

DI particles have been shown to modulate the course of an infection. In cell culture, attenuation of the cytopathic effects is the most frequently described, and DI particles can promote cell survival and establishment of persistent infections. As far as negative-stranded RNA viruses are concerned, copy-back DI RNAs appear to prevent induction of apoptosis through a mechanism which has still to be unravelled, in which a certain category of small leader RNAs may participate. A possible role of the matrix M protein, the concentration of which is decreased to the point where viral assembly and budding at the cell surface is highly diminished in the presence of DI genomes, has also been considered. DI ability selectively allows emergence of St virus (St_2) which escapes interference (**Figure 2(b)**). St_2, resistant to interference, is selectively amplified over St_1, still sensitive to interference, and soon becomes predominant. It loses its ability to support DI_1, which therefore disappears. St_2 will generate its own DI_2, which in turn will favor emergence of a new St variant. Thus, DI viruses serve as mutational drivers favoring virus evolution, through cycles of high and low interference whose periodicity is this time measured in months or in hundreds of viral passages (compare **Figures 2(a)** and **2(b)**).

DI Particles in Experimental Animals

DI particles are generated and amplified in the whole animal, as well as in cell culture. They can change the symptoms of viral infection from rapid death to slow, progressing paralysis. They can sometimes fully protect the animals from an otherwise lethal infection. Interference is likely to be involved in this modulation of symptoms, but other phenomena like increased interferon induction and immune response modulation are reported.

DI Particles in Natural Infections

Involvement of DI particles in natural infections is still poorly documented. This is partly because the experimental results supporting the strong potential for infection modulation of DI particles have not been fully recognized, as detection of DI particles in natural infections is not straightforward. Unpredictable cyclic variations in DI replication, efficiency of DI replication changing with the types of infected tissues, and potent interfering ability associated with poor DI particle replication are all factors which undoubtedly make DI particle detection difficult *in vivo*. Last but not least, virus isolation, which is used to characterize the infectious agent associated with a disease, often represents conditions (low multiplicity of infection) known to impair DI particle replication drastically.

Nevertheless, association of a chicken influenza virus strain, efficiently producing DI particles, with an epidemic of low morbidity and low mortality, and conversely, a high-mortality epidemic associated with a strain free of DI particles, have been reported. Murine and feline leukemia virus strains causing immune deficiency syndromes are shown to contain predominantly replication-defective viral genomes before onset and during the development of the disease. The pathogenicity of some bovine and swine pestiviruses has clearly been associated with presence of DI RNAs in the animals. For the bovine viral diarrhea virus (BVDV), a pestivirus of the same family as hepatitis C virus, the presence of a particular DI RNA can turn noncytopathic virus into a fatal infectious agent. In plants, at least three examples of DI RNAs are described to be involved in infection modulation. Interestingly, depending on the types of viruses, DI RNAs can either attenuate or exacerbate the symptoms.

DI RNAs are identified in stool and blood samples of humans suffering from hepatitis A virus, an infection known to be rather moderate and prolonged. DI particles are identified in measles virus-attenuated vaccine preparations which have been, and are being, widely and successfully used (raising the question of DI particle participation in vaccine attenuation). Measles viruses defective in viral assembly are currently found associated with human subacute sclerosing panencephalitis (SSPE). The brain cells of SSPE patients were, moreover, shown to harbor many species of measles virus copy-back DI RNAs. Direct amplification of a portion of the HIV *tat* gene from infected patients demonstrates that about a third of the sequences correspond to defective *tat* function. Moreover, human immunodeficiency virus (HIV)-1 sequences isolated from a cohort of six blood or blood product recipients infected with one donor all contained a similar deletion in the *nef* gene. Remarkably, all the patients harboring this deleted viral genome remained free of HIV-related diseases 12–16 years after infection, suggesting that this defective species of HIV genome is responsible for this decreased pathology. Epstein–Barr virus (EBV) replicative infections developing in human epithelial lesions involve a deleted rearranged form of EBV DNA (het DNA). This het DNA is associated, in experimental infections, with disruption of latency and persistent productive infections. Specific identification of viral hepatitis B (HBV) genomes containing an interrupted precore antigen (HBeAg) coding sequence in patients dying from fulminant hepatitis suggests that such defective genomes may be responsible for the exacerbation of the disease. This contrasts with more recent data reporting experimental evidence for the existence of DI-like viruses in HBV human chronic carriers; fluctuations between these naturally occurring core internal deletion variants and helper HBV in three chronic carriers were reminiscent of the cycling phenomenon in other DI viral systems.

Future Perspectives

Defective interfering particles are ubiquitous in the realm of animal and plant viruses. In experimental conditions they appear as necessary companions of their nondefective homologs. Capable of affecting the extent of viral growth, the course of viral infections, and serving as selective pressure to drive mutational changes, they can be seen as natural regulators of virus evolution. The demonstration of their participation in natural infections, and of their ability to affect the course of diseases, constitutes a challenge for the years to come. As pointed out by the few examples listed earlier, their direct detection in infected tissues will certainly be needed to assert their involvement in natural infections. The availability of sensitive detection techniques (like polymerase chain reactions), allowing direct observation of viral genomes without the possible distortion of virus isolation, bears great hope. More than giving increased insights into the physiopathology of viral infections, in times where the modifications of the viral genomes represent an imperative step in the generation of viral recombinant vaccines or of appropriate vectors for gene targeting, DI viral genomes represent natural versions of defective genomes that can serve as model tools for creation of more adapted vectors.

See also: Orthomyxovirus Antigen Structure; Orthomyxoviruses: Molecular Biology.

Further Reading

Barrett ADT and Dimmock NJ (1986) Defective interfering viruses and infection of animals. *Current Topics in Microbiology and Immunology* 128: 55.

Perrault J (1981) Origin and replication of defective interfering particles. *Current Topics in Microbiology and Immunology* 93: 151.

Roux L, Simon AE, and Holland JJ (1991) Effects of defective interfering viruses on virus replication and pathogenesis. In: Shatkin A (ed.) *Advances in Virus Research*, vol. 40, 181pp. New York: Academic Press.

Interfering RNAs

K E Olson, K M Keene, and C D Blair, Colorado State University, Fort Collins, CO, USA

Glossary

dsRNA Double-stranded RNA, the trigger for RNAi and intermediate in RNA virus replication.

miRNA MicroRNA; small RNAs, about 21–23 bases in length, that post-transcriptionally regulate the expression of genes by binding to the 3′ untranslated regions (3′ UTR) of specific mRNAs.

RISC RNA-induced silencing complex; a multiprotein complex that brings the guide strand of the siRNA duplex and the cellular mRNA together and cleaves the mRNA with associated endonuclease activity. mRNA is then degraded.

RNAi An evolutionarily conserved, sequence-specific antiviral pathway triggered by double-stranded RNA (dsRNA) that leads to degradation of both the dsRNA and mRNA with homologous sequence.

siRNA Small interfering RNAs; 21–25 bp duplexes that provide guides for sequence-specific cleavage of mRNA. siRNAs are considered the hallmark of RNAi.

Introduction

Antiviral innate immune responses in vertebrate hosts restrict viral invasion until humoral and cell-mediated acquired immune responses specifically clear virus infection. In mammals, innate immune responses are induced when infected cells recognize viral components such as nucleic acids through host pattern-recognition receptors. Toll-like receptors are important pattern-recognition sensors that recognize viral components (e.g., double-stranded RNA (dsRNA)) and signal induction of type I interferons and inflammatory cytokines that lead to an antiviral state. Organisms such as insects probably depend entirely on antiviral innate immune responses to overcome viral infection, since acquired immune responses have not been identified in invertebrates. Interferon-like molecules also have not been found in insects, making it unclear how insects cope with viral infections. In 1998, researchers first described an intracellular response in the worm *Caenorhabditis elegans* that could be triggered by dsRNA and efficiently silenced the expression of genes having sequence identity with the dsRNA trigger. This response, termed RNA interference or RNAi, is now known to be an ancient antiviral innate immune response in eukaryotic organisms including worms, plants, insects, and mammals and is a common evolutionary link among these organisms in their fight against virus invasion.

RNA Interference

RNAi, post-transcriptional gene silencing (PTGS), quelling, and sense suppression are terms for related pathways that have been described in different organisms. All are RNAi responses triggered by dsRNA that result in degradation of both dsRNA and mRNA with cognate sequence. These pathways are highly conserved evolutionarily and exist in many organisms including plants, fungi, and animals. RNAi and related pathways have several functions, including regulation of development, silencing and regulation of gene expression, and defense against viruses and transposable elements. In this article, we focus on RNAi as an antiviral, innate immune pathway.

RNAi Mechanism of Action

The RNAi pathway is divided into an initiator phase and an effector phase. The initiator phase consists of the recognition and processing of long dsRNA molecules by the RNaseIII enzyme Dicer into small interfering RNAs (siRNAs) of 21–25 bp. siRNAs, considered the hallmark of the RNAi response, are duplexes with 3′ overhangs of 2 nt. Each strand has a 5′-PO_4 and 3′-OH. In the fruitfly *Drosophila melanogaster* (drosophila), the siRNAs are unwound and incorporated into the RNA-induced silencing complex, or RISC, with the assistance of Dicer-2 and R2D2, to start the effector phase of the pathway. In the effector phase, one strand of the siRNA duplex acts a as guide sequence to target the RISC to complementary mRNAs and determine the cleavage site on the mRNA. The RISC is known to include products of the following genes: *Argonaute2* (*Ago2*), *Vasa intronic gene* (*VIG*), *fragile X mental retardation* (*FXR*), and *Tudor Staphylococcal nuclease* (*Tudor-SN*). Other genes in drosophila encode proteins having RNA helicase activity associated with RNAi and include *spn-E*, *Rm62*, and *armi*. The latter gene products have been implicated with heterochromatin and transposon silencing.

RNAi Components and Function: Dicers, the Sensors of dsRNA and Initiators of siRNA Production

The majority of biochemical information related to Dicer proteins has been elucidated from studies in drosophila, showing that Dicer enzymes are intracellular sensors of dsRNA that initiate RNAi in drosophila cultured cell and embryo lysates. Candidate genes from three families encoding RNase III motifs were expressed in drosophila S2 cells, immunoprecipitated, and tested *in vitro* for their ability to transform long dsRNA molecules (>30 bp) into small RNAs of ~21 nt (**Figure 1**). The enzyme capable of producing siRNAs was termed Dicer and contained a helicase domain as well as two RNase III domains. The production of siRNAs by Dicer-2 in drosophila was ATP-dependent and the enzyme was inactive in degrading single stranded RNAs. Dicer depletion by immunoprecipitation from drosophila cell lysates resulted in decreased siRNA production. Dicer is a ~200 kDa protein with an N-terminal RNA helicase domain, a PAZ (PIWI/Argonaute/Zwille) domain, a conserved domain of unknown function (DUF283), two C-terminal RNase III domains, and an RNA binding motif. In its active form, Dicer is a homodimer. Dicer is evolutionarily conserved, and is found in drosophila, *C. elegans*, humans, mice, trypanosomes, zebrafish, the fungi *Magnaporthe oryzae* and *Neurospora crassa*, budding yeast (*Schizosaccharomyces pombe*), plants, and many other organisms. The number of *dicer* (*dcr*) genes varies in different organisms. In drosophila there are two *dcr* genes (*dcr1* and *dcr2*); in the plant *Arabidopsis thaliana* there are four, each of which processes siRNAs from different dsRNA sources; however, in the genomes of humans and *C. elegans*, there is only one *dcr* gene.

Dicers and miRNA Biogenesis

The RNAi pathway has two branches leading to production of either siRNA or microRNA (miRNA). The miRNAs *lin-4* and *let-7* were first discovered in *C. elegans* and play a crucial role in development. Subsequently, hundreds of different miRNAs have been found in most eukaryotic organisms and exploring their expression and function is now an important topic of research in a wide range of invertebrates and vertebrates. miRNAs are produced from endogenous pre-miRNA transcripts that form

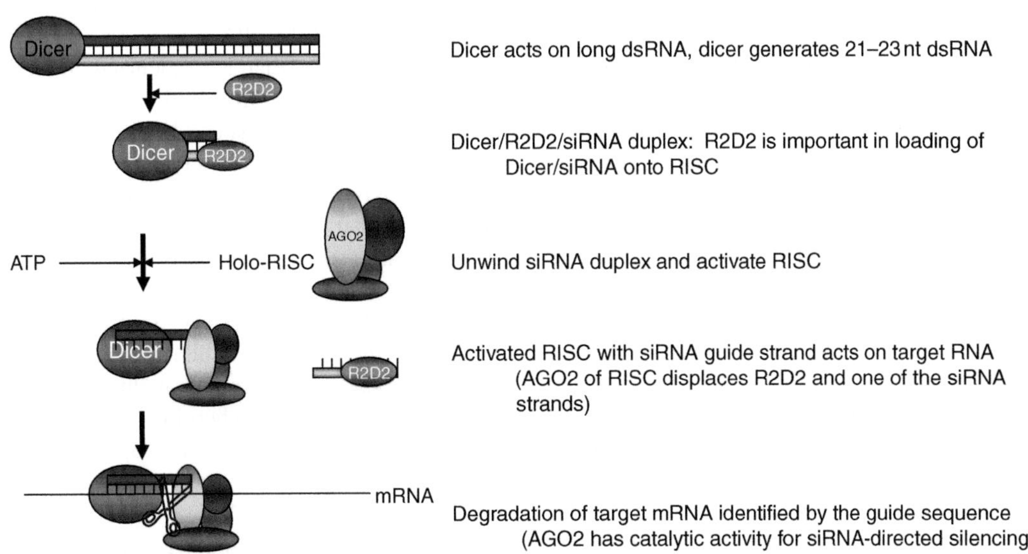
Dicer
R2D2
Dicer acts on long dsRNA, dicer generates 21–23 nt dsRNA
Dicer/R2D2/siRNA duplex: R2D2 is important in loading of Dicer/siRNA onto RISC
AGO2
ATP
Holo-RISC
Unwind siRNA duplex and activate RISC
Activated RISC with siRNA guide strand acts on target RNA (AGO2 of RISC displaces R2D2 and one of the siRNA strands)
mRNA
Degradation of target mRNA identified by the guide sequence (AGO2 has catalytic activity for siRNA-directed silencing)

Figure 1 General scheme for RNAi leading to degradation of a specific mRNA.

stem–loop precursors. Silencing of genes by the miRNA pathway occurs not by degradation of the mRNA, but rather by translational arrest during protein synthesis. Also, unlike the siRNA pathway, miRNA silencing does not require complete base pairing between the miRNA and the target sequence to be silenced. Distinct RISCs process small RNAs for the effector phase of the two gene-silencing mechanisms. It is not known how the distinction is made between siRNA production and miRNA production with a single Dicer enzyme in *C. elegans.* In humans, interferon and other innate immune responses are induced by long dsRNA, so Dicer-like activity generating siRNAs may not be as crucial. In organisms such as drosophila that encode two Dicer proteins, Dicer-1 is the enzyme that produces miRNAs from endogenous transcripts.

R2D2 Protein and RNAi

R2D2 is a 36 kDa protein with two dsRNA binding domains that bridges the initiator and effector stages of the RNAi pathway. R2D2 co-purifies with Dicer-2 from an siRNA-generating extract of drosophila S2 cells. R2D2 association does not affect the enzymatic activity of Dicer-2, but is required to load the newly formed siRNAs into RISC. Dicer-2 forms a heterodimer complex with R2D2 and the siRNA duplex that appears to detect the thermodynamic asymmetry of the siRNA duplex. The passenger strand of the siRNA duplex is separated from the guide strand, which interacts at its 5′ and 3′ ends with the PIWI (possessing RNase H-like activity) and PAZ domains of Argonaute-2 (AGO2), respectively. Strand selection (passenger strand vs. guide strand) depends on the thermodynamic stability of the first four nucleotides of the 5′ terminus of an siRNA duplex. The siRNA strand whose 5′ end has lower base-pairing stability becomes the guide strand, leaving the more stable strand as the passenger strand. R2D2 binds to the thermodynamically more stable end of an siRNA whereas Dicer-2 binds on the opposite end. Release of the degraded passenger strand is believed to be an ATP-dependent reaction. The guide siRNA strand remains associated with RISC and guides AGO2 to the target mRNA containing the complementary sequence. After hybridization, AGO2 cleaves the phosphodiester backbone of the target mRNA. Target RNA cleavage occurs between the 10th and 11th nucleotides of the guide siRNA measured from its 5′ phosphate group. The phosphate group of the 5′ end of the guide siRNA influences the fidelity of the cleavage position as well as the stability of RISC. Recently it was confirmed that the RNAi pathway of drosophila functions as an antiviral immunity mechanism. Drosophila with null mutations for genes encoding Dicer-2, Argonaute-2, or R2D2 were highly susceptible to RNA viruses such as Flock house virus (FHV) (*Nodaviridae*), Drosophila C virus (*Dicistroviridae*), or Sindbis virus (*Togaviridae*) which in some cases were lethal for the flies.

Argonaute Proteins and RNAi

Another family of genes that has already been discussed in this review as important to RNAi is the *argonaute* gene family. In drosophila, there are five *argonaute* genes. Two of the genes, *argonaute1* (*ago1*) and *argonaute2* (*ago2*), have been implicated in RNAi. Argonaute-2 is an important component of the RISC complex. AGO2 has been termed 'Slicer' and is the only component of human RISC that is required for the degradation of mRNA molecules, and its PIWI domain may contain the endonuclease activity. Drosophila embryo mutants lacking *ago2* are unable to perform siRNA-directed mRNA cleavage, although miRNA-directed cleavage is still possible. The embryos also lack the capacity to load siRNAs into RISC.

Drosophila Argonaute-1 (AGO1) is not involved in siRNA-directed cleavage as is AGO2. AGO1 is believed to function downstream of the production of the siRNAs and not as a component of RISC. These studies also demonstrated that *Drosophila* mutants lacking AGO1 are embryonic lethals, implicating AGO1 in *Drosophila* development. AGO1 is required for miRNA-directed cleavage and dispensable for siRNA-directed cleavage, showing divergent roles for different Argonaute proteins. This functional differentiation of the Argonaute proteins has also been noted in plants.

Argonaute proteins are characterized by the PAZ domain and the PIWI domain. The crystal structures of the PAZ domain of AGO2 from *Drosophila* and the thermophile *Pyrococcus furiosus* were determined and shown to have structural properties similar to proteins that bind single-stranded nucleic acid. The PAZ domain recognizes the 3′ overhangs of the siRNA duplexes. Interestingly, this is the same region that is recognized by certain viral suppressors of RNAi. In addition to its endonuclease motif, the PIWI domain is involved with protein–protein interactions between Argonaute and Dicer and may play a role in siRNA loading onto RISC. Recently, researchers have shown that both human AGO1 and AGO2 proteins reside in intracellular structures known as 'cytoplasmic bodies'. These areas of the cell are believed to be sites of regulation of cellular mRNA turnover. Several other AGO proteins from drosophila are associated with the RNAi pathway. These proteins include Piwi, implicated in transposon silencing, Aubergine, associated with germline gene repression, and AGO3, an argonaute-like protein of unknown function. As already mentioned, a number of other proteins have now been associated with the RNAi pathway, including Vig, Tudor-SN, and Fmr-1, all associated with RISC.

Unique Properties of RNAi in Plants and Animals

In *C. elegans* and plants, an amplification of the RNAi response occurs. In these systems, the original siRNAs act as primers for synthesis of new intracellular dsRNA by the endogenous RNA-dependent RNA polymerase (RdRP). These newly produced dsRNAs are processed by Dicer and generate an additional and potentially more diverse pool of siRNAs. This phenomenon is termed 'transitive RNAi' and allows for the degradation of a full-length mRNA even when the initial trigger sequence represents only a portion of the gene or genome. Amplification of the siRNA signal is bidirectional along the transcript in plants; however, in *C. elegans* the signal can only spread 3′–5′ along the transcript. Transitive RNAi does not occur in species such as *D. melanogaster*, where RdRP genes are absent from the genome. Gene knockdown experiments in organisms lacking transitive RNAi require design of dsRNA for specific disruption of gene expression.

One unique aspect of PTGS in plants is the ability of the silencing signal to spread throughout the organism. The siRNAs, complexed with host proteins, are part of a silencing complex that can move to other tissues in the plant through phloem tissues. The distance that the siRNAs travel is dependent on their exact length and the Dicer enzyme by which they were generated. The exact mechanism for long-distance movement has yet to be elucidated. RNAi in *C. elegans* can spread from the point of induction, especially if the dsRNA is introduced into the intestine either by injection or by feeding.

RNAi has become an important reverse genetics tool for studying gene function. Systematic knockdown of all genes has been accomplished in both drosophila and *C. elegans* with great success, identifying previously unknown functions of genes in various biological pathways and revealing differences in the RNAi pathways of each organism. In drosophila, the siRNA signal appears not to be amplified (transitive RNAi) and does not spread as seen in *C. elegans* and plants. This means that RNAi activity is most likely confined to those cells in which the dsRNA trigger is detected by Dicer. RNA silencing has been used in many other invertebrates, including mosquito disease vectors. Several studies have shown that RNAi can be used to knock down endogenous and exogenous gene expression in mosquitoes that transmit medically important pathogens to animals and humans. Studies of RNAi in mosquitoes have also shown that it is possible to silence RNAi complex genes using RNAi. As an example, the ability to silence the RNAi pathway in a hemocyte cell line of *Anopheles gambiae* was tested by transfecting dsRNA derived from exon sequences of the *A. gambiae dcr1* and *dcr2* and *argonaute 1–5* (*ago1–5*) genes. RNAi in *A. gambiae* cells required expression of Dicer-2, AGO2, and AGO3 proteins. This study also demonstrated that RNAi in the mosquito, as in drosophila, does not spread from the target cells, suggesting that RdRP-mediated transitive amplification is absent in the mosquito.

Viruses and RNAi

Since the discovery of RNA silencing, a number of researchers have hypothesized that RNAi plays an important role in antiviral defense. RNA viruses with positive-sense genomes, in particular, form dsRNA intermediates as they replicate in host cells. Whether this is the general viral trigger of RNAi in infected cells is still controversial, since viruses with negative-strand genomes generate little detectable dsRNA during their replication. Studies to elucidate a mechanism termed pathogen-derived resistance (PDR) in plants were among the first to show that viruses could be targeted by RNA silencing. Transgenic plants were engineered to express a portion of the tobacco etch virus (TEV) coat protein. Upon challenge with TEV, the transgenic plants were found to be resistant to the virus. No overt symptoms were seen, and no virus could be recovered from the leaf tissue. PDR was shown to be virus specific, as the transgenic plants could be infected with a genetically unrelated virus; however, any RNA that was introduced into plant cells that shared homology with the RNA in the transgene was degraded. Plant virologists also observed that the delivery of a fragment of viral RNA to plant cells by a heterologous virus expression vector made the cells resistant to challenge by the first virus, and termed this related phenomenon virus-induced gene silencing. Soon after the discovery that plants used an RNA-mediated defense mechanism against viruses, it was shown that many plant viruses encoded protein suppressors of RNA silencing. This would be expected as the virus and host develop a biological arms race to express countermeasures that limit the ability of one to overcome the other.

Viral Encoded Suppressors of RNAi

Several suppressors have been found that interrupt the RNA-silencing pathway at different steps, indicating that evasion of RNA silencing has evolved more than once. One protein, helper component-proteinase, or HC-Pro, from potyviruses of plants is thought to be the most potent suppressor of RNA silencing found to date. HC-Pro functions at a step that prevents the accumulation of siRNAs by interacting with the RNase III enzyme Dicer. HC-Pro also has the capability to reverse established silencing of a transgene, suggesting that the protein inhibits a mechanism required for the maintenance of silencing. The 19 kDa protein (p19) from tombusviruses acts as a suppressor of PTGS in plants in a different manner from HC-Pro. p19 does not block production of the 21–25 nt siRNAs; rather it binds them via the 2-nt 3′ overhangs. Notably, the protein will bind only double-stranded 21 nt sequences; single-stranded RNAs of the same length are not recognized by p19. In binding the siRNAs, p19 forms a homodimer and sequesters the guide sequences that are required for RISC incorporation and targeting mRNA degradation. Another virus, cucumber mosaic virus, encodes a protein 2b that interferes with the spread of siRNA signal in the plant host, allowing systemic spread of the virus.

In addition to the PTGS suppressors encoded by plant viruses, the insect virus Flock house virus encodes a protein, B2, which can suppress RNAi activity in both drosophila S2 and plant cells. In transgenic plants containing a green fluorescent protein (GFP) gene, in which transient expression of siRNAs targeting GFP mRNA had silenced its expression, the presence of B2 protein reversed silencing of GFP. siRNAs were still detected in the tissues, indicating that the suppression of gene silencing did not occur before the production of siRNAs. It is likely that this protein sequesters the siRNAs from the RNAi machinery, possibly in a manner similar to p19. Finally, two other animal viruses encode suppressors of interferon induction that also have apparent RNAi suppressor activity. These are the influenza virus NS1 protein and the vaccinia virus E3L protein.

Viruses may employ other strategies to evade the RNAi pathway. These strategies include (1) sequestration of the viral dsRNA replicative intermediates in viral cores or in double-membrane structures in the host cell formed during viral replication; (2) viral replication and spread outpacing the RNAi pathway; and (3) replication in tissues that are resistant to RNAi.

miRNAs and Viruses

In the last couple of years, a number of viral-encoded miRNAs have been discovered. The functions of most viral-derived miRNAs are unknown; however, DNA viruses such as polyomaviruses and herpesviruses transcribe miRNAs in infected vertebrate cells that appear to regulate expression of critical viral and host genes during infection. The location of miRNAs within different virus genomes are not conserved, suggesting that miRNAs are likely to be recent acquisitions in viral genomes that help adapt the virus to the host during the virus lifecycle. For instance, a viral miRNA in SV40 virus was recently discovered that regulates the viral T antigen. The SV40 miRNA accumulates in late stages of infection and targets the early T antigen mRNA for degradation, thus reducing its expression. Studies of many families of RNA viruses have failed to identify miRNAs in RNA genomes and this finding is consistent with the prominent role of the cellular DNA-dependent RNA polymerase II in the biogenesis of miRNA precursors.

Use of RNAi for Disease Control

Virus diseases of plants cause extensive economic losses. Control of plant virus diseases has usually been associated with control of insects or nematodes that transmit viruses, or through sanitation and quarantine of infected plants. The transformation of plants with effector genes designed to transcribe inverted-repeat RNA (dsRNAs) that target plant viruses can provide novel virus-resistant varieties. Transgenic plants expressing inverted repeats of viral sequences exhibit varying degrees of resistance to the virus or viruses with genome sequences closely related to the source of the transgene. This resistance is due to PTGS wherein viral mRNA is degraded in the cytoplasm soon after synthesis.

In a similar genetic approach to that described previously in plants, we have genetically modified *Aedes aegypti* mosquitoes to exhibit impaired vector competence for dengue type 2 virus (DENV-2) transmission. DENVs are normally transmitted by *A. aegypti* mosquitoes to humans during epidemic outbreaks of dengue diseases. If a DENV-derived dsRNA trigger is expressed in the cell prior to viral translation and replication, an antiviral state can be induced in the mosquito that blocks virus infection. To do this, mosquitoes were genetically modified to express an inverted-repeat (IR) RNA derived from the premembrane protein coding region of the DENV-2 RNA. The IR RNA formed a 560 bp dsRNA in infected midgut epithelial cells of the mosquitoes to induce the RNAi pathway. A transgenic family, Carb77, was selected that expressed IR RNA in the midgut after a blood meal. Carb77 mosquitoes ingesting an artificial blood meal with 10^7 pfu ml^{-1} of DENV-2 exhibited marked reduction of viral envelope antigen in midguts and salivary glands after infection. Transmission of virus by the Carb77 line was significantly diminished when compared to control mosquitoes. As evidence that the resistance was RNAi mediated, DENV-2-derived siRNAs were readily detected in RNA extracts from midguts following ingestion of a blood meal with no virus. In addition, loss of the resistance phenotype was observed when the RNAi pathway was interrupted by injecting *ago2* dsRNA 2 days prior to induction of IR-RNA transgene, confirming that DENV-2 resistance was caused by an RNAi response.

Targeting of replicating animal viruses using RNAi has prompted discussion about whether RNAi can be used as an antiviral therapy. In mammalian cells, siRNAs, rather than long dsRNAs, are required to induce RNAi as an antiviral therapy because these cells possess interferon and other antiviral innate immune pathways that are triggered by dsRNA >30 bp in length. Numerous studies in cell culture have shown that HIV replication can be halted when cells are treated with siRNAs that target the viral genome. West Nile virus (WNV) replication also can be reduced in cultured cells by treatment with siRNAs targeting the virus RNA. These studies showed a significant reduction in levels of WNV RNA if cells were pretreated with siRNAs; however, the cells that were treated subsequent to the establishment of viral replication did not show the same reduction in viral RNA, suggesting that the RNA may be sequestered from the RNAi machinery after replication is established in the cell. Studies investigating RNAi as a therapy for hepatitis C virus (HCV) infection have used siRNAs to effectively target HCV replicon RNAs in cultured human cells as well as in a mouse model.

Model Systems for Studying Role of RNA in Virus Infections

While the genetics and biochemistry of RNAi in *C. elegans* and *D. melanogaster* have been investigated in detail, there have been no virus infection models of these animals until very recently. RNAi-based innate immunity has been detected in *C. elegans*-derived cultured cells infected with vesicular stomatitis virus (VSV; *Rhaboviridae*). FHV replication in *C. elegans* triggered potent antiviral silencing that required RDE-1, an argonaute protein essential for RNAi mediated by siRNAs. This antiviral innate immunity was capable of rapid virus clearance in *C. elegans* in the absence of FHV RNAi suppressor protein B2. Two recent papers have shown that successful infection and killing of *Drosophila* by FHV was strictly dependent on expression of the viral suppressor protein B2. *Drosophila* with a knockout mutation in the gene encoding Dicer-2 showed enhanced susceptibility to infection by FHV, cricket paralysis, and *Drosophila* C viruses (*Dicistroviridae*) and Sindbis virus (*Alphavirus*; *Togaviridae*). These data demonstrate the importance of RNAi for controlling virus replication *in vivo* and establish *dcr2* as a drosophila susceptibility locus for virus infections. *C. elegans* and drosophila are important models for studying virus–RNAi interactions. The drosophila model system allows advanced genetic approaches such as generation of null mutants of RNAi components in a system that has few other dsRNA-triggered defensive responses to complicate mechanistic studies. The availability of the annotated drosophila genome sequence and established genetic approaches is critical for understanding RNAi mechanisms and continues to make the drosophila model important to our understanding of innate immune responses to viruses.

Mosquitoes, RNAi, and Arboviruses

Mosquitoes and arboviruses provide an important naturally occurring insect–virus system to study the potential role of RNAi in host defense. Arboviruses are RNA viruses that must replicate in their arthropod vector for amplification before they can be transmitted to a vertebrate host, such as humans. Obviously, arboviruses must somehow evade the RNAi pathway to successfully replicate in the mosquito prior to transmission. There are several advantages to studying RNAi in mosquitoes. First, the complete genome sequence is now available for at least two medically important vectors, *A. gambiae* and *A. aegypti*. Second, the RNAi pathway of mosquitoes is similar in structure and function to the pathway in drosophila and many of the component genes of RNAi have been identified. Third, new genetic approaches are allowing researchers to manipulate genes that affect innate immune responses in the mosquito. Fourth, infectious cDNA clones of arbovirus genomes from at least three virus families are available to allow manipulation of the viral genes and identify determinants of RNAi modulation. Fifth, RNAi–virus interactions studies can occur in systems directly relating to medically important pathogens.

Mosquitoes, like drosophila, do not have responses comparable to interferon induction, so they can be injected with long dsRNAs (300–500 bp) to trigger the RNAi pathway and efficiently silence specific vector genes that may participate in innate immune pathways, including RNAi. As an example, to determine whether RNAi conditions the vector competence of *A. gambiae* for O'nyong-nyong virus (ONNV; *Alphavirus*), a genetically modified ONNV expressing GFP (eGFP) was developed to readily track virus infection. After intrathoracic injection, ONNV-eGFP slowly spread to other *A. gambiae* tissues over a 9 day period. Mosquitoes were co-injected with virus and dsRNA derived from the ONNV nsP3 gene. Treatment with nsP3 dsRNA inhibited virus spread significantly, as determined by GFP expression patterns. ONNV-GFP titers from mosquitoes co-injected with nsP3 dsRNA also were significantly lower at 3 and 6 days after injection than in mosquitoes co-injected with non-virus-related β-galactosidase (β-gal) dsRNA. However, mosquitoes co-injected with ONNV-GFP and dsRNA derived from the *A. gambiae ago2* gene displayed widespread GFP expression and virus titers 16-fold higher than β-gal dsRNA controls at 3 or 6 days after injection. These observations provided direct evidence that RNAi is an antagonist of ONNV replication in *A. gambiae* and suggest that this innate immune response plays a role in conditioning vector competence. These types of experiments could be vital to understanding why some mosquito species are excellent vectors of disease and others are not.

The Biological Arms Race between Viruses and Hosts

RNA virus–host interactions are often characterized as an escalating arms race between two mortal enemies. The host evolves innate and acquired immune pathways to counter the destructive effects of virus invasion and the virus adapts to these defense measures by rapidly evolving new ways of evading the host's attempts at pathogen control. RNA–virus interactions with the host's RNAi pathway exemplify this struggle for dominance. Many RNA viruses evolve rapidly because their RNA-directed RNA polymerases are error prone, providing significant variation in virus genome populations needed to probe weaknesses in the host's defense and allow selection of new virus variants that have an advantage in their replication. As we have described earlier, a number of families of plant RNA viruses have evolved RNAi suppressors. Each family has developed a different strategy to downregulate RNAi, as shown by the fact that their suppressors attack different steps in the pathway. Still other viruses may have adapted to host RNAi by sequestering their replicative intermediate dsRNA triggers in double-membrane structures derived from host endoplasmic reticulum. Rapid evolution of RNA viruses also produces significant challenges to developing therapeutic strategies for humans using siRNAs to target and destroy viruses. Strategies that use siRNAs to trigger RNAi can be thwarted by point mutations in the target sequence. This has prompted development of siRNAs that target multiple regions of the viral RNA, highly conserved regions of the viral genome, or host genes essential for virus infection. However, RNA viruses are constrained in the amount of genetic variation they can tolerate and remain genetically fit for cell entry, replication, packaging, and egress.

Viruses that have complex life cycles, such as arthropod-borne viruses that must replicate in both vertebrate and invertebrate cells, are further constrained in their evolutionary potential. Finally, there is evidence that the host can evolve to counter the threat posed by RNA viruses. A recent finding emphasized the critical role of RNAi as an innate immune mechanism in *Drosophila* when researchers showed that RNAi pathway genes (*dcr2, ago2, r2d2*) are among the 3% fastest evolving genes among drosophila species, confirming the biological arms race between viruses and insects. The antiviral role of RNAi has been studied for less than a decade and is only now being exploited as a mechanism to fight viral diseases. Knowledge gained since the discovery of RNAi in 1998 should allow researchers to fully exploit RNAi as a means of controlling a number of infectious agents that cause disease in plants, animals, and humans.

Acknowledgments

Our research is funded by NIH grants AI34014 and AI48740 and the Grand Challenges in Global Health through the foundation for NIH.

See also: Satellite Nucleic Acids and Viruses.

Further Reading

Akira S, Uematsu S, and Takeuchi O (2006) Pathogen recognition and innate immunity. *Cell* 124: 783–801.

Baulcombe D (2004) RNA silencing in plants. *Nature* 431: 356–363.

Franz AWE, Sanchez-Vargas I, Adelman ZN, *et al.* (2006) Engineering RNA interference-based resistance to dengue virus type 2 in genetically modified *Aedes aegypti*. *Proceedings of the National Academy of Sciences, USA* 103: 4198–4203.

Galiana-Arnoux D, Dostert C, Schneemann A, Hoffmann JA, and Imler JL (2006) Essential function *in vivo* for dicer-2 in host defense against RNA viruses in *Drosophila*. *Nature Immunology* 7: 590–597.

Hammond SM, Boettcher S, Caudy AA, Kobayashi R, and Hannon GJ (2001) Argonaute2, a link between genetic and biochemical analyses of RNAi. *Science* 293: 1146–1150.

Kavi HH, Fernandez HR, Xie W, and Birchler JA (2005) RNA silencing in *Drosophila*. *FEBS Letters* 579: 5940–5949.

Keene KM, Foy BD, Sanchez-Vargas I, Beaty BJ, Blair CD, and Olson KE (2004) RNA interference as a natural antiviral response to O'nyong-nyong virus (*Alphavirus; Togaviridae*) infection of *Anopheles gambiae*. *Proceedings of the National Academy of Sciences, USA* 101: 17240–17245.

Leonard JN and Schaffer DV (2006) Antiviral RNAi therapy: Emerging approaches for hitting a moving target. *Gene Therapy* 13: 532–540.

Li WX, Li H, Lu R, *et al.* (2004) Interferon antagonist proteins of influenza and vaccinia viruses are suppressors of RNA silencing. *Proceedings of the National Academy of Sciences, USA* 101(5): 1350–1355.

Voinnet O (2005) Induction and suppression of RNA silencing: Insights from viral infections. *Nature Reviews Genetics* 6: 206–220.

Zamore PD (2002) Ancient pathways programmed by small RNAs. *Science* 296: 1265–1269.

Viruses as Vectors in Gene Therapy

K I Berns and T R Flotte, University of Florida College of Medicine, Gainesville, FL, USA

Glossary

Biodistribution Tissues infected by a viral vector.

Episomal form Existence of the transgene as an extrachromosomal element.

Gene therapy Introduction of exogenous genes into cells to correct genetic or physiologic defects.

Nonenveloped virus A virus particle in which the viral coat is not surrounded by a lipid membrane.

Nonhomologous recombination Recombination between two unrelated DNA sequences.
Persistent infection An infection in which the viral genome is retained for extended periods, often for life, in host cells.
Replication competent A vector which can self-replicate and potentially spread from cell to cell.
Transgene A gene carried by a vector into a target cell.
Vector The means by which is a transgene is introduced into a target cell.

Introduction

Gene therapy represents the ultimate application of the evolution of molecular genetics from the elucidation of the structure of the double helix to patient treatment. There are only two requirements: the gene or DNA to be used to correct a genetic defect and a vehicle to introduce the DNA into the patient's cells. With the advent of cloning and recombinant DNA technology and the determination of the sequence of the human genome, a large number of genes are available for the therapy of genetic defects. Yet the clinical application of gene therapy remains in its infancy, albeit it has probably moved beyond the neonatal period. The major reason for the lack of more extensive progress has been the challenge of developing effective vectors. Successful vectors must be able to get the transgene (the new gene to be introduced) to the appropriate cell, transport the transgene into the cell, and then into the nucleus. Once present in the nucleus, a successful vector will express the transgene at the desired level and most often for an extended period of time. A major concern is toxicity, which may reflect either a host reaction to the vector itself, a commonly observed problem for early-generation vectors, or to the transgene, if it is perceived as nonself, which could occur in the case of a person with a null mutation. A further consideration is the ability of any DNA molecule introduced into the nucleus to potentially integrate into the host cell genome, most often by nonhomologous recombination. Nonhomologous recombination, which can occur at many sites in the genome, has the potential of disrupting normal gene expression by either altering the gene product or the regulation of the expression of the gene. Thus, important cellular functions may be lost which can impair the normal biology of the cell and in some cases lead to oncogenic transformation. Because of these considerations or hurdles, development of safe, effective vectors has proved to be much more challenging than initially appreciated.

Introduction of the transgene into patients can be done either *in vivo* or *ex vivo*. A variety of routes have been employed for administration *in vivo*; intramuscular, intravenous, via the airway, intraocular (either subretinal or into the vitreous humor), etc. The alternative approach is to remove the target cells from the body, introduce the transgene, and return the cells to the body. This method has been used primarily with the bone marrow. A second consideration is whether the vector will preferentially persist as an extrachromosomal element or will integrate into the host genome, either randomly or at a specific location. If the target consists of cells which divide rarely, if at all, then persistence as an extrachromosomal element may be an advantage. However, if the cell is a stem or progenitor cell which will undergo a sizable number of divisions, the vector is likely to be diluted out if it is not integrated into the genome.

A number of types of vectors have been tried, ranging from purified DNA administered either IV or via a gene gun, to liposomes, to vectors derived from viruses. The latter have been a favorite because viruses have evolved in nature to function by delivering and expressing genetic material within cells. All DNA viruses which replicate in the nucleus (including retroviruses) cause persistent infections which usually last for the life of the host. However, most of the viruses we know about also cause disease and thus the challenge is to engineer the virus so that it can function as a vector without causing disease. Interestingly, two of the viruses to be discussed below are not known to cause human disease. In any event, what has become clear is that to develop safe and effective vectors requires a detailed knowledge of the fundamental biology of the viruses to be used.

Despite the various challenges to be surmounted, gene therapy has developed to the point where there have been numerous clinical trials in which various levels of toxicity have been observed. In the most paradoxical example 12 children in France with severe combined immunodeficiency disease were cured by gene therapy; however, two of the children developed leukemia caused by the vector (happily, they have been successfully treated for this problem, as well). Clearly, much developmental work remains to be done, yet the promise is great. Several of the types of viral vectors which have been developed are described in the sections below.

There are several general considerations in the design of viral vectors. One is the host range or tissue targeting of the vector. A broad host range means that the vector can potentially be used to target many organs; however, if target specificity is desirable, a broad host range is not desirable. In other cases the desired target normally may not be infected by the vector. Thus, in many instances there have been attempts to modify the original host range. This has been achieved naturally by the use of different serotypes with differing host ranges, by modifying the genes encoding the capsid so that new epitopes are present on the surface of the virion, or, in some cases, by chemical linking of ligands to surface proteins. Another approach to altering host range is by creation of pseudotypes, enveloped viruses in which the envelope glycoprotein of a second virus is substituted for the normal constituent, for example, vesicular stomatitis glycoprotein has been substituted for a retrovirus envelope protein. Vector host range may also be affected by the alteration of the regulation of gene expression, so that it can occur or not in specific tissues.

Retrovirus Vectors

Retroviruses have long been considered to be good candidates to be developed as vectors for gene therapy. The viral capsid is surrounded by a lipoprotein envelope and contains two identical copies of a linear, single stranded RNA of plus polarity, in addition to at least two enzymes, reverse transcriptase and an integrase. As a consequence the viral life cycle passes through a double stranded DNA intermediate which integrates at a variety of sites in the cellular genome through the action of the virion integrase. Three different types of retroviruses have been used as vectors; the oncogenic retroviruses, lentiviruses, and spumaviruses. The oncogenic retroviruses have a wide host range (which can be extended by pseudotyping, in which the envelope proteins can be replaced by glycoproteins from other viruses), but can only integrate their genome in dividing cells after the nuclear membrane is removed before mitosis occurs. This property has limited their utility on the one hand; on the other, it has some desirable features because it limits the potential for horizontal spread since most cells in the body do not divide. Since the oncogenic retroviruses cause cancer, it is important that vector preparations not contain replication competent retrovirus (RCR). The viral genome contains three genes, *gag*, *pol*, and *env*. All of these are deleted from vector constructs; the only original sequences that are retained are the long terminal repeats (ltr), the packaging signal, and an ori. Vector production requires the normal gene products and the challenge is to make the probability of recombination to produce RCR very small so that quality checks can be passed. Since the viral genome integrates at a variety of sites, another potential concern is insertional mutagenesis in which either an essential gene is disrupted or a normally silent gene under tight regulation gains the ability to be expressed constitutively. The former seems relatively remote since most genes are present in at least two copies in a diploid genome. However, the latter could be potentially a one-hit phenomenon and lead to cancer. Indeed, this appears to have happened in a clinical trial, which otherwise was one of the great success stories of gene therapy. Twelve French infants with an X-linked form of severe combined immunodeficiency were treated with a vector derived from Moloney murine leukemia virus; nine of the children were 'cured' of this otherwise lethal disease. However, two children developed leukemia; in both cases the disease was clonal and the cells in each case contained the vector inserted next to the LMO gene. The conclusion was that the inserted vector had turned on gene expression, leading to the disease. Fortunately, it has been possible to successfully treat the children for this additional problem. Interestingly, the French government decided to continue the trial since the disease was otherwise lethal; that is, the benefit outweighed the proven risk.

Lentivirus vectors are derived from human immunodeficiency virus (HIV) and, thus, there is the real and psychological challenge of using a vector based upon a dangerous human pathogen. In addition to the three genes contained in the genomes of all retroviruses, the lentivirus genome contains six additional genes which are the products of splicing and function in a variety of regulatory roles. Again the approach has been to generate vectors from which all the viral genes have been deleted in producer cells where the helper genes required have a minimal amount of sequence homology so that there is very little chance of recombination to generate RCR. Lentivirus vectors are easier to produce in higher titers than the oncogenic retrovirus vectors and have the great potential advantage that they can successfully infect both dividing and nondividing cells and integrate the genome in both cases to establish persistence. Whether the psychological question can be successfully overcome remains to be seen. An alternative possibility is the development of spumavirus vectors. These have many of the same desirable features as lentivirus vectors without the problem of being derived from a serious pathogen. Spumavirus is a common human infection but has never been convincingly associated with human disease. They have been used successfully in a variety of animal models and indeed seem promising.

Adenovirus Vectors

Certain DNA viruses offer the potential to transduce both dividing and nondividing cells, with a much lower risk of integration-related carcinogenesis. Recombinant adenovirus (rAd) and recombinant adeno-associated virus (rAAV) have both been used in clinical gene therapy trials, as have recombinant herpes simplex virus (rHSV). Adenoviruses are nonenveloped, double stranded DNA viruses with a 36 kb genome that contains both early genes (encoding regulatory proteins expressed prior to DNA replication during a lytic life cycle) and late genes (encoding structural proteins expressed after DNA replication). Adenoviruses are present in over 60 serotypes in humans, and they commonly cause acute, self-limited infections of the respiratory and gastrointestinal tracts of humans. rAd vectors based on the group C adenoviruses, Ad2 and Ad5, were initially developed by deleting portions of the early genes E1a and E3 and inserting a therapeutic gene of interest. These first-generation rAd vectors are relatively easy to propagate and mediate robust short-term expression in a wide range of cell types, *in vitro* and *in vivo*. rAd vectors also trigger innate and adaptive immune responses in the host under most circumstances, which can limit both the duration of expression and the safety of *in vivo* gene therapy. An acute inflammatory response to a first-generation rAd vector resulted in the death of one patient with partial ornithine transcarbamylase (OTC) deficiency in a well-publicized incident. Nonetheless, clinical trials of rAd vectors continue in cancer patients and as recombinant vaccines.

Recognition of the immunogenicity of rAd vectors has led to the generation of later-generation versions of the vector, in which more viral coding sequences have been deleted. Second-generation vectors were produced by deletion of other early genes such as E2a and E4, in addition to the E1a and E3 deletions described above. The second-generation vectors mediate less expression of viral proteins, especially the late proteins, and trigger less adaptive immunity than first-generation vectors. The latest version of the rAd technology, known as the high-capacity or helper-dependent adenoviral (HD-Ad) vector, has a greater payload for therapeutic genes and expression elements, a lower risk of adaptive immune responses, and a longer duration of effect. HD-Ad vectors are currently beginning early-stage clinical trials.

Adeno-Associated Virus Vectors

Adeno-associated viruses (AAVs) are nonpathogenic, and include among others, human and primate parvoviruses, with a nonenveloped icosahedral capsid and a 4.7 kbp single stranded DNA genome. The genome contains genes required for replication (*rep*) and for the capsid components (*cap*). The use of internal promoters and alternate splicing allows for the production of a total of four Rep proteins (Rep78, Rep68, Rep52, and Rep40) and three capsid proteins (VP1, VP2, and VP3). These two genes are flanked by two palindromic inverted terminal repeats (ITRs), which contain all of the *cis*-acting elements required for replication and packing of the AAV genome.

AAVs are not adenoviruses, but were originally discovered as contaminants of adenovirus cultures. Wild-type AAV requires a helper virus (usually an adenovirus or herpesvirus) for efficient replication. In the absence of helper virus co-infection, AAV enters into the latent stage of its life cycle, either by maintaining itself as a stable episome or by integrating its DNA into the host genome, often into a specific site on human chromosome 19, the AAVS1 site. This site-specific integration is dependent upon the AAV Rep protein, which is deleted from rAAV vectors. Thus, the vectors have been found to persist primarily as episomal form, with a low frequency of non-site-specific integration. rAAV vectors are also capable of transducing nondividing cells and have been shown to persist long term *in vivo* in nondividing cells in animal models. One of the major limitations of the rAAV system is the small packaging capacity of the virion, with a payload of approximately 4.5 kb of exogenous DNA. This may be overcome in certain instances by taking advantage of the propensity of rAAV vector genomes to form hetero-multimers, thus allowing for *trans*-splicing between two different vector genomes within a single target cell.

rAAV vectors have been used in clinical trials in cystic fibrosis, hemophilia B, limb-girdle muscular dystrophy, Parkinson's disease, Batten's disease, and alpha-1 antitrypsin (AAT) deficiency. Results of these early phase trials confirm the fact that rAAV is safe over a wide dose range when administered to the airways, muscle, and central nervous system. Delivery to the liver in patients with hemophilia B resulted in transient elevations of liver enzymes that appeared to be associated with cell-mediated immune responses to rAAV capsid proteins retained in hepatocytes. While these immune responses may have limited the duration of therapeutic effect of the vector, no clinically important toxicity was observed.

All of the previous clinical trials experienced with rAAV have been using the type strain, AAV serotype 2. In recent years, over 100 new serotypes and genomic variants of AAV have been identified. Many other serotypes have been developed as rAAV vectors and some have substantially higher efficiency for transduction of specific tissues than rAAV2. One trial of an alternative serotype has just been initiated in AAT deficient patients with AAV serotype 1 capsids. The preclinical rAAV1 experience demonstrates that switching to a different AAV serotype capsid can still retain the basic property of long-term persistence of vector DNA, yet capsid–receptor interactions may lead to a distinct pattern of vector biodistribution to distant organs and to increased innate immunity as the distinct capsids interact with pattern recognition receptor proteins. Altered biodistribution could be advantageous if it leads to increased transduction of a target cell population. However, it could also lead to an increased risk of inadvertent germ line transmission of vector DNA, a phenomenon that is generally to be avoided. The clinical utility of alternative rAAV serotype vectors may become clearer in the near future as more clinical trials with these vectors are undertaken.

Both rAd and rAAV vector capsids have also been genetically engineered to create receptor-targeted versions of their respective vectors. Essentially, this approach involves insertion of specific peptides that may serve as ligands for specific cell-surface receptors. These 'designer' vectors have the potential for highly efficient and specific cell targeting. One key to the use of targeting technology is the identification of sites within the capsid which can be modified without a loss of the integrity of the capsid. Another important factor is the size of the peptide insert that may be tolerated within the insertion site. In the case of rAAV, one of the three capsid proteins, VP2, can either be completely deleted from the vector without loss of infectivity (and with a modest increase in packaging capacity), or can be modified at its N-terminus to include very large inserts, including single-chain antibodies. This form of rationally designed vector appears to hold much promise for future vector development, as does another variant of rAAV capsid alteration, an *in vitro* evolution, or DNA shuffling approach. In these types of approaches libraries of rAAV capsid variants are developed by mutagenesis and/or recombination of domains between distinct AAV serotypes. Serial passage on relatively nonpermissive cell types allows for positive selection for novel AAV variants capable of transducing those cell types, thus further expanding the repertoire of AAV capsids available as vehicles for therapeutic gene transfer.

Herpesvirus Vectors

Human herpesviruses are large double stranded DNA viruses with a complex structure. An envelope is separated from the nucleocapsid by the tegument which is composed of a large number of proteins which subserve various functions during viral infection, including turning off cell protein synthesis and initiating viral transcription. The herpes simplex virus (HSV) genome is composed of two parts, a long unique sequence and a short unique sequence, both bounded by terminal repeats. There are three classes of cellular genes: (1) immediate early, which control gene expression; (2) early, involved in DNA replication; and (3) late, which encode structural proteins. HSV preferentially infects cells of the central nervous system; after initial entry through epithelial cells of the skin or mucosa the virions travel retrograde through axons to cell bodies, particularly in sensory ganglia, where the virus most commonly establishes a latent infection. The viral genome is maintained as an extrachromosomal circle. Because the neurons which are persistently infected do not divide, the latent state is maintained unless a variety of environmental stimuli which can stress the cell activate viral replication. HSV vectors of two types have been developed and most often are intended to be lytic and cidal for tumors of the brain. The first type of vector has had several viral genes which are normally associated with virulence removed (immediate early genes and the gene for the tegument protein which shuts off cell protein synthesis) and replaced with genes specifically toxic to tumor cells. The second type is termed an amplicon, which consists of a cassette containing the transgene with its regulatory sequences, the HSV origin of replication and the packaging signal. The former type has a transgene capacity of almost 30 kbp while the latter can accommodate almost the full 160 kbp in the full genome. Since the latter expresses no viral genes, toxicity associated with wild-type virus and an inflammatory immune response can be avoided (unless caused by the transgene). HSV vectors of the first type have been used in several clinical trials and in phase I trials have been found to be nontoxic and to indicate the possibility of beneficial effects on the target tumors, although the number of patients in completed trials has been too small for definitive results.

Clinical Trials

The future of clinical gene therapy may lie with the appropriate matching of vector properties with the properties of the cell target and the desired timing of transgene expression. Situations in which transient, high-level expression is needed may be best addressed with rAd or liposomal vectors. First-generation rAd vectors may be particularly well suited for recombinant vaccine approaches since the innate responses to Ad may serve as an adjuvant for adaptive immune responses to the transgene product. Longer-term expression in nondividing cells might be better addressed with lentivirus or rAAV vectors. Finally, long-term expression in rapidly dividing cells might be best addressed with onco-retrovirus or lentivirus vectors since episomal rAAV vector genomes may be lost in such cells. The choice of vector and route of administration is often dictated by the disease process itself. While many applications of gene therapy have been for recessive genetic disorders, which usually require long-term gene expression within a particular organ, others have aimed to induce immune responses to malignant tumors, to stimulate a new vascular supply to ischemic limbs or segments of myocardium, to augment wound healing, or to relieve inflammatory diseases, such as rheumatoid arthritis. Each of these examples would require a different route of administration, as well as different timing and level of gene expression in order to achieve the desired therapeutic effect.

As more gene therapy vectors are developed into clinical products for specific diseases, a regulatory pathway to the clinic has emerged in USA and in other industrialized countries. The paradigms established for biological agents by the US Food and Drug Administration (FDA) provide the model for such regulatory approval. The pathway begins with proof-of-principle studies in which an appropriate vector is chosen to transfer the therapeutic gene into a cell culture or animal model. These studies generally address the efficiency and duration of gene expression, the biological effects of the transgene product, and functional correction of the defect. The gene transfer approach used in the proof-of-principle study generally mimics the ultimate clinical use of the vector, in terms of whether the vector is injected directly *in vivo* into a target organ, or is used for *ex vivo* transduction of a cell population that is amenable to reimplantation.

If the proof-of-principle studies demonstrate a potentially therapeutic effect, additional preclinical safety testing is usually undertaken in animals. These studies often are done in a larger number of animals, so that important potential toxicities can be understood and their frequency predicted. In some cases, both small animal (usually rodent) models and larger animal (occasionally nonhuman primate) models will be used. Formal safety studies should cover a fairly broad dose range and should include evaluations at both early and late time points. Safety studies in gene therapy have special considerations, including studies of the biodistribution of vector genomes to distant organs, particularly to germ cells within the gonads, studies of the immune responses to vector components and transgene products, and possibly a survey for vector-related tumorigenesis. The complete toxicology package should address potential toxicity of the vector over a specific dose range, and thus guide both the dosing and the safety studies within the early-phase clinical trials.

Traditionally, clinical trials have included phase I safety studies, phase II studies that focus on safety and biological effects, and phase III studies to prove efficacy and document lower frequency toxicities. In recent years, there has been a growing emphasis on phase IV studies, which occur after licensure of a drug or biological product for distribution and sale. In addition, there is a new concept of phase 0 studies in gene transfer, which are used as an experimental context to learn about properties of a given vector or vector class, without necessarily being part of the regulatory process for a specific gene therapy product. Once again, clinical gene therapy studies have certain special considerations, including sampling of semen (as an indicator of potential for inadvertent germ line transmission in males), studies of the immune responses to the vector and transgene product, and long-term follow-up of gene therapy study participants. The latter is designed to monitor the long-term effects of vector-mediated mutagenesis, that might result in secondary malignant changes in target cells at a later time. There are also special regulatory processes for gene therapy trials, particularly those that receive federal funding in USA. These include the involvement of the Office of Biotechnology Activities at the National Institutes of Health, and its Recombinant DNA Advisory Committee, as well as the use of Data and Safety Monitoring Boards by sponsoring NIH institutes for ongoing monitoring of clinical trials.

Given the fairly broad range of available vectors and the rapidly developing systems for clinical testing of vector candidates, it seems likely that a growing number of gene therapy agents will reach clinical application in the coming years. However, the long-term impact of gene therapy will likely take much longer to be fully realized. It remains to be seen just how broad that impact will be.

Further Reading

Baum C, Schambach A, Bohne J, and Galla M (2006) Retrovirus vectors: Toward the plentivirus? *Molecular Therapy* 13: 1050–1063.

Epstein AL, Marconi P, Argnani R, and Manservigi R (2005) HSV-1-derived recombinant and amplicon vectors for gene transfer and gene therapy. *Current Gene Therapy* 5: 445–458.

Ghosh SS, Gopinath P, and Ramesh A (2006) Adenoviral vectors: A promising tool for gene therapy. *Applied Biochemistry and Biotechnology* 133: 9–29.

Hibbitt OC and Wade-Martins R (2006) Delivery of large genomic DNA inserts >100 kb using HSV-1 amplicons. *Current Gene Therapy* 6: 325–336.

Loewen N and Poeschla EM (2005) Lentiviral vectors. *Advances in Biochemical Engineering Biotechnology* 99: 169–191.

Mergia A and Heinkelein M (2003) Foamy virus vectors. *Current Topics in Microbiology and Immunology* 277: 131–159.

Morris KV and Rossi JJ (2006) Lentiviral-mediated delivery of siRNAs for antiviral therapy. *Gene Therapy* 13: 553–558.

Schambach A, Galla M, Modlich U, *et al.* (2006) Lentiviral vectors pseudotyped with murine ecotropic envelope: Increased biosafety and convenience in preclinical research. *Experimental Hematology* 34: 588–592.

Shen Y and Nemunaitis J (2006) Herpes simplex virus 1 (HSV-1) for cancer treatment. *Cancer Gene Therapy* 13: 975–992.

Wu Z, Asokan A, and Samulski RJ (2006) Adeno-associated virus serotypes: Vector toolkit for human gene therapy. *Molecular Therapy* 14: 316–327.

VIRUS INFECTION

Viral Receptors

D J Evans, University of Warwick, Coventry, UK

Introduction

The majority of viruses have an extracellular phase in the life cycle, which is a necessary part of virus transmission and dissemination. To initiate the replication cycle for the production of progeny virions, the virus must first enter a cell. Cell entry is mediated by the specific interaction of the virus with molecules on the cell surface – the receptor – resulting in attachment, a process that leads to internalization of either part or all of the virus particle, or the genome. A cell that expresses the cognate cell-surface receptor for a virus is termed susceptible. Not all susceptible cells to which the virus binds are capable of supporting infection and replication, perhaps due to the absence of necessary intracellular components. Cells that support virus binding, entry, replication, and progeny virion release are termed permissive.

With certain notable exceptions, such as viruses of fungi that exhibit no extracellular stage in the life cycle or plant viruses introduced mechanically by insect vectors, the virus receptor therefore forms the primary determinant of virus tropism. The identification and characterization of the receptor therefore provides important insights into the early stages of the virus replication cycle, and has been a major research focus of molecular virologists worldwide.

A Definition of Virus Receptor and Co-Receptors

Our enhanced understanding of virus–receptor interactions and the early events in cell entry have demonstrated that this process is, in many cases, considerably more complicated than originally thought, involving multiple cell-surface or intracellular molecules with different roles in the entry events. A classical definition of a virus receptor would be the cell-surface molecule that mediates virus attachment and cell entry. Our current understanding of these processes demonstrates that many viruses use different molecules to mediate attachment and post-attachment entry events. For these reasons the term co-receptor is often used to indicate an accessory molecule implicated in cell entry. Both receptors and co-receptors will be considered within the scope of this article. Rather than aim to provide a complete catalog of viruses and receptors, the focus of this article is to use key examples to illustrate specific types of virus–receptor interactions and introduce the general concepts involved in receptor identification and characterization.

The Identification of Virus Receptors

Since a working definition of a receptor is a specific cell-surface molecule involved in virus attachment, the identification of the receptor requires a means of detecting virus binding, and a requirement to demonstrate the specificity of the process.

Direct detection of a virus bound to the cell requires some form of labeling – whether metabolic labeling with a radioisotope or labeling with a conjugated fluorescent marker – and subsequent purification of the virus particle, often by density gradient centrifugation. Indirect methods of detecting virus can also be used, for example, the hemagglutination phenotype demonstrated by sialic acid binding viruses such as influenza A virus, or by detection of a bound virus with antibodies to the virus particle.

The specificity of the observed binding event typically requires the inhibition of the process using a defined block. The routine availability of monoclonal antibodies has hugely facilitated this approach to receptor identification, though alternate methods such as the use of enzymatic pretreatment of cells with heparinase or neuraminidase (NA) have also been used to demonstrate interactions with nonprotein cell-surface molecules.

Finally, in cases where the receptor is a single protein molecule, the specificity of the virus interaction can be further demonstrated by transformation of a nonsusceptible cell with DNA encoding the protein. Following expression at the cell surface, the acquisition of a virus binding phenotype is tested, perhaps combined with the demonstration of specificity by the inhibition with monoclonal antibodies. For example, the molecular cloning of the poliovirus receptor (PVR) involved expression of the protein in murine fibroblasts, which conferred susceptibility to an already permissive cell type.

Typically, a combination of the approaches outlined above is necessary to unequivocally demonstrate the identity of a virus receptor. Multiple confirmatory approaches are required as each technique has inherent uncertainties. For example, monoclonal antibodies may cross-react with closely related membrane proteins, or transformation of DNA may induce receptor expression, rather than encode the receptor *per se*. Similar techniques, and the associated caveats, are subsequently used in the initial stages of characterization of the virus–receptor interaction, for example, the receptor protein domain(s) to which the virus attaches, or the cell or tissue specificity of attachment.

Specific Examples of Virus Receptors

The examples used below illustrate a range of different types of virus–receptor interactions – those involving a single type of target molecule on the cell surface, either proteins or carbohydrate moieties, and those in which multiple receptors and co-receptors have been implicated. Due to their pivotal role in determining tissue and host tropism, and hence influence on pathogenesis, particular effort has been made in identifying receptors for important human and animal virus pathogens. However, the broad concepts illustrated below apply equally to the cell-surface proteins of bacteria used by bacteriophages – where the exquisite specificity of the interaction is used in classification and typing of *Salmonella*.

Sialic Acid

Influenza A is an enveloped virus that causes epidemic and pandemic respiratory tract infections in humans, but is naturally an enteric pathogen of aquatic birds. Both of the external virus glycoproteins are involved in interaction with the cellular receptor, sialic acid, the first identified receptor for a human virus. The trimeric hemagglutinin (HA) molecule, consisting of a stalk and globular head, carries a conserved cleft or pit at the membrane-distal end into which the sialic acid binds. Sialic acids are a family of negatively charged monosaccharides derived from neuraminic acid; those used as virus receptors are usually attached to a penultimate galactose by either an (α2,3)- or (α2,6)-linkage. The structure of an HA–sialyllactose complex has been determined at atomic resolution and provides insight into the specificity of binding, which in human strains involves (α2,6)-linked sialic acid in contrast to the binding of avian strains which exclusively interact with (α2,3)-linked sialic acid. Interaction of influenza with sialic acid is a low-affinity process, attachment being facilitated by multimeric binding to many HA molecules. The tetrameric NA virus glycoprotein has enzymatic activity that hydrolyses the glycosidic linkage of the receptor sialic acid, so enabling the release of progeny virus from the cell surface. NA-inhibitors have been developed as successful antiviral therapies.

Although influenza A is the best-characterized sialic acid-binding virus, there are many other examples, including human parainfluenza virus (hPIV) types 1–3 (and the murine hPIV analog Sendai virus), picornaviruses (rhinovirus type 67, enterovirus type 70), polyomaviruses (JC and BK virus), type 3 reoviruses, group C rotaviruses and adeno-associated virus. As with influenza A virus, predominant bias to either (α2,6)- or (α2,3)-linked sialic acid has been detected in each of these examples.

Heparan Sulfate and Other Glycosaminoglycans

Glycosaminoglycans (GAGs) (including heparan, chondroitin, and dermatan sulfate) are negatively charged sulfated sugars which are widely implicated in attachment of a range of viruses as diverse as herpes simplex virus (HSV), enteroviruses, flaviviruses, and respiratory syncytial virus. However, despite the isolation of clinical virus isolates with GAG-binding capacity, definitive evidence that GAGs have a role in the cell infection process *in vivo* is yet to be obtained. The relevance of GAGs has been further questioned by the observation that foot and mouth disease virus (FMDV), a picornavirus that binds $\alpha_v\beta_3$ integrins *in vivo*, can readily adapt to use GAGs in cell culture – this process is discussed further below.

GAG binding is routinely demonstrated by inhibiting virus binding to the cell surface, or virus infection, with soluble analogs such as heparin sulfate, by heparinase pretreatment of cells, or by analyzing binding to cells deficient in GAG synthesis. The interaction of FMDV and heparin has also been determined crystallographically and shown to involve a region of the capsid distinct from the surface-exposed loop that binds the *in vivo* receptor, $\alpha_v\beta_3$ integrin. GAG-binding domains are usually linear or conformational regions of the virus surface that are rich in basic residues, and adaptation to GAG binding often involves the acquisition of basic (lysine or arginine) amino acid substitutions.

Most viruses shown to bind GAGs are also known to interact with at least one additional cell surface receptor. Glycoprotein C (gC) of HSV-1 and HSV-2 binds HS, and the gD interaction with certain forms of GAGs appears necessary for postbinding prefusion events in cell infection. However, gD also interacts with HVEM (herpes virus entry mediator, a member of the tumor necrosis factor receptor family) and nectins 1 and 2, suggesting that HSV may either exhibit considerable redundancy in cell-surface interactions or, interestingly, that the virus carries a number of binding activities tailored to particular cell or tissue types. Similarly, GAG-binding echoviruses appear to also require the primary receptor for entry, decay accelerating factor (DAF) (Williams and Evans, unpublished data).

The cell–receptor interactions of human immunodeficiency virus (HIV) have received particular attention. In addition to the primary receptor, CD4, and the chemokine co-receptors (see below), syndecans with GAG chains have been implicated in macrophage infection by HIV. Similarly, infection of CD4-negative microvascular endothelial cells, a probable route via which the virus gains access to the central nervous system (CNS), has been shown to be a lipid raft-independent process involving cell-surface proteoglycans. Again, as with HSV, the complexity of interaction with multiple cell receptors, and the *in vivo* function, is yet to be satisfactorily dissected.

Histo-Group Antigens as Receptors

The wide range of nonprotein molecules used for attachment by viruses also includes the histo-blood group antigens (HBGAs), which serve as receptors for the noroviruses, positive strand RNA viruses that cause acute gastroenteritis. HBGAs are complex carbohydrates present both as free oligosaccharides and attached to the surface of mucosal epithelia. The HBGA phenotypes of humans are polymorphic and complex, and are associated with multiple gene families. This was consistent with the demonstration that norovirus infection was influenced by both prior acquired immunity and the genetic background – patients with blood group O were significantly more susceptible to infection than those with blood group B. Subsequent binding studies with norovirus virus-like particles (VLPs) – which retain the receptor-binding capacity of the virus which cannot be cultured *in vitro* – and specific antibodies to blood-group antigens, soluble blood-group antigens or glycosidases, have confirmed a role for these molecules in virus attachment.

Immunoglobulin Superfamily Proteins

Our understanding of the range of cell-surface proteins that function as virus receptors is influenced by the methods utilized to identify the protein. Since these methods are predominantly geared to isolating single-gene traits, for example, cDNA library screening, it is unsurprising that the majority of proteins definitively identified as receptors are single proteins, rather than one component of a multiprotein complex. Perhaps reflecting their relative abundance at the cell surface, proteins belonging to the immunoglobulin superfamily are implicated in the attachment of many viruses.

Poliovirus, the causative agent of paralytic poliomyelitis, binds the membrane-distal Ig-like domain of CD155, the PVR. Receptor binding initiates a series of irreversible conformational changes in the virus particle, which results in the externalization of capsid proteins usually hidden in the mature infectious particle, and the delivery of the virus genome to the cell cytoplasm. Data are accumulating to suggest that this process involves formation of a pore traversed by the genome, though other models exist. That PVR is sufficient to mediate these processes has been elegantly demonstrated by several methods; PVR expression in murine cells renders them fully permissive for poliovirus infection, and is the basis for the generation of PVR-transgenic mouse lines for neurovirulence studies, the conformational changes observed at the cell surface can be recapitulated using poliovirus and soluble PVR (sPVR) *in vitro*, and – most recently – the bound virus–receptor complex has been visualized by crystallography and, using cryo-EM, studied at the surface of PVR-loaded membrane vesicles.

Human rhinoviruses are very closely related to members of the enterovirus genus to which poliovirus belongs. The rhinoviruses can be divided into two predominant groups based upon their receptor usage; of these, the major receptor group bind to the membrane-distal Ig-like domain of intracellular adhesion molecule type 1 (ICAM-1). Like poliovirus, receptor binding initiates conformational changes in the rhinovirus particle that are considered essential for infection. A further similarity is the region of the capsid involved in the receptor interaction. In both poliovirus and the major group rhinoviruses, the receptor binds to a canyon or cleft surrounding the fivefold axis of icosahedral particle symmetry. Two decades ago, Rossmann proposed that this canyon is too narrow to allow the binding of virus neutralizing antibodies, thereby allowing the virus to retain a relatively invariant receptor-binding domain in the presence of immunoselection. More recent crystallographic studies have demonstrated that some neutralizing antibodies penetrate deep into the rhinovirus canyon. This suggests that the canyon has less to do with evasion of immune surveillance, but has instead evolved in response to receptor binding requirements and to accommodate the structural changes that occur post-binding.

The Ig-like protein that has received some of the most intense scrutiny is CD4, the primary receptor for HIV. The identification of CD4 as a receptor for HIV went some way to explain the tropism and pathogenesis of the virus; CD4+ T cells were often depleted in HIV-infected individuals. Antibodies to CD4 blocked virus attachment, fusion and infection, and expression of CD4 in murine cells permitted virus binding. Detailed analysis using monoclonal antibodies specific for each Ig-like domain of CD4, and subsequent site-directed mutagenesis of the receptor, demonstrated that the attachment protein of HIV (the virus surface glycoprotein gp120) binds to a surface-exposed loop of the most membrane-distal of the four Ig-like domains of CD4. Gp120 carries the conserved CD4 attachment site which is protected, though not completely hidden, from interaction with neutralizing antibodies by the diversity, conformational flexibility, and heterogeneous glycosylation of the surrounding protein. One characteristic exhibited by HIV-infected long-term nonprogressors is the presence of a potent neutralizing antibody response to the CD4 binding site – because the virus must retain a largely invariant receptor-binding domain, these antibodies are active against divergent virus isolates, and significant effort has been expended in generating vaccine candidates capable of inducing a similar immune response. Other therapeutic approaches based upon the conserved interaction of gp120 and CD4 have been investigated. Although soluble CD4 receptors, lacking the transmembrane domains, block HIV infection *in vitro*, they have been largely ineffective in clinical trials. These disappointing results have been attributed to the need to maintain high serum concentrations of a protein with a short half-life, and the ability of the virus to spread directly from cell to cell by fusion, rather than by extracellular virus. Similarly, approaches involving toxins fused to soluble CD4 that are used to target gp120 expressed at the surface of infected cells have only been promising in *in vitro* assays.

Complement Control Proteins as Virus Receptors

A number of viruses have been shown to bind to complement control proteins, also termed the regulators of complement activity (RCA) family proteins. These proteins, characterized by a two-disulfide protein fold termed a short consensus repeat (SCR), are involved in protecting autologous cells from complement-mediated lysis.

Epstein–Barr virus (EBV) attachment to B lymphocytes occurs by the interaction of the virus gp350/220 envelope glycoprotein with the cell-surface C3dg receptor designated CR2 (CD21). CR2 was identified as the receptor following the observed similarity between gp350/220 sequences and C3dg, the complement protein target for CR2 activity. Further supporting evidence included the demonstration that multimerized peptides of the proposed gp350/220 or C3dg binding domains inhibited attachment, and the receptor identity was confirmed by expression studies in CR2-deficient target cells. CR2 possesses 16 extracellular SCR domains, the two most membrane-distal being the site of gp350/220 binding, as demonstrated by the construction of murine-human CR2 chimeric receptors, and using SCR-specific monoclonal antibodies in blocking studies.

Laboratory-adapted measles virus binds another RCA family protein member, CD46, a four-SCR domain protein that acts as a cofactor in the proteolytic inactivation (and hence cell protection) of C3b/C4b. Antisera to CD46 blocks virus binding and infection, and transfection of rodent cells with CD46 renders them susceptible to infection by the Edmonston and Hallé strains of measles virus. Further studies demonstrated that clinical isolates of measles cultured in human B-cell lines exhibited a cell tropism inconsistent with the use of CD46 as the receptor – further details are provided in the section on laboratory adaptation.

DAF, a 55 kDa RCA-family protein, is the receptor for a significant number of human enteroviruses, including the coxsackie B viruses and many echoviruses. In contrast to many of the other examples presented here, where binding is to the membrane-distal domain of the receptor, the majority of the known DAF-binding enteroviruses interacts with the third SCR, together with the second and/or fourth SCR domains. Recent completion of the atomic structure of DAF shows it to be a rod-shaped protein, with the four SCR domains projecting in a broadly linear fashion from the top of a heavily glycosylated Ser/Thr-rich stalk, which distances the complement-control functions from the cell membrane. Unsurprisingly, considering the receptor structure, interaction of DAF with echoviruses does not involve the canyon on the virus surface, but instead occurs across the twofold axis of symmetry, with the receptor interface occupying a surface-exposed region of the capsid. In the case of the coxsackieviruses, DAF binding serves both to attach the virus to the cell surface and initiate a cascade of signaling events which results in the transport of the virus to the co-receptor for infection – CAR, the coxsackie and adenovirus receptor, which is located at tight junctions and not normally accessible to the virus.

Co-Receptors for Infection

The distinction between a receptor and a co-receptor for infection is largely semantic. As our understanding of virus binding and entry improves, it is clear that this is often a multistage process involving a cascade of interactions between the virus and cell-surface proteins. The initial interaction is usually defined as being to the receptor, with subsequent interactions being mediated by co-receptors. This order can be reversed if the first interaction is a relatively nonspecific event – for example, the binding of HSV to HS – in which case co-receptor binding may be considered to temporally precede interaction with a more specific molecule that perhaps defines the cell or tissue tropism of the virus. Therefore, the discrimination between nonspecific attachment molecules, receptors, and co-receptors – in which all are sometimes required for infection – is open to interpretation. What is clearly important for many viruses is the requirement for a range of interactions between the virus and the cell surface. Indeed, as our understanding of virus infection improves, the number of examples in which a single receptor is implicated – such as poliovirus – is dwindling, and may represent the exception, rather than the rule.

Although the coxsackie B viruses bind DAF at the cell surface, infection also requires CAR, an Ig-superfamily protein with two extracellular domains, which interacts with the canyon of coxsackie B viruses or the fiber of adenovirus types 2 and 5. Like the poliovirus–PVR interaction, binding of coxsackie B viruses to CAR – at the cell surface or in solution – triggers the irreversible conformational changes associated with particle uncoating. However, CAR is naturally located in tight junctions where it mediates homotypic cell adhesion, and is therefore not immediately accessible for virus attachment at the apical cell surface. Recent elegant studies have demonstrated that virus binding to DAF at the latter site results in Abl-kinase mediated signaling events which cause actin rearrangements that allow virus movement to the tight junctions where uncoating occurs.

DAF binding also activates Fyn-kinase, which results in the phosphorylation of caveolin and subsequent virus entry in caveolin vesicles. The coxsackie B viruses therefore exploit the signaling capacity of their low-affinity receptor (DAF) to traffic to the high-affinity, uncoating receptor (CAR) and thereby cross the epithelial barrier.

Like poliovirus and rhinovirus, binding of HIV to CD4 initiates conformational changes in the trimeric gp120 that are a prerequisite for the subsequent fusion events. However, unlike the two picornavirus examples, HIV must also interact with a co-receptor as a necessary precursor to the fusion of viral and cellular membranes. This requirement for a co-receptor explains why HIV can bind to human CD4 expressed on murine cells – which lack the co-receptor – without undergoing membrane fusion. A mouse-CD4 cell line was transformed with a human cDNA library and used to functionally screen for additional proteins which would enable gp120-mediated membrane fusion. This, or analogous, screens were used to define the co-receptor for HIV as one of a related series of receptors for chemotactic cytokines such as RANTES, MIP-1α and MIP-1β. The chemokine receptors identified, CXCR4 or CCR5, belong to the seven-transmembrane-domain G-protein-coupled receptor superfamily. Co-receptor specificity appears to be determined by the third hypervariable domain (V3) of gp120 – a target for potently neutralizing antibodies *in vitro* – with an additional conserved domain of gp120 contributing to generic features of the chemokine receptor attachment. The critical conserved residues of the latter domain remain hidden from immune surveillance until exposed by the structural changes in gp120 that occur as a consequence of CD4 binding. The importance of the chemokine co-receptor in HIV infection is supported by the observed genetic resistance to HIV of individuals with a deletion and frameshift within CCR5. Similarly, co-receptor identification provided an explanation for the observed suppression of HIV infection by the chemotactic cytokines released by CD8+ T cells. Finally, the identification of chemokine receptors as critical co-receptors of HIV infection has also contributed to our understanding of virus tropism and evolution *in vivo*. CCR5-tropic strains of HIV are predominant for transmission and during the asymptomatic phase of virus infection. CCR5 is expressed at high levels in activated/memory cells and in gut-associated lymphoid tissue, the primary site of HIV replication. As infection progresses, CXCR4-tropic viruses evolve, able to exploit the wider tissue expression of the CXCR4 receptor. The, perhaps logical, assumption that this broadening of cell and tissue tropism is the cause of disease progression is contradicted by the failure to observe a switch to CXCR4-tropic viruses in a significant proportion of patients with acquired immune deficiency syndrome (AIDS).

Laboratory Adaptation and Receptor Usage by Clinical Virus Isolates

Due to the possession of error-prone replication strategies, many viruses exhibit an exquisite ability to adapt to the environment in which they are grown – for example, by the generation of mutations that confer resistance to neutralizing antibodies. The same variation can confound the correct identification of virus receptors in laboratory studies, due to the adaptation to alternate receptors for cell attachment and entry.

The example of heparan sulfate binding by FMDV has already been used to illustrate this process. HS-adapted FMDV isolates are apathogenic in animals, presumably reflecting the evolutionary adaptability of the virus to an environment in which receptor usage is not a rate-limiting process in infection, transmission, or dissemination. The virus interacts with the *in vivo* receptor ($\alpha_v\beta_3$ integrin) using a conserved Arg-Gly-Asp amino acid triplet (RGD motif) on a surface-exposed loop of the VP1 capsid protein. Laboratory-adapted HS-binding FMDV strains can dispense with $\alpha_v\beta_3$ integrin binding altogether, as demonstrated by the use of reverse genetics to mutate the RGD motif to another sequence. Being acid-labile, it is likely that uncoating of FMDV is mediated by entry of the virus–receptor complex into an endosome. This is supported by the further demonstration that it is possible to engineer anti-FMDV antibodies onto the cell surface to which the virus is bound, and via which infection can occur in the absence of either HS or integrin binding.

CD46, like most other complement-control proteins, is expressed ubiquitously on serum-exposed cells. Clinical isolates of measles virus, isolated in either B95a cells or human B-cell lines, do not infect all CD46-expressing cell lines, suggesting that they may use an alternate receptor. Subsequent functional expression cloning studies resulted in the isolation of a cDNA encoding human SLAM (signaling lymphocyte-activation molecule; also known as CDW150) that conferred susceptibility to B95a-cell cultured measles virus. SLAM is an Ig-superfamily protein expressed on a range of immune system cells (macrophages, dendritic cells, some B and T cells) that, when liganded to SLAM on an adjacent cell, transduces a signal resulting in T_H2 cytokine (IL13 and 4) production. All measles virus isolates tested can infect cells via SLAM, though some – including the lab-adapted Edmonston strain – can also use CD46. However, studies suggest that CD46 binding is irrelevant *in vivo* and arises upon culture by the substitution of as few as a single amino acid in the HA protein of the virus.

Summary

Virus receptors are key components of the early events involved in cell infection. The examples outlined above should illustrate that the definition of receptors acting solely as attachment molecules is overly simplistic. In addition to attachment, receptors also actively contribute to entry by initiating conformational changes in the virus that lead to uncoating. In addition, they provide mechanisms for internalization – which increasingly appear to involve signaling events that occur upon receptor binding – in which the virus may subvert a natural receptor cycling process. Furthermore, the identification of virus receptors contributes significantly to our understanding of host, tissue, and cell tropism, and helps explain aspects of virus pathogenesis. The selected bibliography provides further examples illustrating aspects of virus receptor identification and function.

See also: Cytokines and Chemokines; Viral Pathogenesis.

Further Reading

Belnap DM, McDermott BMJ, Filman DJ, *et al.* (2000) Three-dimensional structure of poliovirus receptor bound to poliovirus. *Proceedings of the National Academy of Sciences, USA* 97: 73–78.

Bergelson JM, Cunningham JA, Droguett G, *et al.* (1997) Isolation of a common receptor for coxsackie B viruses and adenoviruses 2 and 5. *Science* 275: 1320–1323.

Coyne CB and Bergelson JM (2006) Virus-induced Abl and Fyn kinase signals permit coxsackievirus entry through epithelial tight junctions. *Cell* 124: 119–131.

Evans D and Almond J (1998) Cell receptors for picornaviruses as determinants of cell tropism and pathogenesis. *Trends in Microbiology* 6: 198–202.

Hogle JM (2002) Poliovirus cell entry: Common structural themes in viral cell entry pathways. *Annual Review of Microbiology* 56: 677–702.

Lindahl G, Sjobring U, and Johnsson E (2000) Human complement regulators: A major target for pathogenic microorganisms. *Current Opinion in Immunology* 12: 44–51.

Lusso P (2006) HIV and the chemokine system: 10 years later. *EMBO Journal* 25: 447–456.

Maddon PJ, Dalgleish AG, McDougal JS, Clapham PR, Weiss RA, and Axel R (1986) The T4 gene encodes the AIDS virus receptor and is expressed in the immune system and the brain. *Cell* 47: 333–348.

Mendelsohn CL, Wimmer E, and Racaniello VR (1989) Cellular receptor for poliovirus: Molecular cloning, nucleotide sequence, and expression of a new member of the immunoglobulin superfamily. *Cell* 56: 855–865.

Olofsson S and Bergstrom T (2005) Glycoconjugate glycans as viral receptors. *Annals of Medicine* 37: 154–172.

Pettigrew DM, Williams DT, Kerrigan D, Evans DJ, Lea SM, and Bhella D (2006) Structural and functional insights into the interaction of echoviruses and decay-accelerating factor. *Journal of Biological Chemistry* 281: 5169–5177.

Rossmann MG (1989) The canyon hypothesis. Hiding the host cell receptor attachment site on a viral surface from immune surveillance. *Journal of Biological Chemistry* 264: 14587–14590.

Tan M and Jiang X (2005) Norovirus and its histo-blood group antigen receptors: An answer to a historical puzzle. *Trends in Microbiology* 13: 285–293.

Ward T, Pipkin PA, Clarkson NA, Stone DM, Minor PD, and Almond JW (1994) Decay accelerating factor (CD55) identified as the receptor for echovirus 7 using CELICS, a rapid immuno-focal cloning method. *EMBO Journal* 13: 5070–5074.

Yanagi Y, Takeda M, and Ohno S (2006) Measles virus: Cellular receptors, tropism and pathogenesis. *Journal of General Virology* 87: 2767–2779.

Virus Entry to Bacterial Cells

M M Poranen and A Domanska, University of Helsinki, Helsinki, Finland

Glossary

Adsorption Initial interaction between a virus particle and a cellular receptor molecule.

Bacteriophage (or phage) A virus that infects bacteria.

Capsid The protective protein coat of a virus particle.

Cell envelope Plasma membrane and cellular structures located outside the plasma membrane.

Cell wall Cellular structures located outside the plasma membrane.

Lipopolysaccharide A unique glycolipid of the outer membrane of Gram-negative bacteria.

Peptidoglycan A polymer consisting of long glycan chains cross-linked via peptide bridges and forming homogenous layer outside the plasma membrane of eubacteria component of the cell wall.

Receptor A specific molecule or molecular assembly exposed on the surface of a cell to which a virus entering the cell attaches.

Receptor-binding protein A virion protein responsible for the interaction of a virion with a specific cellular receptor molecule.

Vertex Fivefold symmetry position of the icosahedra; one icosahedral particle has 12 fivefold symmetry positions.

Viral envelope An outer lipid–protein bilayer of a virus.

Virion A virus particle, the extracellular form of a virus.

Introduction

Viruses are intracellular parasites that are dependent on the metabolic apparatus of the cell. Unlike other parasitic self-replicating systems, like plasmids and viroids, viruses possess an extracellular phase that allows spread from one infected cell or organism to another. Consequently, viruses have to have means to infect new host cells. The entry into a suitable host cell is a key event for the viral reproduction and survival.

Host Cell Barriers

The nature of the host cell wall has a great influence on the viral entry strategy. Gram-positive bacteria have a single internal lipid bilayer and a thick cell wall made of peptidoglycan while Gram-negative cells are covered by an internal membrane, a thin layer of peptidoglycan, and an outer membrane. In addition, bacterial cells may secrete polysaccharides that make a protective extracellular capsule on the surface of the cell.

The relatively strong and inert cell wall of eubacteria efficiently restricts the passage of macromolecules. In addition, bacterial cells do not have endocytic-like uptake systems, which are commonly utilized by eukaryotic viruses to gain access into the host cell. These features of the host strongly influence the mechanism employed by bacterial viruses to gain access into the host cytoplasm. In fact, the capsids of most bacteriophages are never internalized into the host cell; only the viral genome with some necessary protein factors is delivered across the host envelope.

Virions as a Genome Delivery Devise

Virions represent an extracellular form of the virus. It is a vehicle, which allows the virus to resist the harsh environment outside the cell. In addition to its protective nature, the main task of the virion is to recognize the host and to deliver the viral genome with necessary accessory factors to the new host cell. The mechanism of the genome delivery is typically reflected in the structure of the capsid (see below).

The nature of the viral genome influences the mechanisms of virus entry. Viruses that have genomes which cannot be expressed using the enzymatic apparatuses of the host, need to also bring viral polymerases into the host cell. This applies to all viruses having dsRNA or negative-sense ssRNA genomes. Regardless of the type of the genome, many viruses deliver some accessory protein factors inside the host. These are required in the early stage of the infection, prior to the viral genome expression, either to complete the entry process or for successful genome replication and expression.

Host Recognition

Host cell recognition by viruses is a highly specific process. Basically, all bacteriophages have, on their exterior surface, some protein that binds to a receptor molecule exposed on the surface of a susceptible cell. The receptor-binding proteins are often localized in the vertices of the icosahedral virions or the tips of the helical virions. In tailed bacteriophages the initial recognition is carried out by the fibers that are connected to the distal end of the tail (**Figure 1(a)**). The specific recognition leads to irreversible structural rearrangements in the virion components; the viral receptor-binding complexes overcome an energy barrier and fold into a minimal energy state. These conformational changes lead to more tight attachment and eventually trigger the entry process. The rigid structure of the virion is destabilized so that the genome delivery can be accomplished. The icosahedral capsids of bacteriophages seldom disassemble completely; only the structures at the vertices become labile making openings for genome release.

Many bacteriophages, including icosahedral ssRNA bacteriophages (e.g., MS2, Qβ), filamentous ssDNA bacteriophages (e.g., Ff, as M13, fd, and fl), and enveloped dsRNA phage (ϕ6), utilize a bacterial pilus as their primary receptor. The pilus enables efficient capture of the virion at a distance from the cell surface, and the retraction of the pilus translocates the bound phages to the host envelope (**Figure 1(b)**). This allows phages to get access to the cell surface regardless of the polysaccharide capsule, which could restrict the easy access to the cell surface. Some phages (e.g., K1-5, K5, K1E), however, may also use the polysaccharide capsule as the initial site of recognition and binding. Typically, these phages have enzymatic activities associated with the virion for capsule degradation.

Other receptor sites for bacteriophages are lipopolysaccharides, various cell envelope and flagellar proteins, as well as cell-wall carbohydrates. The tailed phages infecting Gram-negative bacteria use either the lipopolysaccharide moieties (e.g., T2, T4, T7) or envelope proteins, such as porins and transporters (e.g., PP01, T1, T5, λ) as their receptors, while phages infecting Gram-positive bacteria typically attach to the cell-wall teichoic acids. Membrane proteins or peptidoglycan moieties rarely are used as attachment sites for bacteriophages infecting Gram-positive bacteria.

Cell-Wall Penetration

The virions of many bacteriophages contain peptidoglycan-hydrolyzing enzymes. These are specialized proteins that locally and temporarily disrupt the peptidoglycan network, thus allowing the penetration of the cell wall.

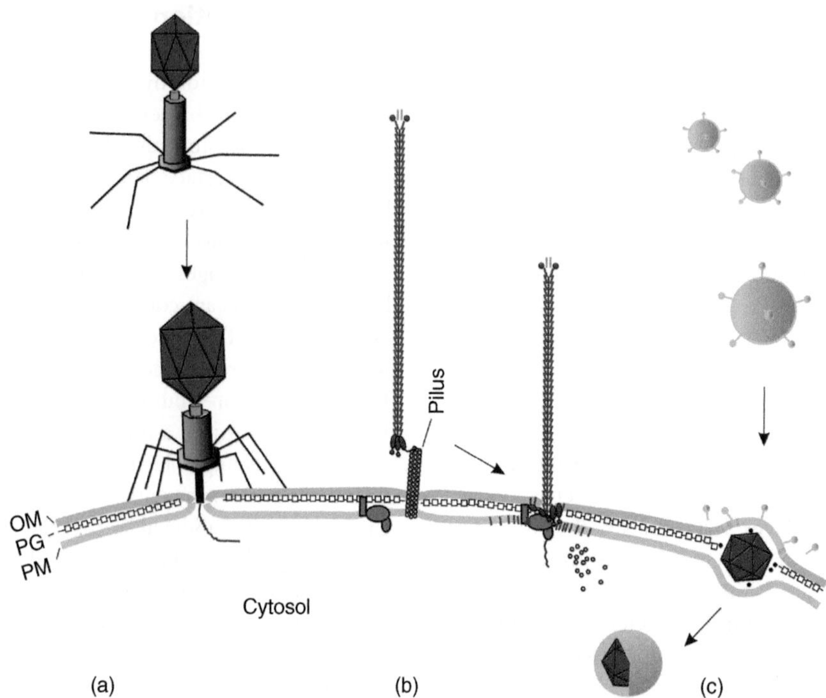

Figure 1 Schematic presentation of the main entry strategies utilized by bacterial viruses. OM, outer membrane; PG, peptidoglycan; PM, plasma membrane. (a) Genome delivery through an icosahedral vertex; given example is a phage with long-contractile tail from the family *Myoviridae* (e.g., T4). Primary interaction between the phage and the host cell is mediated by the tail fibers (black). Contraction of the tail sheath (blue) facilitates the penetration of the cell envelope. The viral genome (purple) is released from the head (red) of the phage virion through the tail tube (black) into the host cytosol. The protein capsid remains outside the cell. (b) Dissociation of filamentous phage capsid at the cell envelope. The receptor binding protein located at the tip of the helical virion interacts with the bacterial pilus. Pilus retraction brings the virion on the cell surface enabling interaction with the co-receptor molecule (green). The viral genome (purple) is released into the host cytosol as the virion proteins are inserted into the plasma membrane. (c) Penetration of the nucleoprotein complex of an enveloped dsRNA virion. The viral spike proteins (green) on the virion surface mediate the interaction between the host and the virion. Fusion between the host outer membrane and phage envelope (blue) takes place leading to the mixing of host and viral lipids. The nucleoprotein assembly (nucleocapsid; red) released into the bacterial periplasm penetrates the peptidoglycan network with the aid of a lytic enzyme (black) located on the nucleocapsid surface. Subsequently, the host plasma membrane is penetrated via an endocytic-like process.

Similar enzymes are also involved in the release of progeny phages from the host cell at the end of the infection cycle. However, when assembled into virions, they likely have an active and crucial role in the entry process.

The lytic activity of virion-associated peptidoglycan-hydrolyzing enzymes is stringently controlled to ensure localized openings in the cell wall. This allows maintaining cell integrity, which is essential for successful production of progeny viruses during the infection cycle. For this reason such enzymes are tightly incorporated into the virus particle. The peptidoglycan-hydrolyzing enzymes of tailed bacterial viruses are often associated with the tail structures or, in the case of some short-tailed phages (e.g., T7), with internal head proteins that are ejected at the beginning of the infection cycle. Icosahedral dsDNA phages with internal membrane and no tail (e.g., PRD1, Bam35) as well as the enveloped dsRNA phage ϕ6 also possess peptidoglycan-hydrolyzing enzymes in their virions (see below; **Figure 1(c)**).

The most common cell-wall-degrading enzymes found in bacteriophage virions are lysozymes and lytic transglycosylases. Both lysozymes and lytic transglycosylases cleave the same glycosidic bond between *N*-acetylmuramic acid and *N*-acetylglucosamine units of the glycan strands. The end product of lysozyme is a disaccharide with reducing end of *N*-acetylmuramic acid, whereas that of lytic transglycosylase is a disaccharide containing a ring structure (anhydromuramic acid).

Not all phages have specialized enzymes in their virions for the penetration of the host cell wall during entry. These phages rely on preexisting channels within the host envelope. This applies to bacterial viruses with ssDNA and ssRNA genomes. For example, infection of filamentous ssDNA phages absolutely depends on a protein complex assembled in the bacterial plasma membrane spanning the peptidoglycan layer (**Figure1(b)**).

Genome Delivery Mechanisms of Phages

In general, entry strategies of bacterial viruses fall into three main categories: (1) genome delivery through an icosahedral vertex, (2) virion dissociation at the cell envelope, and (3) virion penetration via membrane fusion and endocytic-like event (**Figure 1**). All known tailed dsDNA phages (e.g., T4, T5, T7), dsDNA phages with an internal membrane (tectiviruses), as well as icosahedral ssDNA and ssRNA phages deliver their genome through a genome delivery apparatus located at one of the capsid vertices (**Figure 1(a)**). Capsids of filamentous phages and membrane-containing phage PM2 disassemble completely at the cell envelope (**Figure 1(b)**), whereas the enveloped dsRNA bacterial viruses utilize a unique membrane fusion-type uncoating at the outer membrane and virus subparticle internalization through the plasma membrane (**Figure 1(c)**).

Icosahedral Tailed dsDNA Bacterial Viruses

The tailed dsDNA bacteriophages use the tail structure as a genome delivery devise (**Figure 1(a)**). The tail is a protein tube of various length and complexity attached at one of the vertices of the icosahedral head. During the entry process the viral genome enclosed within the head travels through the tail into the host cell. Despite the numerous types of tailed bacterial viruses, they all possess one of the following three types of tails: long contractile, long noncontractile, or short noncontractile tail. The type of tail serves as the main criterion for classification of tailed bacteriophages and also influences the genome delivery mechanism.

The contractile tail which is common for members of the family *Myoviridae* is a complex, long tube assembly surrounded by a sheath that is terminated by a base plate typically decorated with terminal tail fibers (**Figure 1(a)**). The tail fibers act as sensor to detect the host bacterium. Interaction between the tail fibers and their surface receptors leads to conformational changes in the base plate and subsequent irreversible adsorption of the virus on bacterial cells. This triggers contraction of the tail sheath, which then drags the base plate along the tail tube making the tube protrude from the base plate (**Figure 1(a)**). Subsequently, the tail tube penetrates the cell envelope and phage DNA is released into the cell through a transmembrane channel made from viral proteins, host proteins, or a combination of both. The force required for penetration is associated with the contraction of the tail sheath. The peptidoglycan-hydrolyzing activity present in the tail tube facilitates the penetration of the peptidoglycan layer. Some viral proteins enter the cell with the phage DNA in order to protect it from host exonuclease activities and are implicated in the early stages of phage genome transcription.

The bacteriophages of the family *Syphoviridae* have long noncontractile tails. The noncontractile tail is composed of a relatively flexible tube ending with a tip (or straight fiber) and few tail fibers attached to the tube. In contrast to myoviruses, tails of syphoviruses do not possess a contractile sheath and therefore do not contract during DNA ejection. Nevertheless, the long noncontractile tail undergoes considerable conformational changes, which occur after primary interaction with the host cell. These conformational changes mostly concern the distal part of the tail (straight fiber). The straight fiber penetrates the cell envelope and concomitantly shrinks in length while it increases in diameter. This signals the DNA to be released from the head of the virus. The straight fiber of the tail provides the channel for DNA penetration through the host cell envelope. Alternatively, viral DNA travels through a pre-existing pore within the host envelope, as many bacteriophages with noncontractile tails adsorb to channel forming protein complexes such as porins and transporters.

Members of the family *Podoviridae* have short noncontractile tails decorated with tail fibers. Tail fibers are responsible for host recognition and binding. Most of these viruses bind to the lipopolysaccharide of the outer membrane of Gram-negative bacteria. In contrast to the long tails of myo- and syphoviruses, the tails of podoviruses are too short to span the host cell envelope. After tail fiber attachment to lipopolysaccharide, several proteins are ejected from the virion. In the intact virion these proteins either form the internal core structure of the head or are located in the tail. Proteins ejected from the virion are involved in the formation of the channel across the cell envelope, thus extending the short tail and allowing the phage genome to travel from the virion into the cytoplasm.

Icosahedral dsDNA Bacterial Viruses with an Internal Membrane

The internal membrane of membrane containing dsDNA bacteriophages plays an important role during the entry process. According to the genome, the icosahedral membrane containing dsDNA phages fall into two families, *Tectiviridae* with the type member PRD1 and *Corticoviridae* with the only member PM2. The genome of tectiviruses is a linear dsDNA molecule with terminal proteins attached to the 5′-end, whereas the PM2 genome is a highly supercoiled circular dsDNA molecule. Obviously, these viruses use different strategies to deliver their genomes into the host cell.

Icosahedral dsDNA bacterial viruses with an internal membrane from the family *Tectiviridae* (e.g., PRD1) attach to the host cell receptor via a spike protein. The receptor structure for tectiviruses infecting Gram-negative cells is encoded by a multiple drug resistance conjugative plasmid. Specific interactions between the phage and the host induce dissociation of the spike complex from one of the capsid vertices and subsequent formation of a pore in the viral capsid. The internal membrane is transformed into a tubular tail-like structure, which protrudes from the capsid through the pore and penetrates the outer layers of the host cell. The rigid protein capsid stays intact outside the cell. Two lytic enzymes are associated with the virion (membrane) and function during the peptidoglycan penetration. A current model suggests that the viral integral membrane fuses with the host plasma membrane, thus connecting the cytoplasm with the interior of the viral membrane leading to delivery of the viral genome into the host cytoplasm. The members of the family *Tectiviridae* that infect Gram-positive bacteria (e.g., Bam35) attach to the host cell peptidoglycan structure. The internal membrane of these viruses also transforms into a tubular structure suggesting a similar mechanism for DNA entry into the host cell. The membrane tube of tectiviruses forms a channel for DNA translocation and thus is a counterpart to the tails of dsDNA bacteriophages discussed above.

Bacteriophage PM2 (*Corticoviridae*) infects Gram-negative marine bacteria from the genus *Pseudoalteromonas*. It has a circular dsDNA genome. The internal membrane of PM2 does not transform into a tubular structure in contrast to the internal membranes of tectiviruses. Instead, the capsid of the virus completely dissociates on the cell surface and the lipid core faces the host's outer membrane. The viral membrane most likely fuses with the bacterial outer membrane delivering the highly supercoiled dsDNA genome into the periplasm, from where it travels to the host cytosol.

Pleomorphic dsDNA Bacterial Viruses

The quasi-spherical virions of pleomorphic dsDNA bacterial viruses (*Plasmaviridae*) are nucleoprotein complexes within a lipid-protein membrane. The only classified member of the family *Plasmaviridae* is Acholeplasma phage L2. Acholeplasma are wall-less bacteria; thus, the phage adsorbs directly on the plasma membrane. It is assumed that adsorption leads to fusion of viral and host cell membranes resulting in entry of the nucleoprotein complexes into the cell.

Icosahedral ssDNA Bacterial Viruses

The icosahedral ssDNA bacterial viruses of the family *Microviridae* infect either free-living enterobacteria (e.g., ϕX174, G4, and α3) or obligate intracellular parasites lacking a cell wall, like chlamydia (e.g., Chp2) and spiroplasma (SpV4).

Icosahedral ssDNA bacterial viruses infecting free-living enterobacteria contain large spike complexes at the vertices of the virion. The spike complex is composed of five major spike proteins (G in ϕX174) and one minor spike protein (H in ϕX174) located at the spike apex. These spikes are responsible for the phage binding to the host cell, as well as for ejection of the ssDNA genome into the host cytoplasm. Bacteriophages infecting free-living enterobacteria adsorb to the lipopolysaccharide structures of the cell's outer membrane. Although microviruses are tail-less, they may follow an entry pathway similar to that of tailed phages (e.g., T4). Initially, ϕX174-like viruses adsorb reversibly to the cells. This is analogous to the interaction of the tail fibers of tailed phages with the host cell. After reversible adsorption, which allows detection of the host bacterium, follows the irreversible adsorption. Virus binding to a suitable lipopolysaccharide induces conformational changes in the spike proteins and leads to DNA release from the viral capsid. Similarly to the tail of tailed phages, the major spike proteins constitute a channel through which the ssDNA travels into the host cytosol. The minor spike protein penetrates through the host plasma membrane along with the viral DNA. The viral capsid lacking DNA and one of 12 minor spike proteins remains outside the cell.

Large spike complexes are not present in the virions of phages that infect parasitic bacteria. Instead, they have elaborate viral coat protein protrusions on the threefold axis of symmetry not seen in the ϕX174-like viruses. These protrusions are likely responsible for the receptor recognition. Bacteriophages infecting parasites mostly use protein receptors present in the plasma membrane of wall-less bacteria.

Filamentous ssDNA Bacterial Viruses

Filamentous ssDNA bacterial viruses (family *Inoviridae*) mostly infect Gram-negative bacteria. Usually, they utilize the sites of bacterial envelope where outer and inner membranes are in close proximity. The protein complexes functioning as co-receptors are preferentially localized at these sites of the cell envelope.

The receptor-binding protein of filamentous bacteriophages is a minor coat protein (pIII in Ff phages) located at one end of the extended phage particle (**Figure 1(b)**). At the beginning of the infection process this protein binds to the tip of the bacterial pilus. Bacterial pilus is a complex assembly immobilized in the bacterial plasma membrane that spans the periplasmic space and protrudes through the outer membrane (**Figure 1(b)**). Pili are able to retract when they bind to the solid surface and are used for bacterial motility. Upon phage adsorption the pilus rapidly retracts bringing the phage particle through the outer membrane close to its co-receptor, a protein complex (TolQRA for Ff phages) located in the inner membrane and periplasm of the bacterial cell envelope (**Figure 1(b)**). Virus adsorption to the primary receptor (pilus) uncovers the co-receptor-binding domain in the receptor-binding protein of the virion. This allows the interaction between the receptor-binding protein of the virus and the co-receptor to proceed, thereby anchoring the virion to the bacterial plasma membrane. This triggers insertion of adjacent major coat proteins of the filamentous phage particle into the bacterial plasma membrane leading to uncoating of the virion and ssDNA translocation into the host cytoplasm (**Figure 1(b)**). The uncoating of the virion and genome translocation both require the functional co-receptor complex, which is involved in the formation of the channel for DNA traveling through the peptidoglycan and the plasma membrane.

Icosahedral Enveloped dsRNA Bacterial Viruses

The dsRNA bacteriophages (family *Cystoviridae*) have a unique entry mechanism among bacteriophages. As other dsRNA viruses, bacteriophages with a dsRNA genome deliver their genome within a large protein capsid, which protects the enclosed dsRNA genome throughout the infection cycle. Internalization of large nucleoprotein complexes into bacterial cells is rare. The protein capsid of dsRNA bacterial viruses, however, penetrates the host plasma membrane via a mechanism similar to endocytic entry of animal viruses (**Figure 1(c)**).

The virion of dsRNA bacteriophages contains three structural layers that sequentially assist in the penetration of the host outer membrane, the peptidoglycan layer, and the plasma membrane. The spikes protruding from the virion surface are involved in the receptor recognition and binding (**Figure 1(c)**). The dsRNA bacteriophages adsorb either to a bacterial pilus (e.g., ϕ6), which then retracts, or to polysaccharide (e.g., ϕ13). After phage adsorption at the outer membrane, the viral envelope fuses with the host membrane, thus uncoating the virion and placing the resulting viral nucleoprotein complex (or nucleocapsid) into the periplasm (**Figure 1(c)**). The fusion between the viral envelope and bacterial outer membrane is driven by the transmembrane proteins of the viral envelope whose fusogenic properties are activated after virus adsorption. The lytic enzyme of the virus located between the envelope and nucleocapsid locally digests the host cell-wall peptidoglycan, thereby allowing the nucleocapsid to reach the plasma membrane (**Figure 1(c)**). Eventually, the nucleocapsid penetrates the plasma membrane via an endocytic-like route. This involves the formation of the plasma membrane curvature at the contact site resulting in an intracellular vesicle, which then pinches off from the plasma membrane (**Figure 1(c)**). The transcriptionally active viral particle is released from the vesicle into the host cytosol.

Icosahedral ssRNA Bacterial Viruses

All of the known icosahedral ssRNA phages (e.g., MS2, Qβ, GA, family *Leviviridae*) are pilus specific and utilize bacterial pili as attachment site. The virions of icosahedral ssRNA phages contain a single attachment protein (or maturation protein), which carries out the initial interaction with the host pilus. The attachment protein is covalently linked to both ends of the genomic ssRNA. In addition, this protein is partially exposed on the capsid surface likely forming one of the capsid vertices. Virus adsorption initiates the cleavage of the attachment protein into two fragments and later dissociation of those fragments from the viral capsid. As the pilus retracts into the cell, the attachment protein fragments together with the genomic ssRNA are pulled inside the host cell. The empty capsid remains outside the cell.

Bacteriophages with ssRNA genomes do not possess the peptidoglycan-hydrolyzing enzymes present in many other bacterial viruses which facilitate the peptidoglycan penetration during the entry process. Instead, ssRNA phages rely on the host cell pilus assembly to reach the cytoplasm.

Energetics

The viral genome transport across the host membrane is not a passive process. The force required for cell envelope penetration comes from different sources. As discussed earlier, phages with long contractile tails gain energy from the contraction of the tail sheath. This powerful process allows viral DNA to pass the host membrane in a very short time (e.g., 30 s for T4). Another type of energy comes from the pressure inside the viral capsid, which often is very high: In the capsids of dsDNA bacteriophages, the pressure may reach 50 atm (~5MPa)! This internal pressure of the virion, built up during DNA packaging, likely facilitates DNA ejection from the capsid during the initial stages of genome delivery.

The viral genome translocation may also be dependent on different host cell activities. The genome of T7 bacteriophage is pulled into the cytoplasm first by the host RNA polymerase and then by the RNA polymerase encoded from the phage genome. The retraction of bacterial pili and the transfer of the virus particles along the pili also depend on the cellular energy. Also, in many cases viral genome translocation requires membrane potential, which might provide energy to move DNA or RNA, stabilize a transmembrane channel, or regulate the genome translocation process.

Further Reading

Bennett NJ and Rakonjac J (2006) Unlocking of the filamentous bacteriophage virion during infection is mediated by the C domain of pIII. *Journal of Molecular Biology* 356: 266–273.

Fane BA, Brentlinger KL, Burch AD, *et al.* (2006) ϕX174 *et al.*, the *Microviridae*. In: Calendar R (ed.) *The Bacteriophages,* 2nd edn., pp. 129–145. New York: Oxford University Press.

Grahn AM, Daugelavicius R, and Bamford DH (2002) Sequential model of phage PRD1 DNA delivery: Active involvement of the viral membrane. *Molecular Microbiology* 46: 1199–1209.

Kivela HM, Daugelavicius R, Hankkio RH, Bamford JKH, and Bamford DH (2004) Penetration of membrane-containing double-stranded-DNA bacteriophage PM2 into *Pseudoalteromonas* hosts. *Journal of Bacteriology* 186: 5342–5354.

Letellier L, Boulanger P, de Frutos M, and Jacquot P (2003) Channeling phage DNA through membranes: From *in vivo* to *in vitro*. *Research in Microbiology* 154: 283–287.

Ponchon L, Mangenot S, Boulanger P, and Letellier L (2005) Encapsidation and transfer of phage DNA into host cells: From *in vivo* to single particles studies. *Biochimica et Biophysica Acta* 1724: 255–261.

Poranen MM, Daugelavicius R, and Bamford DH (2002) Common principles in viral entry. *Annual Review of Microbiology* 56: 521–538.

Rydman PS and Bamford DH (2002) Phage enzymes digest peptidoglycan to deliver DNA. *ASM News* 68: 330–335.

Van Duin J and Tsareva N (2006) Single-stranded RNA phages. In: Calendar R (ed.) *The Bacteriophages,* 2nd edn., pp. 175–196. New York: Oxford University Press.

Membrane Fusion

A Hinz and W Weissenhorn, UMR 5233 UJF-EMBL-CNRS, Grenoble, France

Glossary

Fusion pore Small opening at the site of two merged lipid bilayers, which allows the exchange of fluids. Fusion pores expand gradually to complete membrane fusion.

Hemifusion Membrane fusion intermediate state with the two proximal leaflets of two opposed bilayers merged to one.

Lipid raft Small membrane microdomain enriched in cholesterol and glycosphingolipids. These domains are resistant to solubilization by Triton X-100.

Type 1 TM protein A glycoprotein composed of an N-terminal external domain and a single transmembrane region followed by a cytoplasmic domain.

Introduction

Enveloped viruses contain a lipid bilayer that serves as an anchor for viral glycoproteins and protects the nucleocapsid containing the genetic information from the environment. The lipid bilayer is derived from host cell membranes during the process of virus assembly and budding. Consequently, infection of host cells requires that enveloped viruses fuse their membrane with cellular membranes to release the nucleocapsid and accessory proteins into the host cell in order to establish a new infectious cycle. Glycoproteins from enveloped viruses evolved to combine two main features. Firstly, they contain a receptor-binding function, which attaches the virus to the host cell. Secondly, they include a fusion protein function that can be activated to mediate fusion of viral and cellular membranes. Both tasks can be encoded by a single glycoprotein or by separate glycoproteins, which act in concert.

Three different classes of viral fusion proteins have been identified to date based on common structural motifs. These include class I fusion proteins, characterized by trimers of hairpins containing a central alpha-helical coiled-coil structure, class II fusion proteins, characterized by trimers of hairpins composed of beta structures, and class III proteins, forming trimers of hairpins by combining structural elements of both class I and class II fusion proteins (**Table 1**).

Viral glycoproteins interact with distinct cellular receptors by initiating conformational changes in the fusion protein leading to membrane fusion. Fusion occurs either at the plasma membrane, where receptor binding triggers conformational changes in the glycoprotein, or in endosomes upon virus uptake by endocytosis. In the latter case the low pH environment of the endosome leads to protonation (key histidine residues have been specifically implicated in the process), which induces conformational changes that lead to fusion of viral and cellular membranes.

The biophysics of membrane fusion is dominated by the stalk hypothesis. According to this view, fusion of two lipid bilayers in an aqueous environment requires that they come into close contact associated with a significant energy barrier. This process involves local membrane bending creating a first site of contact. Complete dehydration of the initial contact site induces monolayer rupture that allows mixing of lipids from the two outer leaflets, resulting in a hemifusion stalk. In a next step, the model predicts that radial expansion of the stalk leads to either direct fusion pore opening or to the formation of another intermediate, the hemifusion diaphragm, an extended bilayer connecting both membranes. The hemifusion diaphragm may also expand into a fusion pore. Fusion pore formation, which is characterized by an initial opening and closing ('flickering') of the pore may be mediated by several factors such as lateral tension in the hemifusion stalk or bilayer and the curvature at the

Table 1 Crystal structures of viral fusion proteins

Virus family	*Virus species*	*PDB code*
Class I		
Orthomyxoviridae	*Influenza A virus HA*	1HA0, 3HMG, *1HTM, 1QU1*
	Influenza C virus HEF	1FLC
Paramyxoviridae	*Simian parainfluenza virus* 5 F	2B9B,*1SVF*
	Human Parainfluenza virus F	1ZTM
	Newcastle disease virus F	*1G5G*
	Respiratory syncytial F	*1G2C*
Filoviridae	*Ebola virus* gp2	*1EBO, 2EBO*
Retroviridae	*Moloney Murine leukemia virus* TM	*1AOL*
	Human immunodeficiency virus 1 gp41	*1ENV, 1AIK*
	Simian immunodeficiency virus gp41	*2SIV, 2EZO*
	Human T cell leukemia virus 1 gp21	*1MG1*
	Human syncytin-2 TM	*1Y4M*
	Visna virus TM	1JEK
Coronaviridae	*Mouse hepatitis virus* S2	*1WDG*
	Sars corona virus E2	*2BEQ, 1WYY*
Class II		
Flaviviridae	*Tick-borne encephalitis virus* E	*1URZ,* 1SVB
	Dengue 2, and 3 virus E	*1OK8 IUZG, 1OAN, 1TG8*
Togaviridae	*Semliki forest virus* E1	1E9W *1RER*
Class III		
Rhabdoviridae	*Vesicular stomatitis virus G*	2GUM
Herpesviridae	*Herpes simplex virus* gB	*2CMZ*

edges of the hemifusion state. Finally the fusion pore extends laterally until both membranes form a new extended lipid bilayer (**Figure 1**).

The applicability of the stalk model to viral membrane fusion processes is supported by a number of observations. Labeling techniques allow to distinguish between merging of lipid bilayers and content mixing thus visualizing intermediate steps in membrane fusion. This has been applied to several liposome fusion systems demonstrating that membrane fusion steps can be arrested at different stages. Furthermore, certain lipids such as inverted cone-shaped lysophospholipids induce spontaneous positive bilayer curvature and inhibit hemifusion, while cone-shaped phosphatidylethanolamines induce negative curvature and promote hemifusion. In contrast, the lipid effect on the opening of the fusion pore is the opposite. Finally, electron microscopy images of influenza virus particles fused with liposomes reveal structures resembling stalk intermediates.

These observations are consistent with the hypothesis that viral fusion proteins generate initial contacts between two opposing membranes and their extensive refolding regulates and facilitates fusion via lipidic intermediate states by lowering the energy to form stalk-like intermediate structures.

Class I Fusion Glycoproteins

Biosynthesis of Fusion Proteins

Class I fusion proteins are expressed as trimeric precursor glycoproteins that are activated by proteolytic cleavage with subtilisin-like enzymes such as furin. This produces a receptor-binding subunit that is either covalently or non-covalently attached to the membrane fusion protein subunit, which anchors the heterotrimer to the viral membrane. The endoproteolytic cleavage positions a hydrophobic fusion peptide at or close to the N-terminus of the fusion domain. Subtilisin-like proteases recognize a conserved multibasic recognition sequence R-X-K/R-R or a monobasic cleavage site present in various glycoproteins. The nature of the cleavage site and its efficient cleavage (e.g., influenza virus hemagglutinin) has been associated with pathogenicity. The multibasic recognition sequences present in influenza virus HA, *SV5* F protein, *HIV*-1 gp160, and Ebola virus GP lead to mostly intracellular processing, whereas monobasic cleavage sites in Sendai virus F protein or influenza virus HA are efficiently cleaved extracellularly, resulting in a more tissue-restricted distribution of these viruses.

Cleavage activates the fusion potential of the viral - glycoproteins and is required for most class I glycoprotein-mediated fusion events. Although some evidence suggests Ebola virus processing by furin is not required for entry, it still requires the activity of endosomal cysteine proteases for efficient entry. Proteolytic cleavage thus generates in most cases a metastable glycoprotein structure that can switch into a more stable structure upon cellular receptor interaction including proton binding in the acidic environment of endosomes. This metastability was first recognized to play an important role in influenza virus hemagglutinin-mediated entry and has since been associated with all class I glycoproteins.

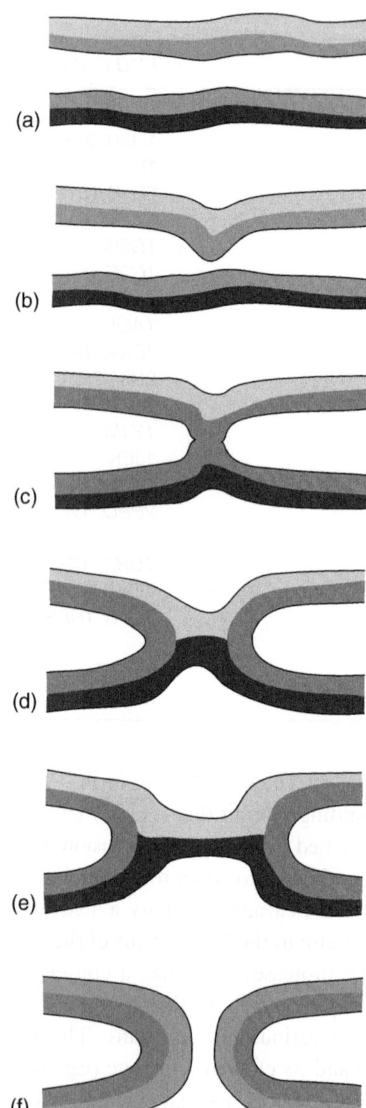

Figure 1 Fusion of two lipid bilayers. (a) Two parallel lipid bilayers do not approach closely. (b) Close contact mediated by local membrane bending. Hemifusion stalks with contact of outer leaflets (c) and inner leaflets (d). Fusion pore opening (f) may proceed directly from the stalk structure (d) or via a hemifusion diaphragm (e).

Structure of Native Influenza Virus Hemagglutinin

Since the structure solution of influenza virus hemagglutinin (HA), HA has served as the prototype of a class I fusion protein. The HA_1 domain, which contains the receptor-binding domain, folds into a beta structure that binds sialic acid-containing cellular receptors at the top of the molecule. In addition, both N- and C-termini of HA_1 interact with the stem of fusion domain HA_2 in an extended conformation. HA_2 anchors hemagglutinin to the viral membrane and folds into a central triple-stranded coiled-coil structure that is followed by a loop region and an antiparallel helix, which extends towards the N-terminal fusion peptide that is buried within the trimer interface (**Figure 2**).

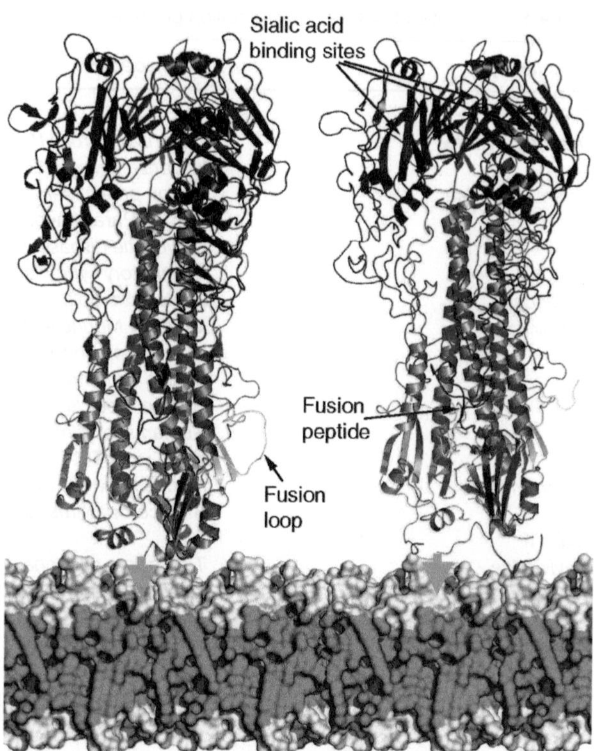

Figure 2 Ribbon diagram of trimeric conformations of influenza A virus hemagglutinin before (left, pdb code 1HA0) and after (right, pdb code 3HMG) proteolytic processing. The HA_1 receptor-binding domain is shown in blue and the positions of the sialic acid-binding sites are indicated. The fusion protein subunit HA_2 is shown in orange. The orientation towards the lipid bilayer is indicated by the orange triangle.

Structure of the Precursor Influenza Virus Hemagglutinin HA_0

The structure of uncleaved influenza virus HA_0 shows that only 19 residues around the cleavage site are in a conformation, which is different from the one seen in the native cleaved HA structure. This difference entails an outward projection of the last residues of HA_1 (323–328) and the N-terminal residues of HA_2 (1-12), resulting in the exposure of the proteolytic cleavage site (**Figure 2**). Upon cleavage, HA_2 residues 1–10 fill a mostly negatively charged cavity adjacent to the cleavage site, which leads to the sequestering of the fusion peptide within the trimeric structure (**Figure 2**).

Structure of the Low pH-Activated Conformation of Hemagglutinin HA_2

Low pH destabilizes the HA_1 trimer contacts, which causes the globular head domains to dissociate. This movement facilitates two major conformational changes. (1) A loop region (residues 55–76) refolds into a helix (segment B in HA) and extends the central triple-stranded coiled-coil in a process that projects the fusion peptide approximately 100 Å away from its buried position in native HA. (2) Another dramatic change occurs toward the end of the central triple-stranded coiled-coil, where a short fragment unfolds to form a reverse turn which positions a short helix antiparallel against the central core. This chain reversal also repositions a β-hairpin and the extended conformation that leads to the transmembrane region (**Figure 3**). Although its orientation changes, the core structure of the receptor-binding domain HA_1 does not change upon acidification.

Since both the neutral pH structure and the core of the low pH structure from hemagglutinin have been solved, a number of class I fusion protein structures have been determined and a common picture has emerged for their mode of action (**Table 1**). The characteristic of all class I fusion protein cores is their high thermostability suggesting that they represent the lowest energy state of the fusion protein. Secondly, they all contain a central triple-stranded coiled-coil region with outer C-terminal antiparallel layers that are either mostly helical or adopt extended conformations, thus forming trimers of helical hairpins. Since they resemble the low pH form of HA, it is assumed that they all represent the postfusion conformation. Although there is only structural evidence for extensive conformational rearrangement of the fusion protein subunit in case of hemagglutinin HA and the paramyxovirus F protein (**Figure 4**), it is assumed that all known class I fusion protein core structures are the product of conformational rearrangements induced by receptor binding.

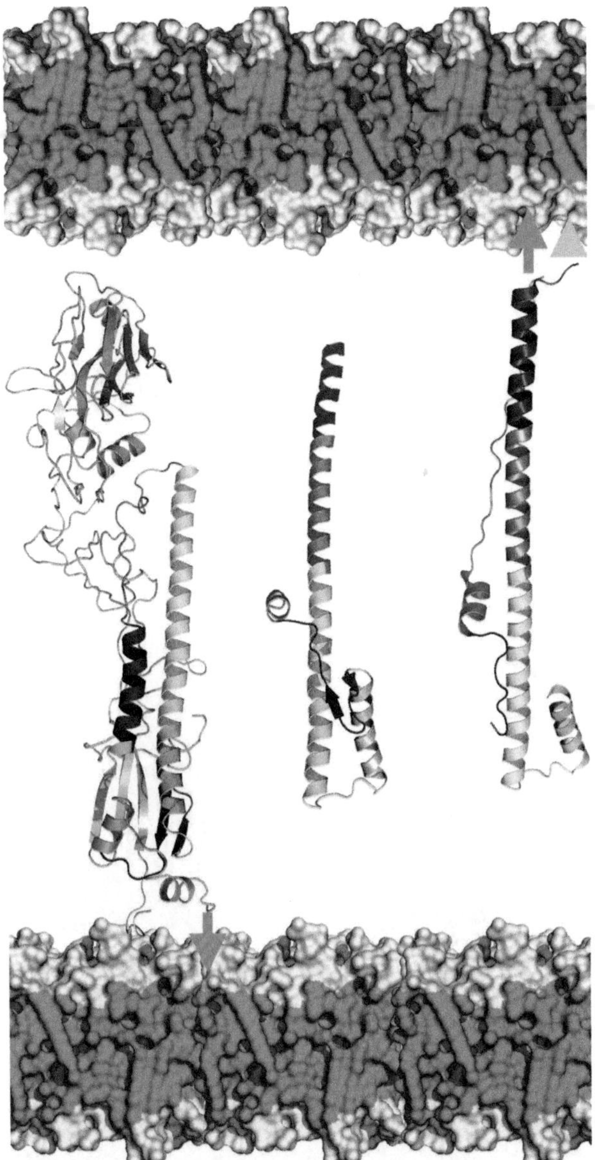

Figure 3 Ribbon representations of the conformational changes in HA_2 upon low pH exposure. Only one monomer is shown for clarity. Left panel: Native cleaved HA (pdb code 3HMG), HA_1 in gray, the secondary structure elements for HA_2 that change are shown in different colors. Middle panel: Low pH HA_2 (pdb code 1HTM) projecting the N-terminus leading to the fusion peptide towards the target cell membrane. Right panel: The C-terminal region has completely zipped up against the N-terminal coiled-coil domain (pdb code 1QU1). The membrane orientations of the TM region and the fusion peptide are indicated by green arrows and yellow triangles, respectively.

Class I-Mediated Membrane Fusion

The positioning of the N- and C-terminal ends containing the fusion peptide and the transmembrane region at the same end of a core structure, which was first established for the *HIV*-1 gp41 core structure, led to the proposal of the following general fusion model. (1) Proteolytic cleavage activation transforms the glycoprotein into a metastable conformation. (2) Receptor binding induces conformational changes in the fusion protein that exposes the fusion peptide and allows fusion peptide interactions with the target membrane (**Figures 5(a)** and **5(b)**). This generates a prehairpin intermediate structure that can be targeted by fusion inhibitors such as the HIV-1-specific T-20 peptide (**Figure 5(b)**). (3) Extensive refolding of the fusion domain most likely requires the dissociation of the C-terminal regions (**Figures 5(b)** and **5(c)**, indicated by blue lines) and leads to the apposition of the two membranes, concomitantly with the zipping up of the C-terminal region against the N-terminal coiled-coil domain ultimately forming the hairpin structure (**Figures 5(c)** and **5(d)**). The complete refolding process is thought to pull the two membranes into close-enough proximity to concomitantly allow membrane fusion. The extensive rearrangement of the fusion protein is thought to control the formation of different intermediate bilayer structures such as the hemifusion stalk (**Figure 5(d)**), and or the hemifusion

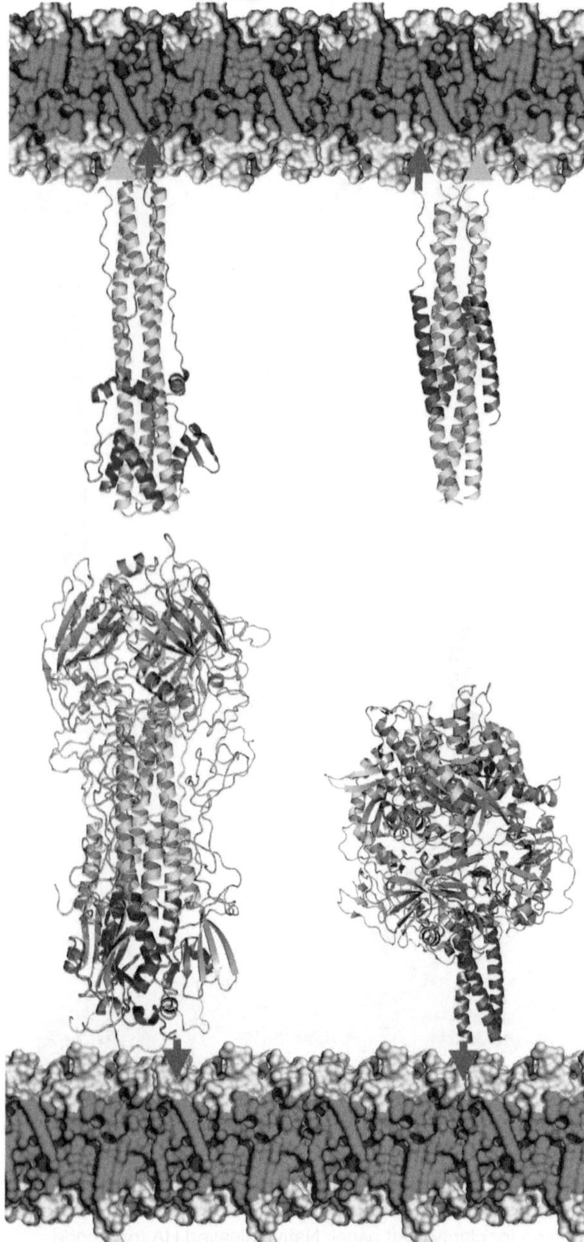

Figure 4 Comparison of the conformational changes induced upon receptor binding of two class I fusion proteins, influenza A virus hemagglutinin (left panel, pdb codes 3HMG and 1QU1) and paramyxovirus F protein (right panel, Simian parainfluenza virus 5 F, pdb codes 2B9B and 1SVF). The lower panel shows ribbon diagrams of native HA and F with secondary structure elements that change conformation upon activation highlighted in two colors (the inner triple-stranded coiled-coil region of the postfusion conformation in yellow and the C-terminal layers in orange). Although both native structures differ quite substantially, the conformational changes result in similar hairpin structures (upper panel) orienting both membrane anchors toward the target cell membrane. The membrane orientations of the TM region and the fusion peptide are indicated by orange arrows and yellow triangles, respectively.

diaphragm, followed by fusion pore opening and expansion (**Figure 5(e)**). It is generally assumed that membrane fusion occurs while the helical hairpin structure is formed and the core fusion protein structures represent postfusion conformations. Refolding of the fusion protein might produce defined stable intermediate structures, as suggested by the two low pH structures of influenza virus HA_2. One indicates that most of the outer layer has not yet zipped up to form the hairpin structure (**Figure 3**, middle panel: the C-terminal ends could extend back to the transmembrane region), while the other one reveals the extended conformation of the outer layer which forms together with the N-terminal coiled-coil a stable N-capped structure (**Figure 3**, right panel). Stepwise refolding may thus lock the fusion process irreversibly at distinct steps in agreement with a general irreversibility of class I-mediated fusion processes. The two membrane anchors, which are not present in the fusion conformation structures, also play an active role in the fusion reaction. Replacement of the transmembrane region by a glycophosphatidylinositol anchor leads to a hemifusion phenotype in case of hemagglutinin-driven membrane fusion, highlighting the role of the transmembrane region. Furthermore the C-terminal membrane proximal region plays an important role in fusion as shown in the case of *HIV*-1 gp41-mediated fusion.

Fusion Peptide

Class I fusion peptide sequences vary between virus families and are usually characterized by their hydrophobicity and a general preference for the presence of glycine residues. Although most fusion peptides of class I fusion proteins locate to the very N-terminus of the fusion protein, a few are found within internal disulfide-linked loops (e.g., filovirus Gp2 and the Avian sarcoma virus fusion proteins). NMR studies on the isolated influenza virus hemagglutinin fusion peptide revealed a kinked helical arrangement, which was suggested to insert into one lipid bilayer leaflet. This mode of bilayer interaction was proposed not only to mediate membrane attachment but also to destabilize the lipid bilayer. A further important function of fusion peptide sequences might be their specific oligomerization at the membrane contact site, which might constitute sites of initial membrane curvature.

Cooperativity of Fusion Proteins

A number of studies suggest that more than one class I fusion protein trimer is required to promote class I-driven membrane fusion. It has been suggested that activated hemagglutinin glycoproteins interact with each other in a synchronized manner and cooperativity of refolding allows the synchronized release of free energy required for the fusion process. This implies that activated fusion proteins

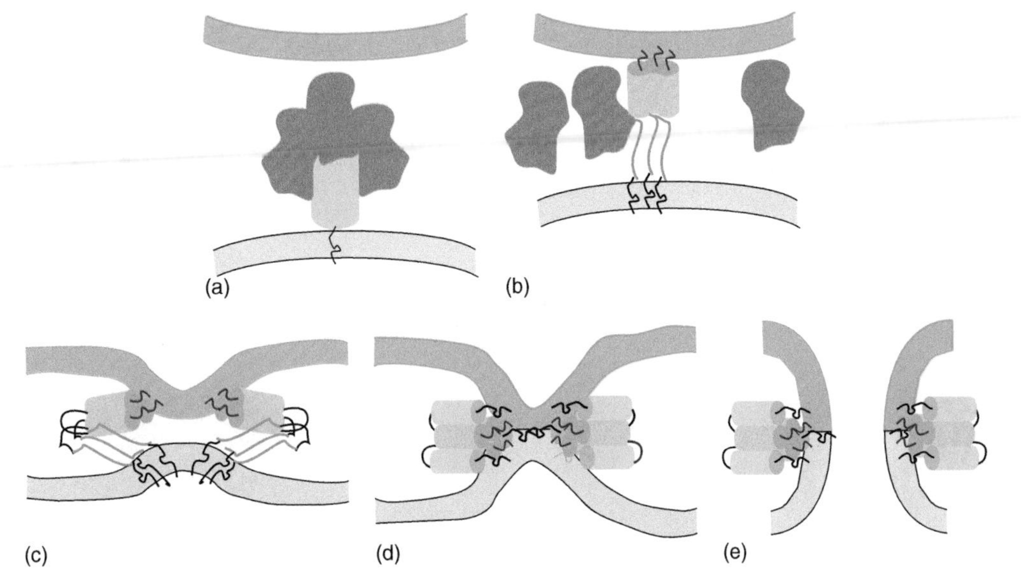

Figure 5 Model for class I glycoprotein-mediated membrane fusion. See text for explanation. The receptor-binding domains are indicated in brown and the fusion protein as cylinders. Note that some fusion proteins such as F from paramyxoviruses associate with an attachment protein (HN, H, or G). The latter interacts with F and cellular receptors triggering F-mediated fusion at the plasma membrane.

assemble into a protein coat-like structure that helps to induce membrane curvature, possibly also by inserting the fusion peptides into the viral membrane outside of the direct virus–cell contact site. However, it should be noted that no clearly ordered arrays of activated class I glycoproteins have yet been observed experimentally.

Role of Lipids in Fusion

The lipid content of a viral envelope such as that of *HIV*-1 was shown to contain mostly lipids normally present in lipid raft microdomains at the plasma membrane. Lipid rafts are small ordered lipid domains that are enriched in cholesterol, sphingomyelin, and glycosphingolipids. A number of enveloped viruses (e.g., influenza virus, HIV-1, Ebola virus, measles virus) use these platforms for assembly and budding, and some evidence suggest that raft platforms are also required for virus entry. Since the fusion activity of viral glycoproteins such as *HIV*-1 Env is affected by cellular receptor density as well as Env glycoprotein density, it has been suggested that both ligands have to be clustered efficiently to cooperatively trigger productive Env-mediated membrane fusion. This observation is consistent with the sensitivity of *HIV*-1 entry to cholesterol depletion.

Class II Fusion Proteins

Biosynthesis of Fusion Proteins

Class II fusion proteins comprise the fusion proteins from positive-strand RNA viruses such as the *Togaviridae* family, genus *Alphaviruses* (e.g., *Semliki Forest virus* (SFV)), and the *Flaviviridae* (e.g., *Dengue, Yellow fever, and Tick-borne encephalitis* virus (TBE)) (**Table 1**). Flaviviruses express the glycoprotein E that associates with a second precursor glycoprotein prM, while alphaviruses express two glycoproteins, the fusion protein El and the receptor-binding protein E2. E1 associates with the regulatory precursor protein p62. Both E-prM and E1-p62 heterodimerization are important for folding and transport of the fusion proteins. Cleavage of the fusion protein chaperones p62 and prM by the cellular protease furin in the secretory pathway is a crucial step in the activation of E and E1 fusion proteins.

Structure of the Native Fusion Protein

The native conformations of the flavivirus E glycoproteins and that of the alphavirus E1 glycoproteins are similar and fold into three domains primarily composed of β-sheets, with a central domain I, flanked by domain III connecting to the transmembrane region on one side and domain II on the other side (**Figure 6**, lower panel). Domain II harbors the fusion loop that is stabilized by a disulfide bridge and mostly sequestered within the antiparallel flavivirus E glycoprotein homodimer. In analogy, the fusion loop might be sequestered within the *SFV* E1–E2 heterodimer. Dimeric E–E and E1–E2 interactions keep the glycoproteins in an inactive, membrane-parallel conformation that covers the viral membrane. *SFV* E1–E2 heterodimers form an icosahedral scaffold with $T=4$ symmetry. Similarly, flavivirus E homodimers completely cover the viral membrane surface. The arrangement of the class II glycoproteins is thus

Figure 6 Ribbon diagram of the structures of *SFV* E1 (left panel, pdb codes 2ALA and 1RER) and of the *TBE* E (right panel, pdb codes 1URZ and 1SVB) in their native dimeric state (lower panel) and low-pH-activated trimeric conformation (upper panel). The three main domains of E1 and E are colored differently: domain I in blue, domain II in orange, and domain III in yellow. In both cases, activation of the conformational changes leads to trimeric hairpin structures. The membrane orientations of the TM region and the fusion peptide are indicated by blue arrows and red triangles, respectively.

completely different from the appearance of class I glycoprotein spikes, which do not form a specific symmetrical protein coat. In addition to forming the outer protein shell, flavivirus E and alphavirus E2 interact with cellular receptors, which direct the virion to the endocytotic pathway.

Structure of the Activated Fusion Protein

There are only minimal changes in secondary structure during the low pH-induced rearrangement of *TBE* E and *SFV* E1. However, the conformational changes result in an approximate 35–40 Å movement of domain III and a rotation of domain II around the hinge axis connecting domains I and II. This rearrangement produces a hairpin-like structure with a similar functional architecture as class I fusion proteins (**Figure 6**, upper panel). The outside of the trimer reveals a groove that was suggested to accommodate the segment, which connects to the transmembrane anchor and thus positions the fusion loops next to the membrane anchors. One significant difference between the *TBE* E and *SFV* E1 low pH conformations are the orientations of the fusion loops. *TBE* E fusion loops undergo homotrimer interactions, while *SFV* E1 fusion loops do not interact within trimers (**Figure 6**).

Class II-Mediated Membrane Fusion

At the low pH of endosomes E and E1 undergo conformational rearrangements that involve three major steps. Firstly, the homo- or heterodimers dissociate from the membrane-parallel conformation in a reversible manner assuming monomeric fusion proteins that expose their fusion loop to the target membrane (**Figures 7(a)** and **7(a′)**). This seems to be a main difference between class I and class II fusion, since trimer dissociation into

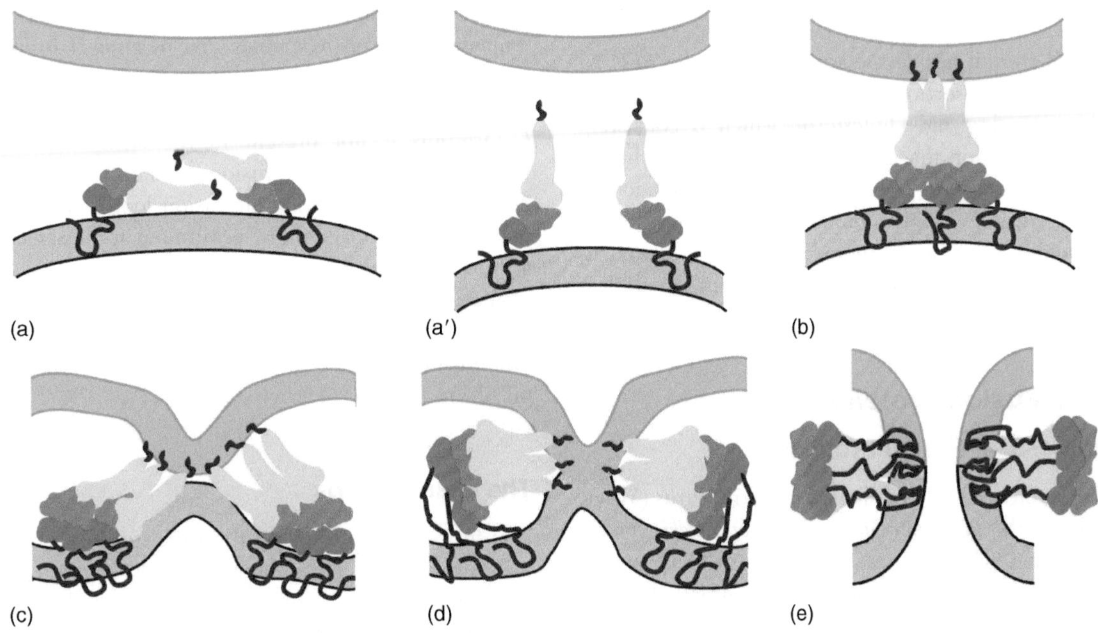

Figure 7 Model for membrane fusion of class II fusion proteins. See text for explanation. The three domains are colored as in **Figure 6**.

monomers has not been implicated in any class I fusion pathway. Secondly, fusion loop membrane interaction leads to the formation of homotrimers with an extended conformation. Trimerization is irreversible and tethers the fusion protein to the target membrane (**Figure 7(b)**). It is comparable to the postulated prehairpin structure of class I fusion proteins such as *HIV*-1 gp41 (**Figure 7(b)**). Notably, both fusion intermediates can be targeted by either fusion protein peptides (T-20 in case of *HIV*-1 gp41) or recombinant fusion protein domains (such as the E3 domain in case of *TBE*), to block membrane fusion. Further refolding, namely the reorientation of domain I, then pulls the two membranes into closer apposition that ultimately leads to the formation of a hemifusion stalk-like structure (**Figures 7(c)** and **7(d)**). Finally complete zipping up of the C-terminal ends against the N-terminal core domains allows fusion pore opening and its expansion (**Figure 7(e)**). Similar to the case of class I fusion protein-driven fusion, refolding is thought to provide the energy for fusion (**Figure 7**).

Fusion Peptide

The native and low-pH-induced crystal structures of the *TBE* virus E, dengue fever virus E, and *SFV* E1 proteins reveal that the conformation of the fusion loop changes upon acidification. The low-pH structures indicate that only hydrophobic side chains of the loop insert into the hydrocarbon chains of the outer leaflet of a target membrane. This is sufficient to anchor the fusion protein to the host cell membrane. Further oligomerization of fusion loops, as shown in the case of the low pH form of *SFV* E1, where crystal packing analysis revealed fusion loop interactions between trimers, was suggested to induce local membrane deformation, such as induction of a nipple-like membrane deformation (**Figure 7(c)**). This has been predicted in the stalk model to play an important role in the generation of lipidic intermediates during membrane fusion. Therefore the *SFV* E1 conformation might reflect an intermediate fusion state preceding the suggested postfusion conformation of flavivirus E trimers with homotrimeric fusion loop interactions. *In vivo*, the latter conformation might be induced by the final refolding of the C-terminal membrane proximal region and thus determining 'open' and 'closed' conformations of *SFV* E1 trimers and *TBE virus* E trimers, respectively (**Figure 6**).

Fusion Protein Cooperativity in Membrane Fusion

Homo- or heterodimeric class II fusion proteins already form a protein shell covering the complete viral membrane in the native state. Upon activation *in vitro*, both, soluble *SFV* E1 protein and flavivirus E protein insert their fusion loops into liposomes and form arrays of trimers organized in a lattice composed of rings of five or six. The E protein lattice on liposomes contains preferentially rings of five, which seems to affect the curvature of coated liposomes. In contrast rings of six form mostly flat hexagonal arrays *in vitro*. E1 pentameric rings can also be reconstructed from the crystal packing of E1 trimers. This strongly suggests that formation of a distinct fusion protein lattice might exert a cooperative effect on the fusion process.

Role of Lipids in Fusion

Although heterodimer dissociation exposes the *SFV* E1 fusion loop, its insertion into a lipid bilayer requires low pH triggering and cholesterol, which is consistent with the observation that *SFV* fusion depends on cholesterol and sphingolipids. The lipids required for E1 activation and fusion imply indirectly that lipid raft microdomains might be targeted for fusion. Flavivirus fusion, however, seems to be less dependent on cholesterol than alphavirus fusion.

Class III Fusion Proteins

Biosynthesis of Fusion Proteins

The glycoprotein G from vesicular stomatitis virus (VSV), a member of the *Rhabdoviridae* (e.g., *VSV* and *Rabies virus*), negative-strand RNA viruses, and gB from Herpesvirus, a member of the *Herpesviridae*, double-stranded DNA viruses, constitute a third class of viral fusion proteins based on the structural similarity of the postfusion conformation of their respective glycoproteins. Unlike class I and II envelope proteins both, *VSV* G and herpesvirus gB, are neither expressed as precursor proteins nor are they proteolytically activated.

Rhabdoviruses express a single trimeric glycoprotein G, which acts as receptor-binding domain to induce endocytosis and as the fusion protein that controls fusion with endosomal membranes upon acidification. However, different from class I and class II fusion machines the conformational changes induced by low pH are reversible. Changes in pH can easily revert the three proposed conformations of G, the native conformation as detected on virions, an activated state that is required for membrane interaction, and an inactive postfusion conformation.

Herpesvirus entry and fusion is more complex since it requires four glycoproteins, namely gD, gH, gL, and gB. Glycoprotein gD contains the receptor-binding activity and associates with gB as well as gH and gL. While gB seems to constitute the main fusion protein, the others are thought to be required for activation of the fusion potential of gB, which is pH independent.

Structure of the Low-pH-Activated VSV Glycoprotein and Herpesvirus gB

VSV G is composed of four domains that, interestingly, show similarities to both class I and class II fusion protein structures. It contains a β-sheet-rich lateral domain at the top, a central α-helical domain that mediates trimerization, and resembles the α-helical hairpin structure of class I fusion molecules, a neck domain containing a pleckstrin homologous (PH) domain, and the fusion loop domain that builds the trimeric stem of G. This stem-like domain exposes two loops at its very tip containing aromatic residues constituting the membrane-interacting motif of G. The stem domain resembles that of class II fusion proteins, albeit its different strand topology, which could be the result of convergent evolution. Although the complete C-terminus is not present in the structure, it points towards the tip of the fusion domain, indicating that both membrane anchors, the fusion loops and the transmembrane region, could be positioned at the same end of an elongated hairpin structure (**Figure 8**). The overall similarity of the structural organization of *VSV* G with that of herpesvirus gB indicates a strong evolutionary relationship between the rhabdovirus G and herpesvirus gB fusion proteins (**Figure 8**).

The Fusion Loops

The fusion loops extending from the stem-like domain of *VSV* G is similar to those observed in class II fusion proteins. The architecture is such that only few hydrophobic side chains can intercalate into one lipid bilayer leaflet, potentially up to 8.5 Å. Intercalation of side chains into one leaflet may induce curvature of the outer leaflet with respect to the inner leaflet, which would satisfy the stalk

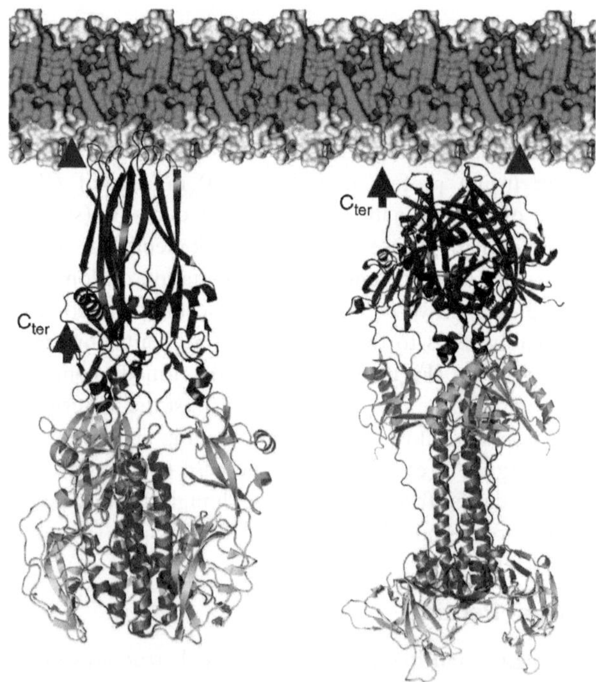

Figure 8 Ribbon diagram of class III fusion glycoproteins from *VSV* G (left panel, pdb code 2CMZ) and Herpesvirus gB (right panel, pdb code 2GUM). The individual domains are colored as follows: domain I in yellow, domain II in orange, domain III in green, and domain IV in blue. Their orientation toward the target membrane is indicated and shows the attachment of the putative fusion loops to one leaflet of the bilayer (red triangles) and the putative orientation of the TM region toward the same end of the fusion loop region (blue arrow).

model. The role of lipidic intermediate states, including the hemifusion state, has been confirmed experimentally in case of rhabdovirus G-mediated fusion.

Since both *VSV* G and herpesvirus gB resemble class I and class II fusion proteins which adopt a hairpin conformation with both membrane anchors at the same end of the molecule, it is most likely that they follow very similar paths in membrane fusion as suggested for class I and class II fusion proteins.

Summary

Although accumulating structural evidence suggests that the structural motifs used by viral fusion proteins and the mode of their extensive refolding varies substantially, the final product, namely the generation of a hairpin-like structure with two membrane anchors at the same end of an elongated structure, is maintained in all known postfusion conformations of viral glycoproteins. Thus the overall membrane fusion process is predicted to be the same for class I, II, and III fusion proteins, although the kinetics of refolding and fusion might vary to a large extent due to the involvement of different structural motifs to solve the problem of close apposition of two membranes.

Further Reading

Chernomordik LV and Kozlov MM (2005) Membrane hemifusion: Crossing a chasm in two leaps. *Cell* 123(3): 375–382.

Earp LJ, Delos SE, Park HE, and White JM (2005) The many mechanisms of viral membrane fusion. *Current Topics in Microbiology and Immunology* 285: 25–66.

Gallo SA, Finnegan CM, Viard M, *et al.* (2003) The HIV Env-mediated fusion reaction. *Biochimica Biophysica Acta* 1614: 36–50.

Harrison SC (2005) Mechanism of membrane fusion by viral envelope proteins. *Advances in Virus Research* 64: 231–261.

Kelian M and Rey FA (2006) Virus membrane-fusion proteins: More than one way to make a hairpin. *Nature Reviews Microbiology* 4(1): 67–76.

Lamb RA, Paterson RG, and Jardetzky TS (2006) Paramyxovirus membrane fusion: Lessons from the F and HN atomic structures. *Virology* 344(1): 30–37.

Roche S and Gaudin Y (2003) Pathway of virus-induced membrane fusion studied with liposomes. *Methods in Enzymology* 372: 392–407.

Skehel JJ and Wiley DC (2000) Receptor binding and membrane fusion in virus entry: The influenza hemagglutinin. *Annual Review of Biochemistry* 69: 531–569.

Replication of Viruses

A J Cann, University of Leicester, Leicester, UK

Glossary

(+)-sense RNA (plus-sense RNA) A virus with a single-stranded RNA genome of the same polarity ('sense') as mRNA.

(−)-sense RNA (minus-sense RNA) A virus with a single-stranded RNA genome of the opposite polarity ('sense') as mRNA.

Assembly The stage of replication during which all the structural components come together at one site in the cell and the basic structure of the virus particle is formed.

Attachment The binding of a virus particle to a specific receptor on the surface of a host cell.

Capsid A protein shell comprising the main structural unit of a virus particle.

Envelope A lipid membrane enveloping a virus particle.

Fusion protein The protein(s) on the surface of a virus particle responsible for fusion of the virus envelope with cellular membranes.

Gene expression An important stage of viral replication at which virus genetic information is expressed: one of the major control points in replication.

Genome replication The stage of viral replication at which the virus genome is copied to form new progeny genomes.

Matrix protein A structural protein of a virus particle which underlies the envelope and links it to the core.

Maturation The stage of viral replication at which a virus particle becomes infectious.

Molecular epidemiology The use of nucleotide sequence information to study the diversity and distribution of virus populations.

mRNA Messenger RNA, translated on ribosomes to produce proteins.

Nucleocapsid The core of a virus particle consisting of the genome plus a complex of proteins.

Penetration The stage of viral replication at which the virus genome enters the cell.

Polyprotein A long polypeptide encoding several mature proteins which are subsequently released by protease cleavage.
Receptor A specific molecule on the surface of a cell which is used by a virus for attachment.
Release The stage of viral replication at which virus particles escape the infected cell.
Tropism The ability of a virus to infect specific cell or tissue types.
Uncoating The stage of viral replication at which structural proteins are lost and the virus genome is exposed to the replication machinery.
Virions Structurally mature, extracellular virus particles.
Virus attachment protein The protein on the surface of a virus particle responsible for binding the receptor.

Unlike cellular organisms, which 'grow' from an increase in the integrated sum of their components and reproduce by division, virus particles are produced from the assembly of preformed components. Once manufactured, virus particles (virions) do not grow or undergo division. This alone makes the process of virus replication distinct from the growth of all other biological agents, and although the term 'grow' is sometimes used in the vernacular to refer to propagation of viruses, it is best to avoid this word when referring to the processes of virus replication.

Although this article will attempt to paint a general picture of the process of virus replication, the type of host cell infected by the virus has a profound effect on the replication process. There are many examples of viruses undergoing different replicative cycles in different cell types. However, the coding capacity of the genome determines the basic replication strategy used by different viruses. This strategy may involve heavy reliance on the host cell, in which case the virus genome can be very compact and need only encode the essential information for a few proteins, for instance, in parvoviruses. Alternatively, large and complex virus genomes, such as those of poxviruses, encode most of the information necessary for replication, and the virus is only reliant on the cell for the provision of energy and the apparatus for macromolecular synthesis, such as ribosomes. Viruses with RNA genomes have no apparent need to enter the nucleus, although during the course of replication, some do. DNA viruses, as might be expected, mostly replicate in the nucleus where host cell DNA is replicated and the biochemical apparatus necessary for this process is located. However, some viruses with DNA genomes (e.g., poxviruses) have evolved to contain sufficient biochemical capacity to be able to replicate in the cytoplasm, with minimal requirement for host cell functions.

Virus replication can be divided into eight stages, as shown in **Figure 1**. It should be emphasized that these are arbitrary divisions, used here for convenience in explaining the replication cycle of a theoretical, 'typical' virus. Regardless of their hosts, all viruses must undergo each of these stages in some form to successfully complete their replication cycle. Not all the steps described here are detectable as distinct stages for all viruses; often they blur together and appear to occur almost simultaneously. Some of the individual stages have been studied in great detail and a considerable amount of information is known about them. Other stages have been much harder to study, and less information is available.

Attachment

The attachment phase of replication comprises specific binding of a virus-attachment protein (or 'antireceptor') to a cellular receptor molecule. Virus receptors on cell surfaces may be proteins (usually glycoproteins) or carbohydrate residues present on glycoproteins or glycolipids. Some complex viruses (e.g., in the *Poxviridae* or *Herpesviridae*) use more than one receptor and therefore have alternative routes of uptake into cells. Most bacteriophage receptors are on the bacterial cell wall, although certain phages use cellular appendages (pili, flagella) as primary adsorption sites. Attachment is an automatic docking process and the kinetics of receptor binding are controlled by the chemical and thermodynamic characteristics of the molecules involved, that is, their relative concentrations and availability.

In most cases, the expression (or absence) of receptors on the surface of host cells determines the tropism of a particular virus, that is, the types of cell in which it is able to replicate. The attachment phase of infection therefore has a major influence on viral pathogenesis and in determining the course of a virus infection. Plant viruses must overcome different problems to animal viruses in initiating infection. The outer surfaces of plants are composed of protective layers of waxes and pectin, and each cell is surrounded by a thick wall of cellulose overlying the cytoplasmic membrane. No known plant virus uses a specific cellular receptor of the type that animal and bacterial viruses use to attach to cells and plant viruses must rely on mechanical breaks in the cell wall to directly introduce a virus particle into a cell.

Some virus receptors consist of more than one protein and multiple interactions are required for virus entry. An example of this is human immunodeficiency virus-1 (HIV-1), the primary receptor for which is the T-cell antigen, CD4. The binding site for the HIV-1 attachment protein (antireceptor), gp 120, has been mapped to the first variable region of CD4, although additional amino acids of the second variable domain also contribute toward binding. The sequences important for CD4 binding have also been

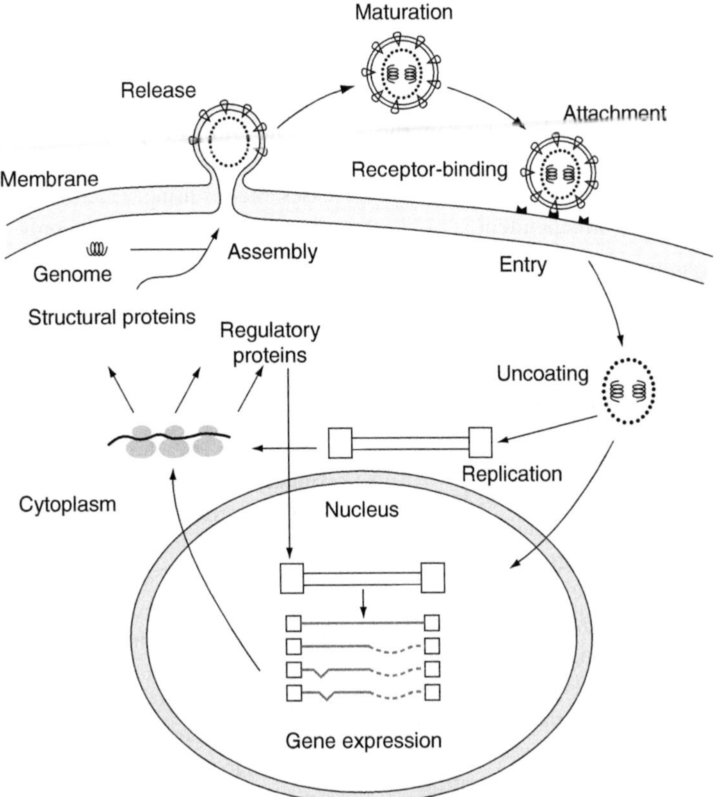

Figure 1 Schematic overview of a generalized scheme of virus replication. Reproduced from Cann AJ (2004) *Principles of Molecular Virology*, 4th edn. Amsterdam: Elsevier, with permission from Elsevier.

mapped in gp120. Deletions in this region or site substitutions abolish CD4 binding. In addition to CD4, there is at least one accessory factor which is necessary to form a functional HIV-1 receptor. These factors have now been identified as a family of proteins known as β-chemokine receptors. Multiple members of this class of proteins have been shown to play a role in the entry of HIV-1 into cells, and their distribution in the body is the primary control for the tropism of HIV-1 for different cell types.

Occasionally, the specificity of receptor binding can be subverted by nonspecific interactions between virus particles and host cells. Virus particles may be taken up by cells by pinocytosis or phagocytosis. However, without some form of physical interaction which holds the virus particle in close association with the cell surface, the frequency of these events would be very low. In addition, the fate of viruses absorbed into endocytic vacuoles is usually to be degraded, except in cases where the virus particle enters cells by this route. On occasion, binding of antibody-coated virus particles to Fc receptor molecules on the surface of monocytes and other blood cells can result in virus uptake. The presence of antiviral antibodies can result in increased virus uptake by cells and increased pathogenicity, rather than virus neutralization, as would normally be expected. The significance of such mechanisms *in vivo* is not known.

Entry

Entry of the virus particle into the host cell normally occurs a short time after attachment of the virus to the receptor. Unlike attachment, cell entry is generally an energy-dependent process, that is, the cell must be metabolically active for this to occur. Three main mechanisms are observed:

1. Translocation of the entire virus particle across the cytoplasmic membrane of the cell. This process is relatively rare among viruses and is poorly understood. It is mediated by proteins in the virus capsid and specific membrane receptors.
2. Endocytosis of the virus into intracellular vacuoles. This is probably the most common mechanism of virus entry into cells. It does not require any specific virus proteins (other than those already utilized for receptor binding) but relies on the normal formation and internalization of coated pits (term to be explained) at the cell membrane. Receptor-mediated endocytosis is an efficient process for taking up and concentrating extracellular macromolecules.
3. Fusion of the virus envelope (where present) with the cell membrane, either directly at the cell surface or following endocytosis in a cytoplasmic vesicle. Fusion

requires the presence of a specific fusion protein in the virus envelope, for example, influenza A virus hemagglutinin or the transmembrane glycoproteins of retroviruses. These proteins promote the joining of the cellular and virus membranes which results in the nucleocapsid being deposited directly in the cytoplasm. There are two types of virus-driven membrane fusions: pH-dependent and pH-independent.

The process of endocytosis is almost universal in animal cells and requires the formation of clathrin-coated pits which results in the engulfment of a membrane-bounded vesicle by the cytoplasm of the cell. At this point, any virus contained within these structures is still cut off from the cytoplasm by a lipid bilayer and therefore has not strictly entered the cell. As endosomes fuse with lysosomes, the environment inside these vessels becomes progressively more hostile as they are acidified and the pH falls, while the concentration of degradative enzymes rises. This means that the virus must leave the vesicle and enter the cytoplasm before it is degraded. There are a number of mechanisms by which this occurs, including membrane fusion and rescue by transcytosis. The release of virus particles from endosomes and their passage into the cytoplasm is intimately connected with (and often impossible to separate from) the process of uncoating.

Uncoating

Uncoating describes the events which occur after host cell entry, during which the virus capsid is partially or completely degraded or removed and the virus genome exposed, usually still in the form of a nucleic acid–protein complex. Uncoating occurs simultaneously with or immediately after entry and is thus difficult to study. In bacteriophages which inject their genome directly into the cell, entry and uncoating are the same process.

The removal of a virus envelope during membrane fusion is the initial stage of the uncoating process for enveloped viruses. Uncoating may occur inside endosomes, being triggered by the change in pH as the endosome is acidified, or directly in the cytoplasm. Entry into the endocytic pathway is a hazardous process for viruses because if they remain in the vesicle too long, they will be irreversibly damaged by low pH or lysosomal enzymes. Hence, some viruses have evolved proteins to control this process; for example, the influenza A virus M2 protein is a membrane channel which allows entry of hydrogen ions into the nucleocapsid, facilitating uncoating. The M2 protein is multifunctional, and also has a role in virus uncoating. In the picornaviruses, penetration of the cytoplasm by exit of virus from endosomes is tightly linked to uncoating. The acidic environment of the endosome causes a conformational change in the particle at around pH 5 that reveals hydrophobic domains not present on the surface of mature virus capsids. These hydrophobic patches interact with the endosomal membrane and form pores through which the RNA genome passes into the cytoplasm of the host cell.

The ultimate product of uncoating depends on the structure of the virus genome/nucleocapsid. In some cases, the resulting structure is relatively simple; for example, picornaviruses have only a small basic protein of approximately 23 amino acids covalently attached to the 5′ end of the RNA genome. In other cases, the virus core which remains is highly complex; for example, in the poxviruses uncoating occurs in two stages – removal of the outer membrane as the particle enters the cell and in the cytoplasm, followed by further uncoating as the core passes into the cytoplasm. In this case, the core still contains dozens of proteins and at least 10 distinct enzymes.

The structure and chemistry of the nucleocapsid determines the subsequent steps in replication. Reverse transcription can only occur inside an ordered retrovirus core particle and does not proceed to completion with the virus RNA free in solution. Eukaryotic viruses which replicate in the nucleus, such as members of the *Herpesviridae*, *Adenoviridae*, and *Polyomaviridae*, undergo structural changes following penetration, but overall remain largely intact. This is important because these capsids contain nuclear localization sequences responsible for attachment to the cytoskeleton and this interaction allows the transport of the entire capsid to the nucleus. At the nuclear pores, complete uncoating occurs and the nucleocapsid passes into the nucleus.

Transcription and Genome Replication

The replication strategy of a virus depends, in large part, on the structure and composition of its genome. For viruses with RNA genomes in particular, genome replication and transcription are often inextricably linked, and frequently carried out by the same enzymes. Therefore, it makes most sense to consider both of these aspects of virus replication together.

Group I: Double-Stranded DNA

This class can be further subdivided into two as follows:

1. Replication is exclusively nuclear or associated with the nucleoid of prokaryotes. The replication of these viruses is relatively dependent on cellular factors. In some cases, no virus-encoded enzymes are packaged within these virus particles as this is not necessary, whereas in more complex viruses numerous enzymatic activities may be present within the particles.
2. Replication occurs in cytoplasm. These viruses have evolved (or acquired from their hosts) all the necessary

factors for transcription and replication of their genomes and are therefore largely independent of the cellular apparatus for DNA replication and transcription. Because of this independence from cellular functions, these viruses have some of the largest and most complex particles known, containing many different enzymes.

Group II: Single-Stranded DNA

The replication of these virus genomes occurs in the nucleus, involving the formation of a double-stranded intermediate which serves as a template for the synthesis of new single-stranded genomes. In general, no virus-encoded enzymes are packaged within the virus particle since most of the functions necessary for replication are provided by the host cell.

Group III: Double-Stranded RNA

These viruses all have segmented genomes, as each segment is transcribed separately to produce individual monocistronic messenger RNAs. Replication occurs in the cytoplasm and is largely independent of cellular machinery, as the particles contain many virus-encoded enzymes essential for RNA replication and transcription since these processes (involving copying RNA to make further RNA molecules) do not normally occur in cellular organisms.

Group IV: Single-Stranded (+)-Sense RNA

These viruses can be subdivided into two groups.

1. *Viruses with polycistronic mRNA such as flaviviruses and picornaviruses.* As with all the viruses in this group, the genome RNA represents mRNA which is translated after infection, resulting in the synthesis of a polyprotein product, which is subsequently cleaved to form the mature proteins.
2. *Viruses with complex transcription such as coronaviruses and togaviruses.* In this subgroup, two rounds of translation are required to produce subgenomic RNAs which serve as mRNAs in addition to the full-length RNA transcript which forms progeny virus genomes. Although the replication of these viruses involves copying RNA from an RNA template, no virus-encoded enzymes are packaged within the genome since the ability to express genetic information directly from the genome without prior transcription allows the virus replicase to be synthesized after infection has occurred.

Group V: Single-Stranded (–)-Sense RNA

The genomes of these viruses can also be divided into two types.

- *Segmented.* The first step in the replication of these viruses (e.g., orthomyxoviruses) is transcription of the (–)-sense RNA genome by the virion RNA-dependent RNA polymerase packaged in virus particles to produce monocistronic mRNAs, which also serve as the template for subsequent genome replication.
- *Nonsegmented.* Monocistronic mRNAs for each of the virus genes are produced by the virus transcriptase in the virus particle from the full-length virus genome. Subsequently, a full-length (+)-sense copy of the genome is synthesized which serves as a template for (–)-sense progeny virus genomes (e.g., paramyxoviruses and rhabdoviruses).

Group VI: Single-Stranded RNA with DNA Intermediate

Retrovirus genomes are composed of (+)-sense RNA but are unique in that they are diploid and do not serve directly as mRNA but as a template for reverse transcription into DNA. A complete replication cycle involves conversion of the RNA form of the virus genetic material into a DNA form, the provirus, which is integrated into the host cell chromatin. The enzyme reverse transcriptase needs to be packaged into virus particles to achieve this conversion, as virus genes are only expressed from the DNA provirus and not from the RNA genome found in retrovirus particles of retroviruses.

Group VII: Double-Stranded DNA with RNA Intermediate

This group of viruses also relies on reverse transcription, but unlike the retroviruses, this occurs inside the virus particle during maturation. On infection of a new cell, the first event to occur is repair of the gapped genome, followed by transcription. As with group VI viruses, a reverse transcriptase enzyme activity is present inside virus particles, but in this case, the enzyme carries out the conversion of virus RNA into the DNA genome of the virus inside the virus particle. This contrasts with retroviruses where reverse transcription occurs after the RNA genome has been released from the virus particle into the host cell.

Assembly

During assembly, the basic structure of the virus particle is formed as all the components necessary for the formation of the mature virion come together at a particular site in the cell. The site of assembly depends on the pattern of virus replication and the mechanism by which the virus is eventually released from the cell and so varies for different viruses. Although some DNA virus particles form in the nucleus, the cytoplasm is the most common site of particle assembly. In the majority of cases, cellular membranes are used to anchor virus proteins, and this initiates the process of assembly.

For enveloped viruses, the lipid covering is acquired through a process known as budding, where the virus particle is extruded through a cell membrane. Lipid rafts are membrane microdomains enriched in glycosphingolipids (or glycolipids), cholesterol and a specific set of associated proteins. Lipid rafts have been implicated in a variety of cellular functions, such as apical sorting of proteins and signal transduction, but they are also used by viruses as platforms for cell entry (e.g., for HIV-1, SV40, and the rotaviruses), and as sites for particle assembly, budding and release from the cell membrane (e.g., in influenza A virus, HIV, measles virus, and rotaviruses).

As with the earliest stages of replication, it is often not possible to identify the assembly, maturation, and release of virus particles as distinct and separate phases. The site of assembly has a profound influence on all these processes. In general terms, rising intracellular levels of virus proteins and genomes reach a critical concentration and this triggers assembly. Many viruses achieve high levels of newly synthesized structural components by concentrating these into subcellular compartments known as inclusion bodies. These are a common feature of the late stages of infection of cells by many different viruses. Alternatively, local concentrations of virus structural components can be boosted by lateral interactions between membrane-associated proteins. This mechanism is particularly important in enveloped viruses which are released from the cell by budding (see above).

Maturation

Maturation is the stage of the replication cycle at which virus particles become infectious. This often involves structural changes in the newly formed particle resulting from specific cleavages of virus proteins to form the mature products or from conformational changes in proteins which occur during assembly (e.g., hydrophobic interactions). Protein cleavage frequently leads to substantial structural changes in the capsid. Alternatively, internal structural alterations, for example, the condensation of nucleoproteins with the virus genome, often result in changes visible by electron microscopy.

Proteases are frequently involved in maturation, and virus-encoded enzymes, cellular proteases or a mixture of the two may be used. Virus-encoded proteases are usually highly specific for particular amino acid sequences and structures, only cutting a particular peptide bond in a particular protein. Moreover, they are often further controlled by being packaged into virus particles during assembly and only activated when brought into close contact with their target sequence by the conformation of the capsid, for example, by being placed in a local hydrophobic environment, or by changes of pH or cation cofactor concentrations inside the particle as it forms. Retrovirus proteases are good examples of enzymes involved in maturation which are under tight control. The retrovirus core particle is composed of proteins from the *gag* gene and the protease is packaged into the core before its release from the cell on budding. During the budding process, the protease cleaves the gag protein precursors into the mature products – the capsid, nucleocapsid, and matrix proteins of the mature virus particle. Other protease cleavage events involved in maturation are less closely controlled. Influenza A virus hemagglutinin must be cleaved into two fragments (HA_1 and HA_2) to be able to promote membrane fusion during infection. Cellular trypsin-like enzymes are responsible for this process, which occurs in secretory vesicles as the virus buds into them prior to release at the cell surface; however, this process is controlled by the virus M2 protein, which regulates the pH of intracellular compartments in influenza virus-infected cells.

Release

For lytic viruses (most nonenveloped viruses), release is a simple process – the infected cell breaks open and releases the virus. The reasons for lysis of infected cells are not always clear, but virus-infected cells often disintegrate because viral replication disrupts normal cellular function, for example, the expression of essential genes. Many viruses also encode proteins that stimulate (or in some cases suppress) apoptosis, which can also result in release of virus particles.

Enveloped viruses acquire their lipid membrane as the virus buds out of the cell through the cell membrane, or into an intracellular vesicle prior to subsequent release. Virion envelope proteins are picked up during this process as the virus particle is extruded. This process is known as budding. As mentioned earlier, assembly, maturation, and release are usually simultaneous processes for viruses which are released by budding. The release of mature virus particles from their host cells by budding presents a problem in that these particles are designed to enter, rather than leave, cells. Certain virus envelope proteins are involved in the release phase of replication as well as in receptor binding. The best-known example of this is the neuraminidase protein of influenza virus. In addition to being able to reverse the attachment of virus particles to cells via hemagglutinin, neuraminidase is also believed to be important in preventing the aggregation of influenza A virus particles and may well have a role in virus release. In addition to using specific proteins, viruses which bud have also solved the problem of release by the careful timing of the assembly-maturation-release pathway. Although it may not be possible to separate these stages by means of biochemical analysis, this does not mean that spatial separation of these processes has not evolved as a means to solve this problem.

Further Reading

Cann AJ (2004) *Principles of Molecular Virology*, 4th edn. Amsterdam: Elsevier.

Freed EO (2004) HIV-1 and the host cell: An intimate association. *Trends in Microbiology* 12: 170–177.

Kasamatsu H and Nakanishi A (1998) How do animal DNA viruses get to the nucleus? *Annual Review of Microbiology* 52: 627–686.

Lopez S and Arias CF (2004) Multistep entry of rotavirus into cells: A Versaillesque dance. *Trends in Microbiology* 12: 271–278.

Moore JP, Kitchen SG, Pugach P, and Zack JA (2004) The CCR5 and CXCR4 coreceptors – central to understanding the transmission and pathogenesis of human immunodeficiency virus type 1 infection. *AIDS Research and Human Retroviruses* 20: 111–126.

Rossmann MG, He Y, and Kuhn RJ (2002) Picornavirus–receptor interactions. *Trends in Microbiology* 10: 324–331.

Schneider-Schaulies J (2000) Cellular receptors for viruses: Links to tropism and pathogenesis. *Journal of General Virology* 81: 1413–1429.

Persistent and Latent Viral Infection

E S Mocarski and A Grakoui, Emory University School of Medicine, Atlanta, GA, USA

Concepts

By nature of severity and ready association of symptoms with infection, acute viral diseases have played a dominant role in developing concepts of viral pathogenesis. Acute viral diseases were the first to succumb to vaccines that were based largely on the knowledge that initial exposure to a pathogen leaves the host with immune memory, reducing levels of secondary infection and disease. As a result, vaccination now controls a variety of acute viral diseases. Chronic viral infection underlies a wide variety of other medically important diseases, that either follow directly from primary infection or that require months, years, or decades to develop. Chronic viral infections are often widespread or even universal within a host species; however, disease rarely occurs in more than a small fraction of the infected population. Also, acute as well as chronic disease can follow infection with a single pathogen. For chronic disease, proof of etiology may only become evident once vaccination has successfully controlled the acute disease. Thus, measles vaccination provides protection from acute measles as well as the chronic disease, subacute sclerosing panencephalitis (SSPE). Vaccination against hepatitis B virus (HBV) prevents acute and chronic hepatitis as well as hepatocellular carcinoma arising from chronic infection. Similarly, vaccination against varicella-zoster virus (VZV) prevents chickenpox (varicella) as well as shingles (zoster) and the neuropathic pain that often follows disease recurrence. Most recently, a human papillomavirus (HPV) vaccine specifically targeting chronic disease states, condyloma and cervical carcinoma, has been developed. Other pathogens associated with significant disease, such as human immunodeficiency virus (HIV), hepatitis C virus (HCV), and a number of herpesviruses (**Table 1**), have not yet surrendered to vaccination. Antiviral therapies have succeeded in controlling some of these; however, the chronic nature of the underlying infection complicates antiviral strategies and favors selection of drug-resistant progeny within the chronically infected host. The list of chronic diseases associated with long-term viral infection has grown, with experimental models suggesting that chronic virus infection contributes to certain cancers, as well as to chronic diseases such as diabetes and atherosclerosis. Many of these chronic diseases remain uncontrolled, and a link to viral (or microbial) pathogens might open avenues to therapy as well as vaccination.

Acute, Persistent, and Latent Infection

Chronic viral infection is distinguished from acute infection by timing and completeness of clearance. Initial exposure to any virus results in primary infection that may be resolved in any of the three general infection patterns (**Figure 1**). One common pattern for viruses is 'acute infection', which may be accompanied by acute disease (**Figure 2**), and is typically controlled by the host adaptive immune response such that the pathogen is completely eliminated by the host response to infection. Immunological memory that initiates following primary infection provides protection from a subsequent, secondary infection. Immunological memory is the reason why vaccines substitute for primary infection and provide long-lived immunity. Exceptions where immunity is not sufficient to protect from reinfection (**Figure 2**), such as the well-known example of annual influenza virus

Table 1 Persistent and latent human viral infection and disease

Virus	*Genome maintenance*	*Cell and tissue tropism*	*Chronic disease syndrome*
Flavivirus			
HCV	Persistent infection	Hepatocyte	Chronic hepatitis[b] Hepatocellular carcinoma
Retrovirus			
HIV-1 and -2	Integrated provirus-persistent infection	$CD4^+$ T lymphocyte Mono/Macs/DCs	AIDS[b]
HTLV-1 and -2	Integrated provirus-persistent infection	T lymphocyte	Leukemia
Hepadnavirus			
HBV	Persistent infection	Hepatocyte	Chronic hepatitis[a,b] and hepatocellular carcinoma[a]
Herpesvirus			
HSV-1 and -2	Latent episome	Sensory neuron	Recurrent vesicular lesions[b]
VZV	Latent episome	Sensory ganglion	Zoster (shingles)[a,b]
EBV	Latent episome	B lymphocyte	Lymphoproliferative disease, nasopharyngeal carcinoma, African Burkitt's lymphoma and Hodgkin's lymphoma
CMV	Latent episome	Myeloid progenitor	Congenital disease, opportunistic disease[b]
HHV-6A,-6B, and -7	Latent episome	$CD4^+$ T Lymphocyte	Opportunistic disease
KSHV (HHV-8)	Latent episome	B lymphocyte (?)	Kaposi's sarcoma
Adenovirus			
Adenoviruses	Persistent infection	Adenoid lymphocyte and other sites	(None)
Papovavirus			
JC	Latent episome	Ductal epithelial, lymphoid, astrocyte	Progressive multifocal leukencephalopathy[b]
BK	Latent episome	Ductal epithelial	(None)
HPV	Latent episome	Basal epithelial	Warts, condyloma[a], and cervical cancer[a]
Poxvirus			
Molluscum contagiousum	Persistent infection	Basal epithelial	Skin lesion
Parvovirus			
AAV	Integrated-persistent infection	Epithelial cell and lymphocyte	(None)
B19	Persistent infection	Erythroid progenitor	Erythroblast crisis and persistent anemia
Paramyxovirus			
Measles	Persistent infection	CNS neuron	Subacute sclerosing panencephalitis and inclusion body encephalitis
Prion	Autocatalytic protein	CNS neuron	Transmissible spongiform encephalopathies

[a]Controlled by vaccination.
[b]Antiviral.

epidemics, may be due to genetic changes in the virus (so-called antigenic 'drift' and 'shift') that allows this RNA virus to escape from existing immunity. Due to the nature of vaccines, immunity from vaccination often wanes with time, such as has been observed with measles, mumps, and, most recently, VZV. A booster vaccination can raise the levels of immunity. Continuous replication, or 'persistent infection' following primary infection (**Figure 1**), follows from incomplete immune clearance, sometimes occurring because infection occurs at an immune-privileged host site, and may continue without disease or in a chronic or progressive disease pattern (**Figure 2**). A biological reservoir of quiescent virus is called 'latent infection' (**Figure 1**). Latent infection is associated with maintenance of viral genome without active viral replication, and occurs in a cellular lineage specific for the particular virus. Latent infection is accompanied by host immune control that completely suppresses continuous replication; however, this surveillance does not prevent latent infection or the sporadic reactivation and recurrent virus replication that is a hallmark of latency (**Figure 1**). Such patterns may be associated with recurrent, chronic, or opportunistic disease patterns (**Figure 2**). Persistence and latency/recurrence patterns sustain pathogens within individuals and improve transmission within populations. The propensity of a particular virus to initiate chronic infection in either of the two general patterns (persistent or latent infection) depends on the type of virus as well as

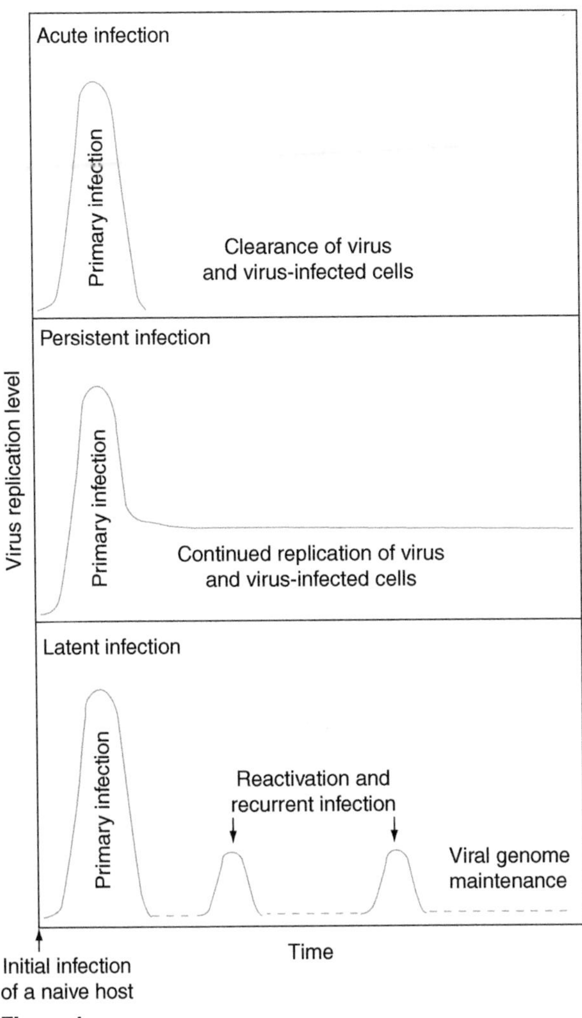

Figure 1

host cell tropism and immune response determinants. The replication scheme and particular cellular niche of the virus are important. A wide range of immunomodulatory viral functions have been identified in latent viruses, and these functions contribute to escape from host immunity. Immunological memory to latent viruses may be dysregulated as has been shown for persistent viruses leaving the host response ineffective in preventing recurrent infection.

Lifelong latent infection depends upon the participation of host cells that are not susceptible to productive replication but that form a biologic reservoir for reactivation, and reactivation can be dictated by changes in status or differentiation state of host cells. It is easy to imagine that latent infection may be established as soon as a virus reaches the appropriate cell lineage, and that this occurs while active replication continues at other body sites. This view of latency focuses on cells rather than the intact host, and follows from the fact that host cell populations may be permissive or nonpermissive. While a reservoir of latently infected cells may certainly be established concurrent with primary infection, the events of acute infection may leave many cells that are abortively infected or that are transiently nonpermissive and do not survive to become a biological reservoir. The term 'latency' best describes the period when actively replicating virus is no longer present anywhere in the host. Latency is a property of some virus groups (**Table 1**) and may be determined through the way the viral genome is maintained in host cells. The nature of genome maintenance allows precise distinction between latent and persistent infection. Finally, it is sometimes difficult to distinguish abortive infection from latent infection. Abortive infection leaves either the complete viral genome or genome fragments deposited in host cells but these do not constitute a biological reservoir for reactivation, although abortive infection of nonpermissive cells may underlie pathogenesis such as the contribution of some DNA viruses to cancer.

Pathogenesis of Chronic Infection

Pathogenesis refers to the complex biological process of infection in the host animal and is influenced by the levels and distribution of virus during primary infection, just as characterized for acute viral infections. The diversity of viruses is immense; however, every virus in Nature infects only a defined set of susceptible host animals, which represents its species range. Although exceptions exist, human chronic viral pathogens tend to be species-restricted and exhibit evidence of a long co-evolution with the host species. This co-existence and co-evolution as the host species evolve enables these pathogens to reach a balance (or *détente*) with host-clearance mechanisms (cell-intrinsic, innate immune, and adaptive immune responses). The pathogenesis of chronic virus infection depends on the interplay of genetic and environmental determinants that are specific to a particular virus type as well as to the host animal species. Further, there may also be susceptibility determinants that vary between individuals within a particular species. Reactivation and recurrent viral replication are important components of pathogenesis of latent virus infections. Long-term persistent replication or latency may have a dramatic impact on the physiology of cells, altering normal cellular function. Thus, chronic viral damage underlies a variety of known conditions, including immunosuppression, organ dysfunction, and certain cancers, and has been suggested to play a role in many others based on experimental models.

Every member of some animal virus groups, including retroviruses, papovaviruses, adenoviruses, hepadnaviruses, and herpesviruses, relies on persistent and/or

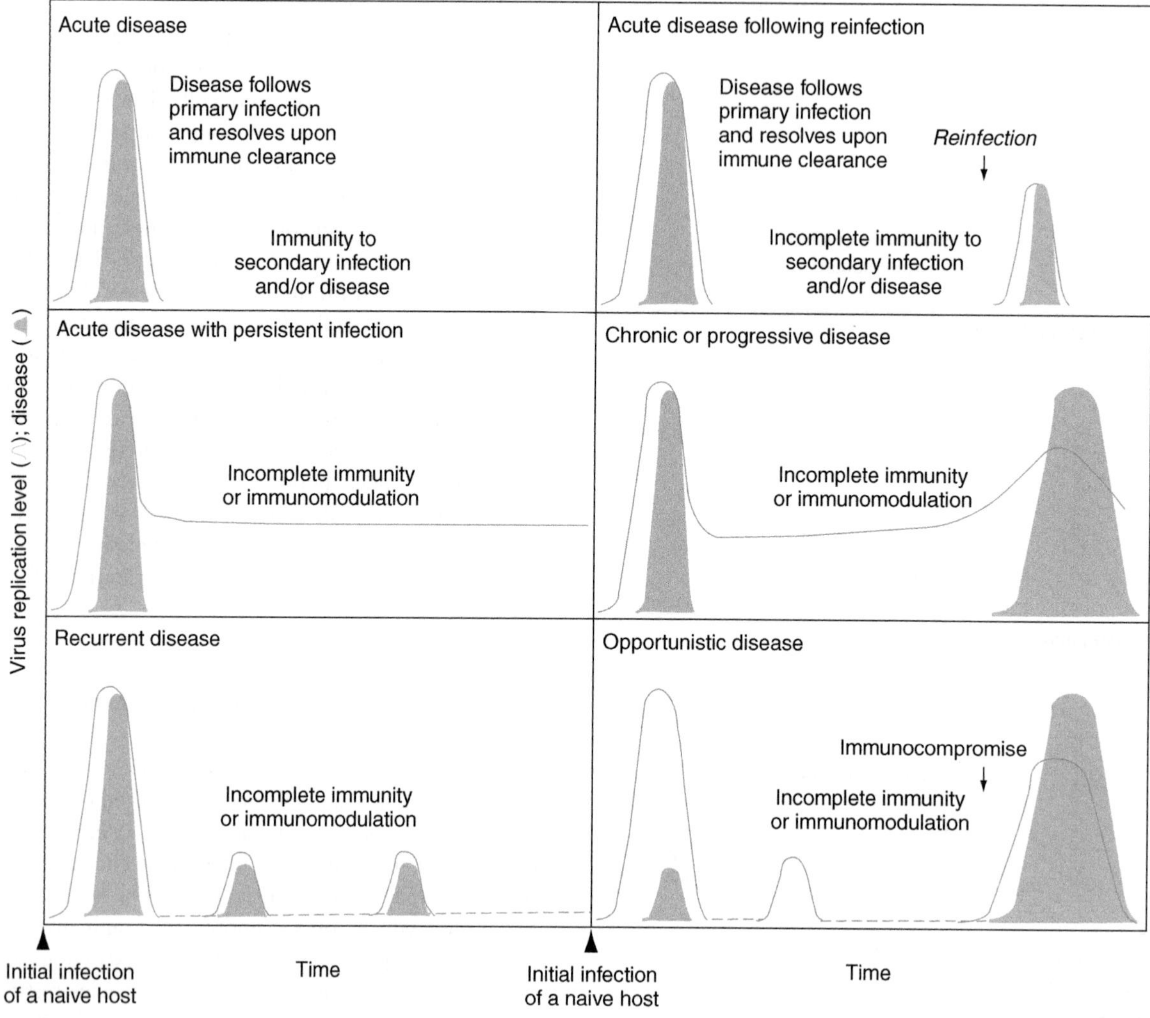

Figure 2

latent infection patterns. Some members of other virus groups rely only on persistent infection, sometimes in a host-dependent fashion, including flaviviruses, iridoviruses, and a range of others. In other cases, properties of the host dictate outcome, as with the regulatory mechanisms first characterized in lysogenic bacteriophages, or the persistent infection patterns that are intrinsic to plant and insect viruses. Disease pathogenesis (**Figure 2**) may follow various patterns. (1) Persistent infection may drive disease, such as with HIV-associated acquired immune deficiency syndrome (AIDS) or HCV-associated chronic liver damage and hepatocellular carcinoma. (2) Latent viral infection may alter host cell behavior such as with HPV-associated cervical cancer or Epstein–Barr virus (EBV) lymphoproliferative disease. (3) Reactivation of latent infection may lead to disease such as in herpes simplex virus (HSV)-recurrent cold sores or cytomegalovirus (CMV)-associated opportunistic infections in immunocompromised hosts. With various acute or chronic viral pathogens, reinfection may also contribute to disease patterns; however, existing immunity generally reduces the likelihood of significant disease. Common principles of chronic viral pathogenesis as well as in the diseases that arise from chronic infection have become more widely appreciated.

Immune Control

Immune control of chronic infections relies on cell-intrinsic, innate, and adaptive immunity in mammalian hosts, just the same as acute infections, although the biology of chronic infection relies on pathogen functions that can deflect or modulate host clearance. Cell-intrinsic and innate immunity reduce or delay acute replication levels, and are sometimes even sufficient to prevent disease through well-known mediators such as interferons and natural killer (NK) lymphocytes. The innate immune system is also responsible for recognition and processing of antigens to prime the adaptive immune response to infectious agents. Immune control of infection is influenced by the way in which a pathogen interacts with antigen-presenting cells (APCs): dendritic cells (DCs) and macrophages (Macs). This may occur through direct infection and antigen presentation, or indirectly through pathogen phagocytosis and cross-presentation to lymphocytes in secondary lymphoid tissues to dictate the ultimate breadth and effectiveness of the adaptive immune response.

All host cells sense and mount intrinsic responses to intracellular pathogens such as viruses, triggering cellular signaling cascades, programmed cell death, induction of interferon responses, and repression of viral genome transcription or replication. DCs and Macs, classically considered part of the reticuloendothelial system, produce an abundance of interferons, other cytokines, and a range of physiologically active mediators such as prostaglandins, vasoregulators, and hormones that may all be considered part of the innate response to infection. Cytokines and interferons initially drive the activation of NK lymphocytes, cytotoxic cells that recognize infected cells through reduced major histocompatibility complex (MHC) class I antigen levels, and increased MHC-like stress protein expression at the cell surface. This process is independent of the particular pathogen, and leads to the production of important cytokines that regulate the behavior of other leukocytes. DCs and Macs are considered crucial to innate responses because they acquire virus in the periphery and migrate to secondary lymphoid organs (lymph nodes, tonsils, and spleen) where an adaptive immune response unfolds. As the major producers of interferons and cytokines during viral infection, various subsets of DCs (and likely Macs) carry out the range of pattern recognition, antigen presentation, and co-stimulation activities with lymphocytes to guide both innate and adaptive immune responses. The antigen-specific adaptive immune response plays a crucial role in control of primary infection, as well as in establishing a reservoir of primed lymphocytes that constitute a memory response to secondary infection and prevents disease upon re-exposure to pathogens. Antibodies, produced by B lymphocytes following antigen-driven differentiation into plasma cells with the help of CD4+ T lymphocytes, form one critical arm of the adaptive response. Antibodies bind to and neutralize viruses, trigger complement-dependent lysis of virus-infected cells, and facilitate phagocytosis. Passive antibodies, or vaccines that only induce humoral immunity, are sufficient to control various acute viral pathogens, including influenza virus, paramyxoviruses, rubella virus, picornaviruses, rotaviruses, and vector-borne viruses. The B-lymphocyte response is central to control and clearance and occurs via neutralization of viral infectivity. Adaptive cellular immunity carried out by CD8+ T lymphocytes functioning as cytotoxic cells, often in collaboration with CD4+ cells, form the other critical arm. Control of persistent and latent viral infection, which is highly cell associated, relies on cytotoxic T-cell as well as antibody clearance mechanisms. In these viruses, cellular immunity is typically critical. CD4+ and CD8+ T lymphocytes also produce cytokines that contribute to suppressing replication and may induce immunopathology. There is sufficient evidence to say that both arms of the adaptive immune response facilitate long-lived immunity to reinfection, sometimes dependent on transmission route or virus type, so generalities may not apply to particular viral pathogens. For example, there is a strong interest in antibody-based immunity in control of HIV infection.

Immune Regulation and Immunopathology

The immune response is regulated, distinguishing self from non-self, while responding to infectious agents. Immune response processes may contribute to disease, by the direct recognition of viral antigen or by breaking the natural tolerance to self. Both antigen-specific and nonspecific damage to tissues as well as collateral damage to uninfected tissues may underlie disease. The damage to liver from chronic HBV (or chronic HCV) replication as well as the stromal keratitis due to recurrent HSV infection is triggered by viral infection but proceeds as a result of active antiviral immunity driving immunopathology. Although antibodies and lymphocytes are important effectors of immune control, dysregulation of the immune response during chronic infection may miscue processes that are normally beneficial to the host. Chronic infection can confound one important property of the adaptive immune response and cause a breakdown in the ability of the immune response to discriminate self from non-self. Thus, a balance of immunologic memory and the regulation of immunity, currently believed to be the job of regulatory CD4+ T lymphocytes as well as cell surface proteins that regulate memory T-lymphocytes, can be achieved. The capacity of persistent and latent viruses to deflect or avoid host immune clearance through immunomodulatory functions likely also contributes to infection and disease patterns. Immunologic memory associated with B or T lymphocytes contributes to secondary (anemnestic) responses upon subsequent exposures to the same or closely related viruses. Incompleteness in the immune response opens the way to reactivation as well as reinfection. Chronic viruses (papillomaviruses, retroviruses, herpesviruses, adenoviruses) rely on reactivation, but often benefit from transmission patterns that include reinfection with closely related strains. Immunopathology may be promoted by chronic infection but is by no means a requisite of chronic infection; this depends on the type of virus. Even acute infections that are typically cleared by adaptive immune effector mechanisms may trigger immunopathology upon re-exposure and this may manifest as chronic disease.

Immunomodulation

Like the immune response that the host mounts to control chronic infection, viruses carry an arsenal of defensive functions that undermine the effectiveness of immune clearance. Some of these functions go beyond defense and actively subvert or even exploit intrinsic and innate responses to infection. Although the main challenge to vaccination is often viewed as antigenic variation, such as characterizes HIV, or the number of viral strains in circulation, such as characterizes HCV, viral functions that deflect, subvert, or exploit host responses themselves may undermine vaccine strategies to control chronic infections. When viewing the large number of immunomodulatory functions encoded by large DNA viruses (poxviruses and herpesviruses), or the more restricted numbers encoded by smaller viruses (retroviruses and flaviviruses), the overwhelming impression is that viruses have found diverse ways to deal with a set of common challenges for chronic infection (**Table 2**). In many cases, the pathways that viruses target are critical to clearance, so neither the pathogen nor the host ultimately wins. The standoff that ensues is what characterizes chronic infection. The presence of immunomodulatory functions in a particular virus reveals the importance of specific immune pathways in control of virus infection. Remarkably, viruses with small genomes that encode fewer than 20 genes (retroviruses, papovaviruses, hepadnaviruses, flaviviruses, paramyxoviruses) have succeeded as well as viruses encoding greater than 20 genes (adenoviruses, poxviruses, herpesviruses), suggesting that there are many successful ways to reach this balance and achieve persistent or latent infection. Here, we will discuss a subset of these processes to provide a perspective on this growing field.

Certain cell-intrinsic and innate responses to infection are used throughout invertebrate and vertebrate hosts to detect and eliminate pathogens. Cell stress resulting from

Table 2 Immunomodulation in chronic viral infection

Virus	*Cell-intrinsic*	*Innate*	*Adaptive*
Flavivirus			
HCV	TLR, RIGI, IRF3, APOP	IFN, NK, APC	VAR, CD8,
Retrovirus			
HIV-1 and -2	TLR, APOP, AUTO, APOBEC	IFN, NK, PKR, CYTK, APC	VAR, MHC-I, CD4
Hepadnavirus			
HBV	APOP	IFN, APC	
Herpesvirus			
HSV-1 and -2	TLR, IRF3, APOP, AUTO, HDAC	IFN, NK, PKR, APC	MHC-I, MHC-II, Ab, C′, CD8, CD4
VZV	TLR, IRF3, APOP, HDAC	IFN, NK, APC	MHC-I, MHC-II, Ab, CD8, CD4
EBV	TLR, IRF3, IRF7, APOP, AUTO, HDAC	IFN, NK, CYTK, APC	MHC-I, MHC-II, Ab, CD8, CD4
CMV	TLR, IRF3, APOP, AUTO, HDAC	IFN, NK, PKR, CYTK, APC	MHC-I, MHC-II, Ab, CD8, CD4
HHV-6A, -6B, and -7	TLR, IRF3, APOP, HDAC	IFN, NK, PKR, CYTK	MHC-I, Ab, CD8
KSHV (HHV-8)	TLR, IRF3, IRF7, APOP, HDAC	IFN, NK, PKR, CYTK	MHC-I, MHC-II, Ab, C′, CD8, CD4
Adenovirus			
Adenoviruses	TLR, IRF3, APOP, AUTO, HDAC	IFN, NK, PKR, CYTK	MHC-I, CD8, CD4
Papillomavirus			
HPV	APOP, HDAC	IFN, NK, CYTK, APC	MHC-I
Poxvirus			
Molluscum contagiousum	APOP	IFN, NK, PKR, CYTK	MHC-I

Cell-intrinsic: TLR, Toll-like receptor; RIG-I, retinoic acid-inducible gene-I; IRF3/IRF7, interferon-regulated factor-3 and -7; APOP, apoptosis; AUTO, autophagy; APOBEC, family of cytosine deaminases; HDAC, histone deacetylases. Innate: IFN, interferon; PKR, protein kinase R; CYTK, cytokines; NK, natural killer lymphocytes; APC, antigen-presenting cell. Adaptive: MHC-I, major histocompatability class I; MHC-II, major histocompatability class II; Ab, antibody; C′, complement; CD8, CD8+ T lymphocytes; CD4, CD4+ T lymphocytes; VAR, variation in genome sequence.

sensing an intruder and initiating pattern recognition and signaling cascades are initially important. Intrinsic stress leading to programmed cell death (apoptosis, autophagy, and other forms) is an important and ancient mechanism to get rid of a host of intracellular pathogens. Extrinsic apoptosis due to signaling via tumor necrosis factor (TNF) family ligands is a development of vertebrates to expand the ways that apoptosis can be induced. The induction of apoptosis stops viruses before they have a chance to replicate and disseminate. Virus-encoded cell death suppressors are common (**Table 2**) in persistent and latent virus infection, likely delaying cell death.

The sensing of virus infection by cellular receptors that detect pattern recognition, such as Toll-like receptors (TLRs), are important to induce interferons and subversion of TLR signaling is common in persistent and latent viruses. Various components of TLR signaling pathways, including IRF-3, IRF-7, RIG-I, and NF-κB, may be targeted.

A variety of cellular enzymes may block steps in viral replication. These need to be overcome by viral or cellular functions in order for replication to proceed, so these have been found to play roles in establishing latency. The impact of histone deactylases (HDACs) on herpesviruses is one well-established example, and these likely also play into papillomavirus and adenovirus infection. APOBEC, a cytoplasmic cytidine deaminase, is a potent inhibitor of HIV and restricts viral replication by introducing mutations into the viral genome.

Persistent and latent infections are controlled over the course of the first few days by innate clearance, making subversion of innate clearance mechanisms an important area. Inflammatory cytokines, the most ancient of which are in the interleukin-1 (IL-1) family, but joined by a range of pro-inflammatory cytokines (IL-6, TNF family, interferons), are induced by viral infection with a range of specific as well as nonspecific inhibitors of these pathways.

In addition to the impact on cytokines, NK cells effectively reduce levels of virus-infected cells within the first few days post infection, and some viruses, particularly in the herpesvirus family, have a number of NK subversion functions.

A variety of these subversion pathways can be imagined to be active in infected APCs. For example, some proteins produced by replicating EBV directly inhibit the antigen-presentation pathway by specifically preventing their own presentation on the cell surface, thereby influencing how well the adaptive immune response to the virus is primed. However, many immune-evasion strategies that have been studied in detail have been found to be effective only in somatic cells and are more important in reducing the potency of adaptive immune clearance mechanisms that are induced.

Persistent and latent viruses have found myriad ways to subvert the effectiveness of host immune effector functions, ranging from deflecting antibody and complement to reducing MHC-I and MHC-II levels to make infected cells less recognizable by T lymphocytes. Many viruses make cytokine/cytokine receptor and chemokine/chemokine receptor mimics that likely subvert both innate and adaptive phases of the immune response by altering cell differentiation or migration.

Chronic Disease Manifestations

Latent and persistent viral infections contribute to chronic disease manifestations in a variety of ways including causing (1) congenital infection with long-term sequelae, (2) transformation of infected cells into malignant cells, (3) virus-mediated organ damage, (4) immune-mediated organ damage. Viruses that are able to infect a newborn during delivery or cross the placenta during pregnancy pose both acute and chronic disease risks. Such pathogens may initiate spontaneous abortion or cause progressive diseases in newborns. The herpesvirus, HSV, causes a rare but fatal (if untreated) systemic infection with neurological damage in newborns exposed at delivery. Cytomegalovirus (CMV) is a common congenital viral infection worldwide that may cause severe neurological disease, but more often causes subtle, progressive hearing, eyesight, and learning disabilities. This virus causes congenital infection with chronic progressive disease consequences. Neonates born to mothers acutely infected with CMV acquire the infection through transplacental transmission. Although the mother's infection is often asymptomatic, the infant with severe congenital CMV disease (also called symptomatic infection) may have microencephaly (an abnormally small head), hepatosplenomegaly (an enlarged spleen and liver), a bruise-like rash, retinitis, and progressive central nervous system complications including deafness, blindness, and psychomotor retardation. More than half of those infants with symptomatic CMV infections at birth will have neurologic complications later in life. Immunocompromised patients are susceptible to reactivation of a wide variety of latent viruses, with CMV most frequently observed. Other herpesviruses, adenoviruses, and papovaviruses may also cause chronic disease manifestations.

In some viruses, proliferative diseases and malignancies may result directly during latent infection, without any requirement for reactivation or replication. Viruses such as EBV and HPV directly transform cells in which they reside latently. Thus, EBV specifically targets naive B lymphocytes, converts them to a memory phenotype as if they had encountered antigen, and immortalizes these cells such that they remain latent hosts for life. EBV has several distinct viral latency 'programs' characterized by differential viral protein expression and associated chronic lymphoproliferative diseases as well as malignancies including endemic Burkitt's lymphoma, some varieties of Hodgkin's lymphoma, NK/T-cell lymphomas, and nasopharyngeal lymphoma. HPV is a second example of a virus causing both transformation of infected cells into malignant cells and virus-mediated organ damage depending on the viral genotype. For example, HPVs 16 and 18 are considered high-risk genotypes with propensity to lead to malignant transformation and are implicated in the development of cervical cancer. The E6 and E7 proteins from these high-risk genotypes are oncogenic, whereas the E6 and E7 proteins from lower-risk genotypes are not. Some HPV types have virtually no oncogenic potential but cause chronic damage to the skin in the form of plantar warts (HPV 1) or common warts (HPV 2). Similarly, HSV can be a chronic problem as a cause of recurrent aphthous stomatitis or painful genital mucosal ulcerations. Although the chronic damage by low-risk HPV or mucosal ulcerations of HSV may not be life threatening, other viruses can directly and indirectly cause grave chronic illnesses. Although the mechanism is unclear, parvovirus B19 has been associated with cessation of red blood cell, white blood cell, and platelet production by the bone marrow. This aplastic anemia can be fatal. Certainly HIV, known to rapidly deplete the host of its important CD4+ T lymphocytes, progresses to life-endangering AIDS when untreated. Even when the virus itself is not implicated in causing cytopathic damage to host organ systems, the immune response to the viral infection may instead wreak havoc on critical tissue. For example, HBV and HCV, while not strictly cytopathic themselves, lead to persistent inflammation in the liver with the host immune response largely responsible for the eventual development of cirrhosis and end-stage liver disease. HCV-related end-stage liver disease is now the single leading indication for liver transplantation in the United States illustrating the impact of persistent infection on public health.

Many viruses are controlled adequately in the immunocompetent host only to cause significant disease in immunocompromised hosts, and in this instance are referred to as opportunistic infections. There are many reasons why a host may be immunocompromised including HIV disease in which there is a progressive loss of CD4+ T-cell function, iatrogenic immunosuppression to maintain organ viability following transplantation, or chemotherapy to suppress cancerous cell growth. HIV patients with very low CD4+ T-cell counts, rendering them severely immunocompromised with AIDS, often manifest with multiple opportunistic infections. For example, whereas HSV usually causes self-limited disease processes in most immunocompetent hosts, in immunosuppressed hosts, HSV can cause an AIDS-related encephalitis. Similarly, HHV 8 has been implicated in the development of Kaposi's sarcoma in AIDS patients. CMV is an example of a virus that is usually well controlled by immunocompetent hosts but presents life-threatening infection in immunosuppressed transplant patients. HPV can similarly lead to development of verruca and condyloma recalcitrant to treatment in the immunosuppressed host with more frequent development of HPV-related carcinoma; and the poxvirus responsible for molluscum contagiosum generally causes skin lesions only in children, whose immune systems are still developing, and adults who are immunocompromised. Finally, hosts co-infected with HIV and HCV demonstrate more rapid development of HCV-related liver disease than immunocompetent patients.

In summary, the interplay between viruses and the infected host determines whether a virus is acutely resolved, becomes latent or persistent, and causes minimal or significant disease manifestations. Many challenges lie ahead to determine strategies sufficient to alter viral pathogenesis, augment host immunity, and change the course of virus-induced disease manifestations. These strategies may ultimately be in the form of prophylactic or therapeutic vaccination or may capitalize on host immune response modulation.

Acknowledgments

The authors acknowledge support from the Cancer Research Institute Investigator Award (A. Grakoui), Woodruff Health Sciences Fund (A. Grakoui and E. S. Mocarski), the Georgia Cancer Coalition (E. S. Mocarski), and the USPHS (A. Grakoui and E. S. Mocarski).

See also: Apoptosis and Virus Infection.

Further Reading

Biron C and Sen G (2006) Innate immune responses to viral infection. In: Knipe DM, Howley PM Griffin DE, *et al.* (eds.) *Fields Virology,* 5th edn., pp. 249–278. Philadelphia: Lippincott Williams and Wilkins.

Braciale T, Hahn YS, and Burton D (2006) Adaptive immune responses to viral infection. In: Knipe DM, Howley PM, Griffin DE, *et al.* (eds.) *Fields Virology,* 5th edn., 279–325. Philadelphia: Lippincott Williams and Wilkins.

Virgin S (2006) Pathogenesis of viral infection. In: Knipe DM, Howley PM, Griffin DE, *et al.* (eds.) *Fields Virology,* 5th edn., pp. 2701–2772. Philadelphia: Lippincott Williams and Wilkins.

Apoptosis and Virus Infection

J R Clayton and J M Hardwick, Johns Hopkins University Schools of Public Health and Medicine, Baltimore, MD, USA

Glossary

Bcl-2 Founding member of anti-apoptotic Bcl-2 protein family; homologs found in cells and viruses.
Caspase Family of cysteine proteases that cleave after aspartate residues.
DD/DED/CARD Structurally related protein–protein interaction domains found in a variety of different proteins (e.g., death receptors, adaptor molecules, and caspases).
FLICE Caspase-8 (original name: Fas-associated death domain protein interleukin-1beta-converting enzyme (ICE)-like).
FLIP Family of caspase-8 (FLICE)-like inhibitory proteins encoded by viruses (vFLIPs) and cells (cFLIPs).
IAP (inhibitor of apoptosis) Family of proteins with BIR (baculovirus IAP repeats) and a RING finger; homologs found in cells and viruses.
Micro-RNA Small hairpin RNAs that regulate mRNAs through RNAi or translational inhibition.
NF-κB Transcription factor that when activated translocates to the nucleus and regulates a variety of cellular and viral functions.
RNAi (RNA interference) Mechanism by which RNA is targeted for degradation in a sequence-specific manner.
TRAILR TRAIL (tumor necrosis factor-related apoptosis inducing ligand) receptor.

Introduction

Programmed cell death is any process by which cells participate in their own death. Cell suicide programs are facilitated by the actions of gene products encoded by the cell destined to die. These death-promoting genes evolved, at least in part, for the purpose of orchestrating cell autonomous death. The term apoptosis describes the morphology of naturally occurring programmed cell death during development and following certain pathological stimuli. In contrast to this deliberate cell death, the term necrosis describes cell morphology resulting from accidental cell death caused by acute injury. However, necrosis is less well defined, and is also used to describe the appearance of cells undergoing nonapoptotic programmed cell death. In addition, the term apoptosis can be used as a mechanistic term that refers to all types of programmed cell death. In multicellular organisms, programmed cell death is essential for development, tissue homeostasis, and regulating immune responses. In humans, it is estimated that billions of cells normally die per day by programmed death. Insufficient or excessive cell death characterizes most human disease states, including virus infections.

Classical apoptosis is characterized by a number of morphological and biochemical changes, including condensation of chromatin, cleavage of DNA between nucleosomes (DNA laddering), plasma membrane blebbing, exposure of phosphatidyl-serine on the outer leaflet of the plasma membrane (detected by annexin V binding), and fragmentation/division of mitochondria. These classic apoptotic events are caused by the activity of a subset of intracellular proteases known as caspases. Caspases are a family of cysteine proteases that cleave after specific aspartate residues in selected cellular proteins. A second subset of caspases includes the proteases that activate inflammatory mechanisms and may promote nonapoptotic programmed cell death. Other nonapoptotic programmed cell death pathways involve other types of proteases such as lysosomal enzymes, but less is known about these pathways, and little is known about their role in virus infections.

Any perturbation of the cell can potentially lead to activation of programmed cell death as a host defense mechanism to eliminate aberrant or damaged cells. Similarly, virus infection of a cell usually triggers the activation of programmed cell death. The ability of cells to recognize intruding viruses and activate cell suicide provides an important host defense mechanism for eliminating viruses by eliminating virus-infected cells prior to mounting host immune responses. However, viruses have developed myriad mechanisms to adapt to and regulate cellular death processes. As a consequence, many viruses cause disease primarily by inducing host cell death.

Cellular Death Pathways

Of the multiple pathways leading to cell death, caspase-dependent apoptosis is the best characterized. There are two general pathways for activating caspases, the receptor-activated (extrinsic) pathway, and the mitochondrial (intrinsic) pathway. Different viruses regulate multiple steps in both of these pathways.

Receptor-mediated pathways for the activation of caspases can be further subdivided into two groups: cell surface death receptors of the tumor necrosis factor receptor (TNFR) superfamily, and the intracellular pattern recognition receptors that activate primarily inflammatory caspases. The cytoplasmic portions of death receptors recruit and activate long pro-domain caspases (e.g., caspase-8). In turn, caspase-8 cleaves and activates the short pro-domain caspases (e.g., caspase-3). Caspase-3 is responsible for cleaving most of the known cellular substrates to facilitate cell death. Death receptor pathways are regulated by both viral and cellular factors that bind the receptor or other associated components in a manner that prevents caspase activation (**Figure 1**). The second receptor subgroup includes the pattern recognition receptors (PRRs), which can be further subdivided into three groups: the plasma membrane toll-like receptor family and two intracellular receptor families, the NOD-like receptors (NLRs), and the RIG-like helicases (RLHs), which are capable of detecting intracellular invaders. NLRs are activators of inflammatory caspases. Of particular relevance to virus infections, the intracellular RLH receptors recognize viral nucleic acids and trigger host defense mechanisms.

The mitochondrial pathway for activating caspases is triggered when cytochrome *c* is released from the intermembrane space of mitochondria into the cytoplasm. Released cytochrome *c* and ATP bind to Apaf-1 resulting in oligomerization of Apaf-1 and its binding partner caspase-9. Like the oligomerization-induced activation of caspase-8 by death receptors, oligomerization of the DED-like caspase-recruitment domain (CARD) motifs of caspase-9 also serves to activate the protease. Thus, initiator caspases (e.g., caspases-8 and-9) are activated by oligomerization, while effector caspases (e.g., caspases-3 and -7) are activated by proteolytic processing. Like caspase-8, activated caspase-9 also cleaves and activates caspase-3 to mediate apoptotic cell death. The mitochondrial pathway is regulated by pro- and anti-apoptotic cellular Bcl-2 family proteins that promote or inhibit cytochrome *c* release. Pro-apoptotic Bcl-2 family members Bax and Bak form homo-oligomers in the mitochondrial outer membrane that directly or indirectly release cytochrome *c* and other mitochondrial factors. This function of Bax and Bak is inhibited directly or indirectly by anti-apoptotic Bcl-2 family members Bcl-2, Bcl-x_L, and others. A more diverse subclass of Bcl-2-related proteins, the BH3-only proteins, inhibits antideath family members and/or promotes the functions of Bax and Bak. Cross-talk between the mitochondrial pathway and the receptor-activated pathway occurs when receptor-activated caspase-8 cleaves and activates the BH3-only protein Bid, leading to Bax- or Bak-mediated cytochrome *c* release. Cross-talk serves to amplify the apoptotic death pathway. Viruses also modulate multiple steps in the mitochondrial cell death pathway (**Figure 1**). For example, many viruses encode homologs of cellular anti-apoptotic Bcl-2 family proteins or other antagonists of Bax/Bak-mediated cell death. Still other viruses encode direct inhibitors of caspases, both pro-apoptotic and pro-inflammatory caspases.

Viruses also cause nonapoptotic cell death, but the mechanisms are less well characterized. Nonprogrammed necrotic cell death may simply be a consequence of extensive virus-induced cell injury after the virus has successfully inhibited apoptosis. Nonapoptotic cytopathologies induced by viruses include the formation of membranous vacuoles (often the sites of virus replication), activation of autophagy (lysosome-mediated degradation), and plasma membrane rupture/lysis. Viruses also trigger inflammation, which is mediated in part by caspases that cleave and activate pro-inflammatory cytokines, such as caspase-1 cleavage and activation of IL-1β. However, even if nonapoptotic death mechanisms are facilitated by cellular gene products, they may or may not cause programmed cell death. For example, cell death caused by lysosomal proteases that are released accidentally in damaged cells is not programmed if the pro-death function of lysosomal proteases did not evolve, at least in part, for the purpose of promoting cell suicide (to benefit the whole organism). However, this pathway is programmed if cathepsin cleavage sites in the pro-apoptotic BH3-only protein Bid, or other substrates, were selected during evolution to facilitate purposeful cell death during infection. Both programmed and non-programmed cell death mechanisms induced by viruses can be useful targets for therapeutic intervention.

Death Receptor Signaling

Death-inducing receptors on the cell surface can be placed into two subgroups, one containing Fas/CD95 and the other containing tumor necrosis factor receptor 1 (TNFR1). Fas recruits caspase-8 to directly and potently activate cell death. TNFR1 recruits different factors to form a distinct complex to inefficiently induce apoptosis or to activate other cellular signaling pathways.

Binding of Fas ligand (FasL) to Fas, or binding of TRAIL to death receptors 4 and 5 (DR4 and DR5), triggers the cytoplasmic tails of these receptors to form a complex known as the death-inducing signaling complex (DISC). The death domain (DD) in the cytoplasmic tail of the receptor binds directly to the DD domain of the small adaptor protein Fas-associated death domain (FADD). The DD-related death-effector domain (DED) of FADD binds directly to the DED domains in the N-terminus of caspase-8 to oligomerize caspase-8 in the DISC. In the other subgroup of death-inducing receptors, tumor necrosis factor-alpha (TNF-α) binds to TNFR1, allowing TNFR1 to recruit intracellular adaptor molecules receptor-interacting protein (RIP) kinase, TNFR-associated factors (e.g., TRAF2), and TNFR1-associated death domain (TRADD) protein to assemble a complex

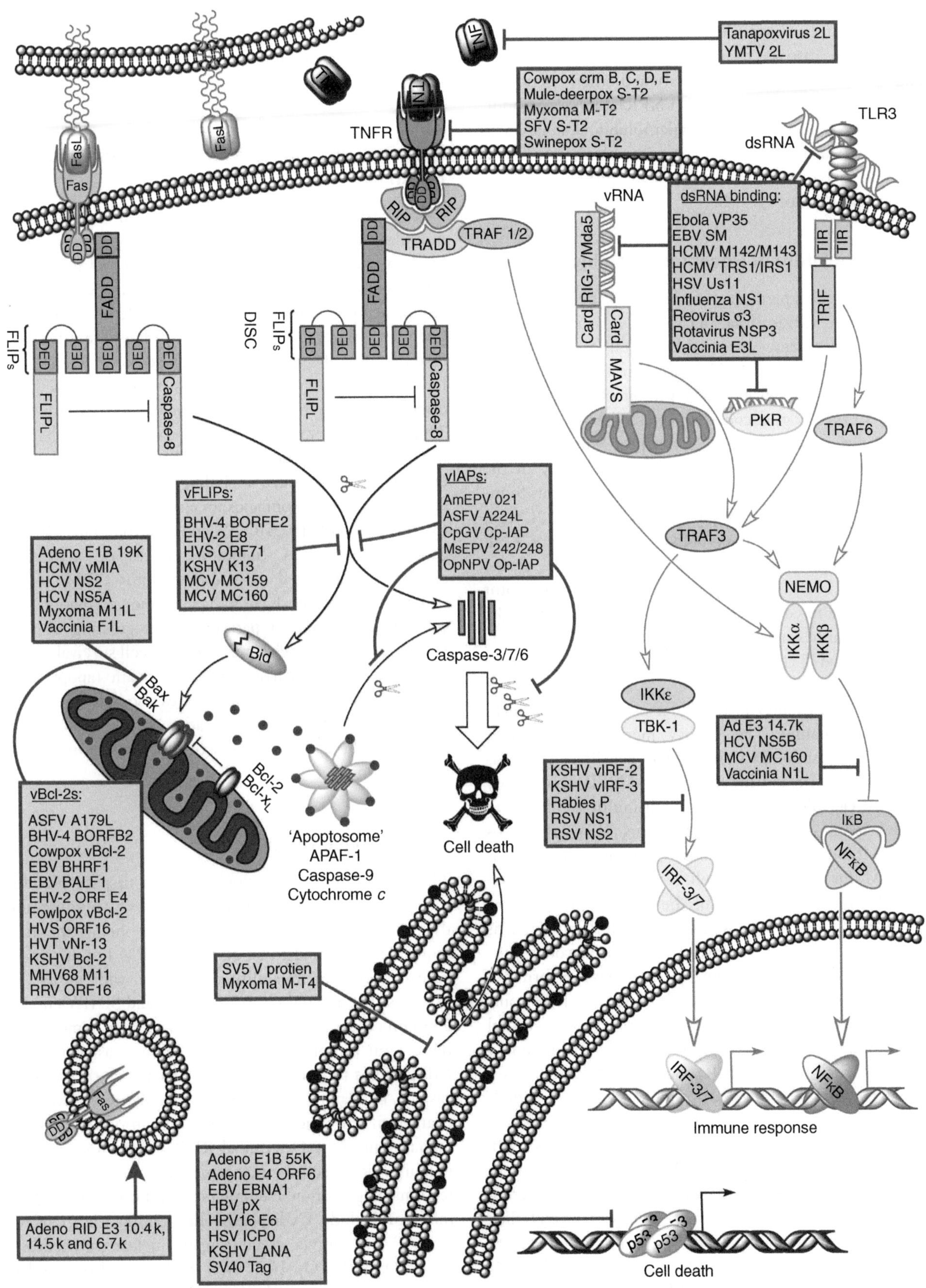

Figure 1 Cell death signalling during virus infection. Abbreviations are as in **Table 1**. Grayed areas indicate innate immunity signalling. Viral cell death regulators (yellow boxes, red arrows); cellular death regulators (black arrows); proteolytic cleavage events (scissors); ribosomes (black circles); cytochrome *c* (red circles).

that dissociates from the membrane, recruits FADD and caspase-8 that in turn cleaves and activates the effector caspases (e.g., caspase-3,-7) to cause cell death. Both cellular and viral factors regulate cell death and other functions of TNFR1 and related receptors. TNFR1 can recruit additional factors, including the cellular inhibitors of apoptosis (e.g., cIAP1 and cIAP2), and can cause pleiotropic effects on cells, including the activation of the pro-inflammatory transcription factor NF-κB.

Viral Regulators of Receptor Signaling

Viruses encode a plethora of molecules that interfere with signaling by extracellular ligands and their respective receptors. In some cases, these viral factors mimic TNFR1, such as the cytokine response modifier (CRM) proteins of cowpox virus, M-T2 protein of myxoma virus, and S-T2 orthologs encoded by several poxviruses. These soluble or membrane-bound receptor mimics act as TNFα-sinks, thereby diminishing the frequency of the interaction between TNFα and the host receptor. Similarly, the viral TNF-binding protein (vTNF-BP) 2L from tanapox (and several other poxviruses) sequesters TNF directly, but bears amino acid similarity to MHC class I rather than TNFR. Another TNFR-superfamily receptor mimic found in cowpox and mousepox viruses is vCD30, which diminishes signaling to the host cell and the proliferation of B and T cells via CD30L, dampening the immune response to infection (**Table 1**).

Other viruses alter intracellular mechanisms of death receptor signaling. The latency membrane protein 1 (LMP1) of Epstein–Barr virus (EBV) acts like a constitutively active (ligand-independent) TNFR-superfamily receptor capable of interacting with multiple TRAFs to inhibit apoptosis. Herpes simplex virus (HSV) glycoprotein D binds a TNFR-related protein as a receptor (HVEM) for entry into the cell, resulting in the activation of NF-κB. Wild strains of human cytomegalovirus (HCMV) encode a HVEM-like molecule UL144 that is missing in laboratory strains. Adenovirus encodes proteins E3–10.4k, E3–14.5k, and E3–6.7k that can form the 'receptor internalization and degradation' or RID complex that binds to the cytoplasmic portion of several different death receptor-superfamily members (including TNFR1, Fas, and TRAILR). Several different viruses target the adaptor TRAF2 to redirect cell death signals toward activation of NF-κB. Herpesvirus saimiri (HVS) StpC, KSHV K15, and rotavirus VP4 also activate NF-κB. Similarly, hepatitis C virus (HCV) nonstructural protein 5A (NS5A) can redirect the death signal by interacting with TRADD to inhibit its association with FADD, thus inhibiting death signaling, while HCV core protein selectively recruits FADD to TNFR1 to promote cell death. NS5A is also known to inhibit TRAF2, further directing the death signal away from NF-κB activation and survival signaling. These are only a few of the many examples where viral factors alter death receptor signaling apparently to alter programmed cell death and other signaling pathways (**Figure 1**).

FLIP Proteins Interfere with Receptor-Induced Death

Viral FLIP proteins (vFLIPs), identified in herpesviruses, modulate the activation of caspase-8 by competitively binding with the DED domain of FADD and preventing the oligomerization of preformed caspase-8-containing complexes that otherwise activate cell death (**Figure 1**). Cellular counterparts of cFLIPs were later identified and take the form of catalytically inactive caspase-8 or shorter forms that, like viral FLIPs, consist of only two DED domains, thereby mimicking the pro-domain of caspase-8. Thus, vFLIPs function more as suppressors than as modulators of caspase activation. Many different viruses encode FLIP proteins, including several herpesviruses and the molluscum contagiosum virus (MCV) that encodes two distinct vFLIPs, MC159 and MC160. The HCMV gene product vICA (inhibitor of caspase activation) encodes an unrelated caspase-8 inhibitor (**Table 1**). In contrast, adenovirus E1A induces tumor formation and cell death, perhaps in part by downregulating $cFLIP_S$. However, just as cytochrome *c* has both cell survival (mitochondrial electron carrier) and cell death (apoptosome formation) functions, cellular caspases also have alternative functions, including cell proliferation, at least at low activity levels. Therefore, caspase activation by E1A may have dual functions.

Viral Caspase Regulators

The P35 protein of baculoviruses is a caspase pseudosubstrate that effectively inhibits caspases by binding into the caspase active site. In addition to being the first molecule of its kind to be discovered, P35 has proved to be exceptional in its ability to inhibit a wide variety of caspases. The related viruses BmNPV and SlNPV encode P35 homologs. To date, there are no cellular homologs of these viral proteins that directly inhibit caspases. Viral IAP proteins related to cellular cIAPs are another class of caspase regulators discovered in baculoviruses. IAPs are characterized by their baculovirus inhibitory repeat (BIR) domains and by a C-terminal RING finger domain with ubiquitin ligase activity that targets its substrates for proteasomal degradation. Exemplars from this class of caspase inhibitor include Cydiapomonella granulovirus (CpGV) Cp-IAP and Orgyia pseudotsugata nuclear polyhedrosis virus (OpNPV) Op-IAP. Viral IAPs are also encoded by poxviruses; these include Amsacta moorei entomopoxvirus (AmEPV) gene 021 and Melanoplus sanguinipes

Table 1 Virus-encoded cell death modulators, listed alphabetically by virus family

Family	*Name*	*Viral gene*	*Function*
Adenoviridae	Adenovirus	E1A	Downregulates cFLIP$_S$, induces TNF-mediated caspase-8 activation, inhibits NF-κB
		E1B-19k	Binds and inhibits Bax/Bak
		E1B-55k	Ubiquitination of p53
		E3 (10.4k, 14.5k, 6.7k)	Internalization of TNF-related death receptors and their subsequent degradation
		E3-14.7k	Binds IKK and p50, suppresses NF-κB; inactivates caspase-8
		E4-ORF6	Ubiquitination of p53
		VAI	Inhibits Dicer as pseudo-substrate
		VAI	Inhibits PKR as pseudo-substrate
Asfarviridae	ASFV[a]	A179L	Bcl-2 homolog, inhibits MMP and cell death
		A224L	vIAP, activator of IKK
		A238L	Non-phosphorylatable IκB ortholog
Baculoviridae	AcMNPV[b]	P35	Broad spectrum caspase inhibitor
	BmNPV[c]	P35	Broad spectrum caspase inhibitor
	CpGV[d]	Cp-IAP	Binds caspases; Ub-ligase
	OpNPV[e]	Op-IAP	Binds caspases, Ub-ligase
	SlNPV[f]	P49	P35-related caspase inhibitor
Bunyaviridae	LAC[g]	NSs	Suppresses RNAi
	SAV[h]	NSs	Inhibits transcription/translation; *Drosophila* reaper-like
Filoviridae	Ebola virus	VP35	Binds dsRNA
Flaviviridae	HCV[i]	Core	Enhances Bid cleavage and TRAIL signaling, binds TNFR1
		E2	PKR pseudo-substrate, interacts with PKR
		IRES	vRNA; competitive inhibitor of PKR
		NS2	Interacts with CIDE-B and prevents MMP
		NS3/4A	Cleaves MAVS, inhibits interferon response; cleaves TRIF, inhibits recognition by TLR3
		NS5A	Direct interaction with PKR; inhibits TRADD/FADD association; interacts with TRAF2, inhibits NF-κB; binds Bax
		NS5B	IKKα-specific inhibitor of IKK complex
Hepadnaviridae	HBV[j]	pX	Direct binding to and inhibition of p53, activation of Src, MAPK, and NF-κB
Herpesviridae	BHV-4[k]	BORFB2	Bcl-2 homolog, inhibits MMP and cell death
		BORFE2	vFLIP, modulates caspase-8 activation
	EBV[l]	BALF1	Suppresses BHRF1 indirectly
		BHRF1	Bcl-2 homolog; inhibits cell death
		BZLF1	Downregulates TNFR1 expression
		EBER1,2	PKR pseudoactivator
		EBNA1	Transcriptionally downregulates HAUSP, leads to p53 poly-Ub
		LMP1	Constitutively active CD40, important for transformation
		miR-BHRF1	Latency-related
		SM	Binds dsRNA, interacts with PKR
	EHV-2[m]	E8	vFLIP, modulates caspase-8 activation
		ORF E4	Bcl-2 homolog, inhibits MMP
	HCMV[n]	IE1	Transcriptional regulator of NF-κB effectors
		IE1/IE2	Affects TNFR trafficking
		M142/M143	Binds dsRNA
		TRS1, IRS1	Binds dsRNA
		UL144	TNFR Superfamily; NF-κB activation
		UL36/vICA	Non-FLIP caspase-8 inhibitor
		UL37/vMIA	Binds ANT, GADD45; promotes mitochondria fragmentation but inhibits cell death
	HSV[o]	γ(1)34.5	eIF2α phosphatase
		ICP0	Sequesters the deubiquitinating enzyme HAUSP, inactivates p53
		LAT	Micro-RNA of TGF-β and SMAD3, suppresses stress response
		Us11	Binds dsRNA, interacts with PKR
		gD	Binds TNFR-related HVEM, activates NF-κB
		US3	vSer/Thr Kinase; modifies Bad, blocks Bad induced death
	HVS[p]	ORF 16	Bcl-2 homolog, inhibits MMP
		ORF 71	vFLIP, modulates caspase-8 activation
		StpC	Interacts with TRAF2, activates NF-κB

Continued

Table 1 Continued

Family	*Name*	*Viral gene*	*Function*
		Tip	Lck Adaptor; NF-κB activation
	HVT[q]	vNr-13	Bcl-2 homolog, inhibits MMP
	KSHV[r]	K1	vTyrosine kinase, constitutive Lyn activation
		K13	vFLIP, modulates caspase-8 activation
		K15	Interacts with TRAF2, activates NF-κB
		K7	Complexes with Bcl-2/caspase-3; regulates Ca^{2+} levels
		KSBcl-2	Bcl-2 homolog, inhibits MMP
		LANA	Suppresses p53 transcription
		vIRF-2	Direct interaction with PKR
		vIRF-3	Inhibits IKKβ and NF-κB activation
	Marek's disease virus	MEQ	Transcriptional inhibitor of apoptosis through TNFR, ceramide, UV and serum withdrawal
	MCMV[s]	Unknown	Disregulates TNFR levels
	MHV68[t]	M11	Bcl-2 homolog, inhibits MMP
Iridoviridae	RRV[u]	ORF16	Bcl-2 homolog, inhibits MMP
	ATV[v]	RelF2H	v-eIF2α; not phosphorylatable
Nodaviridae	FHV[w]	B2	Binds dsRNA
			Orthomyxoviridae
Influenza virus	NS1	Binds dsRNA	
		NS2	Triggers RIG-I
		PB1-F2	Mitochondrial fragmentation, MMP loss and cytochrome *c* release
Papillomaviridae	HPV16[x]	E6	eIF2α phosphatase; targets p53 for ubiquitination and degradation
		E6/E7	Interferes with DISC formation
Paramyxoviridae	Hendra virus	V	Inhibits Mda5
	Mumps virus	V	Inhibits Mda5
	RSV[y]	M2-1	Associates with RelA, activates NF-κB
		NS1, NS2	Interferes with IRF3 activation
	Sendai virus	C	Abolishes IFN/STAT signaling
		V	Inhibits Mda5
	SV5[z]	V	Inhibits Mda5; suppressor of ER-stress-mediated cell death
Picornaviridae	Coxsackie B3	Unknown	Degradation of Cyclin D1, p53, and β-catenin
	Poliovirus	3A	Affects TNFR trafficking
		3C	Cleaves RelA
Polydnaviridae	MdBV[aa]	H4, N5	vIκB, inhibit NF-κB translocation
Polyomaviridae	SV40[bb]	miR-S1	Suppresses TAg and early transcripts in late stages; immune evasion
		TAg	eIF2α phosphatase; binds and inactivates p53
Poxviridae	AmEPV[cc]	021	vIAP; IAP-binding motif antagonist
	Cowpox virus	CrmA/B12R	Serpin and caspase-1, 8, 10 inhibitor
		CrmB,C,D,E	TNFR/death receptor mimic
		vBcl-2	Bcl-2 homolog, inhibits MMP
		vCD30	CD30 homolog that dampens proliferative signal to B and T Cells
	DPV[dd]	S-T2-like	TNF-receptor decoy; sequesters TNF
	Fowlpox virus	vBcl-2	Bcl-2 homolog, inhibits MMP
	MCV[ee]	MC066L	vGlutathione-peroxidase; protects from UV and peroxide stress
		MC159	vFLIP, interacts with FADD, TRAF3 and caspase-8; promotes NF-κB activation
		MC160	IKKα-specific inhibitor of IKK complex and NF-κB activation; vFLIP; inhibits Fas signaling; binds FADD and caspase-8
	Mousepox virus	P28/012	Anti-apoptotic ubiquitin ligase
		vCD30	CD30 homolog that dampens proliferative signal to B and T Cells
	MsEPV[ff]	242	vIAP; IAP-binding motif antagonist
		248	vIAP; IAP-binding motif antagonist
	Myxoma virus	E13L	Interacts with ASC; inhibits caspase-1 activation
		M11L	Inhibits MMP and cell death
		MNF	vIκB, inhibits NF-κB translocation
		M-T2	TNF-receptor decoy; sequesters TNF
		M-T4	Inhibits apoptosis, localizes to ER
		Serp2/M151R	Caspase and granzyme B inhibitor
	SFV[gg]	P28/N1R	Anti-apoptotic ubiquitin ligase
		S-T2	TNF-receptor decoy; sequesters TNF

Continued

Table 1 Continued

Family	*Name*	*Viral gene*	*Function*
	Swinepox virus	S-T2-like	TNF-receptor decoy; sequesters TNF
	Tanapox virus	2L	vTNF-BP; sequesters TNF
	Vaccinia virus	A46R	TLR3 decoy receptor, TIR domains only
		CrmA/B13R	Serpin and caspase-1, 8, 10 inhibitor
		E3L	Binds dsRNA, independently inhibits PKR
		F1L	Interacts with Bim and Bak to inhibit MMP
		K1L	Prevents IκB degradation, inhibits NF-κB
		K3L	v-eIF2α, inhibits PKR function
		N1L	Inhibits TRAF2/6 and IKKα/β
	YMTV[hh]	2L-like	vTNF-BP; sequesters TNF
Reoviridae	Reovirus	σ3	Binds dsRNA
		Unknown	Prevents IκB degradation
	Rotavirus	NSP3	Binds dsRNA
		VP4	TRAF2 binding, IKK activating impairs caspase-8 activation
Retroviridae	BLV[ii]	Unknown	Disregulation of TNFR cell surface expression
	HIV[jj]	gp120	Interacts with CXCR4, causes apoptosis in CTLs; cross-links CD4, induces Fas expression in uninfected cells
		miR-N367	Suppresses Nef, suppresses its own replication
		Nef	Enhances NF-κB signaling and TNF-mediated apoptosis; reduces MHC-I, CD4 expression; phosphorylation and inhibition of Bad
		Protease	Cleaves caspase-8, degradation of Bcl-2
		Tar	vRNA binds TRBP; suppresses RISC
		Tat	PKR pseudo-substrate; enhances NF-κB signaling, TNF-apoptosis
		Vpr	G2/M arrest; glucocorticoid-like; Bax-dependent apoptosis
		Vpu	Suppresses NF-κB activation
	HTLV-1[kk]	Tax	IKKγ adaptor ; enhancer of Bcl-x_L, repressor of Bax,
	REV-T[ll]	v-Rel	Activated c-Rel
Rhabdoviridae	Rabies virus	P	Interferes with IRF3 activation

[a]African swine fever virus; [b]*Autographa californica* multiple nucleopolyhedrovirus; [c]*Bombyx mori* nucleopolyhedrovirus; [d]*Cydiapomonella* granulovirus; [e]*Orgyia pseudotsugata* multiple nucleopolyhedrovirus; [f]*Spodoptera litura* nucleopolyhedrovirus; [g]La Crosse virus; [h]San Angelo virus; [i]Hepatitis C virus; [j]Hepatitis B virus; [k]Bovine herpesvirus-4; [l]Epstein Barr virus; [m]Equid herpesvirus-2; [n]Human cytomegalovirus; [o]Herpes simplex virus; [p]Herpesvirussaimiri; [q]Herpesvirus of turkeys; [r]Kaposi's sarcoma-associated herpesvirus; [s]Mouse cytomegalovirus; [t]Mouse herpesvirus strain 68; [u]Regina ranavirus; [v]*Ambystoma tigrinum stebbensi* virus; [w]Flock house virus; [x]Human papillomavirus 16; [y]Respiratory syncytial virus; [z]Simian virus 5; [aa]*Microplitis demolitor* bracovirus; [bb]Simian virus 40; [cc]*Amsacta moorei* entomopoxvirus; [dd]Mule-deer poxvirus; [ee]*Molluscum contagiosum* virus; [ff]*Melanoplus sanguinipes* entomopoxvirus 'O'; [gg]Shope fibroma virus; [hh]Yaba monkey tumorvirus; [ii]Bovine leukemia virus; [jj]Human immunodeficiency virus; [kk]Human T-lymphotropic virus 1; [ll]Reticuloendotheliosis virus strain T.

entomopoxvirus genes 242 and 248. Viral and cellular IAP proteins were suggested to directly inhibit caspases, but this model is currently challenged by evidence suggesting more indirect mechanisms to explain their inhibitory effects. Other poxviruses encode RING finger proteins that are suggested to have E3 ubiquitin ligase activity, though they lack overall amino sequence similarity to the IAP family.

Another important poxvirus caspase inhibitor produced by vaccinia and the closely related cowpox virus is CrmA, a member of the serine protease inhibitor (serpin) superfamily. CrmA inhibits both serine and cysteine proteases, including caspases-1,-8, and -10. Myxoma virus (Serp-2) can inhibit caspases as well as the related cysteine protease granzyme B. Myxoma virus E13L also acts at the level of caspase activation by binding and inhibiting the apoptosis-associated speck-like protein containing a CARD (ASC) protein, a component of the caspase-1-activating inflammasome complex (**Table 1**).

Regulatory Viral RNAs and RNA-Binding Proteins

Virus Recognition by Cellular PRRs

Early recognition of several viruses is mediated by PRRs and results in the activation of interferon and NF-κB. Recent advances have been made in this area through the discovery of RIG-I and Mda5 as well as their downstream signaling partner, the CARD-containing mitochondrial antiviral signaling protein (MAVS). The protease NS3/4A from the flavivirus hepatitis C virus (HCV) cleaves and inactivates MAVS, thereby short-circuiting virus recognition and signaling through the CARD. Meanwhile, the same protease also cleaves the TIR domain-containing adaptor-inducing IFN-β (TRIF) and short-circuits TLR signaling as well. Vaccinia virus uses a clever trick to dampen TLR signaling; A46R contains a TIR domain to sequester TRIF away from TLR3, thus preventing the recognition of viral antigens. Other virus factors modulate RIG-I and Mda5 by alternative mechanisms, including NS2 from Influenza virus and V proteins of paramyxoviruses. B2 protein of the betanodavirus flock house virus (FHV) is an important determinant of virulence that acts by sequestering double-strand RNA and preventing its recognition by the host. RNA-binding proteins encoded by both RNA viruses (influenza, Ebola virus, and reovirus) and DNA viruses (vaccinia, CMV, and HSV) have also been implicated in modulating cellular strategies for controlling RNA levels.

The nonstructural protein (NSs) encoded on the small genome segment of a subset of mosquito-borne bunyaviruses (e.g., La Crosse virus and San Angelo virus) shares limited sequence similarity with the *Drosophila* pro-death molecule Reaper, and suppresses host cell gene expression, perhaps through an RNA interference (RNAi) mechanism, in both mammalian and cell extracts. These related proteins are also known to inhibit translation, suggesting that they act at multiple levels to direct cell death. The virus-associated interfering RNA (VAI) encoded by adenovirus acts at least in part by forming secondary structures that competitively inhibit the dsRNA binding by the endonuclease Dicer. Similarly, the RNA element TAR of HIV can bind TRBP (TAR RNA-binding protein), a cellular component of the Dicer-containing dsRNA-induced silencing complex (RISC), to inhibit RNA degradation activity. Therefore, post-transcriptional gene silencing through RNAi is involved in regulating host cell responses to RNA virus infections.

Virus-Encoded Micro-RNAs

Micro-RNAs (miRNAs) are small hairpin-forming RNAs that are processed to mature forms both in the nucleus by the nuclease Drosha and in the cytoplasm by Dicer. Just as endogenous cellular miRNAs regulate cellular protein expression, the number of known virus-encoded miRNAs is growing rapidly. HIV miR-N 367 targets and suppresses the translation of Nef and negatively regulates virus replication. The polyomavirus simian virus 40 (SV40) produces an miRNA that suppresses the T-antigen (TAg), effectively evading immune surveillance. The Bcl-2 homolog BHRF-1of EBV is regulated by three distinct miRNAs that are encoded in the 3′-UTR of BHRF1 on the opposite strand. The HSV latency-associated transcript (LAT) also encodes an miRNA, whose targets were recently identified as components of the TGF-β signaling pathway, including TGF-β itself and the downstream adapter SMAD3, the regulation of which results in the suppression of cellular stress responses.

Viral Regulators of the Mitochondrial Pathway

The mitochondrial cell death pathway is regulated by viral and cellular Bcl-2 proteins. Cellular Bcl-2 proteins are generally divided into three functional subgroups, anti-apoptotic (Bcl-2, Bcl-x_L, Mcl-1, Bcl-w), pro-apoptotic Bcl-2 family members (Bax and Bak), and the more distantly related pro-apoptotic BH3-only proteins (Bad, Bid, and others). While these classifications generally apply, there are definitive examples where the endogenous functions of these proteins are opposite to expected. For example, Bax, Bak, and Bad protect neurons from Sindbis virus-induced apoptosis, while Bcl-2 and Bcl-x_L can promote cell death when they are cleaved by caspases.

Viral Bcl-2 Family Proteins

Viral homologs of the cellular anti-apoptotic protein Bcl-2 are encoded by the γ-herpesviruses, including Epstein–Barr virus (EBV) and Kaposi's sarcoma-associated herpesvirus (KSHV/HHV8), and several different poxviruses including the myxoma virus M11L protein and fowlpox virus-Bcl-2. Viral Bcl-2 proteins have the same three-dimensional structure as their cellular counterparts, but differ from the cellular factors in that vBcl-2 proteins are resistant to normal cellular regulatory mechanisms, such as cleavage by caspases, which can convert cellular Bcl-2 into a pro-death factor. Like cellular Bcl-2, vBcl-2 prevents permeabilization of the outer mitochondrial membrane by Bax and Bak, thereby preventing cytochrome *c* release and caspase activation (**Figure 1**).

Other Viral Proteins That Target Pro-Death Cellular Bcl-2 Proteins

Adenovirus E1B-19K, vaccinia virus F1L, cytomegalovirus vMIA (viral mitochondrial inhibitor of apoptosis) and vICA (viral inhibitor of caspase-8 activation) proteins also function at the same step as Bcl-2 as they bind and regulate cellular Bax and/or Bak. vMIA is encoded by exon 1 of immediate early gene UL37. vMIA localizes to mitochondria and causes mitochondrial fragmentation/division that is generally associated with apoptotic cell death, yet vMIA potently suppresses apoptosis triggered by diverse stimuli (**Table 1**).

Yeast Viruses Induce Yeast Programmed Cell Death

Programmed cell death was assumed to arise during evolution with the origin of multicellular organisms. However, unicellular species such as yeast and bacteria also undergo programmed cell death. In mammals, virus-induced apoptosis can be responsible for the pathogenesis caused by viruses such as HIV, mosquito-borne encephalitis viruses, and many others. Most wild yeast strains are persistently infected with viruses, such as the dsRNA L-A virus and associated satellite dsRNA M viruses. These viruses are transmitted to a new host cell by cell–cell fusion. Yeast viruses are more analogous to sexually transmitted human viruses such as retroviruses and herpesviruses that remain in latently infected cells throughout the life of the host organism.

The M1 and M2 viruses of *Saccharomyces cerevisiae* encode a preprotoxin that is cleaved by the Kex proteases to produce the α- and β-chains of active M1- and M2-encoded toxins. Interestingly, yeast M1 virus preprotoxin is organized analogously to mammalian preproinsulin. The M1-encoded K1 toxin induces programmed cell death in neighboring susceptible cells by activating the cell death function of mitochondrial factors encoded by the cells destined to die. Therefore, mitochondrial factors may regulate an evolutionarily conserved cell death pathway that provides single cell organisms with a type of host defense or innate immunity. This virus–host relationship in yeast is analogous to pathogenic human viruses that persist throughout life and can occasionally reactivate to cause disease.

Concluding Remarks

There are nearly as many strategies by which viruses interface with the cell death/survival machinery of host cells as there are viruses, from initial recognition of a virus to the ultimate proteolytic dismantling of the cell by proteases. A large number of virus-encoded cell death modulators have been reported in recent years (partial list in **Table 1**), and many more are likely to be discovered. The importance of viral regulators of cell death is clear from the number of these regulators encoded by a single virus, particularly DNA viruses with large genomes, while small RNA viruses must develop additional strategies. Common themes uncovered through the study of virus-modulated cell death pathways have proved to be broadly applicable to biology, as well as to the study of virus infection, pathogenesis, and evolution.

Further Reading

Benedict CA, Norris PS, and Ware CF (2002) To kill or be killed: Viral evasion of apoptosis. *Nature Immunology* 3(11): 1013–1018.

Chen YB, Seo SY, Kirsch DG, Sheu T-T, Cheng W-C, and Hardwick JM (2006) Alternate functions of viral regulators of cell death. *Cell Death and Differentiation* 13: 1318–1324.

Flint SJ, Enquist LW, Racaniello VR and Skalka AM (eds.) (2004) *Principles of Virology: Molecular Biology, Pathogenesis, and Control of Animal Viruses,* 2nd edn. Washington, DC: ASM Press.

Hiscott J, Nguyen T-LA, Arguello M, Nakhaei P, and Paz S (2006) Manipulation of the nuclear factor-κB pathway and the innate immune response to viruses. *Oncogene* 25: 6844–6867.

Rahman MM and McFadden G (2006) Modulation of tumor necrosis factor by microbial pathogens. *PLoS Pathogens* 2(2): e4 0066–0077.

Viral Pathogenesis

N Nathanson, University of Pennsylvania, Philadelphia, PA, USA

Introduction

Viral pathogenesis deals with the interaction between a virus and its host. Included within the scope of pathogenesis are the stepwise progression of infection from virus entry through dissemination to shedding, the defensive responses of the host, and the mechanisms of virus clearance or persistence. Pathogenesis also encompasses the disease processes that result from infection, variations in viral pathogenicity, and the genetic basis of host resistance to infection or disease. A subject this broad cannot be treated in a single entry, and this article focuses on the dissemination of viruses and their pathogenicity.

Sequential Steps in Viral Infection

One of the cardinal differences between viral infection of a simple cell culture and infection of an animal host is the structural complexity of the multicellular organism. The virus must overcome a number of barriers to accomplish the stepwise infection of the host, beginning with entry, followed by dissemination, localization in a few target tissues, shedding, and transmission to other hosts.

Entry

Some viruses (such as papillomaviruses, poxviruses, and herpes simplex viruses) will replicate in cells of the skin or mucous membranes. Following infection of skin, viruses may spread further by passage of intact virions or virus-infected macrophages and dendritic cells to the regional lymph nodes. A considerable number of viruses cause sexually transmitted infection, which is usually acquired by either by penetration through breaks in the skin or by direct invasion of the superficial epithelium of the mucous membranes. Contaminated blood, semen, or secretions may introduce hepatitis B virus directly into the circulation via breaks in skin and membranes, while human immunodeficiency virus (HIV) may be trapped on the surface of dendritic cells (Langerhans cells) in the epidermis and then transported to draining lymph nodes. Although the skin is a formidable barrier, mechanical injection that breaches the barrier is a 'natural' route of entry for many viruses. More than 500 individual arboviruses are maintained in nature by a cycle that involves a vector and a vertebrate host. When the infected vector takes a blood meal, virus contained in the salivary gland is injected. Also, virus may be transmitted accidentally by contaminated needles, injection of a virus-contaminated therapeutic, transfusion with virus-contaminated blood or blood products, or tattooing.

The oropharynx and gastrointestinal tract are important portals of entry for many viruses, and enteric viruses may invade the host at a variety of sites from the oral cavity to the colon. Some viruses produce localized infections that remain confined to the gastrointestinal tract whereas others disseminate to produce systemic infection. Most enteric viruses, such as rotaviruses, infect the epithelium of the small or large intestine, but some enteroviruses, such as poliovirus, replicate in the lymphoid tissue of the gut and nasopharynx. There are numerous barriers to infection via the enteric route. First, much of the ingested inoculum will remain trapped in the luminal contents and never reach the wall of the gut. Second, the lumen constitutes a hostile environment because of the acidity of the stomach, the alkalinity of the small intestine, the digestive enzymes found in saliva and pancreatic secretions, and the lipolytic action of bile. Third, the mucus that lines the intestinal epithelium presents a physical barrier protecting the intestinal surface. Fourth, phagocytic scavenger cells and secreted antibodies in the lumen can reduce the titer of infectious virus. Viruses that are successful in using the enteric portal tend to be resistant to acid pH, proteolytic attack, and bile, and some may actually exploit the hostile environment to enhance their infectivity. The importance of anal intercourse as a risk factor for hepatitis B and HIV infection has led to the recognition that some viruses can gain entry through the lower gastrointestinal tract, but the exact mechanism of HIV infection through the anocolonic portal remains to be determined.

Respiratory infection may be initiated either by virus contained in aerosols that are inhaled by the recipient host, or by virus that is contained in nasopharyngeal fluids and is transmitted by hand-to-hand contact. Aerosolized droplets are deposited at different levels in the respiratory tract depending upon their size: those over 10 μm diameter are deposited in the nose, those 5–10 μm in the airways, and those <5 μm in the alveoli. Once deposited, the virus must bypass several effective barriers, including phagocytic cells in the respiratory lumen, a covering layer of mucus, and ciliated epithelial cells that clear the respiratory tree of foreign particles. The temperature of the respiratory tract varies from 33 °C in the nasal passages to 37 °C (core body temperature) in the alveoli. Viruses (such as rhinoviruses) that can replicate well at 33 °C but not at 37 °C are limited to the upper respiratory tract and, conversely, viruses that replicate well at 37 °C but not at 33 °C (such as influenza virus) mainly infect the lower respiratory tract. Most respiratory viruses initiate infection by replicating in epithelial cells lining the alveoli or the respiratory tree, but some viruses will also replicate in phagocytic cells located either in the respiratory lumen or in subepithelial tissues.

Spread

Once infection is established at the site of entry, most viruses will spread locally by cell-to-cell transmission of infection. Some viruses remain localized near their site of entry while others disseminate widely. Viruses that are released only at the apical surface of infected cells tend to remain localized, while those that are released at the basal surface tend to disseminate. Usually, the first step in dissemination of the infection is the transport of virus via the efferent lymphatic drainage from the initial site of infection to the regional lymph nodes, either as free virions or in virus-infected phagocytic cells.

The single most important route of dissemination is the circulation, which can potentially carry viruses to any site in the body. There are several sources of viremia. Virus may enter with the efferent lymphatic flow, or may be shed from infected endothelial cells or circulating mononuclear leukocytes. Viruses can circulate either free in the plasma phase of the blood ('plasma' viremia) or associated with formed elements ('cell-associated' viremia), and these two types of viremia have quite different characteristics. An 'active' viremia is caused by the active replication of virus in the host, which occurs after a lag period required for tissue replication and shedding into the circulation. Termination of viremia is often quite abrupt and coincides with the appearance of neutralizing antibody in the serum.

Plasma viremia reflects a dynamic process in which virus continually enters the circulation and is removed. The rate of turnover of virus within the plasma compartment is best expressed as transit time, the average duration of a virion in the blood compartment. Typically, transit times range from 5 to 60 min and tend to decrease as the size of the virions increases. Circulating viruses are removed by the phagocytic cells of the reticuloendothelial system, principally in the liver (Kupffer cells), and to a lesser extent in the lung, spleen, and lymph nodes. Once the host has developed circulating antibody, plasma virus is very rapidly neutralized in the circulation and transit time is reduced by several fold. Plasma viremias are usually short-lived (about a week), but there are notable exceptions.

A number of viruses replicate in cells that are found in the circulation, particularly monocytes, B or T lymphocytes, or (rarely) erythrocytes, and produce a 'cell-associated' viremia. Cell-associated viremia may be of short duration, as in the case of ectromelia and other poxviruses. However, in many instances, cell-associated virus titers are low, and viremia persists for the life of the host rather than terminating when neutralizing antibody appears.

There are several ways in which a virus might cross the vascular wall, but the precise mechanism for penetrating the blood–tissue barrier is unknown in most instances. Potential routes into tissues include passage between endothelial cells at sites where there are no tight junctions, translocation of virus-containing endosomes through endothelial cells, replication within endothelial cells, and tissue invasion of virus-infected lymphocytes or monocytes.

Some viruses can disseminate by spreading through the axons of peripheral nerves. Although less important than viremia, neural spread plays an essential role for certain viruses (such as rabies viruses and several herpesviruses) while other viruses (such as poliovirus and reovirus) can utilize both mechanisms of spread.

Localization or Tropism

One of the salient features that distinguishes viruses is their localization or tropism within the animal host. The names of individual viral diseases often reflects the organs or tissues that are involved; thus, smallpox, poliomyelitis, and hepatitis each have their characteristic features. Tropism is regulated both by viral dissemination and cellular susceptibility. Disease localization may not correspond to the distribution of infection since it reflects both the spread of the virus and the host response to infection. The localization of a virus is determined at several phases of infection, including the portal of entry, systemic spread by viremia or the neural route, and the invasion of local organs or tissues.

Most viruses are quite selective in the cell types that are infected *in vivo* and this selectivity plays a significant role in localization. One major determinant of cellular susceptibility is the presence of viral receptors, namely, molecules on the cellular surface that act as receptors for the specific virus, usually by binding a protein on the virion surface (the viral attachment protein). For instance, poliovirus can only replicate in primate tissues, because the poliovirus receptor is expressed only by primate cells. Transgenic mice expressing the the poliovirus receptor can be infected and develop typical paralytic poliomyelitis after intracerebral injection.

Cellular susceptibility also may be determined at post-entry steps in the replication cycle. An example is Newcastle disease virus, a paramyxovirus of birds. Virulent isolates of Newcastle disease virus encode a fusion protein that is readily cleaved by furin, a proteolytic enzyme present in the Golgi apparatus, so that the virion is activated during maturation prior to reaching the cell surface, and before budding of nascent virions. This makes it possible for virulent strains of the virus to infect many avian cell types, thereby increasing its tissue host range, and causing systemic infections that are often lethal. Avirulent strains of Newcastle disease virus cannot be cleaved by furin and depend upon extracellular proteases, which restricts the spread of the virus beyond the respiratory tract.

For oncogenic viruses, the tumors may represent a distant echo of virus localization. Epstein–Barr virus (EBV) infects epithelial cells and B cells. Latently infected B cells are immortalized, and, rarely, these B cells undergo further transformation into lymphomas by virtue of additional genetic events such as chromosomal transposition (t8:14).

Shedding and Transmission

Acute viral infections are characterized by brief periods (days to a few weeks) of intensive virus shedding into respiratory aerosols, feces, urine, or other bodily secretions or fluids. Persistent viruses are often shed at relatively low titers, but this may be adequate for transmission over the prolonged duration of infection (months to years). Although many viruses replicate in the skin, relatively few are spread from skin lesions. Several viruses are shed in the semen, including hepatitis B and human immunodeficiency virus. A number of viruses are excreted in colostrum and milk, including cytomegalovirus, mumps, rubella, and HIV. Blood is also an important source of transmitted virus, particularly for those viruses that produce persistent plasma or cell-associated viremias, such as hepatitis B, C, and D viruses, HIV, HTLV-I, HTLV-II, and cytomegalovirus, and which can be transmitted by contaminated needles and transfusions. A number of viruses have been isolated from the urine but viruria is probably not important for transmission of most viruses.

Once shed, there are a several means by which a virus is transmitted from host to host in a propagated chain of infection. Probably the most important mechanism is contamination of the hands of the infected transmitter from feces, oral fluids, or respiratory secretions expelled during coughing or sneezing. The virus is then passed by hand-to-hand contact leading to oral, gastrointestinal, or respiratory infection. A second common route is inhalation of aerosolized virus. A third significant mechanism involves direct person-to-person contact (oral–oral, genital–genital, oral–genital, or skin–skin). Finally, indirect person-to-person transmission can occur via blood or contaminated needles. Common source transmissions are less frequent but can produce dramatic outbreaks, via contaminated water, food products, or biologicals such as blood products and vaccines.

Viral Virulence

Virulence refers to the ability of a virus to cause illness or death in an infected host, relative to other isolates or variants of the same agent. The study of virulence variants can provide important insights into pathogenesis, and carries the potential for development of attenuated live virus variants that can be used as vaccines. One approach to virulence is to map the genetic determinants that underlie the virulence phenotype; the other approach is a study of pathogenesis, which describes the differences in infections with viral variants of different virulence. Measures of virulence may be quantitative, or qualitative, since variants of a single virus may differ markedly in their cellular tropism, their mode of dissemination in the infected host, or the disease phenotype that they produce.

Viral Genetic Determinants of Virulence

Viral virulence is encoded in the viral genome and expressed through structural proteins, nonstructural proteins, or noncoding sequences. A large body of information about virus variants has established several principles.

1. The use of mutants has made it possible to identify the role of individual genes and proteins as determinants of the biological behavior of many viral variants.
2. There is no 'master' gene or protein that determines virulence, and attenuation may be associated with changes in any of the structural or nonstructural proteins, or in the noncoding regions of the genome.
3. The virulence phenotype can be altered by very small changes in the genome, if they occur at 'critical sites'. At such sites, a single point mutation leading to the substitution of a specific amino acid or base is often sufficient to alter virulence. For most viruses with small genomes (<20 kb), only a few discrete 'critical sites' have been discovered, usually fewer than 10 per genome.
4. It is possible to create variants with attenuating mutations at several critical sites and these may be more attenuated than single point mutants, and less prone to reversion.
5. Attenuating mutations are often host range mutations that affect replication in some cells but not in others.
6. Although many attenuating mutations have been sequenced, relatively few have been characterized at a biochemical or structural level as to their mechanism of action.

Genetic determinants of viral virulence and attenuation may be illustrated with examples from a few selected virus groups. The attenuation of the vaccine strains of poliovirus is due to critical sites, both in the 5′ nontranslated region (NTR) upstream from the long open reading frame and in selected structural and nonstructural proteins. Each of the three strains of oral polio vaccine (OPV) carry attenuating point mutations in the 5′ NTR (at positions 480, 481, or 472, in types 1, 2, and 3, respectively). Attenuating mutations at these sites are associated with reduced neurovirulence in monkeys and reduced ability to replicate in the central nervous system of transgenic mice bearing the poliovirus receptor. Each of the OPV strains carries at least one mutation that is associated with an alteration in a viral structural or nonstructural protein and confers temperature sensitivity and reduced neurovirulence, and these mutations involve different proteins for the three OPV strains. It has been suggested that mutations in the structural proteins produce attenuation via structural transitions that occur either during virion uncoating or assembly.

Bunyaviruses are negative stranded RNA viruses with a trisegmented genome. The large (L) RNA segment encodes the viral polymerase; the middle (M) RNA segment encodes two glycoproteins (G1 and G2) and a nonstructural protein, Ns_m; the small (S) segment encodes the nucleoprotein (N) and a nonstructural protein, Ns_s. Attenuation can involve either the ability of the virus to replicate in the central nervous system (neurovirulence) or its ability to cause viremia and reach the central nervous system (neuroinvasiveness). Reduced neuroinvasiveness of one bunyavirus virus strain is associated with reduced ability to replicate in striated muscle and consequent low viremogenicity. Attenuation maps to the M RNA segment encoding the viral glycoproteins suggesting that there is an alteration in the infection of myocytes (but not of neurons). Another attenuated bunyavirus showed a striking reduction in its ability to replicate in the central nervous system of adult mice, and attenuation maps to the L RNA segment. The attenuated virus replicates poorly in C1300 NA neuroblastoma and other murine cell lines, and presumably has a mutant polymerase that restricts viral replication in neurons and certain other cell types.

Reoviruses are double-stranded 10-segmented RNA viruses. Most of the virulence determinants represent qualitative differences between type 1 Lang and type 3 Dearing in tissue and organ tropism rather than quantitative differences in disease severity

1. Reovirus type 1 Lang disseminates through the blood and causes an ependymitis in the brain while type 3 Dearing disseminates through the neural route and causes a neuronotropic encephalitis in the brain. The S1 segment encodes the $\sigma 1$ protein, which is the viral attachment protein, and is the major determinant of dissemination and brain cell tropism.
2. The M1 segment encoding the $\mu 2$ protein influences replication in cardiac myocytes and myocarditis.
3. The M2 segment encoding the $\mu 1$ protein affects the protease sensitivity of the virion, and quantitative neurovirulence.

4. The L1 segment encoding λ3 protein influences replication in cardiac myocytes and myocarditis.
5. The L2 segment encoding λ2 protein influences replication levels in the intestine, titers of shed virus, and horizontal transmission between mice.

A new class of virus-encoded proteins has been recognized that contribute to the virulence of viruses by mimicking normal cellular proteins. This group of 'cell-derived' genes has been identified primarily within the genomes of large DNA viruses that have a greater capacity to maintain accessory genes than do viruses with small genomes. 'Virokines' are secreted from virus-infected cells and mimic cytokines, thereby perturbing normal host responses. 'Viroceptors' resemble cellular receptors for cytokines (including antibodies or complement components) that are thereby diverted from their normal cellular targets. Some virus-encoded proteins interfere with antigen presentation and immune induction, while others prevent apoptosis, or interrupt intracellular signaling initiated by cytokines or interferons.

Pathogenic Mechanisms of Virulence

There are many sequential steps in a viral infection and a difference in the comparative replication of two virus variants at any one of these steps can account for their relative degrees of virulence. Attenuated variants usually will replicate or spread less briskly in one or several tissues associated with the pathogenic process. A reduction in viremia can be an important mechanism for the reduction in virulence. For instance, the Mahoney strain of type 1 poliovirus causes a much higher viremia and paralytic rate than several other poliovirus strains that are less viremogenic. Different variants of a single virus can exhibit differences in their tropism for organs, tissues, and cell types, which confers a multidimensional character upon virulence. The three attenuated strains in the oral poliovirus vaccine provide an example. Both attenuated and wild-type polioviruses are able to replicate in the gastrointestinal tract to similar levels, judging from the shedding of virus in the feces, but the attenuated strains show a reduction of about 10 000-fold (relative to the wild-type strains) in their ability to replicate in the spinal cord after intraspinal injection. In this instance, virulence for the target organ (the spinal cord), rather than viremogenicity, appears to be the most important determinant of virulence.

Host Determinants of Susceptibility and Resistance

Studies with inbred animals have documented that the outcome of infections often varies in different strains of mice, and genetic analyses have been conducted to determine the form of inheritance of susceptibility, and to identify the responsible gene. Several generalizations may be made:

1. Most genetic loci control susceptibility to a specific family of viruses and not to all viruses.
2. Susceptibility can often be mapped to a single autosomal locus, but multiple loci have been identified in some instances.
3. Susceptibility may be either dominant or recessive.
4. Where loci have been mapped, they usually are distant from the major histocompatibility locus, and in most such instances, the mechanism of susceptibility is not immunological.
5. The exact mechanism of susceptibility has not been well defined in most instances, although there are some examples that map to the major histocompatibility complex (MHC), implying an immunological explanation.

It is difficult to investigate genetic determinants of host susceptibility in outbred populations of animals or humans. However, recent studies of HIV-1 have provided substantial evidence of a number of human genetic determinants of susceptibility to infection or to the rate of progression to overt AIDS. The most striking of these is a polymorphism in the gene that encodes the chemokine receptor, CCR5, which is the major co-receptor for HIV-1. Some individuals have a deletion in the gene (the Δ32 deletion) such that no CCR5 protein is expressed. Those who are homozygous for the Δ32 mutation are at very low risk of HIV infection, and those who are heterozygous for the Δ32 deletion can be infected but have prolonged incubation periods to onset of AIDS by comparison with CCR5+ homozygous individuals.

The host response to a viral infection may also be influenced by a variety of physiological variables such as age, sex, stress, and pregnancy, all of which have been shown to influence the outcome of infection with some viruses. In general, very young animals are more susceptible than adult hosts to acute viral infections, and are more likely to undergo severe or fatal illness. The innate high susceptibility of infants born of nonimmune mothers is revealed only in those rare instances where a virus has disappeared from an isolated community so that all age groups lack immune protection. A classical example is measles, which can be a devastating disease with a mortality of up to 25% in nonimmune infants. Less commonly, advanced age can also be associated with increased susceptibility to viral infection. St. Louis encephalitis and West Nile encephalitis are examples of this phenomenon.

See also: Defeating Innate Immunity; Innate Immunity; Persistent and Latent Viral Infection.

Further Reading

Barber DL, Wherry EJ, Casopust D, *et al.* (2006) Restoring function in exhausted CD8 T cells during chronic viral infection. *Nature* 439: 682–687.

Martin MP and Carrington M (2005) Immunogenetics of viral infections. *Current Opinion in Immunology* 17: 510–516.

Nathanson N, Ahmed R, Brinton MA, *et al.* (eds.) (2007) *Viral Pathogenesis and Immunity,* 2nd edn. London: Academic Press.

Nathanson N and Tyler K (2005) The pathogenesis of viral infections. In: Mahy BWJ and ter Meulen V (eds.) *Topley and Wilson's Microbiology*, pp. 236–269. London: Hodder Arnold.

Racaniello VR (2006) One hundred years of poliovirus pathogenesis. *Virology* 344: 9–16.

Tumpey TM, Basler CF, Aguilar PV, *et al.* (2005) Characterization of the reconstructed 1918 Spanish influenza pandemic virus. *Science* 310: 77–80.

Antiviral Agents

H J Field, University of Cambridge, Cambridge, UK
R A Vere Hodge, Vere Hodge Antivirals Ltd., Reigate, UK

Introduction

There have been two previous articles in *Encyclopedia of Virology,* the latter being in 1999. That review, by A. K. Field and C. A. Laughlin, gave a good update, especially in the treatment of HIV. In order to avoid repetition, we have focused on antiviral targets. We have not attempted to cover vaccines nor immunomodulating agents, such as interferons, except when these are mentioned briefly in those cases in which they are the therapy of choice. These approaches which utilize a cellular target are outside the scope of this article but the reader is referred to a recent review by L. Schang.

The term 'virucidal' has long been used to describe an agent that destroys virions while outside the host cell. A few compounds inactivate viruses by highly specific mechanisms but many, such as alcohol wipes, have a broad spectrum of activity against many infectious entities. Virucidal agents have an important role, especially in hospitals, in preventing the transmission of viruses and may be useful in reducing the risk of virus transmission between individuals, for example, during sexual contact. These types of agent will not be considered further in this article.

The focus of this article is those antiviral compounds which target virus replication selectively. After a brief historical overview, we discuss the concepts and challenges which have guided and stimulated progress with antiviral therapy – selectivity, spectrum, viral targets, prodrugs, resistance and drug combinations, viral fitness, latency. A summary of the most important antiviral agents is given in **Table 1**. This article outlines the principles underlying the successful development of antiviral compounds for the many clinical applications available today. Some prospects for the future are discussed.

Historical Perspective

The First Uncertain Steps

Following the discovery of antibiotics for treating bacterial infections, for several decades it was thought that 'safe' antiviral chemotherapy would be difficult if not impossible. The earliest antiviral compounds only emphasized the problems. Marboran was introduced in the early 1960s to treat smallpox and vaccinia; its effectiveness was equivocal and its use short-lived. Amantadine and rimantadine were used to treat influenza with relatively little side effects but the influenza virus became resistant within a few days of treatment and the resistant virus spread readily to contacts. The first nucleoside analogs, with antiviral activity, emerged during the search for drugs to treat cancer. The first of these, idoxuridine (IDU), was discovered to be active against herpes viruses by Dr. William (Bill) Prusoff in 1959. However, its toxicity limited its use to topical treatments only (e.g., infections of the eye). Vidarabine (Ara A) was slightly more selective; its main systemic use was to treat herpes encephalitis. No convincing selective antiviral compound had yet been discovered.

The Major Breakthrough

The discovery of the antiherpes drug, acyclovir (ACV) in 1978, was the major milestone in antiviral therapy. For the first time, it was demonstrated that an effective, nontoxic, antiviral drug is an achievable aim. Moreover, chronic suppressive acyclovir therapy for several years, to prevent the misery of recurrent herpes, is possible without adverse effects. More than two decades of worldwide use has proved that acyclovir is one of the safest drugs in clinical therapy. Acyclovir proved that not all nucleoside analogs had to be mutagenic and/or

Table 1 List of important current antiviral agents

(a) Primarily active against Herpes viruses

Generic name (abbreviation) Trade name (company)	*Structural information*	*Mechanism viral target*	*Fig 1 stage*	*Target viruses*
Valaciclovir (VACV) Valtrex® (GSK)	Prodrug of acyclovir (ACV) Zovirax®	Activated by viral TK, inhibits viral polymerase	VI	HSV-1 and -2 VZV (CMV)

Notes: ACV is extremely safe and, apart from some problems related to high doses and crystallization in the kidney, no serious side effects have been encountered. Drug-resistance in immunocompetent patients has remained rare (< 1%) in over two decades of clinical experience. In immunocompromised patients, resistant variants (generally in TK and occasionally in DNA polymerase) occur in up to 5% patients. Fortunately, TK-deficient HSV appears to be less pathogenic and less able to reactivate from latency than wt HSV.

VACV is a valine ester of ACV. In addition to the main target viruses, VACV is sometimes used to prevent CMV infections in transplant patients. During suppressive therapy for genital herpes, VACV reduced the HSV-transmission rate by about 50%.

Generic name (abbreviation) Trade name (company)	*Structural information*	*Mechanism viral target*	*Fig 1 stage*	*Target viruses*
Famciclovir (FCV) Famvir® (Novartis)	Prodrug of penciclovir (PCV) Denavir®/ Vectavir®	Activated by viral TK, inhibits viral polymerase	VI	HSV-1 and -2 VZV

Notes: This was the first antiviral agent to be developed as the prodrug and stimulated the search for an ACV prodrug. FCV was shown to be highly effective, vs. placebo, against HSV-1 and -2 and VZV. In the only clinical trial comparing FCV and VACV for the treatment of shingles, they appeared to be equally effective. Interestingly, PCV has a higher affinity than ACV for HSV TK but is a less potent inhibitor of herpes DNA polymerase. However, the intracellular half-life of PCV triphosphate is significantly longer than ACV triphosphate and this may help to explain its potent antiviral activity. The dynamics of resistance selection may differ subtly from ACV but clinical resistance in immunocompetent patients remains extremely rare.

Generic name (abbreviation) Trade name (company)	*Structural information*	*Mechanism viral target*	*Fig 1 stage*	*Target viruses*
Foscarnet (PFA) Foscavir® (Astra Zeneca)	Pyrophosphate analog	Polymerase inhibitor	VI	HSV-1 and -2 VZV

Notes: PFA is an effective inhibitor of HSV replication. However, the compound must be administered iv and suffers from toxicity (e.g., tendency to accumulate in bone). Therefore it is used in immunocompromised patients with herpes infections resistant to ACV and PCV.

Generic name (abbreviation) Trade name (company)	*Structural information*	*Mechanism viral target*	*Fig 1 stage*	*Target viruses*
Valganciclovir (VGCV) Valcyte® (Roche)	Prodrug of ganciclovir (GCV) Cymmevene®	Activated by kinase encoded by UL 97, polymerase inhibitor	VI	CMV

Notes: VGCV has become the drug of choice for the prevention and treatment of CMV in immunocompromised patients.

(b) Primarily active against RNA viruses

Generic name Trade name (company)	*Structural information*	*Mechanism viral target*	*Fig 1 stage*	*Target viruses*
Zanamivir Relenza® (GSK)	Sialic acid analog	Neuraminidase inhibitor	X	Influenza A and B

Notes: Zanamivir prevents the influenza virions leaving the infected cell. It can be used either for the treatment or prophylaxis of influenza infections. It has a good safety record but its use is limited by the need for inhalation. It is currently being stockpiled, as the second line agent, in preparation for a possible outbreak with H5N1 virus.

Generic name Trade name (company)	*Structural information*	*Mechanism viral target*	*Fig 1 stage*	*Target viruses*
Oseltamivir Tamiflu® (Roche)	Sialic acid analog	Neuraminidase inhibitor	X	Influenza A and B

Notes: Like zanamivir, oseltamivir has good, selective activity against influenza viruses. In contrast to zanamivir, it has good bioavailability which has led to it becoming the first line agent for treatment of influenza. In adult patients, resistance is rare and it appears that those resistant strains are partially disabled. It is the main agent for stockpiling for an H5N1 outbreak.

An inhibitor of influenza RNA polymerase, T-705, has demonstrated excellent activity in cell culture assays and animal models. It is expected to be entering clinical trials soon. With the threat of an H5N1 outbreak, it will be useful to have another treatment option, possibly used in combination with oseltamivir.

Continued

Table 1 Continued

(b) Primarily active against RNA viruses

Generic name Trade name (company)	*Structural information*	*Mechanism viral target*	*Fig 1 stage*	*Target viruses*
Ribavirin Virazole® (Schering-Plough)	Nucleoside analog	Possibly, no direct viral target	?	HCV RSV

Notes: HCV – Current therapy is with one of several interferon forms, usually with ribavirin which seems to enhance the chance for long-term response. Being immune modulators, these are outside the scope of this review. However, antiviral agents, targeting either HCV protease or polymerase, are being developed.

RSV – Although ribavirin has been used to treat serious cases of RSV, its efficacy has been questioned. Palivizumab and RSV immune globulin (RSV-IGIV) are FDA approved.

(c) Primarily active against Hepadnaviruses (HBV)

Generic name (abbreviation) Trade name	*Structural information*	*Mechanism viral target*	*Fig 1 stage*	*Company*
Lamivudine (3TC) Zeffix®, Heptovir®	NA inhibitor	Polymerase	VI	GSK

Notes: Lamivudine has become the first line choice for therapy of chronic HBV although viral resistance appears in about 50% of patients after two years monotherapy. Lamivudine is unusual as it is in routine use for both HBV and HIV.

Adefovir dipivoxil (ADV) Hepsera®	Prodrug of adefovir NA inhibitor	Polymerase	VI	Gilead

Notes: In comparison with lamivudine, the rate of appearance of resistance is much slower, in about 2.5% patients after 2 years monotherapy.

Entecavir Baraclude®	NA inhibitor	Polymerase	VI	BMS

Notes: Entecavir was approved by FDA for HBV in 2006. It is very active with tablet sizes of 0.5 and 1 mg.

Telbivudine Tyzeka®	NA inhibitor	Polymerase	VI	Idenix/Roche

Notes: Telbivudine is the most recent (2007) of FDA approved agents for HBV. Telbivudine has an excellent safety record in animal toxicity tests. Although the tablet size (600 mg) is larger than with the other compounds, it is perhaps the most effective in reducing HBV DNA levels but it should not be used to treat patients failing on lamivudine therapy. It may replace lamivudine as the first line therapy.

(d) Primarily active against Retroviruses (HIV)

Generic name (abbreviation) Trade name	*Structural information*	*Mechanism viral target*	*Fig 1 stage*	*Company*
Zidovudine/Lamivudine (AZT/3TC) Combivir®	Two NRTIs	Polymerase	VI	GSK

Notes: As single agents, the trade names are Retrovir®/Epivir® respectively.

AZT/3TC/abacavir Trizivir®	Three NRTIs	Polymerase	VI	GSK

Notes: The trade name for abacavir as a single agent is Ziagen® and, when combined with lamivudine, Epzicom® or Kivexa®.

Emtricitabine/tenofovir/ efavirenz Atripla®	Two NNRTIs and one NNRTI	Polymerase	VI/VI/VII	Gilead & BMS (Jointly)

Notes: The trade name for the combination emtricitabine/tenofovir is Truvada®. As single agents, the trade names are Emtrive®/ Viread®/Sustiva® respectively.

Continued

Table 1 Continued

(d) Primarily active against Retroviruses (HIV)

Generic name (abbreviation) Trade name	*Structural information*	*Mechanism viral target*	*Fig 1 stage*	*Company*
Nevirapine Viramune®	NNRTI	Polymerase	VII	Boehringer
Notes: Should be used with at least two other agents, one of which should be an NRTI.				
Fosamprenavir Lexiva®	PI	Protease	V	GSK
Notes: Fosamprenavir is the oral prodrug of amprenavir (Agenerase®) and is replacing it in clinical use.				
Saquinavir mesylate Invirase® Fortovase® (Roche)	PI	Protease	V	Roche
Notes: Invirase capsules and tablets should be used only with ritonavir which significantly inhibits the metabolism of saquinavir to provide plasma saquinavir levels at least equal to those achieved with Fortavase capsules which should be use, if at all, as the sole PI in a regimen.				
Lopinavir/ritonavir Kaletra®	2 PIs	Protease	V	Abbott Lab
Notes: Lopinavir is not marketed separately but combined with ritonavir (ratio 4:1) which enhances the plasma levels of lopinavir. The trade name for ritonavir is Norvir®.				
Indinavir Crixivan®	PI	Protease	V	Merck
Notes: Indinavir should be used only with other agents.				
Darunavir Prezista®	PI	Protease	V	Tibotec
Notes: Darunavir was given fast track status and was approved in 2006. Darunavir has less than tenfold decreased susceptibility in cell culture against 90% of 3309 clinical isolates resistant to other PIs.				
Efuviritide (T-20) Fuzeon® (Roche)	Fusion inhibitor	gp41 envelope protein	II	Roche
Notes: Administered sc bid. Mainly used in salvage therapy. It is the only FDA approved fusion inhibitor.				

Further information, structures, and references may be obtained from AVCC FactFile and company web sites.

carcinogenic. The early doubts, about the potential for discovering safe antiviral agents, were dispelled for ever.

The selectivity of acyclovir for the herpes viruses is dependent on it being activated only in herpes virus-infected cells. The critical initial step in that activation is its phosphorylation to ACV-monophosphate by the viral thymidine kinase. Cellular enzymes convert the monophosphate to the triphosphate which inhibits selectively herpes viral DNA polymerase. Following the discovery of acyclovir, other selective antiherpes virus compounds were discovered, including BVDU, BvaraU, and penciclovir (PCV). Currently, ACV and PCV, and their prodrugs valaciclovir (VACV) and famciclovir (FCV), respectively, are the most commonly used drugs to treat infections with HSV-1 and -2 and VZV. Another compound, ganciclovir (GCV), was discovered to have good activity against CMV. Like ACV and PCV, GCV is activated specifically in virus-infected cells but by the kinase encoded by the CMV UL97 gene. The prodrug of GCV, valganciclovir (VGCV), is the preferred therapy for CMV prevention and treatment in immunocompromised patients.

Impetus from the Challenge of HIV

Following the emergence of AIDS, an enormous research effort worldwide was directed at the search for inhibitors of HIV, resulting in a large number of new drugs. Initially, these inhibitors were nucleoside analogs. These had to be phosphorylated (activated) to their triphosphate entirely by cellular enzymes; HIV does not encode a thymidine kinase. Therefore, the selectivity was due entirely to the greater inhibition of HIV reverse transcriptase than the cellular polymerases. Because the medical need was so great, several inhibitors were brought quickly into clinical use even though they were associated with long-term toxicity problems, for example, mitochondrial toxicity with didanosine (ddI) and zalcitibine (ddC) and nephrotoxicity with cidofovir (HPMPC), used to treat CMV in AIDS patients. More recently, great progress has been made in selecting nucleoside and nucleotide analogs (**Table 1**) with much less potential for toxicity and so long-term therapies became normal practice. Once HIV protease had been shown to be essential for HIV replication, protease inhibitors (PIs) were discovered. Although these inhibitors have a high degree of selectivity for the HIV protease, they are not without some side effects, including the development or redistribution of fatty masses. However, one of the PIs is commonly included in the combination therapy now known as HAART (highly active anti-retroviral therapy).

Expansion into Therapies for Other Viruses

A spin-off from this work with HIV has led to inhibitors of other families of viruses, particularly hepatitis B virus (HBV). Lamivudine and adefovir have become the treatments of choice for HBV. Meanwhile, there seemed to be progress discovering compounds active against picornaviruses (which include rhinoviruses causing the common cold). These compounds bind into a pocket within the viral capsid. The best example is pleconaril although unacceptable side effects stopped its development. However, the viruses quickly became resistant and some strains even became dependent on the 'antiviral' compound. Therefore, this approach has not resulted in any clinically useful drugs.

Once the structure of influenza virus neuraminidase was known, new inhibitors of influenza viruses were discovered. Recent advances with replication systems for hepatitis C virus (HCV) have allowed the discovery of anti-HCV compounds. Currently, the threat of bio-terrorism has prompted the successful search for drugs active against poxviruses and, with the prospect of the next influenza pandemic looming, there is renewed research effort directed to anti-influenza drugs. There are at least 25 (**Table 1**) widely used licensed antiviral agents and that number looks set to increase rapidly in the coming years.

Key Concepts

Definition of Selective Activity Index

Viruses are obligate intracellular parasites. Thus, inhibitors of virus replication must do so without toxic effects on the cells, tissues, and organs of the host. This is the concept of 'selective activity' against viruses.

For those viruses which form plaques in a cell monolayer, reduction in plaque formation has long been regarded as the 'gold standard' assay. Antiviral activity is usually defined as the concentration of the inhibitor which reduces viral plaques by 50% (50% effective concentration; EC_{50}). Alternatively, the reduction in the yield of infectious virus can give useful information by using different multiplicities of infection (MOI); in such assays, the concentration of inhibitor to reduce replication by 99% (EC_{99}) is usually reported. For some viruses, for example, HBV, reduced production of a measurable virus product (e.g., nucleic acid or protein) is the only way to assess viral replication. For inhibition of a viral enzyme, the 50% inhibitory concentration (IC_{50}) is commonly reported.

Irrespective of the antiviral assay, it is necessary to assess cellular toxicity of the test compound. This is essential to eliminate the possibility that the lack of virus replication is due to the destruction of the cells. However, the ratio of the concentration to inhibit the replication of the virus to the concentration to destroy nondividing cells in a monolayer is *not* an indication of selective activity. To determine selective activity, it is essential to compare like with like, replicating virus with replicating cells. To test for an effect on replicating cells, it is necessary to aim for a tenfold increase (just over three doublings) so that at the 50% cytotoxic concentration (CC_{50}), at least one doubling has occurred and the second round of replication has started. The ratio of 50% viral inhibitory concentration to 50% cytotoxic concentration is defined as the selective index (SI).

In the literature, authors may not abide by the above definition of selective index. Therefore, published values for selectivity should be considered with some skepticism unless the methodology is fully described. We urge editors of antiviral articles to be aware of this difficulty.

The determination of CC_{50} values in a single assay may be sufficient to select compounds in a primary screen. For assessing compounds worthy of progression into development, it is prudent to assess the compounds in a variety of cytotoxicity tests using different cells. For example, inhibition of mitochondrial DNA can be indicated by using granulocyte-macrophage (CFU-GM) and erythroid (BFU-E) cells. Human bone marrow stem cells in primary culture are a good predictor for potential hematoxicity. A particular class of compounds may be known to have a potential for a particular toxicity, such as phosphonates for nephrotoxicity. In this case, the established pattern of nephrotoxicity in the clinic was reflected in a cell culture assay using primary human renal proximal tubule cells (RPTCs).

When antiviral agents are evaluated against virus infections *in vivo*, the ratio of the effective dose (ED_{50}) to the minimum toxic dose is sometimes referred to as the 'therapeutic ratio'. These measurements are widely used during the preclinical development of new antiviral agents and are often helpful in making comparisons between alternative compounds.

Toxicity of Antiviral Compounds

As mentioned above, SIs provide a numerical estimate of the degree of selectivity for antiviral compounds and may be useful in proceeding with structure–activity relationships (SARs) from a lead compound. However, these results must be treated with great caution when extrapolating to man. Although a high (>1000) SI value may be encouraging, a low SI should not be ignored. During preclinical evaluation, many potential toxic effects can be assessed. For example, pharmacokinetic studies can identify drug–drug interactions. For nucleoside analogs, it is not sufficient to study the possible interactions in plasma; it is also necessary to study drug–drug interactions on phosphorylation within cells. There are well-established tests for indicating potential genetic toxicity of nucleosides and nucleotides and toxicity for mitochondrial enzymes. However, there always remains the possibility for unexpected effects such as lipodystrophy which may not become apparent until the drug is used in man – possibly for several years. In some cases, these complications once identified can be avoided or managed. In other cases, for example, ddI and ddC, the development of safer options eclipses the older compounds. For those infections requiring long-term therapies, over a year for HBV, several years for suppression of recurrent genital herpes, over 10 years for HIV, the test for clinical safety is very exacting. It is a credit to research workers in antiviral chemotherapy that compounds, which have stood the test of time, have been brought into clinical practice.

Spectrum of Antiviral Action

There are few, if any, truly broad spectrum antivirals. Given the highly specific nature of the mechanism of action of most antivirals, this is unsurprising. Most antiviral compounds affect a narrow range of viruses – often restricted to a single virus family or particular subfamily. Telbivudine is highly specific for the HBV (and the related duck and woodchuck viruses) and is inactive against all other viruses tested. Acyclovir is typical in being active against some viruses within one family, the herpes viruses. Lamivudine is unusual in becoming a common treatment for both HIV and HBV.

Broad-spectrum activity against all enveloped viruses does seem to be a possibility. Rep 9 is a phosphorothioate oligonucleotide with a random order of bases (A, C, G, and T), the antiviral activity being dependent not on the base sequence but on the size, the optimal being about 40–50 bases. This oligonucleotide has both hydrophobic (along the backbone) and hydrophilic (bases) surfaces which seem to be essential for activity. Various enveloped viruses have a surface protein (e.g., HIV gp41, influenza hemagglutin) with an alpha-helix which matches the length of the 40 nucleotides in REP 9. The viruses that have been shown to be sensitive to REP 9 include members of the *Herpes-*, *Orthomyxo-*, *Paramyxo-*, and *Retroviridae.*

Virus Replication Cycle Presents Antiviral Targets

The Virus Replication Cycle

All viruses are dependent upon a host cell for protein synthesis. Thus, all viruses replicate via a broadly similar sequence of events (**Figure 1**). The virus must first attach ('adsorb') to the cell. The virus or virus nucleic acid genome then enters ('penetrates') the cytoplasm. The genome is liberated from the protective capsid ('uncoats'). In some cases the viral genome enters the cell nucleus; in all cases the genome is transcribed and thus the viral mRNA directs protein synthesis. Virus products include proteins that regulate further transcription; enzymes including those involved in genome replication; proteases that process other virus-induced proteins; and many proteins that subvert host-defense mechanisms within and without the infected cell. The virus undergoes 'genome replication' and together with virus structural proteins form new virions ('assembly') which are 'released' from the cells. In some cases release is associated with budding through cellular membranes that have been decorated with virus glycoproteins enabling the virus to acquire its envelope. All stages are highly regulated both temporarily and quantitatively, and vary for each particular virus.

Important Classes of Antiviral Targets and Compounds

The major targets and compounds are illustrated in **Figure 1**.

Virus attachment

Currently, there are no major drugs which prevent virus attaching to the host cell. The problem has been illustrated by the picornavirus capsid-binding compounds such as pleconaril. Other compounds are being tried for HIV but with little success so far. It seems that many sites on the surface of a virus can undergo changes without great loss of viral viability; the compounds present too low a genetic barrier (see below).

HIV fusion inhibitors

To date, only one entry inhibitor has been approved by the US Food and Drug Administration (FDA). This is efuviritide (T-20), a synthetic peptide that targets the HIV gp41 envelope protein to prevent fusion. The broad-spectrum antiviral agent, Rep 9, is progressing through development.

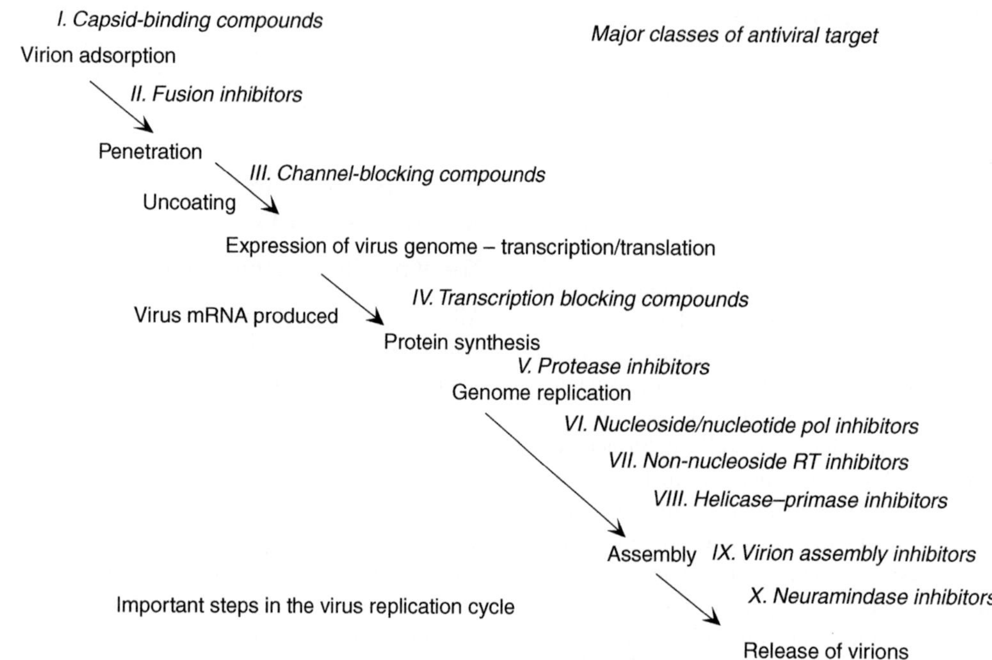

Figure 1 Key steps in the virus replication cycle that provide antiviral targets.

Influenza channel-blocking compounds

Amantadine and rimantadine act by binding to M2 and thus interfering with the penetration of hydrogen ions into the virion, a process essential for uncoating the single-strand RNA genome. Key resistant mutations map to the *M2* gene (and to a lesser extent, HA gene). Such resistant variants are selected quickly and are fully viable. This has severely limited the clinical usefulness of these drugs. During the 2005/06 season in the USA, 109/120 (91%) of H3N2 isolates were resistant to amantadine and rimantadine. Combination with other drugs now seems unlikely to be a viable option and so their future role may be minimal.

Antisense binding to viral mRNA

The antisense approach has the potential for great selectivity due to the sequence of bases in the oligonucleotide being complementary to a particular viral sequence. Conceptually, it may be possible to interfere with latent virus, even when the HIV genome has integrated into the DNA of a long-lived human cell. In practice, there have been many difficulties in progressing antisense compounds to the clinic. So far, only formivirsen has been licensed by the FDA. Its use is limited to intravitreal inoculation in patients with CMV retinitis.

Protease inhibitors

Many viruses encode one or more enzymes involved in the cleavage of virus or cellular proteins. HIV induces an enzyme that specifically cleaves protein precursors in the maturation of gag and pol polyproteins. Peptide mimetics were designed specifically to target HIV protease. The resulting compounds, several of which are now standard HIV drugs, are among the most active antiviral compounds known and inhibit HIV in the nanomolar range. Like other HIV antivirals, drug-resistant mutants arise and so the compounds are used in combination with reverse transcriptase inhibitors. Other families of viruses encode proteases; except for HCV, the search for effective inhibitors has yielded few useful compounds.

Viral DNA/RNA polymerase inhibitors

Nucleoside and nucleotide analogs play a pivotal role in antiviral chemotherapy. The ultimate target is the viral DNA/RNA polymerase which is well conserved within different strains of a particular virus and sometimes within a virus family. For those nucleosides which are activated initially to their monophosphates by a viral enzyme, there is an additional potential for selectivity. In general, the viral enzymes are less stringent in their structural requirements than the corresponding host (e.g., human) equivalents. Because the viral polymerase is crucial to the replication of that virus, there are few potential mutations which allow the polymerase to retain full function but reject the antiviral compound. This results in a relatively high genetic barrier (see below).

Nucleoside/nucleotide analogs are the treatment of choice for herpes viruses and HBV. They are included as two or three components of HAART for HIV. For influenza, T-705 is a good candidate progressing through development. For HCV, viral polymerase inhibitors, such as R1626, are being developed.

Non-nucleoside RT inhibitors

In the search for new inhibitors of retrovirus reverse transcriptase (RT), several chemically unrelated compounds emerged from screens that bind to RT at sites other than the nucleotide binding site. Some are highly selective and inhibit HIV-1 RT at nanomolar concentrations. Enzyme studies showed that these compounds inhibit RT noncompetitively. The compounds are generally thought to be allosteric inhibitors. An early problem associated with these compounds was the rapid emergence of drug-resistant strains. Because the drug-resistant mutations do not directly involve the active sites on the enzyme, they are less likely to compromise enzyme function. Therefore, the genetic barrier is lower than that corresponding to nucleosides targeting the catalytic site. However, adding non-nucleoside RT inhibitor (NNRTI) to NRTI(s) raises the genetic barrier and so the NNRTIs play an important role in HIV combination therapy (see below).

Helicase inhibitors

A new target for effective inhibition of HSV has been discovered. These compounds are being developed by Bayer AG Pharma (e.g., BAY 57–1293) Boehringer-Ingelheim (e.g., BILS 22 BS) and Tularik Inc. (e.g., T157602), and others. To date, they are all aminothiazole derivatives and interact with the helicase–primase complex. This comprises a group of three proteins concerned with unwinding the dsDNA and priming the daughter strand during DNA replication. The specific target is thought to be the product of the HSV UL5 gene, the helicase proteins. The compounds appear to be highly potent in tissue culture and in animal models where they appear to be superior to the nucleosides, ACV, and PCV, especially their efficacy upon delayed therapy. However, it remains to be seen whether toxicity will be an issue; furthermore, preliminary evidence suggests that the development of drug resistance may impede their development.

Virus assembly

As yet, there are no FDA-approved drugs targeting virus assembly but ST-246 is progressing through development. It was discovered as part of the program to prepare for bioterrorism, pox viruses being one of the perceived threats. Investigations with ST-246 resistant virus indicated that the mechanism involves the F13L gene which codes for a major virus envelope protein, p37. Intracellular enveloped viruses (IEVs) are formed from intracellular mature viruses (IMVs) and p37 participates in this wrapping process. ST-246 inhibits this stage and so IEVs are not available to be transported to the cell surface to produce extracellular virus. Exactly how ST-246 acts is being elucidated. Deleting F13L from vaccinia resulted in a virus (delta F13L-Vac) which replicates in cell culture although producing smaller plaques than wild-type (wt) virus. In mice infected with delta F13L-Vac, no lesions were produced but there was a good immune response which protected mice from subsequent challenge with wt virus.

Neuraminidase inhibitors

Mature virions of influenza virus comprise enveloped particles. The envelope membrane is decorated with two glycoproteins which form morphologically distinct 'spikes'. The most prominent of these are hemagglutinin (HA) molecules. Much less numerous but with a distinctive long-stalked 'mushroom' shape are the neuraminidase (NA) molecules. The function of the latter is to cleave sialic acid from sialylated proteins. Sialic acid is the receptor for the influenza virus on mammalian cells. Thus, NA is thought to help the release of virus from the host cell, preventing re-adsorption to the same cell; the enzyme may also aid the passage of virus through mucus to facilitate the colonization of respiratory or intestinal tissues.

During the early 1990s, the crystal structures of NA and its binding to sialic acid were solved. The NA active site was shown to contain well-formed large and relatively rigid pockets. Molecules were designed to interact with this pocket leading to a series of potent inhibitors. Two compounds with this mechanism of action are zanamivir and oseltamivir. Both are potent inhibitors of influenza A virus replication attesting to the important functional role of NA in virus replication. Other compounds targeting this function are currently in development.

Challenges for Antiviral Therapies

Antiviral Prodrugs

The efficacies of several important antiviral drugs (including the nucleoside analogs acyclovir and penciclovir) are severely limited by poor oral bioavailability. One approach to this has been the synthesis of chemically modified derivatives that are rapidly converted to the nucleosides by the host metabolic enzymes. Valaciclovir, the prodrug for acyclovir, and famciclovir, the prodrug for penciclovir, are readily absorbed yielding high blood levels of the parent compound. Both are used widely for treating herpes simplex and varicella zoster infections. Similarly, other oral prodrugs have been developed: the influenza drug oseltamivir, the HIV drugs tenofovir disoproxil fumarate and fosamprenavir, the HBV drug adefovir dipivoxil and the HCV drug R1626. When new compounds are identified with antiviral activity, this kind of approach may be considered at an early stage of development. For example, famciclovir, rather than penciclovir, was used in the clinical trials and currently R1626, the oral prodrug of R1479, is being developed for HCV therapy.

Besides considering prodrugs for improving oral bioavailability, there are many other potential aims. In order to improve cell permeability of GS9148, a prodrug (GS9131) was selected. GS9131 showed potent anti-HIV activity with mean EC_{50} of 50 nM against multiple HIV-1 isolates. After oral administration to dogs at $3\,mg\,kg^{-1}$, GS9131 effectively delivered the active metabolite, GS9148-DP, at high levels for a prolonged period into PBMCs ($C_{max} = 9.2\,\mu M$, $T_{1/2} > 24$ h).

A new approach was reported in the *Prusoff Award Lecture* by Dr. Tomas Cihlar (Gilead Sciences, Foster City, CA, USA) at ICAR 2006. The concept is well established that nucleoside analogs, as their triphosphates, are trapped within the cell and therefore give prolonged activity against virus replication. Could the same strategy be applied to HIV protease inhibitors (PI)? Is it possible to make a phosphonate prodrug of a known PI so that, after entering the cell, it is converted to a charged active phosphonate metabolite with enhanced intracellular retention? This approach is currently under investigation.

Potential for Mutations Leading to Drug Resistance

Most antiviral compounds are highly selective and target a single virus protein. A natural consequence is that point mutations in the virus genome can result in drug-resistant variants. Double-stranded DNA viruses have a relatively low intrinsic error rate (1×10^{-7} to 1×10^{-8} errors per nucleotide) and a 3′–5′ exonuclease proofreading function can edit out errors. In contrast, some RNA viruses have very high intrinsic mutation rates with no proofreading. This high rate of mutation ($\geq 1 \times 10^{-4}$), coupled with a large population of virions, can quickly lead to enormous genetic diversity within a single infected host. For example, HIV has a single-strand RNA genome of approximately 9000 nucleotides. The replication rate in an infected individual has been estimated to be approximately 10^9 daily; thus, $10^{-4} \times 9000 \times 10^9 = 9 \times 10^8$ mutants occur each day. This means that, in theory, every point mutation occurs 10^5 times per day in an HIV-infected individual and every double mutant 10 times per day! As a result, HIV actually exists as a 'quasispecies' or 'swarm' around a particular consensus sequence. Similarly, HCV exists as 'quasispecies'; it has the fastest known daily replication rate of 10^{12} virions daily. HBV, which has a reverse transcriptase step within its replication cycle, mutates readily but it has overlapping reading frames so that a change in the gene encoding the surface protein may also change the gene for the polymerase. This may limit the number of viable mutations. Influenza virus has an additional mechanism for the spread of resistance. The influenza virion contains eight separate segments of single-stranded RNA genome. When two influenza strains co-infect a single cell, the segments can re-assort and pass a resistance mutation from one strain to another.

Based upon virus mutation rates alone, viruses would evolve to resist any specific inhibitor if it were not highly effective in reducing virus replication, ideally to zero. In practice, this is, almost, achieved with drug combinations but patients have a responsibility for their own therapy – any missed dose gives the virus a chance. Also, resistance mutations usually have a biological price; virus fitness may be compromised.

Mutations, that give rise to amino acid substitutions at the interaction site and cause reduced binding of the inhibitor, are termed 'primary mutations'. Such mutations usually arise early upon exposure to the inhibitor. Further mutations, termed 'secondary mutations', may accumulate and contribute to the level of resistance. Yet further mutations may appear which are apparently unrelated to the interaction site and may have no effect on the resistance level. However, these can increase enzyme efficiency so as to compensate for the deleterious effects of the primary and secondary mutations. Many compensating mutations are suspected but often their precise role has yet to be elucidated.

Drug Resistance

Genetic barrier

When a virus is being inhibited by an antiviral compound, resistance mutations are selected but the ease with which this is done depends upon how many potential mutations can give resistance without compromising viral fitness. This has become known as the genetic barrier. With monotherapies, resistance appears quickly with NNRTIs but more slowly with NRTIs which target the same reverse transcriptase (RT) but at the catalytic site. For PIs, which target the protease catalytic site, the rate of appearance of resistance is about comparable to that for NRTIs. It was thought that combining one NRTI and one PI would delay the appearance of resistance greatly, but clinical practice showed that the delay was modest. It is more effective to combine two or three compounds which target the same HIV enzyme but have differing mutation patterns. One high genetic barrier is more effective in delaying resistance than two low genetic barriers.

The anti-influenza M2 channel blockers and anti-picornarvirus capsid binding compounds are examples of agents which appear to create too low a genetic barrier to become useful clinical therapies.

Mechanism of Antiviral Action

The most important classes of virus inhibitors are shown in **Figure 1**. This defines the mechanism of action which may be determined by various approaches. The time-of-addition of the compound in relation to its activity may give important clues. Patterns of cross-resistance to previously characterized agents give useful information. Drug-resistant mutants may be selected by passage of virus in suboptimal concentrations of the inhibitor. The location of drug resistance, found by virus genome sequencing followed by site-directed mutagenesis or marker transfer, provides direct evidence for the mechanism and may give information about the precise interaction between compound and the target protein; this may be at or near the active site of an enzyme. In the search for new antiviral compounds, inhibition of a viral enzyme may be the test screen. Computer modeling of compounds binding to their target proteins may aid further optimization and define the mechanism more precisely.

Virus Fitness

Drug-resistant mutant viruses are often attenuated or 'less fit'. The efficiency of virus replication can be reduced leading to smaller plaque size in cell cultures or reduced pathogenicity when tested in suitable animal infection models. The TK^{-} strains of herpes viruses resistant to ACV and PCV are usually less pathogenic and less able to reactivate from latency than wt HSV. The His274Tyr influenza virus, resistant to oseltamavir, is less pathogenic in ferrets than wt influenza. In chronic HBV patients, continued therapy with lamivudine, even after the appearance of the YMDD resistant strain, used to be better than no treatment although now patients would be switched to alternative therapy. In HIV patients being treated with PIs, the initial resistant mutations give HIV protease with reduced catalytic activity. Later, a constellation of mutations gives rise to more successful drug-resistant viruses that will then dominate the population.

Antiviral Combination Therapy

The problem of antiviral resistance led directly to the introduction of antiviral combination therapy as a crucial feature to control chronic virus infections, notably those caused by HIV and probably will be with HBV and HCV. Initially, there was much opposition to the introduction of antiviral combinations, probably best explained by an underlying fear of enhanced toxicity: this was the problem that dominated the early phase of HIV therapy but eased as newer, better tolerated drugs became available. HIV presented a new urgency; very high levels of HIV genome RNA turnover in the patient and the high mutation rate in the HIV genome meant that antiviral resistance inevitably led to the failure of monotherapies. The concept of genetic barrier was developed at this time. The aim is to create the highest possible genetic barrier and stop all virus replication so that the virus has no chance to overcome that barrier by mutation. This ultimately led to HAART involving the use of triple and quadruple combinations of active compounds. Experience has shown that sequential use of monotherapies is undesirable. Switching from one drug to another enables the sequential development of multiple resistant strains. Current guidelines recommend that at least two RT inhibitors are included in any HAART regimen. This has been very successful in controlling the disease but put a heavy burden on patients who had to take many pills at varying times through the day. So companies developed combinations of their drugs, the first being Combivir (**Table 1(d)**). Recently, two major pharmaceutical companies have been cooperating in the development of a joint formulation in which three antivirals are combined within a single pill (Atripla) leading to a greatly improved convenience and, hopefully, compliance. This first example has started a trend. A similar approach is considered applicable to HBV. For example, telbivudine and valtorcitabine together are being evaluated in clinical trials.

Favorable drug/drug interaction is another rationale for combining drugs. Nearly all current selective antiviral compounds are considered to be virustatic agents. This concept has had major implications in the treatment of HIV. Continual antiviral blood levels of drugs, with intervening troughs, are not sufficient for control of HIV replication. There must be continuous high levels which well exceed the antiviral concentration for that drug. Ritonavir, by blocking the elimination of other PIs, enhances both the concentration and duration of the PI in patients. Lopinavir is co-formulated with ritonavir in a 4:1 ratio primarily to enhance the pharmacokinetics of lopinavir (**Table 1(d)**).

When the symptoms of an infection require treatment, then combining an antiviral agent with an immunomodulator, anesthetic, or anti-inflammatory substance may give added benefit. Thus, acyclovir has been combined with cortisone to simultaneously reduce pain and irritation of the herpes lesion at the same time reducing the replication of infectious virus which would otherwise be prolonged by the anti-inflammatory component. This concept is likely to be widened to other cases in which the inflammatory response plays an important role in the disease, for example, influenza.

Virus Latency

Virus latency remains an obstacle to successful antiviral chemotherapy for which no solution appears to be in sight. Members of the *Herpesviridae* are notable for being able to establish latent infections; with HSV and VZV, latency is established in neuronal cells. During latent HSV-1 infection, the latency-associated transcript (LAT) of HSV-1 is the only known viral gene expressed. A micro-RNA (miR-LAT), which is generated from the exon 1 region of the LAT gene, reduces the induction of apoptosis and so contributes to the survival of that latently infected neuron. Few, if any, HSV proteins are expressed, so affording little chance for either the immune system or antiviral drugs to control the infection. Latency can be established despite ongoing antiviral chemotherapy although there is some evidence from animal models that famciclovir may be more effective than valaciclovir in reducing the load of latent virus when given very early during experimental infections. It still remains to be demonstrated whether or not this has any practical value in patients where therapy shortly after primary exposure is unlikely to be possible. While long-term therapy using nucleoside analogs such as valaciclovir or famciclovir are very successful in suppressing recrudescent lesions in a latently infected individual, on termination of suppressive therapy reactivation leading to recurrent disease is prone to occur. Other viruses, notably HIV, also establish latency in a few long-lived cells, thus making the complete eradication of these infections especially difficult to achieve.

Future Prospects

Herpes

Some bicyclic nucleoside analogs (BCNAs) had been found to be exceptionally active and selective for VZV. A lead compound is Cf 1743. Like ACV and PCV, Cf 1743 is activated by VZV TK but, unlike ACV and PCV, it seems not to act via its triphosphate. Therefore, it will be particularly interesting to evaluate this highly active compound for its ability to reduce postherpetic neuralgia (PHN) in patients with shingles. Cf 1743 as its valyl ester prodrug (FV 100) has recently entered development and progress has been reported on large-scale synthesis and initial toxicity testing.

Helicase-primase inhibitors (see above) represent a new class of compounds active against the herpes viruses. During the last decade, there has been a notable lack of new classes of antiherpes compounds. It will be interesting to see how they progress.

HBV

Since 1999, four drugs have received FDA approval, lamivudine, adefovir dipivoxil, entecavir, and telbivudine (**Table 1(c)**). Although entecavir is the most active compound on a weight basis, the greatest reductions in HBV DNA levels have often been achieved with the most recently (2007) approved compound, telbivudine. In addition, some interesting compounds are progressing through development.

Telbivudine, in phases IIb and III, has been consistently more active than lamivudine against HBV. Because HBV strains (with rtM204I and rtL180M mutations in the polymerase) are cross-resistant to both telbivudine and lamivudine, telbivudine should not be used to treat patients with HBV resistant to lamivudine. However, it may be expected that the greater control of HBV replication with telbivudine than with lamivudine would lead to a lower proportion of patients with resistant virus. Telbivudine, in both pre-clinical toxicological tests and clinical trials, has shown a remarkably good safety record. Therefore, telbivudine may become the first line drug of choice, replacing lamivudine. Since telbivudine and valtorcitabine showed synergy in both cell culture assays and in the woodchuck hepatitis model, these two compounds are being evaluated together in phase IIb trials. Only time will tell if this combination is also able to reduce the rate of appearance of resistance. Telbivudine preferentially inhibits the second strand of HBV DNA synthesis and valtorcitabine inhibits both the first and second strand of HBV DNA synthesis; HBV would have to overcome two stages in its replication cycle and the genetic barrier (especially for the second strand synthesis) would be increased.

With most HBV therapies, although there may be a short lag period after treatment, HBV DNA levels return to baseline. This is typical for drugs which are virustatic. In contrast, clevudine showed a prolonged effect. In a phase I/II trial, at a daily dose of 100 mg, HBV DNA levels were reduced by 3.0 $\log_{10}$ at the end of 4 weeks of therapy and were still 2.7 $\log_{10}$ below baseline at 6 months. However, until an antiviral mechanism has been elucidated for this prolonged effect, one is not sure if this is a great benefit or a warning sign. If there is an antiviral mechanism, then clevudine clearly differs from other anti-HBV drugs and cannot be regarded as a virustatic drug. We suggest that an antiviral compound, which acts via inhibition of the replication cycle and destroys the ability of the virus to continue replication, should be described by a new term, a virureplicidal drug.

HCV

Progress has been severely restricted by the lack of suitable test systems. It seems ironic that, in patients, the production rate of HCV is 10^{12} virions/day, the highest known rate of any virus to date and yet it has taken so long to discover a replication system in cell culture. In 1999, an HCV replicon system became available. No infectious particles are formed but viral RNA synthesis can be followed. Alternatively, HCV pseudoparticles can be obtained by adding the HCV genes for E1(GP35) and E2(gp66) to a partial HIV. The resulting hybrid virus can generate pseudoparticles which incorporate E1 and E2 as heterodimers. These pseudoparticles can be used to raise neutralizing antibody against HCV.

Only in 2005 was full HCV virus replication achieved by Wakita *et al.* JHF1 cDNA was cloned from an individual with fulminant HCV, strain 2a. After transfection into human hepatoma cell line (Huh7), the JFHI genome replicates and leads to the secretion of viral particles. These virions were then used to infect Huh7 cells and chimpanzees. Essentially, only this combination of JHF1 and Huh7 cells has been successful although some chimeric viruses, with the E1 and E2 genes of JFH1 replaced by those from another strain, can also replicate in Huh7 cells.

The availability of the HCV replicon system allowed screening for HCV RNA polymerase inhibitors. One example is R1479 which has progressed through to phase I/II trials. Its activity in HCV-infected patients was sufficient to encourage the development of the oral prodrug, R1626. An alternative approach is the inhibition of the HCV NS3/NS4A protease. Two compounds in phase II studies are VX-950 and SCH 503034. Both gave reductions in HCV RNA levels when used as monotherapies and both appear to have additive or synergistic activity with interferons. A potential limitation of these compounds is the requirement to be dosed 3 times daily but the addition of ritonavir improves the pharmacokinetics such that less frequent dosing may be possible, just as for the HIV protease inhibitors.

Influenza

Since 1999, two neuraminidase inhibitors, zanamivir and oseltamivir, have received FDA approval. Further development of other compounds became slow due to lack of interest but the threat of avian influenza A (H5N1) has stimulated renewed interest. Zanamivir dimer has a long duration of activity, about a week. Peramivir failed to lessen the duration of symptoms in a phase III trial but an oral prodrug may be worth trying.

T-705 (6-fluoro-3-hydroxy-3-pyrazine carboxamide), as its ribo-triphosphate (T-705RTP), inhibits the influenza virus polymerase in a GTP-competitive manner. In cell culture assays, it is active against many strains of A, B, and C influenza viruses. In mice infected with an H5N1 strain, T-705 was highly effective and, on rechallenge with 100-fold more virus on day 21, all the T-705 treated mice survived. T-705 seems to be a likely future therapy.

HIV

Over the last decade, there have been some big gains (reductions in morbidity and mortality) and harsh realities (replication rate 10^{10} virions/day, long-lived latent pool, side effects and viral resistance). Prospects for a vaccine remain poor but new compounds are being developed, including entry and integrase inhibitors. Great efforts have been made to make HIV therapies available to resource-limited areas. Because so many factors differ between areas, outcomes are most likely to differ. Most importantly, research, studying clinical outcomes in these resource-limited settings, should be continued. Otherwise resistant virus may be unknowingly spread around. Several firms have agreed to support the creation of centers of excellence. A great challenge is to investigate the role that therapeutics might play in preventing transmission.

A novel therapeutic concept was described by Jan Balzarini (Rega Institute for Medical Research). HIV gp120 is heavily glycosylated with about 11 'high-mannose' type glycans/gp120 molecule. Pradimicin A (PRM-A) can be considered as a prototype. A high genetic barrier is created by several PRM-A molecules binding to each HIV gp120 protein; resistance can occur in cell culture but only when there have been a large number (>5) of mutations leading to the loss of glycosylation sites. In HIV-infected patients, every loss of a glycosylation site presumably exposes part of the HIV gp120 protein surface to the immune system. Thus, the more resistant the HIV becomes toward PRM-A, the more immunogenic it may become. This is a 'hard choice' which has not yet been presented to HIV with any current therapy. Could the same approach be successful with HBV and HCV?

Conclusions

Niche Targets for Antivirals

Many viruses are well controlled by vaccines; the most outstanding instances are smallpox and polio. Cases of rubella, munps, and measles have been greatly reduced by routine vaccination. Vaccines, mainly used to protect travellers, are available for hepatitis A, rabies, yellow fever, and others. New vaccines can prevent those cervical cancers caused by papillomavirus types 16 and 18. Although these two types account for the majority of cervical cancers, activity against many other types would be required to give full protection. An antiviral could possibly be active against many strains and be effective in an established infection. However, the papillomavirus genome is so small, not even encoding a DNA polymerase, there are few targets. New infections with VZV and HBV are preventable with vaccines but there remains a large pool of infected patients for whom antiviral therapy is required. The threat of avian influenza has led to the stockpiling of anti-influenza drugs. The lack of effective vaccines for HSV, HIV, and HCV has encouraged antiviral therapies. West Nile virus is an example of a newly emerging virus for which there is neither vaccine nor specific antiviral.

Broad Spectrum Antivirals?

Antiviral chemotherapy has come a long way since the first faltering steps with marboran, amantadine and rimantadine, idoxuridine (IDU), and vidarabine (Ara A). Acyclovir set the standard for effectiveness together with clinical safety. For over two decades, clinical safety has been associated with high selectivity for a single virus or a few closely related viruses. So far, there are no FDA-approved antiviral drugs, targeting viral replication, which are active against a broad spectrum of viruses. However, Rep 9 may be the first example.

Further Reading

Anderson RM and May RM (1992) *Infectious Diseasese of Humans Dynamics and Control.* New York: Oxford University Press.

De Clercq E (2005) Recent highlights in the development of new antiviral drugs. *Current Opinion in Microbiology* 5: 552–560.

De Clercq E (2005) Antiviral drug discovery and development where chemistry meets with biomedicine. *Antiviral Research* 67: 56–75.
De Clercq E, Branchale A, Vere Hodge A, and Field HJ (2006) Antiviral chemistry and chemotherapy's current antiviral agents FactFile 2006. *Antiviral Chemistry and Chemotherapy* (1st edition) 17: 113–166.
De Clercq E and Field HJ (2006) Antiviral prodrugs – the development of successful prodrug strategies for antiviral chemotherapy. *British Journal Pharmacology* 147: 1–11.
Field AK and Loughlin CA (1999) Antivirals. In: Granoff A and Webster RG (eds.) *Encyclopedia of Virology*, 2nd edn., pp. 54–68. New York: Academic Press.
Field HJ and De Clercq E (2004) Antiviral drugs: A short history of their discovery and development. *Microbiology Today* 31: 58–61.
Field HJ and Whitley RJ (2005) Antiviral chemotherapy. In: Mahy BWJ and ter Meulen V (eds.) *Topley and Wilson's Microbiology and Microbial Infections,* 10th edn., vol. 2, pp. 1605–1645. London: Hodder Arnold.
Schang LM, St. Vincent MR, and Lacasse JJ (2006) Five years of progress on cyclin-dependent kinases and other cellular proteins as potential targets for antiviral drugs. *Antiviral Chemistry and Chemotherapy* 17(6): 293–320.
Wakita T, Pietschmann T, Kato T, *et al.* (2005) Production of infectious hepatitis C virus in tissue culture from a cloned viral genome. *Nature Medicine* 11(7): 791–796.
Wutzler P and Thust R (2001) Genetic risks of antiviral nucleoside analogues – a survey. *Antiviral Research* 49: 55–74.

Diagnosis Using Serological and Molecular Approaches

R Vainionpää, University of Turku, Turku, Finland
P Leinikki, National Public Health Institute, Helsinki, Finland

Glossary

EIA Enzyme immunoassays are methods used to estimate virus-specific IgG and IgM antibodies or virus antigens by enzyme-labeled conjugates.
PCR By the polymerase chain reaction (PCR) and with specific primers, DNA sequences can be multiplied.
RT-PCR For RNA, nucleic acid has to be transcribed with reverse transcriptase (RT) enzyme to complementary DNA prior to PCR.

Introduction

Specific virus diagnostics can be used to determine the etiology of acute viral infection or the reactivation of a latent infection. Two approaches can be used: demonstration of a specific antibody response or of the presence of the virus itself. Serological methods are used for measuring the antibody response while the presence of virus can be demonstrated by cultivation or demonstration of specific antigens or gene sequences. For the latter, molecular diagnostic methods have become more and more widely applied.

In this article, we briefly describe the principles of the most important serological methods and molecular applications that are used to provide information about the viral etiology of the clinical condition presumed to be caused by a viral infection.

The diagram of the course of acute virus infection (**Figure 1**) indicates the optimal methods for viral diagnosis. Following transmission, the virus starts to multiply and after an incubation period clinical symptoms appear with simultaneous shedding of infectious virus. Virus-specific antibodies appear somewhat later (from some days to weeks, called a window period). When the virus-specific antibody production reaches the level of detection, at first immunoglobulin M(IgM) antibodies and some days later immunoglobulin G(IgG) antibodies appear, and the amount of infectious virus starts to decrease. If this is the first encounter with this particular virus, that is, a primary immune response, IgG antibody levels can stay at a relatively low level, whereas in a later contact with the same antigen, that is, in secondary response, IgG levels increase rapidly and reach high levels while IgM response may not be detectable at all. Antibodies are usually investigated from serum samples taken at acute and convalescent phase of the infection. In selected cases other materials such as cerebrospinal fluid and other body fluids can also be analyzed.

The presence of infectious virus or viral structural components can be investigated directly from various clinical specimens either by virus isolation, nucleic acid detection assays, or antigen detection assays. In order to reach the best diagnosis for each patient, it is important to select the most suitable method using the right sample collected at the right time.

Principles of Serological Assays

During most primary infections IgM antibody levels peak at 7–10 days after the onset of illness and then start to decline, disappearing after some weeks or months. An IgM response is usually not detected in reactivated infections or reinfections. The production of IgG antibodies

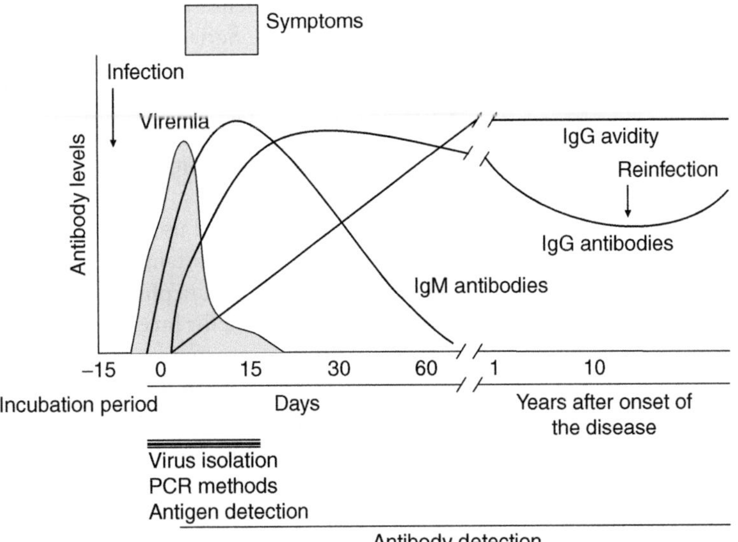

Figure 1 The course of virus infection. The shedding of infectious virus after incubation period and typical antibody response. Recommended diagnostic laboratory methods have been marked.

starts a few days after IgM response and these antibodies often persist throughout life.

Serological diagnosis is usually based on either the demonstration of the presence of specific IgM antibodies or a significant increase in the levels of specific IgG antibodies between two consecutive samples taken 7–10 days apart. The antigen for the test can be either viable or inactivated virus or some of its components prepared by virological or molecular methods. Isotype-specific markers or physical separation are used to demonstrate the isotype of the reacting antibody. In some cases, even IgG subclass specificities are determined although they have limited value in diagnostic work.

During the early phase of acute infection the specific avidity of IgG antibodies is usually low but it increases during the maturation of the response. Diagnostic applications of the measurement of the avidity of IgG antibodies against specific antigens have been developed to help distinguish serological responses due to acute infections from those of chronic or past infections.

Serological assays are useful for many purposes. In primary infections they often provide information about the etiology even after the acute stage when infectious virus or its components can no longer be demonstrated in the samples. They are widely used for screening of blood products for the risk of certain chronic infections, evaluation of the immune status, and need for prophylactic treatments in connection with certain organ transplantations. They are also widely used for epidemiological studies, determination of vaccine-induced immunity, and other similar public health purposes.

Serological assays have their limitations. In some infections the antibody response is not strong enough or the limited specificity of the antigens used in the assay does not allow unambiguous interpretation of the results. In infections of newborns the presence of maternal antibodies may render the demonstration of the response in the baby impossible. In immunocompromised patients the serological response is often too weak to allow the demonstration of specific responses. In these cases other virological methods should be considered.

Other clinical specimens than sera can be used for antibody assays. IgM and IgG antibody determinations from cerebrospinal fluid are used for diagnosis of virus infections in the central nervous system although new molecular methods are increasingly replacing them. Recently, increasing attention has been given to the use of noninvasive sample materials such as saliva or urine. They are becoming important for public health purposes but their value for diagnosing individual patients is still limited.

Principles of the Most Common Serological Tests

Neutralizing Antibody Assay

Antibodies that decrease the infectious capacity of the virus are called neutralizing antibodies. They are produced during acute infection and often persist during the entire lifetime. They are also useful as an indication of immunity. Both IgM and IgG antibodies participate in the neutralization.

In the assay, known amounts of infectious virus are mixed with the serum sample and incubated for a short period after which the residual infectivity is measured using cell cultures or test animals. This infectivity is then compared with the infectivity of the original virus and the neutralizing capacity is calculated from this result. Today, neutralizing antibody assays are often done by plaque reduction assays with better accuracy but with somewhat more complex technical requirements.

Neutralizing antibody assay is specific and sensitive, but time-consuming and laborious, and therefore it is not widely used in routine diagnostic services.

Hemagglutination Inhibition Test

Many viruses bind to hemagglutinin molecules found at the surface of red blood cells of various animal species and this can cause aggregation of red cells in suitable conditions. Prevention of this aggregation, called hemagglutination inhibition, by specific antiviral antibodies in the patient's serum has been widely used for diagnostic purposes. The test, known as hemagglutination inhibition test, has important diagnostic and public health applications in certain infections, most notably in influenza where antibodies measured by this test show additional specificity compared to other tests and therefore provide more detailed information about the immunity and past infections of individuals. However, for the diagnosis of individual patients, the assay is no longer widely used and is replaced by more modern immunoassays.

In the test, a virus preparation with a predetermined hemagglutinating capacity is mixed with the serum sample and after proper incubation the residual hemagglutination capacity is measured. Both IgM and IgG antibodies are able to inhibit hemagglutination.

Complement Fixation Test

The complement fixation test (CFT) is a classical laboratory diagnostic test, which is still used for determination of virus antibodies in patient sera or cerebrospinal fluid samples during an acute infection. The test mainly measures IgG antibodies.

The test is based on the capacity of complement, a group of heat-labile proteins present in the plasma of most warm-blooded animals to bind to antigen–antibody complexes. When the complexes are present on the surface of red blood cells, complement causes their lysis which can be visualized by a suitable experimental setup.

In the actual test, the complement in the patient's serum is first destroyed by heating; the serum is then mixed with appropriate viral antigen and after incubation; when the antigen–antibody complexes are formed, exogenous complement (usually from fresh guinea pig serum) is added. This complement then binds to the complexes and having been 'fixed', it is then no longer able to cause lysis of added indicator red cells. Usually, sheep red cells coated with antisheep red cell antibodies are used as indicator to measure the presence of any residual complement. The effect is measured by a suitable test protocol. Serial dilutions of the patient serum are used and the highest dilution where the serum can still prevent complement activity in the indicator system is taken as the CFT titer of the sample. The tests are usually carried out on microtiter plates and the results are observed by eye.

CFT is still used for diagnosis of acute virus infection. It measures certain types of antibodies which occur only during the acute phase of the infection. Therefore, CFT is not suitable for investigation of immune status. The assay procedure is quite complex, because the test is dependent on several biological variables, which have to be standardized by pretesting. The method is less sensitive than many other immunoassays. In addition, the method is very labor intensive and is not amenable to automation. The use of CFT in virus diagnostics is increasingly replaced by modern immunoassays.

Immunoassays

In immunoassays, antibodies binding to specific immobilized antigens can directly be observed using bound antigens and proper indicators such as labeled anti-immunoglobulin antibodies. The antigens can be immobilized to plastic microtiter plates, glass slides, filter papers or any similar material. Different immunoassays are nowadays widely used to measure virus-specific IgM and IgG antibodies. The most recent formats of immunoassays make it possible to detect simultaneously both antigens and antibodies decreasing significantly the window period between infection and immune response. Numerous commercial kits with high specifity and sensitivity are available. Automation has made immunoassay techniques more rapid, accurate, and easier to perform.

In the basic format of solid-phase immunoassays, virus-infected cells, cell lysates, purified or semipurified, recombinant viral antigens or synthetic peptides are immobilized to a solid phase, usually plastic microtiter wells or glass slides. Patient's serum is incubated with the antigen and the bound antibody, after washing steps, is visualized using labeled anti-immunoglobulin antibodies ('conjugate') (**Figure 2(a)**). If the label used is an enzyme, the test is called enzyme immunoassay (EIA) or enzyme-linked immunosorbent assay (ELISA) and the bound antibody is detected by an enzyme-dependent color reaction. If a fluorescent label is used, the method is called immunofluorescent test (IFT). The enzyme labels most commonly used are horseradish peroxidase (HRP) and alkaline phosphatase (AP). In HRP-EIA the color-forming system consists of ortho-phenyldiamine (OPD) as a chromogen and hydrogen peroxidase (H_2O_2) as a substrate. If the HRP-conjugate is bound to antibody–antigen complexes, the colorless chromogen becomes yellow and color intensity is measured with a photometer at a wavelength of

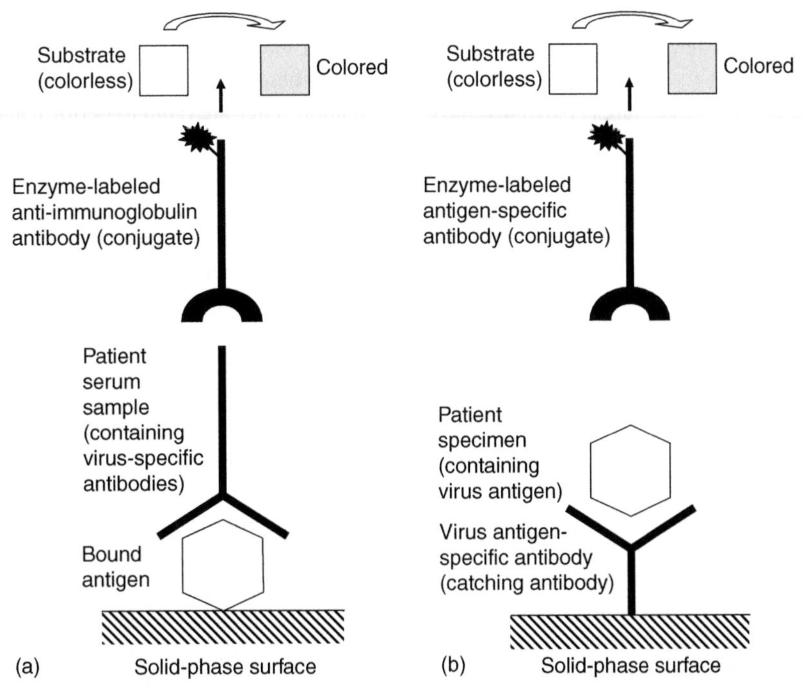

Figure 2 Enzyme immunoassay. (a) Detection of virus-specific antibodies. (b) Detection of virus antigens.

490–492 nm. The intensity of the color is proportional to the amount of bound conjugate and to the amount of specific antibodies in a patient serum sample. If the serum contains no specific antibodies, the conjugate is not bound and no color reaction occurs. By using either anti-IgG or anti-IgM conjugates it is possible to determine separately immunoglobulin subclasses.

The specificity and sensitivity of these immunoassays are high. The sensitivity can be improved further, by using an additional incubation step where IgM antibodies are first enriched ('captured') in the sample by using anti-IgM immunoglobulin. Modifications to improve assay specificity by various methods of antigen handling and by using monoclonal antibodies or synthetic peptides have been developed.

Immunofluorescent tests were used in the past for measuring virus-specific antibodies, but are now replaced by EIA techniques. The principle of the method is similar to EIAs. In IFT infected cells are placed on a glass slide and bound antibodies are detected by fluorescein-labeled anti-immunoglobulin antibodies. The glass slides are examined under a fluorescence microscope. The method is specific and sensitive, but quite labor intensive and reading the test demands considerable experience.

Immunoblotting

In some infections (e.g., that caused by human immunodeficiency virus (HIV)), antibodies against certain components of the virus are more informative than other less-specific antibodies and they are detected by immunoblotting assays. Different virus antigens, prepared by gel diffusion or other techniques, are absorbed as discrete bands on a solid strip of cellulose or similar material and the strip is incubated with the patient's serum. Antibodies present in the serum bind to specific antigens and are detected using an HRP-conjugate and nitroblue tetrazolium as the precipitating color chromogen. The color reaction is observed and compared to positive and negative control samples assayed on separate strips.

Lateral-Flow and Latex Tests

A technique known as lateral-flow technology has also been used to identify antibodies or antigens. These tests involve application of serum or other samples directly on a strip of suitable material such as cellulose, where the antibodies are diffused laterally and eventually reach a site in the strip where appropriate antigen has been applied and chemically fixed. Specific antibodies become bound to the site while nonreacting antibodies diffuse out from the area. The presence of antibodies is visualized using labeled conjugates.

Although such tests are not quantitative, they are valuable for infections where the presence of specific antibodies is indicative, such as HIV infection. Performance of the test is often very simple and the result is available in a few minutes or a few hours, making such tests suitable for bed-side screening. In more advanced tests, several different antibodies can be detected by a single assay and the test conditions can be modified further so that antigens can also be detected. Many such tests have become commercially available in recent years.

For some applications, coated latex particles have replaced strips with fixed antigen as the solid phase. Binding of specific antibodies can be visualized with chromogenic or otherwise labeled indicator antibodies or a positive reaction can be detected by agglutination of the latex particles.

Point-of-Care Tests

Point-of-care tests (POC tests) are becoming increasingly common in clinical practice. Most of them are based on easy-to-use lateral-flow or latex particle technology and are able to give the result in a few minutes. POC tests are nowadays available for antibody screening of an increasing number of virus infections (HIV, hepatitis C virus (HCV), varicella-zoster virus (VZV), cytomegalovirus (CMV), Epstein–Barr virus (EBV)). Some authorities still question the validity of POC tests for clinical use although there is considerable evidence that many of the commercially available kits give reliable results.

Detection of Viral Antigens

The presence of viral antigens in clinical specimens, such as nasopharyngeal aspirates, fecal specimens, vesicle fluids, tissue specimens, as well as serum samples can be demonstrated by antigen detection assays.

In immunofluorescence tests, cells from a clinical specimen are fixed on a glass slide and viral antigens present in the cells are detected by fluorescein-labeled virus-specific antibodies. More reliable results can be obtained using enzyme immunoassay or time-resolved fluoroimmunoassay (TR-FIA). Europium-labeled monoclonal antibodies can be used as a conjugate. Solubilized antigens in clinical specimens are first captured using specific monoclonal antibodies bound to a solid phase, and are then detected with enzyme- or europium-labeled virus-specific antibodies (**Figure 2(b)**).

Antigen detection methods are especially recommended in the case of virus reactivation, for example, for herpes simplex and varicella zoster virus diagnosis where the serological response can be very weak. Antigen detection assays are also widely used in respiratory tract infections like influenza and respiratory syncytial virus infections. A simple test for the demonstration of rotavirus and adenovirus antigens in children with gastroenteritis is also available.

Nucleic Acid Detection Assays

Direct demonstration of viral nucleic acids in clinical samples is an increasingly used technique for virus diagnosis. Using the polymerase chain reaction (PCR) with specific primers, viral sequences can be rapidly multiplied and identified. These techniques are largely replacing classical virus isolation. They are rapid to perform and in many cases more sensitive than virus isolation or antigen detection methods making earlier diagnosis possible. They have proved particularly valuable for the diagnosis of viruses that cannot be cultivated such as papillomaviruses, parvoviruses, and hepatitis viruses. Semiquantitative and quantitative applications have been developed allowing monitoring of viral load during antiviral treatment. These tests cannot distinguish between viable and replication-incompetent virus, warranting caution in the interpretation of the results in certain cases. Also, sensitivity to cross-over contamination in the laboratory has caused some problems in clinical laboratory settings.

The specificity of these tests is based on the extent of pair-matching sequences between the viral nucleic acids and the primers. Extremely high sensitivity is typical for PCR methods; 1–10 copies of viral nucleic acid can be detected in a few hours. PCR methods are available for both RNA and DNA viruses. For RNA viruses viral nucleic acid has to be transcribed with reverse transcriptase (RT) enzyme to complementary DNA (RT-PCR).

Viral nucleic acid is extracted from the sample material and amplified in three successive steps. The double-stranded DNA is first heat-denaturated and separated into single strands. The specific target fragment of DNA strand is then amplified (**Figure 3**) by pairs of target-specific oligonucleotide primers, each of which hybridize to one strand of double-stranded DNA. The hybridized primers act as an origin for heat-stable polymerase enzyme and a complementary strand is synthesized via sequential addition of deoxyribonucleotides. After annealing of the primers, extension of the DNA fragment will start. These cycles are repeated 35–40 times, each cycle resulting in an exponentially increasing numbers of copies.

After the amplification is completed, the products can be detected by several methods. Agarose gel electrophoresis combined with ethidium bromide staining of the products is a classical method (**Figure 4**). The size of the amplified product is compared to control amplicons and other standards in the same gel. Various hybridization assays, based on labeled complementary oligonucleotides (probes), are also used to improve the sensitivity and specificity of the detection.

The amplified fragments can also be sequenced giving additional information about the virus. Comparison of the sequences with known virus sequences allows identification of species, strains, or subtypes that may be important for public health or medical purposes. Sequencing after RT-PCR is also the current method-of-choice for investigating the emergence of antiviral drug resistance among HIV-infected patients.

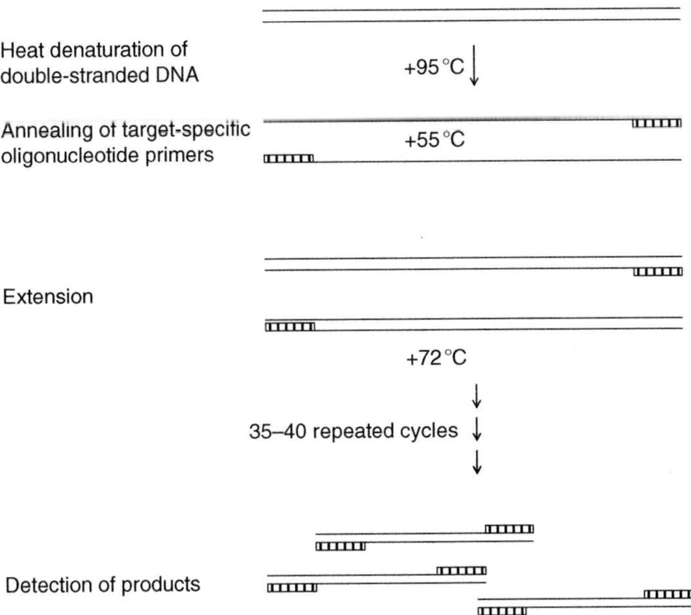

Figure 3 Polymerase chain reaction.

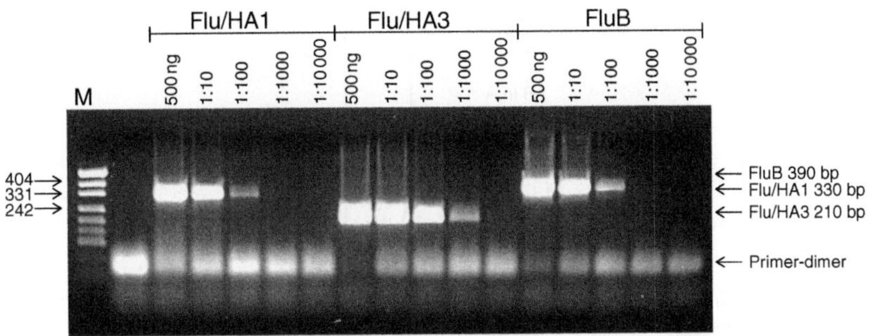

Figure 4 Detection of PCR products (amplicons) by an agarose gel electrophoresis after ethidium bromide staining.

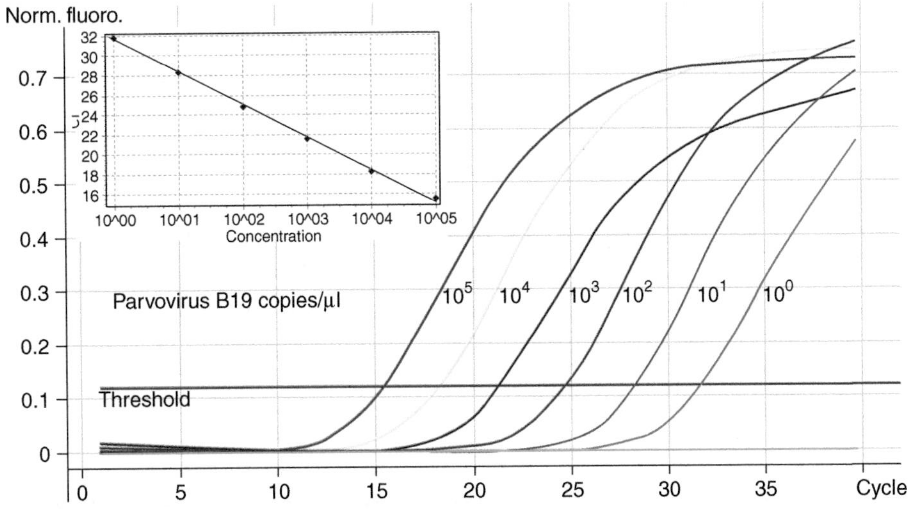

Figure 5 Quantitative real-time PCR with fluorescent-labeled probes for parvovirus B19.

Real-time PCR instruments monitor accumulation of amplicons by measuring the fluorescence continuously in each cycle of the reaction. The earlier the amplification product becomes detectable over the background, the higher is the amount of virus in the sample (**Figure 5**). One application, based on the use of melting temperatures, allows simultaneous detection and analysis of several different nucleic acids. It also allows testing for more than one virus from the same sample (**Figure 6**).

The PCR assays are extremely sensitive and can therefore be influenced by inhibitors of the polymerase enzyme that are sometimes present in clinical samples. Internal controls can be included into reaction mixtures. Nucleases present in samples or in reagents can also cause

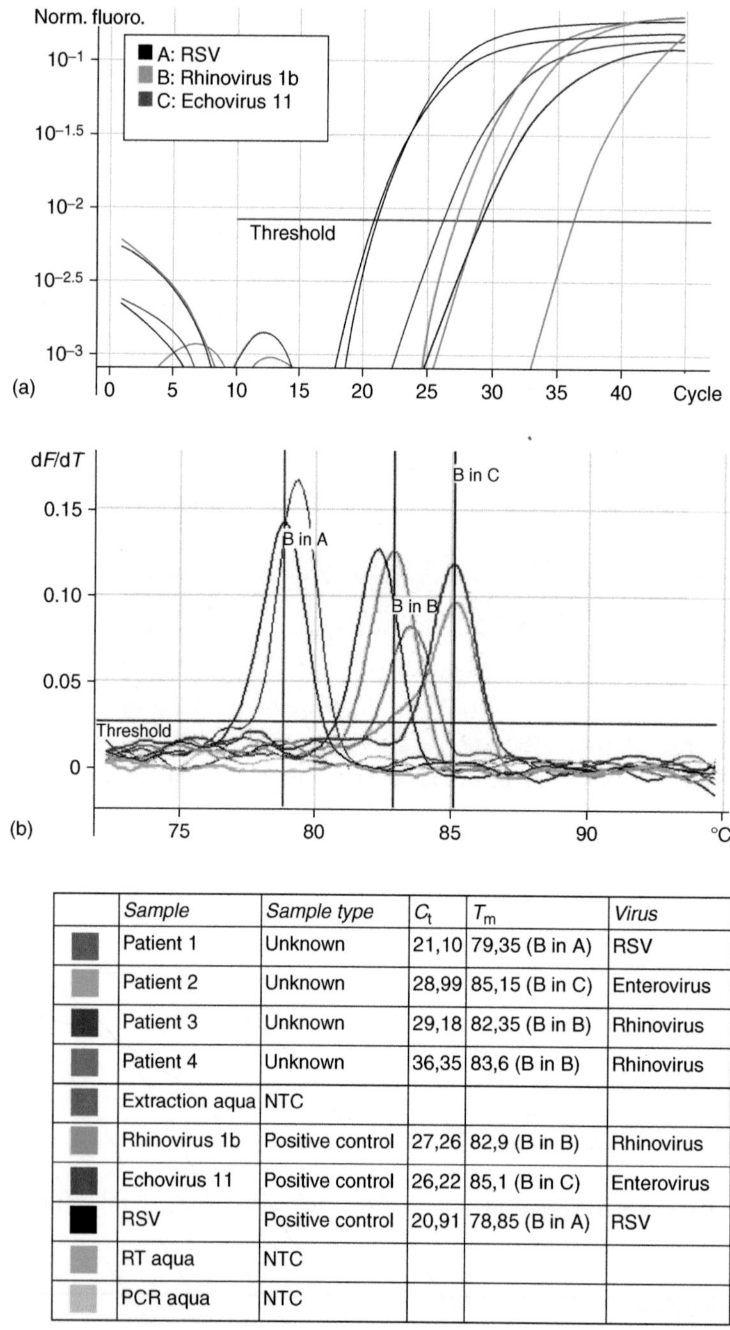

	Sample	Sample type	C_t	T_m	Virus
	Patient 1	Unknown	21,10	79,35 (B in A)	RSV
	Patient 2	Unknown	28,99	85,15 (B in C)	Enterovirus
	Patient 3	Unknown	29,18	82,35 (B in B)	Rhinovirus
	Patient 4	Unknown	36,35	83,6 (B in B)	Rhinovirus
	Extraction aqua	NTC			
	Rhinovirus 1b	Positive control	27,26	82,9 (B in B)	Rhinovirus
	Echovirus 11	Positive control	26,22	85,1 (B in C)	Enterovirus
	RSV	Positive control	20,91	78,85 (B in A)	RSV
	RT aqua	NTC			
	PCR aqua	NTC			

(c)

Figure 6 RT-PCR with real-time detection (a) and with melting curve analysis (b) for the detection of respiratory syncytial virus (RSV), rhinovirus, and enterovirus in respiratory secretions (c). C_t is a threshold cycle number, T_m is a melting temperature, and NTC is a nontemplate control. Unpublished results by Waris M, Tevaluoto T, and Österback R.

false negative results by degrading viral nucleic acids. Furthermore, amplicons may also cause product carry-over and false positive results. Extremely high care has to be applied in handling the clinical specimens, the reagents, as well as the reaction products.

One of the great advantages of the PCR technology is its potential to detect new emerging viruses. By using primers from related viruses or so-called generic primers important information regarding the new virus can be obtained for further development of more specific tests. A good example is the severe acute respiratory syndrome (SARS) virus, for which specific diagnostic tests became available soon after the taxonomic position of the virus became known. The technology also allows safe handling and transport of virus samples, since extraction buffers added to the samples inactivate virus infectivity.

Future Perspectives

Driven by public health, scientific and commercial interests, new diagnostic tests for the laboratory diagnosis of viral infections are continuously being developed. The main area for development will probably be new molecular detection methods, where automation will provide rapid, well-standardized, and easy-to-use technology.

Use of multianalyte methods is becoming a practical reality and they might significantly change diagnostics of infectious diseases in future. They provide an opportunity to screen simultaneously for a wide range of viruses increasing the rapidity of the diagnostic procedure. A single microarray test (microchip) can contain thousands of virus-specific oligonucleotide probes spotted on a glass slide. Several kit applications for detecting viral nucleic acids and antigens or virus-specific antibodies already exist. Microarrays based on random PCR amplification can be used to detect a variety of viruses belonging to different families. Screening of some other infection markers can be also included in the same test format. Microarrays are not widely used for clinical purposes because of limited sensitivity and the difficulties of developing analytical instruments suitable for diagnostic laboratories. Another line of development is the increasing number of POC tests, which may form an important part of future diagnostic testing of infectious diseases.

See also: Antigenicity and Immunogenicity of Viral Proteins; Antibody - Mediated Immunity to Viruses; Diagnosis Using Microarrays.

Further Reading

Halonen P, Meurman O, Lövgren T, *et al.* (1983) Detection of viral antigens by time-resolved fluoroimmunoassay. In: Cooper M, *et al.* (eds.) *Current Topics in Microbiology and Immunology,* vol. 104, pp. 133–146. Heidelberg: Springer.

Hedman K, Lappalainen M, Söderlund M, and Hedman L (1993) Avidity of IgG in serodiagnosis of infectious diseases. *Review of Medical Microbiology* 4: 123–129.

Hukkanen V and Vuorinen T (2002) Herpesviruses and enteroviruses in infections of the central nervous system: A study using time-resolved fluorometry PCR. *Journal of Clinical Virology* 25: S87–S94.

Jeffery K and Pillay D (2004) Diagnostic approaches. In: Zuckerman A *et al.* (eds.) *Principle and Practise of Clinical Virology*, pp. 1–21. Chichester, UK: Wiley.

Wang D, Coscoy L, Zylberberg M, *et al.* (2002) Microarray-based detection and genotyping of viral pathogens. *Proceedings of the National Academy of Sciences, USA* 99: 15687–15692.

Diagnosis Using Microarrays

K Fischer, A Urisman, and J DeRisi, University of California, San Francisco, San Francisco, CA, USA

Motivation for New Diagnostic Methods

Numerous human diseases exist for which viral etiologies are suspected, yet specific causal agents are not known. Among these are up to 20% of cases of acute hepatic failure, up to 35% of cases of acute aseptic meningitis and acute encephalitis, up to 50% of cases of acute respiratory infections, and numerous other conditions. In addition, infectious agents may be involved in the pathogenesis of a number of chronic conditions, most notably disorders such as chronic inflammation, autoimmune and degenerative conditions, as well as some forms of cancer. While it is unlikely that viruses cause all of these diseases, identifying causative agents in even a modest number of disorders will have profound implications for understanding, diagnosis, and treatment of these conditions.

New approaches to viral diagnostics and discovery are needed to overcome the shortcomings of existing methods. These methods include viral culture, electron microscopy, serology, specific polymer chain reaction (PCR)-based methods, and techniques based on subtractive hybridization. While these methods have been critical for identifying many important human and nonhuman pathogens, each of these methods has intrinsic limitations. For example, many viruses are refractory to culture. Inspection by electron microscopy may prove difficult depending on the titer and morphological features of the virus. Serology- and PCR-based techniques are highly specific methods, but their specificity frequently renders them ineffective for the detection of variant or novel viral species. In the case of PCR, there exist dozens if not hundreds of variations on the method which may extend the scope of the assay, usually through multiplexing or by the use of degenerate primers, yet the number of possible targets that can be interrogated remains small relative to the number of known viral pathogens. Finally, subtractive hybridization techniques, while unbiased, are difficult to troubleshoot and essentially impossible to scale up for high throughput.

DNA Microarrays

While there exist many different forms of DNA microarrays, produced by both researchers and corporations, they fundamentally share the same properties. All microarrays exist as some form of solid substrate, typically glass or silicon, to which is bound different species of nucleic acid. Each microarray may contain 10^5–10^6 different species of DNA, arranged in a grid. Fluorescently labeled nucleic acid derived from any biological sample may be interrogated by hybridization to the microarray. In this manner, the abundance of thousands of different nucleic acid species may be simultaneously measured. The most common application of microarray technology is the measurement of relative gene expression, often for entire genomes. While gene expression profiling has enjoyed tremendous success over the last decade, it has long been realized that the microarray format would be amendable to the detection of exogenously derived nucleic acids for purpose of identifying the presence of pathogens in a background of host material. The general concepts and methodologies are similar to those required for expression profiling; however, there are key differences to consider for the design of the array and the mechanics of sample processing. Furthermore, most implementations of virus-detection microarrays strive only to determine the presence of viral sequences, rather than attempting to quantify the amount of virus in a particular sample. In practical terms, this allows more aggressive amplification strategies to be used during sample preparation.

Microarray Design for Viral Detection

Algorithms to guide the process of sequence selection for expression microarrays are well developed, and many of the design principles apply directly to microarrays for viral detection. The parameters shared in common include physical properties of the oligonucleotide itself, such as the propensity of the sequence to form hairpins, melting temperature, and sequence complexity. Beyond this, the design considerations for viral detection and expression profiling differ substantially. First, the majority of microarrays intended for viral detection are designed to specifically detect the products of a multiplex PCR reaction using specific primers. In this role, the array simply sorts the product of amplification. The design specifications for these types of arrays are straightforward and are primarily guided by the choice of flanking PCR primers for virus amplification. This configuration of array does not exploit the full potential of the microarray for panviral detection. However, the design parameters for a generalized virus detection chip that does not rely on specific multiplex PCR primers must take into account additional factors.

While it is essential that probes for expression arrays are unique with respect to target sequence (to prevent cross hybridization from other mRNA species), the same does not necessarily apply to viral probes. In fact, to maximize the probability of detecting any viral species from a known family, it is often desirable to choose sequences that are the most conserved among a group of viruses. For example, while there exists a large range of sequence diversity among human rhinoviruses, sequences in the 5′ UTR are highly conserved, even among more distant picornaviruses. These sequences may thus serve as a type of universal 'hook' to capture both existing and new variants of these RNA viruses. In case of novel pathogens, such as the SARS coronavirus, the use of conserved sequences was a key determinant for successful detection by microarray. Clearly, unique species-specific or even genotype-specific oligonucleotides can further augment the discrimination of the microarray. A logical extension of this strategy takes into account features of viral taxonomy. Rather than choosing all conserved, or all unique sequences, one may attempt to cover, by design, each level of the taxonomic tree for each family. Thus, some sequences would be chosen to be species specific (terminal nodes on the tree), some would be genus specific, and so on. Various bioinformatic tools, including online databases, are currently available to assist in such design efforts. In all cases, it is critical to prescreen each choice for matches within the human genome to prevent inappropriate cross-hybridization to host material.

A more simple approach is to simply tile overlapping oligonucleotides spanning the entire genome of the viruses in question. This approach is appropriate for relatively small panels of viruses since each species will result in large numbers of sequences, depending on genome length. In general, this approach becomes impractical when extended beyond a few viral species.

After satisfying the basic design requirements, more sophisticated considerations may also contribute to choice of viral sequence for representation on a microarray. For example, to enhance detection of latent herpesviruses, it may be advantageous to overrepresent sequences specific for genes specific for latent phase expression, rather than those involved in lytic processes. In this case, it is assumed that the RNA rather than DNA will be analyzed, which highlights the importance of sample processing and amplification considerations.

Sample Processing and Amplification

The protocol by which nucleic acids are isolated from specimens, and the subsequent amplification of the material, if needed, is also important to consider, as this may also affect both microarray and experimental design. In general, isolation of total RNA is the preferred and more conservative route. While this may seem biased toward RNA viruses, all DNA viruses produce mRNA as part of their lifecycle, so this choice does not exclude them. However, if viral particles are collected, or host material is removed from the specimen by filtration or other size-selection techniques, it may be advisable to isolate total nucleic acid (both RNA and DNA) to maximize sensitivity. The origin of the sample also bears on this issue. When processing a relatively acellular material, such as cerebral spinal fluid, a total nucleic acid extraction would be appropriate, whereas in the case of a solid organ, such as liver or brain, an RNA extraction would avoid the unnecessary complexity brought by co-purifying massive quantities of the host genomic DNA.

After nucleic acid has been isolated, an amplification step is typically employed to generate sufficient quantities for successful microarray hybridization. In certain situations, where large amounts of primary material are available, the yield of nucleic may be such that an amplification step may be bypassed altogether, but these situations are the exception. The choice of an amplification strategy is closely linked to the design of the array itself. Several virus-detection microarrays have been designed to serve as detectors for the products of multiplex PCR amplification strategies, but in these cases, the broad spectrum and unbiased nature of the microarray is not realized.

For panviral microarrays, where no assumptions are made as to the probable identity of the target, a general, randomized amplification strategy is required. Numerous random amplification strategies exist, but all begin with a priming step using a randomized oligo of various lengths, ranging from 6 to 15 bp. At this point, PCR adapters may be added, either by ligation, or through priming via a common sequence linked to the random primer. Alternatively, various RNA polymerase promoter sequences may be appended which then allow linear amplification. For all these methods, contamination with previously isolated material is a critical concern and good laboratory practices and appropriate controls, such as nontemplated amplification reactions, must be an integral part of the protocol.

The overall complexity of the sample, with respect to the viral species, is a critical factor for the success of random amplification strategies. As previously noted, the ideal samples are those that contain relatively low amounts of host cellular material and high titers of virus. For many biological specimens, it may be possible to reduce the complexity by filtration, centrifugation, or by pretreatment of the raw material with various nucleases. In the latter case, free host material such as genomic DNA and ribosomal RNA may be degraded, yet viral-packaged nucleic acid will escape destruction. Such 'preprocessing' can significantly enhance the signal to noise of the final microarray assay, although these steps add both cost and complexity to the overall protocol. Potentially, the use of novel microfluidic techniques for particle size discrimination may serve as a rapid and reproducible way to deterministically reduce complexity in biological samples.

Bioinformatics

Analysis of hybridization results depends greatly on the design of the experiment and of the DNA microarray in particular. Microarrays that are narrowly targeted, for the purposes of distinguishing between strains of a particular virus species, for example, human influenza or smallpox, are subjected to different analysis techniques than are multigenus, multifamily, or even panviral arrays. The utility of DNA microarrays in all these fields is demonstrable when investigators use appropriate analysis techniques.

In strain and species typing applications there are usually ample controls available to researchers. By use of control hybridizations, characteristic species or strain hybridization patterns can be classified manually. Often these patterns are the result of iterative microarray design where features with the desired specificity are spatially clustered on the chip. These patterns can be used as templates for visual inspection of the experimental arrays where different classes of microarray features are enumerated and the input sample is thereby placed into one of the known groups or unclassified as ambiguous. Machine-learning techniques, such as probabilistic neural networks, are also sometimes used in these studies, but to date have not been show to be more accurate than simple enumerative methods.

Broader explorations of viral populations, for example, panviral studies, must face the problem that the number of control samples available is small when compared to the number of distinct viral genomes that may be encountered. Many viral targets are extremely mutable, introducing one or more mutations per virus genome for every cycle of replication. Such diversity quickly outstrips the resources of any effort to perform exhaustive control hybridizations.

In panviral studies standard methods such as hierarchical clustering and several particular bioinformatic approaches can be employed. First the microarray should be designed to include conserved regions of the target viruses to minimize the chance that a divergent but related virus will escape detection. Second, estimates of the hybridization patterns expected from the possible targets for which sequence data is available can be generated using biophysical models of hybridization. Experimental hybridization patterns can then be compared to the model profiles using a correlation metric. If a virus present in a hybridization shares homology with one of the sequences used to estimate the hybridization profiles, this similarity will be reported as a significant correlation between the profile and the experimentally observed pattern. In the last method, the intensity history of the microarray's features can be compared to an experimental hybridization. Extraordinarily strong signal at a particular feature can indicate *bona fide* virus, especially when signal at taxonomically related features is similarly elevated. During the design stage of the experiment, it is important to consider that in multifamily, metagenomic studies it is critical to have negative controls to characterize the background of microarray designed with broad specificity, while in studies with narrower focus a significant number of positive controls are needed for each class of virus being considered.

Real-World Applications: Research and Clinical

To date, only a few studies have examined the use of viral-detection microarrays using actual prospectively collected patient samples, as opposed to viruses cultured in the lab or previously characterized retrospective samples. As of the time of this writing, no large-scale study has been published comparing the use of virus microarrays (using a random amplification) to traditional laboratory medicine diagnostics, such as commercial DFA kits and PCR assays. Preliminary data and studies using modest numbers of clinical samples suggest that a microarray-based viral diagnosis outperforms conventional DFA assays, both in terms of sensitivity and specificity. In comparison to a PCR assay using specific primer pairs, microarray assays are likely to have comparable specificity, yet the sensitivity is anticipated to be somewhat lower, depending on amplification strategy. However, microarray-based assay coupled to a random amplification protocol provides much broader detection capabilities, often including essentially every known viral pathogen. When considering the potential of DNA microarrays for clinical diagnostics, it may be the case that the microarray will serve a complementary role to specific PCR assays, especially when the latter fails to yield positive results.

For clinical cases where conventional diagnostic assays have failed, the use of a panviral microarray assay may allow identification of new, or unanticipated pathogens. In a recent case report, Chiu and colleagues reported the diagnosis of a previously healthy 28-year-old woman suffering from a severe respiratory tract infection of unknown etiology. Extensive panels of diagnostics for bacteria, fungi, and viruses failed to reveal any positive results, and all viral cultures remained negative. DNA microarray analysis of RNA isolated from endotracheal aspirate revealed the presence of human parainfluenza virus-4, a virus that is not normally included on standard DFA panels or PCR tests. Viral sequence was recovered directly from the patient sample confirming the identity of the virus, and it was further shown that the patient seroconverted during the time of the illness. While it is generally believed that human parainfluenza virus-4 causes only mild, self-limiting infections, this data taken together suggests that the spectrum of disease may extend to respiratory failure in an otherwise healthy adult. This example demonstrates an attractive feature of the microarray assay, namely, the power to detect the unexpected.

In addition to the detection of previously known pathogens, DNA microarrays have also been effective for the detection of novel viral species. In the case of SARS, total nucleic acid from a supernatant from an infected vero cell culture revealed a coronavirus signature consisting of oligonucleotides originating from avian infectious bronchitis, human and bovine coronaviruses, and, interestingly, several astroviruses. At first glance, the hybridization signal from astrovirus-derived oligonucleotides would seem to be aberrant. In fact, this is expected, since several astroviruses and coronaviruses share conserved sequences at the 3′ end of their genomes. These particular sequences were represented on the microarray since the panviral design algorithm purposely selected conserved sequences within and among viral families.

The same principles applied to a separate study in which a novel xenotropic gamma retrovirus was detected in prostate tumor biopsies of men with a mutant variant of the *RNASEL* gene. Integration sites and full-length genomes were subsequently cloned and the virus was demonstrated to be replication competent, thus validating the microarray result. Again, the broad-spectrum nature of the DNA microarray was critical to the success of the project, since there were no preconceptions that such a virus might be a candidate, given that no xenotropic gamma retrovirus had been previously observed in a human subject.

Limitations

Several important limitations of using microarrays for viral detection and discovery should not be overlooked. The most important limitation of the approach is its reliance on known viral sequences. Although most of the novel viruses discovered in the last decade share homology with previously known viruses, viruses lacking even short regions of homology cannot be detected by any hybridization-based method. In the case of profoundly divergent viruses, more brute-force approaches, such as shotgun sequencing, are likely to be applicable. Other existing and likely surmountable limitations of the microarray-based methods are their cost, the need for specialized equipment, and access to computational resources. It is likely that these limitations will become less pronounced as streamlined versions of the technology become available through academic and commercial efforts. It may also be the case that the utility of DNA microarrays may be surpassed by next-generation, massively parallel shotgun sequencing technologies, which would permit cheap, fast, and unbiased analysis of clinical samples.

Scientific and Public Health Implications

Currently, a substantial fraction of human disease, with a presumed viral cause, goes without a clinical diagnosis. This is especially true for common ailments, such as upper respiratory tract infections, where despite advances in PCR assays, the etiology of 30–60% of infections remains unidentified. Without considering the complexities of diagnostic regulatory approvals, an unbiased DNA microarray approach to the detection of viral pathogens should substantially increase the number of successful diagnoses, and as a consequence, may lead to improved therapeutics and supportive care. It should be noted that use of a virus microarray extends beyond the analysis of clinical samples. The wide scope of detection and the power to discover new pathogens has broad application to agriculture, veterinary medicine, ecology, and environmental genomics. While it is impossible to accurately predict the number of undiscovered viral pathogens remaining on this planet, tools such as virus-detection DNA microarrays permit rapid inroads into this fascinating and important aspect of virology.

See also: Diagnosis Using Serological and Molecular Approaches.

Further Reading

Chiu CY, Rouskin S, Koshy A, *et al.* (2006) Microarray detection of human parainfluenzavirus 4 infection associated with respiratory failure in an immunocompetent adult. *Clinical Infectious Diseases* 43: e71–e76.

Lin B, Wang Z, Vora GJ, *et al.* (2006) Broad-spectrum respiratory tract pathogen identification using resequencing DNA microarrays. *Genome Research* 16: 527–535.

Lin FM, Huang HD, Chang YC, *et al.* (2006) Database to dynamically aid probe design for virus identification. *IEEE Transactions on Information Technology in Biomedicine* 10: 705–713.

Urisman A, Fischer KF, Chiu CY, *et al.* (2005) E-Predict: A computational strategy for species identification based on observed DNA microarray hybridization patterns. *Genome Biology* 6: R78.

Urisman A, Molinaro RJ, Fischer N, *et al.* (2006) Identification of a novel gammaretrovirus in prostate tumors of patients homozygous for R462Q RNASEL variant. *PLoS Pathogen* 2: e25.

Wang D, Coscoy L, Zylberberg M, *et al.* (2002) Microarray-based detection and genotyping of viral pathogens. *Proceedings of the National Academy of Sciences, USA* 99: 15687–15692.

Wang D, Urisman A, Liu YT, *et al.* (2003) Viral discovery and sequence recovery using DNA microarrays. *PLoS Biology* 1: e2.

Wang Z, Daum LT, Vora GJ, *et al.* (2006) Identifying influenza viruses with resequencing microarrays. *Emerging Infectious Diseases* 12: 638–646.

Disease Surveillance

N Noah, London School of Hygiene and Tropical Medicine, London, UK

Glossary

Case–fatality rate Number of persons dying of an infection divided by total number of persons with the disease. Thus, CFR of 5% means that 5 of 100 persons with the infection died.

Epidemic and outbreak Most infectious disease epidemiologists do not distinguish between these terms and use them interchangeably. Generally, an outbreak is defined as a localized increase in cases, whereas an epidemic is more widespread, perhaps affecting a whole country.

GUM clinics The term GenitoUrinary Medicine clinics is often used for special clinics for sexually transmitted infections (STIs).

Incidence The rate of new infections (number by population) in a given time – for example, 5 cases of influenza per 1000 per year. Good for short-term infections.
Outbreak When an infection occurs at a frequency higher than expected for that time or place. It is basically an increased incidence which is usually unexpected. Two linked cases is also theoretically an outbreak.
Pandemic This term should be restricted to infections affecting many countries, though it does not have to be worldwide. SARS spared many countries and indeed some continents, but it was a pandemic. AIDS is undoubtedly a pandemic.
Prevalence The number of infections at any one time in a given population, expressed as a rate. Good for chronic infections and serological studies (e.g., prevalence of varicella antibody at age 15 in a given population is 95%).

Introduction

Surveillance is undoubtedly an essential – indeed critical – ingredient of any disease control program. It is used to monitor the impact of an infection, the effect of an intervention or health promotion strategy, health policy, planning, and delivery. Surveillance is the ongoing and systematic collection of routine data which are then analyzed, interpreted, and acted upon. It is essentially a practical process, which nevertheless can be useful in other ways. Its main purpose is to analyze time trends – but these can include not simply fluctuations in overall numbers, but also changes in age and sex distributions, geographical locations, and even possibly, in some of the more sophisticated established surveillance systems, at-risk groups (such as particular social, ethnic and occupational groups). Surveillance is essential for evaluating the impact of an intervention, such as mass vaccination, on a population. It can be a fairly sensitive system for the detection of outbreaks earlier than they would have been recognized otherwise.

The word ongoing in the description of surveillance helps to distinguish it from a survey, which is usually finite, tends to focus more on one or more groups of persons, and involves a questionnaire. Nevertheless, surveillance does not have to continue forever – when it is no longer useful, it should be stopped.

To be systematic is another important ingredient of surveillance. If reporting centers are not consistent in what they report, nor regular, the data they send in will be uninterpretable, and probably useless. Defining what needs to be reported and agreeing on the criteria for making a reportable diagnosis are necessary to make sense of the data.

The collection of routine data is another important characteristic of surveillance, especially with surveillance of laboratory infections. Generally, testing of samples is done for diagnosis, not primarily for surveillance. As laboratory testing is expensive, making further use of the results by contributing to surveillance makes for more efficient use of information, and the contribution to surveillance itself can often justify further testing. Typing echoviruses or coxsackie viruses can seldom be justified on clinical grounds alone, but the surveillance of serotypes can provide valuable information on the epidemiology of these viruses, their clinical characteristics, seasonality, and age and sex distributions. Echovirus 4 for example tends to be rare, fairly localized even within a country, but, when it occurs, it has a high aseptic meningitis rate, and causes a short and sharp autumn outbreak. Echovirus 9 on the other hand is far more common, more widespread geographically when it causes an epidemic, more benign, with a macular rash, and with meningitis not an especially common manifestation.

Although surveillance is essentially a practical exercise, this article attempts to show that surveillance can also be useful in giving us clues about an infection, whether it be its natural history, etiology, severity, or outcome.

Collection of Data: Sources of Data in Surveillance of Viruses

Death Certification

In most developed and middle-income countries, deaths are certified primarily for legal reasons, but have proved to be an important data source to use for surveillance. Clearly, they tend to be useful mainly for serious infections, and may suffer from inaccuracies, but remain a useful basic data source. Diseases with a high mortality rate and a short duration of illness (the viral hemorrhagic fevers for example) will obviously be better represented by death certification, than those with a low mortality. HIV/AIDS, before the era of HAART, had a high mortality but long periods of symptomless and symptomatic infection, so that death certification data needed careful interpretation. Laboratory reporting systems can be another useful source of information on mortality from infection.

Notifications

Surveillance systems such as statutory notification tend to be based mainly on clinical features. These can be very useful for common diseases with distinctive clinical syndromes, such as measles and mumps. It is important however not to make a disease notifiable unless there is a good reason for it: 'good' reasons include a mass vaccination program (when surveillance is virtually mandatory), any other mass control program, serious diseases for which contact tracing, mass or close contact prophylaxis, or investigation into source, is necessary. Serious less common viral infections such as poliomyelitis are notifiable in most countries. This is because contact tracing and preventive measures can be taken. Broader clinical diagnoses, such as aseptic (viral) meningitis, may on the face of it be less useful to notify, as it is usually impossible at the bedside to distinguish such causes of it as, for example, the coxsackie B viruses, echoviruses, and mumps. Nevertheless, notification of aseptic meningitis can be useful because a rapid rise in notified cases may need to be investigated. Moreover, the timing of any epidemic may give a clue to etiology – mumps meningitis tends to increase in spring and is usually accompanied by a concurrent outbreak of clinical mumps, while enterovirus epidemics are more likely to occur in autumn.

Other Clinical Sources of Data

Specific general practitioner (GP) surveillance systems are useful to provide data of epidemiological value for infections that are not notifiable, such as the common cold or chickenpox (in UK). They are of course clinically based, but are nevertheless useful, and often surprisingly accurate, possibly because those GPs who subscribe to a surveillance system are motivated to do so. GP surveillance systems are often sentinel-based, that is, based on a sample of GPs in a country, region, or area. Thus, they are good for common infections which to make notifiable would possibly be wasteful, for example, chickenpox. Moreover each sentinel would normally provide complete reporting. In the English system, GPs provide data on the base populations of their practices, so that rates of infection can be provided as a routine, a feature that is almost unique among surveillance systems. GP surveillance systems also tend to be good for timeliness and completeness.

Laboratory Data

It is useful to think of laboratory data as being of qualitative rather than quantitative value as they add quality and detail to disease surveillance. Thus, in the example already used for aseptic meningitis, a precise diagnosis of mumps, echovirus, or coxsackie type is particularly useful in clustering and outbreaks. Indeed, as the enteroviruses exhibit a strong late summer/autumn seasonal pattern with each enterovirus returning to its baseline in winter, when numbers of a particular type continue to be reported at a level higher than the winter baseline, the return of this virus to cause another epidemic in the following summer can usually be safely predicted. Laboratory data are also essential in qualifying food poisoning and gastroenteritis. Separating viral from bacterial causes is a useful first step, as their management tends to be very different. Norovirus gastroenteritis can be food-borne, but also spreads very easily from person-to-person because an extremely small dose is necessary for infection to occur, and the virus being fairly resistant to the environment will survive for some time. Management therefore must concentrate on hygiene. With most bacterial causes of food poisoning, especially the salmonellas, management often depends on the removal of the offending food. Laboratory surveillance is particularly essential for unraveling the mass of respiratory viral infections that inflict humans – respiratory syncytial virus, adenoviruses, parainfluenza viruses, rhinoviruses, etc. Indeed, it is particularly useful for influenza – not only separating it from influenza-like illness, but also in identifying influenza A and B, and if A, the subtype and variant.

Surveillance of Outbreaks

Surveillance of outbreaks (as opposed to individual infections) can be revealing, and important to allow public health measures. The existence of the noroviruses was suspected in the UK many years before the organisms were identified.This was because some outbreaks which did not fit the characteristics of known infections but had characteristics of their own had occurred.

Hospital Admissions

These can be useful for certain more serious infections, such as hepatitis and encephalitis. They can be unwieldy, generally lack detail, and often are published a year or more after the events have occurred.

Serological Surveillance

Serological surveillance has become increasingly important, and is a useful tool in assessing the immunity of a population, though it can also be used to identify vulnerable individuals. The immunity of a population can be vaccine-induced, and serological surveillance is a valuable adjunct to the methods available to monitor a mass immunization program. Vulnerable age groups can be identified, and booster doses of the vaccine introduced. An example of the importance of serological surveillance in determining public health policy is included below (analysis by person).

Surveillance of Viruses in Nonhuman and Environmental Sources

To build a picture of an infection, animals, birds, and the environment have been placed under surveillance. Rabies in foxes and other wildlife, influenza in birds, pigs, and other animals, are examples of important and fairly successful surveillance systems.

Other Sources of Data

Records of sickness absence, absence from school, calls to an emergency room, if available, can provide speedy information that something has happened, but tend to be nonspecific. Surveillance of antiviral resistance will become important.

Surveillance of HIV/AIDS

The association of HIV/AIDS with stigma makes surveillance of this infection especially difficult. The uniqueness and seriousness of this infection warrants a separate section. It is an example of the importance of tailoring surveillance to a specific serious infection if it becomes necessary to do so.

In the UK and some other countries with data protection acts, HIV infection, as with other STIs, is not notifiable. Special confidential surveillance systems through clinicians and GUM clinics, as well as laboratories, are in place. These are especially important for assessing risk factors. Inclusion of risk factors is essential for targeted intervention – for example, the proportions and rates of new diagnoses attributed to men who have sex with men, heterosexual sex, mother-to-infant, blood transfusion, IVDUs, and other needlestick injury.

Laboratory reporting is essential. Death certification is useful, though it has been shown that men who have sex with men, and probably those with other risk factors, are under-represented. In the UK, matching reported cases with death certificates is very important, as it allows for detection of deaths due to AIDS (such as pneumonia) as well as deaths associated with AIDS, and which are seen now in HIV-infected individuals – these include liver and cardiovascular disease, overdoses, and malignancies.

A surveillance system, based on unlinked anonymous testing of samples of blood taken routinely from certain at-risk population or occupational groups has been shown to provide valuable information on HIV infection in these populations. Specific screening systems for blood donors, military recruits, commercial sex workers, and family planning/termination of pregnancy clinics are also useful if these populations are to be targeted. Behavioral surveillance should also be seriously considered, to assist in identifying future trends, healthcare planning, as well as for specific health promotion efforts.

Attributes of Surveillance Systems

Completeness

Incompleteness is an almost universal drawback of most notification systems. They should never be dismissed for this reason alone. Statutory notification systems can be essential for surveillance and control. For common infections completeness may not be worth striving for, because notifications (assuming consistency in reporting) will generally provide information on trends, as well as fairly accurate information on age, sex and seasonal distributions, and possibly on place. The effect of mass vaccination programs can also be monitored fairly closely with statutory notification systems, as with measles (**Figure 1**) and acute paralytic poliomyelitis (**Figure 2**) in the UK. When a mass vaccination or other universal control program reduces the incidence of an infection to low levels, completeness becomes much more essential. For serious infections also, such as SARS or Lassa fever, for which contact tracing or other control measure is necessary, completeness is essential.

In active surveillance, reporters make negative returns if they have had no cases during the reporting period, to ensure completeness. In some countries, enhanced surveillance has been used to assess more accurately the true incidence of an infection – regions or districts are chosen to report all cases of a particular infection or infections. It is a hybrid of active and sentinel surveillance.

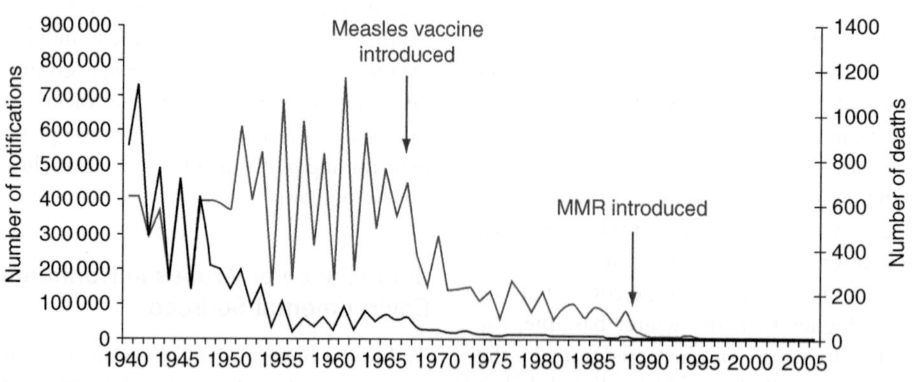

Figure 1 Measles notifications: cases and deaths, England and Wales 1940–2006. Reproduced from the Health Protection Agency (www.hpa.org.uk).

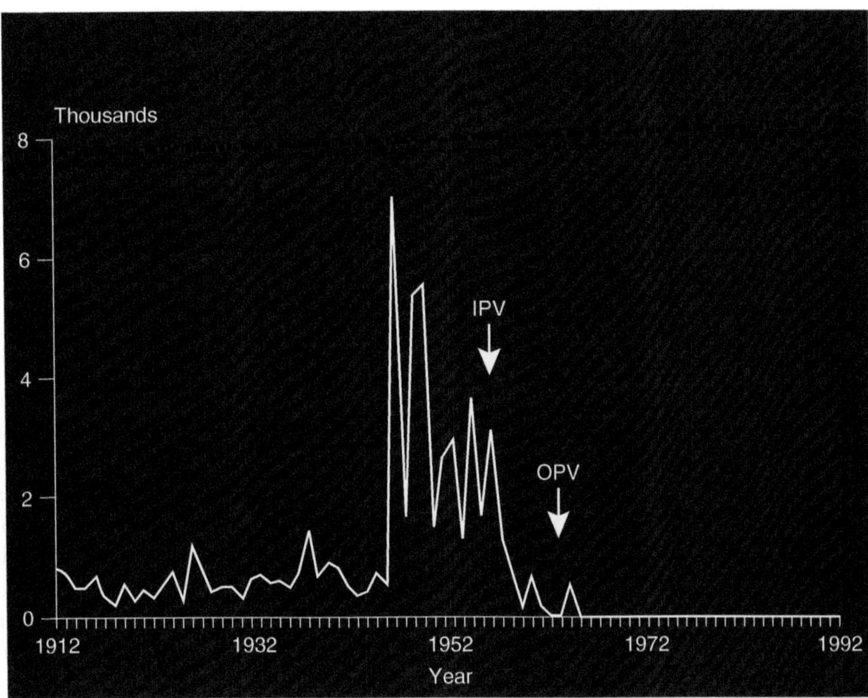

Figure 2 Notifications of acute poliomyelitis in England and Wales 1912–93. Reproduced from the Health Protection Agency (www.hpa.org.uk).

Timeliness

Timeliness is important for infections for which urgent public health measures have to be undertaken, such as poliomyelitis and viral hemorrhagic fever, as well as any outbreak. In some instances, infections, not normally urgent, can become so as an elimination program progresses. In a country with elimination of measles as its goal, a case of indigenous or imported measles needs to be dealt with urgently, as it may lead to an outbreak if not controlled immediately. Laboratory data are often not timely, and hospital data generally even less so, but can make up in accuracy what they lose in timeliness.

Accuracy

Accuracy is clearly important, though some minor degrees of inaccuracy can be tolerated in some common infections. Clinical data are most liable to have some inaccuracies, though even laboratory data can be inaccurate. Case definitions and quality control systems can be useful to improve accuracy.

Representativeness

For surveillance to provide an accurate picture of the impact of a particular infection, representativeness is essential. It is perhaps the most important quality for any surveillance system. Having a wide coverage of reporting clinicians and laboratories, or a well-chosen sample of sentinel sites, is necessary for the data collected to be representative of an infection in a country. Sometimes it may be necessary to assess data from various sources, such as notifications/GPs (clinical), hospital, laboratory, and death certificates.

Consistency

Consistency is another crucial basic attribute of any surveillance system. Reporters must know what to report (case definition) and how often. Otherwise it will not be possible to interpret trends.

Analysis of Data

Time

The three basic analyses by time, place, and person should be routine. Computer programs have made analyses of data quicker but somewhat less flexible. There is no standard period for analysis by time. Depending on what the surveillance intends to show, yearly, quarterly, monthly, four-weekly, or weekly time intervals can be used. In surveillance, time intervals shorter than this are rarely used. Monthly intervals have the disadvantage of having unequal number of days in each month, and are difficult to use when reporting is weekly; for seasonal trends four-weekly periods are better but cannot be divided into quarterly periods. In viral surveillance, four-weekly rather than weekly intervals tend to be most useful in showing seasonal changes. There is more likely to be more variation ('noise') in weekly intervals, making for less smooth changes. For secular trends, quarterly or annual intervals are generally used. Analyzing by time can reveal regular changes in the periodicity of viruses, enabling some of them to be predicted.

A basic knowledge of seasonal and secular patterns makes it easier to detect changes that signify a possible epidemic, and to differentiate these from a random variation. It is important to remember when analyzing laboratory data that there is often an interval, which can be 2 weeks or more, between date of onset and date of reporting.

Person

Analysis by age and sex is another basic analysis in surveillance. It can identify those most affected, and vulnerable groups. Changes in age distributions may provide important clues about a changing viral infection, and the effect of mass interventions on the age distribution of an infection can be monitored. Changes in the age distribution of measles in 1994 in the UK signified that an epidemic in older children was imminent, and the vaccine schedule was changed to include an extra booster injection (MR) to children aged 5–16 years. This averted the outbreak and the booster dose became a permanent feature of the routine immunization schedule in the UK. Indeed the changes in age distribution following mass vaccination could be considered an epidemiological side effect of mass vaccination. Requests for occupational groups and travel histories should be selective. For poliomyelitis, SARS, dengue, and the viral hemorrhagic fevers, travel histories are required. Occupational group may be useful for norovirus, and hepatitis types A, B, or C. Specific risk factors may be worthwhile for HIV, hepatitis B and C.

Place

Analysis by place can pinpoint local outbreaks. Some echoviruses (e.g., echovirus type 4) can cause rare short local outbreaks, other types (e.g., types 9 and 11) are more common and more widespread. Food-borne outbreaks of hepatitis A are generally picked up locally through routine surveillance, but sometimes more extensive outbreaks caused by a more widely distributed foodstuff, including shellfish or frozen soft fruit, may be identified.

Interpretation

Collection and analysis are generally routine functions; skill is required in interpretation of the data. No statistic is perfect, and surveillance data, like all data, must be interpreted with caution. One must take into account the origins of the data – clinical, laboratory, or hospital. Not only must the reliability, or otherwise, of the data be evaluated, but what the data signify in the natural history of the infection must also be recognized.

Every viral infection has its stages and these must be recognized before surveillance data can be sensibly interpreted. At what stage are the data being collected important to understanding and interpretation? Using influenza or hepatitis A as examples, and a defined population (**Table 1**), only a proportion of persons in the defined population will be infected with the virus. They can only be comprehensively detected by screening, and serological surveillance will identify these persons, or assess population immunity (B1 in **Table 1**). In HIV/AIDS surveillance, unlinked anonymous testing of samples of blood taken routinely at say, an antenatal clinic, can give vital information on the prevalence of HIV infection, since in this infection, presence of antibody denotes infection, not immunity. A smaller proportion will be ill (B2), but only some of these will visit a doctor (B3). Surveillance systems based on GP consultation rates have now been recognized as an important addition to the spectrum of a disease, and many countries have excellent systems. Of those patients that do visit their family doctor, only some will be admitted to hospital (B4). Finally, only some will die (B5).

For laboratory data, the stages are slightly different (**Table 2**), but still important to understanding what the reported data mean. As before, only a proportion of persons will be infected (B1), a smaller proportion will be ill (B2), and a smaller proportion still visit their doctors (B3). Not all doctors will send specimens to a laboratory (B4), and only a proportion of these specimens (B5), depending on accuracy of the identification process, the method of transport, the fragility of the organism, and the swabbing or other sampling technique, will be positive. Finally, depending on the level of consistency of reporting,

Table 1 Stages of a viral infection

Clinical
A. Uninfected
B. Infected
1. Asymptomatic
2. Symptomatic unreported
3. Symptomatic, sees a doctor
4. Symptomatic, admitted to hospital
5. Symptomatic, dies/survives

Table 2 Stages in laboratory diagnosis

A. Uninfected
B. Infected
1. Asymptomatic
2. Symptomatic, unreported
3. Symptomatic, sees a doctor
4. Symptomatic, specimens submitted
5. Symptomatic, specimens positive
6. Symptomatic, specimens reported to surveillance system

only some of these will be reported (B6). It is important to recognize these stages in interpreting surveillance data.

It is important moreover to recognize the biases that will inevitably occur between these stages. Collection of data in routine surveillance is not normally a scientific process as one has to rely on readily available data – data obtained mostly for other reasons, such as to make a definitive diagnosis. Only the most severe cases die, and death certification thus provides, at best, a limited view of any disease. Similarly, only certain types and severity of cases will be admitted to hospital (some admissions are for social reasons for example) or even visit their family doctor. Certain age, sex, and perhaps social or occupational groups are more likely to seek medical help, be investigated and be reported. In laboratory data, more severe cases, or children, are perhaps much more likely to be investigated in detail.

These shortcomings of surveillance data do not make them useless – but their strengths and limitations must be recognized.

Feedback

If interpretation is turning statistics into information, feedback is getting the information across to those that matter, and those that need to know, so that action – the objective of surveillance – can be taken. Without feedback, surveillance is pointless. Feedback is most likely to be informative if undertaken by those most closely involved in the surveillance cycle, and who understand the significance of the data they are receiving.

Feedback should be aimed at contributors and those in public health. Contributors will then be aware of which viruses are circulating and this will help them to know what to look for in their own tests (e.g., what echovirus or adenovirus types are in circulation). Moreover, routine surveillance will undoubtedly uncover outbreaks of infection, which will need further investigation and control at local, national, and even international level. An interesting side effect of a flourishing microbiological/feedback surveillance system is that it often stimulates better quality control within reporting laboratories.

Regular feedback is helpful, not only to contributors, but also to those who can act for the public health. Generally, the periodicity of feedback should reflect the frequency of reporting – weekly feedback for weekly reports for example. Regular topic-based reviews are important.

Evaluation of Surveillance

A surveillance system is like a country's train system. Once the rail lines have been built, the goods that will be carried along those lines can be changed according to need. Similarly, in a surveillance network, once the lines of communication have been laid down, the data being reported can be changed according to what is most important at the time (though probably not too frequently). Nevertheless, surveillance systems should ideally be frequently evaluated for usefulness, as well as for accuracy, efficiency, and effectiveness.

They should also be sufficiently flexible, so that 'new' or emerging infections can be included in an emergency or when the need arises. The successful implementation of international surveillance for SARS was instrumental in controlling it. Emergency surveillance was also essential following the tsunami of 2004, and is also necessary for the successful management of other disasters following earthquakes, hurricanes, and floods.

Surveillance systems should be evaluated before they are set up, and again at regular intervals thereafter. Before implementing a surveillance system, is there an adequate public health and administrative infrastructure in place to take action? Are the data to be collected representative and sufficiently timely for the specific infection? Are they useful, and is action being taken on the information? If not, is the feedback inadequate?

Global and International Surveillance

The ease of modern travel, the distribution of goods (especially foodstuffs) across increasingly wide parts of the world, and the uncontrollable spread of birds and other wildlife across boundaries has made global and international surveillance essential for outbreak and infection control. Surveillance of influenza now requires the expertise of many professionals – epidemiologists, virologists, and vaccinologists, clinicians, statistical modelers, veterinarians, managers, and planners – in many different countries so that information can be exchanged, and attempts made on a global basis, to prevent the next pandemic. Only recently, a new variant of chikungunya virus jumped from Kenya, where it seems to have started, to islands in the Indian Ocean and hence to India. It has now, in 2007, even reached Italy. In Reunion alone, it affected 265 000 people, an astoundingly high incidence of 34%, and an estimated case-fatality rate of 1/1000. In India 1.3 million persons are thought to have been affected (so far, to February 2007).Two species of mosquito have been involved, *Aedes aegypti* and *A. albopictus.* Epidemics of dengue and West Nile virus have also spread widely recently. AIDS/HIV was destined to become a global problem almost from the time of its first discovery. On a more positive note, SARS was contained through the use of international surveillance; and surveillance was the backbone of the smallpox eradication program.

International surveillance can also be used for the detection of international outbreaks of food poisoning caused by the distribution of foodstuffs across a wide number of countries. An outbreak of hepatitis A in England was caused by frozen raspberries grown and frozen in another country and another outbreak of hepatitis A, this time in Czechoslovakia (before it became separate republics) was caused by strawberries used to make ice cream; the strawberries had been imported from another Eastern European country. There are now well-established trans-European surveillance systems for salmonella infections and legionnaires' disease.

The need for surveillance will never diminish or disappear. Surveillance systems will only improve, become increasingly sophisticated, and become increasingly relied upon and used. Control of infection will not be possible without it.

Further Reading

Charrel RM, de Lamballerie X, and Raoult D (2007) Chikungunya outbreaks – The globalization of vectorborne diseases. *New England Journal of Medicine* 356: 769–771.

Chin J (ed.) (2002) *Control of Communicable Diseases Manual,* 17th edn. Washington, DC: APHA.

Chorba TL (2001) Disease surveillance. In: Thomas JC and Weber DJ (eds.) *Epidemiologic Methods for the Study of Infectious Diseases*, ch. 7. Oxford, UK: Oxford University Press.

Communicable Disease Report(1994) National measles and rubella immunisation campaign. *Communicable Disease Report Weekly* 4(31): 146–150.

Heymann D and Rodier GR (1998) Global surveillance of communicable diseases. *Emerging Infectious Diseases* 4: 362–365.

Noah N (2006) *Controlling Communicable Disease*, chs. 1–4, pp. 14–19. Maidenhead, UK: Open University Press.

UNAIDS (2003) Introduction to second generation HIV surveillance. http://www.data.unaids.org/Publications/IRC-pub03/2nd_generation_en.ppt (accessed September 2007).

Thomas MEM, Noah ND, and Tillett HE (1974) Recurrent gastroenteritis in a preparatory school caused by *Shigella sonnei* and another agent. *Lancet* 1: 978–981.

Relevant Websites

http://www.cdc.gov – CDC (Centers for Disease Control and Prevention).

http://www.hpa.org.uk – Communicable Disease Report Weekly, Communicable Disease Report, Health Protection Agency (HPA).

http://www.hpa.org.uk – Enter-net(Gastrointestinal), International, Health Protection Agency.

http://www.who.int – Epidemiologic Surveillance, WHO.

http://www.who.int – Epidemiology, WHO.

http://www.eiss.org – European Influenza Surveillance Scheme (EISS).

http://www.eurosurveillance.org – Eurosurveillance is Freely Available Weekly and Monthly, European Centre for Disease Prevention and Control (ECDC).

http://www.hpa.org.uk – Health Protection Agency (HPA).

http://www.hpa.org.uk – HIV and Sexually Transmitted Infections, Health Protection Agency.

http://www.who.int – National Influenza Pandemic Plans, WHO.

http://www.hpa.org.uk – Response, Co-ordination and Specialist Support for Outbreaks and Incidents, Centre for Infections, HPA.

http://www.who.int – Surveillance of Noncommunicable Disease Risk Factors, WHO.

http://www.EWGLI.org – The European Working Group for Legionella Infections, EWGLI.

IMMUNE RESPONSES

Innate Immunity

F Weber, University of Freiburg, Freiburg, Germany

Glossary

Apoptosis A form of programmed cell death.

Apoptotic bodies Remnants of cells which underwent apoptosis.

Complement system A pathogen-triggered cascade of biochemical reactions involving more than 20 soluble and cell-bound proteins. Complement activation results in opsonization, priming of humoral immune responses, and perforation of membranes.

Cytokines Proteins which mediate cell–cell communication related to pathogen defense. Secreted by immune cells or tissue cells.

Innate immunity Physical and chemical barriers, cells, cytokines, and antiviral proteins which exclude, inhibit, or slow down infection with little specificity and without adaptation or generation of a protective memory.

Interferons (IFNs) Cytokines mediating antiviral activity. Distinguished into type I (IFN-α/β), type II (IFN-γ), and type III (IFN-λ). Type I and type III IFNs directly mediate antiviral activity in responding cells, whereas type II IFN is more immunomodulatory.

Interferon-stimulated response element (ISRE) A promoter element common to all type I IFN-stimulated genes.

Opsonization Tagging of infected cells or pathogens for destruction by phagocytic cells.

Pathogen-associated molecular patterns (PAMPs) Molecular signatures of pathogens used by the innate immune system to distinguish self from non-self. Often highly repetitive patterns.

Pattern recognition receptors (PRRs) Intracellular and extracellular receptors recognizing specific PAMPs.

Phagocytosis Uptake of particles by cells.

Introduction

Viruses attempting to conquer a mammalian body are faced with an array of problems. 'Innate immunity' in a wider sense means all sorts of factors which exclude, inhibit, or slow down infections in a rapid manner but with little specificity and without adaptation or generation of a protective memory. Many of these efficient and not at all primitive defenses are evolutionarily old and can be found in all metazoans. For the sake of brevity, however, the discussion in this article is restricted to mammals as these are the best investigated organisms in that respect. RNA interference, the innate immune system of plants and nonvertebrates, is not covered here.

Mammalian innate immune defenses against virus infections can be divided into several distinct parts such as mechanical and chemical barriers (not further mentioned here), complement system, phagocytic/cytolytic cells of the immune system which act in a nonspecific manner, and cytokines (most prominently the type I interferons).

The Complement System

The complement system (which 'complements' the adaptive immune system in the defense against pathogens) primes the adaptive immune response and is also directly effective against pathogens. Complement activation is achieved by specific receptors recognizing pathogens or immunocomplexes. Three different pathways are being distinguished which are termed the classical pathway (triggered by antigen–antibody complexes), the mannan-binding lectin pathway (triggered by lectin binding of pathogen surfaces), and the alternative pathway (triggered by complement factor C3b-coated pathogen surfaces). They all activate a cascade of reactions involving more than 20 soluble and cell-bound proteins, thus resulting in a rapid and massive response. The complement system is able to (1) tag infected cells and pathogens for destruction by phagocytic cells (opsonization), (2) prime humoral immune responses, and (3) perforate membranes of infected cells by the membrane-attack complex. In response, viruses have evolved effective countermeasures such as incorporation of cellular complement-regulatory proteins into particles or expressing specific inhibitors in infected cells.

Cellular Innate Immunity

Macrophages/monocytes, granulocytes, natural killer cells, and dendritic cells belong to the cellular branch of the innate immune system. Monocytes circulate in the bloodstream for several hours before they differentiate into macrophages. These potent phagocytic cells either continue patrolling or they permanently settle in particular tissues (i.e., the Kupffer cells of the liver), being able to rapidly remove viral particles and apoptotic bodies. Activated macrophages also synthesize inflammatory cytokines such as interferon (IFN)-γ and tumor necrosis factor (TNF)-α, thus triggering an adaptive immune response. Granulocytes are also able to remove viral particles and apoptotic bodies by phagocytosis. They are rapidly attracted to inflammatory sites and enter the tissue by transendothelial migration. Both macrophages and granulocytes cleave the ingested viral proteins into fragments and present them to T lymphocytes.

Natural killer (NK) cells are able to recognize infected cells in an antigen-independent manner and destroy them by their cytotoxic activity. Also, they rapidly produce large amounts of IFN-γ to activate the adaptive immune system. NK cells are regulated by a fine balance between stimulatory and inhibitory receptors. One of their prominent features is their ability to destroy cells which lack MHC I molecules on their surface. As many viruses downregulate MHC expression in order to avoid an adaptive immune response, NK surveillance represents an important early warning and attack system against virus infections.

A key connection between the innate and the adaptive immune system is provided by dendritic cells (DCs). These specialized immune cells sample antigen at the site of infection, activate themselves and the surrounding tissue cell by cytokine synthesis, and then migrate to secondary lymphatic organs in order to mobilize T cells against the presented antigen. The differentiation into efficient antigen-presenting cells (APCs) is achieved by cytokine production which, in turn, is triggered by stimulation of receptors recognizing pathogen-specific molecular patterns (PAMPs). Two main types of DCs are being distinguished: myeloid DCs (mDCs) and plasmacytoid DCs (pDCs). mDCs are an early split-off of the myeloid bone marrow precursors, that is, the stem cells which are also giving birth to macrophages/monocytes and granulocytes, among others. Depending on the location, several subsets of mDCs such as Langerhans cells or interstitial cells are being distinguished. pDCs, which are not segregated into subpopulations, are thought to be derived from lymphatic precursor cells. Both mDCs and pDCs can sense viral infection by several intra- and extracellular PAMP receptors (see below). Depending on the DC type, high levels of interleukins or interferons are being produced which coin the subsequent immune reaction. pDCs are potent producers of the main antiviral cytokines, the type I interferons.

Antiviral Cytokines: The Type I Interferons

Isaacs and Lindenmann discovered in 1957 that cells which had been in contact with virus particles secrete a soluble factor which confers resistance to influenza viruses, a phenomenon called 'interference'. In the subsequent years, it became more and more clear that the so-called type I interferon (IFN-α/β) system is our primary defense mechanism against viral infections. In fact, humans with genetic defects in the IFN signaling pathway have a bad prognosis as they die at an early age of viral diseases which would otherwise pose little problems. Similarly, knockout mice with a defective IFN system quickly succumb to viral pathogens of all sorts although they have an intact adaptive immune system.

In response to virus infection, pDCs are particularly well equipped to synthesize and secrete IFN-α/β, but in principle all nucleated cells are able to do so. In an autocrine and paracrine manner, IFNs trigger a signaling chain leading to the expression of potent antiviral proteins which limit further viral spread. In addition, IFNs initiate, modulate, and enhance the adaptive immune response. The signaling events which culminate in the direct IFN-dependent restriction of virus growth can be divided into three steps, namely (1) transcriptional induction of IFN synthesis, (2) IFN signaling, and (3) antiviral mechanisms.

Interferon Induction

A number of pattern recognition receptors (PRRs) recognize conserved PAMPs of viruses and initiate induction of IFN genes (see **Figure 1**). PRRs can be divided into the extracellular/endosomal toll-like receptors (TLRs) and the intracellular receptors RIG-I, MDA-5, and PKR. The main PAMPs of viruses appear to be nucleic acids, namely double-stranded RNA (dsRNA), single-stranded RNA (ssRNA), and double-stranded DNA (dsDNA).

dsRNA is an almost ubiquitous transcriptional by-product of RNA and DNA viruses. It is recognized by TLR3, the related RNA helicases RIG-I and MDA-5, and the protein kinase PKR. A third dsRNA-binding member of the RIG-I helicase family, LGP2, acts as a negative-feedback inhibitor.

Viruses with a negative-strand ssRNA genome (e.g., influenza virus) are unique in that they do not produce substantial amounts of intracellular dsRNA. Their genomic ssRNA is recognized in the endosome by TLR7 and -8. Interestingly, in the cytoplasm, RIG-I recognizes the influenza virus genome in a 5′-triphosphate-dependent manner. The question how much the well-documented dsRNA-binding and unwinding activity of RIG-I contributes to its 5′-triphosphate-dependent recognition of viral genomes remains to be solved.

The third important PAMP, viral dsDNA, is again recognized both by an endosomal receptor, TLR9, and an unknown intracellular receptor. Thus, for all three nucleic acid-based PAMPs of viruses, there are specific PRRs present both in the endosomal and the intracellular compartment.

Besides nucleic acids, some viral proteins can provoke a TLR response such as the envelope proteins of respiratory syncytial virus and measles virus by activating TLR4 and TLR2, respectively.

All PRR-triggered signaling pathways eventually culminate in a strong activation of type I IFN transcription.

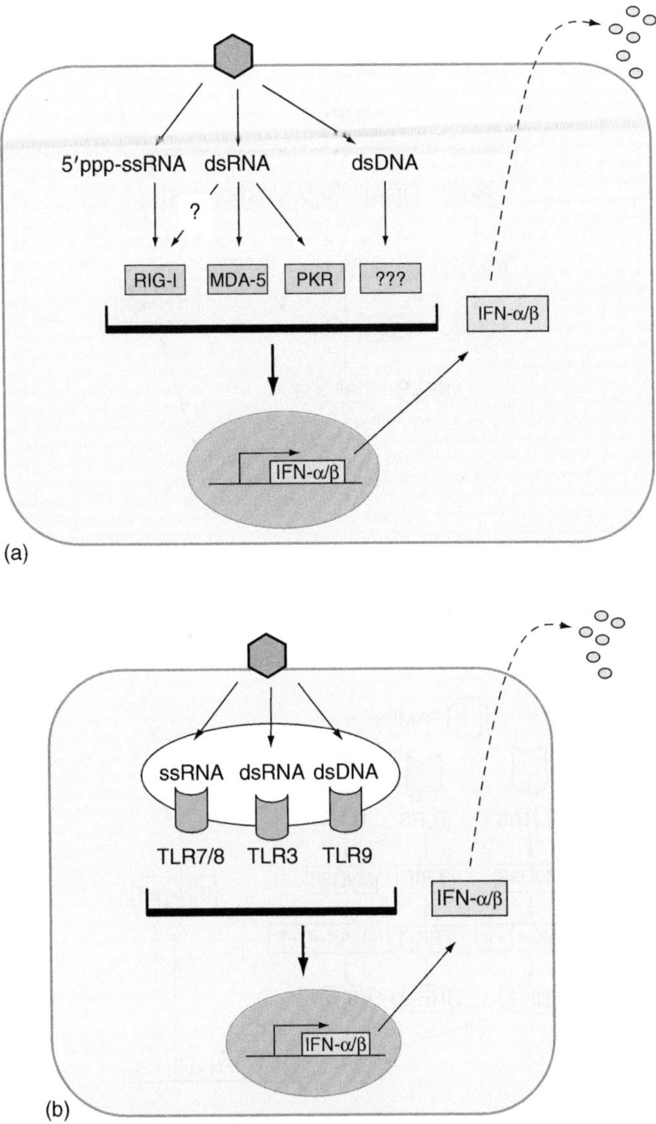

Figure 1 Depending on the virus, ssRNA, dsRNA, dsDNA, or combinations thereof represent characteristic by-products of infection which lead to induction of IFN-α/β genes. (a) These signature molecules are recognized by the intracellular PRRs RIG-I, MDA-5, PKR, and an unknown receptor for viral dsDNA. RIG-I recognizes dsRNA, but was shown to be important for recognition of 5′-triphosphate-containing ssRNA *in vivo* (see text). (b) Intracellular PAMP recognition is mirrored by the endosomal TLR pathways recognizing the same characteristics, except that ssRNAs do not need to be 5′-triphosphorylated.

The 'classic' intracellular pathway of IFN-β gene expression involves RIG-I and MDA-5 which both contain two N-terminal caspase-recruiting domain (CARD)-like regions and a C-terminal DExD/H box RNA helicase domain (**Figure 2(a)**). RNA binding to the helicase domain induces a conformational change which liberates the CARD domain to interact with the signaling partner IPS-1 (also called Cardif, MAVS, or VISA). This adaptor mediates RIG-I and MDA-5 signaling and needs to be located at the mitochondrial membrane. IPS-1 has a CARD-like domain which binds to RIG-I and MDA5 and a C-terminal region which activates the kinases IKKε and TBK-1. These kinases are known to phosphorylate the transcription factor IFN regulatory factor (IRF)-3, a member of the IRF family. Phosphorylated IRF-3 homodimerizes and is transported into the nucleus. In addition, the transcription factor nuclear factor-kappa B (NF-κB) is recruited in a PKR/TRAF- and IPS-1-dependent way. Together, IRF-3 and NF-κB strongly upregulate IFN gene expression. This leads to a 'first wave' of IFN production (IFN-β and IFN-α4 in mice) which triggers the expression of the transcription factor IRF-7. Recent evidence has shown that IRF-7 is a master regulator of IFN gene expression and that IRF-3 seems to cooperate with IRF-7 for full activity. IRF-7 can be activated in the same way as IRF-3 and is responsible for a positive-feedback loop that initiates the synthesis of several IFN-α subtypes as the 'second-wave' IFNs.

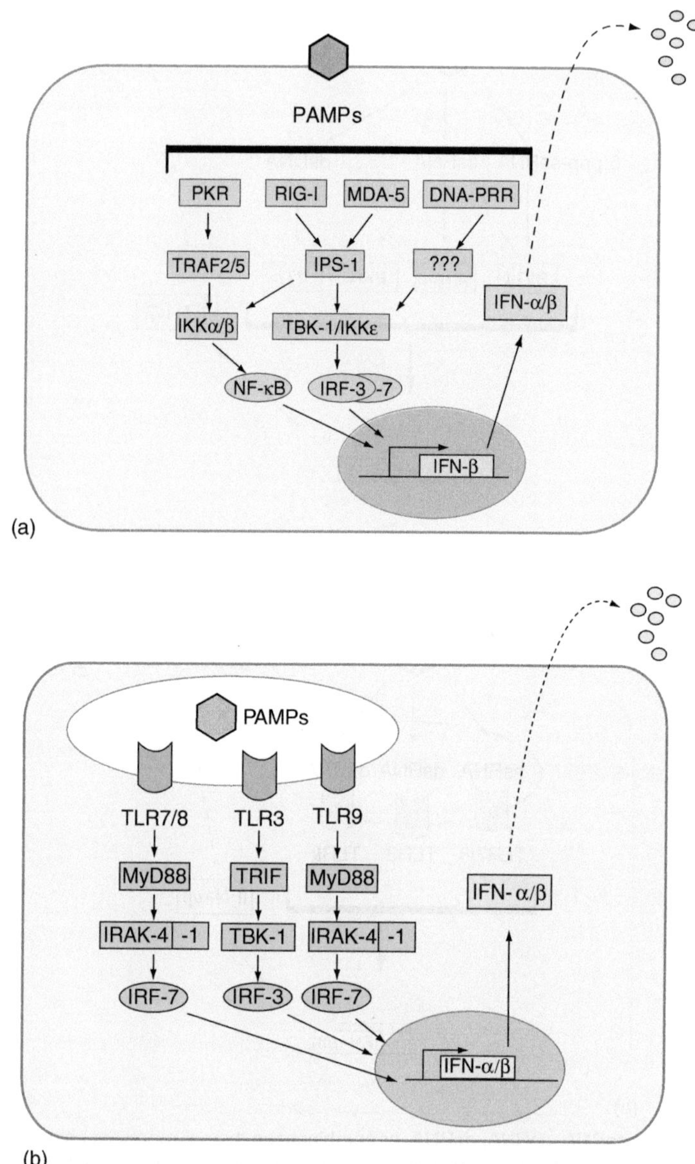

Figure 2 (a) PAMP recognition by intracellular PRRs leads to activation of the transcription factors NF-κB, IRF-3, and AP-1 (not shown). The cooperative action of these factors is required for full activation of the IFN-β promoter. IRF-3 is phosphorylated by the kinases TBK-1 and IKKε which in turn are activated by RIG-I and MDA5 via IPS-1. NF-κB is activated by the PKR pathway as well as by IPS-1. The IFN-induced IRF-7 later enhances IFN gene transcription, but is also essential for immediate early IFN-β transcription. (b) PAMP recognition by endosomal PRRs. IRF-7 is activated by IRAK-1, which in turn is phosphorylated by IRAK-4 in an MyD88-dependent manner. Both TLR7/8 and TLR9 use the MyD88 adaptor, whereas TLR3 activates IRF-3 via TRIF and TBK1.

mDCs can sense dsRNA by the classic intracellular pathway and, in addition, by TLR3 (**Figure 2(b)**). dsRNA-induced triggering of endosomal TLR3 proceeds via TRIF and TRAF3 which activate the kinase TBK-1, leading to phosphorylation of IRF-3 and, subsequently, to the activation of IFN-β gene expression.

pDCs sense the presence of viral ssRNA or dsDNA by TLR7, TLR8, and TLR9 (**Figure 2(b)**). Upon activation, TLR7, -8, and -9 signal through their adaptor molecule MyD88, the IRAK kinases, and IRF-7 to transcriptionally activate multiple IFN-α genes. In contrast to other cell types, pDCs contain considerable amounts of constitutively expressed IRF-7. IRF-7 is further upregulated in response to IFN and generates a positive-feedback loop for high IFN-α and IFN-β production. Furthermore, TLR7 and TLR9 are retained in the endosomes of pDCs to allow prolonged IFN induction signaling.

Type I IFN Signaling

IFN-β and the multiple IFN-α subspecies activate a common type I IFN receptor (IFNAR) which signals to

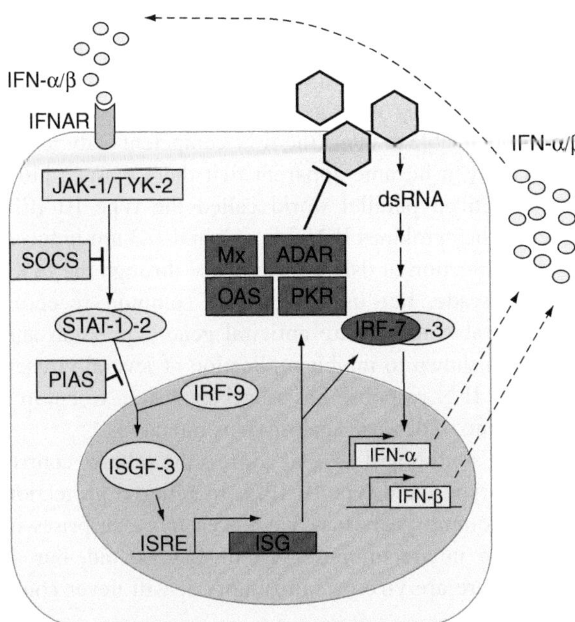

Figure 3 IFN-α and IFN-β bind to the type I IFN receptor (IFNAR) and activate the expression of numerous ISGs via the JAK/STAT pathway. IRF-7 amplifies the IFN response by inducing the expression of several IFN-α subtypes. Mx, ADAR, OAS, and PKR are examples of proteins with antiviral activity. Modified from Haller O, Kochs G, and Weber F (2006) The interferon response circuit: Introduction and suppression by pathogenic viruses. *Virology* 344: 119–130, with permission from Elsevier.

the nucleus through the so-called JAK–STAT pathway (**Figure 3**). The STAT proteins are latent cytoplasmic transcription factors which become phosphorylated by the Janus kinases JAK-1 and TYK-2. Phosphorylated STAT-1 and STAT-2 recruit a third factor, IRF-9, to form a complex known as IFN-stimulated gene factor 3 (ISGF-3) which translocates to the nucleus and binds to the IFN-stimulated response element (ISRE) in the promoter region of interferon-stimulated genes (ISGs). Specialized proteins serve as negative regulators of the JAK–STAT pathway. The suppressor of cytokine signaling (SOCS) proteins prevent STAT activation whereas protein inhibitor of activated STAT (PIAS) family members function as small ubiquitin-like modifier (SUMO) E3 ligases and inhibit the transcriptional activity of STATs.

Direct Antiviral Effects of Type I IFNs

Type I IFNs activate the expression of several hundred IFN-stimulated genes (ISGs) with multiple functions. To date, five antiviral pathways have been studied in great detail, namely the protein kinase R (PKR), the RNA-specific adenosine deaminase 1 (ADAR 1), the 2–5 OAS/RNaseL system, the product of the ISG56 gene (p56), and the Mx proteins. PKR, ADAR1, and 2–5 OAS are constitutively expressed in normal cells in a latent, inactive form. Basal mRNA levels are upregulated by IFN-α/β and these enzymes need to be activated by viral dsRNA. PKR is a serine-threonine kinase that phosphorylates – among other substrates – the α-subunit of the eukaryotic translation initiation factor eIF2. As a consequence, translation of cellular and viral mRNAs is blocked. PKR also plays a role in virus-induced NF-κB activation, as described above. ADAR 1 catalyzes the deamination of adenosine on target dsRNAs to yield inosine. As a result the secondary structure is destabilized due to a change from an AU base pair to the less stable IU base pair and mutations accumulate within the viral genome. The 2–5 OAS catalyzes the synthesis of short 2′–5′ oligoadenylates that activate the latent endoribonuclease RNaseL. RNaseL degrades both viral and cellular RNAs, leading to viral inhibition. P56 binds the eukaryotic initiation factor 3e (eIF3e) subunit of the eukaryotic translation initiation factor eIF3. It functions as an inhibitor of translation initiation at the level of eIF3 ternary complex formation and is likely to suppress viral RNA translation. Mx proteins belong to the superfamily of dynamin-like large GTPases and have been discovered as mediators of genetic resistance against orthomyxoviruses in mice. They most probably act by enwrapping viral nucleocapsids, thus preventing the viral polymerase from elongation of transcription.

The antiviral profiles of the IFN effectors listed above are distinct but often overlapping. Mx proteins, for example, mainly inhibit segmented negative-strand RNA viruses and also Semliki Forest virus, whereas the 2–5 OAS/RNaseL system appears more important against positive-strand RNA viruses. Moreover, only rarely the presence of one particular IFN effector determines host resistance. Rather, it is the sum of antiviral factors affecting, for example, genome stability, genetic integrity, transcription, and translation that confers the full antiviral power of IFN.

Indirect Antiviral Effects of Type I IFNs

Besides the effector proteins listed above, several ISGs contribute in a more indirect manner to the enhancement of both innate and adaptive immune responses. Virus-sensing (and in part antiviral) PRRs such as TLR3, PKR, RIG-I, and MDA5 are by themselves upregulated in a type-I-IFN-dependent manner. Similarly, IRF-7 and and STAT1, the key factors of type I IFN and ISG transcription, respectively, are ISGs. The strong positive-feedback loop mediated by the upregulation of these PRRs and transcription factors is counterbalanced by several negative regulators such as LGP2, SOCS, and PIAS, which are either ISGs or depend on IFN signaling for their suppressive action.

Type I IFNs can directly enhance clonal expansion and memory formation of $CD8^+$ T cells. Also, IFNs promote NK cell-mediated cytotoxicity and trigger the synthesis of cytokines such as IFN-γ or IL-15 which modulate the adaptive immune response, enhance NK cell proliferation, and support $CD8^+$ T-cell memory. Moreover, by upregulating TLRs, MHCs, and costimulatory molecules, IFNs enable APCs (most prominently DCs) to become competent in presenting viral antigens and stimulating the adaptive immune response.

Good Cop–Bad Cop

Given their massive impact on the cellular gene expression profile, it is quite expected that type I IFNs have not only antiviral, but also antiproliferative and immunomodulatory effects. Treatment with IFNs is an established therapy against several viral and malignant diseases such as hepatitis B, hepatitis C, Kaposi's sarcoma, papillomas, multiple sclerosis, and several leukemias and myelomas. However, the strong and systemic effects of IFNs do not come without a price. Administration of IFN can locally produce inflammation, and systemically cause fever, fatigue, malaise, myalgia, and anemia. It is no coincidence that these latter are 'flu-like' symptoms, since in many acute infections IFNs play a dominant role. The effects of IFN which are desired and beneficial if restricted to the site of first infection can turn into a life-threatening 'cytokine storm' if it becomes systemic. Severe acute respiratory syndrome (SARS) and human infections with H5N1 influenza viruses are examples of such out-of-control innate immune responses. Another 'dark side' aspect is that patients with autoimmune diseases have chronically elevated levels of IFNs, and that IFN therapy can aggravate autoimmune disorders. It is thought that pDCs (and in part B cells) are autostimulated by self-DNA via TLR9 and by small nuclear RNA complexes (snRNPs) via TLR7. Chronic production of IFNs causes maturation of mDCs, which in turn activate autoreactive T and B cells.

Concluding Remarks

The concept of innate immunity certainly comprises more than the IFN system (see above), but type I IFNs represent a central part. These cytokines not only have direct antiviral effects but also orchestrate the first defense reactions and the subsequent adaptive immune response, thus determining the course of infection. The recent findings that basically every virus appears to have evolved one or several countermeasures for controlling the IFN response is testament to its importance. In addition, IFNs are not only antiviral, but also effective tumor suppressors. Tumor cells often eliminate the IFN system during the transformation process. The payoff is an increased susceptibility to infection, an Achilles heel which is exploited by the therapeutic concept of oncolytic viruses. Tumor selectivity of such viruses can be even more increased by using IFN-sensitive mutants. The inability of those mutants to fight the IFN response is complemented by the mutations of the tumor cells, thus allowing virus growth. At the same time, these viruses are unable to infect the IFN-competent body cells.

Recently, it became apparent that there exists a hitherto unnoticed parallel world called the type III IFN system. The cytokines IFN-λ1, -λ2, and -λ3 are induced by virus infection or dsRNA and signal through the JAK/STAT cascade, but use a separate common receptor. They are able to activate antiviral gene expression and have been shown to inhibit replication of several viruses. Thus, the IFN response has a backup system to enforce the first line of defense against virus infections.

Future studies will have to address the relative contribution of type I and type III IFNs to antiviral protection and the coming years may have even more surprises in stock. The innate immune system may be old, but as long as there are viruses and tumors, it will never come out of fashion.

See also: Antibody - Mediated Immunity to Viruses; Defeating Innate Immunity; Interfering RNAs.

Further Reading

Akira S and Takeda K (2004) Toll-like receptor signalling. *Nature Reviews Immunology* 4: 499–511.

Ank N, West H, and Paludan SR (2006) IFN-λ: Novel antiviral cytokines. *Journal of Interferon and Cytokine Research* 26: 373–379.

Colonna M, Trinchieri G, and Liu YJ (2004) Plasmacytoid dendritic cells in immunity. *Nature Immunology* 5: 1219–1226.

Diefenbach A and Raulet DH (2003) Innate immune recognition by stimulatory immunoreceptors. *Current Opinion in Immunology* 15: 37–44.

Garcia MA, Gil J, Ventoso I, *et al.* (2006) Impact of protein kinase PKR in cell biology: From antiviral to antiproliferative action. *Microbiology and Molecular Biology Reviews* 70: 1032–1060.

Garcia-Sastre A and Biron CA (2006) Type 1 interferons and the virus–host relationship: A lesson in detente. *Science* 312: 879–882.

Haller O and Kochs G (2002) Interferon-induced mx proteins: Dynamin-like GTPases with antiviral activity. *Traffic* 3: 710–717.

Haller O, Kochs G, and Weber F (2006) The interferon response circuit: Induction and suppression by pathogenic viruses. *Virology* 344: 119–130.

Hoebe K, Janssen E, and Beutler B (2004) The interface between innate and adaptive immunity. *Nature Immunology* 5: 971–974.

Honda K, Takaoka A, and Taniguchi T (2006) Type I interferon gene induction by the interferon regulatory factor family of transcription factors. *Immunity* 25: 349–360.

Reis e Sousa C (2006) Dendritic cells in a mature age. *Nature Reviews Immunology* 6: 476–483.

van Boxel-Dezaire AH, Rani MR, and Stark GR (2006) Complex modulation of cell type-specific signaling in response to type I interferons. *Immunity* 25: 361–372.

Vilcek J (2006) Fifty years of interferon research: Aiming at a moving target. *Immunity* 25: 343–348.

Volanakis JE (2002) The role of complement in innate and adaptive immunity. *Current Topics in Microbiology and Immunology* 266: 41–56.

Defeating Innate Immunity

C F Basler, Mount Sinai School of Medicine, New York, NY, USA

Glossary

Complement A system of serum proteins that function to promote the phagocytosis or lysis of pathogens or pathogen-infected cells.

Conventional dendritic cell (cDC) A phagocytic cell type that serves as a sentinel for the immune system and that links innate and adaptive immune responses. Upon encountering an antigen, cDCs can be induced to undergo maturation, such that they produce cytokines, upregulate to their surface co-stimulatory molecules, and promote T-cell activation.

Cytokine A secreted protein that promotes or modulates immune responses.

Interferon α/β (IFN-α/β) A family of structurally related cytokines the expression of which can be induced by virus infection and other stimuli. IFN-α/β proteins bind to the IFN-α/β receptor activating a signaling pathway that modulates the expression of hundreds of genes and induces in cells an antiviral state.

Natural killer (NK) cell A lymphocyte that can kill target cells in a non-antigen-specific manner. Killing by NK cells is tightly regulated by positive and negative signals.

Opsonization The coating of a particle so as to promote its phagocytosis.

Pathogen-associated molecular patterns (PAMPs) Molecular patterns found on pathogens that are recognized as 'foreign' and that trigger innate immune responses.

Pattern recognition receptor (PRR) A host molecule that recognizes PAMPs and signals to stimulate innate immune responses.

Toll-like receptor (TLR) A member of a family of type I transmembrane proteins that serve as pattern-recognition receptors. TLRs signal through association with cytoplasmic adaptor proteins.

Innate Immunity

Immune responses have evolved, in part, to eliminate or to contain infectious agents such as viruses. Innate immunity consists of a variety of relatively nonspecific and short-lived responses typically triggered soon after the appearance of an antigen (such as a virus or other pathogen). Innate immune mechanisms can be contrasted with adaptive immune responses in which responses are antigen specific and are characterized by immunological memory.

Innate immunity can be mediated by proteins, for example, interferons (IFNs), other cytokines, or complement; and by specific cell types such as macrophages, dendritic cells (DCs), and natural killer (NK) cells. These innate responses act rapidly to suppress infection. Given that viruses have co-evolved with their hosts, it is perhaps not surprising that viruses have evolved ways to defeat innate immune responses. For virtually any effector of the innate immune response, an example can be found whereby a virus overcomes or counteracts this host response. The strategies employed by viruses to evade these responses will vary. In general, however, viruses with large genomes, such as are found in many DNA viruses, may produce 'accessory proteins' that specifically carry out immune-evasion functions. In contrast, viruses with small genomes, such as are common among many RNA viruses, often encode multifunctional proteins. These may carry out both immune-evasion functions and functions essential for virus replication. However, regardless of the general strategy employed, suppression of innate immunity is generally critical for viral pathogenesis.

The IFN-α/β System

The IFN-α/β response serves as a major component of the innate immune response to virus infection, can be activated in most cell types, and triggers in cells an 'antiviral state'. It should be noted that there is also an IFN-γ which is structurally distinct from IFN-α/β and which is produced mainly by immune cells. Despite these differences, IFN-γ can activate signaling pathways similar to those activated by IFN-α/β and can also induce in cells an antiviral state. The IFN-α/β system can be activated in most cell types and can be viewed, in simplified terms, in two phases (**Figure 1**). The first is an induction phase in which virus infection or another stimulus activates latent transcription factors. These induce expression of the IFN-α/β genes which encode multiple IFN-α proteins and, in humans, a single IFN-β protein. In the second phase, the secreted, structurally related IFN-α/β proteins bind to the IFN-α/β receptor. This activates a Jak-STAT signaling pathway that induces expression of numerous genes, some of which have antiviral properties. There is only a single IFN-β gene

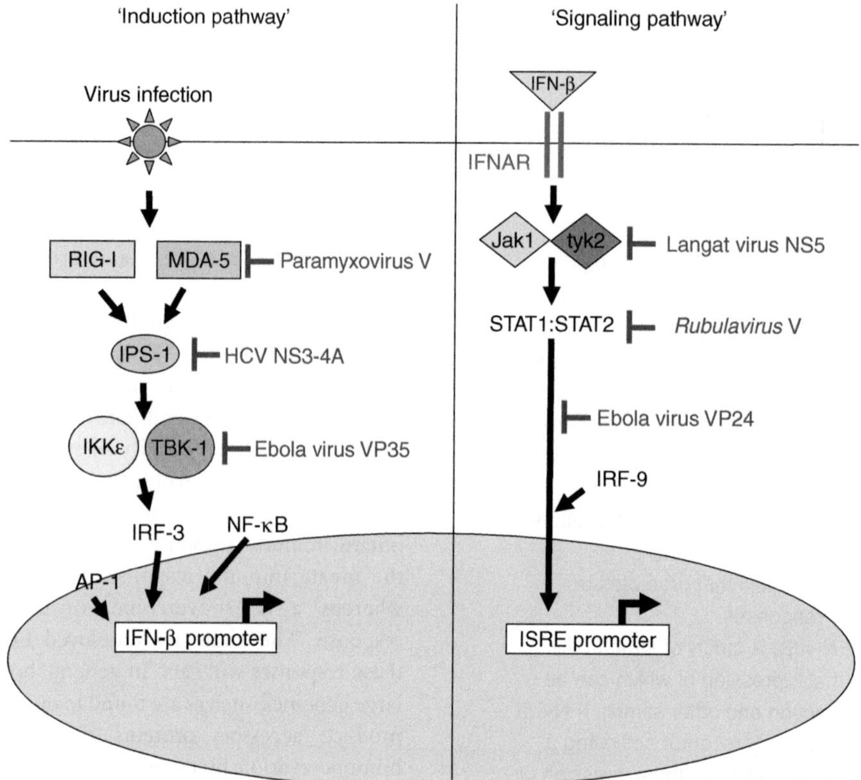

Figure 1 A simplified, schematic diagram of the IFN-α/β system and examples of viral proteins that disrupt the IFN-α/β response. Here the IFN-α/β system is depicted in two phases, each subject to virus intervention. First, the 'induction pathway' is activated by virus infection. Products of the infection activate signaling pathways, either through cytoplasmic sensors of virus infection such as retinoic acid inducible gene I protein (RIG-I) or melanoma differentiation antigen 5 protein (MDA-5) or through select Toll-like receptor (not shown). Signaling through RIG-I or MDA-5 requires IPS-1 and results in activation of the interferon regulatory factor 3 (IRF-3) kinases IKKε and TBK-1. These signaling pathways trigger activation of the transcription factors IRF-3, nuclear factor-kappa B (NF-κB), and AP-1. These cooperate to induce IFN-β gene expression. Examples of viral proteins that target the induction phase include paramyxovirus V proteins that bind and inhibit MDA-5, the hepatitis C virus (HCV) NS3-4A protease that cleaves IPS-1, and the Ebola virus VP35 protein that may act at the level of the IRF-3 kinases. Second, in the 'signaling pathway', secreted IFN-α/β binds to the IFN-α/β receptor, activating a Jak-STAT signaling cascade, resulting in the expression of numerous genes. Examples of viral proteins that target the IFN signaling pathways are the NS5 protein of Langat virus that prevents Jak1 and tyk2 kinase activation, the V proteins of the paramyxovirus genus *Rubulavirus*, which target STAT proteins for proteasome-dependent degradation, and the Ebola virus VP24 protein that prevents nuclear accumulation of activated STAT1. A third aspect of the response is the production of IFN-induced genes that encode antiviral gene products (not shown). Viral inhibitors of specific IFN-induced antiviral effector proteins also exist. Examples of inhibitors described in the text are depicted in red font in the figure.

but multiple IFN-α genes, and, in most cell types, it is primarily IFN-β that is produced after initial virus infection. However, in a positive-feedback loop, IFN-α/β can prime cells to produce numerous IFN-α types and larger overall amounts of IFN-α/β. The consequence is an amplification of the antiviral response.

Pathogen-Associated Molecular Patterns and Pattern-Recognition Receptors

Innate immune mechanisms, including IFN-α/β responses, may be triggered by diverse stimuli. Therefore, mechanisms have evolved to recognize as foreign structures and molecules that are conserved among groups of pathogens but distinct from host molecules. The pathogen structures and molecules recognized by the host are generically referred to as pathogen-associated molecular patterns (PAMPs), while the host molecules that recognize such structures are referred to as pattern-recognition receptors (PRRs). PRRs important for detection of virus infection include select members of the Toll-like receptor (TLR) family of transmembrane PRRs, including TLR3, TLR4, TLR7, TLR8, and TLR9, and particular cellular RNA helicases including the retinoic acid inducible gene I protein (RIG-I) and the melanoma differentiation antigen 5 protein (MDA-5).

TLRs are expressed primarily on cells, such as DCs and macrophages, which function to trigger innate immune responses and to act as antigen-presenting cells, initiating adaptive immune responses. Several TLRs have the capacity to recognize viral nucleic acids. TLR3 recognizes double-stranded RNA (dsRNA), such as polyI:polyC experimentally added to cells or dsRNA produced during the course of virus infection. TLR7 and TLR8 can be activated by virus derived single-stranded RNAs (ssRNAs) rich in guanosine or uridine, such as those derived from influenza virus or human immunodeficiency virus (HIV). TLR9, in contrast recognizes dsDNA; and TLR9 signaling can be stimulated by CpG DNA motifs or by virus-derived dsDNA such as that from herpes simplex virus. There are also examples of other, non-nucleic acid, viral products signaling through TLRs, although such observations are often viewed as controversial because it is difficult to completely exclude the possibility that the virus preparations used contain some low level contamination with a bacterial TLR ligand. Examples of viral protein recognition by TLRs include the recognition of respiratory syncytial virus (RSV) fusion (F) protein by TLR4 and measles virus hemagglutinin protein activation of TLR2 signaling.

In general, activation of TLR signaling would seem to be a disadvantage for a virus, as it will promote innate antiviral responses. However, in some cases, viruses may take advantage of the innate response. For example, mouse mammary tumor virus (MMTV) infects B cells early during the course of infection *in vivo*. This virus can also activate TLR4 signaling. Because replication of this retrovirus requires that its host cell be dividing, the activation of B cells via TLR4 may facilitate establishment of virus infection. Similarly, activation of DCs, another early target of MMTV infection, via TLR4 signaling increases DC numbers and increases expression of CD71 the viral receptor on these cells, again promoting the infection process.

Different TLRs have different distributions among cell types, and the consequences of TLR activation will vary, depending both upon the specific TLR in question as well as the cell type on which it is activated. For example, conventional dendritic cells (cDCs) function as sentinels which, in an immature state, constitutively sample their environment for foreign antigen. Upon encountering antigen and activation of PRRs, the cDCs mature and become potent activators of T-cell responses. Of the IFN-α/β-inducing TLRs, human cDCs express TLRs 3 and 4. Their activation results in IFN-β production and the maturation of the cDC, thereby providing a link between the relatively nonspecific innate response to infection and the antigen-specific adaptive immunity of T-cell responses. In contrast, human plasmacytoid DCs (pDCs) express TLRs 7, 8, and 9, and activation of pDCs results in the production of copious amounts of IFN-α.

Many cell types do not express TLRs. However, such cells typically have the capacity to mount an IFN-α/β response to virus infection. In such cells, intracellular PRRs detect and signal in response to virus infection. Two such intracellular PRRs are RIG-I and MDA-5. RIG-I binds to and is activated by RNA molecules possessing 5′ triphosphate groups, although activation of signaling may be influence by other properties of the RNAs such as secondary structure or modifications to the RNAs. MDA-5 can be activated by intracellular dsRNA, such as transfected polyI:polyC. These properties appear to fit with data obtained from mice in which either RIG-I or MDA-5 were knocked out. Thus, viruses that produce RNAs with 5′ triphosphates during the course of their replication, including influenza viruses, paramyxoviruses, and flaviviruses, trigger IFN-β production through RIG-I. In contrast, RIG-I was not essential for detection of the picornavirus encephalomyocarditis virus in MDA-5 knockout cells, presumably because picornaviruses produce RNAs where the 5′ end is occupied by the covalently linked viral protein, VPg. MDA-5, in this case, probably activates IFN-α/β production due to recognition of viral dsRNA. It is notable however that expression from viruses of inhibitors of RIG-I or MDA-5 might influence the outcome of such analyses. For example, multiple paramyxoviruses have been found to encode 'V' proteins (discussed further below) that inhibit MDA-5 function. Thus, it is possible that paramyxovirus infection yields products that activate not only RIG-I but also MDA-5. However, because the presence of the V proteins blocks MDA-5 function, it may appear that paramyxoviruses can only activate RIG-I. The evolutionary pressure to retain an inhibitor of MDA-5 may in fact suggest a role for detection of negative-strand RNA viruses by MDA-5.

Signaling Leading to IFN-α/β Production

The cellular molecules immediately downstream of TLRs differ, depending upon which TLR is engaged. Similarly, RIG-I and MDA-5 share common downstream signaling molecules but these differ to some extent from the TLR-activated pathways.

TLR signaling is mediated by specific downstream 'TIR-domain'-containing adaptor proteins. TLR3 signaling to induce IFN-α/β expression appears to require the adapter TRIF, while TLR4 signaling appears to require TRIF and a second adaptor, TRAM. For the RIG-I and MDA-5 helicases, recognition of the appropriate nucleic acid molecule is thought to induce structural rearrangements in the helicase, exposing an otherwise repressed caspase recruitment domain (CARD, a protein:protein interaction domain), allowing the helicase to interact with downstream signaling molecules. A mitochondria-localized protein IPS-1 (also called MAVS, Cardif, or VISA) lies downstream of RIG-I and MDA-5; it is essential for the signaling from RIG-I and MDA-5. Each of these pathways leads to the activation of the cellular kinases IKKε and TBK-1 which phosphorylate interferon regulatory factor 3 (IRF-3) and also activate nuclear factor-kappa B (NF-κB), transcription factors critical for IFN-β gene expression. In contrast, TLRs 7, 8, and 9 utilize signaling pathways that are dependent upon the adapter MyD88, which involve activation of IRF-7 and which leads to the production of IFN-α.

IRF-3 and IRF-7

In most cell types, IFN-β is the primary form of IFN that is first produced. The IFN-β promoter transcription occurs following the coordinated activation of IRF-3, NF-κB, and the ATF-2/c-Jun form of AP1. The initial production of IFN-β, as well as select IFN-α genes, induces expression of a similar IRF family member, IRF-7. IRF-7, like IRF-3, is activated by phosphorylation in response to virus infection. However, unlike IRF-3, IRF-7 is able to participate in the activation of numerous IFN-α genes, thus providing a mechanism by which IFN-α/β production can be greatly amplified in response to infection. In contrast to this model, in pDCs, IRF-7 appears to be constitutively expressed and this contributes to the copious production of IFN-α by these cells in response to TLR 7, 8, or 9 agonists.

IFN Signaling

IFN-α/β and IFN-γ bind to two distinct receptors but activate similar signaling pathways. For both pathways, ligand binding activates receptor-associated Jak family tyrosine kinases. These undergo auto- and transphosphorylation and phosphorylate the cytoplasmic domains of the receptor subunits. The receptor-associated phosphotyrosine residues then serve as docking sites for the SH2 domains of STAT proteins. The receptor-associated STATs then undergo tyrosine phosphorylation and form homo- or heterodimers via reciprocal SH2 domain–phosphotyrosine interactions. Signaling from the IFN-α/β receptor results predominately in the formation of STAT1:STAT2 heterodimers which additionally interact with IRF-9. IFN-γ signaling results predominately in the formation of STAT1:STAT1 homodimers. Upon dimerization, the STAT1:STAT2 heterodimer or the STAT1:STAT1 homodimer interacts with a specific member of the karyopherin α (also known as importin α) family of nuclear localization signal receptors, karyopherin α1 (importin α5). This interaction with karyopherin α1 mediates the nuclear accumulation of these STAT1-containing complexes. The consequence of the activation and nuclear accumulation of these complexes is the specific transcriptional regulation of numerous genes, some of which have antiviral properties.

IFN-Induced Genes Encoding Antiviral Proteins

Treatment of cells with interferon induces expression of more than 100 genes and induces in cells an 'antiviral state', meaning that the cells become resistant to virus infection and/or replication. Our understanding of the IFN-induced antiviral state is very incomplete. However, the products of several IFN-induced genes have been demonstrated to exert antiviral properties. Examples include the dsRNA-activated protein kinase, PKR, which upon activation, phosphorylates the translation initiation factor eukaryotic initiation factor 2α (eIF2α). This leads to a general inhibition of translation which can suppress virus replication. Similarly, 2′-5′ oligoadenylate synthetases (OASs) are a family of IFN-induced, dsRNA-activated enzymes that catalyze formation of 2′-5′ oligoadenylates. These activate RNase L, an enzyme that destroys mRNAs and cellular RNAs and suppresses translation of proteins. Mx proteins, including Mx1 in mice or MxA in humans, belong to the dynamin family of GTPases. Their expression is IFN-inducible, and members of this family have been shown to inhibit replication of specific virus families. For example, human MxA can inhibit replication of bunyaviruses, orthomyxoviruses, paramyxoviruses, rhabdoviruses, togaviruses, picornaviruses, and hepatitis B virus. At least part of the antiviral effect of these proteins is exerted by their ability to alter the intracellular trafficking of viral proteins; for example, human MxA can sequester orthomyxovirus nucleocapsids in the cytoplasm, preventing entry into the nucleus where virus RNA synthesis takes place.

Viral Evasion of the Host IFN Response

It is likely that most, if not all, viruses successfully maintained in nature have evolved mechanisms to counteract IFN responses.

Examples of the mechanisms by which viruses defeat the IFN system include inactivation of the PRR signaling pathways, inactivation of the IFN-induced signaling pathways, and the targeting of IFN-induced antiviral 'effectors'.

Targeting Components of the PRR Signaling Pathways

Numerous examples now exist of viruses targeting the signaling pathways activated by TLRs, RIG-I, or MDA-5. Mechanisms range from targeting of the PRRs themselves, to the targeting of the transcription factors that these pathways activate, to the targeting of kinases that activate these transcription factors. The RIG-I pathway is targeted by several viruses, including Ebola virus and hepatitis A virus. In poliovirus-infected cells, the cytoplasmic PRR MDA-5 is cleaved in a caspase- and proteosome-dependent manner, presumably due to poliovirus activation of apoptosis. This observation is consistent with the idea that MDA-5 is particularly relevant to picornavirus infection. Hepatitis C virus serves as an example of a virus that can target both a TLR (TLR3) and intracellular PRR (RIG-I) pathways. HCV is an enveloped, positive-strand RNA virus of the flavivirus family associated with persistent liver infections, cirrhosis, and liver cancer.

The HCV genome encodes NS3-4A, a noncovalent complex with protease and RNA helicase activity. NS3-4A has been found to inhibit, in a protease-dependent manner, virus-induced activation of IRF-3, which, as noted above, plays a major role in activating IFN-α/β responses. TLR3 signaling requires the function of the adaptor TRIF, and the protease activity of NS3-4A cleaves and inactivates TRIF, to prevent TLR3-induced IFN-α/β expression. NS3-4A was also found to cleave IPS-1, the mitochondria-localized adaptor downstream of both RIG-I and MDA-5, similarly preventing RIG-I-induced IFN-α/β expression.

Targeting Components of the IFN Signaling Pathways

Numerous examples also exist of virus-mediated inhibition of the IFN-activated Jak-STAT signaling cascades. There are many examples among the members of the paramyxovirus family itself. It should be noted that several of these also inhibit IFN-α/β production and might therefore also be classified as proteins that target multiple aspects of the IFN system (see the section titled 'Viruses targeting multiple aspects of the IFN system'). Viruses of the genus *Rubulavirus* of the paramyxovirus family encode V proteins which target STAT proteins for ubiquitin-dependent degradation. However, the specific STAT targeted by a particular V protein will vary. For example, the simian virus 5 (SV5) V protein targets STAT1 while human parainfluenza virus type 2 (HPIV2) encodes a V protein that promotes STAT2 degradation. In these cases, the V proteins appear to direct the STAT protein to a ubiquitin ligase complex that includes the proteins DDB1 and cullin 4A. In contrast, Nipah virus, an emerging zoonotic paramyxovirus that causes highly lethal encephalitis in humans, encodes proteins (P, V, and W) that share a common N-terminal domain. This domain contains sequences that mediate an interaction with STAT1, and for the V protein, also STAT2. These Nipah virus proteins inhibit STAT1 not by targeting it for degradation; rather, the P, V, and W appear to sequester STAT1 preventing the tyrosine phosphorylation that would otherwise activate it.

Upstream and downstream steps in the IFN signaling pathways are also subject to interruption by virus infection. The tyrosine kinases that activate STAT1 and STAT2 typically undergo tyrosine phosphorylation following exposure of cells to IFN. However, in cells infected with Langat virus, a flavivirus of the tick-borne encephalitis virus complex, neither STAT1/2 nor tyk2/Jak1 phosphorylation is seen. This inhibition is mediated by as association of the viral polymerase protein NS5 with both the IFN-α/β receptor and the IFN-γ receptor. In contrast, Ebola virus infection does not prevent STAT1 tyrosine phosphorylation. However, in Ebola virus-infected cells, tyrosine-phosphorylated STAT1 fails to enter the nucleus. The failure of otherwise activated STAT1 to traffic to the nucleus appears to be due to the action of the Ebola virus VP24 protein. Tyrsoine-phosphorylated STAT1 has been reported to enter the nucleus through an interaction with the nuclear localization signal receptor karyopherin alpha 1 (importin α5). VP24 interacts with this specific karyopherin alpha, and in doing so, inhibits STAT1–karyopherin alpha 1 interaction, thus explaining the defect in STAT1 nuclear import.

Targeting of Antiviral Effector Proteins

Among the IFN-induced genes with demonstrated antiviral activity, PKR has been the most heavily studied, and antagonists of PKR function were among the earliest described viral inhibitors of the IFN response. Upon activation by dsRNA, PKR phosphorylates the translation factor eIF2α, a subunit of the translation initiation factor eIF2 that brings methionyl (Met) initiator tRNA to the ribosome. Phosphorylation of eIF2α on serine 51 inhibits a GDP-to-GTP exchange reaction required for the continuing function of eIF2, and therefore this phosphorylation arrests translation at the initiation phase. As a consequence, numerous viruses have devised diverse ways to preserve the translation of their mRNAs in the face of PKR activation. Examples include the adenovirus VA_I RNA and the vaccinia virus E3L protein, which impede PKR activation. By inhibiting PKR function, these molecules not only help preserve translation of viral mRNAs but also facilitate virus replication in the presence of an IFN response. Another notable counterbalance to PKR function is the herpes simplex virus $\gamma_1$34.5 protein. In this case, the viral protein recruits cellular phosphatase 1 alpha (PP1α) to dephosphorylate eIF-2α. A functional $\gamma_1$34.5 has been found to be required for HSV-1 neurovirulence, demonstrating its importance *in vivo*, although subsequent studies have identified additional herpes simplex virus proteins, including US11, that also facilitate translation of viral mRNAs.

Viruses Targeting Multiple Aspects of the IFN System

As was noted above, it has become apparent that viruses will often target multiple components of the IFN response. Influenza viruses serve as such an example. The influenza A virus NS1 protein is a multifunctional inhibitor of innate immune responses. Evidence suggests that NS1 directly targets RIG-I, the PRR that detects influenza viruses in most cell types. This has the consequence of preventing signaling that would lead to the activation of IRF-3 and NF-κB and to IFN-α/β production. However, NS1 proteins also can affect, in a more global way, the expression of cellular gene expression. This occurs because NS1 can interfere with host cell pre-mRNA processing and can inhibit the nuclear export of cellular mRNAs. The ability of NS1 to suppress gene expression in this way has the potential to inhibit expression of IFN-α/β genes. In addition, this function may also suppress expression of genes induced by IFN thus preventing induction of an antiviral state. Finally, NS1 is able to inhibit the activation of PKR and therefore can directly target the function of at least one IFN-induced, dsRNA-activated antiviral protein. That the NS1 protein is critical for viral evasion of IFN responses is highlighted by *in vivo* experiments. Mutant influenza viruses either lacking or encoding altered NS1s display attenuated phenotypes in mice able to mount an IFN response. However, this attenuation can be largely reversed in animals lacking a fully competent IFN system, demonstrating that the attenuation in the wild-type mice is directly related to the absence of an IFN-antagonist protein.

Defeating IFN Responses – Consequences for Adaptive Immune Responses

It should be recognized that innate cytokine responses, including IFN-α/β responses, not only serve to suppress virus replication, but also act to promote adaptive immune responses. IFN-α/β can, for example, affect T-cell responses, influencing the ability of CD8+ T-cell populations to expand and generate memory cells, although the magnitude of the IFN-effect depends upon the pathogen administered. Similarly, pathogen-specific effects of IFN-α/β upon clonal expansion of CD4+ T cells have also been described. Connections between the IFN-α/β response and the function of antigen-presenting cells have also been identified. For example, the same stimuli that activate IFN-α/β production can also activate DC maturation, promoting the upregulation of major histocompatibility complex molecules and co-stimulatory molecules, such as CD80 and CD86, and the production of cytokines. DC maturation then promotes activation of T-cell responses. Therefore, the virus-encoded mechanisms that tend to suppress IFN-α/β responses, particularly those that inhibit IFN-α/β production, may serve to suppress adaptive immune responses as well. This hypothesis is supported in part by the observation that expression of the influenza virus NS1 protein, a suppressor of IFN-α/β production, rendered viruses less able to promote human DC maturation and function compared with viruses that lacked the IFN-antagonist NS1 protein.

Other Cellular Antiviral Proteins

Other cellular proteins may also serve as innate defense against virus infection. Prime examples are members of the APOBEC3 family of cytidine deaminases which function as a defense against retrovirus infection. APOBEC3 proteins deaminate cytidines to uridines in single-stranded DNA. Family member APOBEC3G can incorporate into budding HIV-1 particles and, during reverse transcription in the subsequently infected cell, deaminate cytidines. Because this results in introduction of dC to dU changes on minus-strand DNA, the result is the dG to dA hypermutation of the positive, coding-strand of the virus genome. This can result in the introduction of lethal (to the virus) mutations as well as susceptibility to cleavage by uracil DNA glycosylase and apurinic-apyrimidinic (AP) endonuclease enzymes. However, HIV-1 can defeat the antiviral effect of APOBEC3G through the function of its protein, viral infectivity factor (Vif). Vif is required for HIV-1 replication in many cell types. This is due, in part, to the ability of Vif to bind APOBEC3G and target it for ubiquitin-dependent degradation via a ubiquitin ligase complex. The ability of Vif to eliminate APOBEC3G from cells of a particular species can determine its ability to replicate in that species. Thus, HIV-1 Vif can target APOBEC3G in humans and chimpanzees. However, the inability of HIV-1 to target APOBEC3G in African green monkeys or rhesus macaques restricts the ability of HIV-1 to infect these species, thus providing a demonstration of the critical role of Vif in overcoming the innate antiviral effect of APOBEC3G.

Cellular Components of Innate Immunity to Virus Infection

As noted above, cellular components of the innate immune response also exist. Included among such cell types are DCs (discussed above), macrophages, and NK cells. NK cells function to produce cytokines or to lyse cells in a non-antigen-specific manner. Cytokines released by NK cells may exert antiviral effects or regulate other aspects of innate or adaptive immune responses. NK killing of target cells can occur through exocytosis of cytotoxic granules or through engagement of death receptors. A role for NK function in controlling virus replication has been described for several virus types including arenaviruses, paramyxoviruses, HIV, and, most notably, herpes viruses. NK activity is regulated via a series of positive and negative regulatory signals, and it is the balance of these regulatory signals that appears to determine NK cell activation. The best-studied examples of virus evasion of NK responses are those used by cytomegalovirus (CMV). One classic mechanism of NK cell activation is recognition of a cell with low levels of class I major histocompatibility complex (MHC-I) on its surface. Because NK cells express upon their surface receptors that recognize MHC-I and negatively regulate their function, targets lacking MHC-I drive this balance toward NK cell activation. CMV-infected cells typically express low levels of MHC-I upon their surface. This is due, at least in part, to virus-encoded proteins designed to downregulate MHC-I expression, so as to permit evasion of antigen-specific T-cell responses. CMV must therefore also evade NK responses that would otherwise be activated by low MHC-I. One such mechanism is the production by CMV of MHC-I-like proteins. In addition, CMV infection downregulates ligands on the infected cell that would trigger activating signals in NK cells.

Complement

Complement is a group of proteins, both soluble and cell associated, that serve multiple innate immune functions. Antiviral properties of the complement system include its ability to enhance activation (priming) of adaptive immune responses, to lyse membrane-bound viruses and infected cells, and to promote phagocytosis of viruses and infected cells. Three major complement pathways exist and are referred to as the classical, the alternate, and the lectin-binding pathways. Although activated by different means, each pathway can, through a cascade of events, lead to the deposition of the complement protein C3b on a target. This process, termed opsonization, promotes phagocytosis. Deposition of C3b can also lead to the formation of a 'membrane attack complex' which can lyse the coated particle or cell. The classical pathway is activated by binding of a complement protein, C1q, which attaches to the Fc portion of an antibody. Thus, virus or virus-infected cells can be targeted by complement and subject to C3b deposition and subsequent lysis or phagocytosis when they become bound by antibody. Activation of the alternative pathway, in contrast, occurs in the absence of antibody and occurs when an activated form of the complement protein C3 recognizes a (usually) foreign surface. In the lectin pathway, lectins may bind to carbohydrates on antigens such as viruses, triggering a pathway similar to the classical pathway.

Because complement is strongly pro-inflammatory and cytolytic, it must be tightly regulated to protect the host, and some of these host-encoded protections can either be mimicked or co-opted by viruses to evade the antiviral effects of complement. For example, several either encode 'regulators of complement activation (RCA)' proteins or package host-encoded RCA proteins. For example, gamma herpes viruses (γHVs), large DNA viruses that establish lifelong infection, encode RCA homologs. One well-studied example is the murine γHV68 RCA which can inhibit C3 and C4 deposition *in vitro*. Deletion of this γHV68-encoded RCA decreased virulence during acute infection, but virulence was restored by deletion of host C3, demonstrating the *in vivo* importance of the viral RCA as an antagonist of the complement system. An alternate strategy can be employed by HIV-1 which is able to incorporate into virus particles two glycosylphosphatidylinositol (GPI)-linked complement control proteins, CD55 and CD59. These make virus resistant to complement-mediated lysis. Finally, viruses may also produce complement inhibitors with no obvious homology to the host complement system. West Nile virus, for example, produces a secreted glycoprotein NS1. NS1 can bind to cell surfaces but it also accumulates in serum. NS1 binds to and recruits a complement regulatory protein, factor H, and the ability of NS1 to interact with factor H correlates with its ability to impair complement activation and to suppress deposition of C3 and formation of membrane attack complexes on infected cells.

Conclusion

As is illustrated by the selected examples provided above, innate mechanisms targeting virus infection are numerous, and the manner in which viruses overcome such responses are equally diverse. Defeating such mechanisms is also critical for maintenance of viruses in nature and for viral pathogenesis. It should be recognized, however, that it may not be evolutionarily advantageous for a virus to absolutely defeat host immune responses, as this might result in the rapid demise of the host with reduced opportunity for virus amplification and transmission. Therefore, viruses do not necessarily fully block innate immune responses. Rather many viruses will have evolved mechanisms to attenuate innate immunity only to such a degree that their spread to new hosts is facilitated.

See also: Antibody - Mediated Immunity to Viruses; Cell-Mediated Immunity to Viruses; Innate Immunity.

Further Reading

Andoniou CE, Andrews DM, and Degli-Esposti MA (2006) Natural killer cells in viral infection: More than just killers. *Immunological Reviews* 214: 239–250.

Basler CF and Garcia-Sastre A (2002) Viruses and the type I interferon antiviral system: Induction and evasion. *International Reviews of Immunology* 21: 305–337.

Chiu YL and Greene WC (2006) Multifaceted antiviral actions of APOBEC3 cytidine deaminases. *Trends in Immunology* 27: 291–297.

Finlay BB and McFadden G (2006) Anti-immunology: Evasion of the host immune system by bacterial and viral pathogens. *Cell* 124: 767–782.

Gale M, Jr., and Foy EM (2005) Evasion of intracellular host defence by hepatitis C virus. *Nature* 436: 939–945.

Hilleman MR (2004) Strategies and mechanisms for host and pathogen survival in acute and persistent viral infections. *Proceedings of the National Academy of Sciences, USA* 101(supplement 2): 14560–14566.

Horvath CM (2004) Weapons of STAT destruction. Interferon evasion by paramyxovirus V protein. *European Journal of Biochemistry* 271: 4621–4628.
Kawai T and Akira S (2007) Antiviral signaling through pattern recognition receptors. *Journal of Biochemistry (Tokyo)* 141(2): 137–145.
Levy DE and Garcia-Sastre A (2001) The virus battles: IFN induction of the antiviral state and mechanisms of viral evasion. *Cytokine and Growth Factor Reviews* 12: 143–156.
Lopez CB, Moran TM, Schulman JL, and Fernandez-Sesma A (2002) Antiviral immunity and the role of dendritic cells. *International Reviews of Immunology* 21: 339–353.
Mastellos D, Morikis D, Isaacs SN, Holland MC, Strey CW, and Lambris JD (2003) Complement: Structure, functions, evolution, and viral molecular mimicry. *Immunologic Research* 27: 367–386.

Antigen Presentation

E I Zuniga, D B McGavern, and M B A Oldstone, The Scripps Research Institute, La Jolla, CA, USA

Introduction

The immune system is responsible for the tremendous task of fighting a wide range of pathogens to which we are constantly exposed. This system can be broadly subdivided in innate and adaptive components. The innate immune system exists in both vertebrate and invertebrate organisms and represents the first barrier against microbial invasion. This arm of the immune system rapidly eliminates the vast majority of microorganisms that we daily encounter and is responsible for limiting early pathogen replication. The adaptive response is a more sophisticated feature of vertebrate animals involving a broad repertoire of genetically rearranged receptors that specifically recognize microbial antigens (antigen is a generic term for any substance that can be recognized by the adaptive immune system). The hallmark of the adaptive response is the generation of a potent and long-lasting defense specifically directed against the invading pathogen.

B and T lymphocytes represent the effector players of adaptive immunity and carry on their surface antigen-specific receptors, B-cell receptors (BCRs) and T-cell receptors (TCRs), respectively. There are two major classes of T lymphocytes: CD8 cytotoxic and CD4 helper T cells. Upon antigen encounter, lymphocytes undergo clonal expansion and differentiation of their unique functional features. B cells differentiate into plasma cells and secrete antibodies that specifically bind the corresponding antigen. CD8 T cells directly kill infected cells or release cytokines that interfere with viral replication, while CD4 T cells activate other cells such as B cells and macrophages. Unlike B cells, which can directly bind native free antigen, T cells only recognize antigen-derived peptides displayed on cell surfaces in the context of major histocompatibility complex (MHC) class I (MHC-I, CD8 T cells) or class II (MHC-II, CD4 T cells) molecules.

Different pathogens preferentially replicate in distinct cellular compartments. While viruses and intracellular bacteria replicate in the cytosol, microbes such as mycobacterium and protozoan parasites are intravesicular and colonize the endosomal and/or lysosomal compartments. In addition, extracellular bacteria release antigens, such as toxins, that are engulfed by antigen-presenting cells (APCs) to also reach the endosomal pathway. Antigenic peptides derived from these sources are exhibited on cell surfaces by MHC molecules. This process, which represents the major focus of this article, is named 'antigen presentation' and is a fundamental pillar of antimicrobial host defense.

Antigen-Presenting Cells

For initiation of an immune response, naive T cells need to be activated or 'primed'. For that, they require both the recognition of the specific MHC–peptide complex (signal 1) and simultaneous co-stimulation (signal 2). Although all nucleated cells express MHC-I and can potentially display MHC-I –microbial peptide complexes after infection, only a specialized group of leukocytes, named APCs, expresses both MHC-I and MHC-II as well as co-stimulatory molecules. The best-characterized co-stimulatory molecules are B7-1 and B7-2, which bind to the CD28 molecule on the T-cell surface. In addition, T cells express CD40 ligand, which interacts with CD40 on APC further enhancing co-stimulation and enabling T-cell response. Finally, there is another group of adhesion molecules such as lymphocyte function-associated antigen-1 (LFA-1) on APCs which binds to ICAM-1 on T cells that seal the APC–T-cell interface. During APC–T-cell interactions, all these molecules cluster together forming a highly organized supramolecular adhesion complex (SMAC), enabling the intimate contact between the two cells that is referred to as the immunological synapse (**Figure 1**).

APCs are composed of macrophages, B cells, and dendritic cells (DCs). They differ in location, antigen uptake,

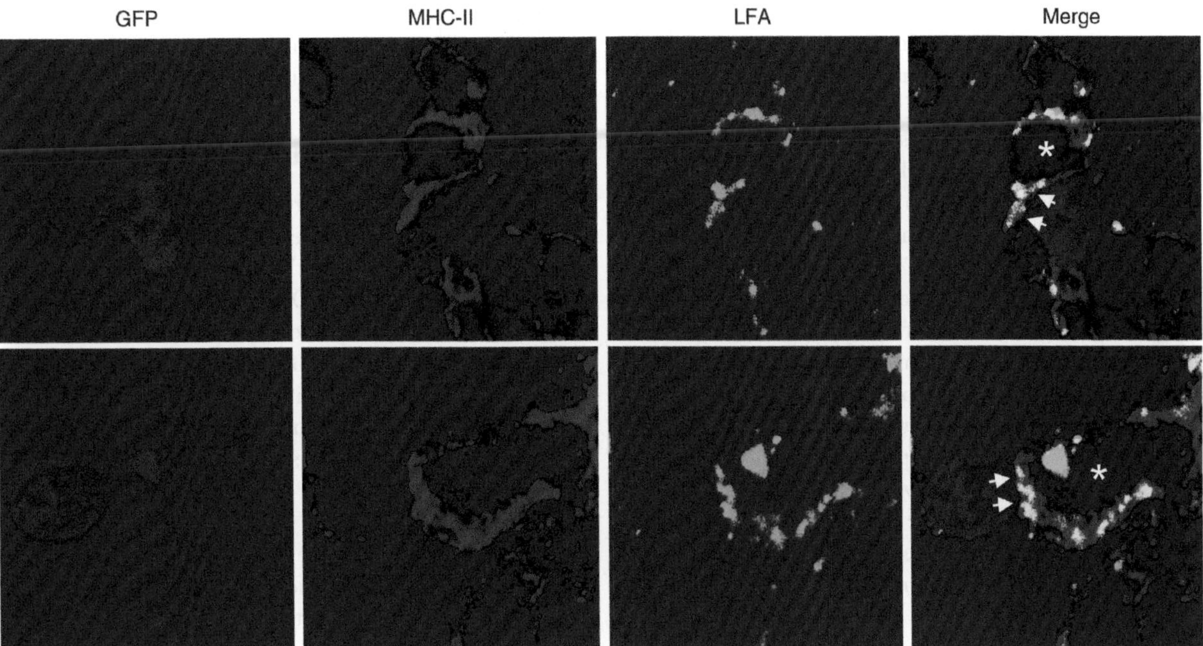

Figure 1 Interactions between virus-specific T cells and APCs. Three-color confocal microscopy was used to demonstrate immunological synapse formation between lymphocytic choriomeningitis virus-specific T cells (blue) and MHC-II$^+$ APCs (red) in the central nervous system. Immunological synapses were indicated by the polarization of the adhesion molecule LFA-1 (green) between the CTL and APC. Asterisks denote the engaged APC, and arrows denote the contact point between the two cells. LFA-1 is expressed on both CTLs and APCs, but note that all of the CTL-associated LFA-1 is focused toward a contact point at the CTL–APC interface. Reproduced from Lauterbach H, Zúñiga EI, Truong P, Oldstone MBA, and McGavern DB (2006) Adoptive immunotherapy induces CNS dentritic cell recruitment and antigen presentation during clearance of a persistent viral infection. *Journal of Experimental Medicine* 203 (8): 1963–1975, with permission from Rockefeller University Press.

and expression of antigen-presenting and co-stimulatory molecules. Macrophages are localized in connective tissues, body cavities, and lymphoid tissues. Within the secondary lymphoid tissues, macrophages are mainly distributed in the marginal sinus and medullary cords. They specialize in phagocytosis and engulf particulate antigens through scavenger germline receptors such as the mannose receptor. On the other hand, B cells form follicular structures within secondary lymphoid organs and recirculate through the blood stream and lymph seeking their specific antigen. B cells recognize antigens specifically through a rearranged BCR. DCs are the most professional and robust of the APCs. They are widely distributed through the body at an 'immature' stage of development, acting as sentinels in peripheral tissues. They continuously sample the antigenic environment by both phagocytosis and macropinocytosis, which is the engulfment of large volume of surrounding liquid. Within the secondary lymphoid organs, some DCs strategically localize within T-cell areas where they can optimally encounter circulating naive T lymphocytes that actively scan the DC network.

APCs are able to detect components of invading pathogens which trigger their activation/maturation. Specifically, pathogen-associated molecular patterns (PAMPs), as these components are termed, range from lipoproteins to proteins to nucleic acids carried by potential invaders. These PAMPs are recognized by evolutionary conserved 'pattern recognition receptors' (PRRs) on APCs. Among PRRs, the Toll-like receptors (TLRs) have emerged as critical players in determining APC imprinting on the ensuing immune response. TLR triggering has pleiotropic effects on APCs, promoting survival, chemokine secretion, expression of chemokine receptors, migration, cytoskeletal and shape changes, and/or endocytic remodeling. After interacting with these pathogen signatures, the microbial antigens are processed and presented as peptides associated with MHC molecules and activated APCs upregulate both antigen-presenting and co-stimulatory molecules initiating a 'maturation' process. As part of this process, APCs in peripheral tissues change their chemokine receptors and initiate migration to secondary lymphoid organs where the adaptive immune response is initiated.

The strategic migration and location of DCs into T-cell areas of the secondary lymphoid organs coupled to their superior antigen-presenting capacity make them the most powerful APCs. Indeed, DCs are about 1000 more efficient than B cells or macrophages in stimulating naive T cell. This has been shown by several experiments in which elimination of DCs prevented the initiation of antigen-specific T-cell responses. Interestingly, DCs are a heterogeneous cell population composed of different subtypes which present unique and overlapping functions. As many as six different subsets of DCs occupy the lymph nodes. Three major defined populations of DCs have been recognized in mouse spleen and humans: CD8+ conventional DCs (cDCs), CD11b+ cDCs, and plasmacytoid DCs. These subpopulations differ not only in surface phenotype but also in functional potential and localization. In this regard, cDCs are potent activators of naive T lymphocytes, as CD8+ DCs are believed to be specialized in cross-presentation of exogenous antigens. A recent study suggests that CD8+ and CD11b+ cDCs differ from each other in their intrinsic antigen-processing capacity being specialized in MHC-I and MHC-II antigen presentation, respectively. In contrast, plasmacytoid DCs are poorer activators of T cells, even after stimulation *in vitro*. They likely play a more protagonist role during innate immunity by secreting specific cytokines and chemokines, such as type 1 interferons (IFNs an important antiviral mediator), and activation of a broad range of effector cells, such as natural killer (NK) cells. Thus, the heterogeneity inherent to DC populations significantly influences the varieties of immune responses to different pathogens, which are subsequently amplified by cross talk between the various subsets.

Major Pathways of Antigen Presentation

Although in all healthy individuals MHC molecules play the same crucial role of antigen presentation, these molecules are highly polymorphic. There are hundreds of different alleles encoding the MHC molecules in the whole population and each individual exhibits only few of them. The major allelic variants of MHC are found in key amino acids forming the peptide-binding cleft. Thus, although a given MHC-I molecule can bind several different peptides; particular amino acids are preferred in certain positions of the peptide resulting in differential peptide sets for particular MHC variants. Importantly, T-cell specificity involves co-recognition of a particular antigenic peptide together with a particular MHC variant, a feature known as T-cell MHC restriction.

Like other polypeptide chains of proteins destined to arrive at the cell surface, MHC molecules are translocated to the lumen of the endoplasmic reticulum (ER) during synthesis. In this compartment, the subunits of MHC molecules are assembled together and the peptide-binding groove or cleft is formed. However, MHC molecules are unstable in the absence of bound peptide. In the following sections, we will consider how MHC molecules are folded and generated peptides are bound to MHC-I or MHC-II molecules. After binding, MHC–peptide complexes travel to the cell surface where they are recognized by antigen-specific T cells. Although not discussed in this article, it should be noted that other MHC-like molecules (i.e., CD1) also display peptide and lipid antigens contributing to antigen presentation, especially during mycobacteria infections.

MHC-I Antigen Presentation

MHC-I molecules are expressed in most if not all nucleated cells. MHC-I molecules are heterodimers of a highly polymorphic α-chain (43 kDa) that binds noncovalently to β2-microglobulin (12 kDa), which is nonpolymorphic. The α-chain contains three domains. The $\alpha 3$ domain crosses the plasma membrane while the $\alpha 1$ and $\alpha 2$ domains constitute the antigen-binding site. The peptides that bind the MHC-I molecule are usually 8–10 amino acids long and contain key amino acids at two or three positions that anchor the peptide to the MHC pocket and are called anchor residues.

As mentioned above, the peptide-binding site of MHC molecules is formed in the ER. However, all proteins, including viral-derived antigens, are synthesized in the cytosol. Numerous studies in the last years outlined the molecular events connecting the antigen generation in the cytosol with the peptide binding to the MHC-I molecule in the ER. A highly conserved multicatalytic proteasome complex is in part responsible for cytosolic protein degradation into small peptides. The proteasome contains 28 subunits forming a cylindrical structure composed of four rings, each of seven units. Under normal conditions, the proteasome complex exists in a constitutive form. During viral infections, IFNs released by cells of the innate immune system induce the synthesis of three different proteasome subunits, which replace their constitutive counterparts to form the immunoproteasome. This inflammatory form of the proteasome favors the production of peptides with a higher chance of MHC binding. Moreover, IFNs can also enhance the rate of proteasome peptide degradation increasing the availability of peptides and reducing their excessive cleavage. It is important to highlight that other cytosolic proteases also contribute to MHC-I peptide generation and further cleavage can occur within the ER before MHC binding. The source of peptides for MHC-I complexes still holds its secrets. Proteasome substrates may encompass *de novo* synthesized, mature stable, and/or defective proteins. It is believed that defective ribosomal products (DRiPs), which are proteins targeted for degradation due to premature termination or misfolding, constitute an important source of MHC-I peptides.

Peptides available in the cytosol are transported into the lumen of the ER by ATP-dependent transporters-associated antigen-processing 1 and 2 (TAP-1 and TAP-2) proteins. TAP proteins are localized in the ER membrane forming a channel through which peptides can pass. Within the ER, the newly synthesized MHC-I α-chain binds to a chaperone molecule called calnexin, which retains the incomplete MHC molecule in the ER. After binding to the β2-microglobulin, calnexin is displaced and the emerging MHC molecule binds to a loading complex composed by the chaperone protein calreticulin, TAP, the thiol oxidoreductase Erp57, and tapasin, which bridges MHC-I molecule and TAP. After peptide binding, the fully folded MHC-I molecule and its bound peptide are released from the complex and transported to the cell membrane. Importantly, under steady-state conditions, the MHC-I molecules in ER are in excess with respect to peptides allowing the rapid appearance of microbial peptides onto the cell surface during infection. However, since MHC-I molecules are unstable without bound peptide, they also present self antigens under normal conditions. Because of the absence of microbial signatures, antigen presentation of self peptides by inactivated/immature APCs leads to T-cell tolerance rather than activation. This is one of the important ways anti-self or autoimmune responses are controlled.

For several years, intracellular peptides were believed to be the only source of MHC-I molecules. However, it is now clear that exogenous proteins also have access to the cytosolic compartment and bind MHC-I in the ER. This mechanism is known as cross-presentation and is believed to be particularly important for enabling MHC-I presentation by cells that are not directly infected by the virus but instead are engulfing viral particles by phagocytosis or micropinocytosis. The molecular mechanism by which MHC-I molecules access exogenous peptides is of considerable interest. Different nonexclusive possibilities have been proposed, including sampling of phagosome-generated peptides by MHC-I molecules, transference of ER molecules (including MHC-I and its loading complex) into phagosomes, re-entry of plasma membrane MHC-I into recycling endosomes with the subsequent peptide exchange, and finally the acquisition of peptides from other cells through GAP junctions.

MHC-II Antigen Presentation

The MHC-II molecule is composed by two noncovalently bound transmembrane glycoprotein chains, α (34 kDa) and β (29 kDa). Each chain has two domains and altogether form a four-domain heterodimer similar to the MHC-I molecule. α1 and β1 domains form the peptide-binding cleft resulting in a groove which is open at the ends, which is different from the MHC-I groove in which the extremes of the peptide are buried at the ends. Peptides that bind to MHC-II are larger than those that bind to MHC-I molecules, being 13–17 amino acids long or even much longer.

Since MHC-I is a surface protein, its biosynthesis is initiated in the ER. To prevent newly synthesized MHC-II molecules from binding cytosolic peptides that are abundant in the ER, its peptide-binding cleft is covered by a protein known as MHC-II-associated invariant chain (Ii). Through a targeting sequence in its cytoplasmic domain, the Ii also directs MHC-II molecules to acidified late endosomal compartments, where Ii is cleaved leaving only the Ii pseudopeptide (CLIP) covering the peptide-binding groove. MHC-II molecules bound to CLIP cannot bind other peptides, indicating that CLIP must be dissociated or displaced by the antigenic peptide.

Proteins that enter the cell through endocytosis or are derived from pathogens that replicate in vesicles are degraded by endosome proteases. These proteases become activated as the endosome pH progressively decreases. The final set of peptides available in the endosomal compartment is a result of antigen processing by several acid proteases that exist in endosomes and lysosomes. For instance, the cathepsin S is a very predominant acid protease and mice deficient in this enzyme have a compromised antigen-processing capacity. Vesicles carrying peptides fuse with the vesicles carrying MHC-II molecules, achieving CLIP dissociation and the incorporation of antigenic peptides to MHC-II molecules. An MHC-II-like molecule that is predominant in the endosome facilitates this process. This molecule contributes to 'peptide editing', removing weakly bound peptides and assuring that the emerging MHC-II–peptide complexes are stable enough to be scanned by CD4 T cells.

MHC-II molecules seem to be in excess and are rapidly degraded unless microbial peptides become available to fill the groove. This excess is important to permit MHC-II availability upon infection. However, during infection, APCs are exposed to both self and microbial peptides. How APCs discriminate between self and nonself represents a fundamental question in immune biology. Recent evidence suggests that the efficiency of presenting antigens from phagocytosed cargo is dependent on the presence of a TLR ligand within the cargo. Thus, TLR signaling would mark a particular phagosome for an inducible mode of maturation dictating the fate of the cargo-derived peptides and favoring their presentation by MHC-II molecules in a phagosome autonomous fashion.

Because they travel through the endocytic pathway, which can be considered a topological continuation of the extracellular space, MHC-II molecules were believed to be specialized in the presentation of exogenous antigens. However, the analysis of MHC-II peptidome revealed many peptides of cytosolic or even nuclear origin. Autophagy or 'self-eating' explains MHC-II access to cytosolic peptides. This highly conserved pathway could be accomplished by several mechanisms including microphagy (when lysosomal invagination sequesters cytosolic componets), macrophagy (when a double membrane structure that encloses and isolates cytoplasmic components and eventually fuses with lysosome), and chaperone-mediated autophagy (when cytosolic proteases generate peptides that are transported into lysosomes).

Viral Subversion of Antigen Presentation

Considering the crucial role of antigen presentation for host defense, it is not surprising that many viruses have evolved maneuvers to evade or divert this process. Particularly, the essential role played by APCs in host defense to pathogens makes them an ideal target for viruses to suppress the immune response, thereby maximizing their chances of survival, replication, and transmission. Indeed, many viruses that cause major health problems are able to interfere with the ability of APCs to prime an efficient and effective antiviral immune response. In fact, many viruses have developed different mechanisms to subvert each stage of APC biology. Furthermore, with the greater understanding of antigen presentation pathways comes the discovery of novel viral immune-evasion strategies. In this section, we illustrate selective viral strategies to subvert antigen presentation by describing particular cases.

An interesting example of virus blockade of antigen presentation from very initial steps is the ability of the prototypic arenavirus lymphocytic choriomeningitis virus (LCMV), to dramatically block DC development from early hematopoietic progenitors. Fms-like tyrosine kinase 3 ligand (Flt3L) is known to induce the expansion of undifferentiated progenitors into DCs within the spleen and bone marrow (approximate 20-fold increase), both in mice and humans. In contrast, LCMV-clone (CL)-13 that suppresses the immune response and causes a persistent infection in mice is associated with DC early progenitors that become refractory to the stimulatory effects of Flt3L.

TLRs function in APCs as an early sensor against pathogens; therefore, impairment in TLR signaling confers another selective advantage to certain infectious agents. As an example, vaccinia virus (VV) blocks TLR signaling and the subsequent maturation of APCs. Specifically, two proteins of VV suppress intracellular signaling of interleukin-1 (a potent pro-inflammatory host factor) and TLR-4.

Migration of DCs is a crucial step in initiating the adaptive immune response. Examples of viruses that have developed mechanisms to prevent migration of infected DCs to lymphoid organs are herpes simplex virus (HSV) 1 and human cytomegalovirus (HCMV). In both cases, there is an inhibition of complete DC maturation and subsequent expression of chemokine homing receptors. In addition, HCMV inhibits DC migration one step further by preventing APCs from arriving at a site of infection by producing homologs of chemokines that interfere with host pro-inflammatory chemokine gradients.

Another effective immune-evasion strategy used by viruses to disrupt APCs is the prevention of or interference with antigen-specific T-cell activation. The ability to disrupt MHC–peptide binding has evolved in many different virus species including adenovirus and human immunodeficiency virus (HIV). Herpesviruses have also evolved to block host cell antigen presentation. Some mechanisms utilized by herpesviruses to disrupt the antigen-presentation pathway include blocking peptide transport to the ER through interference with TAP proteins (HSV ICP47, HCMV US6), transport of particular MHC-I heavy chains from the ER to the cytosol where they are destroyed (HCMV US11, HCMV US2), retention of specific MHC-I heavy chains in the ER (HCMV US3, murine CMV–MCMV-m152), and disruption of T-cell recognition of MHC-I on the cell surface (MCMV m04). That viruses have independently evolved numerous mechanisms to disrupt MHC–peptide presentation indicates the effectiveness and importance of this strategy to the survival of viruses with different infectious life cycles.

Maturation of APCs results in upregulation of co-stimulatory molecules and expression of cytokines that enable them to stimulate naive T cells. Viruses that can impair T-cell stimulation by preventing the upregulation of co-stimulatory molecules include Ebola virus, Lassa fever virus (LFV), HSV-1, and HIV. Additionally, a number of viruses (hepatitis C virus (HCV), HIV, measles virus (MV), and dengue virus (DV)) are also able to inhibit interleukin (IL)-12 production, which is often required for effective T-cell response. HCV does this through the action of its core and nonstructural protein 3 (NS3), which induces production of IL-10. DV, on the other hand, is able to inhibit IL-12 production through an IL-10 independent mechanism. In addition, compelling evidence showed that *in vivo* persistent infection of mice with LCMV, as well as persistent HCV infection in humans, induces IL-10 production by APCs resulting in the blunting of the CD8 T-cell response and chronic viral infection. Remarkably, antibodies blocking IL-10/IL-10R interactions correct T-cell exhaustion by restoring T-cell function, which results in purging of virus from mice persistently infected with LCMV.

Finally, a novel immunosuppressive molecule, programmed cell death-1 (PD-1), is upregulated in nonfunctional CD8 T cells during chronic infections (LCMV, HIV, HCV). Interaction of PD-1 with PD-ligands on APCs (or parenchymal cells) inhibits lymphocyte activation. As for IL-10 blockade, antibodies interfering with PD-1/PDL interactions also promote viral clearance from a persistently infected host.

The fact that not all viruses are able to block APC maturation does not necessarily represent a failure of the pathogen or a success for the host. A good example of this is observed following MV infection that exploits the ability of DCs to mature and migrate to lymphoid organs in response to infection. MV benefits greatly by having infected DCs home to lymphoid compartments where the infected cells are able to actively suppress T-cell proliferation (mediated through T-cell contact with surface viral glycoproteins) and also facilitate virus spread to more lymphoid cells. Therefore, the full understanding of the virus–host relationship requires not only studying the active mechanisms that viruses use to disable the immune system, but by also asking how a virus benefits by not altering a particular immune function.

Concluding Remarks

Co-evolution of certain hosts and pathogens for millions of years has resulted in a fine-tuned equilibrium that enables survival of both. Antigen presentation is one of the critical elements in this balance. While antigen presentation is an essential process for long-term effective host defense, targeting APCs represents a common maneuver of many viruses to avoid host surveillance and establish a chronic or persistent infection. A major challenge in biomedical research is to thwart microbial APC subversion to promote eradication of the pathogen. A better understanding of the mechanisms used by APC to display microbial antigens as well as the virus strategies to subvert APC functions during immune responses will provide new tools for designing novel vaccination approaches and immunotherapeutic treatments for human infectious diseases.

Acknowledgments

This is publication no. 18909 from Molecular and Neuroscience Integrative Department, The Scripps Research Institute (TSRI). This work was supported by NIH grants AI 45927, AI 05540, and AI 09484.

E. I. Zuniga is a Pew Latin American Fellow.

See also: Antibody - Mediated Immunity to Viruses; Cytokines and Chemokines; Cell-Mediated Immunity to Viruses; Immunopathology; Persistent and Latent Viral Infection; Vaccine Strategies.

Further Reading

Bevan MJ (2006) Cross-priming. *Nature Immunology* 7: 363–365.
Blander JM and Medzhitov R (2006) On regulation of phagosome maturation and antigen presentation. *Nature Immunology* 7: 1029–1035.
Dudziak D, Kamphorst AO, Heidkamp GF, *et al.* (2007) Differential antigen processing by dendritic cell subsets *in vivo*. *Science* 315: 107–111.
Itano AA and Jenkins MK (2003) Antigen presentation to naive CD4 T cells in the lymph node. *Nature Immunology* 4: 733–739.
Janeway CA, Travers P, Walport M, and Shlomchik MJ (2005) *Immunobiology: The Immune System in Health and Disease.* New York: Garland Science Publishing.
Lauterbach H, Zúñiga EI, Truong P, Oldstone MB, and McGavern DB (2006) Adoptive immunotherapy induces CNS dendritic cell recruitment and antigen presentation during clearance of a persistent viral infection. *Journal of Experimental Medicine* 203(8): 1963–1975.
Menendez-Benito V and Neefjes J (2007) Autophagy in MHC class II presentation: Sampling from within. *Immunity* 26: 1–3.
Norbury CC and Tewalt EF (2006) Upstream toward the 'DRiP'-ing source of the MHC class I pathway. *Immunity* 24: 503–506.
Oldstone MB (2007) A suspenseful game of 'hide and seek' between virus and host. *Nature Immunology* 8: 325–327.
Reis e Sousa C (2006) Dendritic cells in a mature age. *Nature Reviews Immunology* 6: 476–483.
Shen L and Rock KL (2006) Priming of T cells by exogenous antigen cross-presented on MHC class I molecules. *Current Opinion in Immunology* 18: 85–91.
Shortman K and Liu YJ (2002) Mouse and human dendritic cell subtypes. *Nature Reviews Immunology* 2: 151–161.
Strawbridge AB and Blum JS (2007) Autophagy in MHC class II antigen processing. *Current Opinion in Immunology* 19: 87–92.
Yewdell JW and Nicchitta CV (2006) The DRiP hypothesis decennial: Support, controversy, refinement and extension. *Trends in Immunology* 27: 368–373.

Antigenicity and Immunogenicity of Viral Proteins

M H V Van Regenmortel, CNRS, Illkirch, France

Glossary

B-cell epitope Surface region of a native protein antigen that is specifically recognized by the binding sites of free and membrane-bound antibody molecules. The membrane-bound antibodies are the B-cell receptors that recognize the antigen during the immunization process.

Cryptotope Antigenic site or epitope hidden in polymerized proteins and virions because it is present on the surfaces of subunits that become buried. Cryptotopes are antigenically active only after dissociation of protein aggregates and virions.

Mimotope A peptide possessing similar binding activity as that of a peptide epitope but showing little or no sequence similarity with it. Originally, mimotopes were defined as peptides able to bind an antibody but showing no sequence similarity with the protein antigen used to induce the antibody, usually because the antibody was directed to a discontinuous epitope.

Neotope Antigenic site or epitope specific for the quaternary structure of polymerized proteins or virions. Neotopes arise from the juxtaposition of residues from neighboring subunits or through conformational changes in the monomers resulting from intersubunit interactions.
Paratope Binding site of an antibody molecule that binds specifically to an epitope of the antigen. Paratopes are constituted of residues from six complementarity-determining regions (CDRs) located on the heavy and light chains of immunoglobulins. The CDRs vary greatly in sequence and in length in individual antibodies.

Virus Antigenicity

The antigenic reactivity or antigenicity of viruses corresponds to their capacity to bind specifically to the functional binding sites of certain immunoglobulin molecules. Once such binding has been observed experimentally, the particular immunoglobulin becomes known as an antibody specific for the virus.

The antigenicity of nonenveloped viruses resides in the viral proteins that form the viral capsid, whereas the antigenicity of enveloped viruses resides mostly in the exposed proteins and glycoproteins that are anchored in the viral, lipid membrane. The oligosaccharide side chains of viral glycoproteins contribute significantly to the antigenic properties of enveloped viruses.

The antigenic sites or B-cell epitopes of viral proteins correspond to those parts of viral capsids and envelope proteins that are specifically recognized by the binding sites or paratopes of free and membrane-bound antibody molecules. Antibody molecules anchored in the outer membrane of B cells correspond to the receptors of B cells that recognize the antigen when it is administered to a vertebrate host during immunization. The B-cell receptors recognize the native tertiary and quaternary structure of viral proteins and the antibody molecules that are released when the B cells have matured into plasmocytes, also recognize the native conformation of the proteins.

During a natural viral infection or after experimental immunization, the immune system of the host may also encounter dissociated viral protein subunits and it will elicit antibodies specific for these components. In addition, during a viral infection, nonstructural viral proteins that are not incorporated into the virions will also induce the production of specific antibodies in the infected host. Diagnostic tests that detect antibodies to nonstructural viral proteins are useful for differentiating animals infected with, for instance, foot-and-mouth disease virus from vaccinated animals that possess only antibodies directed to the capsid proteins of the virus. In this case, the ability to differentiate vaccinated animals from infected animals by a suitable immunoassay is an important prerequisite for convincing trading partners that a cattle-exporting country is free of foot-and-mouth disease.

Viral Antigenic Sites

In the absence of further qualification, the term epitope used in the present article refers to an antigenic site of a protein recognized by B cells. Immune responses are also mediated by T cells, that is, by lymphocytes that recognize protein antigens through T-cell receptors, after the antigen has been processed into peptide fragments. T-cell responses to viral antigens which involve the so-called T-cell epitopes will not be discussed here.

Because of the additional antigenic complexity that arises in viral proteins as a result of the quaternary structure of virions, it is useful to distinguish two categories of viral epitopes known as cryptotopes and neotopes. Cryptotopes are epitopes that are hidden in the intact, assembled capsid. They are located on the surfaces of viral subunits that become buried when the subunits associate into capsids, and they are antigenically active only after dissociation or denaturation of virions. Neotopes are epitopes that are specific for the quaternary structure of virions and which are absent in the dissociated protein subunits. Neotopes arise either through conformational changes in the monomers induced by intersubunit interactions or result from the juxtaposition of residues from neighboring subunits. These terms are useful when it is important to distinguish between the epitopes carried by different aggregation states of viral proteins. The availability of monoclonal antibodies has made it easy to identify neotopes and cryptotopes in many viruses whose antigenic structure has been analyzed in detail. There is evidence, for instance, that the trimeric form of the gp160 protein of human immunodeficiency virus (HIV) possesses neotopes that are not present in the monomeric form but are important for the induction of neutralizing antibodies.

Another important category of viral epitope is the neutralization epitopes that are recognized by antibodies able to neutralize viral infectivity. Since it is the antibodies that bring about neutralization, it is appropriate to talk of neutralizing antibodies and not of neutralizing epitopes. These epitopes are best referred to as neutralization epitopes.

Infectivity neutralization depends on properties of the virus, the antibody, and the host cell. The ability of an antibody to interfere with the process of infection always takes place in a specific biological context and it cannot be described adequately only in terms of a binding reaction between virus and antibody. In many cases loss of infectivity occurs when the bound antibody molecules inhibit the ability of the virus to attach to certain receptors of the host cell. It can happen that the antibody prevents the virus from infecting one type of host cell but not another type.

Some picornaviruses such as the rhinoviruses occur in the form of as many as 100 different variants known as serotypes, each serotype being neutralized by its own antibodies but not by antibodies specific for other serotypes. In contrast, another picornavirus, that is, poliovirus, exists only as three serotypes. The structural basis for this difference in the number of neutralization serotypes between different picornaviruses has not yet been clarified.

Antibodies to Viral Antigens

The capacity of antibodies to recognize viral epitopes resides in their functional binding sites or paratopes. The most common type of antibody is an immunoglobulin known as IgG which possesses two identical binding sites. These paratopes are constituted of three complementarity-determining regions (CDRs) located in a heavy chain and three CDRs in a light chain. The CDRs comprise a total of about 50–70 amino acid residues and form six loops that vary greatly not only in sequence but also in length from one antibody to another. Each individual paratope is made up of atoms from not more than 15–20 CDR residues which make contact with a specific epitope. This means that as many as two-thirds of the CDR residues of an antibody molecule do not participate directly in the interaction with an individual epitope. These residues remain potentially capable of binding to other epitopes that may have little structural resemblance with the first epitope, a situation which gives rise to antibody multispecificity. The relation between an antibody and its antigen is thus never of an exclusive nature and antigenic cross-reactivity will be observed whenever it is looked for. The so-called antibody footprint corresponds to an area on the surface of the protein antigen of about 800 $Å^2$.

Continuous and Discontinuous Epitopes

Protein epitopes are usually classified as either continuous or discontinuous depending on whether the amino acids included in the epitope are contiguous in the polypeptide chain or not (**Figure 1**). This terminology may give the impression that the elements of recognition

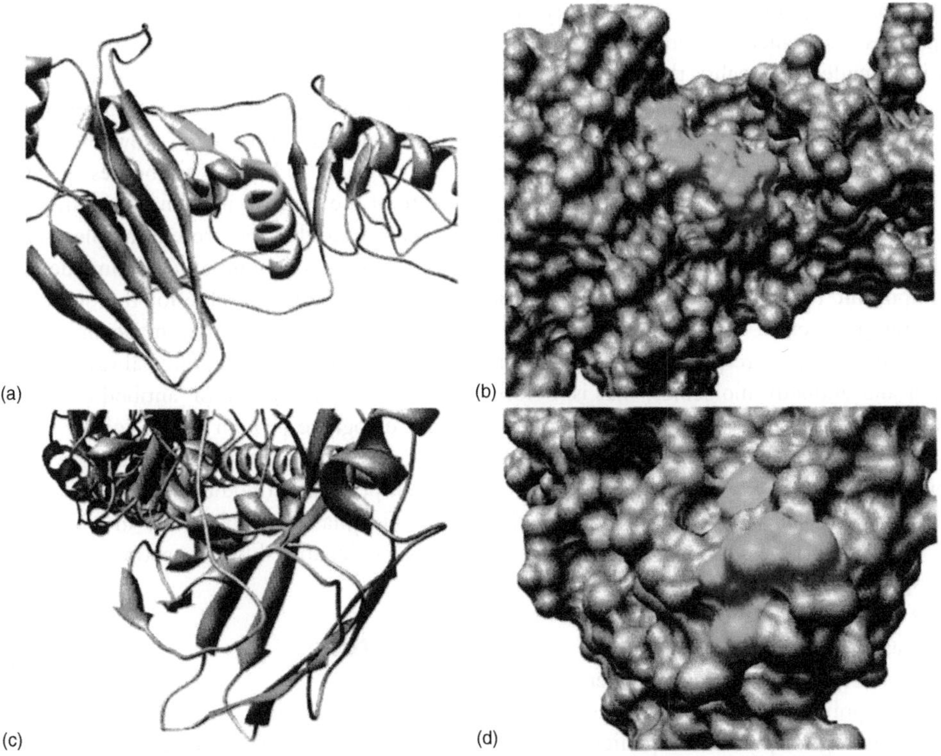

Figure 1 Illustration of the difference between a continuous (a, b) and discontinuous epitope (c, d) in the hemagglutinin protein of influenza virus. The residues that are part of epitopes are colored in blue ; the remaining residues are gray. (a) Ribbon representation of a continuous epitope. (b) Surface representation of the epitope shown in (a). (c) Ribbon representation of a discontinuous epitope. Residues distant in the sequence are brought close together by the folding. (d) Surface representation of the epitope in (c). Reproduced from Greenbaum JA, Andersen PH, Blythe M, *et al.* (2007) Towards a consensus on datasets and evaluation metrics for developing B cell epitope prediction tools. *Journal of Molecular Recognition* 20: 75–82, with permission from Wiley.

operative in epitope–paratope interactions are individual amino acids, whereas it is in fact at the level of individual atoms that the recognition occurs. The distinction between continuous and discontinuous epitopes is not clear-cut since discontinuous epitopes often contain stretches of a few contiguous residues that may be able, on their own, to bind to antibodies directed to the cognate protein. As a result, such short stretches of residues may sometimes be given the status of continuous epitopes. On the other hand, continuous epitopes often contain a number of indifferent residues that are not implicated in the binding interaction and can be replaced by any of the other 19 amino acids without impairing antigenic activity. Such continuous epitopes could then be considered to be discontinuous.

Since discontinuous epitopes consist of surface residues brought together by the folding of the peptide chain, their antigenic reactivity obviously depends on the native conformation of the protein. When the protein is denatured, the residues from distant parts of the sequence that collectively made up the epitope are scattered and they will usually no longer be individually recognized by antibodies raised against the discontinuous epitope.

Discontinuous epitopes are often called conformational epitopes because of their dependence on the intact conformation of the native protein. However, this terminology may be confusing since it seems to imply that continuous epitopes, also called sequential epitopes, are conformation independent. In reality the linear peptides that constitute continuous epitopes necessarily also have a conformation which, however, is mostly different from that of the corresponding region in the intact protein.

Most of our knowledge of the structure of discontinuous epitopes has been obtained from a small number of X-ray crystallographic studies of antibody–antigen complexes. It is important to realize that the structure of epitopes and paratopes seen in the complex may be different from the structure of the respective binding sites in the free antigen and antibody molecules, that is, before they have been altered by the mutual adaptation or induced fit that occurs during the binding interaction. For this reason, the structure of epitopes after complexation tends to be an unreliable guide for identifying the exact epitope structure that was recognized by the B-cell receptors during the immunization process. Crystallographic studies of antigen–antibody complexes have shown that the vast majority of protein epitopes are discontinuous and consist of residues from between two and five segments of the polypeptide chain of the antigen.

Most of our knowledge of protein antigenicity has not been obtained by X-ray crystallography but was derived from studies of the antigenic cross-reactivity between the intact protein and peptide fragments. In such studies, antibodies raised against the virus or isolated viral proteins are tested for their ability to react in various immunoassays with natural or synthetic 6–20-residue peptides derived from the protein sequence. Any linear peptide that is found to react in such an assay is labeled a continuous epitope of the protein. It is customary to test peptides of decreasing size and to give the status of epitope to the smallest peptide that retains a measurable level of antigenic reactivity. Usually this leads to the identification of continuous epitopes with a length of 5–8 residues, although the lower size limit tends to remain ill-defined. It is not unusual for certain di- or tripeptides to retain a significant binding capacity in particular types of solid-phase immunoassays.

On the other hand, increasing the length of peptides does not always lead to a higher level of cross-reactivity with antiprotein antibodies since longer peptides may adopt a conformation that is different from the one present in the intact protein. There is also no reason to assume that short peptides will have a unique conformation mimicking that of the corresponding region in the protein. Cross-reactivity of peptides with antiprotein antibodies is commonly observed because of antibody multispecificity and of the induced fit and mutual adaptation capacity of the two partners.

Many investigators believe that the majority of so-called continuous epitopes described in the literature and listed in immunological databases actually correspond to unfolded regions of denatured protein molecules, that is, that they are not genuine epitopes of native proteins. They argue that there is usually little experimental evidence to show that short peptides are actually able to bind to antibodies specific for the native state of the cognate protein. It is indeed very difficult to demonstrate that the protein sample used in an immunoassay does not contain at least some denatured molecules that could be responsible for the observed binding reaction. Furthermore, claims made in the early 1980s that immunization with peptides always leads to a high frequency of induction of antibodies that recognize the native cognate protein are no longer considered valid. It is now accepted that such claims arose because the ability of antipeptide antibodies to recognize proteins was tested in solid-phase immunoassays in which the proteins had become denatured by adsorption to plastic.

Mimotopes

The multispecificity of antibody molecules is illustrated by the existence of so-called mimotopes. The term mimotope was coined in 1986 by Mario Geysen and was originally defined as a peptide that is able to bind to a particular antibody, but is unrelated in sequence to the protein antigen used to induce that antibody, usually because the antibody is directed to a discontinuous epitope. Currently, the term mimotope is applied to any epitope mimic, irrespective of whether the epitope being mimicked is continuous or discontinuous. It is indeed possible to mimic a continuous epitope with a cross-reactive mimotope peptide that shows little sequence similarity with the original epitope. This cross-reactivity of mimotopes is due to the fact that dissimilar amino acid residues may actually contain a sufficient number of identical atomic groups to allow the peptide to cross-react with atoms of the antibody-combining site.

To qualify as a mimotope, a peptide should not only be able to bind a particular antibody but it should also be capable of eliciting antibodies that recognize the epitope being mimicked. This requirement stems from the fact that a single immunoglobulin always harbors a number of partly overlapping or nonoverlapping paratopes, each one capable of binding related or unrelated epitopes. When different subsites in the immunoglobulin-binding pocket partly overlap, binding to one epitope may prevent a second unrelated epitope from being accommodated at a nearby location. Therefore, when a peptide is labeled a mimotope of epitope A because of its capacity either to bind to an anti-A antibody or to inhibit the binding of epitope A to the antibody, it cannot be excluded that the putative mimotope actually binds to a different subsite from the one that interacts with epitope A. This is why it is necessary to show that a putative peptide mimotope is also able to induce antibodies that cross-react with epitope A, thereby demonstrating that it really is a mimotope of epitope A.

Nowadays, mimotopes are often identified by testing combinatorial peptide libraries, obtained by chemical synthesis or phage display, for their capacity to bind monoclonal antibodies specific for viral proteins. It is also possible to screen a phage library by means of sera collected from individuals that recovered from a viral infection and have seroconverted. Mimotopes identified in this manner may find applications in peptide-based diagnostic assays. It is also believed that mimotopes could be used for developing peptide-based vaccines.

The antigenic reactivity of viral proteins discussed so far is based on chemical reactions between epitopes and paratopes and it can be described using parameters such as the structural and chemical complementarity of the two partners, electrostatic and hydrogen bond interactions between them, the kinetics and equilibrium affinity constants of the interaction, the discrimination potential of individual antibody molecules, etc. Such immunochemical descriptions take the existence of antibodies for granted and do not ask questions about the biological origin, synthesis, or maturation of antibody molecules. The situation is different when it comes to investigations of the immunogenicity of viral proteins since this property cannot be analyzed outside of the biological context of an immunized host.

Immunogenicity

Whereas the antigenicity of proteins is a purely chemical property, their immunogenicity is a biological property that has a meaning only in the context of a particular host. Immunogenicity is the ability of a protein to give rise to an immune response in a competent host and it depends on extrinsic factors such as the host immunoglobulin repertoire, self-tolerance, the production of cytokines, and various cellular and regulatory mechanisms definable only in a given biological context.

When a number of continuous epitopes of a protein have been identified, this knowledge does not provide information on which particular immunogenic structure, present in the antigen used for immunization, was recognized by B-cell receptors and initiated the production of antibodies. Our ignorance of the exact immunogenic stimulus is often referred to as the black box conundrum and this makes the study of immunogenicity very much an empirical endeavor.

Although any peptide identified as a continuous epitope will readily elicit antipeptide antibodies, it is only rarely able to induce antibodies that also recognize the cognate, native protein antigen. In an immunoassay, an antibody raised against the native protein may be able to select one conformation in a peptide or it may induce a reactive conformation by an induced fit or mutual adaptation mechanism, the result in both cases being the occurrence of a cross-reaction between the peptide and the antiprotein antibody. On the other hand, when the same peptide meets a variety of B-cell receptors during the immunization process, it may not be able to bind preferentially to those rare paratopes in the receptors that, in addition to recognizing the peptide, also cross-react with the native protein. As a result most elicited antipeptide antibodies will only recognize the peptide and will not cross-react with the cognate protein.

As far as the immunogenicity of discontinuous epitopes is concerned, the situation is even more uncertain since the residues belonging to the epitope were identified by crystallographic analysis at the end of a process of mutual adaptation and conformational change in the two binding partners. The discontinuous epitope with its unique conformational features cannot be dissected out of the three-dimensional assembly of residues in the native protein and its immunogenic potential cannot therefore be studied independently of the rest of the protein antigen. As a result, the exact three-dimensional structure of the immunogen which was recognized by B-cell receptors and initiated the immune response cannot be known with certainty. Furthermore, reconstituting a discontinuous epitope in the form of a linear peptide that would include all the epitope residues from distant parts of the antigen sequence and would assemble them in the correct conformation appears to be an extremely difficult task.

Implications for Vaccine Development

Our increasing knowledge of the structure of viral epitopes has given rise in some quarters to the expectation that it should be possible to develop peptide-based viral vaccines. However, the results so far have been disappointing. In spite of the hundreds of viral epitopes that have been identified by studying the antigenicity of peptide fragments of viral proteins, no commercial peptide vaccine has yet reached the marketplace.

Although most peptide fragments are immunogenic in the sense that they readily induce antibodies that react with the peptide immunogen, this type of immunogenicity is irrelevant for vaccination purposes. What is required is the induction of antibodies which, on the one hand, recognize the cognate, native viral antigen (so-called cross-reactive immunogenicity) and which in addition also neutralize the infectivity of the virus (so-called cross-protective immunogenicity). Unfortunately, very few of the continuous epitopes of viruses that have been described possess the required cross-reactive and cross-protective immunogenicity. Many attempts have been made to increase the conformational similarity between peptide and intact protein, for instance, by constraining the peptide conformation by cyclization but this has not resulted in peptides possessing adequate cross-protective immunogenicity. There is also some evidence that the intrinsic disorder in certain loop regions of viral proteins may be responsible for the finding that peptides corresponding to such disordered regions are sometimes better vaccine immunogens than peptides with a constrained conformation.

The difficulties that must be overcome to transform a continuous epitope into an effective vaccine immunogen are illustrated by the many studies of the peptide ELDKWAS corresponding to residues 662–668 of the gp41 protein of HIV-1. This peptide which is recognized by the anti-HIV-1 broadly cross-reactive neutralizing monoclonal antibody (Mab) 2F5 has, for a long time, been regarded as a promising vaccine candidate because it is located in a conserved region of gp41 necessary for envelope-mediated fusion of the virus.

The ELDKWAS peptide has been incorporated into a variety of immunogenic constructs in an attempt to have it elicit antibodies with the same neutralizing capacity as Mab 2F5. When additional gp41-derived flanking residues were added to the peptide or when its conformation was constrained, peptides were obtained which had a tenfold higher affinity for the 2F5 antibody than the unconstrained peptide. However, in spite of their improved antigenicity, the peptide constructs, when used as immunogens, were still unable to induce antibodies with detectable neutralizing capacity.

In an attempt to ascertain which structural elements close to the ELDKWAS residues in the gp41 immunogen may have influenced the induction of the neutralizing 2F5 antibody, the crystal structure of the 2F5 antibody in complex with various gp41 peptides was determined. The conformation of the bound peptides was found to differ significantly from the corresponding region in the gp41 protein, indicating that the putative ELDKWAS epitope was able to assume various conformations depending on the fusogenic state of gp41. However, it was not clear which conformation should be stabilized in the peptide constructs intended for vaccination.

It seems that the viral epitopes involved in the immunogenic stimulus may in fact be dynamic structures with variable conformations and it has been suggested that such epitopes should be referred to as transitional epitopes. Such a label is an appropriate reminder that it is necessary to include the fourth dimension of time in the description of antigenic specificity.

The findings obtained with the ELDKWAS peptide suggest that these residues are part of a more complex discontinuous epitope that elicited the neutralizing Mab 2F5. There is also evidence that the hydrophobic membrane environment close to the ELDKWAS sequence played a role in the induction of the neutralizing antibody 2F5.

Unfortunately, only some of the antibodies induced by viral antigens possess neutralizing activity and it is not known which structural features in the immunogen are responsible for the appearance of neutralizing rather than non-neutralizing antibodies. In recent years, attempts to find out have relied mainly on the crystallographic analysis of viral proteins complexed with neutralizing Mabs. Although such studies reveal the structure of the epitopes present in the complexes, they do not provide information on which transitional immunogenic epitopes were able to induce neutralizing antibodies. It seems likely that this type of information will only be obtained by systematic empirical trials in which numerous candidate immunogens are tested for their ability to induce protective immune responses.

See also: Antigenic Variation; Antibody - Mediated Immunity to Viruses; Neutralization of Infectivity; Vaccine Strategies.

Further Reading

Clavijo A, Wright P, and Kitching P (2004) Developments in diagnostic techniques for differentiating infection from vaccination in foot-and-mouth disease. *Veterinary Journal* 167: 9–22.

Geysen HM, Rodda SJ, and Mason TJ (1986) *A priori* delineation of a peptide which mimics a discontinuous antigenic determinant. *Molecular Immunology* 23: 709–715.

Greenbaum JA, Andersen PH, Blythe M, *et al.* (2007) Towards a consensus on datasets and evaluation metrics for developing B cell epitope prediction tools. *Journal of Molecular Recognition* 20: 75–82.

Klenk H-D (1990) Influence of glycosylation on antigenicity of viral proteins. In: Van Regenmortel MHV and Neurath AR (eds.) *Immunochemistry of Viruses II*, pp. 25–37. Amsterdam: Elsevier.

Sheppard N and Sattentau Q (2005) The prospects for vaccines against HIV-1: More than a field of long-term nonprogression? *Expert Review Molecular Medicine* 7(2): 1–21.
Uversky VN, Oldfield CJ, and Dunker AK (2005) Showing your ID: Intrinsic disorder as an ID for recognition, regulation and cell signaling. *Journal of Molecular Recognition* 18: 343–384.
Van Regenmortel MHV (1996) Mapping epitope structure and activity: From one dimensional prediction to four-dimensional description of antigenic specificity. *Methods: Companion to Methods in Enzymology* 9: 465–472.
Van Regenmortel MHV (1998) From absolute to exquisite specificity. Reflections on the fuzzy nature of species, specificity and antigenic sites. *Journal of Immunological Methods* 216: 37–48.
Van Regenmortel MHV (1999) Molecular design versus empirical discovery in peptide-based vaccines: Coming to terms with fuzzy recognition sites and ill-defined structure-function relationships in immunology. *Vaccine* 18: 216–221.
Van Regenmortel MHV (2006) Immunoinformatics may lead to a reappraisal of the nature of B cell epitopes and of the feasibility of synthetic peptide vaccines. *Journal of Molecular Recognition* 19: 183–187.
Van Regenmortel MHV and Muller S (1999) In: *Synthetic Peptides as Antigens*, pp. 1–381. Amsterdam: Elsevier.
Wyatt R and Sodroski J (1998) The HIV-1 envelope glycoproteins: Fusogens, antigens and immunogens. *Science* 280: 1884–1888.
Yuan W, Bazick J, and Sodroski J (2006) Characterization of the multiple conformational states of free monomeric and trimeric human immunodeficiency virus envelope glycoproteins after fixation by cross-linker. *Journal of Virology* 80: 6725–6737.
Zolla-Pazner S (2004) Identifying epitopes of HIV-1 that induce protective antibodies. *Nature Reviews Immunology* 4: 199–210.
Zwick MB, Jensen R, Church S, *et al.* (2005) Anti-human immunodeficiency virus type 1 (HIV-1) antibodies 2F5 and 4E10 require surprisingly few crucial residues in the membrane-proximal external region of glycoprotein gp41 to neutralize HIV-1. *Journal of Virology* 79: 1252–1261.

Cytokines and Chemokines

D E Griffin, Johns Hopkins Bloomberg School of Public Health, Baltimore, MD, USA

Glossary

Chemokine Chemotactic cytokine that regulates trafficking of cells during immune responses.
Cytokine Soluble protein produced by a variety of cells that mediates cell-to-cell communication important for innate and adaptive immune responses.
Inflammasome A multiprotein complex of more than 700 kDa that is responsible for the activation of caspases 1 and 5, leading to the processing and secretion of the pro-inflammatory cytokines IL-1β and IL-18.
Lymphotoxin A cytokin secreted by activated T_H1 cells, fibroblasts, endothelial and epithelial cells.

Introduction

Cytokines are small soluble proteins produced by a wide variety of cells and are important mediators of cell-to-cell communication. Production of these biologically potent proteins is particularly associated with generation of the innate and acquired immune responses. Cytokines regulate both the initiation and maintenance of immune responses and some also induce production of cellular antiviral molecules. Chemokines (chemotactic cytokines) regulate trafficking of leukocytes during immune responses and infiltration of leukocytes into infected tissues. Both cytokines and chemokines share the properties of pleiotropy and redundancy. Each cytokine and chemokine has more than one function and most functions can be performed by more than one of these soluble factors.

Cytokines can act at a distance, but their effects are usually most potent in the areas in which they are produced. They mediate their effects by binding to small numbers of high-affinity receptors on responding cells. Cytokine receptors have a modular design consisting of two or more chains, each with a single transmembrane-spanning domain. Individual chains may participate in formation of the receptor for more than one cytokine. Chemokine receptors are single G-protein-coupled molecules with seven-transmembrane domains. The biologic function of many cytokines and chemokines is regulated by a short half-life for both the protein and the mRNA and by the production of circulating cytokine inhibitors such as soluble forms of the cytokine receptor or biologically inactive forms of the cytokine itself. A number of viruses, particularly in the herpesvirus and poxvirus families, encode inhibitory homologs of cytokines, chemokines, or their receptors, attesting to their importance for antiviral immunity.

Overview of Cytokine Networks

Cytokines associated with the innate immune response are those induced early after infection by virus replication *per se*. One of the most important of these early immunologically nonspecific responses are type 1 (α and β) interferons (IFNs) that are generally synthesized and released by infected cells. Some cells, for example, plasmacytoid dendritic cells, produce IFN rapidly after infection. IFN-α/β can induce an antiviral state in surrounding cells and can also influence the characteristics of the subsequent adaptive immune response. In addition, phagocytic cells, responsible for processing and presentation of viral antigens to initiate the antigen-specific immune response, produce a variety of cytokines, including IFN-α/β. These antigen-presenting cells, generally dendritic cells and monocyte/macrophages, may or may not be infected or produce infectious virus. Important cytokines and chemokines produced by antigen-presenting cells in the initial phases of the response to many viral infections are tumor necrosis factor (TNF), monocyte chemotatic proteins (MCP), and interleukin (IL)-1, IL-6, IL-12, and IL-18. These soluble factors promote the activation of T cells and recruitment of inflammatory cells into areas of virus replication. IL-12 stimulates natural killer (NK) cells to produce IFN-γ, promotes differentiation of $CD8^+$ cytotoxic T lymphocytes, and influences the early development and differentiation of CD4 T cells toward a delayed type hypersensitivity cellular immune response (type 1).

$CD4^+$ and $CD8^+$ T lymphocytes are the effectors of the specific cellular immune response and function primarily through synthesis of cytokines. T cells do not produce cytokines constitutively, but upregulate expression in response to stimulation through the T-cell receptor. $CD8^+$ T cells are stimulated by viral antigens presented in association with class I major histocompatibility (MHC) antigens. This type of stimulation is most likely to occur if the virus replicates in the antigen-presenting cell because antigens presented in association with MHC class I must be processed by proteasomes in the cytosol. $CD8^+$ T cells often have cytotoxic activity through production of granzymes and perforin or Fas ligand, but also produce substantial amounts of IFN-γ and lymphotoxin (LT). Viral antigens are presented to $CD4^+$ T cells as peptides in association with class II MHC antigens. Processing of antigens for presentation in association with MHC class II occurs through an endosomal pathway; therefore, virus replication is not required. Activated $CD4^+$ T cells produce IL-2, a T-cell growth factor that supports proliferation of both $CD4^+$ and $CD8^+$ T cells.

With further development of the virus-specific immune response, $CD4^+$ T cells tend to differentiate into different types of T-helper (Th) cells. Type 1 $CD4^+$ T (Th1) cells develop under the influence of IFN-γ and produce primarily IFN-γ and TNF-β (LT-α) and are associated with classic delayed type hypersensitivity or cellular immune responses. Type 2 $CD4^+$ T cells develop under the influence of IL-4 and produce primarily IL-4, IL-5, and IL-13 and are associated with strong antibody responses to viral antigens. A third type of $CD4^+$ effector cells (Th17) develop under the influence of transforming growth factor (TGF)-β and produce IL-17 and IL-6 and are associated with inflammation and autoimmune disease. Cytokines produced by $CD4^+$ T cells are necessary for B-cell development and for the switch from B cells producing IgM to B cells producing more mature forms of immunoglobulin such as IgG, IgA, and IgE and for B cell maturation into long-lived antibody-secreting cells. Both $CD4^+$ and $CD8^+$ T-cell responses are associated with mononuclear infiltrates into areas of virus replication. Regulatory T cells control and dampen ongoing immune responses, in part, through production of IL-10 or TGF-β.

Brief Description of Cytokine Families

TNF

The TNF family has three members: TNF-α, TNF-β/LT-α, and LT-β. These cytokines have approximately 30% homology and interact with cellular receptors in the TNF receptor (TNFR) superfamily. Activated macrophages are the major source of TNF-α and T cells are the major source of LT. TNF-α is synthesized as a proprotein (26 kDa) which is membrane bound and cleaved by a specific multidomain cell-surface metalloprotease to yield a 17 kDa protein. The active form of the soluble protein is a cone-shaped trimer that can bind to either of two TNFRs (TNFR1, p55 or TNFR2, p75). Binding of TNF to TNFR1 induces nuclear factor kappa B (NF-κB) and can trigger apoptosis. Homotrimers of LT-α also bind to TNFR1 and TNFR2, but LT-β reacts with a separate member of the TNFR family, the LT-β receptor. TNF interacts with a wide variety of cells systemically and locally and induces pro-inflammatory cellular responses such as expression of adhesion molecules, release of cytokines, chemokines and procoagulatory substances, synthesis of acute phase proteins, and production of fever. TNF also has some antiviral activity and is an important early participant in the immune response to infection. In bacterial infections, TNF is one of the mediators of endotoxin-induced (septic) shock. TNF and LT also play an important role in development of lymphoid tissues because lymphoid architecture in mice with deletions of these genes is abnormal.

IFN-γ

IFN-γ is a glycosylated protein of 25 kDa that is produced by NK cells and by type 1 CD4 and CD8 T cells (immune IFN). Although IFN-γ induces antiviral activity, it is structurally unrelated to IFN-α/β and uses a distinct receptor. However, IFN-γ shares some components of the intracellular signaling pathways for type 1 IFN. The biologically active form of IFN-γ is a dimer. A primary role for IFN-γ is the activation of macrophages to increase phagocytosis, tumoricidal properties, and intracellular killing of pathogens, particularly bacteria and fungi. IFN-γ induces macrophage production of a variety of inflammatory mediators and reactive oxygen and nitrogen intermediates. IFN-γ increases expression of high-affinity immunoglobulin F_c receptors on phagocytes, which increases recognition of opsonized microorganisms by these cells. IFN-γ also increases expression of MHC antigens by macrophages and this facilitates antigen presentation to T cells. The IFN-γ cell-surface receptor is composed of two chains, IFN-γR1 (α) and IFN-γR2 (β). The α (R1) chain is sufficient for binding, but the β (R2) chain is required for signaling and receptor complex formation. T-cell responsiveness is regulated by receptor expression. Both Th1 and Th2 cells express IFN-γR1, but only Th2 cells express IFN-γR2. Therefore, Th1 cells produce, but do not respond to, IFN-γ while Th2 cells respond to, but not produce this cytokine. Macrophages express both receptor chains and are a primary target of IFN-γ activity.

IL-1

The IL-1 family has three members: IL-1α, IL-1β, and IL-1 receptor antagonist (IL-1RA). IL-1α and IL-1β have only distantly related amino acid sequences, but similar structure, and recognize the same receptors. IL-1RA binds the IL-1 receptor without transducing a signal and blocks the activities of IL-1α and IL-1β. Activated monocytes and macrophages are the major source of IL-1, although many other cells can produce this cytokine. IL-1β is synthesized as a precursor protein of 31 kDa that is processed by the caspase interleukin-1 β-converting enzyme (caspase-1) within the 'inflammasome' to the active secreted 17 kDa form. IL-1 acts systemically, as well as locally, and can produce fever, sleep, and anorexia, frequent symptoms of viral infection. Hepatocytes are among the cells that produce IL-1RA as a part of the acute-phase response to inflammation and infection, presumably to control the effects of IL-1. There are two IL-1 receptors but only IL-1R1 transduces an activation signal to responding cells. IL-1R2 appears to function as a decoy receptor that regulates IL-1 activity by binding the cytokine without transducing a signal.

IL-18, also known as IFN-γ-inducing factor, is an 18 kDa cytokine produced by macrophages that shares structural features with the IL-1 family of proteins. It is synthesized as a proprotein that is cleaved by caspase 1 to its biologically active form. IL-18 stimulates IFN-γ production by T cells, NK cytotoxicity, and T-cell proliferation.

IL-2

IL-2 is a unique 15 kDa cytokine produced primarily by activated type 1 $CD4^+$ T cells. IL-2 plays an important role in supporting proliferation of activated helper T cells, cytotoxic T cells, B cells, macrophages, and NK cells. For stimulated $CD4^+$ T cells, IL-2 acts in an autocrine fashion because Th1 cells can produce IL-2 and upregulate high-affinity IL-2 receptors as well. The high-affinity IL-2R consists of three chains – IL-2Rα, β, and γ – which are expressed on the surface of activated T cells. IL-2Rα is a unique protein, whereas IL-2Rβ and IL-2Rγ are members of the hematopoietic growth factor receptor superfamily. Members of this superfamily share a 210-amino-acid domain in the extracellular region of the molecule that contains a distinct cysteine and tryptophan motif at the N-terminus and a Trp-Ser-X-Trp-Ser motif at the C-terminus. The whole domain is folded into two barrel-like structures with the ligand-binding region between. In addition to the IL-2 receptor, this receptor superfamily includes components of receptors for cytokines IL-3, IL-4, IL-5, IL-6, IL-7, IL-9, granulocyte macrophage-colony stimulating factor (GM-CSF), and granulocyte-colony stimulating factor (G-CSF). Signal transduction is through the IL-2Rβ chain. Intermediate-affinity IL-2R βγ receptors are expressed on monocytes and NK cells. IL-2Rγ (γ constant) also participates in the formation of the multichain IL-4, IL-7, IL-9, and IL-15 receptors. Defects in expression of this molecule lead to severe combined immunodeficiency in humans. Binding of IL-2 to the high-affinity receptor leads to T-cell proliferation.

IL-4

There are five members of the IL-4 family of cytokines: IL-3, IL-4, IL-5, IL-13, and GM-CSF. These cytokines have approximately 30% protein sequence homology, have similar α-helical structures, are linked in the same chromosomal region, and are produced by type 2 $CD4^+$ T cells. IL-3 and GM-CSF are two of the four major human myeloid growth factors that also include G-CSF and M-CSF. IL-3 is produced by activated T cells and binds to a heterodimeric receptor. The β subunit of the IL-3R is shared with the receptors for IL-5 and GM-CSF.

IL-4 and IL-13 have similar effects on B cells and monocytes. Both can serve as a costimulating factor for B-cell proliferation, drive activated B cells into immunoglobulin secretion, and, in human B cells, induce class switching to IgG4 and IgE. IL-4 is regarded as an anti-inflammatory cytokine because it antagonizes the effects of IFN-γ on macrophages and downregulates type 1 $CD4^+$ T-cell responses. IL-13 induces monocyte proliferation and monocytosis *in vivo*. IL-4 is important for promoting the differentiation of $CD4^+$ T cells to type 2 cytokine production. IL-4 and IL-13 are produced by type 2 $CD4^+$ T cells. IL-4 is also produced by a subset of $CD4^+$ T cells that are $NK1.1^+$ (NK T cells), by type 2 $CD8^+$ T cells and by mast cells, basophils, and eosinophils.

IL-5

IL-5, along with IL-3 and GM-CSF, acts as an eosinophil stimulating factor and its receptor shares a common β chain with these cytokines. The high-affinity IL-5R has a unique α chain. IL-5 induces proliferation and differentiation of eosinophil progenitors, whereas IL-3 and GM-CSF probably act at earlier stages of development. IL-5 is produced by type 2 CD4$^+$ and CD8$^+$ T cells and by NK cells. IL-5 also induces antigen-stimulated B cells to differentiate into immunoglobulin-secreting plasma cells and enhances secretion of IgA.

IL-6

IL-6 is a member of a family of cytokines and neuronal differentiation factors that includes leukemia inhibitory factor, oncostatin M, IL-11, and ciliary neurotrophic factor. The receptors for these cytokines share the transmembrane protein gp130 that transduces the cytokine signal. IL-6 is a functionally diverse cytokine produced by macrophages and monocytes early in the response to infection. It induces hepatic synthesis of acute phase proteins, supports proliferation of B cells, and has an important role in production of IgA.

IL-7

IL-7 is a 15 kDa protein that is produced primarily by epithelial cells and supports the growth and development of immature B and T cells. IL-7 is produced by bone marrow stromal cells, thymic epithelial cells, keratinocytes, and intestinal epithelial cells and supports lymphocyte development in these locations. In the thymus, IL-7 induces rearrangement of germline T-cell receptor genes by upregulating expression of the RAG-1 and RAG-2 recombinases. The α chain of the IL-7 receptor is a member of the hematopoietic growth factor family and associates with the IL-2Rγ chain to form a functional high-affinity IL-7R.

IL-10

IL-10 is an 18 kDa polypeptide that forms homodimers. It is predicted to be a member of the 4 α–helix bundle family of cytokines. It is produced by a wide variety of cells, including type 2 CD4$^+$ and CD8$^+$ cells, B cells, mast cells, macrophages, and keratinocytes. Epstein–Barr virus encodes a viral homolog of IL-10 that is functionally active. The activities of IL-10 are diverse, but its principal function appears to be limitation of inflammatory responses. IL-10 inhibits cytokine production by Th1 cells, NK cells, and macrophages and has been associated with viral persistence. It suppresses the synthesis of TNF-α, IL-1α, IL-6, IL-8, GM-CSF, and G-CSF by macrophages while increasing the production of IL-1RA. IL-10 stimulates proliferation and differentiation of activated B cells and T cells and of mast cells and is important for differentiation of regulatory T cells. The functional IL-10R consists of two subunits, IL-10R1 and IL-10R2, both of which belong to the class II cytokine receptor family. Expression of both subunits is required for signal transduction.

IL-12

IL-12 is a structurally unique heterodimeric cytokine composed of two disulfide-linked 35 and 40 kDa polypeptides. The 40 kDa protein conveys receptor-binding activity, is shared with IL-23 and is constitutively produced by many cells. Only the heterodimer is biologically active, so regulation of synthesis of the 35 kDa chain controls IL-12 activity, while synthesis of the IL-23 19 kDa chain controls IL-23 activity. IL-12 is produced by macrophages and dendritic cells and stimulates the production of IFN-γ by NK and T cells. IL-12 is important for differentiation of CD4$^+$ T cells into Thl cells and IL-23 promotes differentiation into Th17 cells. At least two classes of the IL-12 receptor exist. The IL-12R β1 chain is expressed in both Th1 and Th2 cells while the β2 chain is expressed only in Th1 cells. The IL-23 receptor is composed of IL-12β1 and a unique IL-23R subunit. For IL-12 responsiveness and for T-cell differentiation, both β1 and β2 subunits are necessary so only potential Th1 cells respond to this cytokine. Both IL-12R β1 and β2 subunits are related to the gp130 group of cytokine receptors. Homodimers of p40 bind with high affinity to the IL-12R, transduce no signal and block the effect of the heterodimer.

IL-15

IL-15 is a novel cytokine functionally related to IL-2, but without significant IL-2 sequence homology. It binds to a receptor with the β and γ chains of the IL-2R, but has its own unique α chain. IL-15 induces T-cell proliferation, enhances NK cell function, is a potent T-lymphocyte chemoattractant, and stimulates production of IL-5. Unlike IL-2, it is produced primarily by monocyte/macrophages and endothelial cells.

IL-17

IL-17 (IL-17A) is the founding member of a family of six cytokines (IL-17A-F). IL-17 is a 155-amino acid glycoprotein that functions as a disulfide-linked 35 kDa homodimer. IL-17 is produced by a separate lineage of activated CD4$^+$ (Th17) and CD8$^+$ T cells that have been implicated in autoimmune disease.

Chemokines and Chemokine Families

The chemokine superfamily includes a large number of structurally related, basic, heparin-binding, small molecular weight (8–10 kDa) proteins that have potent chemotactic and immunoregulatory functions. Chemokines attract specific, but overlapping, populations of inflammatory cells, induce integrin activation, and can be divided into four families with different biologic activities based on the spacing of the first two of four conserved disulfide-forming cysteine residues in the amino-terminal part of the molecule. The largest groups are the CXC or α-chemokines and the CC or β-chemokines. In general, CXC chemokines are the primary attractants for neutrophils, whereas CC chemokines modulate responses of monocytes, macrophages, lymphocytes, basophils, and eosinophils. C chemokines have only one set of cysteines and are chemotactic for lymphocytes. The CX3C chemokine fractalkine is unusual in that it is a type 1 transmembrane protein with the chemokine domain on top of an extended mucin-like stalk. Herpesviruses and poxviruses encode viral chemokine binding proteins (vCKBPs) that are important regulators of the host immune response.

α-Chemokines

The first CXC chemokine to be identified was platelet factor 4 (PF4, CXCL4) that is stored in the granules of platelets that also contain neutrophil-activating protein (NAP-2, CXCL7) ready to be released at sites of tissue injury. The best-characterized α-chemokine is IL-8 (CXCL8) that can be rapidly induced in most cells and is also chemotactic for neutrophils. Other α-chemokines with chemotactic activity for neutrophils include GRO-α (CXCL1), GRO-β (CXCL2), GRO-γ (CXCL3)γ, epithelial cell-derived neutrophil activating protein (CXCL5), and granulocyte chemotactic protein-2 (CXCL6). Sequence identities between these various family members range from 24% to 46%. IP-10 (CXCL10) is a structurally similar, but functionally distinct, CXC chemokine that is an attractant for monocytes and T cells rather than granulocytes.

β-Chemokines

The CC chemokines have two large subgroups: the MCPs and the macrophage inflammatory proteins (MIPs). MCP-1 (CCL2), produced by endothelial cells in response to IL-1β and TNF-α, functions as a dimer, attracts and activates monocytes, NK cells, $CD4^+$, and $CD8^+$ T cells, promotes Th2 responses and is a potent histamine-releasing factor for basophils. MCP-2 (CCL8) and MCP-3 (CCL7) are structurally and functionally related chemokines important in Th2 responses, whereas MIP1α (CCL3) and MIP1β (CCL4) are important in Th1 responses. Other CC chemokines commonly increased during viral infection are RANTES (CCL5) and I-309 (CCL1). Sequence identities between family members are 29–71%. These chemokines activate integrins, thereby inducing or increasing adhesion to endothelial cells or extracellular matrix proteins to promote development of an inflammatory response. Two novel, related β chemokines, MCP-4 (CCL13) and eotaxin (CCL11), are preferential chemoattractants for eosinophils. CCL11 is produced by epithelial and endothelial cells, often in concert with IL-5.

Chemokine Receptors

Chemokine receptors are seven-transmembrane, G-protein-coupled molecules that signal through the pertussis toxin-sensitive Gαi subfamily of hetereotrimeric G proteins. In general, the α chemokine receptors (CXCRs) are distinct from the β receptors (CCRs). Each receptor can bind multiple chemokines within its class and each chemokine can bind multiple receptors. Therefore, the specificity of cellular attraction to a site and subsequent activation is probably achieved by selective mixtures of chemokines and cellular receptors. These molecules serve as secondary receptors for entry of human immunodeficiency virus into T cells and monocytes.

CXCR5 and CCR7 and their ligands (CXCL13, CCL19, and CCL21) are major regulators of dendritic cell and lymphocyte trafficking to secondary lympoid tissue. Th1 cells more frequently express CXCR3, CXCR6, CCR2, CCR5, and CX3CR1 than Th2 cells which more frequently express CCR3, CCR4, and CCR8. Tissue-specific homing is chemokine dependent. For instance, T cells that home to skin preferentially express CCR4 and CCR10, whereas their ligands CCL2 and CCL27 are produced by dermal macrophages, dendritic cells, and keratinocytes. B cells producing IgG upregulate CXCR4 which promotes homing.

See also: Antibody - Mediated Immunity to Viruses; Cell-Mediated Immunity to Viruses; Immunopathology.

Further Reading

Boehm U, Klamp T, Groot M, and Howard JC (1997) Cellular responses to interferon-gamma. *Annual Review of Immunology* 15: 747–795.

Brown MA and Hural J (1997) Functions of IL-4 and control of its expression. *Critical Reviews in Immunology* 17: 1–32.

Del Prete G (1998) The concept of type-1 and type-2 helper T cells and their cytokines in humans. *International Reviews of Immunology* 16: 427–455.

Dinarello CA (1996) Biologic basis for interleukin-1 in disease. *Blood* 87: 2095–2147.

Kolls JK and Linden A (2004) Interleukin-17 family members and inflammation. *Immunity* 21: 467–476.

Moore KW, de Waal Malefyt R, Coffman RL, and O'Garra A (2001) Interleukin-10 and the interleukin 10 receptor. *Annual Review of Immunology* 19: 683–765.

Murphy PM (2003) Chemokines. In: Paul WM (ed.) *Fundamental Immunology,* 5th edn., pp. 801–840. Philadelphia: Lippincott Williams and Wilkins.

Sugamura K, Asao H, Kondo M, *et al.* (1996) The interleukin-2 receptor gamma chain: Its role in the multiple cytokine receptor complexes and T cell development in XSCID. *Annual Review of Immunology* 14: 179–205.
Taga T and Kishimoto T (1997) Gp130 and the interleukin-6 family of cytokines. *Annual Review of Immunology* 15: 797–819.
Trinchieri G (1998) Immunobiology of interleukin-12. *Immunologic Research* 17: 269–278.
Weaver CT, Harrington LE, Mangan PR, Gavrieli M, and Murphy KM (2006) Th17: An effector CD4 T cell lineage with regulatory T cell ties. *Immunity* 24: 677–688.

Cell-Mediated Immunity to Viruses

A J Zajac and L E Harrington, University of Alabama at Birmingham, Birmingham, AL, USA

Glossary

Antigen presentation The process by which proteins are degraded into peptides that are loaded onto MHC molecules and these complexes are targeted to the cell surface.
Central-memory T cell A population of memory T cells which primarily reside in secondary lymphoid organs; characterized by the expression of CD62L and CCR7.
Chemokines Chemotactic cytokines which stimulate the migration of cells.
Cytokines Secreted proteins that regulate cellular actions by signaling via specific receptors.
Cytotoxic T lymphocyte (CTL) T cells which can kill virus-infected cells upon activation.
Effector T cell Cells capable of immediate functional activity resulting in pathogen removal.
Effector-memory T cell A population of memory T cells poised for immediate effector function that primarily resides outside the lymphoid organs.
Immune homeostasis Maintenance of lymphocyte populations at steady-state levels.
Immunodominance The hierarchy of T-cell responses to the array of individual epitopes which are presented during any given viral infection.
Immunological memory The ability of the host to mount rapid recall responses upon re-exposure to the inducing antigen.
Immunopathology Tissue damage that results from the actions of the host's immune response.
Major histocompatibility complex (MHC) A cluster of genes involved in immune recognition and regulation; MHC class I molecules couple with β2 microglobulin to present peptides to CD8 T cells; MHC class II molecules present peptides to CD4 T cells.
T-cell exhaustion The progressive loss of antiviral T-cell functions which can culminate in the complete deletion of specific T-cell populations during chronic viral infections.
T-cell receptor (TCR) Heterodimeric receptor expressed by T cells that binds specific peptide–MHC complexes.
T helper 1 (Th1) cell Effector CD4 T-cell subset characterized by the production of IFN-γ; associated with immune responses to intracellular bacteria and viruses.
T helper 2 (Th2) cell Effector CD4 T-cell subset characterized by the secretion of IL-4, IL-5, and IL-13; important for helminth infections; linked to allergies and asthma.

General Overview

Cell-mediated immune responses play a critical role in combating viral infections. They are comprised of T-cell responses, which fundamentally differ from antibody (humoral) responses in the way they bring about infection control. The cardinal trait of cell-mediated responses is that the physical presence of reactive T cells is required for immunity, whereas humoral responses are conferred by the presence of soluble antibodies. T cells, together with B cells, form the adaptive immune response to viral infections. The hallmarks of adaptive immunity include antigen specificity and memory. These features allow T cells to elaborate responses which specifically target the numerous viruses which may infect the host. The ability to establish long-lived immunological memory provides a unique mechanism to better protect the host during subsequent viral exposures.

Due to their importance in controlling pathogens, cell-mediated immune responses are widely studied.

Significantly, much of our understanding of cell-mediated immunity, including the fundamental concepts of major histocompatibility complex (MHC) restriction, tolerance, T-cell diversity, and immunological memory, has been determined by analyzing immune responses to viruses. Cell-mediated immune responses are dynamic, diverse, and display a broad range of phenotypic and functional properties.

T-Cell Recognition

T cells differ from antibodies (humoral responses) in the way they recognize viral antigens. Antibodies are capable of binding to intact viral proteins, including structural components of viral particles and also viral proteins present at the surface of infected cells. By being able to bind to conformationally complex structures, antiviral antibodies have the unique ability to neutralize the infectivity of viral particles present in the circulation or at mucosal surfaces, a function that cannot be performed by T cells. T cells cannot recognize intact viral proteins and therefore play no role in directly neutralizing whole viral particles. Instead, T cells recognize short peptide fragments presented at the cell surface in a noncovalent association with MHC molecules (see **Figure 1**). Thus, T-cell recognition is referred to as being MHC-restricted, since an individual T cell will only bind strongly to one particular MHC molecule, and is also peptide specific, as a T cell will predominately only recognize one specific short antigenic peptide.

T-Cell Receptors

CD4 and CD8 T cells express a unique surface receptor, the T-cell receptor (TCR) that determines the MHC restriction and peptide specificity of an individual T cell. In the vast majority of T cells, this is a heterodimeric receptor comprised of the TCR α- and β-chains. A smaller population of T cells (~5% of circulating T cells in humans) express an alternative form of the TCR, formed by the noncovalent association of TCR γ- and δ-chains; however, the roles of γδ T cells in controlling viral infections are not well defined. Each T cell expresses only one unique version of the TCR whose precise sequence and structure represents the end result of a series of gene rearrangements. This recombinatorial process, which occurs during T-cell ontogeny in the thymus, generates a massive repertoire of T cells with tremendous diversity between their individual TCR sequences. Estimates of the size of the T-cell repertoire suggest that 2.5×10^7 different TCRs

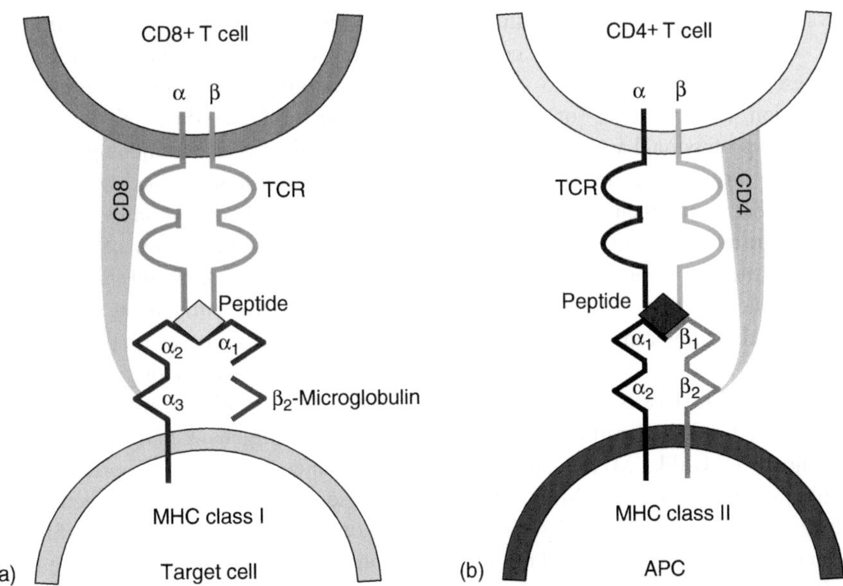

Figure 1 Similarities and differences between CD4 and CD8 T-cell recognition. Both CD4 and CD8 T cells express unique T-cell receptors at the cell surface which determine their antigen specificity and MHC restriction. (a) CD8 T cells recognize MHC class I molecules together with a non-covalently associated antigenic peptide, typically of 8–10 amino acids in length. MHC class I complexes are widely expressed and usually present endogenously synthesized antigens, including peptides derived from the degradation of viral proteins. (b) CD4 T cells recognize peptide antigens presented by MHC class II molecules. These antigenic peptides are typically 13–17 amino acids in length. These peptides are usually derived from extracellular antigens which have been endocytosed into professional antigen-presenting cells where they are proteolytically processed and re-presented at the cell surface bound to MHC class II molecules. MHC class II complexes have a much more limited tissue distribution than MHC class I molecules and are primarily expressed by macrophages, dendritic cells, and B cells.

are detectable in human blood. This large ensemble of individual T-cell clones is collectively capable of recognizing and responding to the vast array of antigens, including virally encoded antigens, which may be encountered during the lifespan of the host.

Antigen Processing and Presentation

CD8 T cells recognize MHC class I complexes (**Figure 1**). MHC class I molecules are expressed on virtually all cell types and usually present peptides derived from endogenously synthesized proteins. Antigen processing occurs continuously as newly synthesized proteins become degraded into peptide fragments by proteasomes. These fragments, typically of 8–10 amino acids in length, enter the endoplasmic reticulum and, if they have sufficient binding affinity, associate with MHC class I heavy chains together with the nonpolymorphic protein β2-microglobulin. These assembled MHC peptide complexes are then transported to the cell surface. This process allows MHC class I molecules to sample peptide fragments derived from proteins which are produced within the cell, including normal cellular proteins as well as virally encoded proteins, and present them for inspection by CD8 T cells. This ongoing immunological surveillance allows CD8 T cells to detect, respond to, and remove host cells which express 'non-self' viral proteins.

Although the endogenous pathway of antigen presentation provides a valuable mechanism for revealing the presence of virally infected cells to CD8 T cells, an alternative 'cross-presentation' pathway also operates. During cross-presentation, viral particles (or other antigens) are endocytosed by professional antigen-presenting cells and then undergo proteolytic degradation. The resulting peptide fragments can then bind to MHC class I molecules, and traffic to the cell surface. This enables antigen-presenting cells, such as dendritic cells, to display virally derived peptides to CD8 T cells even if the antigen-presenting cell itself is not capable of supporting virus replication.

CD4 T cells differ from CD8 T cells as they recognize peptides presented by MHC class II rather than MHC class I complexes (**Figure 1**). Unlike MHC class I complexes, which are ubiquitously expressed, MHC class II molecules are only presented by certain cell types such as dendritic cells, macrophages, and B cells. This limits CD4 T-cell recognition to professional antigen-presenting cells, since it is these specialized cells that have the capacity to display peptide–MHC class II complexes to CD4 T cells. Due to structural differences in the peptide-binding groove of MHC class I and MHC class II complexes, the viral peptides that are presented to CD4 T cells are typically longer (~13–17 amino acids in length) compared to those displayed by MHC class I molecules. In addition, whereas MHC class I complexes primarily present endogenously synthesized antigens, MHC class II molecules usually present antigens derived from extracellular sources. These exogenous antigens can include viral particles and also remnants of virally infected cells. Following uptake by professional antigen-presenting cells, these antigens are degraded into peptide fragments within acidified endosomes. Alternatively, if the antigen-presenting cell is actively infected with the virus, then intracellular vesicles containing viral proteins can serve as a source of peptides for associating with MHC class II complexes. Once at the cell surface these presented antigens can be detected by, and activate, CD4 T cells which express TCRs that are capable of specifically recognizing the peptide–MHC class II combination.

Immunodominance

Individual viruses encode multiple potential T-cell epitopes; therefore T-cell responses elicited during viral infections are not monoclonal or monospecific. Instead, oligoclonal subsets of cells are induced and, although each individual T cell is responsive to only one particular peptide–MHC combination, the overall pool of cells is sufficiently diverse to ensure that numerous epitopes can be detected. The kinetics, magnitudes, phenotypic and functional traits, as well as the stability of T responses to each individual virally encoded epitope can differ. Consequently, an ordered hierarchy can emerge as certain epitopes elicit more abundant, or immunodominant responses, whereas others are less prevalent and give rise to subdominant responses (**Figure 2**).

Our understanding of the precise determinants of immunodominance is incomplete; however, the hierarchy of T-cell responses is likely to be shaped by many factors. The magnitude of responses to individual viral epitopes is influenced by properties of the host's T cells including precursor frequencies and the avidity of the T cells for the presented viral antigen, as well as viral related factors including the ability of viral peptides to bind MHC complexes, the abundance of presented antigen, the kinetics of viral protein synthesis, and the types of cells which present the viral antigens. Changes in the patterns of immunodominance have been reported most notably during the course of persistent viral infections, as well as following secondary exposures to viruses which have been previously controlled (**Figure 2**). In the case of lymphocytic choriomeningitis virus (LCMV) infection of C57BL/6 mice, the NP396 epitope is co-dominant following well-controlled acute infections, and CD8 T cells specific for this epitope respond most vigorously during secondary exposures to LCMV. By contrast, during persistent LCMV infections, responses to this usually dominant epitope can become completely undetectable. During Epstein–Barr virus (EBV) infections shifting patterns of immunodominance are observed as responses to immediate early and early

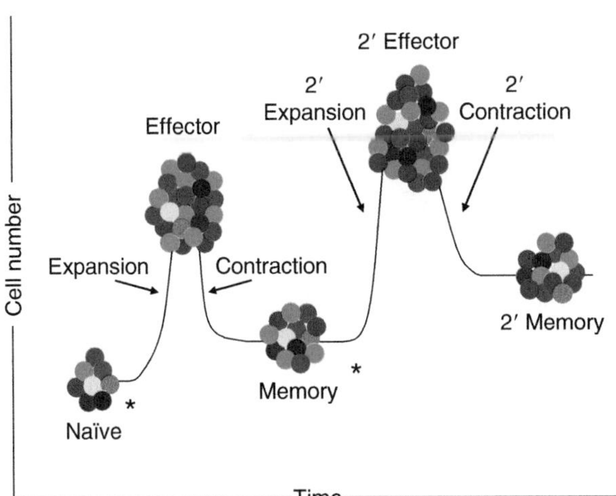

Figure 2 Changes in T-cell immunodominance can occur during primary and secondary immune responses following acute viral infections. Viruses elicit T-cell responses to a range of individual viral epitopes. These responses are not necessarily equal and in the example depicted the T cells specific for the 'green' epitope are immunodominant following the primary infection. This pattern of immunodominance is maintained during the memory phase but secondary exposure to the virus results in an anamnestic recall response during which the 'red' epitope-specific T cells predominate. Asterisk indicates when exposure to the virus occurred.

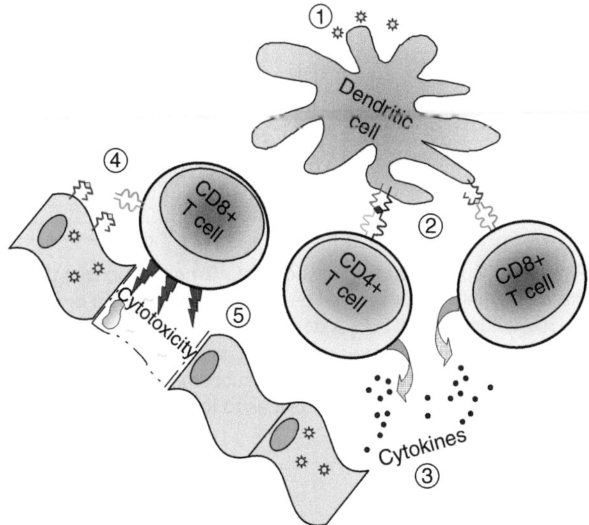

Figure 3 The induction and function of cell-mediated immune responses. (1) The presentation of viral antigens by professional antigen-presenting cells is a critical step in initiating cell-mediated immune responses. (2) Recognition of cognate peptide–MHC complexes by CD4 and CD8 T cells stimulates their proliferation and differentiation into effector cells. (3) Antigen-activated effector CD4 and CD8 T cells can express cytokines such as IL-2, TNF-α, and IFN-γ. These cytokines have important roles in coordinating the antiviral immune response and can also have direct antiviral effects. (4) Once activated antiviral T cells disperse into tissues where can they respond locally, at the sites of viral infection. (5) If viral peptide–MHC complexes are recognized, then effector CD8 T cells can elaborate cytotoxic effector functions which kill the infected cells. CD4 T cells can also become cytotoxic during certain infections; however, their impact is less promiscuous as they recognize the MHC class II complexes which are only expressed by professional antigen-presenting cells.

viral proteins are initially detected, but as viral latency becomes established responses to lytic cycle proteins decline and responses to latent viral proteins predominate. These observations demonstrate that not all antiviral T cells respond equally and suggest that certain specificities of T cells may be more effective at combating particular viral infections.

Induction of Cell-Mediated Immunity during Viral Infections

As a virus infection becomes established in the host, a series of molecular and cellular signals are initiated which activate cell-mediated immune responses. These signals include the production of interferons, other cytokines, and inflammatory mediators, in addition to the mobilization of local dendritic cells. Dendritic cells are thought to provide a critical cellular link for priming naive CD4 and CD8 T cells (**Figure 3**). It has been proposed that these cells are especially prone to infection by viruses which facilitate their role as cellular sensors for signaling the occurrence of an infection. Even if dendritic cells are not permissive to active infection with particular viruses, they can also present antigens through cross-priming to CD8 T cells, as well as to CD4 T cells via the classical exogenous antigen processing pathway.

The primary activation events which induce cell-mediated immunity predominately occur in secondary lymphoid organs including regional lymph nodes and the spleen. During the early stages of viral infections dendritic cells residing at the initial sites of infection take up viral antigens, become activated, and migrate to regional lymph nodes. Within the lymph nodes these dendritic cells encounter naïve T cells which are circulating through these organs as part of their normal immunosurveillance protocol. Engagement of TCRs on the naive T cells with viral–peptide MHC complexes presented by the dendritic cells results in sequestration of the T cells and launches the antiviral T-cell response (**Figure 2**). The ensuing proliferation and differentiation of virus-specific T cells also occur in conjunction with inflammatory mediators such as interferons and other danger signals. Many parameters, including the duration and strength of antigenic stimulation, co-stimulatory interactions, the presence of cytokines, and the provision of CD4 T cells help guide the developing response. These early events play a critical role in driving the generation of both the effector T cells

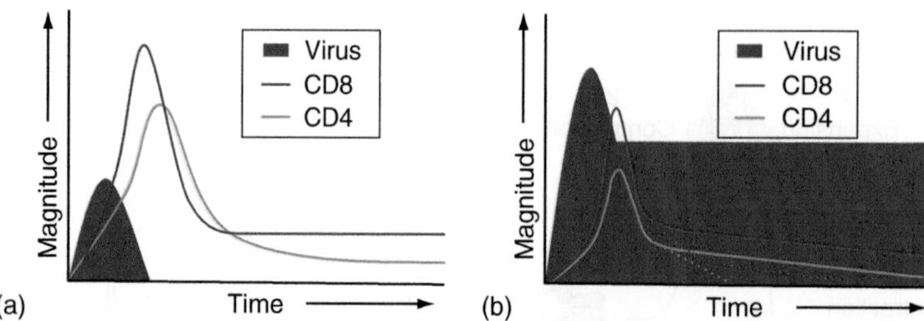

Figure 4 Successful and unsuccessful T-cell responses during acute and chronic viral infections. (a) During acute viral infections massive T-cell responses can be induced which play a principal role in clearing the infection. Following the resolution of the infection, the responding T-cell pool is downregulated but a long-lived pool of memory T cells becomes established which helps protect against subsequent viral exposures. (b) During chronic viral infections T-cell responses are elicited but a variety of phenotypic and functional defects manifest as these responses succumb to exhaustion. A gradation of exhausted phenotypes is often observed, ranging from an inability to produce effector cytokines to the complete deletion of virus-specific T cells.

as well as the subsequent establishment of the memory T cell pool.

CD8 T Cells

One of the most impressive aspects of CD8 T-cell responses is the massive proliferation of these cells which occurs during the initial phase of many viral infections (**Figures 2** and **4**). Experimental studies of acute LCMV infection of mice have demonstrated that antiviral CD8 T cells can increase over 10 000-fold during the first week of infection; over 50%, and perhaps even more, of the host's CD8 T cells are LCMV-specific at the peak of the response! Marked expansions of virus-specific CD8 T cells are a common feature of many virus infections including influenza, vaccinia virus, EBV, yellow fever virus, and early following human immunodeficiency virus (HIV) infection. During this expansion phase the patterns of gene expression change promoting the synthesis of cytokines and cytotoxic effector molecules as well as alterations in surface molecules including cytokine receptors and adhesion molecules. This results in an expanded pool of virus-specific effector cells with functional attributes necessary to control the infection. The ensemble of virus-specific effector cells which emerge during the acute phase of the infection is remarkably heterogeneous and comprises of subsets which differ in their epitope specificity, clonal abundance, effector potential, expression of adhesion molecules and cytokine receptors, and ultimate fate. Although the initial activation of T-cell responses occurs in secondary lymphoid organs, the effector cells become dispersed throughout the host. In this way the T cells are available locally, at the sites of infection, where they operate to eliminate the host of virally infected cells.

CD8 T cells are potent antiviral effector cells due to their ability to produce both inflammatory mediators as well as cytotoxic effector molecules (**Figure 3**). CD8 T cells are commonly referred to as cytotoxic T lymphocytes (CTLs), which emphasizes their ability to kill virally infected target cells. These killing functions are triggered as the effector T cell become activated following engagement with a virally infected target cell displaying an appropriate peptide–MHC complex. The subsequent release of perforin and granzyme molecules by the T cells ensures the swift destruction of the infected cell. Ideally, this targeted removal of the infected cell occurs before progeny virus is released. Alternative Fas and TNF-dependent cytotoxic mechanisms have been reported but their *in vivo* significance in killing virus infected cells is not well defined. In addition to their direct killing functions, CD8 T cells also produce a range of cytokines and chemokines. In the laboratory the production of these effector molecules is often used to detect the presence of antiviral T cells. Most importantly, within the infected host the production of these soluble mediators, such as IFN-γ and TNF-α, can also help clear viral infections without causing death of infected cells. This cytokine-mediated purging of infected cells has been most convincingly shown during viral hepatitis.

CD4 T Cells

CD4 T cells are traditionally known as helper T cells because of their ability to provide help to B cells and CD8 T cells, resulting in antibody production, class switching, cytotoxic T cell activity, and memory development. In addition to assisting cells of the adaptive immune system, CD4 T cells produce an array of cytokines and chemokines that stimulate cells of the innate immune system, such as macrophages and neutrophils, to traffic to the sites of infection and elaborate their effector activities. It has also been demonstrated that CD4 T cells are directly capable of antiviral functions, through the production of IFN-γ and, in some circumstances, by inducing lysis of virally infected cells (**Figure 3**). Thus, CD4 T cells are critical constituents of the cell-mediated immune response to viral infections; however, it should be emphasized that innate immunity as well as humoral immune responses, which are helped by CD4 T cells, are key components of the host overall antiviral response.

Like CD8 T cells, naive CD4 T cells circulate through secondary lymphoid organs in a relatively quiescent state. Following recognition of antigen in the context of MHC class II, a cascade of signaling events is initiated within the CD4 T cell which results in activation, proliferation, and differentiation into an effector CD4 T cell (**Figure 3**). Classically effector CD4 T cells have been divided into two polarized subsets based on their cytokine production profile. T helper 1 (Th1) cells primarily produce IFN-γ and are critical for the immune responses to various viral infections, as well as infections with intracellular bacteria. This subclass of effector cells is typically associated with antiviral cell-mediated immunity. Conversely, T helper 2 (Th2) cells predominantly secrete the cytokines IL-4, IL-5, and IL-13, assist with the eradication of helminth infections, and have historically been linked with the production of antibodies and humoral immune responses.

In recent years, the definition of CD4 T-cell subsets has expanded beyond Th1 and Th2 cells, with the importance of unique populations of regulatory CD4 T cells and also IL-17 producing 'Th17' cells becoming evident. Regulatory cells are pivotal for preventing autoimmunity by suppressing the activation of autoreactive T cells. Regulatory T cells are typically characterized by the expression of the transcription factor Foxp3 and by the production of the suppressive cytokines IL-10 and TGF-β. Relatively little is known regarding the significance of these CD4 T-cell populations during viral infections; however it has been proposed that regulatory T cells are both beneficial to the host, by limiting immunopathology, and detrimental, by dampening effector functions.

T-Cell Memory

Ideally, the primary immune response overwhelms the infection and results in the complete eradication of the virus from the host. If the infection is successfully resolved then the expanded pool of effector T cells does not remain constitutively activated. Instead, a downregulation phase ensues during which typically the majority (<90%) of the virus-specific T cells present at the peak of the immune response die by apoptosis. The remaining 5–10% of T cells survive the contraction phase and constitute a long-lived pool of memory T cells (**Figures 2** and **4**). In this way, a beneficial memory of past infections is established as, by comparison with naïve hosts, an increased number of virus-specific T cells are maintained which are tuned to rapidly respond if they re-encounter infected cells. The population of memory T cells which emerges following the resolution of the infection and restoration of homeostasis is not uniform, as phenotypic and functional diversity is apparent even within subsets of memory T cells which recognize the same viral epitope. This is well illustrated by the classification of memory T cells into broad categories termed effector- and central-memory T cells. These subsets have been defined based upon their anatomical location, functional quality, proliferative potential, and expression of surface molecules.

Since T cells neither recognize nor neutralize the infectivity of viral particles they do not confer sterilizing immunity and cannot prevent secondary infections. Nevertheless, as the host cells become infected, preexisting memory T cells which developed following prior viral exposures can mount robust recall responses. These anamnestic responses are characteristically more rapidly induced, greater in magnitude, and possibly more functionally competent than primary T-cell responses. Such pronounced secondary responses help protect the host by contributing to the swift control of the infection thereby reducing the morbidity and mortality. Memory T-cell responses are not the only components of secondary immune responses as these cells act in conjunction with antiviral antibodies to protect the host during viral reexposures.

Analysis of both clinical specimens and experimental animal models has demonstrated that acute viral infections can induce very long-lived memory T-cell responses (**Figure 4**). Studies using experimental mice have demonstrated that memory CD8 T cells reactive against various infections such as LCMV, vaccinia virus, and influenza are maintained at remarkably stable levels for over 2 years following infection. By contrast, CD4 T-cell responses are not as consistent and have been reported to gradually decay over time. Although natural exposures to viruses lead to the formation of immunological memory, these advantageous responses are also induced following vaccinations. Vaccines are successful in protecting the host against subsequent infections because of their ability to promote long-lived memory responses. In humans the longevity of T-cell responses has been investigated in detail following smallpox vaccination. Smallpox-specific T-cell responses are detectable in individuals who received a single dose of the vaccine 75 years previously! Notably, the findings suggested that the responses do decline slowly with predicted half-lives of 8–15 years.

Persistent Viral Infections

Although T-cell responses can be highly effective at controlling acute infections and contribute to protective secondary responses, persistent viral infections do arise and are often associated with the development of phenotypically and functionally inferior responses (**Figure 4**).

These types of infections include many viral pathogens which are of significant public health importance such as HIV and hepatitis C virus (HCV). A common feature of these infections is that T-cell responses are initially induced but qualitative and quantitative defects become apparent as the generation of robust sets of effector cells, as well as the progression of memory T-cell development are subverted. By comparision with successful T-cell responses elaborated during acute viral infections, a spectrum of phenotypic and functional defects have been detected during persistent infections. The production of cytokines including IL-2, TNF-α, and IFN-γ, as well as cytotoxic effector molecules such as perforin may be diminished or abolished, and decreased proliferative potential has also been observed. The severe loss of effector activity as well as the physical deletion of antiviral T cells which can occur during persistent infections has been termed exhaustion (**Figure 4**).

The parameters which contribute to T-cell exhaustion are not fully understood. Comparative analysis of T-cell responses to viral infections which result in different levels of antigenic exposure, such as influenza, cytomegalovirus, EBV, HCV, and HIV, indicate that antiviral T cells may adopt different preferred phenotypic and functional set points. Experimental studies suggest that many factors including, but not limited to, viral targeting and destruction of dendritic cells, the production of immunosuppressive cytokines such as IL-10, the depletion of CD4 T cell subsets, and the induction of weak neutralizing antibody responses can all contribute inferior cell-mediated immune responses. Changing viral loads may also impact the functional quality of the T-cell response. During acute HCV infection antiviral CD8 T cells transiently lose the ability to produce IFN-γ, but recover from this 'stunned' state as the viral loads are brought under control. Importantly, this suggests that under certain conditions the exhaustion of virus-specific T cells may be prevented or even reversed. Several reports have now demonstrated that during persistent LCMV, HIV, and HCV infections, antiviral T cells can express the inhibitory receptor PD-1. Antibody-based therapeutic treatments to block this receptor have been shown to promote proliferation of previously exhausted T cells and restore their functional activities. This is a promising experimental observation; however, the jury is still out on whether this approach will be a beneficial treatment for persistent infections of humans.

Immunopathology

Since viruses are obligate intracellular pathogens they must infect permissive host cells in order to replicate. Infected cells die as a direct result of the virus' lytic replication cycle or are killed as a consequence of the actions of the antiviral immune response. Although immune-mediated destruction of virally infected cells is necessary to contain the infection, it can also result in immunopathology, which represents collateral damage to the host caused by the actions of the immune response. A classical example of immunopathology occurs following intracranial infection of adult mice with LCMV. Mice infected by this route succumb to a characteristic lethal disease and expire approximately 1 week following infection. Death can be prevented by immunosuppression of the mice, which has shown that the disease is a consequence of a vigorous CD8 T-cell response rather than due to the infection *per se*. HBV-associated viral hepatitis is another instance where anti-viral CD8 T-cell responses, which are attempting to clear the infection, cause liver damage in the infected individual.

Most viral infections are associated with the development of an IFN-γ-producing Th1-CD4 T cell response. In various animal models, the absence of CD4 T cell help during viral infection results in impaired clearance of the infectious agent. However, not all CD4 T-cell responses are beneficial as the induction of inappropriate types of CD4 T-cell responses can be deleterious to the host, due to immunopathology. In the 1960s, a group of young children were administered a formalin-inactivated vaccine for respiratory syncytial (RS) virus and following exposure to live RS virus, these children exhibited enhanced infection rates and immunopathology linked to increased frequencies of eosinophils and neutrophils within the airways. Animal studies have indicated that the vaccine was associated with a Th2-biased virus-specific immune response (increased levels of IL-4 and IL-13, as well as eosinophil recruitment to the lungs) that upon live infection displayed many of the signatures of immunopathology which manifested in these vaccinated children. Supporting experiments suggest that immunization to promote the Th1 responses or ablation of Th2 responses prevents the development of these pathological effects following live viral infection.

Immune Evasion

Arguably, one of best indications of the importance of cell-mediated immunity in controlling viral infections is the observation that many viruses have evolved strategies to evade the actions of the host immune response. There is, however, no one universal evasion mechanism; instead, viruses have adopted various preferred approaches to escape antiviral T-cell responses. A common strategy is to change the amino acid sequence of T-cell epitopes or nearby flanking residues that impede the recognition or processing of the antigenic peptide. Many viruses rely on error-prone polymerases in order to replicate, which favor the incorporation of mutations in progeny viral genomes. The resulting variant viruses will have a selective advantage if the amino acid substitution abolishes the ability of the epitope to associate with MHC molecules, negatively impacts recognition of the epitope by T cells, or prevents antigen processing.

In addition to mutating epitope sequences, many viruses encode specific molecules which function to interfere with the antiviral immune response. Both MHC class I and class II antigen-presenting pathways are targeted by several viral proteins. These immune evasion molecules block antigen presentation in various ways, ranging from preventing the transport of antigenic peptides into the endoplasmic reticulum to inhibiting the egress of peptide-loaded MHC complexes. The end result of these inhibitory strategies is to impair immunological surveillance. Although there is much anecdotal evidence that interference with antigen processing diminishes cell-mediated immune responses, experimental studies using murine cytomegalovirus (MCMV) question this notion. Infection of mice with mutants which lack several viral genes known to block antigen processing did not effect the ability of the host to elaborate an anti-MCMV CD8 T-cell response.

See also: Antigenicity and Immunogenicity of Viral Proteins; Antibody - Mediated Immunity to Viruses; Antigen Presentation; Antigenic Variation; Cytokines and Chemokines; Immunopathology; Defeating Innate Immunity; Innate Immunity; Persistent and Latent Viral Infection; Vaccine Strategies; Viral Pathogenesis.

Further Reading

Alcami A and Koszinowski UH (2000) Viral mechanisms of immune evasion. *Immunology Today* 9: 447–445.

Doherty PC and Christensen JP (2000) Accessing complexity: The dynamics of virus-specific T cell responses. *Annual Review of Immunology* 18: 561–592.

Ertl HC (2003) Viral immunology. In: Paul WE (ed.) *Fundamental Immunology,* 5th edn., pp. 1021–1227. Philadelphia, PA: Lippincott Williams and Wilkins.

Finlay BB and McFadden G (2006) Anti-immunology: Evasion of the host immune system by bacterial and viral pathogens. *Cell* 124: 767–782.

Frelinger JA (2006) *Immunodominance – The Choice of the Immune System.* Weinheim, Germany: Wiley-VCH.

Klenerman P and Hill A (2005) T cell and viral persistence: Lessons from diverse infections. *Nature Immunology* 6: 873–879.

Oldstone MBA (2006) Viral persistence: Parameters, mechanisms and future predictions. *Virology* 344: 111–118.

Seder RA and Ahmed R (2003) Similarities and differences in CD4+ and CD8+ effector and memory T cell generation. *Nature Immunology* 4: 835–842.

Yewdell JW and Bennink JR (1999) Immunodominance in major histocompatibilty complex class I-restricted T lymphocyte responses. *Annual Review of Immunology* 17: 51–88.

Zinkernagel RM and Doherty PC (1997) The discovery of MHC restriction. *Immunology Today* 18: 14–17.

Antibody-Mediated Immunity to Viruses

A R Neurath, Virotech, New York, NY, USA

Milestones in History

The primeval notion about immunity to disease arose in Greece about 25 centuries ago from the observation that those who recovered from an apparently contagious disease became resistant to a subsequent similar sickness. The earliest known attempts to intentionally transfer immunity to an infectious disease occurred in China in the tenth century. It involved exposing uninfected people to material from lesions caused by smallpox. This not always successful practice was introduced in the seventeenth century to the Ottoman Empire and subsequently to England and its colonies in North America.

This approach was revolutionized by replacing material from human lesions by that derived from cowpox lesions first in 1774 in England by a farmer, Benjamin Jesty, who used it on his family, and 22 years later by Edward Jenner. Thereafter, the widely used vaccination has been considered as the first example of immunization with a life-attenuated virus. Nevertheless, both the cause of the disease and the success of vaccination remained unexplained.

An understanding started to emerge about a century later when (1) Koch and Pasteur set forth the germ theory of infectious diseases, and Pasteur developed attenuated vaccines against anthrax and rabies; and (2) Von Behring and Kitasato demonstrated that immunity could be transferred by a soluble serum component(s). Several decades later, these components were shown to be antibody immunoglobulins (Heidelberger and Kabat) whose fundamental structure was elucidated by Porter and Edelman in the late 1950s, and further characterized by numerous X-ray crystallography and sequencing studies.

However, several findings indicated that antibodies are not the only mediators of specific immunity. Immune responses corresponding to delayed hypersensitivity and allograft rejections appeared unrelated to the presence of serum antibodies. Immunity could also be transferred by cells from immunized to naive animals (Landsteiner and Chase). It was shown by Gowans and colleagues in 1962 that lymphocytes are essential for immune responses.

Antibodies to very many distinct antigens can be produced upon immunization. Therefore, it was presumed that the diverse antibodies were not preformed but would be generated on demand following antigenic stimuli. In theory, there were two possibilities: that an antigen either directs or selects for the formation of a specific antibody. Further studies fully supported the clonal selection theory (Burnet, Jerne, and Talmage). The repertoire of diverse antibodies undergoing a process of maturation leading to high-affinity antibodies is generated by somatic rearrangements and hypermutation of immunoglobulin genes (Tonegawa).

Although it became evident that both immunoglobulins and cells are essential for immunity, the details of the process whereby antigens and viruses elicit immune responses were contributed to by results of studies in transplantation biology and notably the discovery of the major histocompatibility complex (MHC; HLA molecules on human cells) (Benaceraff, Dausset, and Snell). Their role in presentation of antigen fragments and interactions of specific antibody producing B-cells with T-cells, as well as the function of cytokines, chemokines, and adhesion molecules will be discussed in later sections of this article.

Immunoglobulins, lymphocytes, MHC molecules, and antigen receptors are all components of adaptative (acquired) immunity. Components of innate immunity, responding rapidly to an invading pathogen, play a key role in initiating and orchestrating the adaptive response (Janeway).

The major milestones in the history of immunology and their time lines are summarized in **Table 1**.

Immunoglobulin Phylogeny

The adaptive immune system, based on clonally diverse repertoires of antigen recognition molecules, arose in jawed vertebrates (gnasthostomes) about 500 million years ago. Homologs of human immunoglobulins, T-cell antigen receptors (TCRs), MHC I and MHC II molecules, and recombination activating genes (RAGs) have been identified in all classes of gnasthostomes. This has provided evolutionary advantages allowing the recognition of potentially lethal pathogens, including viruses, and the initiation of protective responses against them (**Figure 1**), and represented an add-on to a preexisting innate immune system. For immunoglobulins, somatic variation was further expanded through class switching, gene conversion, and somatic hypermutation. Jawless vertebrates (agnathans) were shown to also have an adaptive immune system based on recombinatorial assembly of genetic units different from those of gnasthostomes, to generate a diverse repertoire of lymphocytes, each with distinct receptors (**Figure 2**). The assembly relies on highly variable leucine-rich repeat (LRR) protein modular units.

Most gnasthotomes generate a large part of their immuglobulin diversity in ways similar to those in humans. However, exceptions have occurred. Two of these, relevant to passive immunotherapy and development of diagnostics using the corresponding immunoglobulins, are mentioned here. (1) Only three classes of H-chain genes, corresponding to IgM, IgY, and IgA, exist in chicken. Class switching occurs from IgM to either IgY or IgA. IgY is considered to be the ancestor of 'present-day' IgG and IgE. Unlike human IgG, IgY does not bind to cellular Fc receptors and to proteins A and G, respectively, and does not activate the complement system. (2) H-chain antibodies devoid of light chains occur in the Camelidae species (camels, dromedaries, and llamas). Single-domain fragments from such antibodies react with specific antigens with high affinity constants (nanomolar range). The binding involves longer than average third variable loops of the antibody molecules. These unique antibodies and their antigen-specific fragments are expected to become important for biotechnological and medical applications, including intrabodies (intracellular antibodies).

Innate Immunity

The innate immune system provides an early defense against pathogens and alerts the adaptive immune system when initial invasion by a pathogen has occurred. The innate system predates evolutionarily the adaptive system and relies on antimicrobial peptides and on germline-encoded pattern-recognition receptors (PRRs). Antimicrobial peptides can damage enveloped viruses and interfere with processes involved in fusion of viruses with target cell membranes. PRRs recognize microbial (viral) components defined as pathogen-associated molecular patterns (PAMPs). Different PRRs are specific for distinct PAMPs, have distinct expression patterns, activate different signaling pathways, thereby eliciting distinct responses against invading pathogens. PRRs can be subdivided into two classes: Toll-like receptors (TLRs) and non-Toll-like innate immune proteins.

Toll was originally identified as a differentiation protein in *Drosophila*, and later (1996) shown to play a role in defense against fungal infection. Searching the entire DNA sequence database for similarities with the coding sequence for *Toll* resulted in identification of Toll-like sequences, and ultimately in discovery of at least 10 functional TLRs in humans. TLRs are type 1 integral membrane glycoproteins the extracellular domain of which contains several LRR motifs and a cytoplasmic

Table 1 History of immunology timeline

1798	Smallpox vaccination	Edward Jenner
1876	Validation of germ theory of disease by discovering bacterial basis of anthrax	Robert Koch
1879	Chicken cholera vaccine development	Louis Pasteur
1890	Discovery of diphtheria 'antitoxin' in blood	Emil Von Behring and Shibasaburo Kitasato
1882	Isolation of the tubercle bacillus	Robert Koch
1883	Delayed type hypersensitivity	Robert Koch
1884	Phagocytosis; cellular theory of immunity	Elie Metchnikoff
1891	Proposal that antibodies are responsible for immunity	Paul Ehrlich
1891	Passive immunity	Emil Roux
1894	Complement and antibody activity in bacteriolysis	Jules Bordet
1900	A, B, and O blood groups	Karl Landsteiner
1901	Cutaneous allergic reaction	Maurice Arthus
1903	Opsonization by antibody	Almroth Wright and Stewart Douglas
1907	Discipline of immunochemistry founded	Svante Arrhenius
1910	Viral immunology theory	Peyton Rous
1917	Haptens discovered	Karl Landsteiner
1921	Cutaneous reactions	Carl Prausnitz and Heinz Kustner
1924	Reticuloendothelial system	Ludwig Aschoff
1939	Discovery that antibodies are gamma globulins	Elvin Kabat
1942	Adjuvants	Jules Freund and Katherine McDermott
1942	Cellular transfer of sensitivity in guinea pigs (anaphylaxis)	Karl Landsteiner and Merill Chase
1944	Immunological hypothesis of allograft rejection	Peter Medawar
1948	Demonstration of antibody production in plasma B-cells	Astrid Fagraeus
1949	Distinguishing self vs. nonself and its role in maintaining immunological unresponsiveness (tolerance) to self	Macfarlane Burnet and Frank Fenner
1952	Discovery of agammaglobulinemia (antibody immunodeficiency)	Ogden Bruton
1953	Immunological tolerance hypothesis	Rupert Billingham, Leslie Brent, Peter Medawar, and Milan Hasek
1955–59	Clonal selection theory	Niels Jerne, David Talmage, and Macfarlane Burnet
1957	Discovery of interferon	Alick Isaacs and Jean Lindenmann
1958	Identification of first autoantibody and first recognition of autoimmune disease	Henry Kunkel
1959–62	Elucidation of antibody structure	Rodney Porter and Gerald Edelman
1959	Lymphocytes as the cellular units of clonal selection; discovery of lymphoid circulation	James Gowans
1961	Discovery of thymus involvement in cellular immunity	Jacques Miller
1968	Recognition of B- and T-cells in immunodeficiencies	Robert Good
1968	Distinction of bone marrow- and thymus-derived lymphocyte populations; discovery or T- and B-cell collaboration	Jacques Miller and Graham Mitchell
1965	Demonstration of allelic exclusion in B-cells	Benvenuto Pernis
1968–70	Elaboration of two-signal model of lymphocyte activation	Peter Bretscher and Melvin Cohn
1970	Discovery of membrane immunoglobulins	Benvenuto Pernis and Martin Raff
1971	Recognition of hypervariable regions in Ig chains	Elvin Kabat
1972	Elucidation of the major histocompatibility complex	Baruj Benacerraf, Jean Dausset, and George Snell
1974	Discovery of MHC restriction	Rolf Zinkernagel and Peter Doherty
1974	HLA-B27 predisposes to an autoimmune disease	Derek Brewerton
1975	Monoclonal antibodies used in genetic analysis	Georges Kohler and Cesar Milstein
1976	First demonstration of cross-priming	Michael Bevan
1978	Direct evidence for somatic rearrangement in immunoglobulin genes	Susumu Tonegawa
1978	Recognition that dendritic cells are distinctive and highly potent antigen-presenting cells	Ralph Steinman
1979	Discovery of leukocyte adhesion molecules and their role in lymphocyte trafficking	Eugene Butcher
1984–87	Identification of genes for the T-cell antigen receptor	Mark Davis, Leroy Hood, Stephen Hedrick, and Gerry Siu
1987	Crystal structure of MHC peptide solved	Pam Bjorkman, Jack Strominger, and Don Wiley
1989	Emerging field of innate immunity, infectious nonself model of immune recognition ('stranger hypothesis')	Charlie Janeway

Continued

Table 1 Continued

1994	Danger hypothesis of immune responsiveness	Polly Matzinger
1986	Discovery of T-helper subsets	Tim Mossmann and Bob Coffman
1989	Discovery of first chemokines	Edward Leonard, Teizo Yoshimura, and Marco Baggiolini
1991	Discovery of the first costimulatory pathway (CD28/B7) for T-cell activation	Kevin Urdahl and Mark Jenkins
1992	Cloning of CD40 ligand and recognition of its role in T-cell-dependent B-cell activation	Armitage, Spriggs, Lederman, Chess, Noelle, Aruffo, *et al.*
1996–7	Discovery of the role of Toll and Toll-like receptors in immunity	R. Medzhitov, CA. Janeway, Jr., and J. Hoffmann

Reprinted with permission from Steven Greenberg, MD, from http://www.columbia.edu/itc/hs/medical/pathophys/immunology/readings/ConciseHistoryImmunology.pdf.

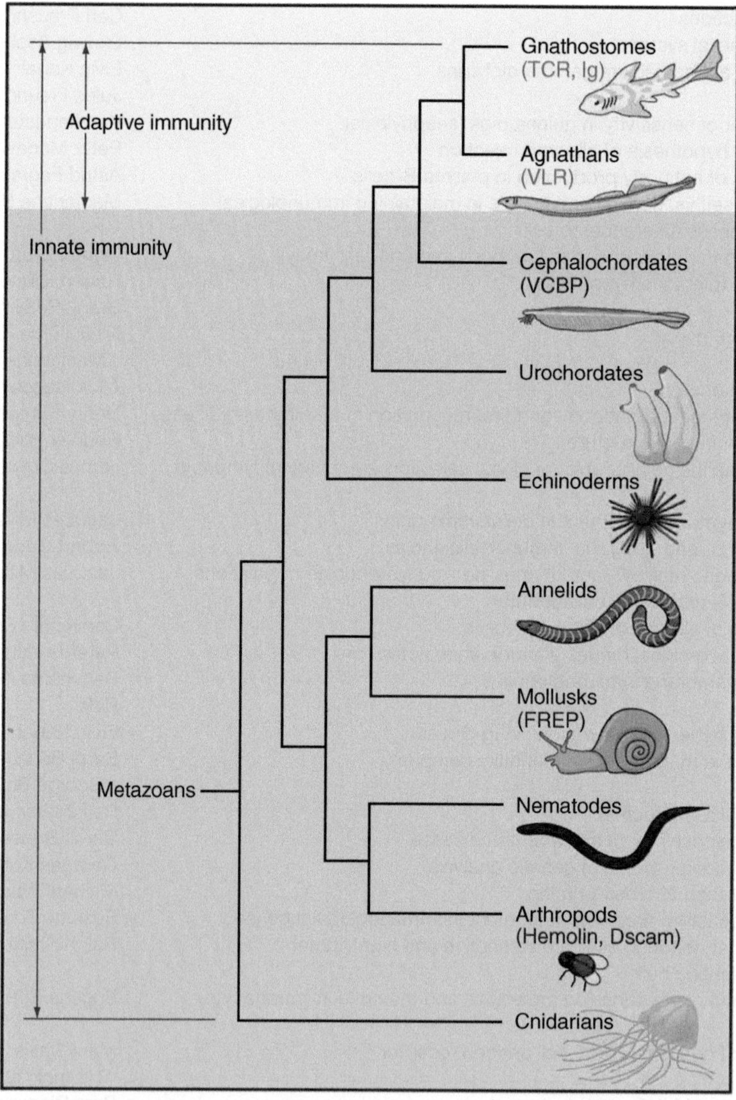

Figure 1 Phylogenetic tree indicating theoretical evolutionary relationships of metazoans and the emergence of adaptive immunity in conjunction with innate immunity. Families of immune molecules, other than Toll-like receptors (TLRs), are indicated in blue: chitin-binding domain-containing proteins (VCBPs), fibrinogen-related proteins (FREPs), hemolin and Down's syndrome cell adhesion molecule (Dscam). The recombinatorial based immune receptors are indicated in green: T-cell receptors (TCRs), immunoglobulins (Ig's), and variable lymphocyte receptors (VLRs). Reprinted from Cooper MD and Alder N (2006) The evolution of adaptive immune systems. *Cell* 124: 815–822, with permission from Elsevier.

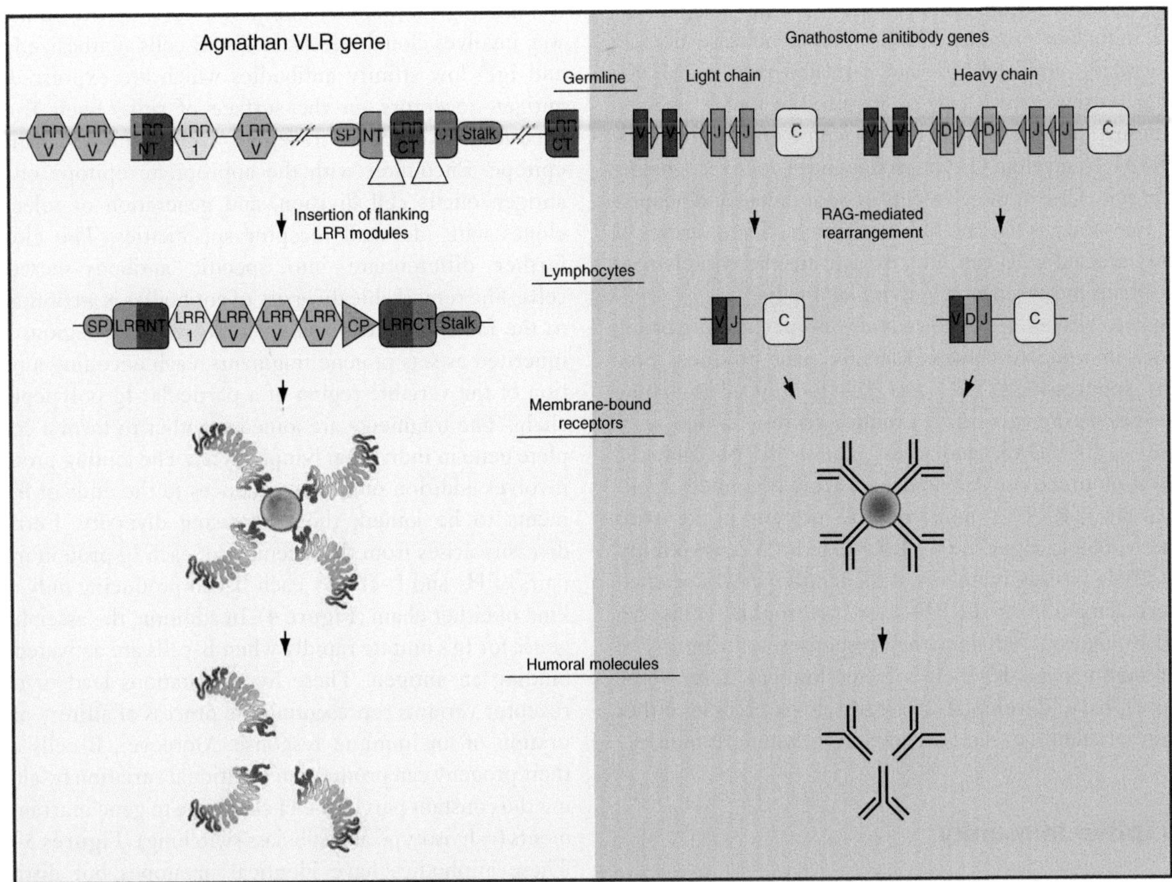

Figure 2 Two recombinatorial systems used for generating diverse antigen receptors in vertebrates. The figure compares the assembly of leucine-rich repeat (LRR) modular genetic units in agnathan lymphocytes to generate variable lymphocyte receptor (VLR) genes vs. the rearrangement of Ig gene segments in gnathostome B-lymphocytes to generate diverse antibody genes. Variable 24-amino-acid LRR (LRRV, green); N-terminal capping LRR (LRRNT, blue); variable 18-amino-acid LRR (LRR1, yellow); signal peptide (SP, orange); first six amino acids of LRRNT (NT, light blue); C-terminal capping LRR (LRRCT, red); last nine amino acids of LRRCT (CT, pink); variable 13 amino acid connecting peptide (CP, orange); and invariant VLR stalk (Stalk, purple). The small orange triangles adjacent to the representative V (blue), D (green), and J (light blue) gene segments represent recombination signal sequences (RSSs). The C (yellow) indicates the immunoglobulin constant region. Reproduced from Cooper MD and Alder N (2006) The evolution of adaptive immune systems. *Cell* 124: 815–822, with permission from Elsevier.

signaling domain homologous to that of the interleukin-1 receptor (IL-1R). TLRs 2, 3, 4, 7, 8, and 9 recognize PAMPs characteristic for viruses. Viral DNAs rich in CpG motifs are recognized by TLR9, leading to activation of pro-inflammatory cytokines and type 1 interferon (IFN) secretion. TLR7 and TLR8, expressed within the endosomal membrane, are specific for viral single-stranded RNA (ssRNA). Double-stranded RNA (dsRNA), generated during viral infection as an intermediate for ssRNA viruses or during transcription of viral DNA, is recognized by TLR3. TLR3 is expressed in dendritic cells (DCs) and in a variety of epithelial cells, including airway, uterine, corneal, vaginal, cervical, biliary, and intestinal cells. Cervical mucosal epithelial cells also express functional TLR9, suggesting that TLR3 and TLR9 provide an antiviral environment for the lower female reproductive tract. Some viral envelope glycoproteins are recognized by TLR2 and TLR4, each expressed at the cell surface, leading to production of pro-inflammatory cytokines. In general, the engagement of TLRs by microbial PAMPs triggers signaling cascades leading to the induction of genes involved in antimicrobial host defenses. This includes the maturation and migration of DCs from sites where infection occurs to lymphoid organs where DCs can initiate antigen-specific immune responses. Triggering distinct TLRs elicits different cytokine profiles and different immune responses. Engagement of TL3 and TRL 4, respectively, upregulates polymeric immunoglobulin receptor expression on cells. Thus, bridges between innate and adaptive immune responses are established.

TLRs are expressed either at the cell surface or in lysosomal/endosomal membranes. Therefore, they would not recognize pathogens that succeeded in invading the cytosolic compartment. These pathogens are detected by a variety of cytoplasmic PRRs. They include retinoic acid inducible protein (RIG-1) with a helicase domain recognizing viral dsRNA, and a related protein, MDA5. Other proteins which may be involved in innate antiviral immunity include a triggering receptor on myeloid cells (TREM-1), myeloid C-type lectins and siglecs, recognizing sialic acid. One of the elements of both innate and adaptative immunity is the complement system. Some viruses or virus-infected cells can directly activate the complement cascade in the absence of antiviral antibodies.

Many viruses are endowed with properties subverting innate immune responses. Vaccinia virus produces proteins suppressing TLR- and IL-1R-induced signaling cascades. Paramyxoviruses produce proteins which associate with MDA5, and thus inhibit dsRNA-induced activation processes. Adenoviruses avoid immune surveillance by TRL9. A nonstructural protein of hepatitis C virus blocks signaling by RIG-1 and MDA5. Marburg and Ebola viruses, members of the family *Filoviridae*, elicit direct activation of TREM-1 on neutrophils. This can lead to vigorous inflammatory responses contributing to fatal hemorrhagic fevers in infected humans. Thus, some viruses have developed strategies to overcome either innate or adaptive (see next section) immune responses.

Adaptive Immunity

Adaptive immunity is a complex anticipatory system triggered by exposure to antigens, including viruses. Its hallmarks are selectivity, diversification, specificity, and memory. The principal effector molecules of the system are antigen-binding receptors (immunoglobulin (Ig) and T-cell receptors (TCR)). The following simplified overview (see **Figure 3**) will be limited to Ig's and humoral immune responses.

Differentiation of B-cells into antibody-generating plasma cells occurs through distinct pathways. Two pathways lead to rapid IgM and IgA antibody production by B-1 cells against T-cell-independent antigens. These antibodies have low antigen-binding constants, and immunological memory does not evolve. The third pathway involves clonal selection. The B-cells synthesize IgM and IgD low affinity antibodies which are expressed as antigen receptors on the surface of cells. Each B-cell produces a different receptor recognizing a distinct epitope. Encounter with the appropriate epitope on an antigen elicits cell division, and generation of selected clones with identical receptor specificities. The clones further differentiate into specific antibody-secreting cells. The remarkable diversity of antibodies is attributable to the fact that genes coding for Ig variable regions are inherited as sets of gene fragments, each encoding a portion of the variable region of a particular Ig polypeptide chain. The fragments are joined together to form a complete gene in individual lymphocytes. The joining process involves addition of DNA sequences to the ends of fragments to be joined, thus increasing diversity. Further diversity arises from the assembly of each Ig protein from pairs of H- and L-chains, each B-cell producing only one kind of either chain (**Figure 4**). In addition, the assembled genes for Ig's mutate rapidly when B-cells are activated by binding an antigen. These hypermutations lead to new receptor variants representing the process of affinity maturation of an immune response. Moreover, B-cells and their progeny can produce an additional variation by altering the constant part of the H chain due to gene rearrangements (=Ig isotype and subclass switching) (**Figures 5–7**). These antibodies have identical paratopes but distinct effector functions (complement activation, phagocytosis, transcytosis, etc.). As a consequence of the aforementioned processes, there are B-lymphocytes of at least 100 million distinct specificities in every human individual at any given time.

The majority of Ig is produced in mucosa-associated tissues, predominantly in the intestine, rather than in the bone marrow, spleen, and lymph nodes, in the form of IgA. This harmonizes with the fact that mucosal surfaces represent the predominant sites for entry of pathogens, including viruses. Coincidentally, it has been demonstrated that the gut is the major site for HIV-1 replication and depletion of CD4+ cells. There are two subclasses of

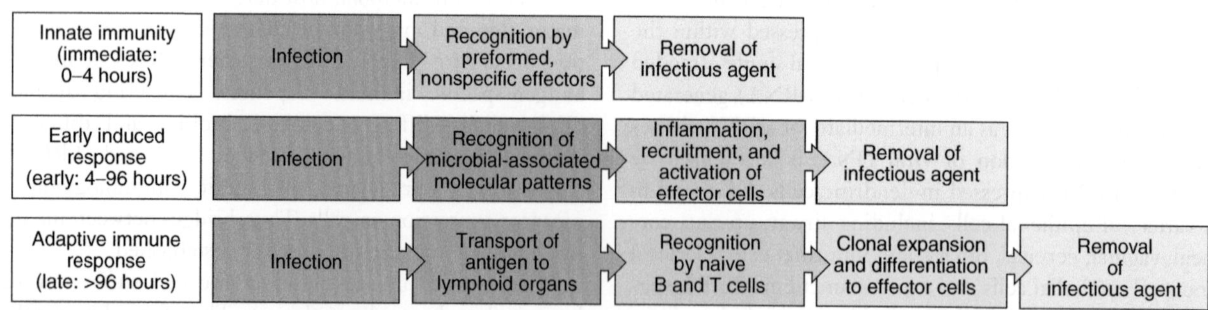

Figure 3 Three phases of a response to an initial infection. Copyright 2005 from Janeway CA, Jr., Travers P, Walport M, and Shlomchik MJ (2005) *Immunobiology. The Immune System in Health and Disease*, 6th edn, figure 2.1, p. 37. New York: Garland Science. Reproduced by permission of Garland Science/Taylor & Francis LLC.

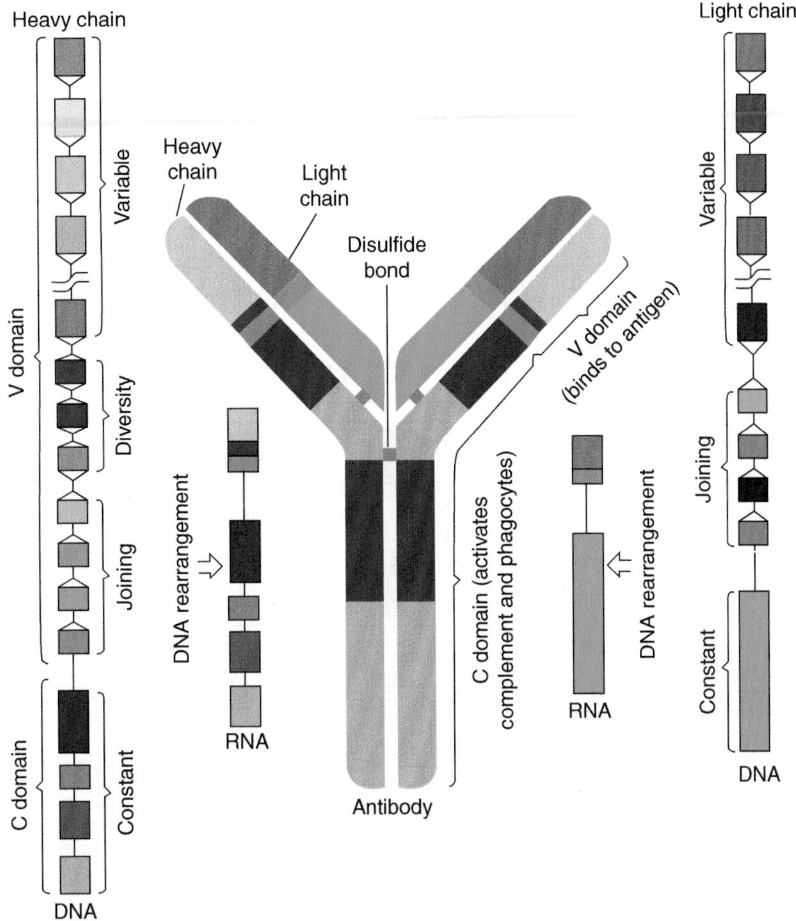

Figure 4 Antibody molecule consisting of pairs of H- and L-chains each encoded by genes assembled from different DNA segments. The segments rearrange to generate genes for chains that are distinct in each B-cell. The joining is variable so that the gene segments can encode the estimated 100 million specific antibodies each human is capable of producing. Reprinted from Janeway CA, Jr. (1993) How the immune system recognizes invaders. *Scientific American* (Sep.): 73–79, with permission of Scientific American and Ian Worpole (artist).

IgA, IgA1 and IgA2, which occur in monomeric, dimeric, tetrameric, and polymeric (pIgA) forms. A distinguishing feature of secretory IgA is its association with another glycoprotein, the secretory component (SC) (**Figure 8**). SC is also the extracellular portion (which can be generated by proteolytic cleavage) of an integral epithelial cell membrane protein, the polymeric Ig receptor (pIgR) mediating transcytosis of pIgA and IgM. The latter process allows neutralization of pathogens within intracellular vesicular compartments.

An essential aspect of adaptive immunity is its ability to recall past encounters with a pathogen for decades or even an entire lifetime. This fundamental feature is the foundation of successful vaccination. Germinal center-derived memory B-cells have the following attributes: antigen specificity; hypermutated Ig variable gene segments; and ability to bestow immunological memory following their adoptive transfer to immunologically naïve recipients. The CD27 surface antigen is a marker for memory B-cells.

The multifaceted performance of B-cells is the result of a thoroughly orchestrated ensemble involving CD4+ helper T-cells, DCs, MHC class II antigens, cell-differentiation antigens (CDs), cytokines, etc. In summary, in addition to occupancy of the Ig receptor, B-cells must interact with antigen-specific T-cells. The T-cells, through specific TCRs, recognize peptide fragments generated from the antigen internalized by the B-cell, and displayed on the surface of the B-cell as a peptide–MHC class II complex. Helper T-cells stimulate the B-cell following binding of the CD40 ligand on the T-cell to CD40 on the B-cell; the interaction of tumor necrosis factor (TNF)–TNF receptor family ligand pairs; and the release of specific cytokines. Further details of these interactions are shown in **Figures 9–11**.

The intricacies of all these tightly coordinated events seem to minimize the possibility that 'rationally designed' synthetic vaccines will be able to successfully recreate the specificity of virus neutralization B-cell epitopes or neotopes.

	Immunoglobulin								
	IgG1	IgG2	IgG3	IgG4	IgM	IgA1	IgA2	IgD	IgE
Heavy chain	γ_1	γ_2	γ_3	γ_4	μ	α_1	α_2	δ	ε
Molecular weight (kDa)	146	146	165	146	970	160	160	184	188
Serum level (mean adult mg ml^{-1})	9	3	1	0.5	1.5	3.0	0.5	0.03	5×10^{-5}
Half-life in serum (days)	21	20	7	21	10	6	6	3	2
Classical pathway of complement activation	++	+	+++	−	+++	−	−	−	−
Alternative pathway of complement activation	−	−	−	−	−	+	−	−	−
Placental transfer	+++	+	++	−/+	−	−	−	−	−
Binding to macrophage and phagocyte Fc receptors	+	−	+	−/+	−	+	+	−	+
High-affinity binding to mast cells and baseophils	−	−	−	−	−	−	−	−	+++
Reactivity with staphylococcal protein A	+	+	−/+	+	−	−	−	−	−

Figure 5 Properties of human immunoglobulin isotypes. The molecular mass of IgM corresponds to that of a pentamer. IgE is associated with immediate hypersensitivity. When attached to mast cells, it has a much higher half-life than in plasma. Copyright 2005 and permission as shown in legend for **Figure 3** (source = figure 4.17).

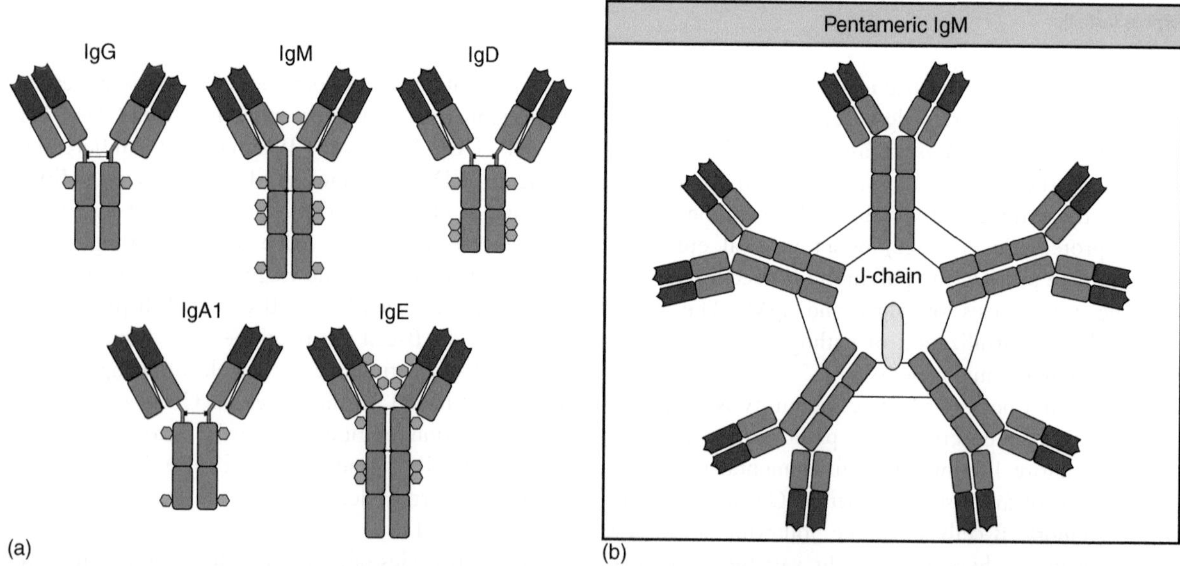

Figure 6 (a) Structural organization of human immunoglobulin isotype monomers. Both IgM and IgE lack a hinge region and each contains an additional heavy chain domain. Disulfide bonds linking the chains are indicated by black lines. N-linked glycans are shown as turquoise hexagons. (b) Pentameric IgM is associated with an additional polypeptide, the J-chain. The monomers are cross-linked by disulfide bonds to each other and to the J-chain. Copyright 2005 and permission as shown in legend for **Figure 3** (source figures 4.18 and 4.23).

Functional activity	IgM	IgD	IgG1	IgG2	IgG3	IgG4	IgA	IgE
Neutralization	+	–	++	++	++	++	++	–
Opsonization	+	–	+++	*	++	+	+	–
Sensitization for killing by NK cells	–	–	++	–	++	–	–	–
Sensitization of mast cells	–	–	+	–	+	–	–	+++
Activates complement system	+++	–	++	+	+++	–	+	–

Distribution	IgM	IgD	IgG1	IgG2	IgG3	IgG4	IgA	IgE
Transport across epithelium	+	–	–	–	–	–	+++ (dimer)	–
Transport across placenta	–	–	+++	+	++	+/–	–	–
Diffusion into extravascular sites	+/–	–	+++	+++	+++	+++	++ (monomer)	+
Mean serum level (mg ml^{-1})	1.5	0.04	9	3	1	0.5	2.1	3×10^{-5}

Figure 7 Functions and distribution of human immunoglobulin isotypes. Copyright 2005 and permission as shown in legend for **Figure 3** (source figure 9.19).

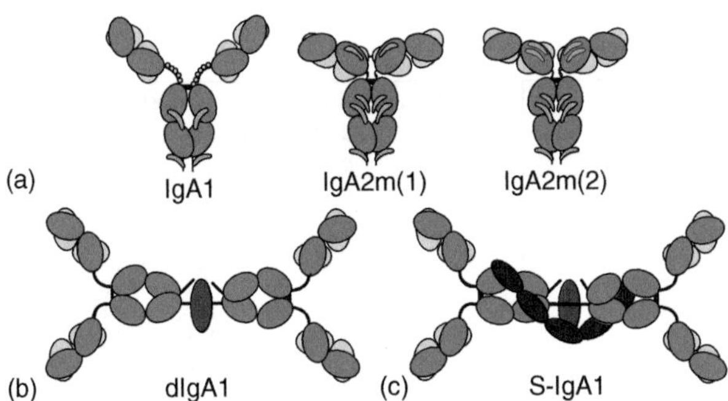

Figure 8 Structural organization of different molecular forms of human IgA. Heavy chains are shown in mid-blue, light chains in yellow, J-chain in red, and the secretory component (SC) in navy blue. (a) Monomeric mIgA. N-linked glycans are shown in orange, and O-linked glycans by small green circles. (b) Dimeric IgA1. (c) Dimeric secretory IgA1 (S-IgA1). For clarity, glycans are not shown for (b) and (c). Reprinted from Woof JM and Mestecky J (2005) Mucosal immunoglobulins. *Immunological Reviews* 206: 64–82, figure 1, with permission from Blackwell Publishing.

Biological Functions of Antiviral Antibodies

The surface of viruses is represented by a mosaic cluster of protein or glycoprotein subunits. The subunits correspond to a single or two or more species. The pattern of repetitiveness is usually a key factor responsible for the efficiency of early and rapid B-cell responses, potent IgM antibody production, and efficient downstream antibody class switching. However, the immune response is not restricted to antigenic sites (epitopes and neotopes) on the surface of viruses. Virus particles, following initial infection or provided as a vaccine, also separate into constituent parts. Consequently, unassembled surface subunits (their epitopes and cryptotopes) and internal virus components become exposed to the immune system, ultimately resulting in the production of antibodies having multiple specificities. Only some of these are directed against intact viruses, and may have virus-neutralizing properties. The formation of antibodies with distinct specificities may not be simultaneous but rather sequential. Especially in case of some not directly cytopathic viruses (hepatitis B and C, lymphocytic choriomeningitis, and HIV-1, prone to elicit a chronic carrier state), virus-neutralizing antibodies (VNAbs) appear with a delay after non-neutralizing antibodies. The latter may function as a 'decoy' if they are targeted like VNAb to virus surface epitopes.

Virus-Neutralizing and Protective Antibodies

VNAbs are crucial for protection against reinfection by a virus the VNAbs are specific for. Protection by efficacious

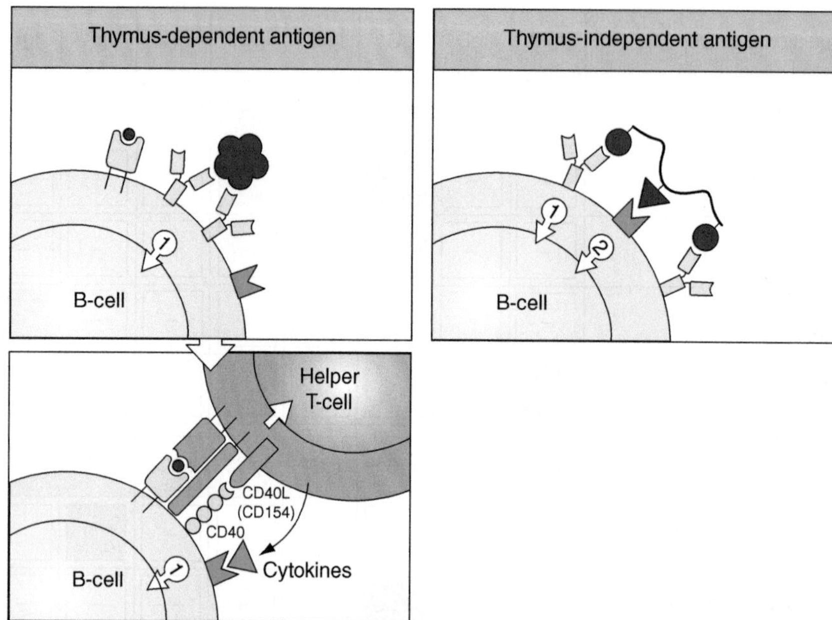

Figure 9 T-helper-cell-dependent initiation of the humoral immune response (two left panels). The first signal required for B-cell activation is delivered by binding of antigen (virus) (large red particle) to Ig cell receptors corresponding to monomeric IgM. Internalization and degradation of the antigen, and complex formation of the resulting peptide(s) (small red circle) with MHC class II molecules on the B-cell, allow the second signal to be delivered, that is, by the interaction between CD40 on the B-cell and the CD40 ligand (=CD154) on the CD4+ helper T-cell, and the engagement of the T-cell receptor (TCR) with the peptide-MHC class II complex on the B-cell. The activation is promoted by binding of cytokines to their specific receptors (see **Figure 10**). For comparison (right panel), in case of T-helper-cell-independent antigens, the second signal can be delivered by the antigen itself, either through binding of a part of the antigen to a receptor of the innate immune system (e.g., TLRs; green), or by extensive cross-linking of the membrane IgM by a polymeric antigen. Copyright 2005 and permission as shown in legend for **Figure 3** (source figure 9.2).

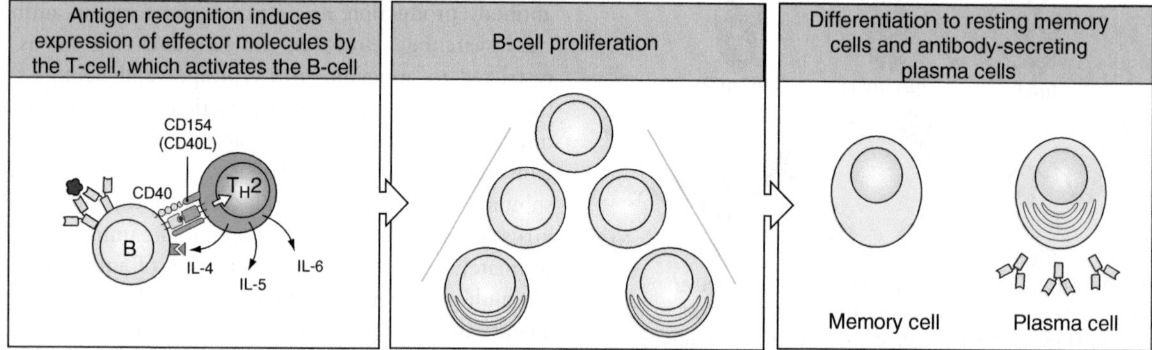

Figure 10 Antigen recognition induces the expression of B-cell stimulatory interleukins IL-4, IL-5, and IL-6 (and/or others) by the T-cell, driving the proliferation and differentiation of B-cells into antibody-secreting plasma cells. Activated B-cells can alternatively become memory B-cells. Copyright 2005 and permission as shown in legend for **Figure 3** (source figure 9.3).

vaccines correlates closely with *in vitro* determined VNAb titers of sera from immunized individuals. Protection by passive immunization relies on VNAb recognizing neutralization epitopes (or neotopes) on the virus surface. Coating of virus particles by antibodies is necessary but not always sufficient for virus neutralization. The effectiveness of virus neutralization correlates with the rate of antibody binding to critical epitopes and is augmented by slow dissociation of the formed antigen–antibody complexes. These kinetic parameters can be determined experimentally.

Spatial adaptive complementarity, electrostatic interactions, hydrogen bonds, and van der Waals forces contribute to the binding. Experimentally determined virus neutralization can depend on the target cells used. Virus neutralization is a multihit process and is successful when the number of unencumbered viral molecules, essential

Role of cytokines in regulating Ig isotype expression							
Cytokines	IgM	IgG3	IgG1	IgG2b	IgG2a	IgE	IgA
IL-4	Inhibits	Inhibits	Induces		Inhibits	Induces	
IL-5							Augments production
IFN-γ	Inhibits	Induces	Inhibits		Induces	Inhibits	
TGF-β	Inhibits	Inhibits		Induces			Induces

Figure 11 Role of cytokines in regulating Ig isotype switching. The individual cytokines either induce (violet) or inhibit (red) the production of particular Ig isotypes. IFN, interferon; TGF, transforming growth factor. Copyright 2005 and permission as shown in legend for **Figure 3** (source figure 9.7).

for initiation of the virus replicative cycle, is brought below a minimum threshold level. The mechanism of neutralization depends on processes obligatory for reproduction of a particular virus, and may involve the following steps: attachment to cell receptors; post-attachment events, internalization (endocytosis); fusion with cell membranes or endosomal vesicles; uncoating and/or intracellular localization; and enzymatic activities (e.g., transcription). Direct occupancy by VNAb of cell receptor binding sites on the virus surface might not be obligatory for neutralization. Steric hindrance or induction of deleterious conformational changes may be sufficiently effective. A unique feature of several anti-HIV-1 human monoclonal VNAb, having distinct specificities, is a very long finger-like third complementarity determining region of the immunoglobulin heavy chain allowing access to a recessed critical site on HIV-1 gp120. Such feature is rare in the human immunoglobulin repertoire but is common in the Ig's of Camelidae.

The mechanism and magnitude of VNAb neutralizing effects are influenced by their immunoglobulin isotype and subtype which affect interactions with complement, Fc receptors, and transcytosis through mucosal epithelia. Non-neutralizing virus-surface-binding antibodies sometimes enhance the effectiveness of VNAb, limit viral spread in the early phases of infection, and contribute to its suppression through antibody-mediated cellular cytotoxicity (ADCC). Antibodies directed to epitopes (or neotopes) on distinct surface proteins may act synergistically in virus neutralization.

The principle that VNAbs specific for virus surface components provide protection against disease is not absolute. The flavivirus nonstructural protein NS1 elicits a protective immune response against yellow fever, dengue, and tick-borne encephalitis viruses. The paratope binding site containing F(ab′)2 fragments are ineffective. Thus the immunoglobulin Fc portion is obligatory for the protective effect.

Antibody-Dependent Enhancement of Viral Diseases

Some viruses make use of antiviral antibodies to gain entry into target cells thus widening cell receptor usage to initiate infection. The infectious virus–antibody complexes rely upon the Fc portion of IgG antibodies to gain entry into monocytes/macrophages and granulocyte through Fc (FcR) or complement receptors (CR) expressed on these cells.The characteristic feature of the viruses is their propensity to establish persistent infections and their antigenic diversity. Antibody-dependent enhancement (ADE) has been demonstrated to occur *in vitro* with members of the families *Bunyaviridae*, *Coronaviridae*, *Flaviviridae*, *Orthomyxoviridae*, *Paramyxoviridae*, *Retroviridae*, *Rhabdoviridae*, and *Togaviridae*. A link between *in vitro* ADE and clinical manifestations cannot be always established. A relationship between ADE and disease exacerbation has been observed for dengue, measles, yellow fever, and respiratory syncytial viruses (RSVs). ADE may occur in children infected at a time when the level of transferred maternal antiviral antibodies declines to insufficient levels. ADE could represent an obstacle for development of vaccines, as has been the case for anti-RSV vaccines. A vaccine consisting of formaldehyde-treated measles virus hemagglutinin (ineffective to elicit antibodies to the virus fusion protein) induced antibodies causing ADE and led to aggravated atypical disease following infection with measles virus.

FcR- and CR-independent ADE was shown to occur following binding to HIV-1 of antibodies eliciting conformational changes in the gp120 envelope glycoprotein, allowing direct virus binding to cellular co-receptors while bypassing the primary binding to the primary CD4 cell receptor.

Immune System Evasion by Viruses

Persistence in an infected host and repeated reactivation of many viruses rely on several specific evasion strategies of adaptive immunity. Thus common protective responses are redirected or altered to the advantage of the infectious agent. This includes antiviral antibody responses and involves (1) specific paratope–epitope interactions (Fab fragments) and (2) effector mechanisms mediated by the Fc portion of antibodies.

The first mechanism is provided by genetically determined amino acid replacements leading to changes of virus epitopes (or neotopes) involved in virus neutralization. The sites of these escape mutations are usually on the viral surface, result in structural changes in antibody/virus contact sites, and lead to much less favorable kinetic parameters for antibody binding or completely abrogate binding. Presentation of new glycan chains on enveloped viruses or elimination of these chains due to mutations of N-glycosylation sites may cause substantial epitope alterations. The rate of escape mutation appearance is promoted by error-prone replication of the viral genome. Antibodies with new specificities must be produced to bring the mutant viruses under control. The process is repeated, and if not successful, persistent infection is established. A similar process leads to evasion from T-cell-mediated protective responses.

Secondly, several viruses bypass clearance processes facilitated by the Fc portion of bound antibodies by encoding and expressing Fc receptor analogs. Subversion of the complement cascade provides another way how to block clearance of cell-free virus and infected cells.

Additional scenarios for escaping immune surveillance include: interference with MHC class I restricted antigen presentation involving also inhibition of MHC class I cell surface expression or synthesis of viral MHC class I homologs; blocking MHC class II restricted antigen presentation; downregulation of cellular CD4 or its degradation; interference with cytokine effector functions; and other strategies.

Immunoglobulins for Passive Immunization against Human Viruses

Transfer of immunity from an immune donor to an unprotected recipient by serum is one of the early landmarks in the history of immunology (Von Behring). The active serum components have later been identified as immunoglobulins. The half-life of immunoglobulin isotypes in serum is 6–21 days (**Figure 5**). Consequently, administered antivirus immunoglobulins can provide only short term prophylactic and therapeutic benefits, respectively, unless they are administered repeatedly or incorporated into a slow-release medical device. The most common applications are pre-exposure (travel, protection against community-wide infection(s), medical professionals, immunosuppressed individuals, combination with live vaccines to minimize their potential side effects) and post-exposure prophylaxis (passive immunization may provide immediate protection while the benefits from vaccination are delayed). Local mucosal applications of immunoglobulins 'as needed' appear promising against perinatal and sexual transmission, respectively, of several viruses (e.g., herpesviruses and HIV-1).

The immunoglobulins are isolated from serum of individuals pre-screened for high levels of antibodies against a particular virus or from vaccinated individuals. The immunoglobulins are purified, treated to remove or inactivate infectious agents which might be present in the pooled serum source, notwithstanding rigorous screening of the individual sera entering the pool. The products are further processed depending on their intended intramuscular or intravenous applications. All these immunoglobulins are polyclonal with respect to the 'indicated' virus and contain other antibodies originally present in the pooled sera. Oral administration of antibodies produced in bovine colostrums or chicken yolk has been recently suggested.

Alternatively, monoclonal antibodies (mAbs) specific for epitopes, known to elicit virus-neutralizing and protective immune responses, are used. They are prepared using hybridoma technologies, immortalized human peripheral B-cells, transgenic mice and bacteriophage expression libraries. If derived from animal species, the mAbs are 'humanized' using recombinant DNA techniques by replacing amino acid sequences outside the antigen-binding sites with sequences corresponding to human immunoglobulins. By an *in vitro* directed evolution process allowing manipulation of antigen-binding kinetics, mAb variants having much faster antigen association rates and much slower dissociation rates can be produced. Such antibodies have a much improved capacity to neutralize the target virus and may have a great clinical potential. Their production in high yield in plants offers an economically advantageous approach applicable to both IgG and secretory IgA antibodies.

While polyclonal antibodies from human sera provide immunological diversity, mAbs are highly specific for a single virus epitope. Potential alterations of such epitopes may generate virus neutralization escape mutants and decrease or eliminate the effectiveness of mAb prophylactics/therapeutics. Polyclonal antibody preparations prepared from pooled sera are not uniform and vary with the source of serum pools.

This problem can be overcome by the development of human recombinant antigen-specific polyclonal antibodies by a novel Sympress technology (Symphogen, Lyngby, Denmark).

Immunoglobulins, either already in clinical use or in development, are directed against one of the following viruses: hepatitis A and B; cytomegalovirus; rabies; respiratory syncytial virus; smallpox; vaccinia; varicella zoster; measles; mumps; rubella; parvovirus B19; Epstein–Barr virus; herpes simplex; tick-borne encephalitis; poliovirus; Hantavirus; West Nile virus; rotavirus; poliovirus; HIV-1; Ebola virus; and severe acute respiratory syndrome-associated coronavirus.

Unlike vaccines, passive immunization can rapidly deliver protective levels of antibodies directly to susceptible mucosal sites where many virus infections are initiated. Secretory IgA because of its polyvalency and relative stability may have advantages over IgG for passive immunization at these sites.

Antibodies function also as immunomodulators which can bridge innate and acquired, and cellular and humoral immune responses, respectively. Infected host cells can be targeted by linking anticellular toxins to antiviral antibodies or by bispecific antibodies in which one Fab fragment of the antibody is virus specific and the other one recognizes a host cell component or receptor.

Research on antibody-mediated immunity and understanding of antibody-based prophylaxis and therapies for virus diseases have provided a foundation for research on and development of antivirus vaccines.

Vaccines against Human Viral Diseases

Vaccination is the most successful medical intervention against viral diseases. Vaccines prevent or moderate illnesses caused by virus infection in an individual and prevent or diminish virus transmission to other susceptible persons, thus contributing to herd immunity. This effect is expected to be long term, depends on establishment of immunological memory at both the B- and T-cell levels, and may require consecutive or repeated vaccinations. The effectiveness of vaccination might be diminished or compromised for viruses occurring in the form of simultaneous quasi-species or undergoing time-dependent changes of antigenic properties (antigenic drift and antigenic shift). However, in some cases, vaccination predisposes to aggravated disease elicited by infection with a virus identical or related to that used for vaccination, that is, antibody enhancement of virus infection occurs (dengue and respiratory syncytial viruses and HIV-1).

The following categories of vaccines can be distinguished (**Table 2**): (1) live attenuated; (2) whole virus (inactivated); (3) glycoprotein subunit vaccines; and (4) protein vaccines based on recombinant DNA technologies. One vaccine in category (3), hepatitis B surface antigen (HBsAg), is derived from plasma of hepatitis B virus carriers. It is remarkable that vaccines in categories (3) and (4) correspond to multi-subunit self-assembled particles having antigenic specificities closely similar or identical to those expressed on the surface of virus particles. On the other hand, individual virus protein or glycoprotein subunit molecules or their peptide fragment have been less suitable candidates for vaccine development because of insufficient similarities with intact viruses.

Table 2 Past and present vaccines against human viral diseases

Live attenuated	*Killed whole virus*	*Glycoprotein subunit*	*Genetically engineered*
Influenza (nasal)	Influenza	Influenza	Papillomavirus
Rabies	Rabies	Hepatitis B	Hepatitis B
Poliovirus	Poliovirus		
Yellow fever	Japanese encephalitis		
Measles	Tick-borne encephalitis		
Mumps	Hepatitis A		
Rubella			
Adenovirus			
Varicella zoster			
Rotavirus			
Smallpox (vaccinia)			

Vaccine formulations require the incorporation of an immunological adjuvant to enhance their immunogenicity. Adjuvants are designed to optimize antigen delivery and presentation, enhance the maturation of antigen-presenting dendritic cells, and induce immunomodulatory cytokines.

Currently, most vaccines are administered parenterally using syringes with needles.The procedure is disliked by many, and is questionable for mass vaccination programs in developing countries. Therefore, efforts are being made to produce vaccines which can be delivered onto mucosal surfaces, that is, mostly orally or nasally. In addition, high-workload needle-free injection devices are being developed. At this juncture, the bifurcated needle, developed by Benjamin Rubin over 40 years ago, must be mentioned. It proved to be essential for the successful campaign to eradicate smallpox worldwide.

Veterinary Vaccines

The development and use of veterinary vaccines has the following aims: cost-effective prevention and control of virus diseases in animals; induce herd immunity; improve animal welfare and food production for human

Table 3 Current veterinary vaccines

Species	*Live*	*Killed*	*Recombinant*	*DNA*
Avian	Encephalomyelitis		Encephalomyelitis (fowl pox vector)	
	Influenza		Influenza (fowl pox vector)	
	Pneumovirus	Paramyxovirus		
		Polyomavirus		
		Reovirus		
	Bursal disease	Bursal disease	Bursal disease (Marek's disease vector)	
	Marek's disease		Marek's disease (Marek's disease vector)	
	Fowl pox		Fowl pox	
	Newcastle disease	Newcastle disease	Newcastle disease (fowl pox vector)	
		Bronchitis		
	Anemia			
	Laryngotracheitis		Laryngotracheitis (fowl pox vector)	
	Duck enteritis			
	Duck hepatitis			
	Canary pox			
Feline	Calicivirus	Calicivirus immunodeficiency virus		
		Leukemia virus	Leukemia virus (canarypox vector)	
	Infectious peritonitis			
	Rhinotracheitis	Rhinotracheitis		
	Panleukopenia	Panleukopenia		
		Rabies	Rabies (canary pox vector)	
Canine	Adenovirus 2			
	Parvovirus	Parvovirus		
	Coronavirus	Coronavirus		
	Parainfluenza		Parainfluena (canary pox vector)	
	Canine distemper		Canine distemper (canary pox vector)	
	Measles			
	Hepatitis			
	Rabies	Rabies		
Sheep and goat	Bluetongue			
	Ovine ecthyma			
	Poxviruses			
		Louping ill		
Equine	Influenza	Influenza	Influenza (canary pox vector)	
	Rhinopneumonitis	Rhinopneumonitis		
		Rotavirus		
	Arteritis			
	African horse sickness	African horse sickness		
		Encephalomyelitis		
	West Nile virus	West Nile virus	West Nile virus (canary pox vector)	West Nile virus
	Flavivirus chimera			
Bovine	Respiratory syncytial virus	Respiratory syncytial virus		
	Rhinotracheitis	Rhinotracheitis		
	Diarrhea	Diarrhea		
	Bronchitis	Bronchitis		
	Parainfluenza 3			
	Rotavirus	Rotavirus		
	Coronavirus	Coronavirus		
	Herpes 1	Herpes 1		
	Foot-and-mouth disease	Foot-and-mouth disease		
	Rinderpest			

Continued

Table 3 Continued

Species	Live	Killed	Recombinant	DNA
Porcine	Pseudorabies	Pseudorabies		
	Enterovirus	Influenza		
	Parvovirus	Circovirus		
	Rotavirus	Rotavirus		
	Transmissible gastroenteritis	Transmissible gastroenteritis		
	Reproductive and respiratory syndrome	Reproductive and respiratory Syndrome		
	Hog cholera	Hog cholera		

Several of the described vaccines are being administered as combination vaccines.

consumption; decrease the usage of veterinary drugs, thereby minimizing their environmental impact and food contamination; and decrease the incidence of zoonoses (e.g., infections by avian influenza, rabies, West Nile, Rift Valley fever viruses, respectively).

Research, development, and production of some veterinary vaccines have been on the forefront of the general field of vaccinology. The foot-and-mouth disease virus vaccine was the first one produced at an industrial scale (Frenkel method). A vaccinia-rabies virus G protein recombinant vaccine was among the first biotechnology-based vaccines licensed. The world's first DNA vaccine (against West Nile virus in horses) was approved by the US Department of Agriculture in July 2005. DIVA (Differentiating Infected from Vaccinated Animals; also termed marker) veterinary recombinant vaccines and companion diagnostic tests have been developed. They can be applied to programs to control and eradicate virus infections. These are examples to be considered for the development of human vaccines. The latter will require rigorous evaluations for safety and efficacy which are more difficult to obtain than in veterinary settings. A list of licensed veterinary vaccines is shown in **Table 3**.

See also: Antigenicity and Immunogenicity of Viral Proteins; Antigen Presentation; Antigenic Variation; Cell-Mediated Immunity to Viruses; Cytokines and Chemokines; Diagnosis Using Serological and Molecular Approaches; DNA Vaccines; Neutralization of Infectivity; Vaccine Strategies.

Further Reading

Ahmed R (ed.) (2006) Immunological memory. *Immunological Reviews* 211, 5–337.

Burton DR (ed.) (2001) Antibodies in viral infection. *Current Topics in Microbiology and Immunology* 260, 1–300.

Casadevall A, Dadachova E, and Pirofsky LA (2004) Passive antibody therapy for infectious diseases. *Nature Reviews Microbiology* 2: 695–703.

Cooper MD and Alder N (2006) The evolution of adaptive immune systems. *Cell* 124: 815–822.

Frank SA (2002) *Immunology and Evolution of Infectious Disease.* Princeton: Princeton University Press.

Hangartner L, Zinkernagel RM, and Hengartner H (2006) Antiviral antibody responses: The two extremes of a wide spectrum. *Nature Reviews Immunology* 6: 231–243.

Janeway CA, Jr. (1993) How the immune system recognizes invaders. *Scientific American* (Sep.): 73–79.

Janeway CA, Jr., Travers P, Walport M, and Shlomchik MJ (2005) *Immunobiology. The Immune System in Health and Disease,* 6th edn. New York: Garland Science.

Levine MP, Kaper JB, Rappuoli R, Liu M and Good MF (eds.) (2004) *New Generation Vaccines* 3rd edn. New York: Dekker.

O'Neill LA (2005) Immunity's early-warning system. *Scientific American* (Jan.): 38–45.

Paul WE (ed.) (2003) *Fundamental Immunology* 5th edn. Philadelphia: Lippincott Williams and Wilkins.

Plotkin SA and Orenstein WA (2004) *Vaccines,* 4th edn. Philadelphia: Saunders–Elsevier.

Pulendran B and Ahmed R (2006) Translating innate immunity into immunological memory: Implications for vaccine development. *Cell* 124: 849–863.

Van Regenmortel MHV (2002) Reductionism and the search for structure–function relationships in antibody molecules. *Journal of Molecular Recognition* 15: 240–247.

Weissman IL and Cooper MD (1993) How the immune system develops. *Scientific American* (Sep.): 65–71.

Woof JM and Mestecky J (2005) Mucosal immunoglobulins. *Immunological Reviews* 206: 64–82.

Orthomyxovirus Antigen Structure

R J Russell, University of St. Andrews, St. Andrews, UK

Glossary

Hemagglutinin The surface glycoprotein responsible for receptor binding and cell fusion.

Influenza The prototypical member of the family *Orthomyxoviridae* of viruses.

Neuraminidase The surface glycoprotein necessary for viral release.

Introduction

The family *Orthomyxoviridae* is defined by viruses that have a negative-sense, single-stranded, and segmented RNA genome. There are five different genera in the family: *Influenzavirus A, B,* and *C, Thogotovirus,* and *Isavirus.*

Influenza viruses are classified into three types: A, B, and C. Type A influenza viruses are further classified into subtypes based on the antigenic properties of their surface antigens, hemagglutinin (HA) and neuraminidase (NA). There are 16 different HA subtypes (H1–H16) and nine different NA subtypes (N1–N9) of influenza A. Viruses containing HAs from each of the 16 HA subtypes and NAs of the nine NA subtypes, in different combinations, have been isolated from avian species. Human viruses that caused pandemics in the last century were of the combination H1N1 in 1918, H2N2 in 1957, and H3N2 in 1968. Viruses that are currently circulating are of the H1N1, H3N2 combination and B. Influenza viruses also infect a range of other animals including swine and equine, with H1N2 and H3N2 currently infecting the former and H3N8 the latter.

Different influenza virus strains are named according to their type, the species from which the virus was isolated (omitted if human), place of virus isolation, the number of the isolate and the year of isolation, and in the case of the influenza A viruses, the HA and NA subtypes. For example, the 282nd isolate of an H5N1 subtype virus from chickens in Hong Kong in 2006 is designated: influenza A virus (A/chicken/Hong Kong/282/2006(H5N1)).

The genus *Thogotovirus* contains two different species, *Dhori virus* and *Thogoto virus.* Viruses from both these species were isolated from ticks, and therefore are different from influenza viruses with respect to their host range. The genus *Isavirus,* with the type species *Infectious salmon anemia virus,* is also distinct from influenza viruses A, B, and C, although numerous studies identify these isolates as members of the family *Orthomyxoviridae.*

The majority of the research into the family *Orthomyxoviridae* of viruses has been concerned with influenza viruses and therefore this article will focus exclusively on the structures of the influenza viral antigens.

Influenza Virus Surface Glycoproteins

The surface of type A and B influenza viruses has two major surface glycoproteins, HA and NA, whereas influenza C viruses have a single glycoprotein, hemagglutinin–esterase-fusion (HEF). Antibodies are raised to both HA and NA, although HA elicits the dominant antigenic response, and it is the natural variation in amino acid sequence (antigenic drift) and subsequent structural changes that limits the long-term effectiveness of an anti-influenza vaccine. Consequently, annual updates to the vaccine composition are required to maintain a match to the viruses that are currently circulating.

Influenza A and B Hemagglutinin

In type A and B influenza viruses, HA mediates attachment of the virus to the host cell via binding to sialic acid residues on the termini of cell surface glycoproteins. HA is also responsible for the fusion of viral and host cell membranes.

HA is synthesized as a single-chain polypeptide (HA0) of approximately 550 amino acids, which is subsequently proteolytically cleaved into two chains, HA1 and HA2. An interchain disulfide bond exists between cys-14 (HA1) and cys-137 (HA2). The monomers then associate non-covalently to form the functional trimer. The highly conserved and hydrophobic N-terminal residues of HA2, known as the fusion peptides (see below), are buried from solvent upon trimerization. It is known that these residues insert into the host membrane to facilitate virus membrane–cell membrane fusion. The C-terminal end of HA2 has a transmembrane anchor that tethers the molecule to the viral membrane.

HA is a highly N-glycosylated protein. For example, the H1 HA from the virus that caused the pandemic of 1918 has five glycosylation sites in HA1 and one in HA2. The number of glycosylation sites varies between virus species and has a marked impact on the antigenicity of the molecule (see below). H1 HA isolated from viruses circulating in 2002 had accumulated an additional six potential glycosylation sites with respect to the 1918 H1 HA. In addition to their influence on the antigenicity of HA, glycosylation sites that are conserved in all 16 HAs play a role in the co-translational folding of HA via binding to host cell chaperones.

To facilitate structural studies of HA, the protein is proteolytically cleaved from the surface of the virus, which removes the hydrophobic transmembrane anchor from the molecule. To date, no structure has been elucidated of a full-length HA.

The first crystal structure of HA to be determined was from the A/Aichi/2/68 (H3N2) virus, and showed the protein to be composed of a membrane-proximal, triple-stranded α-helical stem-like structure, that is composed of residues predominantly from HA2, that supports a membrane-distal globular multidomain structure, that is composed of residues solely from HA1 (**Figure 1(a)**). The membrane-distal part of each monomer can be subdivided into a vestigial esterase domain and a receptor-binding domain, the latter located at the very tip of each monomer. The receptor-binding domain consists of three secondary structure elements – the 190 helix (residues 190–198), the 130 loop (residues 135–138), and the 220 loop (residues 221–228) – that form the sides of each site, with the base made up of the conserved residues Tyr-98, Trp-153, His-183, and Tyr-195.

HA mediates the first stage of virus infection via binding to terminal sialic acid sugars on host glycoproteins and glycolipids. Sialic acids are usually found in either α-2,3- or α-2,6-linkages to galactose, the predominant penultimate sugar of N-linked carbohydrate side chains. The binding preference of an HA for one of these linkage types correlates with the species that the virus infects. The avian enteric tract has predominantly sialic acid in the α-2,3-linkage and all HAs found in the 16 antigenic subtypes found in avian influenza viruses bind preferentially to this linkage. In cells of the human upper respiratory tract, however, sialic acid in the α-2,6-linkage predominates. Indeed, HAs from human influenza viruses display a binding preference for sialic acid in the α-2,6-linkage. An avian origin has been proposed for human influenza viruses and therefore a change in binding specificity is required for cross-species transfer. The precise details of the amino acid changes underlying this switch in specificity seem to be HA subtype dependent but it has been shown that for both H1 and H3 HAs, only a small number of mutations are necessary. For example, a single mutation in H3 HA of Gln-226 to Leu-226 causes a switch in specificity from α-2,3- to α-2,6-binding preference. Insights into the amino acids involved in avian and human receptor binding have also been gained from elucidating the structure of HA in complex with receptor analogs.

The structure of HA0 has also been elucidated via crystallization of a protein in which Arg-329 had been

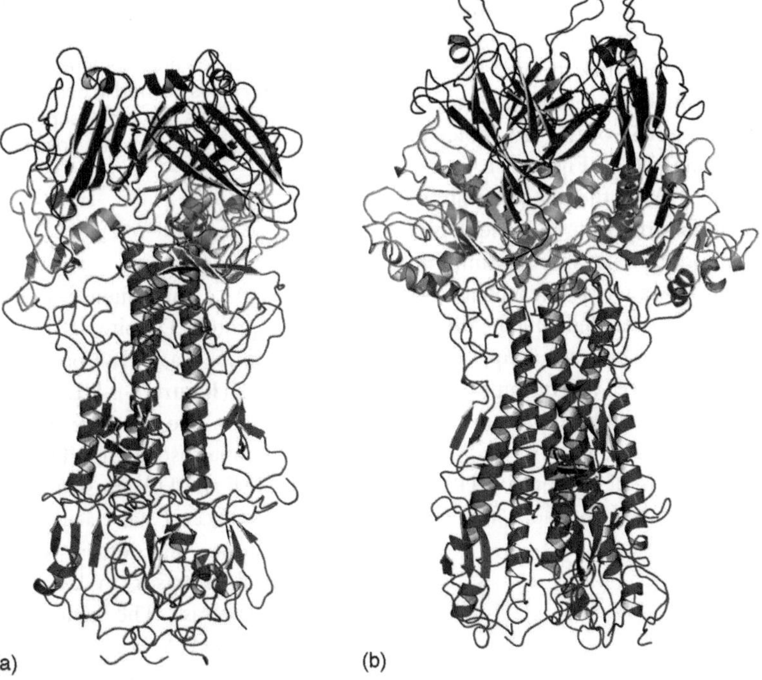

Figure 1 (a) Cartoon representation of influenza A HA. The receptor-binding domain of HA1 is colored blue and the vestigial esterase domain green. HA2, which forms the central α-helical stem, is colored red. (b) Cartoon representation of influenza C HEF. The receptor-binding domain of HEF1 is colored blue and the esterase domain green. HEF2, which forms the central α-helical stem, is colored red.

mutated to a glutamine to prevent proteolytic cleavage of the polypeptide. The structure revealed that the cleavage site is located in a prominent surface loop which lies adjacent to a cavity which is not present in the cleaved-HA structure. There are three ionizable residues in this cavity which are buried from solvent after cleavage due to the burial of the nearly formed N-terminus of HA2 (fusion peptide). Thus it has been proposed that cleavage of HA0 results in a metastable form of the protein in which a low-pH trigger has been set due to the burial of ionizable residues. A characteristic of highly pathogenic influenza viruses such as the currently circulating H5N1 viruses is an insertion of a series of basic residues adjacent to the cleavage site. This polybasic insertion would create a much more extended surface loop which would facilitate enhanced intracellular cleavage.

In addition to sialic acid receptor binding, HA mediates virus membrane–cell membrane fusion. Subsequent to receptor binding, the influenza virus is internalized via endocytosis, exposing the virus to a low-pH environment which triggers a dramatic and irreversible conformational change in HA. Exposure of HA to low pH causes an alteration in the protonation state of residues that are buried by the fusion peptide, resulting in its removal from its buried site. Subsequently, the middle of the original long α-helix of HA2 unfolds to form a reverse turn, jack-knifing the C-terminal half backward toward the N-terminus, and results in HA2 adopting a rod-like coiled-coil structure. HA1, which was missing from the crystal structure, de-trimerizes and swings away from the HA2 but the interchain disulfide bond is maintained. As a consequence of these molecular rearrangements, the N-terminal fusion peptide, which inserts into the host membrane, and the C-terminal transmembrane anchor of HA2 are placed at the same end of the HA molecule, thereby facilitating membrane fusion (**Figure 2**).

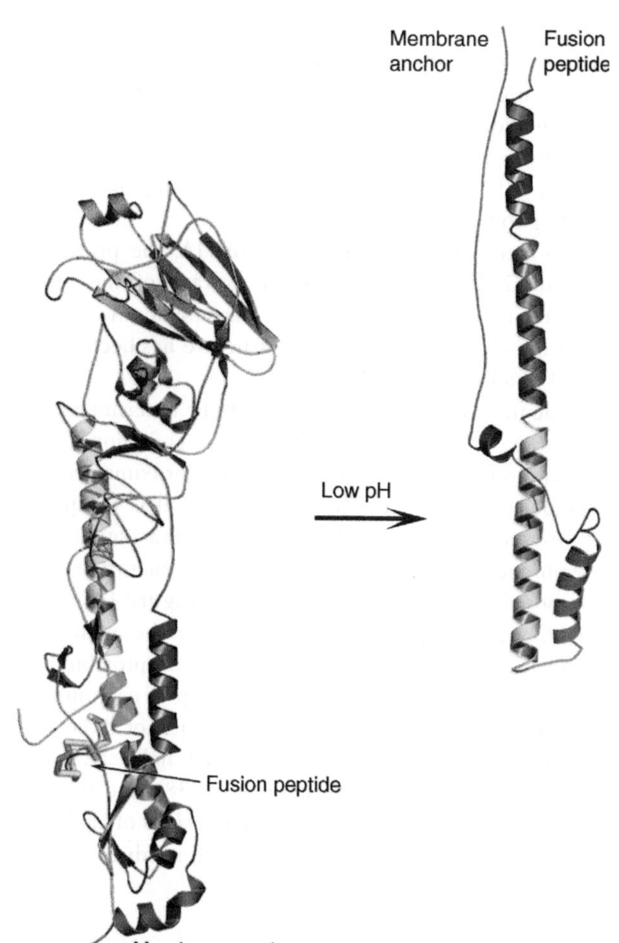

Figure 2 Schematic representation of the molecular rearrangements of HA upon exposure to low pH. The location of the fusion peptide and membrane anchor is shown in both structures. HA1 is colored in gold and HA2 is colored to highlight equivalent regions of the polypeptide. Note that only the yellow part of HA2 adopts the same position and conformation in both forms.

The structure adopted by HA2 at the pH of membrane fusion shares a number of features with equivalent parts of the membrane fusion proteins of human immunodeficiency virus (HIV) and Ebola virus. In each case, the molecules contain central triple-stranded coiled coils that are surrounded by three α-helices that pack antiparallel to the core helices. As a consequence, the fusion peptides and membrane anchor regions of all of the proteins are at the same end of a rod-shaped molecule.

At present, no crystal structure of influenza B HA has been elucidated but it is expected to adopt the same conformation as influenza A HAs.

Influenza C HEF

In contrast to the HA of influenza A and B viruses, the major glycoprotein, HEF, of the C viruses has a receptor-destroying activity, as well as receptor-binding and fusion activities. As a consequence, influenza C viruses lack the NA glycoprotein. Despite sharing only 12% sequence identity with influenza A HAs, both HEF and HA are structurally similar (**Figure 1**). HEF has a membrane-proximal, α-helical stem-like structure and a membrane-distal globular multidomain structure. The receptor-binding domain of both HEF and HA are structurally similar despite HEF binding 9-*O*-acetyl sialic acid rather than sialic acid. The receptor-destroying domain shows structural homology to bacterial esterases, in keeping with its activity as a 9-*O*-acetylesterase. The stem region of HEF is similar to that observed in HA except that the triple-stranded, α-helical bundle diverges at both its ends and that the fusion peptide is partially exposed to solvent. The receptor-binding domain is inserted into a surface loop of the esterase domain which itself is inserted into a surface loop of the stem. Thus all three functions of the HEF are segregated into structurally distinct domains.

Influenza A and B Neuraminidase

After virus replication, NA removes sialic acid from virus and cellular glycoproteins to facilitate virus release and the spread of infection to new cells, otherwise viral aggregation would occur. Thus NA is commonly known as a receptor-destroying enzyme.

NA is synthesized as a single polypeptide chain but unlike HA no post-translational cleavage occurs. NAs have a highly conserved short cytoplasmic tail and a hydrophobic transmembrane region that provides the anchor for the stalk and the head domains. Like HA, NA is a highly N-glycosylated protein and serves to influence the antigenicity of NA.

NA suitable for structural studies is made via proteolytic cleavage from the virus surface which removes the membrane anchor and stalk domain. A deletion in the stalk domain is characteristic of highly pathogenic influenza viruses but as yet no structural information is available for this domain of NA.

The first crystal structure of the head domain of NA was of a N2 subtype, and showed that the molecule was homotetrameric with circular fourfold symmetry. Each monomer is composed of six topologically identical four-stranded antiparallel β-sheets that are themselves arranged like the blades of a propeller (**Figure 3**).

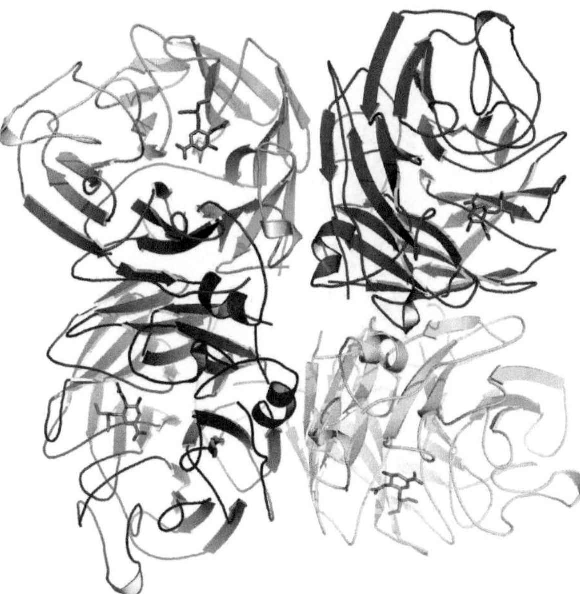

Figure 3 Cartoon representation of influenza A NA. Each monomer of the tetramer is colored differently, with the monomer at the top-left colored to emphasize the six-bladed β-propeller structure. Oseltamivir is bound to the active site and shown in stick representation and colored blue.

Sialic acid, the product of catalysis, binds in a deep pocket on the surface of the molecule, roughly in the middle of each monomer. No conformational changes occur upon binding of sialic acid. The amino acids in the active site that interact with sialic acid are highly conserved across all NA subtypes. The active sites of all influenza NAs contain three arginine residues, Arg 118, Arg 292, and Arg 371, that bind the carboxylate of the substrate sialic acid; Arg 152 that interacts with the acetamido substituent of the substrate; and Glu 276 that forms hydrogen bonds with the 8- and 9-hydroxyl groups of the substrate.

These properties of the active site made NA an attractive target for structure-based drug design programs, and resulted in the synthesis of oseltamivir (Tamiflu) and zanamivr (Relenza), two clinically licensed anti-influenza drugs.

The crystal structure of influenza B NA has also been elucidated and is essentially identical to influenza A NA, with the structure having the same homotetrameric arrangement and a high degree of conservation of the active site.

Influenza A Subtype Variation in Antigen Structure

To date, the crystal structures of H1, H3, H5, H7, and H9 HAs have been elucidated, which represent the four phylogenetic clades of influenza A HAs. Sequence identity between clades is between 40% and 60%. All of the HAs have the same molecular fold but HAs representative of each clade are distinguished by differences in the orientation of their membrane distal globular domains relative to the central trimeric coiled coil. For example, the receptor-binding domain of H3 HA is rotated clockwise by 24° relative to H1 HA. Also, the conformation of the interhelical loop of HA2 is different in HA from each clade. The clade-specific differences between HAs are clustered in regions that undergo conformational changes in membrane fusion, for example near to the fusion peptide, and it has been suggested that differences in the stability of HAs to pH and temperature may represent selection pressures in the evolution of influenza A HAs.

Influenza A NAs form two distinct phylogenetic groups. Crystal structures of N1, N4, and N8 have been elucidated as representatives of group 1 and N2 and N9 as representatives of group 2. All NAs have the same homotetrameric conformation but group-specific differences occur in the active site. Relative to group 2 NAs, a large cavity exists in group 1 NAs adjacent to the 4-amino group of oseltamivir. Upon binding of inhibitors to group 1 NAs, a conformational change of the 150-loop occurs which results in the active site of the two groups of NA being essentially identical. Chemical exploitation of this cavity is currently being undertaken in the development of a new generation of anti-influenza drugs to address the problem of mutations in NA that cause resistance to oseltamivir and zanamivir.

Antigenic Variation and Antibody Binding

Influenza viruses undergo a process known as 'antigenic drift' whereby natural variants of HA and NA occur, due to the error-prone nature of the viral RNA polymerase. HA is the major molecule recognized by the adaptive immune system of the host, although NA does elicit an immune response. Upon infection of a cell, the virus causes an immune response that commonly results in the production of neutralizing antibodies, in the case of HA, which acts to prevent the virus from binding to cells. Antibodies against NA, while they do not inhibit virus entry, prevent spread of the virus and afford some protection against challenge with the same or similar virus.

In the last century, three viruses were introduced into the human population from the avian reservoir of influenza viruses, and the subsequent pandemics caused millions of deaths. Following them, immune pressure causes antigenic variation that has the potential to continue until the fitness of the virus is compromised via deleterious mutations in HA which affects its receptor-binding properties.

Examination of the positions of amino acid substitutions of natural variants of HAs show that they are scattered throughout the molecule. Of the changes that are retained by circulating viruses, so-called 'fixed' substitutions, the majority are located on the surface of the molecule. Whereas, in the case of H3 HA subtype virus isolated between 1968 and 2005, two-thirds of the residues that are not retained are buried. This suggests that the 'fixed' substitutions have been selected because they alter the local structure of HA and prevent antibody binding.

This is supported by the fact that the location of amino acid substitutions in antigenic variants of HAs that have been selected by growing virus in the presence of anti-HA monoclonal antibodies (mAbs) map to the same location as the 'fixed' substitutions. Thus sites of mAb-selected mutations indicate the sites at which selecting antibodies bind. These mutants escape neutralization by the selecting antibody. The antibodies nevertheless still bind but with a much reduced affinity. There appears to be no preference in the type of amino acid change that occurs in selected variants.

Prevention of antibody recognition frequently occurs through the introduction of new glycosylation sites in HA. Oligosaccharide attachment at these sites prevents recognition first by covering of the protein surface and thus no antibodies being selected and second, because the sugars in the oligosaccharides are made by cellular enzymes, they are deemed by the immune system to be antigenically 'self'.

Sites of amino acid substitutions that occur in natural and antibody-selected variants are predominantly on the

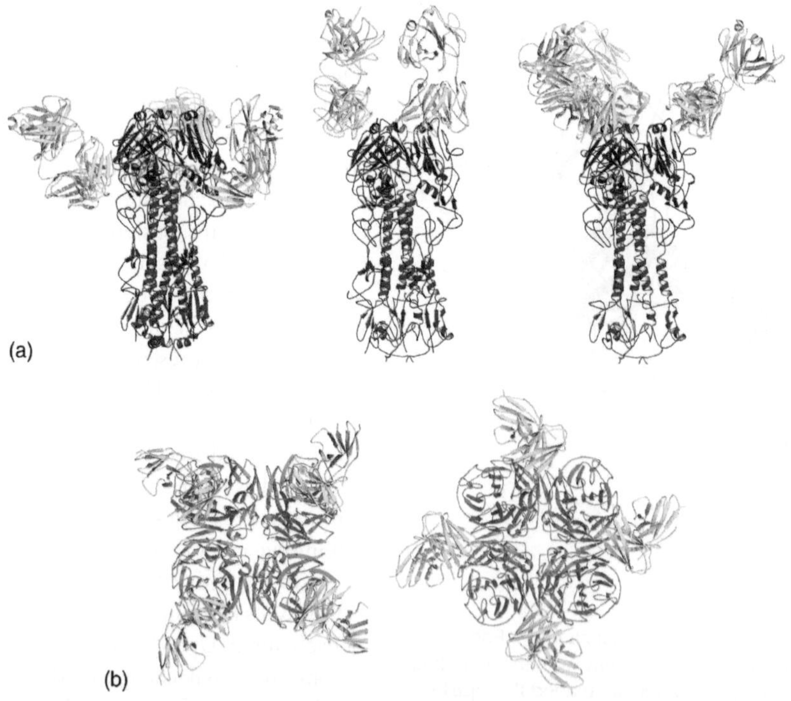

Figure 4 Antibody binding to influenza antigens. (a) Three different crystal structures of influenza HA bound to Fab fragments. (b) Two different crystal structures of influenza NA bound to Fab fragments.

surface of the membrane distal part of HA, and commonly surround the receptor-binding site. This suggests that there is a link between the neutralization of viral infectivity and the prevention of virus binding to cells.

Crystal structures have been elucidated for HA and NA in complex with antibodies (**Figure 4**), and has shown that fragment antigen binding (Fab) antibody fragments do indeed bind to different parts of the molecules, supporting the observation that natural variants are scattered throughout the molecule. It is interesting to note that the stoichiometry of Fab binding is not always three molecules of Fab to one trimeric spike of HA. Of the three anti-HA antibodies that have been studied structurally, two of the binding sites overlap. All three of the antibodies have been shown to neutralize infectivity by the prevention of virus binding to cells. One of the antibodies, however, also blocks the conformational change of HA that is necessary for fusion.

Conclusion

Knowledge of the three-dimensional structures of the surface antigens of influenza viruses have had a major impact on the understanding of their role in the life cycle of influenza viruses. Insights into receptor binding, fusion mechanisms, and drug discovery have been gained and the study of these antigens from a range of influenza viruses has given insights into their structural evolution in relation to immune recognition.

See also: Orthomyxoviruses: Molecular Biology.

Further Reading

Lamb RA and Krug RM (2001) *Orthomyxoviridae*: The viruses and their replication. In: Knipe DM, Howley PM, Lamb RA, *et al.* (eds.) *Fields Virology*, 4th edn., p. 1487. Philadelphia: Lippincott Williams and Wilkins.

Palese P and Shaw ML (2006) *Orthomyxoviridae*: The viruses and their replication. In: Knipe DM, Howley PM, Griffin DE, *et al.* (eds.) *Fields Virology*, 5th edn., pp. 1647–1689. Philadelphia: Lippincott Williams and Wilkins.

Skehel JJ and Wiley DC (2000) Receptor binding and membrane fusion in virus entry: The influenza hemagglutinin. *Annual Review of Biochemistry* 69: 531.

Wright PF, Neumann G, and Kawaoka Y (2006) Orthomyxoviruses. In: Knipe DM, Howley PM, Griffin DE, *et al.* (eds.) *Fields Virology*, 5th edn., pp. 1691–1740. Philadelphia: Lippincott Williams and Wilkins.

Vaccine Strategies

I Kusters and J W Almond, sanofi pasteur, Lyon, France

Introduction

To develop a new viral vaccine, get it licensed, and bring it to market is a lengthy, complex, and very expensive task. It requires a detailed knowledge of all aspects of the virus, especially its structure, epidemiology, pathology, and immunobiology, and demands a close collaboration between fundamental scientists, regulatory authorities, and industrial scientists and engineers. Completely new vaccines against human viruses appear infrequently on the market and the cost and complexity of their development has escalated with time, mainly due to the increased regulatory pressures to have highly defined products and to ensure complete clinical safety and high efficacy. Over the past 20 years or so, six new human virus vaccines have been developed for licensure in major markets; these are hepatitis B, Japanese encephalitis, hepatitis A, varicella (recently with a zoster formulation), and, over the past year, rotavirus, and human papilloma virus (HPV). The introduction of HPV vaccines is a technological triumph that offers to dramatically reduce the incidence of cervical cancer worldwide in the long term. The newest rotavirus vaccines seem, so far, to be free from the complications that led to the withdrawal of the earlier Rotashield vaccine in 1999 and promise to be very effective in reducing the burden of rotavirus diarrhea worldwide. In spite of this impressive progress, however, virus vaccine development has not accelerated over recent decades, and there remains a significant list of human virus diseases of widespread prevalence for which there are no vaccines available. These include human immunodeficiency virus (HIV), hepatitis C, hepatitis E, Eptein–Barr virus, herpes simplex virus (HSV), cytomegalovirus (CMV), respiratory syncytial virus (RSV), and parainfluenza viruses (PIV) (see **Table 1**).

In the veterinary world, the list of successes is significantly more impressive, and both the time cost of developing a new vaccine can be substantially reduced by the ability to reach proof of concept through direct challenge experiments, often with the wild-type virus. In this arena, the requirements are often very different to those for human vaccines, especially for vaccines of

Table 1 Human viruses causing disease with important medical need for which no vaccines are available

Adenoviruses	Human immunodeficiency virus (HIV)
Chikungunya	Human metapneumovirus (hMPV)
Cytomegalovirus (CMV)	
Dengue	Norwalk virus
Enterovirus 71 (EV71)	Parainfluenza virus (PIV)
Epstein–Barr virus (EBV)	Parvovirus B19
Hantavirus	Respiratory syncytial virus (RSV)
Hepatitis C (HCV)	Rhinovirus
Hepatitis E (HEV)	SARS
Herpes simplex virus (HSV)	West Nile virus

production animals where ease of administration and cost per dose are paramount. Recognizing that many approaches and outstanding fundamental challenges are similar for human and veterinary virus vaccines, this article focuses principally on the former with occasional reference to veterinary vaccines where they illustrate particular concepts.

Fueled by the new technologies of genomics, proteomics, and molecular immunology, the past 20 years have seen an impressive increase in our knowledge of all aspects of virology, providing insights to guide new vaccine concepts. The biological properties of viruses influencing choice of strategy include pathogenesis, serotype diversity, antigenic variation, immune evasion mechanisms, latency, and route of transmission. New vaccine candidates have been described for a good number of the viruses listed in **Table 1**, and although many of these are in an early, pre-proof-of-concept stage, some are substantially developed and offer realistic prospects of licensure over the next decade. The most promising of these, based on pre-clinical and clinical results obtained so far, are dengue, hepatitis E, and HSV and CMV. Significant challenges remain however, and really promising candidate vaccines against pathogens such as HIV, hepatitis C, and infant RSV remain elusive. This article focuses on the strategies available to develop new viral vaccines and discusses some of the challenges posed by the more difficult targets.

Types of Vaccines

There are multiple possible approaches to the development of a viral vaccine that can be generally described as follows:

1. Killed whole or split virus vaccines. This approach requires that the virus can be grown to high titer in cell culture or other scalable medium such as hens eggs; that the virus can be successfully and completely inactivated using an agent such as formaldehyde or B-propiolactone without destroying immunogenicity; that, from an industrialization perspective, the immunogenic dose is low to modest with respect to virus yield (in the 10 μg range) and that the killed whole or split particle elicits protective immunity. This approach has had excellent successes in the form of vaccines such as inactivated polio vaccine (IPV), hepatitis A (HAV) vaccine, and influenza.
2. Subunits or single proteins prepared by recombinant DNA methods and fermentation processes in cell culture. This approach may work well when a single protein can provide immunity and where the expression system allows appropriate folding and processing of the viral protein.
3. Live-attenuated vaccines. There are several approaches possible.
   a. Use of a closely related animal virus that is not well adapted for efficient and widespread replication in humans and therefore does not cause disease, but nevertheless provokes an immune response that protects against the corresponding human virus. The best known example of this is the use of vaccinia virus to vaccinate against smallpox, but a similar approach has been used for rotaviruses, with genetic reassortment to confer appropriate antigenicity.
   b. Development of an empirically attenuated human virus by multiple passages in tissue culture, typically of nonhuman origin, and/or passage in animals. Attenuation is usually achieved by the accumulation of a number of mutations that affect the efficient functioning at normal body temperature of various genes or gene products, thereby reducing virulence. Replication competence, however, is maintained at a sufficient level to stimulate a protective immune response. Although there are some drawbacks to this approach from a safety perspective (e.g., a risk of reversion), this method has provided a bedrock of vaccinology over many decades and has worked well for viruses such as polio – oral polio vaccine (OPV), mumps, measles, rubella, and yellow fever.
   c. Live-attenuated vaccines prepared by knowledge-based manipulation of the viral genome. There are several examples of candidate vaccines in this category including HSV and influenza.
4. Vectored or chimeric virus approaches. This is where an existing virus vaccine can be modified genetically to carry genes encoding antigens from a foreign virus. The chimeric vaccine should retain the attenuation and growth characteristics of the parent vaccine strain but stimulate immunity against the foreign virus.
5. Naked DNA. This is where a DNA encoding viral antigens plus appropriate expression control sequences is administered directly to the recipient. Expression of the DNA leads to an immune response against the antigens encoded.

Principles of Vaccine Development and Examples

When developing a new vaccine, the choice of approach is made very much on a case-by-case basis, and for a given virus is driven by knowledge of its biology, structure, antigenic diversity, and pathogenesis. High importance should be given to what type of immunity arises as a result of natural infection and whether the pathogen can cause persistent and/or repeated infections in a single host. Experiments in animal models may also allow the dissection of the immune response to identify correlates of protection. The use of primates in particular can be useful if the disease produced is similar to that observed in humans. However, many viruses are highly host specific and may have evolved strategies to evade immune responses that may also be host specific (such as recruitment of downregulators of complement fixation). Care must therefore be taken as results in animal models may not be entirely reproducible in the natural host.

Killed Vaccines

Evidence that circulating antibodies are sufficient to provide immunity may come, for example, from the observation that the disease is modified or exacerbated in immune deficiencies such as hypo-gammaglobulinaemia and/or that passive immune globulin can protect against infection and disease. The latter observation was made for several viruses prior to vaccine development, including hepatitis A, suggesting that the key to vaccine development in these cases would be the stimulation of a strong humoral immune response. Such responses can often be adequately provided by killed vaccines, and so this was an obvious choice of approach. The strategy will also be influenced by the successes and failures with closely related viruses (either human or animal) that have similarities in epidemiology, pathogenesis, and mode of transmission. Thus, for HAV, the successful paradigm of the inactivated polio virus vaccine from the same virus family (*Picornaviridae*) provided further confidence that a killed whole virus particle approach would be effective. Indeed, inactivated HAV vaccines were developed successfully on this basis by several companies in the early to mid-1990s. Current HAV vaccines are prepared by propagating the virus in an approved cell substrate, human fibroblasts or human diploid cell culture MRC-5, purification, inactivation using formalin, and adjuvanted with aluminum hydroxide. The vaccines are given parenterally, as a two-dose series, 6–18 months apart.

A further example of a killed vaccine is rabies vaccine. Several WHO-recommended inactivated rabies vaccines are available currently. They are all similar in being whole virus, used after inactivation by β-propiolactone, and purification and concentration by ultracentrifugation and/or ultrafiltration. The vaccines are required to have a protective potency defined as >2.5 international units (IU). The potency is determined by the National Institute of Health (NIH) potency test which is based on assays using intra-cerebral challenge of previously immunized mice. The main differences between the rabies vaccines currently available lies in the cell culture used for production. The cell cultures used for the WHO-recommended vaccines are MRC-5 cells for purified human diploid cell vaccine (HDCV), Vero cells for purified Vero cell culture rabies vaccine (PVRV), and primary duck or chicken embryo fibroblasts for purified duck embryo vaccine (PDEV) and purified chick embryo culture vaccine (PCECV), respectively.

When a new virus emerges that poses a severe threat to human health such as the human coronavirus which caused severe acute respiratory syndrome (SARS) in 2002, it is necessary to start working immediately to develop a vaccine. The huge challenge in such a scenario is time. Generally, to develop a vaccine from basic research through animal studies, clinical lot development, analytical test development, clinical trials, industrial scaleup, and licensure takes 8–12 years. These timelines can be compressed in case of extreme urgency, but this compression is not unlimited. Although experience on existing vaccines can be exploited, all processes and procedures need to be evaluated, validated, and implemented. Working with a BMBL Section III Laboratory Biosafety Level 3 (BSL 3) agent, such as human SARS coronavirus, is not exceptional for vaccine manufacturers, but many precautions need to be taken in regards to equipment and laboratory practices, as little was known about the SARS virus in the early stages. The choice of the vaccine approach was indeed influenced by the time factor. If the virus grows well in cell culture, an inactivated viral vaccine is usually the option of choice as it is the fastest to accomplish. Fortunately, in case of the human SARS coronavirus, the virus did grow well on Vero cells, was efficiently inactivated either by formol or β-propiolactone, and appeared to be very immunogenic in several animal models as well as in human beings. After 2003, no more human SARS cases were observed and the development of such vaccines has generally been put on hold for the present.

Recombinant Protein Vaccines

As discussed above, a further simple strategy, particularly when only antibodies are required, is to use recombinant DNA methods to express a single surface structural protein of the virus in a host–vector system such as *Escherichia coli* or one of several yeast species. This approach may be adopted when the virus cannot be propagated efficiently in culture, making a killed vaccine approach impossible, when the inactivation process may diminish immunogenicity, or when a focused immune response against a specific protein is required. This approach has proved very successful for hepatitis B, where the vaccine is composed of particulate complexes of the virus surface glycoprotein HBsAg produced in yeast. These particles mimic virus-like particles produced during natural infection and induce a highly protective and long-lasting immune response. This vaccine has been on the market since 1992 as a three-dose series of injection, each containing 5–20 μg HBsAg. A new version of this vaccine has recently been licensed with a formulation containing 20 μg HBsAg, adjuvanted with monophosphoryl lipid A (MPL) and alum, which reduces the incidence of nonresponders compared to a population vaccinated with the licensed vaccine.

The recently introduced human papilloma virus vaccines have also been developed using the recombinant protein approach (see below). But as with killed vaccine approaches, one size does not fit all, and for many viruses the approach of recombinant protein expression has not proved successful for a variety of reasons. For example, many viruses have a complex structure that cannot easily be reproduced in foreign hosts at high yield, particularly when the final structure is formed from several conformationally interdependent proteins. Incorrectly folded or immature proteins may not elicit functional, protective antibodies. Second, for some viruses, immune responses to several proteins together may be necessary to provide complete protection against disease. Third, recombinant proteins administered conventionally are generally poor at providing cellular responses of a Th1 profile that may be necessary for protection against viruses such as HIV, hepatitis C virus (HCV), and members of the herpes family. Consequently, for many viruses, live-attenuated approaches or more complex production systems and/or methods of delivery are required as discussed below.

Live-Attenuated Vaccines

Live-attenuated vaccines are used to prevent diseases such as yellow fever, polio, mumps, measles, and rubella. They are based on viral strains that have lost their virulence, but are still capable of replicating sufficiently well to provoke a protective immune response. They cause infection but without inducing the clinical manifestations, eliciting a humoral as well as cellular immune response. Historically, the attenuation was obtained by passage in animals. The first demonstration of attenuation of a virus in cell culture was that of the yellow fever virus by Lloyd and Theiler. The attenuation resulted from a prolonged passage on cultures of chick embryo tissue. Another example is that of the development of the oral polio vaccines by Albert Sabin in the 1950s. Wild-type strains of each of the three serotypes were passaged in monkey testicular tissue both *in vitro* and *in vivo*, while testing in monkeys was performed by intracerebral inoculation at various stages of passage. Eventually, strains were selected that were unable to induce paralysis in animals. The number of passages and cloning steps required to achieve the desired level of attenuation varied between the serotypes. These vaccines are still routinely used in many countries of the world and have been the principal tool with which the WHO has pursued its campaign of global polio eradication.

The advent of genome sequencing and recombinant DNA techniques in the 1980s allowed the key mutations conferring attenuation, empirically introduced by Sabin's passages, to be identified. In addition to temperature sensitivity mutations affecting protein structure, all three attenuated strains had in common, point mutations in the 5′ noncoding regions which affected the stability of RNA secondary structure believed to be important for interaction with host factors and for internal entry of ribosomes.

Rotavirus

As discussed above, live-attenuated vaccines may also be based on a closely related animal virus. The rotavirus vaccine licensed by Merck in 2006 is based on the bovine WC3 rotavirus. The original monovalent bovine strain is naturally attenuated for human beings, but does not induce protective immunity. To improve the effectiveness of the strain for human use, reassortant strains were prepared which contained genes encoding capsid proteins from the most common human serotypes on a background of the bovine strain. The present vaccine contains five single gene reassortants, each containing a gene for a capsid protein from human serotypes G1, G2, G3, G4, and PIA (**Figure 1**). A three-dose regimen with $2.0–2.8 \times 10^6$ infectious units per reassortant, administered orally beginning at age 6–12 weeks with a 4–10 week interval between doses, provides 70% protection against both mild and severe rotavirus diarrhea.

Nowadays, for some viruses, attenuated viral vaccines can be designed on a more rational basis, by specifically targeting virulence factor functions that may not be essential for virus replication, especially in cell culture, but are necessary *in vivo* to counter host innate defense mechanisms. An example here is the NS1 gene of influenza A viruses. This protein is able to downregulate interferon production in the virus-infected cell and some deletion mutants of NS1 lack this function and are therefore much more easily controlled by the host interferon response and are thereby attenuated. The augmented interferon (IFN) response provoked by such viruses may also have the advantage of providing stronger immune stimulation resulting in increased immune responses. So far, such strains have only been tested in animal models but they offer promise as future influenza vaccine strains.

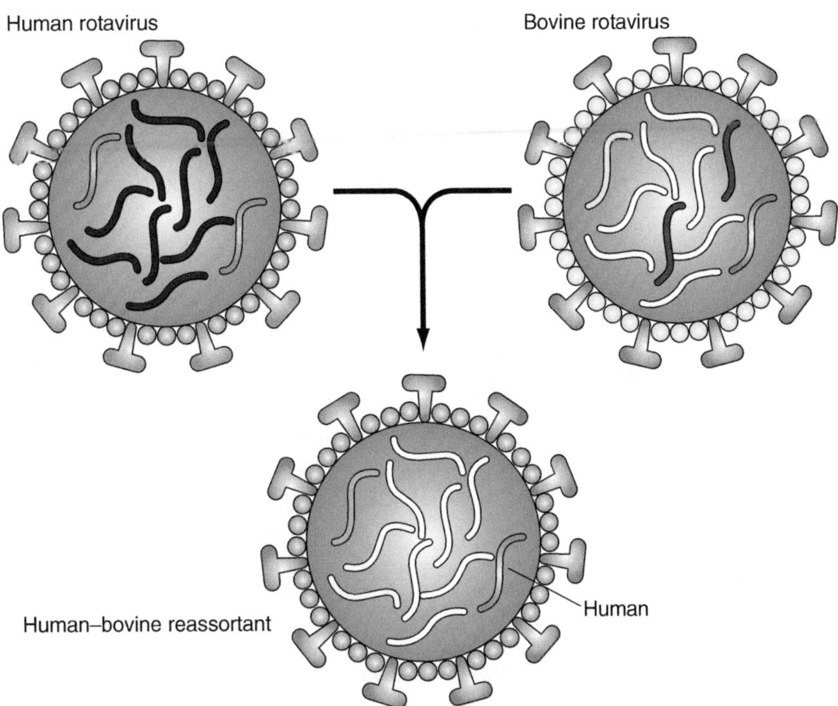

Figure 1 Rotavirus reassortant to generate oral live virus vaccine. RotaTeq is a polyvalent vaccine consisting of five human-bovine reassortants: four G serotypes (G1, G2, G3, G4) representing 80% of the G strains circulating worldwide, and one P serotype representing >75% of the P strains circulating worldwide. Reprinted by permission from Macmillan Publishers Ltd: *Nature Medicine* (Buckland BC (2005) The process development challenge for a new vaccine. *Nature Medicine* 11: S16–S19.), copyright (2005).

A drawback of many live-attenuated vaccines is that they require a cold chain from point of production to point of use and this may pose logistical difficulty, especially in developing countries. Also, the safety of live-attenuated viral vaccines is under constant scrutiny because of the risk, albeit small, that the mutations conferring attenuation will revert to wild type, allowing the virus to become virulent again. This is the reason why some live-attenuated vaccines are not recommended for immunosuppressed patients.

Viral Vectors

If neither the killed nor the attenuated vaccine approach is appropriate or feasible, one can consider the use of a viral vector. In this case, an attenuated virus is used as a backbone carrying immunogenic proteins of the virus of interest. In general, the viral proteins chosen are the membrane and/or envelope proteins as these proteins are presented on the outside of the virus particle and recognized by the immune system. An example of a viral vector is the yellow virus vaccine strain 17D. This vaccine strain was developed in the 1930s, since which time over 400 million people have been immunized with this vaccine. The strategy here is to use the 17D vaccine as a vector to deliver the two structural proteins, the premembrane PRE-M and envelope proteins from closely related flaviviruses. The resulting chimeric virus needs to be viable and to replicate efficiently in an acceptable cell substrate for vaccine production. The chimerivax dengue virus approach (by Acambis in collaboration with sanofi pasteur) has been developed using the PRE-M and envelope genes of wild-type clinical isolates. The technology involved is illustrated in **Figure 2**. The yellow fever virus genome has been cloned as an infectious cDNA. This infectious cDNA is manipulated to remove the PRE-M and E-Genes of yellow fever virus and exchange them for the coat protein genes of each of the dengue virus serotypes. Thus four individual chimerivax cDNAs are constructed. Transcribing these cDNAs to RNA provides infectious RNA with which to transfect cells in culture. Thus, the resulting virus is a heterologous virus containing the immunizing antigens of dengue virus with the replicative engine of the yellow fever 17D vaccine. Chimeric dengue viruses are expected to mimic the biological properties of yellow fever 17D which has the excellent characteristics of providing minimal reactogenicity and lifelong immunity. The candidate vaccines have been characterized in preclinical models, neurovirulence in mice, viremia and immunogenicity in monkeys, and shown to have desirable characteristics. Moreover, when the four constructs are mixed and administered to monkeys, seroconversion against all four dengue serotypes appears to occur simultaneously in most infected

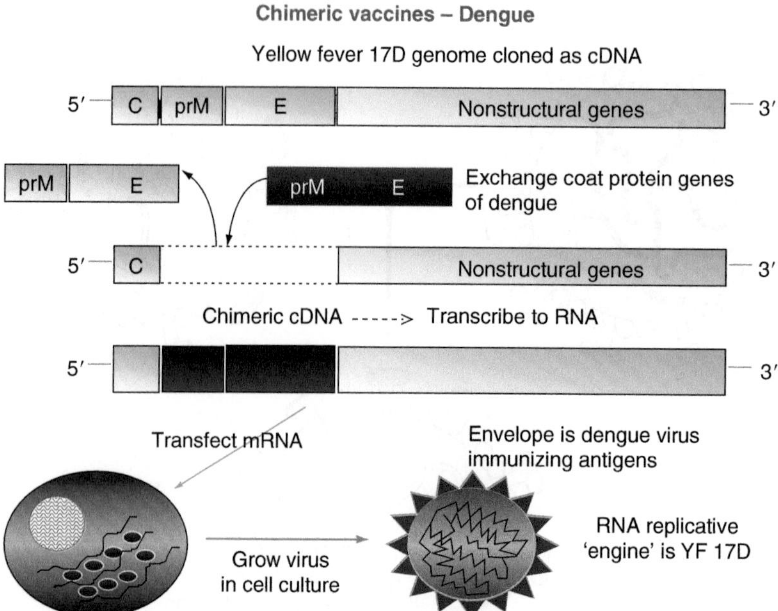

Figure 2 A DNA copy of the genome of yellow fever virus (blue) is manipulated to replace the prM and E genes by those of a related flavivirus such as Dengue (red). Transfection of mRNA transcribed from the resulting cDNA produces a 'chimeric virus' in cell culture.

animals. Moreover, the antibodies generated seem to be functional in that they neutralize dengue viruses in plaque reduction tests. The chimerivax dengue viruses grow well in culture and are well suited to industrial scaleup. In human volunteers, the chimeric viruses are safe and well tolerated and elicit specific immune responses against the different dengue serotypes. These strains therefore provide excellent candidates for further development.

DNA

Since the early 1990s, there has been considerable interest in the possible use of naked DNA as a vaccine delivery method. Naked DNA has the advantage that it can be taken up by cells and express the viral protein encoded. Depending on the conditions, this expression can be mid- to long term, thereby providing a substantial stimulation of the immune response. DNA vaccinology has apparently worked well in mice, but, so far, results in humans have been mainly disappointing requiring milligram amounts of DNA. Delivery of the DNA on colloidal gold, however, seems to offer a better prospect of success, as reported recently for hepatitis B virus and influenza. Regulatory issues concerning the use of DNA vaccines and its possible insertion integration into chromosomal genes are potential drawbacks to this type of approach, especially for use in prophylactic vaccination in infants. Further work is needed on safety issues before it can be seriously considered as a means to vaccinate populations.

The Challenge of Antigenic Variation

Antigenic Variation

Antigenic variation is displayed by a number of important pathogenic viruses and poses a particular problem for vaccine developers. The variation may be manifest in different ways depending on the virus' natural biology. Thus, for some viruses such as foot-and-mouth disease virus (FMDV), rhinoviruses, and HPV, multiple antigenic variants or serotypes co-circulate, sometimes with particular geographical patterns or ecological niches. Individual strains may show antigenic drift, presumably generating new serotypes over time. The rate of drift and the generation of new serotypes are not well understood for these viruses but may involve genetic recombination as well as cumulative mutational change. Other viruses may show a different pattern of antigenic variation: for example, influenza A viruses circulate as a limited number of subtypes (currently two in humans, H1 and H3), and each of these accumulate antigenic changes over several seasons (antigenic drift), escaping the most recently generated population-based immunity as they evolve. Occasionally, a new subtype may emerge (antigenic shift) either through genetic reassortment of a human strain with an avian strain, or possibly through direct evolution from an avian strain. Generally new subtypes displace existing subtypes as was the case when H2 emerged to displace H1 in 1957 and when H3 displaced H2 in 1968. For reasons that are not clear, this displacement did not occur when H1 reemerged in 1976 and since then there have been two influenza A subtypes co-circulating in the human population. This type of 'longitudinal' antigenic variation is clearly different from that of the multiple serotype viruses discussed above, and is generally more tractable in terms of vaccine development. Yet a different pattern is observed with HIV and to some extent HCV, where a limited number of genetic clades may contain very many different antigenic variants and where longitudinal variation to escape recently generated immune responses occurs within a single persistently infected individual. This type of natural biology generates a plethora of antigenic variants that can co-circulate. Providing immunological protection against all of these is an immense challenge for vaccine developers.

Multiple Serotype Vaccines

So what are the strategies available to develop vaccines against antigenically variable viruses? Most straightforwardly, one can generate simple killed vaccines against the currently circulating strains as discussed above and use these where the virus is prevalent. This strategy has had some success in the case of FMDV, where the geographical range of the virus may be (at least partially) restricted by regulations on movement of susceptible farm animals, and the vaccines are cheap and quick to prepare. However, even though simple killed vaccines have been shown to work for individual serotypes of rhinoviruses, it is difficult to imagine that this approach would be effective for this virus where, presumably because of widespread human contact and international travel, there seems to be a freer global circulation of multiple and unpredictable serotypes each winter. Moreover, for most people, rhinovirus infections are relatively trivial and therefore the balance of medical need versus industrial feasibility/commercial attractiveness of preparing vaccines in advance against multiple strains does not favor such a strategy. It would perhaps be a different matter if it were possible to design a simple rhinovirus vaccine that provided cross-protection against all serotypes.

The case of HPV is more manageable because of the small number of serotypes associated with severe disease. Thus, in this case, a significant impact on disease can be made by vaccinating against just a few of the many different genotypes. Both currently available vaccines contain HPV types 16 and 18 to vaccinate against cervical cancer, and the Merck vaccine contains, in addition, HPV types 6 and 11 to vaccinate against genital warts (condylomata accuminata). The absence of efficient cell culture systems for papillomaviruses has required the development of eukaryotic expression systems to produce virus-like particles composed of the L1 capsid protein, which are highly immunogenic. Future, second-generation vaccines will likely incorporate one or more of the additional highly or moderately oncogenic serotypes, such as 31, 33, and 35.

Influenza Annual Vaccination

For influenza, the limited number of circulating subtypes in any particular season makes it possible to adopt a strategy of annual vaccination. Thus, current influenza vaccines are trivalent, containing strains of H1 and H3 of influenza A and an influenza B strain. The strategy of annual vaccination is not without risk and requires a high level of international cooperation on disease surveillance and strain isolation, construction of high-yielding seed viruses, preparation of reagents for formulation, and industrial production. Following strain selection and recommendation by WHO, vaccine production has to occur over a very tight time schedule to ensure that vaccine is ready for the following winter (**Figure 3**). Occasionally problems arise such as a mismatch between the selection of a particular vaccine strain and the virus that eventually circulates during the following winter. This may compromise the effectiveness of that particular component of the vaccine. Other potential problems include less than optimal growth of the high growth reassortant seed at the industrial scale, leading to less vaccine being produced, and the lateness of seeds or reagents impacting on prompt delivery of vaccine for the flu season. The vast majority of influenza vaccine used currently is partially purified, killed whole or split virus prepared in embryonated hen's eggs, formulated to 15 μg of hemagglutinin (HA) of each strain, and provides generally good protection that correlates well with the induction of virus neutralizing or hemagglutination-inhibiting antibodies. However, vaccines prepared using different technologies are arriving on the market or are in advanced stages of development. These include live-attenuated (cold-adapted) strains, licensed by MedImmune in the USA in 2003, influenza-recombinant surface protein (hemagglutinin) produced in a baculovirus expression system from Protein Sciences, and inactivated virus vaccines prepared in cell cultures such as Vero, Madin–Darby canine kidney (MDCK), or PerC-6. This exploration of alternative technologies in recent years has been fueled by several criteria, perhaps the most important of which is greater and more flexible scaleup capability. The recent concerns over the possible emergence of an H5 pandemic has focused attention on the present worldwide limits in capacity, most notably in a situation of 'surge' demand, and governments have responded by providing the industry with incentives to increase capacity and diversify methods of production. A further response to this concern has been 'dose sparing' clinical studies on influenza vaccines adjuvanted with alum and other, proprietary adjuvants. These studies, using an H5N1 strain, suggest that it may be possible to reduce the vaccine dose from 2 × 90 μg HA nonadjuvanted, to 2 × 30 μg HA adjuvanted

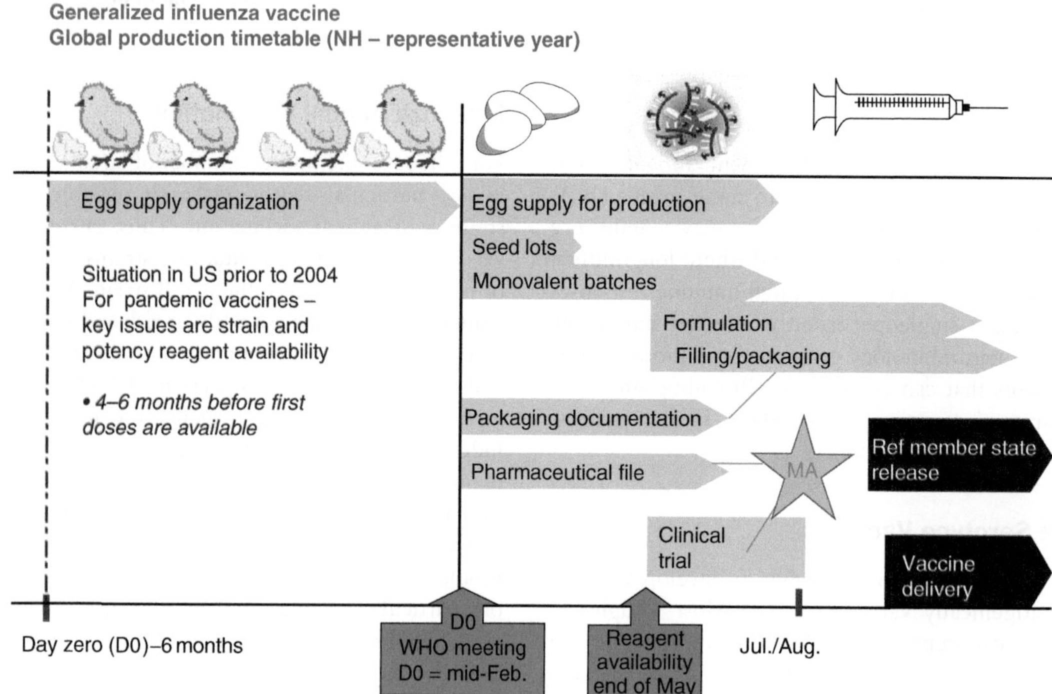

Figure 3 Approximate time schedule for the production of annual influenza vaccine from embryonated hens eggs.

with aluminum hydroxide or even as low as 3.75 μg HA adjuvanted with new proprietary adjuvants. Moreover, there have been renewed suggestions that it may be possible to develop vaccines with a broader and perhaps multiseasonal protective effect by stimulating cellular immune responses, particularly against nucleoprotein (NP), a claim currently made for the live-attenuated approach, and even a universal flu vaccine, for example, based on the well-conserved M2 virus surface protein. Animal challenge experiments, especially using multiple arrays of the M2 protein, have been encouraging to date and suggest it may be possible to provoke a much stronger response against M2 than that induced by natural infection. Whether such a response will provide solid protection in humans however remains to be established.

HIV Approaches

The vast array of antigenic variants of HIV and HCV renders the approaches discussed above extremely difficult for these types of viruses. For HIV in particular, many strategies have been tried, so far without significant success. Early on in the AIDS pandemic the focus was on the use of simple recombinant surface glycoproteins, aimed at inducing neutralizing antibodies in the hope that even limited protection against homologous or closely related strains would provide a proof of concept that could be built upon. Unfortunately, this type of approach was not successful and a large phase III clinical trial carried out by Vaxgen using *E. coli*, which produced gp120, was not able to provide convincing evidence of protection even against strains closely related to that present in the vaccine formulation. Subsequently, there has been considerable effort on the induction of cellular responses, especially CD8 cytotoxic T lymphocytes, and more recently on a balanced cellular response to include CD4 effector mechanisms. The objective here is to provide the means for the immune system to launch an immediate attack on the first cells to become infected following exposure to the virus. Ideally, such an attack would prevent the primary viremia by eliminating the virus before it becomes established in the body. However, more realistically, there is evidence from primate studies that strong preexisting T-cell responses can control the primary viremia and reduce the viral set point during the asymptomatic phase. A low viral set point is associated with slow or no progression to AIDS. In addition, the HIV evades immune surveillance by actively downregulating the major histocompatibility complex I (MHCI) molecules on the surface of infected cells by the HIV nef protein. Vaccination strategies to produce cellular responses have mainly used vectors such as the vaccinia virus strains MVA and NYVAC, canarypox, adenoviruses, or naked DNA, either with multiple doses of a single type of construct or in heterologous prime-boost strategies. Antigens delivered have ranged from substantial regions of gag-pol-env of

HIV to multiple copies in a 'string of beads' format of defined T-cell epitopes presented by common human leukocyte antigen (HLA) haplotypes. In general, these studies have not delivered T-cell responses of sufficient magnitude to be strongly encouraging, although one such strategy based on canarypox delivery of gag-pol-env antigens, followed by boosting with a recombinant env protein, has progressed to a clinical phase III study. Although many commentators have expressed doubt about whether this approach will show efficacy, it may generate useful information on the role of cellular responses in controlling HIV loads.

Most recently, HIV vaccine efforts have again turned to the induction of neutralizing antibodies, this time aimed specifically at epitopes that have been defined by studying unusual but highly informative broadly neutralizing monoclonal antibodies. The fact that such antibodies exists is highly encouraging from a vaccine perspective. The concepts here are based on the notion that certain conserved but crucial regions of gp120 or gp41 are naturally poorly immunogenic, either because they are relatively hidden in the conformationally folded protein or are shielded by strongly immunogenic noncritical domains or by glycosylation, or because they are only transiently exposed during the structural rearrangements that accompany cell binding and virus penetration. It is argued that because antibodies against these regions are neutralizing they will be protective if they can be generated prophylactically with sufficient avidity and at sufficient titer. This 'cryptic epitope' idea has been discussed for several viruses over many years and is akin to that mentioned above for influenza M2, in that the objective is to generate a far stronger response against a particular antigen or epitope than that resulting from natural infection. Such antibodies, once induced, will need to have the kinetic properties necessary to effectively neutralize the virus *in vivo*. So far there are no examples among virus vaccines that prove this concept. For HIV gp160, the particular construction, presentation, and formulation of molecules able to raise high-titer antibodies against these conserved regions (many of which are imprecisely defined) are far from obvious. Nevertheless, the induction of prophylactic immune responses of this type is certainly worth detailed investigating in detail, given the magnitude of the HIV problem. The challenge of developing a vaccine against HIV however remains immense.

Conclusions and Perspectives

The easy viral vaccine targets of significant medical importance have been done. The viruses against which we do not have vaccines today are either of regionalized or sporadic importance medically and the incentives to develop them have not been sufficiently large or they are viruses that pose significant challenges in terms of their biological characteristics. Thus, HIV and HCV pose challenges because of their antigenic variation and the fact that natural immune responses are unable to protect and/or eliminate the virus. RSV poses challenges because of the immunopotentiation of pathogenesis that, for infants, must be avoided at all costs.

Nevertheless, there are grounds for optimism. A new generation of adjuvants, making it possible to selectively orientate immune responses toward Th1 or Th2 as necessary, promises the possibility of being able to 'improve on nature' in terms of immune response provoked by the viral antigens. New developments in vectors and virus 'chimeras' offer promise for vaccines such as dengue and perhaps RSV and parainfluenza viruses and targeted modification of immunomodulatory genes may offer prospects of new vaccines against herpes family viruses. Fundamental studies on virus pathogenesis, epidemiology, and immunobiology are greatly aided by new technologies such as genomics and proteomics and it is likely that the improved understanding will increase the technical and scientific feasibility of developing new viral vaccines in the years ahead.

See also: Antibody - Mediated Immunity to Viruses; Antigenicity and Immunogenicity of Viral Proteins; Antigenic Variation; DNA Vaccines; Neutralization of Infectivity; Vaccine Safety.

Further Reading

Buckland BC (2005) The process development challenge for a new vaccine. *Nature Medicine* 11: S16–S19.

Garber DA, Silvestri G, and Feinberg MB (2004) Prospects for an AIDS vaccine; Three big questions, no easy answers. *Lancet* 4: 379–413.

Koelle D and Corey L (2003) Recent progress in herpes simplex virus immunobiology and vaccine research. *Clinical Microbiological Reviews* 16(1): 69–113.

Plotkin SA, Rupprecht CE, and Koprowski H (2005) Rabies vaccine. In: Plotkin SA and Orenstein WA (eds.) *Vaccines,* 4th edn., pp. 1011–1038. Philadelphia: Saunders.

DNA Vaccines

S Babiuk, National Centre for Foreign Animal Disease, Winnipeg, MB, Canada
L A Babiuk, University of Alberta, Edmonton, AB, Canada

Glossary

Electroporation The perturbution of cell membranes by very weak electrical impulses to allow the uptake of plasmid DNA.

Th1 immune response The quality of the immune response represented by a more active cell-mediated immune response with the concomitant secretion of specific cytobines such as interferon gamma.

Th2 immune response The quality of the immune response represented by a larger humoral antibody response, with the secretion of specific cytobines such as interleukin-4.

Introduction

Although the era of vaccinology has begun over 200 years ago, some of the early advances continue to form the foundation of our most successful vaccines. For example, even though it is over 200 years since Jenner introduced the concept of vaccination, using a live virus vaccine to protect people from smallpox, the concept of vaccinating with live agents continues. Indeed, some of the most effective vaccines are live-attenuated or live-heterologous vaccines. These vaccines have been successful in eliminating smallpox, or dramatically reducing the economic impact of diseases such as measles, polio, and many animal viral infections such as rinderpest. The reason for this success is that as the agent replicates, it resembles a natural infection, thereby stimulating the appropriate immune responses, including humoral and cell-mediated immunity. Furthermore, the persistence of the infecting agent, over an extended period of time, also provides long-term memory. Approximately 100 years after Jenner, Pasteur introduced a new approach to vaccination by killing the infectious agent and introducing the killed agent into the body as a nonreplicating vaccine. Similar to the original live vaccines, killed vaccines continue to be used for many infectious agents and are an integral component of our current armamentarium for disease control. Indeed, in some cases, live vaccines, which, in many cases, are more effective than killed vaccines, can be used initially to reduce the level of circulating agent in the environment, followed by killed vaccines to 'mop up' the remaining agent. The best example of such an approach is the use of live polio virus vaccine to provide broad immunity and reduce the viral load in the environment or country of interest and then use killed polio virus vaccine as a final step in disease eradication. The reason for such a strategy is that live vaccines can revert to virulence and may have certain risks associated with their use, which are worth taking when the disease burden is high but less so when the disease burden is low.

A century after Pasteur, we embarked on a new era of vaccination, primarily driven by advances in molecular biology and biotechnology, as well as a much deeper understanding of the host immune responses and antigens of each pathogen that are involved in providing protection from infection. These advances are the underpinning of the era of vaccination employing genetic engineering. This resulted in the development of vaccines against hepatitis B and human papillomavirus, which not only prevent infection but also reduce the development of tumors. These vaccines were genetically engineered to contain all of the critical epitopes involved in inducing protective immunity. However, they do not contain any nucleic acid, thereby making them relatively safe and they are considered to be killed vaccines. Unfortunately, they require adjuvants and generally do not induce the broadest range of immune responses often seen with live vaccines.

Since production and purification of genetically engineered subunit vaccines are generally very expensive, require strong adjuvants, and generally induce a skewed systemic Th2 response, the quest continues for the development of more effective vaccines resembling live vaccines. To combine the benefits of a subunit vaccine such as safety with the broad immune responses produced by live vaccines, the concept of introducing the gene encoding putative protective proteins in a plasmid into the host was proposed. Indeed, this was heralded as a 'third generation of vaccinology', genetic vaccination, or DNA vaccination. The major advantage of such an approach, especially for viral infections, is that the antigen is expressed endogenously and the antigens are, therefore, processed in a manner resembling a viral infection. As a result, the antigen is presented by both major histocompatibility complex (MHC) I and MHC II pathways, inducing a more balanced immune response. This article describes the progress made to date in the area of DNA vaccination, including a description of the DNA vaccines that are already licensed as well as some of the challenges faced by this technology.

Furthermore, the opportunities to overcome these challenges will also be discussed.

Concept of DNA Vaccines

The major reason for continued interest in DNA-based vaccines is their simplicity in concept, ease of production, potential to develop a broad range of immune responses, as well as their perceived safety and ability to induce immunity in neonates in the absence or presence of maternal antibodies. With regards to simplicity, a DNA vaccine is comprised of a plasmid containing various regulatory elements to ensure efficient production of the plasmid in bacterial systems, such as an origin of replication and a selectable marker as well as an expression cassette containing the gene of interest under a eukaryotic promoter usually human cytomegalovirus for efficient expression of the gene inserted into mammalian cells. Since the general features of a plasmid are identical for all vaccines, the single platform makes DNA vaccines very attractive from the prospective of manufacturing. Thus, the only difference between different vaccines would be the gene insert. Thus, if a company establishes a process for manufacturing one plasmid-based vaccine, they can use the same process for production and purification of a variety of different vaccines.

Recent advances in our understanding of pathogenesis, comparative biology, molecular biology, bioinformatics, and immunology make identification of putative protective antigens to most pathogens relatively easy. Second, identifying and isolating the gene encoding a protective antigen, combined with gene sequencing to ensure the cloned gene contains the correct sequence and is correctly positioned in the plasmid, is relatively straightforward. Furthermore, the gene sequence can also be modified to optimize the codon biases for expression in the host cell of interest. Once the plasmid is constructed, the production of the plasmid is relatively simple, using well-established fermentation processes in *Escherichia coli*. Since *E. coli* grows to very high densities, the quantity of plasmid produced in this manner is very high. Combining fermentation with well-established downstream processing currently developed for plasmid purification makes this process very attractive to commercial companies. Indeed, processes for plasmid purification are considered easier to perform than protein purification used in subunit vaccine production. Currently, a number of companies have established the downstream processing steps for purifying high-quality plasmids suitable for DNA vaccine production. These processes can easily be transferred to developing countries to provide secure supplies of vaccine for the developing world, should that be necessary. Indeed, all of the processes including fermentation, plasmid purification, and quality control can occur in a time frame as short as 1 month. This is a significant advantage over current subunit vaccine protein production.

The process of inserting a gene into the plasmid is routine and once the process is developed for one vaccine, the same process can be used for inserting any gene of interest. The development of a new vaccine would not require additional skills, except knowledge of the gene of interest. As a result, extensive trials and procedures required to demonstrate the absence of reactogenic components often required for subunit or conventional vaccines is reduced to a minimum. This not only reduces the concerns for safety, but also reduces the time for reaching the market.

Safety Issues with DNA-Based Vaccines

DNA vaccines are considered to be safe since they can be highly purified to remove extraneous materials and are noninfectious. The infectious nature is a common concern with live-attenuated vaccines where reversion to virulence has occurred. Preliminary trials in animals and humans demonstrated low inherent toxicity and since the plasmids are often administered in physiological saline, in liposome-based formulations, or various other carriers which have already been tested and used over extended periods in humans, these vaccines are considered relatively safe. Initially, there was concern regarding the potential of the introduced plasmids to integrate into somatic or germline cells. However, studies have suggested that the probability of this occurring is 3 orders of magnitude lower than spontaneous mutations. Finally, there was concern that introduction of large amounts of DNA might lead to the generation of anti-DNA antibodies. Fortunately, a number of studies designed to test this possibility have demonstrated that this is likely to be a rare event if it would occur at all. Thus, it is currently widely believed that DNA-based vaccines are relatively safe.

Attractiveness of DNA-Based Vaccines

One desirable feature of any vaccine is the ability to protect individuals against a variety of disease agents following single administration of a vaccine. Thus, multicomponent vaccines are gaining popularity, not only because of the breadth of protection they induce, but also because of improved compliance since individuals do not need to return for numerous vaccinations. One limitation of combining multiple conventional vaccines is the possibility of interference between various components in the vaccine. With DNA vaccines this is also a possibility, but it should be possible to identify which component is interfering in induction of immunity. For example, in the study of a vaccine containing nine different plasmids encoding nine different malarial antigens, reduced immune responses to various components were observed compared to responses induced by individual plasmids. Similarly, we observed interference between plasmids encoding the genes for bovine herpesvirus-1 gD, parainfluenza-3 HA, and influenza virus HA. We also showed that co-administration of BHV-1 gD reduced the immune responses to BHV-1 gB. Since in these last two instances, the gD gene was responsible for the interference, it will be necessary to re-engineer this gene to remove the interference. In contrast, the use of genes encoding the four different dengue serotypes did not result in any interference. Indeed, combining the four serotypes resulted in higher antibody levels against dengue-4 than if animals were immunized with a monovalent dengue-4 plasmid-based vaccine. Similarly, broadening of the immune response was seen with multiple DNA vaccine components of various HIV proteins. Thus, it appears that each plasmid combination may need to be investigated independently for maximal benefit.

In addition to combining plasmids encoding different genes of interest, it is also possible to introduce two different genes encoding two different proteins from either the same or different pathogens in a single plasmid, thus further reducing the cost of production since one vaccine would protect against two different agents. This can be achieved by generating a plasmid which contains two gene expression cassettes or by simply linking the genes from two pathogens and inserting them into the plasmid. In this case, the two genes would be expressed as a single protein from a single promoter. Since the conformational changes imposed by linking two proteins together may alter their immunogenicity and the stability of the plasmid may be compromised as a result of its size, a number of plasmids have been constructed with each gene driven by its own promoter or by insertion of an encephalomyocarditis virus (EMCV) internal ribosome entry site (IRES) to translate the second gene. In some of these cases, the level of expression of the second gene was not as effective as that of the first gene, whereas in other instances there was no reduction in gene expression of either gene. Using such manipulations, it should be possible to broaden the protection to induce immunity to a number of different diseases simultaneously.

Neonatal Immunization

Since many diseases occur during the first few weeks or months of life, it is critical to induce immunity from vaccines as early as possible. Unfortunately, most vaccines given to neonates are less effective than in adults. There are many reasons for this. In the case of live vaccines, the presence of maternally derived antibodies generally limit the degree of replication of the vaccine, resulting in poor immune responses due to reduced antigenic mass produced *in vivo.* In the case of killed vaccines where there is less interference by maternally derived antibodies, the immune response is generally skewed to a Th2 bias response, thought to be due to the immaturity of the immune system. Using DNA vaccines, it has been shown that not only was it possible to induce immunity in neonates, but immunity could even be induced *in utero.* Thus, sheep immunized *in utero* during the third trimester were born fully immune. More importantly, the animals developed long-term memory. Indeed, the induction of memory by DNA vaccines is a very attractive feature of these vaccines since immune memory and duration of immunity are desirable for many vaccines. Numerous other studies have also shown the induction of long-term memory of T cells and B cells following DNA vaccination. This has been shown in a variety of models including pigs, sheep, and cytomologous monkeys. Memory could be further increased by incorporating plasmids encoding for IL-12 or IL-15 with the DNA vaccine.

It is also interesting that DNA vaccines appear to be able to induce immune responses in the presence of high levels of antibody. Thus, the fact that DNA vaccines can induce immunity in neonates in the presence or absence of maternal antibodies makes this approach to vaccination extremely attractive, especially for diseases such as herpes simplex virus-2, human immunodeficiency virus (HIV), hepatitis B, group B strep, and chlamydia which often infect children during birth or shortly thereafter.

Limitations and Clinical Applications of DNA Vaccines

Although there are over 1000 publications demonstrating the effective induction of immunity using DNA vaccines in mice which protect mice from subsequent infection, DNA vaccines have not been as successful in larger animals or humans. Possibly the greatest challenge to adopting DNA vaccination as a routine in large animals and humans is the poor efficiency of transfection leading to suboptimal induction of immunity. This has limited the introduction of vaccines into the market. However, recently two DNA vaccines were licensed for commercial use. The first was a vaccine to immunize fish against infectious hematopoietic necrosis virus, followed shortly after with one for West Nile virus in horses. Although there have been numerous clinical trials in humans, the greatest success has been when DNA vaccines were used to prime the individual, followed by booster with a recombinant or subunit vaccine. Indeed, the DNA prime, followed by protein boost is currently the method of choice for induction of the immune responses using DNA vaccines in humans. Thus, it is our contention that if better delivery mechanisms could be introduced to enhance the transfection efficiency in large animals and humans, DNA vaccines have the potential to become a critical component in our armamentarium against infectious diseases.

Vaccine Delivery

Once the plasmids are purified, they are introduced into the host (animals or humans) and the DNA is taken up by specific cells where the gene is expressed and protein is produced. Unfortunately, only a very small percentage of the DNA is internalized, which is one of the drawbacks of DNA vaccination. Even for the DNA molecules that are internalized, not all are transported to the nucleus where they express their encoded antigen. However, the few molecules that enter the nucleus are transcribed and translated to produce protein, making the animal cells a bioreactor. Thus, the vaccine antigen is produced in the animal. Even though the quantity of protein produced by the transfected cell is very low, the protein is presented to the immune system and induces a broad spectrum of immune responses including antibody- and MHC-I-restricted CTL responses. Clearly, the magnitude of the immune response as well as the quality of the immune response can be influenced by the route of delivery; intradermal delivery is generally better than intramuscular since the numbers of antigen-presenting cells are more numerous in the skin than in muscle. Indeed, muscle cells are not generally capable of presenting antigen to the immune system, although immunity occurs as a result of antigen release from myocytes leading to cross-priming. The quality of the immune response can also be modulated by the background of the plasmid. A number of groups have modified the plasmid background to incorporate sufficient and appropriate CpG motifs to stimulate the innate immune response. By stimulating the innate immune response, it is possible to recruit the appropriate cells to the site of antigen expression and create a cytokine microenvironment conducive to developing an adaptive immune response. To improve the cytokine microenvironment conducive to induction of immunity, a number of investigators have co-administered DNA vaccines with plasmids expressing cytokines, defensins, or other immune stimulator molecules to recruit dendritic cells to the vaccination site. Other constructs have contained both the antigen and the cytokine/chemokines within the single plasmid. In most cases, the immune response was enhanced by the co-administration of cytokines such as granulocyte-macrophage colony-stimulating factor (GMCSF) and interleukins IL-2, IL-12, and IL-15. In addition, the cytokine–chemokine combination plasmids could bias the immune response. For example, macrophage inflammatory protein-1 (MIP-1)β biased the immune response of Th2-like response, whereas MIP-2 and MIP-1α favored a Th1 like response. Currently, a number of studies are ongoing using IL-12 and IL-15 for enhancing immune response to HIV vaccines.

Since intracellular uptake of DNA is not a very efficient process, numerous investigators have employed a variety of methods to enhance DNA uptake into cells. One very efficient delivery system is encapsulation of DNA in liposomes or polylactide-L-glycolide (PLG) microparticles. DNA delivery by these microparticles serves two purposes. First, the encapsulated DNA is protected from degradation, and, second, the liposomes or microparticles facilitate uptake of the DNA into antigen-presenting cells. Indeed, it is even possible to deliver the microparticles orally to induce mucosal immunity as well as systemic immunity. Since encapsulation of DNA in PLG is thought to cause DNA damage due to the process of encapsulation, other investigators have attached the DNA to the surface of PLG microspheres. This process has resulted in excellent immune responses regardless of whether the PLG microparticles were given mucosally or delivered systemically, suggesting the microspheres play a crucial role in DNA uptake in the cells.

Improving uptake of DNA is thought to be critical for enhancement of immunity, and since the response is better when more cells are transfected, significant efforts have been focused on enhancing DNA uptake by various physical methods. The earliest approach to enhancing cell uptake was the use of a gene gun. In this instance, the DNA is coated on gold particles and is propelled into the skin. Some of these particles directly penetrate cells, whereas others are taken up by antigen-presenting cells. Although this approach has been shown to be effective, the quantity of DNA that can be coated onto gold beads is generally low, thereby reducing the effectiveness of such an approach.

The limitation of the quantity of DNA that can be taken up by cells by a gene gun can be overcome by jet injection of DNA. Depending on the pressure used to deliver the DNA, the DNA can be deposited either in the epidermis, subcutaneously, or even intramuscularly. However, an even more efficient method to deliver DNA into cells is the use of electroporation. Various companies have developed devices to both inject and electroporate the DNA simultaneously. Electroporation does not only enhance DNA uptake, but it also causes localized inflammation and induces the various mediators of innate immunity, resulting in enhanced immune responses. These studies have demonstrated the importance of modifying the local environment at the site of injection since the mere enhancing of expression does not always increase the immune responses. These studies demonstrate that one requires all of the elements of the immune response to be present at the site, as well as the specific antigen for the immunity to be induced.

Future Prospects

DNA vaccines represent an exciting addition to an already impressive track record of vaccines, which have saved millions of lives and improved the quality of life of almost every individual on the planet. Furthermore, vaccines have added significant economic benefits to society. Unfortunately, individuals still suffer from many infections. The reasons for this are varied, including the cost of vaccines, politics, as well as distribution issues. One of the major challenges to distribution is the need for a cold chain. If this could be overcome, the vaccines could be distributed to even remote regions of the world. DNA vaccines offer an advantage in this regard. Furthermore, it should be possible to develop these vaccines more economically, ensuring that the poor in remote regions could be vaccinated. Unfortunately, for this dream to become a reality, it will be necessary to improve the transfection efficiency of DNA vaccines to ensure that sufficient antigen is produced to induce an immune response. Currently, improvements in DNA vaccines often have been marginal. What is required is a 10–50-fold improvement to make the vaccines economical in most veterinary species and for humans in the developing world. However, the fact that two different DNA vaccines have already been licensed provides hope that within the next decade we will see a number of new DNA-based vaccines licensed and shown to be effective in reducing both morbidity and mortality. Whether these vaccines will be comprised solely of DNA-based vaccines or will be comprised of DNA-based vaccines which will prime the immune response, followed by boosting with conventional or recombinant vaccines, remains to be determined. Indeed, we will probably see both types of vaccine configurations in the future.

See also: Antibody - Mediated Immunity to Viruses; Cell-Mediated Immunity to Viruses; Vaccine Strategies.

Further Reading

Babiuk S, Babiuk LA, and van Drunen Littel-van den Hurk S (2006) DNA vaccination: A simple concept with challenges regarding implementation. *International Reviews of Immunology* 25: 51–81.

Gerdts V, Snider M, Brownlie R, Babiuk LA, and Griebel P (2002) Oral DNA vaccination *in utero* induces mucosal immunity and immune memory in the neonate. *Journal of Immunology* 168: 1877–1885.

O'Hagan DT, Singh M, and Ulmer JB (2004) Microparticles for delivery of DNA vaccines. *Immunological Reviews* 199: 191–200.

Ulmer JB, Wahren B, and Liu MA (2006) DNA vaccines: Recent technological and clinical advances. *Discovery Medicine* 6: 109–112.

van Drunen Littel-van den, Hurk S, Babiuk SL, and Babiuk LA (2004) Strategies for improved formulation and delivery of DNA vaccines to veterinary target species. *Immunological Reviews* 199: 113–125.

Vaccine Safety

C J Clements, The Macfarlane Burnet Institute for Medical Research and Public Health Ltd., Melbourne, VIC, Australia

G Lawrence, The Children's Hospital at Westmead, Westmead, NSW, Australia and University of Sydney, Westmead, NSW, Australia

Glossary

Adjuvant A chemical additive to vaccines that improves the presentation of the antigen to the immune system of the recipient, thereby increasing the effectiveness of the vaccine. The commonest adjuvant is an aluminum salt.

Adventitious agent A contaminating virus or other infectious organism.

Adverse event following immunization Any medical incident that follows immunization and is believed to be caused by the immunization.

Anaphylaxis A hypersensitivity reaction characterized by sudden onset and rapid progression of signs and symptoms in at least two organ systems. It results in release of histamine from mast cells and may follow administration of drugs and vaccines. The severity of the reaction varies from mild symptoms of flushing to severe bronchospasm, collapse, and death. It is treatable with adrenaline.

Auto-disable syringe A form of disposable syringe that automatically locks after one use, preventing its reuse. It is the syringe design of choice for immunization.

Causality assessment A scientific method for assessing the relationship between an adverse event and the receipt of a dose of vaccine.

Clinical trial A research activity that involves the administration of a test regime to humans to evaluate its efficacy and safety. There are classically four phases in clinical trials assessing vaccines.
Guillain–Barré syndrome A polyradiculoneuritis resulting in various degrees of motor and sensory disturbance. The condition frequently follows an acute infection or administration of vaccine.
High-titer measles vaccine A measles vaccine that contains at least one log greater (>log 10^4) numbers of live viral particles than a standard vaccine preparation (log 10^3).
Hypotonic hyporesponsive episode (HHE) The sudden onset of limpness, reduced responsiveness, and change in skin color in an infant following the administration of a vaccine. Typically (though not exclusively), this may follow administration of pertussis-containing vaccines.
Immunization The process of presenting an antigen to the immune system with the purpose of inducing an immune response. Immunization is synonymous with vaccination.
Injection equipment Any material needed to introduce a vaccine into the body of the recipient. Classically refers to needles and syringes, but new technologies for administering vaccines will widen its usage.
Neurovirulence test One of several tests that is required of live viral vaccines before release to ensure that the vaccine virus does not negatively affect the central nervous system.
Presentation How a vaccine is presented for use (e.g., as a liquid, freeze-dried powder, one-dose vial, 10-dose vial, etc.).
Preservative A chemical (most frequently thiomersal/thimerosal) that is included in a liquid vaccine preparation that inhibits the growth of organisms that inadvertently contaminate the vaccine.
Program error A mistake made by vaccination staff that occurs during administration of a vaccine.
Protective immunity Immunity that protects against actual disease, not simply evoking an immune response.
Regulatory authority An institution/laboratory that takes responsibility for quality control of vaccines used in a country.
Safety box A tough cardboard box into which can be placed used injection equipment. It is generally then transported to an incinerator or can be burnt on the spot, thus safely disposing of medical waste that might otherwise be a hazard.
Stabilizer Proteins or chemicals that are added to a liquid vaccine to maintain it in a stable or unchanging state.
SV40 Simian virus 40 is a virus found in certain monkeys and cell cultures derived from monkey tissues.
Vaccine diluent A liquid (usually sterile water or saline) that is used to reconstitute freeze-dried vaccine.
Vaccine safety The discipline of ensuring that vaccines are manufactured, transported, and administered with minimum risk to the recipient, vaccinator, or general public.

Introduction

Vaccines are arguably the most effective weapon in the modern fight against infectious diseases. They avert millions of deaths and cases of disease every year. But there is no such thing as a 'perfect' vaccine that protects everyone who receives it and is entirely safe for everyone. The scientist attempts to create a vaccine that protects nearly everyone, is extremely effective, and is as safe as humanly possible. That still leaves some vaccine recipients who are not fully protected; and worse, a very small number who suffer negative consequences from the vaccine itself.

There are many aspects that can be considered in the design phase which influence the ultimate safety and public acceptability of a vaccine (**Figure 1**). Vaccine safety can be categorized into aspects related to the manufacture of the vaccine, delivery of the vaccine to the recipient through an immunization programme, and individual and population responses to the vaccine. This article describes the efforts taken to ensure that a vaccine is manufactured, tested, and administered as safely as possible, and identifies how vaccine safety is measured and monitored after a vaccine is licensed for use in the population.

Vaccines are designed to evoke a protective response from the immune system as if it were the actual infectious disease. While a vaccine is designed to maximize the body's reaction and minimize any pathological effects, effective vaccines (i.e., capable of inducing protective immunity) may also produce some undesirable side effects. Clinical trials determine whether side effects are tolerable or not, and detect certain adverse events before the vaccine reaches general distribution. Most adverse reactions are mild and clear up quickly.

Design factors can help minimize programmatic errors in the administration of the vaccine by reducing the chance of human error. The most common types of programmatic errors are unsafe or contaminated injections. It is sobering to realize that up to one-third of injections (not just for vaccination) are not sterile. New technologies have been introduced in the last decade that make injections much safer, including an auto-disable syringe that can only be used once and must be discarded because the plunger mechanism locks after use. This device helps minimize transmission of diseases that result from the reuse of contaminated injection equipment.

Vaccine adverse reactions can be due to the vaccine itself, or to immunological factors in the individual such as anaphylaxis or hypersensitivity reactions. However, the majority of adverse events thought to be related to the administration of a vaccine are actually not due to the vaccine itself – many are simply coincidental events. For example, many children have a cough around the time of vaccination, but this is independent of the vaccine, not caused by it. An appropriate causality assessment, using internationally consistent methods, is necessary to fully determine whether a particular vaccine causes a specific reaction. Even when scientists have been able to show that a particular event is unrelated to vaccine administration, the public may still perceive there to be a link.

Vaccine Design

Manufacture

A vaccine consists of many parts, only one of which is the antigen by which it is known. Other components may include an adjuvant, a preservative, antibiotics, stabilizers, and certain ingredients that remain from the manufacturing process. There may be components not stated on the information sheet that are classified as proprietary that the manufacturers are not obliged to declare. The safety of the vaccine can be affected by each and all of its components.

How a vaccine is manufactured will also determine its content. For instance, thiomersal may be used in the manufacture of pertussis vaccines to maintain sterility during the manufacturing process. While the bulk of the chemical is removed, the final product will still contain minute amounts of thiomersal. Influenza vaccines are grown on chicken eggs – trace amounts of egg proteins not removed in the purification process can cause severe allergic reactions in susceptible individuals.

Safeguards in vaccine manufacture now make it highly unlikely that currently available vaccines are contaminated with infectious agents that cause disease. Manufacturers supplying the international market are required to submit vaccines to extensive tests to detect known

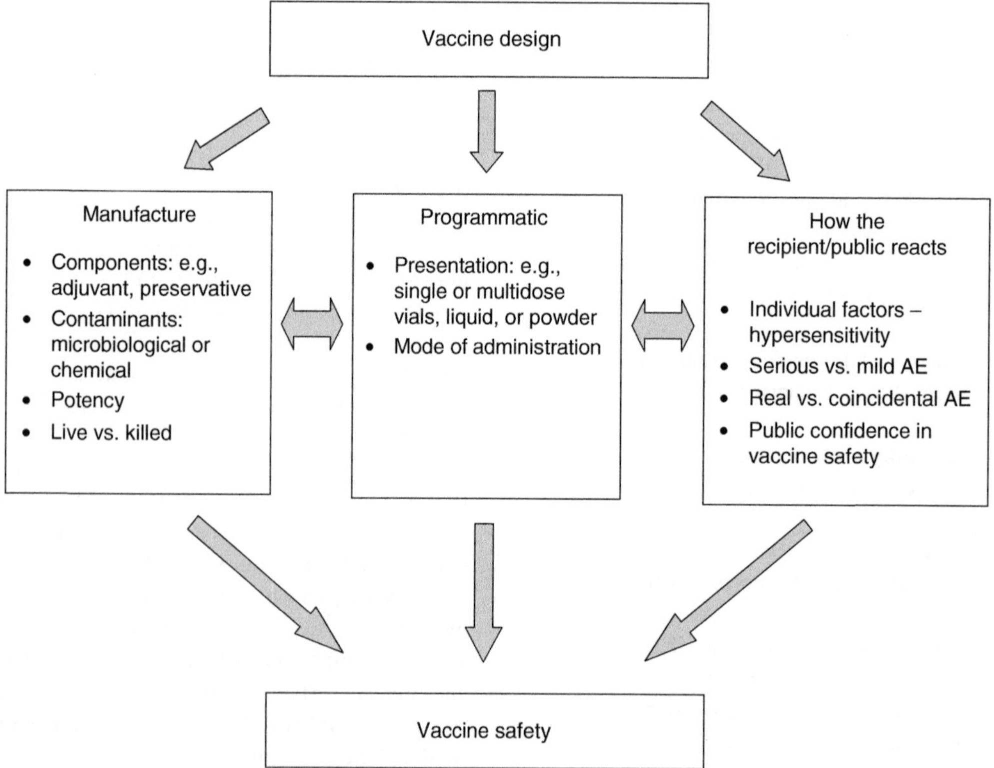

Figure 1 Components of vaccine safety.

potential pathogens. As well, they must test cell lines used in the production of vaccines for the presence of a variety of infectious agents. Once manufactured, vaccines must undergo rigorous clinical trials for safety and effectiveness before being licensed by regulatory authorities.

Procedures designed to detect and prevent the presence of adventitious agents in vaccines include characterization of the cell substrates used for vaccine production as well as testing the products by specified methods. For example, manufacturers are required to test their cell lines used for the production of poliovirus vaccines (generally monkey kidney cells) for a variety of infectious agents such as tuberculosis, SV40, herpes viruses (including simian cytomegalovirus), Coxsackie virus, and lymphocytic choriomeningitis virus. Cell tissue culture tests must show an absence of cytopathogenic effects. Adult and suckling animals are also used to screen for viable microbial agents. A neurovirulence safety test is conducted for the live oral poliovirus vaccine.

The advent of bovine spongiform encephalitis (BSE) in the 1990s generated concern that the etiological agent might be contaminating vaccines through the use of calf serum during manufacture. Industry has taken steps to ensure that the risk to vaccines from BSE is virtually zero.

New technologies can also give rise to new concerns during manufacture. For instance, some modern vaccines are grown on chick embryo fibroblasts. These vaccines may test positive by a polymerase chain reaction assay for avian leukocytosis virus, a retrovirus not unlike the HIV/AIDS virus. Because the assay is so sensitive, it detects nonreplicating fragments of the virus that have no pathological significance. The cell substrates used for measles/mumps/rubella (MMR) vaccine are derived from flocks free of avian leukosis virus. MMR vaccine is also tested extensively for adventitious viral activity. As technology improves, regulatory authorities are likely to require the application of new methods to screen for infectious agents in vaccines.

Potency

The potency (or viral titer) of a vaccine can influence both its safety and its ability to induce protective immunity. When developing new rotavirus vaccines, the safety and efficacy of low-, medium-, and high-viral-titer formulations of human–bovine reassortant vaccines were assessed in randomized controlled clinical trials of infants. The medium-potency formulation was selected for use in phase III trials as it was found to be more efficacious than the high- and low-titer formulations and had a slightly lower, but not clinically significant, incidence of fever. In another example, a high-titer formulation of measles vaccine was used in some African countries in 1989–92 to protect infants from as early as 6 months of age. Unexpected results showed that the survival of infants who received the high-titer vaccine before 9 months of age was substantially lower than those who received the standard titer vaccine at 9 months of age. The high-titer vaccine was withdrawn from use.

Live or killed vaccine

There are three main types of viral vaccines currently in widespread use: live attenuated virus (e.g., MMR, yellow fever, varicella vaccines), killed virus (e.g., influenza vaccine), and recombinant protein (e.g., hepatitis B vaccine). Safety issues differ to some extent for the three types of vaccine. For all, the risk of an adverse event is highest in the immediate post-vaccination period due to reactions to vaccine components and to programmatic errors. Live-attenuated viral vaccines must undergo viral replication in the recipient to produce sufficient antigen to generate a protective immune response. The vaccine recipient may experience a mild form of the disease, particularly at the time of peak viral load. Febrile seizure in young children 5–14 days after receiving MMR vaccine is a known adverse event. There is also the potential risk of reversion to virulence if the vaccine strain of virus undergoes genetic mutations either *in vitro* during manufacture or in the vaccinee, that can cause significant illness. Killed vaccines usually contain trace amounts of the inactivating agent (often formaldehyde) that can cause reactions in some individuals. There is also the potential risk of incomplete inactivation with the effect that the vaccine could contain trace amounts of live virus capable of replication in the host to cause disease. An atypical variant of measles infection has been reported in some recipients of killed measles vaccine who were at risk of developing severe delayed hypersensitivity reaction following exposure to wild measles virus. As well as containing no viral genetic material, recombinant protein vaccines have the advantage of presenting fewer antigens to the immune system, thus minimizing the risk of hypersensitivity reaction. However, they are often produced in yeast cells, traces of which may remain after purification and cause allergic reactions in some individuals.

Programmatic Factors, Presentation, and Administration

A 'programmatic error' is caused by an error or errors in the handling or administration of a vaccine. The error is usually the fault of a person rather than the fault of the vaccine or other technology. It can generally be prevented through staff training and an adequate supply of safe injection equipment although vaccine design factors can help minimize the chance of human error.

A programme error may lead to a cluster of events. Improper immunization practice can result in abscesses or other blood-borne infections. Freeze-dried vaccines (such as measles-containing vaccines and yellow fever) are presented as a dry powder in a vial that needs reconstituting with liquid. Errors can occur in reconstituting the powder, such as the use of inappropriate or contaminated liquid. Once a vaccine is reconstituted, it is open to the environment and makes an ideal culture medium for organisms such as *Staphylococcus.* Fatal toxic shock syndrome has been reported following administration of a multidose vial of reconstituted measles vaccine that had become contaminated. In such circumstances, a number of infants immunized from the same vial may die within a short time of injection. To avoid this catastrophe, field staff are required to discard reconstituted vaccine after 6 h, or at the end of the vaccination session, whichever is sooner.

There are options for the way a vaccine is presented that can affect its safety in the field. To reduce cost, vaccines are often presented in multidose vials. It is common in developing countries for diphtheria–tetanus–pertussis (DTP) and measles vaccines to be in 10-dose vials, while BCG is often in 20-dose vials. The downside of multidose vials is that they have the potential for contamination (see above). In contrast, single-dose presentations are unlikely to become contaminated as the vial is discarded once the single dose has been used. Prepackaged mono-dose vials reduce the need for human handling (and hence reduce the risk of contamination) but are more costly per unit. Industrialized countries tend to favor mono-dose presentations because of their increased safety and convenience.

There are a number of different modes of administration of vaccines, and these affect safety in different ways. Aerosol presentations are attractive because they do not need a needle or syringe, thus reducing the risk of contamination. However, administering an inhaled vaccine to a 9-month-old infant is not without its practical problems. Nasal delivery bypasses the intestinal barrier, but recipients with purulent rhinitis (a common situation in developing countries) may interfere with successful seroconversion. Edible vaccines and vaccines presented using skin patches are also attractive for their apparent freedom from these safety concerns.

The injectable vaccine is by far the commonest presentation in the world today. Injections are one of the most common healthcare procedures. Five to ten percent of some 16 billion injections administered worldwide each year are given for immunization. Unsafe injections or unsafe practices in relation to immunization are not only responsible for cases of hepatitis B and C, HIV, and other serious and potentially lethal side effects suffered by vaccine recipients, but may also pose an occupational hazard to health providers and an environmental hazard to communities. Furthermore, unsafe injection practices can seriously impede the progress made by immunization programmes, and have a substantial effect on global immunization coverage. Fortunately, a number of technologies are at hand that help make injections safer.

The reuse of standard single-use disposable syringes and needles becomes unsafe when they are not disposed of after use but are reused multiple times. The auto-disable syringe, which is now widely available at low cost, presents the lowest risk of person-to-person transmission of blood-borne pathogens (such as hepatitis B or HIV) because it cannot be reused. The auto-disable syringe is the equipment of choice for administering vaccines, both in routine immunization and mass campaigns. In 1999, the World Health Organization (WHO) and other United Nations organizations issued a joint statement promoting the use of auto-disable syringes in immunization services. 'Safety boxes', puncture-proof containers – for the collection and disposal of used disposable and auto-disable syringes, needles, and other injection materials – reduce the risk from contaminated needles and syringes. The safe disposal of used needles and syringes is a critical component of any vaccination program if infection is to be prevented. Poor management of injection-related waste exposes patients, healthcare workers, waste handlers, and, indeed, the wider community, to infections, toxic effects, and injuries.

Program errors may occur when a new vaccine is integrated into an existing immunization programme that has important (and sometimes subtle) differences in its presentation, preparation, or administration. Errors such as the omission of a component, injection of oral vaccines, and the use of incorrect diluent to reconstitute freeze-dried vaccines are examples of this type of errors. While thermostabilty of a vaccine is not strictly an issue of vaccine safety, failure to maintain the cold chain is a significant component of programmatic error. This can lead to vaccine failure and the need to revaccinate populations, thus re-exposing them to the risk, albeit small, of vaccine adverse reactions.

How the Recipient/Public Reacts

While the recipient of a dose of vaccine will hopefully react by developing immunity to the challenging antigen, it is common for there to be various other less desirable reactions following vaccine administration. The WHO describes an adverse vaccine reaction as either 'common' or 'rare'. Common reactions are generally mild, settle without treatment, and have no long-term consequences. More serious reactions are very rare – usually of a fairly predictable (albeit extremely low) frequency.

Rates of common, mild vaccine reactions

The purpose of a vaccine is to induce immunity by causing the recipient's immune system to react to the vaccine. It is not surprising that vaccination commonly results in certain mild side effects (**Table 1**). Local reaction, fever, and systemic symptoms can result as part of

Table 1 Summary of common minor viral vaccine reactions

Vaccine	Local reaction (pain, swelling, redness)	Fever	Irritability, malaise, and nonspecific symptoms
Hepatitis B	Adults up to 30% Children up to 5%	1–6%	
Measles/MMR	Up to 10%	Up to 5%	Up to 5%
Oral polio (OPV)	None	Less than 1%	Less than 1%[a]
Influenza	>10%	Up to 10%	Up to 10%

[a]Diarrhoea, headache, and/or muscle pains.
Note: the rates due to the vaccine administration will be lower as these symptoms occur independently as part of normal childhood.
Source: Supplementary information on vaccine safety. Part 2: Background rates of adverse events following immunization. WHO/V&B/00.36. http://www.who.int/vaccines-documents/DocsPDF00/www562.pdf.

the normal immune response. In addition, some vaccine components (e.g., aluminum adjuvant, antibiotics) can lead to reactions. A successful vaccine reduces these reactions to a minimum while inducing maximum immunity.

Rates of rare, more severe vaccine reactions

Most of the rare vaccine reactions (e.g., seizures, thrombocytopenia, hypotonic hyporesponsive episodes, persistent inconsolable screaming) are self-limiting and do not lead to long-term problems (**Table 2**). Anaphylaxis, while potentially fatal, is treatable without leaving any long-term effects.

Public perceptions of vaccine safety

While vaccine manufacture and administration have been getting progressively safer over the last 30 years, the perception of the public has been changing. Certain reports of adverse events following immunization (AEFIs) published in the medical literature over the past few years have resulted in controversy. While generating provocative hypotheses, the studies on which these reports are based have generally not fulfilled the criteria needed to draw conclusions about vaccine safety with any degree of certainty. Yet these reports have had a major influence on public debate and opinion making. Sadly, health professionals have not always handled public announcements about such matters very well and the debate has often spilt over into the political arena and policymaking. At times, mishandling of public information has resulted in a reduced public acceptance of a vaccine. Even when scientific evaluation has been correct, the public has often been swayed by other factors. This again underlines the importance of a correct assessment of causality.

Until very recently, communication with the public by health professionals and vaccine manufacturers has often been on a paternalistic basis: "We know what is best for you." The level of education of the general public has risen, helped by the Internet; freedom-of-information laws in many countries have allowed the public to access what was previously privileged information. As a result, members of the public are often well informed on a given public health issue such as vaccine safety.

Discussions about vaccine safety now must contend with this new era. It is no longer sufficient for professionals to be factually correct – they must present information in ways that are perceived to be correct. A key component of this is representing themselves in ways that generate the confidence and trust of the public. In Jordan, a vaccine scare followed hard on the heels of a scare about water contamination, with major consequences. The public lost confidence in the truthfulness of government spokespersons, even though their claims that the vaccines in question were safe later proved to be correct. A similar situation occurred in the United Kingdom when there was a large decline in MMR immunization rates after publication of a now-disproved hypothesis that MMR vaccine caused autism. One of the factors attributed to this was loss of public confidence in the government and health spokespersons because of the way the BSE situation had been handled some years earlier.

Monitoring and Assessment of Vaccine Safety

The clearest and most reliable way to determine whether an adverse event is causally related to vaccination is by comparing rates of the event in a vaccinated and nonvaccinated group in a randomized clinical trial. Such trials, however, are unlikely to be large enough to assess very rare events, and postmarketing surveillance systems are required to identify events potentially related to vaccination. Postmarketing surveillance capability is improving; more countries now have monitoring systems, and more importance is attached to the reporting of suspected links between vaccination and adverse events. These systems have been successful in bringing to light serious adverse events after vaccines have been approved for use by regulatory authorities. A recent example is intussusception after administration of reassortant rhesus rotavirus vaccine.

Table 2 Summary of rare, serious viral vaccine reactions, onset interval, and rates

Vaccine	*Reaction*	*Onset interval*	*Rate per million doses*
Hepatitis B recombinant vaccine	Anaphylaxis	0–1 h	1–2
	Guillain–Barré syndrome (plasma derived)	1–6 weeks	5
Measles/MMR[a]	Febrile seizures	5–12 days	333
	Thrombocytopenia (low platelets)	15–35 days	33
	Anaphylaxis	0–1 h	1–50
Oral polio (OPV)	Vaccine-associated paralytic poliomyelitis (VAPP)	4–30 days	1.4–3.4[b]
Japanese encephalitis	Serious allergic reaction		10–1000
	Neurological event		1–2.3
Yellow fever	Post-vaccination encephalitis	7–21 days	500–4000 in infants less than 6 months[c]
	Allergic reaction/anaphylaxis	0–1 h	5–20
	Catastrophic system collapse resembling naturally acquired infection	2–7 days	2–32 (higher rates in the elderly or history of thymoma)[d]

[a]Reactions (except anaphylaxis) do not occur if already immune (~90% of those receiving a second dose); children over 6 years unlikely to have febrile seizures.
[b]VAPP risk is higher for first dose (1 per 1.4–3.4 million doses) compared to 1 per 5.9 million for subsequent doses and 1 in 6.7 million doses for contacts.
[c]Isolated cases with no denominator make it difficult to assess the rate in older children and adults, but it is extremely rare (less than four cases per 8 million doses).
[d]Studies have reported >10-fold higher risk of this rare adverse event among elderly vaccine recipients. There is a suggestion that reduced thymus gland activity (through age, immunosuppression, or surgical removal) increases the risk.
Source: Supplementary information on vaccine safety. Part 2: Background rates of adverse events following immunization. WHO/V&B/00.36. http://www.who.int/vaccines-documents/DocsPDF00/www562.pdf.

Evaluating Adverse Events – Causality Assessment

Clinical events such as anaphylaxis or rash may occur subsequent to vaccination. But are they 'caused' by vaccination? It would be highly likely that collapse within minutes of receiving a dose of vaccine was due to the vaccine. But a rash 3 weeks later may not be – more likely it was coincidental. It is the area between these two scenarios that needs clarification.

Assessment of whether a given vaccine causes a particular adverse reaction varies from the casual observation to the carefully controlled study. The public frequently forms a decision about a vaccine's safety based on the information available to them – often a report based on unscientific observations or analyses that fail to stand the scrutiny of rigorous scientific investigation. Submitting a study to a scientific process rather than to partially informed opinion is crucial in determining whether a vaccine actually causes a given reaction. If undertaken carelessly or without scientific rigor, the study results will be inconclusive at best, may result in the inappropriate withdrawal of a valuable vaccine from use, or at worst may result in the exposure of a population to a dangerous vaccine.

When the Global Advisory Committee on Vaccine Safety (GACVS) undertakes causality assessment, the criteria it looks for in arriving at conclusions include consistency of observations, strength of the association, specificity, temporal relation, and biological plausibility. Clearly, not all these criteria need to be present, and neither does each carry equal weight for a causal relationship between an adverse event and the vaccine to be determined. Biological plausibility is a less robust criterion than the others described.

An association between vaccine administration and an adverse event is most likely to be considered strong when the evidence is based on well-conducted human studies that demonstrate a clear association in a study specifically designed to test the hypothesis of such association. Such studies may be randomized controlled clinical trials, cohort studies, case-control studies, or controlled case-series analyses. Case reports by themselves, however numerous and complete, do not fulfill the requirements for testing hypotheses.

Surveillance of AEFIs

Vaccines are generally administered to healthy individuals, so any report of vaccine administration causing harm needs to be taken seriously. Although the vaccines in general used in national immunization programmes are extremely safe and effective, adverse events can occur following vaccine administration and no vaccine is perfectly safe. In addition to the vaccines themselves, the process of immunization is a potential source of adverse events. An AEFI is any adverse event that follows immunization that is believed to be caused by the immunization. Using this terminology allows description and analysis of the event without prejudging causality. Immunization will tend to be blamed for any event that happens after immunization. AEFIs can be classified into five categories:

- vaccine reaction: a reaction due to the inherent properties of the vaccine;
- programme error: an error in the immunization process;
- coincidental event: unrelated to the immunization, but has a temporal association;
- anxiety-related: a reaction can arise from the pain of the injection rather than the vaccine; and
- unknown: in some cases the cause of the AEFI remains unknown.

It is important to investigate a report of an AEFI to determine into which of the above categories it fits. The response to an AEFI will vary depending on the cause. However, many countries have ineffective reporting systems or no system at all. Currently, only 35% of 192 countries, and only 25% of 165 nonindustrialized countries, have an adequately functioning system for monitoring adverse events following immunization. The benefits of good surveillance of AEFIs are shown in **Table 3**.

Benefits of monitoring AEFIs

As vaccine-preventable infectious diseases continue to decline, people have become increasingly concerned about the risks associated with vaccines. Furthermore, technological advances and continuously increasing knowledge about vaccines have led to investigations focused on the safety of existing vaccines which have sometimes created a climate of concern. Allegations regarding vaccine-related adverse events that are not rapidly and effectively dealt with can undermine confidence in a vaccine and ultimately have dramatic consequences for immunization coverage and disease incidence.

Table 3 Benefits of monitoring AEFIs

Improves vaccine safety by creating awareness among health professionals of the risks of vaccination
Identifies urgent problems that need investigation and action
Improves capacity to respond to AEFI reports
Improves the quality of the vaccination programme by monitoring performance and increasing staff confidence
Stimulates other aspects of programme monitoring, e.g., case reporting
Public perceives that health authorities understand and monitor vaccine safety, building public confidence in immunization
Detects signals for potential follow-up and research
Estimates rates for serious AEFIs
For comparison between products
To determine risks and benefits of immunization
To validate prelicensure data
Identifies programmatic errors and batch problems

Not only is public awareness of vaccine safety issues rising, there are increasing opportunities for AEFIs to occur due to various factors. For instance, an increasing number of vaccine doses is now given in campaigns than ever before (**Table 4**). Polio virus is targeted for eradication and measles elimination is being implemented in many parts of the world. As a result, vast numbers of doses of vaccine are being given over short time intervals. Thus, even if the AEFI rates stay the same, the actual number of cases is likely to be much higher than would be expected over the same time period, simply because so many doses have been administered.

Classical Safety Events

SV40

Between 1959 and 1963, both inactivated and live attenuated poliovirus vaccines were inadvertently contaminated with simian virus 40 (SV40), a monkey virus known to be oncogenic for newborn hamsters. SV40 was found to be present in monkey kidney tissue used for propagation of the poliovirus during manufacture. From 1961, manufacturers were required by control authorities to test for SV40 using infectivity assays and different cell cultures, and lots positive for SV40 were not released. A few years later, the source of monkeys used for production was changed to species that do not harbor SV40. There is no evidence that polio vaccine administered since the early 1960s has contained SV40, and large epidemiological studies have not identified an elevated cancer risk in persons who received SV40-contaminated vaccines, but fragments of SV40 DNA have recently been identified in certain human tumors.

Swine flu

Influenza vaccine has been the source of a number of issues related to safety over the years. Mass immunization against swine-like influenza was carried out in 1976–77 using the A/New Jersey swine-influenza vaccine. Initially it appeared that this vaccine generated an increased risk of acquiring vaccine-related Guillain–Barré syndrome (GBS). In

Table 4 Possible effect of immunization campaigns on vaccine safety

A real rise in program errors due to pressure on staff
Increased workload on staff can increase human error
Extra or new staff may not be fully trained in administration technique
Because large numbers of doses are administered there may be a rise in the absolute number of AEFIs even though rates may not change
Campaigns can generate popular opposition and rumors about vaccine safety
Rate of AEFIs prior to campaign may be unknown, so that any report of an AEFI appears to be due to the campaign

contrast, the 1978–79 influenza vaccine was not associated with a statistically significant excess risk of GBS. Although the original Centers for Disease Control study of the relation between A/New Jersey/8/76 (swine flu) vaccine and GBS demonstrated a statistical association and suggested a causal relation between the two events, controversy has persisted. At least one recent review of the data seemed to confirm the link. Another review suggested that there was no increased risk of acquiring GBS associated with the influenza vaccines administered during these seasons and that the causative 'trigger agent' in the A/New Jersey (swine) influenza vaccine administered in 1976 has not been present in subsequent influenza vaccine preparations.

There was no increase in the risk of vaccine-associated GBS from 1992–93 to 1993–94. A study by Lasky suggested only slightly more than one additional case of GBS per million persons vaccinated against influenza. More recently, an oculorespiratory syndrome has been described following influenza vaccine.

Recent Safety Events

A number of events in recent years has provoked concern that certain vaccines might be unsafe. Most of these concerns have proved to be unfounded, and some remain unproven or incompletely understood.

Thiomersal

Thiomersal is a mercury-containing preservative used in certain liquid vaccines such as DTP and hepatitis B vaccines. It came into the public gaze in 1999 as the result of a study being undertaken to estimate how much mercury was being administered to children in North America through vaccines. It transpired that the amount exceeded the allowed maximum according to environmental health permitted limits. Scientists assumed that ethyl mercury in thiomersal would have the same toxicology as methyl mercury, a chemical about which a great deal was known. Over the ensuing years, evidence was accumulated that showed it actually had a different metabolic pathway in the human body compared with methyl mercury. Both were metabolized in the liver and excreted in the gut, but methyl mercury was reabsorbed and accumulated in the body. Ethyl mercury, on the other hand, was found to be passed in the stool and lost from the body, thus avoiding a cumulative effect. Overwhelming scientific evidence indicates no clear link between neurodevelopmental disorders or autism and the receipt of vaccines containing thiomersal. One set of authors clearly flies in the face of the weight of scientific evidence, yet support a widely held public belief that thiomersal (in the quantities present in vaccines) is bad for children. Live viral and live bacterial vaccines do not contain thiomersal, largely because the preservative would kill the vaccine organism, rendering the vaccine useless.

Hepatitis B vaccine and multiple sclerosis

Reports of multiple sclerosis developing after hepatitis B vaccination have led to the concern that this vaccine might be a cause of multiple sclerosis in previously healthy subjects. A number of reliable studies have been consistent in showing that there is no association between hepatitis B vaccination and the development of multiple sclerosis.

MMR autism and bowel disease

Reports have proposed a link between the administration of the MMR vaccine and juvenile autism, Crohn's disease, and other forms of inflammatory bowel disease. The weight of scientific evidence strongly favors the conclusion that there is no direct association between measles virus, measles vaccine, autism, or inflammatory bowel disease.

Intranasal influenza vaccine and Bell's palsy

After the introduction of an inactivated intranasal influenza vaccine that was used only in Switzerland, 46 cases of Bell's palsy were reported. The relative risk of Bell's palsy in one study was found to be 19 times the risk in the controls, corresponding to 13 excess cases per 10 000 vaccinees within 1–91 days after vaccination. As a result, this vaccine is no longer in clinical use.

Yellow fever vaccine

Until recently, yellow fever vaccine was considered to be very safe. It has been given to many millions of people over more than 60 years to protect against yellow fever virus infection, which can be fatal in 5% or more of cases. From 1996 onward, a previously unrecognized adverse event of multiorgan failure and death was detected through passive surveillance in the United States, Australia, and other countries. Subsequent studies found that the risk of this rare adverse event was approximately 3 per 1 million doses, and vaccine recipients aged 65 years and older had a 10-fold higher risk than younger vaccinees, particularly following the first dose of the vaccine.

HIV Infection

The era of the HIV epidemic has had an impact on vaccine safety in a number of ways.

- It may be transmitted by injection (not only during vaccination). It could thus be transmitted by incorrect injection practices during routine immunization to a small number of infants attending a clinic at the same time. And it could theoretically be transmitted during mass campaigns where many more children might be at potential risk.
- Staff may be infected because of needle-stick injury or other improper handling of infected body fluids.
- It may alter the immune response of the recipient, either making him/her vulnerable to some negative

effect of the vaccine, or diminishing the effect the vaccine has on protecting the infant. Some vaccines are not as effective in more immune-suppressed individuals, and severely immune-depressed infants must avoid vaccines such as BCG as they are at risk of disseminated BCG, which may be lethal.

- The course of the disease might be altered. The activation of CD4+ T lymphocytes following immunization could potentially augment HIV replication and result in accelerated progression to disease. Several, but not all, investigators have described increased HIV RNA plasma levels lasting several days following immunization.

On the positive side, the HIV epidemic has alerted health workers to the potential dangers of contamination during vaccine administration and has resulted in the safer use of syringes and proper disposal of injection equipment.

Conclusions

The design and manufacture of vaccines are of such high standards today that they have become extremely safe. But it will never be possible to say that everything is known about the subject and that vaccines are totally safe. Nor can the unpredictable human element in administration ever be totally excluded from the equation. But new methods of safety testing will emerge as each new threat is discovered. And new technologies will complement existing safety measures. However, communicating to vaccine recipients or their parents that the vaccine they are about to receive is neither totally effective nor totally safe makes health professionals uncomfortable. But the reality is that vaccines are many orders of magnitude safer than contracting the disease. Through constant vigilance, we can anticipate that vaccines will remain extremely safe – as safe as humanly possible. Any lesser alternative is unacceptable.

See also: Antigenicity and Immunogenicity of Viral Proteins; DNA Vaccines; Antibody - Mediated Immunity to Viruses; Neutralization of Infectivity; Vaccine Strategies.

Further Reading

Afzal MA, Minor PD, and Schild GC (2000) Clinical safety of measles, mumps and rubella vaccine. *Bulletin of the World Health Organization* 78(2): 199–204.

Bellaby P (2003) Communication and miscommunication of risk: Understanding UK parents' attitudes to combined MMR vaccination. *British Medical Journal* 327: 725–728.

Causality assessment of adverse events following immunization(2001) *Weekly Epidemiology Record, 23 March, 12,* 76: 85–92. http://www.who.int/vaccine_safety/reports/wer7612.pdf(accessed February 2007).

Clements CJ and McIntyre PB (2006) When science is not enough – A risk/benefit profile of thiomersal-containing vaccines. *Expert Opinion on Drug Safety* 5(1): 17–29.

Cutts FT, Clements CJ, and Bennett JV (1997) Alternative routes of measles immunization: A review. *Biologicals* 25: 323–338.

Dellepiane N, Griffiths E, and Milstein JB (2000) New challenges in assuring vaccine quality. *Bulletin of the World Health Organization* 78: 2155–2162.

Halsey N (2003) Vaccine safety: Real and perceived issues. In: Bloom BR and Lambert P-H (eds.) *The Vaccine Book*, pp. 371–385. Amsterdam: Academic Press.

Kharabsheh S, Al Otoum H, Clements CJ, *et al.* (2001) Mass psychogenic illness following Td vaccine in Jordan. *Bulletin of the World Health Organization* 79(8): 764–770.

Madsen KM, Lauritsen MB, Pedersen CB, *et al.* (2003) Thimerosal and the occurrence of autism: Negative ecological evidence from Danish population-based data. *Pediatrics* 112(3, Part 1): 604–606.

Moss WJ, Clements CJ, and Halsey NA (2003) Immunization of children at risk for infection with human immunodeficiency virus. *Bulletin of the World Health Organization* 81(1): 61–70.

Safranek TJ, Lawrence DN, Kurland LT, *et al.* (1991) Reassessment of the association between Guillain-Barré syndrome and receipt of swine influenza vaccine in 1976–1977: Results of a two-state study. Expert Neurology Group. *American Journal of Epidemiology* 133(9): 940–951.

Shah K and Nathanson N (1976) Human exposure to SV40: Review and comment. *American Journal of Epidemiology* 103: 1–12.

Simonsen L, Kane A, Lloyd J, Zaffran M, and Kane M (1999) Unsafe injections in the developing world and transmission of bloodborne pathogens: A review. *Bulletin of the World Health Organization* 77(10): 789–800.

Supplementary information on vaccine safety. Part 2: Background rates of adverse rates of adverse events following immunization. WHO/V&B/00.36. http://www.who.int/vaccines-documents/DocsPDF00/www562.pdf(accessed February 2007).

Thompson NP, Montgomery SM, Pounder RE, and Wakefield AJ (1995) Is measles vaccination a risk factor for inflammatory bowel disease? *Lancet* 345: 1071–1074.

Relevant Websites

http://www.cioms.ch – Council for International Organizations of Medical Sciences (CIOMS) – WHO Working Group on Vaccine Pharmacovigilance.

http://www.brightoncollaboration.org – The Brighton Collaboration.

http://www.who.int – The WHO Expert Committee on Biological Standardization; WHO Programme for International Drug Monitoring; WHO's Global Advisory Committee on Vaccine Safety (GACVS).

http://www.who-umc.org – Uppsala Monitoring Centre (UMC), A WHO Collaborating Center.

Neutralization of Infectivity

P J Klasse, Cornell University, New York, NY, USA

Glossary

Affinity The strength of an intermolecular interaction, for example, of antibody (Ab) binding to antigen (Ag).

Dissociation constant K_d [M], in the law of mass action for a bimolecular association, it corresponds to the concentration of an unbound molecule that yields half-maximal binding to the other molecule. It quantifies affinity: the lower the K_d, the higher the affinity.

Epitope The site on an antigen that makes contact with the paratope of an Ab.

Molecularity The number of molecules involved in the rate-limiting step of a chemical reaction; here it refers to the number of Abs that must bind to a virus particle in order to neutralize it.

Occupancy θ, the degree to which one molecule is ligated by another, for example, the percentage of Ag sites bound by Ab.

Paratope The part of an Ab molecule that makes contact with the epitope of an Ag.

Stoichiometry In the context of neutralization, it is the study of integer ratios of molecules in Ab–virion complexes.

Valency The number of separate, usually identical, areas of contact in bimolecular association.

Introduction

Miscellaneous agents inhibit viral infection. Among them are antibodies (Abs), which have disparate effects on viral infection both *in vivo* and *in vitro*. One such effect is virus neutralization, which can be stringently defined as the reduction in infectivity by interference before the first biosynthetic step in the viral replicative cycle, through the binding of antibodies to epitopes on the surface of the virion. The definition delimits neutralization from other effects of Abs on viral replication and on infected cells. For example, Abs to neuraminidase of influenza virus inhibit viral release from the cell surface, a late step in the replicative cycle. That does not qualify as neutralization. Another example is Abs to viral receptors. Such Abs can block infection, but they do not fulfill the criterion of binding to the virion.

Maybe the definition too restrictively specifies Abs as the neutralizing agents. Obviously, some loosening of this criterion will allow a comparison with the neutralizing effects of fragments of Abs, peptide mimics of paratopes, fragments of receptors and their mimics, or even other ligands, as long as they bind to the viral surface.

What the definition implicitly does include is also noteworthy. Enveloped viruses carry cellular passenger antigens (Ags). If Abs to these inhibit infection, that counts as neutralization. Under certain circumstances, Abs aggregate virions. If that diminishes infectivity, the definition includes it as neutralization – although some researchers have chosen not to. Similarly, the definition can accommodate for effects of complement, in addition to that of the Abs, as enhancement of neutralization. Whether or how such enhancement occurs is a matter for investigation. As a case in point, a report found neutralization through virolysis to be more complement-enhanced when mediated by Abs to a retroviral transmembrane protein than by Abs to the surface unit of the same envelope glycoprotein complex. Notably, the definition stipulates little about mechanism. That is left to hypotheses and their testing by experiment.

A closer look at the defining terms 'reduction', 'first biosynthetic step', and 'binding' raises some questions. First, does any degree of 'reduction' count as neutralization? Does neutralization entail the complete abrogation of the infectivity of at least some virions? Or can it be partial, a mere reduction in the propensity to infect? Second, the 'first biosynthetic step' occurs after viral entry. Why does the definition put the limit at so late a step? How could Abs affect events occurring after the viral genome or core has entered the cytoplasm from the cell surface or from an endosomal vesicle? Controversial cases of postentry neutralization have been promulgated. Therefore, the definition should not exclude them; they must be refuted or corroborated by experiment. Empirically then, which replicative steps does neutralization block? Third, the definition makes Ab 'binding' a necessary condition for neutralization. But is it also sufficient? Or are there Abs that can bind to surface epitopes on virions but still not neutralize? And if so, why?

Experimental Measurement

Passive immunization, that is, transfer of neutralizing antibodies (NAbs), can protect an organism against some viral infections. Many successful vaccines induce high titers of NAbs and neutralization is, quite plausibly, often crucial to immune defenses *in vivo*. But it is measured *in vitro*. Classically, a neutralization assay consists of four steps: First, virus is incubated with test and control Abs. Then these reaction mixes are added to target cells, and the virus is left to adsorb. After free virus is washed off, virus that has entered the cells is allowed to replicate. Finally, the degree of replication is measured.

This basic scheme comes in different forms. A key distinction is between single- and multicycle assays. Early studies measured inhibition of replication in plaque- and focus-forming assays. These quantal assays were adapted for both bacteriophage and animal viruses. Such tools have a theoretical advantage: if the inoculum is dilute, a primary plaque or focus most likely will represent the local spread from a single infectious event. Furthermore, if progeny virus is prevented from forming secondary plaques and foci, the number of infectious units measured will correspond to the first cycle of replication. Then the degree of neutralization can be expressed as the proportion of initial infectious units prevented from infecting.

Today neutralization studies often employ defective viruses: a viral gene has been deleted and replaced with a reporter gene. Envelope glycoprotein genes can be supplied in *trans*. Hence, one can make pseudotype particles with neutralization targets of choice. Like quantal assays, this method limits the infectivity to the first replicative cycle.

Other neutralization assays measure the reduction in viral Ag production during the culture. If multiple replicative cycles are allowed, though, the reduction in Ag may not be proportional to the fraction of virus that was originally inactivated by the binding of Abs.

Antibody–Antigen Binding and Epitope Occupancy on Virions

The binding of Abs to epitopes on virions is the *sine qua non* of neutralization. The concentration of the antibody, together with its affinity for the epitope, determines the occupancy, θ, of Ab on the virion, that is, the proportion of specific epitopes that the Ab ligates. If the occupancy is too low, the virion will not be neutralized. Instead, the infectivity may even be enhanced by subneutralizing occupancies.

The occupancy can be estimated from the law of mass action, provided the molar excess of Ab over Ag is vast. Usually, this is the case since NAbs tend to neutralize in the nM, or higher, concentration range, which viral Ags in inocula rarely reach. This is reflected in the percentage law: the proportion neutralized by an Ab at a certain concentration is approximately constant when the amount of virus is varied.

Suppose we know the number of epitopes on the virion, and the K_d of the Ab for binding to those epitopes. In theory, we can then estimate how many Abs per virion are bound at any given Ab concentration. In practice, however, this is often impossible. Notably, virions do not always have a constant number of epitopes. The virions of some viruses vary in size. Also, enveloped viruses can incorporate varying numbers of envelope glycoprotein molecules; subunits of these can be lost from the virions after budding. Furthermore, the relevant K_d is seldom known. Often the K_d is measured for Ab binding to a form of the neutralization Ag that differs from the functional one on the infectious virion: the binding studies may have used an unprocessed precursor or a truncated, immobilized monomeric subunit of an oligomer. In addition, the oligomeric context of the epitope strongly affects its antigenicity. The spacing of Ag sites on virions, as well as in binding assays, can also affect the avidity, and therefore the functional affinity, of Ab binding. Clearly, the relevant affinity of NAbs is central to an understanding of neutralization.

The IC_{50} of a NAb, that is, the concentration at which it neutralizes half of the inoculum, is also a notable variable. According to the reasoning above, it should stand in a definite relation to the K_d for Ab binding to virions. If the minimum neutralizing occupancy happens to be 50%, the IC_{50} and the K_d will be similar. With higher occupancies required, we get $IC_{50} > K_d$; with lower, $IC_{50} < K_d$ – all according to the formula $\theta = ([Ab]/K_d)/(([Ab]/K_d) + 1)$.

The maximal extent of neutralization is intriguing. As mentioned, some Abs fail entirely to neutralize. Others have only marginally inhibitory effects. Yet others may give a plateau of neutralization at around, say 75% or 90%. We can imagine a whole spectrum. At one extreme end of the spectrum, where we find the most effective NAbs, no residual infectivity is detectable with low-infectivity inocula. But with higher viral doses, even a strong neutralization leaves a detectable residual infectivity. Furthermore, when neutralization is studied over time, it may reach a plateau. Such residual infectivity, called the 'persistent fraction', has been studied extensively. Usually, the persistent and the neutralized fractions do not differ genetically. Nor does spontaneous aggregation of virus explain all persistence. No comprehensive explanation has been found, but some contributing factors deserve mention. Because of both steric hindrance and the stochastic nature of binding, the Ab may not yield a sufficient occupancy for a complete abrogation of the infectivity of all virions. Conformational, and therefore antigenic, heterogeneity among epitopes may leave some unoccupied. Moreover, the Ab binding is in principle reversible, and at the cell surface Abs and receptors may compete for binding to the virion. Finally, NAbs act in parallel with, and may differentially affect, spontaneous decay of the virus. A better understanding of these dynamics may suggest which properties of NAbs render them most effective. Suffice it to say that the maximal extent of neutralization is not predictable from the IC_{50} value of an Ab. And *in vivo*, that maximum may be more decisive than a titration midpoint.

Mechanism

The first replicative step prevented by the Ab is one aspect of the neutralization mechanism. Another is how, at the intermolecular level, the Ab effects this block.

The initial step in the replicative cycle is attachment to a susceptible cell. NAbs to widely different viruses block this step. For example, Abs to distinct epitopes of influenza virus hemagglutinin (HA), when occupying one in four HA trimers, neutralize by blocking viral attachment. Some viruses, however, interact with ancillary attachment molecules on the surface of cells before docking onto their major receptors. In such cases, the NAb may not interfere with the initial attachment but with the subsequent receptor interactions. Moreover, assortments of NAbs to HIV-1, rotavirus, and papilloma virus show widely different capacities to block the attachment of each virus. These differences may stem from how distally on a viral spike or other protrusion the epitope is located, and at what angle to the virus particle the Ab binds; other contributing factors may be the maximum number of epitopes that can simultaneously be occupied, the size of the Ab, and the valency of its binding.

After attachment to the cell, the next obligatory step for many viruses is internalization by endocytosis. Hence, if endocytic uptake is blocked by the Ab, the virus is neutralized. Failing that, neutralization can still occur inside the endosome. By delaying fusion or penetration, the Ab will promote the eventual degradation of the virus in lysosomes.

We can also hypothesize the converse. Some viruses cannot use the endosomal route for productive entry. Therefore shunting the virion–Ab complex onto that route, for example, through Fc–receptor interactions, would instead lead to neutralization.

After docking onto a major receptor, many viruses interact with co-receptors as a necessary entry step. If that interaction is blocked by an Ab, the virus is neutralized. At this stage, however, steric constraints may prevent access to the relevant epitopes, as multimolecular complexes form. Indeed, a Fab binding adjacently to the co-receptor binding site of HIV-1 neutralizes more efficiently than the corresponding immunoglobulin G (IgG).

We can study post-attachment neutralization (PAN) by letting the virus adsorb to cells at a temperature too low to allow entry. After that, the Ab is allowed to bind to the adsorbed virus. As the cells are warmed up, the non-neutralized virus will infect. At which steps, then, can PAN intervene? In spite of the name and experimental setup, PAN does not necessarily affect only pure post-attachment events. In fact, Abs that interfere directly with virus–receptor interactions often mediate PAN. That may be explained by the necessity of recruiting multiple receptors into a fusion or entry complex. The lower temperature prevents collateral diffusion of receptors; the Ab covers remaining sites not engaged by the receptors; then, at the warm-up, the Ab precludes the requisite additional receptor interactions. Or else PAN occurs because the Ab induces the dissociation of the virus particle. Most importantly, we cannot declare that PAN involves major neutralization mechanisms just because we can make it happen. For when Abs are instead first allowed to bind to virions, attachment may be blocked as the complexes later encounter the cells, which pre-empts any further replicative steps. It also pre-empts PAN.

Particularly the neutralization of picornaviruses, such as poliovirus and rhinovirus, was long thought to interfere with a late penetration step and to depend on the induction of conformational changes in their antigens. Many NAbs can change the isoelectric point, pI, of the picornavirus particle. But their capacity to do so correlates poorly with their neutralizing efficacy. Besides, many of these Abs effectively block virus attachment.

Structural studies have shown that sometimes the paratope of a NAb changes more than the epitope upon binding. An extreme example is one of the most potent NAbs against rhinovirus, which remolds its paratope but leaves the Ag conformationally intact. Furthermore, if neutralization depended on conformational changes in the Ag, there ought to be Abs that block neutralization by competing with NAbs, thereby preventing conformational changes. Such antibodies, though, are not readily found.

Of course, viral proteins, while mediating entry, change conformation or even refold. Therefore, hampering such changes may promote neutralization at the postattachment stage. Or blocking interaction with a receptor, the role of which is to trigger conformational changes, will prevent them by default.

The block of receptor interactions is not necessarily direct. On the contrary, because of the relative sizes of surface viral Ags and Abs, indirect steric interference is potentially strong. Furthermore, when an irrelevant epitope is tagged onto an attachment- and entry-mediating viral protein, at least in some cases, Abs to that epitope can neutralize the virus. What seems most crucial, therefore, is simply the Ab occupancy on functional viral proteins.

Kinetics

The dissociation constant, K_d, is equal to the ratio of the two rate constants for the binding reaction, k_{off}/k_{on}. Yet how the kinetics of NAb binding influences the potency and extent of neutralization is largely unexplored. It is known, though, that the higher the density of target cells and receptors, the greater the demands on neutralization. Thus, the kinetic constants for the Ab, as well as for the receptor binding may influence neutralization in the competitive situation at the cell surface.

One kind of kinetic study takes advantage of PAN. When cells with attached virus are warmed up, NAbs differ in the length of time during which they can still neutralize. Several factors may be responsible for such differences: how fast the on-rate is, whether the epitopes are only transiently exposed, and precisely when the Ab interferes in the chronology of molecular entry events.

Kinetic studies of a different sort have influenced the theory of neutralization profoundly. If the reaction between Ab and virus can be stopped, and the mixture rapidly transferred to target cells, the decline in infectivity can be monitored as a function of reaction time. Much debate has focused on the first part of such kinetic plots of declining relative infectivity. The absence of a shoulder on the curve was interpreted as evidence of neutralization through the very first Ab binding events. It has been questioned, though, whether Ab binding is sufficiently slow to allow the recording of the infectivity of virions with single Abs bound. In any case, at low temperatures, with low concentrations of low-affinity Abs, such kinetic neutralization curves do evince shoulders. Even if the first few Abs reduce the infectivity, they may not abrogate the entire infectious potential of a virion. Perhaps they only dent it. Such interpretations depend on whether neutralization and infectivity are all-or-nothing phenomena or more gradual and incremental. Another complication is the need to estimate stochastic, rare multihits when on average only few Abs have bound.

In a seminal study, Dulbecco *et al.* demonstrated that the rate of neutralization of poliovirus and Western equine encephalitis virus, expressed as a function of Ab concentration, follows approximate first-order kinetics. They took this as evidence that each virion is neutralized by a single Ab, so-called 'single-hit neutralization'. The order of a reaction, however, does not have to be an integer; it is not equivalent to the number of Abs required for neutralization of a virion (see section 'Molecularity'). In addition, the discussion above of the molar ratio of Ab to Ag in an ordinary neutralization reaction makes it clear that what Dulbecco and followers have observed are reactions of 'pseudo'-first order. This is still of practical value. For it allows a calculation of the 'neutralization rate constant'. But the kinetic study of reaction order cannot determine whether neutralization requires single or multiple hits.

Molecularity

Molecularity is the minimum number of molecules involved in the rate-determining step of a reaction. In the case of neutralization, the two reactants are Ab and virus. The molecularity is then the minimum number of Abs a virion must bind in order to be neutralized. Stoichiometry describes only relative amounts, but since the goal is to find the number of Abs on each virion, that term is also often used in this context.

The average number of Abs bound to virus particles, at different recorded degrees of neutralization, has been ascertained through radio labeling and electron microscopy. Such data allow us to calculate the minimal number of Abs required for neutralization, if we assume that Ab binding follows a Poisson distribution (**Figure 1**). For example, if a virus were neutralized by the single hit of an Ab, then with an average of 1 Ab per virion, 37% of the initial viral infectivity would remain. That is merely a special case. It therefore makes no sense to identify the number of Ab molecules on the x-axis at 37% residual infectivity as the number of hits required for neutralization, except on the single-hit curve. Instead, a proper Poisson analysis involves the plotting of many relative infectivities as a function of the number of bound Abs per virion. Then these empirical data should be compared with the Poisson-derived molecularity curves. The closest one best represents the molecularity.

An additional premise of this Poisson analysis is that a virion is either infectious or not. In other words, there is an absolute threshold of neutralization: with a minimum number of Abs bound, the virion loses all its infectivity at one fell swoop. A radical alternative to this view is that every Ab bound diminishes the infectious propensity of the virion by an equal amount. In fact, there is much evidence of partial reductions in infectivity by NAbs. But a plausible model of neutralization might locate the effects of Abs somewhere in between the two models: there may be a zone of occupancies over which the virion suffers drastic but not total losses in infectivity.

Much direct evidence supports multihit molecularities of neutralization. Four to five Abs are required to give significant neutralization of poliovirus, 36–38 for papillomavirus, 70 for influenza virus, and 225 for rabies virus. Accordingly, Burton and co-workers found an approximately linear relationship between surface area of the virion and the number of Abs required for neutralization. Neutralization could then be seen as the coating of virus particles by Abs, which could interfere with attachment or entry directly or indirectly. The theory can be refined. For example, two enveloped viruses with the same surface area but different numbers of envelope glycoprotein spikes may not require the same number of Abs for neutralization. There is much evidence to support this theory and little to refute it.

Escape and Resistance

Viral genes vary in sequence to widely different extents. The most variable ones encode neutralization antigens; within such antigens surface-exposed elements, often loops containing some of the neutralization epitopes, vary most (**Figure 2**). If a mutation in such a region reduces the affinity of a NAb, then the virus gains a replicative advantage in the presence of the Ab, *in vivo* or *in vitro*. The selection of such mutations constitutes viral escape from neutralization.

Suppose neutralization did not result directly from Ab binding, but indirectly from the induction of conformational

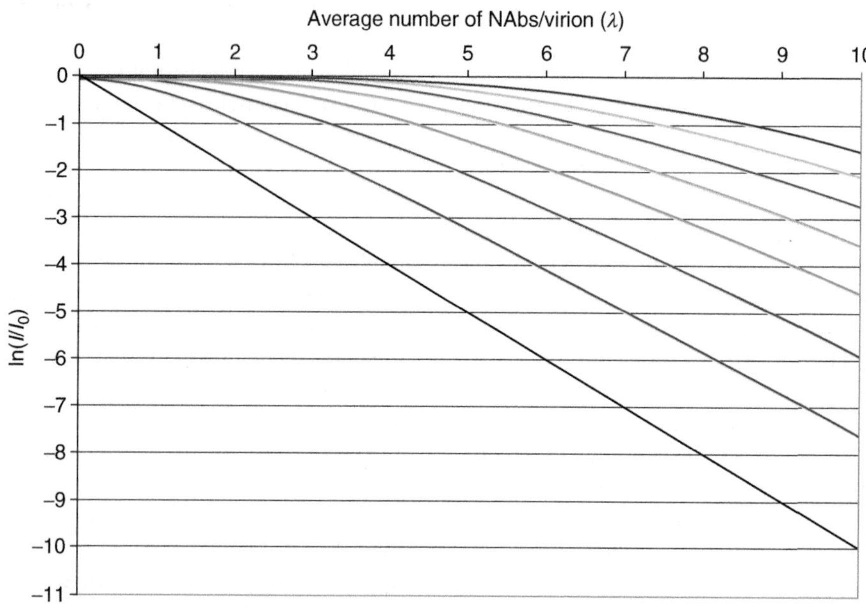

Figure 1 Molecularity of neutralization. The diagram shows the virus infectivities that can be predicted from a Poisson distribution of low numbers of Abs bound to virions. Different absolute thresholds for neutralization are assumed. The natural logarithm of the still infectious fraction of virus, $\ln(I/I_0)$ is plotted on the *y*-axis as a function of the average number of NAbs per virion, λ, on the *x*-axis. The latter can be determined, for example, by radio-labeling techniques or electron microscopy. If the minimum number of NAbs per virion required for neutralization is *L*, then the infectious fraction of virions will be equal to the cumulated fractions with fewer than *L* NAbs bound to them: $I/I_0 = \Sigma_{r=0}^{L-1}(\lambda^r e^{-\lambda})/r!$ Thus with $L = 1$, the contested single-hit molecularity, and at an average of 1 NAb per virion, $\lambda = 1$, $I/I_0 = e^{-1}$ so that $\ln(I/I_0) = -1$: if the single-hit hypothesis were true, approximately 37% of the infectivity would remain when the virions have on average one NAb bound to them. As can be seen in the diagram only the curve for $L = 1$ (black) goes through the point with the coordinates [1,−1]. The other curves represent other molecularities: $L = 2$ (red), $L = 3$ (blue), $L = 4$ (green), $L = 5$ (orange), $L = 6$ (magenta), $L = 7$ (cyan), and $L = 8$ (purple). These curves do not go through [*L*,−1]. It has been a common error in the literature to read the x value at $y = -1$, or 37% infectivity on a nonlogarithmic scale, and assume that to be the value of *L*, the number of hits required for the neutralization. Instead, the diagram shows that the best fitting molecularity can be obtained by comparing the empirical with the theoretical curves.

changes. Then viral escape from neutralization might sometimes be mediated by mutations blocking those secondary effects of the Abs rather than diminishing the Ab binding itself. Yet, a study of rhinovirus found that all mutations conferring neutralization escape also reduced Ab binding. The routes of escape taken by viruses thus shed light on the mechanisms of neutralization.

Some NAbs bind to more conserved epitopes, which are so intricately involved in attachment or entry functions that the virus can accommodate no great changes there. To produce such Abs has a fitness value for the host; to minimize their elicitation has a fitness value for the virus. A result of these evolutionary forces pitted against each other can be seen in influenza virus. As influenza epidemics rage, the viral HA protein keeps varying residues located on a rim surrounding a receptor-binding pocket. The mutations abrogate the binding of, and hence the neutralization by, Abs that the previous serotype of the virus elicited. But the receptor-binding capacity of the virus, guarded from Ab access in the depth of a pocket, remains intact.

The abrogation of NAb binding does not have to be so direct; mutations outside the epitope can confer resistance; they do not even have to be close in space to the epitope surface. When a neutralization-escape mutant of foot-and-mouth-disease virus was compared with the wild type structurally and serologically, the resistant phenotype was attributed to residues well outside the epitope in the VP1 protein. The mutated residues seem to affect the crucial orientation of the epitope loop from a distance in space. An escape mutant of HIV-1 provides another intriguing example. The subtle substitution of a Thr for an Ala in a highly conserved region of the transmembrane protein confers resistance to NAbs directed to the CD4-binding site on the other, noncovalently linked, moiety of the envelope glycoprotein. Perhaps most intriguing of all, substitutions in the cytoplasmic tail of the HIV-1 transmembrane protein can reduce the sensitivity to NAbs that bind on the other side of the viral membrane, to epitopes on the outer subunit of the envelope glycoprotein complex. This antigenic change must involve intricate conformational effects across the membrane and from subunit to subunit. Even so, the end result points to the most straightforward mechanism of resistance: the affinities of the NAbs are reduced.

Although the molecular basis thus varies for how NAb binding is reduced, such reduction is the route of escape generally taken. One provocative exception has been found. An escape mutant of rabies virus can still infect when coated by NAb; hypothetically, the virus has switched to a new

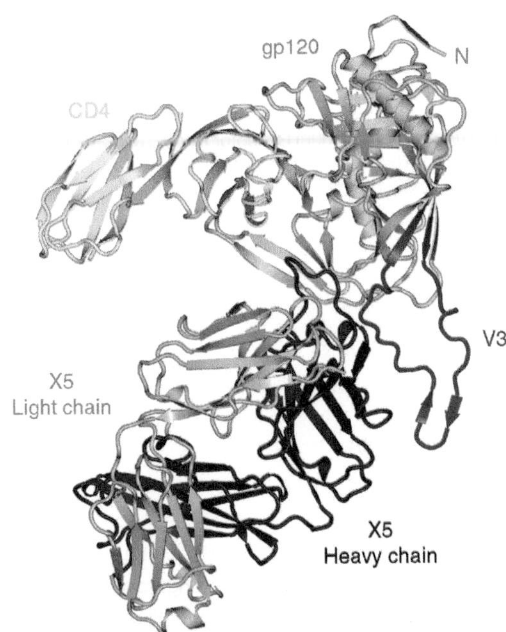

Figure 2 NAbs to HIV-1 are directed to its trimeric envelope glycoprotein, both to the surface unit, gp120, and to the transmembrane protein, gp41. Here, a three-dimensional model of the core of gp120 (gray) is shown. The structure was obtained by high-resolution crystallography of a trimolecular complex. The second molecule in the complex is a two-domain fragment of CD4 (yellow), the major receptor for HIV. Conserved neutralization epitopes overlap the site where CD4 binds. Abs to such epitopes compete with CD4 for binding to gp120 and can thereby directly block a necessary docking event in the viral attachment and entry process. The third molecule in the complex is a neutralizing Fab of the Ab X5. This binds to an epitope generated when CD4 interacts with the full-length gp120; it is located adjacent to the co-receptor binding site. Interestingly, the Fab of X5 neutralizes more efficiently than the whole Ab because its epitope is not induced until the envelope glycoprotein complex docks onto CD4, at which stage access to the epitope is scarce for the larger whole Ab but not for the Fab. Lastly, a variable region, V3, of the gp120 molecule is shown jutting out like a hook (red). This region also contains important neutralization epitopes, but because of the high sequence variability of V3, Abs to these epitopes generally do not cross-neutralize strains. V3 and more conserved adjacent surfaces make contacts with the co-receptor. Reproduced from Huang CC, Tang M, Zhang MY, *et al.* (2005) Structure of a V3-containing HIV-1 gp120 core. *Science* 310 (5750): 1025–1028. Reprinted with permission from AAAS.

entry mechanism, independent of its highly NAb-occupied envelope glycoprotein.

How does tissue culture affect sensitivity to neutralization? The isolation of a virus removes it from the immune system of the infected host; it entails the lifting of any selective neutralization pressure exerted *in vivo.* Since resistance *in vivo* may have come at a price in basic fitness, when there is no longer any gain from such sacrifices, the replicative functions will evolve anew toward optimum. When propagated *in vitro* the virus may therefore become artificially sensitive to neutralization. That seems to have happened to some strains of HIV.

HIV and many other viruses that survive the onslaught of the host immune system have developed a panoply of phenotypic traits protecting them specifically against neutralization: much of the Ab response is misdirected to epitopes that are exposed only on disassembled oligomers and denatured or degraded protein. Neutralization may require a high occupancy. Glycan shields can render surfaces on envelope glycoproteins nonimmunogenic; the glycan moieties may shift from site to site as a result of escape mutations, providing a malleable shield. Essential interactions with receptors may be limited to a subset of pocket-lining residues, so that the virus can tolerate variation not just around the receptor-binding site but also within it.

As mentioned, some potent NAbs induce barely any changes in the epitopes. Indeed, less potent Abs to overlapping sites sometimes generate greater changes. When such changes are entropically unfavorable, the adverse effect on affinity and occupancy favor the virus. Thus, entropic masking constitutes yet another viral defense.

Few viruses may exploit the entire repertoire of potential defenses. But these traits attest to the importance of neutralization in natural infection. When combined and richly developed, they also posit a formidable challenge to the induction of NAbs through vaccination.

See also: Antigenic Variation; Antibody - Mediated Immunity to Viruses; Persistent and Latent Viral Infection; Vaccine Strategies; Viral Receptors.

Further Reading

Andrewes CH and Elford WJ (1933) Observations on anti-phage sera. I: The percentage law. *British Journal of Experimental Pathology* 14(6): 367–376.

Burnet FM, Keogh EV, and Lush D (1937) The immunological reactions of the filterable viruses. *Australian Journal of Experimental Biology and Medical Science* 15: 227–368.

Burton DR (ed.) (2001) *Antibodies in Viral Infection. Vol. 260. Current Topics in Microbiology and Immunology.* Berlin: Springer.

Burton DR, Desrosiers RC, Doms RW, *et al.* (2004) HIV vaccine design and the neutralizing antibody problem. *Nature Immunology* 5(3): 233–236.

Della-Porta AJ and Westaway EG (1978) A multi-hit model for the neutralization of animal viruses. *Journal of General Virology* 38(1): 1–19.

Dulbecco R, Vogt M, and Strickland AGR (1956) A study of the basic aspects of neutralization. Two animal viruses. Western equine encephalitis virus and poliomyelitis virus. *Virology* 2: 162–205.

Fazekas de St. GS (1961) Evaluation of quantal neutralization tests. *Nature* 191: 891–893.

Gollins SW and Porterfield JS (1986) A new mechanism for the neutralization of enveloped viruses by antiviral antibody. *Nature* 321 (6067): 244–246.

Huang CC, Tang M, Zhang MY, *et al.* (2005) Structure of a V3-containing HIV-1 gp120 core. *Science* 310(5750): 1025–1028.

Jerne NK and Avegno P (1956) The development of the phage-inactivating properties of serum during the course of specific immunization of an animal: Reversible and irreversible inactivation. *Journal of Immunology* 76: 200–208.

Klasse PJ and Moore JP (1996) Quantitative model of antibody- and soluble CD4-mediated neutralization of primary isolates and T-cell line-adapted strains of human immunodeficiency virus type 1. *Journal of Virology* 70(6): 3668–3677.
Klasse PJ and Sattentau QJ (2002) Occupancy and mechanism in antibody-mediated neutralization of animal viruses. *Journal of General Virology* 83(Pt 9): 2091–2108.
Mandel B (1978) Neutralization of animal viruses. *Advances in Virus Research* 23: 205–268.
Nybakken GE, Oliphant T, Johnson S, Burke S, Diamond MS, and Fremont DH (2005) Structural basis of West Nile virus neutralization by a therapeutic antibody. *Nature* 437(7059): 764–769.
Skehel JJ and Wiley DC (2000) Receptor binding and membrane fusion in virus entry: The influenza hemagglutinin. *Annual Review of Biochemistry* 69: 531–569.

Immunopathology

M B A Oldstone, The Scripps Research Institute, La Jolla, CA, USA
R S Fujinami, University of Utah School of Medicine, Salt Lake City, UT, USA

Glossary

Adaptive immune system Comprises T cells, B cells, and antibodies. Elements of the innate immune system are required for initiation of the adaptive immune response. The adaptive immune system recognizes specific antigenic epitopes. Immunopathology is mainly mediated by the adaptive immune response to viral infection.

Antibody Antibodies are secreted by plasma cells and can neutralize viruses, bind to the surface of infected cells, activate the complement system, and form immune complexes. Antibodies are an important part of the adaptive immune response.

B cell Each B cell has a specific type of antibody molecule on its surface (B cell receptor). Viral proteins are recognized by B cell receptors. B cells mature into plasma cells that secrete large amounts of antibodies.

Complement A protein cascade system mainly found in serum. The cascade can be activated by immune complexes of viral proteins and antibodies. The protein cascade generates chemotactic peptides for phagocytic cells, proteins that can coat viral particles, and lyse viral membranes.

Cytokines Proteins that can be produced by various cells within the body. These proteins are produced in response to infection. Particularly, immune cells but other cell types have cytokine receptors. Cytokines mediate inflammation and can modulate the immune response through cytokine receptors.

Immune complex A complex between antibody molecules and antigens. These complexes can activate the complement system through the Fc receptor on the antibody molecules and can lodge in glomeruli and small vessels causing immunopathology.

Immunopathology Disease resulting from the immune response to infection causing tissue injury and damage.

Innate immune system Comprising phagocytic cells (macrophages and neutrophils), complement system, natural killer (NK) cells, and various cytokines that can act within minutes to hours after infection. Recognition is to patterns of microbial structures, not specific epitopes.

Major histocompatibility complex (MHC) A region encoding highly polymorphic proteins that are involved in immune recognition of T cells, some cytokines and proteins of the complement cascade. This region is coded for by genes on human chromosome 6 and on mouse chromosome 17.

Natural killer (NK) cells A population of lymphocytes that can kill virus-infected cells. NK cells can secrete large amounts of the antiviral cytokine IFN-γ. IFN-γ can also activate macrophages. NK cells are part of the innate immune system.

T cell T cells are mature in the thymus. These lymphocytes recognize viral peptides in the context of MHC molecules through the T-cell receptor on the surface of T cells. T cells can kill virus-infected cells and release potent antiviral cytokines such as interferons.

Introduction

When a virus infects a vertebrate host, an immune response is rapidly generated against the virus. This immune response will often determine the survival or death of the host in acute infections or the establishment of a persistent infection. Tissue damage or pathology can result due to the antiviral immune response. This is different from direct viral effects where virus infection of cells can lead to tissue damage and apoptosis. A vigorous antiviral immune response is responsible for the immunopathology associated with acute and persistent/chronic virus infections. There is a necessary balance between viral clearance by the immune system and pathology induced by the immune system in attempting to clear the virus (**Figure 1**).

One important component of the adaptive immune system is the lymphocyte. Due to specialized receptors on the surface of lymphocytes (T cell receptor (TCR), B cell receptor (BCR)/antibody molecule), these cells are able to recognize or discriminate between different antigenic peptides derived from microbes or self-proteins. Lymphocytes can be divided into several broad populations. These cells can act directly to lyse or kill virus-infected cells (discussed below) or produce antiviral substances such as cytokines (interferons (IFNs), tumor necrosis factor (TNF)) that have antiviral activity but also can disturb the host's cell function or by themselves cause cellular and tissue injury. Lymphocytes in collaboration with cells of the innate immune system can also kill virus-infected cells or secrete antibodies with antiviral neutralizing capacity (discussed below).

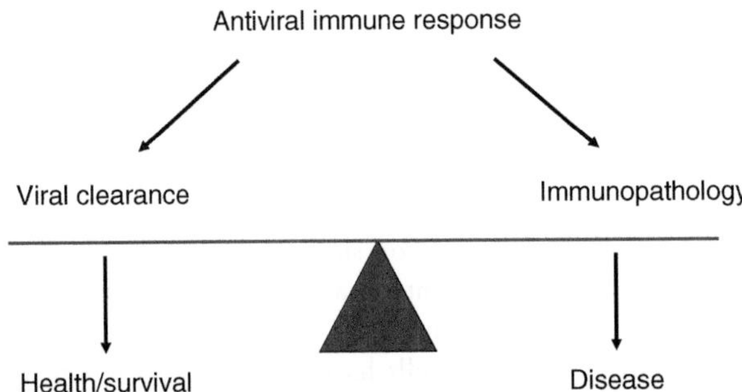

Figure 1 The antiviral immune response is a balance between viral clearance and immunopathology. One direction leads to survival of the host, whereas the other can lead to disease.

One population of lymphocytes are the B cells which are bursa or bone marrow-derived lymphocytes. These lymphocytes express BCRs on their surfaces called immunoglobulins or antibodies (**Figure 2**). The antibodies are secreted from differentiated B cells called plasma cells. Different antibodies can bind to specific viral proteins or antigens; but in general, one specific antibody is originally derived from one B cell and can recognize a unique epitope. Antibodies can neutralize viruses by attaching to the surface of the virion, thus inhibiting the attachment, penetration, or uncoating phases of viral replication (**Figure 2**, arrow 1). Antibodies can also bind to virions in the blood (viremia) or to viral proteins (such as hepatitis B virus surface antigen), and form antigen–antibody complexes (immune complexes) (**Figure 2**, arrow 2). If the complexes are of the appropriate size, glomerulonephritis

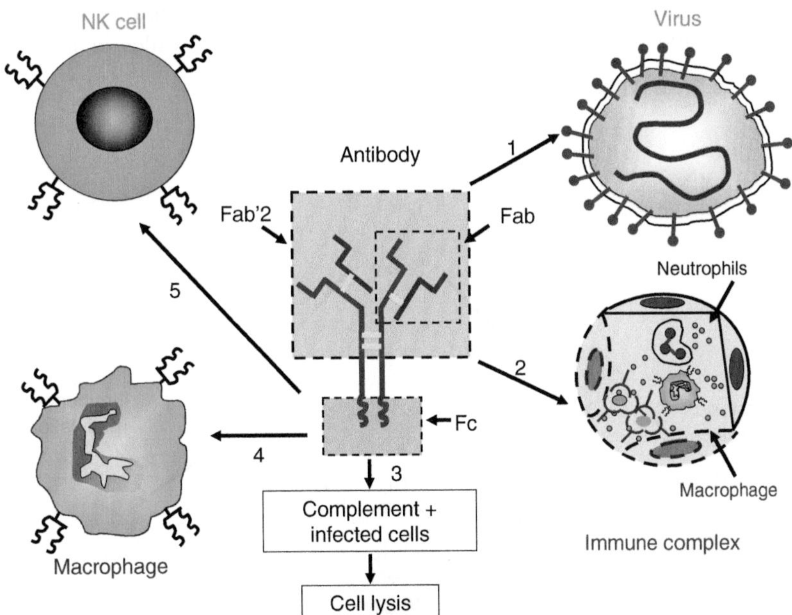

Figure 2 Antiviral antibodies can play an important role in viral clearance and/or disease. The Fab portion of the antibody molecule can bind and neutralize the virus (arrow 1). Antibodies binding to viral proteins or virions can form immune complexes that can lead to immune complex disease (arrow 2). Antibody binding to virus-infected cells through the Fc portion of the antibody molecule can activate the complement system leading to lysis of the infected cell (arrow 3). The Fc portion of the antibody molecule can also bind to Fc receptors in macrophages and NK cells. The antibody-coated macrophages and NK cells can kill infected cells via antibody-dependent cell-mediated cytotoxicity (arrows 4 and 5).

can occur due to immune complex deposition (discussed below). Immune complexes are also engulfed by Fc receptor bearing cells of the innate immune system found in the liver and spleen. However, since immune complexes are often infectious, by this means, Fc or complement expressing cells can also be infected. However, this process aids in clearance of virus from the blood, thereby limiting the viremia. Infected cells can be killed by antibodies specific for viral proteins expressed on the surface of virus-infected cells. This requires the presence of complement proteins. Complement proteins binding to the Fc portion of the immunoglobulin (**Figure 2**, arrow 3) activate the complement cascade leading to the formation of the complement membrane attack complex on the cell membrane and eventually cell lysis. Antibodies via the Fc portion of the immunoglobulin molecule can bind to Fc receptor positive cells (macrophages and NK cells – two cell types of the innate immune system) (**Figure 2**, arrows 4 and 5). The Fab portion of the immunoglobulin (antigen recognition site) gives specificity for the killing of the virus-infected cells. Therefore, antibodies can participate in clearing/neutralizing circulating free virus, as well as killing virus-infected cells with the help of complement and cells of the innate immune system. However, the number of antibody molecules required to kill a cell is usually in excess of 1×10^6, which makes this an inefficient process.

T cells represent another population of lymphocytes that participates in viral clearance and immunopathology. T cells can be further divided into two subpopulations, $CD4^+$ and $CD8^+$ T cells (**Figure 3**). T cells with the surface molecule CD4 are called T-helper cells (Th) due to their ability to secrete certain cytokines that 'help' B cells to differentiate into antibody producing plasma cells and 'help' the subpopulation of T cells expressing CD8 molecules to acquire cytotoxic or killer cell activity. $CD4^+$ T cells are 'restricted' by major histocompatibility complex (MHC) class II molecules. The MHC class II molecule presents viral peptides as a complex that is recognized by TCRs on $CD4^+$ T cells. The $CD4^+$ T cells can be further subdivided, depending on the types of cytokines they secrete, into $CD4^+$ Th1 T cells that are involved in delayed-type hypersensitivity immune responses and $CD4^+$ Th2 T cells that help B cells. When activated $CD4^+$ Th1 T cells secrete IFN-γ that has antiviral activity and can also in turn activate macrophages. These activated macrophages, cells of the innate immune system, can kill virus-infected cells indirectly by releasing reactive oxygen species and toxic cytokines, such as TNF or lymphotoxin (LT) that lyse cells by what is known as bystander killing. The $CD4^+$ T-cell population is also involved in the maintenance of antiviral T-cell memory. T-cell memory is key for immune individuals to respond to subsequent infections with the same or similar viruses. The $CD4^+$ T-cell population also contains a subset of T cells that also express a surface protein, CD25, and an internal protein called FoxP3. These T cells have the ability to modulate the adaptive immune response, and therefore are known as T-regulatory cells.

The $CD8^+$ T-cell population is intimately involved in viral clearance and immunopathology. $CD8^+$ T cells kill virus-infected cells by two mechanisms and are very efficient usually requiring recognition of ten or fewer viral peptide molecules in association with the MHC

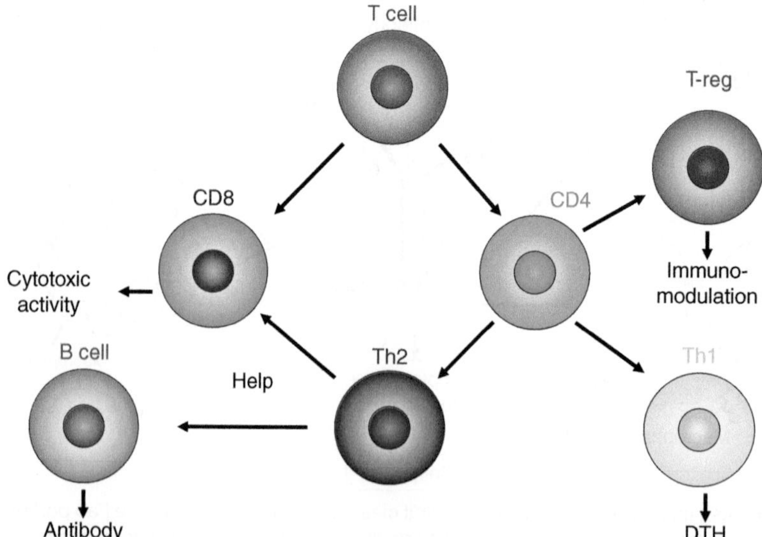

Figure 3 T cells can be divided into different subpopulations depending on their function. $CD8^+$ T cells have the ability to kill virus-infected cells and are restricted by MHC class I molecules. $CD4^+$ T cells can be further subdivided. Some $CD4^+$ T cells have the ability to provide 'help' to other T cells and B cells. Others can act as regulatory cells modulating the function of other T cells. $CD4^+$ T cells can also be directly involved in inflammatory responses through the production of various pro-inflammatory cytokines.

class I complex (see below). The first mechanism is by release of perforin molecules that form pores in the target cell's membrane similar to the complement membrane attack complex. The second mechanism involves the interaction of FasL-Fas on the T cell and the infected cell, respectively. Fas is a member of the TNF receptor family. Fas binding to its ligand leads to activation of caspase 8, a death signal in the infected cells that results in apoptosis. The cytotoxic T-lymphocyte (CTL) killing of virus infected cells occurs when infected cells are killed in a MHC class I restricted manner (TCR on $CD8^+$ T cells recognizing viral peptide complexed to class I molecules) prior to the maturation of infectious virus thus reducing the virus' ability to replicate and disseminate within the body. This supplements the action of antiviral antibodies that can neutralize infectious virus. Memory $CD8^+$ T cells form during the antiviral immune response and are long-lived T cells that can protect animals in the event of subsequent infections with the same virus.

Thus B cells, $CD4^+$ and $CD8^+$ T cells, in conjunction with complement and cells (NK, macrophage, neutrophil) of the innate immune system, are responsible for viral clearance. These are also the cellular participants that are involved in immunopathology associated with virus infections.

Immunopathology

Antiviral immune responses are responsible for the immunopathology associated with acute and persistent/ chronic virus infection. Immunopathology is determined by several factors. These involve contributions from both the host and the virus. Host factors include the age of the individual, genetic makeup of the host (MHC and non-MHC genes), route and/or site of infection, and whether the immune system is intact or immunocompromised. Characteristics of the virus consist of the strain or type and the size of the inoculum/dose.

Age of the host can play an important role in determining whether an individual survives, becomes immune, or acquires a persistent infection. For example, infection with lymphocytic choriomeningitis virus (LCMV), an arenavirus, has different outcomes depending on the age of the host. The natural host of the virus is the mouse. Its genome is comprised of two segments of RNA and is ambisense. The two RNA segments are encased by a nucleocapsid protein forming the nucleocapsid that is surrounded by a lipid envelope embedded with two viral glycoproteins. For the most part LCMV is noncytolytic and does not induce cell death after infection. Neonatal infection, infection of newborn mice with LCMV, or of adult mice with immunosuppressive variant(s) results in a persistent infection where many tissues are infected and mice survive. At this early age, the mouse's immune system is not fully developed and the mouse is not able to mount a functional CTL response. It was first proposed by Burnet that mice persistently infected with LCMV were immunologically tolerant to the virus (the host's immune system is not able to recognize the virus and mount an immune response). Accordingly, this purported immune tolerance to LCMV was why the virus was able to persist. However, seminal studies by Oldstone and colleagues demonstrated that mice persistently infected with LCMV were not tolerant. These mice produced antivirus antibodies and routinely developed glomerulonephritis. The inflammatory lesions in the kidneys were due to LCMV–anti-LCMV immune complex deposition in the glomeruli activating the complement cascade and recruiting of cells of the innate immune system leading to disease (to be discussed below) (**Figure 2**, arrow 2). Similar findings followed for congenital murine retrovirus infections.

In contrast, if mice are infected as adults with a moderate dose of LCMV (nonimmunosuppressive variant) by an extraneural route of infection (outside the central nervous system (CNS)) little or no immunopathologic disease is observed. Anti-LCMV CTLs are induced that are able to clear virus-infected cells. A portion of these LCMV-specific T cells go on to become memory T cells. Subsequent intracerebral (ic) infection of an adult immunocompetent immune mouse results in far less inflammation and immunopathology in the CNS and the animal survives. Interestingly, after ic LCMV infection of adult nonimmune mice, virus is able to replicate in certain epithelial cells. After about 5–7 days post infection the infected mouse mounts a robust anti-LCMV CTL immune response which is responsible for inflammation in the choroid plexus, ependyma, and leptomeninges causing choriomeningitis. Death is due to the action of the anti-LCMV CTL immune response altering the blood–brain barrier and initiating brain edema. This infection illustrates several of the factors involved in immunopathology. The age of the host is critical, since in very young animals the immune systems are not fully developed. These animals are immunocompromised, that is, unable to mount an effective CTL response allowing a noncytolytic virus-like LCMV to persist. Other studies have shown the importance of CTL in initiating immunopathology and death. Adult mice whose T cells are ablated or inhibited and then infected with LCMV via the ic route of infection develop a persistent infection and do not die of the choriomeningitis. Adoptive transfer of immune spleen cells, immune T cells, or LCMV-specific $CD8^+$ T cells into LCMV carrier mice leads to clearance of viral infection. Thus, this experiment complements studies of acute infection and substantiates the role of the antiviral T cells in clearance of virus and termination of infection.

Mice that are infected as newborns or earlier with LCMV develop a persistent infection, but are not without immunopathology. As mice infected at birth or earlier develop into adults, they produce anti-LCMV antibodies that can bind to viral antigens (present due to the persistent infection) forming immune complexes. Depending on the size of the immune complexes and numbers formed, the complexes deposit in the glomeruli of the kidneys due to the filtering action within the glomeruli. As a consequence of immune complex deposition in the glomeruli, the complement cascade is initiated releasing chemotactic peptides that recruit and activate polymorphonuclear leukocytes (PMNs) and macrophages into the glomeruli. These activated cells release inflammatory cytokines, such as TNF and interleukin (IL)-1, and reactive oxygen intermediates, as well as lysosomal enzymes that induce inflammation and immunopathology resulting in glomerulonephritis. The extent of glomerulonephritis induced by persistent LCMV infection varies and depends both on the strain of mouse (genetic influences discussed below) and strain of virus.

The genetic composition of the host can play an important role in whether immunopathology occurs. For example, infection with Theiler's murine encephalomyelitis virus (TMEV), a picornavirus related to the cardioviruses, has different outcomes depending on the strain of the host. As with LCMV the mouse is the natural host. Because of the availability of inbred mice and mapping of their molecules, particularly MHC and background genes involved in innate immunity, the mouse has been an exquisite model to measure not only host genetics but influence of viral genetics as well. TMEV is a positive stranded RNA virus. The genome is approximately 8000 bases in length. The icosohedral capsid is comprised of four viral proteins. The virion is not encapsidated by a membrane. TMEV can be divided into two groups depending on neurovirulence. The GDVII subgroup is comprised of viruses that are highly neurovirulent. Members of the TO subgroup are less neurovirulent. Ic infection with the DA strain of TMEV (TO subgroup) into adult C57BL/6 mice leads to an antiviral CD8$^+$ T-cell response that clears the virus from the CNS and other tissues of the body. There is limited inflammation in the CNS but mice survive and are immune to subsequent infections with TMEV.

In contrast, ic infection of SJL/J mice with the DA strain of TMEV induces an antiviral CD8$^+$ T-cell response which is not able to clear the virus. Therefore, DA virus is able to establish a persistent infection in the CNS. Immunopathology results as a consequence of a chronic anti-TMEV immune response attempting to clear virus-infected cells in the CNS. Infected SJL/J mice eventually develop large areas of inflammation and demyelination in the spinal cord with a spastic paralysis. Genetic differences in the MHC complex between the two strains of mice are implicated in whether one mouse strain is able to clear the virus and the other mouse strain is not able to clear the virus resulting in disease. In addition, innate immune responses are also genetically determined. Therefore, the early inflammation in response to infection is encoded within the genome and varies from individual to individual.

The initial site of infection is important. A peripheral infection with LCMV in an adult mouse generates a robust anti-LCMV CTL response and immunity. Since the virus is not present in the CNS or only a few cells are infected, the CTL response clears the virus and little pathology ensues. Infection by the ic route also induces a vigorous CTL response. However, a significant number of meningial cells are infected with the virus. CTLs recognize virus-infected cells in the meninges lysing the cells via a perforin-mediated mechanism. This contributes to the death of the mouse. Therefore, the site or route of infection and, which cells are infected by the virus, are important factors.

The strain of the virus can play a major role in immunopathology, as shown by the following two examples. As mentioned previously TMEV can be divided into two groups depending on neurovirulence. The GDVII subgroup comprises viruses that are highly neurovirulent. An intravenous (IV) or ic infection of mice with the GDVII virus will result in the virus killing the mouse in 7–10 days. The death is due to the virus' ability to infect large numbers of neurons and directly kill these cells. In contrast, infection of mice (SJL/J) with viruses from the TO subgroup with 100 000–1 000 000 times more virus leads to mild acute disease with TMEV establishing a persistent infection in the CNS. The chronic anti-TMEV immune response results in extensive demyelination and inflammation. The pathology is caused by the antiviral immune response comprising antibodies capable of neutralizing the virus, CD4$^+$ and CD8$^+$ T cells recognizing virus-infected cells and either directly killing the infected cells, as in the case of antiviral CD8$^+$ T cells or TMEV antibody and complement, or killing the infected cells via bystander mechanisms which are mediated by CD4$^+$ T cells and the effector/killer cells are macrophages (as in delayed-type hypersensitivity responses).

CNS infection of adult mice with LCMV leads to the generation of antiviral CD8$^+$ T cells and infected mice die about 7–10 days later as the CD8$^+$ T-cell response develops. As mentioned above there is extensive infiltration of T cells and mononuclear cells in the meninges, alterations in the blood–brain barrier, and edema. In contrast, infection of adult mice with a variant of LCMV, clone 13, that differs in two amino acids from wild-type Armstrong LCMV, induces an immunosuppression due to infection of dendritic cells which are then not able to arm and expand T- and B-viral-specific cells. The variant LCMV is able to persist in adult mice. These persistently infected mice also develop immune complexes in their kidneys. Different strains of virus, such as clone 13, can initiate different types of disease leading to immunopathology.

The amount of virus the host first encounters can set the stage for later disease. For example, high doses of certain strains of LCMV when given by the IV route results in dissemination to the various lymphoid tissues. Virus can replicate to high titers at these sites constantly stimulating the antiviral T cells leading to 'CTL exhaustion'. This is likely due to both the continual stimulation of the TCR by antigen presenting cells leading to the apoptosis of LCMV-specific CTLs and to infection of dendritic cells. With the depletion or failure to generate virus-specific CTLs, LCMV is able to establish a persistent infection. This is in contrast when low or moderate amounts of LCMV are used to infect mice. The infected mice are able to mount an effective CTL response, virus is cleared, and animals are immune.

In summary, immunopathology results from an imbalance between the immune system's ability to clear the virus and resulting tissue damage due to the antiviral immune response. These can take the form of antibody and complement, $CD4^+$ T cells and $CD8^+$ CTLs causing tissue damage in the process of eliminating virus and virus-infected cells.

See also: Antigen Presentation; Antibody - Mediated Immunity to Viruses; Cell-Mediated Immunity to Viruses.

Further Reading

Borrow P and Oldstone MBA (1997) Lymphocytic choriomeningitis virus. In: Nathanson N, Ahmed R, Gonzalez-Scarano F, *et al.* (eds.) *Viral Pathogenesis*, pp. 593–627. Philadelphia: Lippincott-Raven Publishers.

Buchmeier MJ and Oldstone MBA (1978) Virus-induced immune complex disease: Identification of specific viral antigens and antibodies deposited in complexes during chronic lymphocytic choriomeningitis virus infection. *Journal of Immunology* 120: 1297–1304.

Henke A, Huber S, Stelzner A, and Whitton JL (1995) The role of $CD8^+$ T lymphocytes in coxsackievirus B3-induced myocarditis. *Journal of Virology* 69: 6720–6728.

Oldstone MBA (1982) Immunopathology of persistent viral infections. *Hospital Practice (Office Education)* 17: 61–72.

Oldstone MBA and Dixon FJ (1969) Pathogenesis of chronic disease associated with persistent lymphocytic choriomeningitis viral infection. I. Relationship of antibody production to disease in neonatally infected mice. *Journal of Experimental Medicine* 129: 483–505.

Selin LK, Cornberg M, Brehm MA, *et al.* (2004) CD8 memory T cells: Cross-reactivity and heterologous immunity. *Seminars in Immunology* 16: 335–347.

Tsunoda I and Fujinami RS (1996) Two models for multiple sclerosis: Experimental allergic encephalomyelitis and Theiler's murine encephalomyelitis virus. *Journal of Neuropathology and Experimental Neurology* 55: 673–686.

DESCRIPTION AND MOLECULAR BIOLOGY OF SOME VIRUSES

Adenoviruses: Molecular Biology

K N Leppard, University of Warwick, Coventry, UK

Introduction and Classification

Adenoviruses were first discovered during the 1950s, in studies of cultures from human adenoids and virus isolation from respiratory secretions. Their study accelerated dramatically in the 1960s after it was demonstrated that a human adenovirus could cause tumors in experimental animals, leading to the hope that such work would lead to a better understanding of human cancer. What has been learned subsequently, as well as making a significant contribution in this area, has also had a major influence on our understanding of fundamental eukaryotic cell processes such as RNA splicing and apoptosis.

The adenoviruses are classified in the family *Adenoviridae*. Historically, this family was divided into two genera, *Aviadenovirus* and *Mastadenovirus*, containing viruses that infected avian and mammalian hosts, respectively, but the advent of large-scale DNA sequencing has led to two further genera being established, based on comparisons of genome sequences. Thus, for example, the genus *Atadenovirus* contains two of five species of bovine adenovirus and one of four species of sheep adenovirus, the others being members of the genus *Mastadenovirus*. The fourth genus, *Siadenovirus*, so far contains a frog adenovirus and a turkey adenovirus.

There are six human adenovirus species, *Human adenovirus A* through *F*, all of which are classified in the genus *Mastadenovirus*. These species correspond to the subgroups of human adenovirus established earlier on the basis of the hemagglutination characteristics of their virions, their genome nucleotide composition, and their oncogenicity in rodents. There are 51 known human adenovirus serotypes that are distributed between these species. Each causes characteristic disease symptoms in man. Although there has been an increasing level of interest in other adenoviruses in recent years, the majority of what is known about adenovirus molecular biology comes from the study of human adenoviruses 2 and 5, two very closely related members of species *Human adenovirus C*. The following account is therefore based on these two viruses; important differences from this picture that pertain to other adenoviruses are pointed out where appropriate.

Particle Structure

Adenovirus particles have icosahedral capsids containing a core that is a complex of proteins and DNA (**Figure 1**). There is no lipid envelope and no host proteins are found in the virion; thus all components of the particle are encoded by the virus (see the section of titled 'Gene expression'). The structural proteins of the particle are known by roman numerals in order of decreasing apparent size on polyacrylamide gels.

The capsid shell is built of 252 capsomers. Twelve of these are known as pentons. They occupy the vertices of the shell and are formed of two components, penton base, which is a pentamer of the penton polypeptide

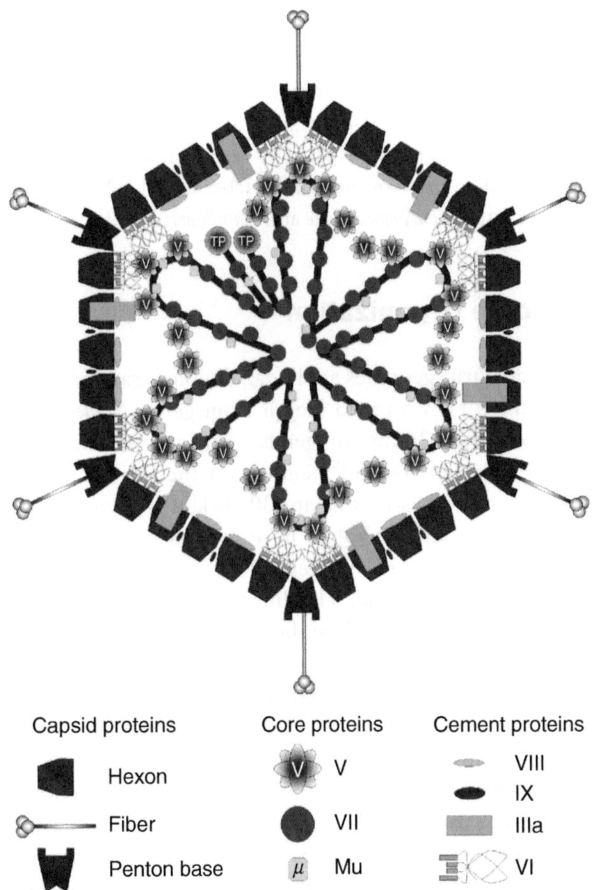

Figure 1 A cut-through diagram of a mature adenovirus particle. The positions of proteins within the shell of the particle are defined based on cryoelectron microscopy analyses. However, the indicated structure of the DNA core is largely hypothetical; the core appears disordered in structural analyses. Several of these proteins (IIIa, VI, VII, μ, and TP) are synthesized in precursor form (pIIIa, pVI, pVII, pX, and pTP, respectively) and assembled as such into immature particles before processing by the adenovirus protease. Reproduced from Russell WC (2000) Update on adenovirus and its vectors. *Journal of General Virology* 81: 2573–2604, with permission from Society for General Microbiology.

(protein III), and fiber, which is a trimer of the fiber polypeptide (protein IV). This fiber unit projects outward from each vertex of the particle; it has a long shaft, formed of a repeating β-sheet module, and a distal knob domain. Both the penton base and fiber components play specific roles in attachment and internalization (see the section of titled 'Attachment and entry'). Each of the remaining 240 capsomers, which are known as hexons, are trimers of the hexon polypeptide (protein II), which is thus the major capsid protein of the virus.

Inside the shell, the viral core comprises the genome in complex with three proteins: VII, V and μ, the latter being a processed form of polypeptide X. The capsid shell is stabilized by proteins IIIa, VI, VIII, and IX, some of which bridge between the shell and the core. Protein IX trimers occupy crevices on the external faces of the shell; when IX is absent, the particle is less stable and hence its capacity for DNA is reduced. Finally, the particle carries a virus-coded, sequence-specific protease of 23 kDa that is essential both for the maturation of newly formed particles to an infectious state and for particle uncoating after infection. Thus, adenovirus particles are considerably more complex than those of the smaller icosahedral DNA viruses of the families *Polyomaviridae* and *Papillomaviridae*.

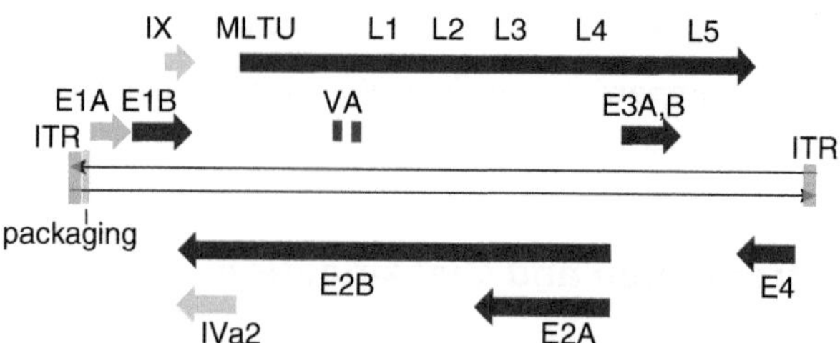

Figure 2 A schematic representation of the genes and major genome features of human adenoviruses (not to scale). The duplex DNA is represented by black lines, with arrowheads indicating 3′-ends. Inverted terminal repeats (ITRs) and the packaging sequence are indicated as green and gray boxes, respectively. Protein-coding genes are indicated as colored arrows, the color indicating the phase of infection when expression commences (see also **Figure 3(a)**), and RNA pol III transcripts as red boxes.

Genome Organization

All adenovirus genomes are linear, double-stranded DNA molecules. They vary in length from genus *Siadenovirus* members at ~26 kbp to genus *Aviadenovirus* members at ~45 kbp; genomes of members of the genera *Mastadenovirus* and *Atadenovirus* are 30–36 kbp in length. The genomes are all thought to have a virus-coded protein, known as the terminal protein (TP), covalently attached to each genome 5′-end. This linkage is a consequence of the mechanism of DNA replication (see below). Also at each end of the linear genome is an inverted terminal repeat sequence varying in length from ~40 to 200 bp; the terminal repeats of human adenovirus 5 are 103 bp.

Among the human adenoviruses, there is a high level of conservation of genome organization. Both genome strands are used to encode proteins from genes that are designated E1A, E1B, E2, E3, E4, IX, IVa2, and the major late transcription unit (MLTU), while two small noncoding RNAs are expressed from the VA genes (**Figure 2**). The E2 and E3 genes are divided into regions A and B while the MLTU is divided into five regions, L1–L5, each with a distinct polyadenylation site. Within each gene/region there are typically multiple proteins encoded. This genome organization applies to all human adenoviruses with only subtle variations, such as the presence of one rather than two VA genes, or the presence of a second fiber coding region in L5, in a few types. Considering adenoviruses in general, the organization of the center of the genome (where the structural proteins and replication proteins are coded) is conserved but the regions closer to the genome termini do not necessarily contain recognizable E1A, E1B, and E4 sequences. However, it is presumed that the products encoded here perform the same types of function as these genes supply in human adenoviruses. Unlike the other adenovirus genes, E3 is highly divergent in both sequence and the number and nature of its protein products, even between different human adenoviruses. This may be responsible for aspects of viral pathogenicity and tropism that differ between adenovirus types.

Attachment and Entry

Primary attachment of particles to the target cell surface is thought to be mediated, for all adenoviruses, by the knob domains at the distal ends of the fibers that project from the virion vertices. For human viruses in species *Human adenovirus A, C, E,* and *F,* the cell receptor for attachment appears to be a molecule designated CAR (coxsackie adenovirus receptor), a cell-surface protein of the immunoglobulin superfamily that is present on a variety of cell types. CAR mediates cell-to-cell interactions, being present at tight junctions and other types of intercellular contact, and is also used as a receptor by members of the unrelated human coxsackie viruses (members of the genus *Enterovirus*, family *Picornaviridae*). The location of CAR has consequences for virus entry and exit (see the section titled 'Assembly and release'). Other adenoviruses, such as members of species *Human adenovirus D* and certain animal adenoviruses, may also use this receptor. However, members of species *Human adenovirus B* clearly use a distinct receptor, the CD46 molecule. CD46 is involved in regulating complement activation and is widely distributed on human cells. Other cell surface molecules may also serve as receptors for some or all of these viruses under particular circumstances.

Following primary attachment, there is a secondary interaction between the penton base component of the particle and cell-surface integrins. These are a family of heterodimeric cell adhesion molecules, several of which can bind the penton base polypeptide via a conserved RGD peptide sequence motif in the viral protein. The interaction between fiber and CAR or CD46 and between the penton base and integrin together lead to internalization of the particle in an endocytic vesicle. Acidification of the endosome then causes conformational changes in the virion that lead to the loss of several proteins from the particle vertices, including fiber and VI. It is this latter protein that is now thought to cause lysis of the endosomal membrane, permitting the residual particle to enter the cytoplasm. The ability of the particle to undergo these uncoating steps depends crucially on the prior action of the encapsidated viral protease, which cleaves several particle proteins during final maturation (see the section titled 'Assembly and release'). The function of protein VI in particular is controlled by sequential specific cleavage events. Cleavage of particle proteins is thought to render the particle metastable, meaning that it is primed for uncoating during the subsequent round of infection.

Once the residual virion has escaped from the vesicle, it is rapidly transported along microtubules toward the nucleus. During this process, there is further loss of virion proteins. Ultimately, the core of the particle is delivered into the nucleus where all the remaining steps of the replication cycle, excepting actual protein synthesis, take place.

Gene Expression

Once the incoming genome has reached the nucleus, it is used as template for transcription. The principal enzyme involved is host RNA polymerase II, although the VA genes are transcribed by RNA polymerase III. A program of viral gene expression is established, during which essentially the entire length of each genome strand is transcribed. However, not all the regions of the genome are expressed at the same time during infection. Instead, gene expression is tightly regulated, resulting in a temporal cascade of gene activation and hence protein production (**Figure 3(a)**). Key to this is the action of a combination of viral and host proteins as specific activators of viral genes. Thus, initially upon infection, the E1A gene is the only one to be transcribed. If protein production from these mRNAs is prevented with metabolic inhibitors, little further gene expression is seen, indicating that E1A proteins are important activators of the remainder of viral gene expression. One of the E1A proteins, known as the 13 S RNA product or the 289 residue protein, has a potent transactivation function although it lacks specific DNA-binding activity. Instead, its recruitment to promoters is thought to be through protein–protein interactions. The action of this protein, together with host transcription factors that E1A activates indirectly, leads to activation of the remaining early gene promoters. Among the proteins produced from these genes are the ones required for DNA replication. Other early proteins modulate the host environment (see the section titled 'Effects on the host cell'). The production of large numbers of replicated DNA templates, together with the action of the intermediate gene product IVa2 and L4 33 kDa/22 kDa proteins working in a feed-forward activation mechanism, dramatically activates the MLTU, ultimately resulting in high levels of transcription of the genes that encode the structural proteins needed to form progeny particles.

Figure 3(b) shows a detailed transcript map of human adenovirus 5. There are only nine promoters for RNA polymerase II in the genome, one for each of the eight genes excepting E2, which has distinct promoters for the early and late phases of infection. Of these eight protein-coding genes, only the one encoding protein IX has a single product. Extensive use of alternative splice sites and polyadenylation sites within the primary transcripts of the other genes leads to multiple mRNAs, each with distinct coding potential; in total, at least 50 different proteins are made. The MLTU, for example, encodes around 18 proteins, each with a unique and essential role as either a virion component, assembly factor, or late gene expression regulator.

As well as allowing the expression of a wide range of proteins, the use by adenovirus of differential RNA processing also opens up the possibility of further controls on the temporal pattern of gene expression during the course of infection. For all the genes that show alternative splicing, there is a trend toward the removal of larger or greater numbers of introns as infection proceeds. A well-documented example is the MLTU L1 region. This limited segment of the MLTU is actually expressed at low level even in advance of replication beginning, but only the 5′-proximal reading frame (encoding the 52/55 kDa proteins) is accessed by splicing. Later in infection, the production of L1 mRNA switches to access the distal reading frame encoding protein IIIa. This is achieved through virus-directed modification of host-splicing components and the direct involvement of L4 33 kDa protein; this protein is also needed to generate the full pattern of protein expression from other regions of the MLTU.

DNA Replication

Adenoviral DNA replication follows a mechanism that is completely distinct from that of other animal DNA viruses,

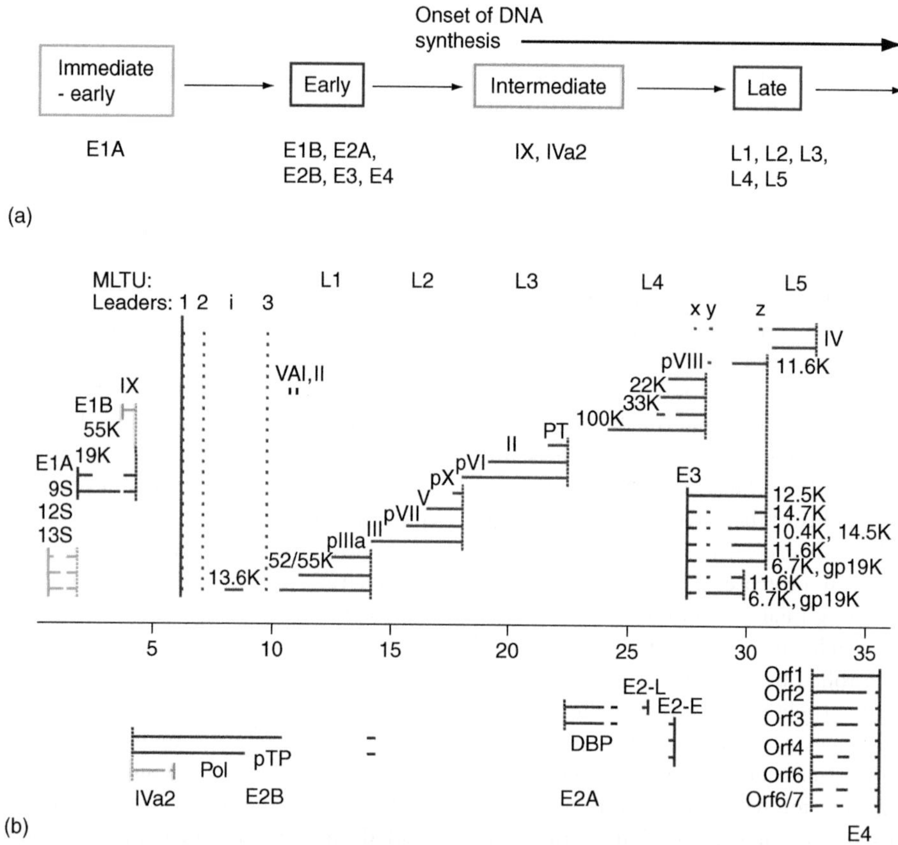

Figure 3 Gene expression by human adenovirus 5. (a) The phases of gene expression. The numbers E1A, L1, etc. refer to regions of the viral genome from which transcription takes place. (b) A transcription map of the genome. The genome is represented at the center of the diagram as a line scale, numbered in kbp from the conventional left end, with rightward transcription shown above and leftward transcription below. Genes or gene regions are named in boldface. Promoters of RNA polymerase II transcription are shown as solid vertical lines and polyadenylation sites as broken vertical lines. VAI and VAII are short RNA polymerase III transcripts. Individual mRNA species are shown as solid lines, color-coded according to the temporal phase of their expression in (a), with introns indicated as gaps. The protein(s) translated from each mRNA is indicated above or adjacent to the RNA sequence encoding it. Structural proteins are shown by roman numerals; PT, 23 kDa virion proteinase; DBP, 72K DNA-binding protein; pTP, terminal protein precursor; Pol, DNA polymerase. (a) Reproduced from Dimmock NJ, Easton AJ, and Leppard KN (2006) *Introduction to Modern Virology*, 6th edn., figure 9.6. Oxford: Blackwell Scientific, with permission from Blackwell Publishing. (b) Adapted from Leppard KN (1998) Regulated RNA processing and RNA transport during adenovirus infection. *Seminars in Virology* 8: 301–307. Academic Press, with permission from Elsevier.

and from that of its host organism. To achieve this, it encodes three proteins that function directly in replication: a DNA polymerase (DNApol), the terminal protein precursor (pTP), and a single-stranded DNA-binding protein (DBP), all from the E2 gene. Three host proteins also contribute to the replication process: two transcription factors and topoisomerase I. Aside from this, none of the host enzymes that function in DNA replication are needed.

The requirements for adenovirus replication were established using a cell-free replication system in which accurate initiation and elongation were obtained from the viral origin of replication. This is a sequence of 51 bp that is present at the two ends of the genome as part of the inverted terminal repeat. Within this sequence, only the terminal 18 bp (the core origin) is essential for initiation of replication; these bind a complex of DNApol and pTP. The remainder of the 51 bp origin is accessory sequences that bind the two host transcription factors. This binding alters the conformation of the DNA so as to facilitate binding of the DNApol/pTP complex to the adjacent core origin. Once bound at the origin, DNApol uses a serine residue hydroxyl side chain in pTP as a primer for DNA synthesis. The 5′-residue is almost invariably a cytosine, templated by the guanidine residue at the fourth position from the 3′-end of the template strand. After initiation has occurred, the complex then slips back to pair with the residue at the 3′-end of the template strand (also a guanidine), and synthesis then proceeds 5′ to 3′ away from the origin. DBP and topoisomerase I are not needed for initiation of replication, but are essential for elongation.

A scheme for the complete replication of adenoviral DNA is shown in **Figure 4**. A key feature of the process is that there is no synchronous replication of the leading and lagging strands of the template duplex at a single

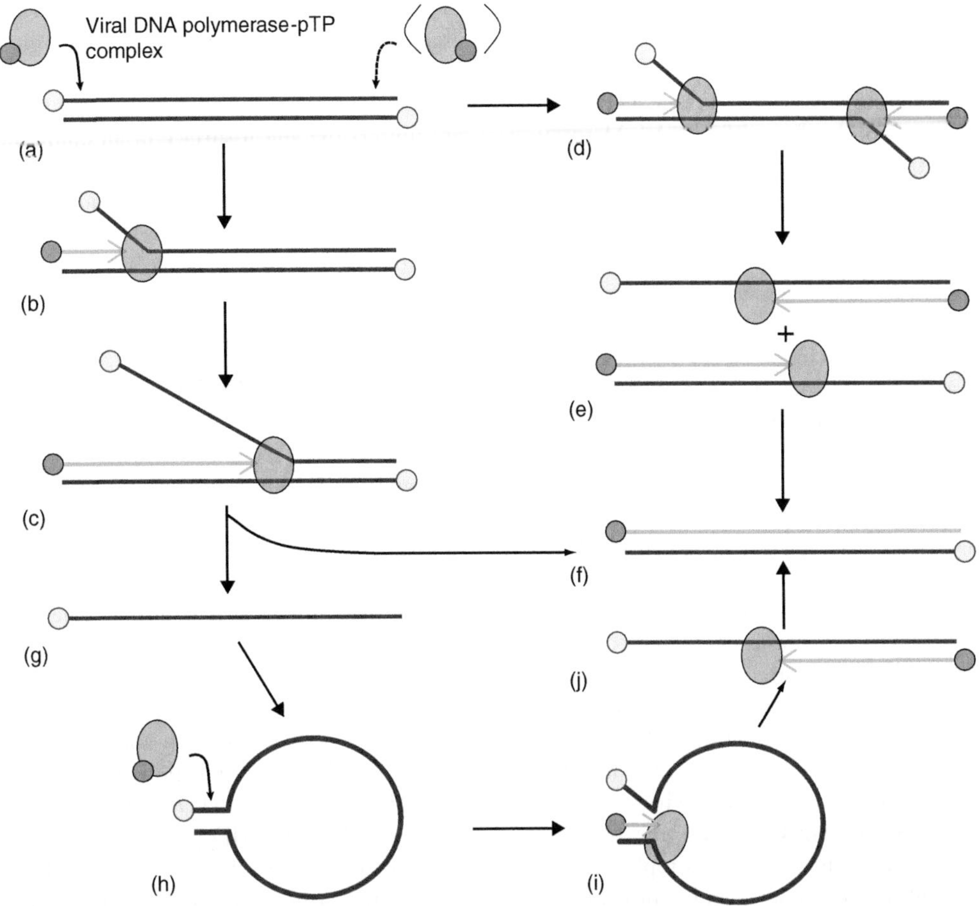

Figure 4 A general scheme for adenovirus DNA replication. Dark blue, parental DNA; light blue, new DNA. Arrowheads on new DNA strands represent 3′-ends. Yellow circles represent the terminal protein (TP) attached to parental DNA 5′-ends and brown circles the terminal protein precursor molecules (pTP) which prime new DNA synthesis. The viral DNA polymerase is represented in pink. Reproduced from Dimmock NJ, Easton AJ, and Leppard KN (2006) *Introduction to Modern Virology*, 6th edn., figure 6.7. Oxford: Blackwell Scientific, with permission from Blackwell Publishing.

replication fork, as is seen in eukaryotic DNA replication. Instead, the complementary strand of the template DNA duplex is displaced until it can be replicated later. This requires large amounts of DBP to stabilize it. Replication of this displaced single strand is achieved in one of two ways. Either its replication will already have been initiated, via the second origin at the other genome end, prior to its displacement from the duplex by the advancing replication fork, or else it can form a pan-handle structure in which its two ends come together to form a functional origin on which its replication can be initiated subsequently.

Assembly and Release

Electron microscopy performed on nuclei of cells late in the infectious cycle reveals the formation of paracrystalline arrays of adenovirus particles; this is the site of adenovirus assembly. The initial step in particle formation is the bringing together of the necessary components. Progeny genomes are generated within prominent replication centers in the nucleus while the proteins necessary to package this DNA are produced in the cytoplasm and then transported back to the nucleus. Hexon polypeptides require the assistance of a large 100 kDa protein, encoded in the MLTU L4 region, and protein pVI, to form trimers and reach the nucleus efficiently. It is generally believed that immature adenovirus particles assemble with the aid of scaffolding proteins (proteins that appear in intermediate virion forms but not the mature virion), prior to the insertion of the DNA genome. DNA packaging into these particles requires specific sequences in the genome that lie between the left-end copy of the inverted terminal repeat and the E1A gene (**Figure 2**). Packaging also involves the viral IVa2, L1 52/55 kDa, and L4 22 kDa proteins, and probably other host factors, to achieve efficient encapsidation of the DNA, which is accompanied by the ejection of scaffolding proteins. Sequence-specific binding of IVa2 to elements in the packaging sequence is important in this process. The final step in particle maturation is the cleavage of a number of the component proteins, including the genome-attached pTP, by the virion 23 kDa protease, an L3 gene product. The protease is stimulated by a peptide cleaved from pVI. Maturation cleavage is essential for the particles to acquire infectivity (see the section titled 'Attachment and entry').

In immortalized cell lines such as HeLa, infected cells show severe cytopathic effect, with loss of cytoskeleton leading to rounding and detachment from the monolayer within 1–2 days at high multiplicity of infection. However, the release of virus particles is rather inefficient, with considerable amounts of virus remaining cell associated. One of the E3 gene products, an 11.6 kDa protein (ADP, an abbreviation for adenovirus death protein) that is produced late in infection, facilitates loss of cell viability and particle release. In its absence, although cytopathic effect occurs normally, cell death is delayed and virus release is even less effective. However, this protein is not widely conserved, even among human adenoviruses, so the general significance of this function is uncertain.

The time taken from the initial addition of virus to the point of maximum progeny yield in cultures of HeLa cells is 20–24 h. However, in cultures of normal fibroblasts, this period is considerably longer, at around 72 h. The precise factors that determine the rate of adenovirus replication in different cell types are not known, but the availability of host functions needed for adenovirus replication may be higher in established cell lines than in normal cells.

The tissues that are targeted by adenovirus *in vivo* are epithelial cell sheets. Individual cells are linked by tight junctions that define distinct apical and basolateral membrane surfaces (externally facing and internally facing, respectively; **Figure 5**). In a culture model of such epithelia, virus infection leads to initial virus release via the basolateral surface. However, fiber–CAR interactions (see the section titled 'Attachment and entry') break down the CAR–CAR adhesion between adjacent cells, leading to increased permeability across the epithelium and escape of virus to the apical surface through gaps between the cells. The location of CAR on the basolateral surface of the cell means that when virus first enters the body and encounters the apical surface of an epithelium, it cannot access the receptors necessary to initiate infection. The initial round of infection must depend either on a break in epithelial integrity exposing the basolateral surface or on an alternative receptor that the virus can use to enter the first target cell in the epithelium.

Effects on the Host Cell

The interaction between adenovirus and its host is complex and subtle. Infection leads to a variety of effects on the molecular and cell biology of the host. These effects can usually be rationalized as favoring the production of progeny virus. A very large number of such effects have been studied; this section mentions some of those that are best understood.

One area of intervention by viral proteins is in reacting to aspects of the host response to infection. Several viral proteins act to block the pro-apoptotic signaling that is activated early in the infectious process. These include the E1B 19 kDa and 55 kDa proteins, the E3 14.7 kDa protein, and the E3 RID protein complex. The adaptive immune response is also impaired by the E3 gp19 kDa protein, which prevents export to the cell membrane of mature major histocompatibility complex class I antigens. A third intervention is by the VA RNA, which blocks activation of protein kinase R, an aspect of the interferon response that is triggered by double-stranded RNA produced during the infection by hybridization between transcripts produced from the two genome strands; this would otherwise stop protein synthesis. Susceptibility of adenovirus infections to interferon is also apparently blocked via the action of E4 Orf3 protein on PML nuclear

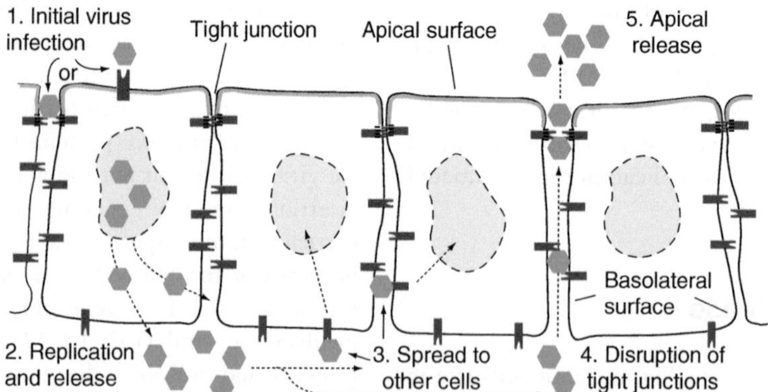

Figure 5 A schematic diagram of a polarized epithelial cell sheet, showing the sequence of events (1–5) by which adenovirus is thought to infect the epithelium and then spread both locally and to distant sites. CAR receptor molecules are shown in red, and a hypothetical alternative receptor in pink. See text for further details.

bodies (promyelocytic leukemia protein-containing structures also known as ND10). Finally, the host cell stress and DNA damage sensor, p53, is inactivated and targeted for degradation, through the action of E1B 55 kDa and E4 Orf6 proteins, while these same viral proteins (as well as E4 Orf3 in species *Human adenovirus C*) also block the double-stranded DNA break repair pathway, which otherwise leads to end-to-end joining of progeny genomes and a substantial failure of replication.

Adenovirus proteins also have effects on the host gene expression apparatus that impact on viral gene expression. First, through the action of E1A and E4 Orf6/7 proteins, the host transcription factor E2F is released from inhibitory complexes with Rb (retinoblastoma protein) family members so that it can activate the E2 promoter; this also has the effect of activating cell-cycle progression. Second, the activity of protein phosphatase 2A is altered through the action of E4 Orf4 protein, leading to a variety of effects including changes in the activity of SR splicing proteins and consequent changes in viral mRNA splicing pattern during the course of infection. Third, the E1B 55 kDa and E4 Orf6 proteins cause a block to the export of host cell mRNA from the nucleus during the late phase of infection and instead favor the export of viral mRNA. Finally, the L4 100 kDa protein is involved in allowing efficient translation of viral mRNA while the normal process of cap-dependent recognition of mRNA by ribosomes is inhibited.

Adenoviruses as Vectors

Several features of human adenoviruses have led to interest in them as potential gene delivery vectors for human cells. Applications are under development in which such vectors are designed to elicit a specific immune response to a delivered gene product (i.e., a recombinant vaccine), to give tumor-specific cell killing for cancer therapy, or to give long-term persistence and expression of a delivered gene (a gene therapy approach). Initial attention focused on human adenovirus 5 because of its experimental tractability (ease of making recombinants, ability to grow virus to very high titers in cell culture, and high level of genetic characterization), its ability to deliver genes into a variety of cell types whether or not they were actively dividing, and its mild disease course in a natural context. Difficulties experienced with these vectors, including the relative nonaccessibility of the CAR receptor on target tissues (see the section titled 'Assembly and release'), have since led to the development of vectors based on several other adenoviruses.

To convert human adenovirus 5 into a passive gene delivery vector, genome modifications are required to render it replication-defective and to create sufficient spare capacity for the insertion of foreign DNA; otherwise, only an additional 1–1.5 kbp of DNA can be accommodated. The E3 region is usually deleted, as its functions are dispensable for growth in cell culture, and the E1A and E1B genes are also removed. This latter deletion renders the virus largely replication defective, but the deficiency can be readily complemented in widely available cell lines to allow growth of the vector in the laboratory. Adding further deletions in other essential genes, such as E2 and E4, further impairs replication, giving even lower levels of late gene expression. Ultimately, all the viral genes can be stripped from the vector genome to leave only the replication origins and packaging sequence from the termini. These vectors have given the best performance among adenovirus vectors for long-term gene delivery but are more difficult to complement for growth in culture and hence much lower yields of vector particles are achieved.

See also: Viruses as Vectors in Gene Therapy.

Further Reading

Akusjärvi G and Stévenin J (2003) Remodelling the host cell RNA splicing machinery during an adenovirus infection. *Current Topics in Microbiology and Immunology* 272: 253–286.

Berk AJ (2005) Recent lessons in gene expression, cell cycle control, and cell biology from adenovirus. *Oncogene* 24: 7673–7685.

Berk AJ (2006) *Adenoviridae*: The viruses and their replication. In: Knipe DM and Howley PM (eds.) *Fields Virology,* 5th edn., ch. 63. Philadelphia: Lippincott Williams and Wilkins.

Davison AJ, Benkö M, and Harrach B (2003) Genetic content and evolution of adenoviruses. *Journal of General Virology* 84: 2895–2908.

Dimmock NJ, Easton AJ, and Leppard KN (2006) *Introduction to Modern Virology*, 6th edn. Oxford: Blackwell Scientific.

Fabry CMS, Rosa-Calatrava M, Conway JF, *et al.* (2005) A quasi-atomic model of human adenovirus type 5 capsid. *EMBO Journal* 24: 1645–1654.

Leppard KN (1998) Regulated RNA processing and RNA transport during edenovirus infections. *Seminars in Virology* 8: 301–307.

Liu H, Naismith JH, and Hay RT (2003) Adenovirus DNA replication. *Current Topics in Microbiology and Immunology* 272: 131–164.

Meier O and Greber UF (2003) Adenovirus endocytosis. *Journal of Gene Medicine* 5: 451–462.

Ostapchuk P and Hearing P (2005) Control of adenovirus packaging. *Journal of Cellular Biochemistry* 96: 25–35.

Russell WC (2000) Update on adenovirus and its vectors. *Journal of General Virology* 81: 2573–2604.

Zhang Y and Bergelson JM (2005) Adenovirus receptors. *Journal of Virology* 79: 12125–12131.

Baculoviruses: Molecular Biology of Granuloviruses

S Hilton, University of Warwick, Warwick, UK

Glossary

Defective interfering particle Virus particles that are missing part of their genome. These deletions in their genome mean that they cannot sustain an infection by themselves and depend on coinfection with a suitable helper virus.

Occlusion body A crystalline protein matrix which surrounds the virions of some insect viruses.

Origin of replication A unique DNA sequence at which DNA replication is initiated.

Palindrome A sequence of DNA equal to its complementary sequence read backwards.

Introduction

The *Baculoviridae* are a family of invertebrate viruses with large circular, double-stranded DNA genomes (80–180 kbp). The genomes are packaged into nucleocapsids, which are enveloped and embedded in proteinaceous occlusion bodies. Baculoviruses are divided taxonomically into two genera, *Nucleopolyhedrovirus* (NPV) and *Granulovirus* (GV). GVs and NPVs show major differences, not only in the morphology of their occlusion bodies but also in their tissue tropism and cytopathology. NPVs have large polyhedral-shaped occlusion bodies measuring 0.15–15 μm, comprised predominantly of a single protein called polyhedrin and typically have many virions embedded. GVs have small ovicylindrical occlusion bodies, which average about 150 nm × 500 nm. These are comprised predominantly of a single protein called granulin, which is related in amino acid sequence to polyhedrin. GVs normally have a single virion consisting of one nucleocapsid within a single envelope (**Figure 1(a)**). They infect both agricultural and forest pests, making them important as potential biological insecticides. To date, eight complete GV genomes have been sequenced (**Table 1**).

Infection Cycle

Baculoviruses produce two distinct virion phenotypes, occlusion-derived virus and budded virus, which are responsible for infection of insects (*per os* infection) and insect cells (cell-to-cell spread), respectively. Infection for both GVs and NPVs begins with the ingestion and solubilization of the occlusion bodies in the host larval midgut. The released occlusion-derived virus then fuses with the microvilli of midgut epithelial cells, and nucleocapsids pass through the cytoplasm to the nucleus. At the nuclear pore, the GV DNA is thought to be injected into the nucleus leaving the capsid in the cytoplasm. As replication proceeds, the nucleus enlarges and the nucleoli and chromatin move to its periphery (**Figure 1(b)**). The nuclear membrane ruptures early in infection in GVs, following the production of limited numbers of nucleocapsids. This is followed by extensive virogenesis, nucleocapsid envelopment and the formation of occlusion bodies in the mixed nuclear-cytoplasmic contents. Some nucleocapsids continue to form and pass out of the cell, forming budded virus which initiates secondary infection (**Figure 1(c)**). Baculoviruses encode two different major budded virus envelope glycoproteins, GP64 for group I NPVs and F protein for group II NPVs and GVs, which mediate membrane fusion during viral entry. Other nucleocapsids are enveloped and occluded within granulin-rich sites throughout the cell. As the number of occlusion bodies increase, the cell greatly enlarges and eventually lyses (**Figure 1(d)**). When the larvae die, the remaining cells lyse and the occlusion bodies are released back into the environment.

Taxonomy and Classification

GVs have been isolated only from the insect order Lepidoptera (butterflies and moths). They have been isolated from over 100 species belonging to at least 10 different host families within Lepidoptera, mainly Noctuidae and Tortricidae. There has been no formal classification of subgroups within the GVs. However, it is recognized that there are three types of GVs based on their tissue tropism, although these groups are not supported by phylogenetic analyses.

The first type of GV is slow-killing and predominantly infects larvae belonging to the family Noctuidae, but also includes the Torticidae-specific virus adoxophyes orana GV (AdorGV). They infect only the fat body of the larvae after the virus has passed through the midgut epithelium. Slow-killing GVs tend to kill in the final instar regardless of the instar in which the larvae are infected. These infected larvae tend to take longer to die than other GV- or NPV-infected larvae, typically from 10 to 35 days. A possible explanation for this is that important tissues such as the tracheal matrix and the epidermis are not infected.

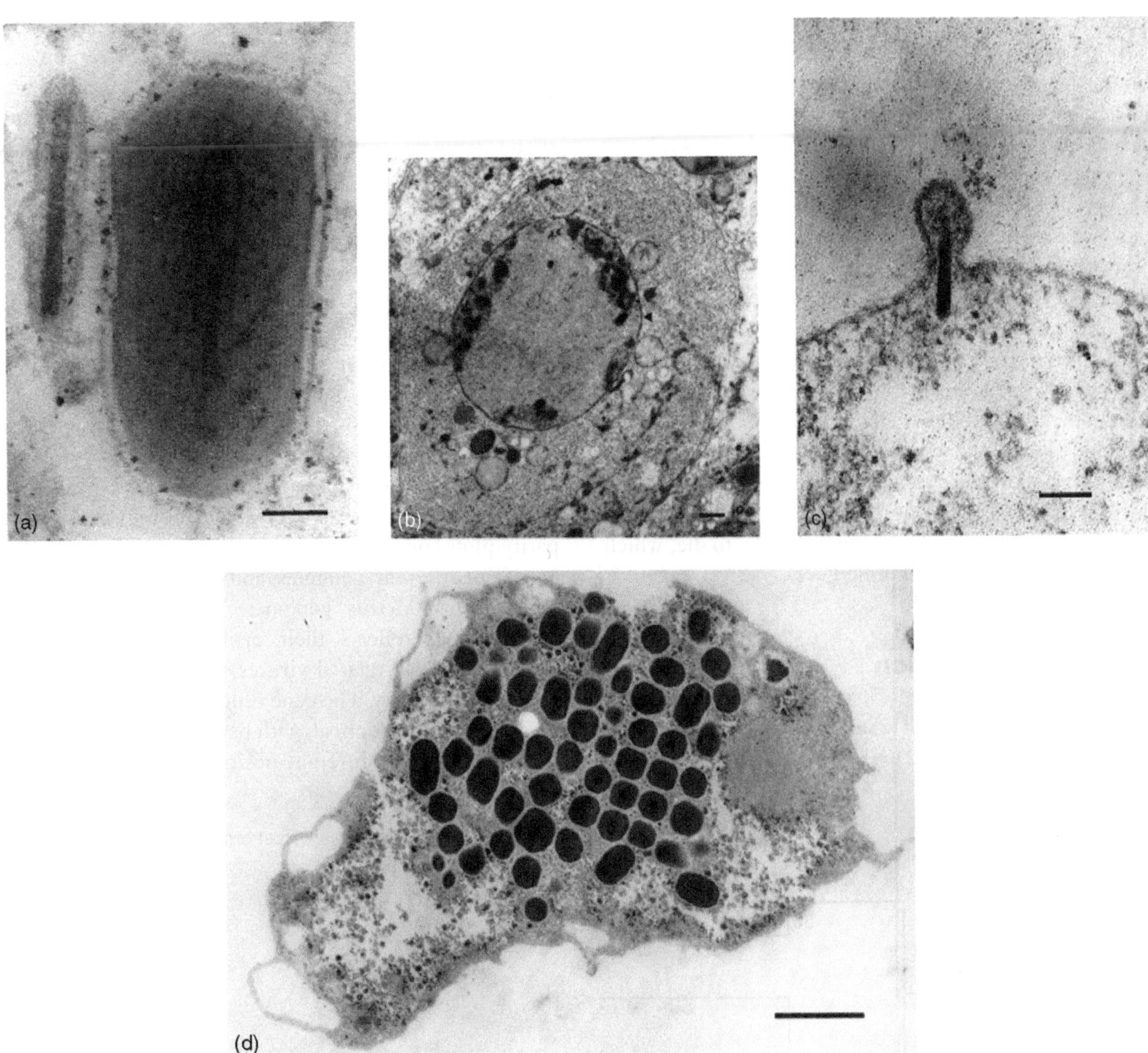

Figure 1 GV morphology. (a) Enveloped GV virion and adjacent occluded virion. (b) Early stages of GV replication. Nucleocapsids are present throughout the nucleus prior to disintegration of the nuclear membrane. The nuclear membrane is indicated by an arrowhead. (c) virus budding from plasma membrane. (d) *In vitro* replication of CpGV. *C. pomonella* cell late in infection containing occluded virus particles. Scale = 100 nm (a, c); 1 μm (b, d). (a–c) Reprinted from Winstanley D and Crook NE (1993) Replication of *Cydia pomonella* granulosis virus in cell cultures. *Journal of General Virology* 74: 1599–1609, with permission from Society for General Microbiology.

Table 1 Features of the eight completely sequenced granuloviruses

Granulovirus	*Abbreviation*	*Length (nt)*	*AT content (%)*	*Number of ORFs*	*Host family*
Adoxophyes orana granulovirus	AdorGV	99 657	65.5	119	Tortricidae
Agrotis segetum granulovirus	AgseGV	131 680	62.7	132	Noctuidae
Choristoneura occidentalis granulovirus	ChocGV	104 710	67.3	116	Tortricidae
Cryptophlebia leucotreta granulovirus	CrleGV	110 907	67.6	128	Tortricidae
Cydia pomonella granulovirus	CpGV	123 500	54.8	143	Tortricidae
Phthorimaea operculella granulovirus	PhopGV	119 217	64.3	130	Gelechiidae
Plutella xylostella granulovirus	PlxyGV	100 999	59.3	120	Plutellidae
Xestia c-nigrum granulovirus	XecnGV	178 733	59.3	181	Noctuidae

The second type of GV includes the GV type species, *Cydia pomonella* GV (CpGV). They are relatively fast-killing, taking only 5–10 days to kill. The fat body is infected along with the epidermis, malpighian tubules, tracheal matrix, hemocytes, and many other tissues to a lesser extent. The faster speed of kill could be due to the infection of a wider range of body tissues. Fast-killing GVs tend to kill in the instar in which the larvae are infected or within the following instar. These viruses infect larvae from a variety of families of Lepidoptera including Tortricidae, Pieridae, Yponomeutidae, Pyralidae, and Gelechiidae.

The third type of GV contains only one species to date, *Harrisina brillians* GV (HabrGV). HabrGV replicates only in the midgut epithelium cells of *H. brillians* larvae, from the family Zygaenidae. The larvae develop diarrhoea which consists of a discharge containing occlusion bodies. This can lead to a rapid spread of infection. The younger the larvae are infected, the longer they take to die, which is also typical of the slow-killing GVs.

Genome Organization

Phylogenetic analysis of the completely sequenced baculoviruses (**Figure 2**) shows that there is one clade which contains all of the Tortricidae-specific GVs. Although there are no other clearly defined clades, it is clear that the GVs tend to group based on the family of their host insect and not on their tissue tropism type. The only known type 3 GV, HabrGV, which infects insects from the family Zygaenidae, has not been completely sequenced and so was not included in these analyses. Previous phylogenetic analysis using the granulin gene placed HabrGV in a clade with the Torticidae-specific viruses. However, baculovirus relationships based solely on occlusion body protein sequence can disagree with other gene phylogenies, and further HabrGV genes need to be analyzed before evolutionary conclusions can be made.

The genomic organization of the GVs has been compared using gene parity plot analysis (**Figure 3**). Gene parity plots compare the relative positions of homologous genes in different genomes and illustrate conservation between baculovirus genomes. Gene arrangement in baculoviruses reflects their evolutionary history, with the more closely related viruses sharing a higher degree of gene collinearity. The gene order among the sequenced GVs is virtually identical, with between only one and five common genes in different positions along the genome.

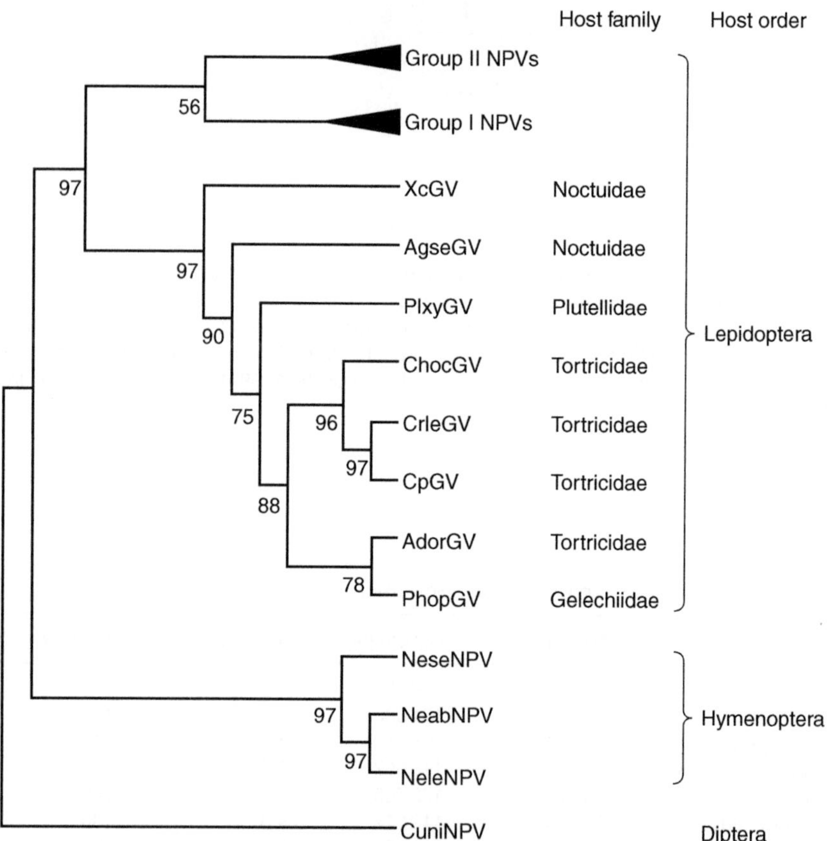

Figure 2 Baculovirus phylogeny (maximum pasimony) of completely sequenced baculoviruses based on the LEF-8 and LEF-9 concantenated sequences. Bootstrap percentage support values are indicated.

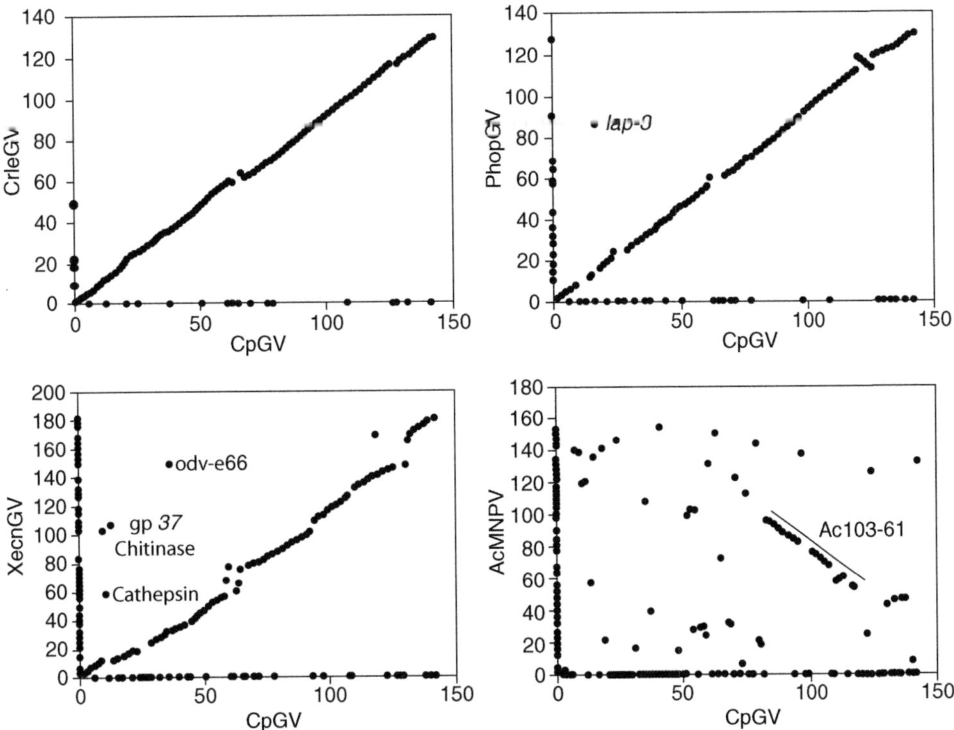

Figure 3 Gene parity plots of CpGV gene organization versus CrleGV, PhopGV, XecnGV, and AcMNPV.

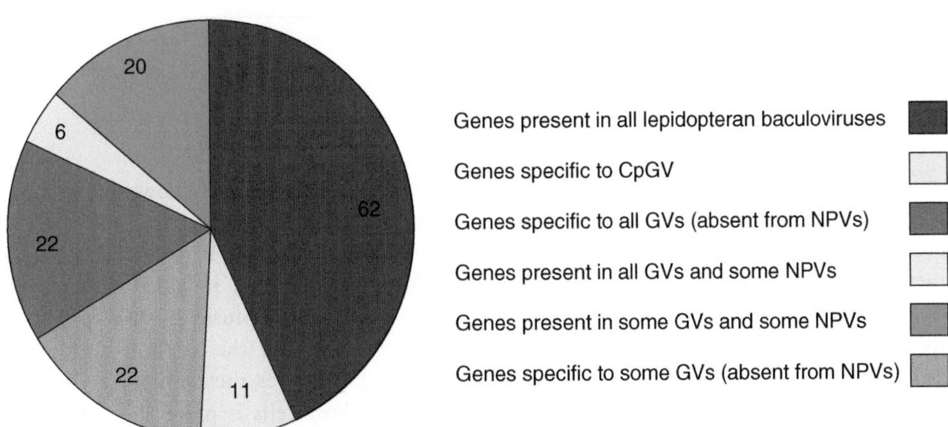

Figure 4 CpGV gene content. The number of CpGV genes (from a total of 143) are shown in each section of the pie chart.

One of these, the *iap-3* gene, is in three different positions in the five genomes which contain it. This suggests that this gene may be a more recent acquisition. The GVs appear to be far more collinear than the NPVs. There are many inversions and rearrangements among the NPV genomes relative to each other. The only GV genome to show any inversion of a block of genes is phthorimaea operculella GV (PhopGV) which has a block of six genes (Phop113–118) inverted relative to the other GVs. This block of genes flanks a putative homologous repeat region, which in NPV genomes, are sites where most major rearrangements, insertions, and deletions are found. The gene order among GVs is quite different from that of NPVs (**Figure 3**). GVs share only one main region of collinearity with NPVs, although this is inverted with respect to the granulin/polyhedrin gene. This region is from autographa californica NPV (AcMNPV) open reading frames (ORFs) Ac103-61.

The gene content of CpGV is shown in **Figure 4** and **Table 2**. There are 62 genes common to all Lepidopteran baculoviruses sequenced to date. There are a further 28 genes common to all GVs, 22 of which are absent from NPVs. A further 22 GV genes are found in more than one GV, but not in NPVs. The GV-specific genes

Table 2 Genes present in all lepidopteran baculoviruses and those present in all GVs

	Genes present in all lepidopteran baculoviruses	*Additional genes present in all GVs (genes in bold italic are absent from NPVs)*
Transcription	*39K, p47, lef-5, lef-4, lef-6, lef-8, lef-9, lef-11, vlf-1*	
Replication	*lef-1, lef-2, dnapol, helicase, dbp1, lef-3, ie-1, me53,*	*Cp120 (dnaligase), Cp126 (helicase-2)*
Structural	*F protein, gp41, odv-ec27, odv-e56, p6.9, p74, vp91, vp39, vp1054, pif, fp25K, odv-e18, odv-e25, odv-e66, pk1, polh*	***pep/p10 (Cp22)****, Cp71(p24capsid)*
Auxiliary	*alk-exo, fgf, ubiquitin*	***mp-nase (Cp46), fgf-1 (Cp76), iap-5 (Cp116), fgf-3 (Cp140)****, Cp59(sod)*
Unknown	*38K, ac22, ac29, ac38, ac53, ac66, ac68, ac75, ac76, ac78, ac81, ac82, ac92, ac93, ac96, ac106, ac109, ac110, ac115, ac142, 38.7K, ac145, ac146, p40, p12, p45*	***Cp2, Cp4, Cp5, Cp20, Cp23, Cp29, Cp33, Cp39, Cp45****, Cp47(p13),* ***Cp50, Cp56****, Cp79(Ac150),* ***Cp99, Cp100, Cp115, Cp122, Cp135, Cp136***
Total	62	28 (**22**)

Table 3 Auxiliary gene content of GVs

	egt	*enhancin*	*rr1*	*rr2a*	*pe38*	*ptp-2*	*chitinase*	*cathepsin*	*Cp94 (iap)*	*iap-3*	*bro*	*lef-10*
AdorGV	✓	×	×	×	×	×	×	×	×	✓	×	×
AgseGV	✓[a]	✓	×	×	×	×	✓	✓	×	✓	×	✓
ChocGV	✓	×	×	×	×	×	×	×	×	✓	×	✓
CrleGV	✓	×	×	×	✓	✓	✓[b]	✓	✓	✓	×	✓
CpGV	✓	×	✓	✓	✓	✓(2)	✓	✓	✓	✓	✓[a]	✓
PhopGV	✓	×	×	×	×	×	×	×	✓	×	✓	✓
PlxyGV	✓	×	×	×	×	×	×	×	×	×	×	×
XecnGV	×	✓(4)	×	×	×	×	✓	✓	×	×	✓(7)	✓

Numbers in parenthesis indicate gene copy number if greater than one.
[a]Likely not functional due to truncation.
[b]Likely not functional due to chitinase active site absent.

could be responsible for the biological differences between these two baculovirus groups. The relationship between biological differences and gene content of GVs and NPVs awaits the characterization of further genus specific genes. CpGV also contains 11 CpGV-specific genes, which may encode host-specific factors. There are also 20 genes which are present in some GVs and some NPVs, these are likely to be auxiliary genes, which are not essential for viral replication, but provide the virus with some selective advantage. These include *chitinase, cathepsin, iap-3, pe38, bro, ptp-2, rr1* and *rr2a, lef-10,* and egt. The GV auxiliary genes which have been assigned putative functions are shown in **Table 3.**

GV Replication *In Vitro*

Several NPVs replicate well in cell culture and some cell lines are commercially available. In contrast, there are only a few laboratories worldwide that have managed to maintain GV-permissive cell lines, which yield low virus titers. As a result, there are very few gene expression studies and little biochemical characterization of the GV genes. Currently, little is known regarding the molecular basis for phenotypic differences between GVs and NPVs. Consequently, some GV genes have been studied using NPV systems. These include the CpGV *cathepsin* and *iap-3* genes, trichoplusia ni GV (TnGV) *helicase* and *enhancin* genes and the xestia c-nigrum GV (XecnGV) *mmp* gene. However, the study of GV genes in their native virus and host cells is preferable. Only three GVs have replicated in cell culture. TnGV was reported to infect *Trichoplusia ni* cells but their permissive character was not maintained. Complete viral replication of PhopGV has been reported in a *P. operculella* cell line but has not led to any molecular studies. A *C. pomonella* embryonic cell line permissive for CpGV has enabled molecular studies on GVs to progress and recombinant CpGV viruses to be produced.

Gene Expression

The gene expression of NPVs has been found to be sequential and coordinated. It is essentially split into two stages. Early gene expression precedes DNA replication and late/very late gene expression occurs after the onset of DNA replication. The successive stages of viral gene expression are dependent on the previous gene products. Early genes are thought to be transcribed using the host RNA polymerase II. Therefore, promoters of NPV early genes tend to mimic promoters of eukaryotic genes. Late and very late genes are transcribed by virus-induced RNA polymerase that recognizes the late promoter motif (A/T/G)TAAG. Temporal studies on GV gene expression are limited due to a lack of highly permissive cell lines. However, GV genes possess promoters similar to the early and late promoters described above, and transcription of a small number of genes has been shown to initiate from them. It is therefore likely that GVs follow the same cascade of gene expression as NPVs.

AT Content

The sequenced GV genomes are AT-rich, ranging from 54.8% in CpGV, to 67.6% in cryptophlebia leucotreta GV (CrleGV), with an average of 62.6% (**Table 1**). CpGV and CrleGV are the most closely related sequenced GVs, although their AT content is highly divergent. The difference in AT content is mainly due to the base composition of the third nucleotide position within the codon of the coding regions. The host-cell machinery has to provide the virus with sufficient suitable t-RNA to replicate. The viruses may therefore have adapted to the base composition of the host. CrleGV is only infective to *C. leucotreta*, whereas CpGV is infective to both hosts, *C. pomonella* and (to a lesser degree) *C. leucotreta*. However, the AT content of these hosts is not yet known. CpGV is the only sequenced GV to contain the genes encoding the large and small subunits of ribonucleotide reductase (*rr1* and *rr2a*), which are also found in several NPVs and other DNA virus genomes. These enzymes are involved in nucleotide metabolism and catalyze the reduction of host cell rNTPs to dNTPs. It may be advantageous for CpGV to maintain these genes in order to alter the composition of dNTPs, allowing for replication in hosts with a divergent AT composition.

GV Specific Genes

Metalloproteinase

Baculovirus genomes have been shown to contain a number of proteases. Those studied so far are classified as nonessential auxiliary genes. Enhancin is a metalloproteinase which digests the peritrophic membrane in the insect midgut to facilitate virus infection. It has so far been found in all Noctuidae-specific GVs and one Tortricidae-specific GV (choristoneura fumiferana GV) and also in several group II NPVs. A cysteine proteinase (cathepsin) is involved in postmortem degradation of the infected host, in conjunction with a viral chitinase homolog. Cathepsin is found in most lepidopteran NPVs and four of the sequenced GVs (**Table 3**). A third group of proteases has been identified exclusively in the GVs, with homology to the matrix metalloproteinases (MMP). MMPs are active against components of the extracellular matrix and are usually secreted. All eight GV genomes possess an MMP homolog, suggesting it plays an important role in GV infection. Very little is known about MMP activity in infected larvae. A role for MMP in the breakdown of infected host tissue, potentially facilitating virus dispersal, and stimulation of postmortem melanization, has been proposed for XecnGV MMP. This study was performed using XecnGV MMP, overexpressed in occlusion-negative bombyx mori NPV (BmNPV), and may not be representative of its function in GV infection. However, results suggested that XecnGV MMP did encode a functional MMP. Putative signal peptide sequences are found in all of the GV MMP sequences except XecnGV. This suggests that these proteins, in keeping with MMPs, may enter the secretory pathway. Analysis of the predicted GV MMP amino acid sequences shows that residues predicted to be essential for metalloproteinase activity are conserved between all of the GV MMP sequences. The greatest conservation is in the region of the predicted zinc-binding active site of the enzyme. The conservation of MMP in all sequenced GVs implies that MMP plays an important role during GV infection, possibly in the breakdown of the basement membrane. As MMP is found exclusively in the GVs, its activity may lead to a difference in the infection process between GVs and NPVs.

Inhibitors of Apoptosis

Baculoviruses possess two families of genes that suppress apoptosis, the P35/P49 family and the inhibitor of apoptosis (IAP) family. The IAP-3 protein of CpGV was the first member of the baculovirus IAP family of proteins to be identified. It has been shown to block apoptosis in diverse systems and substitute for the *p35* gene in blocking AcMNPV-induced apoptosis in SF21 cells. Since then all of the GVs (and lepidopteran NPVs) have been found to contain IAP homologs. The GVs contain between one and three IAP homologs. Phylogenetic analyses have demonstrated that there are three clades of GV IAPs suggesting that they are not gene duplications. The first group contains the previously characterized IAP-3, which is found in CpGV, CrleGV, choristoneura occidentalis GV (ChocGV), AdorGV, and agrotis segetum GV (AgseGV). A second group includes Cp94 which has homologs, in the same genome position, in CrleGV and PhopGV. This IAP has so far been shown not to have any anti-apoptotic activity. A third group contains IAP-5 which is specific to all of the GVs. All GV genomes sequenced to date contain

IAP-5, which is absent from NPV genomes. The IAP-5 genes are located in the same positions on the GV genomes suggesting an ancient acquisition before the GVs diverged. This is in contrast to IAP-3 which is one of the only GV genes found in different positions within the GV genomes, suggesting a more recent acquisition. The CpGV IAP-5 has not been shown to have anti-apoptotic activity by itself, but it can stimulate the anti-apoptotic activity of CpGV IAP-3. However, its function in genomes lacking IAP-3 is unknown. The P35/P49 family was thought to be NPV specific but recently a P35/P49 homolog has been identified in ChocGV, although functionality has yet to be ascertained.

Fibroblast Growth Factors

Fibroblast growth factors, or FGFs, are a family of growth factors involved in tissue repair and cell proliferation and differentiation. It has been suggested that baculovirus FGFs may be required for efficient spread of infection beyond the midgut. FGFs are conserved among vertebrate and invertebrate organisms, but within the viruses they have only been identified in baculoviruses. All of the GV genomes contain three *fgf* genes, whereas there is only one *fgf* gene present in the NPV genomes (homologs of AcMNPV Ac32). Ac32 is transcribed both early and late in infection and is secreted when expressed in insect cells, which suggests that it acts as an extracellular ligand. An Ac32-null mutant was constructed and compared to wild-type virus in cell culture. No obvious differences were observed in protein or DNA synthesis, although its pathogenesis *in vivo* has not yet been evaluated.

All of the GV FGF proteins show homology to the FGF domain in Ac32. Phylogenetic analyses have demonstrated that there are three clades of GV FGFs suggesting that they are not gene duplications. The NPV FGFs did not group clearly with any GV FGF clade, but FGF-2 is usually annotated as the Ac32 homolog. It has not been determined whether all copies are functional and the advantage of three GV *fgf* genes with respect to pathogenesis is not yet known.

DNA Ligase and Helicase-2

All GV genomes contain the genes, *DNA ligase* and *helicase-2*. The *helicase-2* gene is part of the helicase superfamily I which includes PIF1 from eukaryotes and RecD from *Escherichia coli*. These enzymes are typically involved in DNA metabolism, such as replication, recombination, and repair. The only NPVs to contain either of these genes are lymantria dispar (LdMNPV) which contains a *DNA ligase* and a *helicase-2*, and mamestra configurata NPV (MacoNPV) and spodoptera litura NPV (SpltNPV) which both contain a *helicase-2*. The LdMNPV DNA ligase displays catalytic properties of a type III DNA ligase. However, in LdMNPV, neither the *helicase-2* nor the *DNA ligase* gene stimulates DNA replication in transient assays. The roles of *DNA ligase* and *helicase-2* in viral replication are unknown but based on their homology to genes from other organisms, it is likely that they are involved in DNA recombination or repair systems.

PEP/P10

Three GV genes form part of a highly conserved GV gene family. These are the homologs of the CpGV genes, Cp20, Cp22, and Cp23. Two of the genes, Cp20 and Cp23, are 35% identical to each other. Phylogenetic analyses of Cp20 and Cp23 and their homologs in other GVs clearly suggest that they are likely to be paralogous genes which were duplicated in a common ancestor before the GVs differentiated. The three genes show a significant similarity to domains of the baculovirus polyhedron envelope/calyx protein (PEP). PEP is thought to be an integral part of the polyhedron envelope. It is concentrated at the surface of polyhedra, and is considered important for the proper formation of the periphery of polyhedra. It is thought that PEP may stabilize polyhedra and protect them from fusion or aggregation. Electron microscopic evidence exists for a GV calyx and PEP may play a similar role in GVs.

The GV homologs of Cp22 share a number of motifs with P10, including a proline-rich domain and a heptad repeat sequence. In NPV-infected cells, P10 forms fibrillar structures in the nucleus and cytoplasm. These structures have yet to be identified in electron micrographs of GVs. It is possible that GV fibrous bodies are smaller, more granular structures, which may have been overlooked. The P10 protein is implicated in occlusion body morphogenesis and disintegration of the nuclear matrix, resulting in the dissemination of occlusion bodies. P10 is also crucial for the proper formation of the polyhedron envelope. The Cp22 homologs are over twice the size of most NPV P10 proteins and much of the sequence identity is between sequences of low complexity. The large size of the Cp22 homologs may be due to the presence of domains similar to the calyx/pep protein of NPVs. It has been suggested that in GVs the PEP and P10 may be conserved in a single protein (Cp22) and that the other members of the gene family (Cp20 and Cp23) may also be involved in the formation of the occlusion body envelope.

Origins of Replication

Within the genomes of NPVs, sequences have been identified that function as origins of replication (*oris*) when cloned into plasmids and transfected into infected cells. In some NPVs, these also function as enhancers of early gene transcription. These regions, called homologous regions (*hrs*), typically contain one or more copies of an imperfect

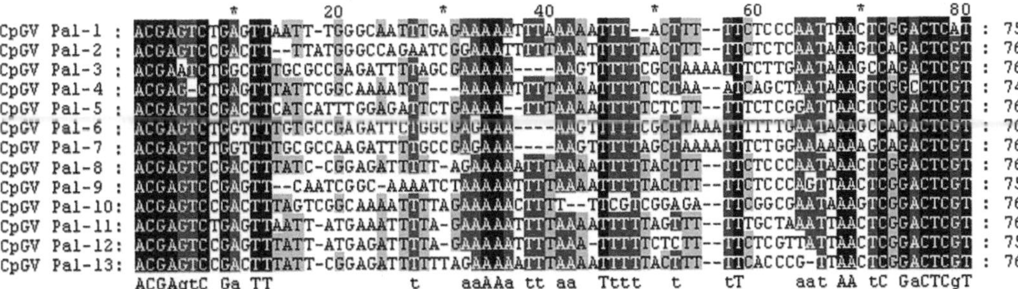

Figure 5 The 13 CpGV imperfect palindromes that acts as origins of replication.

palindrome sequence and are located in several positions on the genome. They have been identified in most NPVs sequenced to date and have been found to act as *oris* in several NPVs using infection-dependent DNA replication assays. In addition, complex structures containing multiple direct and inverted repeats of up to 4000 bp have also been identified that act as *oris*. These have similar structural characteristics to eukaryotic *oris* and are found only once per genome; they are therefore called non-*hr oris*. The discovery that serial passage of some NPVs results in the appearance of defective interfering particles (DIs) containing reiterations of a non-*hr ori*, is further evidence that non-*hr oris* may be involved in genome replication. It has also been suggested that the expansion of non-*hr ori*-like regions is not restricted to DI particles. A 900 bp region within the non-*hr ori*-like region of CrleGV was amplified in submolar populations in CrleGV *in vivo*, resulting in the expansion of the hypervariable region.

The CpGV genome contains 13 imperfect palindromes of 74–76 bp (**Figure 5**), which are found in 11 regions of the genome. Six of these palindromes are within putative genes. The 13 palindromes have been characterized using an infection-dependent replication assay in *C. pomonella* cells and all have been found to replicate. The palindromes are most conserved at their ends and have a highly AT-rich center. The entire 76 bp palindrome is required for replication, with no replication occurring when 10 bp from each end of the palindrome are removed. The specific flanking DNA of each palindrome is required for optimal replication, even though there was no homology between the flanking sequences. A region reminiscent of the non-*hr* type *ori* in NPVs has also been identified in the CpGV genome. However, this does not replicate in the infection-dependent replication assay and its function in GVs is not yet known.

Singleton imperfect palindromes are also found in CrleGV, ChocGV, PhopGV, and AdorGV. These GVs group in a clade with CpGV and it is likely that these palindromes descended from a common viral ancestor. The palindromes range in size from 63 bp in CrleGV to 320 bp in PhopGV and are often found in putative *hrs* which also consist of short direct repeats. The number of palindromes in each putative *hr* ranges from typically one up to four, which is less than the number of repeated palindromes in most NPV *hrs*. The ends of the palindromes are conserved in all of these GVs having a consensus of 13 bp (A(C/T)GAGTCCGANTT). The centers of the palindromes differ in sequence but all are AT-rich. These palindromes may have the potential to form secondary structures, such as cruciform (stem-loop or hairpin) structures through intrastrand base pairing. These structures may allow the initiation of DNA replication via the binding of specific protein factors. They may also remain in the linear state, acting as binding sites for protein dimers. This appears to be the case for the AcMNPV *hr5* palindrome and the regulatory protein IE-1. NPV *hrs* have been demonstrated to be *cis*-acting enhancers of transcription of baculovirus early genes including *39K*, *ie-2*, *p35*, and *helicase*. It is not yet known if CpGV *hrs* act as enhancers of early gene transcription.

The other GVs which do not group in the CpGV clade (PlxyGV, XecnGV, and AgseGV) have quite different putative *hr* regions. PlxyGV has large *hr* regions of up to 2383 bp consisting of direct repeats of 101–105 bp, repeated up to 23 times with a 15 bp palindrome near the center. XecnGV contains direct repeats of about 120 bp, repeated 3–6 times without any palindromes. So far been no putative *hrs* have been identified in the AgseGV genome. The only shared feature between all of the GV *hrs* is the relative positions of some of the *hrs* within the genome. For example, there is often a putative *hr* between the *desmoplakin* and *lef-3* genes, between *sod* and *p74* and flanking or within the *vp91* gene. Within the NPVs, *hr* regions are also often found flanking *sod* and *vp91*, which suggests that some of the *hrs* may have originated before the GVs and NPVs diverged.

See also: Baculoviruses: Molecular Biology of Nucleopolyhedroviruses.

Further Reading

Escasa SR, Lauzon HAM, Mathur AC, Krell PJ, and Arif BM (2006) Sequence analysis of the *Choristoneura occidentalis* granulovirus genome. *Journal of General Virology* 87: 1917–1933.

Hashimoto Y, Hayakawa T, Ueno Y, Fujita T, Sano Y, and Matsumoto T (2000) Sequence analysis of the *Plutella xylostella* granulovirus genome. *Virology* 275: 358–372.
Hayakawa T, Ko R, Okano K, Seong S-I, Goto C, and Maeda S (1999) Sequence analysis of the *Xestia c-nigrum* granulovirus genome. *Virology* 262: 277–297.
Herniou EA, Olszewski JA, Cory JS, and O'Reilly DR (2003) The genome sequence and evolution of baculoviruses. *Annual Review of Entomology* 48: 211–234.
Hess RT and Falcon LA (1987) Temporal events in the invasion of the codling moth *Cydia pomonella* by granulosis virus: An electron microscope study. *Journal of Invertebrate Pathology* 50: 85.
Hilton SL and Winstanley D (2007) Identification and functional analysis of the origins of DNA replication in the *Cydia pomonella* granulovirus genome. *Journal of General Virology* 88: 1496–1504.
Jehle JA (2002) The expansion of a hypervariable, non-*hr ori*-like region in the genome of *Cryptophlebia leucotreta* granulovirus provides *in vivo* evidence for the utilization of baculovirus non-*hr oris* during replication. *Journal of General Virology* 83: 2025–2034.
Jehle JA (2006) Molecular identification and phylogenetic analysis of baculoviruses from lepidoptera. *Virology* 346: 180–193.
Lange M and Jehle JA (2003) The genome of the *Cryptophlebia leucotreta* granulovirus. *Virology* 317: 220–236.
Luque T, Finch R, Crook N, O'Reilly DR, and Winstanley D (2001) The complete sequence of *Cydia pomonella* granulovirus genome. *Journal of General Virology* 82: 2531–2547.
Miller LK (ed.) (1997) *The Baculoviruses.* New York: Plenum.
O'Reilly DR and Vilaplana L (2003) Functional interaction between *Cydia pomonella* granulovirus IAP proteins. *Virus Research* 92: 107–111.
Winstanley D and Crook NE (1993) Replication of *Cydia pomonella* granulosis virus in cell cultures. *Journal of General Virology* 74: 1599–1609.
Wormleaton S, Kuzio J, and Winstanley D (2003) The complete sequence of the *Adoxophyes orana* granulovirus genome. *Virology* 311: 350–365.

Baculoviruses: Molecular Biology of Nucleopolyhedroviruses

D A Theilmann, Agriculture and Agri-Food Canada, Summerland, BC, Canada
G W Blissard, Boyce Thompson Institute at Cornell University, Ithaca, NY, USA

Glossary

Budded virus The type of baculovirus virion that is formed by budding from the cell plasma membrane and mediates the systemic spread of infection in the infected insect.
Homologous repeat element A DNA sequence that is typically comprised of a series of imperfect palindromes and repeated at various locations around a baculovirus genome. Homologous repeat elements function as origins of DNA replication and as transcriptional enhancers.
Occlusion derived virus The type of baculovirus virion that is assembled in the nucleus and becomes occluded with the paracrystalline matrix of the occlusion body.
Occlusion body A crystalline protein matrix which surrounds the virions of some insect viruses.
Polyhedrin The major occlusion body protein of NPV occlusion bodies.

Introduction

The family *Baculoviridae* currently consists of two genera, *Nucleopolyhedrovirus* and *Granulovirus.* The nucleopolyhedroviruses (NPVs) currently include all baculoviruses that are not classified as a granulovirus (GV), including those isolated from hymenopterans (sawflies) and dipterans (mosquitoes). In recent years, it has been recognized that the hymenopteran and dipteran baculoviruses are distinct from all other baculoviruses and it has been proposed that the *Baculoviridae* be subdivided into four new genera: the alphabaculoviruses (which include the current lepidopteran NPVs), betabaculoviruses (which include the current GVs), gammabaculoviruses (which include the current hymenopteran NPVs), and deltabaculoviruses (which include the current dipteran NPVs). This article focuses primarily on the lepidopteran NPVs (the proposed alphabaculoviruses).

Lepidopteran NPVs are the most widely studied of the *Baculoviridae*, primarily due to the availability of insect cell culture systems that are permissive for infection. A number of lepidopteran NPVs have been developed as bioinsecticides, and in addition lepidopteran NPVs are used widely as protein expression vectors and more recently as vectors for mammalian cell transduction. Also, there is a great deal of recent interest in developing certain lepidopteran NPVs as mammalian gene therapy vectors.

Life Cycle

Like all baculoviruses the lepidopteran NPVs contain genomes of double-stranded circular DNA, with genomes

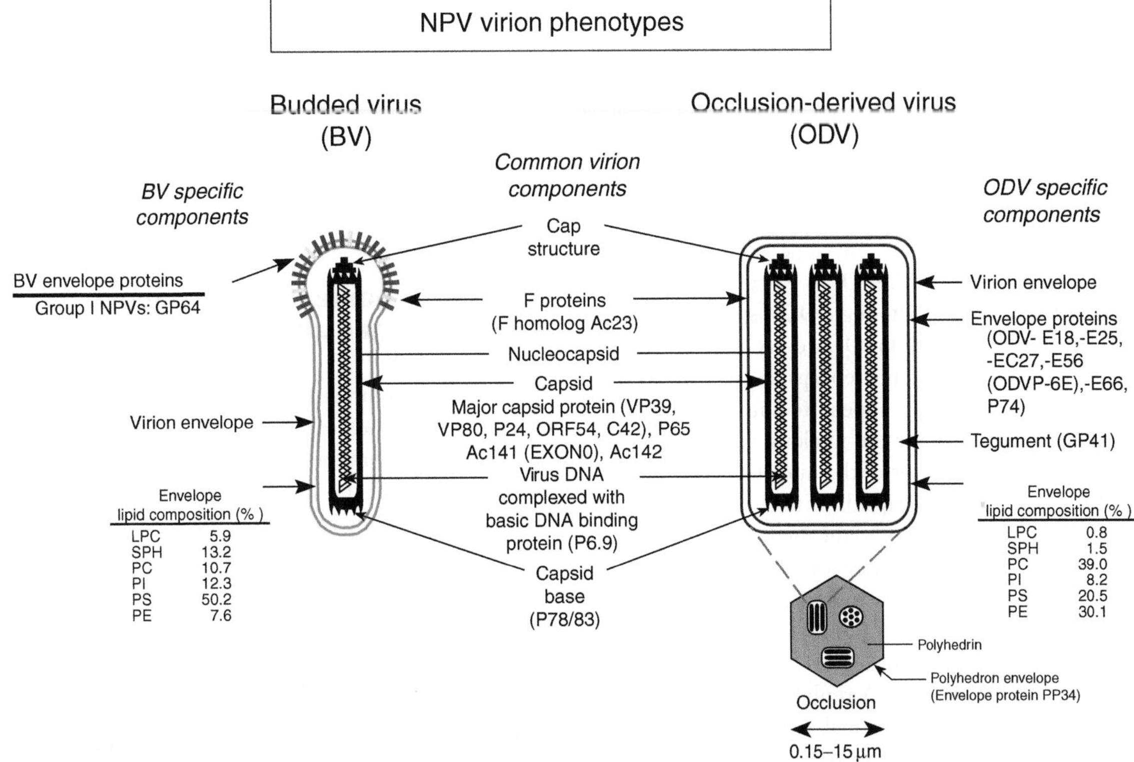

Figure 1 Schematic diagram of NPV virion phenotypes showing BV and ODV structures. Components common to both virion phenotypes are shown in the center and components unique to each phenotype are indicated on the left and right. The lipid compositions of the BV and ODV envelopes are indicated. LPC, lysophosphatidylcholine; SPH, sphingomyelin; PC, phosphatidylcholine; PI, phosphatidylinositol; PS, phosphatidylserine; PE, phosphatidylethanolamine. Adapted from Cytotechnology, vol. 20(1), 1996, pp. 73–93, Baculovirus–insect cell interactions, Blissard GW, Copyright 1996. With kind permission from Springer Science and Business Media.

ranging in size from ~111 to 168 kbp. Two distinct types of virions are produced in the life cycle: the budded virus (BV) and the occlusion-derived virus (ODV) (**Figure 1**). BVs generally contain a single nucleocapsid and obtain an envelope by budding from the plasma membrane of the infected cell. In contrast, ODVs are formed in the nucleus, acquire a membrane that is derived from the inner nuclear membrane, and virions consist of either single (S) or multiple (M) nucleocapsids per envelope. The S and M designation (SNPV and MNPV) does not appear to hold any taxonomic significance above the species level but this characteristic is clearly associated with certain viral species. As the infection progresses, ODVs are assembled in the nucleus and become embedded in a paracrystalline proteinaceous matrix to form occlusion bodies (OBs) (also known as polyhedra) that can range in size from 0.15 to 15 μm. A single matrix protein called polyhedrin makes up the majority of the mass of the polyhedra or OB. The OB as a whole is surrounded by a carbohydrate-protein layer known as the polyhedral envelope.

Because the ODVs and BVs have lipid bilayer envelopes that are derived from different sources (nuclear vs. plasma membranes, respectively), the envelopes of ODV and BV have a significantly different protein composition. Proteomic analyses of ODV from the archetype NPV, autographa californica MNPV (AcMNPV), and from helicoverpa armigera NPV (HearNPV) have resulted in the identification of up to 46 viral proteins of which eight or more may be found in the ODV envelope (summarized in **Figure 1**). ODV-specific proteins include P74, PIF-1 and -2, ODV-E18, ODV-EC27, ODV-E56 (ODVP-6E), ODV-E66, and P96. Such detailed proteomic analyses have not been performed on BV but the BV virion phenotype is known to contain BV-specific genes such as GP64. The NPV BV envelope appears to contain fewer viral proteins than ODV envelopes.

Infection Cycle

The infection cycle of NPVs is initiated when insects ingest polyhedral OBs (**Figure 3(a)**). The OBs pass through the foregut, enter the midgut, and dissolve or disassemble in response to the alkaline environment of the lepidopteran insect midgut. This process releases the ODVs which then traverse the peritrophic membrane (a protein–chitin structure that lines the gut) and infect the columnar epithelial cells of the lepidopteran midgut. This infection process is aided by virally encoded metalloproteases called enhancins that degrade the peritrophic membrane, thus helping virions cross this barrier and interact with the microvilli of the midgut epithelial cells. At the microvillar surface ODVs bind to the cell surface and are believed to enter the cell by direct fusion of the ODV envelope with the plasma membrane at the cell surface. ODV entry has not been examined in great detail although ODV-specific protein P74 appears to be necessary for ODV entry.

Upon entry into the cell, nucleocapsids are transported to the nucleus where they appear to interact with the nuclear pore complex enter the nucleus, and deliver the DNA genome to the interior of the nucleus. The interaction of baculovirus nucleocapsids with the nuclear pore complex and genome delivery to the nucleus is not clearly understood and requires further study. After the initial uncoating of the viral genome, transcription of viral early genes is initiated by host RNA polymerase II, resulting in early transcription and the subsequent production of proteins associated with viral DNA replication and late gene expression. A variety of additional regulatory and even structural protein genes are also transcribed in the early phase. Production of early gene products results in the initiation of viral DNA replication, and the assembly and activity of a virus-encoded RNA polymerase. This viral 'late' RNA polymerase recognizes unique viral late promoters resulting in the transcription of viral structural and other genes. With only rare exceptions, NPV late promoters contain the sequence 'DTAAG' at the late transcription start site. The DTAAG motif and the sequences immediately flanking it comprise the late promoter. Viral DNA replication and nucleocapsid assembly occurs within the nucleus in a viral structure known as the virogenic stroma (**Figure 2(a)**). Nucleocapsids are transported from the cell nucleus to the plasma membrane, where they bud from the cell surface to form the BVs, which are responsible for the systemic spread of the viral infection to other tissues within the insect (**Figure 3(b)**). Nucleocapsids are also retained in the nucleus and become singly or multiply enveloped to form the ODV, which become embedded in polyhedrin to form the OBs (**Figure 2**). Many or most cells and tissues of the host insect are permissive for viral infection and become infected and filled with OBs, which are released when the cells lyse. At least two viral proteins, chitinase and cathepsin, aid in the disruption of infected cells and tissues and the release of OBs into the environment. OBs released into the environment can remain stable for months or even many years if protected from ultraviolet (UV) light. Over 1×10^8 OBs can be produced from an individual infected larva depending on the host species, the virus, and the larval instar. The dissemination of OBs is believed to be aided by viral modification of host behavior. Many NPV-infected insects will exhibit a behavior of climbing to the upper regions of plants during the late stages of infection. Release of the OBs from such elevated locations is believed to enhance the spread of the virus by making the released OBs more accessible to other larvae feeding on lower regions of the plant.

Genome Organization (Genomes, Gene Content, Organization, Evolution)

To date the complete genomic sequences have been determined for 30 lepidopteran NPVs. The smallest genome is that from the maruca vitrata NPV (MaviNPV) at 111 953 bp and the largest from the leucania separata NPV (LeseNPV) at 168 041 bp. The archetype and best-studied lepidopteran NPV, AcMNPV, has a genome size of 133 894 bp (**Figure 4**). The numbers of predicted genes (open reading frames (ORFs) of 50 amino acids or greater) range from 126 (MaviNPV) to 169 (LeseNPV). Taxonomic analysis of lepidopteran NPV genomes has led to the subdivision of the NPVs into two distinct clades: group I and group II NPVs (**Figure 5**). The group I clade appears to represent a more cohesive and well-defined phylogeny, whereas the group II clade represents a less-homogenous group of viruses (**Table 1**).

A distinct feature of nearly all lepidopteran NPVs is regions of homologous repeats (*hr*'s) distributed throughout the genome. The *hr* sequences can vary significantly in size. Each *hr* region contains a series of repeated palindromic sequences. In the AcMNPV genome, for example, seven *hr*'s are present and they all contain repeats of an imperfect palindrome that contains an *Eco*RI restriction site at its core (**Figure 4**). Transient transcription and plasmid replication assays have shown that the *hr* sequences function as transcription enhancers and as replication origins. Single copy non-*hr* replication origins have also been identified in the genomes of AcMNPV, orgyia pseudotsugata MNPV (OpMNPV), and spodoptera exigue MNPV (SeMNPV). The non-*hr* origins appear more similar to eukaryotic cellular replication origins and contain multiple short repeated elements. The AT content of the lepidopteran NPVs can be quite variable, ranging from 42.4% (lymantria dispar MNPV; LdMNPV) to 64.4% (adoxophyes hommai NPV; AdhoNPV) and the significance of this variation is not currently known.

DNA Replication

Baculovirus DNA replication has been best characterized in the lepidopteran NPVs. Transient assays or the use of gene knockout viruses have shown that a number of genes are

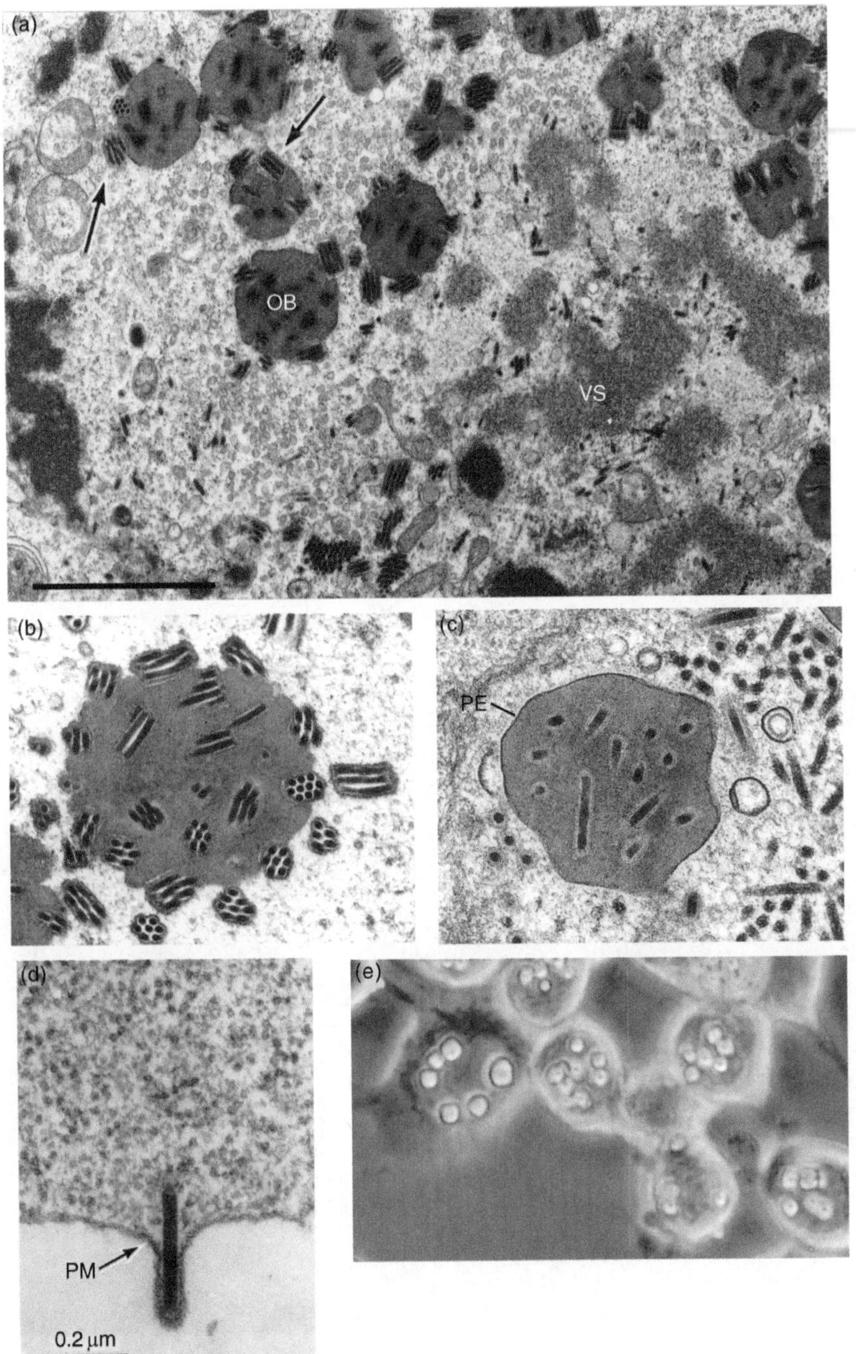

Figure 2 Electron and light micrographs of NPV-infected insect cells. (a) An insect cell infected with AcMNPV showing the nuclear virogenic stroma (VS) that contains developing nucleocapsids and the OBs showing ODV in the occlusion process (arrows). Scale = 1 μm. (b) An OB of the MNPV type, showing the process of occlusion of ODV-containing multiple nucleocapsids in each virion. Many virions are found in an OB. (c) A mature OB of the SNPV type, showing a single nucleocapsid in each virion (ODV) and many ODVs per occlusion body. Also note the formation of the polyhedral envelope (PE). (d) An NPV nucleocapsid budding from the plasma membrane (PM) of an infected cell. (e) Tn5b1–4 cells infected with AcMNPV producing large relatively uniform OBs. (a–d) Courtesy of R. Granados.

involved in viral DNA replication and viral genome processing. This includes *lef-1*, *lef-2*, *lef-3*, *lef-11*, *DNA polymerase*, *helicase*, *vlf-1*, *dna binding protein* (*dbp*), *alkaline exonuclease* (*ae*), *me53*, and *ie0* or *ie1*. *lef-1* has been shown to have primase activity and potentially to interact with *lef-2* which has an unknown function. The LEF-3 protein has a single-stranded DNA-binding (SSB) activity, forms homotrimers, and is required for transport of helicase to the nucleus. Helicase is believed to be involved in unwinding of viral DNA. DNA polymerase has homology to other known polymerases and is

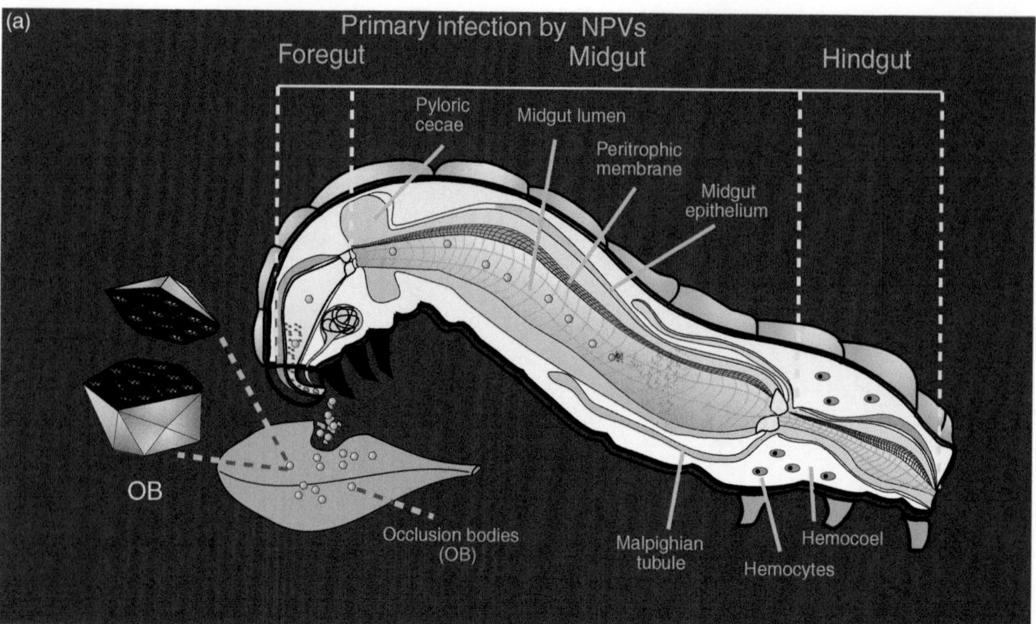

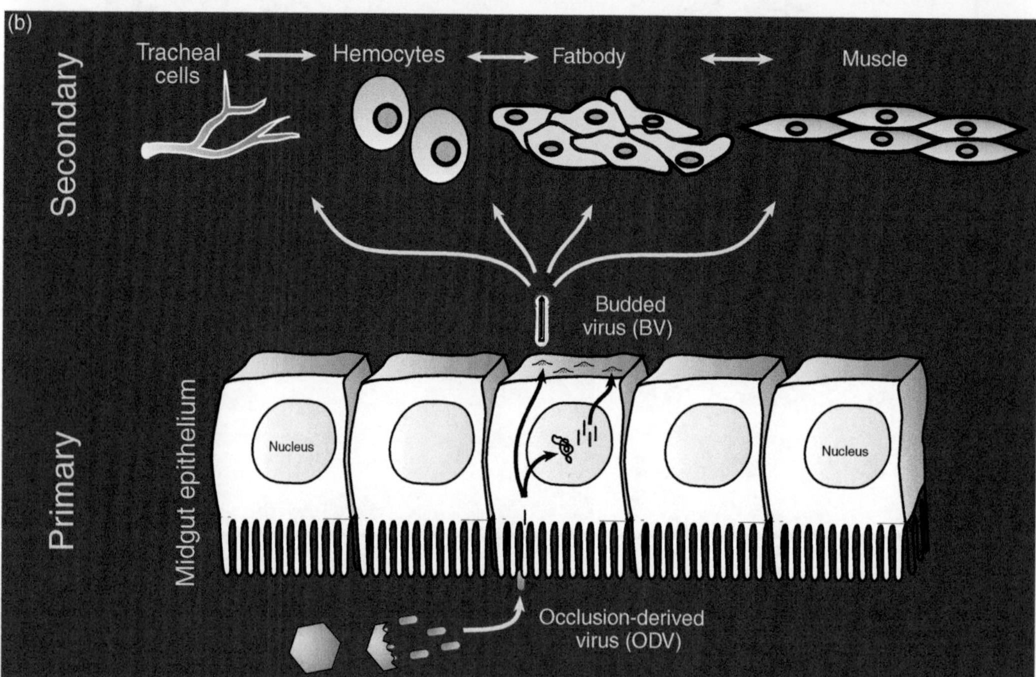

Figure 3 Schematic diagrams illustrating the infection process in a lepidopteran larva after ingestion of NPV OBs from a contaminated food source. (a) Within the midgut, OBs dissolve or 'disassemble' in the alkaline environment and the embedded ODVs are released into the lumen of the midgut. ODVs traverse the lining of the midgut, the peritrophic membrane, aided by virally encoded proteins and establish primary foci of infection in the midgut epithelial cells. (b) Upon infection of midgut epithelial cells by ODV interactions with microvilli at the apical surfaces of midgut epithelial cells, nucleocapsids are transported to the nucleus and initiate a primary round of viral replication, producing BVs that bud from the basal surface and subsequently infect cells of secondary tissues including tracheal cells, hemocytes, fatbody, and muscle. Alternatively, there is evidence that nucleocapsids can traverse the midgut cell and bud directly from the basal surface without uncoating or replicating in the nucleus, potentially avoiding cellular defense mechanisms associated with the midgut and accelerating the infection of secondary tissues. (a) Reproduced from Slack J and Arif BM (2006) The baculoviruses occlusion-derived virus: Virion structure and function. *Advances in Virus Research* 69: 99–165, with permission from Elsevier.

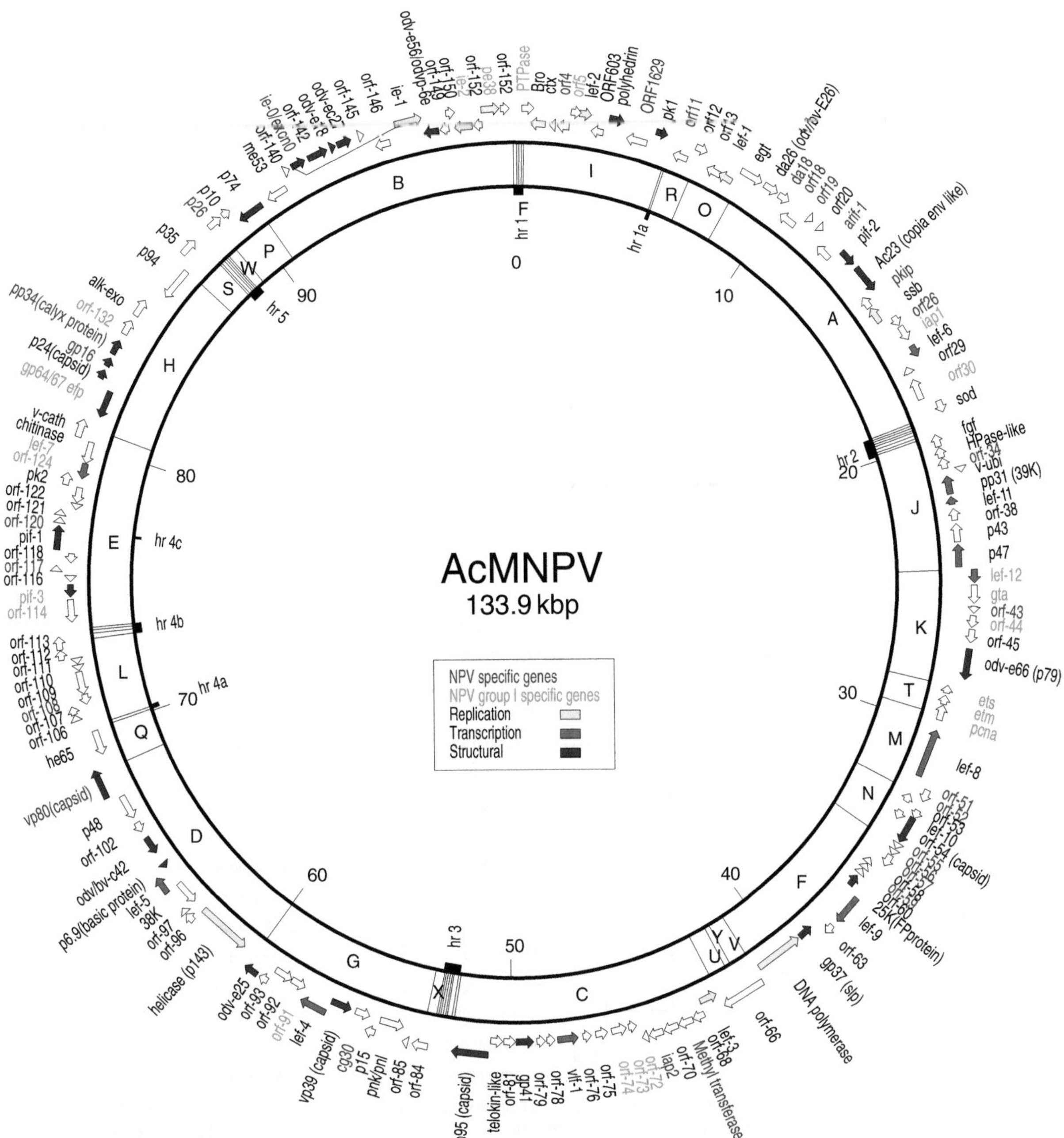

Figure 4 Genomic map of the archetype lepidopteran NPV, AcMNPV, which contains 154 open reading frames that encode predicted proteins of 50 or more amino acids. Functional groups of genes are highlighted (colored arrows) as well as genes specific to lepidopteran NPVs and group I lepidopteran NPVs (text colors). The inner circle shows the *Eco*RI restriction map of the C6 strain of AcMNPV. Locations of homologous repeat or *hr* regions (which contain repeats of *Eco*RI restriction sites) are indicated.

essential when assayed within the context of a viral infection. In addition to LEF-3, DBP has also been shown to have SSB activity and localizes at sites of viral DNA replication. In the absence of DBP, viral DNA is not processed into full-length genomes and is not packaged correctly into nucleocapsids. In transient replication assays, LEF-11 is not required for viral DNA replication; however, a *lef-11* knockout virus is unable to replicate DNA. LEF-11 has been shown to be a nuclear protein but its role in NPV DNA replication is unknown. Similar to *lef-11*, the deletion of *me53* also results in a virus that is unable to replicate viral DNA. VLF-1 was originally identified due to its impact on baculovirus very late gene transcription. However, in the absence of VLF-1, viral DNA is not packaged into nucleocapsids correctly. In transient assays or recombinant viruses, either of the viral transactivators IE0 or IE1 can support viral DNA replication but both

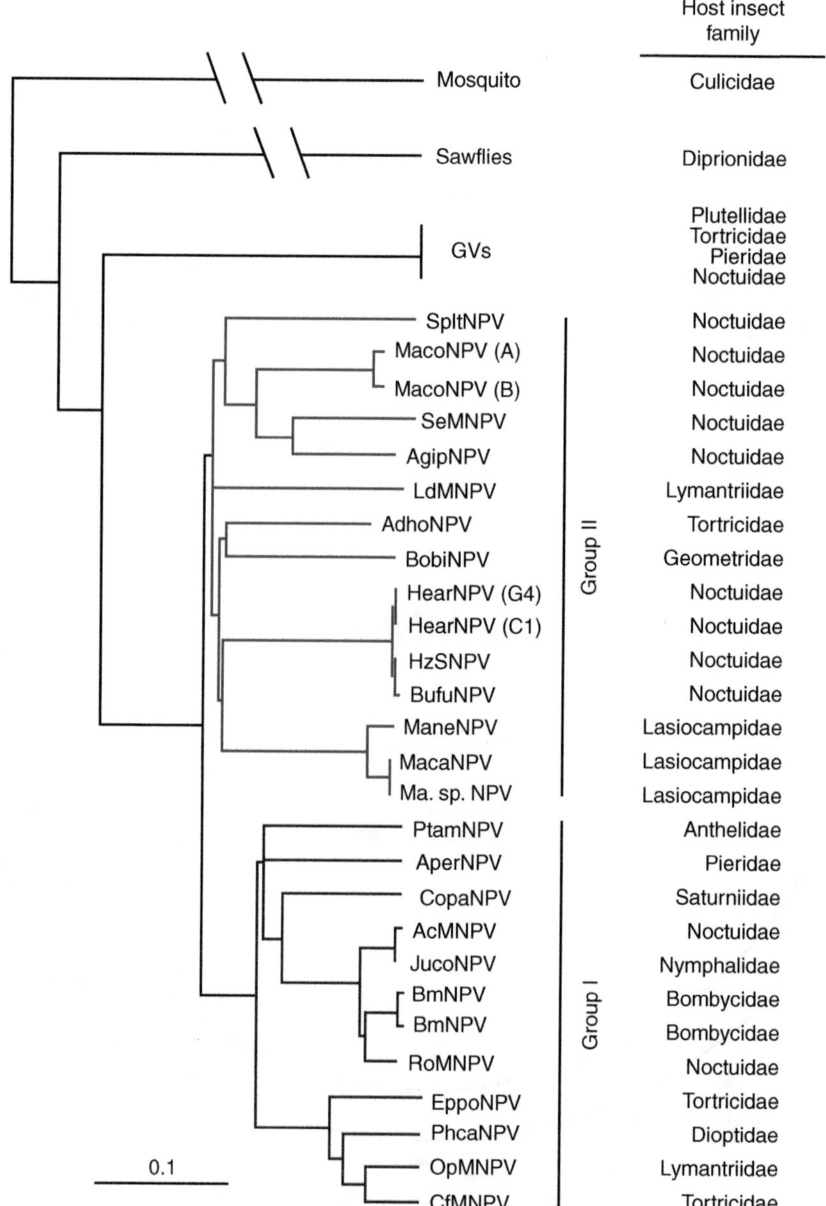

Figure 5 Phylogram showing the evolutionary relationships among lepidopteran NPVs and their lepidopteran hosts. A concatenated partial amino acid sequence of the common proteins polyhedrin/granulin, lef-8, and lef-9 was used to generate the neighbor-joining (NJ) distance tree and CuniNPV was used as outgroup. Blue branches identify group I NPVs, red identify group II NPVs, and green identify GVs. The lepidopteran family which is the known host of each respective baculovirus is shown on the right. Reprinted from Lange M, Wang H, Zhihong H, and Jehle JA (2004) Towards a molecular identification and classification system of lepidopteran-specific baculoviruses. *Virology* 325: 36–47, with permission from Elsevier.

are required to achieve full levels of viral replication. IE-1 and presumably IE-0 bind *hr* sequences and therefore may function as origin binding proteins.

A number of viral proteins have been shown to augment viral DNA replication but are not essential. They include *iap-1* and the group I specific NPV genes *pe38* and *ie2*. All of these genes are RING finger proteins and PE38 and IE2 have ubiquitin ligase activities. However, it is not known whether this enzymatic activity plays a role in DNA replication.

The mechanism by which the viral genome replicates is not known but high molecular weight DNAs that are suggestive of genomic concatemers have been reported. It has therefore been suggested that NPV DNA may replicate as a rolling circle. More recently however, due to presence of high molecular weight non-unit lengths of

Table 1 Summary of genes specific to lepidopteran NPVs, and the group I and group II NPV clades

Gene	*Gene function*
Lepidopteran NPV specific	
arif-1 (ac21)	Rearrangement of cellular actin
iap-2 (ac71) iap3	RING domain protein-possible apoptosis inhibition
orf1629 (ac9)	Essential structural gene of the nucleocapsid
pkip (ac24)	Stimulates the activity of the viral protein kinase-1 *in vitro*
pp34 (ac131)	Polyhedral envelope-associated phosphoprotein
vp80 (ac104)	Capsid protein
cg30 (ac88)	RING domain protein
exon0 (ac141)	BV nuclear egress
ie0 (ac141 and ac147)	Transcriptional transactivator, DNA replication
ac4, ac11, ac17, ac18, ac19, ac26, ac34, ac43, ac51, ac52, ac55, ac56, ac57, ac59, ac69, ac108, ac117, ac120, gp16(ac130), p26 (ac136)	Unknown function
Group I NPV specific	
ptp1 (ac1)	Protein tyrosine/serine phosphatase, affects OB formation
odve26 (ac16)	ODV envelope protein
iap1 (ac27)	RING domain protein-possible apoptosis inhibition
ie2 (ac151)	RING domain protein, transcriptional activation
lef7(ac125)	Late gene expression
vp80a (ac87)	Capsids protein
lef12 (ac41)	Late gene expression
gp64 (ac128)	BV specific glycoprotein required for virion entry
pe38 (ac153)	RING domain protein, transcriptional activation
ac5, ac30, ac72, ac73, ac114, ac124, ac132, gta (ac42), etm (ac48), ets (ac47), ac44, ac74, ac91	Unknown function
Group II NPV specific	
orf4PE	Polyhedral envelope protein
rr2b (ld120)	Ribonucleotide Reductase, R2/beta subunit
rr1 (ld148)	Ribonucleotide Reductase small subunit
parg (ld141)	Possible poly (ADP-ribose) glycohydrolase
ld55, ld129, ld111, ld124, ld127, ld142, ld144 , ld138	Unknown function

viral DNA in infected cells, it has been proposed that baculoviruses may use a recombination-dependent mechanism of viral DNA replication.

Temporal Regulation of Transcription

Transfected NPV DNA is infectious indicating that no viral proteins are required to initiate or mediate early transcription from the viral genome. Baculovirus genes are expressed in a temporal cascade beginning with early gene expression and followed by late gene expression. Late gene transcription requires prior or concomitantly with the onset of DNA replication. Very late gene expression occurs at the terminal part of the replication cycle and includes hyperexpressed viral genes such as polyhedrin.

Early Gene Expression

Early NPV expression is dependent upon the host cell RNA polymerase II complex and is sensitive to α-amanitin. NPV early genes have been divided into two categories, immediate early (IE) and delayed early (DE) genes. IE genes require only cellular factors for expression whereas DE genes are either dependent on, or substantially upregulated by prior viral gene expression. Early genes are primarily involved in gene regulation, host modification, DNA replication, and factors required for late gene expression.

Many IE genes have a common motif, TATA-N_{24-26}-CAGT, at their transcription start site similar to many motifs found in insect genomes. The promoters of IE and DE genes resemble typical eukaryotic RNA Pol II promoters and contain host cell transcription factor binding sites.

The primary viral regulatory protein for early transcription is the IE-1/IE-0 complex. The *ie0* gene is the only known spliced baculovirus gene that produces two viral protein products. IE0 contains the entire IE1 coding sequence but in addition has an N-terminal extension that is variable in length depending on the NPV species. IE0 has peak expression prior to viral DNA replication whereas IE1 continues to increase in steady-state levels until the final stages of the infection cycle. IE0 appears to be specific to the lepidopteran NPV as bioinformatics analysis has been unable to identify homologs in non-lepidopteran NPVs or GVs.

Viral gene knockouts have shown that either IE1 or IE0 is essential for viral replication and in the absence of this gene no viral infection is initiated. IE1/0 is an acidic domain transcriptional transactivator, similar to the herpesvirus VP16 protein; and activates viral gene transcription by both enhancer-dependent and independent mechanisms. IE1/0 forms dimers that bind to the *hr* sequences which serve as the transcription enhancers. IE1/0 is also essential for the viral replication complex where it is believed to play the role of an origin binding protein. In support of this, it has been shown that a transcriptionally inactive OpMNPV IE1 is able to support transient viral DNA replication.

Additional IE genes include *ie2* and *pe38*, which have been shown to augment viral transcription and, as indicated previously, viral DNA replication. Unlike IE1/0, both *ie2* and *pe38* can be deleted from the genome and the virus remains capable of replicating. In both cases however, dependent on the cellular environment, BV production and viral DNA replication are reduced. Some of the most highly expressed viral IE genes are *he65*, *me53*, and *gp64*. The localization of G-actin within the nucleus has been shown to be in part due to HE65 function, whereas ME53 is essential for DNA replication. Overall the IE genes play crucial regulatory roles that coordinate the NPV infection cycle.

DE expression requires additional viral factors to achieve nonbasal levels of expression. An example of a DE promoter is the AcMNPV *pp31* promoter which contains a TATA-CAGT sequence at the transcription start site and is dependent on IE1 for activated transcription (above basal-level) expression. The promoters of DE genes have not been clearly defined and it is not clear whether IE1/0 or other regulatory proteins may bind directly to the DE gene promoter.

Late Gene Expression

Baculovirus late genes are transcribed by a viral-encoded DNA-dependent α-amanitin-insensitive RNA polymerase that recognizes baculovirus late promoters. With only very rare exceptions, the baculovirus late promoters contain the sequence DTAAG and late transcription initiates within this seqeunce. The DTAAG sequence is considered to be at the core of the late promoter. In model late promoters that have been examined, only very short sequences immediately up- and downstream of the DTAAG sequence were identified as components of the late promoter. Regions farther downstream have been shown to affect levels of very late gene expression. Using transient assays 20 NPV genes or late expression factors (*lef*'s) were identified as important or necessary for late gene expression (*lef-1 to -12*, *pp31*, *p47*, *DNApol*, *vlf-1*, *helicase*, *p35*, *ie1/0*, *ie-2*). Many of the *lef* genes have been shown to be required for DNA replication, a prerequisite for late gene expression. Three of these genes are only found in the group I lepidopteran NPVs (*lef-7*, *lef-12*, and *ie-2*). Studies of purified NPV RNA polymerase have shown that it is comprised of four major proteins, LEF-4, LEF-8, LEF-9, and P47, and homologs of all these genes have been identified in all baculovirus genomes sequenced to date. LEF-4 has been shown to have enzymatic activity for RNA capping and LEF-8 and LEF-9 contain conserved motifs found in known RNA polymerases.

At very late times post infection, transcription of the hyperexpressed late genes (*polyhedrin* and *p10*) is upregulated, resulting in expression of those gene products at extremely high levels. The burst of very late gene transcription is mediated by the viral protein VLF-1, which binds to an A/T-rich region called a burst sequence located downstream of the DTAAG motif. This hyperexpression during the very late phase forms the basis of the baculovirus expression system. The *vlf-1* gene is a core baculovirus gene found in all sequenced baculovirus genomes.

Morphogenesis

Viral Structures – Nucleocapsids

The NPV life cycle is biphasic and produces two virion phenotypes, the BV and the ODV. Nucleocapsids of both virion phenotypes are believed to be structurally identical. NPV nucleocapsids are rod-shaped and approximately 30–60 nm by 250–300 nm (**Figures 1** and **2**) and are assembled in the ring zone of the nuclear virogenic stroma. DNA is densely packed within the nucleocapsid and small highly basic protein P6.9 is associated with the viral DNA. P6.9 is a core baculovirus protein that contains *c.* 40% arginine and 30% serine/threonine. The highly basic nature of this protein functions in neutralizing the positive charges of the nucleic acid and aids in the condensation and packaging of the viral DNA. Surrounding the viral DNA is the major capsid protein, VP39 (AcMNPV *orf89*). In the virus AcMNPV, the following proteins are associated with the nucleocapsid: P24, VP80, ORF54, BV/ODV-C42, ORF142, P95, ORF1629, and EXON0. ORF1629 is a phosphoprotein that is essential and is found associated with the basal end of the nucleocapsid. EXON0 is required for efficient egress of nucleocapsids from the nucleus. Interestingly, the very late gene transcription factor VLF-1 has also been shown to associate with the ends of the nucleocapsids and the data suggest that it facilitates the packaging of viral DNA and the formation of correctly sized nucleocapsids. Deletion of BV/ODV-C42 showed that this gene is essential and BV is not produced though viral DNA replication is not affected.

Viral Structures - BV and ODV Envelopes

The virion envelopes of BV and ODV are obtained by different mechanisms and contain arrays of proteins that are generally specific to each virion type. Nucleocapsids are believed to acquire an initial envelope when they egress from the nucleus to the cytoplasm. By an unknown mechanism, the envelope is lost and nucleocapsids are transported to the cell surface, potentially utilizing cytoskeletal structures such as microtubules. At the cell surface, the nucleocapsids bud from the plasma membrane acquiring the modified plasma membrane as the BV envelope. Nucleocapsids destined to be incorporated into ODV are retained in the nucleus and are enveloped by a membrane that is derived from the inner nuclear membrane. Analysis of the lipid composition of AcMNPV ODV and BV envelopes show that they differ significantly, reflecting the differences of their origins.

The BV envelope of group I NPVs contains the major envelope glycoprotein, GP64, which is essential for viral attachment and membrane fusion during entry (see below). GP64 is not present in ODV and is therefore a BV-specific protein. GP64 proteins are not found in group II NPVs but a protein that is functionally equivalent, the F (fusion) protein, is present in group II NPVs and GVs. The best-studied GP64 proteins are those of AcMNPV and OpMNPV. F proteins have been studied most extensively from SeMNPV (ORF8), HaNPV (ORF133), and LdMNPV (ORF133). All viruses that encode a GP64 protein (group I NPVs) also encode a homolog of the F protein, although the F homolog in group I NPVs is not essential for virion entry or viral replication (see below). Unlike GP64, F proteins (and homologs in the group I NPVs) may not be specific to BVs. Proteomic analyses have identified F proteins in the ODVs of AcMNPV, HearNPV, and CuniNPV. More detailed studies will be necessary to understand any potential role of F proteins in the ODV.

Viral entry by budded virions of the NPVs is mediated primarily by the major envelope protein, GP64 or F. The GP64 proteins are highly conserved, with approximately 80% amino acid sequence identity among all GP64 protein ectodomains examined. Structurally, GP64 is a type I integral membrane glycoprotein that is phosphorylated, palmytoylated, and heavily glycosylated. It is found on the cell surface and on the virion as a disulfide-linked trimer of GP64 monomers. GP64 functions in both viral entry and exit. In viral entry, GP64 is important for host cell receptor binding and membrane fusion. Little is known of the details of GP64 interactions with host cell receptor(s) and the specific molecule(s) that serves as the host cell receptor for virion binding is unknown. However, GP64 binding appears to be highly promiscuous and this feature has been exploited in the use of baculoviruses as mammalian transduction vectors, as potential gene therapy vectors, and in the use of GP64 for pseudotyping other viruses such as retroviruses and paramyxoviruses. F proteins from group II NPVs presumably serve a similar role in host receptor binding although they may recognize a different host cell receptor. After host cell binding, BVs of viruses such as AcMNPV are internalized via endocytosis and the low pH of the endosome triggers a conformational change in the GP64 protein, resulting in activation of membrane fusion. Unlike many other membrane fusion proteins, the GP64 protein does not require a prior internal cleavage (within the ectodomain) for maturation or activation of the functional fusion protein. Studies of GP64-mediated membrane fusion indicate that large short-lived complexes of approximately 10 or more GP64 trimers form immediately prior to membrane merger and are likely the unit structure of the membrane fusion mechanism. In addition, the opening of the fusion pore occurs rapidly after triggering.

F proteins from group II NPVs such as SeMNPV, HaNPV, and LdMNPV are also low-pH-activated membrane fusion proteins. These F proteins require an internal cleavage by a cellular proprotein convertase, for fusion activity. F proteins from group II NPVs share general structural features with paramyxovirus F proteins.

Studies of a *gp64* gene knockout AcMNPV virus showed that GP64 is also necessary for efficient budding of progeny BV. Because GP64 is found on the surface of infected cells, concentrated in discrete areas, it is thought that these concentrations of GP64 represent the sites of virion budding. GP64 is found on the virion in a polarized manner with the GP64 spikes found concentrated at the end of the virion that corresponds to the end where budding initiated (**Figure 2**). An important question in the biology of this virus is whether GP64 accumulation determines the sites of BV budding, or whether it is simply targeted to the same sites. Like the VSV G protein, GP64 is required for efficient budding. However, the precise domains required for virion assembly or budding are not yet known.

Group I NPVs encode both GP64 and an F homolog (Ac23 in AcMNPV, Op21 in OpMNPV). In AcMNPV, GP64 is essential whereas the F homolog (Ac23) is not and Ac23 can be deleted with no substantial effect in cell culture infections. However, the conservation of F homologs in the genomes of the group II NPVs suggests an important function. Deletion of the Ac23 gene from the AcMNPV genome results in delayed mortality of infected larvae. Thus, while F proteins found in group II NPVs are essential entry proteins, the F homologs in group I NPVs are not essential but appear to serve an accessory role that may be important in the pathogenicity or virulence of the virus.

The GP64 protein has been used for several biotechnological applications, including peptide display on baculovirus particles, and pseudotyping by replacing the envelope protein of another virus with GP64. Gene therapy vectors derived from retroviruses can be effectively pseudotyped with AcMNPV GP64 or other GP64 proteins. Further study of these viral envelope proteins should yield important and useful new tools for applications in biotechnology, agriculture, and medicine.

ODV envelopes have been shown to contain a number of specific proteins that are not present in BV. They include the following AcMNPV proteins (or their homologs): ODV-E18, ODV-E25, ODV-EC27, ODV-E56 (ODVP-6E), ODV-E66, P74, and ORF142. Viruses containing gene knockouts have shown that AcMNPV ORF142 and ODV-EC27 are essential for successful virus assembly and production of infectious virus.

Four genes encode ODV envelope proteins that are required for oral infectivity. *Per os infectivity factor-1, -2* and *-3* (*pif-1, pif-2, pif-3*) and *p74* have been shown to be required for oral infectivity of SeMNPV and AcMNPV. ODVs produced from viruses that do not express P74 are not orally infectious and evidence suggests that P74 is required for receptor binding on the midgut cell surface. Similar results have been obtained with PIF-1, -2, and -3. None of these proteins is required for infection of cells by BV in tissue culture.

Viral Structures – Tegument Proteins and Polyhedra

GP41 is an ODV-specific glycoprotein that does not fractionate with either the nucleocapsid fraction or the envelope fraction and is therefore believed to be a so-called tegument protein. GP41 is a core baculovirus gene, conserved in all baculovirus genomes examined. A temperature-sensitive mutant of the AcMNPV GP41 protein shows that it plays a critical role in viral development. When grown at the nonpermissive temperature, *ts* mutants of *gp41* fail to produce ODVs and polyhedra, and in addition nucleocapsids fail to egress from the nucleus to form BV. GP41 therefore appears to play a key role in the assembly of both virion phenotypes even though it is a component of only the ODV. Tegument proteins appear to be acquired within the nucleus when ODV nucleocapsids are enveloped. BVs do not contain GP41 as it is believed that all nuclear proteins surrounding nucleocapsids are lost when the nucleocapsids migrate from the nucleus to the plasma membrane.

OBs represent a feature that is common to all the currently classified baculoviruses (**Figure 2**) but the shape and size of the OB can vary substantially. Lepidopteran NPVs produce some of the largest OBs (polyhedra) and, unlike the GVs or the dipteran or hymenopteran NPVs, the embedded ODV can contain single or multiple nucleocapsids per envelope (**Figure 2**). The major component of the OBs of NPVs is the OB matrix protein, polyhedrin, which forms the bulk of the paracrystalline array (**Figure 2**). Surrounding the OB is a structure known as the calyx or envelope which is thought to be comprised of carbohydrate and protein. A lepidopteran NPV-specific phosphoprotein, PP34, is the major protein associated with this structure. To date no specific function has been attributed to the polyhedra calyx in the infection cycle of NPV. However, the OBs of an AcMNPV virus with a *pp34* deletion were found to have increased sensitivity to alkali disruption and enhanced virulence in fourth instar *Spodoptera exiguae* larvae, suggesting that the polyhedra calyx may stabilize the OB.

NPV-Specific Genes

The lepidopteran NPVs have 28 genes that are specific to all members of this genus and a further 13 genes specific to group II NPVs and 21 genes specific to group I NPVs (**Figure 4**). A number of these genes have been characterized and are known to impart specific functionality on these viruses but many remain to be investigated to determine their role in NPV biology.

See also: Baculoviruses: Molecular Biology of Granuloviruses.

Further Reading

Blissard GW (1996) Baculovirus–insect cell interactions. *Cytotechnology* 20: 73–93.

Clem RJ (2005) The role of apoptosis in defense against baculovirus infection in insects. *Current Topics in Microbiology and Immunology* 289: 113–129.

Guarino LA, Xu B, Jin J, and Dong W (1998) A virus-encoded RNA polymerase purified from baculovirus-infected cells. *Journal of Virology* 72: 7985–7991.

Herniou EA, Olszewski JA, Cory JS, and O'Reilly DR (2003) The genome sequence and evolution of baculoviruses. *Annual Review of Entomology* 48: 211–234.

Jehle JA, Blissard GW, Bonning BC, *et al.* (2006) On the classification and nomenclature of baculoviruses: A proposal for revision. *Archives of Virology* 151: 1257–1266.

Lange M, Wang H, Zhihong H, and Jehle JA (2004) Towards a molecular identification and classification system of lepidopteran-specific baculoviruses. *Virology* 325: 36–47.

Miller LK (ed.) (1997) *The Baculoviruses*. New York: Plenum.

Okano K, Vanarsdall AL, Mikhailov VS, and Rohrmann GF (2006) Conserved molecular systems of the *Baculoviridae*. *Virology* 344: 77–87.

Oomens AGP and Blissard GW (1999) Requirement for GP64 to drive efficient budding of autographa californica multicapsid nucleopolyhedrovirus. *Virology* 254: 297–314.

Slack J and Arif BM (2006) The baculoviruses occlusion-derived virus: Virion structure and function. *Advances in Virus Research* 69: 99–165.

Smith GE, Fraser MJ, and Summers MD (1983) Molecular engineering of the *Autographa californica* nuclear polyhedrosis virus genome: Deletion mutations within the polyhedrin gene. *Journal of Virology* 46: 584–593.

Stewart TM, Huijskens I, Willis LG, and Theilmann DA (2005) The *Autographa californica* multiple nucleopolyhedrovirus *ie0–ie1* gene complex is essential for wild-type virus replication, but either IE0 or IE1 can support virus growth. *Journal of Virology* 79: 4619–4629.

Theilmann DA, Blissard GW, Bonning B, *et al.* (2005) The *Baculoviridae*. In: Fauquet CM, Mayo MA, Maniloff J, Desselberger U, and Ball LA (eds.) *Virus Taxonomy: Eighth Report of the International Committee on Taxonomy of Viruses*, pp. 177–185. San Diego, CA: Elsevier Academic Press.

Coronaviruses: Molecular Biology

S C Baker, Loyola University of Chicago, Maywood, IL, USA

Glossary

Cell tropism Process that determines which cells can be infected by a virus. Factors such as receptor express can influence the cell type that can be infected.

Discontinuous transcription Process by which the coronavirus leader sequence and body sequence are joined to generate subgenomic RNAs.

Double membrane vesicles (DMVs) Vesicles that are generated during coronavirus replication when viral replicase proteins sequester host cell membranes. These vesicles are the site of coronavirus RNA synthesis.

Transcriptional regulatory sequences (TRSs) Sequences that are recognized by the coronavirus transcription complex to generate leader-containing subgenomic RNAs.

Introduction

Coronaviruses (CoVs) were first identified during the 1960s by using electron microscopy to visualize the distinctive spike glycoprotein projections on the surface of enveloped virus particles. It was quickly recognized that CoV infections are quite common, and that they are responsible for seasonal or local epidemics of respiratory and gastrointestinal disease in a variety of animals. CoVs have been named according to the species from which they were isolated and the disease associated with the viral infection. Avian infectious bronchitis virus (IBV) infects chickens, causing respiratory infection, decreased egg production, and mortality in young birds. Bovine coronavirus (BCoV) causes respiratory and gastrointestinal disease in cattle. Porcine transmissible gastroenteritis virus (TGEV) and porcine epidemic diarrhea virus (PEDV) cause gastroenteritis in pigs. These CoV infections can be fatal in young animals. Feline infectious peritonitis virus (FIPV) and canine coronavirus (CCoV) can cause severe disease in cats and dogs. Depending on the strain of the virus and the site of infection, the murine CoV mouse hepatitis virus (MHV) can cause hepatitis or a demyelinating disease similar to multiple sclerosis. CoVs also infect humans. Human coronaviruses (HCoVs) 229e and OC43 are detected worldwide and are estimated to be responsible for 5–30% of common colds and mild gastroenteritis. Interestingly, HCoV-OC43 and BCoV share considerable sequence similarity, indicating a likely transmission across species (either from cows to humans or vice versa) and then adaptation of the virus to its host. In contrast to the relatively mild infections caused by HCoV-229e and HCoV-OC43, the CoV responsible for severe acute respiratory syndrome (SARS-CoV) causes atypical pneumonia with a 10% mortality rate. Two additional HCoVs, HCoV-NL63 and HCoV-HKU1, have been recently identified using molecular methods and are associated with upper and lower respiratory tract infections in children, and elderly and immunosuppressed patients. CoVs are grouped according to sequence similarity. CoVs that infect mammals are assigned to group 1 and group 2, whereas CoVs that infect birds are in group 3.

To date, the most infamous example of zoonotic transmission of a CoV is the outbreak of SARS in 2002–03. We now know that the outbreak started with cases of atypical pneumonia in the Guangdong Province in southern China in the fall of 2002. The infection was spread to tourists visiting Hong Kong in February, 2003, resulting in the dissemination of the outbreak to Hong Kong, Vietnam, Singapore, and Toronto, Canada. After attempting to treat cases of atypical pneumonia in Vietnam and acquiring the infection himself, Dr. Carlo Urbani alerted the World Health Organization (WHO) that this disease of unknown origin may be a threat to public health. The WHO rapidly organized an international effort to identify the cause of the outbreak, and within months a novel CoV was isolated from SARS patients and identified as the causative agent. Sequence analysis revealed that the virus was related to, but distinct from, all known CoVs. This led to an intensive search for an animal reservoir for this novel CoV. Initially, the masked palm civet and raccoon dog were implicated in the chain of transmission, since a SARS-CoV-like virus could be isolated from some animals found in wild animal markets in China. However, SARS-CoV-like viruses were never detected in animals captured from the wild, indicating that the civets may have only served as an intermediate host in the chain of transmission. Further investigation revealed that the likely reservoir for SARS-CoV is the Chinese horseshoe bat (*Rhinolophus* spp.), which is endemically infected with a virus, named bat-SARS-CoV, that is closely related to SARS-CoV. The existence of an animal reservoir presents the possibility of re-emergence of this significant human pathogen. By improving our understanding of the molecular aspects of CoV replication and pathogenesis, we may facilitate development of appropriate antiviral agents and vaccines to control and prevent diseases caused by known and potentially emerging CoV infections.

Molecular Features of CoVs

CoV virions (**Figure 1(a)**) are composed of a large RNA genome, which combines with the viral nucleocapsid protein (N) to form a helical nucleocapsid, and a host cell-derived lipid envelope which is studded with virus-specific proteins including the membrane (M) glycoprotein, the envelope (E) protein, and the spike (S) glycoprotein. CoV particles vary somewhat in size, but average about 100 nm in diameter. The genomic RNA (gRNA) inside the virion, which ranges in size from 27 to 32 kb for different CoVs, is the largest viral RNA identified to date. CoV gRNAs have a broadly conserved structure which is illustrated by the SARS-CoV genome shown in **Figure 1(b)**. The gRNA is capped at the 5′ end, with a short leader sequence followed by two long open reading frames (ORFs) encoding the replicase polyprotein. The remaining part of the genome encodes the viral structural and so-called accessory proteins. The structural protein genes are always found in the order S–E–M–N, but accessory protein genes may be interspersed at various sites between the structural genes. SARS-CoV has the most complex genome yet identified, with eight ORFs encoding accessory proteins. The expression of these ORFs is not required for viral replication, but they may play a role in the pathogenesis of SARS. In addition, the products of accessory genes may be incorporated into the virus particle, potentially altering the tropism or enhancing infectivity. For SARS-CoV, the proteins encoded in ORFs 3a, 6, 7a, and 7b have been shown to be incorporated in virus particles, but the exact role of these proteins in enhancing virulence is not yet clear.

The features of CoV structural proteins are shown in **Figure 2**. For each structural protein, a schematic diagram of the predicted structure of the protein is shown on the left and a linear display of the features is shown on the right. The CoV spike glycoprotein is essential for attachment of the virus to the host cell receptor and fusion of the virus envelope with the host cell membrane. CoV spike glycoproteins assemble as trimers with a short cytoplasmic tail and hydrophobic transmembrane domain anchoring the protein into the membrane. The spike glycoprotein is divided into the S1 and S2 regions, which are sometimes cleaved into separate proteins by cellular proteases during the maturation and assembly of virus particles. S1 contains the receptor-binding domain (RBD) and has been shown to provide the specificity of attachment for CoV particles. The cellular receptors

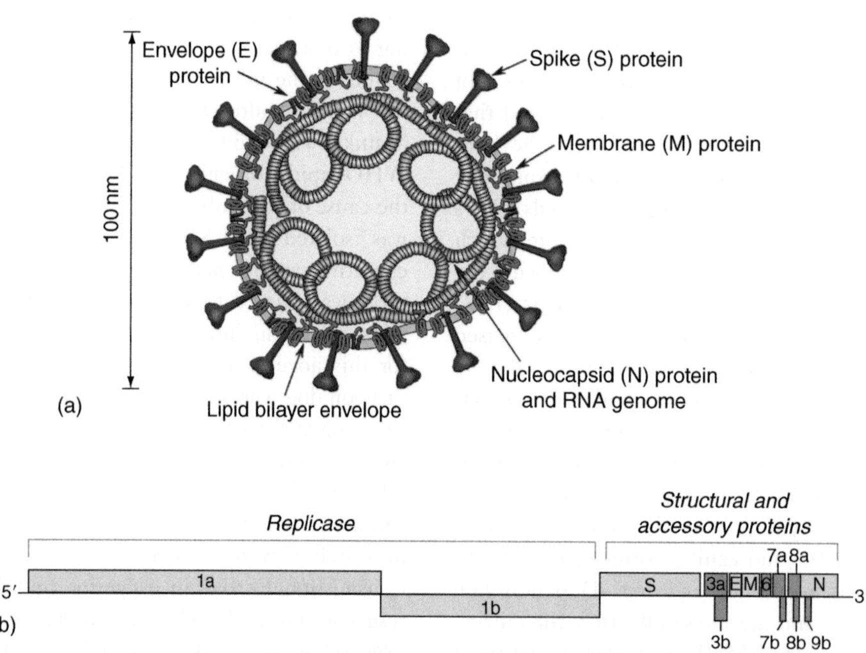

Figure 1 CoV virion and the genome of SARS-CoV. (a) Schematic diagram of a CoV virion with the minimal set of four structural proteins required for efficient assembly of the infectious virus particles: S, spike glycoprotein; M, membrane glycoprotein; E, envelope protein; and N, nucleocapsid phosphoprotein which encapsidates the positive-strand RNA genome. (b) Schematic diagram of the gRNA of SARS-CoV. Translation of the first two open reading frames (ORF1a and ORF1b) generates the replicase polyprotein. ORFs encoding viral structural and accessory (orange) ORFs are indicated at the 3′ end of the genome. (a) Reprinted from Masters PS (2006) The molecular biology of coronaviruses. *Advances in Virus Research* 66: 193–292, with permission from Elsevier.

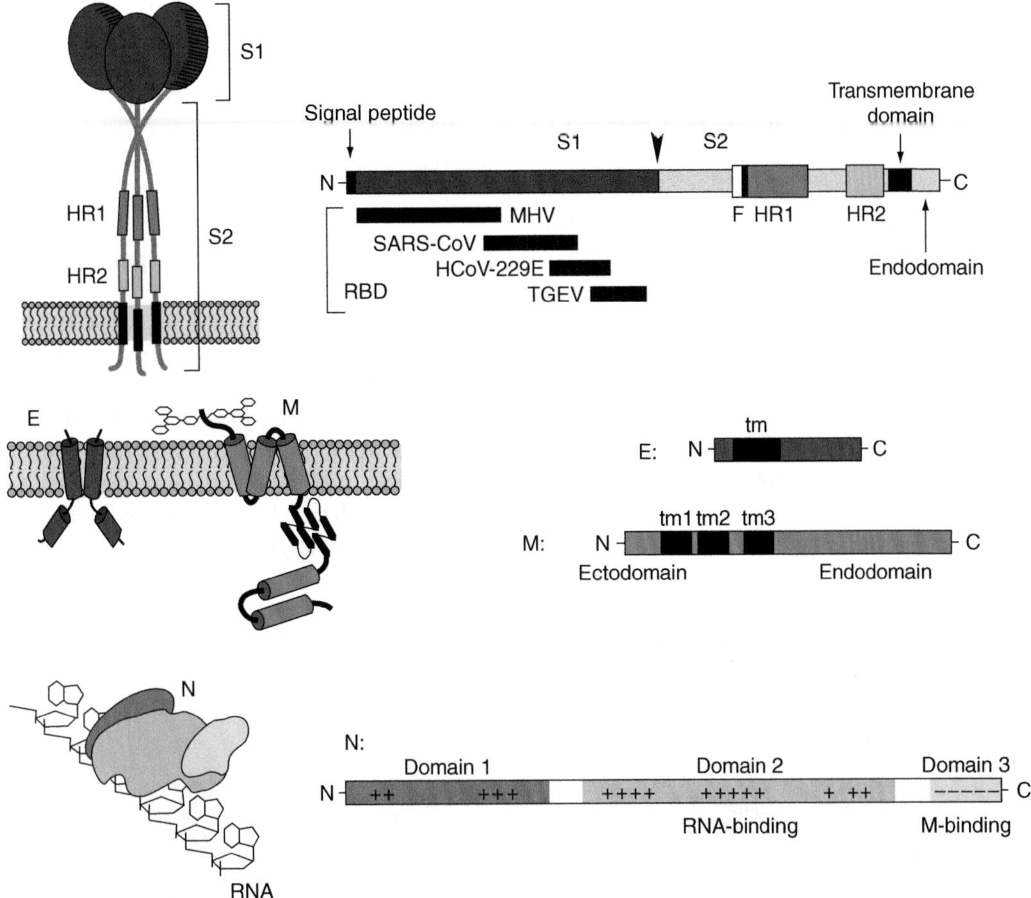

Figure 2 Diagrammatic representation of the spike trimer assembled on membranes, with the S1 receptor binding domain (RBD) and S2 fusion domain indicated. The linear map of spike indicates the location of the RBDs for three CoVs, and the relative location of the heptad repeat domains 1 and 2 (HR1 and HR2) which mediate the conformational changes required to present the fusion peptide (F) to cellular membranes. The membrane (M), envelope (E), and nucleocapsid (N) proteins represented in association with membranes or viral RNA. The linear map of each protein highlights the transmembrane domains of M and E and the RNA-binding and M protein-binding domains of N. Domains 1 and 2 of N are rich in arginine and lysine (indicated by +). Reprinted from Masters PS (2006) The molecular biology of coronaviruses. *Advances in Virus Research* 66: 193–292, with permission from Elsevier.

and corresponding RBDs in S1 have been identified for several CoVs. MHV binds to murine carcinoembryonic antigen-related cell adhesion molecules (mCEACAM1 and MCEACAM2); TGEV, FIPV, and HCoV-229e bind to species-specific versions of aminopeptidase N. Interestingly, both HCoV-NL63 and SARS-CoV have been shown to bind to human angiotensin-converting enzyme 2 (ACE2). ACE2 is expressed in both the respiratory and gastrointestinal tracts, consistent with virus replication at both these sites.

Once the S1 portion of the spike has engaged the host cell receptor, the protein undergoes a dramatic conformational change to promote fusion with the host cell membrane. Depending on the virus strain, this can occur at the plasma membrane on the surface of the cell, or in acidified endosomes after receptor-mediated endocytosis. The critical elements in the conformational change are the heptad repeats, HR1 and HR2, and the fusion peptide, F. After engaging the receptor, there is a dissociation of S1 which likely triggers the rearrangement of S2 so that HR1 and HR2 are brought together to form an antiparallel, six-helix bundle. This new conformation brings together the viral and host cell membranes and promotes the fusion of the lipid bilayers and introduction of the nucleocapsid into the cytoplasm. During infection, the spike glycoprotein is also present on the surface of the infected cell where it may (depending on the virus strain) promote fusion with neighboring cells and syncytia formation. The spike glycoprotein is also the major antigen to which neutralizing antibodies develop. The spike protein is a target for development of therapeutics for treatment of CoV infections. Monoclonal antibodies directed against the spike neutralize the virus by blocking binding to the receptor; synthetic peptides that block HR1-HR2 bundle formation have also been shown to block CoV infection.

The membrane (M) and envelope (E) proteins are essential for the efficient assembly of CoV particles. M is a triple-membrane-spanning protein that is the most abundant viral structural protein in the CoV virion. The ectodomain of M is generally glycosylated, and is followed by three transmembrane domains and an endodomain which is important for interaction with the nucleocapsid protein and packaging of the viral genome. The E protein is present in low copy numbers in the virion, but is important for efficient assembly. In the absence of E protein, few or no infectious virus particles are produced. The exact role of the E protein in the assembly of virus particles is still unknown, but recent studies suggest that E may act as an ion channel. The nucleocapsid protein (N) is an RNA-binding protein and associates with the CoV gRNA to assemble ribonucleoprotein complexes. The N protein is phosphorylated, predominantly at serine residues, but the role of phosphorylation is currently unknown. The N protein has three conserved domains, each separated by highly variable spacer elements. Domains 1 and 2 are rich in arginine and lysine residues, which is typical of many RNA-binding proteins. Domain 3 is essential for interaction with the M protein and assembly of infectious virus particles. The N protein has been shown to be an important cofactor in CoV RNA synthesis and is proposed to act as an RNA chaperone to promote template switching, as described below.

Replication and Transcription of CoV RNA

The replication and transcription of CoV RNA takes place in the cytoplasm of infected cells (**Figure 3**). The CoV virion attaches to the host cell receptor via the spike glycoprotein and, depending on the virus strain, the spike mediates fusion directly with the plasma membrane or the virus undergoes receptor-mediated endocytosis and spike-mediated fusion with endosomal membranes to release the viral gRNA into the cytoplasm. Once the positive-strand RNA genome is released, it acts as a messenger RNA (mRNA) and the 5′ end (ORF1a and ORF1b) is translated by ribosomes to generate the viral

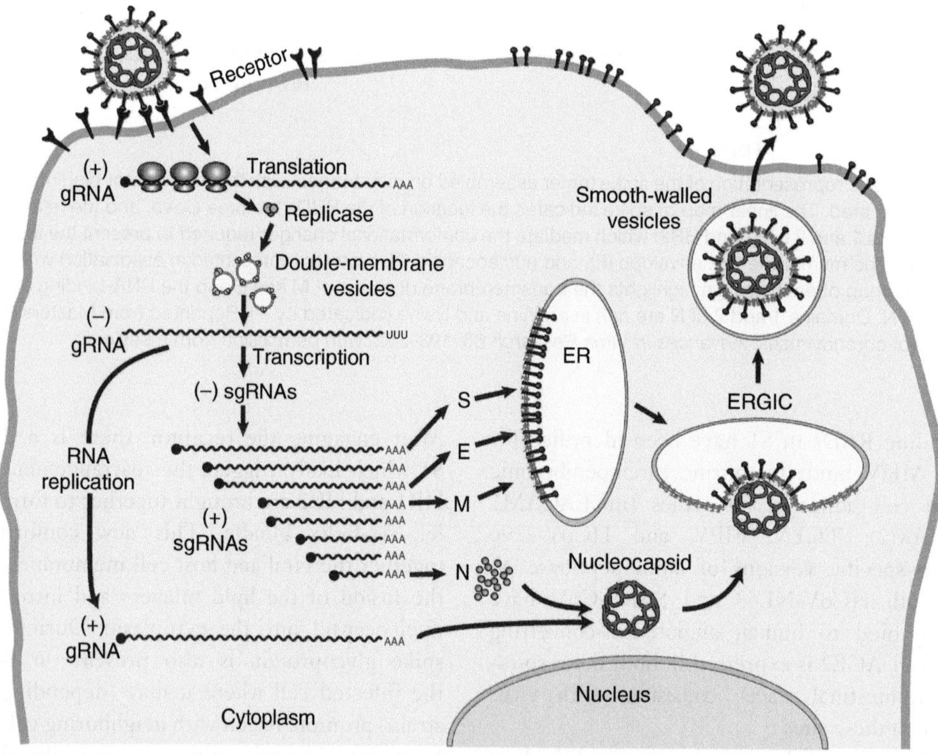

Figure 3 Replication cycle of CoVs. The spike glycoprotein on the virus particle interacts with host cell receptors to mediate fusion of the virus and host cell membranes and release of the positive-strand RNA genome into the cytoplasm. The 5′-proximal open reading frames (ORF1a and ORF1b) are translated from the gRNA to generate the replicase polyprotein. The replicase polyprotein is processed by viral proteases into 16 nonstructural proteins which assemble with membranes to generate double-membrane vesicles (DMVs) where RNA synthesis takes place. A nested set of 3′ co-terminal subgenomic (sg) RNAs is generated by a discontinuous transcription process. The sgRNAs are translated to generate the viral structural and accessory proteins. Viral gRNA is replicated and associates with nucleocapsid protein and viral structural proteins in the endoplasmic reticulum-Golgi intermediate compartment (ERGIC), where virus particles bud into vesicles before transport and release from the cell. Reprinted from Masters PS (2006) The molecular biology of coronaviruses. *Advances in Virus Research* 66: 193–292, with permission from Elsevier.

RNA-dependent RNA polymerase polyprotein, termed the viral replicase. Translation of ORF1b is dependent on ribosomal frameshifting, which is facilitated by a slippery sequence and RNA pseudoknot structure present in all CoV gRNAs. The replicase polyprotein is processed by replicase-encoded proteases (papain-like proteases and a poliovirus 3C-like protease) to generate 16 mature replicase products. These viral replicase proteins sequester host cell membranes to generate distinctive double-membrane vesicles (DMVs) that have been shown to be the site of CoV RNA synthesis. The replicase complex on the DMVs then mediates the replication of the positive-strand RNA genome to generate full-length and subgenomic negative-strand RNAs, and the subsequent production of positive-strand gRNAs and sgmRNAs. The sgmRNAs are translated to generate viral structural and accessory proteins, and virus particles assemble with positive-strand gRNA in the endoplasmic reticulum-Golgi intermediate compartment (ERGIC) and bud into vesicles, with subsequent release from the cell. Depending on the virus strain, this replication can be robust and result in destruction of the host cell or a low-level, persistent infection that can be maintained in cultured cells or infected animals.

A hallmark of CoV transcription is the generation of a nested set of mRNAs, with each mRNA having the identical 'leader' sequence of approximately 65–90 nt at the 5′ end (**Figure 4(a)**). The leader sequence is encoded only once at the 5′ end of the gRNA. Each subgenomic mRNA (sgmRNA) has the identical leader sequence fused to the 5′ end of the body sequence. How are the leader-containing mRNAs generated during CoV transcription? Current evidence supports a model of discontinuous transcription, whereby the replicase complex switches templates during the synthesis of negative-strand RNA (**Figure 4(b)**). The key sequence element in this process is the transcriptional regulatory sequence (TRS). The TRS is a sequence of approximately 6–9 nt (5′-ACGAAC-3′ for SARS-CoV) which is found at the end of the leader sequence and at each intergenic region (the sites between the open reading frames encoding the viral structural and accessory proteins). Site-directed mutagenesis and deletion analysis has revealed the critical role of the TRS in mediating transcription of sgmRNAs. Deletion of any intergenic TRS results in loss of production of the corresponding sgmRNA. In addition, the CoV leader TRS and the intergenic TRS sequences must be identical for optimal production of the sgmRNAs. A three-step

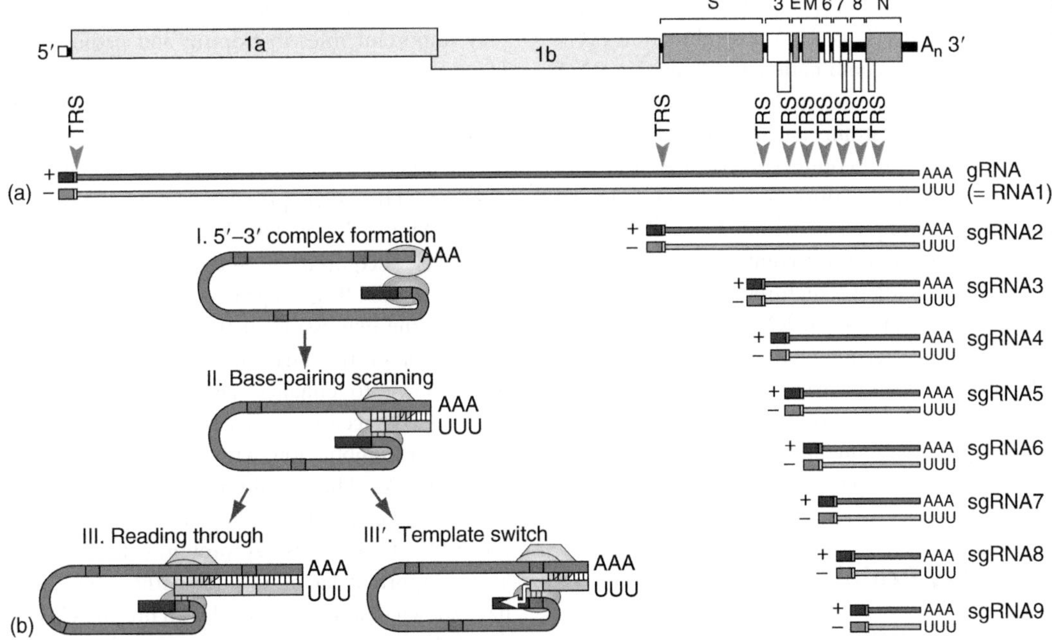

Figure 4 Model of SARS-CoV gRNA and sgRNAs, and a working model of discontinuous transcription. (a) Diagram of gRNA and the nested set of sgRNAs of SARS-CoV. The 5′ leader sequence, the transcriptional regulatory sequences (TRSs), and the positive- and negative-sense sgRNAs are indicated. (b) A working model of CoV discontinuous transcription. I. 5′–3′ complex formation. Binding of viral and cellular proteins to the 5′ and 3′ ends of the CoV gRNA is represented by ellipsoids. The leader sequence is indicated in red, the TRS sites are in orange. II. Base-pair scanning step. Minus-strand RNA (light blue) is synthesized from the positive-strand template by the viral transcription complex (hexagon). At the TRS site, base-pairing may occur between the template, the nascent negative-strand RNA, and the leader TRS sequence (dotted lines). III. The synthesis of negative-strand RNA can continue to make a longer sgRNA III, or a template switch can take place III′ to generate a leader-containing subgenomic negative-strand RNA, which could then serve as the template for leader-containing positive-strand sgRNAs. Modified from Enjuanes L, Almazán F, Sola I, and Zunia S (2006) Biochemical aspects of coronavirus replication: A virus–host interaction. *Annual Reviews in Microbiology* 60: 211–230.

working model for template switching during negative-strand RNA synthesis has been proposed to describe the process for the generation of CoV leader-containing sgmRNAs (**Figure 4(b)**). In this process, the 5′ end and 3′ end of the gRNA form a complex with host cell factors and the viral replication complex. The 3′ end of the positive strand is used as the template for the initiation of transcription of negative-strand RNA. Negative-strand RNA synthesis continues up to the point of the TRS. At each TRS, the viral replicase may either read through the sequence to generate a longer template, or switch templates to copy the leader sequence. The template switch allows the generation of a leader-containing sgmRNA. In this model, alignment of the leader TRS, the newly synthesized negative-strand RNA, and the genomic TRS is critical for the template switching to occur. Disruption of the complex, or loss of base-pairing within the complex, will result in the loss of production of that sgmRNA. Further studies of the CoV replication complex may yield new insights into the role of the viral helicase and endoribonuclease in the generation of the leader-containing CoV RNAs.

Another hallmark of CoV replication is high-frequency RNA recombination. RNA recombination occurs when a partially synthesized viral RNA dissociates from one template and hybridizes to similar sequences present in a second template. Viral RNA synthesis continues and generates a progeny virus with sequences from two different parental genomes. This RNA recombination event is termed copy-choice recombination. Copy-choice RNA recombination can be demonstrated experimentally when two closely related CoV strains (such as MHV-JHM and MHV-A59) are used to coinfect cells. Recombinant viruses with cross-over sites throughout the genome can be isolated, although sequences within the spike glycoprotein may be a 'hot spot' for recombination due to the presence of RNA secondary structures that may promote dissociation and reassociation of RNA. It has been proposed that copy-choice recombination is also the mechanism by which many CoVs have acquired accessory genes, and it has been exploited experimentally for the deletion or insertion of specific sequences in CoV genomes to assess their role in virus replication and pathogenesis.

CoV Accessory Proteins

Sequence analysis of CoVs isolated from species ranging from birds to humans has revealed that all CoVs encode a core canonical set of genes, replicase (rep), spike (S), envelope (E), membrane (M), and nucleocapsid (N), and additional, so-called accessory genes (**Table 1**). The canonical genes are always found in the same order in the genome: rep-S-E-M-N. Reverse genetic studies (see below) have shown that this is the minimal set of genes required for efficient replication and assembly of infectious CoV particles. However, the genomes of all CoVs sequenced to date encode from one to eight additional ORFs, which code for accessory proteins. As the name implies, these accessory proteins are not required for CoV replication in tissue culture cell lines, but they may play important roles in tropism and pathogenesis *in vivo.* How were these additional genes acquired? Current evidence indicates that these additional sequences may have been acquired by RNA recombination events between co-infecting viruses. For example, the hemagglutinin-esterase (HE) glycoprotein present in four different CoVs (MHV, BCoV, HCoV-OC43, and HCoV-HKU-1) was likely acquired by recombination of an ancestral CoV with the HE glycoprotein gene of influenza C. Interestingly, the expression of the HE gene has no effect on replication of the virus in cultured cell lines, but has been shown to enhance virulence in infected animals. Other CoV accessory genes may have been acquired through recombination with host cell mRNA or other viral mRNAs. The specific role of the accessory proteins in CoV replication and pathogenesis is under investigation. For SARS-CoV, accessory protein 6 has been implicated as an important factor in viral pathogenesis. Researchers have shown that mice infected with murine CoV expressing SARS-CoV protein 6 rapidly succumb to the infection, indicating that the protein 6 enhances virulence. In addition, recent studies suggest that SARS-CoV accessory proteins may play a role in blocking host cell innate immune responses, which may enhance viral replication and virulence. Other accessory proteins, such as SARS-CoV 3a and 7a, have been shown to be packaged into virus particles, where they may enhance infectivity or alter cell tropism. Future studies will be aimed at elucidating

Table 1 Coronavirus **canonical** and *accessory* proteins

Virus	*Proteins:* **canonical (rep-S-E-M-N)** *and accessory*
Group 1	
TGEV	**rep-S**-*3a,3b*-**E-M-N**-*7*
FIPV	**rep-S**-*3a,3b,3c*-**E-M-N**-*7a,7b*
HCoV-229E	**rep-S**-*4a,4b*-**E-M-N**
PEDV	**rep-S**-*3*-**E-M-N**
HCoV-NL63	**rep-S**-*3*-**E-M-N**
Group 2	
MHV	**rep**-*2a, HE*-**S**-*4-5a*,**E-M-N**,*7b*
BCoV	**rep**-*2a, HE*-**S**-*4a,4b-5*,**E-M-N**,*7b*
HCoV-OC43	**rep**-*2a, HE*-**S**-*5*,**E-M-N**,*7b*
HCoV-HKU1	**rep**-*HE*-**S**-*4*-**E-M-N**,*7b*
SARS-CoV	**rep-S**-*3a,3b*-**E-M**-*6-7a,7b-8a,8b*-**N**,*9b*
Bat-SARS-CoV	**rep-S**-*3*-**E-M**-*6-7a,7b-8*-**N**,*9b*
Group 3	
Avian IBV	**rep-S**-*3a,3b,3c*-**E-M**-*5a,5b*-**N**

Reprinted from Master PS (2006) The molecular biology of coronaviruses. *Advances in Virus Research* 66: 193–292, with permission from Elsevier.

how CoV accessory proteins may modulate the virulence of CoV infection.

Manipulating CoV Genomes Using RNA Recombination and Reverse Genetics

Genetic manipulation of CoV sequences is challenging because of the large size (27–32 kbp) of the RNA genomes. However, two approaches have been developed to allow researchers to introduce mutations, deletions, and reporter genes into CoV genomes. These approaches are (1) targeted RNA recombination and (2) reverse genetics using infectious cDNA constructs of CoV. The first approach exploits high-frequency copy-choice recombination to introduce mutations of interest into the 3′ end of the CoV gRNA. In the first step of targeted RNA recombination, a cDNA clone encoding the region from the spike glycoprotein to the 3′ end of the RNA is generated. These sequences can be easily manipulated in the laboratory to introduce mutations or deletions, or for the insertion of reporter or accessory genes, into the plasmid DNA. Next, RNA is transcribed from the plasmid DNA and the RNA is transfected into cells coinfected with the CoV of interest. RNA recombination occurs between the replicating CoV and the transfected substrate RNA, and viruses with the 3′ end sequences derived from the transfected substrate RNA will be generated. The recombinant viruses are generated by high-frequency copy-choice recombination, but the challenge is to sort or select for the recombinant virus of interest from the background of wild-type virus. To facilitate selection of recombinant viruses, Masters and Rottier introduced the idea of host range-based selection. They devised a clever plan to use a mouse hepatitis virus (MHV) that encodes the spike glycoprotein from a feline CoV as the target for their recombination experiments. This feline-MHV, termed fMHV, will infect only feline cell lines. Substrate RNAs that encode the MHV spike and mutations of interest in the 3′ region of the genome can be transfected into feline cells infected with fMHV, and progeny virus can be collected from the supernatant and subsequently selected for the ability to infect murine cell lines. Recombinant CoVs that have incorporated the MHV spike gene sequence (and the downstream substrate RNA with mutations of interest) can be selected for growth on murine cells, thus allowing for the rapid isolation of the recombinant virus of interest. This host range-based selection step is now widely used by virologists to generate recombinant viruses with specific alterations in the 3′ end of the CoV genome.

The second approach for manipulating CoV sequences, generating infectious cDNA constructs of CoV, has been developed in several laboratories. Full-length CoV sequences have been cloned and expressed using bacterial artificial chromosomes (BACs), vaccinia virus vectors, and from an assembled set of cDNA clones representing the entire CoV genome. The generation of a full-length cDNA and subsequent generation of a full-length CoV gRNA allows for reverse genetic analysis of CoV sequences. Successful reverse genetics systems are now in place to study the replication and pathogenesis of SARS-CoV, MHV, HCoV-229e, and IBV. These reverse genetics systems have allowed researchers to introduce mutations into the replicase gene and identify sites that are critical for enzymatic activities of many replicase products such as the helicase, endoribonuclease, and the papain-like proteases. Reverse genetic approaches are also being used to investigate the role of the TRSs in controlling the synthesis of CoV mRNAs. Interestingly, the SARS-CoV genome can be 're-wired' using a novel, nonanonical TRS sequence, which must be present at both the ends of the leader sequence and at each intergenic junction. This 're-wired' SARS-CoV may be useful for generating a live-attenuated SARS-CoV vaccine. An important feature of this 're-wired' virus is that it would be nonviable if it recombined with wild-type virus, since the leader TRS and downstream TRS would no longer match in a recombinant virus. The development of reverse genetics systems for CoVs has opened the door to investigate how replicase gene products function in the complex mechanism of CoV discontinuous transcription, and provides new opportunities to generate novel CoVs as potential live-attenuated or killed virus vaccines to reduce or prevent CoV infections in humans and animals.

Vaccines and Antiviral Drug Development

Because of the economic importance of CoV infection to livestock and domestic animals, a variety of live-attenuated and killed CoV vaccines have been tested in animals. Vaccines have been developed against IBV, TGEV, CCoV, and FIPV. However, these vaccines do not seem to provide complete protection from wild-type virus infection. In some cases, the wild-type CoV rapidly evolves to escape neutralization by vaccine-induced antibodies. In studies of vaccinated chickens, a live-attenuated IBV vaccine has been shown to undergo RNA recombination with wild-type virus to generate vaccine escape mutants. Killed virus vaccines may also be problematic for some CoV infections. Vaccination of cats with a killed FIPV vaccine has been shown to exacerbate disease when cats are challenged with wild-type virus. Therefore, extensive studies will be required to carefully evaluate candidate vaccines for SARS-CoV. A variety of approaches are currently under investigation for developing a SARS-CoV vaccine, including analysis of killed virus vaccines, live-attenuated virus vaccines, DNA immunization, and viral vector vaccines (such as modified vaccine virus Ankara, canarypox, alphavirus, and adenovirus vectors). The development of improved animal models for SARS will be essential for evaluating SARS-CoV candidate vaccines. Transgenic mice expressing human ACE-2 may be an appropriate small animal model. Initial studies suggest that Syrian hamsters and ferrets develop pneumonia and lung pathology similar to that seen in humans after infection with SARS-CoV, and therefore may be appropriate animal models for viral pathogenesis. CoV vaccine studies will benefit from an improved understanding of conserved viral epitopes that can be targeted for vaccine development.

The use of neutralizing monoclonal antibodies directed against the SARS-CoV spike glycoprotein is another approach that may provide protection from severe disease. The success in the development and use of humanized monoclonal antibodies against respiratory syncytial virus (family *Paramyxoviridae*) to protect infants from severe disease indicates that this approach is certainly worth investigating. Preliminary studies have indicated that patient convalescent serum and monoclonal antibodies directed against the SARS-CoV spike glycoprotein efficiently neutralize infectious virus. Further studies are essential to evaluate any concerns about potential antibody-mediated enhancement of disease and to determine if neutralization escape mutants arise rapidly after challenge with infectious virus. Studies evaluating monoclonal antibodies directed against a variety of structural proteins, and monoclonal antibodies directed against conserved sites in the spike glycoprotein will provide important information on the efficacy of passive immunity to protect against SARS.

Currently, there are no antiviral drugs approved for use against any human CoV infection. With the potential for the emergence or re-emergence of pathogenic CoV from animal reservoirs, there is considerable interest in identifying potential therapeutic targets and developing antiviral drugs that will block viral replication and reduce the severity of CoV infections in humans. Two promising targets for antiviral drug development are the SARS-CoV protease domains, the papain-like protease (PL^{pro}) and the 3C-like protease ($3CL^{pro}$, also termed the main protease, M^{pro}) (**Figure 5**). These two protease domains are encoded within the replicase polyprotein gene and protease activity is required to generate the 16 replicase nonstructural proteins (nsp1–nsp16) that assemble to generate the viral replication complex. The crystal structure of the $3CL^{pro}$ was determined first from TGEV and then from SARS-CoV. Rational drug design, much of which was based on our knowledge of inhibitors directed against the rhinovirus 3C protease, has provided promising lead compounds for $3CL^{pro}$ antiviral drug development. Interestingly, these candidate antivirals have been shown to inhibit the replication of SARS-CoV and other group 2 CoVs such as MHV, and the less related group 1 CoV, HCoV-229e. This indicates that the active site of $3CL^{pro}$ is highly conserved among CoVs and that antiviral drugs developed against SARS-CoV $3CL^{pro}$ may also be useful for inhibiting the replication of more common human CoVs such as HCoV-229e, HCoV-OC43, HCoV-NL63, and HCoV-HKU1. Further studies are needed to determine if these inhibitors can be developed into clinically useful antiviral agents.

Analysis of SARS-CoV papain-like protease led to the surprising discovery that this protease is also a viral deubiquitinating (DUB) enzyme. The SARS-CoV PL^{pro} was shown to be required for processing the amino-terminal end of the replicase polyprotein and to recognize conserved cleavage site (-LXGG). The LXGG cleavage site is also the site recognized by cellular DUBs to remove polyubiquitin chains from proteins targeted for degradation by proteasomes. Analysis of the X-ray structure of the SARS-CoV PL^{pro} has revealed that it has structural similarity to known cellular DUBs. These studies suggest that

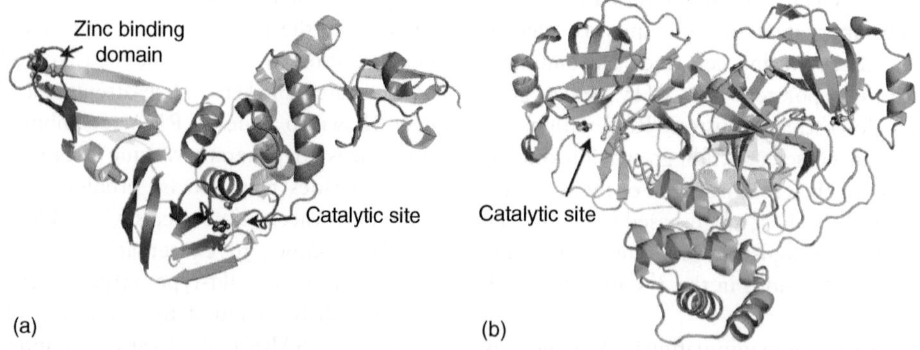

Figure 5 CoV proteases are targets for antiviral drug development; X-ray structures of the two SARS-CoV protease domains encoded in the replicase polyprotein. (a) The SARS-CoV papain-like protease (PL^{pro}) with catalytic triad cysteine, histidine, and aspartic acid residues, and zinc-binding domain indicated. (b) The 3C-like protease ($3CL^{pro}$, also termed main protease, M^{pro}) dimer with catalytic cysteine and histidine residues indicated.

CoV papain-like proteases have evolved to have both proteolytic processing and DUB activity. The DUB activity may be important in preventing ubiquitin-mediated degradation of viral proteins, or the DUB activity may be important in subverting host cell pathways to enhance viral replication. PL^{pro} inhibitors are now being developed using structural information and by performing high-throughput screening of small molecule libraries to identify lead compounds. Additional CoV replicase proteins, particularly the RNA-dependent RNA polymerase, helicase, and endoribonuclease, are also being targeted for antiviral drug development.

Future Perspectives

The development of targeted RNA recombination and reverse genetics systems for CoVs has provided new opportunities to address important questions concerning the mechanisms of CoV replication and virulence, and to design novel CoV vaccines. In the future, improved small animal models for testing vaccines and antivirals, and the availability of additional X-ray crystallographic structure information for rational drug design will be critical for further progress toward development of effective vaccines and antiviral drugs that can prevent or reduce diseases caused by CoVs.

See also: *Nidovirales*.

Further Reading

Baker SC and Denison M (2007) Cell biology of nidovirus replication complexes. In: Perlman S, Gallagher T, and Snijder E (eds.) *The Nidoviruses*. Washington, DC: ASM Press.

Baric RS and Sims AC (2005) Development of mouse hepatitis virus and SARS-CoV infectious cDNA constructs. *Current Topics in Microbiology and Immunology* 287: 229–252.

Enjuanes L, Almazán F, Sola I, and Zuñiga S (2006) Biochemical aspects of coronavirus replication: A virus–host interaction. *Annual Reviews in Microbiology* 60: 211–230.

Lau YL and Peiris JS (2005) Pathogenesis of severe acute respiratory syndrome. *Current Opinion in Immunology* 17: 404–410.

Li W, Wong SK, Li F, *et al.* (2006) Animal origins of the severe acute respiratory syndrome coronavirus: Insights from ACE2-S-protein interactions. *Journal of Virology* 80: 4211–4219.

Masters PS (2006) The molecular biology of coronaviruses. *Advances in Virus Research* 66: 193–292.

Masters PS and Rottier PJM (2005) Coronavirus reverse genetics by targeted RNA recombination. *Current Topics in Microbiology and Immunology* 287: 133–160.

Perlman S and Dandekar AA (2005) Immunopathogenesis of coronavirus infections: Implications for SARS. *Nature Reviews Immunology* 5: 917–927.

Ratia K, Saikatendu K, Santarsiero BD, *et al.* (2006) Severe acute respiratory syndrome coronavirus papain-like protease: Structure of a viral deubiquitinating enzyme. *Proceedings of the National Academy of Sciences, USA* 103: 5717–5722.

Shi ST and Lai MMC (2005) Viral and cellular proteins involved in coronavirus replication. *Current Topics in Microbiology and Immunology* 287: 95–132.

Stadler K, Masignani V, Eickmann M, *et al.* (2003) SARS – Beginning to understand a new virus. *Nature Reviews Microbiology* 1: 209–218.

Thiel V and Siddell S (2005) Reverse genetics of coronaviruses using vaccinia virus vectors. *Current Topics in Microbiology and Immunology* 287: 199–228.

Wang L, Shi Z, Zhang S, Field H, Daszak P, and Eaton BT (2006) Review of bats and SARS. *Emerging Infectious Diseases* 12: 1834–1840.

Yang H, Xie W, Xue X, *et al.* (2005) Design of wide-spectrum inhibitors targeting coronavirus main proteases. *PLoS Biology* 3: 1742–1751.

Yount B, Roberts RS, Lindesmith L, and Baric RS (2006) Rewiring the severe acute respiratory syndrome coronavirus (SARS-CoV) transcription circuit: Engineering a recombinantion-Resistant genome. *Proceedings of the National Academy of Sciences, USA* 103: 12546–12551.

Relevant Websites

http://patric.vbi.vt.edu – Coronavirus bioinformatics resource, PATRIC (PathoSystems Resource Integration Center).

http://www.cdc.gov – Severe acute respiratory syndrome (SARS) resource, Centers for Disease Control and Prevention.

Hepatitis B Virus: Molecular Biology

T J Harrison, University College London, London, UK

Glossary

Pseudogene Nonfunctional DNA sequence that is very similar to that of a known gene.

Tolerogen A substance that produces immunological tolerance.

Introduction

Hepatitis B virus (HBV) is the prototype of the family *Hepadnaviridae*, a group of viruses that infect mammals (primates and rodents) and birds. These viruses are distantly related to the retroviruses and, although their genomes are DNA, they are replicated by the reverse transcription of an RNA pregenome.

Structure and Replication of HBV

The 42 nm hepatitis B virion is composed of the DNA genome packaged together with a copy of the virus-encoded polymerase in an icosahedral nucleocapsid made up of dimers of the hepatitis B core antigen, HBcAg. In turn, the nucleocapsid is covered by an envelope composed of a lipid bilayer derived from internal cellular membranes and embedded with the hepatitis B surface protein, HBsAg. The open reading frame (ORF) encoding HBsAg has three in-frame initiation codons (**Figure 1**) which are used for the translation of the large (L, pre-S1 + pre-S2 + S), middle (M, pre-S2 + S), and small (S or major) surface proteins. All three proteins, which share the same C-terminus, are found in the virions.

The early events in infection of the hepatocyte are not well understood, but begin with the virion binding to an unidentified receptor (and, most likely, co-receptors) on the plasma membrane. The first contact seems to involve a domain located near to the N-terminus of L (and not present in M or S), and other interactions involving S also seem to be important. Virus entry likely is via the endosomal route and results in the delivery of the genome, with or without the nucleocapsid, to the nucleus.

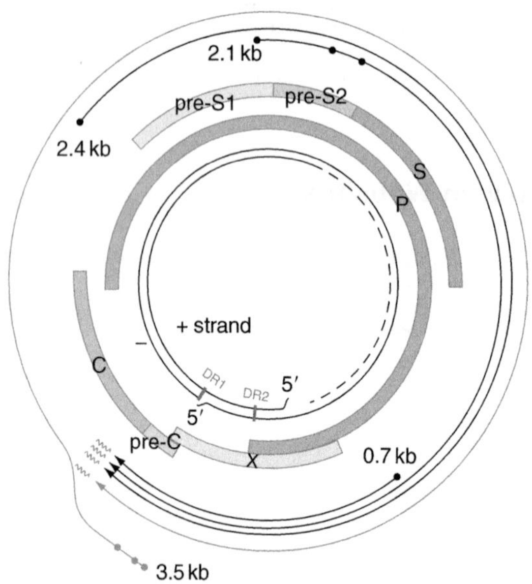

Figure 1 Organization of the HBV genome. The inner circles depict the complete minus strand and incomplete plus strand, and the positions of the direct repeats (DRs) are indicated. The blocks surrounding the genome show the locations of the four overlapping ORFs; C and S contain two and three in-frame initiation codons, respectively. The outer arrows show the viral transcripts, solid circles indicate the positions of the 5′ ends, and the arrowheads indicate the common polyadenylation site. Reproduced from Kidd-Ljunggren K, Miyakawa Y, and Kidd AH (2002) Genetic variability in hepatitis B viruses. *Journal of General Virology* 83: 1267–1280, with permission from Society for General Microbiology.

The 3.2 kbp genome is composed of two linear strands of DNA held in a circular configuration by base pairing of a short region ('cohesive end region') where the 5′ ends overlap (**Figure 1**). The plus strand is incomplete so that the circle is partially single-stranded. After delivery to the nucleus, the plus strand is completed (probably by a host enzyme), the primers that remain attached to the 5′ ends of both strands are removed, and the ends are ligated with the introduction of superhelical turns to yield covalently closed circular (ccc)-DNA. This cccDNA associates with histones and other host proteins and resides in the nucleus as a 'minichromosome', the template for transcription of the viral RNAs.

The genome encodes four overlapping ORFs, specifying the core and surface proteins, the polymerase, and a small protein known as HBx (**Figure 1**). All of the *cis*-acting elements, including two enhancers, four promoters, and the single polyadenylation signal, are embedded within the ORFs. Two families of RNAs (2.1 and 3.5 kb in size) are sufficiently abundant to be detectable by northern blotting of RNA from HBV-infected human liver. The 2.1 kb family is translated from a promoter (surface promoter 2, SP2) in the pre-S1 region and encodes M and S. The 3.5 kb RNAs, transcribed from the core promoter, include the precore RNA, which is translated to yield hepatitis B e-antigen (HBeAg), described in detail below, and the pregenomic RNA, which also encodes HBcAg and the polymerase. Surface promoter 1 (SP1), upstream of the surface ORF, transcribes a less-abundant 2.4 kb RNA that encodes L, and the X promoter transcribes an RNA of ~0.7 kb, HBx. All of the RNAs are 3′ co-terminal, being polyadenylated in response to a polyA signal in the core (or C) ORF (**Figure 1**).

The pregenomic RNA is the 3.5 kb transcript that is reverse-transcribed to the HBV genome and also acts as the mRNA for HBcAg and the polymerase. HBcAg forms dimers and has arginine-rich motifs at the C-terminal end that are believed to interact with the viral nucleic acid in the nucleocapsid. Less frequently, the downstream ORF is translated to yield the polymerase, probably following a ribosomal translocation that bypasses the core initiation codon. There is a stem–loop structure, termed epsilon (ε), near to the 5′ end of the pregenomic RNA (**Figure 2(a)**), and the polymerase binds to a bulge on the stem, priming first strand synthesis and signaling packaging of the pregenome and polymerase into 'precores', as the dimers of HBcAg self-assemble around the RNA–protein complex. The primer is the N-terminal domain of the polymerase, in which a tyrosine residue is covalently linked to guanosine, and this is extended by a further three residues on the bulge. The extended primer then translocates to a complementary sequence near to the 3′ end of the RNA (**Figure 2(b)**).

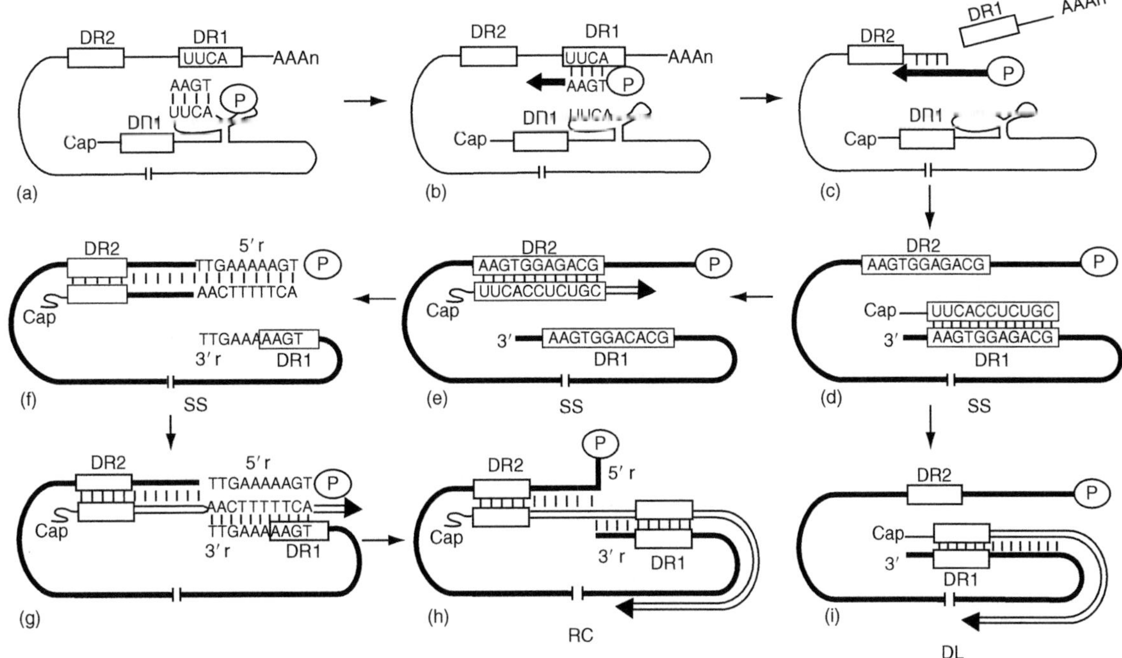

Figure 2 Replication of the HBV genome. See text for details. Reproduced from Liu N, Ji L, Maguire ML, and Loeb DD (2004) *cis*-Acting sequences that contribute to the synthesis of relaxed-circular DNA of human hepatitis B virus. *Journal of Virology* 78: 642–649, with permission from American Society for Microbiology.

The direct repeats (DR1 and DR2) are motifs of 11 nt that occur twice in the HBV genome (**Figure 1**). Located at the positions of the 5′ ends of the two strands, they play an important role in template switches during viral DNA synthesis. The translocated primer binds within DR1, and minus-strand synthesis commences with concomitant degradation of the template by an RNaseH activity in the C-terminal domain of the polymerase (**Figure 2(c)**). Completion of minus-strand synthesis leaves a capped oligoribonucleotide (which was the 5′ end of the pregenome) containing the DR1 sequence (**Figure 2(d)**). This translocates to the copy of DR2 near to the 5′ end of the minus strand and primes plus-strand synthesis, which then proceeds to the 5′ end of the minus strand (**Figures 2(e)** and **2(f)**). The short (~8 nt) terminal redundancy of the minus strand allows another template switch and circularization of the genome. Completion of the nucleocapsid during plus-strand synthesis starves the polymerase of the substrate, leaving the characteristic partially single-stranded structure that typifies the HBV genome (**Figures 2(g)** and **2(h)**).

Mature cores then bud through cellular membranes containing HBsAg (S, M, and L) and are exocytosed. A vast excess of subviral particles, which lack nucleocapsids and are noninfectious, are also secreted from the hepatocyte and are presumed to overwhelm the immune response of the host. These comprise 22 nm spheres, composed of S and M, and tubular forms of the same diameter, composed of S, M, and L (**Figure 3**). An endoplasmic-retention signal within the pre-S1 domain of L is involved in the formation of the virions and tubular subviral particles.

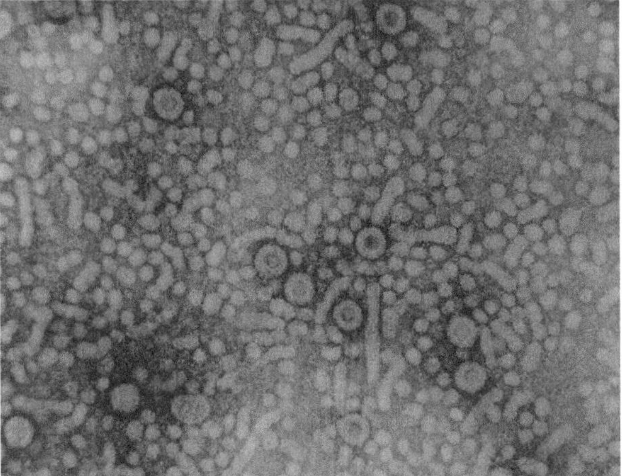

Figure 3 Electron micrograph of particles from the plasma of an infectious carrier. The 42 nm virions (Dane particles) are outnumbered by 22 nm spherical and tubular noninfectious, subviral particles.

The Surface Protein, HBsAg

The 226-amino-acid (aa) residue S protein is quite hydrophobic and structural models suggest that it contains four membrane-spanning regions (**Figure 4(a)**). The major

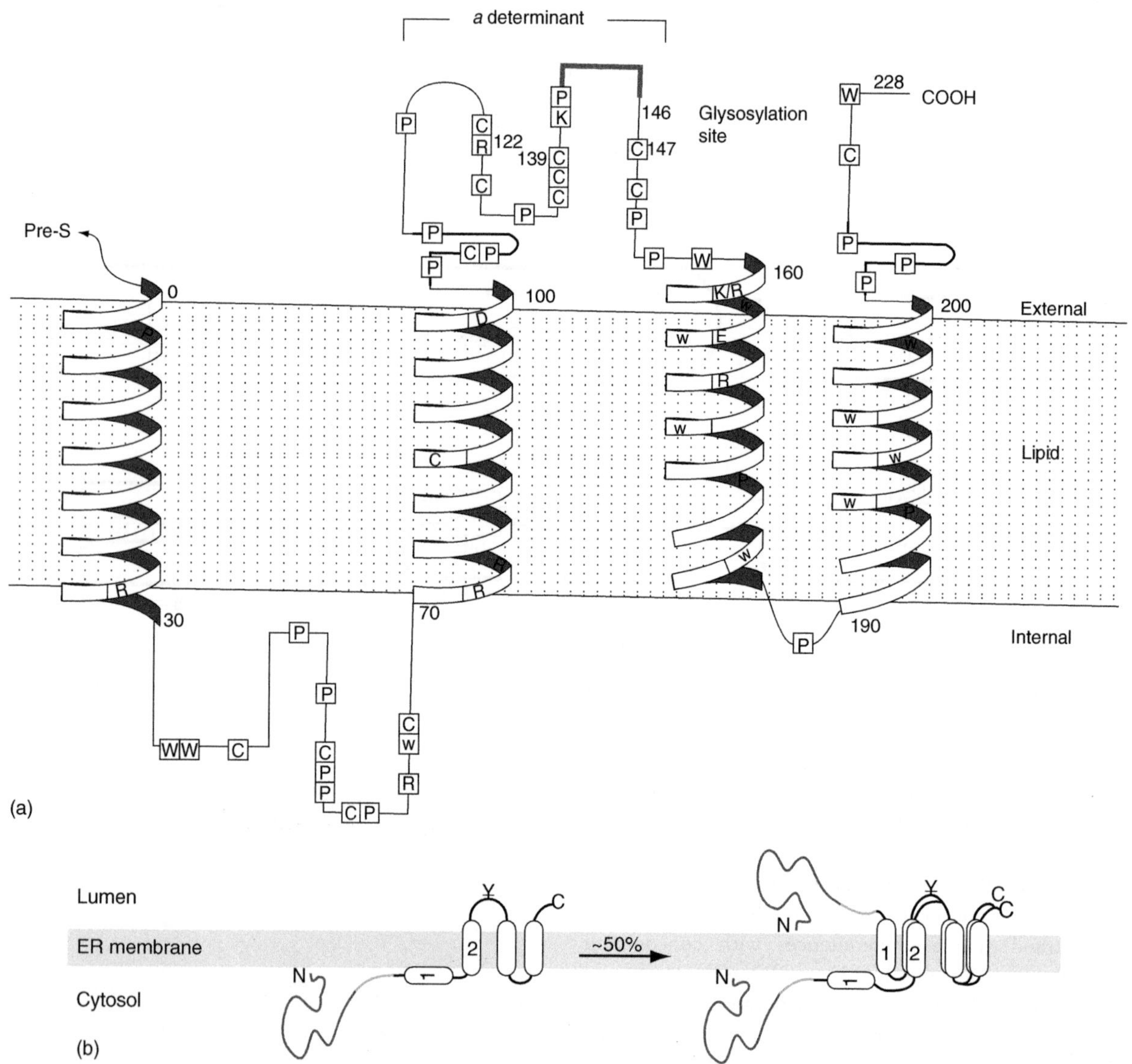

Figure 4 (a) Predicted structure of the major surface protein, S. Note the abundance of cysteine residues in the major hydrophilic region. (b) Dual topology of the large surface protein, L. (a) Reproduced from Stirk HJ, Thornton JM, and Howard CR (1992) A topological model for hepatitis B surface antigen. *Intervirology* 33: 148–158, with permission from Karger Publishers. (b) Reproduced from Lambert C, Mann S, and Prange R (2004) Assessment of determinants affecting the dual topology of hepadnaviral large envelope proteins. *Journal of General Virology* 85: 1221–1225, with permission from Society for General Microbiology.

hydrophilic loop is exposed on the surface of viral particles and is the main target of the immune response to HBsAg. This domain is rich in cysteine residues, which are believed to be linked by disulfide bridges, forming a complex structure that includes the *a* determinant and *d/y* and *w/r* subdeterminants. Antibody (anti-HBs) responses to this domain give immunity to infection in convalescence or following immunization with hepatitis B vaccine. Approximately half of the S residues are N-glycosylated at aa 146 (in the *a* determinant) so that the protein exists in two forms, p24 and gp27.

The N-terminus of S is exposed on the surfaces of the virions and subviral particles and, in M, is extended by 55 aa (the pre-S2 region). The pre-S2 region is glycosylated in M to give two proteins, gp33 and gp36, depending upon the glycosylation of asparagine 146 in the S domain. M is much less abundant than S in the virion and subviral particles. The SP2 promoter lacks a 'TATA box' and the 5′ ends of the 2.1 kb RNAs are heterogeneous, most being located downstream of the pre-S2 initiation codon (**Figure 1**). Furthermore, that codon is in a poor 'Kozak' context for recognition by the ribosome; consequently, much more S than M is produced.

The large surface protein, L, has a further extension of the N-terminus by the pre-S1 region. Again, there are two forms, p39 and gp42, depending on glycosylation at aa 146 of S (the pre-S2 region is not glycosylated in L). The glycine at aa 2 of the pre-S1 region is myristylated and embedded in the membrane. L has a dual topology: upon expression, the pre-S loop may translocate into the endoplasmic reticulum (ER) lumen, so that it is exposed on the exterior of the particle and presents the receptor-binding domain. However, in around 50% of the L molecules, the loop does not translocate but remains in the cytosol and is believed to interact with the nucleocapsid during budding and maintain this contact in the virion (**Figure 4(b)**).

The Hepatitis B Vaccines

The hepatitis B vaccines contain HBsAg and are given via an intramuscular injection to stimulate a protective anti-HBs response. The first-generation (so-called 'plasma derived') vaccine was produced by purifying HBsAg from donated blood. Despite some worries that blood-borne viruses (particularly human immunodeficiency virus (HIV)) potentially might contaminate particular batches, this method of vaccine production is safe provided that manufacturing protocols, including steps that inactivate potential infectivity, are adhered to strictly. Even today, plasma-derived vaccines constitute the majority of doses given worldwide. Currently, most doses of hepatitis B vaccine given in the West are produced by expression of HBsAg in yeast (*Saccharomyces cerevisiae*), and several vaccines based on expression in mammalian cells also have been licensed.

Most individuals produce a protective immune response with the standard, three-dose regimen of immunization, but there is a problem of nonresponsiveness in up to 5% of recipients, particularly the immune-suppressed and older individuals, and especially males. The vaccines contain only S and there has been interest in whether vaccines containing pre-S epitopes may circumvent this problem of nonresponsiveness. Up to 50% of nonresponders may sero-convert following immunization with a vaccine containing pre-S epitopes and the nonresponsiveness of many of the remainder seems to be genetically based.

Most countries worldwide have now introduced universal immunization of infants. Although the vaccine may be incorporated into the Expanded Programme of Immunization, it is essential that it is given as soon as possible after birth to the infants of infectious mothers, in order to break the chain of mother-to-infant transmission (see below). Immunization within 12 h of birth protects around 70% of the infants of carrier mothers and up to 90% success may be achieved when passive immunization with hepatitis B immune globulin (HBIG) also is given, at a contralateral site, to neutralize any maternal virus that reaches the newborn infant. Rarely, however, the antibody may select variants with mutations that alter the amino acid sequence of the *a* determinant. Such 'antibody escape' variants may also be selected in liver-transplant recipients given HBIG in an attempt to prevent HBV infection of the graft and, indeed, in the natural course of infection, with seroconversion to anti-HBs. The most common mutations affect the highly conserved codons 144 and 145 (for aspartic acid and glycine, respectively) of S. Fortunately, there is no evidence that these 'antibody escape' variants can be transmitted to individuals with vaccine-induced immunity.

Hepatitis B e-Antigen, HBeAg

This antigen circulates as a soluble protein in the plasma of some persistently infected individuals and is recognized as a marker of infectivity. The synthesis of HBeAg is nonessential for virus replication and its function is thought to be as a tolerogen; it may be especially important in perinatal transmission. HBV does not cross the placenta, but HBeAg is believed to do so and to circulate in the fetus. The infants of viremic mothers almost invariably become infected at or around the time of birth. At least 90% of such infants become persistently infected and often are extremely immune-tolerant of the virus, with very high viral loads that may persist for decades or even for life. When they are females, they infect their children in turn; thus, HBeAg seems to be important in maintaining HBV in the human population through the generations.

As noted above, the pregenomic RNA is translated from the second initiation codon in the core ORF to produce the nucleocapsid protein. The precore RNA has its cap site further upstream, so that it contains the precore initiation codon and the entire core ORF is translated (**Figure 5**). There are an additional 29 aa, encoded by the precore region, at the N-terminus of the precursor to HBeAg (p25) and these include a signal sequence that targets the protein to the ER. Here, the cellular signal peptidase cleaves this sequence and the protein is secreted with further proteolysis, removing the arginine-rich domain at the C-terminus.

Seroconversion from HBeAg positivity to anti-HBe, with clearance of HBV replication, is often a feature of recovery from persistent infection. However, a significant proportion of patients with circulating anti-HBe remain viremic, infected with HBV variants that do not synthesize, or synthesize reduced quantities of, HBeAg. The most common mutation affects codon 28 in the precore region, changing it from a tryptophan to a termination codon and preventing the synthesis of p25. Because this mutation affects base-pairing in ε, it is constrained in some genotypes and subgenotypes of HBV. Other, less common 'precore mutations' introduce termination codons elsewhere in the precore region, destroy the precore initiation codon, or modify the signal peptidase cleavage site.

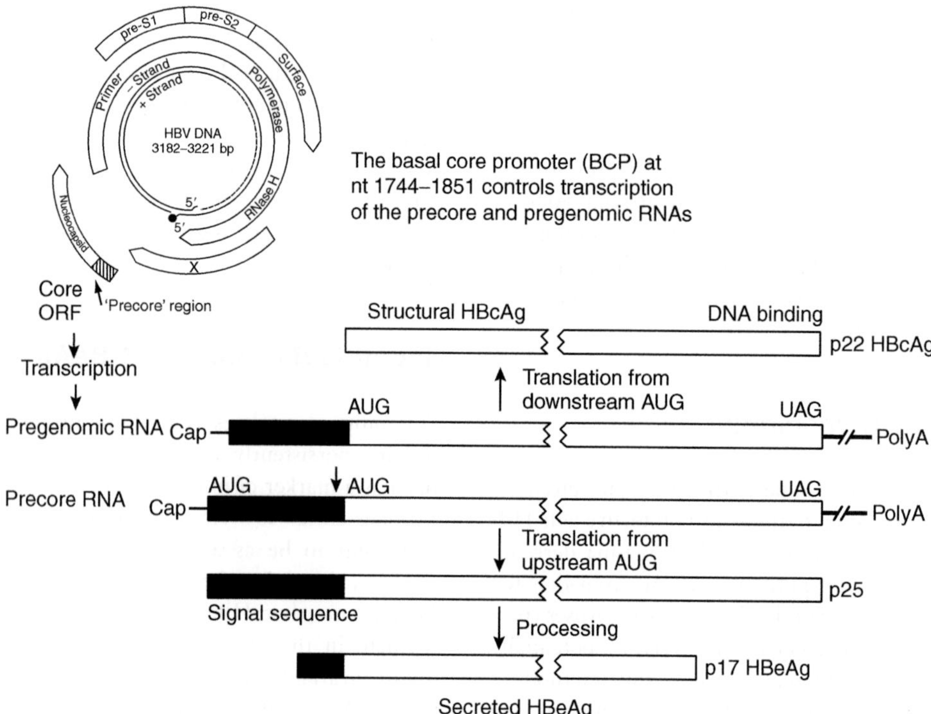

Figure 5 Expression of the HBV core ORF. The pregenomic RNA is translated from the second initiation codon in the ORF to yield HBcAg. The precore RNA is translated from the upstream initiation codon to p25, which is processed to HBeAg. The arrow indicates the position of the most common precore mutation, changing a tryptophan to a termination codon. Reproduced from Harrison TJ (2006) Hepatitis B virus: Molecular virology and common mutants. *Seminars in Liver Disease* 26: 87–96, with permission from Thieme International (Stuttgart).

Another class of mutations found in HBV from anti-HBe-positive individuals affects the core promoter. These mutations decrease the transcription of the precore RNA (and may increase the transcription of the pregenomic RNA), resulting in reduced synthesis of HBeAg. Core promoter mutations may be associated with a particularly high risk of developing hepatocellular carcinoma (HCC), especially in genotype C.

Treatment of Chronic Hepatitis B

Prior to the introduction of specific antiviral therapeutics, persistent HBV infection was treated with interferon, with mixed success. In some studies, up to one-third of treated individuals cleared the virus, but the treatment was rather less successful in certain populations, particularly individuals from the Far East. Interferon seems to work by modulating the immune response of the host, rather than a direct antiviral effect, and was found to be most successful in individuals who had evidence of hepatitis or, in other words, already were mounting a cellular immune response to the virus.

Because the HBV genome is replicated via an RNA intermediate, treatment with nucleoside and nucleotide analogs is an option. The predicted amino acid sequence of the HBV polymerase is similar to those of retroviruses and a motif, Tyr-Met-Asp-Asp (YMDD), in the active site is highly conserved. The HBV polymerase has not been crystallized but its structure is inferred from those of other RNA-dependent DNA polymerases, particularly that of HIV type 1 (HIV-1). Several conserved regions, designated A–F, are recognized within the reverse transcriptase domain and the YMDD motif is located within conserved region C (**Figure 6**).

Despite the similarity between the HIV and HBV polymerases, many of the nucleoside analogs first used against HIV, such as zidovudine, were found not to be effective against HBV. Lamivudine, a deoxycytidine analog which acts as a chain terminator during reverse transcription of the pregenome and can result in a 4–5 $\log_{10}$ suppression of viral load, was the first nucleoside analog to be used successfully and licensed for therapy of hepatitis B. However, prolonged monotherapy with lamivudine results in the emergence of resistant virus in approximately 40% of patients after 2 years of therapy (and 65% after 5 years). Resistance to lamivudine is associated with point mutations which result in substitution of the methionine residue in the YMDD motif by either valine or isoleucine, changes which parallel lamivudine resistance in HIV-1. The valine (but usually not the isoleucine) substitution is associated with a second, leucine-to-methionine substitution

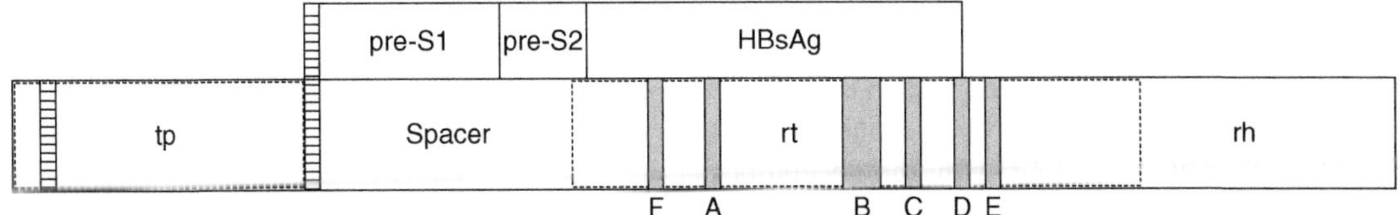

Figure 6 Conserved domains in the HBV reverse transcriptase. The polymerase and overlapping surface ORFs are shown. tp, terminal protein (primase); rt, reverse transcriptase; rh, RNaseH. Reproduced from Stuyver L, Locarnini S, Lok A, *et al.* (2001) Nomenclature for antiviral-resistant human hepatitis B virus mutations in the polymerase region. *Hepatology* 33: 751–757, with permission from the American Association for the Study of Liver Diseases.

in the upstream, conserved region B and this seems to compensate partially for the adverse effect of the former on replication efficiency.

Another nucleoside analog, adefovir dipivoxil, inhibits the replication not only of wild-type HBV but also of lamivudine-resistant mutants. Resistance to adefovir emerges at a slower rate than to lamivudine, reaching around 22% after 2 years for patients initially treated with lamivudine, and perhaps less frequently for nucleoside-naive individuals. Adefovir resistance is associated particularly with substitutions in conserved regions D or B and these mutants are susceptible to lamivudine. The absence of substitutions in conserved region C is consistent with the concept that adefovir inhibits the priming of reverse transcription, rather than chain elongation, and suggests that combination therapy with lamivudine and adefovir may inhibit the emergence of resistant mutants.

In vitro, the deoxyguanosine analog entecavir inhibits priming of reverse transcription as well as chain elongation of both the minus and plus strands. The drug is active against lamivudine- and adefovir-resistant HBV. Resistance to entecavir has been described in patients who had been treated previously with lamivudine and in whom lamivudine-resistant mutants had already been selected, but resistance seems to arise in otherwise treatment-naive patients at a rate of less than 1% per year. Other analogs, including telbivudine, which is similar in structure to lamivudine but seems to have better efficacy, have also been licensed recently for use in hepatitis B but the design of regimens for combination therapy lags way behind that for HIV.

HBV and HCC

When tests for HBsAg were first developed, it became clear that regions of the world with a high prevalence of chronic hepatitis B and a high annual incidence of HCC were coincident and that, in those regions, most individuals with HCC were HBsAg-positive. In Taiwan, a large prospective study confirmed the increased risk of HCC for carriers of HBsAg, and there is now evidence that the hepatitis B vaccine will reduce considerably the risk of developing the tumor. Although the importance of hepatitis C virus as a cause of HCC has increased considerably over recent decades, HBV remains a critical and preventable cause of one of the most common cancers of humans.

Cloning of the HBV genome enabled the use of virus-specific DNA sequences as probes in Southern hybridization assays, leading to evidence of chromosomal integration of viral sequences in at least 80% of HBV-associated HCCs. The chromosomal sites of integration of viral DNA seem to be random. Nevertheless, the tumors are clonal with respect to the integrated HBV DNA (often with multiple copies or partial copies of the genome), suggesting they arise from a single cell and that integration is the first step in the oncogenic process. In fact, integrated HBV DNA often can be detected by Southern hybridization of DNA from liver biopsies of infected patients who do not have HCC, suggesting that clonal expansion of hepatocytes containing integrated HBV DNA may be common in persistent infection.

Although the chromosomal sites of integration are random, there are hot spots in the viral genome, particularly around the direct repeats, for recombination with human DNA. A double-stranded linear form of HBV DNA, an aberrant product (and not a replicative intermediate) generated when the plus-strand primer fails to translocate and primes *in situ* (**Figure 2(i)**), may be that which recombines with the host DNA. Integration seems to be of no value to the virus, it is not a part of the replication cycle, and the integrants often are rearranged and never have sufficient redundancy to act as templates for the transcription of the 3.5 kb pregenome. On the other hand, the surface ORF often is expressed from integrants and many HBV-associated HCCs secrete HBsAg.

Despite considerable study, there is no consensus regarding the mode of HBV oncogenesis. Interest has focused on the product of the X ORF; this acts as a transactivator of transcription and seems to be important in 'kick-starting' the infection by stimulating transcription from the viral promoters. The protein seems to have pleiotropic effects in the cell, including stimulating various signal pathways and perhaps also preventing apoptosis. Transgenic mice with liver-specific expression of HBx develop HCC. Truncated middle surface proteins, produced following transcription of incomplete surface ORFs in integrated HBV DNA, also may act as transactivators, perhaps by locating in the plasma membrane and interacting with the receptors that initiate various signaling cascades. Some mice with the entire HBV genome as transgenes show accumulation of the large surface protein (L) in the ER of the hepatocytes and the resultant ER stress also seems to lead to the development of HCC.

Researchers have also investigated the hypothesis that HBV acts as an insertional mutagen, perhaps by inactivating a tumor suppressor gene or causing the inappropriate expression of a normally quiescent gene. Early studies, based on the analysis of genomic libraries of tumor DNA, identified insertions in or next to human genes, such as for cyclin A and the retinoic acid receptor β gene. However, these were isolated instances and no common targets were identified. More recently, amplification of virus–host junctions using polymerase chain reaction (PCR)-based techniques has identified further examples, particularly the human telomerase reverse transcriptase (hTERT), which seems to be a target of integration in several tumors. It is worth noting that, while the initial sites of HBV DNA integration are random, the integrants in tumors are a selected subset that may be implicated in the oncogenic process.

Woodchuck hepatitis virus (WHV) also causes HCC in its natural host. In contrast to HBV, in woodchuck tumors WHV DNA often is integrated within or adjacent to myc family oncogenes (c-myc or N-myc). In fact, woodchucks possess an additional copy of N-myc, a retrotransposed pseudogene that lacks introns. Insertion of the WHV enhancer 5′ or 3′ of the myc-coding region activates transcription from the myc promoter.

Further Reading

Bertoletti A and Gehring AJ (2006) The immune response during hepatitis B virus infection. *Journal of General Virology* 87: 1439–1449.

Bouchard MJ and Schneider RJ (2004) The enigmatic X gene of hepatitis B virus. *Journal of Virology* 78: 12725–12734.

Harrison TJ (2006) Hepatitis B virus: Molecular virology and common mutants. *Seminars in Liver Disease* 26: 87–96.

Kidd-Ljunggren K, Miyakawa Y, and Kidd AH (2004) Genetic variability in hepatitis B viruses. *Journal of General Virology* 83: 1267–1280.

Kremsdorf D, Soussan P, Paterlini-Brechot P, and Brechot C (2006) Hepatitis B virus-related hepatocellular carcinoma: Paradigms for viral-related human carcinogenesis. *Oncogene* 25: 3823–3833.

Lambert C, Mann S, and Prange R (2004) Assessment of determinants affecting the dual topology of hepadnaviral large envelope proteins. *Journal of General Virology* 85: 1221–1225.

Liu N, Ji L, Maguire ML, and Loeb DD (2004) *cis*-Acting sequences that contribute to the synthesis of relaxed-circular DNA of human hepatitis B virus. *Journal of Virology* 78: 642–649.

Norder H, Courouce A-M, Coursaget P, *et al.* (2004) Genetic diversity of hepatitis B virus strains derived worldwide: Genotypes, subgenotypes, and HBsAg subtypes. *Intervirology* 47: 289–309.

Stirk HJ, Thornton JM, and Howard CR (1992) A topological model for hepatitis B surface antigen. *Intervirology* 33: 148–158.

Stuyver L, Locarnini S, Lok A, *et al.* (2001) Nomenclature for antiviral-resistant human hepatitis B virus mutations in the polymerase region. *Hepatology* 33: 751–757.

Zoulim F (2006) Antiviral therapy of chronic hepatitis B. *Antiviral Research* 71: 206–215.

Herpes Simplex Viruses: Molecular Biology

E K Wagner[†] **and R M Sandri-Goldin,** University of California, Irvine, Irvine, CA, USA

The Virion

The Capsid

Like all herpesviruses, herpes simplex virus (HSV) type 1 (HSV-1) and the closely related HSV type 2 (HSV-2) have enveloped, spherical virions. The genome of HSV-1 is densely packaged in a liquid-crystalline, phage-like manner within a 100 nm icosahedral capsid. The capsid comprises 162 capsomeres of 150 hexons and 12 pentons. Hexons contain six molecules of the 155 kDa major capsid protein (VP5 or UL19) with six copies of the vertex protein VP26 (UL35) at the tips. Pentons contain five copies of VP5. Hexons are coordinated in threefold symmetry with a triplex structure made up of two other proteins: one copy of VP19C (UL38) and two copies of VP23 (UL18) per triplex. Small amounts of other viral proteins are also associated with the capsid; these include VP24 (derived from UL26), a maturational protease, and the UL6 gene product. The capsid portal formed from the UL6 gene product is now thought to occupy one of the penton positions, which leaves the capsid with 11 rather than 12 VP5-containing pentons.

The Tegument

The capsid is surrounded by a layer consisting of approximately 20 tegument or matrix proteins. These include the

[†] Deceased

α-*trans* inducing factor (αTIF, VP16, or UL48), and a virion-associated host shutoff function (vhs or UL41).

The Envelope

The trilaminar viral lipid envelope forms the outer surface of the virion. This envelope has a diameter of 170–200 nm, although exact dimensions vary depending upon the method of visualization. The envelope is derived from the host cell nuclear membrane and contains at least 10 virally encoded glycoproteins. One glycoprotein (gG–US4) has sufficient differences in amino acid sequence to serve as a type-specific immune reagent to differentiate HSV-1 and HSV-2. At least four (gL–UL1, gH–UL22, gB–UL26, and gD–US6) are involved in virus penetration into the host cell. The sequence of gB is highly conserved among a broad range of herpesviruses. Another glycoprotein (gC–UL44) facilitates the initial attachment of the virion to glycosaminoglycans (GAGs) on the cell surface. Two others (US7–gI and US8–gE) function together as a heterodimer that binds the Fc region of immunoglobulin G (IgG) and influences cell-to-cell spread of virus.

The Genome

The complete genomic sequences of prototype strains of HSV-1 and HSV-2 are available in standard databases. Homology between HSV-1 strains is >99% for most regions of the genome while overall homology between HSV-1 and HSV-2 approximates 85% within translational reading frames and is significantly less outside them. The HSV-1 genome is shown schematically in **Figure 1**. It is a linear, double-stranded (ds) DNA duplex, 152 000 bp in length, with a base composition of 68% G+C. The genome circularizes upon infection. Because the genome circularizes, the genetic/transcription map is conveniently shown as a circle. The genome contains at least 77 open reading frames (ORFs) translated into functional proteins. Therefore, the map is complex.

The organization of the genome can be represented as $a_n bU_L b'a'_m c'U_S ca$. The genome is made up of five important regions. These are:

1. The ends of the linear molecules, with the left end comprised of multiple repeats of the 'a' sequence (a_n), which are important in circularization of the viral DNA upon infection and in packaging the DNA in the virion. Variable numbers of the 'a' sequence in inverted form (a'_m) are present internally in the genome.
2. The 9000 bp long inverted repeat (R_L – 'b' region). This region encodes an important immediate early regulatory protein (ICP0), and most of the 'gene' for the latency-associated transcript (LAT), including the promoter. It also encodes the ORF for the ICP34.5 gene, which is important in neurovirulence, and two partially overlapping ORFs antisense to that of ICP34.5, namely ORF-O and ORF-P. Two other short translational reading frames 5′ of the LAT cap site, termed ORF-X and ORF-Y, are also present, but have unknown functions.
3. The long unique region (U_L), which is 108 000 bp long, and encodes at least 58 distinct proteins. It contains an origin of DNA replication (ori_L) and genes encoding DNA replication enzymes, including the DNA polymerase. A number of capsid proteins and glycoproteins are encoded in U_L as well as many other proteins.
4. The 6600 bp short inverted repeats (R_S – 'c' region) encode the very important immediate early transcriptional activator, ICP4. R_S also encodes another origin of DNA replication (ori_S) and the promoters for two other immediate early genes.
5. The 13 000 bp short unique region (U_S) contains 13 ORFs, a number of which encode glycoproteins important in viral host range and response to host defense.

The arrangement of the inverted repeat elements allows the unique regions of the viral genome to invert relative to each other into four 'isomers': prototype (P), inversion of U_L relative to U_S (I_L), inversion of U_S relative to U_L (I_S), and inversion of both U_L and U_S (I_{LS}). Inversion is facilitated by the 'a' sequence elements noted above.

Viral Replication

Entry

Initial association of HSV with a host cell is mediated by the association with the envelope glycoproteins gB (UL27) and gC (UL44) and glycosamminoglycans (GAGs), such as heparin sulfate, on the cell surface. Virus internalization involves the mediation of four essential glycoproteins, gB (UL27), gH (UL22), gD (US6), and gL (UL1). In many cells, this is mediated by interaction of gD with a cellular surface protein named the herpesvirus entry mediator (HVEM). HVEM is a member of the TFN/NGF family of proteins. Other cellular surface proteins may also function in a similar capacity since some cells lacking this receptor still allow efficient viral entry. Following entry, HSV exists as a de-enveloped virion in the cytoplasm. This particle must then be transported to the nucleus by a process that requires the cell microtubule machinery, and then the viral DNA released into the nucleus (uncoating). The details of these processes are poorly understood.

Transcription and DNA Replication

Once in the nucleus, the viral DNA becomes available for transcription by host RNA polymerase II into mRNAs.

The viral genome is accompanied into the nucleus by the VP16 or αTIF protein (UL48), which functions in enhancing immediate early viral transcription via cellular transcription factors. Gene expression occurs in three phases – immediate early or α (the first phase), early or β (which requires the immediate early protein ICP4), and late or γ (which requires the immediate early proteins ICP4 and ICP27 and viral DNA replication for maximal expression). DNA replication has a significant influence on viral gene expression. Early expression is reduced following the start of DNA replication, while late genes begin to be expressed at high levels. Immunofluorescence studies show that DNA replication occurs at discrete sites or replication compartments in the nucleus. Many of the early proteins are involved in viral DNA replication. Some of these early proteins, such as the two-subunit DNA polymerase and the three-subunit helicase–primase, participate directly in DNA synthesis, while others, such as thymidine kinase and the two-subunit ribonucleotide reductase, are involved indirectly by increasing the pools of deoxynucleoside triphosphates.

RNA Export, Translation, and Post-Translational Processing

HSV mRNAs are capped at the 5′ end and processed to form the 3′ end by polyadenylation, but the majority of viral transcripts do not undergo splicing. HSV transcripts are exported to the cytoplasm using the cellular TAP/NXF1 mRNA export pathway, and this process is facilitated by the HSV immediate early protein ICP27, which binds to viral RNAs as well as to the TAP/NXF1 receptor. HSV mRNA is translated on polyribosome structures using the host translational machinery.

Extensive proteolytic processing of HSV proteins does not occur, but the maturation of the capsid absolutely requires the activity of the maturational protease encoded by the UL26 gene product. Many viral proteins are extensively phosphorylated, including the immediate early regulatory proteins ICP4, ICP0, and ICP27, as well as a number of structural proteins. Other modifications such as arginine methylation of ICP27 have also been described. The precise role of such post-translational modifications upon the function of these proteins has not been well characterized.

Assembly and Release

Following DNA replication in the nucleus, the completed viral genomes are packaged into preformed capsids as the scaffolding proteins (the products of UL26.5 and the 3′ portion of UL26) are displaced. This assembly process entails several proteins that comprise the DNA encapsidation/cleavage machinery. The packaging process also involves the ‘a’ sequence at the other end of the genome-sized DNA fragment, leading to cleavage of the growing DNA chain and resulting in encapsidation of a full-length genome. Assembly is accompanied by a major change in capsid structure (maturation), which entails the action of the UL26 viral protease. This is followed by a very complicated process of egress. While there is controversy on how this process occurs, in the currently favored model, the nucleocapsid first buds through the inner nuclear membrane acquiring an envelope. The virion then enters the cytoplasm by fusing with the outer nuclear membrane, thereby losing its initial envelope. Subsequently, it becomes re-enveloped by budding into intracellular membranous compartments and proceeds through the cytoplasm. Whatever the exact mechanism, infectious virus can be recovered from cells many hours before cellular disintegration and release of virions into the extracellular medium. Virus spread by cell to cell contact and/or cell fusion is probably an important feature of the pathogenesis of infection in the host and can be readily observed with infections in cultured cells.

Viral Transcripts and Proteins

Each viral protein is expressed from its own independently regulated transcript. The viral transcription map is included in **Figure 1**. Independent starts of overlapping transcripts, (rare) splicing, and temporally differentiated polyadenylation site utilization result in the number of independent transcripts expressed exceeding 100.

General Properties of Transcripts and Promoters

Generally, the ORFs of HSV-1 are expressed as unique, intronless mRNA molecules that are controlled by temporal class-specific promoters. Nested, partially overlapping transcripts, each encoding a unique ORF but utilizing a shared polyadenylation site, are common (e.g., transcripts encoding UL24, UL25, and UL26 and those encoding US5, US6, and US7). Some complementary reading frames also exist, such as the mRNAs encoding ORF-P and ORF-O with that encoding ICP34.5 and transcripts encoding reading frames complementary to UL43.

Most HSV promoters are recognizable as eukaryotic RNA polymerase II promoters with obvious ‘TATA’ box homologies 20–30 nucleotides (nt) 5′ of the mRNA cap sites. The *cis*-acting elements mediating transcription of immediate early (α) transcripts include upstream enhancer elements at sites distal to and extending to several hundred nucleotides upstream of the cap sites, as well as several transcription factor binding sites located within the region from −120 to −50. The enhancers contain multiple copies of

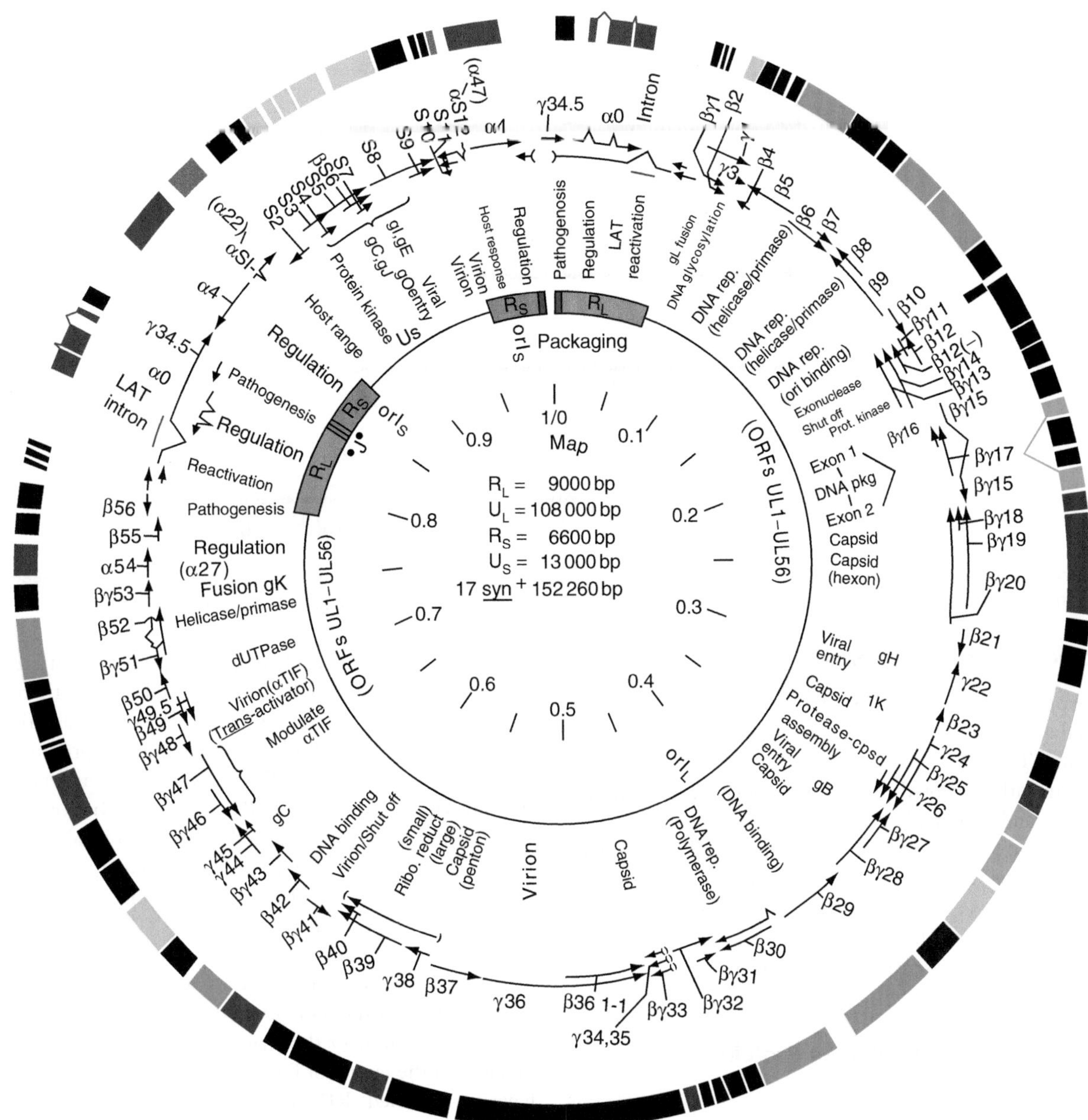

Figure 1 The genetic and transcriptional map of HSV-1 (strain 17*syn*+). The map is shown as a circle to indicate the fact that the linear genome circularizes upon infection. ORFs predicted from sequence data are named numerically according to their position on the viral genome, and transcripts expressing all or part of them are indicated with the arrowheads showing the transcription termination/polyadenylation sites. Known kinetic classes of transcripts are indicated; the LATs in the R_L regions are the only transcripts expressed during latent infection. Also shown are the location of the origins of replication and the 'a' sequences at the genome ends which are involved in encapsidation. In addition, and where known, the functions of individual genes are shown. In the literature, gene names sometimes contain a subscript (e.g., U_L19 rather than UL19).

a 'TAATGARAT' sequence, which interacts with cellular transcription factors of the POU family (notably, Oct-1). Enhancement of transcription occurs through the action of virion-associated tegument protein VP16 or αTIF (UL48), which is a powerful transcriptional activator. The *cis*-acting elements mediating transcription of early (β) promoters include several transcription factor binding sites and the TATA box at –30, but the 'TAATGARAT' enhancer elements are not present. Leaky-late (βγ) and late transcript (γ) promoters contain critical sequence elements mediating full levels of transcription at or near the cap site, and at least some contain transcription factor binding sites in the proximal part of the leader sequence downstream of the start site of transcription. HSV-1 promoters representing the temporal classes of productive infection are shown in **Figure 2**.

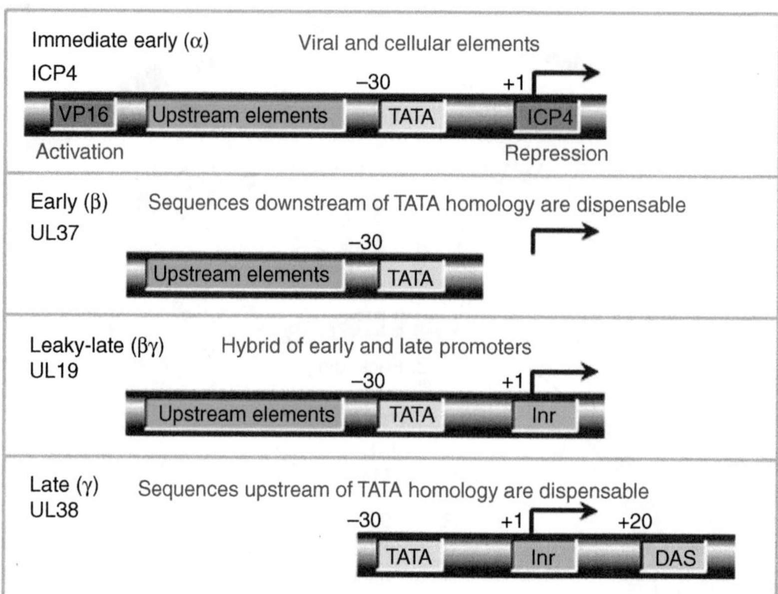

Figure 2 The architecture of HSV promoters. Features of the promoters controlling representative members of the kinetic classes of viral transcripts expressed during productive infection are shown. Inr, initiator element; DAS, downstream activator sequence.

Nomenclature of Proteins

The proteins encoded by HSV-1, their functions (if known), and the locations of their coding regions on the genome are given in **Table 1**.

Currently accepted protein nomenclature is a mix of historic and systematic systems. Proteins encoded in the repeats and a number of other proteins of well-established function are referred to by their historic or functional names. Proteins encoded by ORFs in the U_L and U_S regions are often referred to by the location of the ORF. Both nomenclatures are used in **Table 1**.

Functional Classification of Proteins

Regulatory proteins

The cascade of viral gene expression begins with the expression of immediate early or α proteins. There are five immediate early genes and their expression is highly activated by the action of VP16 or αTIF (UL48), which interacts with cellular transcription factors, Octl and HCF. The resulting complex binds to the 'TAATGARAT' sequences upstream of immediate early gene promoters. Three of the five immediate early genes, ICP4, ICP0, and ICP27, have roles in regulating viral gene expression at the level of transcription. ICP4 is a transcriptional activator that is required for the transcription of early and late viral mRNAs. Multiple phosphorylated forms of the protein have been identified in the infected cell. ICP4 apparently stabilizes the formation of the transcription initiation complexes at the TATA box of viral promoters.

The essential ICP27 protein is multifunctional, acting at both the transcriptional and post-transcriptional levels to regulate viral and cellular gene expression. Different functional domains of ICP27 are required for its activities, which include recruiting cellular RNA polymerase II to sites of viral DNA, inhibiting cellular splicing, promoting export of viral intronless mRNAs, and enhancing translation of viral transcripts. ICP27 interacts with myriad cellular and viral proteins, and it binds to viral RNAs. ICP27 is phosphorylated and methylated on some arginine residues, and these modifications presumably help regulate its various activities.

ICP0 is also a multifunctional protein that is, first and foremost, a transcriptional activator. ICP0 appears to play a central role in regulating whether HSV infection goes into the lytic phase or the virus becomes latent. ICP0 is a member of the family of E3 ubiquitin ligase enzymes that have a so-called RING finger zinc-binding domain. This domain confers on ICP0 the ability to induce the proteasome-dependent degradation of a number of cellular proteins. This process results in multiple consequences, including the disruption of cellular nuclear substructures known as ND10 or PML nuclear bodies. ICP0 interacts with or forms complexes with a large number of proteins. It also undergoes extensive post-translational modification, including phosphorylation by viral and cellular kinases, and ubiquitination. The phosphorylation status of ICP0 alters as infection progresses, suggesting that the change in ICP0 localization from nuclear to predominantly cytoplasmic at later times of infection may in part be regulated by phosphorylation.

DNA replication proteins

Seven early proteins are required for the formation of the viral DNA replication complex: The DNA polymerase

Table 1 Genetic functions encoded by HSV-1

Location	*Name*	*Function*
R_L	'a'	*cis* genome cleavage, packaging signal
	ICP34.5	RL1, neurovirulence, inhibits cellular apoptosis
	ORF-O	Modulates ICP0, ICP22 – inhibits splicing
	ORF-P	Modulates ICP4
	ICP0	RL2 (α0, IE1) immediate early *trans*-activator, E3 ubiquitin ligase, disrupts nuclear ND10 structures
	LAT-intron	Stable accumulation in nuclei of latently infected neurons
	LAT	*c.* 600 nt in 5′ region facilitates reactivation
	ORF-X	Unknown
	ORF-Y	Unknown
U_L	UL1	gL, essential glycoprotein that functions as a heterodimer with gH – involved in viral entry
	UL2	Uracil DNA glycosylase, DNA repair
	UL3	Nuclear protein
	UL4	Nuclear protein
	UL5	Part of helicase–primase complex, essential for DNA replication
	UL6	Capsid portal protein, capsid maturation, DNA packaging in capsid
	UL7	Unknown, dispensable in cell culture
	UL8	Part of helicase–primase complex, essential for DNA replication
	UL9	Ori-binding protein, essential for DNA replication
	UL10	gM, glycoprotein of unknown function
	UL11	Tegument protein, capsid egress, and envelopment
	UL12	Alkaline exonuclease, DNA packaging (?), capsid egress
	UL12.5	C-terminal 2/3 of UL12, expressed by separate mRNA – specific function unknown
	UL13	VP18.8, tegument-associated protein kinase
	UL14	Unknown
	UL15	DNA packaging, spliced mRNA, exons flank UL16 and UL17
	UL16	Tegument protein
	UL17	Cleavage and packaging of DNA
	UL18	VP23, capsid protein, triplex
	UL19	VP5, major capsid protein, hexon
	UL20	Membrane-associated, virion egress
	UL21	Tegument protein, auxiliary virion maturation function (?)
	UL22	gH, glycoprotein involved in viral entry as heterodimer with gL
	UL23	TK, thymidine kinase
	UL24	Nuclear protein
	UL25	Tegument protein, capsid maturation, DNA packaging
	UL26	VP24 (N-terminal half), protease, minor scaffolding protein (C-terminal half)
	UL26.5	Scaffolding protein
	UL27	gB, glycoprotein required for virus entry, mediates binding through interaction with GAGs in plasma membrane
	UL28	ICP18.5, capsid maturation, DNA packaging
	UL29	ICP8, single-stranded DNA-binding protein, essential for DNA replication
	ori_L	Origin of DNA replication
	UL30	DNA polymerase
	UL31	Nuclear lamina protein involved in nuclear egress
	UL32	Capsid maturation, DNA packaging
	UL33	Capsid maturation, DNA packaging
	UL34	Nuclear membrane protein involved in nuclear egress
	UL35	VP26, capsid protein
	UL36	VP1/2, tegument protein
	UL37	Tegument phosphoprotein
	UL38	VP19C, capsid protein, triplex
	UL39	Large subunit of ribonucleotide reductase
	UL40	Small subunit of ribonucleotide reductase
	UL41	vhs, virion host shutoff protein, degrades mRNA
	UL42	Polymerase accessory protein, essential for DNA replication
	UL43	Multiple membrane spanning protein
	UL43.5	Antisense to UL43
	UL44	gC, glycoprotein involved in initial stages of virion-cell association with GAGs, complement binding protein
	UL45	Membrane associated
	UL46	VP11/12, tegument associated

Continued

Table 1 Continued

Location	*Name*	*Function*
	UL47	VP13/14, tegument associated
	UL48	αTIF, VP16, virion-associated transcriptional activator, enhances immediate early transcription
	UL49	VP22, tegument protein
	UL49.5	Membrane protein associated with gM, glycosylated in other herpesviruses (gN) but not in HSV-1
	UL50	dUTPase, nucleotide pool metabolism
	UL51	Unknown
	UL52	Part of helicase–primase complex, essential for DNA replication
	UL53	gK, glycoprotein involved in virion egress
	UL54	ICP27, immediate early regulatory protein, multifunctional regulator, inhibits splicing, mediates viral RNA export, enhances translation initiation
	UL55	Unknown
	UL56	Tegument protein, affects pathogenesis
R_L		See R_L above
	R_L/R_S junction	Joint region, site of inversions of long and short segments, contains 'a' sequences
R_S	LAT-poly(A) site	Polyadenylation site for the primary latency associated transcript and the transcripts encoding ORF-O and ORF-P
	ICP4	RS1, essential immediate early transactivator, required for early and late gene expression
	ori_S	Origin of DNA replication
U_S	ICP22	US1, nonessential immediate early protein, affects host range
	US2	Unknown
	US3	Tegument-associated protein kinase, phosphorylates UL34 and US9
	US4	gG, glycoprotein of unknown function
	US5	gJ, glycoprotein of unknown function
	US6	gD, glycoprotein involved in virus infectivity and entry, binds HVEM
	US7	gI, glycoprotein, which as heterodimer with gE binds IgG-Fc and influences cell-to-cell spread of virus
	US8	gE, glycoprotein, which as heterodimer with gI binds IgG-Fc and influences cell-to-cell spread of virus
	US8.5	Nucleolar protein
	US9	Tegument-associated phosphoprotein (type II membrane protein)
	US10	Tegument-associated protein
	US11	Tegument-associated phosphoprotein, RNA binding, post-transcriptional regulation
	US12	ICP47, immediate early protein, inhibits major histocompatibility complex type 1 antigen presentation in human and private cells

(UL30) functions in a complex with UL42, a processivity factor. The ori binding protein (UL9) binds to the critical core sequence 'GTTCGCAC' in ori_S and may function as a dimer. Equimolar amounts of proteins UL5, UL8, and UL52 make up the helicase–primase complex; and ICP8 (UL29), the major single-stranded DNA-binding protein functions in a manner analogous to phage T_4 gene 32 protein in keeping the replication fork open.

A number of other early proteins are involved in genome replication by altering the pool of deoxyribonucleoside precursors and for repair and proofreading functions. These include thymidine kinase (UL23), the large and small subunits of ribonucleotide reductase (UL39 and UL40), and deoxyuridine triphosphatase (UL50). An enzyme with a potential repair function is uracil-DNA glycosylase (UL2).

Structural proteins

More than 30 HSV late proteins are associated with mature virions. These include five capsid proteins, of which VP5 is the major capsid protein. A large number of tegument proteins and 10 or more glycoproteins are also found in mature virions.

Capsid assembly can be readily accomplished both *in vitro* and in insect cells with proteins expressed from a panel of recombinant baculoviruses expressing the four principal capsid proteins (UL19–VP5, UL38–VP19C, UL18–VP23, and UL35–VP26), the scaffolding protein (UL26.5), and the maturational protease (UL26). Other proteins have been implicated genetically in the capsid assembly process, but are dispensable in minimal *in vitro* systems.

Packaging of the viral DNA into the mature capsid requires a large number of viral proteins, including UL6, UL15, UL17, UL25, UL28, UL32, and UL33. The protein encoded by UL15, in conjunction with that encoded by UL28, may function as a terminase, cleaving genome-sized fragments from the growing concatemeric replication complex. The UL15–UL28 complex also appears to mediate the initial entry of viral DNA into the capsid, that is, it serves a 'docking' function. The alkaline exonuclease gene (UL12) also plays a role in DNA packaging.

Proteins involved in pathogenesis and cytopathology

After the onset of viral DNA synthesis, a large fraction of the viral genome is represented in abundant amounts of complementary mRNA that can form stable dsRNAs, which serve to activate protein kinase R (PKR). Activated PKR phosphorylates the α subunit of the translation initiation factor 2 (eIF-2α), which shuts off protein synthesis. To combat this response, HSV encodes a protein named $\gamma_1$34.5 (ICP34.5), which binds phosphatase 1α and redirects it to dephosphorylate eIF-2α, thereby preventing the shutoff of protein synthesis by activated PKR. One of the immediate early proteins, ICP47, appears to have a role in modulating host response to infection by specifically interfering with the presentation of viral antigens on the surface of infected cells by major histocompatibility complex class I (MHC-I), precluding immune surveillance by CD8$^+$ T lymphocytes at the very onset of infection.

A number of the envelope glycoproteins found in the intact virion appear to have a significant role in controlling host response. The complex of glycoproteins gE (US9) and gI (US8) functions as an IgG Fc receptor blocking the presentation of this region of the antibody molecule and precluding activation of the complement cascade. In addition, gC (UL44) binds several components of the complement complex, and viruses with mutations in gD (US6) display altered pathogenesis in mice.

Other viral gene products also have a role in cytopathology and pathogenesis. While dispensable for productive infection in some cells, the immediate early protein ICP22 is required for HSV replication in others – perhaps by mediating the expression of a set of late transcripts. Viruses lacking thymidine kinase (UL23) have significantly altered pathogenic patterns in experimental animals while certain mutations in the DNA polymerase gene (UL30) have altered patterns of neurovirulence. Viral mutants defective in the virion host-shutoff protein (vhs; UL41) also have altered virulence in mice.

Latent Infection

HSV-1 latency is characterized by the persistence of viral genomes as episomes in the nuclei of sensory neurons. During latency, lytic functions are repressed and only one region of the genome is transcribed abundantly: the region encoding the LATs. The LAT domain is transcriptionally complex, and while the predominant species that accumulates during latency is a 2.0 kb stable intron, other RNA species are transcribed from this region of the genome, including a number of lytic or acute-phase transcripts. The LAT promoter (**Figure 3**) has not been characterized in quite as much detail as those promoters controlling productive cycle transcripts. It appears to have a number of regulatory elements important in neuronal expression over extended periods of time.

Latent infection with HSV can be viewed as having three separable phases: establishment, maintenance, and reactivation. In the establishment phase, the virus must enter a sensory neuron, and, following entry, there must be a profound restriction of viral gene expression so that the cytopathic results of productive infection do not occur. Thus, productive cycle genes are quiescent transcriptionally and functionally and only LAT is expressed. The maintenance of the HSV genome in latently infected neurons requires no viral gene expression apart from the LATs. Latent HSV genomes are harbored within the nucleus of a nondividing sensory neuron and do not need to replicate. HSV DNA is maintained as a nucleosomal, circular episome in latent infections and low levels of genome replication might occur or be necessary for the establishment or maintenance of a latent infection from which virus can be efficiently reactivated.

Reactivation of HSV results in the appearance of infectious virus at the site of entry in the host. Expression of only a 350 bp region of LAT near its 5′ end is both necessary and sufficient to facilitate reactivation in several animal models. Whatever the specific mechanism for this facilitation, it does not involve expression of a protein;

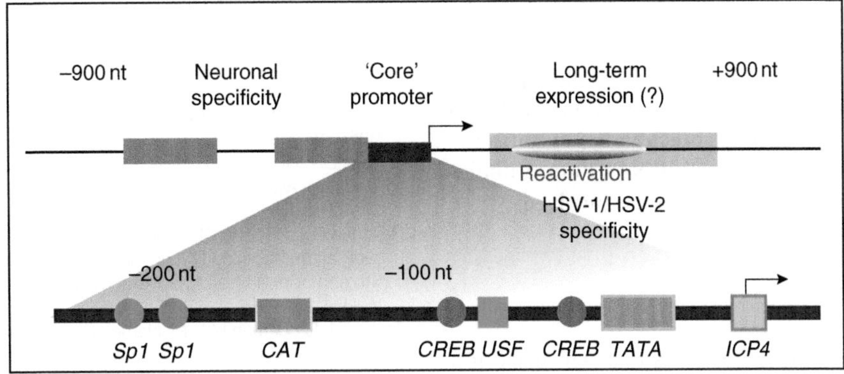

Figure 3 The architecture of the HSV LAT promoter. The transcriptional initiation site (arrow) and binding sites for transcriptional factors are shown in the expansion of the 'core' promoter.

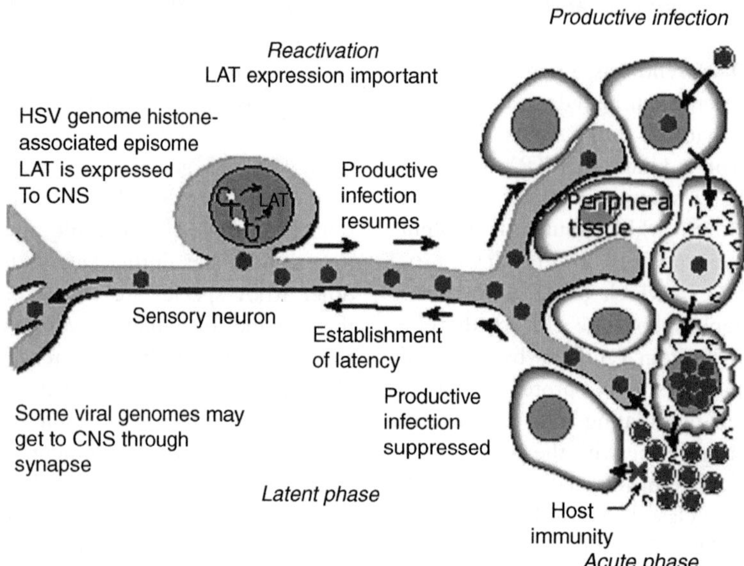

Figure 4 Schematic representation of HSV latent infection and reactivation. CNS, central nervous system.

rather, the active region appears to function by a *cis*-acting mechanism, the details of which are currently obscure. The process of reactivation is triggered by stress as well as other signals that may transiently lead to increased transcriptional activity in the harboring neuron, which may achieve a threshold that leads to lytic infection. The latency and reactivation cycle is shown in **Figure 4**.

Latent Phase Transcripts

Latently infected neurons represent 1–10% of the total neurons in the ganglia. In some neurons, an 8.5 kb polyadenylated primary LAT is expressed, which extends through the joint region (across the 'a' sequences) to a polyadenylation signal next to that of the ICP4 gene. The primary LAT is extensively processed, forming a stable 2 kb poly(A-) intron in lariat form, which accumulates in the latently infected neuronal nucleus. The mechanism by which LATs contribute to the establishment of and reactivation from latency has not been elucidated; in part, this is because LAT mutants behave somewhat differently in the different animal models used to study HSV latency. Because LAT does not become translated into protein, the RNA is believed to be the regulatory factor. It has been proposed that LAT may function as an antisense RNA or a small interfering RNA (siRNA). Future studies will reveal whether these possibilities are involved.

Future Trends

HSV possesses a number of features that will make it a continuing research model in the future. Among the most obvious are: (1) the neurotropism of the virus, along with the ability to generate recombinants containing foreign genes, recommends HSV as a vector for introducing specific genes into neural tissue; (2) the fact that replication of HSV is restricted in terminally differentiated neurons suggests its potential as a therapeutic agent against neural tumors, and several studies have suggested real promise; (3) the abundance of viral genes dedicated to the efficient replication of viral DNA in the productive cycle provides a number of potential targets for antiviral chemotherapy; and (4) the number of multifunctional viral proteins that interact with many different cellular proteins makes HSV a useful model for detailed analysis of specific aspects of eukaryotic gene expression and cellular functions.

Further Reading

Advani SJ, Weichselbaum RR, Whitley RJ, and Roizman B (2002) Friendly fire: Redirecting herpes simplex virus-1 for therapeutic applications. *Clinical Microbiology and Infection* 89: 551–563.
Bloom DC (2004) HSV LAT and neuronal survival. *International Reviews of Immunology* 23: 187–198.
Campadelli-Fiume G, Cocchi F, Menotti L, and Lopez M (2000) The novel receptors that mediate the entry of herpes simplex viruses and animal alphaherpesviruses into cells. *Reviews in Medical Virology* 10: 305–319.
Enquist LW, Husak PJ, Banfield BW, and Smith GA (1998) Infection and spread of alphaherpesviruses in the nervous system. *Advances in Virus Research* 51: 237–347.
Roizman B and Knipe DM (2001) Herpes simplex viruses and their replication. In: Knipe DM and Howley PM (eds.) *Fields Virology* 4th edn., pp. 2399–2459. Philadelphia: Lippincott Williams and Wilkins.
Taylor TJ, Brockman MA, McNamee EE, and Knipe DM (2002) Herpes simplex virus. *Frontiers in Bioscience* 7: 752–764.
Wagner EK and Bloom DC (1997) The experimental investigation of herpes simplex virus latency. *Clinical Microbiology Reviews* 10: 419–443.

Human Cytomegalovirus: Molecular Biology

W Gibson, Johns Hopkins University School of Medicine, Baltimore, MD, USA

Introduction

This chapter provides a general description of the molecular biology of human cytomegalovirus (HCMV) and highlights a number of specific features concerning the virus and its replication cycle. It is necessary to point out that most studies on the molecular biology of HCMV have utilized highly passaged, genetically deficient laboratory strains grown in human fibroblast cells. Although the findings are anticipated to apply in large part to wild-type strains grown in fibroblast and other cell types, and therefore to be relevant to the clinical situation, there are probably important differences to be discovered in the future.

Properties of the Virion

Typical of the herpesvirus group, the virion of HCMV (species *Human herpesvirus 5*, subfamily *Betaherpesvirinae*, family *Herpesviridae*) is approximately 230 nm in diameter and is composed of a DNA-containing capsid, surrounded by a less structured tegument layer, and bounded by a trilaminate membrane envelope. The capsid is isosahedral, has a diameter of approximately 110 nm, and is made up of four principal protein species that are organized into 162 capsomeres (150 hexamers plus 12 pentamers) and 320 triplexes located between the capsomeres. By analogy with herpes simplex virus type 1 (HSV-1), one of the pentamer positions is occupied by a portal complex (a dodecamer of the portal protein) through which DNA enters and leaves the capsid. The capsomeres are approximately 20 nm in length and 15 nm in diameter, have a channel about 3 nm in diameter that is open at the exterior end, possess favored cleavage planes along the longitudinal axis, and have short, spicule-like protrusions, probably representing the triplexes, extending out symmetrically and resulting in a pinwheel appearance of the capsomere viewed end on. The tegument region is approximately 50 nm thick and includes seven relatively abundant virus-encoded protein species, five of which are phosphorylated. The virion envelope is estimated to be 10 nm thick and contains at least 10 abundant protein species. Both the tegument and envelope contain a substantial number of less abundant virus-encoded and host-cell proteins. In addition, the virion has been reported to contain phospholipids (phosphatidylcholine, phosphatidylethanolamine, and phosphatidylinositol) and polyamines (spermidine and spermine). The properties of the virion DNA and proteins are described below.

Properties of the Genome

The genome of HCMV is composed of a linear, double-stranded DNA molecule with a size of 236 kbp in wild type virus. Thus, it is over 50% larger than that of HSV-1, and the largest among the human herpesviruses. It is a class E genome, like that of HSV-1, consisting of a long unique (U_L) and a short unique (U_S) sequence, both of which are flanked by much shorter inverted repeat sequences that enable U_L and U_S to invert relative to each other and give rise to the four structural isomers of the genome found in virions. The origin of DNA replication (*ori*-Lyt) has been localized to a 3–4 kbp region near the center of U_L. Like *ori*-L, one of three duplicated origin sequences in the HSV-1 genome, the HCMV *ori*-Lyt region is located adjacent to the promoter for the early, single-stranded DNA-binding protein (encoded by gene UL57). HCMV *ori*-Lyt includes a 2.4 kbp sequence that has no homology with other described virus DNA replication origins and contains direct and inverted repeat sequences. This region contains consensus cyclic AMP (cAMP) response elements and other transcription factor-binding sites, and 23 copies of the sequence 'AAAACACCGT that are conserved near the homologous *ori*-Lyt of simian cytomegalovirus (SCMV).

The complete nucleotide sequence of the high passage strain AD169 was the first to be determined for HCMV, with a size of 229345 bp. It has a G + C content of 57.2 mol%, and was assessed as containing 189 protein-coding open reading frames (ORFs), some of which are spliced and some of which are duplicated in the inverted repeats. The low passage strain Toledo was found to contain an additional ~15 kbp at the right end of U_L containing 19 additional ORFs. Analysis of the complete 235645 bp sequence of the low passage strain Merlin and a reevaluation of the genetic content of strain AD169 have indicated that wild-type HCMV contains ~165 protein-coding genes. The wild-type genome contains 13 families of related genes, many with members clustered into blocks and encoding predicted or recognized glycoproteins (e.g., the RL11, US6, and US27 families). Taken together, these gene families represent ~40% of the total gene number and in part account for the comparatively large size of the HCMV genome. It is also worth noting that the HCMV genome contains homologs of several cellular genes, including surface receptors such as class I major histocompatability (MHC) antigens, G protein-coupled receptors (GPCRs), and a tumor necrosis factor receptor.

Properties of the Proteins

HCMV proteins share the general pattern of expression characteristic of the herpesvirus group. Immediate early (IE or α), early (E or β), and late (L or γ) proteins are synthesized sequentially from corresponding mRNAs whose transcription is regulated in a temporal cascade. IE proteins are required to regulate transcription from their own promoters and those of subsequently expressed genes. E proteins include many of the enzymes and regulatory factors needed to carry out the synthesis of progeny DNA and proteins. L proteins include most of the virion structural proteins. Members of all three classes have been described for HCMV, though only a small number of the proteins encoded by HCMV have been identified. Those that have been reported include both nonvirion and virion species and their properties are briefly described below.

Nonvirion Proteins

Many of the recognized nonvirion proteins are made at early times after infection and localize to the nucleus. The first of these to appear is the major IE protein (IE1, encoded by UL123), a 72 kDa phosphoprotein that, together with products of a second IE gene, IE2 (UL122), can transactivate HCMV E promoters. Both IE1 and IE2 are expressed from spliced transcripts generated using the major IE promoter. In addition, the IE2 gene gives rise to a family of proteins ranging in size from 23 to 86 kDa that are generated by differential splicing and translational start sites. IE2 proteins can repress the major IE promoter by acting on a sequence immediately upstream of the transcription initiation site. Other IE proteins are encoded by genes IRS1, TRS1, US3, and UL36–38, and can act synergistically with IE1 and the US3 protein to regulate both cellular and virus gene transcription.

The E proteins include the 140 kDa virus DNA polymerase (encoded by UL54), a 140 kDa nuclear single-stranded DNA-binding protein (DB140; UL57), and a set of closely related nuclear phosphoproteins (UL112). Like its homologs in other herpesviruses, the DNA polymerase has an associated 3′-specific nuclease activity. Like its homolog (UL29) in HSV-1, UL57 is located adjacent to *ori*-Lyt, and its product DB140 localizes to discrete foci within the nuclei of infected cells and accumulates in the cytoplasm under conditions of inhibited virus DNA synthesis. UL112 encodes a set of four E, nuclear, DNA-binding phosphoproteins (34, 43, 50, and 84 kDa) that appear to be produced by alternate splicing, are related to each other by their common N-terminal sequence, and are differentially phosphorylated. Expression of these proteins appears also to be regulated at both the transcriptional and post-transcriptional levels. The functions of these proteins are not known, and they may have a role in DNA replication or gene regulation.

Another nonvirion nuclear DNA-binding phosphoprotein is distinguished by its high abundance and kinetics of synthesis. The ~52 kDa species, DB52, is the product of gene UL44. On the basis of its reduced but not eliminated synthesis under conditions of inhibited virus DNA replication, DB52 is classified as a delayed-early (DE or γ_1) protein. Expression of this gene is under different promoter control at early and late times of infection, and there is evidence that its expression is also post-transcriptionally regulated. DB52 and its SCMV homolog bind to DNA *in vitro* and *in vivo* and associate with the virus DNA polymerase to enhance its activity, analogous to the HSV-1 UL42 protein. DB52, like the HSV-1 UL42 protein, has the structure of a homotrimeric sliding clamp similar to that of the cellular processivity factor, proliferating cell nuclear antigen. Three other proteins that function in virus DNA replication or processing are a phosphotransferase (encoded by UL97), and the ATPase (encoded by UL89) and its partner subunit (encoded by UL56) that constitute an apparent homolog of the bacteriophage DNA cleavage/packaging complex, terminase.

Several other nonvirion HCMV proteins have recently been shown to interfere with the MHC class I antigen presentation system (i.e. the US3, US6, and US11 proteins). These proteins are thought to help the virus-infected cell escape recognition and destruction by immune surveillance, and provide a mechanism for virus persistence in the immune-competent host. It is viewed as likely that wild-type HCMV encodes many proteins involved in modulating the host response to infection.

Virion Proteins

Estimates of the number of protein species in HCMV virions depend on the techniques used. Classical approaches utilizing sodium dodecyl sulfate polyacrylamide gel electrophoresis (SDS-PAGE) have detected a total of 30–35 protein species. The characteristics of the most abundant of these proteins, including size, charge, carbohydrate and phosphate content, and deduced locations in the particle, are summarized in **Table 1** and described below. Virions have also been reported to contain several enzymatic activities (including protein kinase, DNA polymerase, DNase, and topoisomerase II), and several host proteins (including PP1 and PP2A phosphatases and a 45 kDa cellular actin-related protein), although

Table 1 Principle proteins of the HCMV virion

Descriptive name	*Predicted M_r*[a]	*Observed M_r*[b]	*Modification*[c]	*Charge*[d]	*Gene*[e]
Capsid					
Major capsid protein (MCP)	153.9	153		N	UL86
Portal protein	78.5			B	UL104
Precursor assembly protein (pAP)[f]	38.2	50	P, C	B	UL80.5
Minor capsid protein (mCP)	34.6	34		B	UL85
mCP-binding protein (mCP-BP)	33.0	35		N	UL46
Protease assemblin (A, A_N, A_C)[g]	28.0	28		A	UL80a
Smallest capsid protein (SCP)	8.5	8		B	UL48A
Tegument					
High mol. wt. protein (HMWP)	253.2	212		N	UL48
Basic phosphoprotein (BPP; pp150)	112.7	149	G, P	B	UL32
HMWP-binding protein (HMWP-BP)	110.1	115		N	UL47
80 K		80	P		
Upper matrix protein (UM; pp71)	61.9	74	P	N	UL82
Lower matrix protein (LM; pp65)	62.9	69	P	B	UL83
24 K (pp28)	28.0	24	P	B	UL99
Envelope					
Glycoprotein B (gB)	102.0	102	G, P, C	N	UL55
N-terminal portion (gB_N)	52.6	130	G	A	UL55
C-terminal portion (gB_C)	49.5	62	G, P	A	UL55
Glycoprotein H (gH)	84.4	84	G	N	UL75
Glycoprotein M (gM)	42.9	43			UL100
Glycoprotein N (gN)	14.5	57	G	A	UL73
Glycoprotein O (gO)	54.2	125	G		UL74
GPCR27	42.0	42	G	B	US27
GPCR33	46.3	44	G	B	UL33

[a]Molecular masses (kDa).
[b]Molecular masses (kDa) given as determined by SDS-PAGE.
[c]G, glycosylation; P, phosphorylation; C, proteolytic cleavage.
[d]Relative net charges were determined by 2-D PAGE (charge/size separation) with respect to the major capsid protein, which is approximately neutral in such separations. N, neutral; A, acidic; B, basic.
[e]Gene locations are shown in figure 1 of the article on Human Cytomegalovirus: General Features.
[f]Present in immature capsids and noninfectious enveloped particles but absent from virions.
[g]Assemblin (A) is derived from a 74 kDa precursor (encoded by UL80a), and is autoproteolytically converted to amino (A_N) and carboxyl (A_C) portions.

it is not known whether these are specific associations. A larger number of proteins have been detected by a mass spectrometric analysis of virions, amounting to species encoded by 59 recognized HCMV genes and over 70 host proteins. The sensitivity of this technique is sufficiently high to raise questions about whether all these proteins are specifically incorporated into virions, and whether they are present in all virions.

The capsid shell is composed of four abundant protein species referred to as the major capsid protein (MCP), the minor capsid protein (mCP), the mCP-binding protein (mCP-BP), and the smallest capsid protein (SCP). Formation of the capsid shell is coordinated by precursors of an internally located maturational protease (assemblin) and assembly protein (pAP), which serve a scaffolding role and are present in nascent intranuclear capsids (e.g., B capsids) and in noninfectious enveloped particles (see below), but not in virions. These internal scaffolding proteins are the only abundant capsid proteins known to be post-translationally modified (**Table 1**).

The bulk of the protein mass of the virion tegument is contributed by its seven principal protein species. The largest of these, called the high molecular weight protein (HMWP), is the homolog of HSV-1 VP1/2 (encoded by HSV-1 UL36), and is an active ubiquitin-specific cysteine protease. HMWP forms heterooligomers with the HMWP-binding protein (HMWP-BP) encoded by the adjacent gene. Neither HMWP nor HMWP-BP is detectably phosphorylated or glycosylated. The basic phosphoprotein (BPP or pp150) and the upper matrix (UM or pp71) and lower matrix (LM or pp65) proteins account for the greatest protein mass of the tegument constituents and are highly phosphorylated, both *in vivo* and *in vitro*, by the virion-associated protein kinase. BPP is distinguished among the virion proteins by having the highest density of O-linked *N*-acetylglucosamine residues, which are attached at a single site or tight cluster of sites. UM is a transacting inducer of IE gene transcription that may be functionally analogous to the α-*trans*-inducing factor of HSV-1. LM appears to be an important component in cellular immunity and there are reports that it has an associated protein kinase activity. The '24K protein' (pp28) is also phosphorylated and, like its HSV-1 homolog UL11, forms oligomers (with the protein encoded by UL94). HMWP and BPP are thought to make direct contact with the capsid.

The virion envelope contains at least 10 protein species. The high degree of size and charge heterogeneity among this group has complicated their analysis. The most abundant species are distributed among three glycoprotein complexes (gCI, gCII, and gCIII), each of which is disulfide linked. The complex gCI consists of three forms of glycoprotein B (gB). HCMV gB is synthesized as a 130–160 kDa, cotranslationally N-glycosylated precursor that undergoes at least four further modifications: (1) its 17–19 predicted N-linked oligosaccharides are converted from nontrimmed to trimmed high-mannose forms, and then a portion is further modified to complex structures; (2) O-linked carbohydrates are added; (3) the C-terminal region of the molecule is phosphorylated; and (4) the protein is cleaved between Arg460 and Ser461 to yield an intramolecularly disulfide-linked heterodimer composed of a highly glycosylated 115–130 kDa N-terminal fragment (gB_N), and a less extensively modified 52–62 kDa C fragment (gB_C). This gB_N–gB_C heterodimer can form intramolecular disulfide crosslinks, yielding a complex with the composition $(gB_N–gB_C)_2$.

Early studies of the complex gCII identified proteins with M_rs of 50–52 and >200 kDa representing forms of the integral membrane protein glycoprotein M (gM). These proteins have similar peptide maps indicating that they are closely related, and the 52 kDa component is comparatively highly O-glycosylated and sialylated. Subsequently, the 50–60 kDa glycoprotein N (gN, UL73) was recognized as a second component of gCII.

The complex gcIII was initially characterized as composed of glycoprotein H (gH) and glycoprotein L (gL; not listed in **Table 1**). HCMV gH is an 86 kDa species that is less extensively N-glycosylated than gB and its oligosaccharides are not processed beyond the high-mannose structure. It has been proposed to mediate virus–host membrane fusion during the initial steps of virus infection, and its intracellular trafficking associates with gL. A third component, the highly sialylated, O-glycosylated virion glycoprotein O (gO), was later shown to be specified by UL74. Recently, using strains of HCMV that more closely reflect wild-type virus, an alternative complex consisting of gH, gL, and two small proteins encoded by UL128 and UL130 has been implicated in tropism for nonfibroblast cell types.

Physical Properties

Infectivity of HCMV is eliminated by exposure to 20% ether for 2 h, by heating at 65 °C for 30 min, by exposure to ultraviolet light (4000 erg s^{-1} cm^{-2}; 400 μJ s^{-1} cm^{-2}) for 4 min, by acidic pH (<5), and by treatment with low concentrations of either ionic or nonionic detergents. Infectivity is also reduced following sonication, pelleting by ultracentrifugation, and repeated cycles of freezing and thawing.

Replication

General features of the HCMV replication cycle are typical of the herpesvirus group. Following virus adsorption and entry into the cell, the DNA is transported to the nucleus where it is replicated and packaged into preformed capsids. Maturation of the nucleocapsid involves acquisition of the tegument constituents followed by envelopment, which has been observed at both nuclear and cytoplasmic membranes.

DNA Replication

HCMV DNA is replicated in the nucleus. Parental DNA appears to circularize soon after entering the cell, but maximal rates of progeny DNA synthesis do not occur until approximately 3 days after infection. The 'endless' concatemeric nature of progeny DNA is consistent with a rolling circle mechanism of DNA replication. However, the finding that virus DNA synthesis proceeds bidirectionally from *ori*-Lyt under conditions inhibition of replication by ganciclovir ([9-[2-hydroxy-1-(hydroxymethyl)-ethoxy]methyl] guanine) is not easily reconciled with this model, and perhaps indicates an intermediate stage of replication based on a theta mechanism. Unit-length genomic DNA is released from the concatemers by cleavage at a specific site located in the *a* repeat elements, 30–35 bp from a herpesvirus group-common cleavage/packaging region referred to as the herpes pac homology sequence.

Characterization of Transcription

The pattern and control of HCMV gene expression shares many features with that of HSV-1. As during translation, transcription of the HCMV genome is temporally regulated into three major kinetic classes. The first class of mRNAs made are transcribed from input DNA and are referred to as IE (or α) RNAs. Their expression requires no preceding virus protein or DNA synthesis, and they are operationally defined as those RNAs that can be made in the presence of inhibitors of protein synthesis. The most abundant IE RNAs are transcribed from the major IE promoter. These transcripts are related to each other by alternate splicing patterns and encode IE1 and multiple forms of IE2. The promoter region for this transcription unit is one of the strongest known, presumably due to its high content of consensus binding sites for host-cell transcription factors such as nuclear factor 1, CAAT-binding protein, and SP1, and the large number of repeat sequences that contain cyclic AMP response elements and that may interact with additional DNA-binding proteins. Transcription from this region is enhanced by UM, whose function appears to be analogous to that of the HSV α-*trans*-induction factor. Lower abundance IE RNAs arise from genes IRS1, TRS1, US3, and UL36–38.

Expression of E (or β) RNAs requires the preceding synthesis of IE proteins, but does not require virus DNA synthesis. Thus, E RNAs are made in the presence of inhibitors of virus DNA synthesis, but not in the presence of inhibitors of protein synthesis. Both IE and E genes appear to be transcribed by host-cell RNA polymerase II. Expression of L (or γ) RNAs requires the preceding synthesis of both IE and E virus proteins, as well as synthesis of virus DNA. The possibility that post-transcriptional events may be involved in controlling HCMV RNA processing and translation is suggested by the presence of some E RNAs that are not translated until late times of infection.

Characterization of Translation

HCMV RNAs are translated by the host-cell protein-synthesizing system. Virus-specific proteins that may augment the host system (e.g., initiation or elongation factors) have not yet been identified. Each virus protein species appears to be encoded by a unique mRNA. Cells infected with HCMV do not show the generalized early shut-off of cell protein synthesis that occurs in HSV-1-infected cells. There are, nevertheless, very early changes in the metabolism of HCMV-infected cells that include stimulated transcription of some host genes (e.g., heat shock protein 70, ornithine decarboxylase, thymidine kinase, creatine kinase, and cyclooxygenase 2), decreased transcription of other host genes (e.g., fibronectin), and changes similar to those induced by the GPCR signaling pathway (e.g., decreased intracellular Ca^{2+} stores and increased levels of intracellular cAMP).

Post-Translational Modifications

Modifications of HCMV proteins include phosphorylation, glycosylation, and proteolytic cleavage. Phosphorylation has been detected on IE (e.g., IE1 and IE2), E (e.g., UL112 protein) and L (e.g., DE nonvirion UL44 protein; late virion tegument (UL32, UL82, and UL83) proteins), and capsid (AP) proteins, all of which localize to the infected cell nucleus. At least some of these phosphorylations are likely to be catalyzed by reported virus-coded protein kinases. Glycosylation typifies many of the virion envelope glycoproteins and includes both N-linked (high-mannose and complex structures) and O-linked oligosaccharides. The only nonenvelope virion protein that has been demonstrated to be glycosylated is BPP, which has O-linked *N*-acetylglucosamine. Glycosylation of HCMV proteins is most likely carried out by host cell systems. The third modification that has been described for HCMV proteins is proteolytic cleavage. There are two maturational endoproteolytic cleavages, in addition to glycoprotein signal sequence cleavages. The first of these cuts between Arg460 and Ser461 in envelope gB and is catalyzed by a calcium-dependent cellular, furin-like proteinase. The other is a cleavage between residues Ala308 and Ser309 of the capsid assembly protein precursor (pAP). This latter cleavage is catalyzed by the genetically related, virus-encoded proteinase, assemblin (A), and is essential for the production of infectious virus.

Cytopathology

The mechanisms of cell recognition, binding, and penetration by HCMV are not yet understood, but there are at least five envelope proteins that may be involved: (1) a 'disintegrin-like' domain in virion envelope gB interacts with cellular integrins; (2) β_2-microglobulin is present on virions and may interact with cell-surface MHC proteins; (3) virion envelope gH interacts with a 90 kDa cell protein that may also be involved in binding or penetration; and (4) a highly O-glycosylated envelope glycoprotein complex (gcII) is able to bind heparin, and this interaction could promote early virus–cell interaction as proposed for cellular heparin and HSV-1. Different entry mechanisms may operate for different cell types.

Assembly of the capsid and DNA replication and packaging take place in the nucleus. As with HSV-1, some tegument constituents are presumed to be added to the maturing nucleocapsid in the nucleus, with envelopment and de-envelopment occurring as the tegument-coated capsid transits the two leaflets of the nuclear membrane into the cytoplasm. Further addition of tegument proteins is then thought to precede secondary envelopment at vesicle membranes in the infected cell cytoplasm.

The cytopathic effects of HCMV are distinctive. Early after infection, a transient cell rounding occurs, followed by overall enlargement (cytomegalia) and by the appearance of basophilic intranuclear inclusions, which gradually enlarge to fill and distort the nucleus into an elongated or kidney-shaped form late in infection. These intranuclear inclusions contain virus DNA, proteins, and capsids, and are thought to be nucleocapsid assembly sites. Late in infection, the cytoplasm contains numerous enveloped and nonenveloped capsids, large (>500 nm 'black holes') and smaller (250–500 nm 'dense bodies') electron-dense aggregates, and a large eosinophilic spherical region adjacent to the nucleus that exhibits strong F_C binding.

Release of HCMV from the cell results in three kinds of enveloped particles being freed into the growth medium of cultured fibroblasts: (1) virions; and two aberrant particles referred to as (2) noninfectious enveloped particles, which closely resemble virions in structure and composition but have no DNA and contain the capsid assembly protein that is absent from virions; and (3) dense bodies, which are solid, 250–600 nm spherical aggregates of LM that contain no DNA and are surrounded by an envelope layer.

Acknowledgments

I am grateful to Andrew Davison for his advice and kind assistance in updating the previous edition of this article.

Further Reading

Brignole EJ and Gibson W (2007) Enzymatic activities of human cytomegalovirus maturational protease assemblin and its precursor (pUL80a) are comparable: Maximal activity of pPR requires self-interaction through its scaffolding domain. *Journal of Virology* 81: 4091–4103.

Chee MS, Bankier AT, Beck S, *et al.* (1990) Analysis of the protein coding content of the sequence of human cytomegalovirus strain AD169. *Current Topics in Microbiology and Immunology* 154: 125–169.

Dolan A, Cunningham C, Hector RD, *et al.* (2004) Genetic content of wild type human cytomegalovirus. *Journal of General Virology* 85: 1301–1312.

Mocarski ES and Tan Courcelle C (2001) Cytomegaloviruses and their replication. In: Knipe DM and Howley PM (eds.) *Fields Virology*, 4th edn., pp. 2629–2673. Philadelphia: Lippincott Williams and Wilkins.

Varnum SM, Streblow DN, Monroe ME, *et al.* (2004) Identification of proteins in human cytomegalovirus (HCMV) particles: The HCMV proteome. *Journal of Virology* 78: 10960–10966.

Human Immunodeficiency Viruses: Molecular Biology

J Votteler and U Schubert, Klinikum der Universität Erlangen-Nürnberg, Erlangen, Germany

Glossary

Pandemic An outbreak of an infectious disease that spreads across countries, continents, or even worldwide.

Zoonosis Transmission of a pathogen from one species to another (e.g., humans).

Introduction

Since the beginning of the 1980s, approximately 40 million people have been infected with the human immunodeficiency virus (HIV), the causative agent of the acquired immune deficiency syndrome (AIDS). To date, this multisystemic, deadly, and so far incurable disease has caused more than 20 million deaths. AIDS was first described in 1981 in a group of homosexual men suffering from severe opportunistic infections. Two years later, a retrovirus was isolated from lymphocytes of AIDS patients and was later termed HIV. The two types, HIV-1 and -2, together with the simian immunodeficiency viruses (SIVs) found in nonhuman primates, delineate the genus of primate lentiviruses. Viruses of the genus *Lentivirus* preferentially replicate in lymphocytes and differentiated macrophages and often cause long-lasting and mostly incurable chronic diseases. In contrast to other retroviruses that require the breakdown of the nuclear membrane during mitosis for integration of the viral genome into host cell chromosomes, lentiviruses are uniquely capable of infecting nondividing cells, preferentially terminally differentiated macrophages and resting T cells.

Despite anti-retroviral therapy becoming more effective, infections with HIV remain one of the most devastating pandemics that humanity has ever faced. Unfortunately and despite enormous efforts, there is only limited hope that an effective vaccine will be developed in the near future and there are no other mechanisms available to stimulate the natural immunity against HIV. Another major drawback is the fact that the virus continues to persist even during prolonged therapy in latently infected host cells, classified as the viral reservoirs *in vivo*. Current anti-retroviral treatment is based on drugs that target either the viral enzymes, protease (PR) and reverse transcriptase (RT), or the envelope (Env) protein-mediated entry of the virus into the target cell.

Although introduction of highly active anti-retroviral therapy (HAART) in the mid-1990s, which involves the combination of different classes of both, PR and RT inhibitors, has led to a significant reduction in morbidity and mortality, an eradication of the virus from HIV-1-infected individuals has never been achieved. In addition, these antiviral drugs can induce severe adverse effects, particularly when administered in combination and over prolonged medication periods. A drawback to these treatments is that with the high mutation rate of HIV and replication dynamic, drug-resistant mutants are evolving. Cellular genes have much lower mutation rates, and a potential solution to this problem is to target cellular factors, enzymes, or complex mechanisms that are essential for replication of HIV in the host cells.

Genomic Organization and Replication of HIV

The *Retroviridae* belong to the huge group of eukaryotic retro-transposable elements. Retroviruses are distinguished from other viruses by their ability to reverse-transcribe their RNA genomes into DNA intermediates by using the enzyme RT. The provirus DNA genome is then integrated into the host cell chromosomes by action of the viral enzyme integrase (IN). All retroviruses contain at least three major genes encoding the main virion structural components that are each synthesized as three polyproteins that produce either the inner virion interior (Gag, group specific antigen), the viral enzymes (Pol, polymerase), or the glycoproteins of the virion Env. The genomic organization of HIV-1 is outlined in **Figure 1**. In addition to virus structural proteins and enzymes, some retroviruses code for small proteins with regulatory and auxiliary functions. In the case of HIV-1, these proteins comprise the two essential regulatory elements: Tat, which activates transcription, and Rev, which modulates viral RNA transport. In contrast, Nef, Vpr, Vif, and Vpu are nonessential for replication in certain tissue culture cells and are generally referred as accessory proteins (**Figure 1**).

According to the nomenclature defined by the International Committee on Taxonomy of Viruses (ICTV) convention, members of the *Orthoretrovirinae* contain RNA inside the virus particle and are released from the plasma membrane. However, the assembly of orthoretroviruses follows two different strategies: alpha-, gamma-, delta-, and lentiviruses (ICTV nomenclature) assemble on the inner leaflet of the plasma membrane and are released as immature virions. In contrast, betaretroviruses and spumaviruses first assemble in the cytoplasm to form immature particles that are then transported to the plasma

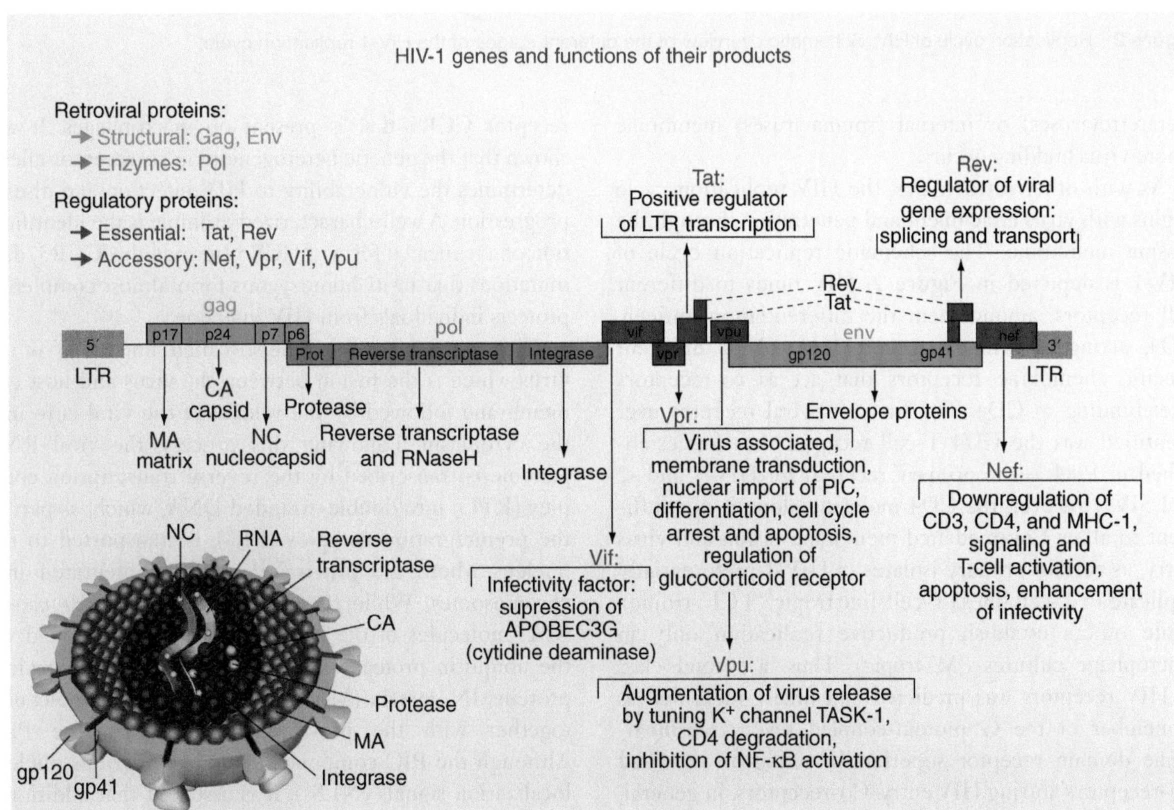

Figure 1 Genomic organization of HIV. This figure illustrates the complex organization of the HIV-1 genome comprising the standard retroviral elements for virus structure, *gag* and *env* (blue), the enzymes *pol* (green), and the six additional genes. These additional genes are either essential as *tat* and *rev*, or accessory as *vpr*, *vpu*, *vif*, and *nef*. The schematic model of the structure of the mature HIV-1 virus particle is given underneath.

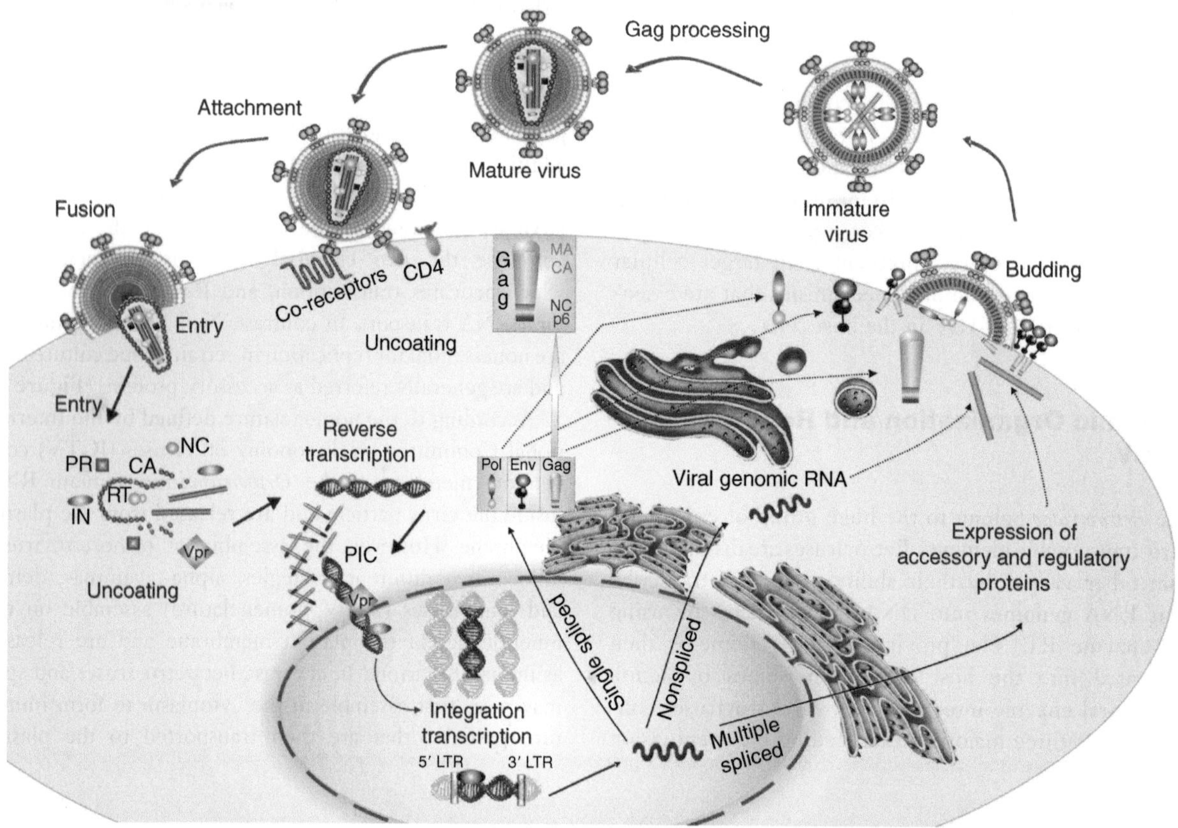

Figure 2 Replication cycle of HIV. Schematic overview of the different stages of the HIV-1 replication cycle.

(betaretroviruses) or internal (spumaviruses) membrane where virus budding occurs.

As with other retroviruses, the HIV replication cycle begins with virus attachment and penetration through the plasma membrane. The schematic replication cycle of HIV-1 is depicted in **Figure 2**. HIV binds to different cell receptors, among them the differentiation antigen, CD4, acting as primary receptor, as well as different specific chemokine receptors that act as co-receptors after binding to CD4. The first retroviral receptor ever identified was the CD4 T-cell receptor that was established in 1984 as the primary receptor for HIV-1 and -2 and SIV. However, the CD4 molecule alone is not sufficient to allow Env-mediated membrane fusion and virus entry, as certain primary isolates of HIV-1 preferentially replicate in T-cell lines (T-cell-line tropic, 'TCL'-tropic), while others establish productive replication only in macrophage cultures ('M'-tropic). Thus, a second class of HIV receptors was predicted and finally identified as a member of the G protein-coupled seven transmembrane domain receptor superfamily acting as so-called 'co-receptors' during HIV entry. Co-receptors, in general, function as cellular receptors for α- and β-chemokines. Two types of co-receptors arbitrate the disparity in cell-type tropism: the α-chemokine receptor CxCR4 that is typically present in T-cell lines, and the β-chemokine receptor CCR5 that is present on macrophages. It was shown that the genetic heterogeneity in co-receptor alleles determines the vulnerability to HIV infection and disease progression. A well-characterized example is the identification of a truncated form of CCR5 (termed the 'CCR5/d32' mutation) that in its homozygous form almost completely protects individuals from HIV infection.

Virus entry starts with the so-called 'uncoating' of the virus which is the fusion between the virus and host cell membrane followed by the release of the viral core into the cytoplasm. Following this process, the viral RNA genome is transcribed by the reverse transcription complex (RTC) into double-stranded DNA, which, as part of the preintegration complex (PIC), is transported to the nucleus where the proviral genome is integrated into chromosomes. While there is evidence that the capsid (CA) molecules of the incoming virus are degraded via the ubiquitin proteasome system (UPS), the three virus proteins IN, matrix (MA), and accessory Vpr molecules together with the provirus genome form the PIC. Although the PIC components comprise various nuclear localization signals (NLSs), it is assumed that additional cellular and viral factors, including a central DNA flap that is formed during reverse transcription, facilitate the import of the PIC into the nucleus. It is worth noting that virus entry involves a number of consecutive and highly organized multistep events that might explain why approximately only 1 out of a 1000 mature HIV-1 particles are capable of establishing a productive infection.

After reverse transcription of the viral genome and nuclear import of the PIC, the proviral DNA is integrated into chromosomal DNA. Activation of HIV-1 long terminal repeat (LTR) promoter-driven retroviral gene expression, followed by export, transport, and splicing of the viral mRNA, the so-called late steps of the HIV-1 replication cycle are initiated at this point. These steps involve membrane targeting of Gag and Gag–Pol polyproteins, assembly, as well as budding and maturation of progeny virions. Upon activation of integrated provirus, viral mRNAs are processed and transported into the cytosol for translation of newly synthesized structural proteins that assemble at the plasma membrane into budding particles. The Gag polyprotein in HIV-1 and HIV-2 consists of different functional domains that mediate the recognition and binding of viral RNA, the membrane targeting of Gag and Gag–Pol polyproteins, virion assembly, and efficient particle release from the plasma membrane as the final step of virus budding. In general, the Gag polyprotein constitutes viral components that are both sufficient and strictly required for virus particle assembly and budding, although further viral components such as the genomic RNA, the envelope, and the viral enzymes are required for production of infectious virions. The processing of the HIV-1 Gag polyprotein Pr55 by the viral PR generates the MA, CA, nucleocapsid (NC), as well as p6 proteins and the two spacer peptides, p2 and p1. The HIV-1 *pol*-encoded enzymes, PR, RT, and IN, are proteolytically released from a large polyprotein precursor, Pr160, the expression of which results from a rare frameshift event occurring during translation of Pr55. Like the Gag proteins, the Env glycoproteins are also synthesized from a polyprotein precursor protein. The resulting surface (SU) gp120 and transmembrane (TM) gp41 glycoproteins are produced by the cellular *protease* during trafficking of gp160 to the cell membrane. While gp120 contains the domains that mediate virus binding to CD4 and co-receptors, the gp41 anchors the trimetric complexes of TM/SU in the virus membrane and includes the determinants that regulate fusion between cellular and virus membranes during virus entry. The N-terminus of the extracellular ectodomain of gp41 harbors the so-called 'fusion peptide' that consists of a hydrophobic domain in concert with two helical motifs and regulates membrane fusion by formation of a six-helix bundle. The highly polymorphic SU protein gp120 is organized into five conserved domains (C1 to C5), and five highly variable domains that in most of the known SU sequences are concentrated near loop structures that are stabilized by disulfide bond formation.

In general, Gag proteins of different retroviruses exhibit a certain structural and functional similarity: MA mediates the plasma membrane targeting of the Gag polyprotein and lines the inner leaflet of the lipid bilayer of the mature virus particle, CA regulates assembly of Gag and forms the core shell of the infectious virus, and NC regulates packaging and condensation of the viral genome. In addition to these canonical mature retrovirus proteins, other Gag domains have been described, such as the HIV-1 p6 region that directs the incorporation of the regulatory protein Vpr into budding virions and governs efficient virus budding.

HIV particles bud from the plasma membrane as immature noninfectious viruses consisting predominantly of uncleaved Gag polyproteins. After virus release and in concert with PR activation which is autocatalytically released from the Gag–Pol polyprotein, processing of Gag and Gag–Pol polyproteins into its mature proteins and condensation of the inner core structure occurs that ultimately results in the formation of a mature infectious virus. Besides PR and Env, at least two other viral factors are known to promote efficient virus release: the HIV-1-specific accessory protein Vpu and the p6 domain. While Vpu supports virus release by an ion channel activity, p6 contains at least two distinct late assembly (L) domains that are required for efficient separation of assembled virions from the cell surface by a yet-undefined mechanism that somehow involves the cellular multivesicular body (MVB) as well as the UPS.

Role of Cellular Factors in HIV Replication

From entry to release and maturation into infectious progeny virions, each individual step in the HIV replication cycle exploits cellular pathways. As for most other intracellular parasites, the replication of HIV-1 depends on the interaction with specific host cell factors, and some of these proteins are specifically incorporated into progeny virions. Conversely, replication of HIV-1 is blocked in cells of the infected host by the action of restriction factors that function as barriers to retroviral replication. These factors are part of the so-called 'innate immune system' for which several mechanisms are known to interfere with replication at different stages of the viral life cycle. HIV-1, however, has evolved strategies to undermine these antiviral responses and, as a consequence, successfully propagates in the specific host environment.

Well-characterized examples for host–virus protein interaction are the virus receptors (CD4 and chemokine receptors) which enable virus entry into specific host cells, the role of the chromatin-remodeling system, and the HMG I family proteins for proviral DNA integration, as well as the requirement of different factors of the endosomal protein trafficking and ubiquitination systems for virus release.

An important case of host–virus interaction is the role of the UPS in virus budding. Recent work has provided intriguing insight into the mechanism of how virus budding exploits the cellular machinery that is normally involved in vacuolar lysosomal protein sorting and MVB-biogenesis. In the case of HIV-1, the recruitment of these cellular factors to the virus assembly site is facilitated by the interaction between the primary L-domain of p6 with at least one important host factor, the tumor susceptibility gene product 101 (Tsg101), an E2-type ubiquitin ligase-like protein. The second L-domain at the C-terminus of p6 mediates the binding of Gag to AIP1/ALIX, a class E vacuolar protein sorting factor that also interacts with Tsg101. AIP1/ALIX also binds to late-acting components of the endosomal sorting complexes required for transport (ESCRTs) and is necessary for the formation of MVB at endosomal membranes. Further, in more recent studies, VPS37B was identified as a new component of the ESCRT that binds to Tsg101. There has been accumulating evidence that HIV-1 recruits the components of the MVB system to the budding machinery which follows two separate and cell-type-specific pathways: in T cells HIV-1 buds primarily from the cell surface, while in monocytes/macrophages the virus buds into vacuoles of the MVB system.

An example for the ability of HIV-1 to escape host cell restriction is the recently discovered relationship between the HIV-1 accessory protein Vif (virus infectivity factor), and APOBEC3G (apolipoprotein B mRNA editing enzyme catalytic polypeptide-like 3G). APOBEC3G is a member of the cellular cytidine deaminase DNA/RNA-editing enzyme family that has the unique capability of hypermutating retroviruses, including HIV-1, with terminal consequences. However, the antiviral effect of APOBEC3G does not correlate with the frequency of mutation induced by APOBEC3G, suggesting that cytidine deaminase activity is not the only underlying mechanism of the antiviral activity of APOBEC. Nevertheless, the resulting innate block in virus replication is counteracted by Vif, which in the virus producer cell binds to APOBEC3G, induces its polyubiquitination by an SCF-like E3 ubiquitin ligase (Cullin5-ElonginB+C), and finally initiates proteasomal degradation of APOBEC3G. It is assumed that by this mechanism Vif is precluding the presence of APOBEC3G in progeny virions.

A second family of proteins with anti-retroviral activity are proteins of the tripartite interaction motif (TRIM) family, exemplified by TRIM5α. TRIM proteins contain a series of three motifs comprising a RING-finger (really interesting new gene), a B-box, and a coiled-coil (CC) domain. Only TRIM5α contains an additional SPRY domain which mediates interaction with the viral CA and is mainly responsible for the species-specific restriction activity. It has been shown that TRIM5α blocks incoming retroviruses at an early step of virus replication, occurring sometime after virus entry and before reverse transcription is initiated, and this antiviral activity of TRIM5α is clearly CA dependent. The RING-finger acts as an E3-type ubiquitin ligase, suggesting that ubiquitination of incoming CA molecules leading to proteasomal degradation might be responsible for the antiviral activity. However, the molecular mechanisms employed by TRIM5α to restrict retroviral infection are still poorly understood and a matter of intensive debate.

HIV Regulator Proteins

While retroviruses share the same fundamental replication cycle and have the same basic genomic organization (e.g., the canonical *gag*, *pol*, and *env* genes), they vary in the content of additional small regulatory genes. These proteins, except for Rev and Tat, dispensable for HIV-1 replication in certain cell lines *in vitro*, contribute enormously to pathogenesis and spread of HIV-1 *in vivo*.

The *trans*-Activator of Transcription (Tat) of HIV-1

It is now generally accepted that the HIV *trans*-activator (Tat) plays an important role in the pathogenesis of AIDS. Although originally described as an activator of the HIV-1 LTR promoter, Tat was later shown to regulate reverse transcription, to affect the expression of various cellular and viral genes, and to be released from infected cells. This so-called extracellular Tat, which acts as a cell membrane transducing peptide in the sense of a so-called 'trojan molecule', can affect neighboring cells, that are both uninfected and infected target cells. Indeed, there is accumulating evidence that Tat in its extracellular form plays a major role in AIDS-associated diseases like Karposi's sarcoma and HIV-associated dementia.

Tat can be expressed in two forms, as the 72-amino-acid one-exon Tat and as the 86–101-amino-acid (depending on the HIV-1 isolate) two-exon Tat expressed primarily early during infection. The 14–15 kDa Tat binds to an RNA stem–loop structure forming the Tat-responsive element (TAR) at the 5′ LTR region. It activates transcriptional elongation by stimulating the protein kinase TAK (Tat-associated kinase) resulting in hyperphosphorylation of the RNA polymerase II. In general, Tat stimulates the production of full-length HIV transcripts and is, therefore, essential for HIV replication.

The Regulator of Expression of Virion Proteins (Rev)

The compact organization of the HIV-1 genome and expression of all structural and regulatory proteins from a single promoter in the 5′-LTR requires a complex splicing regime of the primary transcript. The fully spliced mRNAs encoding Tat, Rev, and Nef are readily transported from the nucleus to the cytoplasm and, in consequence, these proteins are synthesized early in infection. However, the nuclear export of unspliced or single spliced mRNAs encoding structural proteins Gag and Env, the viral enzymes, as well as the accessory proteins Vif, Vpr, and Vpu, all require the activity of Rev. Hence, these proteins are expressed later during viral infection as the Rev concentration in the infected cell increases.

The Rev protein binds to the viral mRNAs via an arginine-rich RNA-binding motif that additionally serves as NLS required for the transport of Rev from the cytosol to the nucleus. Rev recognizes a stem–loop structure in the viral transcripts known as the Rev response element (RRE) that is located in *env.* After binding to RRE, Rev forms multimeric complexes of up to 12 monomers that are exported from the nucleus by interaction with nuclear export factors. These factors are recruited by a leucine-rich nuclear export signal (NES) located in the C-terminal region of Rev and promote the shuttling of the Rev–RNA complexes to the cytosol. Hence, Rev is required for the synthesis of viral proteins and is therefore essential for HIV replication.

The Lentivirus Protein R (Vpr)

Vpr is a virion-associated, nucleocytoplasmatic shuttling regulatory protein that is encoded by (and conserved among) primate lentiviruses, HIV-1, HIV-2, and the SIVs. Although dispensable for growth of HIV-1 in activated and dividing T cells, Vpr appears to play an important role in virus replication *in vivo*, since deletion of *vpr* and the related *vpx* genes in SIV severely compromises the pathogenic properties in experimentally infected rhesus macaques. Furthermore, HIV-2 and SIV also encode an additional Vpr-related protein, Vpx, that is believed to function synergistically with Vpr. The Vpr of HIV-1 is reported to exhibit numerous biological activities, including nuclear localization (based on the presence of at least two NLS), ion channel formation, transcriptional activation of HIV-1 and heterologous promoters, co-activation of the glucocorticoid receptor, regulation of cell differentiation, induction of apoptosis, cell cycle arrest, and transduction through cell membranes. Although significant amounts of Vpr (approximately 0.15-fold molar ratio to viral core proteins) are packaged into budding HIV-1 particles in a process dependent on Vpr's interaction with the C-terminal p6 domain of the Gag, the biological role(s) of virion-associated Vpr still remains to be fully elucidated.

The highly conserved 96-amino-acid Vpr has received considerable attention, and a number of biological functions have been attributed to its presence in various cellular and extracellular compartments. The most intensively investigated biological functions of Vpr are those affecting the translocation of the PIC of the incoming virus from the cytoplasm to the nucleus, and the arrest in the G_2 phase of the cell cycle. The nuclear targeting function of Vpr has been associated with productive infection of terminally differentiated macrophages by mediating integration of the proviral DNA into the host genome. It is assumed that Vpr causes dynamic disruptions in the nuclear envelope architecture that enables transport of the PIC across the nuclear membrane. Regarding the second function of Vpr, which leads to G_2 cell cycle arrest in HIV-1-infected T cells, it was suggested that this activity provides an intracellular milieu favorable for viral gene expression. Numerous cellular binding partners of Vpr have been identified, and for some of these a specific role in G_2 arrest has been proposed. Nevertheless, the precise molecular mechanism underlying the Vpr-induced G_2 arrest remains unclear. A potential explanation for the obvious paradigm that Vpr prevents proliferation of infected T cells by arresting them in the G_2 phase was provided by the observation that viral gene expression is optimal in the G_2 phase and that Vpr can increase virus production by delaying cells at this stage of the cell cycle. Interestingly, there are sufficient amounts of Vpr in incoming virus particles to induce G_2 cell cycle arrest even prior to the initiation of *de novo* synthesis of viral proteins. The secondary structures in Vpr emerging from several analyses indicate the presence of an α-helix–turn–α-helix motif at the N-terminus and an extended amphipathic helical region at the C-terminus which might play a key role in self-association and the interaction of Vpr with heterologous proteins, such as p6, NC, Tat of the virus and the adenine nucleotide translocator of the mitochondrial pore.

Other studies suggest that the prolonged G_2 arrest induced by Vpr ultimately leads to apoptosis of the infected cell. Conversely, early anti-apoptotic effects of Vpr have been described which are superseded by its pro-apoptotic effects. These pro-apoptotic effects of Vpr may result from either effects on the integrity of the nuclear envelope or direct mitochondrial membrane permeabilization, perhaps involving Vpr-mediated formation of ion channels in biological membranes.

The HIV-1-Specific Virus Protein U (Vpu)

Vpu is exclusively encoded by HIV-1, with one exception – the HIV-1 related isolate SIV_{cpz} that encodes a Vpu-like protein similar in length and predicted structure to the Vpu from HIV-1 isolates. Vpu of HIV-1 represents an integral membrane phosphoprotein with various biological functions: first, in the endoplasmic reticulum (ER), Vpu induces degradation of CD4 in a process involving the ubiquitin proteasome pathway and casein kinase 2 (Ck-2) phosphorylation of its cytoplasmic tail (Vpu_{CYTO}). Second,

Vpu augments virus release from a post-ER compartment by a cation-selective ion channel activity mediated by its transmembrane anchor (Vpu_{TM}). It was shown previously that Vpu can regulate cationic current when inserted into planar lipid bilayers and in *Xenopus* oocytes. Recent results indicate that the virus release function of Vpu_{TM} involves the mutually destructive interaction between Vpu and the K^+ channel TASK-1. Thus, the virus release function of Vpu might be mediated by tuning the activity of host cell ion channels at the cell membrane.

Structurally, Vpu is a class I oligomeric membrane-bound phosphoprotein composed of an amphipathic sequence of 81 amino acids comprising a hydrophobic N-terminal membrane anchor proximal to a polar C-terminal cytoplasmic domain. The latter contains a highly conserved region with the CK-2 phosphorylation sites that regulate Vpu's function in the ER. The current model of the structure and the orientation in the membrane of the N-terminal hydrophobic Vpu_{TM} provides evidence for water-filled five-helix bundles. However, more work is required to understand Vpu in its functional forms *in vivo* where it exists as the phosphoprotein in mulitprotein complexes involving CD4 and β-TrCP, a component of the SkpI, Cullin, F-box protein (SCF^{TrCP}) E3 ubiquitin ligase complex that regulates the ubiquitination and degradation of cellular proteins by the proteasome.

The Viral Infectivity Factor Vif

The 23 kDa phosphoprotein Vif represents a viral infectivity factor that is highly conserved among primate lentiviruses. Vif regulates virus infectivity in a cell-type-dependent fashion. Permissive cells like Hela, Cos, 293T, SupT1, CEM-SS, and Jurkat cells support HIV replication independent of Vif. In contrast, primary lymphocytes, macrophages, and certain cell lines like H9 cells require Vif for production of fully infectious viruses.

Until recently, the mechanism of how Vif supports formation of infectious virions remained mostly enigmatic. Several findings now support the intriguing hypothesis that at least one main function of Vif is to neutralize cellular cytidine deaminase APOBEC3G, which, in the absence of Vif, is encapsidated into progeny virions. It has been a long-known fact that Vif can act *in trans* when it is expressed in the virus-producer cell, but not in the target cell, indicating that Vif must be situated at the virus-assembly site. This model of a post-entry function for Vif is further supported by the observation that Vifs from diverse lentiviruses function in a species-specific mode. Several hypotheses have been put forward to explain the mechanism of Vif function: the activity of Vif might regulate reverse transcription and proviral DNA synthesis, Vif can support the formation of stable virus cores, or Vif might bind to viral RNA in order to support post-entry steps of virus replication. The discovery of the APOBEC3G–Vif interaction provides at least some explanation for these previous observations.

The Multifunctional Nef Protein

Originally, the 27 kDa N-terminal myristoylated membrane-associated phosphoprotein Nef was described as a 'negative factor' that suppresses the HIV LTR promoter activity. As one of the most intensively investigated HIV-1 accessory proteins, Nef has been shown to fulfill multiple functions during the viral replication cycle. In general, it is now widely accepted that inconsistent with its originally 'negative' nomenclature, Nef plays an important stimulatory role in the replication and pathogenesis of HIV-1. The broad spectrum of Nef-associated activities so far described can be summarized in distinct classes: (1) modulation of cell activation and apoptosis; (2) change in intracellular trafficking of cellular proteins, particularly of cell receptor proteins like CD4 and major histocompatibility complex I (MHC-I); and (3) increase of virus infectivity.

Well characterized and one of the earliest observations in the course of studying the biology of Nef is the Nef-induced downregulation of cell surface CD4. This mechanism occurs on a post-translational level by augmentation of receptor internalization, where Nef binds to the adapter protein complex in clathrin-coated pits, followed by lysosomal degradation. Another influence on immune receptors is the Nef-mediated block in the transport of MHC-I antigen complexes to the cell surface, leading to the disturbance in the recognition of HIV-1 infected cells by cytotoxic T_{CD8}^{+} lymphocytes (CTLs). More recent studies also described downregulation of the co-stimulatory protein CD28 induced by Nef. It appears that the mechanisms by which Nef blocks cell surface expression of CD4, MHC-I, and CD28 are independent and can be genetically separated.

Nef is expressed relatively early in the HIV replication cycle, and there are reports that Nef is even incorporated into budding virions where it can be cleaved by the viral protease. Nef is present during almost all steps of the viral life cycle, which might explain the multiple functions deployed by this accessory protein. Major attention has been focused toward the Nef-mediated modulation of the T-cell signal transduction and activation pathways. Several lines of evidence supported the hypothesis that, by affecting T-cell receptors, Nef may support several aspects of the virus replication cycle, including regulation of apoptosis and immune evasion. Earlier studies in transgenic mice indicated Nef-dependent activation in T-cell signaling. Later studies indicate that Nef imitates T-cell receptor signaling motifs. The Nef-induced T-cell activation requires the tyrosine kinase Zap70 and the ζ-chain of the T-cell receptor. Further, Nef was shown to activate

T cells by binding to the Nef-associated kinase, also known as the p21-activated kinase 2 (PAK2). In addition to the activation of T cells, Nef also induces cell death, since it was shown that expression of the death receptor Fas ligand (FasL) is activated by Nef, and that this upregulation might increase killing of lymphocytes attacking HIV infected cells.

A remarkable turn to our understanding of Nef function was offered recently when it was demonstrated that Nef proteins from the great majority of primate lentiviruses, including HIV-2, efficiently downregulate the T-cell receptor CD3 (TCD3), thereby suppressing activation and apoptosis in T cells. In contrast, *nef* alleles from HIV-1 and its closest related SIV isolates do not encode this activity. Hyperactivation of the immune system as caused by T-cell activation is now generally accepted as an important clinical marker of AIDS progression. Hence, downregulation of TCD3 and the resulting avoidance preceding T-cell activation yields viral persistence without damaging the host immune system and disease progression to AIDS. Thus, in contrast to HIV-1, SIV-derived Nef proteins are better described as persistence factors than as pathogenesis factors.

HIV Subtypes

Analysis of viral sequences allows HIV-1 to be classified into three distinctive groups, representing the HIV-1 lineages main (M), outlier (O), and non-M-non-O (N). It is assumed that each group resulted from an independent zoonotic transfer from chimpanzees, which were infected with SIV_{CPZ} into humans. In addition, eight HIV-2 lineages arose from separate zoonotic transfers from sooty mangabeys infected with SIV_{SM}. Most intriguingly, of these 11 zoonotic events that transferred primate lentiviruses from nonhuman primates to humans, only those resulting in the HIV-1 M-group finally led to the current global AIDS pandemic. Most of group O and group N strains originate from West Africa, especially Cameroon. However, the propagation of these strains was relatively restricted. HIV-2, predominantly A and B groups, is also for the most part endemic in West Africa, particularly in Côte d'Ivoire. In comparison to HIV-1, HIV-2 led to a relatively small number of infections in humans. It is conceivable that more transmissions from nonhuman primates to humans occurred, which, however, were not able to spread efficiently in the human population and therefore have never been detected. At least one reason for the low frequency of successful zoonotic events is that transmissions of primate lentiviruses across the species barrier can be controlled by the existence of species-specific restriction factors (e.g., TRIM5α, as detailed above) that act as barriers against retroviral transfers.

Further Reading

Anderson JL and Hope TJ (2004) HIV accessory proteins and surviving the host cell. *Current HIV/AIDS Reports* 1(1): 47–53.

Coffin JM, Hughes SH, and Varmus HE (1997) *Retroviruses.* New York: CSHL Press.

Flint SJ, Enquist LW, Krung RM, Racaniello VR, and Skalka AM (2000) *Principles of Virology.* Washington, DC: ASM Press.

Freed EO (2002) Viral late domains. *Journal of Virology* 76(10): 4679–4687.

Freed EO and Mouland AJ (2006) The cell biology of HIV-1 and other retroviruses. *Retrovirology* 3: 3–77.

Goff SP (2004) Genetic control of retrovirus susceptibility in mammalian cells. *Annual Reviews of Genetics* 38: 61–85.

Klinger PP and Schubert U (2005) The ubiquitin–proteasome system in HIV replication: Potential targets for antiretroviral therapy. *Expert Review of Anti-Infective Therapy* 3(1): 61–79.

Li L, Li HS, Pauza CD, Bukrinsky M, and Zhao RY (2005) Roles of HIV-1 auxiliary proteins in viral pathogenesis and host–pathogen interactions. *Cell Research* 15(11–12): 923–934.

Morita E and Sundquist WI (2004) Retrovirus budding. *Annual Reviews of Cell and Developmental Biology* 20: 395–425.

Schubert U and McClure M (2005) Human immunodeficiency virus. In: Mahy BWJ and Ter Meulen V (eds.) *Virology,* 10th edn., pp. 1322–1346. London: Topley and Wilson.

Iridoviruses

V G Chinchar, University of Mississippi Medical Center, Jackson, MS, USA
A D Hyatt, Australian Animal Health Laboratory, Geelong, VIC, Australia

Glossary

Anuran An amphibian of the order Salientia (formerly Anura or Batrachia) which includes frogs and toads. Also called, salientians.

Introduction

Members of the family *Iridoviridae*, hereafter referred to as iridovirids to distinguish them from members of the genus *Iridovirus*, are large (120–200 nm), double-stranded DNA viruses that utilize both the nucleus and the cytoplasm in the synthesis of viral macromolecules, but confine virion formation to morphologically distinct, cytoplasmic assembly sites. Virus particles display icosahedral symmetry but, unlike other virus families, infectious virions can be either nonenveloped (i.e., naked) or enveloped, although the latter show a higher specific infectivity. The viral capsid is composed primarily of the major capsid protein (MCP), a ~50 kDa protein that is highly conserved among all members of the family. An internal lipid membrane, that is essential for infectivity, underlies the capsid and encloses the viral DNA core. Approximately 30 virion-associated proteins have been identified by gel electrophoresis, but the functions of most of these proteins are unknown. The viral genome is linear, double-stranded DNA and ranges in size from 103 to 212 kbp, depending upon the viral species. As a likely consequence of its mode of packaging, viral DNA is terminally redundant and circularly permuted. The size of the repeat regions range from 5% to 50% of the genome and, like the genome size, appears to vary with the specific viral species.

Iridovirus Taxonomy

Members of the family *Iridoviridae* are classified into five genera, two of which infect invertebrates (*Iridovirus*, *Chloriridovirus*), and three that infect ectothermic vertebrates (*Ranavirus*, *Lymphocystivirus*, and *Megalocytivirus*). In addition to differences in host range, viruses within the three vertebrate iridovirus genera, with one known exception, contain highly methylated genomes in which every cytosine present within a CpG motif is methylated. Viral isolates within the genus *Megalocytivirus* show high levels of sequence identity and it is not clear if they represent strains of the same viral species, or a small number of closely related species. To a lesser extent, the same is true within the genus *Ranavirus*, although here differences in host range and clinical presentation, along with a higher level of sequence variation, allows identification of individual viral species (**Table 1**).

Recently, genomic analysis validated taxonomic divisions made earlier on the basis of physical characteristics, clinical presentation, and host range. A phylogenetic tree constructed using the inferred amino acid sequences of the major capsid proteins (MCPs) of 16 iridovirids, representing all five genera, supports the division of the family into three genera of vertebrate iridoviruses (*Megalocytivirus*, *Ranavirus*, and *Lymphocystivirus*), and a phylogenetically diverse cluster of invertebrate iridescent viruses (IIVs) composed of the existing *Chloriridovirus* and *Iridovirus* genera (**Figure 1**). Recently, a maximum likelihood (M-L) tree, constructed using a concatenated set of 11 protein sequences conserved among all iridoviruses, confirmed this finding. Moreover, a M-L tree constructed using

Table 1 Taxonomy of the family *Iridoviridae*

Genus	*Viral species [strains]*[a]	*Tentative species*
Iridovirus	*Invertebrate iridescent virus 6* (IIV-6) [Gryllus iridovirus, Chilo iridescent virus] *Invertebrate iridescent virus 1* (IIV-1) [Tipula iridescent virus]	Anticarsia gemmatalis iridescent virus (AGIV), IIV-2, -9, -16, -21, -22, -23, -24, 29, -30, -31
Chloriridovirus	*Invertebrate iridescent virus-3* (IIV-3) [Aedes taeniorhynchus iridescent virus, mosquito iridescent virus]	
Ranavirus	*Frog virus 3* (FV3), [tadpole edema virus, TEV; tiger frog virus, TFV] *Ambystoma tigrinum virus* (ATV), [Regina ranavirus, RRV] *Bohle iridovirus* (BIV) *Epizootic haematopoietic necrosis virus* (EHNV) *European catfish virus* (ECV), [European sheatfish virus, ESV] *Santee-Cooper ranavirus,* [largemouth bass virus, LMBV; doctor fish virus, DFV; guppy virus 6, GV-6]	Singapore grouper iridovirus (SGIV), grouper iridovirus (GIV), Rana esculenta iridovirus, Testudo iridovirus, Rana catesbeiana virus Z (RCV-Z)
Megalocytivirus	*Infectious spleen and kidney necrosis virus* (ISKNV) [Red Sea bream iridovirus, RSIV; African lampeye iridovirus, ALIV; orange spotted grouper iridovirus, OSGIV; rock bream iridovirus, RBIV]	
Lymphocystivirus	*Lymphocystis disease virus 1* (LCDV-1) [flounder lymphocystis disease virus, flounder virus]	LCDV-2, LCDV-C, LCDV-RF; [Dab lymphocystis disease virus]
Unclassified		White sturgeon iridovirus (WSIV) Erythrocytic necrosis virus (ENV)

[a]Recognized viral species are italicized and likely virus strains or isolates are enclosed within brackets. Tentative species are listed, as are commonly used abbreviations for species, strains, isolates, and tentative species.

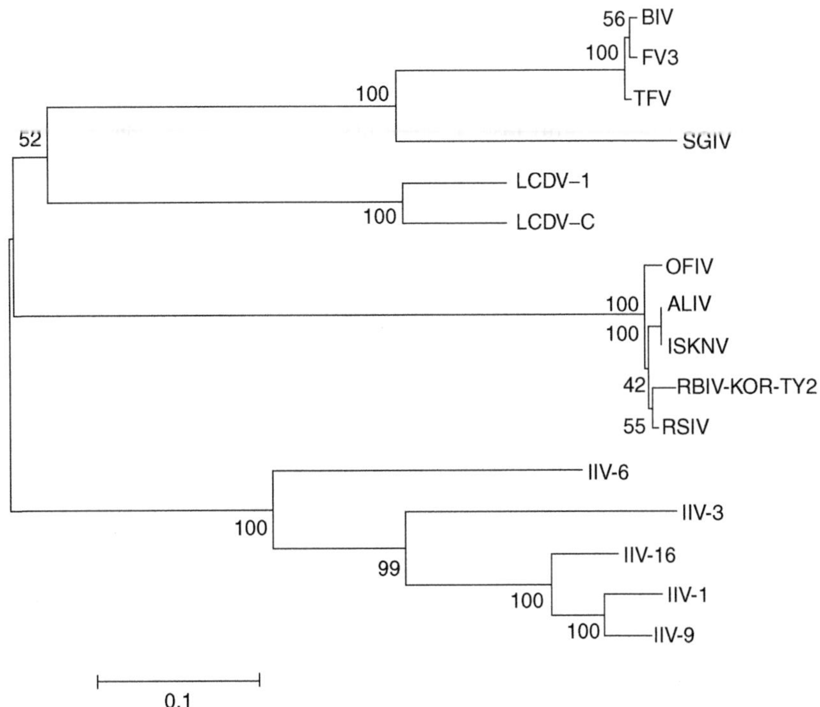

Figure 1 Phylogenetic relationships among iridovirids. The inferred amino acid sequences of the MCP of 16 iridovirids, representing all five currently recognized genera, were aligned using the CLUSTAL W program and used to construct a phylogenetic tree using the neighbor-joining algorithm and Poisson correction within MEGA version 3.1. The tree was validated by 1000 bootstrap repetitions. Branch lengths are drawn to scale. The number at each node indicates bootstrapped percentage values. The sequences used to construct the tree were obtained from the following viruses: genus *Megalocytivirus* – ISKNV, infectious skin and kidney necrosis virus (AF370008); ALIV, African lampeye iridovirus (AB109368); OFIV, olive flounder iridovirus (AY661546); RSIV, red sea bream iridovirus (AY310918); RBIV, rock bream iridovirus (AY533035); genus *Ranavirus* – SGIV, Singapore grouper iridovirus (AF364593); TFV, tiger frog virus (AY033630); BIV, Bohle iridovirus (AY187046); FV3, frog virus 3 (U36913); genus *Lymphocystivirus* – LCDV-1, lymphocystis disease virus (L63545); LCDV-C, lymphocystis disease virus (China) (AAS47819.1); genus *Iridovirus* – IIV-6, invertebrate iridescent virus 6 (AAK82135.1); IIV-16 (AF025775), IIV-1 (M33542), and IIV-9 (AF025774); genus *Chloriridovirus* – IIV-3 (DQ643392).

MCP sequences from 14 invertebrate iridescent viruses (many of which had been tentatively classified as members of the genus *Iridovirus*) has shown that IIV-6 clusters with two tentative members of the genus *Iridovirus* on one branch, whereas IIV-3 is linked with the remaining 12 isolates on a separate branch. Given this admixture of putative chloriridovirus and iridovirus species, it has been suggested that the status of the existing invertebrate genera may require revision. However, the lack of collinearity between IIV-3 and IIV-6, the low levels of amino acid identity between IIV-3 and IIV-6, along with differences in host range, virion size, and GC content, indicate that IIV-3 and IIV-6 are likely members of different viral genera. On a larger taxonomic scale, recent work has linked the family *Iridoviridae* with other large DNA virus families such as the *Poxviridae*, *Asfarviridae*, *Phycodnaviridae*, the newly discovered mimiviruses, and the *Ascoviridae*. However, although it is clear that these large DNA viruses share a set of common genes, it is not known if iridoviruses are more closely related to ascoviruses, the recently discovered mimivirus, or another virus family.

Viral Genome

The 12 completely sequenced iridovirid genomes range in size from 103 to 212 kbp (**Table 2**). The genomes of vertebrate iridovirids are found at the lower end of this size range, whereas the genomes of IIV-3 and IIV-6 (invertebrate iridovirids) occupy the high end. Consistent with the differences in size, vertebrate iridovirids encode ~100 putative ORFs, whereas invertebrate iridovirids code for 126–234 putative proteins. BLAST analysis of the various viral genomes indicates that about a quarter to a third of the viral ORFs share sequence identity/similarity to eukaryotic and viral proteins of known function including the two largest subunits of RNA polymerase II, a viral DNA polymerase, the two subunits of ribonucletide reductase, dUTPase, thymidylate synthase, thymidylate kinase, a major capsid protein, and other proteins likely to be directly involved in viral replication. In addition to these replicative proteins, iridovirids also encode one or more putative 'immune evasion' proteins such as a viral homolog of eukaryotic translational initiation factor 2α (vIF-2α), a

Table 2 Coding potential of iridovirid genomes

Genus	*Virus*	*bp*[a]	*No. of genes*[b]	*%GC*	*GenBank Acc. No.*
Iridovirus	IIV-6	212 482	234	29	AF303741
Chloriridovirus	IIV-3	191 132	126	48	DQ643392
Lymphocystivirus	LCDV-1	102 653	110	29	L63545
	LCDV-C	186 250	176	27	AY380826
Megalocytivirus	ISKNV	111 362	105	55	AF371960
	OSGIV	112 636	121	54	AY894343
	RBIV	112 080	118	53	AY532606
Ranavirus	FV3	105 903	98	55	AY548484
	TFV	105 057	105	55	AF389451
	ATV	106 332	96	54	AY150217
	SGIV	140 131	162	49	AY521625
	GIV	139 793	120	49	AY666015

[a]The value shown represents the unique genome size in base pairs (bp) minus the length of the terminal repeats.
[b]The value shown is an estimate of the total number of nonoverlapping genes encoded by a given virus. It is generally lower than the total number of putative ORFs which includes putative genes encoded on opposing DNA strands and overlapping genes.

steroid oxidoreductase, a caspase activation and recruitment domain (CARD)-containing protein, and a homolog of the TNF receptor. Dot plot analyses have shown that virus species within the genus *Ranavirus* display some degree of collinearity in their gene order, but that inversions occur even between closely related viruses such as Ambystoma tigrinum virus (ATV) and tiger frog virus. Moreover, between more distantly related ranaviruses (e.g., Singapore grouper iridovirus and FV3) and between viruses from different genera, collinearity appears to break down completely, indicating that, although iridovirids share many genes in common, their precise arrangement within the viral genome does not appear to be critical for successful virus replication. Furthermore, although viral genes are temporally expressed, there is no clustering of immediate early (IE), delayed early (DE), or late (L) viral genes. These observations suggest that the marked differences in gene order between different viral species may be due to the propensity of iridovirids to undergo high levels of recombination.

DNA repeats are present within the intergenic regions of all iridovirids, but differ between genera in their extent, arrangement, and sequence motifs. For example, 15 distinct repeats, comprising two distinct groups, 0.8–4.6 kbp in length, are found within IIV-3. The repeats are composed of a ~100 bp sequence that is present in 2–10 copies/genome. They show 80–100% identity within a group and ~60% identity between groups. The function(s) of the repeat regions is unknown but, by analogy to similar regions in other DNA viruses, they may play roles in genome replication and gene expression. Among members of individual vertebrate iridovirid genera, sequence identity within particular coding regions, for example, the MCP, is relatively high (>70%). However, there is considerable diversity among the MCPs of invertebrate viruses, with some members of the genus *Iridovirus* showing only 50% identity to others. As expected, between genera amino acid identity/similarity is only ~50%.

Virus Replication Cycle

Most of what is known about the replication of iridovirids is based on studies conducted with *Frog virus 3* (FV3), the type species of the genus *Ranavirus*. Thus, any discussion of iridovirid life cycles will have a strongly 'ranacentric' orientation. By necessity, that bias will continue here, but with the caveat that other members of the family, especially viruses from different genera, may not follow the FV3 pathway. **Figure 2** summarizes the salient features of FV3 replication.

Virus Entry

The cellular receptor for FV3 is apparently a quite common and highly conserved molecule as a range of mammalian, avian, piscine, and amphibian cell lines can be infected *in vitro*. However, since the host range *in vivo* is restricted mainly to anurans, the receptor molecule may not be expressed as widely in whole animals, at least in cells in which infection may be initiated. In addition, since FV3 does not replicate at temperatures above 32 °C, mammals and birds are nonpermissive hosts. The host range varies with the particular virus. Some iridovirids, such as RSIV and LCDV, possess a broad host range and infect a wide variety of different fish species, whereas other members of the family, such as ATV, infect a limited number of host species. Although both naked and enveloped virions are infectious, early work suggests that these two forms of the virus enter target cells via different mechanisms. Non-enveloped virions are thought to bind to the plasma membrane and inject their viral cores into the cytoplasm whence it is

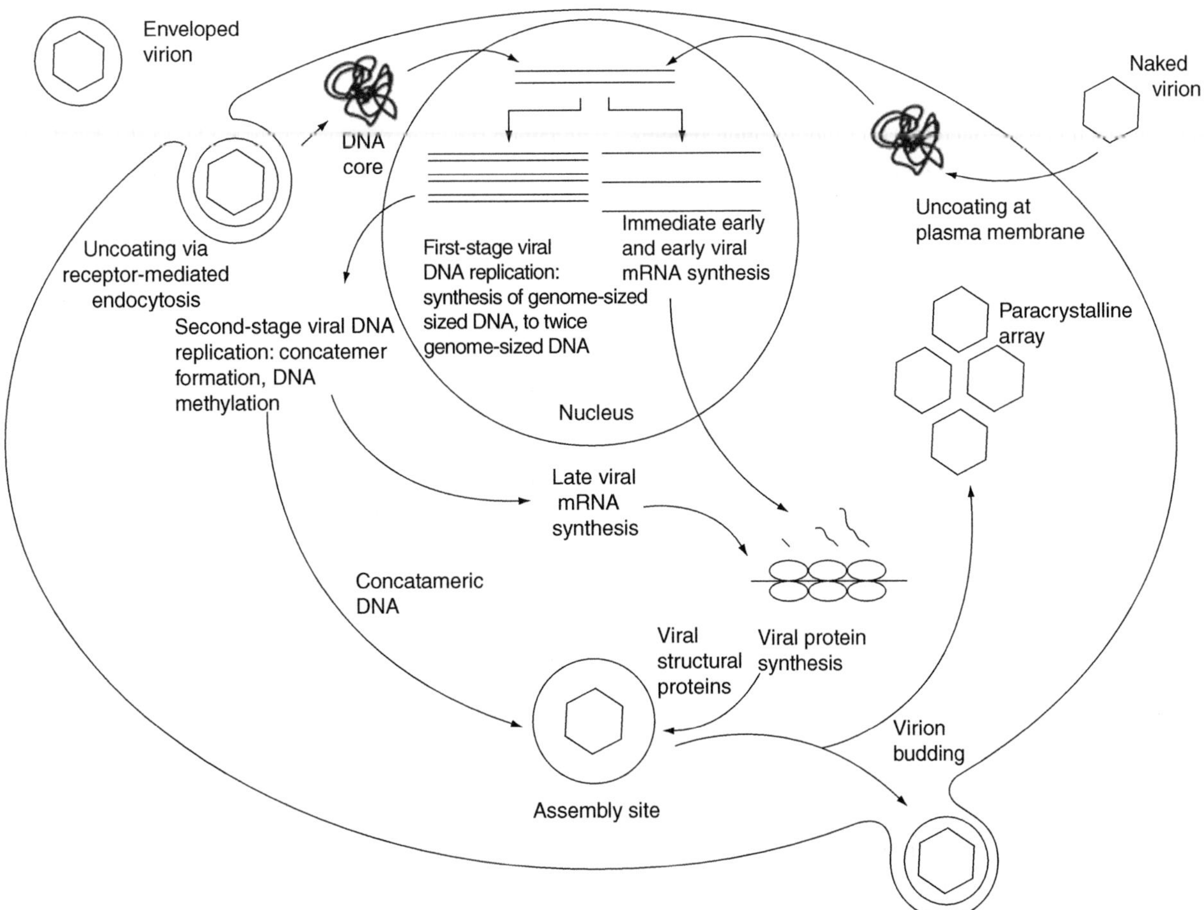

Figure 2 FV3 replication cycle. See text for details. Reproduced from Chinchar VG (2002) Ranaviruses (family *Iridoviridae*): Cold-blooded killers. *Archives of Virology* 147: 447–470, with permission.

transported to the nucleus. In contrast, enveloped virions enter the cell by receptor-mediated endocytosis.

Early Viral RNA Synthesis

Following entry of the viral DNA into the nucleus, IE viral RNA synthesis begins in a process catalyzed by host cell RNA polymerase II (Pol II). Naked viral DNA is not infectious and IE transcription is dependent upon one or more virion-associated proteins, referred to as virion-associated transcriptional transactivators (VATTs). However, it is not known if VATTs modify the viral DNA template or interact with cellular RNA polymerase II (Pol II) itself. Subsequently, one or more newly synthesized IE proteins, designated virus-induced transcriptional transactivators (VITTs), are required for the transcription of delayed early (DE) viral genes. Although its mechanism of action is unknown, VITTs are likely required to overcome the inhibitory effect of high levels of DNA methylation on Pol II-catalyzed transcription. Like other DNA viruses, transcription of both invertebrate and vertebrate iridovirid genes is a highly regulated process in which the three classes of viral genes (IE, DE, and late) are synthesized in a coordinated manner. In general, IE transcripts likely encode regulatory and catalytic proteins required for the synthesis of DE and late viral gene products, whereas late mRNAs are translated to yield viral structural proteins.

Viral DNA Synthesis and Methylation

Following its synthesis in the cytoplasm and translocation to the nucleus, viral DNA polymerase synthesizes genome-sized DNA molecules. Subsequently, viral DNA is transported to the cytoplasm where the second stage of viral DNA replication takes place. In this stage, large, concatameric structures are generated that contain more than 10 times the unit length of viral DNA. The mechanics of concatamer formation are not known, but likely involve multiple recombination events and may occur within viral assembly sites.

Following transport to the cytoplasm, viral DNA is methylated by a virus-encoded cytosine DNA methyltransferase. It is not known whether unit length or concatameric DNA is the template for methylation, or whether both serve equally well. The precise role of DNA methylation in the virus life cycle has not yet been elucidated. It has been postulated that methylation protects viral DNA from degradation mediated by a virus-encoded endonuclease similar to those found in bacterial restriction-modification systems. While this scenario is plausible, the rationale for encoding the endonuclease is not clear, unless one postulates that breakdown products of host cell DNA are used to synthesize viral DNA. Regardless of the reason, methylation is important for successful virus replication since treatment of cells with 5-azacytidine, an inhibitor of DNA methylation, reduces virus yields 100-fold. Analysis of viral DNA following sucrose gradient centrifugation suggests that the reduction in viral yields is likely due to the increased sensitivity of unmethylated DNA to nucleolytic attack and the inability of nicked DNA to be properly packaged into virions. An alternative hypothesis that is currently being tested suggests that viral DNA is methylated to block activation of a Toll-like receptor 9 (TLR-9) response. According to this hypothesis, unmethylated FV3 DNA, like that of herpes simplex virus DNA, binds TLR9 and triggers a pro-inflammatory response. In contrast, methylated DNA may not trigger as rapid an immune response and thus permit higher levels of virus replication *in vivo*.

Late Viral RNA Synthesis

Although early viral transcription is catalyzed by host Pol II, late viral transcription is likely catalyzed by a virus-encoded or virus-modified enzyme, designated vPol II. Supporting this view is the observation that all iridovirids sequenced to date encode homologs of the two largest subunits of Pol II which either form a unique, wholly virus-encoded polymerase, or a chimeric polymerase composed of host and viral components. It is likely that vPol II catalyzes the synthesis of late viral messages within viral assembly sites. However, it is not known if some vPol II returns to the nucleus where it contributes to the synthesis of early transcripts, or if viral transcription remains spatially and temporally separated into nuclear (i.e., early transcription catalyzed by host Pol II) and cytoplasmic (i.e., late transcripts synthesized by vPol II) compartments. Recent studies indicate that cells treated with an antisense morpholino oligonucleotide (asMO) targeted to the largest subunit of vPol II show ~80% reduction in late, but not early, viral message expression, and >95% drop in virus yields. These results support the hypothesis that late viral gene transcription is likely catalyzed by a virus-encoded or virus-modified Pol II-like molecule.

Virion Assembly

Following synthesis, late viral proteins, which are likely structural components of the virion, are transported into assembly sites where they participate in virion formation. Unfortunately, the mechanics of virion assembly are poorly understood. Packaging of viral DNA appears to occur via a headful mechanism in which a unit length of viral DNA, plus an additional variable amount (depending upon the virus species), is inserted into the developing virion. As a consequence of this mechanism, viral DNA is terminally redundant and circularly permuted. Completed (but not yet enveloped) virions are often seen scattered within assembly sites or present within paracrystalline arrays (**Figure 3**). It is unclear whether virions are transported from assembly sites to the arrays, or whether, as virion morphogenesis continues, the original assembly site is transformed into a paracrystalline array. In contrast to the accumulation of naked virions within the cytoplasm, enveloped virions form by budding at the plasma membrane. However, most virions remain cell associated and are only released from the cell upon lysis.

Virus assembly sites are one of the most striking features seen in virus-infected cells following visualization by either fluorescent antibody staining or transmission electron microscopy (**Figure 3**). In ranavirus-infected cells, assembly sites are electrolucent areas within the cytoplasm that are devoid of cellular organelles such as mitochondria, endoplasmic reticulum, ribosomes, or cytoskeletal elements. Assembly sites contain viruses in various stages of formation ranging from partially formed capsids to complete, nonenveloped capsids with electron-dense cores. The matrix is granular and frequently contains smooth-surfaced vesicles and diffuse electron-dense material which may be nucleic acid. FV3 assembly sites are surrounded by intermediate filaments. It has been suggested that the filaments may play a direct role in virus replication, perhaps by anchoring assembly sites and facilitating entry of viral components, or by excluding cellular organelles that might interfere with virion morphogenesis. Furthermore, the association of ranaviruses with the cytoskeleton suggests that intermediate filaments may be involved in virion envelopment and release. Disruption of intermediate filaments by treatment of FV3-infected cells with taxol or microinjection of anti-vimentin antibodies has been shown to interfere with the ability of intermediate filaments to encompass assembly sites. It has also been shown to lead to intrusion of cellular components into the assembly site, reduced accumulation of viral proteins, and 70–80% reduction in virus yields. Moreover, assembly sites are observed in the absence of late gene expression, suggesting that one or more early viral gene products and the viral DNA are sufficient for their formation.

The morphological appearance of cells from fish infected *in vivo* with megalocytiviruses such as RSIV is markedly different from the assembly site formation observed in ranavirus-infected cells. RSIV-infected splenocytes contain large inclusion bodies that are

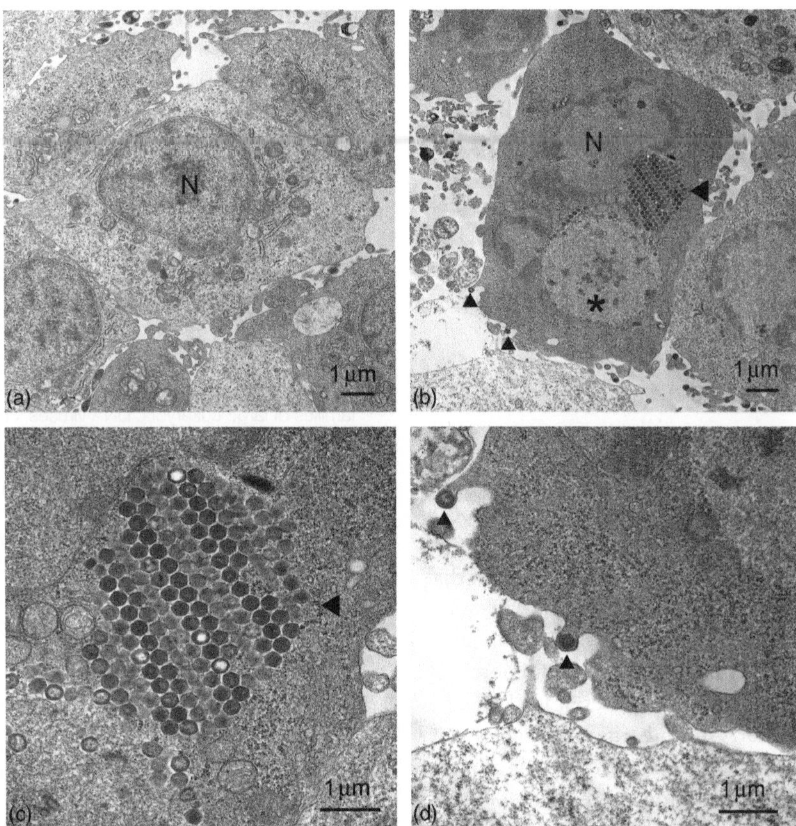

Figure 3 Electronmicrographic analysis of RCV-Z-infected FHM cells. FHM cells were infected with RCV-Z, a tentative member of the genus *Ranavirus*, at an MOI ~20 PFU/cell and fixed for electron microscopy at 9 h post infection. (a) Mock-infected FHM cells; (b) RCV-Z-infected cells showing chromatin condensation within the nucleus (N), a single large paracrystalline array within the cytoplasm (arrowhead), and a viral assembly site (asterisk); (c) enlargement of the paracrystalline array seen within (b); (d) enlargement of a virion in the process of budding from the plasma membrane. Arrowheads in (c) and (d) indicate virions within a paracrystalline array (c) or in the process of budding (d). Reproduced from Majji S, LaPatra S, Long SM, *et al.* (2006) Rana catesbeiana virus Z (RCV-Z): A novel pathogenic ranavirus. *Diseases of Aquatic Organisms* 73: 1–11, with permission from Inter-Research Science Center.

surrounded by a membrane and enclose not only the viral assembly site, but also mitochondria, rough and smooth endoplasmic reticulum, and electron-dense amorphous structures that likely contain viral DNA. Inclusion body-bearing cells are also found within the kidney, liver, heart, and gills, and are a hallmark of megalocytivirus infection. Interestingly, *in vitro* infection of cultured cells with RSIV does not result in the formation of membrane-bound inclusions, and the resulting histology is similar to that seen following ranavirus infection.

Effects of Virus Infection on Host Cell Function and Viability

FV3-infection results in a marked inhibition of cellular DNA, RNA, and protein synthesis and culminates in apoptotic cell death. The mechanisms responsible for these outcomes are slowly being resolved. Whereas the inhibition of cellular DNA synthesis appears to be a secondary consequence of the prior inhibition of host RNA and protein synthesis, the latter are likely to be the direct effect of virus infection. Inhibition of translation appears to be due to a series of events that include the synthesis of abundant amounts of highly efficient viral mRNAs, degradation of preexisting host transcripts, inhibition of host transcription, and the induction of a dsRNA-activated protein kinase (PKR) that phosphorylates and thereby inactivates eIF-2α. However, as viral translation persists in the face of host translational shutdown, FV3, like many other viruses, may encode factors that antagonize PKR-mediated effects. Most ranaviruses encode an eIF-2α homolog (vIF-2α) that, like its vaccinia virus counterpart (K3L), is thought to act as a pseudosubstrate that binds PKR, preventing phosphorylation of eIF-2α. Surprisingly, in at least one strain of FV3, vIF-2α is truncated by a deletion that removes the region homologous to eIF-2α and K3L, rendering it unable to bind to PKR. However, these strains maintain high levels of viral protein synthesis, suggesting that, as in the case of vaccinia virus, other viral proteins also play key roles in maintaining viral translation.

Apoptosis, as evidenced by DNA fragmentation, chromatin condensation, and membrane reversal, takes place in FV3-infected cells from ~6 to 9 h post infection. The immediate trigger for apoptosis is not known, but may involve PKR activation, translational shutdown, or another viral insult. Interestingly, both host cell shutdown and apoptosis can be triggered by productive infections as well as nonproductive infections with heat- or ultraviolet (UV)-inactivated virus, suggesting that a virion-associated protein is responsible. Receptor binding by virions may be sufficient to trigger these effects.

Concluding Remarks

For most of the 40 years since its discovery, FV3 has been studied, not for its impact on the infected host, but because it is the type species of a novel virus family that possesses certain features that can best (or only) be studied in this system. However, as other iridovirids are now emerging as important agents of disease in commercially and ecologically important animal species, it is likely that we will see concerted efforts to study their life cycles and, in the process, uncover novel aspects of iridovirus biogenesis. It is anticipated that the pioneering studies of the 1970s and 1980s will be extended using contemporary techniques such as asMOs and small interfering RNAs to elucidate the function of key viral proteins. In addition, study of the role of the host immune system in combating iridovirus infections in fish and amphibians may provide a way, not only to uncover potential virus-encoded immune evasion proteins, but also to elucidate key elements in antiviral immunity in lower vertebrates.

Further Reading

Allen MJ, Schroeder DC, Holden MTG, and Wilson WH (2005) Evolutionary history of the *Coccolithoviridae*. *Molecular Biology and Evolution* 23: 86–92.

Chinchar VG (2002) Ranaviruses (family *Iridoviridae*): Cold-blooded killers. *Archives of Virology* 147: 447–470.

Chinchar VG, Hyatt AD, Miyazaki T, and Williams T (in press) Family *Iridoviridae*: Poor viral relations no longer. *Current Topics in Microbiology and Immunology*.

Delhon G, Tulman ER, Afonso CL, *et al.* (2006) Genome of invertebrate iridescent virus type 3 (mosquito iridescent virus). *Journal of Virology* 80: 8439–8449.

Hyatt AD, Gould AR, Zupanovic Z, *et al.* (2000) Comparative studies of piscine and amphibian iridovirues. *Archives of Virology* 145: 301–331.

Iyer LM, Aravind L, and Koonin EV (2001) Common origin of four diverse families of large eukaryotic DNA viruses. *Journal of Virology* 75: 11720–11734.

Jancovich JK, Mao J, Chinchar VG, *et al.* (2003) Genomic sequence of a ranavirus (family *Iridoviridae*) associated with salamander mortalities in North America. *Virology* 316: 90–113.

Lua DT, Yasuike M, Hirono I, and Aoki T (2005) Transcription program of Red Sea bream iridovirus as revealed by DNA microarrays. *Journal of Virology* 79: 15151–15164.

Majji S, LaPatra S, Long SM, *et al.* (2006) Rana catesbeiana virus Z (RCV-Z): A novel pathogenic ranavirus. *Diseases of Aquatic Organisms* 73: 1–11.

Murti KG, Goorha R, and Chen M (1985) Interaction of frog virus 3 with the cytoskeleton. *Current Topics in Microbiology and Immunology* 116: 107–131.

Sample RC, Bryan L, Long S, *et al.* (2007) Inhibition of iridovirus protein synthesis and virus replication by antisense morpholino oligonucleotides targeted to the major capsid protein, the 18 kDa immediate early protein, and a viral homologue of RNA polymerase II. *Virology* 358: 311–320.

Stasiak K, Renault S, Demattei M-V, Bigot Y, and Federici BA (2003) Evidence for the evolution of ascoviruses from iridoviruses. *Journal of General Virology* 84: 2999–3009.

Tan WGH, Barkman TJ, Chinchar VG, and Essani K (2004) Comparative genomic analyses of frog virus 3, type species of the genus *Ranavirus* (family *Iridoviridae*). *Virology* 323: 70–84.

Williams T, Barbosa-Solomieu V, and Chinchar VG (2005) A decade of advances in iridovirus research. *Advances in Virus Research* 65: 173–248.

Kaposi's Sarcoma-Associated Herpesvirus: Molecular Biology

E Gellermann and T F Schulz, Hannover Medical School, Hannover, Germany

History

Kaposi's sarcoma (KS) is an endothelial cell-derived tumor. It is common in acquired immunodeficiency syndrome (AIDS) patients, and is known to have occurred before the arrival of human immunodeficiency virus (HIV). Today, it is one of the most frequently occurring tumors in Africa.

In 1994, following epidemiological studies that suggested the involvement of a sexually transmissible agent other than HIV in the pathogenesis of epidemic (AIDS-associated) KS, Y. Chang, and P. S. Moore used representational difference analysis to identify two sequence fragments of a then unknown rhadinovirus (or γ_2-herpesvirus). Epidemiological studies have now established KS-associated herpesvirus (KSHV, also known as human herpesvirus 8, HHV-8) as a necessary and indispensable causative factor in the pathogenesis of all forms of KS. In addition to 'epidemic' KS, these include the 'classic' form that is more common mainly in HIV-negative elderly men in some Mediterranean countries, the 'endemic' form known to have existed in East and Central Africa long before the spread of HIV, and 'iatrogenic' KS as seen, for example, in transplant recipients. KSHV is also essential in the pathogenesis of primary effusion lymphoma (PEL) and frequently found in the plasma cell variant of multicentric Castleman's disease (MCD). PEL is a rare lymphoma characterized by effusions in the pleural or abdominal cavity. MCD is a lymphoproliferative disorder characterized by multiple lesions that involve lymph nodes and spleen and featuring expanded germinal centers with B-cell proliferation and vascular proliferation.

KSHV (species *Human herpesvirus 8*) is a member of the genus *Rhadinovirus* in subfamily *Gammaherpesvirinae* of the family *Herpesviridae*. It was the first rhadinovirus discovered in humans and thereby also the first in an Old World primate. In the 12 years since this discovery, it has become apparent that most, if not all, Old World primate species, including the great apes (chimpanzee, gorilla and orangutan) harbor related rhadinoviruses. Of these, the genomes of two distinct rhesus macaque viruses have been sequenced completely (or nearly completely), and found to represent two major rhadinovirus lineages. Short-sequence fragments of rhadinovirus genomes representing these lineages have been found in most Old World primates, with the exception of humans, who appear to harbor only KSHV.

Virion Structure

KSHV has the characteristic morphological appearance of a herpesvirus, with a core containing the double-stranded, linear viral DNA, an icosahedral capsid (100–110 nm in diameter) containing 162 capsomers, the tegument that surrounds the capsid, and the envelope, in which glycoproteins are embedded. The virion contains at least 24 proteins, among which are five capsid proteins, eight envelope proteins, six tegument proteins, and five other proteins such as the regulator of transcriptional activation (RTA) and replication-associated protein (RAP or K-bZIP) that are not usually considered as structural virion proteins. The presence of RTA may account for the transient lytic gene expression program that occurs soon after virus entry into the cell. Among the envelope glycoproteins are gB and gpK8.1, which interact with cellular receptors like heparan sulphate or integrin molecules, thereby mediating entry of the virus into the host cell.

Genome Organization

The KSHV genome consists of a long unique region of approximately 140 kbp flanked by multiple, variable numbers of a tandemly repeated 801 bp terminal repeat (TR) (**Figure 1**). All the protein-coding open reading frames (ORFs) are located in the long unique region, and around 90 have been described so far. These are named ORF4 through ORF75 and K1 through K15. In addition, a noncoding RNA and an alternatively spliced RNA that is processed into 12 microRNAs (miRNAs) has been reported. miRNAs are a class of evolutionarily conserved, noncoding RNAs of about 22 nucleotides that are presumed to have regulatory functions.

The majority of KSHV genes are homologous to genes in the alpha- and betaherpesviruses. Their functions as structural virion proteins or nonstructural players during the lytic (productive) replication cycle are therefore assumed to be similar to those of their counterparts. These genes are not discussed in further detail in this article.

In addition, several genes are only found in KSHV or closely related rhadinoviruses of Old World primates. These are shaded gray in **Figure 1** and include genes encoding latency-associated nuclear antigen 1 (LANA-1; encoded by ORF73) and several homologs of cellular proteins, including D-type cyclin (vcyc; encoded by ORF72), a homolog of a cellular FLICE-inhibitory protein (vFLIP; encoded by ORF71), interleukin 6 (vIL6; encoded by K2), macrophage inflammatory proteins (vMIP1, vMIP2, and vMIP3; encoded by K6, K4.1, and K4.2), interferon regulatory factors (vIRF1, vIRF2, vIRF3, and vIRF4; encoded by K9, K11, K10.5, and K10), complement regulatory protein (KCP; encoded by ORF4), and two members of a class of membrane-associated ubiquitin E3 ligases involved in the down-regulation of major histocompatability class I (MHC I) and adhesion molecules from the cell surface (MIR1 and MIR2; encoded by K3 and K5). Ten of the miRNAs are located in the intergenic region between K12 and ORF71 and two are within K12.

The KSHV genome contains two lytic origins of DNA replication (Ori-Lyt), which show extensive sequence similarity to each other but are oriented in opposite directions. Each copy of TR contains a latent origin of DNA replication (Ori-Lat) that consists of two binding sites for LANA-1 and an adjacent region that is essential for replication.

Evolution

Rhadinoviruses closely related to KSHV appear to exist in many, if not all, Old World primate species. The phylogenetic relationships between DNA fragments from these viruses suggest the co-evolution of primate rhadinoviruses with their host species. Modern KSHV isolates can be grouped into five subtypes (A–E), three of which are still associated with particular human populations (B, Africa; D, Australasia; E, Native Americans), while

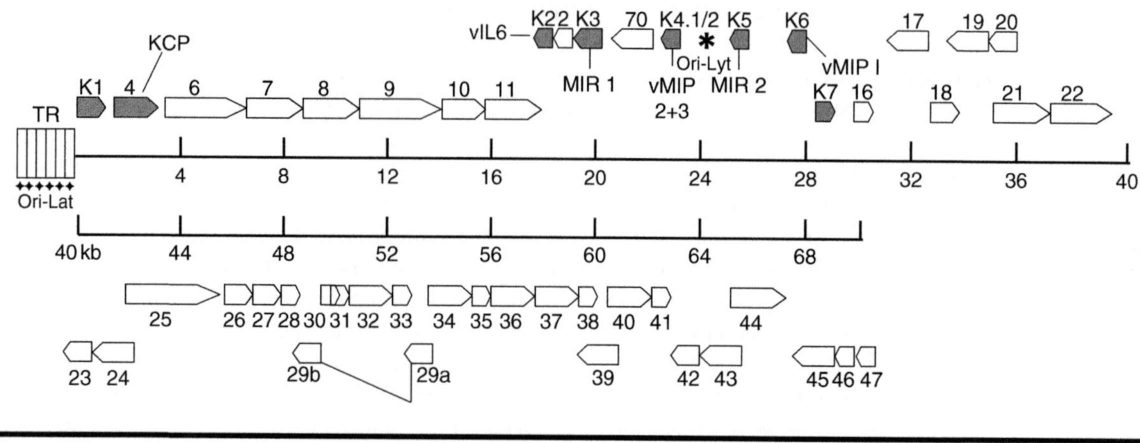

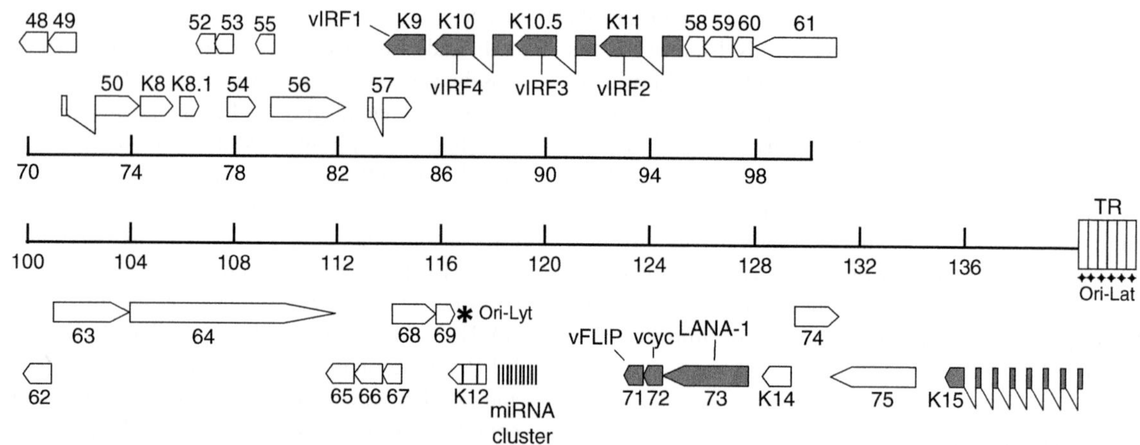

Figure 1 Schematic map of the KSHV genome. Protein-coding ORFs are denoted ORF4–ORF75 and K1–K15, and the names of encoded proteins are given for some (see RefSeq NC_003409 and NC_009333). ORFs typical of Old World or New World Primate rhadinoviruses are shaded gray and other marked features are described in the text. The scale is in kbp.

for two (A and C) such an association is no longer easily discernable, presumably as a result of extensive population mixing in Europe and in Near and Middle Eastern countries. However, one variant of subtype A (A5) is often seen in populations with links to Africa. With the exception of the K1 gene and the right end of the viral genome, variability between subtypes does not exceed a few percent.

Extensive recombination has evidently occurred between KSHV subtypes, and many modern isolates have chimeric genomes. At the right end of the genome, which includes the K15 gene, two types of very divergent and slowly evolving sequences have been found, suggesting that recombination events with rhadinoviruses other than KSHV could have taken place. By contrast, at the left end of the genome, the K1 gene shows evidence of extensive variation and evolution under selective pressure. It is possible that selective pressure exerted by T cells recognizing epitopes may have shaped the marked variability of the encoded membrane glycoprotein, but other explanations are also conceivable.

Tropism and Cell Entry

In vivo, KSHV has been found in human peripheral blood B and T cells, monocytes/macrophages, and endothelial cells. *In vitro*, a wide range of cell lines and primary cell cultures have been infected with KSHV, including human primary and immortalized endothelial cell cultures, epithelial cell lines and primary epithelial cell cultures, fibroblast cultures, monocytes/macrophages, and dendritic cells. Cell cultures from some other host species can also be infected *in vitro*, as can a strain of non-obese diabetic/severe combined immunodeficiency (NOD/SCID) mouse. These observations indicate that the restriction of KSHV for humans in the context of 'natural' infections is not determined at the level of cell entry.

The viral membrane glycoproteins gB (encoded by ORF8) and gpK8.1 (K8.1) are involved in entry of KSHV into target cells. Both interact with heparan sulfate, and gB also binds to $\alpha_3\beta_1$ integrin via an arginine-glycine-aspartate (RGD) motif. In addition, DC-SIGN (CD209) serves as a receptor for KSHV on macrophages and dendritic cells. xCT, the 12-transmembrane light chain of the human cystine/glutamate exchange transporter system x-c, serves as a receptor for KSHV fusion and entry.

The interaction of KSHV with heparan sulfate is thought to mediate the first, 'low-affinity' contact with the target cell, and subsequent binding of individual viral proteins to particular receptors then initiates the next stages of fusion and endocytosis of viral particles. Different entry routes may predominate, depending on the lineage of the infected cell. The importance of the interaction of gB with $\alpha_3\beta_1$ integrin may lie in the integrin-mediated activation of focal adhesion kinase (FAK), whose phosphorylation has been shown to play an important role during the early stages of KSHV entry, such as cellular uptake of viral DNA. The related kinase Pyk2 may also contribute to this process. The subsequent activation of Src, phosphatidylinositol 3-kinase (PI-3K), protein kinase C (PKC), RhoGTPase, mitogen-activated protein kinase (MEK), and extracellular signal-regulated kinase 1/2 (ERK1/2) facilitates virus uptake and immediate-early gene expression. KSHV-activated RhoGTPase in turn contributes to microtubular acetylation and leads to the modulation of microtubule dynamics required for movement of KSHV in the cytoplasm, and for delivery of viral DNA into the cell nucleus.

Latent Persistence and Replication

One of the characteristic features of KSHV is its propensity of quickly establishing a nonproductive, persistent state of infection following cell entry. Akin to its gammaherpesvirus (lymphocryptovirus) cousin, Epstein–Barr virus (EBV), but in contrast to alpha- and betaherpesviruses, this latent state of infection appears to be the default option for KSHV in most cell types, and presumably reflects its mode of long-term persistence *in vivo*. Latency is characterized by a restricted pattern of viral gene expression. Latent viral genes are depicted in **Figure 2** and include those for LANA-1, vcyc, vFLIP, and kaposin A and B, as well as the miRNAs. Several promoters, indicated in **Figure 2**, direct expression of cognate transcripts, of which two (P1 and P3) are strictly latent whereas two others (P2 and P4) become active during the lytic replication cycle. In addition, vIRF3 (also known as LANA-2) is part of the latent gene expression pattern in latently infected B cells. *In vivo*, expression of many of these latent genes has been demonstrated in the majority of KSHV-infected endothelial and spindle cells of KS, and the neoplastic B cells of PEL and MCD, using immunohistochemical or *in situ* hybridization techniques.

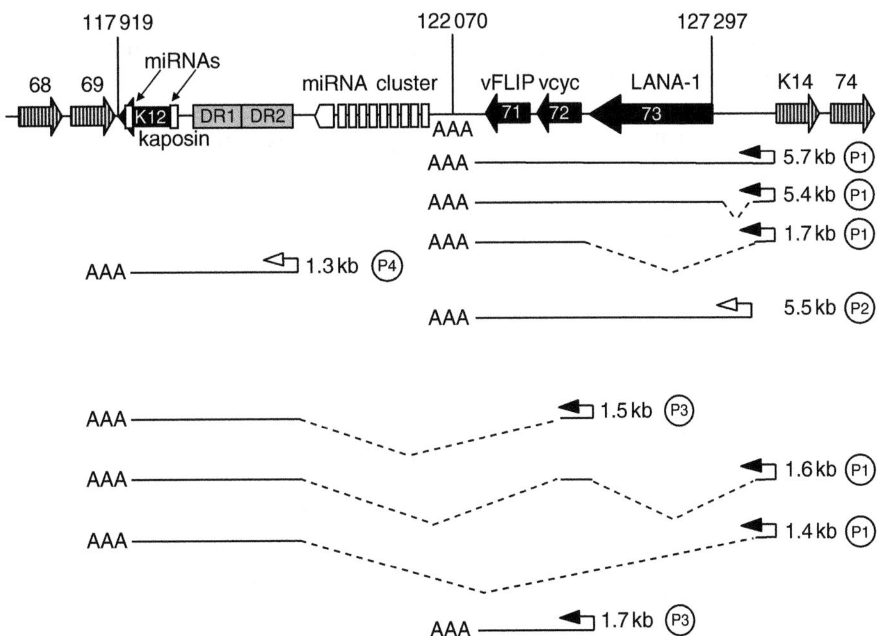

Figure 2 The latency gene cluster in the KSHV genome. The KSHV latency locus encodes four proteins indicated by black arrows (K12/kaposin, ORF71/vFLIP, ORF72/vcyc, and ORF73/LANA-1), as well as 12 miRNAs (white boxes), and is flanked by lytic genes (striped arrows). Lytic promoters are indicated by white arrows and latent promoters by black arrows. Readthrough of a viral polyadenylation signal located at nucleotide 122 070 leads to miRNA expression. P1 (127 880/86) gives rise to a precursor RNA that is spliced to produce 5.7 and 5.4 kb tricistronic (ORF71, ORF72, and ORF73) and 1.7 kb bicistronic (ORF71 and ORF72) mRNAs. Additionally, P1 drives two transcripts that could function as precursors for the miRNAs. As soon as RTA is expressed, P2 (127 610) is induced and gives rise to a 5.5 kb mRNA encompassing all three ORFs. Lytic transcription from P4 (118 758) leads to a 1.3 kb transcript coding for K12/kaposin. P3 (123 751/60) controls a 1.7 kb transcript encoding vFLIP and vcyc and a 1.5 kb transcript that has the potential to encode the miRNAs. Sequence coordinates are derived from GenBank NC_003409.

During latency, the KSHV genome persists as a circular episome, of which multiple copies (up to 100 or more) have been detected in latently infected PEL cell lines, although it is likely that the copy number varies widely in other cell types. Replication of the genome during S-phase is ensured by LANA-1, which binds to viral episomes via its C-terminal region and recruits a range of cellular factors, including the 'origin-recognition complex', which is involved in replication of cellular DNA. Two binding sites for LANA-1, spaced 39 bp apart, are located in each of the multiple 801 bp TRs that flank the viral genome. In addition, LANA-1 attaches viral episomes to mitotic chromosomes via its N-terminal domain, thus ensuring their segregation to daughter cells during mitosis as well as their replication during S-phase.

Possibly linked to its role in the replication of viral episomes is the ability of LANA-1 to promote the transition of the G1/S cell cycle checkpoint, thereby enabling viral DNA synthesis. The physical interaction of LANA-1 with the retinoblastoma protein pRB and with GSK-3β is thought to contribute to this property. In addition, LANA-1 binds to p53 and may thereby inactivate cell cycle checkpoint functions that could interfere with successful viral DNA replication, although this can also lead to genomic instability in the infected cell.

A further facet of the role of LANA-1 during latent persistence is its inhibition of the central activator of the lytic replication cycle, RTA, thereby contributing to maintenance of the latent infection. The ability of LANA-1 to interact with components of silenced and transcriptionally active chromatin, such as MeCP2 (which binds to methylated CpG motifs and thereby promotes chromatin silencing) or brd2/RING3, brd4/MCAP, brd4/HUNK, and brd3/orfX (which bind to acetylated histones in transcriptionally active chromatin regions), may be related to its role as a transcriptional repressor and activator, but such a link is currently still tenuous.

Among the other latent viral proteins is vcyc, which, although a homolog of a D-type cyclin, has a more pleiotropic range of functions. The ability of vcyc to promote the transition of the G1/S cell cycle checkpoint is linked to its phosphorylation of pRB (in concert with cdk4/6) and perhaps its inactivation of p27. This property may be important in the context of establishing S-phase conditions required for viral episome replication. However, recent findings clearly indicate that vcyc also contributes to the enhanced proliferation of virus-infected cells.

Translated from the same bicistronic mRNA as vcyc, vFLIP has anti-apoptotic functions and is a potent activator of the NFκB pathway. It is currently thought that the contribution of this protein to maintaining latency may involve protection against apoptosis, induced either by intracellular surveillance mechanisms detecting the replication of an extraneous agent or by cytotoxic T cells directed against latent viral proteins. Another role could be the repression of the lytic cycle via inhibition of the RTA promoter, which has been shown in several rhadinoviruses to be downregulated by the NFκB pathway. In PEL-derived cell lines, continuous activation of the NFκB pathway is required for the survival of these latently infected cells, and in endothelial cells NFκB activation may contribute to the formation of spindle cells, one of the histological hallmarks of KS lesions.

The fourth strictly latent protein is vIRF-3 (or LANA-2), which is expressed in a strictly B-cell-specific manner by K10.5 (**Figure 1**). Its contribution to latency is incompletely understood, but may involve its ability to interact with p53.

Unlike the latency proteins described above, expression of the transcript encoding kaposin A and B increases after initiation of the lytic replication cycle. It is currently not clear whether these two proteins play a role during latent persistence or whether they contribute to early stages of the lytic cycle. Both proteins influence intracellular signaling pathways, kaposin A by associating with cytohesin-1, a guanine nucleotide exchange factor for ARF GTPases, and regulator of integrin signaling that triggers the Erk/MAPK signaling cascade, and kaposin B by binding to MK2 and increasing the stability of cellular cytokine mRNAs that are regulated by the p38 MAPK pathway. Likewise, the roles of the KSHV miRNAs during latency, if any, are not known precisely, but may involve regulation of cellular and viral mRNAs.

Lytic Replication

Lytic replication results in the production of new KSHV virions. The central regulator of the lytic replication cycle is RTA, which is encoded by ORF50. Overexpression of RTA in latently infected cells is sufficient to trigger reactivation from latency. However, in many experimental systems additional inhibition of histone deacetylases will augment this process. As in other herpesviruses, lytic replication involves a 'rolling circle' mechanism to synthesize new viral DNA concatamers, from which individual linear viral genomes are cleaved during packaging into newly assembled capsids. This cleavage occurs in the TR region, thus ensuring the packaging of the complete long unique region of the genome flanked by a variable number (within limits) of copies of TR. Synthesis of new viral DNA occurs in nuclear replication compartments in the vicinity of ND10/POD domains, which undergo morphological changes during this process. As in other herpesviruses, the exact role of ND10/POD domains in productive herpesviral replication is not fully understood. KSHV proteins recruited to lytic replication compartments include the single-stranded DNA-binding protein (encoded by ORF6), the polymerase processivity factor (ORF59), the DNA polymerase (ORF9), the primase-associated factor (ORF40/41), the primase (ORF56), and the DNA helicase (ORF44), which together mediate viral DNA replication. Other viral proteins are involved in aspects of this process. Thus RAP (K-bZIP; encoded by K8), a structural but not functional homolog of EBV ZTA, is also associated with viral replication compartments and blocks the cell cycle at the G1/S transition by increasing expression levels of p21 and C/EBPα as well as by associating with and downmodulating the activity of cyclin-dependent kinase 2 (cdk-2). The ability to block the cell cycle at G1/S appears to be an important early step in the lytic cycle for most herpesviruses, presumably because it facilitates diversion of the nucleotide pools from cellular to viral DNA replication.

Immune Evasion

To establish an infection in higher vertebrates, viruses must interfere with the host cell immune system. Mechanisms of immune evasion are especially important for viruses that establish long-term infections (e.g., herpesviruses). Like other herpesviruses, KSHV employs several mechanisms to protect infected cells from attacks by the immune system.

Latency represents one strategy to escape from the immune system. During latency only a minimal number of viral proteins are expressed, thus reducing the number of antigens that are presented to the immune system. Another strategy favored by herpesviruses is to downregulate MHC I proteins on the surface of infected cells, thereby decreasing antigen presentation and the recognition of infected cells by cytotoxic T lymphocytes.

KSHV encodes two proteins, modulator of immune recognition 1 and 2 (MIR1 and MIR2; encoded by K3 and K5), which are involved in protecting virus-infected cells against natural killer (NK) cells or cytotoxic T lymphocytes. They are both expressed immediately after viral reactivation. While K3 efficiently modulates multiple MHC alleles, K5 affects the expression not only of HLA-A, but also of ICAM-1 and B7.2. These membrane proteins downregulate MHC I molecules by increasing their endocytosis and degradation rate. This mechanism differs from that of other known viral inhibitors of MHC I expression, which either interfere with the synthesis of MHC I chains or retain them in the endoplasmic reticulum. MIR1 and MIR2 are members of the family of enzyme type 3 (E3) ubiquitin ligases, which regulate the last step of ubiquitination, and thereby increase the degradation rate of MHC I proteins.

The complement system represents a potentially important antiviral mechanism. KSHV encodes a lytic protein called KSHV complement-control protein (KCP; encoded by ORF4), which is incorporated into virions and inhibits activation of the complement cascade, thereby protecting virions and virus-infected cells from complement-mediated opsonisation or lysis.

Kaposin B, noted above for its ability to modulate the p38 MAPK pathway through its interaction with MK2, has an immunomodulatory function. It is able to increase the expression of certain cytokines by stabilizing their mRNAs. The turnover of some cytokine transcripts is regulated via AU-rich elements (AREs) present in the 3′ nontranslated region, which target the RNAs for degradation. The p38 pathway, via MK2, regulates this turnover. By binding to MK2, kaposin B inhibits the decay of cytokine mRNAs containing AREs. This leads to the enhanced production of certain cytokines, such as IL6 and granulocyte macrophage colony-stimulating factor (GM-CSF).

The viral homolog of IL6, vIL6, binds directly to gp130, the signal-transducing chain of the cellular IL6 receptor and a variety of other cellular cytokine receptors. Unlike cellular IL6, which requires binding to the IL6 receptor α-chain specifically to trigger only the α-chain/gp130 receptor complex, vIL6 has a much broader range of cellular targets and more pleiotropic effects. It is a potent stimulant for B-cell proliferation and also acts on other bone-marrow-derived cells.

Finally, three other KSHV genes show significant homologies to cellular chemokines. Their encoded proteins (vMIP1, vMIP2, and vMIP3) modulate the immune system. They are able to bind to chemokine receptors, act as chemo-attractants, and presumably affect the composition of leukocyte infiltrates around KSHV-infected cells. vMIP1 and vMIP2 have also been shown to have angiogenic properties and may therefore contribute to angiogenesis in KS lesions.

Inhibition of Apoptosis

Apoptosis is necessary for the elimination of cells that are no longer required or have become damaged. Many viruses modulate apoptotic pathways to prevent the premature death of an infected cell, which must be avoided in order to complete the replication cycle.

In PEL cells, pharmacological inhibition of the NFκB pathway leads to apoptosis, suggesting that NFκB, which is activated by several viral proteins, plays an important role in preventing apoptosis in KSHV-infected cells. Several viral proteins have been shown to have an anti-apoptotic function, including vFLIP, which blocks Fas- and Fas-ligand-induced apoptosis at the level of procaspase 8 activation and which is also a potent inducer of NFκB (see above).

Among the KSHV vIRFs, vIRF-1 and vIRF-2 both inhibit interferon signaling and subsequently prevent induction of apoptosis. Both vIRF-1 and vIRF-3 inhibit the activation of p53-dependent promoters.

The product of K7, which is the viral inhibitor of apoptosis (vIAP), is thought to operate as a molecular adaptor, bringing together Bcl-2 and effector caspases. The viral homolog of human Bcl-2, vBcl-2 (encoded by ORF16) is expressed only in late-stage KS lesions. It inhibits virus-induced apoptosis and the pro-death protein BAX.

KSHV-Associated Diseases

KSHV is responsible for KS, a tumor of endothelial origin, as well as for PEL, a rare B-cell lymphoma. It is also found in the plasma cell variant of MCD and thought to play an essential role in the pathogenesis of this B-cell-derived lymphoproliferative disorder. Additional sightings of KSHV in a variety of other proliferative or neoplastic disorders have not been confirmed. However, KSHV may play a role in the occasional case of hemophagocytic syndrome and bone marrow failure in immunosuppressed individuals.

Kaposi's Sarcoma

With very few exceptions, KSHV sequences have been detected by PCR in all KS cases. The few exceptions are held to reflect technical problems or the variable load of KSHV DNA in this pleomorphic tumor, which, in addition to KSHV-infected spindle and endothelial cells, contains many other (uninfected) cellular lineages. KSHV adopts a latent gene-expression pattern in the majority of infected endothelial or spindle cells. As shown by immunohistochemistry and *in situ* hybridization studies, the latent genes discussed above (i.e., LANA-1, vcyc, vFLIP, and kaposin) are expressed in the majority of cells, whereas immediate-early, early, or late genes or proteins are expressed only in a small percentage. Most KS tumors also show a uniform pattern of TR lengths in the circular viral episome, consistent with viral latency. This could suggest that viral latent proteins play a key role in the atypical differentiation of endothelial cells into spindle cells, shown by gene expression array studies to represent an intermediate endothelial cell differentiation stage between vascular and lymphatic endothelial cells. However, it is conceivable that certain viral proteins of the lytic cycle, although apparently only expressed in a few cells, could also contribute to pathogenesis by either exerting a paracrine effect (if secreted, as in the case of vIL6 or vMIPs) or inducing the expression of cellular growth factors that could then act on neighboring latently infected cells (e.g., the viral chemokine receptor homolog encoded by ORF74 induces the secretion of VEGF). Given the inefficient persistence of KSHV in latently infected endothelial cell cultures *in vitro*, it is also conceivable that periodic lytic reactivation of KSHV is required to infect new endothelial cells and thereby maintain KSHV in this population. On the other hand, *in vitro* experiments suggest that, in spite of being lost rapidly from most infected dividing cells in culture, a small percentage of such cells can remain stably infected, possibly as a result of epigenetic modifications of the viral genome. It is therefore equally conceivable that the latently infected spindle cells in KS tumors are derived from cells that harbor such epigenetically modified latent viral genomes.

Primary Effusion Lymphoma

The presence of KSHV DNA is a defining feature of PEL, a rare lymphoma entity characterized by a lymphomatous effusion in the pleural or abdominal cavity containing malignant B cells that have a rearranged immunoglobulin gene and often express syndecan-1/CD138 and MUM1/IRF4, two markers characteristic for postgerminal center B cells. The majority of PEL samples are dually infected with KSHV and EBV, but 'KSHV-only' cases also occur. KSHV adopts a latent gene-expression pattern in the majority of PEL cells and in permanent cell lines derived from PEL samples. In addition to the latent proteins found in KS (LANA-1, vcyc, vFLIP; see above and **Figure 2**), vIRF3 is expressed in latently infected PEL cells, indicating a B-cell-specific pattern of latent gene expression. A minority of PEL cells, or PEL-derived cell lines, can spontaneously undergo productive viral replication, or be induced to do so by treatment with phorbol esters, histone deacetylase inhibitors such as sodium butyrate, or overexpression of RTA, the central regulator of the productive viral replication cycle. Such PEL-derived cell lines have for a long time been a source of virus for *in vitro* experiments.

Multicentric Castleman's Disease

KSHV is frequently found in the B-cell population of the plasma cell variant of MCD. As in KS and PEL, the majority of virus-infected cells are in a latent stage of infection, expressing characteristic latency proteins such as LANA-1, but also the B-cell-specific latent protein vIRF3. However, unlike KS and PEL, in which only a very small proportion of virus-infected cells appear to undergo productive viral replication at any point in time (see above), a sizable number of KSHV-infected B cells in MCD spontaneously express viral proteins that are thought to reflect productive viral replication. In addition, case reports suggest that MCD activity is reflected by the level of peripheral blood viral load, indicating that MCD may be the manifestation of, or be accompanied by, productive viral reactivation in B cells. Among the 'lytic' viral proteins expressed in MCD B cells is vIL6, a potent stimulator of B-cell growth (see above).

Further Reading

Ablashi DV, Chatlynne LG, Whitman JE, Jr., and Cesarman E (2002) Spectrum of Kaposi's sarcoma-associated herpesvirus, or human herpesvirus 8, diseases. *Clinical Microbiology Reviews* 15: 439–464.

Brinkmann MM and Schulz TF (2006) Regulation of intracellular signalling by the terminal membrane proteins of members of the *Gammaherpesvirinae*. *Journal of General Virology* 87: 1047–1074.

Damania B (2004) Oncogenic gamma-herpesviruses: Comparison of viral proteins involved in tumorigenesis. *Nature Reviews Microbiology* 2: 656–668.

Dourmishev LA, Dourmishev AL, Palmeri D, Schwartz RA, and Lukac DM (2003) Molecular genetics of Kaposi's sarcoma-associated herpesvirus (human herpesvirus-8) epidemiology and pathogenesis. *Microbiology and Molecular Biology Reviews* 67: 175–212.

Edelman DC (2005) Human herpesvirus 8 – A novel human pathogen. *Virology Journal* 2: 78.

Kempf W and Adams V (1996) Viruses in the pathogenesis of Kaposi's sarcoma – A review. *Biochemical and Molecular Medicine* 58: 1–12.

Moore PS and Chang Y (2001) Kaposi's sarcoma-associated herpesvirus. In: Knipe DM and Howley PM (eds.) *Fields Virology*, 4th edn., pp. 2803–2833. Philadelphia, PA: Lippincott Williams and Wilkins.

Rezaee SA, Cunningham C, Davison AJ, and Blackbourn DJ (2006) Kaposi's sarcoma-associated herpesvirus immune modulation: An overview. *Journal of General Virology* 87: 1781–1804.

Schulz TF (2006) The pleiotropic effects of Kaposi's sarcoma herpesvirus. *Journal of Pathology* 208: 187–198.

Mononegavirales

A J Easton and R Ling, University of Warwick, Coventry, UK

Glossary

Antigenome An RNA molecule complementary in sequence to the single-stranded genomic nucleic acid of a virus.

Nucleocapsid Internal structure of a virus containing its genetic material and one or more proteins. It may exist in infected cells independently of virus particles.

Introduction

The order *Mononegavirales* comprises four families of viruses, the *Bornaviridae*, *Rhabdoviridae*, *Filoviridae*, and *Paramyxoviridae*. This order of viruses includes the agents of a wide range of diseases affecting human, animals, and plants. The disease may be highly characteristic of the virus (e.g., measles, mumps) or of a more general nature (e.g., respiratory disease caused by various paramyxoviruses). Disease and host range vary to some extent between the viral families; these and other differences between the families are summarized in **Table 1**. The feature which unites these viruses is the presence of a genome consisting of a single RNA molecule which is of opposite polarity to the mRNAs that they encode. The genomic RNA, which has inverted terminal complementary repeats, ranges in size from about 9 to 19 kbp and is usually present in a helical nucleocapsid. Typically, 93–99% of the genomic sequence encodes proteins with the genes encoding the structural proteins common to all members of the order arranged in the same relative position in the genome with respect to each other.

The genome RNA is present in the form of a nucleocapsid complex with at least one structural protein and this complex in turn is enclosed in a lipid membrane which in all cases except the bornaviruses has spikes of about 5–10 nm comprised of the viral glycoproteins. The nucleocapsids are incorporated into virions via interactions mediated by a viral matrix protein which acts as a bridge between the complex and the virus glycoprotein(s). The lipid membrane of the virus is derived from the host cell plasma membrane or internal membranes during a budding process. The shape of the viral particles varies although certain forms predominate in different families of viruses.

Taxonomy

The order *Mononegavirales* is divided into four families on the basis of features such as those listed in **Table 1** and common aspects of the genetic organization, shown in **Figures 1** and **2**. There is little conserved sequence between viruses in different families except for regions of the L (polymerase) protein. A phylogenetic tree based on part of the L protein sequence is shown in **Figure 3** in which the different families can be separated. The taxonomic structure of the order is based on a range of attributes including morphological features of the virion and shape of nucleocapsid rather than the sequence similarities. Consequently, the pneumoviruses are classified as a subfamily of the *Paramyxoviridae* despite the L protein sequence apparently being more similar to that of the filoviruses. A large number of viruses that belong to the

Table 1 Distinguishing features of viruses belonging to the indicated families

	Bornaviridae	*Filoviridae*	*Paramyxoviridae*	*Rhabdoviridae*
Genome size (approximate, nt)	8900	19 000	13 000–18 000	11 000–15 000
Virion morphology	90 nm, spherical	Filamentous, circular, 6-shaped	Pleomorphic, spherical, filamentous	Bullet-shaped or bacilliform
Replication site	Nucleus	Cytoplasm	Cytoplasm	Cytoplasm except nucleorhabdoviruses
Host range	Horses, sheep, cats, ostriches	Primates, (bats?)	Vertebrates	Vertebrates, vertebrates + invertebrates, plants + invertebrates
Growth in cell culture	Noncytopathic	Noncytopathic	Lytic, often with syncytium formation	Lytic for animal viruses, rapid for vesiculoviruses
Pathogenic potential	Behavioral disturbances to severe encephalomyelitis	Hemorrhagic fever	Respiratory or neurological illness	Mild febrile illness to fatal neurological disease

order have not been classified as yet in a particular genus. Examples of these are tupaia rhabdovirus and Flanders virus in the *Rhabdoviridae*, various paramyxoviruses isolated from rodents such as tupaia paramyxovirus, Mossman virus, Beilong virus, and J virus, whose L gene sequences cluster near the henipaviruses and morbilliviruses, and fer-de-lance virus with an L gene sequence most similar to that of the respiroviruses.

Virus Structure

All mononegaviruses have an enveloped virus particle which contains glycoprotein spikes projecting from the surface, though Borna disease virus has a less well defined structure with no spikes and an electron-dense center. The shapes of the viruses vary considerably. Viruses in the family *Paramyxoviridae* are highly pleomorphic with the major form usually being roughly spherical, with filamentous and irregular shaped forms also being present. Rhabdoviruses have a more defined shape than the viruses in the other families, being bullet shaped or bacilliform and with a uniform width and length determined by the genome length. Filoviruses such as Ebola virus have a basic bacilliform shape of relatively uniform width (~80 nm) but are often folded into branched, circular, and 6-shaped forms and may form long filaments. Internally, several layers are apparent, the helical outer layer surrounding an electron-dense layer and an axial channel (**Figure 4**).

The nucleocapsid complex structures of the mononegaviruses are all helical which in viruses in the subfamily *Paramyxovirinae* give rise to a herringbone pattern. In contrast, rhabdoviruses and pneumoviruses have a more flexible and less regular structure. Even the more regular structures are nevertheless believed to exist in dynamic states allowing structural changes during RNA synthesis. In the pneumoviruses the width of the nucleocapsid is generally narrower and the helical pitch greater than in other paramyxoviruses, enabling these viruses to be distinguished prior to their molecular characterization. Rhabdovirus nucleocapsids are assembled into a more structured skeleton form in virus particles due to association with a matrix protein.

Features of the Viral Genome RNA

The precise details of the genome RNA vary between viruses with a wide range of lengths possible. An unusual feature of some paramyxoviruses is that the genome length is always a multiple of six nucleotides, referred to as the 'rule of six'. This is due to the nucleoprotein which binds the RNA associating with units of six nucleotides. However, most viruses in the order do not conform to this rule.

Promoter Sequences

The virus RNA contains three functional promoter sequences, the genomic promoters at the 3′ ends of the genome and the antigenome which directs replication of the virion RNA, and the transcriptional promoter which directs transcription from the genome. The genomic and antigenomic promoters share at least a partial sequence similarity due to the inverse complementarity of the virion RNA ends. To date the precise sequence requirements for the genomic and transcriptional promoters have not been defined precisely for most viruses and it is likely that they overlap. In some viruses, a short leader RNA transcript is produced from the genomic promoter. These transcripts, like the full-length antigenome and genome, are neither capped nor, other than for the

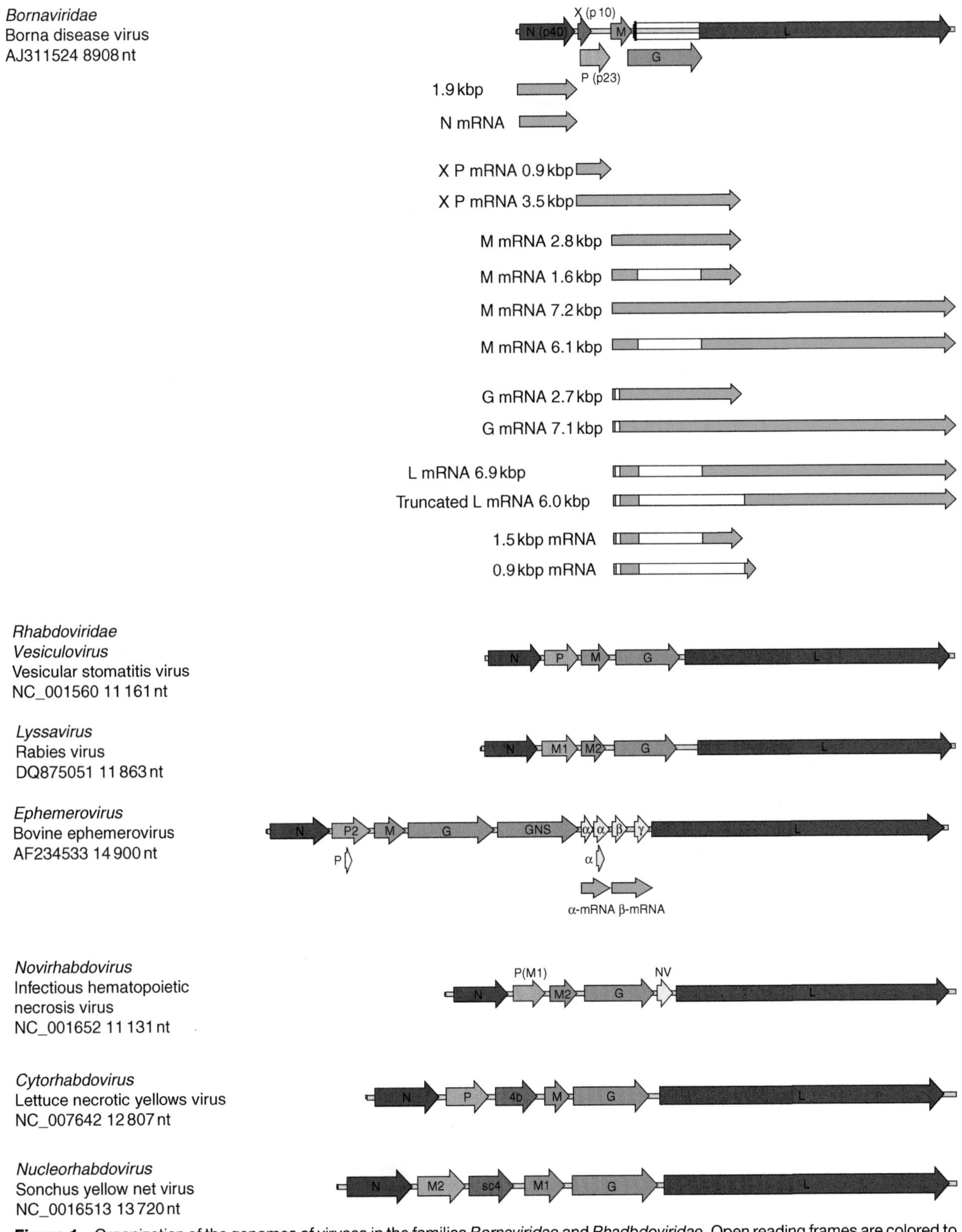

Figure 1 Organization of the genomes of viruses in the families *Bornaviridae* and *Rhadbdoviridae*. Open reading frames are colored to indicate proteins that are believed to have similar functions in the different viruses. Messenger RNAs are only shown where they differ from the monocistronic transcripts normally observed for viruses in the order *Mononegavirales*. Introns in Borna disease virus are indicated by unshaded regions. All the rhabdoviruses have potential second open reading frames in the second (phosphoprotein) gene although this is only shown for bovine ephemerovirus. The P gene was formely known as NS in vesiculoviruses and is sometimes known as M1 in lyssaviruses and novirhabdoviruses and M2 in nucleorhabdoviruses.

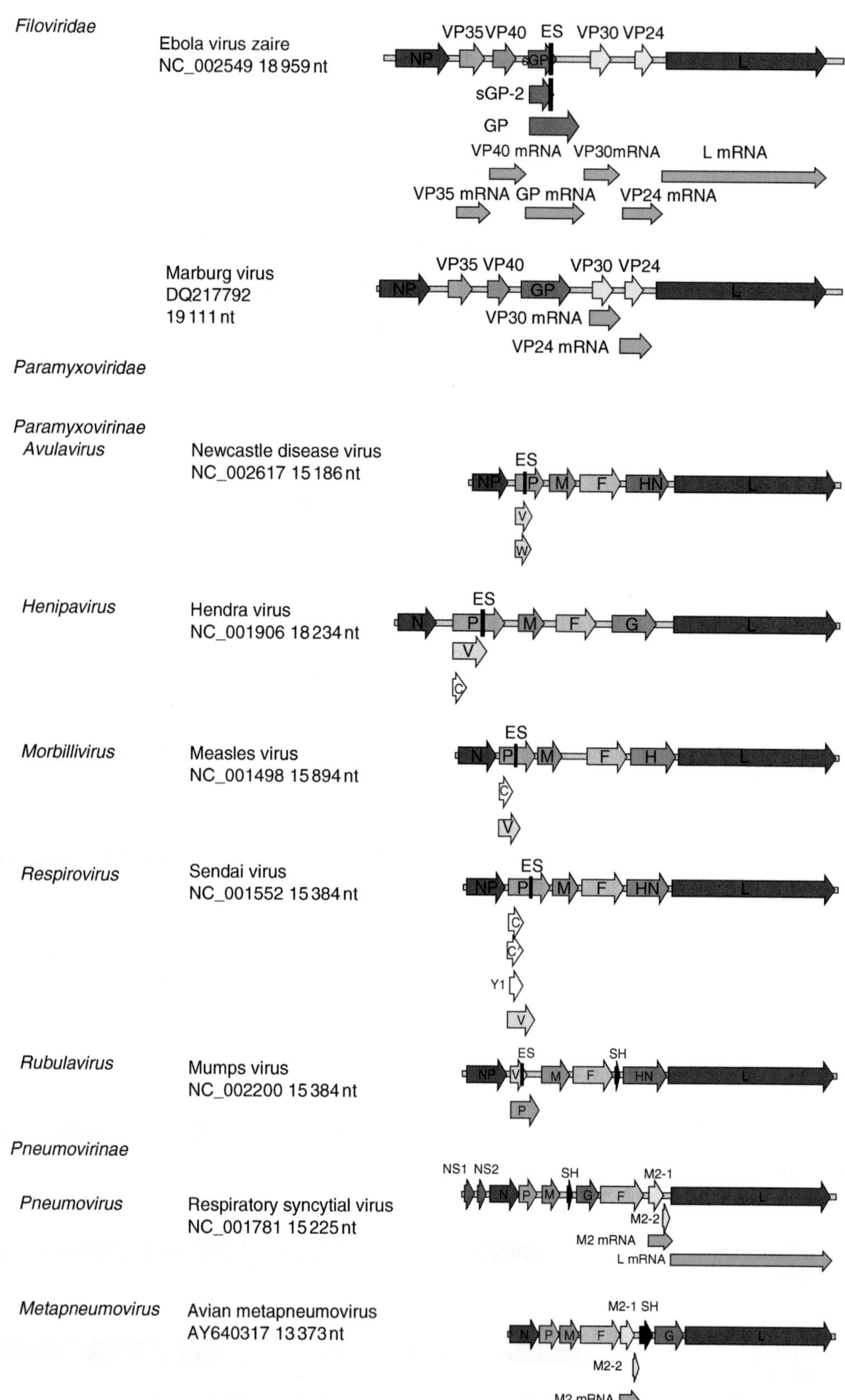

Figure 2 Genomic organization of viruses in the families *Filoviridae* and *Paramyxoviridae*. Open reading frames (ORFs) are colored as in **Figure 1**. Messenger RNAs are again indicated only where they have unusual features such as overlaps between adjacent genes or encoding two ORFs. However, all P genes of viruses in the *Paramyxovirinae* subfamily encode more than one protein, so these mRNAs are not shown. ES indicates an editing site where nontemplated residues are added to generate additional proteins. The ORF in the same line as the ORFs from other genes is the one encoded by the unedited RNA.

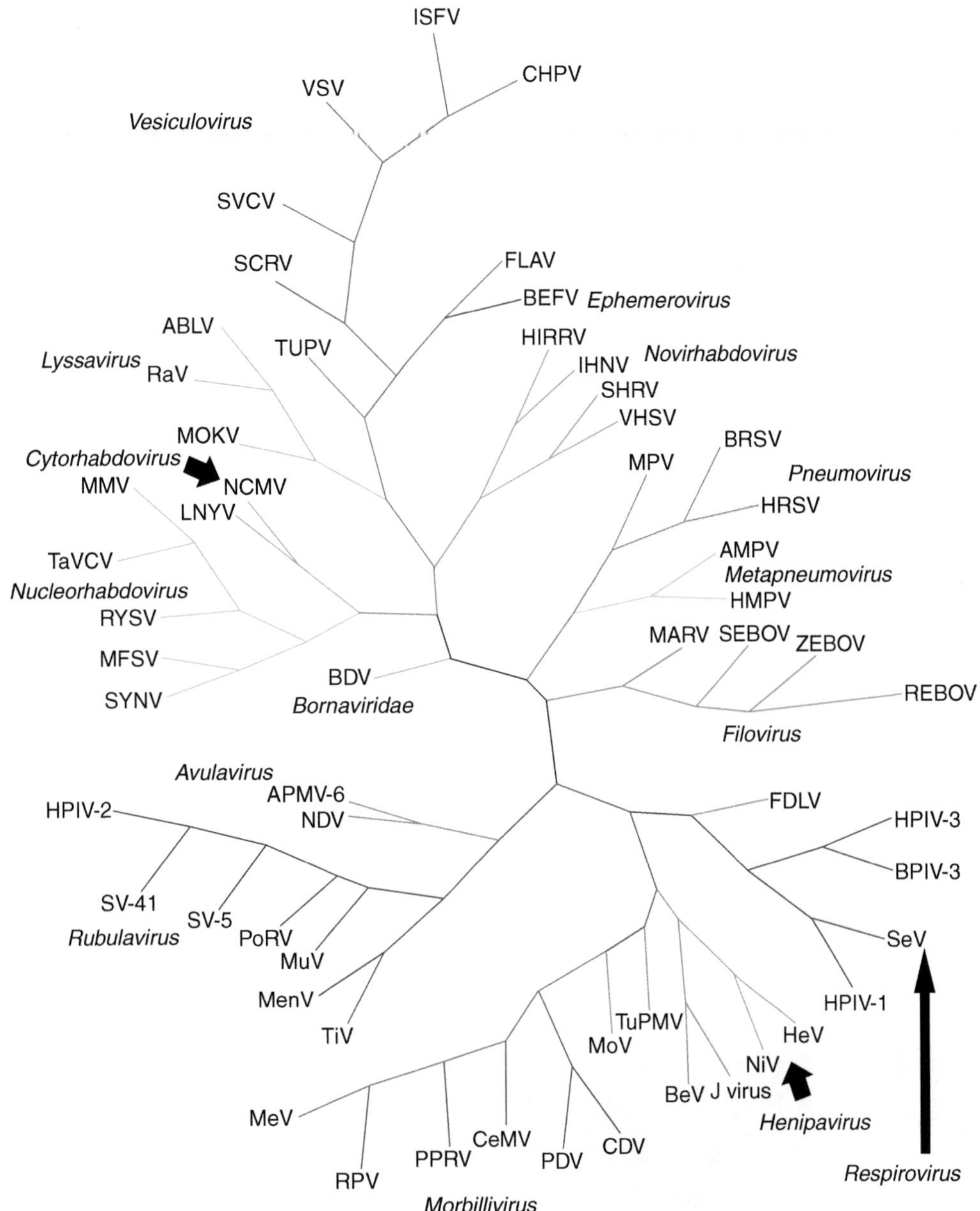

Figure 3 Phylogenetic tree of the largest region of the L protein shared by viruses in each of the mononegavirus genera. Tree branches of members in the same subfamily or family are shaded in similar colors. Virus abbreviations are those in the *Eighth Report of the International Committee on the Taxonomy of Viruses*. Borna viruses: BDV, Borna disease virus. Filoviruses: MARV, Lake Victoria marburgvirus; REBOV, Reston ebolavirus; SEBOV, Sudan ebolavirus; ZEBOV, Zaire ebolavirus Metapneumoviruses: AMPV, avian metapneumovirus; HMPV, human metapneumovirus. Pneumoviruses: BRSV, bovine respiratory syncytial virus; HRSV, human respiratory syncytial virus; MPV, murine pneumonia virus. Vesiculoviruses: CHPV, Chandipura virus; ISFV, Isfahan virus; SVCV, spring viremia of carp virus; SCRV, Siniperca chuatsi virus; VSV, vesicular stomatitis virus (San Juan). Lyssaviruses: ABLV, Australian bat lyssavirus; MOKV, Mokola virus; RaV, rabies virus. Novirhabdoviruses: HIRRV, hirame rhabdovirus; IHNV, infectious hematopoietic necrosis virus; SHRV, snakehead rhabdovirus; VHSV, viral hemorrhagic septicaemia virus. Ephemerovirus: BEFV, bovine ephemeral fever virus. Unassigned animal rhabdoviruses: FLAV, Flanders virus; TUPV, Tupaia rhabdovirus. Cytorhabdoviruses: LNYV, lettuce necrotic yellows virus; NCMV, northern cereal mosaic virus. Nucleorhabdoviruses: MFSV, maize fine streak virus; MMV, maize mosaic virus; RYSV, rice yellow stunt virus; SYNV, Sonchus yellow net virus; TaVCV, taro vein chlorosis virus. Henipaviruses: HeV, Hendra virus; NiV, Nipah virus. Avulaviruses: NDV, Newcastle disease virus; APMV-6, avian paramyxovirus 6. Rubulaviruses: HPIV-2, human parainfluenza virus 2; MenV, menangle virus; MuV, mumps virus; PoRV, porcine rubulavirus; SV-5, simian virus 5; SV-41, simian virus 41; TiV, Tioman virus. Respiroviruses: BPIV-3, bovine parainfluenza virus 3; HPIV-1(3), human parainfluenza virus 1(3); SeV, Sendai virus. Morbilliviruses: CDV, canine distemper virus; CeMV, cetacean morbillivrus; MeV, measles virus; PPRV, peste-des-petit-ruminants virus; PDV, phocine distemper virus; RPV, rinderpest virus. Unclassified paramyxoviruses: BeV, Beilong virus; FDLV, fer-de-lance virus; MoV, Mossman virus; TuPMV, Tupaia paramyxovirus.

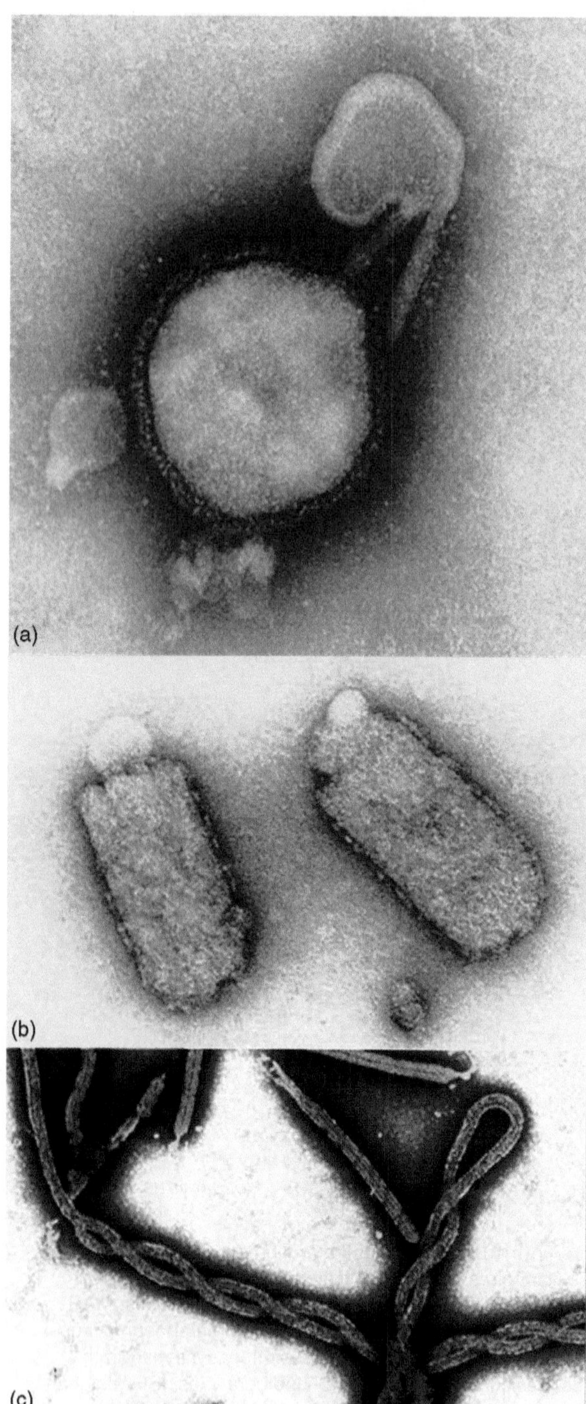

Figure 4 Electron micrographs showing the differing virus particle morphologies seen within the order *Mononegavirales*. (a) Electron micrograph of human parainfluenza virus type 3, a member of the family *Paramyxoviridae*. (b) Electron micrograph of rabies virus, a member of the family *Rhabdoviridae*. (c) Electron micrograph of Ebola virus, a member of the family *Filoviridae*. (a) Reproduced from Henrickson KJ (2003) Parainfluenza viruses. *Clinical Microbiology Reviews* 16: 242–264, with permission from American Society for Microbiology. (b) Reproduced from Mebatsion T, Weiland F, and Conzelmann K-K (1999) Matrix protein of rabies virus is responsible for the assembly and budding of bullet-shaped particles and interacts with the transmembrane spike glycoprotein G. *Journal of Virology* 73: 242–250, with permission from American Society for Microbiology. (c) Reprinted by permission from Macmillan Publishers Ltd: Nature Medicine, vol. 10, pp. S110–S121, Geisbert TW and Jahrling PB, Exotic emerging viral diseases: Progress and challenges, copyright (2004).

nucleorhabdoviruses, polyadenylated. Originally it was believed that the transcriptional promoter initiated transcription from the genomic 3′ terminus producing the short leader RNA and then reinitiated transcription from the first gene to produce the first capped mRNA with a termination and reinitiation cycle at each gene junction to produce the remaining mRNAs as described below. More recently this has become less certain and the transcriptional promoter may cause initiation of transcription to occur directly from the first mRNA start signal.

Gene Start and Gene End Signals

Each gene is flanked by a transcription start and termination sequence. The transcription start and transcription termination/polyadenylation signals, though differing between viruses, are conserved between different genes of the same virus, each being *c.* 10 nt in length. The gene end/polyadenylation signals have a series of uridine residues at the end of the gene (the 5′ end in the genomic sense) that are copied by a 'stuttering' mechanism to give a poly(A) tail.

Intergenic Regions

Most viruses of the order *Mononegavirales* have untranscribed regions between the gene end and gene start signals. These occur between all genes in most of the viruses. However, in the *Bornaviridae*, most of the transcription start sites are located upstream of the transcription end/polyadenylation site of the upstream gene so the only intergenic region is between the N and P genes. In the filoviruses and respiratory syncytial viruses, some transcripts also overlap. In the latter case, the overlap is confined to the L gene transcript starting upstream of the M2 gene, in Marburg virus the mRNAs for the VP30 and Vp24 proteins overlap, and in Ebola virus the VP35/VP40, GP/VP30, and VP24/L transcripts overlap. In some viruses, the size of the intergenic regions is conserved between all the genes, for example, two nucleotides in vesicular stomatitis virus (VSV) and three nucleotides in respiroviruses and henipaviruses, whereas in most viruses the length of the intergenic regions is variable.

Transcription

Transcription is initiated from the single promoter near the 3′ end of the genome and requires the N, P, and

L proteins in addition to the genome as a minimal requirement. In pneumoviruses, transcription of longer mRNAs is enhanced by the M2-1 protein and in nucleorhabdoviruses, nucleocapsid complexes isolated from nuclei, but not virus particles, are transcriptionally active, thus suggesting a requirement for cellular proteins. The involvement of cellular proteins in virus transcription has been suggested for some other viruses whereas transcription can be achieved *in vitro* with nucleocapsids isolated from VSV and cytorhabdoviruses.

The precise start point of transcription is not known. It is now considered possible that transcription is initiated directly from the start of the first gene, though it is possible for some viruses that transcription begins at the $3'$ terminus of the genome with production of a short non-polyadenylated leader RNA. The polymerase initiates mRNA synthesis at the first transcription initiation signal sequence. Having transcribed the gene the polymerase encounters the conserved transcription termination sequence at which point it iteratively adds a polyadenylate tail at the sequence of U residues and ceases transcription. The polymerase then either detaches from the template and has to reinitiate at the $3'$ transcriptional promoter or reinitiates transcription from the next gene start signal. In the case of the L gene of respiratory syncytial virus and most genes in Borna disease virus this gene start signal may be upstream of the gene end signal. This overlapping organization may reduce the level of expression of the downstream gene. More typically an intergenic sequence separates the two genes and the polymerase reinitiates transcription from the downstream gene. In a small proportion of cases the polymerase fails to terminate transcription as normal and continues to the end of the next gene generating a polycistronic transcript. The overall effect of this transcription process is to produce a gradient of transcription with those genes closer to the promoter being expressed at higher levels than those located further away. The steepness of this gradient varies considerably between viruses and between the cell infected (e.g., in measles vs. subacute sclerosing panencephalitis (SSPE)). The attenuation may vary between different genes, so the decline in transcript levels may be in a series of steps rather than a smooth gradient.

Splicing of mRNA

Most viruses in this order replicate in the cytoplasm and produce only unspliced RNA molecules. Nucleorhabdoviruses replicate in the nucleus but splicing has not been observed in these viruses. However, Borna disease virus replicates in the nucleus and three introns have been identified (**Figure 1**). The smallest and most $5'$ terminal (mRNA sense) is spliced out to remove the part of the M open reading frame (ORF) and allow translation of the RNA to give the G glycoprotein. A doubly spliced RNA with both the first intron and a second one removing most of the G ORF produces the L mRNA. A third longer intron shares the splice donor site with the second one but has a different splice acceptor site resulting in larger intron that could result in a truncated L protein or proteins in either of the other two ORFs but these have not been observed. The use of spliced RNAs results in a larger number of transcripts beyond that normally observed for a virus in the order *Mononegavirales* with five gene start sites.

Replication of the Viral RNA

Replication of the viral RNA occurs via a positive-sense antigenome intermediate. This RNA, like the genome, is encapsidated as it is synthesized. The factors determining whether a genomic template is used for transcription or replication are not well understood and may vary between viruses. In some cases, the stoichiometry of the N, P, and L proteins constituting the polymerase are known to differ between the two processes. The antigenome RNA is subsequently used as the template for the synthesis of multiple genomic RNA molecules. The inverse complementarity of the terminal sequences is presumably related to their functions in synthesis of full-length antigenomic and genomic RNAs and their encapsidation.

Viral Proteins

A set of five proteins have counterparts in all the viruses of the order whereas others are unique to viruses in specific families or genera.

Nucleocapsid (N) Proteins

All members of the order possess a protein that coats the genomic and antigenomic RNAs to form a nucleocapsid complex along with the phosphoprotein and polymerase protein. In paramyxoviruses conforming to the 'rule of six', the N protein binds six nucleotides of RNA. The N protein is essential for transcription and replication and is usually encoded by the promoter proximal gene transcribed at the highest level (the exception is in the pneumoviruses where there are two short genes encoding nonstructural proteins). In the nucleorhabdoviruses there is a nuclear localization signal near the C-terminus and the N protein is directed to subnuclear structures by association with the P protein. In Borna disease virus, two forms of N protein, p40 and p38, are translated from a single mRNA using an internal translation initiation event. Both are localized in the nucleus, the p38 protein lacking a nuclear localization signal and being translocated in association with other virus proteins.

Phosphoprotein (P)

This protein is encoded by the gene following that encoding the nucleocapsid protein in all the viruses. It is phosphorylated and required for transcription and replication. In some viruses, the ratio of the N and P proteins is believed to influence whether the viral polymerase carries out transcription or replication. The nucleorhabdovirus P protein has a nuclear export signal and is expressed throughout the cell when expressed alone but localized to the nucleus in the presence of N protein. In Borna disease virus, the P protein has a nuclear localization signal and probably has a role in retaining the N and X proteins in the nucleus. The rabies virus P protein inhibits the production of β-interferon by interfering with the function of the TBK kinase that activates IRF3 that in turn activates transcription from the β-interferon promoter. The P protein of Ebola virus (VP35) also inhibits IRF3-mediated activation of the β-interferon promoter but acts upstream of kinases such as TBK.

Other Proteins Encoded by the P mRNA

The P mRNA encodes additional ORFs in the *Bornaviridae* and members of the subfamily *Paramyxovirinae* by use of alternative initiation codons, which are not always the standard AUG. Rhabdoviruses and filoviruses also have additional ORFs that, in the case of some VSV strains, appear to be used to produce C and C′ proteins which may enhance transcription. In Borna disease virus, leaky scanning results in the P protein being produced from an ORF different from that of the first which encodes the X or p10 protein. The X protein is localized to the nucleus with other viral proteins and has a nuclear export signal that is blocked by association with P protein. A second AUG codon in the same frame as the P protein results in an N-terminally truncated form of the P protein P′ or p16 that associates with the other viral proteins.

In the viruses belonging to the subfamily *Paramyxovirinae* there are two mechanisms for expressing of additional genes from the P mRNA. The C proteins (C, C′, Y1, and Y2) are translated in a different ORF from alternative initiation codons that are sometimes not the standard AUG codon. The V proteins (V and W) share a common N-terminus with the P protein but have an editing site where additional G residues are sometimes inserted by the transcriptase (one for V proteins and two for W) into the mRNA giving the protein an alternative C-terminus. In most cases the unedited mRNA transcript encodes the P protein and the edited ones the V or W proteins but in the rubulaviruses the V protein is encoded by the unedited transcript and P by the edited one. These proteins play a role in the inactivation of interferon induction and signaling, although the details of how this occurs vary between viruses.

Matrix Proteins

The matrix protein is typically encoded by the third transcript (except for the pneumoviruses, nucleorhabdoviruses, and cytorhabdoviruses that have additional genes, as described below). It is a major structural component of the viruses and is important for particle morphogenesis. In some cases, virus-like particles can be obtained by expression of M protein alone and the morphology of rabies virus lacking an M gene loses the characteristic bullet shape in addition to there being a greatly reduced yield of virus particles. Different subpopulations of the M protein are believed to be involved in different aspects of morphogenesis (see below). In the case of vesiculoviruses, the M protein inhibits cellular transcription and mRNA export from the nucleus, thus preventing the interferon response. In addition, the M protein is largely responsible for the cytopathic effect seen in cells infected with VSV. In the case of Borna disease virus, there are conflicting reports as to the glycosylation status and subcellular localization of M protein.

Viral Glycoprotein(s)

All the viruses encode between one and three surface glycoproteins that are responsible for attachment of the virus to infected cell membranes and subsequent fusion either with the plasma membrane or in the case of rhabdoviruses with internal membranes. In addition, some rubulaviruses and all pneumoviruses have a less well characterized small hydrophobic protein that is also membrane associated and sometimes glycosylated.

The G proteins of rhabdoviruses and filoviruses and the fusion proteins of the viruses in the family *Paramyxoviridae* form trimers at the surface of infected cells and on virions. These proteins are type I membrane proteins with an N-terminal signal peptide that is cleaved off in the endoplasmic reticulum, a C-terminal membrane anchor, and a number of glycosylation sites on the ectodomain. The Ebola virus G protein is encoded by an edited mRNA with the unedited RNA and a second edited RNA encoding shorter soluble glycoproteins lacking a transmembrane anchor. The fusion proteins of viruses in the families *Paramyxoviridae* and the filovirus G proteins have a protease cleavage site that results in the mature form of each monomer having two disulfide-linked chains comprising a smaller N-terminal chain, F2, and a larger F1 chain. The N-terminus of the F1 chain has the fusion peptide which is located at the end of an extended α-helix following activation by binding to a receptor or exposure to low pH. This peptide inserts into the target membrane and a hairpin-like structure is formed that results in the C-terminal viral membrane anchor being located at the same end of the molecule as the fusion peptide along with their associated heptad repeats.

Viruses in the family *Paramyxoviridae* have a second membrane glycoprotein variously known as H if it has hemagglutination activity, HN if it has hemagglutination and neuraminidase activity, and G if it has neither (although the murine pneumonia virus G protein was named by analogy with that of the other pneumoviruses despite having hemagglutination activity). This protein is a type II membrane protein with an N-terminal, uncleaved signal-anchor and a C-terminal ectodomain that has N-linked glycosylation sites and in pneumoviruses extensive O-linked glycosylation. The proteins form a tetramer in the cases where the structure has been studied and is conventionally regarded as the attachment protein, although in some cases virus lacking this protein can still infect cells. In some viruses (e.g., Newcastle disease virus, human parainfluenza virus type 3, mumps virus, and canine distemper virus), it is also required to obtain full fusion activity, although in others (e.g., simian virus 5, measles virus, and avian metapneumovirus) F protein alone is sufficient.

The rubulaviruses SV5 and mumps along with all the pneumoviruses have a third small type II membrane protein. This is glycosylated in the case of the pneumovirus proteins and in the metapneumoviruses has an additional C-terminal region with conserved cysteine residues. In the pneumoviruses several forms exist, viz. unglycosylated, glycosylated, and a polylactosamine-modified glycosylated form. Little is known about its function although the avian metapneumovirus SH protein suppresses F-protein-mediated cell–cell fusion.

The ephemeroviruses have an additional nonstructural, highly glycosylated protein of unknown function.

The L protein

The L protein is the largest viral protein and is encoded by all viruses in the order *Mononegavirales* at the 5′ end of the genome and is therefore produced from the least abundant mRNA transcript. It contains conserved motifs and is presumed to contain the major enzymatic activities of the RNA polymerase in the nucleocapsid complex in which, together with the N and P proteins (and in pneumoviruses M2-1 protein), it carries out transcription (including capping and polyadenylation) and replication of the viral RNA.

Additional Virus-Encoded Proteins

Several members of the order encode proteins in addition to those described above. Two nonstructural proteins, NS1 and NS2, are encoded by the pneumoviruses and are not found in the metapneumoviruses or any of the other viruses in the order *Mononegavirales*. They are encoded by the most abundant transcripts and appear to have a role in counteracting the host interferon response. Both proteins are required to block the effects of interferon presumably by acting on downstream effectors because they do not interfere with JAK/STAT signaling. They also interfere with IRF3 activation which is required for interferon β-induction.

The pneumoviruses also contain an M2 gene containing two ORFs located immediately upstream of that encoding the L protein. The first ORF encodes the M2-1 protein which shows marked similarity between all these viruses and has been shown to enhance transcription, particularly of longer genes. The second ORF of the M2 gene shows no similarity to other proteins or conservation between viruses. The M2-2 protein is thought to inhibit virus RNA synthesis.

Three plant viruses (the nucleorhabdovirus, rice yellow stunt virus, and the cytorhabdoviruses, northern cereal mosaic virus and strawberry crinkle virus) have poorly characterized ORFs prior to the polymerase gene which may encode additional proteins. There is tentative evidence that these may be related to sequences likely to be involved in RNA replication/transcription. Novirhabdoviruses have an ORF encoding the nonvirion (NV) protein that appears to enhance virus growth.

Cytorhabdoviruses and nucleorhabdoviruses encode between one and four additional proteins at a location between the phosphoprotein and matrix protein genes. These are not well characterized but appear to have similarities to proteins in plants involved in transport between cells via plasmodesmata and are involved in the spread of nucleocapsids between plant cells via the plasmodesmata.

Viral Entry

Mononegaviruses infecting mammalian cells typically attach to receptors on the plasma membrane of cells and fuse either directly with the plasma membrane or, after uptake into endosomal vesicles and exposure to a more acidic environment, with endosomal membranes. These processes are achieved using one or more virus glycoproteins. In viruses of the family *Paramyxoviridae*, these two activities appear to reside on different proteins with the G, H, or HN protein mediating attachment and the F protein mediating fusion. However, in cell culture, viruses lacking the attachment protein gene are often able to grow, suggesting that viral attachment is possible in its absence. The paramyxoviruses generally fuse with the plasma membrane at neutral pH whereas animal rhabdoviruses fuse with endosomal membranes following activation by low pH. Primary infection of plants by the cytorhabdoviruses and nucleorhabdoviruses involves physical injection into plant tissue by invertebrate vectors such as planthoppers, leafhoppers, and aphids, with the latter responsible for viral entry of all but two of these viruses that infect dicotylydenous plants. Fusion of the viral membrane with cellular membranes results in the release of viral nucleocapsids into either the cytoplasm or, in the case of the nucleorhabdoviruses, the nucleus. The animal viruses, with the exception of the bornaviruses, replicate in the cytoplasm where the viral nucleocapsids are released following fusion of the viral membrane with the plasma membrane or endosomal membrane.

Viral Assembly and Budding

The process of viral assembly has been most thoroughly studied in VSV and the general features are thought to be common to all members of the order though the details will differ. The overall process of VSV assembly involves early stages where nucleocapsids are assembled by N protein being released from N–P dimers to bind the genomic RNA as it is being synthesized. At the same time, the G glycoprotein forms microdomains at the plasma membrane. Different populations of M protein are believed to be involved in different steps of assembly. These are, first, binding to regions of plasma membrane enriched in G protein, second, recruitment of the nucleocapsids to these regions, and third, condensation of the nucleocapsids into helical structures which may have M protein at the center as well as between the membrane and the nucleocapsid. While G protein facilitates viral assembly it is not essential, and the M protein is believed to be the key component in assembly with virus-like particles being produced by expression of M protein alone. As the M protein associated with the membrane binds to the condensing nucleocapsid structures, bullet-shaped protrusions occur at the membrane surface. Cellular proteins are believed to associate with the M protein via a so-called late domain (PPPY in the case of VSV) to directly or indirectly cause fusion of the membrane at the base of the protrusions to release virus particles.

The matrix protein is also the main determinant of virus assembly and budding in viruses of the family *Paramyxoviridae* although the precise mechanisms probably vary between viruses. For example, Sendai virus M protein alone can form virus-like particles whereas SV5 M protein cannot. Interaction of viral and cellular proteins probably varies as well since measles and respiratory syncytial virus production are dependent on the presence of cellular actin whereas human parainfluenza virus 3 release is dependent upon the presence of microtubules. These cellular components are required for efficient RNA synthesis as well as virus production. In measles virus, interaction of the M protein and the glycoproteins may occur in lipid rafts to which M protein is recruited independently of the glycoproteins. However recruitment of the hemagglutinin to the rafts may be dependent upon its interaction with the fusion protein. In contrast, the M proteins of Sendai virus and respiratory syncytial virus may require interaction with viral glycoproteins, in particular the fusion protein, for recruitment into lipid rafts. This association is believed to occur during transport to the cell surface.

Budding of most of the animal viruses in the order occurs at the plasma membrane. However, rabies virus buds into internal membrane compartments as do the plant rhabdoviruses. In the case of the nucleorhabdoviruses, there appears to be extensive rearrangement of the intracellular membranes with enlarging of the nuclei due to the insertion of extra membranes through which the virus buds. These membranes are contiguous with the nuclear membrane resulting in the virus ending up in the perinuclear space. Virus is transmitted from here via insect vectors. The matrix protein has an additional role in the polarity of virus release from cells in measles virus where, in contrast to most viruses in the family *Paramyxoviridae*, the HN protein is not transported in a polarized manner and the F protein is directed to the basolateral surface. In the presence of M protein, complexes of M and the glycoproteins are sorted to the apical surface. In other members of this virus family, the glycoproteins are targeted to the apical surface from which the viruses bud, whereas in VSV sorting of G protein and virus budding occur at the basolateral surfaces.

The cytoplasmic tails of the glycoproteins of the viruses in the family *Paramyxoviridae* have roles in enhancing the efficiency and specificity of virus production, recombinant viruses with truncated cytoplasmic tails showing a lack of localization at the membrane surface, lower virus yields, and virus preparations with reduced amounts of viral glycoproteins and increased levels of cellular proteins. VSV G protein also requires a minimal length of cytoplasmic tail to enhance budding efficiency but there does not appear to be any requirement for a specific sequence. Some paramyxovirus fusion proteins can produce virus-like particles.

All the viruses in the order *Mononegavirales* with the possible exception of Borna disease virus therefore seem to have the matrix protein and at least one glycoprotein as major determinants of virus release but the exact details may vary even within members of the same family or subfamily. Budding to release progeny virions of most of the animal viruses occurs at the plasma membrane. However, rabies virus buds into internal membrane compartments as do the plant rhabdoviruses. In the case of the nucleorhabdoviruses, there appears to be extensive rearrangement of the intracellular membranes with enlarging of the nuclei due to the insertion of extra membranes through which the virus buds. These membranes are contiguous with the nuclear membrane resulting in the virus ending up in the perinuclear space, from where they are transmitted via insect vectors.

Further Reading

Feldmann H, Geisbert TW, Jahrling PB, *et al.* (2005) Family *Filoviridae*. In: Fauquet CM, Mayo MA, Maniloff J, Desselberger U, and Ball LA (eds.) *Virus Taxonomy: Eighth Report of the International Committee on Taxonomy of Viruses*, pp. 645–653. San Diego, CA: Elsevier Academic Press.

Geisbert TW and Jahrling PB (2004) Exotic emerging viral diseases: Progress and challenges. *Nature Medicine* 10: S110–S121.

Henrickson KJ (2003) Parainfluenza viruses. *Clinical Microbiology Reviews* 16: 242–264.

Jackson AO, Dietzgen RG, Goodin MM, Bragg JN, and Deng M (2005) Biology of plant rhabdoviruses. *Annual Reviews of Phytopathology* 43: 623–660.

Jayakar HR, Jeetendra E, and Whitt MA (2004) Rhabdovirus assembly and budding. *Virus Research* 106: 117–132.

Lamb RA, Collins PL, Kolakofsky D, *et al.* (2005) Family *Paramyxoviridae*. In: Fauquet CM, Mayo MA, Maniloff J, Desselberger U, and Ball LA (eds.) *Virus Taxonomy: Eighth Report of the International Committee on Taxonomy of Viruses*, pp. 655–668. San Diego, CA: Elsevier Academic Press.

Lamb RA and Parks GD (2006) *Paramyxoviridae*: The viruses and their replication. In: Knipe DM, Howley PM, Griffin DE, *et al.* (eds.) *Fields Virology,* 5th edn., pp. 1449–1496. Philadelphia: Lippincott Williams and Wilkins.

Lipkin WI and Briese T (2006) *Bornaviridae*. In: Knipe DM, Howley PM, Griffin DE, *et al.* (eds.) *Fields Virology,* 5th edn., pp. 1829–1851. Philadelphia: Lippincott Williams and Wilkins.

Lyles DS and Rupprecht CE (2006) *Rhabdoviridae*. In: Knipe DM, Howley PM, Griffin DE, *et al.* (eds.) *Fields Virology,* 5th edn., pp. 1363–1408. Philadelphia: Lippincott Williams and Wilkins.

Mebatsion T, Weiland F, and Conzelmann K-K (1999) Matrix protein of rabies virus is responsible for the assembly and budding of bullet-shaped particles and interacts with the transmembrane spike glycoprotein G. *Journal of Virology* 73: 242–250.

Pringle CR (2005) Order *Mononegavirales*. In: Fauquet CM, Mayo MA, Maniloff J, Desselberger U, and Ball LA (eds.) *Virus Taxonomy: Eighth Report of the International Committee on Taxonomy of Viruses*, pp. 609–614. San Diego, CA: Elsevier Academic Press.

Sanchez A, Geisbert TW, and Feldmann H (2006) *Filoviridae*: Marburg and Ebola viruses. In: Knipe DM, Howley PM, Griffin DE, *et al.* (eds.) *Fields Virology,* 5th edn., pp. 1409–1448. Philadelphia: Lippincott Williams and Wilkins.

Schwemmle M, Carbone KM, Tomongo K, Nowotny N, and Garten W (2005) Family *Bornaviridae*. In: Fauquet CM, Mayo MA, Maniloff J, Desselberger U, and Ball LA (eds.) *Virus Taxonomy: Eighth Report of the International Committee on Taxonomy of Viruses*, pp. 615–622. San Diego, CA: Elsevier Academic Press.

Tomonga K, Kobayashi T, and Ikuta K (2002) Molecular and cellular biology of Borna disease virus infection. *Microbes and Infection* 4: 491–500.

Tordo N, Benmandour A, Calisher C, *et al.* (2005) Family *Rhabdoviridae*. In: Fauquet CM, Mayo MA, Maniloff J, Desselberger U, and Ball LA (eds.) *Virus Taxonomy: Eighth Report of the International Committee on Taxonomy of Viruses*, pp. 623–644. San Diego, CA: Elsevier Academic Press.

Nidovirales

L Enjuanes, CNB, CSIC, Madrid, Spain
A E Gorbalenya, Leiden University Medical Center, Leiden, The Netherlands
R J de Groot, Utrecht University, Utrecht, The Netherlands
J A Cowley, CSIRO Livestock Industries, Brisbane, QLD, Australia
J Ziebuhr, The Queen's University of Belfast, Belfast, UK
E J Snijder, Leiden University Medical Center, Leiden, The Netherlands

Glossary

3CL^{pro} or M^{pro} 3C-like proteinase, or main proteinase.
ADRP ADP-ribose-1″-phosphatase.
CS TRS Core sequence.
ExoN 3′ to 5′ exoribonuclease.
NendoU Nidovirus endoribonuclease.
O-MT Ribose-2′-O-methyltransferase.
PL^{pro} Papain-like cysteine proteinase.
TRS Transcription-regulating sequence.

Taxonomy and Phylogeny

The order *Nidovirales* includes the families *Coronaviridae*, *Roniviridae*, and *Arteriviridae* (**Figure 1**). The *Coronaviridae* comprises two well-established genera, *Coronavirus* and *Torovirus*, and a tentative new genus, *Bafinivirus*. The *Arteriviridae* and *Roniviridae* include only one genus each, *Arterivirus* and *Okavirus*, respectively. All nidoviruses have single-stranded RNA genomes of positive polarity that, in the case of the *Corona*- and *Roniviridae* (26–32 kbp), are the largest presently known RNA virus genomes. In contrast, members of the *Arteriviridae* have a smaller genome ranging from about 13 to 16 kbp. The data available from phylogenetic analysis of the highly conserved RNA-dependent RNA polymerase (RdRp) domain of these viruses, and the collinearity of the array of functional domains in nidovirus replicase polyproteins, were the basis for clustering coronaviruses and toroviruses (**Figure 2**). The more distantly related roniviruses also group with corona- and toroviruses, thus forming a kind of supercluster of nidoviruses with large genomes. By contrast, arteriviruses must have diverged earlier during nidovirus evolution. The current

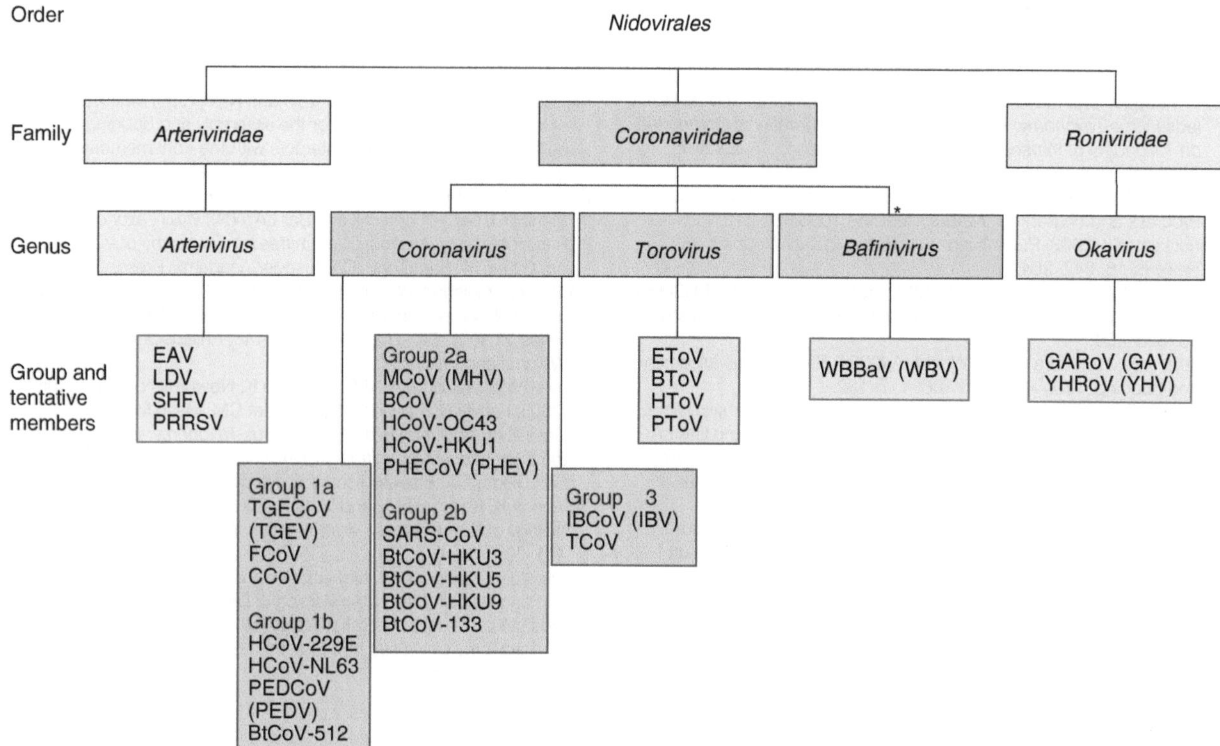

Figure 1 Nidovirus classification and prototype members. The order *Nidovirales* containing the families *Coronaviridae* (including the established genera *Coronavirus* and *Torovirus*, and a new tentative genus *Bafinivirus*), *Arteriviridae*, and *Roniviridae*. Phylogenetic analysis (see **Figure 2**) has confirmed the division of coronaviruses into three groups. In arteriviruses, four comparably distant genetic clusters have been differentiated. To facilitate the taxonomy of the different virus isolates, the types Co, To, Ba, Ro, standing for coronavirus, torovirus, bafinivirus, or ronivirus, respectively, have been included. The following CoVs are shown: human coronaviruses (HCoV) 229E, HKU1, OC43 and NL63, transmissible gastroenteritis virus (TGEV), feline coronavirus (FCoV), porcine epidemic diarrhoea virus (PEDV), mouse hepatitis virus (MHV), bovine coronavirus (BCoV), bat coronaviruses (BtCoV) HKU3, HKU5, HKU9, 133 and 512 (the last two isolated in 2005), porcine hemagglutinating encephalomyelitis virus (PHEV), avian infectious bronchitis virus (IBV), and severe acute respiratory syndrome coronavirus (SARS-CoV); ToV: equine torovirus (EToV), bovine torovirus (BToV), human torovirus (HToV), and porcine torovirus (PToV); BaV: white bream virus (WBV); Arterivirus: equine arteritis virus (EAV), simian haemorrhagic fever virus (SHFV), lactate dehydrogenase-elevating virus (LDV), and three (Euro, HB1, and MLV) porcine reproductive and respiratory syndrome viruses (PRRSV); RoV: gill-associated virus (GAV) and yellow head virus (YHV). Human viruses are highlighted in red. Some nodes are formed by a pair of very closely related viruses (e.g., SARS-CoV and BtCoV-HKU3). Asterisk indicates tentative genus.

taxonomic position of coronaviruses and toroviruses as two genera of the family *Coronaviridae* is currently being revised by elevating these virus groups to the taxonomic rank of either subfamily or family.

A comparative sequence analysis of coronaviruses reveals three phylogenetically compact clusters: groups 1, 2, and 3. Within group 1, two subsets can be distinguished: subgroup 1a that includes transmissible gastroenteritis virus (TGEV), canine coronavirus (CCoV), and feline coronavirus (FCoV), and subgroup 1b that includes the human coronaviruses (HCoV) 229E and NL63, porcine epidemic diarrhoea virus (PEDV), and bat coronavirus (BtCoV) 512 which was isolated in 2005. Within group 2 coronaviruses, two subsets have been recognized: subgroup 2a, including mouse hepatitis virus (MHV), bovine coronavirus (BCoV), HCoV-OC43, and HCoV-HKU1; and subgroup 2b, including severe acute respiratory syndrome coronavirus (SARS-CoV) and its closest circulating bat coronavirus relative, BtCoV-HKU3. A growing number of other bat viruses has been recently identified in groups 1 and 2. It is currently being debated whether some of these viruses (e.g., BtCoV-HKU5, BtCoV-133 (isolated in 2005), and BtCoV-HKU9) may in fact represent novel subgroups or groups. Avian infectious bronchitis virus (IBV) is the prototype of coronavirus group 3, which also includes several other bird coronaviruses. In arteriviruses, there are four comparably distant genetic clusters, the prototypes of which are equine arteritis virus (EAV), lactate dehydrogenase-elevating virus (LDV) of mice, simian hemorrhagic fever virus (SHFV) infecting monkeys, and porcine reproductive and respiratory syndrome virus (PRRSV) which infects pigs and includes European and North American genotypes.

Roniviruses are the only members of the order *Nidovirales* that are known to infect invertebrates. The family *Roniviridae* includes the penaeid shrimp virus, gill-associated virus (GAV), and the closely related yellow head virus (YHV).

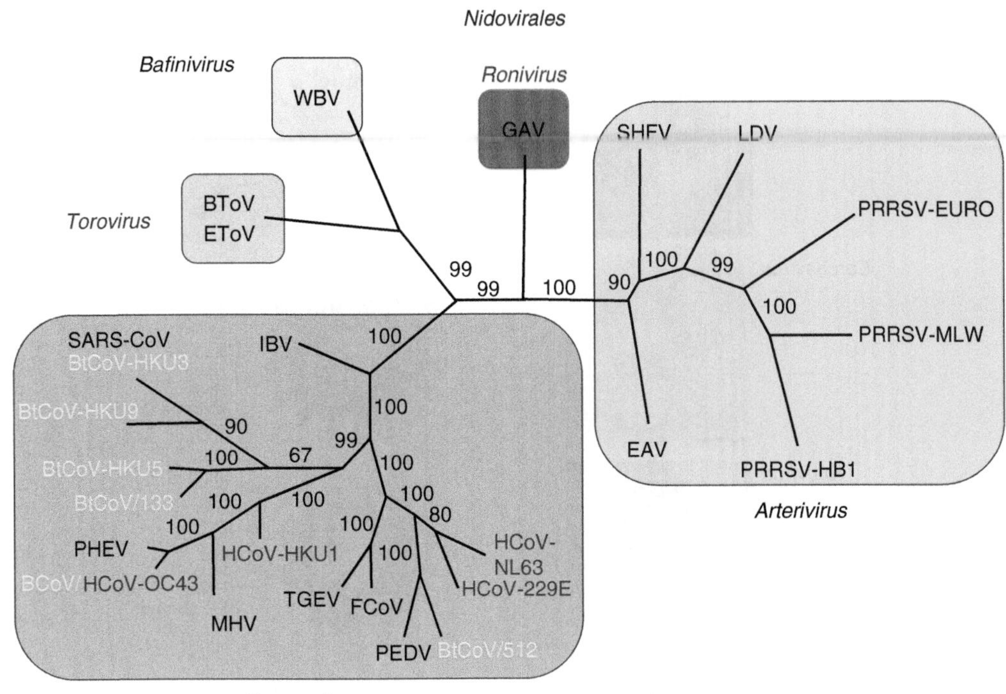

Figure 2 Nidovirus phylogeny. Tree depicting the evolutionary relationships between the five major groups of nidoviruses as shown in **Figure 1** (*Coronavirus, Torovirus, Bafinivirus, Ronivirus*, and *Arterivirus*). This unrooted maximum parsimonious tree was inferred using multiple nucleotide alignments of the RdRp-HEL region of a representative set of nidoviruses with the help of the PAUP*v.4.0b10 software (AEG, unpublished). Support for all bifurcations from 100 bootstraps performed is indicated. The phylogenetic distances shown are approximate. For acronyms, see **Figure 1**.

More than 100 full-length coronavirus genome sequences and around 30 arterivirus genome sequences have been documented so far, whereas only very few sequences have been reported for toroviruses, bafiniviruses, and roniviruses. Therefore, information on the genetic variability of these nidovirus *taxa* is limited.

Diseases Associated with Nidoviruses

Coronavirus infections are mainly associated with respiratory, enteric, hepatic, and central nervous system diseases. In humans and fowl, coronaviruses primarily cause upper respiratory tract infections, while porcine and bovine coronaviruses establish enteric infections, often resulting in severe economic losses. In 2002, a previously unknown coronavirus that probably has its natural reservoir in bats crossed the species barrier and caused a major outbreak of SARS, which led to more than 800 deaths worldwide.

Toroviruses cause gastroenteritis in mammals, including humans, and possibly also respiratory infections in older cattle. Bafiniviruses have been isolated from white bream fish but there is currently no information on the pathogenesis associated with this virus infection. Roniviruses usually exist as asymptomatic infections but can cause severe disease outbreaks in farmed black tiger shrimp (*Penaeus monodon*) and white pacific shrimp (*Penaeus vannamei*), which in the case of YHV can result in complete crop losses within a few days after the first signs of disease in a pond. Infections by arteriviruses can cause acute or persistent asymptomatic infections, or respiratory disease and abortion (EAV and PRRSV), fatal age-dependent poliomyelitis (LDV), or fatal hemorrhagic fever (SHFV). Arteriviruses, particularly PRRSV in swine populations, cause important economic losses.

Virus Structure

In addition to the significant variations in genome size among the three nidovirus families mentioned above, there are also major differences in virion morphology (**Figure 3**) and host range. Nidoviruses have a lipid envelope which protects the internal nucleocapsid structure and contains a number of viral surface proteins (**Figure 3**). Whereas coronaviruses and the significantly smaller arteriviruses have spherical particle structures, elongated rod-shaped structures are observed in toro-, bafini-, and ronivirus-infected cells. The virus particles of the *Corona-* and *Roniviridae* family members carry large surface projections that protrude from the viral envelope (peplomers), whereas arterivirus particles possess only relatively small

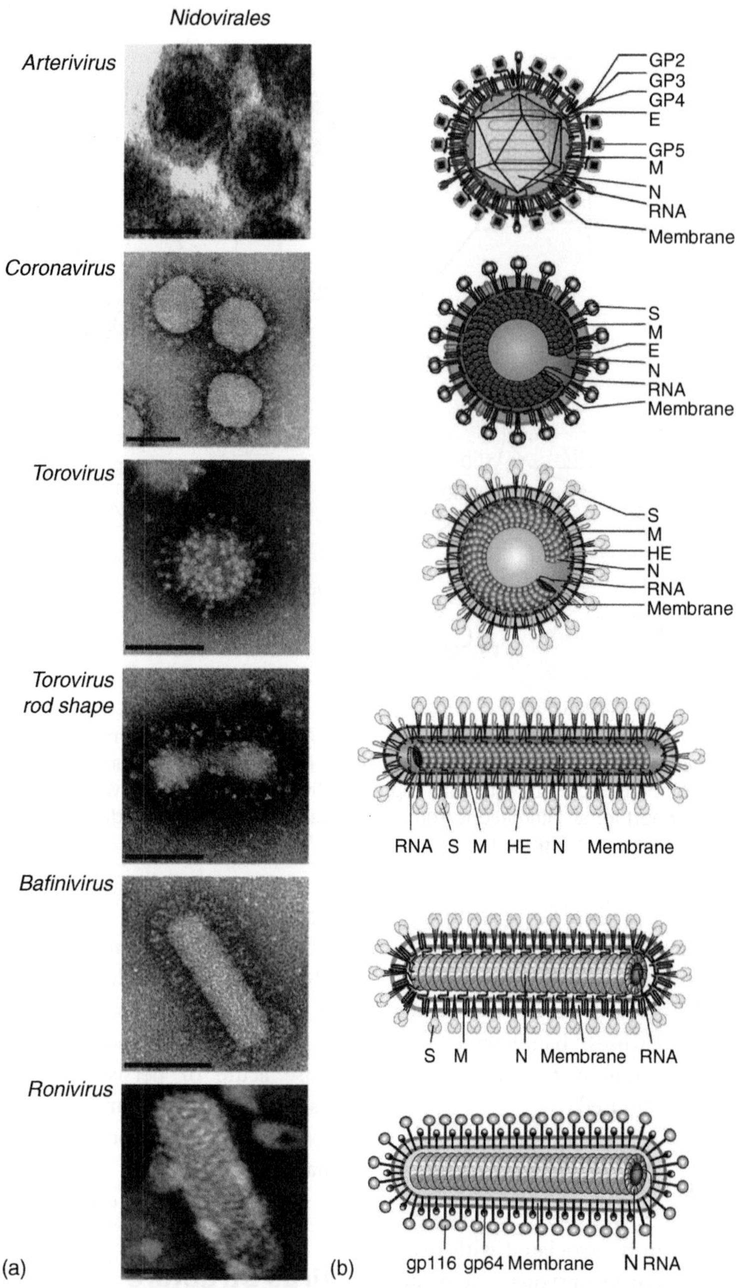

Figure 3 Nidovirus structure. Architecture of particles of members of the order *Nidovirales*: electron micrographs (a) and schematic representations (b) are shown. N, nucleocapsid protein; S, spike protein; M, membrane protein; E, envelope protein; HE, hemagglutinin-esterase. Coronavirus M protein interacts with the N protein. In arterivirus, GP_5 and M are major envelope proteins, while GP_2, GP_3, GP_4, and E are minor envelope proteins. Toro-, bafini-, and roniviruses lack the E protein present in corona- and arteriviruses. Proteins gp116 and gp64, ronivirus envelope proteins. Different images were reproduced with permission from different authors: arterivirus, E. Snijder (Leiden, The Netherlands); ronivirus, P. J. Walker (CSIRO, Australia); bafinivirus, J. Ziebuhr (Queen's University, Belfast) torovirus, D. Rodriguez (CNB, Spain); coronavirus, L. Enjuanes (CNB, Spain).

projections on their surface. Coronaviruses have an internal core shell that is formed by a nucleocapsid featuring a helical symmetry. The nucleocapsid (N) protein interacts with the carboxy-terminus of the envelope membrane (M) protein. The intracellular forms of torovirus, bafinivirus, and ronivirus nucleocapsids have extended rod-shaped (helical) morphology. By contrast, mature (extracellular) toroviruses (but not bafini- and roniviruses) feature a remarkable structural flexibility, which allows them to adopt crescent- and toroid-shaped structures also. Unlike other nidoviruses, arteriviruses have an isometric core shell. In all nidoviruses, the nucleocapsid is formed by only a single N protein that interacts with the genomic RNA.

Both the number and properties of structural proteins vary between viruses of the three families of the *Nidovirales* and may even vary among viruses of the same family. Nidoviruses usually encode at least three structural proteins: a spike (S) or major surface glycoprotein, a trans-membrane (M) or matrix protein, and the N protein (**Figure 3**). Ronivirus particles are unique in that they possess two envelope glycoproteins, gp116 (S1) and gp64 (S2), but no M protein. Coronavirus and arterivirus particles possess another envelope protein called E that is not conserved in toroviruses, bafini-, and roniviruses. Toroviruses and subgroup 2a coronaviruses, such as MHV, have a hemagglutinin esterase (HE) as an additional structural protein, whereas the SARS-CoV has at least four additional proteins that are present in the viral envelope (encoded by ORFs 3a, 6, 7a, and 7b). The proteins may promote virus growth in cell culture or *in vivo*, but they are dispensable for virus replication.

The major envelope proteins are the S and M proteins in coronaviruses and toroviruses, the GP_5 and M proteins in arteriviruses, and the S1 and S2 proteins in roniviruses. Among these, only the corona-, toro-, and bafinivirus S and M proteins share limited sequence similarities, possibly indicating a common origin. Whereas S proteins can differ in size, they share an exposed globular head domain and, with the exception of roniviruses, a stem portion containing heptad repeats organized in a coiled-coil structure. The S proteins of corona- and toroviruses (and most likely those of bafiniviruses) form trimers that bind the cell surface receptor whereas receptor binding in roniviruses is probably mediated by gp116 (S1). The arterivirus envelope proteins form two higher-order complexes: one is a disulfide-linked heterodimer of GP_5 and the M protein; and the other is a heterotrimer of the minor structural glycoproteins GP_2, GP_3, and GP_4. Except for the E and M proteins, all arterivirus structural proteins are glycosylated. By contrast, the M proteins of corona- and toroviruses (and, most likely, bafiniviruses) are glycosylated, and they share a triple-spanning membrane topology with the amino-terminus exposed on the outside of the virions and the carboxy-terminus facing the nucleocapsid. In TGEV, a proportion of the M proteins has a tetra-spanning membrane topology leading to the exposure of both termini on the virion surface.

In the virion, the coronavirus E protein has a low copy number (around 20) and deletion of the E protein gene from the genome of the group 1 coronavirus TGEV blocks virus maturation, preventing virus release and spread. In the group 2 coronaviruses MHV and SARS-CoV, deletion of the E protein results in a dramatic reduction, of up to 100 000-fold, of virus infectivity. The coronavirus E and SARS-CoV 3a proteins are viroporins, that is, they belong to a group of proteins that modify membrane permeability by forming ion channels in the virion envelope.

Genome Organization

Nidovirus genomes contain variable numbers of genes, but in all cases the 5′ terminal two-thirds to three-quarters of the genome is dedicated to encoding the key replicative proteins, whereas the 3′ proximal genome regions generally encode the structural and, in some cases, accessory (group- and virus-specific) proteins (**Figure 4**). Nidovirus genome expression is controlled at the translational and post-translational levels. Thus, for example, ribosomal frameshifting is required for the expression of ORF1b, and the two replicase polyproteins (pp1a and pp1ab) are proteolytically processed by viral proteases. The proteolytic processing occurs in a coordinated manner and gives rise to the functional subunits of the viral replication–transcription complex. By contrast, the expression of the structural and several accessory proteins is controlled at the level of transcription. It involves the synthesis of a nested set of 3′ co-terminal sg mRNAs that are produced in nonequimolar amounts. As in cellular eukaryotic mRNAs, in general only the ORF positioned most closely to the 5′ end of the sg mRNA is translated.

The Replicase

The nidovirus replicase gene is comprised of two slightly overlapping ORFs, 1a and 1b. In corona-, toro-, bafini-, and roniviruses, ORF1a encodes a polyprotein (pp1a) of 450–520 kDa, whereas a polyprotein of 760–800 kDa (pp1ab) is synthesized from ORF1ab. Expression of the ORF1b-encoded part of pp1ab involves a ribosomal frameshift mechanism that, in a defined proportion of translation events, directs a controlled shift into the −1 reading frame just upstream of the ORF1a stop codon (**Figure 4**). In arteriviruses, pp1a (190–260 kDa) and pp1ab (345–420 kDa) are considerably smaller in size. Proteolytic processing of coronavirus pp1a and pp1ab generates up to 16 nonstructural proteins (nsps 1–16), while processing of the arterivirus replicase polyproteins generates up to 14 nsps. It is generally accepted that most of the replicase nsps assemble into a large protein complex, called the replication–transcription complex. The complex is anchored to intracellular membranes and likely also includes a number of cellular proteins. Nidoviruses replicase genes include a conserved array of protease, RNA-dependent RNA polymerase (RdRp), helicase (HEL), and endoribonuclease (NendoU) activities. In contrast to other positive-strand RNA viruses, they employ an RdRp with a characteristic SDD rather than the usual GDD active site.

The vast majority of proteolytic cleavages in pp1a/pp1ab are mediated by an ORF1a-encoded chymotrypsin-like protease that, because of its similarities to picornavirus 3C proteases, is called the 3C-like protease ($3CL^{pro}$). Also the term 'main protease' (M^{pro}) is increasingly used for this enzyme, mainly to refer to its key role in

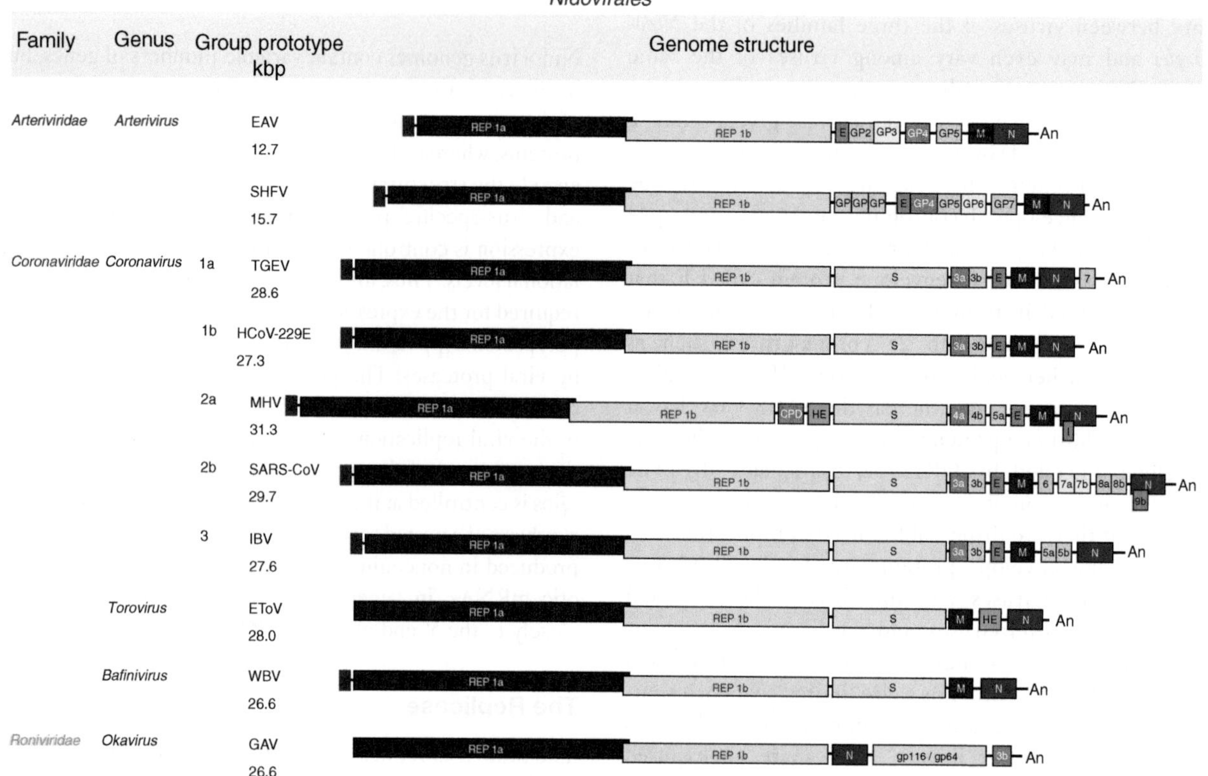

Figure 4 Nidovirus genome structure. Genome organization of selected nidoviruses. The genomic ORFs of viruses representing the major nidovirus lineages are indicated and the names of the replicase and main virion genes are given. References to the nomenclature of accessory genes can be found in the text. Genomes of large and small nidoviruses are drawn to different scales. Red box at the 5′ end refers to the leader sequence. Partially overlapping ORFs have been drawn as united boxes. Spaces between boxes representing different ORFs do not mean noncoding sequences.

nidovirus replicase polyprotein processing (**Figure 5**). Nidovirus–M^{pro}share a three-domain structure. The two N-terminal domains adopt a two-β-barrel fold reminiscent of the structure of chymotrypsin. With respect to the principal catalytic residues, there are major differences between the main proteases from different nidovirus genera. The presence of a third, C-terminal domain is a conserved feature of nidovirus main proteases, even though these domains vary significantly in both size and structure. The C-terminal domain of the coronavirus M^{pro} is involved in protein dimerization that is required for proteolytic activity *in trans*. Over the past years, a large body of structural and functional information has been obtained for corona- and arterivirus main proteases which, in the case of coronaviruses, has also been used to develop selective protease inhibitors that block viral replication, suggesting that nidovirus main proteases may be attractive targets for antiviral drug design.

In arteri-, corona-, and toroviruses, the M^{pro} is assisted by 1–4 papain-like ('accessory') proteases (PL^{pro}) that process the less well-conserved N-proximal region of the replicase polyproteins (**Figure 5**). Nidovirus PL^{pro} domains may include zinc ribbon structures and some of them have deubiquitinating activities, suggesting that these proteases might also have functions other than polyprotein processing. Bafini- and roniviruses have not been studied in great detail and it is not yet clear if these viruses employ papain-like proteases to process their N-terminal pp1a/pp1ab regions (**Figure 5**).

The replicase polyproteins of 'large' nidoviruses with genome sizes of more than 26 kb (i.e., corona-, toro-, bafini-, and roniviruses) include 3′–5′ exoribonuclease (ExoN) and ribose-2′-*O*-methyltransferase (MT) activities that are essential for coronavirus RNA synthesis but are not conserved in the much smaller arteriviruses (**Figure 5**). The precise biological function of ExoN has not been established for any nidovirus but the relationship with cellular DEDD superfamily exonucleases and recently published data suggest that ExoN may have functions in the replication cycle of large nidoviruses that, like in the DEDD homologs, are related to proofreading, repair, and recombination mechanisms.

NendoU is a nidovirus-wide conserved domain that has no counterparts in other RNA viruses. It is therefore considered a genetic marker of the *Nidovirales*. The endonuclease has uridylate specificity and forms hexameric

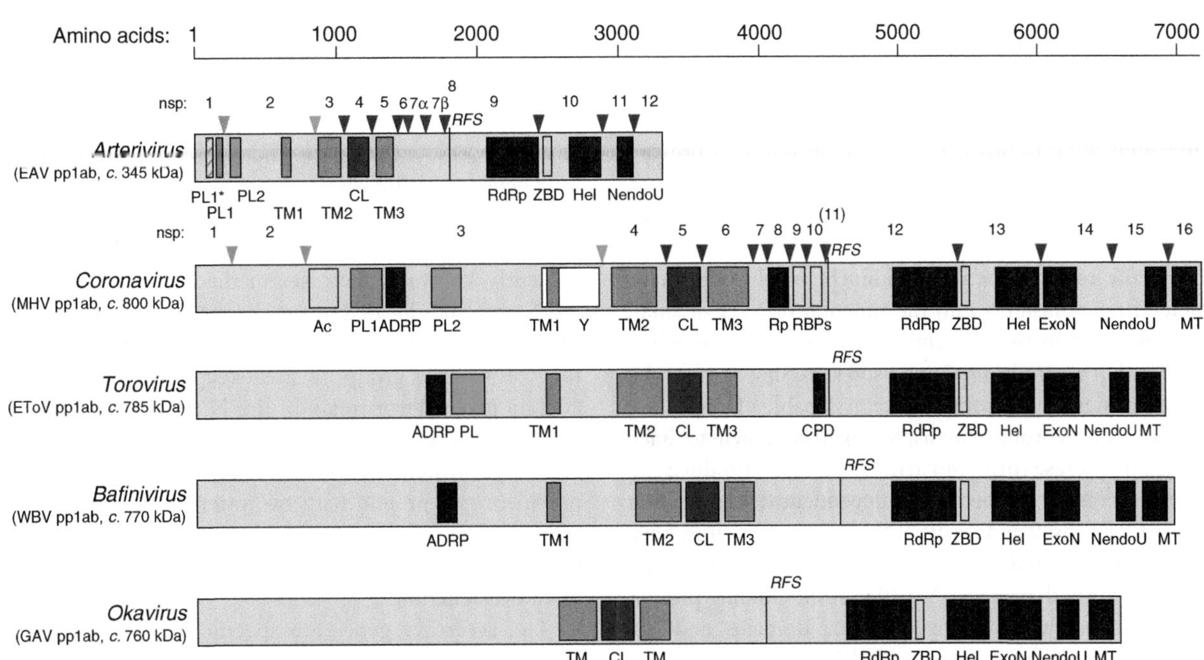

Figure 5 Nidovirus replicase genes. Polyprotein (pp) 1ab domain organizations are shown for representative viruses from the five nidovirus genera. Acronyms as in **Figure 1**. Arterivirus and coronavirus pp1ab processing pathways have been characterized in considerable detail and are illustrated here for EAV and MHV. N-proximal polyprotein regions are cleaved at two or three sites by viral papain-like proteases 1 (PL1) and 2 (PL2), whereas the central and C-terminal polyprotein regions are processed by the main protease, M^{pro}. PL1 domains are indicated by orange boxes and cognate cleavage sites are indicated by orange arrowheads. PL2 domains and PL2-mediated cleavages are shown in green and CL domains and CL-mediated cleavages are shown in red. Note that EAV encodes a second, but proteolytically inactive PL1 domain (PL1*; orange-striped box). For the genera *Torovirus*, *Bafinivirus*, and *Okavirus*, the available information on pp1ab proteolytic processing is limited and not shown here. Other predicted or proven enzymatic activities are shown in blue: ADRP, ADP-ribose 1′-phosphatase; Rp, noncanonical RNA polymerase ('primase') activity; RdRp, RNA-dependent RNA polymerase; HEL, NTPase/RNA helicase and RNA 5′-triphosphatase; ExoN, 3′-to-5′ exoribonuclease; NendoU, nidoviral uridylate-specific endoribonuclease; MT, ribose-2′-O methyltransferase; CPD, cyclic nucleotide phosphodiesterase. Regions with predicted transmembrane (TM) domains are indicated by gray boxes. Other functional domains are shown as white boxes: Ac, acidic domain; Y, Y domain containing putative transmembrane and zinc-binding regions; ZBD, helicase-associated zinc-binding domain; RBPs, RNA-binding proteins. Expression of the C-terminal part of pp1ab requires a ribosomal frameshift, which occurs just upstream of the ORF1a translation stop codon. The ribosomal frameshift site (RFS) is indicated.

structures with six independent catalytic sites. Cellular homologs of NendoU have been implicated in small nucleolar RNA processing whereas the role of NendoU in viral replication is less clear. Reverse genetics data indicate that NendoU has a critical role in the viral replication cycle.

Two other RNA-processing domains, ADP-ribose-1″-phosphatase (ADRP) and nucleotide cyclic phosphodiesterase (CPD), are conserved in overlapping subsets of nidoviruses (**Figure 5**). Except for arteri- and roniviruses, all nidoviruses encode an ADRP domain that is part of a large replicase subunit (nsp3 in the case of coronaviruses). The coronavirus ADRP homolog has been shown to have ADP-ribose-1′-phosphatase and poly(ADP-ribose)-binding activities. Although the highly specific phosphatase activity is not essential for viral replication *in vitro*, the strict conservation in all genera of the *Coronaviridae* suggests an important (though currently unclear) function of this protein in the viral replication cycle. This may be linked to host cell functions and, particularly, to the activities of cellular homologs called 'macro' domains which are thought to be involved in the metabolism of ADP-ribose and its derivatives.

The CPD domain is only encoded by toroviruses and group 2a coronaviruses. In toroviruses, the CPD domain is encoded by the 3′ end of replicase ORF1a (**Figure 5**), whereas in group 2a coronaviruses, the enzyme is expressed from a separate subgenomic RNA. The enzyme's biological function is not clear. Coronavirus CPD mutants are attenuated in the natural host whereas replication in cell culture is normal, suggesting some function *in vivo*. The available information suggests that nidovirus replicase polyproteins (particularly, those of large nidoviruses) have evolved to include a number of nonessential functions that may provide a selective advantage in the host.

ORF1a of all nidoviruses encodes a number of (putative) transmembrane proteins, like the coronavirus nsps 3, 4, and 6 and the arterivirus nsps 2, 3, and 5. These have been shown or postulated to trigger the modification of cytoplasmic membranes, including the formation of unusual double-membrane vesicles (DMVs). Tethering of the replication–transcription complex to these virus-induced membrane structures might provide a scaffold or subcellular compartment for viral RNA synthesis, possibly allowing it to proceed under conditions that prevent or impair detection by cellular defense mechanisms, which are usually induced by the double-stranded RNA intermediates of viral replication.

Finally, recent structural and biochemical studies have yielded novel insights into the function of a set of small nsps encoded in the 3′-terminal part of the coronavirus ORF1a. For example, nsp7 and nsp8 were shown to form a hexadecameric supercomplex that is capable of encircling dsRNA. The coronavirus nsp8 was also shown to have RNA polymerase (primase) activity that may produce the primers required by the primer-dependent RdRp residing in nsp12. For nsp9 and nsp10, RNA-binding activities have been demonstrated and crystal structures have been reported for both proteins. Nsp10 is a zinc-binding protein that contains two zinc-finger-binding domains and has been implicated in negative-strand RNA synthesis.

Structural and Accessory Protein Genes

In contrast to the large genome of *Coronaviridae*, which can accommodate genes encoding accessory proteins (i.e., proteins called 'nonessential' for being dispensable for replication in cell culture (**Figure 6**)), the smaller genomes of arteriviruses only encode essential proteins (**Figure 4**). Coronaviruses encode a variable number of accessory proteins (2–8), while the torovirus genome contains a single accessory gene encoding a hemagglutinin-esterase (HE). Coronavirus accessory genes may occupy any intergenic position in the conserved array of the four genes encoding the major structural proteins (5′-S-E-M-N-3′), or they may reside upstream or downstream of this gene array. Roniviruses are unique among the presently known nidoviruses in that the gene encoding the N protein is located upstream rather than downstream of the gene encoding the glycoproteins. Several members of the coronavirus group 1a are exceptional in that they contain genes downstream of the N protein gene, which has not been reported for other coronaviruses. The ronivirus glycoprotein gene is also unique in that it encodes a precursor polyprotein with two internal signal peptidase cleavage sites used to generate the envelope glycoproteins S1 and S2 as well as an amino-terminal protein with an unknown function.

The accessory genes are specific for either a single virus species or a few viruses that form a compact phylogenetic cluster. Many proteins encoded by accessory genes may function in infected cells or *in vivo* to counteract host defenses and, when removed, may lead to attenuated virus phenotypes. Group 1 coronaviruses may have 2–3 accessory genes located between the S and E genes and up to two other genes downstream of N gene. Viruses of group 2 form the most diverse coronavirus cluster, and they may have between three and eight

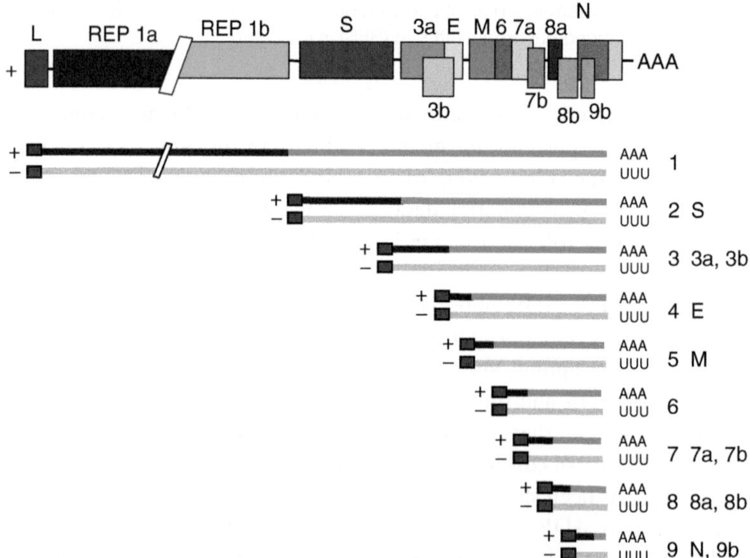

Figure 6 SARS-CoV genome organization and sg mRNA expression. A diagram of CoV structure using SARS-CoV as a prototype. Below the top bar a set of positive- and negative-sense mRNA species synthesized in infected cells is shown. Dark and light blue lines (+), mRNA sequences translated and nontranslated into viral proteins, respectively. Light gray lines (–), RNAs complementary to the different mRNAs. L, leader sequence. Poly(A) and Poly(U) tails are indicated by AAA or UUU, respectively. Rep 1a and Rep 1b, replicase genes. Other acronyms above and below the top bar indicate structural and nonstructural proteins. Numbers and letters to the right of the thin bars indicate the sg mRNAs.

accessory genes. In this cluster, MHV, HCoV-OC43, and BCoV form the phylogenetically compact subgroup, 2a, that is characterized by the presence of (1) two accessory genes located between ORF1b and the S gene encoding proteins with CPD and HE functions, (2) two accessory genes located between the S and E protein genes, and (3) an accessory gene, I, that is located within the N protein gene. Of this set of five accessory proteins, only three homologs are encoded by the recently identified HCoV-HKU1, which is the closest known relative of the cluster formed by MHV, HCoV-OC43, and BCoV. In contrast, the most distant group 2 member, SARS-CoV, has seven or eight unique accessory genes, two between the S and E protein genes, four to five between the M and N protein genes, and ORF9b which entirely overlaps with the N protein gene in an alternative reading frame. In group 3 avian coronaviruses, of which IBV is the prototype, several accessory genes, which are expressed from functionally tri- or bicistronic mRNAs, have been identified in the region between the S and E protein genes (gene 3) and between the M and N protein genes (gene 5).

Some functionally dispensable ORF1a-encoded replicase domains may also be considered as accessory protein functions. For instance, MHV and SARS-CoV nsp2 turned out to be nonessential for replication in cell culture.

Replication

Like in all other positive-stranded RNA viruses, nidovirus genome replication is mediated through the synthesis of a full-length, negative-strand RNA which, in turn, is the template for the synthesis of progeny virus genomes. This process is mediated by the viral replication complex that includes all or most of the 14–16 nsps derived from the proteolytic processing of the pp1a and pp1ab replicase polyproteins of arteriviruses and coronaviruses. The replication complex, which is likely to include also cellular proteins, is associated with modified intracellular membranes, which may be important to create a microenvironment suitable for viral RNA synthesis as well as for recruitment of host factors. Electron microscopy studies of cells infected with arteriviruses (EAV) and coronaviruses (MHV and SARS-CoV) have shown that RNA synthesis is associated with virus-induced, DMVs. The origin of DMVs is under debate and different intracellular compartments including the Golgi, late endosomal membranes, autophagosomes, and the endoplasmic reticulum have been implicated in their formation.

Studies of *cis*-acting sequences required for nidovirus replication have mainly relied on coronavirus defective-interfering (DI) RNAs replicated by helper virus. Genome regions harboring minimal *cis*-acting sequences have been mapped to around 1 kb domains of the genomic 5′ and 3′ ends. Studies with MHV DI RNAs have indicated that both genome ends are necessary for positive-strand synthesis, whereas only the last 55 nt and the poly (A) tail at the genomic 3′ end are required for negative-strand synthesis. It has been postulated that the 5′ and 3′ ends of the genome may interact directly during RNA replication, as predicted by computer-aided simulations of MHV and TGEV genomic RNA interactions in protein-free media. There is, however, some experimental evidence supporting protein-mediated cross-talk between both genome ends in the form of RNA–protein and protein–protein interactions.

Several experimental approaches have implicated, in addition to the nsps encoded by the replicase gene, the N protein in coronavirus RNA synthesis. Early in infection, the coronavirus N protein colocalizes with the site of viral RNA synthesis. In addition, the N protein can enhance the rescue of various coronaviruses from synthetic full-length RNA, transcribed *in vitro* or from cDNA clones. In contrast, arterivirus RNA synthesis does not require the N protein.

Host factors that may participate in nidovirus RNA synthesis have been identified mainly from studies of coronaviruses and arteriviruses. In coronaviruses (MHV and TGEV), heterogeneous nuclear ribonucleoprotein (hnRNP) A1 has been identified as a major protein binding to genomic RNA sequences complementary to those in the negative-strand RNA that bind another cellular protein, polypyrimidine tract-binding protein (PTB). hnRNP A1 and PTB bind to the complementary strands at the 5′ end of coronavirus RNA and could mediate the formation of an RNP replication complex involving the 5′ and 3′ ends of coronavirus genomic RNA. The functional relevance of hnRNP A1 in coronavirus replication was supported by experiments showing that its overexpression promotes MHV replication, whereas replication was reduced in cells expressing a dominant-negative mutant of hnRNP A1. There is also experimental evidence to suggest that the poly(A)-binding protein (PABP) specifically interacts with the 3′ poly(A) tail of coronavirus genomes, and that this interaction may affect their replication. Other cellular proteins found to bind to coronavirus genomic RNA, such as aconitase and the heat shock proteins HS40 and HS70, might be involved in modulating coronavirus replication. Similarly, interactions of cellular proteins such as transcription cofactor p100 with the EAV nsp1, or of PTB or fructose bisphosphate aldolase A with SHFV genomic RNA, suggest that, in arterivirus replication also, a number of cellular proteins may be involved.

Transcription

RNA-dependent RNA transcription in some members of the *Nidovirales* (coronaviruses, bafiniviruses, and arteriviruses), but not in others (roniviruses), includes a discontinuous

RNA-synthesis step. This process occurs during the production of subgenome-length negative-strand RNAs that serve as templates for transcription and involves the fusion of a copy of the genomic 5′-terminal leader sequence to the 3′ end of each of the nascent RNAs complementary to the coding (body) sequences (**Figure 6**). The resulting chimeric sg RNAs of negative polarity are transcribed to yield sg mRNAs that share both 5′- and 3′- terminal sequences with the genome RNA. Genes expressed through sg mRNAs are preceded by conserved 'transcription-regulatory sequences' (TRSs) that presumably act as attenuation or termination signals during the production of the subgenome-length negative-strand RNAs. In arteriviruses and coronaviruses, the TRSs preceding each ORF are presumed to direct attenuation of negative-strand RNA synthesis, leading to the 'jumping' of the nascent negative-strand RNA to the leader TRS (TRS-L). This process is guided by a base-pairing interaction between complementary sequences (leader TRS and body TRS complement) and it has been proposed that template switching only occurs if the free energy (ΔG) for the formation of this duplex reaches a minimum threshold. This process is named 'discontinuous extension of minus strands' and can be considered a variant of similarity-assisted template-switching that operates during viral RNA recombination. The genome and sg mRNAs share a 5′-leader sequence of 55–92 nt in coronaviruses and 170–210 nt in arteriviruses.

Toroviruses are remarkable in that they employ a mixed transcription strategy to produce their mRNAs. Of their four sg mRNA species, the smaller three (mRNAs 3 through 5) lack a 5′ common leader and are produced via nondiscontinuous RNA synthesis. In contrast, sg mRNA2 has a leader sequence that matches the 5′-terminal 18 nt of the genomic RNA and its production requires a discontinuous RNA-synthesis step reminiscent of, but not identical, to that seen in arteri- and coronaviruses.

Synthesis of torovirus mRNAs 3 through 5, and possibly of the two mRNAs in roniviruses, is thought to require the premature termination of negative-strand RNA synthesis at conserved, intergenic, TRS-like sequences to generate subgenome-length negative-strand RNAs that can be used directly as templates for sg mRNA synthesis. In the case of torovirus mRNA2, a TRS is lacking. Fusion of noncontiguous sequences seems to be controlled by a sequence element consisting of a hairpin structure and 3′ flanking stretch of 23 residues with sequence identity to a region at the 5′ end of the genome. It is thought that during negative-strand synthesis, the hairpin structure may cause the transcriptase complex to detach, prompting a template switch similar to that seen in arteri- and coronaviruses.

In addition to regulatory RNA sequences, viral and host components involved in protein–RNA and protein– protein recognition are likely to be important in transcription. For example, the arterivirus nsp1 protein has been identified as a factor that is dispensable for genome replication but absolutely required for sg RNA synthesis. The identification of host factors participating in nidovirus transcription is a field under development and specific binding assays have recently identified a limited number of cellular proteins that associate with *cis*-acting RNA regulatory sequences. For example, differences in affinity of such factors for body TRSs might regulate transcription in nidoviruses by a mechanism similar to that of the DNA-dependent RNA-polymerase I termination system, in which specific proteins bind to termination sequences.

Origin of Nidoviruses

The complex genetic plan and the replicase gene of nidoviruses must have evolved from simpler ones. Using this natural assumption, a speculative scenario of major events in nidovirus evolution has been proposed. It has been speculated that the most recent common ancestor of the *Nidovirales* had a genome size close to that of the current arteriviruses. This ancestor may have evolved from a smaller RNA virus by acquiring the two nidovirus genetic marker domains represented by the helicase-associated zinc-binding domain (ZBD) and the NendoU function. These two domains may have been used to improve the low fidelity of RdRp-mediated RNA replication, thus generating viruses capable of efficiently replicating genomes of about 14 kbp. The subsequent evolution of much larger nidovirus genomes may have been accompanied by the acquisition of the ExoN domain. This domain may have further improved the fidelity of RNA replication through its 3′–5′ exonuclease activity, which might operate in proofreading mechanisms similar to those employed by DNA-based life forms. It has been suggested that the ORF1b-encoded HEL, ExoN, NendoU, and *O*-MT domains may provide RNA specificity, whereas the relatively abundantly expressed CPD and ADRP might control the pace of a common pathway that could be part of a hypothetical oligonucleotide-directed repair mechanism used in the present coronaviruses and roniviruses. The expansion of the replicase gene may have been associated with an increase in replicase fidelity, thus also supporting the further expansion of the 3′-proximal genome region to encode the structural proteins required to form complex enveloped virions.

Effect of Nidovirus Infection on the Host Cell

Compared to other viruses, the interactions of nidoviruses with their hosts have not been studied in great detail. In many cases, information is based on relatively few studies performed on a limited number of viruses from the families *Coronaviridae* and *Arteriviridae.* Also, most studies have been performed with viruses that have been adapted to cell culture and therefore may have properties that differ from those of field strains. Coronaviruses and arteriviruses are clearly the best-studied members of the *Nidovirales* in terms of their interactions with the host.

Coronavirus infection affects cellular gene expression at the level of both transcription and translation. Upon infection, host cell translation is significantly suppressed but not shut off, as is the case in several other positive-RNA viruses. The underlying mechanisms have not been characterized in detail, but data obtained for MHV and BCoV suggest that they may involve the 5′-leader sequences present on coronavirus mRNAs. The viral N protein was reported to bind to the 5′-common leader sequence and it has been speculated that this might promote translation initiation, leading to a preferential translation of viral mRNAs. Furthermore, host mRNAs were reported to be specifically degraded in MHV- and SARS-CoV-infected cells, further reducing the synthesis of cellular proteins. Another mechanism affecting host cell protein synthesis may be based on specific cleavage of the 28S rRNA subunit, which was observed in MHV-infected cells.

Studies on cellular gene expression following nidovirus infections have mainly focused on the coronaviruses MHV and SARS-CoV. For example, SARS-CoV infection was reported to disrupt cellular transcription to a larger extent than does HCoV-229E. Differences in cellular gene expression have been proposed to be linked to differences in the pathogenesis caused by these two human coronaviruses. Apart from the downregulation of genes involved in translation and cytoskeleton maintenance, genes involved in stress response, proapoptotic, proinflammatory, and procoagulating pathways were significantly upregulated. Both MHV and SARS-CoV induce mitogen-activated phosphate kinases (MAPKs), especially p38 MAPK. In addition, activation of AP-1, nuclear factor kappa B (NF-κB), and a weak induction of Akt signaling pathways occur after SARS-CoV infection and the N and nsp1 proteins were suggested to be directly involved in inducing these signaling pathways.

Nidoviruses have also been reported to interfere with cell cycle control. Infection by the coronaviruses TGEV, MHV, SARS-CoV, and IBV was reported to cause a cell cycle arrest in the G0/G1 phase and a number of cellular proteins (e.g., cyclin D3 and hypophosphorylated restinoblastoma protein) and viral proteins (MHV nsp1, SARS-CoV 3b 7a, and N proteins) have been proposed to be involved in the cell cycle arrest in G0/G1.

Many viruses encode proteins that modulate apoptosis and, more generally, cell death, which allows for highly efficient viral replication or the establishment of persistent infections. Infection by coronaviruses (e.g., TGEV, MHV, and SARS-CoV) and arteriviruses (e.g., PRRSV and EAV) have been reported to induce apoptosis in certain cell types. Apoptosis has also been reported in shrimp infected with the ronivirus YHV and is thought to be involved in pathogenesis. Both apoptotic and antiapoptotic molecules have been found to be upregulated, suggesting that a delicate counterbalance of pro- and antiapoptotic molecules is required to ensure cell survival during the early phase of infection, and rapid virus multiplication before cell lysis occurs. Coronavirus-induced apoptosis appears to occur in a tissue-specific manner, which obviously has important implications for viral pathogenesis. For instance, SARS-CoV was shown to infect epithelial cells of the intestinal tract and induce an antiapoptotic response that may counteract a rapid destruction of infected enterocytes. These findings are consistent with clinical observations of a relatively normal endoscopic and microscopic appearance of the intestine in SARS patients. Furthermore, SARS-CoV causes lymphopenia which involves the depletion of T cells, probably by apoptotic mechanisms that are triggered by direct interactions of the SARS-CoV E protein with the antiapoptotic factor Bcl-xL. Also the MHV E protein has been reported to induce apoptosis. The SARS-CoV 7a protein was found to induce apoptosis in cell lines derived from lung, kidney, and liver, by a caspase-dependent pathway. Apoptosis has also been associated with arterivirus infection but information on underlying mechanisms and functional implications is limited.

Coronavirus and arterivirus infections trigger proinflammatory responses that often are associated with the clinical outcome of the infection. Thus, for example, there seems to be a direct link between the IL-8 plasma levels of SARS patients and disease severity, similar to what has been described for pulmonary infections caused by respiratory syncytial virus. In contrast, despite the upregulation of IL-8 in intestinal epithelial cells, biopsy specimens taken from the colon and terminal ileum of SARS patients failed to demonstrate any inflammatory infiltrates, which may be the consequence of a virus-induced suppression of specific cytokines and chemokines, including IL-18, in the intestinal environment.

Innate immunity is essential to control vertebrate nidovirus infection *in vivo.* The induction of type I IFN (IFN-α/β) varies among different coronaviruses and arteriviruses. Whereas some coronaviruses such as TGEV are potent inducers of type I IFN, other coronaviruses (MHV and SARS-CoV) or arteriviruses (PRRSV) do not stimulate its production, thus facilitating virus escape from innate immune defenses. Type I interferon is a key player in innate immunity and in the activation of effective adaptive immune responses. Upon viral invasion, IFN-α/β is synthesized and secreted. IFN-α/β molecules signal through the type I interferon receptor (IFNR), inducing the transcription of several antiviral mediators, including IFN-γ, PKR, and Mx. IFN-γ is critical in resolving coronavirus (MHV and SARS-CoV), and also arterivirus (EAV, LDV, and PRRSV) infections. Like many other viruses, coronaviruses have developed strategies to escape IFN responses. For example, it has been shown that the SARS-CoV 3b, 6, and N proteins antagonize interferon by different mechanisms, even though all these proteins inhibit the expression of IFN by interfering with the function of IRF-3.

In arteriviruses such as PRRSV, IFN-γ is produced soon after infection to promote Th1 responses. However, PRRSV infections or vaccination with attenuated-live PRRSV vaccines cause only limited IL-1, TNF-α, and IFN-α/β responses. This then leads to IFN-γ and Th1 levels that fail to elicit strong cellular immune responses.

See also: Coronaviruses: Molecular Biology.

Further Reading

de Groot RJ (2007) Molecular biology and evolution of toroviruses. In: Snijder EJ, Gallagher T, and Perlman S (eds.) *The Nidoviruses*, pp. 133–146. Washington, DC: ASM Press.

Enjuanes L (ed.) (2005) *Current Topics in Microbiology and Immunology, Vol. 287: Coronavirus Replication and Reverse Genetics.* Berlin: Springer.

Enjuanes L, Almazan F, Sola I, and Zuniga S (2006) Biochemical aspects of coronavirus replication and virus–host interaction. *Annual Review of Microbiology* 60: 211–230.

Gorbalenya AE, Enjuanes L, Ziebuhr J, and Snijder EJ (2006) *Nidovirales*: Evolving the largest RNA virus genome. *Virus Research* 117: 17–37.

Masters PS (2006) The molecular biology of coronaviruses. *Advances in Virus Research* 66: 193–292.

Sawicki SG, Sawicki DL, and Siddell SG (2007) A contemporary view of coronavirus transcription. *Journal of Virology* 81: 20–29.

Siddell SG, Ziebuhr J, and Snijder EJ (2005) Coronaviruses, toroviruses, and arteriviruses. In: Mahy BWJ and ter-Meulen V (eds.) *Virology,* 10th edn. vol.1, pp. 823–856. London: Hoddeer-Arnold.

Snijder EJ, Siddell SG, and Gorbalenya AE (2005) The order *Nidovirales*. In: Mahy BWJ and ter-Meulen V (eds.) *Virology,* 10th edn., vol. 1, pp. 390–404. London: Hodder-Arnold.

Snijder EJ and Spaan WJM (2007) *Arteriviruses*. In: Knipe DM, Howley PM, Griffin DE, *et al.* (eds.) *Fields Virology,* vol. 1, pp. 1205–1220. Philadelphia: Lippincott Williams and Wilkins.

Spaan WJM, Cavanagh D, de Groot RJ, *et al.* (2005) Nidovirales. In: Fauquet CM, Mayo MA, Maniloff J, Desselberger U, and Ball LA (eds.) *Virus Taxonomy: Eighth Report of the International Committee on Taxonomy of Viruses*, pp. 937–945. San Diego, CA: Elsevier Academic Press.

van Vliet ALW, Smits SL, Rottier PJM, and de Groot RJ (2002) Discontinuous and non-discontinuous subgenomic RNA transcription in a nidovirus. *EMBO Journal* 21(23): 6571–6580.

Walker PJ, Bonami JR, Boonsaeng V, *et al.* (2005) *Roniviridae*. In: Fauquet CM, Mayo MA, Maniloff J, Desselberger U, and Ball LA (eds.) *Virus Taxonomy: Eighth Report of the International Committee on Taxonomy of Viruses*, pp. 975–979. San Diego, CA: Elsevier Academic Press.

Ziebuhr J and Snijder EJ (2007) The coronavirus replicase: Special enzymes for special viruses. In: Thiel V (ed.) *Molecular and Cellular Biology: Coronaviruses*, pp. 31–61. Norfolk, UK: Caister Academic Press.

Zuñiga S, Sola I, Alonso S, and Enjuanes L (2004) Sequence motifs involved in the regulation of discontinuous coronavirus subgenomic RNA synthesis. *Journal of Virology* 78: 980–994.

Nodaviruses

P A Venter and A Schneemann, The Scripps Research Institute, La Jolla, CA, USA

Introduction

Viruses belonging to the family *Nodaviridae* are small (28–37 nm), nonenveloped, and isometric. These viruses characteristically package bipartite positive-sense RNA genomes that are made up of RNA1 (3.0–3.2 kb) and RNA2 (1.3–1.4 kb). RNA1 encodes protein A, the RNA-dependent RNA polymerase (RdRp), and RNA2 encodes the capsid protein which is required for formation of progeny virions. The family is subdivided into two genera: *Alphanodavirus*, whose members infect insects, and *Betanodavirus*, whose members infect fish. Alphanodaviruses have become model systems for studies on RNA replication, specific genome packaging, virus structure, and assembly, and for studies on virus–host interactions that are required to suppress RNA silencing in animal cells. Betanodaviruses, on the other hand, cause high mortalities in hatchery-reared fish larvae and juveniles, and are therefore economically important pathogens to the marine aquaculture industry.

Taxonomy

Table 1 lists the definitive and tentative nodavirus species, their natural hosts, and geographic origin. A comparison between the capsid protein sequences of alpha- and betanodaviruses shows that these genera are distantly related with an approximate similarity of only 10%. Viruses within each genus share antigenic determinants but also show distinct immunological reactivities. Within the genus *Alphanodavirus*, the species *Flock house*

Table 1 Natural hosts and geographic origin of viruses belonging to the family *Nodaviridae*

Virus species	*Virus abbreviation*	*Host*	*Geographic origin*
Genus *Alphanodavirus*			
Black beetle virus	BBV	Scarab beetle (*Heteronychus arator*)	New Zealand
Boolarra virus	BoV	Underground grass grub (*Oncopera intricoides*)	Australia
Flock house virus	FHV	Grass grub (*Costelytra zealandica*)	New Zealand
Gypsy moth virus	GMV	Gypsy moth (*Lymantria ninayi*)	Papua New Guinea
Manawatu virus	MwV	Grass grub (*Costelytra zealandica*)	New Zealand
New Zealand virus	NZV	Unknown	Unknown
Nodamura virus	NoV	Mosquitoes (*Culex tritaeniorhynchus*)	Japan
Pariacoto virus	PaV	Southern armyworm (*Spodoptera eridania*)	Peru
Genus *Betanodavirus*			
Atlantic cod nervous necrosis virus	ACNNV	Atlantic cod (*Gadus morhua*)	Canada
Atlantic halibut nodavirus	AHNV	Atlantic halibut (*Hippoglossus hippoglossus*)	Norway
Barfin flounder nervous necrosis virus	BFNNV	Barfin flounder (*Verasper moseri*)	Japan
Dicentrarchus labrax encephalitis virus	DIEV	Sea bass (*Dicentrarchus labrax*)	France
Dragon grouper nervous necrosis virus	DGNNV	Dragon grouper (*Epinephelus lanceolatus*)	Taiwan
Greasy grouper nervous necrosis virus	GGNNV	Greasy grouper (*Epinephelus tauvina*)	Singapore
Grouper nervous necrosis virus	GNNV	Grouper (*Epinephelus coioides*)	Taiwan
Japanese flounder nervous necrosis virus	JFNNV	Japanese flounder (*Paralichthys olivaceus*)	Japan
Lates calcarifer encephalitis virus	LcEV	Barramundi (*Lates calcarifer*)	Israel
Malabaricus grouper nervous necrosis virus	MGNNV	Grouper (*Epinephelus malabaricus*)	Taiwan
Redspotted grouper nervous necrosis virus	RGNNV	Redspotted grouper (*Epinephelus akaara*)	Japan
Seabass nervous necrosis virus	SBNNV	Sea bass (*Dicentrarchus labrax*)	France
Striped jack nervous necrosis virus	SJNNV	Striped jack (*Pseudocaranx dentex*)	Japan
Tiger puffer nervous necrosis virus	TPNNV	Tiger puffer (*Takifugu rubrides*)	Japan
Umbrina cirrosa nodavirus	UCNNV	Shi drum (*Umbrina cirrosa*)	Italy

virus, *Black beetle virus*, and *Boolarra virus* are relatively closely related to each other while *Nodamura virus* (the type species of this genus) and *Pariacoto virus* are evolutionarily the most distant alphanodaviruses. Conversely, available sequences for the RdRp and capsid protein of three betanodaviruses show high levels of identity (*c.* 80% and 90%, respectively). Phylogenetic analysis based on alignments of a variable region within the betanodavirus capsid protein sequence grouped the virus species into four genotypes: *Tiger puffer nervous necrosis virus*, *Striped jack nervous necrosis virus*, the type species of the genus *Betanodavirus*; *Barfin flounder nervous necrosis virus*; and *Redspotted grouper nervous necrosis virus*. Viruses belonging to the *Striped jack nervous necrosis virus* genotype exhibit subtle but distinct serological differences to viruses belonging to the other genotypes.

Host Range and Geographic Distribution

All known alphanodaviruses were originally isolated in Australasia, except for Pariacoto virus (PaV), which was isolated in Peru. Several alphanodaviruses can multiply in a wide range of insect species in addition to their natural insect hosts (**Table 1**) including bees, beetles, mosquitoes, moths, tsetse flies, and ticks. In the laboratory, larvae of the common wax moth (*Galleria mellonella*) are convenient hosts for most of these viruses. Nodamura virus (NoV) is not only infectious to insects, but also has the unique ability to cause hind-limb paralysis and 100% mortality in suckling mice. Flock house virus (FHV), black beetle virus (BBV), and Boalarra virus (BoV) can readily be propagated to very high yields in cultured *Drosophila melanogaster* cells. PaV is not infectious to *D. melanogaster* cells, but a number of insect cell lines that are susceptible to infection by this virus have been identified, including *Helicoverpa zea* FB33 cells. FHV, NoV, and PaV can also be propagated in baby hamster kidney cells when the genomic RNAs or cDNA clones of these viruses are transfected into these cells. Other heterologous expression systems that have been shown to support FHV replication include plant cells, the yeast *Saccharomyces cerevisiae*, and the worm *Caenorhabditis elegans*.

The incidence of virus nervous necrosis, the disease caused by betanodaviruses, has been reported in Asia, Australia, Europe, Japan, and North America. The natural

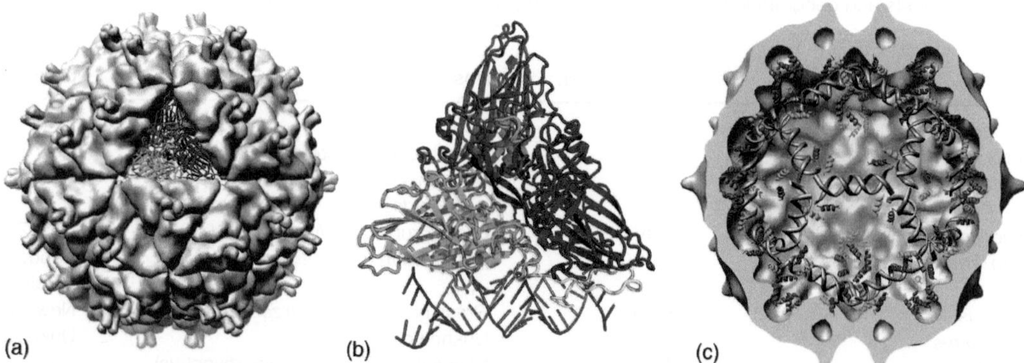

Figure 1 Structure of the alphanodavirus PaV. (a) Molecular surface rendering of PaV based on atomic coordinates. The coat protein subunits or protomers (shown in gray) adopt three different quasi-equivalent positions within the viral capsid. A-subunits are related by fivefold symmetry, while the B- and C-subunits are related by threefold symmetry. The high-resolution structure of one A-, B-, and C-protomer, which together represent one of the 60 icosahedral asymmetric units of the particle, is shown in blue, red, and green, respectively. (b) Top-down view of one icosahedral asymmetric unit showing the interaction between the extended N-terminus (yellow) of the A-subunit (blue) and the duplex RNA (magenta) located at the twofold contacts of the virion. (c) Cut-away view of the PaV virion showing the organization of the internal icosahedrally ordered duplex RNA and the gamma-peptides associated with the A-, B-, and C-subunits. Color coding is as described in panels (a) and (b). The bulk of the coat protein shell (gray) is shown at 22 Å resolution. Panels (a) and (c) were generated with the program Chimera. Courtesy of Dr. P. Natarajan.

hosts are predominantly hatchery-reared larvae and juveniles of fish species, but mortalities in adult fish have also been reported. In the laboratory, some betanodaviruses are infectious to a number of fish cell lines, including cultured cells from striped snakehead fish (SNN-1), sea bass larvae (SBL), and orange spotted grouper (GS). Striped jack nervous necrosis virus (SJNNV) also replicates when *in vitro* synthesized transcripts corresponding to its genomic RNAs are introduced into SSN-1 cells.

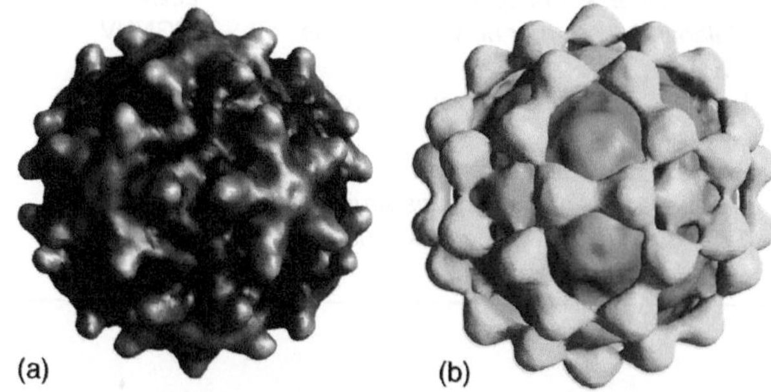

Figure 2 Three-dimensional cryoelectron microscopy reconstruction of (a) native PaV at 22 Å resolution and (b) MGNNV virus-like particle at 23 Å resolution. The inner capsid shell of MGNNV is shown in green, whereas outer protrusions are shown in yellow. Courtesy of Dr. P. Natarajan

Virion Properties

The capsids of nodaviruses consist of 180 copies of a single gene product (or protomer) arranged with $T=3$ icosahedral symmetry. The protomers adopt three slightly different conformations based on the three quasi-equivalent positions within the capsid shell. Sixty of these protomers are arranged into 12 pentamers at the icosahedral fivefold axes, while 120 are arranged into 20 hexamers at the threefold axes. In alphanodaviruses, a significant percentage of the packaged genomic RNA is organized as duplex RNA. Specifically, in the case of PaV, X-ray crystallography visualized 25 bp of double-stranded RNA (dsRNA) at each of the 60 icosahedral twofold contacts of the virion (**Figure 1**). Together, these regions of duplex RNA represent 35% of the genome and they give the impression of a dodecahedral cage in the interior of the particle. RNA cage structures are not observed for betanodaviruses. High-resolution X-ray crystal structures and cryoelectron microscopy image reconstructions of several alphanodaviruses as well as a cryoelectron microscopy image reconstruction of the betanodavirus malabaricus grouper nervous necrosis virus (MGNNV) show that capsid structures are not conserved between the two genera (**Figure 2**).

Alphanodavirus Capsids

Alphanodavirus capsids are approximately 32–33 nm in diameter (**Figure 1(a)**). Each protomer in the viral capsid is initially composed of protein alpha (~44 kDa), which spontaneously cleaves into mature capsid proteins beta (~39 kDa) and gamma (~4 kDa) following assembly of virus particles. Beta represents the N-terminal portion of alpha protein, whereas gamma represents a short C-terminal peptide that remains associated with mature virions. Alpha protein (as well as beta protein) contains a central β-barrel motif that forms the spherically closed shell of the capsid. Loops between the β-strands form the exterior surface of the capsid. The N- and C-termini of alpha protein form the interior surface, which harbors the beta–gamma cleavage site and has predominate α-helical secondary structure. Biophysical studies on the dynamic behavior of FHV particles in solution have shown that the termini are transiently exposed at the exterior surface of the capsid.

The quaternary structure of the termini associated with protomers at the fivefold axes is markedly different from those at the threefold axes:

- Specifically, in PaV, the N-termini of threefold associated protomers are structurally disordered, while the N-termini of fivefold associated protomers are well ordered. Positively charged residues within these termini make extensive neutralizing contacts with the packaged RNA (**Figure 1(b)**). The resultant RNA–protein complexes play an important role in controlling the $T=3$ symmetry of the capsid. This is illustrated by aberrant assembly of FHV capsid proteins with deleted N-termini. These mutants assemble into multiple types of particles, including smaller 'egg'-shaped particles with regions of symmetry that are similar to those of $T=1$ particles.
- The C-terminal α-helices on the gamma-cleavage products of threefold-associated protomers interact with the duplex RNA, while their fivefold-associated counterparts do not contact RNA, but are grouped together into helical bundles along the fivefold axes of the capsid (**Figure 1(c)**). The positioning of these pentameric helical bundles within the capsid and the amphipathic character of each helix makes them ideal candidates for membrane-disruptive agents that are released from the virion during cell entry to facilitate the translocation of genomic RNA into host cells. These helices have in fact been shown to be highly disruptive to artificial membranes *in vitro* as judged by their ability to permeabilize liposomes to hydrophilic solutes.

Betanodavirus Capsids

MGNNV capsids are approximately 37 nm in diameter and thus slightly larger than alphanodavirus capsids. In addition, each of the protomers of the capsid consists of two structural domains as compared to the single domain structures of alphanodavirus protomers. The domains that are more internally located in the capsid form a contiguous shell around the packaged RNA while the outer domains form distinct surface protrusions on MGNNV capsids (**Figure 2**). Each betanodavirus protomer is composed of a single capsid protein (42 kDa) that does not undergo autocatalytic cleavage.

Virion Assembly and Specific Genome Packaging

The assembly pathway of alphanodaviruses is much better characterized than that of betanodaviruses. Alphanodavirus assembly proceeds through a precursor particle, the provirion, while an equivalent assembly intermediate is not evident for betanodaviruses. For alphanodaviruses, 180 copies of newly synthesized capsid precursor protein alpha assemble rapidly in the presence of excess viral RNA into the provirion intermediate, which has identical morphology to that of a mature virion. The assembly of provirions serves as a trigger for a maturation event in which alpha is autocatalytically cleaved into beta and gamma with a halftime of about 4 h. Maturation is required for acquisition of particle infectivity. A dependence on cleavage for infectivity is in agreement with the role of gamma as a membrane disruption agent for RNA translocation during cell entry.

Although mature virions display increased chemical stability compared to provirions, it has been shown that the capsid proteins of mature virions more readily unfold under high pressure. This characteristic combined with the dynamic properties of mature virions could also be critical for their infectivity, because it could accelerate virus uncoating during cell entry.

The two genomic strands of alphanodaviruses are co-packaged into a single virion during assembly. Compelling evidence for co-packaging comes from the discovery that the entire complement of packaged RNA1 and RNA2 forms a stable hetero-complex within FHV virions when these particles are exposed to heat. The mechanism by which FHV co-packages its genome is still unknown, but a number of molecular determinants for specific genome packaging have been elucidated:

- Mutants of capsid protein alpha lacking N-terminal residues 2–31 are unable to package RNA2. The N-terminal residues of this protein are therefore bifunctional for virion assembly, because they control both the specific packaging of RNA2 and the $T=3$ symmetry of capsids.
- The C-terminal residues carried on the gamma-region of alpha protein are not only critical for virus assembly and infectivity, but also for the specific packaging of both viral RNAs.
- A predicted stem–loop structure proximal to the 5′ end of RNA2 (nt 186–217) has been proposed to represent a packaging signal for this genomic segment.

Despite the specificity for genomic RNA under native conditions, random cellular RNAs are readily packaged into virus-like particles (VLPs) when alpha is synthesized in a heterologous expression system. This is also true for the betanodaviruses MGNNV and dragon grouper nervous necrosis virus (DGNNV), whose capsid proteins have been synthesized in insect cells and *Escherichia coli*, respectively. The random RNAs packaged within alphanodavirus VLPs are organized into dodecahedral cages that are very similar to those of wild-type virions. Alpha protein is therefore the sole determinant for RNA cage formation, because specific RNA sequences and lengths are not important for this organization of the packaged nucleic acid.

Genome packaging is not only controlled by specific interactions between alpha protein and the viral RNAs, but also by a coupling between RNA replication and virion assembly. Only alpha proteins synthesized from replicating RNAs, as opposed to nonreplicating RNA, can package viral genomic RNA and partake in the assembly of infectious virions. This is exemplified by an ability to synthesize two distinct populations of FHV particles when alpha is co-synthesized from both replicating and nonreplicating RNAs within the same cell: (1) a population of infectious virions derived from the synthesis of alpha from replicating viral RNA, and (2) a population of VLPs containing random cellular RNA derived from the synthesis of alpha from nonreplicating mRNA.

Genome Structure and Coding Potential

The nodaviral genome consists of RNA1 and RNA2, and both RNAs are co-packaged into single virions. The 5′ ends of these RNAs are capped but their 3′ ends are not polyadenylated. Interestingly, the 3′ ends of alphanodavirus RNAs are unreactive with modifying enzymes to the 3′-hydroxyl groups of RNA molecules, while those of betanodavirus RNAs are marginally more reactive. It therefore appears as if these 3′ ends are blocked by an as-yet unidentified moiety or secondary structure.

A schematic representation of the genomic organization and replication strategy of FHV is shown in **Figure 3**. An open reading frame for the 112 kDa RdRp (protein A) nearly spans the entire length of FHV RNA1. RNA3, a 387 nt subgenomic RNA that is not packaged into virions, is synthesized from RNA1. RNA3 corresponds to the 3′ end of RNA1 and carries two open reading frames. One of these is in the +1 reading frame relative to protein A and encodes B2 (106 residues), which is required for the suppression of RNA silencing in infected host cells. The other matches the protein A reading frame and encodes B1 (102 residues) whose function is unknown. The coding potential of RNA1 and RNA3 is conserved within the family *Nodaviridae* with the exception that BoV RNA3 does not encode B1. RNA2 of all the species within this family has a single open reading frame for the viral capsid protein. In the genus *Alphanodavirus*, this protein is in the form of

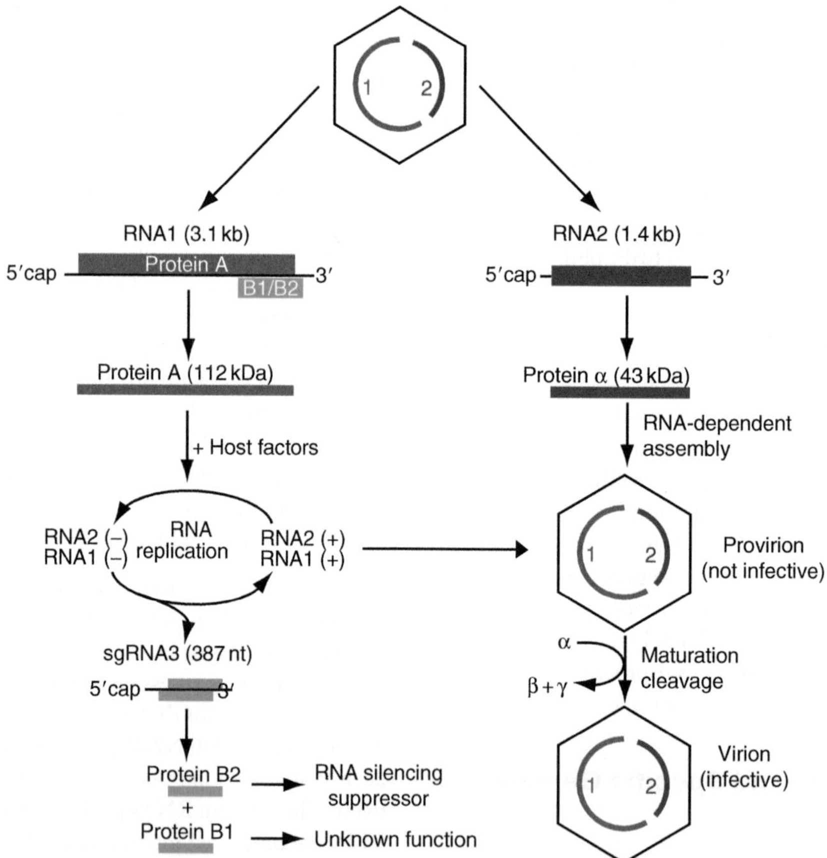

Figure 3 Genomic organization and replication strategy of the alphanodavirus FHV. Adapted from Ball LA and Johnson KL (1998) Nodaviruses of insects. In: Miller LK and Ball LA (eds.) *The Insect Viruses*, pp. 225–267. New York: Plenum.

capsid precursor protein alpha, which undergoes cleavage to produce beta and gamma, while in the genus *Betanodavirus* it is in the form of a capsid protein that is not cleaved.

Regulation of Viral Gene Expression

The information on the replication cycle of nodaviruses in this and the next section pertains primarily to alphanodaviruses, which have been studied in much greater detail than betanodaviruses.

Following virion entry, RNA and protein synthesis are temporally regulated with the net effect of dividing the cellular infection cycle of nodaviruses into two phases. In the early phase, replication complexes are established to enable the amplification of viral RNAs, and in the following phase, the synthesis of capsid protein alpha is boosted to high levels to promote virion assembly. As described in the following section, RNA3 fulfills an important role in regulating this process, which is best characterized for FHV among all the nodaviruses.

RNA Synthesis

RNA1 and RNA2 are detectable at 5 h post infection. Rates of synthesis for these RNAs peak at about 16 h, but the RNAs continue to accumulate to very high levels within the cell throughout the remainder of the infection cycle which lasts 24–36 h. RNA3, on the other hand, is co-synthesized with RNA1 in the early hours of infection, but levels of this RNA decrease dramatically after 10 h. Interactions between RNA3 and RNA2 coordinate the synthesis of RNA1 and RNA2 in two ways. First, RNA3 is required as a transactivator for the replication of RNA2, which is a unique function for a subgenomic RNA. Second, RNA2 suppresses RNA3 synthesis, which accounts for the diminished levels of RNA3 at the later stages of the replication cycle.

Protein Synthesis

During the early stages of virus infection, proteins A and B2 accumulate within the cells to establish replication complexes and to suppress the RNA-silencing defenses within the cell, respectively. The rate of protein A synthesis peaks at 5 h post infection and B2 is optimally synthesized between 6 and 10 h. Subsequent to this, the synthesis of these proteins is suppressed, but both proteins are stable and remain detectable at the very late stages of infection. The shutdown of protein A synthesis is partly due to the increased translation efficiency for RNA2 as compared with RNA1. Conversely, the decreased rate of B2 synthesis is due to the RNA2-mediated suppression of RNA3 synthesis. The synthesis of capsid protein alpha lags 2–3 h behind the synthesis of proteins A and B2. Hereafter, alpha is synthesized at a remarkable rate, which peaks at 16 h, and it accumulates to cellular levels that are significantly greater than those of proteins A and B2.

RNA Replication

The replication cycle of nodaviruses occurs exclusively within the cytoplasm of infected cells and leads to the accumulation of RNA1 and RNA2 to levels that approximate those of ribosomal RNAs. The majority of these RNAs are positive stranded, but *c.* 1% are in the form of negative strands. Protein A is the only virus-encoded protein that is required for RNA replication, because it directs the autonomous replication of RNA1 in the absence of RNA2. In addition, the other proteins encoded on FHV RNA1 (B1 and B2) are redundant for replication in cells that do not inactivate FHV via the RNA-silencing pathway. Consistent with its role in RNA replication, protein A of both alpha- and betanodaviruses contains a GlyAsp Asp motif, as well as a number of other conserved motifs for RdRps.

Replication Complexes

The replication of nodaviral RNA occurs within spherules that have a diameter of 40–60 nm and are formed by the outer membranes of mitochondria in infected cells (**Figure 4**). Within each spherule, protein A functions as an integral membrane protein that is exposed on the cytoplasmic face of the outer membrane. The N-terminus of protein A acts both as a transmembrane domain that-anchors it into the membrane and as a signal peptide for the targeting of this protein to mitochondria. It is presumed that multiple copies of protein A are associated with each spherule, because several copies thereof are visualized in spherules by immunoelectron microscopy, and this protein was shown to self-interact *in vivo* by co-immunoprecipitation and fluorescence resonance energy transfer.

Several alphanodaviruses are able to replicate their RNAs in diverse cellular environments, including those of insects, yeast, plant, worm, and mammalian cells. This shows that the formation of active replication complexes does not rely on factors that are specific to a particular host. The type of intracellular membrane is also not critical, because protein A mutants with N-terminal targeting sequences to endoplasmic reticulum membranes relocate to these membranes and support robust RNA replication. However, membrane association of protein A is important for complete replication of RNA1, RNA2, and RNA3 as gauged by the *in vitro* activities of partially purified preparations of this protein. Apart from its dependence on intracellular membranes, only one cellular host factor that supports RNA replication has

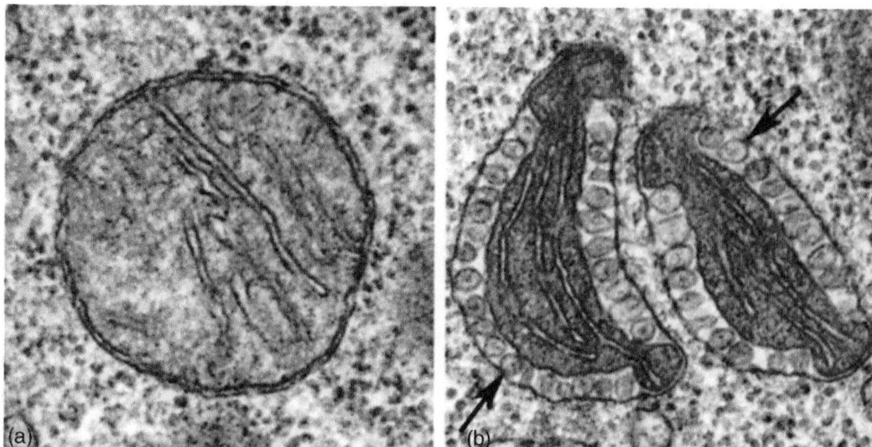

Figure 4 Electron micrographs of mitochondria in (a) uninfected and (b) FHV-infected *D. melanogaster* cells. Arrows point to spherules in which RNA replication occurs. Reproduced from Miller DJ, Schwartz MD, and Ahlquist P (2001) Flock house virus RNA replicates on outer mitochondrial membranes in drosophila cells. *Journal of Virology* 75: 11664–11676.

been identified. This factor, the cellular chaperone heat shock protein 90, does not facilitate RNA synthesis *per se*, but plays an important role in the assembly of FHV RNA replication complexes.

cis-Acting Sequences Controlling Replication

During the replication of RNA1 and RNA2, the synthesis of negative strands is governed by approximately 50–58 nt at the 3′ end of positive-strand templates. Conversely, synthesis of positive strands requires fewer than 14 nt at the 3′ ends of negative-strand templates. Internally located *cis*-acting elements on the positive strands have been identified on both FHV RNA1 (nucleotides 2322–2501) and RNA2 (nucleotides 520–720). Furthermore, it is likely that a predicted secondary structure at the 3′ end of positive-strand RNA2 from FHV, BBV, BoV, and NoV acts as a *cis*-acting signal for the synthesis of RNA2 negative strands. This structure consists of two stem loops and is preceded by a conserved C-rich motif. No conserved sequences or secondary structures are discernible at the 3′ termini of RNA1.

A long-distance base-pairing interaction between two *cis*-acting elements on positive strands of RNA1 controls the transcription of the subgenomic RNA3. The base pairing occurs between a short element (approximately 10 nt) located 1.5 kbp upstream of the RNA3 start site and a longer element (approximately 500 nt) that is located proximal to the start site. It is highly likely that this interaction promotes the premature termination of negative-strand RNA1 synthesis, which would result in the synthesis of negative-strand RNA3. Evidence that the synthesis of negative-strand RNA3 precedes the synthesis of positive-strand RNA3 supports this hypothesis. Following its transcription from RNA1, RNA3 is able to replicate in an RNA1-independent manner.

Genetics

The nodavirus RdRp switches templates to generate two types of minor RNA species by RNA recombination: (1) defective interfering RNAs (DI-RNAs) and (2) RNA dimers. DI-RNAs are internally deleted products of the genomic RNAs and often contain sequence rearrangements. RNA dimers, on the other hand, contain head-to-tail junctions between two identical copies of RNAs 1, 2, or 3, or between a copy each of RNA2 and RNA3. Both positive- and negative-strand versions of these RNAs are generated in infected cells and it is unknown whether dimers play a role in RNA replication.

Reassortment between RNA1 and RNA2 could possibly play a role in the evolution of nodaviruses. However, evidence for this is only experimental in nature, as it was shown that the RdRps from FHV, BBV, and BoV can replicate each other's RNAs and that all six of the possible reassortants are infectious.

Suppression of RNA Silencing by Protein B2

B2 plays a critical role in the life cycle of alphanodaviruses as a suppressor of RNA silencing, which acts as an antiviral defense mechanism in the insect hosts of these viruses. RNA silencing is triggered by dsRNA, which may be present at the onset of an alphanodavirus infection as RNA replication intermediates or as structured regions within single-stranded viral RNAs. During the RNA-silencing pathway, this dsRNA is processed into small (21–25 bp) interfering RNAs (siRNAs) by the endoribonuclease Dicer, and the siRNAs, in turn, guide the RNA-induced silencing complex (RISC) to specifically degrade the viral genome.

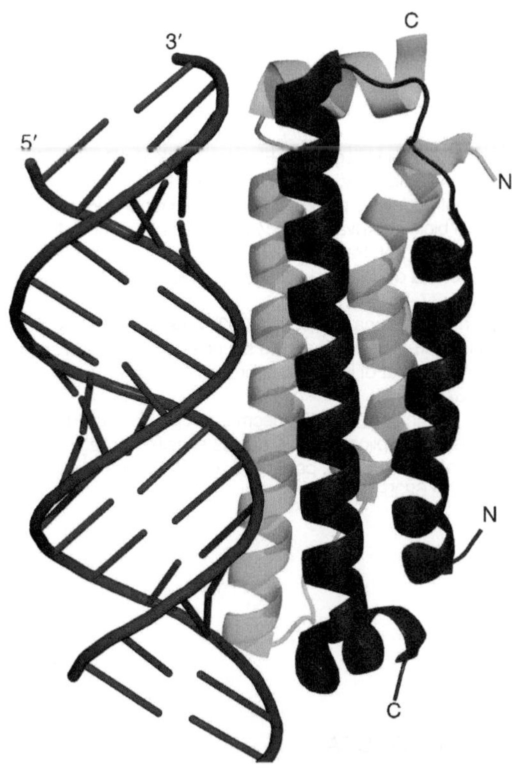

Figure 5 Structure of FHV B2 bound to dsRNA. Blue and green: B2 dimer; red: RNA. B2 forms a homodimer that binds dsRNA in a sequence-independent manner. The interactions occur between the flat face of the B2 molecule and the phosphodiester backbone of the RNA. Courtesy of Dr. P. Natarajan.

B2 has been proposed to suppress RNA silencing at two stages of this pathway. First, it binds to the siRNA cleavage products of Dicer and can therefore prevent their incorporation into RISC. Second, it is able to bind to long dsRNAs and may therefore inhibit the formation of siRNAs from full-length viral RNAs. B2 is a homodimer that forms a four-helix bundle (**Figure 5**). The dimer interacts with one face of the RNA duplex and makes several contacts with two minor grooves and the intervening major groove. These contacts are exclusively with the ribose phosphate backbone of the RNA helix, which explains the sequence-independent binding of B2 to dsRNA.

The role of alphanodaviral B2 as a suppressor of RNA silencing is not limited to insects, but extends to the RNA silencing pathways of plants, worms (*C. elegans*), and mammalian cells. Preliminary studies on betanodaviral B2 suggest that this protein functions as an inhibitor of RNA silencing in fish.

Virus–Host Cell Interactions

A general phenomenon of nodavirus infection is the presence of large cytoplasmic paracrystalline arrays of virus particles. In addition, betanodavirus-infected cells exhibit extensive vacuolation, while alphanodavirus-infected cells characteristically show a clustered distribution of mitochondria. The mitochondria are markedly elongated and deformed, and show the presence of numerous spherules on their surface (see **Figure 4**). In cell culture, alphanodavirus infections are accompanied by a progressive shutdown of host protein synthesis and the accumulation of virus particles to very high yields. Despite the high virus yields, most cell types permissive to infection by these viruses remain intact. The exceptions are *D. melanogaster* line 1 cells infected with either FHV or BBV, because these cells undergo cytolysis at 3 days post infection.

Both FHV and BBV are also able to establish persistent infections in *D. melanogaster* cells, rendering these cells immune to superinfection by FHV and BBV. Persistence is not established by mutations within the viral genes and could therefore only be attributed to cellular changes that affect interactions between the virus and its host.

Epidemiology and Pathogenesis

Disease symptoms for the natural insect hosts of alphanodaviruses have not been characterized, but are presumed to be similar to those of experimentally infected *G. mellonella* larvae, which suffer stunted growth, severe paralysis, and 100% mortality. In general, these viruses show a wide histopathological distribution within insects, and are readily detectable in neural and adipose tissues, midgut cells, salivary glands, muscle, trachea, and fat body. NoV causes severe hind limb paralysis in suckling mice, making it the only alphanodavirus pathogenic to a mammalian host. This disease symptom is caused by the replication of this virus in muscle cells and by the degeneration of spinal cord neurons.

Betanodaviruses are the causative agents of virus nervous necrosis in larvae and juveniles of both marine and freshwater fish, and cause significant problems to the marine aquaculture industry. The ability of these viruses to be transmitted both horizontally and vertically is most probably a contributing factor to the epizootic infections that have been caused by these viruses among hatchery-reared fish. Infected fish larvae show abnormal swimming behavior, which is caused by the development of vacuolating encephalopathy and retinopathy. Like alphanodaviruses, betanodaviruses are also able to establish persistent infections in their fish hosts. This is evident in the detection of these viruses in large populations of marine fish that show no clinical signs of viral nervous necrosis. The development and instatement of screening methods for the detection of these subclinical infections are therefore of great importance to the marine aquaculture industry.

Further Reading

Ball LA and Johnson KL (1998) Nodaviruses of insects. In: Miller LK and Ball LA (eds.) *The Insect Viruses*, pp, 225–267. New York: Plenum.
Chao JA, Lee JH, Chapados BR, *et al.* (2005) Dual modes of RNA-silencing suppression by flock house virus protein B2. *Nature Structural and Molecular Biology* 12: 952–957.
Delsert C, Morin N, and Comps M (1997) Fish nodavirus lytic cycle and semipermissive expression in mammalian and fish cell cultures. *Journal of Virology* 71: 5673–5677.
Li H, Li WX, and Ding SW (2002) Induction and suppression of RNA silencing by an animal virus. *Science* 296: 1319–1321.
Lindenbach BD, Sgro JY, and Ahlquist P (2002) Long-distance base pairing in flock house virus RNA1 regulates subgenomic RNA3 synthesis and RNA2 replication. *Journal of Virology* 76: 3905–3919.
Miller DJ, Schwartz MD, and Ahlquist P (2001) Flock house virus RNA replicates on outer mitochondrial membranes in Drosophila cells. *Journal of Virology* 75: 11664–11676.
Nishizawa T, Furuhashi M, Nagai T, Nakai T, and Muroga K (1997) Genomic classification of fish nodaviruses by molecular phylogenetic analysis of the coat protein gene. *Applied and Environmental Microbiology* 63: 1633–1636.
Schneemann A, Ball LA, Delsert C, Johnson JE, and Nishizawa T (2005) Family *Nodaviridae*. In: Fauquet CM, Mayo MA, Maniloff J, Desselberger U, and Ball LA (eds.) *Virus Taxonomy: Eighth Report of the International Committee on Taxonomy of Viruses*, pp. 865–872. San Diego, CA: Elsevier Academic Press.
Schneemann A, Reddy V, and Johnson JE (1998) The structure and function of nodavirus particles: A paradigm for understanding chemical biology. *Advances in Virus Research* 50: 381–446.
Tang L, Johnson KN, Ball LA, *et al.* (2001) The structure of pariacoto virus reveals a dodecahedral cage of duplex RNA. *Nature Structural Biology* 8: 77–83.
Tang L, Lin CS, Krishna NK, *et al.* (2002) Virus-like particles of a fish nodavirus display a capsid subunit domain organization different from that of insect nodaviruses. *Journal of Virology* 76: 6370–6375.
Venter PA, Krishna NK, and Schneemann A (2005) Capsid protein synthesis from replicating RNA directs specific packaging of the genome of a multipartite, positive-strand RNA virus. *Journal of Virology* 79: 6239–6248.

Orthomyxoviruses: Molecular Biology

M L Shaw and P Palese, Mount Sinai School of Medicine, New York, NY, USA

Glossary

Antigenic drift The gradual accumulation of amino acid changes in the surface glycoproteins of influenza viruses.

Antigenic shift The sudden appearance of antigenically novel surface glycoproteins in influenza A viruses.

Cap snatching The process whereby the influenza virus polymerase acquires capped primers from host mRNAs to initiate viral transcription.

Recombinant virus A virus generated through reverse genetics techniques.

Reverse genetics Techniques that allow the introduction of specific mutations into the genome of an RNA virus.

Subtype Classification of influenza A virus strains according to the antigenicity of their HA and NA genes.

Classification and Nomenclature

Influenza viruses are members of the family *Orthomyxoviridae* which is divided into five genera: *Influenzavirus A, Influenzavirus B, Influenzavirus C, Thogotovirus*, and *Isavirus*. These viruses share many common features but are defined by having a segmented, single-stranded, negative-sense RNA genome that replicates in the nucleus of infected cells. Influenza A and B viruses both have eight genome segments which can encode 11 proteins while influenza C viruses have seven genome segments that encode nine proteins. Division of influenza viruses into the *Influenzavirus A, B*, or *C* genera is based on cross-reactivity of sera to the internal viral antigens, whereas different strains within each genus are distinguished by the antigenic characteristics of their surface glycoproteins. On this basis, 16 different hemagglutinin (HA) subtypes (H1–H16) and nine different neuraminidase (NA) subtypes (N1–N9) have been described for influenza A viruses. In theory, any combination of H and N subtypes is possible but the only subtypes known to have circulated in the human population are the H1N1, H2N2, and H3N2 subtypes. All remaining H and N subtypes have been identified in viruses isolated from animals, particularly avian species. Individual influenza virus strains are named according to their genus (i.e., A, B, or C), the host from which the virus was isolated (omitted if human), the location of the isolate, the isolate number, the year of isolation, and, for influenza A viruses, the H and N subtypes. As an example, the first isolate of a subtype H3N8 influenza A virus isolated from a duck in Ukraine in 1963 is named A/duck/Ukraine/1/63 (H3N8). Influenza A viruses can undergo antigenic shift through reassortment of their glycoprotein genes. In contrast, influenza B and C viruses have only one antigenic subtype and do not undergo antigenic shift; however, all influenza viruses are characterized by minor antigenic differences, known as antigenic drift.

Virion Structure and Composition

Influenza virions are pleomorphic, enveloped particles (**Figure 1**). Spherical influenza virus particles have a diameter of approximately 100 nm but long filamentous particles (from 300 nm to several micrometers in length) have also been observed, particularly in fresh clinical isolates. The viral glycoproteins are embedded in the host-derived lipid envelope and are visible as spikes that radiate from the exterior surface of the virus when particles are viewed under the electron microscope. For influenza A and B viruses, the glycoproteins consist of the HA protein, which is the major surface protein and the NA protein. Influenza C viruses have only one surface glycoprotein, the HEF (hemagglutinin/esterase/fusion) protein. Additional minor components of the viral envelope are the M2 protein (influenza A viruses), NB and BM2 proteins (influenza B viruses), and the CM2 protein (influenza C viruses). The most abundant virion protein, M1, makes up the matrix which lies beneath the lipid membrane and surrounds the ribonucleoprotein (RNP) complexes. RNPs consist of the viral RNAs which are coated with the nucleoprotein (NP) and associated with the heterotrimeric polymerase complex (PB1, PB2, and PA proteins). Finally, small amounts of the nuclear export protein (NEP/NS2) have also been found within influenza A and B virus particles.

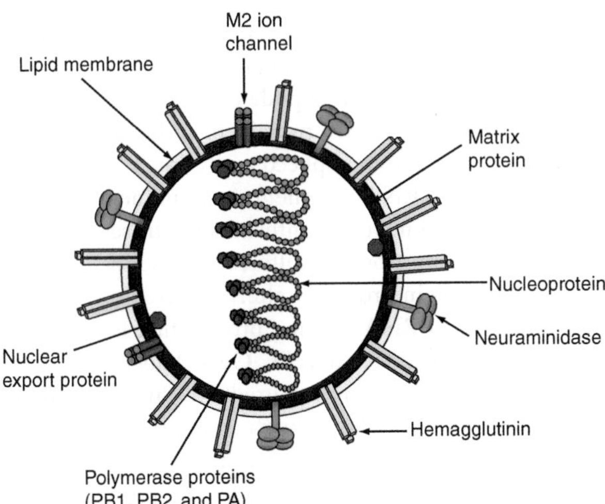

Figure 1 The viral proteins are indicated by arrows. The HA is a trimer while the NA and M2 proteins are both homotetramers. The matrix protein lies beneath the host-derived lipid envelope. Eight ribonucleoproteins form the core of the particle. These consist of the RNA genome segments coated with the nucleoprotein and bound by the polymerase complex. Small amounts of the nuclear export protein are also found within the virus particle.

Physical Properties

Because influenza viruses contain a lipid membrane, they are highly sensitive to delipidating and other denaturing agents. Differences in pH, ionic strength, and ionic composition of the surrounding medium influence the viral resistance to physical and chemical agents. The infectivity can be preserved in saline-balanced fluid of neutral pH at low temperatures. Influenza viruses are also relatively thermolabile, and are rapidly inactivated at temperatures higher than 50 °C. Agents affecting the stability of membranes, proteins, or nucleic acids, such as ionizing radiation, detergents, organic solvents, etc., reduce or completely destroy the infectivity of the virus.

Properties of the Viral Genome and Proteins

The genomes of influenza A and B viruses consist of eight separate negative-sense, single-stranded RNA segments known as vRNA. The influenza C virus genome has seven such segments. Each viral RNA segment exists as an RNP complex in which the RNA is coated with nucleocapsid protein (NP) and forms a helical hairpin that is bound on one end by the heterotrimeric polymerase complex. Noncoding sequences are present at both 5′ and 3′ ends of the viral gene segments; at the extreme termini are partially complementary sequences that are highly conserved between all segments in all influenza viruses (5′ end: AGUAGAAACAAGG and 3′ end: UCG(U/C) UUUCGUCC). When base-paired, these ends function as the viral promoter which is required for replication and transcription. The additional noncoding sequences contain the polyadenylation signal for mRNA synthesis as well as parts of the packaging signals required during virus assembly.

The eight segments of the influenza A virus genome code for the viral proteins (**Table 1**). The three largest segments each encode one of the viral polymerase subunits, PB2, PB1, and PA. The second segment also encodes an accessory protein, PB1-F2, from an alternate open reading frame within the PB1 gene. PB1-F2, which is unique to influenza A viruses, localizes to mitochondria and has pro-apoptotic activity. Segment 4 codes for the HA protein. The mature HA protein is a trimeric type I integral membrane glycoprotein which is found in the lipid envelope of virions and on the surface of infected cells. HA undergoes several post-translational modifications including glycosylation, palmitoylation, proteolytic cleavage, disulfide bond

Table 1 Influenza A virus[a] genes and their encoded proteins

Genome segment	*Length in nucleotides*	*Encoded proteins*	*Protein size in amino acids*	*Function*
1	2341	PB2	759	Polymerase subunit, mRNA cap recognition
2	2341	PB1	757	Polymerase subunit, endonuclease activity, RNA elongation
		PB1-F2[b]	87	Pro-apoptotic activity
3	2233	PA	716	Polymerase subunit, protease activity
4	1778	HA	550	Surface glycoprotein, receptor binding, fusion activity, major viral antigen
5	1565	NP	498	RNA binding activity, required for replication, regulates RNA nuclear import
6	1413	NA	454	Surface glycoprotein with neuraminidase activity, virus release
7	1027	M1	252	Matrix protein, interacts with vRNPs and glycoproteins, regulates RNA nuclear export, viral budding
		M2[c]	97	Integral membrane protein, ion channel activity, virus assembly
8	890	NS1	230	Interferon antagonist activity, regulates host gene expression
		NEP/NS2[c]	121	Nuclear export of RNA

[a]Influenza A/PR/8/34 virus.
[b]Encoded by an alternate open reading frame.
[c]Translated from an alternatively spliced transcript.

formation, and conformational changes. Cleavage of the precursor HA0 molecule into its HA1 and HA2 subunits (which are linked by a disulfide bridge) is mediated by host cell proteases and is essential for the fusion activity of HA. HA is also responsible for binding to host cell-surface receptors (sialic acid) and is the major target of neutralizing antibodies. Segment 4 of the influenza B virus genome also encodes HA but in influenza C virus, this segment codes for the HEF protein. HEF has attachment, fusion, and receptor-destroying activity and therefore incorporates the functions of HA and NA (see below) into one protein, explaining why influenza C viruses possess one less genome segment. HEF also recognizes a different cellular receptor, namely 9-*O*-acetylneuraminic acid.

The NP is encoded by segment 5. This is a highly basic protein whose main function is encapsidation of the viral RNA (an NP monomer binds approximately 24 nt of RNA) which is necessary for recognition by the polymerase. NP also plays a crucial role in transporting the viral RNPs into the nucleus which it achieves through interaction with the host nuclear import machinery. Segment 6 of influenza A and B viruses encodes the NA protein which is the second major viral glycoprotein (**Table 1**). This type II integral membrane protein is a tetramer and its sialidase (neuraminidase) activity is required for efficient release of viral particles from the infected cell. In influenza B viruses, the NA gene also encodes the NB protein from a second open reading frame. Little is known about this protein apart from the fact that it is a membrane protein and is required for efficient virus growth *in vivo*.

Segment 7 of influenza A viruses encodes two proteins, the matrix protein, M1, and the M2 protein. M1 is expressed from a collinear transcript, while M2 is derived from an alternatively spliced mRNA. M1 associates with lipid membranes and plays an essential role in viral budding. It also regulates the movement of RNPs out of the nucleus and inhibits viral RNA synthesis at late stages of viral replication. M2 is a tetrameric type III membrane protein that has ion channel activity. It functions primarily during virus entry where it is responsible for acidifying the core of the virus particle which triggers dissociation of M1 from the viral RNPs (uncoating). The BM2 protein of influenza B virus plays a similar role but it is expressed from the M1 transcript via a 'stop/start' translation mechanism. The role of CM2 in uncoating of influenza C viruses is less well established. This protein is encoded on segment 6 and is produced by signal peptide cleavage of a precursor protein (p42). The matrix protein of influenza C virus, CM1, is expressed from a spliced transcript.

The shortest RNA (segment 8 in influenza A and B viruses and segment 7 in influenza C viruses) encodes the NS1 protein from a collinear transcript and the NEP/NS2 protein from an alternatively spliced transcript. NS1 is an RNA-binding protein that is expressed at high levels in infected cells. Gene knockout studies have shown that NS1 is required for virus growth in interferon-competent hosts but not interferon-deficient hosts, indicating that NS1 acts to inhibit the host antiviral response. Influenza A virus NS1 has also been shown to interfere with host mRNA processing. Influenza B virus NS1 also has interferon antagonist activity and inhibits the conjugation of ISG15 to target proteins. The NEP/NS2 protein mediates the nuclear export of newly synthesized RNPs, corresponding with its expression at late times during viral infection.

Viral Replication Cycle

Virus Attachment and Entry

Influenza viruses bind to neuraminic acids (sialic acids) on the surface of cells via their HA proteins. The receptors are ubiquitous on cells of many species and this is one of the reasons why influenza viruses are able to infect many different animals. There appears to be some specificity associated with certain HAs, which recognize sialic acid bound to galactose in either an α-2,3- or an α-2,6-linkage. The latter linkage is more common in human cells while the α-2,3-linkage is more common in avian cells. Consequently, the HAs of human influenza viruses have a preference for the α-2,6-linkage and the HAs of avian influenza viruses are more likely to bind to α-2,3-linked sialic acids (**Figure 2**).

Entry of influenza viruses into cells involves internalization by endocytic compartments. This process is thought to occur via clathrin-coated pits, but non-clathrin-, non-caveolae-mediated internalization of influenza viruses has also been described. The actual entry process requires a low-pH-mediated fusion of the viral membrane with the endosomal membrane. The fusion activity can only occur after a pH-mediated structural change in the HA of the virus has occurred and if the HA was previously cleaved into an HA1 and an HA2 subunit. The fusion peptide at the N-terminus of the HA2 subunit interacts with the endosomal membrane, and through the acid pH-mediated conformational change the viral and endosomal membranes are brought together and fuse, opening up a pore that releases the viral RNPs into the cytoplasm. This uncoating step also depends on the presence of the viral M2 protein, an ion channel protein. The uncoating can be inhibited by amantadine (and rimantadine), which are FDA-approved anti-influenza drugs targeting the M2 protein. These compounds interfere with the influx of H^+ ions from the endosome into the virus particle, which seems to be required for the disruption of protein–protein interactions, resulting in the release of RNPs into the cytoplasm, free of the other viral proteins.

An important characteristic of the influenza virus life cycle is the dependence on the nuclear functions of the cell. Once the RNP has been released from the endosome into the cytoplasm, it is then transported into the nucleus, where RNA transcription and replication take place (**Figure 2**). This cytoplasmic nuclear transport is an energy-driven process involving the presence of nuclear localization signals (NLSs) on the viral proteins and an intact cellular nuclear import machinery.

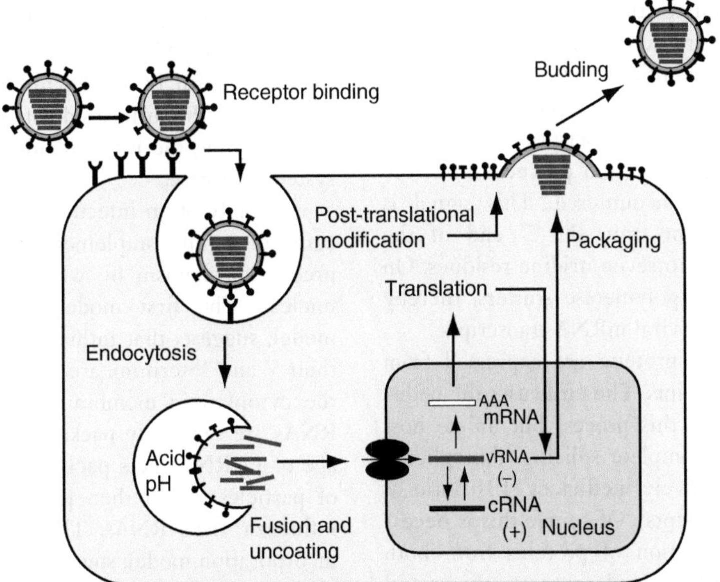

Figure 2 Virus attachment to the host cell is mediated by interaction of the HA protein with sialic acid-containing receptors. During endocytosis the interior of the endocytic vesicle becomes acidified. This induces conformational changes in HA which triggers its fusion activity and leads to fusion between the viral and endosomal membranes. The interior of the virus also becomes acidified due the ion channel activity of the M2 protein. The low pH dissociates the M1 protein from the RNP complexes, which are then transported into the nucleus. Viral RNAs are transcribed into mRNA and are replicated through a cRNA intermediate. Following export into the cytoplasm, the mRNAs are translated into viral proteins. The HA, NA, and M2 proteins are transported to the surface via the endoplasmic reticulum and Golgi (where they undergo post-translational modification). The remaining viral proteins are transported into the nucleus where they are required for either replication or nuclear export of newly synthesized vRNA. Virus assembly takes place at the apical plasma membrane which involves packaging of the eight RNPs into budding virus particles. Efficient release of budding particles requires the activity of the viral neuraminidase.

Viral RNA Synthesis

All influenza virus RNA synthesis occurs in the nucleus of the infected cell. The viral polymerase uses the incoming negative-sense vRNA as a template for direct synthesis of two positive-sense RNA species, mRNA and complementary RNA (cRNA). The viral mRNAs are incomplete copies of the vRNA and are capped and polyadenylated. cRNAs are full-length copies of the vRNA and in turn serve as a template for the generation of more negative-sense vRNA. Both vRNA and cRNA promoters are formed by a duplex of the partially complementary 5′ and 3′ ends, which increases the affinity for polymerase binding. Several studies have also indicated the importance of secondary structure in this region leading to the proposal of a corkscrew configuration. The stem–loop structures in this model have been shown to be critical for transcription.

Transcription

The viral transcription reaction is dependent on cellular RNA polymerase II activity because initiation requires a 5′ capped primer which is cleaved from host cell pre-mRNA molecules. This unique mechanism is known as cap snatching. Initiation of transcription begins with the binding of the vRNA promoter to the PB1 subunit. PB2 then recognizes and binds the cap structure on host pre-mRNAs which stimulates the endonuclease activity of PB1. The host mRNA transcripts are cleaved 10–13 nt from their 5′ caps, and the addition of a G residue which is complementary to the C residue at position 2 of the vRNA 3′ end serves to initiate transcription. Elongation is catalyzed by the PB1 protein and proceeds until the polyadenylation signal is encountered. This signal is located approximately 16 nt from the 5′ end of the vRNA and consists of five to seven uridine residues. On reaching this position the polymerase stutters, thereby adding a poly(A) tail to the viral mRNA transcripts.

Certain influenza virus proteins are expressed from alternatively spliced transcripts. The virus uses the cellular splicing machinery for this process but unlike host mRNAs which undergo complete splicing, the splicing of viral transcripts is relatively inefficient (1:10 ratio of spliced to unspliced transcripts). Of course this is necessary to ensure the production of proteins from both spliced and unspliced mRNAs and consequently several regulatory mechanisms are in place. These include the rate of nuclear export, the activity of host cell splicing factors, as well as *cis*-acting sequences within the NS1 transcript.

Replication

The replication step involves the synthesis of positive-sense cRNA from the incoming vRNA and subsequent generation of new negative-sense vRNA for packaging into progeny virions. In contrast to viral mRNAs, cRNAs are not prematurely terminated (and not polyadenylated) and do not possess a 5′ cap. Therefore the initiation and termination reactions for these two positive-sense RNAs are very different. Initiation of cRNA (and vRNA) synthesis is thought to be primer independent but it is not well understood and we also do not know how the polymerase manages to bypass the polyadenylation signal when it is in replication mode. In contrast to mRNAs, newly synthesized cRNAs and vRNAs are encapsidated and it has been proposed that the availability of soluble NP (i.e., not associated with RNPs) controls the switch between mRNA and cRNA synthesis. Other theories are that there are conformational differences between the transcription- and replication-competent polymerases and also that cellular proteins may be involved.

Packaging, Assembly, and Release of Virus

Influenza viruses assemble their components at (and in) the cytoplasmic membrane. The M1 protein has been shown to play an important role in recruiting the viral RNPs to the site of assembly at the cytoplasmic membrane, and together with the HA, NA, and M2 proteins it is the driving force in particle formation. Several influenza virus proteins, including the HA and NA, have apical sorting signals, which partially explains the observation that influenza viruses bud from the apical side of polarized cells. This asymmetric process may also provide the mechanism by which influenza viruses are largely restricted to the surface side of the respiratory tract rather than reaching the internal side, which would result in a systemic disease.

Assembly of an infectious particle requires the packaging of a full complement of the RNA genome. The precise mechanism by which this is achieved remains unclear. The first model, the random incorporation model, suggests that influenza virus RNAs, by virtue of their 5′ and 3′ termini, are recognized by viral proteins at the cytoplasmic membrane and that the correct eight RNAs are randomly packaged. If, for example, an average of 12 RNAs gets packaged, a reasonable percentage of particles would then possess a full set of the eight influenza virus RNAs. The other model, the selective incorporation model, suggests that the newly made particle has only eight RNA segments because specific signals on each RNA result in the incorporation of one segment each of the eight different RNAs. There is increasing evidence to support the validity of the latter model: there appear to be unique packaging signals on each gene segment which include both coding and noncoding sequences near the 5′ and 3′ ends of the eight different RNAs. However, it is still not clear how (or even whether) such sequences define structural elements required for the packaging of each different RNA.

Infectious particles are only formed after budding is complete. Infectious influenza virus is not found inside of cells, as is the case for poliovirus, adenovirus, and others. As influenza virus particles bud off the cytoplasmic membrane, the viral NA is needed to remove sialic acid from the cell surface as well as from the carbohydrate-carrying viral glycoproteins themselves. In the absence of a functional NA or in the presence of NA inhibitors, virus particles are not released but instead form aggregates on the surface of infected cells. The NA is thus a releasing factor which is needed in order for the newly formed virus to break free and move on to infect neighboring cells. Oseltamivir and zanamivir are two FDA-approved neuraminidase inhibitors which have strong prophylactic and therapeutic activity against influenza A and B viruses.

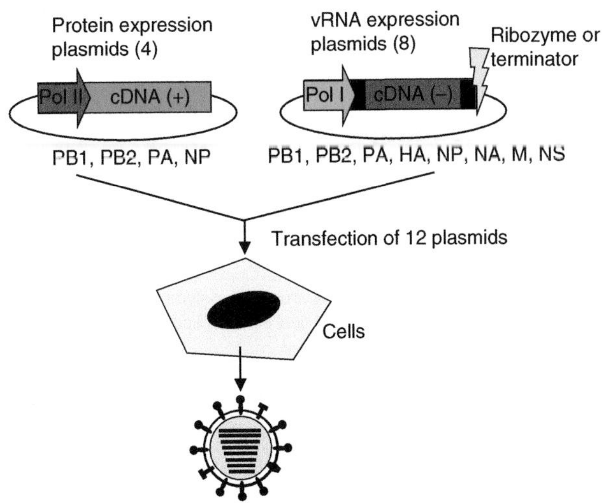

Figure 3 The negative-sense cDNA for each viral segment is cloned between a polymerase I promoter and the hepatitis delta virus ribozyme or polymerase I terminator. These plasmids will give rise to exact copies of the eight viral RNA segments. These eight plasmids are transfected into mammalian cells along with four expression plasmids for the polymerase proteins and NP, which are required for replication of the viral RNAs. The viral RNAs are replicated and transcribed by the reconstituted viral polymerase and result in the formation of infectious recombinant influenza virus.

Reverse Genetics of Influenza Viruses

Influenza viruses are negative-strand RNA viruses, and thus their genomic RNA, when transfected into cells, does not result in the formation of infectious virus. (In contrast, the RNA of poliovirus, a positive-strand RNA virus, after transfection into cells gives rise to infectious poliovirus particles.) Initial experiments involving the transfection of a synthetic influenza virus RNA mixed with purified viral polymerase and superinfection with helper virus resulted in the formation of infectious virus which had at least one cDNA-derived gene (into which mutations could be introduced). More recently, transfection of plasmids expressing the viral RNAs and viral proteins led to a helper-free virus rescue system (**Figure 3**). The term reverse genetics refers to the fact that viral RNA is first reverse-transcribed into DNA, which can be modified. This mutated DNA is then used in the plasmid-only transfection system to create infectious influenza virus. The advances of reverse genetics have been invaluable for the study of the structure/function relationships of the different influenza virus genes and proteins. In many cases, the definitive role of a protein (or domain of a protein) can only be explored by making the appropriate knockout mutant or by changing the genes in a virus and then studying its effect on virus replication.

Reverse genetics has also been used to reconstruct the extinct 1918 influenza virus using sequence data from tissue samples of victims of the 1918 pandemic. The study of this virus showed that the 1918 virus was intrinsically more virulent than other human influenza viruses and that this fact, more than any other, was responsible for the high morbidity and mortality observed in the 1918–19 pandemic.

Another application of reverse genetics techniques relates to making improved influenza virus vaccines. In the case of the avian H5N1 viruses, most strains are highly virulent by virtue of their having a basic cleavage peptide between the HA1 and HA2 subunits of the HA. The presence of this sequence of basic amino acids in the cleavage peptide allows for rapid cleavage of the HA in a variety of cells and has been found to be associated with rapid growth and high virulence. For vaccine-manufacturing purposes, it is advisable to remove this sequence signature of virulence by reverse genetics techniques. Also, the master seed strains for use in killed and live influenza virus vaccines could be constructed by reverse genetics, which would greatly accelerate the time-consuming process of selecting the right strains during interpandemic years.

Reverse genetics also allows the construction of influenza viruses which can express foreign genes. Such constructs are not only helpful in the laboratory for studies of the effects of the foreign genes in the context of an infectious virus, but they may also be useful for the development of combination vaccines. Reverse genetically modified influenza viruses represent a potential platform to express foreign antigens which can induce protective immunity against several disease agents. Genetically modified influenza viruses might also be used in the fight against cancer, as influenza viruses have been shown to be oncolytic and to induce a vigorous T-cell response. The latter could be taken advantage of by expressing tumor antigens from an influenza virus vector, which then would result in the induction of a vigorous cytolytic T-cell response against cancer cells expressing such antigens.

See also: Antiviral Agents; Orthomyxovirus Antigen Structure.

Further Reading

Cros JF and Palese P (2003) Trafficking of viral genomic RNA into and out of the nucleus: Influenza, Thogoto and Borna disease viruses. *Virus Research* 95: 3–12.

Deng T, Vreede FT, and Brownlee GG (2006) Different *de novo* initiation strategies are used by influenza virus RNA polymerase on its cRNA and viral RNA promoters during viral RNA replication. *Journal of Virology* 80: 2337–2348.

Fodor E, Devenish L, Engelhardt OG, *et al.* (1999) Rescue of influenza A virus from recombinant DNA. *Journal of Virology* 73: 9679–9682.

Gamblin SJ, Haire LF, Russell RJ, *et al.* (2004) The structure and receptor binding properties of the 1918 influenza hemagglutinin. *Science* 303: 1838–1842.

Garcia-Sastre A and Biron CA (2006) Type 1 interferons and the virus–host relationship: A lesson in *détente*. *Science* 312: 879–882.

Hayden FG and Palese P (2002) Influenza virus. In: Richman DD, Whitley RJ, and Hayden FG (eds.) *Clinical Virology,* 2nd edn., pp. 891–920. Washington, DC: ASM Press.

Krug RM, Yuan W, Noah DL, and Latham AG (2003) Intracellular warfare between human influenza viruses and human cells: The roles of the viral NS1 protein. *Virology* 309: 181–189.

Neumann G, Watanabe T, Ito H, *et al.* (1999) Generation of influenza A viruses entirely from cloned cDNAs. *Proceedings of the National Academy of Sciences, USA* 96: 9345–9350.

Noda T, Sagara H, Yen A, *et al.* (2006) Architecture of ribonucleoprotein complexes in influenza A virus particles. *Nature* 439: 490–492.

Palese P (2007) Influenza and its viruses. In: Engleberg NC, DiRita V, and Dermody TS (eds.) *Schaechter's Mechanisms of Microbial Disease,* 4th edn., pp. 363–369. Philadelphia, PA: Lippincott Williams and Wilkins.

Palese P and Shaw ML (2007) *Orthomyxoviridae*: The viruses and their replication. In: Knipe DM, Howley PM, Griffin DE, Lamb RA, and Martin MA (eds.) *Fields Virology,* 5th edn., pp. 1647–1689. Philadelphia, PA: Lippincott Williams and Wilkins.

Pinto LH and Lamb RA (2006) Influenza virus proton channels. *Photochemical and Photobiological Sciences* 5: 629–632.

Schmitt AP and Lamb RA (2005) Influenza virus assembly and budding at the viral budozone. *Advances in Virus Research* 64: 383–416.

Stevens J, Blixt O, Tumpey TM, *et al.* (2006) Structure and receptor specificity of the hemagglutinin from an H5N1 influenza virus. *Science* 312: 404–410.

Tumpey TM, Basler CF, Aguilar PV, *et al.* (2005) Characterization of the reconstructed 1918 Spanish influenza pandemic virus. *Science* 310: 77–80.

Antigenic Variation

G M Air and J T West, University of Oklahoma Health Sciences Center, Oklahoma City, OK, USA

Glossary

Envelope protein (Env) The surface spike of HIV-1.

Gp120 The subunit of HIV-1 Env that binds to receptors on cells to initiate infection.

Gp41 The subunit of HIV-1 Env that carries fusion activity.

Hemagglutinin Influenza surface glycoprotein that binds cell surface receptors to initiate infection.

Neuraminidase Influenza surface glycoprotein that cleaves sialic acid receptors to spread virus to new cells.

Neutralizing antibody An antibody that blocks virus infection.

Manifestations of Antigenic Variation

Antigenic variation refers to the observation that different isolates of a single virus species may show variable cross-reactivity when tested with a standard serum. The homologous virus (the isolate that was used to raise the antiserum) usually shows the highest reactivity. Cross-reactivities with other viruses of the same species may vary from high to zero. While zero cross-reaction is due to a different type or subtype of the species, intermediate reactivities define antigenic groups or serotypes.

Immune response to viruses is generally thought of as negative selection – that is, elimination of the pathogen. There are a few examples where immune recognition has a positive influence on the continued proliferation of the recognized pathogen, due to viral mechanisms to exploit immune elimination. Some viruses can use Fc receptors or mannose receptors to internalize antibody-bound virus into a replication mode instead of a destructive one. Alternatively, the virus may sabotage the immune response by changing antigenicity. A primary basis of antigenic variation is selection of virus mutants by antibodies. These are known as escape mutants, and since the escaped virus is resistant to antibody neutralization it possesses a fitness advantage in the presence of antibody. Escape mutants may also be selected by CD4+ or CD8+ T cells. There is certainly variation in the T-cell epitopes, but a lack of definitive examples that these have been selected by T cells as an immune evasion mechanism in the same way as occurs with antibody selection.

There are several mechanisms that allow mutations to occur with sufficiently high frequency to be selected by antibodies. Viruses with an RNA genome show the highest degrees of antigenic variation. The mutations are not induced by antibody but are present in the population and can be selected out by antibodies when wild-type virus particles are neutralized. The origin of this rather high rate of mutation is the viral replicase. RNA polymerases have no 3′ editing function and so insertion of an occasional mismatched base is not corrected and quasispecies with random mutations are present in any virus population. The mystery of RNA viruses is not why they show so much antigenic variation, but why some RNA viruses are antigenically almost invariant. Rhinoviruses exist in over 100 serotypes while poliovirus, another picornavirus, is stable enough that the vaccine did not need updating during the WHO Global Eradication campaign. Dengue virus exists in four serotypes and shows large variations in antigenic cross-reactivity within each serotype, but another flavivirus, yellow fever virus, is antigenically stable. Respiratory syncytial virus exists as two distinct antigenic groups with high antigenic diversity within each group, but other paramyxoviruses, such as measles, mumps, and rubella, are stable and therefore amenable to easier vaccination protocols.

It is clear from the above observations that vaccine success is inversely related to the degree of antigenic variation in the targeted virus. The measles-mumps-rubella vaccine has reduced those diseases almost to zero in the developed world and where vaccine compliance is high, but there is still no vaccine against respiratory syncytial virus. It does not follow that an antigenically stable virus necessarily leads to a successful vaccine, since viruses have evolved many different mechanisms to evade the immune system. Some viruses remain hidden from antibodies – for example, in neural tissues. Some make cytokine mimics that block immune signaling, or code for proteins that inhibit signaling pathways. But it is antigenic variation that causes as yet insurmountable barriers to making effective vaccines for many pathogens where the diversity is high. If all antigenic groups are represented in the vaccine, the dose becomes extremely large with consequent danger of adverse side effects. The recently licensed human papillomavirus vaccine contains the three genotypes considered oncogenic, but difficulties in culturing human papillomaviruses has precluded a serological classification of the more than 100 genotypes that exist and it remains unclear how broadly effective the current vaccine will be in areas where different or unique subtypes co-circulate.

DNA viruses do not generally incorporate mutations during replication. DNA polymerases can only extend from a base that is correctly hydrogen bonded to its partner on the template strand. A mismatched nucleotide causes synthesis to stop until the wrong base is excised by the 3′ nuclease activity. Nevertheless, DNA viruses have accumulated mutations during evolution, and some have a high rate of recombination that leads to antigenic diversity. Some small DNA viruses are quite diverse in sequence (papillomaviruses) and canine parvovirus has evolved different antigenic properties since it first appeared in the 1970s. There is evidence that the Epstein–Barr and Kaposi's sarcoma herpesviruses undergo some degree of antigenic variation.

There are distinct patterns of antigenic diversity among viruses. Most exist in several serotypes that 'rotate' in the human population, resulting in waves of disease. As antibodies accumulate in the population against one serotype, another can move in because it is not neutralized by those antibodies. Presumably, there is little antigenic memory against these viruses, because even with >100 serotypes of rhinoviruses, a person would eventually become exposed to all of them. Alternatively, the lack of immunity against all serotypes could be because the immune response focuses on single or a few immunodominant but noncross-reactive epitopes (a phenomenon known as original antigenic sin) or rapid clearance of the antigen may result in failure to achieve affinity maturation of the antibody response. Influenza viruses, on the other hand, are constantly evolving into new antigenic variants and the old ones never seem to return. It is not clear if other viruses, even very variable RNA viruses, undergo the same style of antigenic drift as influenza: a progressive, unidirectional evolution. The most variable virus of all is human immunodeficiency virus (HIV), the cause of AIDS. During HIV infection each infected individual generates a unique swarm of virus variants known as a quasispecies that diversify in a radial fashion. For many virus species, the pattern of variation is not well understood, due to infrequent epidemics, geographic isolation, or lack of sufficiently extensive molecular analysis. Here we compare and contrast the antigenic variability of the most-studied viruses, influenza virus and HIV.

Antigenic Variation in Influenza Viruses

Influenza viruses are classified as types A, B, or C based on cross-reactivity of internal antigens. Type A influenza viruses, those most commonly associated with human infection, show two distinct mechanisms of antigenic variation: antigenic shift and antigenic drift. Antigenic shift is a replacement of one subtype of surface antigen with another. There are two surface glycoprotein antigens, hemagglutinin (HA) and neuraminidase (NA) that exist in multiple serotypes. To date there are 16 subtypes of HA (H1–H16) and nine subtypes of NA (N1–N9), based on lack of antigenic cross-reactivity. H1, H2, H3, N1, and N2 have been found in human epidemic viruses. All subtypes are found in avian influenza viruses and thus birds are considered the natural viral reservoir, providing new antigens that occasionally are transferred into human viruses. The segmented influenza genome facilitates this replacement, since a new antigen gene can be readily exchanged into an existing human virus during a mixed infection without the need for RNA recombination. However, avian viruses do not easily infect mammals, so one hypothesis is that such mixing can only occur in a species that can host both of the parental viruses, such as the pig. Pigs have not been implicated in the transmission of avian influenza H5N1 to humans, which continues to be a rare event, and so far no reassortant viruses with H5 HA have been found. Antigenic shifts are shown in **Figure 1**. In recent years influenza viruses circulating in humans have been H1N1, H3N2, and type B.

Antigenic drift is a less dramatic form of antigenic variation but is responsible for annual epidemics of influenza. The term refers to a gradual change in serological cross-reactivity when compared to the original pandemic virus. Drift has been continuous in H3 HA since 1968, in N2 NA since 1957, and in H1 and N1 since 1977. Drift is detected using ferret antiserum, which discriminates better than antisera of other species, raised against a particular H3N2 strain. There is some degree of cross-reactivity to viruses isolated before and after the appearance of that strain, typically over a period of about 10 years, but for better protection, the vaccine strain is changed as soon as cross-reactivity is decreased.

The progressive, unidirectional changes in antigenic cross-reactivity correlate with progressive accumulation of changes in amino acid sequence. **Figure 2** shows the sequence changes that have occurred in the H3 HA since it appeared in the human population in 1968. Comparison of isolates in any given year shows the same antigenic properties and virtually the same sequence worldwide. This is commonly attributed to air travelers carrying the virus across and between continents, but it is also possible that the same mutations have been independently selected in different places.

Mechanism of Antigenic Drift

The mechanism of antigenic drift in influenza viruses is relatively well understood, although it does not allow one to predict the next epidemic strains. The development of monoclonal antibodies, and the demonstration that these can be used to select antigenic variants that escape from antibody neutralization by a single amino acid change, led to the mapping of neutralizing epitopes on both HA and NA. Crystal structures of HA and NA bound to neutralizing antibodies show that the amino acids that change in escape mutants are within the epitope as defined by amino acids that contact the antibody. Potentially, escape mutants could have amino acid sequence changes outside the epitope causing a conformational change that is transmitted to the epitope, but this has not been observed. The reason is that a widespread conformational change would almost certainly affect the function of the HA or NA, debilitating the virus, and the selection of escape mutants

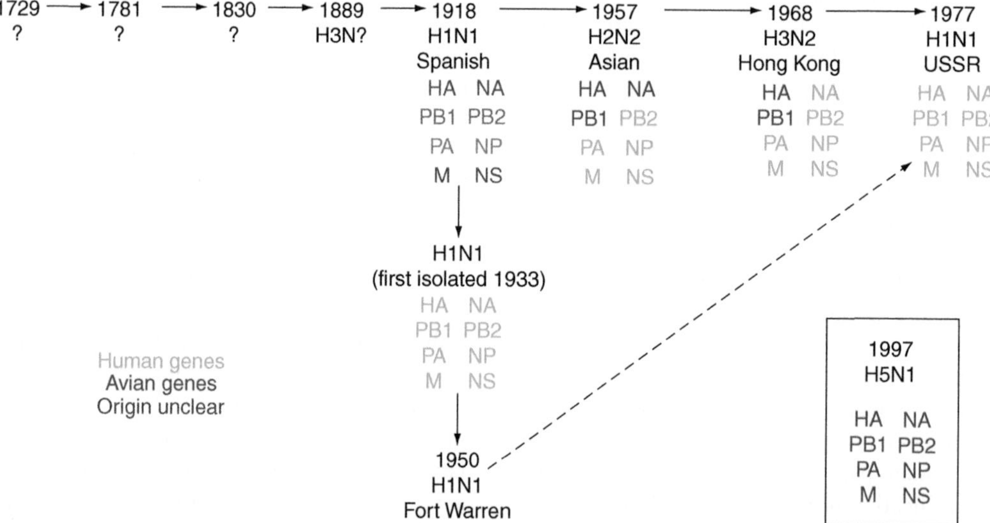

Figure 1 Influenza pandemics are very distinctive in their epidemiology and descriptions can be found back to ancient times. The figure shows some well-described pandemics (antigenic shifts) of more recent times. Retrospective serology of samples taken before the H3N2 pandemic of 1968 from people who had lived through the 1889 pandemic showed the presence of H3 antibodies. The virus was first isolated in 1933 and antibodies raised against it showed no break back to the pandemic of 1918, classified as H1N1. The H2 HA and N2 NA introduced in 1957 were both of avian origin, as was the H3 HA in 1968. The return of H1N1 in 1977 is not regarded as a pandemic since people over the age of 20 still had antibodies. The genes have been sequenced from the 1918 virus and many authors say they were of avian origin, but in the absence of an essentially identical avian influenza sequence from the same time period, this conclusion is not substantiated. The H5N1 virus that has occasionally infected people since 1997 is wholly avian in its genes. Potentially it could provide new antigens to human influenza viruses, or mutate to spread more widely into the human population.

```
N-linked glycosylation: yellow
Antigenic sites: A red, B blue, C green, D magenta, E cyan

                               10         20         30         40         50         60         70         80    85
                                |          |          |          |          |          |          |          |     |
       Conserved:      *  * ***** ********** *** ***** ********* *********  **  ** *** * ******* **** ** ** *  **
A/Aichi/2/68           QDLPGNDNST ATLCLGHHAV PNGTLVKTIT DDQIEVTNAT ELVQSSSTGK ICNNPHRILD GIDCTLIDAL LGDPHCDVFQ NETWD
A/Udorn/307/72         ..F....... .......... .......... N......... .......... .......... .......... .......G.. .....
A/Port Chalmers/1/73   ..F....... .......... .......... N......... .......... .......... ..N....... .......G.. .....
A/Bangkok/1/79         .N........ .......... .......... N......... .........R ..DS...... .KN....... .......G.. ..K..
A/Philippines/2/82     .N........ .......... .......... N......... .........R ..DS...... .KN....... .......G.. ..K..
A/Leningrad/360/86     .K........ .......... .......... N......... .........R ..DS...... .KN....... .......G.. ..K..
A/Beijing/353/89       .K........ .......... .......... N......... .........R ..DS...... .KN....... .......G.. .KE..
A/Beijing/32/92        .K........ .......... .......... N......... .........R ..DS...... .KN....... .......G.. .KE..
A/Wuhan/359/95         .K........ .......... .......... N......... .........R ..DS...... .KN....... .......G.. .KE..
A/Sydney/5/97          .KI....... .......... .......... N......... .........R ..DS...... .EN....... .......G.. .KE..
A/Panama/2007/99       .K........ .......... S......... N......... .........R ..DS..Q... .EN....... .......G.. .KE..
A/Fujian/411/2002      .K........ .......... ...I...... N......... .........G ..DS..Q... .EN....... ....Q..G.. .KK..
A/California/7/2004    .K........ .......... ...I...... N......... .........G ..DS..Q... .EN....... ....Q..G.. .KK..
A/Wisconsin/67/2005    .K........ .......... ...I...... N......... .........G ..DS..Q... .EN....... ....Q..G.. .KK..
A/New York/928/2006    .K........ .......... ...I...... N......... .........E ..DS..Q... .EN....... ....Q..G.. .KK..
A/Oklahoma/1050/06     .K..R..... .......... ...I...... N......... .........E ..DS..Q... .EN....... ....Q..G.. .KK..

                             90        100        110        120        130        140        150        160        170
                              |          |          |          |          |          |          |          |          |
                       ** ***** * ********** ********** *****  * * **** * *  *  * *  ********       ** ** *****
A/Aichi/2/68           LFVERSKAFS NCYPYDVPDY ASLRSLVASS GTLEFITEGF TWTGVTQNGG SNACKRGPGS GFFSRLNWLT KSGSTYPVLN VTMPN
A/Udorn/307/72         .......... .......... .......... ......S... .......... ........D. .........Y .......... .....
A/Port Chalmers/1/73   .......... .......... .......... ......N... .......... ........D. .........Y ....A..... .....
A/Bangkok/1/79         .......... .......... .......... ......N... N......S.. .Y.....SDN S........Y E.E.K..... .....
A/Philippines/2/82     .......... .......... .......... ......N... N......S.. .YT....SNN S........Y E.E.K..... .....
A/Leningrad/360/86     ..I....... .......... .......... ......N... N......S.. .YT....SVN S........Y E.EYK..A.. .....
A/Beijing/353/89       ........Y. .......... .......... ......N.D. N....A.S.E .Y.....SVK S........H E.EYK..A.. .....
A/Beijing/32/92        ........Y. .......... .......... ......N.D. N....A.D.. .Y.....SVN S........H ..EYK..A.. .....
A/Wuhan/359/95         ........Y. .......... .......... .....TN... N....A.D.T .Y.....SVK S........H .LEYK..A.. .....
A/Sydney/5/97          ........Y. .......... .......... .....NN.S. N....A...T .Y....SSIK S........H QLKYK..A.. .....
A/Panama/2007/99       ........Y. .......... .......... .....NN.S. N....A...T .S....RSNN S........H QLKYK..A.. .....
A/Fujian/411/2002      ........Y. .......... .......... .....NN.S. N........T .S....RSNK S......... HLKYK..A.. .....
A/California/7/2004    ........Y. .......... .......... .....NN.S. N........T .S....RSNN S......... HLKFK..A.. .....
A/Wisconsin/67/2005    ........Y. .......... .......... .....ND.S. N........T .S....RSNN S......... HLKFK..A.. .....
A/New York/928/2006    ........Y. .......... .......... .....NN.S. N........T .S....RSNN S......... HLKFK..A.. .....
A/Oklahoma/1050/06     ........Y. .......... .......... .....NN.S. N........T .S....RSNN S......... HLKFK..A.. .....

                                180        190        200        210        220        230        240        250   255
                                  |          |          |          |          |          |          |          |     |
                         ******* * *** *    *        ** * **** *** **  * ***  *  *    *** ** ******* * * *** ** *****
A/Aichi/2/68           NDNFDKLYIW GIHHPSTNQE QTSLYVQASG RVTVSTRRSQ QTIIPNIGSR PWVRGLSSRI SIYWTIVKPG DVLVINSNGN LIAPR
A/Udorn/307/72         .......... .V.....D.. .......... ......K... .......... .......... .......... .I........ .....
A/Port Chalmers/1/73   .......... .V.....D.. ..N....T.. ......K... .......... .......... .......... .I........ .....
A/Bangkok/1/79         .G........ .V.....DK. ..N...R... ......K... .......... .......... .......... .I.L...... .....
A/Philippines/2/82     .GK....... .......DK. ..N..IR... ......K... ..V....... .......... .......... .I.L...T.. .....
A/Leningrad/360/86     .GK....... .V.....EK. ..N...R... ......K... ..V....... .......... .......... .I.L...T.. .....
A/Beijing/353/89       .GK....... .V.....DR. ..K...R... ......K... ..V....... .......... .......... .I.L...T.. .....
A/Beijing/32/92        .GK....... .V.....DRD ......R... ......K... ..VT...... .....Q.... .......... .I.L...T.. .....
A/Wuhan/359/95         ..K....... .V.....DSD ...I...... ......K... ..V....... .....I.... .......... .I.L...T.. .....
A/Sydney/5/97          ..K....... .V.....DSD ...I.A.... ......K... ..V....... .....I.... ..H....... .I.L...T.. .....
A/Panama/2007/99       .EK....... .V.....DSD .I.I.A.... ......K... ..V......I .....V.... .......... .I.L...T.. .....
A/Fujian/411/2002      .EK....... .V...G.DSD .I...A.... .I....K... ..V....... .R..DV.... .......... .I.L...T.. .....
A/California/7/2004    .EK....... .V...G..ND .I...A.... .I....K... ..V....... .R..DIP... .......... .I.L...T.. .....
A/Wisconsin/67/2005    .EK....... .V...G.DND .IF.HA.... .I....K... ..V....... .RI.NIP... .......... .I.L...T.. .....
A/New York/928/2006    .EK....... .V...G.DND .IF..A.... .I....K... ..V....... .R..DIP... .......... .I.L...T.. .....
A/Oklahoma/1050/06     .EK....... .V...G.DND .IF..A.... .I....K... ..V....... .R..NIP... .......... .I.L...T.. .....

                             260        270        280        290        300        310        320        330       340
                               |          |          |          |          |          |          |          |         |
                       ***  * *** *********   * ******* ******** * *** ** *** * ******** ********** ****¯Start HA2
A/Aichi/2/68           GYFKMRTGKS SIMRSDAPID TCISECITPN GSIPNDKPFQ NVNKITYGAC PKYVKQNTLK LATGMRNVPE KQTRGLFGAI AGFIE
A/Udorn/307/72         .......... .........G .......... .......... ......?... .......... .......... .......S.. .....
A/Port Chalmers/1/73   .......... .........G .......... .......... .......... .......... .......... ....------ -----
A/Bangkok/1/79         ....I..... .........G ..S....... ........?. .......... .......... .......... .....I.... .....
A/Philippines/2/82     ....I..... .........G ..S....... .......... .......... .R........ .......... .....I.... .....
A/Leningrad/360/86     ....I..... .........G ..S....... .......... .......... .R........ .......... ....------ -----
A/Beijing/353/89       ....I..... .........G ..S....... .......... ...R...... .R........ .......... .....I.... .....
A/Beijing/32/92        ....I.N... .........G ..S....... .......... ...R...... .R........ .......... .....I.... .....
A/Wuhan/359/95         ....I.S... .........G N.N....... .......... ...R...... .R........ .......... .....I.... .....
A/Sydney/5/97          ....I.S... .........G K.N....... .......... ...R...... .R........ .......... .....I.... .....
A/Panama/2007/99       ....I.S... .........G K.N....... .......... ...R...... .R........ .......... .....I.... .....
A/Fujian/411/2002      ....I.S... .........G K.N....... .......... ...R...... .R........ .......... .....I.... .....
A/California/7/2004    ....I.S... .........G K.N....... .......... ...R...... .R........ .......... .....I.... .....
A/Wisconsin/67/2005    ....I.S... .........G K.N....... .......... ...R...... .R........ .......... .....I.... .....
A/New York/928/2006    ...RI.S... .........G K.N....... .......... ...R...... .R........ .......... .....I.... .....
A/Oklahoma/1050/06     ....I.S... .........G K.N....... .......... ...R...... .R........ .......... .....I...- -----
```

(a)

Figure 2 Continued

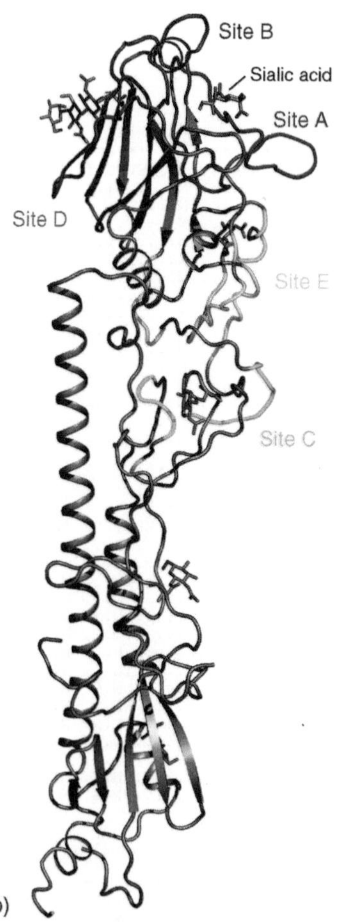

Figure 2 (a) The amino acid sequence of A/Aichi/2/68, representative of the first human H3N2 isolates. The HA precursor peptide is cleaved into two polypeptides called HA1 and HA2. Only HA1 is shown here since there are no known antigenic sites on HA2 and it shows very little sequence drift. Changes in HA1 of later viruses are shown, with dots where there is no change. Mutations that map in the five antigenic regions are colored: red, site A; blue, site B; green site C, magenta, site D; cyan, site E. The assignment of residues to antigenic sites is accessible at the Influenza Sequence Database (ISD). Sequons that predict N-linked glycosylation are shown as yellow bands. The progressive nature of antigenic (and genetic) drift is clearly seen; previous mutations are generally retained in later isolates. The sequences chosen were in most cases used as vaccine components, and essentially the same sequence (and antigenic character) was found worldwide. Two 2006 isolates are included for comparison. (b) Alpha carbon trace of a monomer of HA showing one or more sugars at each site of N-linked glycosylation and a bound sialic acid in the receptor binding site. Each peptide segment of the antigenic sites in (a) are seen to come together in the three-dimensional structure. Antigenic sites A through E are colored as in (a). Site D is partly buried in the trimer interface when the whole molecule is present. Site C is some distance from the receptor binding site but the crystal structure of a complex with antibody shows that the constant domain of the antibody reaches to and obscures the sialic acid binding site.

can only occur if the mutants are close to wild type in fitness.

The neutralizing epitopes, determined crystallographically, consist of 11–18 amino acids that are separated in the linear sequence but come together in the three-dimensional structure to form a binding surface that exists only when the antigen is folded into its native structure (**Figure 2**). The footprint of an antibody on HA and NA is quite large and so it is not surprising that antibodies contact more than one linear segment of the protein. Many years of effort have not yielded a peptide-based vaccine for influenza and it seems unlikely that this will be possible. In contrast, for HIV there is still very active research ongoing to develop peptide-based vaccines or therapeutic antibodies that bind to linear peptides, because the surface antigen gp160 may be more dynamic than influenza HA or NA and linear segments may become transiently accessible to antibodies (see below).

Mechanisms of Neutralization

The basis of antibody-mediated neutralization of influenza virus (and HIV, see below) is generally considered to be blocking of biological function rather than Fc-mediated mechanisms such as phagocytosis or complement activation. The most commonly used assay to assess effectiveness of vaccination is the hemagglutination inhibition (HAI) test, which measures blocking of the receptor-binding site by antibodies. This might seem to be an oversimplification of many potential mechanisms of neutralization, but the correlation between HAI and neutralization or protection has held up for several decades. Similarly, although NA antibodies are not as much studied, monoclonal antibodies do not neutralize or protect from infection unless they inhibit the NA activity. Thus, neutralizing antibodies against influenza bind close enough to the sialic acid binding site (HA) or the enzyme active site (NA) to inhibit binding to receptors or substrates. The binding sites themselves have constraints on tolerating changes because they must retain function, but the polypeptide loops surrounding them are able to mutate so that the neutralizing antibody can no longer bind, or binds with such low affinity that it no longer inhibits function. Site-directed mutagenesis of all side chain contacts of the epitope has shown that a subset contributes the major energy of the interaction. These two or three critical contacts are the ones that are found to change in escape mutants, and are the only changes in site-directed mutants that abolish antibody inhibition. In a few cases, the antibody still binds but does not inhibit function, but most commonly there is no detectable binding of antibody to a mutant that has a change in a critical contact. This explains how a change in only one out of the 11–19 amino acids that make up the epitope can eliminate the antibody inhibition, and greatly reduces the number of changes needed for a new antigenic variant.

How Does Antigenic Drift Occur?

The surface area of the HA accommodates a multitude of epitopes that have been grouped into antigenic sites A through E (**Figure 2**) by determining if binding of one antibody blocks the binding with another. If it does, the two antibodies are considered to belong to the same site. In the face of a human polyclonal antibody response, it would seem necessary to mutate each of the five distinct sites to generate an escape mutant. The early H3 variants show changes in all sites, but more recently the changes have been confined to fewer sites. The difference is likely to be in immunogenicity, and suggests that the human antibody response to recent viruses is not very polyclonal. The possible reasons include previous substitutions rendering a site nonimmunogenic, or blocking of sites by the addition of N-linked carbohydrates. A/Aichi/68 HA1 contained six carbohydrate chains but by 1999 there were 11 sites for carbohydrate addition. Site A may now be completely shielded from antibodies. The mobilization and addition of carbohydrate is a primary mechanism of immune avoidance in HIV (see below). The antigenically significant transition from California/04 to Wisconsin/05 was due to changes only in site B.

Is There Significant Antigenic Drift in NA?

Antibodies against NA have been shown to contribute to protection. They do not inhibit entry of human viruses into cells but they inhibit spread of progeny viruses because sialic acid is not removed from the N linked carbohydrates of the surface glycoproteins, causing aggregation of virus particles by HA–HA interaction. The rate of amino acid substitution in NA is about half that in HA, suggesting a lesser antigenic selection. Ferret antisera that show clear differences in HA antigenicity from 1 year to the next may not discriminate between drifted NAs. However, the crystal structure of the NA of A/Memphis/31/98 complexed with a neutralizing antibody suggests that NA may be contributing significantly to antigenic drift in H3N2 viruses. Of the 11 amino acids that make contact through side chains to the antibody, five have changed in isolates up to 2006 (**Figure 3**).

Antigenic Variation in HIV

Antibody and Antigenic Variation in HIV-1 Infection

Nearly all HIV-1-infected individuals develop antibodies capable of mediating some level of neutralization. The presence of strong, broadly neutralizing responses in some long-term nonprogressors suggests that neutralizing antibodies may be a correlate of protection, but their absence during acute infection, when the escalating HIV-1 viral

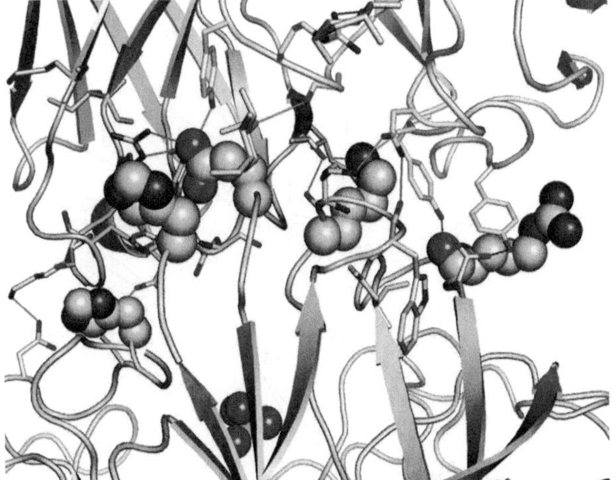

Figure 3 Changes in the epitope of a 1998 N2 NA up to 2006. The NA is brown, Fab H chain is dark blue, and L chain is light blue. Amino acid side chains that are in contact in the interface are shown as stick models, except for those on NA that have changed in later isolates, which are shown in space-filling form. Reproduced from Venkatramani L, Bochkareva E, Lee JT, *et al.* (2006) An epidemiologically significant epitope of a 1998 human influenza virus neuraminidase forms a highly hydrated interface in the NA-antibody complex. *Journal of Molecular Biology* 356(3): 651–663 (cover picture), with permission from Elsevier.

load is reduced, suggests they play a minor role in initial immune control and that cellular immune responses are more effective. HIV-1 neutralizing responses take months or even years to develop in patients but during a long and chronic infection, antibody may be critical in limiting both virus-to-cell and cell-to-cell spread of virus. It is possible that non-neutralizing responses such as antibody-dependent cell-mediated cytotoxicity (ADCC) are important in acute infection.

In contrast to influenza, there seems to be no basis for a serological definition of subtypes in HIV-1. Instead, subtypes are defined based on genetic relatedness. HIV-1 exhibits intra-subtype sequence differences of up to 20% in the envelope glycoprotein gene (*env*) and differences of up to 35% between subtypes. Only antibodies directed against the viral envelope glycoprotein that limit or prevent virus entry into CD4-bearing target cells are effective in reducing viral replication. HIV-1 is also selected by cellular immune mechanisms, and recent studies suggest that CTLs target Env more often than previously appreciated; however, here we will focus on the antigenic variation in Env and its interplay with humoral immune selection.

HIV-1 Evolves to Create Global and Intra-Patient Antigenic Variation

HIV-1, a member of the genus *Lentivirus* of the subfamily *Orthoretrovirinae* of the family *Retroviridae*, is one of the most rapidly evolving pathogens ever studied. Through a combination of rapid production of highly genetically diverse progeny and the extreme plasticity of many of its viral proteins, HIV-1 has managed to elude efforts aimed at control through drug treatment or vaccination. The most recent common ancestor of the HIV-1 pandemic subtypes came into the human population less than a century ago. Over the course of that relatively short interval, the virus has become not only one of the most significant human pathogens, but also one of the most diverse. The pandemic HIV-1 can be segregated into subtypes A–K (minus E, I, and K) (**Figure 4**). Most subtypes co-circulate in West-Central Africa pointing to this region, where the geographical habitat of chimpanzees and humans overlap, as a nexus for the phylogenetically inferred initial transmission events into humans. Circulating recombinant forms (CRFs) are rapidly emerging in areas where more than one subtype co-circulates. These recombinants pose a serious concern, as they can be thought of as analogous to reassortants in influenza and could rapidly obviate the effectiveness of vaccination unless sufficiently broad cross-reactivity can be generated. Antigenic variation within the global HIV-1 population, as well as within each infected individual, is a primary mechanism underlying the success of HIV-1 as a human pathogen.

The HIV-1 Replicase and the Generation of Diversity/Variation

In order to understand the mechanisms that produce variation, it is necessary to become familiar with some of the

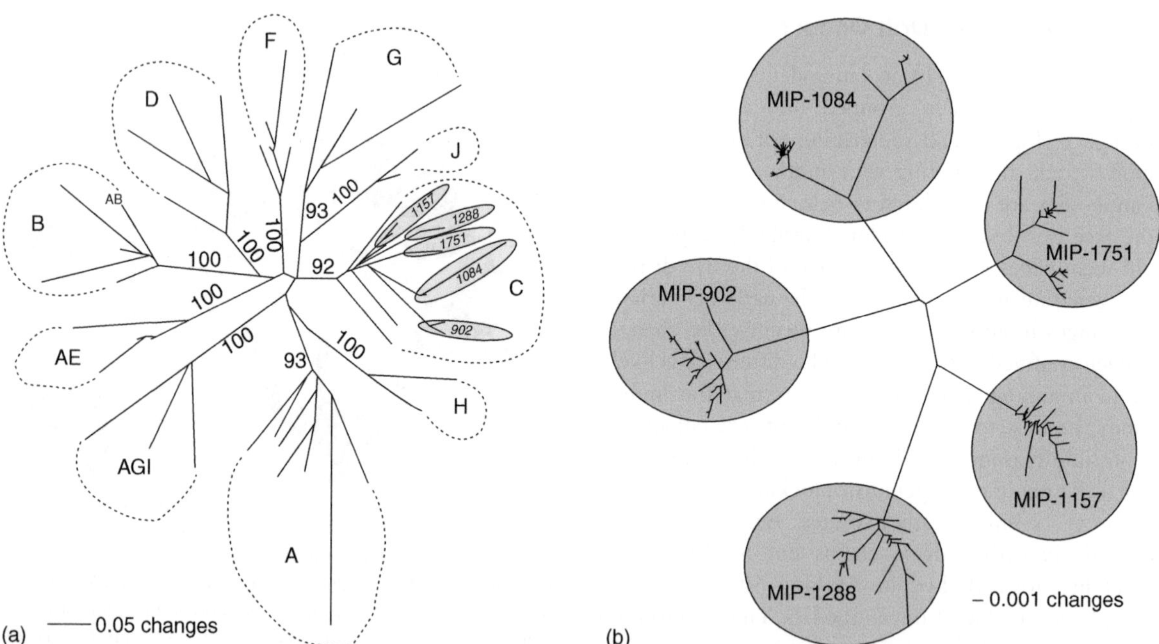

Figure 4 (a) Shows that the various HIV-1 subtypes segregate one from another when compared using envelope glycoprotein sequences and that individual viruses cluster within the subtype into distinct branches. (b) An unrooted phylogenetic tree shows that each of five mother–infant pair's glycoprotein sequences form an independent lineage despite all being subtype C viruses isolated from the same geographic region. Over time, infant and maternal sequences will also segregate as they undergo independent viral antigenic variation and selection. HIV-1 diversification appears perpetual.

basic processes of retroviral replication. HIV-1 contains two copies of message (+)-sense RNA. Replication of the HIV-1 genome proceeds through a dsDNA intermediate that becomes permanently integrated into the target cell genome. The HIV-1 replicase is an RNA-dependent DNA polymerase or reverse transcriptase (RT). The HIV-1 RT lacks any proofreading capacity. The rate of misincorporation of nucleotides by HIV-1 RT is between 10^{-3} and 10^{-4}/position/round of replication. Given a genome size of approximately 10 000 base pairs this suggests that HIV-1 introduces at least one mutation with each genome copy.

HIV-1 Diversity Influences Its Population Genetics

The viral loads in plasma for individuals undergoing acute infection are as high as 10^8 copies ml^{-1}. This means that the virus has the potential to produce all possible single mutations each day, as well as a population containing multiple mutations. The replication rate is accompanied by rapid viral and infected cell turnover, pressure to avoid immune recognition and elimination, all coupled with fitness maximization. The population of closely related but distinct viruses generated by replication is referred to as a quasispecies. For HIV, this term refers to the dynamic group of viral genotypes found in an infected individual at a given sampling. It is the creation of this swarm of subtly different progeny viruses and the resultant selection of those progeny that drives HIV natural selection and antigenic variation.

In addition to single-base mutations, HIV readily undergoes recombination but this does not occur through reassorting genomic segments as in influenza. HIV-1 does not possess a segmented genome, but instead packages two copies of viral genomic RNA. If co-infection of target cells and co-packaging of distinct RNA genomes occurs, reverse transcriptase can switch viral RNA templates during synthesis of DNA. The result is the creation of large insertions and deletions that introduce antigenic changes in a single replication cycle. While co-infection has proved difficult to monitor *in vivo*, the globally increasing number of HIV-1 CRFs suggests that co-infection occurs with relatively high frequency. In the absence of a vaccine the abundance of such recombinants will only increase given the continuation of human cultural, medical, and sexual practices that promote virus transmission, and the occurrence of infection rates in the developing world as high as 35%. Both vaccine development and drug treatment efforts will be further complicated as these recombinants become more widely intermixed and distributed.

The HIV-1 Envelope Glycoprotein: Variable Loops and Decoys

The target of HIV-neutralizing responses is the envelope glycoprotein. The glycoprotein complex exists in the virion membrane as a trimer of noncovalently associated heterodimers of surface protein (SU) and transmembrane protein (TM). The gp120 component is the most exposed component of the complex and contains five hypervariable loops (V1–V5) and five relatively invariant or constant regions (C1–C5). Vaccination with gp120 peptides, glycoprotein, or glycoprotein-expression constructs has met with only limited success, as defined by elicitation of broadly neutralizing antibody responses. Neutralization of HIV-1 correlates with blocking virus entry into cells. The diversity of HIV sequences means that the vast majority of epitopes are not shared and thus the only hope for vaccination is to target conserved epitopes. Much of what we know about Env epitopes that are cross-reactive comes from studies of a handful of human monoclonal antibodies that have broad neutralizing activity for multiple viral strains and subtypes. Structural studies of some of these antibodies in complex with gp120 have revealed their binding sites and it is now clear that they do not bind to linear epitopes but to conserved structural or conformational epitopes that the virus has evolved to rarely reveal to the immune system (**Figures 5** and **6**).

HIV-1 entry is a series of steps mediated through interactions between Env and receptors on the target cell plasma membrane. The first of these steps is a high-affinity interaction between gp120 and the primary receptor CD4. Thus, antibodies that interfere with CD4 binding would be anticipated to be effective at inhibiting infection. The interaction with CD4 induces substantial conformational changes in Env that lead to exposure of the chemokine co-receptor (usually CCR5 or CxCR4) binding site in the V3 loop and other induced structures. The CD4-induced conformation is another step at which it is thought that neutralizing antibodies might exert a selective and protective influence. The interaction with CD4 and chemokine co-receptor induces further changes in the gp120 structure that promote exposure of gp41 and insertion of its N-terminal fusion domain into the target cell membrane. This final step is also potentially a target for recognition of epitopes and the development of fusion-inhibiting antibody responses. However, despite considerable antigenic diversity, the HIV-1 glycoprotein has evolved to be an exceedingly poor neutralization immunogen; at least it is a very poor inducer of broadly neutralizing responses particularly at conserved sites most pertinent to the functional processes of Env defined above.

Some of the hypervariable loops of gp120 tolerate extensive sequence variation as well as insertion and deletion, and are excellent inducers of antibody, but rarely neutralizing antibody. In the event a loop-induced neutralizing response is generated, it is isolate specific and not cross-reactive; the virus utilizes the diversity it generates through replication to escape the neutralizing response and repopulate. Thus such loop-targeted antibodies are unlikely to have durable neutralizing capacity even in the individual. It is therefore likely that at least one function

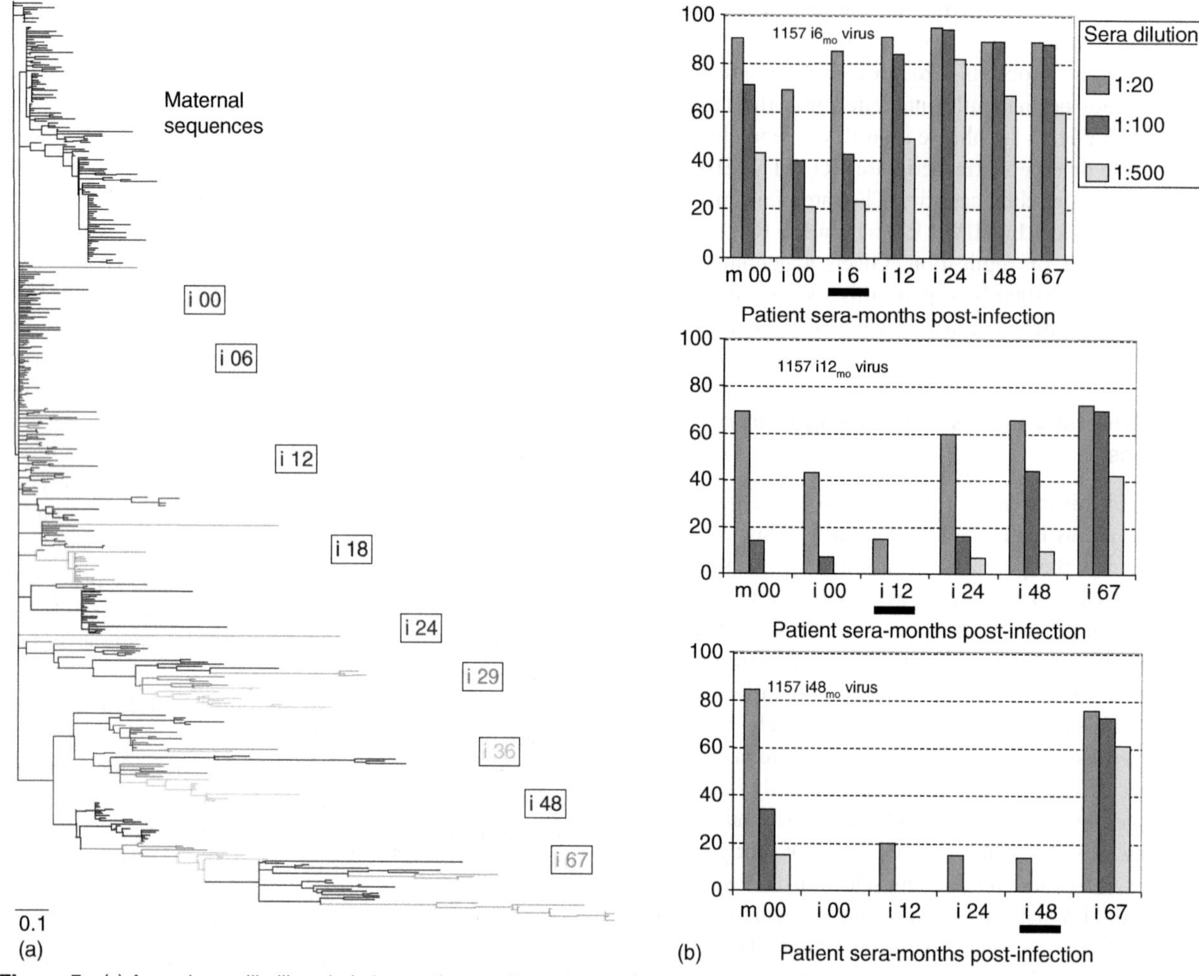

Figure 5 (a) A maximum-likelihood phylogenetic tree displaying the diversifying evolution of HIV-1 Env in a single individual. The numbers in color indicate the months post infection when samples were obtained for sequence analysis. Note that divergence from a maternal root sequence continues with time and that at later time points the virus appears to resample earlier genotypes. (b) Neutralization of viruses derived from some of the time points in (a) by autologous contemporaneous and noncontemporaneous sera suggests that recognition of early (6 month) viruses is maintained even though the genotype is clearly replaced, and that the neutralizing response generated early is incapable of neutralizing virus generated after selection (12 month serum is ineffective on 12 month virus or subsequent time points). In total, these data suggest that diversity generates an antibody response that is never 100% effective, not broadly reactive and perhaps mistargeted by 'original antigenic sin' as discussed in the text.

of the loops is to decoy the immune response, misdirecting it to portions of the protein that can continually support variation to no discernable functional detriment for the virus. In support of this concept, HIV-1 with deletion of V1–V2 is still replication competent but is more susceptible to neutralization.

The CD4 binding site is conserved in all HIV subtypes and is contained within a deep and inaccessible pocket of the protein that is inaccessible to antibodies (the canyon hypothesis). Ab that could interact with Env to prevent CD4 association would be neutralizing. However, it appears that the loops also serve to mask important targets for neutralization such as the CD4 binding site and CD4-induced epitopes such as the co-receptor binding regions by sterically inhibiting the access of antibody to potential epitopes or by limiting the affinity with which such Ab can associate. The loops are also extensively glycosylated and this also impacts recognition as discussed below.

Env Glycosylation and Antibody Recognition

HIV-1 Env gp120 and gp41 have been shown to be extensively modified by an average of ~30 carbohydrate additions for both proteins, and glycosylation of the HIV-1 Env is essential for folding and oligomerization. Moreover, glycosylation participates in protecting HIV-1 from recognition by antibodies. This concept, termed glycan shielding, suggests that the concentration of carbohydrate additions on gp120 block antibody access to the protein domains beneath that are critical for CD4 recognition, co-receptor binding, and exposure of the fusion peptide. Highly conserved glycans often play important structural

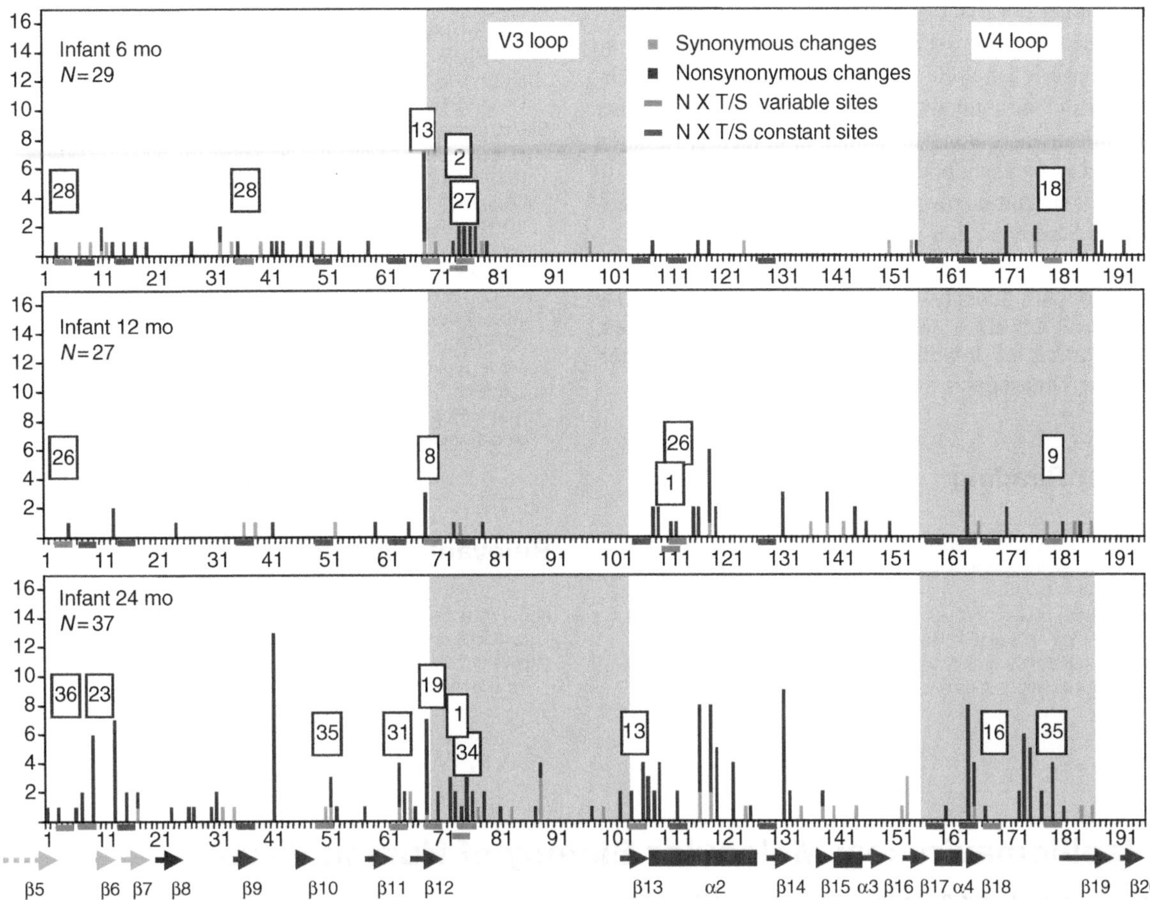

Figure 6 Amino acid sequence changes found in the HIV-1 subtype C Envelope glycoprotein in a newly infected child. C2-V4 diversity (amino acids 228–427 in HXB2 numbering) for sequences in a single patient (N = number of sequences per time point) in relation to potential glycosylation sites that are either constant in the population at that time point (blue) or show variability (pink). The proportion of the glycans maintained, for variable sites, is shown in the boxes above the sites. Red bars indicate the presence of nonsynonymous changes (codon level) and their height indicates the abundance in the population. Green bars indicate synonymous polymorphism. Note that variability increases with time, sites undergoing positive selection (red bars) change with time in some instances but not in others, and that glycan addition sites change both in abundance and position over the course of infection. Designations at the bottom of the figure shows relationships to structural elements identified in the unliganded SIV envelope glycoprotein compared with HIV-1 structures PDB id 2BF1.

and functional roles, but others are highly polymorphic between subtypes and within infected individuals. Thus, variability in the protein sequence in or around glycosylation sites affects immune recognition by subtly altering the primary sequence, the fold of the structure, and/or accessibility to epitopes. The continual diversification leads to a dynamic of appearing and disappearing glycosylation sites followed by humoral selection and outgrowth of the escape mutant.

It is important to realize that the nature of the shield itself is also a component of protection. Unlike the highly immunogenic cell wall components in prokaryotes (e.g., lipopolysaccharide), the sugars in HIV-1 are synthesized and modified as host components and, therefore, indistinguishable from carbohydrate additions to a multitude of other membrane proteins. Vaccine strategies that stimulate recognition of glycans will need to take into consideration targeting what is essentially 'self' carbohydrate. Some broadly neutralizing anti-HIV-1 envelope monoclonal antibodies have polyspecific or self-reactivity to host antigens but are not routinely made due to elimination by host B-cell tolerance mechanisms.

Concluding Remarks

Though influenza virus and HIV both show significant antigenic variation in response to neutralizing antibody, the mechanisms they use to generate variation are both similar and distinct. Influenza virus diversification by drift is unidirectional whereas HIV tends to generate radial diversity. Both viruses undergo shift to induce completely novel antigens; influenza virus by reassorting genomic segments and HIV by RT-mediated recombination.

A new influenza variant spreads through the human population whereas a new HIV variant merely repopulates the infected individual. Both viruses utilize carbohydrate additions to avoid the antibody response, but in HIV the number of modifications is more substantial. In influenza, variation is utilized to avoid antibody binding whereas HIV diversity promotes recognition of epitopes lacking functional significance, and therefore non-neutralizing responses.

See also: Antigenicity and Immunogenicity of Viral Proteins; Antibody - Mediated Immunity to Viruses; Neutralization of Infectivity; Orthomyxovirus Antigen Structure; Quasispecies; Vaccine Strategies.

Further Reading

Air GM, Laver WG, and Webster RG (1987) Antigenic variation in influenza viruses. *Contribution Microbiology and Immunology* 8: 20–59.

Hammond AL, Lewis J, May J, Albert J, Balfe P, and McKeating JA (2001) Antigenic variation within the CD4 binding site of human immunodeficiency virus type 1 gp120: Effects on chemokine receptor utilization. *Journal of Virology* 75: 5593–5603.

Knipe DM, Howley PM, Griffin DE, *et al.* (2001) *Field's Virology*, 4th edn. Philadelphia: Lippincott Williams and Wilkins.

Korber B, Gaschen B, Yusim K, Thakallapally R, Kesmir C, and Detours V (2001) Evolutionary and immunological implications of contemporary HIV-1 variation. *British Medical Bulletin* 58: 19–42.

Skehel JJ and Wiley DC (2000) Receptor binding and membrane fusion in virus entry: The influenza hemagglutinin. *Annual Review of Biochemistry* 69: 531–569.

Venkatramani L, Bochkareva E, Lee JT, *et al.* (2006) An epidemiologically significant epitope of a 1998 human influenza virus neuraminidase forms a highly hydrated interface in the NA-antibody complex. *Journal of Molecular Biology* 356(3): 651–663.

Wilson IA and Cox NJ (1990) Structural basis of immune recognition of influenza virus hemagglutinin. *Annual Review of Immunology* 8: 737–771.

Yusim K, Kesmir C, Gaschen B, *et al.* (2002) Clustering patterns of cytotoxic T-lymphocyte epitopes in human immunodeficiency virus type 1 (HIV-1) proteins reveal imprints of immune evasion on HIV-1 global variation. *Journal of Virology* 76: 8757–8768.

Relevant Websites

http://www.hiv.lanl.gov – A series of freely accessible reviews on HIV-1 diversity, drug resistance and immunology.

http://www.flu.lanl.gov – The Influenza Sequence Database (ISD). To see the sequence diversity in influenza viruses.

Papillomaviruses: Molecular Biology of Human Viruses

P F Lambert and A Collins, University of Wisconsin School of Medicine and Public Health, Madison, WI, USA

Introduction

Papillomaviruses (PVs) are nonenveloped, double-stranded DNA tumor viruses that infect mammalian and avian hosts. There are more than 100 genotypes of human papillomaviruses (HPVs) that can either infect mucosal or cutaneous epithelia. PVs all share the common property of inducing benign proliferative lesions within the epithelial tissues they infect. In these lesions, the virus is replicated in a manner that is exquisitely tied to the differentiation program of the epithelial host cell. Within the mucosal genotypes, HPVs are further subcategorized as low- or high-risk genotypes, reflective of the latter's additional association with frank cancer. High-risk HPVs are best known as the etiologic agent of virtually all cervical cancer, other anogenital cancers, as well as a subset of head and neck cancers. The high-risk HPV most commonly found in these cancers is HPV-16. In this article, the nature of the papillomaviral life cycle as discovered through the study of this high-risk HPV is described. We also compare and contrast its life cycle to that of other PVs, where possible.

Genome

PVs contain an ~8000 bp, circular, double-stranded DNA genome. The encapsidated genome is thought to exist in a chromatinized state within the virus particle; however, there is one report claiming an absence of cellular histones in virus particles. In the infected host cell, the viral genome is delivered to the nucleus where it remains in an extrachromosomal state throughout the viral life cycle. Integration of the viral DNA into the host chromosome is not a normal part of the viral life cycle but can occur in some PV-associated cancers.

The viral genome is organized into three regions: the long control region (LCR), which contains many *cis*-elements involved in the regulation of both DNA replication and transcription; as well as the early (E) and late (L) regions that encode subsets of genes so denoted by their onset of expression in the context of the viral life cycle (**Figure 1**). Transcription of the viral genome occurs unidirectionally and is directed from multiple promoters that are located within the LCR and the 5′ end of the E region. Multiple splicing and polyadenylation signals

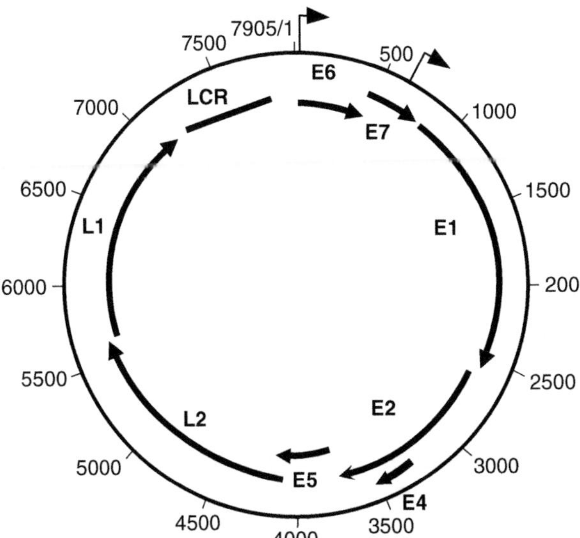

Figure 1 The HPV genome. The locations of the protein-coding sequences of early (E) and late (L) genes, and those of the long control region (LCR) and transcriptional promoters (arrows), are shown.

are used depending on the stage of the viral life cycle, leading to a variety of transcripts with differential coding capacity. The viral genome codes for eight viral genes. The early gene products, which are coded from the E region, contribute to the transcription, replication, and maintenance of the genome as a stable, nuclear plasmid within the poorly differentiated compartment of the infected tissue, as well as its amplification and encapsidation within the more terminally differentiated cellular compartment to make progeny virus. A subset of the early genes (E5, E6, and E7) possesses transforming properties in tissue culture, and tumorigenic properties in experimental animal models. The late structural genes encode for the major capsid protein, L1, and the minor capsid protein, L2. The individual role of each viral gene in the life cycle is described in this article.

Pathogenesis

PVs cause warts: benign, self-regressing lesions of the skin, oral cavity, and anogenital tract. The mucosotropic HPVs are the leading viral cause of sexually transmitted disease in women, with over 50% of sexually active women having been infected in their lifetime. A subset of these mucosotropic HPVs, the so-called 'high-risk' genotypes, are the cause of virtually all cervical cancer, with approximately 500 000 new cases diagnosed yearly, and approximately half as many deaths due to this cancer, making it the second leading cause of death by cancer among women worldwide. Precursor lesions to cervical cancer, called cervical intraepithelial neoplasia (CIN) grades 1–3 (with grade 3 having the most dysplastic characteristics), can be detected by the Pap smear, a cytological screening of cervical scrapes. Pap smears have led to a threefold reduction in the incidence of cervical cancer in those countries employing it as a routine preventative screening technique.

Recently approved prophylactic vaccines that prevent infection by two of the most common high-risk HPVs, HPV-16 and HPV-18, are predicted to lead to an 80% reduction of cervical cancer worldwide beginning in 2040 if the vaccine is universally administered to the world population of women. Socioeconomic conditions may limit the utility of these vaccines.

Other less frequently observed anogenital cancers, of the anus, penis, vulva, and vagina, are also caused by these high-risk HPVs. On average, there is a latency of over a decade between infection and onset of these anogenital cancers. In these cancers, one can find retention of the HPV genome, often in the integrated state, with a subset of viral oncogenes, specifically E6 and E7, upregulated in their expression. In mouse models, the high-risk HPV E6 and E7 oncogenes have been demonstrated to predispose mice to cervical cancer.

The same high-risk HPVs, particularly HPV-16, are etiologically associated with approximately 25% of head and neck squamous cell carcinoma (HNSCC), particularly of the oropharynx where the etiological correlation is 50% or greater. Again, the continued expression of E6 and E7 is found in HPV-positive HNSCC, and, in a mouse model, the HPV-16 E6 and E7 oncogenes have been demonstrated to predispose mice to HNSCC.

Cutaneous HPVs are also associated with a subset of skin cancers. This has been demonstrated most clearly for a subset of cutaneous HPVs that cause epidermodysplasia verruciformis (EV), a skin lesion that appears to arise preferentially in certain families, arguing a genetic predisposition. Patients with EV are highly prone to develop skin cancers. In mouse models, EV-associated HPVs have been experimentally demonstrated to predispose animals to skin cancer.

Life Cycle

With the exception of the so-called fibropapillomaviruses (e.g., bovine papillomaviruses (BPVs) and other PVs that infect ungulate hosts), most PVs are specifically epitheliotropic. Their life cycle occurs within stratified squamous epithelia, which are composed of basal and suprabasal compartments (**Figure 2**). The basal cells within the three-dimensional architecture of such epithelial tissues comprise a single layer sheet of poorly differentiated, proliferative cells that are physically attached to the underlying basement membrane that separates the epithelium from the underlying stroma. The suprabasal compartment

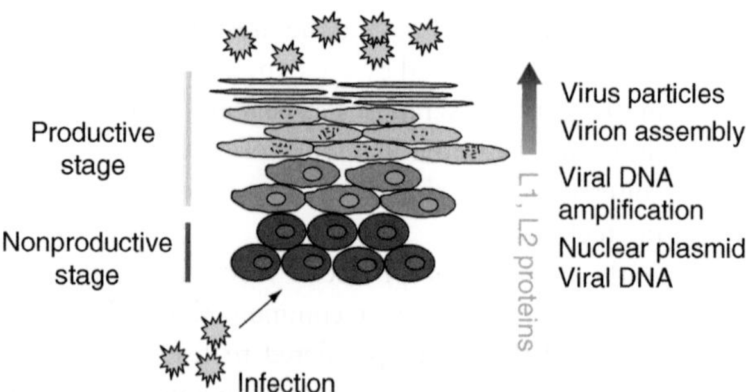

Figure 2 The HPV life cycle. A schematic cross section of stratified squamous epithelium is shown with basal cells at the bottom and terminally differentiated squames at the top. The various processes that arise within the nonproductive and productive stages of the viral life cycle are indicated at relevant positions on the periphery.

comprises the multiple layers of cells positioned above the basal compartment that are quiescent and, as they become more superficial in their position, progressively more differentiated. When basal cells undergo cell division, one daughter cell loses contact with the basement membrane, elevating it to the suprabasal compartment where it exits the cell cycle and begins a program of terminal differentiation. As these cells move upward, they become progressively more terminally differentiated and lose nuclear membrane integrity, leading to the production of terminally differentiated squames, which eventually slough off into the environment. PVs simultaneously exploit and disrupt these normal cellular programs of differentiation and quiescence for their life cycle and the production of progeny virions.

Fibropapillomaviruses differ from other PVs in that they can also infect the underlying dermal fibroblasts, leading to a hyperplasia of the cells. Lesions induced by fibropapillomaviruses can be quite large due to this underlying fibroplasia. The life cycle of these fibropapillomaviruses, however, still relies directly upon the infection of the epithelial cell, where the production of progeny virus specifically occurs.

Entry

It is presumed, though it has never been directly demonstrated, that PVs initiate infection of the host epithelia via entry into cells within the basal compartment. The specific mechanism(s) PVs use to gain access to, bind, and enter basal cells is not clear. The integrin $\alpha6{:}\beta4$ heterodimer has been implicated as a cellular receptor for the virus, allowing for entry into cells upon its expression. However, in the absence of this integrin, some PVs retain the ability to infect cells. Other factors such as heparin sulfate proteoglycans are thought to play a role in cellular binding/entry by certain HPV genotypes at least in some cell types. Both the major (L1) and minor (L2) capsid proteins are required for viral particle infectivity, and both have been postulated to bind to cellular receptors. PVs have been argued to enter cells via clathrin-dependent or caveolin-dependent pathways and to require endosomal acidification. Once inside, L2 is thought to bind to β-actin and this is thought to facilitate movement of the virus particle through the cytoplasm to the nucleus. Alternatively, it has been argued that L2 binds to the microtubule motor dynein, facilitating intracytoplasmic transport. How the encapsidated DNA actually enters the nucleus is currently unknown; however, once there it appears to associate with subnuclear domains called promyelocytic leukemia protein oncogenic domains (PODs).

Establishment of the Nonproductive Infectious State

Once an HPV particle enters the host cell and translocates its DNA to the nucleus, it must rely primarily on cellular machinery to replicate its genome since HPVs only encode one component of the machinery required for initiating DNA replication, E1, a DNA helicase. It is partially for this reason that it is presumed that HPVs initiate their viral life cycle by infecting cells within the basal compartment, as these are the only cells within the epithelia that are normally proliferating and have active cellular replication machinery. In these infected basal cells, the HPV genome becomes established as a low-copy-number nuclear plasmid. The 'early' promoter (designated P_{97} in the case of HPV-16) directs expression of the early genes, some of which are required for the replication of the viral genome, with transcripts being observed as early as 4 h post infection. This early promoter appears to stay active throughout the viral life cycle, whereas the 'late' promoter (designated P_{742} in HPV-16) is only active in cells upon their differentiation. Thus within the basal compartment, early but not late viral genes that encode the structural proteins are expressed, and therefore no progeny virus is produced. Consequently, the infectious state found in the basal cells is referred to as the 'nonproductive' stage of the infectious life cycle.

Maintenance of the Nonproductive Infectious State

A hallmark of HPV infections is their long-term persistence over many years. In the case of high-risk HPVs, this persistence is a prerequisite for the development of cancers associated with these viruses. For the virus to persist long term, it must maintain its genome within the basal compartment over multiple cell divisions, allowing for the continual regeneration of daughter cells that can provide a reservoir of proliferation-competent cells harboring the viral genome. Several early genes have been implicated in the establishment, as well as maintenance of viral genomes (see below). The virus must also affect an amplification of cells within the basal compartment that harbor the viral genome, as seen in warts where one finds a localized expansion of cells within the infected epithelia.

One possible explanation for both persistence and localized expansion of infected cells is that the virus infects epithelial stem cells. From these stem cells are derived the transiently amplifying cell population that populates the local area with HPV-positive cells. This is an attractive hypothesis; however, stem cells are infrequent. Therefore, the necessity to infect stem cells would indicate that productive infection is an inefficient process. An alternative explanation is that HPVs infect the transiently amplifying nonstem cells and extend their lifespan, thereby leading both to viral persistence and an expansion of infected cells at the original site of infection. Perhaps reflective of this alternative explanation is the observation that high-risk viruses have the unique capability among PVs to immortalize epithelial cells. This immortalization potential of high-risk HPVs, in turn, reflects the unique properties of their oncogenes E6 and E7. Whether the high-risk HPVs are unique in their capacity to establish persistent infections without the need to infect stem cells, as might be predicted by their immortalization potential, remains unclear.

Productive Stage

Normally when a basal cell divides, it produces two daughter cells, one of which separates physically from the basement membrane, transits to the suprabasal compartment, and begins a program for terminal differentiation that includes exiting the cell cycle and halting new rounds of DNA synthesis. When a daughter cell that harbors HPV genomes transits to the suprabasal compartment, it begins the so-called 'productive' stage of the HPV life cycle in which progeny viruses are made. In this stage of the viral life cycle, the HPV genome induces a delay and perturbation of the terminal differentiation program, which, importantly, includes the retention of an ability of these suprabasal cells to support DNA synthesis. This DNA synthesis-competent state is critical to allow for the amplification of the viral genome that is necessary for the production of progeny virus. Production of progeny virus within the suprabasal compartment also relies upon the selective expression of the late viral genes that encode the viral capsid proteins. These steps are spatially and temporally regulated. Specifically, induction of suprabasal DNA synthesis and a delay in early steps of epithelial cell differentiation both occur early in the productive stage and within the lower strata. Late gene expression, viral DNA amplification, and progeny virus production arise in the upper strata and are associated with further perturbations in terminal differentiation. Thus, specific life cycle events correspond intimately with different alterations in the terminal differentiation process within certain regions of the infected epithelium.

The multistage mode of replication of viral DNA genomes exemplifies the role that differentiation plays in the viral life cycle. HPV genomes harbored extrachromosomally within the basal compartment are replicated on average once per cell cycle and are stably maintained within this proliferating cell compartment at a low copy number. In contrast, the viral DNA is amplified to a much higher copy number in the upper layers of the epithelia strata. This difference in replication state in basal versus suprabasal cells correlates with a switch in the mode of replication from theta to, seemingly, rolling-circle DNA replication, and is induced by cell differentiation.

Functions of Individual Viral Genes

The Role of E1

PVs rely heavily on cellular machinery to replicate their DNA. However, PVs do encode one gene that is enzymatically involved in viral DNA replication and is required for plasmid replication, E1. E1 binds the viral origin of replication (ori) and possesses DNA helicase activity that permits the unwinding of the viral genomic template for replication. *In vivo* viral replication is mediated by E1, as well as the viral transactivator, E2. E2 is able to bind specific DNA sequences, E2 binding sites (E2BSs), as well as the E1 protein; both of these E2 properties help drive E1-mediated plasmid replication. The current model proposes a multistep, interdependent process for viral replication in which E1 homodimers and E2 homodimers cooperatively bind to ori, then undergo conformational changes that allow E1 to assemble into a helicase-competent hexamer. Many cellular factors contribute to E1-dependent viral replication, including DNA polymerase α, chaperone proteins, histone H1, and replication factor A. In some cases, E1 has been shown to recruit these factors directly.

While E1 is required for the establishment of papillomaviral DNA replication, E1 is dispensable for plasmid maintenance of the BPV genome. In bovine papillomavirus type 1 (BPV-1)-transformed mouse C127 cells, in which that genome was already established as a nuclear plasmid, E1 was not required for stable maintenance of the viral plasmid. An E1-independent form of replication could also be detected in *Saccharomyces cerevisiae* and the *cis*-requirements for this replication were mapped. The ability of PVs to employ an E1-independent replication pathway may be a property of or reflect various modes of replication that the virus utilizes during different stages of its life cycle. In contrast, or in addition, the replication mechanism may reflect the ability of the virus to take advantage of a different host cell environment.

The Role of E2

In addition to a role in viral DNA replication, E2 also regulates viral transcription and the maintenance of the extrachromosomal viral genome. The ability of E2 to regulate transcription is mediated by its binding to E2BSs within the papillomaviral genome. In the PV genome there are multiple E2BSs that have different affinities for E2. Their relative locations are thought to determine the effects of E2 on transcription. If E2 binds an E2BS that is near to, but not overlapping, promoter elements, E2 activates transcription from that promoter. However, if the E2BS overlaps promoter elements, then it can repress transcription by sterically hindering binding of cellular factors necessary for transcription from that promoter.

In the context of the high-risk HPV genomes, E2's ability to suppress transcription from the viral promoter that drives expression of the viral early genes is thought to reflect one reason why integration of the viral genome into the host chromosome, and the consequent disruption of E2, contributes to increased expression of the viral oncogenes E6 and E7. However, these integration events are likely to have pleiotropic effects, as it has been shown that integration results in increased stability of the E6/E7 mRNAs due to the loss of an mRNA instability element located in the 3′ end of the early region. The ability of E2 to repress transcription of the viral oncogenes has been exploited to turn off expression of E6 and E7 in cervical cancer cells lines and thereby demonstrate the requirements for constitutive expression of E6 and E7 in these cancer-derived cell lines. Interestingly, E2 only appears to repress transcription of viral genomes when they are integrated into the host genome. This may reflect the fact that E2BSs in extrachromosomal HPV genomes can be methylated, and this methylation is known to prevent E2 binding. It is likely that epigenetic events also modulate the expression of integrated copies of the viral genome in cervical cancers by altering the chromatin state of the viral genome. In addition to the full-length E2 gene product, E2-mediated transcriptional regulation can also be mediated by N-terminally truncated forms of E2, E8^E2, and E2TR, which are synthesized from alternative transcripts. These E2 proteins can act as dominant negative factors to suppress E2-mediated transcriptional activation either by competing themselves for binding to the E2BSs or through heterodimerization with the full-length E2 protein.

PV genomes are maintained as nuclear plasmids throughout the viral life cycle. Therefore, there is a need for viral plasmids to be segregated efficiently to daughter cells during cell division. For BPV-1, the full-length E2 protein has been shown to mediate this process and is thought to reflect E2's ability to associate with host chromosomes during mitosis. E2 can bind to the chromosomal attachment protein, bromodomain 4 (Brd4), and, in yeast, the ability of BPV-1 E2 to mediate plasmid maintenance relies upon the co-expression of mammalian Brd4. These data suggest that Brd4 contributes to BPV-1 E2's plasmid maintenance function. Further study is required to ascertain whether this proposed mechanism is relevant to HPVs. Of concern is the observation that mutant HPV-31 genomes that contain amino acid substitutions in the E2 protein predicted to disrupt Brd4 binding, based upon prior studies in BPV-1 E2, retain an ability to replicate stably as a nuclear plasmid, though less efficiently than wild-type HPV-31 genomes.

The Role of E1^E4

As with the other PV proteins, numerous functions are ascribed to the E1^E4 gene product. The small (10–20 kDa) regulatory viral protein is generated from a spliced mRNA that fuses the E1 and E4 translational open reading frames (see **Figure 1**). E1^E4 is poorly expressed in undifferentiated monolayer cultures containing HPV, as well as within basal cells of raft cultures. However, E1^E4 expression is increased substantially in the upper strata of papillomas. Within raft cultures the greatest E1^E4 expression is seen in the more superficial layers of epithelium that also contain amplified viral genomes. Owing to colocalization of E1^E4 and amplified viral genomes, it was proposed that E1^E4 functions in the productive stage of the viral life cycle, specifically viral DNA amplification. Indeed, when severe truncation mutations of E1^E4 were made in the context of the entire cottontail rabbit papillomavirus (CRPV), HPV-16, or HPV-31 genomes, defects in viral DNA amplification were observed, indicating that E1^E4 contributes to the amplification of viral genomes. However, effects in viral genome establishment were also seen with some HPV-16 E1^E4 mutants. How E1^E4 mediates viral genome establishment is not clear.

E1^E4 contains three major motifs postulated to contribute to the role of E1^E4 in the viral life cycle. A leucine-rich motif in the N-terminus is required for association with keratins and perhaps to contribute to viral DNA amplification. A proline-rich region in the central section, containing threonine residues, is required for induction of cell cycle arrest at the G2/M phase and sequestering of cyclinB/cdk1 complexes in the cytoplasm. The C-terminus contains a mucosal homology domain, and regulates the ability of E1^E4 to oligomerize, bind the DEAD-box RNA helicases, and induce collapse of keratin filamentous networks. In addition, the association of E1^E4 with cornified cell envelopes (CCEs) and PODs has not been mapped, but both interactions are postulated to contribute to the role of E1^E4 in the PV life cycle.

The association of E1^E4 with both CCEs and keratin filaments is thought to aid egress of the virus. PVs are not lytic; rather, they assemble their progeny virions within the granular and cornified layers of the epithelia. The progeny virions are thought to be released in the context of the natural process of epithelial sloughing. These E1^E4 interactions are proposed to aid the disruption of cellular structure and thus the ability of progeny virions to be released into the environment.

The associations of E1^E4 with cyclinB/cdk1, PODs, and the E1^E4-associated DEAD-box RNA helicase are each postulated to mediate a role for E1^E4 in viral replication or amplification. Overexpression of E1^E4 leads to a cell cycle profile consistent with cells that are arrested at G2/M. It is thought that E1^E4 mediates G2 arrest via decreasing the soluble nature of the cyclin complex and preventing it from translocating to the nucleus. It is hypothesized that these cells retain an ability to support viral DNA amplification. In HPV-1-induced warts, E1^E4 associates with promyelocytic leukemia protein (PML) and the *in vitro* overexpression of E1^E4 induces relocalization of PML to E1^E4 nuclear inclusion bodies. These data suggest that PODs, with E1^E4, may play a role in the life cycle of HPVs.

The Role of E5

E5 is a small, dimeric, hydrophobic membrane protein that localizes to the Golgi membrane and endoplasmic reticulum. Initially, most studies to understand the function and contribution of E5 to transformation, replication, or oncogenesis by PVs were performed within the context of BPV. In the case of BPV-1, E5 is the major transforming protein and can induce immortalization and transformation of rodent fibroblasts and keratinocytes. In contrast, E6 and E7 are the major transforming proteins of HPV. Though, in the context of the HPV genome, E5 is not absolutely required for the induction of DNA synthesis or immortalization, it contributes quantitatively to both of these processes. Furthermore, E5 can independently transform rodent keratinocytes, induce anchorage-independent growth, and override cellular growth arrest signals to promote proliferation. Therefore, E5 appears to play an important yet secondary role in transformation and oncogenesis induced by HPVs, whereas it plays a primary role in BPV-induced transformation.

The ability of E5 to transform cells is attributed to its ability to activate growth factor receptors and to inhibit the 16 kDa vacuolar ATPase (vATPase). BPV-1 E5 can induce the activation of platelet-derived growth factor receptor β (PDGFRβ) in a ligand-independent fashion. This activation occurs through direct binding of BPV-1 E5 to PDGFRβ, which induces receptor oligomerization and activation. In contrast, HPV E5 increases the signaling capacity of the epidermal growth factor receptor (EGFR). This difference in target growth receptors is presumably mediated by tissue tropism specificity. Fibroblasts, which BPVs infect along with keratinocytes, contain abundant amounts of PDGFR. In keratinocytes, which are the normal host cells of HPVs, EGFR is a key growth factor receptor. EGFR phosphorylation, a marker of its activation, is increased in the presence of E5, at least when it is overexpressed. Expression of EGF in concert with E5 leads to greater proliferation of keratinocytes. However, when HPV E5 was studied in the context of the complete viral genome, HPV E5 did not induce a detectable increase in either levels or phosphorylation of EGFR. E5 contributes with E6 and E7 to immortalize keratinocytes and contributes quantitatively to the proliferative capacity of HPV and the ability of HPV to override growth arrest signals. Therefore, while *in vitro* and heterologous system studies predicted that HPV E5 contributes to the viral life cycle through upregulation of EGFR signaling, EGFR-independent pathways may actually be responsible for the proliferative contribution of E5 to the HPV life cycle. Alternatively, E5 may mediate its effects via upregulation of EGFR but the threshold level modified by E5 may not be distinguishable with current technical methodologies.

While E5 does not contribute to the nonproductive stage of the HPV-16 or HPV-31 life cycle, it does contribute quantitatively to the productive stage of both HPV-16 and HPV-31. In the absence of a full-length E5 gene, HPV genomes are defective in their ability to override the normal differentiation program. Normally, HPV-16 and HPV-31 can induce DNA synthesis and proliferation markers in cells induced to undergo differentiation, which are normally quiescent. However, in the absence of E5, DNA synthesis and proliferation marker expression was quantitatively less than the induction of these markers by wild-type HPV genomes. E5 was also shown to be required for efficient viral DNA amplification (HPV-16 and HPV-31), late viral gene expression, and re-entry into the cell cycle (HPV-31). These data illustrate that E5 contributes quantitatively to many aspects of the productive stage of the HPV life cycle. Further support for a role of E5 in productive infection lies in *in vivo* studies that examine papilloma formation by E5 alone or E5 mutant genomes. In the context of CRPV, E5 mutant genomes are less effective at inducing papillomas than the wild-type CRPV genome. In addition, HPV-16 E5 transgenic mice develop spontaneous papillomas and hyperplasia. Thus, E5 contributes to proliferation both in the context of animal models and organotypic raft culture viral life cycle studies.

The Role of E6

E6, an approximately 150 amino acid residue protein, is encoded by the first translational open reading frame (ORF) within the HPV genome and is one of the major HPV oncoproteins. High-risk HPV E6 is able to transform rodent fibroblasts and immortalize keratinocytes, as well as other epithelial cells, in conjunction with E7, the other major HPV oncogene. In addition, *in vivo* E6 can inhibit epithelial differentiation, induce hyperplasia, suppress DNA damage-induced cell cycle arrest, and contribute to carcinogenic promotion and progression. All of these activities of E6 are thought to contribute to its oncogenicity. E6 does not possess any intrinsic enzymatic activity; rather, its oncogenic capabilities are mediated through its ability to act as a scaffold and regulate protein–protein interactions. Within E6 there are two motifs that are generally thought to mediate such interactions: two zinc finger domains and an α-helix-binding domain. High-risk E6 also contains a C-terminal PSD-95/Dlg/ZO-1 (PDZ) domain as well. Some of the protein interactions that have been mapped to these E6 domains include: p53, E6-associated protein (E6AP), E6-binding protein (E6BP), c-myc, p300/CBP, paxillin, PDZ proteins, interferon regulatory factor 3, and Bcl-2 homologous antagonist/killer (Bak).

The most studied of these interactions is the E6–p53 interaction. p53 is a transcription factor and tumor suppressor that is activated upon aberrant DNA replication, cellular stress, or cellular damage signals. As a response to cellular damage signals, the tumor suppressor p53 is upregulated and can induce cell cycle arrest or apoptosis through its transcriptional activity. p53 is a major regulator of DNA synthesis inhibition through cell cycle arrest and induction of cell death through apoptosis, which is presumably the reason that p53 is one of the most commonly mutated proteins in human cancer. Through the association of E6 with E6AP, a ubiquitin ligase, E6 is able to bind and induce the degradation of p53, though recent studies argue that other factors might also mediate E6's ability to induce p53 degradation. The ability of E6 to deregulate p53 correlates, at least in part, with its ability to transform cells in tissue culture and induce tumors in mice.

Clearly, the E6–p53 interaction is important for the oncogenic capabilities of E6. However, *in vivo* studies of the high-risk HPV-16 E6 using HPV transgenic mouse models indicate that an E6 mutant that retains the ability to inactivate p53 is partially defective for oncogenic phenotypes observed in mice transgenic for wild-type HPV-16 E6, including its abilities to induce skin tumors and cervical cancer. Thus, p53 inactivation is not sufficient to account for E6's oncogenic properties. The E6 mutant studied, competent to inactivate p53, is defective in its interaction with PDZ proteins. The ability of E6 to bind PDZ proteins also correlates with its ability to transform baby rat kidney cells. Only E6 proteins encoded by high-risk HPVs can bind PDZ proteins; E6 proteins encoded by low-risk HPVs do not contain the four-residue C-terminal, PDZ interaction domain. PDZ proteins are named for the protein family members that were initially identified to possess copies of a motif of approximately 80–100 residues, termed PDZ domains. These proteins are scaffolding proteins that generally interact with the C-terminus of their protein partners, which possess a charged four-residue motif, the PDZ interaction domain. The evolutionarily conserved PDZ proteins are required for development, cell adhesion, cellular growth and differentiation, and cell cycle processes. The E6 interaction with PDZ proteins is reported to lead to degradation of the PDZ proteins. Data indicate that the E6–PDZ interaction allows for the dysregulation of normal cellular growth controls and E6-induced hyperproliferation, transformation, and carcinogenesis. Therefore, while the E6–p53 interaction is required for E6-induced transformation and proliferation, the E6–PDZ interaction also contributes to E6 oncogenicity in a fashion that is distinct from p53.

E6 contributes to the viral life cycle. In HPV-31, as well as HPV-16, E6 is required for viral episome maintenance. In transient replication assays, HPV-31 E6-null genomes are able to replicate, indicating that E6 is not required for the initial establishment of HPV genomes. However, these E6-null genomes are not able to maintain the viral episome in long-term assays. HPV-31 genomes that are specifically defective in their ability to interact with PDZ proteins are still able to exhibit properties of intact nonproductive and productive stages, yet are slightly defective at each step of the life cycle. In contrast, there are clear defects in the viral life cycle when E6–p53 interactions are disrupted. Although the ability to bind p53 is not required for the establishment of viral episomes, the ability of viral episomes to be maintained correlates with the ability of E6 to induce degradation of p53. How the degradation of p53 contributes to maintenance of the viral episome is not clear. E6-mediated p53 degradation may be needed for viral maintenance to protect against induction of cellular damage signals induced during viral replication, especially as a consequence of the dysregulation of the tumor-suppressor protein, retinoblastoma (pRb), by E7.

The Role of E7

E7 is a 98-residue protein composed of two N-terminal conserved regions and a zinc finger domain in its C-terminus. The C-terminus regulates E7 multimerization. The N-terminus is partially homologous to two other virally coded oncoproteins, adenovirus (Ad) E1A and simian virus 40 T antigen (SV40TAg). E7 contains a pocket protein-binding motif, LXCXE, that confers an ability for E7 to bind pRb, as well as the two other 'pocket protein' family members, p107 and p130. The LXCXE motif is required for the ability of E7 to transform rodent fibroblasts and, in conjunction with E6, to immortalize human fibroblasts and keratinocytes efficiently. In addition, the LXCXE motif is required for E7 repression of pRb-induced senescence of cells harboring HPV genomes; and, in mice, the ability of E7 to bind pRb is essential for E7's induction of epithelial hyperplasia. This conserved region within high-risk E7s, and their interaction with the pocket proteins, correlate with their oncogenic properties. However, E7 is a multifunctional protein that is reported to bind not only the pocket proteins but also as many as 100 other cellular factors. Other potentially relevant partners include: the pocket proteins (pRb, p107, and p130), the cyclin-dependent kinase inhibitors p21 and p27, CK2 (formerly casein kinase II), and histone deacetylase (HDAC).

The pocket proteins are cell cycle regulators that modulate cell cycle progression and DNA synthesis. pRb is a tumor suppressor and is part of one of the most frequently mutated regulatory pathways in human cancers. p107 and p130 are closely related to pRb and possess similar biochemical activities. Together the pocket proteins regulate overlapping and distinct parts of the cell cycle. Both low- and high-risk E7 are able to bind the pocket proteins, although the relative binding preference for the pocket proteins appears to differ between low- and high-risk E7 proteins. In addition, high-risk HPV E7s can induce degradation of the pocket proteins. Two other DNA tumor viruses, Ad and SV40, also bind and inactivate the pocket proteins and affect the cell cycle.

The most studied E7–pocket protein interaction is the E7–pRb interaction. Normally, pRb is hypophosphorylated early in the cell cycle. In its hypophosphorylated state, pRb binds the transcription factor E2F/DP complex. The E2F/DP complex is a transcriptional activation complex that drives the expression of S phase genes. In early G1 phase of the cell cycle, hypophosphorylated pRb binds to E2F/DP and consequently inactivates the transcriptional complex. In late G1, pRb is phosphorylated by G1-specific cyclin/cdk complexes composed of cyclin D and cyclin E, and this leads to release of the E2F/DP complex from pRb. The free E2F/DP complex then becomes transcriptionally active and S-phase promoting genes are transcribed. In the context of HPV infection, E7 expression negates the requirement for pRb phosphorylation to activate the E2F/DP transcriptional complex. Specifically, E7 binding preferentially to hypophosphorylated pRb leads to the release of E2F/DP complexes from pRb and the consequent activation of the E2F complex. Therefore, E7 inactivation of pRb bestows on HPV the ability to override the pRb inhibition of the cell cycle.

In the context of the entire genome, the overall mechanistic requirements for E7 differ between genotypes. In HPV-16, E7 is required for the productive stage of the viral life cycle. Specifically, an HPV-16 genome carrying a null mutation of E7, while able to become established stably as a nuclear plasmid in human keratinocytes, is unable to support many aspects of the life cycle that arise in the differentiating compartment of stratified epithelium, including the inhibition of differentiation, reprogramming of suprabasal cells to support DNA synthesis, and viral DNA amplification. A similar requirement for E7 specifically in the productive stage of the life cycle was reported for HPV-18, and subsequent studies with HPV-16 demonstrated the importance of the E7–pocket protein interactions in mediating its role in the life cycle. An HPV-16 mutant genome carrying a mutation within the E7 ORF that disrupts the ability of E7 to bind pRb and the other pocket proteins displayed all of the defects in the productive life cycle that were observed with the E7-null HPV-16 genome. In contrast, a mutant genome in which E7 is disrupted in its ability to induce the degradation of the pocket proteins retained a partial ability to induce suprabasal DNA synthesis and viral DNA amplification but was completely defective in inhibiting cellular differentiation. Thus, there are differential requirements for E7's binding versus degradation of pRb and the other pocket proteins in E7's contribution to the various stages of the viral life cycle.

In contrast to what was observed with HPV-16 and HPV-18, the intermediate-risk HPV-31 genome appears to require E7 during the nonproductive stage of the viral life cycle as well as during the productive stage. E7-null HPV-31 genomes are unable to be maintained stably as nuclear plasmids in cultures of poorly differentiated human keratinocytes, whereas E7-null HPV-16 and HPV-18 genomes do stably replicate as nuclear plasmids. This difference apparently reflects genotype-specific differences.

The Role of L1 and L2

L1 and L2 are late viral genes encoding the major and minor capsid proteins, respectively. The PV capsid contains 72 pentameric capsomeres arranged on a $T=7$ icosahedral lattice and is approximately 55 nm in diameter, as determined by cryoelectron microscopy. The majority of the viral capsid is composed of the major capsid protein, L1, with L2 composing a minor fraction. Expression of L1 alone allows for the formation of pseudovirions or virus-like particles, VLPs, which are visually indistinguishable from native virions. If L2 is expressed in conjunction with L1, it is also incorporated into VLPs, but L2 is not required for capsid formation. L1 and L2 are specifically expressed in the most superficial layers of stratified keratinocytes, where progeny virus production arises.

Although L2 is not required for capsid formation, it does possess activities that likely contribute to a role in the viral life cycle. Some of these have been described above, under the section on virus entry, and include binding to cell surface receptors, actin, and PML, all of which have been implicated in early steps of virus infection. In addition, L2 is thought to be required for viral DNA encapsidation, although this requirement is obviated in some pseudovirion assembly systems. In life cycle studies carried out with HPV-31, L2-null genomes were able to support the nonproductive stage and most aspects of the productive viral life cycle in transfected human keratinocytes, with the exception that the level of encapsidated DNA in the progeny virions was reduced significantly and these progeny viruses were further reduced in their infectivity. Therefore, L2 seems to have a role in the viral life cycle in at least two distinct steps, encapsidation and infectivity. Whether the role of L2 in infectivity is at the level of receptor recognition or viral trafficking or both is something that should be determined in the future.

Further Reading

Bernard HU (2002) Gene expression of genital human papillomaviruses and potential antiviral approaches. *Antiviral Therapy* 7: 219–237.

Collins A, Nakahara T, and Lambert PF (2005) Contribution of pocket protein interactions in mediating HPV16 E7's role in the papillomavirus life cycle. *Journal of Virology* 79: 14769–14780.

Demeret C, Desaintes C, Yaniv M, and Thierry F (1997) Different mechanisms contribute to the E2 mediated transcriptional repression of human papillomavirus type 18 viral oncogenes. *Journal of Virology* 71: 9343–9349.

Evander M, Frazer IH, Payne E, Qi YM, Hengst K, and McMillan NA (1997) Identification of the α6 integrin as a candidate receptor for papillomaviruses. *Journal of Virology* 71: 2449–2456.

Lambert PF, Ozbun MA, Collins A, Holmgren S, Lee D, and Nakahara T (2005) Using an immortalized cell line to study the HPV life cycle in organotypic "raft" cultures. *Methods in Molecular Medicine* 119: 141–155.

Longworth MS and Laimins LA (2004) Pathogenesis of human papillomaviruses in differentiating epithelia. *Microbiology and Molecular Biology Reviews* 68: 362–372.

Munger K, Baldwin A, Edwards KM, *et al.* (2004) Mechanisms of human papillomavirus-induced oncogenesis. *Journal of Virology* 78: 11451–11460.

Nakahara T, Doorbar J, and Lambert PF (2005) HPV16 E1^E4 contributes to multiple facets of the papillomavirus life cycle. *Journal of Virology* 79: 13150–13165.

Ozbun MA (2002) Human papillomavirus type 31b infection of human keratinocytes and the onset of early transcription. *Journal of Virology* 76: 11291–11300.

Pyeon D, Lambert PF, and Ahlquist P (2005) Production of infectious human papillomavirus independently of viral replication and epithelial cell differentiation. *Proceedings of the National Academy of Sciences, USA* 102: 9311–9316.

Sanders CM and Stenlund A (2001) Mechanism and requirements for bovine papillomavirus, type 1, E1 initiator complex assembly promoted by the E2 transcription factor bound to distal sites. *Journal of Biological Chemistry* 276: 23689–23699.

Schiller JT and Lowy DR (2006) Prospects for cervical cancer prevention by human papillomavirus vaccination. *Cancer Research* 66: 10229–10232.

Skiadopoulos MH and McBride AA (1998) Bovine papillomavirus type 1 genomes and the E2 transactivator protein are closely associated with mitotic chromatin. *Journal of Virology* 72: 2079–2088.

zur Hausen H (2002) Papillomaviruses and cancer: From basic studies to clinical application. *Nature Reviews. Cancer* 2: 342–350.

Partitiviruses

S A Ghabrial, University of Kentucky, Lexington, KY, USA
W F Ochoa and T S Baker, University of California, San Diego, La Jolla, CA, USA
M L Nibert, Harvard Medical School, Boston, MA, USA

Glossary

Cryo-transmission electron microscopy Transmission electron microscopy of unstained, unfixed, frozen-hydrated (vitrified) specimens, preserved as close as possible to their native state.

Hyphal anastomosis The union of a hypha with another resulting in cytoplasmic exchange.

Icosahedral symmetry Arrangement of 60 identical objects or asymmetric units adopted by many isometric (spherical) viruses, with a combination of two-, three-, and fivefold rotations equivalently relating the units about a point in space.

Mycoviruses Viruses that infect and multiply in fungi.

'$T = 2$' symmetry The so-called 'forbidden' triangulation symmetry, in which 120 chemically identical protein monomers, form 60 identical, asymmetric dimers arranged with icosahedral symmetry. Monomers in a '$T = 2$' structure occupy two distinct, nonequivalent positions and therefore differ from the 'allowed' symmetries where monomers are either equivalently or quasi-equivalently related.

Virus capsid Generally a protective shell, composed of multiple copies each of one or more distinct protein subunits, that encapsidate the viral genome.

Introduction

In the early 1960s, interest in the antiviral activities associated with cultural filtrates of *Penicillium* spp. led to the discovery of double-stranded RNA (dsRNA) isometric viruses in these and other filamentous fungi. Fungal viruses, or mycoviruses, are now known to be of common occurrence. The isometric dsRNA viruses isolated from *Penicillium* spp. were among the first to be molecularly characterized and were shown to have segmented genomes. Those with bipartite genomes are currently classified in the genus *Partitivirus* in the family *Partitiviridae.* Interestingly, plant viruses (also called cryptoviruses) with bipartite dsRNA genomes and very similar properties to the fungal partitiviruses were discovered in the late 1970s and are presently classified in the genera *Alphacryptovirus* and *Betacryptovirus* in the family *Partitiviridae.* The fungal partitiviruses and plant partitiviruses are discussed elsewhere in this encylopedia. The goals of this article are to examine the similarities and differences between these two groups of viruses and to discuss future perspectives of partitivirus research.

Biological Properties

Fungal and plant partitiviruses are both associated with latent infections of their respective hosts. There are no known natural vectors for any partitivirus. Fungal partitiviruses are transmitted intracellularly during cell division and sporogenesis (vertical transmission) as well as following hyphal anastomosis, that is, cell fusion, between compatible fungal strains (horizontal transmission). In some ascomycetes (e.g., *Gaeumannomyces graminis*), virus is usually eliminated during ascospore formation. Experimental transmission of fungal partitiviruses has been reported by transfecting fungal protoplasts with purified virions. The plant cryptoviruses, on the other hand, are not horizontally transmitted by grafting or other mechanical means, but are vertically transmitted by ovule and/or pollen to the seed embryo. Thus, whereas sexual reproduction of the host is required for the survival of plant partitiviruses, it is detrimental to the continued existence of the fungal partitiviruses that infect some ascomycetes. Transmission of fungal partitiviruses through asexual spores, however, can be highly efficient, with 90–100% of single conidial isolates having received the virus. In summary, transmission of fungal partitiviruses by asexual spores and plant partitiviruses by seed provide the primary or only means for disseminating these viruses.

Both fungal and plant partitiviruses are generally associated with symptomless infections of their hosts. While cryptoviruses are present in very low concentrations in plants (e.g., 200 μg of virions per kg of tissue for white clover cryptic virus), fungal partitiviruses can accumulate to very high concentrations (at least 1 mg of virions per g of mycelial tissue for penicillium stoloniferum virus F (PsV-F)). Mixed infections of fungal or plants hosts with two distinct partitiviruses are not rare. PsV-S and PsV-F represent one example in which two partitiviruses infect the same fungus, *Penicillium stoloniferum.* Interestingly, a significant increase in cryptovirus concentration has been observed in mixed infections with unrelated viruses belonging to other plant virus families.

Virion Properties

The buoyant densities of virions of members of the partitivirus family range from 1.34 to 1.39 gm cm^{-3}, and the sedimentation coefficients of these virions range from 101S to 145S (S_{20w} in Svedberg units). Generally, each virion contains only one of the two genomic dsRNA segments. However, with some viruses like the fungal partitivirus PsV-S, purified preparations can contain other distinctly sedimenting forms that include empty particles and replication intermediates (see the next section).

All fungal and plant partitiviruses examined to date have been shown to possess virion-associated RNA-dependent RNA polymerase (RdRp) activity, which catalyzes the synthesis of single-stranded RNA (ssRNA) copies of the positive strand of each of the genomic dsRNA molecules. The *in vitro* transcription reaction occurs by a semi-conservative mechanism, whereby the released ssRNA represents the displaced positive strand of the parental dsRNA molecule and the newly synthesized positive strand is retained as part of the duplex.

Genome Organization and Replication

Virions of members of the partitivirus family contain two unrelated segments of dsRNA, in the size range of 1.4–2.3 kbp, one encoding the capsid protein (CP) and the other encoding the RdRp. The two segments are usually of similar size and are encapsidated separately, that is, each particle generally contains only one dsRNA segment. The genomes of at least 16 members of the genus *Partitivirus* have recently been completely sequenced (**Table 1**). In contrast, the complete genome sequences of only three alphacryptoviruses and no betacryptoviruses have yet been determined (**Table 1**). The genomic structure of Atkinsonella hypoxylon virus (AhV-1), the type species of the genus *Partitivirus,* comprising segment 1 (2180 bp, encoding the RdRp) and segment 2 (2135 bp, encoding the CP), is schematically represented in **Figure 1**.

The presence of one or more additional dsRNA segments is common among members of the family *Partitiviridae.* For example, in addition to the two genomic segments, preparations of AhV-1 contain a third dsRNA segment of 1790 bp. With the exception of the termini, this third segment is also unrelated to the other two. The absence of any long open reading frame (ORF) on either strand of segment 3 of AhV-1 suggests that it is a satellite segment, not required for replication. Satellite dsRNAs are often associated with infections by members of the family *Partitviridae* (**Table 1**).

Table 1 List of viruses in the family *Partitiviridae* with sequenced genomic dsRNAs

Virus[a]	*Abbreviation*	*dsRNA segment no. (size in bp; encoded protein, size in kDa)*	*GenBank accession no.*
Genus: *Partitivirus*			
*Atkinsonella hypoxylon virus**	AhV	1 (2180; RdRp, 78)	L39125
		2 (2135; CP, 74)	L39126
		3 (1790; satellite)	L39127
Ceratocystis resinifera virus	CrV	1 (2207; RdRp, 77)	AY603052
		2 (2305; CP, 73)	AY603051
*Discula destructiva virus 1**	DdV1	1 (1787; RdRp, 62)	NC_002797
		2 (1585; CP, 48)	NC_002800
		3 (1181; satellite)	NC_002801
		4 (308; satellite)	NC_002802
*Discula destructiva virus 2**	DdV2	1 (1781; RdRp, 62)	NC_003710
		2 (1611; CP, 50)	NC_003711
*Fusarium poae virus 1**	FpV-1	1 (2203; RdRp, 78)	NC_003884
		2 (2185; CP, 70)	NC_003883
*Fusarium solani virus 1**	FsV-1	1 (1645; RdRp, 60)	D55668
		2 (1445; CP, 44)	D55669
*Gremmeniella abietina virus MS1**	GaV-MS1	1 (1782; RdRp, 61)	NC_004018
		2 (1586; CP, 47)	NC_004019
		3 (1186; satellite)	NC_004020
*Helicobasidium mompa virus**	HmV	V1-1 (2247; RdRp, 83)	AB110979
		V1-2 (1776; RdRp, 63)	AB110980
		V-70 (1928; RdRp, 70)	AB025903
*Heterobasidion annosum virus**	HaV	1 (2325; RdRp, 87)	AF473549
Ophiostoma partitivirus 1	OPV-1	1 (1744; RdRp, 63)	AM087202
		2 (1567; CP, 46)	AM087203
Oyster mushroom virus	OMV	1 (2038; RdRp, 70)	AY308801
Penicillium stoloniferum virus F*	PsV-F	1 (1677; RdRp, 62)	NC_007221
		2 (1500; CP, 47)	NC_007222
		3 (677; satellite)	NC_007223
*Penicillium stoloniferum virus S**	PsV-S	1 (1754; RdRp, 62)	NC_005976
		2 (1582; CP, 47)	NC_005977
Pleurotus ostreatus virus	PoV	1 (2296; RdRp, 82)	NC_006961
		2 (2223; CP, 71)	NC_006960
*Rhizoctonia solani virus 717**	RhsV-717	1 (2363; RdRp, 86)	NC_003801
		2 (2206; CP, 76)	NC_003802
Rosellinia necatrix virus 1	RnV-1	1 (2299; RdRp, 84)	NC_007537
		2 (2279; CP, 77)	NC_007538
Genus: *Alphacryptovirus*			
*Beet cryptic virus 3**	BCV-3	2 (1607; RdRp, 55)	S63913
*Vicia cryptic virus**	VCV	1 (2012; RdRp, 73)	NC_007241
		2 (1779; CP, 54)	NC_007242
*White clover cryptic virus 1**	WCCV-1	1 (1955; RdRp, 73)	NC_006275
		2 (1708; CP, 54)	NC_006276
Unclassified viruses in the family *Partitiviridae*			
Cherry chlorotic rusty spot associated partitivirus	CCRSAPV	1 (2021; RdRp, 73)	NC_006442
		2 (1841; CP, 55)	NC_006443
Fragaria chiloensis cryptic virus	FCCV	1 (1743; RdRp, 56)	DQ093961
Pyrus pyrifolia cryptic virus	PpV	1 (1592; RdRp, 55)	AB012616
Raphanus sativus cryptic virus 1	RasV-1	1 (1866; RdRp, 67)	NC_008191
(or Radish yellow edge virus*)	(RYEV)	2 (1791; CP, 56)	NC_008190
Raphanus sativus cryptic virus 2	RasV-2	1 (1717; RdRp, 55)	DQ218036
		2 (1521; unknown)	DQ218037
		3 (1485; unknown)	DQ218038

[a]An asterisk next to the virus name indicates it is presently recognized by ICTV as a member or a tentative member in the family *Partitiviridae*. Family members or tentative members that have not been sequenced to date are not included.

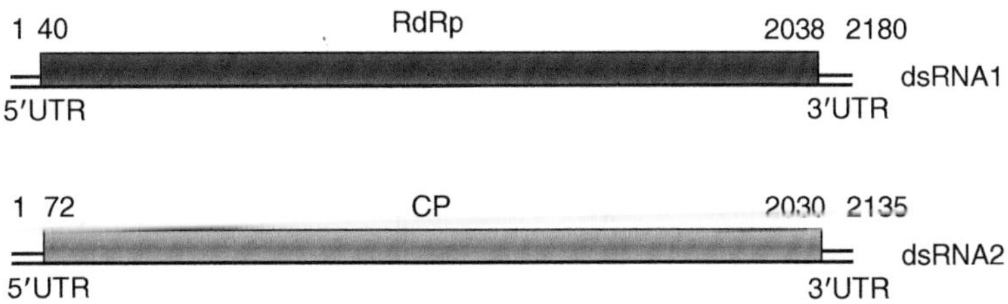

Figure 1 Genome organization of atkinsonella hypxylon virus (AhV), the type species of the genus *Partitivirus*. dsRNA1 contains the RdRp ORF (nt positions 40–2038) and dsRNA2 codes for the CP ORF (nt positions of 72–2030). The RdRp and CP ORFs are represented by rectangular boxes. Reproduced from Ghabrial SA, Buck KW, Hillman BI, and Milne RG (2005) *Partitiviridae*. In: Fauquet CM, Mayo MA, Maniloff J, Desselberger U, and Ball LA (eds.) *Virus Taxonomy: Eighth Report of the International Committee on Taxonomy of Viruses*, pp. 581–590. San Diego, CA: Elsevier Academic Press, with permission from Elsevier.

Limited information on how fungal viruses in the genus *Partitivirus* replicate their dsRNAs is derived from *in vitro* studies of virion-associated RdRp and the isolation from naturally infected mycelium of particles that represent various stages in the replication cycle. The RdRp is believed to function as both a transcriptase and a replicase. The transcriptase activity within an assembled virion catalyzes the synthesis of progeny positive-strand RNA from the parental dsRNA template, accompanied by the displacement of the parental positive strand and its release from the virion. This can presumably occur through repeated rounds, giving rise to multiple positive-strand copies. Each released positive strand can then be consecutively or alternatively (1) used as a template for protein (CP or RdRp) translation by the host machinery or (2) packaged by CP and RdRp into an assembling progeny virion. The replicase activity within this assembling virion then catalyzes the synthesis of negative-strand RNA on the positive-strand template, reconstituting a genomic dsRNA segment. The replication of plant partitiviruses is presumed to mimic that described for the fungal viruses.

Partially purified virion preparations are known to include a small proportion of particles that contain only one ssRNA molecule corresponding to the genomic positive strand. These may be particles in which the replicase reaction is defective or has not yet occurred. In addition, partially purified virion preparations are known to contain a relatively large proportion of a heterogeneous population of particles more dense than the mature virions. These dense particles may represent various stages in the replication cycle including particles containing the individual genomic dsRNAs with ssRNA tails of varying lengths, particles with one molecule of dsRNA and one molecule of its ssRNA transcript and particles with two molecules of dsRNA (**Figure 2**).

Structures of Partitivirus Virions

The isometric dsRNA mycoviruses are classified within the families *Totiviridae*, *Chrysoviridae*, and *Partitiviridae*. The capsid structures of representative members of each of the first two families have been determined, at least one at near atomic resolution using X-ray crystallography, and the others at low to moderate resolutions (~1.5–2.5 nm) using cryo-transmission electron microscopy (cryo-TEM) combined with three-dimensional (3D) image reconstruction. We have recently initiated systematic cryo-TEM and image reconstruction studies of three viruses, PsV-S, PsV-F, and FpV-1, within the genus *Partitivirus*. Based on phylogenetic analysis of partitivirus CPs or RdRps, PsV-S, and PsV-F form sister clades in the same cluster, whereas FpV-1 is placed in a separate and distinct cluster within this genus (see section on 'Taxonomic and phylogenetic considerations'). These viruses are structurally distinguished primarily on the basis of significant differences in the sizes of the respective CPs and percent nucleotide and amino acid sequence identity of their RdRps and CPs. Structures of partitviruses from the genera *Alphacrypto-* and *Betacryptovirus*, all of which are plant viruses, have yet to be reported.

Preliminary studies of PsV-S, PsV-F, and FpV-1 strongly suggest that all members of the genus *Partitivirus* share a number of common features. The structure of PsV-S is used to highlight these features, and significant differences are identified below.

PsV-S represents one of the simplest of the dsRNA viruses. Each virion comprises of one of the two genome segments (S1, 1754 bp and S2, 1582 bp), one or two copies of the 62 kDa RdRp, and 120 copies of the 47 kDa CP. Virions exhibit an overall spherical morphology in negative stain or in vitrified solution (**Figure 3**). Reports of virion diameters based on measurements from stained specimens likely underestimate the true virion size, owing to distortions and other potential artifacts associated with the imaging of negatively stained specimens. Based on cryo-TEM images, the maximum diameter of fully hydrated PsV-S virions is 35 nm. 3D reconstruction of the PsV-S structure at ~21 Å resolution demonstrates that the capsid is a contiguous shell (average thickness ~2 nm) of subunits arranged with icosahedral symmetry (**Figure 4(a)**). The external surface of PsV-S displays 60 prominent protrusions that rise approximately 3 nm

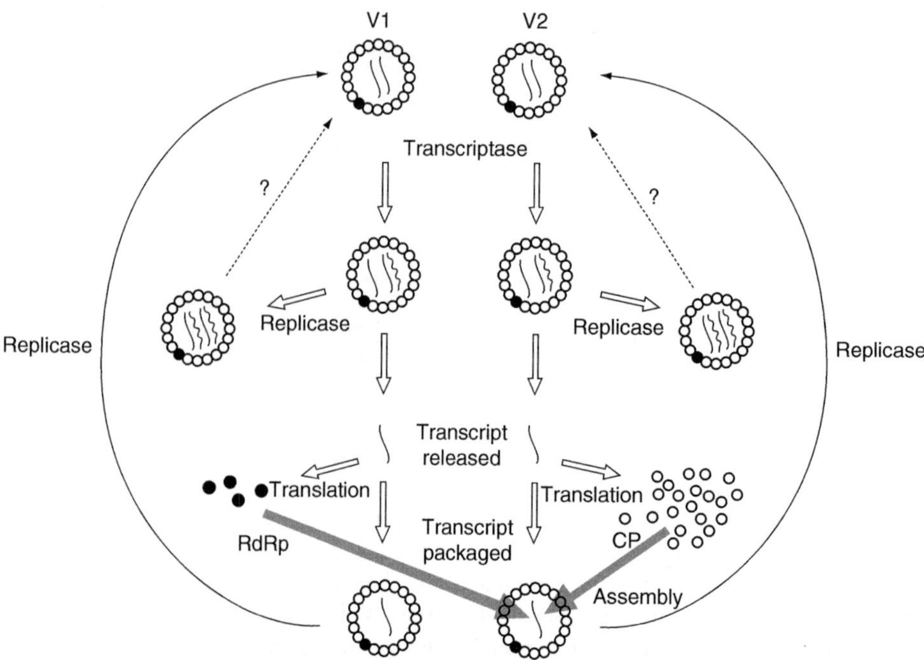

Figure 2 Model for replication of penicillium stoloniferum virus S (PsV-S). The open circles represent capsid protein (CP) subunits and the closed circles represent RNA-dependent RNA polymerase (RdRp) subunits. Solid lines represent parental RNA strands whereas wavy lines represent newly synthesized progeny RNA strands. Reproduced from Ghabrial SA and Hillman BI (1999) Partitiviruses-fungal (*Partitiviridae*). In: Granoff A and Webster RG (eds.) *Encyclopedia of Virology*, 2nd edn., pp. 1477–1151. San Diego: Academic Press, with permission from Elsevier.

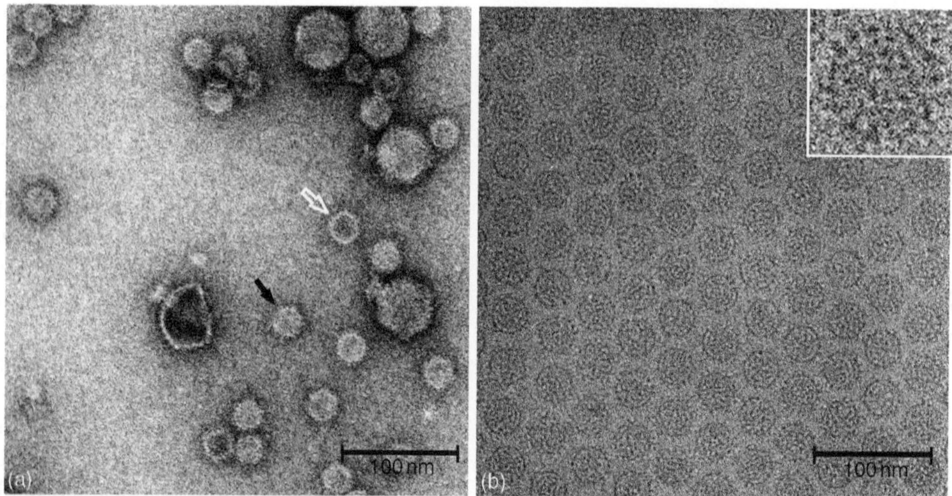

Figure 3 Electron micrographs of penicillium stoloniferum virus S (PsV-S). Samples were negatively stained in 2% uranyl acetate (a) or prepared unstained and vitrified (b). Micrographs were recorded on a CCD detector in an FEI Polara transmission electron microscope operated at 200 keV with samples at (a) room or (b) liquid nitrogen temperatures. In (a) bacteriophage P22 (five largest particles) was mixed with PsV-S to serve as a calibration reference. Heavy metal stain surrounds virions (e.g., black arrow) and contrasts their surfaces against the background carbon support film. Stain penetrates into the interior of 'empty' capsids (e.g., white arrow), resulting in particle images in which only a thin, annular shell of stain-excluding material (capsid) is seen. In (b) the unstained PsV-S sample was vitrified in liquid ethane. Here particles appear dark (higher density) against a lighter background of surrounding water (lower density). The inset shows a three-times enlarged view of an individual particle, in which several knobby surface features are clearly visible.

above an otherwise relatively featureless, spherical shell. Each protrusion and a portion of the underlying shell, constitutes one asymmetric unit (1/60th) of the capsid. The dimeric morphology of each protrusion is consistent with an asymmetric unit being composed of two CP monomers. Hence, there are 120 protein subunits arranged as 60 asymmetric dimers packed in a so-called '$T = 2$' ('forbidden') lattice, which requires

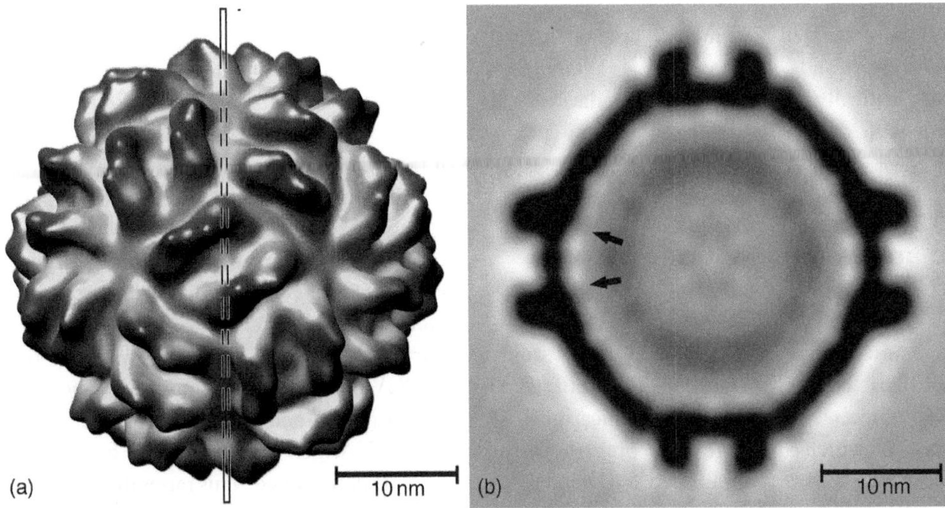

Figure 4 Three-dimensional (3D) structure of PsV-S. (a) Shaded, surface representation of PsV-S 3D reconstruction viewed along an icosahedral twofold axes. The 3D map is color-coded to emphasize the radial extent of different features (yellows and greens highlight features closest to the particle center, and oranges and reds those farthest from the center). A total of 60 prominent protrusions extend radially outward from the capsid surface. Each protrusion exhibits an approximate dyad symmetry, which is consistent with the expectation that the partitivirus capsid consists of 120 capsid protein monomers, organized as 60 asymmetric dimers in a so-called 'T=2' lattice. (b) Density projection image of a central, planar section through the PsV-S 3D reconstruction (from region marked by dashed box in (a)). Darker shades of gray correspond to higher electron densities in the map section and lighter shades represent low-density features such as water outside as well as inside the particles. The capsid shell appears darkest because it contains a closely packed, highly ordered (icosahedral) arrangement of capsid subunits. The genomic dsRNA on the inside appears at lower density, in part because the RNA is not as densely packed and in part because the RNA adopts a less ordered arrangement. The protrusions seen in (a) appear as large 'bumps' in the central section view that decorate the outside of a contiguous, ~2 nm thick, shell. Arrows point to faint density features that appear to form contacts between the inner surface of the protein capsid and the underlying RNA. These contacts occur close to the fivefold axes of the icosahedral shell.

that the monomers do not interact equivalently or quasiequivalently as often occurs, especially in smaller, simpler virus capsids (e.g., with triangulation symmetries such as T= 1, 3, 4).

An additional, symmetric ball of weak density within the capsid is attributed to the genomic dsRNA, and does not appear to adopt any regular structure at the limited resolution of this initial 3D reconstruction (**Figure 4(b)**). Lack of detectable genome organization may, in part, be attributed to averaging effects that occur when images of both PsV-S particle types are combined to produce the reconstructed 3D density map. Potential interactions between the genome and the inner wall of the capsid are suggested by faint lines of density observed in thin, planar sections through the density map (arrows; **Figure 4(b)**).

PsV-F and PsV-S co-infect and are co-purified from the fungal host, *Penicillium stoloniferum*. The coat protein of PsV-F (420 aa; 47 kDa) is very similar in size to that of PsV-S, but they only share 17% sequence identity. PsV-F virions are ~37 nm in diameter and the gross morphology and T= 2 organization of the capsid is remarkably similar to that of PsV-S (not shown). However, the 60 protrusions in PsV-F are narrower, they extend further above the shell (~6.5 nm), and their long axes are rotationally aligned ~18° in a more anti-clockwise orientation compared to those in PsV-S.

Fusarium poae virus (FpV-1) also exhibits a 'T= 2' arrangement of its more massive coat protein subunits (637 aa; 70 kDa) (not shown). These subunits assemble to form ~42 nm diameter virions, significantly larger than both PsV-S and PsV-F. The average thickness of the FpV-1 shell varies considerably and is much less uniform than the shells of the smaller partitiviruses. The dimeric protrusions in FpV-1 extend ~3 nm above the capsid shell and have a wide base that appears to form more extensive interactions or connections to the underlying shell compared with the other two partitviruses. Central sections through a low-resolution density map of FpV-1 exhibit three concentric layers of weak density, with a 3 nm spacing consistent with a relative close packing of nucleic acid as found in other dsRNA virus capsids.

Current studies are aimed at obtaining higher-resolution 3D reconstructions for each of the fungal partitiviruses mentioned above. In the absence of any crystallographic structure determinations, higher-resolution cryo-TEM reconstructions should enable more detailed comparisons to be made, and potentially provide a means to locate the RdRps that are presumed to be fixed inside the capsid shell and possibly associated with a specific recognition site on each of the separately encapsidated genomic dsRNA segments. Such studies will also allow evolutionary links to the 'T= 2' cores of other well-studied dsRNA viruses that infect both fungal and nonfungal hosts to be explored. Structural studies expanded to include members of the genera *Alphacrypto-* and *Betacryptovirus* also have the potential to systematically characterize the similarities as well as differences in the life cycles of fungal and plant partitiviruses.

Taxonomic and Phylogenetic Considerations

Recent phylogenetic analyses based on amino acid sequences of RdRps (conserved motifs) of members of the family *Partitiviridae* led to the identification of four clusters, two of which are large and comprise only members of the genus *Partitivirus* (**Figure 5**). One large cluster with strong bootstrap support includes the partitiviruses DdV1, DdV2, FsV, GaV-MS1, OPV, and PsV-S (see **Table 1** for virus abbreviations). The CPs of these viruses share significant amino acid sequence identities (34–62%), but less than the RdRps (55–70% identity), suggesting that the CPs have evolved at a faster rate. The second partitivirus RdRp cluster consists of AhV, CrV, FpV, PoV, RhsV, and RnV1. These two large clusters were proposed to comprise two subgroups (subgroups 1 and 2) of the genus *Partitivirus* (**Figure 5**). Interestingly, the CPs of these two subgroups differ significantly in size with average sizes of 47 and 74 kDa for subgroups 1 and 2, respectively. Of the remaining two RdRp clusters (**Figure 5**), one consists of BCV (genus *Alphacryptovirus*) and two other, unclassified plant cryptoviruses (FCCV and PpV). The fourth cluster consists of a mixture of plant viruses (WCCV and RYEV, genus *Alphacryptovirus*) and fungal viruses (OMV and HmV, genus *Partitivirus*), together with CCRSAV, which may be either a fungal or a plant virus, as yet to be determined. This raises the interesting possibility of horizontal transfer of members of the family *Partitiviridae* between fungi and plants. This is a reasonable possibility because some of the viruses in these clusters have fungal hosts that are pathogenic to plants. Based on our current

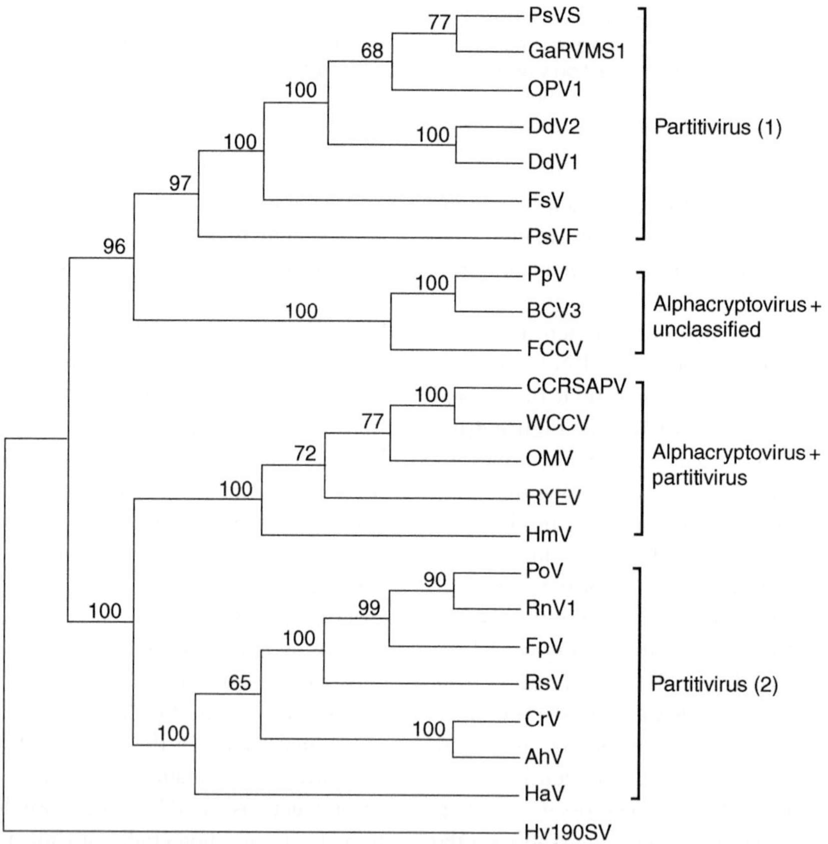

Figure 5 Phylogenetic analysis of the RdRp conserved motifs and flanking sequences derived from aligned, deduced amino acid sequences of members of the family *Partitiviridae* using the program CLUSTAL X. Motifs 3 through 8 and the sequences between the motifs, as previously designated by Jiang and Ghabrial (*Journal of General Virology* 85, 2111–2121, 2004) were used. See **Table 1** for virus name abbreviations. Bootstrap numbers out of 1000 replicates are indicated at the nodes. The tree was rooted with the RdRp of Helminthosporium victoriae 190S virus (Hv190SV), a member of the genus *Totivirus* in the family *Totiviridae* (GenBank accession no. NC_003607), which was included as an outgroup.

knowledge of partitviruses, the taxonomy of the family *Partitiviridae* will probably need to be reconsidered. As additional molecular and structural information from a wider range of plant and fungal partitiviruses is gathered, the need for new or revised taxonomic classifications may become even more apparent.

Future Perspectives

Development of Infectivity Assays for Fungal Partitiviruses

The recent success of using purified RnV-1 virions to infect virus-free isolates of the fungal host *Rosellinia necatrix* is very promising since there is considerable need to advance our knowledge of the biology of fungal partitiviruses (e.g., host-range, virus–host interaction, and molecular basis of latent infection). The natural host range of partitiviruses is restricted to the same or closely related vegetative compatibility groups that allow lateral transmission. At present, there are no known experimental host ranges for fungal partitiviruses because suitable infectivity assays have yet to be developed. With the success of the *Rosellinia necatrix*-RnV-1 infectivity assays, it is expected that future research would explore the potential experimental host range of RnV-1 including other fungal species in the genus *Rosellinia* and related genera. It is also anticipated that comparable infectivity assays using purified virions would be developed for other fungal partitiviruses. Because of the possibility alluded to earlier, of horizontal transfer of partitiviruses between fungi and plants, it is interesting to test the infectivity, using fungal protoplasts, of plant partitiviruses such as the alphacryptoviruses WCCV-1 and RYEV, which are known to be more closely related to certain fungal partitiviruses (e.g., OMV, HaV, and HmV) than to other plant partitiviruses. Alternatively, it would likewise be interesting to explore the infectivity of fungal partitiviruses such as OMV, HaV, and HmV to appropriate plant protoplasts.

Fungal partitiviruses are known to be associated with latent infections of their natural hosts. But, conceivably, some partitiviruses may induce phenotypic changes and/or virulence attenuation in one or more of their experimental host fungi. If true, this would provide excellent opportunities for exploiting fungal partitiviruses for biological control and for basic studies on host–pathogen interactions. Furthermore, transfection assays using fungal protoplasts electroporated with full-length, *in vitro* transcripts of cloned cDNA to viral dsRNAs should also be applicable to partitiviruses. This would allow for the development of partitivirus-based vectors for expressing heterologous proteins in fungi. Because many partitiviruses are known to support the replication and encapsidation of satellite dsRNA, this property would facilitate the construction of recombinant vectors. Genes of interest could be inserted into the satellite molecule between the conserved termini that are presumed to be required for replication and encapsidation. Considering the high level of partitivirus accumulation in their fungal hosts, the use of partitiviral vectors provides an attractive and cost-effective means for the overproduction of valuable proteins in filamentous fungi.

Reconsideration of the Taxonomy of the Family *Partitiviridae*

There is an urgent need to characterize at the molecular level, a broad range of plant partitiviruses that should include representatives of the genus *Betacryptovirus*. Taxonomic considerations of partitiviruses may benefit greatly from elucidating the capsid structure of representative members of each of the four clusters delineated by phylogenetic analysis. It may, however, require concerted efforts to generate purified virions of alphacryptoviruses and betacryptoviruses in quantities needed for structural studies since they generally occur at very low concentrations in their hosts.

Further Reading

Antoniw JF (2002) *Alphacryptovirus (Partitiviridae)*. In: Tidona CA and Darai G (eds.) *The Springer Index of Viruses*, pp. 676–679. New York: Springer.

Antoniw JF (2002) *Betacryptovirus (Partitiviridae)*. In: Tidona CA and Darai G (eds.) *The Springer Index of Viruses*, pp. 680–681. New York: Springer.

Crawford LJ, Osman TAM, Booy FP, *et al.* (2006) Molecular characterization of a partitivirus from ophiostoma himal-ulmi. *Virus Genes* 33: 33–39.

Ghabrial SA (2001) Fungal viruses. In: Maloy O and Murray T (eds.) *Encyclopedia of Plant Pathology*, pp. 478–483. New York: Wiley.

Ghabrial SA (2002) *Partitivirus (Partitiviridae)*. In: Tidona CA and Darai G (eds.) *The Springer Index of Viruses*, pp. 685–688. New York: Springer.

Ghabrial SA and Hillman BI (1999) Partitiviruses-fungal (*Partitiviridae*). In: Granoff A and Webster RG (eds.) *Encyclopedia of Virology*, 2nd edn., pp. 1477–1151. San Diego: Academic Press.

Ghabrial SA, Buck KW, Hillman BI, and Milne RG (2005) *Partitiviridae*. In: Fauquet CM, Mayo MA, Maniloff J, Desselberger U, and Ball LA (eds.) *Virus Taxonomy: Eighth Report of the International Committee on Taxonomy of Viruses*, pp. 581–590. San Diego, CA: Elsevier Academic Press.

Sasaki A, Kanematsu S, Onoue M, Oyama Y, and Yoshida K (2006) Infection of *Rosellinia necatrix* with purified viral particles of a member of *Partitiviridae* (RnPV1-W8). *Archives of Virology* 151: 697–707.

Phycodnaviruses

J L Van Etten, University of Nebraska–Lincoln, Lincoln, NE, USA
M V Graves, University of Massachusetts–Lowell, Lowell, MA, USA

History

Since the early 1970s, viruses or virus-like particles (VLPs) have been been reported in at least 44 taxa of eukaryotic algae, which include members in 10 of the 14 classes of algae. However, most of the early reports described single accounts of microscopic observations. The VLPs were not characterized because they were difficult to obtain in reasonable quantities. Several factors contributed to the low virus concentrations: (1) often only a few algal cells contained particles; (2) usually the cells only contained particles at one stage of the algal life cycle; (3) cells containing particles tended not to lyse; (4) in most cases the particles were not infectious; and (5) some hosts could not be cultured easily.

However, this situation began to change with the discovery of a family of large double-stranded DNA (dsDNA)-containing viruses that infect and replicate in certain strains of unicellular, eukaryotic, exsymbiotic, chlorella-like green algae. The first such 'chlorella viruses' were discovered in 1978 in chlorella symbiotic with *Paramecium bursaria* and in 1981 in chlorella symbiotic with the green coelenterate *Hydra viridis*. The algae from *P. bursaria* can be grown free of the paramecium in culture, and these cultured, naturally endosymbiotic *Chlorella* strains (NC64A and Pbi or their equivalents) serve as hosts for many similar viruses. The lytic chlorella viruses can be produced in large quantities and assayed by plaque formation using standard bacteriophage techniques. Recently, a plaque-forming virus that infects chlorella symbiotic with the heliozoon *Acanthocystis turfacea* was described. This virus does not infect *Chlorella* NC64A or *Chlorella* Pbi. The prototype chlorella virus is PBCV-1, which stands for *Paramecium bursaria* chlorella virus. The genomes (313–370 kbp) of several of the chloroviruses have either been sequenced or are in the process of being sequenced.

Large polyhedral, dsDNA-containing viruses that infect certain marine algae are also under active investigation. These include viruses that infect filamentous brown algae, *Ectocarpus* sp. (EsV viruses) and *Feldmannia* sp. (FsV viruses) (**Table 1**), and viruses that infect *Emiliania huxleyi* (EhV viruses). The genomes of some of these viruses have also been sequenced recently. Although all of these algal viruses arose from a common ancestor, they can have different lifestyles. For example, EsV and FsV viruses have a lysogenic phase in their life cycle and are only expressed as virus particles in sporangial cells of their host. In contrast, the chlorella viruses and EhV viruses are lytic.

The first algal viruses to be discovered were large dsDNA viruses; consequently, it was assumed for several years that algae were only infected by large dsDNA viruses. However, this scenario is changing rapidly. A positive-sense 9.1 kbp single-stranded RNA (ssRNA) virus has been discovered that infects a toxic bloom-forming alga, *Heterosigma akashiwo* (called HaRNAV) that is related to the picorna-like virus superfamily. A dsRNA, reo-like virus that infects a microalga, *Micromonas pusilla*, has been reported and finally a virus (CsNIV) with an unusual genome structure that infects diatoms in the genus *Chaetoceros* has been described. The CsNIV genome consists of a single molecule of covalently closed circular single-stranded DNA (ssDNA) (6005 nucleotides) as well as a segment of linear ssDNA (997 nucleotides). These recently discovered algal viruses are described in other articles in this encyclopedia.

Taxonomy and Classification

Members and prospective members of the family *Phycodnaviridae* constitute a genetically diverse, but morphologically similar group of viruses with eukaryotic algal hosts from both fresh and marine waters. Accumulating genetic evidence indicates that the phycodnaviruses together with the poxviruses, iridoviruses, asfarviruses, and the 1.2 Mbp mimivirus have a common evolutionary ancestor, perhaps, arising at the point of eukaryogenesis, variously reported to be 2.0–2.7 billion years ago. All of these viruses share nine gene products and 33 more gene products are present in members of at least two of these five viral families. Collectively, these viruses are referred to as nucleocytoplasmic large DNA viruses (NCLDV).

Phycodnaviruses are large (mean diameter 160 ± 60 nm) icosahedrons, which encapsidate 160–560 kbp dsDNA genomes. Where known, the viruses have an internal-membrane that is required for infection. Phylogenetic analyses of their δ-DNA polymerases indicate that they are more closely related to each other than to other dsDNA viruses and that they form a monophyletic group, consistent with a common ancestor. However, the phycodnaviruses fall into six clades which correlate with their hosts and each has been given genus status. Often the genera can be distinguished by additional properties, for example, lytic versus lysogenic lifestyles or linear versus circular genomes. Members of the genus *Chlorovirus* infect freshwater algae, whereas, members of the other five genera (*Coccolithovirus*,

Table 1 Taxonomy and general characteristics of some phycodnaviruses

Genus[a]	Type species[a]	Known host range[a]	Source	Particle diameter (nm)	Genome size (kbp) and conformation	Latent period (h)	Burst size
Chlorovirus	*Paramecium bursaria* chlorella virus 1 (PBCV-1)	*Chlorella* NC64A *Chlorella* Pbi *Chlorella* SAG 3.83	FW	190	313–370 Closed linear dsDNA, hairpin termini	6–8	200–350[b]
Coccolithovirus	*Emiliania huxleyi* virus 86 (EhV-86)	*Emiliania huxleyi*	MW	160–200	407–415 Circular	4–6	400–1000
Phaeovirus	*Ectocarpus siliculousus* virus 1 (EsV-1)	*Phaeophycea* *Ectocarpus siliculousus* *Ectocarpus fasciculatus* *Feldmannia simplex* *Feldmannia irregularis* *Feldmannia species* *Hincksia hincksiae* *Myriotrichia clavaeformis* *Pilayella littoralis*	MW	150–170	170–340 Open linear, single stranded regions	ND	$>1\times10^6$
Prasinovirus	*Micromonas pusilla* virus SP1 (MpV-SP1)	*Micromonas pusilla*	MW	115	200	7–14	1800–4100
Prymnesiovirus	*Chrysochomulina brevifilum* virus PW1 (CbV-PW1)	*Haptophyceae (aka Prymnesiophyceae)* *Chrysochomulina brevifilum* *Chrysochomulina strobilus* *Chrysochomulina globosa*	MW	145–170	510	ND	800–1000
Raphidovirus	*Heterosigma akashiwo* virus 01 (HaV01)	*Heterosigma akashiwo*	MW	202	294	30–33	770

[a]Data abstracted from http://www.ncbi.nlm.nih.gov/ICTVdb/Ictv/index.htm.
[b]PFU/cell.
FW, freshwater; MW, marine/coastal water; ND, not determined.
Reproduced from Dunigan DD, Fitzgerald LA, and Van Etten JL (2006) Phycodnaviruses: A peek at genetic diversity. *Virus Research* 117: 119–132, with permission from Elsevier.

Phaeovirus, *Prasinovirus*, *Prymnesiovirus*, and *Raphidovirus*) infect marine algae.

Structure and Composition

Chlorella virus particles are large (molecular weight $\sim 1 \times 10^9$ Da) and complex. The PBCV-1 virion contains more than 100 different virus-encoded proteins. The PBCV-1 54 kDa major capsid protein is a glycoprotein and comprises ~40% of the total protein. The major capsid protein consists of two eight-stranded, antiparallel β-barrel, jelly-roll domains related by pseudo sixfold rotation. This structure resembles the major coat proteins from some other dsDNA viruses that infect all three domains of life including bacteriophage PRD1, human adenoviruses, and a virus STIV infecting the Archaea, *Sulfolobus solfataricus*. This finding led to the suggestion that these three viruses may also have a common evolutionary ancestor with the NCLDVs, even though there is no significant amino acid sequence similarity among their major capsid proteins.

Cryoelectron microscopy and three-dimensional image reconstruction of the PBCV-1 virion (**Figure 1**) indicate that the outer capsid is icosahedral and covers a lipid bilayered membrane. The membrane is required for infection because the virus loses infectivity after exposure to organic solvents. The outer diameter of the virus capsid ranges from 1650 Å along the two- and threefold axes to 1900 Å along the fivefold axis. The capsid shell consists of 1680 doughnut-shaped trimeric capsomers plus 12 pentameric capsomers at each icosahedral vertex. The trimeric capsomers are arranged into 20 triangular facets (trisymmetrons, each containing 66 trimers) and 12 pentagonal facets (pentasymmetrons, each containing 30 trimers and one pentamer at the icosahedral vertices). Assuming all the trimeric capsomers are identical, the outer capsid of the virus contains 5040 copies of the major capsid protein. The virus has a triangulation number of 169. However, PBCV-1 is not the largest phycodnavirus; phaeocystis pouchetti virus (PpV01) has an icosahedral capsid with a triangulation number of 219.

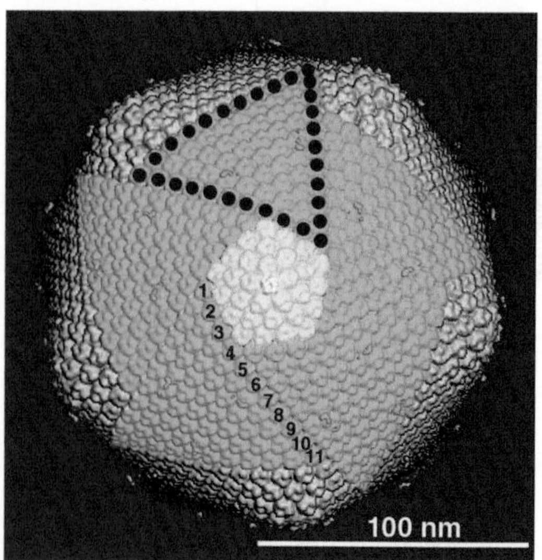

Figure 1 Three-dimensional image reconstruction of chlorella virus PBCV-1 from cryoelectron micrographs. The virion capsid consists of 12 pentasymmetrons and 20 trisymmetrons. Five trisymmetrons are highlighted in the reconstruction (blue) and a single pentasymmetron is colored yellow. A pentavalent capsomer (white) lies at the center of each pentasymmetron. Each pentasymmetron consists of one pentamer plus 30 trimers. Eleven capsomers form the edge of each trisymmetron (black dots) and therefore each trisymmetron has 66 trimers. Reprinted, with permission, from Van Etten JL (2003) Unusual life style of giant chlorella viruses. *Annual Review of Genetics* 37: 153–195, ©2003 by Annual Reviews.

Structural proteins of many viruses, such as herpesviruses, poxviruses, and paramyxoviruses, as well as the chlorella viruses, are glycosylated. Typically, viral proteins are glycosylated by host-encoded glycosyltransferases located in the endoplasmic reticulum (ER) and Golgi and then transported to a host membrane. Nascent viruses acquire the glycoprotein(s) and only become infectious by budding through the membrane, usually as they are released from the cell. Consequently, the glycan portion of virus glycoproteins is host specific.

However, glycosylation of PBCV-1 major capsid protein differs from this paradigm. Accumulating evidence indicates that PBCV-1 encodes most, if not all, of the enzymes involved in constructing the complex oligosaccharides attached to its major capsid protein and that the process occurs independently of the ER and Golgi. Furthermore, five of six putative PBCV-1-encoded glycosyltransferases are predicted to be located in the cytoplasm. PBCV-1 also encodes several additional proteins involved in post-translational modification that may alter virus structural proteins. These include a prolyl-4-hydroxylase and several protein kinases and a phosphatase.

Genomes

The 331 kbp PBCV-1 genome is linear and nonpermuted. The genome termini consist of 35 nucleotide long, incompletely base-paired, covalently closed hairpin loops that exist in one of two forms (flip and flop). Each hairpin loop is followed by an identical 2.2 kbp inverted repeat sequence; the remainder of the genome consists primarily of single-copy DNA. The PBCV-1 genome has ~695 open reading frames (ORFs) that have 65 or more codons, of which ~366 are probably protein encoding. The putative protein-encoding genes are evenly distributed on both strands and intergenic space is minimal, 275 ORFs are separated by less than 100 nucleotides (**Figure 2**). One exception is a 1788 bp sequence near the middle of the

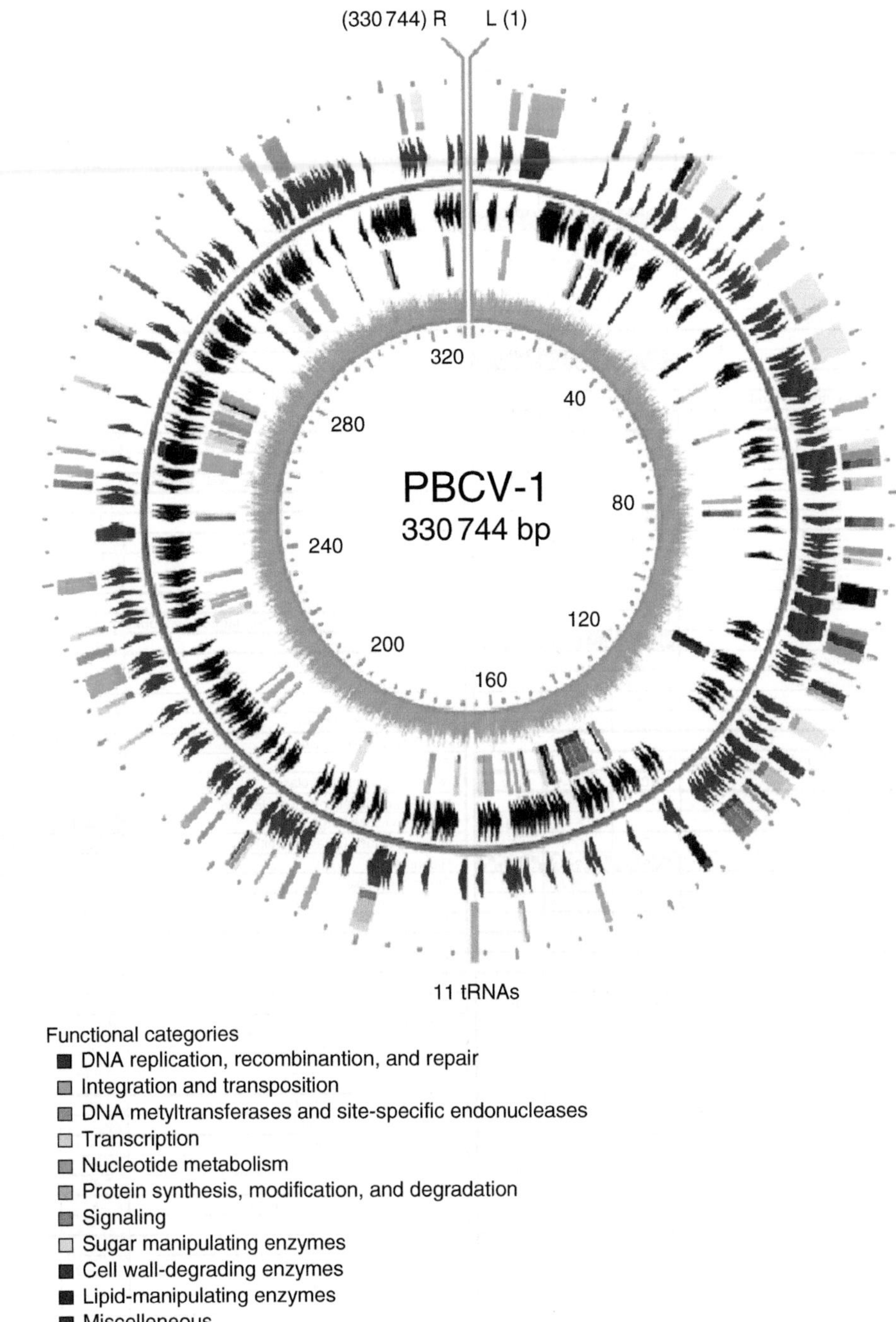

Figure 2 Map of the chlorella virus PBCV-1 genome visualized as a circle using Circular Genome Viewer (Paul Stothard, Genome Canada). However, the genome is a linear molecule and the ends are depicted at the top of the figure as green lines (L and R represent the left and right ends of the genome, respectively). The red and blue arrows represent the 366 protein-encoding genes; red arrows depict genes transcribed in the rightwards direction and blue arrows genes transcribed in the leftwards direction. The two rings that flank the protein-encoding genes show the predicted functions of the proteins, color-coded by function (see insert in the figure). The location of the polycistronic gene encoding the 11 tRNAs is indicated in the outermost ring (i.e., this gene is transcribed in the rightwards direction). The innermost ring (in green) represents the A + T content determined using a 25 bp window. Note that the A + T content is fairly constant over the genome (60% A + T).

genome. This sequence has a polycistronic gene containing 11 tRNAs (**Figure 2**).

Approximately 40% of the 366 PBCV-1 gene products resemble proteins in the databases, many of which have not previously been associated with viruses (some are listed in **Figure 3**). Eighty-four ORFs have paralogs within PBCV-1, forming 26 groups. The size of these groups ranges from two to six members.

DNA replication, recombination, and repair			
Description	PBCV-1	EsV-1	EhV-86
Archaeo-eukaryotic primase			
ATPase[a]	4	2	
DNA ligase	*		
DNA topoisomerase II	*		
DNA-binding protein	#		
Exonuclease[b]	*	2	
Helicase			2
Helicase-Rec BCD-like			
Helicase-Superfamily III			
Nucleic acid-binding protein			
PCNA[c]	2		2
Pyrimidine dimer-specific glycosylase	*		
Replication factor C (lg subunit)			
Replication factor C (sm subunit)		4	
RNaseH			
δ-DNA polymerase			

Protein synthesis, modification, and degradation			
Description	PBCV-1	EsV-1	EhV-86
ATPase (AAA+ class)			
ATP-dependent protease proteolytic subunit			
Esterase			
Hydrolase			
Prolyl 4-hydroxylase	*#		
Protease-Cysteine			
Protease-OTU like cysteine			2
Protease-Serine			5
Protein disulfide isomerase	#		
SKP1 protein			
Thiol oxidoreductase	#		
Translation elongation factor-3			
Ubiquitin C-terminal hydrolase			
Zn metallopeptidase			
tRNAs[d]	11		5

Integration and transposition			
Description	PBCV-1	EsV-1	EhV-86
Antirepressor of lysogeny			
Endonuclease			3
Homing endonuclease GIY-YIG	7		
Homing endonuclease HNH[e]	6	2	
Integrase			
Protelomerase			
Repressor of lysogeny			
Transposase		2	

Sugar and lipid manipulation			
Description	PBCV-1	EsV-1	EhV-86
Alginate mannuronan epimerase			
D-lactate dehydrogenase			
Fatty acid desaturase			2
Fructose-2, 6 bisphosphatase			
Fucose synthetase	*		
GDP-D-mannose dehydratase	*		
Glucosamine synthetase	*		
Glycerophosphoryl diesterase			
Glycosyltransferase	6		
Hyaluronan synthase	*		
Lipase			
Lipid phosphate phosphatase			
Lysophospholipase			
N-acetyltransferase			
Patatin phospholipase			
Salidase			
Serine palmitoyltransferase			
Sterol desaturase			
Sugar lyase			
Transmembrane fatty acid elongation protein			
UDP-glucose dehydrogenase	*		

Signaling			
Description	PBCV-1	EsV-1	EhV-86
Dual specificity phosphatase	#		
Glutamate receptor			
Hybrid histidine kinase		6	
Ligand-gated channel protein			
Phosphoshuttle			
Potassium channel protein	*#		
Ser/Thr protein kinase[f]	8*#	4	2

DNA restriction/modification			
Description	PBCV-1	EsV-1	EhV-86
Adenine DNA methylase	2*		
Cytosine DNA methylase	3*		
DNA restriction endonuclease	2*#		

Transcription			
Description	PBCV-1	EsV-1	EhV-86
DNA-dependent RNA polymerase II largest			
DNA-directed RNA polymerase II subunit			
DNA-directed RNA polymerase subunit			3
Histone H3, Lys 27 methylase	*#		
mRNA guanylyltransferase	*		
Oligoribonuclease			
RNA triphosphatase	*		
RNase III	*		
Superfamily II helicase[g]	3#		
SWI/SNF helicase	#		
Transcription factor TFIIB			
Transcription factor TFIID			
Transcription factor TFIIS			
VLTF2-type transcription factor			

Cell wall degradation			
Description	PBCV-1	EsV-1	EhV-86
Chitinase	2*#		
Chitosanase	*#		
β and α 1, 4-linked glucuronic lyase	*		
β-1, 3-glucanase	*		

Nucleotide metabolism			
Description	PBCV-1	EsV-1	EhV-86
Aspartate transcarbamylase	*		
Cytosine deaminase			
dCMP deaminase	*		
Deoxynucleoside kinase			
dUTP pyrophosphatase	*		
Glutaredoxin	*		
NTP pyrophosphohydrolase			
Nucleic acid-independent nucleoside			
Ribo. reductase (large subunit)			
Ribo. reductase (small subunit)			
Ribonuclease			
Thermonuclease			
Thioredoxin	#		2
Thymidylate kinase			
Thymidylate synthase X	*		
Thymidylate synthase-bifunct. dihydrofolate			

Miscellaneous			
Description	PBCV-1	EsV-1	EhV-86
ABC transporter protein	#		
Agmatine iminohydrolase	*		
Amidase	*#		
Cu/Zn-superoxi dedismutase	*#		
Fibronectin binding protein			
Histidine decarboxylase			
Homospermidine synthase	*#		
Monoamine oxidase	#		
N-carbamoylput amidohydrolase	*		
O-methyltransferase			
Ornithine decarboxylase	*		
Pathogenesis-related protein			
Calcium-binding protein			
Collagen-like protein			
Thaumatin-like protein			
Lectin protein			2
Longevity-assurance family protein			
Major facilitator			
Phosphate permease			
Phosphoglycerate mutase			

Present in all 3 viruses	
PBCV-1 + EsV-1	
PBCV-1 + EhV-86	
EsV-1 + EhV-86	
Unique to one virus	
Functional Enzyme	*
Virion Associated	#

Figure 3 Continued

Some PBCV-1 genes are interrupted by introns; a gene encoding a transcription factor-like protein contains a self-splicing type I intron, whereas the δ-DNA polymerase gene contains a spliceosomal-processed type of intron. In addition, one of the PBCV-1 tRNA genes is predicted to have an intron.

One unusual feature of PBCV-1 DNA, as well as the other chlorella virus DNAs, is that they contain methylated bases. Chlorella virus genomes contain 5-methylcytosine in amounts varying from 0.1% to 48% of the total cytosines. Many viral DNAs also contain N^6-methyladenine with concentrations up to 37% of the total adenines. This led to the discovery that many chlorella viruses encode multiple DNA methyltransferases, as well as site-specific (restriction) endonucleases.

The 407 kbp circular dsDNA genome of one of the EhV viruses is predicted to have 472 protein-encoding genes. Only 66 (14%) of these 472 gene products match a sequence in GenBank. The EhV virus encodes several unexpected genes never found in a virus before, including four gene products involved in sphingolipid biosynthesis and two additional gene products which encode desaturases.

The structure of the 335 kbp EsV genome is unknown. Although several experiments suggest that the genome is circular, DNA sequencing indicates that it has defined ends with inverted repeats. The virus is predicted to contain 231 protein-encoding genes, 48% of which resemble various proteins in the public databases. About 12% of the EsV genome consists of tandem repeats and portions of the genome have ssDNA regions. Collectively, PBCV-1, EsV, and EhV viruses have in excess of 1000 unique ORFs. A total of 123 putative ORFs from these three viruses is organized into metabolic domains (**Figure 3**). Interestingly, these three viruses only have 14 gene products in common. Not surprisingly, several of these common ORFs are involved in DNA replication, such as the δ-DNA polymerase, large and small subunits of ribonucleotide reductase, proliferating cell nuclear antigen (PCNA), superfamily II and III helicases, and the newly recognized archaeo-eukaryotic primases. Another common ORF among the three phycodnaviruses is the major capsid protein.

Virus Replication

PBCV-1 attaches rapidly, specifically, and irreversibly to the external surface of cell walls, but not to protoplasts, of host *Chlorella* NC64A. Attachment always occurs at a virus vertex, followed by degradation of the wall at the attachment point. Following wall degradation, the internal membrane of the virus probably fuses with the host membrane resulting in entry of the virus DNA and virion-associated proteins into the cell, leaving an empty capsid on the surface. Several observations suggest that the infecting DNA, plus associated proteins, is rapidly transported to the host cell nucleus and the host

Figure 3 Selected ORFs in the PBCV-1, EsV, and EhV genomes are arranged by their metabolic domains. If a genome encodes a putative protein more than once, a number in the box indicates the number of genes of this type per genome. Color-coding is indicated on the figure and is used to depict the relationship between viruses. Red indicates proteins that are encoded by all three viruses; yellow indicates proteins that are encoded by PBCV-1 and EsV, but not EhV; green indicates proteins that are encoded by PBCV-1 and EhV, but not EsV; orange indicates proteins that are encoded by EsV and EhV, but not PBCV-1; and blue indicates there are no shared homologs. Solid colored boxes indicate that the putative proteins are homologs. A diagonally stripped box indicates that the putative proteins are nonhomologous, and a checkered box indicates that the putative proteins are a mix of homologous, nonhomologous, or unique ORFs. In this case, a footnote has been added to clarify the specific differences; in parentheses the ORF has been defined by the gene number and any ORF beginning with an 'A' is from the PBCV-1 genome, an 'EsV' is from the EsV-1 genome, and an 'EhV' is from the EhV-86 genome. Proteins known to be functional are indicated with a star (*) and proteins known to be associated with the virion are indicated with a pound sign (#). Superscript 'a' indicates ATPase – One homolog between all three viruses (A392R, EsV-26, and EhV072), one homolog between PBCV-1 and EsV (A565R and EsV-171), and two PBCV-1 ATPases which have no homologs in EsV or EhV (A561L and A554/556/557L). Superscript 'b' indicates Exonuclease – One homolog between PBCV-1 and EsV (A166R and EsV-64) and one unique to EsV (EsV-126). Supercript 'c' indicates PCNA – One homolog between all three viruses (A193L, EsV-132, and EhV020), one homolog between PBCV-1 and EhV (A574L and EhV020). However, EhV-440, another EhV-encoded PCNA, has no homologs in PBCV-1 or EsV. Superscript 'd' indicates tRNAs – PBCV-1 has 11 tRNA genes, encoding AA Leu (2), Ile, Asn (2), Lys (3), Tyr, Arg, and Val. EhV has five tRNA genes, encoding AA Leu, Ile, Gln, Asn, and Arg. Supercript 'e' indicates Homing endonuclease HNH – One homolog between two viruses (A422R and EhV087). PBCV-1 (A87R) and EsV (EsV-119) are homologous. Four other HNH endonucleases are unique to PBCV-1 (A267L, A354R, A478R, and A490L) and one is unique to EsV (EsV-1–16). Superscript 'f' indicates Ser/Thr protein kinase – Four PBCV-1, two EsV-1, and one EhV-86 S/T kinases ORFs are homologous (A248R, A277L, A282L, A289L, EsV-82, EsV-111, and EhV451). The remaining S/T kinase ORFs from the three genomes (A34R, A278L, A614L, A617R, EsV-104, EV-156, and EhV-402) are unique. Suerscript 'g' indicates that superfamily II helicase, PBCV-1 (A153R), and EsV (EsV-66) are homologous. The two additional PBCV-1-encoded helicases (A241R and A363R) and EhV104 are unique. Reproduced from Dunigan DD, Fitzgerald LA, and Van Etten JL (2006) Phycodnaviruses: A peek at genetic diversity. *Virus Research* 117: 119–132, with permission from Elsevier.

transcription machinery is reprogrammed to transcribe virus RNAs. This process occurs rapidly because early PBCV-1 transcripts can be detected within 5–10 min post infection (p.i.). PBCV-1 translation occurs on cytoplasmic ribosomes and early PBCV-1-encoded proteins can be detected within 10 min p.i.

PBCV-1 DNA synthesis and late virus transcription begins 60–90 min p.i. Ultrastructural studies of PBCV-1-infected chlorella suggest that the nuclear membrane remains intact, at least during early stages of virus replication. At approximately 2–3 h p.i., assembly of virus capsids begins in localized regions in the cytoplasm, called virus assembly centers, which become prominent at 3–4 h p.i.. By 5 h p.i., the cytoplasm is filled with infectious progeny virus particles (~1000 particles/cell) and by 6–8 h p.i. localized lysis of the host cell releases progeny. Of the progeny released, 25–50% of the particles are infectious. Mechanical disruption of cells releases infectious virus 30–50 min prior to cell lysis, indicating the virus does not acquire its glycoprotein capsid by budding through the host plasma membrane as it is released from the cell. Other chlorella viruses have longer replication cycles than PBCV-1. For example, virus NY-2A requires approximately 18 h for replication and consequently forms smaller plaques.

Virus EsV initiates its life cycle by infecting free-swimming, wallless gametes of its host. Virus particles enter the cell by fusion with the host plasma membrane and release a nucleoprotein core particle into the cytoplasm, leaving remnants of the capsid on the surface. The viral core moves to the nucleus within 5 min p.i. One important feature that distinguishes the EsV life cycle from the other phycodnaviruses is that the viral DNA is integrated into the host genome and is transmitted mitotically to all cells of the developing alga. The viral genome remains latent in vegetative cells until it is expressed in the algal reproductive cells, the sporangia or gametangia. Massive viral DNA replication occurs in the nuclei of these reproductive cells, followed by nuclear breakdown and viral assembly that continues until the cell becomes densely packed with virus particles. Virus release is stimulated by the same factors that induce discharge of gametes from the host, that is, changes in temperature, light, and water composition. This synchronization facilitates interaction of viruses with their susceptible host cells.

In contrast to the chlorella and EsV viruses, very little is known about the EhV life cycle, including how it infects its host. One property that clearly distinguishes EhV from the other phycodnaviruses is that it encodes six RNA polymerase subunits; in contrast, neither PBCV-1 nor EsV encodes a recognizable RNA polymerase component. Thus like the poxviruses, EhV may carry out its entire life cycle in the cytoplasm of its host. The latent period for EhV is 4–6 h with a burst size of 400–1000 particles per cell.

Virus Transcription

Detailed studies on transcription are lacking for all of the algal viruses and only a few general statements can be made about PBCV-1 transcription:

1. PBCV-1 infection rapidly inhibits host RNA synthesis.
2. Viral transcription is programmed and early transcripts appear within 5–10 min p.i. Late viral transcription first begins about 60–90 min p.i.
3. Some early viral transcripts are synthesized in the absence of *de novo* protein synthesis. As expected, the synthesis of later transcripts requires translation of early virus genes.
4. Early and late genes are dispersed in the PBCV-1 genome.
5. PBCV-1 ORFs are tightly packed on both DNA strands and the coding regions of some of the genes overlap. The largest distance between PBCV-1 ORFs is a 1788-nucleotide stretch in the middle of the genome. This sequence has a polycistronic gene containing 11 tRNAs (**Figure 2**).
6. Consensus promoter regions for early and late genes have not been identified, although the 50 nucleotides preceding the ATG start codon of most functional PBCV-1 genes are at least 70% A+T.
7. Transcription of some PBCV-1 genes appears to be complex. For example, some gene transcripts exist as multiple bands and these patterns change between early and late times in the virus life cycle.

Additional *Chlorella* Viruses

Several hundred plaque-forming chloroviruses have been characterized to various degrees. They infect either *Chlorella* NC64A cells (NC64A viruses), an endosymbiont of *P. bursaria* isolated from North America, or *Chlorella* Pbi cells (Pbi viruses) that are endosymbiotic with a paramecium isolated in Europe. Like PBCV-1, each of these viruses contain many structural proteins, a large (>300 kbp) dsDNA genome, and they are chloroform sensitive. The DNAs of some of these viruses hybridize strongly with PBCV-1 DNA, while others hybridize poorly.

Three additional chlorella virus genomes have been sequenced recently and others are nearing completion. The largest, the 370 kbp genome of virus NY-2A, contains ~400 protein-encoding genes. Most common genes are colinear in viruses PBCV-1 and NY-2A, which infect the same host chlorella. However, almost no colinearity exists between common genes in Pbi virus MT325 and those in PBCV-1, NY-2A, and AR158 suggesting plasticity in the chlorella virus genomes. Additionally, the G + C contents of the three NC64A viruses range from 40% to 41% whereas the G + C contents of the Pbi viruses are approximately 45%. These last two observations suggest that these two virus groups have been separated for considerable evolutionary time. Viruses morphologically similar to the NC64A and Pbi viruses have also been isolated from chlorella symbiotic in the coelenterate *Hydra viridis* and very recently from the heliozoon *Acanthocystis turfacea.* The symbiotic hydra chlorella have not been cultured and so very little is known about these viruses. However, the *A. turfacea* viruses can be isolated by plaque formation.

Other Algal Viruses

Field isolates representing at least six genera of filamentous brown algae contain virus particles that are morphologically similar to EsV and FsV. Virus expression is variable; virions are rarely observed in vegetative cells but often are common in unilocular sporangia (FsV) or both unilocular and plurilocular sporangia (EsV). EsV viruses only infect the free swimming, zoospore stage of *Ectacarpus* sp. All natural isolates of *Feldmania* sp. are infected with virus and so infection studies cannot be conducted.

Viruses that infect *Emiliania huxleyi*, *Micromonas pusilla*, and *Chrysochromulina brevfilum* have been isolated from many marine environments. These viruses can be distinguished by DNA restriction patterns.

Phycodnavirus Genes Encode Some Interesting and Unexpected Proteins

Many chlorella virus-encoded enzymes are either the smallest or among the smallest proteins of their class. In addition, homologous genes in the chloroviruses can differ in nucleotide sequence by as much as 50%, which translates into amino acid differences of 30–40%. Therefore, comparative protein sequence analyses can identify conserved amino acids in proteins as well as regions that tolerate amino acid changes. The small sizes and the finding that many chlorella virus-encoded proteins are 'user friendly' have resulted in the biochemical and structural characterization of several PBCV-1 enzymes. Examples include: (1) The smallest eukaryotic ATP-dependent DNA ligase, which is the subject of intensive mechanistic and structural studies. (2) The smallest type II DNA topoisomerase. The virus enzyme cleaves dsDNAs about 50 times faster than the human type II DNA topoisomerase; consequently, the virus enzyme is being used as a model enzyme to study the mechanism of topoisomerase II DNA cleavage. (3) An RNA guanylyltransferase that was the first enzyme of its type to have its crystal structure resolved. (4) A small prolyl-4-hydroxylase that converts Pro-containing peptides into hydroxyl-Pro-containing peptides in a sequence-specific fashion. (5) The smallest protein (94 amino acids) to form a functional potassium ion channel. These minimalist enzymes may represent evolutionary precursors of contemporary proteins, but it is also possible that they are products of evolutionary optimization during viral evolution.

The chloroviruses are also unusual because they encode enzymes involved in sugar metabolism. Three PBCV-1 encoded enzymes, glutamine:fructose-6-phosphate amindotransferase, UDP-glucose dehydrogenase, and hyaluronan synthase, are involved in the synthesis of hyaluronan, a linear polysaccharide composed of alternating β-1,4-glucuronic acid and β-1,3-*N*-acetylglucosamine residues. All three genes are transcribed early in PBCV-1 infection and hyaluronan accumulates on the external surface of the infected cells.

Two PBCV-1-encoded enzymes, GDP-D-mannose dehydratase and fucose synthase, comprise a three-step pathway that converts GDP-D-mannose to GDP-L-fucose. The function of this putative pathway is unknown. However, fucose, a rare sugar, is present in the glycans attached to the major capsid protein.

PBCV-1 encodes four enzymes involved in polyamine biosynthesis: ornithine decarboxylase (ODC), homospermidine synthase, agmatine iminohydrolase, and *N*-carbamoyl-putrescine amidohydrolase. ODC catalyzes the decarboxylation of ornithine to putrescine, which is the first and the rate-limiting enzymatic step in polyamine biosynthesis. Not only is the PBCV-1-encoded ODC the smallest known ODC, the PBCV-1 enzyme is also interesting because it decarboxylates arginine more efficiently than ornithine.

The genome sequences of other phycodnaviruses have revealed several interesting and unexpected gene products. However, except for an aquaglyceroporin encoded by chlorella virus MT325, none of these products have been expressed and tested for enzyme function.

Ecology

Eukaryotic algae are important components of both freshwater and marine environments; however, the significance of viruses in these systems is only beginning to be appreciated. The chlorella viruses are ubiquitous in freshwater collected throughout the world and titers as high as 100 000 infectious particles per mililiter have been reported in native waters. Typically, the titer is 1–100 infectious particles per milliliter. The titers are seasonal with the highest titers in the spring. It is not known whether chlorella viruses replicate exclusively in algae symbiotic with paramecia or if the viruses have another host(s). In fact, it is not known if paramecium chlorellae exist free of their hosts in natural environments. However, the chlorellae are protected from virus infection when they are in a symbiotic relationship with the paramecium.

The concept that viruses might have a major impact on the marine environment began about 15 years ago with the discovery that seawater contains as many as 10^7 VLPs per milliliter. This huge population consits of both bacterial and algal viruses and is important because phytoplankton, consisting of cyanobacteria and eukaryotic microalgae, fix 50–60% of the CO_2 on Earth. At any one time, 20–40% of these photosynthetic organisms are infected with a virus. Consequently, these viruses contribute to microbial composition and diversity, as well as, nutrient cycling in aqueous environments. Thus viruses, including the phycodnaviruses, have a major impact on global carbon/nitrogen cycles that is only beginning to be appreciated by scientists, including those who model such cycles.

Algal viruses are also believed to play major roles in the termination of marine algal blooms and so there are active research efforts to understand the natural history of these algal/virus systems. For example, the coccolithophorid *Emiliania huxleyi* is a unicellular alga found throughout the world; the alga can form immense coastal and mid-oceanic blooms at temperate latitudes that cover 10 000 km^2 or more. One of the primary mechanisms for terminating *E. huxleyi* blooms is lysis by the EhV viruses described above.

The filamentous brown algae, *Ectocarpus* sp. and *Feldmania* sp. isolated from around the world, are infected with lysogenic EsV and FsV viruses, respectively. Lysogeny is consistent with the observation by early investigators that VLPs appear infrequently in eukaryotic algae and only at certain stages of algal development. The apparent lack of infectivity by many of the previously observed VLPs in eukaryotic algae is also consistent with a lysogenic lifestyle. The VLPs might either infect the host and resume a lysogenic relationship or be excluded by preexisting lysogenic viruses.

Perspectives

Sequence analyses of three phycodnaviruses suggest that this family may have more sequence diversity than any other virus family. These three viruses only have 14 homologous genes, which means that there are in excess of 1000 unique ORFs in just these three viruses. Despite the large genetic diversity in these three sequenced phycodnaviruses, phylogenetic analyses of δ-DNA polymerases and DNA primases indicate that the phycodnaviruses group into a monophyletic clade within the NCLDVs. A recent analysis using eight concatenated core NCLDV genes also indicates that the phycodnaviruses cluster together and are members of the NCLDV 'superfamily'. However, it is obvious that the identification and characterization of phycodnaviruses is in its infancy. Metagenomic studies, such as DNA sequences from the Sargasso Sea samples, indicate that translation products of many unknown sequences are more similar to PBCV-1 proteins than the next known phycodnavirus.

Further Reading

Brussaard CPD (2004) Viral control of phytoplankton populations – A review. *Journal of Eukaryotic Microbiology* 51: 125–138.

Dunigan DD, Fitzgerald LA, and Van Etten JL (2006) Phycodnaviruses: A peek at genetic diversity. *Virus Research* 117: 119–132.

Iyer ML, Balaji S, Koonin EV, and Aravind L (2006) Evolutionary genomics of nucleocytoplasmic large DNA viruses. *Virus Research* 117: 156–184.

Muller DG, Kapp M, and Knippers R (1998) Viruses in marine brown algae. *Advances in Virus Research* 50: 49–67.

Van Etten JL (2003) Unusual life style of giant chlorella viruses. *Annual Review of Genetics* 37: 153–195.

Van Etten JL, Graves MV, Muller DG, Boland W, and Delaroque N (2002) Phycodnaviridae – large DNA algal viruses. *Archives of Virology* 147: 1479–1516.

Wilson WH, Schroeder DC, Allen MJ, *et al.* (2005) Complete genome sequence and lytic phase transcription profile of a Coccolithovirus. *Science* 309: 1090–1092.

Yamada T, Onimatsu H, and Van Etten JL (2006) Chlorella viruses. *Advances in Virus Research* 66: 293–336.

Relevant Websites

http://www.giantvirus.org – Giantvirus.org.
http://greengene.uml.edu – Greengene at UML.

Picornaviruses: Molecular Biology

B L Semler and K J Ertel, University of California, Irvine, CA, USA

Glossary

***Cre* (*cis*-acting replication element)** First found in human rhinovirus genomic RNA and subsequently identified in other picornavirus genomes, the *cre* acts as template for the viral RNA polymerase $3D^{pol}$ to uridylylate VPg to VPg-pU-pU. Evidence suggests that this *cre* can function in *trans* as well.

IRES (internal ribosome entry site) An RNA sequence typically characterized by extensive nucleic acid secondary structure. The 40S ribosomal subunit of the cellular translation machinery interacts with RNA stem loops/ sequence, and subsequently allows translation of downstream RNA sequence of the IRES. Translation, therefore, proceeds without recognition of a 5′ cap. Utilized by some virus families (including *Picornaviridae*) and cellular messenger RNAs.

Polyprotein In the context of the discussion of *Picornaviridae*, this refers to the long protein resulting from translation of the single open reading frame of picornavirus RNA. The polyprotein is processed by viral proteinases to yield mature viral proteins.

Positive-strand RNA A single molecule of picornavirus RNA that encodes functional viral protein when translated in the 5′–3′ direction. This is the 'sense' orientation of picornavirus RNA as it enters the cell that is also encapsidated in progeny virions.

RNP complex Ribonucleoprotein complex. Describing a stable interaction of RNA and protein(s), either *in vivo* or *in vitro*.

Uridylylation Refers to the addition of two uridylate residues to the VPg molecule by the picornavirus RNA-dependent RNA polymerase $3D^{pol}$ (and other viral proteins) using a viral RNA template.

VPg Virus protein, genome-linked. Also known as 3B, the function of VPg is to act as a protein primer for the picornavirus RNA-dependent RNA polymerase. Following uridylylation, VPg-pU-pU is covalently attached to the 5′ end of picornavirus RNAs (positive- and negative-strands).

Overview of Virus Family *Picornaviridae*

Introduction

Viruses belonging to the family *Picornaviridae* are small (Latin *Pico*) RNA (*rna*) viruses whose host range is typically restricted to mammals. Genera associated with *Picornaviridae* include erbovirus, teschovirus, kobuvirus, aphthovirus, cardiovirus, enterovirus, hepatovirus, parechovirus, and rhinovirus. The first three of these genera are relatively recent additions to the picornavirus family, and the last four contain pathogens that are the most extensively studied picornaviruses capable of infecting humans. Particularly, poliovirus of the enterovirus family is widely considered to be the 'prototypical' picornavirus, and perhaps the most feared by humans due to the potential for poliovirus infection to result in paralytic poliomyelitis. This debilitating affliction can result in paralysis of one or more limbs in an infected individual, and in rare cases even death. Human rhinovirus infection in humans results in the common cold, and is one of the most prevalent diseases throughout the world. While in the developed world the common cold is often seen merely as an inconvenience at worst, it is the most important cause of asthma exacerbations, and there is no effective vaccine for the virus nor any effective medical treatment for infection.

Due to the intense interest in *Picornaviridae* based upon the diseases associated with picornavirus infections, the virus family has received extensive scientific attention to understand the mechanisms of gene expression and replication of its members. In turn, insights into the manner in which members of the *Picornaviridae* propagate have allowed a better understanding of how these viruses cause disease and also how other unrelated virus families replicate within their own host cell systems. The aim of this article is to discuss the general mechanisms of gene expression and propagation of the *Picornaviridae* as well as to highlight some of the differences between the individual picornavirus family members.

Classification

From a disease perspective, individual virus species of the family *Picornaviridae* are grouped by genus based on their pathogenic properties and/or route of infection (**Table 1**). For example, the enterovirus genus includes poliovirus and coxsackievirus based upon the natural oral route of entry into the host and replication in gut tissue.

Table 1 Members of *Picornaviridae*

Genus	*Representative species*	*Associated disease*
Aphthovirus	*Foot-and-mouth disease virus (FMDV)*	Foot-and-mouth disease (livestock)
Cardiovirus	*Encephalomyocarditis virus (EMCV), Theiler's murine encephalomyocarditis virus (TMEV)*	Encephalomyelitis, myocarditis (mice and livestock), demyelination in mice
Enterovirus	*Poliovirus, coxsackievirus*	Poliomyelitis; hand, foot, and mouth disease; myocarditis
Erbovirus	*Equine rhinitis virus*	Upper respiratory infection (horses)
Hepatovirus	*Hepatitis A virus*	Acute liver disease
Kobuvirus	*Aichi virus*	Gastroenteritis
Parechovirus	*Human parechovirus*	Gastroenteritis, paralysis, encephalitis, neonatal carditis
Rhinovirus	*Human rhinovirus*	Common cold
Teschovirus	*Porcine teschovirus*	Encephalitis, paralysis (swine)

The genus cardiovirus includes encephalomyocarditis virus (EMCV), which infects heart and nervous tissues in rodents. Based upon the ability of neutralizing antibody to recognize capsid antigens on the surface of the virion the species of *Picornaviridae* are further subclassified on the basis of serotype. This can range from a single serotype in the case of hepatitis A virus (HAV), or, in the case of rhinovirus, to ~100 identified serotypes.

In terms of molecular genetics, picornaviruses are broadly classified according to the internal ribosome entry site (IRES) in the 5′ noncoding region (NCR). Based upon similarity in RNA secondary structure and sequence studies, three distinct types of picornavirus IRES elements have been characterized. The genomes of enteroviruses (poliovirus and coxsackievirus) and rhinovirus contain type I IRESs, while type II IRESs are found in the genomes of cardiovirus, aphthovirus, erbovirus, and parechovirus. The HAV IRES is considered an outlier of the type II IRES and is classified as a type III IRES. The recently identified teschovirus and kobuvirus IRES elements are yet to be examined in enough detail for placement into a specific type. The function and characteristics of picornavirus IRES types are discussed in the section on 'Translation' of this article.

The other primary difference that distinguishes the picornavirus genera relates to structure and function of the proteins encoded by the polyprotein. For example, the genomes of enteroviruses, rhinoviruses, parechoviruses, and hepatoviruses do not code for a 'leader' (L) protein. Instead, many of these viruses encode a proteinase in the P2 region of the genome designated 2A. The 2A proteinase is structurally similar among the enteroviruses and rhinoviruses. Cardiovirus, aphthovirus, erbovirus, teschovirus, and kobuvirus genomes code for an L protein at the amino terminus of the uncleaved polyprotein, but only the aphthovirus L protein has been reported to be a proteolytically active proteinase. Parechovirus and HAV RNA genomes do not encode an L protein and the 2A regions of the genomes of these viruses are not proteolytically active.

Genome Structure and Features

All picornaviruses contain a single molecule of RNA of messenger RNA (mRNA) sense (hereafter referred to as positive-sense) within the capsid (**Figure 1**). The nucleotide (nt) length of the RNA genome found in virion particles is quite short compared to most other viruses, ranging from 6500 to 9500 nt in length. The 5′ end of the viral genome is covalently linked to the viral polypeptide VPg (virus protein, genome linked). The 5′ RNA terminus also contains RNA secondary structures important for viral RNA synthesis (see the section titled 'Viral RNA replication'). In comparison to cellular mRNAs, the picornavirus 5′ NCR is quite long often in excess of 700 nt and up to 1200 nt. Encompassing nearly 10% of the total RNA genome length, the 5′ NCR contains the picornavirus IRES, which allows initiation of translation of the viral genome. The 3′ NCRs of picornaviruses are, in contrast to many cellular mRNAs, quite short, ranging from 50 to 100 nt in length. Lastly, a homopolymeric poly(A) tract encoded by the viral genome follows the 3′ NCR and may participate in the formation of stable RNA secondary structure(s) with the heteropolymeric 3′ NCR. The poly (A) tract likely stabilizes and protects the viral RNA genome from intracellular degradation enzymes, much as the poly(A) tracts of cellular mRNAs protect their cognate RNAs. Despite the genetic and phenotypic differences that distinguish the members of the family *Picornaviridae*, all members utilize similar strategies of genome replication and production of infectious virus particles in their respective host-cell environments. Due to a positive-sense RNA genome, picornavirus RNA can immediately be translated into polyprotein upon entry into the cell cytoplasm, and does not require the inclusion of viral proteins such as polymerase in the capsid (as in the case of negative-strand RNA viruses). Moreover, nearly all picornaviruses inhibit cap-dependent translation of host cell mRNAs while utilizing a cap-independent mechanism to translate their own genome, reducing the competition

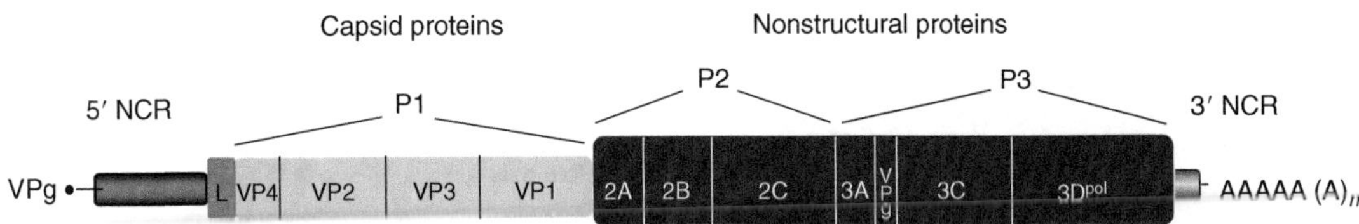

VPg:
- Protein primer for RNA synthesis

5′ NCR:
- Translation from IRES element
- Binding site for host/viral factors for RNA replication

Leader protein (L)

Nonstructural proteins:
- Polyprotein processing
- Host translation shutoff
- RNA replication
- Cellular membrane rearrangement

Poly(A) tract:
- Maintenance of RNA stability
- PABP binding site
- Uridylylation of VPg, and initiation of RNA replication

3′ NCR:
- Replication complex binding site(?)
- RNA stability(?)

Figure 1 Organization of a typical picornavirus RNA genome. VPg is covalently attached to the 5′ terminus of genomic RNA. The 5′ NCR contains RNA sequences forming RNA structural elements used in virus RNA translation and replication. The coding region is divided into three sections consisting of the P1, P2, and P3 regions. The RNA region of P1 is translated and processed to virion capsid proteins. Some of the functions of the nonstructural proteins contained in the P2 and P3 regions are listed above. Following the P3 region, the 3′ NCR and poly(A) tract function in replication and RNA stability.

for translation factors between the virus and host cell mRNAs in favor of the virus.

Picornavirus Entry and Gene Expression

Capsid

Picornavirus genomes are packaged into icosahedral virion structures consisting of 60 individual protomers formed by the structural proteins in the P1 coding region: VP1, VP2, VP3, and VP4. A single copy of VP1, VP2, and VP3 is organized into a triangular subunit on the surface of the capsid. Five of these subunits are organized around a fivefold symmetrical axis to form a pentameric protomer (**Figure 2**). Twelve of these pentameric subunits (a total of 60 triangular protomers) form the icosahedron, with twofold and threefold axes of symmetry located at the connection points. The small structural protein VP4 is not present on the outside of the mature virion; it is located in the interior of the capsid structure and interacts with the encapsidated RNA genome. During assembly, inclusion of the RNA genome coincides with the processing of a precursor polypeptide (VP0) into VP4 and VP2. The mechanism for the maturation of VP0 to VP4 and VP2 is currently unknown. What is known is that the viral proteinases 2A, L, and 3C/3CD do not appear to be involved, suggesting a cellular factor may be required for virion maturation or that viral RNA/capsid proteins effect this cleavage.

The virion capsid proteins VP1, VP2, and VP3 of enteroviruses and rhinoviruses protrude from the surface of the capsid structure at the axes of symmetry, forming the 'mesa' at the fivefold axes and the 'propeller' at the threefold axes (**Figure 2**). Observations of these extrusions using X-ray crystallography give the impression of clefts between the axes of symmetry, termed 'canyons'. The canyon region of *Picornaviridae* was initially proposed to be the site of virion binding to cell receptors prior to entry into the cytoplasm, based upon its predicted inaccessibility to neutralizing antibody. This hypothesis has been supported by cryoelectron micrograph studies of cell receptor bound to the canyon region of rhinovirus and coxsackievirus virion particles. While the canyon region of cardioviruses and aphthoviruses is not nearly as pronounced as in enteroviruses and rhinoviruses, neutralization studies investigating the virion capsid of foot-and-mouth disease virus (FMDV) indicate that the G–H loop of the VP1 capsid protein is important for cell receptor recognition.

A summary of some of the cellular molecules identified as cellular receptors for picornaviruses is listed in **Table 2**. Some of the receptors, such as the rhinovirus major group receptor ICAM-1, have known cellular functions, while the normal cellular function of other virus receptors, such as the HAV receptor HAVcr-1, have not been determined. For poliovirus, entry of the virion RNA into the cell is not fully understood but may occur via a variation of receptor-mediated endocytosis; however, acidification of the endosome may not be necessary as release of the viral RNA is not pH dependent. Whether the mature virion is internalized wholly in a membrane-bound vesicle or whether the RNA exits at the plasma membrane surface is also unknown. Once bound to CD155, the virion capsid induces a conformational change and the internal VP4 capsid protein is extruded to the external surface to contact the plasma membrane along

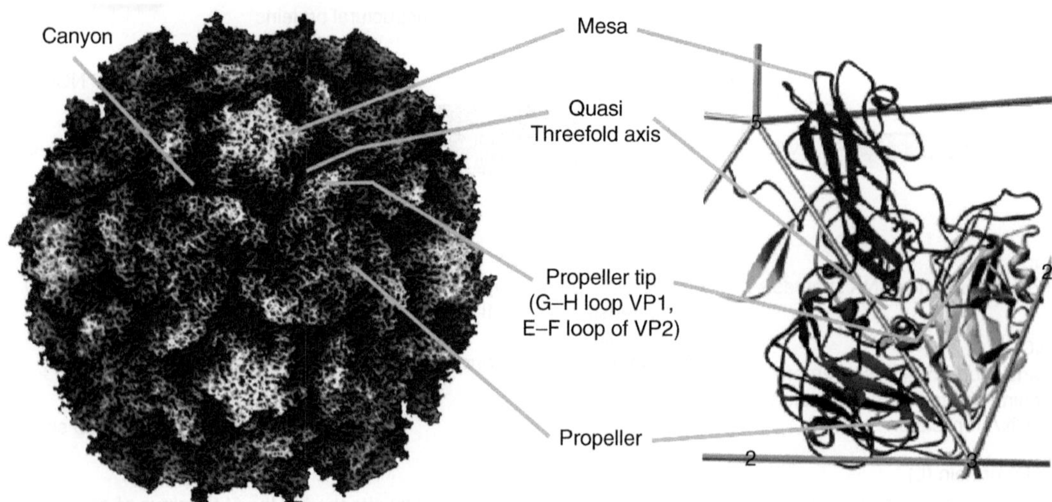

Figure 2 Atomic structure of poliovirus capsid. Left panel: A radially depth-cued rendering of the atomic model of the structure of the 160S poliovirus virion particle. Right panel: An expanded representation of a single protomer showing ribbon diagrams of VP1 (blue), VP2 (yellow), VP3 (red), and VP4 (green) overlaid on an icosahedral framework. The fivefold, threefold, and twofold axes are indicated by numbers. Cyan lines point to prominent surface features in the two panels. Reproduced from Bubeck D, Filman DJ, Cheng N, Steven AC, Hogle JM, and Belnap DM (2005) The structure of the poliovirus 135S cell entry intermediate at 10-angstrom resolution reveals the location of an externalized polypeptide that binds to membranes. *Journal of Virology* 79: 7745–7755, with permission from the American Society for Microbiology.

Table 2 Examples of cell receptors for picornavirus entry

Genus	*Species*	*Cellular receptor*
Aphthovirus	*Foot-and-mouth disease virus*	Integrin (strains A12), heparan sulfate (strain O1)
Cardiovirus	*Encephalomyocarditis virus*	VCAM-1
Enterovirus	*Coxsackievirus B1-6*	CAR, DAF (CD55)
	Poliovirus 1–3	CD155 (Pvr)
Hepatovirus	*Hepatitis A virus*	HAVcr-1
Rhinovirus	*Major rhinovirus group*	ICAM-1
	Minor rhinovirus group	LDL

VCAM, Vascular cell adhesion molecule; CAR, coxsackie-adenovirus receptor; DAF, decay-accelerating factor (CD55); Pvr, poliovirus receptor (CD155); HAVcr-1, hepatitis A virus cell receptor; ICAM-1, intercellular adhesion molecule; LDL, low-density lipoprotein.

with the amino terminus of VP1 around the fivefold axis of symmetry, forming a channel by which the RNA has been proposed to enter the cytoplasm and begin the replicative cycle through translation of the RNA genome.

Translation of Genomic RNA

Once the virus has been uncoated and its RNA released into the host cell cytoplasm, the positive-strand RNA genome is translated. This viral mRNA contains *cis*-acting RNA sequence and structural elements distinct from those of host cell mRNAs. Eukaryotic mRNAs contain a 7-methyl guanosine cap that is attached to the nascent transcript in the nucleus during RNA processing. In addition to protecting the transcript from degradation, recognition of the cap structure by the eIF-4F complex serves as a scaffold for the assembly of other initiation factors and the 40S ribosomal subunit. Picornavirus genomes lack anything resembling a cap structure. Instead, the 5′ end of genomic RNA (as well as negative-strand RNA intermediates) is covalently bound to the viral protein VPg. This protein does not appear to have a role in translation of the polyprotein. Studies carried out with picornavirus mRNAs associated with translating polysomes have not detected VPg bound to viral RNAs, and further evidence indicates an unknown cellular factor as capable of cleaving the RNA–protein bond.

Translation of picornavirus positive-strand RNAs does not follow the conventional scanning model of most cellular mRNAs. Analysis of the first completely sequenced picornavirus, poliovirus, indicated that the 5′ NCR contains numerous AUG start codons in contexts favorable for initiation of translation that were not utilized by the virus. Further, biochemical analysis of the 5′ NCR

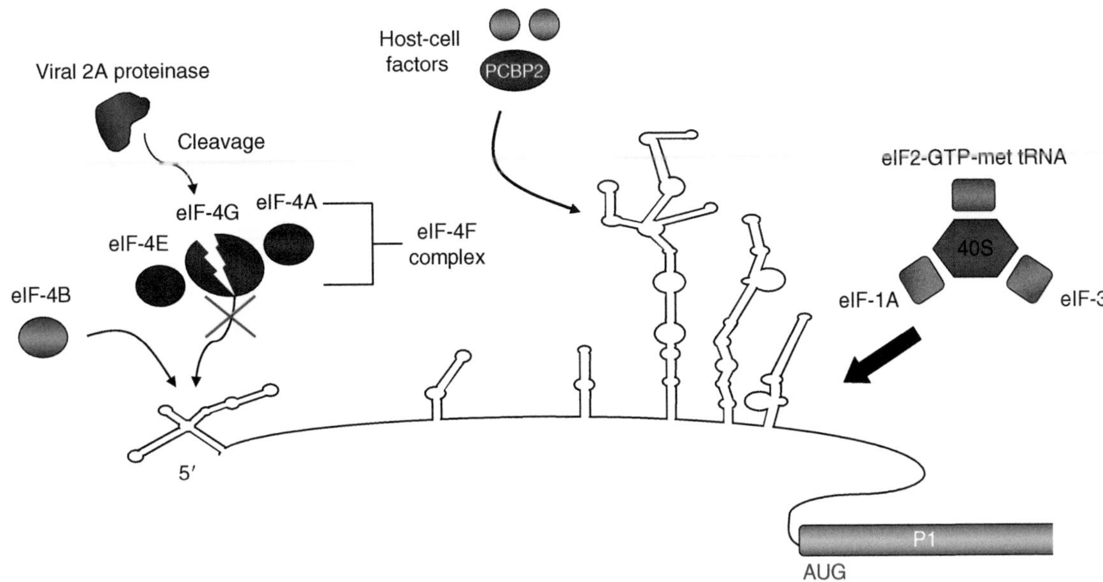

Figure 3 Cap-independent RNA translation of poliovirus RNA. Binding of cellular protein PCBP2 and other host cell factors to RNA secondary structures in the IRES element stabilizes the RNA, allowing association of the 40S ribosomal subunit with eIF-1A, eIF2-GTP-met, and eIF-3 bound. Following binding of the 60S subunit to form the complete 80S ribosome, initiation of RNA translation begins at the AUG start codon preceding the P1 coding region. Translation of the genomic RNA leads to the generation of the 2A proteinase, which will shut off host cell cap-dependent RNA translation through inactivation of the eIF-4F complex via cleavage of eIF-4G. Also note the absence of a cap structure at the 5′ end of the RNA.

determined that this region of viral RNA contains several RNA secondary structures with sufficient thermodynamic stability to most likely inhibit the 40S ribosomal subunit from scanning toward a start codon. Subsequent studies demonstrated that an IRES is utilized to translate the poliovirus RNA genome. All picornaviruses contain an IRES in their 5′ NCRs. IRES elements have also been identified in the genomes of other RNA viruses as well as in cellular mRNAs. An example of the mechanism of picornavirus RNA translation is shown in **Figure 3** for the poliovirus IRES.

In contrast to host cell cap-dependent translation, cap-independent translation initiation from an IRES does not use a cap structure at the 5′ end of the RNA to interact with the eIF-4F complex. In fact, many picornaviruses will actually proteolytically cleave or otherwise sequester eukaryotic initiation factors with high efficiency. The notable exception is HAV; cleavage of the eIF-4G component of the eIF-4F complex inhibits IRES activity, suggesting that an intact eIF-4F complex may interact with a site within the IRES to initiate HAV translation. While cleaved eIF-4G is rendered nonfunctional for cellular mRNA translation, eIF-4A and eIF-4B may participate with the cleaved portion of eIF-4G to bring the eIF-1A/eIF-2-met tRNA/eIF-3/40S ribosome complex to the viral IRES for enteroviruses, rhinoviruses, and aphthoviruses.

By convention, the RNA structure elements of the picornavirus type I IRES (enteroviruses and rhinoviruses) and HAV (denoted as having a type III IRES) are described by roman numerals, while the distinct 5′ NCR secondary structure elements of the type II IRES (including those of cardioviruses and aphthoviruses) are denoted by letters.

Chemical and enzymatic RNA structure probing aided by computer modeling studies have determined that the enterovirus and rhinovirus 5′ NCRs (approximately 700–800 nt in length) contain six stem–loop structures, with the region containing the IRES encompassing stem loops II through VI (**Figure 4**). Extensive genetic analysis of type I stem–loop structures involving point mutations, substitutions, or complete deletions of specific stem–loop structures indicates that specific RNA sequence elements are also important for IRES function. In particular, the poliovirus stem–loop IV contains a short region of single-stranded cytidines and a GNRA tetra-loop sequence important for cellular protein binding and efficient viral genome translation. Similar sequence and structural elements have also been predicted for the IRES of the closely related enterovirus coxsackievirus B.

The type II IRES elements of FMDV, EMCV, and the recently classified IRES of parechoviruses are bounded between stem-loops D through L (**Figure 4**) and are contained within a 5′ NCR of greater length than their enterovirus counterparts, from over 800 nt for EMCV to 1200 nt for FMDV. In addition to a greater number of RNA stem–loop structures within the 5′ NCR, some type II IRES elements also include a poly(C) region preceding stem–loop D that is missing in picornaviruses with a type I IRES (Theiler's murine encephalomyocarditis virus, a

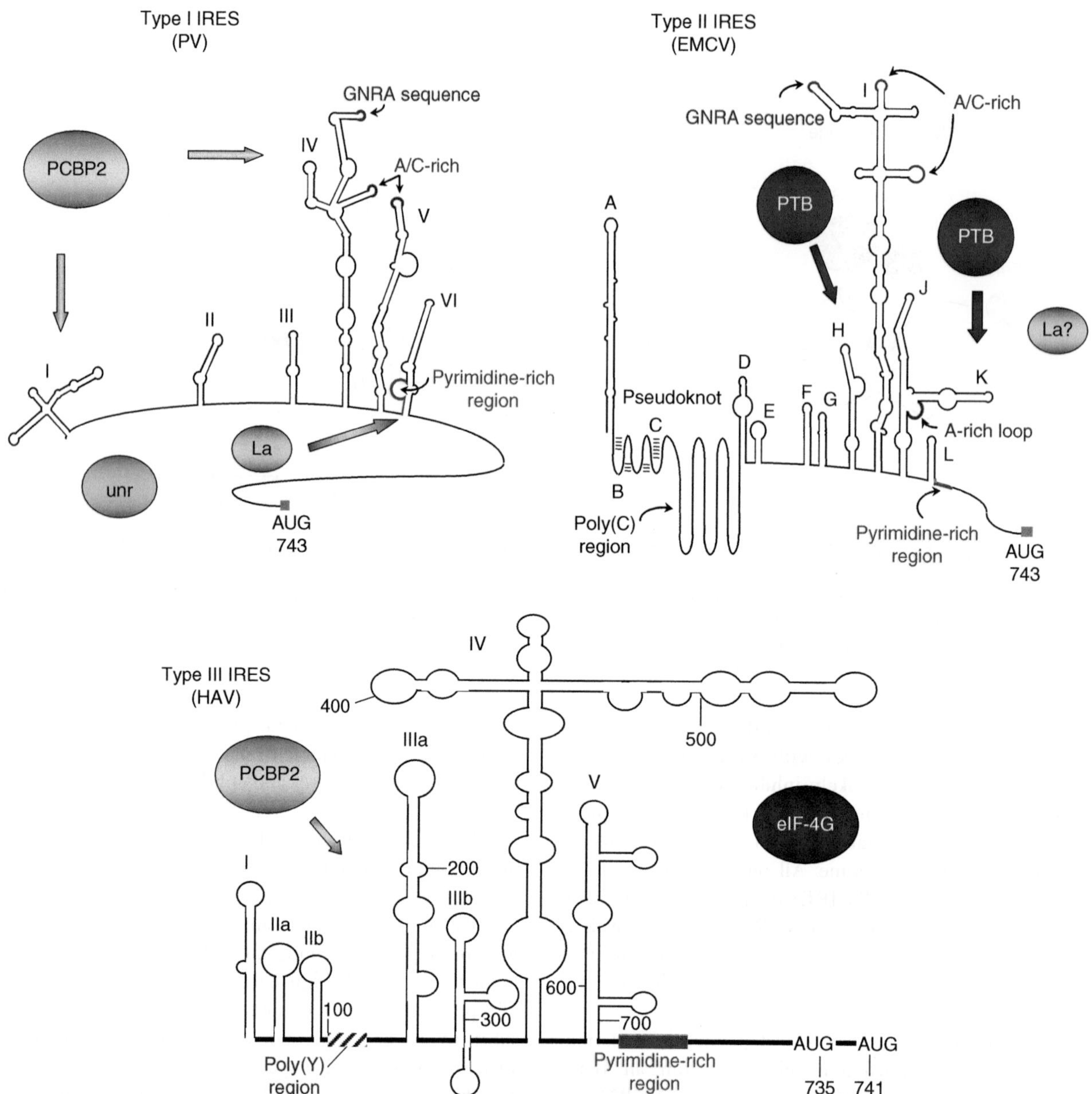

Figure 4 Comparison of IRES elements in the 5′ NCR of picornavirus RNA genomes. The predicted RNA secondary structure motifs associated with type I, II, and III IRES elements are represented by poliovirus, encephalomyocarditis virus, and hepatitis A virus, respectively. The secondary structure of the RNA sequence for each of the 5′ NCRs has been predicted by computer modeling and confirmed by RNA structure probing. Also depicted are some of the cellular factors known to bind to each particular IRES type, and their putative binding sites. Reproduced from Stewart SR and Semler BL (1997) RNA determinants of picornavirus cap-independent translation initiation. *Seminars in Virology* 8: 242–255, with permission from Elsevier.

cardiovirus related to EMCV, does not contain a poly(C) tract). The function of this homopolymeric stretch of cytosine residues (ranging from 60 to over 400 nt) is unknown. Despite the structural differences compared to type I IRES elements, the type II IRES also contains a GNRA tetra-loop sequence in the stem–loop I, as well as A/C-rich regions that are found in type I IRES elements. The IRES of HAV is contained within stem loops IIIa to V, and although its stem–loop structures more closely resemble those of the type II IRES, the requirements for efficient IRES function are different than the type I or type II class of elements.

Picornavirus IRES elements operate to direct the assembly of initiation complexes using the host cell protein synthesis machinery. This assembly is mediated by the interactions between IRES RNA elements and cellular factors, including cellular proteins that do not appear to have a role in cap-dependent cellular translation.

These RNA–protein interactions between viral RNA and host cell proteins have been identified as either being crucial for optimal translation efficiency specifically for one IRES type, or important for translation of more than one IRES type, perhaps acting to stabilize the RNA secondary structure that allows internal ribosome entry. For example, rabbit reticulocyte lysate (RRL) is a commonly used and efficient *in vitro* system for the translation of mRNA. However enterovirus, rhinovirus, and HAV IRES elements initiate translation aberrantly in RRL. In particular, the efficiency and fidelity of poliovirus translation in RRL can be increased significantly by the addition of crude extracts from HeLa cells or by high concentrations of the human La protein. In contrast, the type II IRES elements of EMCV, FMDV, and Theiler's virus RNA direct translation efficiently in RRL without the addition of La or other factors. Another cellular protein, polypyrimidine tract binding protein (PTB), interacts with poliovirus, human rhinovirus, EMCV, and FMDV IRES elements (**Figure 4**). The high-affinity RNA binding of PTB to the type II IRES elements of EMCV and FMDV suggests a role for the protein in directing viral RNA translation. Mutations introduced into the EMCV IRES designed to abrogate binding of PTB to the IRES result in a loss of IRES activity. **Table 3** lists some of the known interactions of cellular proteins with picornavirus IRES elements.

The cellular protein poly(rC) binding protein (PCBP) binds to the 5′ NCR (specifically stem–loop I and IV) of enterovirus and rhinovirus genomic RNAs and is necessary for the activity of the IRES element. Depletion of the PCBP isoform PCBP2 from HeLa cytoplasmic extracts drastically reduces the translation efficiency of poliovirus RNA, which can be restored with the addition of recombinant PCBP2. Intriguingly, while FMDV and EMCV IRES elements bind PCBP2, depletion of PCBP2 does not affect the ability of these IRESs to direct translation *in vitro*. A substantial amount of research has focused on the role of PCBP in poliovirus infection, as binding of PCBP (either as PCBP1 or PCBP2) to stem-loop I of poliovirus is necessary for RNA replication (described below).

Table 3 Examples of host proteins interacting with picornavirus 5′ NCRs

Cell proteins	*Virus RNA binding*
La autoantigen	PV, HRV, CVB, EMCV, FMDV
PCBP2	PV, HRV, CVB, FMDV, HAV
unr	PV, HRV
PTB	PV, HRV, EMCV, FMDV
ITAF-45	FMDV
eIF-4A	EMCV
eIF-4B	EMCV, PV, FMDV
eIF-4E	EMCV
eIF-4G	EMCV, FMDV
eIF-2	EMCV, PV

Polyprotein Processing

Ribosome recognition and initiation within the picornavirus IRES allows the translation of the long polyprotein (approximately 250 kDa) encoded by the single open reading frame (ORF) encoded by picornavirus genomes. The viral polyprotein is proteolytically cleaved by the viral proteinases L (encoded by aphthoviruses), or 2A (encoded by enteroviruses and rhinoviruses), and 3C and 3CD (encoded by all picornaviruses) (**Figure 5**). Viral protein processing yields the precursor molecules P1, P2, and P3. The P1 region of the polyprotein is further processed to three or four distinct proteins, which are the structural components of the viral capsid, and in the case of cardioviruses, aphthoviruses, erboviruses, teschoviruses, and kobuviruses, the nonstructural protein L. The proteolytically processed P2 precursor contains proteins necessary for restructuring of cytoplasmic membranes and vesicles utilized in virus replication, for the shutoff of host cell translation, and for viral RNA replication. The P3 precursor is processed to produce major proteins necessary for RNA replication, including the RNA-dependent RNA polymerase $3D^{pol}$, and the viral proteinases 3C and 3CD, both of which are responsible for the majority of viral polyprotein cleavage events. The translation stop codon for picornavirus genomes terminates synthesis of the polyprotein immediately after synthesis of the P3 polyprotein, at the carboxy terminus of the $3D^{pol}$ sequence.

Substrates of Viral Proteinases

Picornavirus Capsid Proteins

The P1 sequence encodes the capsid proteins VP4, VP2, VP3, and VP1. Initial processing of the viral polyprotein generates VP0 (a precursor to VP4 and VP2), VP3, and VP1. Parechoviruses and kobuviruses do not process the P1 capsid precursor in a manner analogous to how poliovirus processes VP0; therefore, complete processing of the capsid proteins results in only three proteins used in the formation of the virion capsid. For poliovirus, cleavage of the P1 region into the mature capsid components requires the proteolytic activity of 3CD. For all picornaviruses that process VP0 into VP4 and VP2, an unknown host or viral factor may effect this novel cleavage event.

Viral Proteins Involved in RNA Replication

Viral protein 2B is known to alter and increase the permeability of intracellular membranes including the endoplasmic reticulum. The 2B protein of coxsackievirus B3 (CVB3) contains a region of amino acids that resemble a

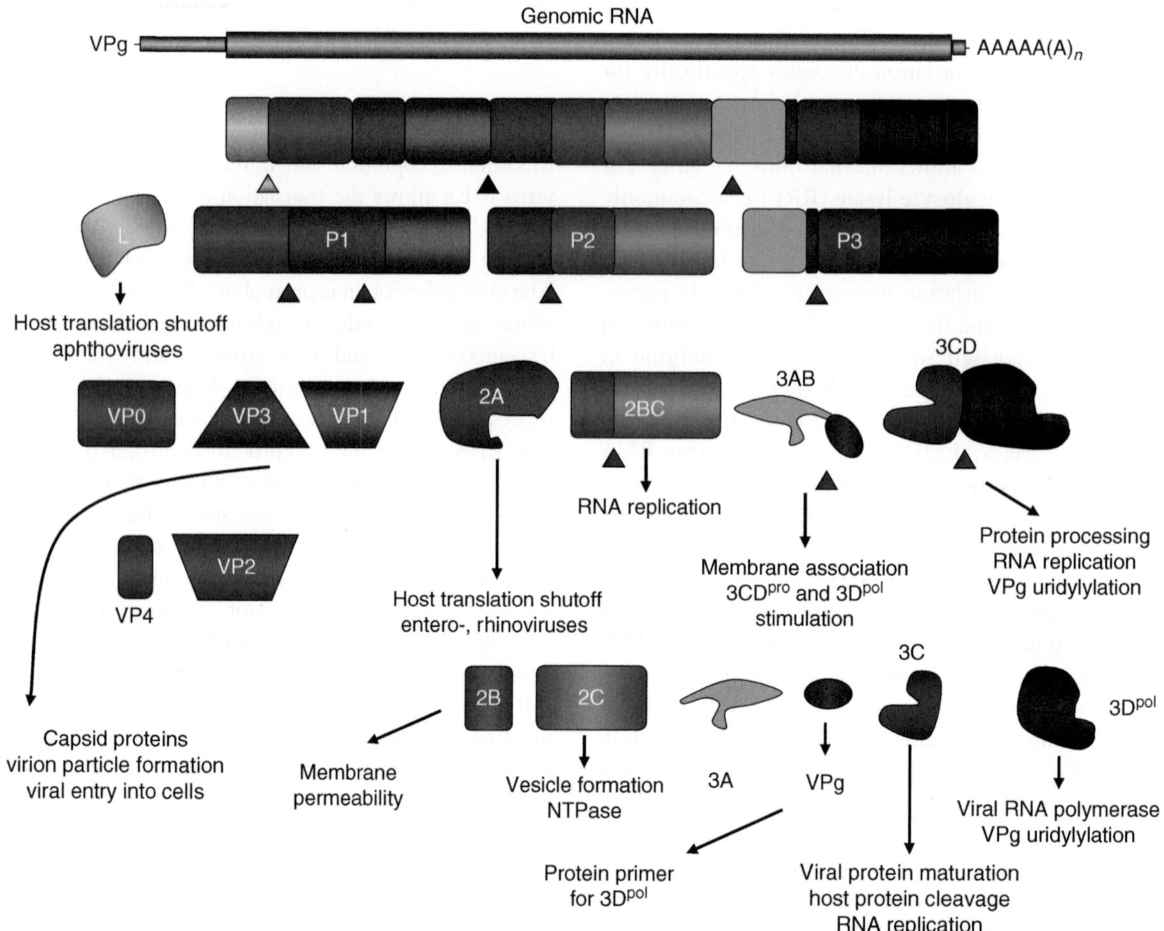

Figure 5 Picornavirus polyprotein processing. As the polyprotein is translated, autocatalysis by viral proteinases 2A/L or 3C/3CD results in the P1, P2, and P3 precursors. (The L protein of cardioviruses is not a proteinase and thus is not released by autocatalysis, and is instead released by 3C/3CD mediated cleavage from the P1 precursor.) These precursors are further processed into the virus structural proteins (P1) or nonstructural proteins (P2 and P3). The functions of the precursor and mature virus proteins are described above. The color-coded triangles correspond to specific cleavage events by the viral proteinases (green, L; black, 2A; red, 3C/3CD).

cationic α-helix. Infection of tissue culture cells with a CVB3-containing mutations in the 2B α-helix predicted to disrupt the cationic character of the domain resulted in a decrease in virion progeny being released into the extracellular space, suggesting that 2B, possibly in conjunction with the viral proteins 2C/2BC, may facilitate virion release during later stages of picornavirus infection. While the precise function of picornavirus 2B is unknown, insertion mutations into the hydrophobic region of the poliovirus 2B produced a virus with defects in viral RNA synthesis that was incapable of rescue by complementation with wild-type virus, suggesting a *cis*-acting role for the 2B protein in picornavirus RNA replication.

During poliovirus infection, 2C and its precursor molecule 2BC are the viral factors responsible for induction of membranous vesicles that are rearranged by the 2B protein (see above) creating a vesicle scaffold for the viral replication complex to begin RNA synthesis. Both proteins are known to interact with the vesicles associated with the virus RNA replication complexes, and 2BC and 2C (but not 2B) induce the formation of these membranous vesicles when expressed alone in tissue culture cells.

The 2C coding region contains several putative structural domains that point to its role as a multifunctional protein in the picornavirus replication pathway. Poliovirus 2C protein is predicted to contain two amphipathic α-helices at its amino- and carboxy-termini and a cysteine-rich motif resembling a DNA zinc-binding finger domain. The middle of the coding region contains a domain with sequence elements found in RNA helicases; while helicase activity is yet to be reported, the 2C protein does have NTP-binding and hydrolysis functions. Although its precise function in RNA synthesis is currently unknown, the 2C protein of picornaviruses clearly has a role in viral RNA replication based upon studies using guanidine hydrochloride (Gu–HCI). In the presence of low concentrations of Gu–HCI (2 mM), picornavirus negative-strand RNA replication is specifically inhibited. Sequence analysis of poliovirus mutants resistant to the addition of Gu–HCI revealed lesions in the 2C coding region, indicating that poliovirus sensitivity to Gu–HCI mapped to the 2C protein. Poliovirus 2C and 2BC viral proteins also bind to the RNA stem–loop structure at the 3′ end of negative-strand RNA, suggesting that these proteins may also have a role in positive-strand RNA synthesis. Recent evidence has also suggested that poliovirus 2C may regulate the catalytic activity of the 3C proteinase to process viral precursor molecules.

The carboxy-terminal hydrophobic domain of the viral 3A protein anchors this small (approximately 130 amino acids) polypeptide and its precursor 3AB into cellular membranes. Mutations within the hydrophobic region predictably disrupt the interaction of the poliovirus 3AB precursor with cellular membranes, but more importantly they also disrupt efficient cleavage of viral P2-P3 or P2-3AB protein precursor molecules by the 3C. Thus, 3AB may provide a means of sequestering viral factors required for RNA synthesis and protein processing in the same cytoplasmic location as the vesicles formed from intracellular membranes, providing at the same time the VPg protein primer for RNA replication.

The viral proteinases 3C and 3CD cleave the polyprotein encoded by the genomic RNA to produce mature viral proteins. These proteinases also cleave cellular proteins, perhaps acting to disrupt host cell translation machinery or an antiviral response mounted by the cell. In addition, 3CD also participates in processes critical to viral RNA replication. Interaction of poliovirus 3CD, the precursor of 3C and the polymerase $3D^{pol}$, is required to form an RNP complex with PCBP2 and the 5′ end of genomic RNA. This complex has been shown to be required for poliovirus RNA replication, possibly by facilitating negative-strand RNA synthesis via communication of the 5′ and 3′ termini of the genomic RNA. In addition, poliovirus 3CD enhances the rate of uridylylation of VPg from the *cre* element *in vitro*, suggesting that 3CD may assist the $3D^{pol}$ through either direct interaction or binding the RNA sequence of the *cre* to alter its conformation in a manner suitable to serve as template.

The $3D^{pol}$ of picornaviruses is the RNA-dependent RNA polymerase responsible for the synthesis of negative- and positive-strand RNAs. *In vitro* studies examining the requirements for poliovirus $3D^{pol}$ RNA synthesis determined that although the viral polymerase cannot initiate RNA synthesis without a primer sequence, it can elongate a nascent RNA strand following initiation without need of additional viral or cellular proteins. Interestingly, it has been proposed that 3AB may have a stimulatory role in the elongation step.

Host Cell Protein Cleavage and Translation Shutoff

In addition to their role in the maturation of viral proteins, the 2A (enteroviruses, rhinoviruses) and L (aphthoviruses) proteinases also have a role in influencing the translation activity of the host cell by cleaving factors involved in cap-dependent translation initiation. Both proteinases shut down host cell protein synthesis by cleaving the eIF-4G component of the eIF-4F initiation factor complex, resulting in an inability of the amino-terminal domain of eIF-4G to recognize and bind eIF-4E that is associated with the 7-methyl cap structure on the 5′ end of cellular mRNAs. Enterovirus and rhinovirus protein synthesis is upregulated as a consequence of this increase of free ribosomes that are available for internal ribosome entry. EMCV uses an alternate method to inhibit host cell translation by activating a cellular translational repressor, 4E-BP1, which binds cellular eIF-4E and therefore inhibits host cap-dependent translation.

Viral RNA Replication

Genomic RNA Elements Involved in Picornavirus RNA Replication

Like all RNA viruses, picornavirus replication uses a viral-encoded RNA-dependent RNA polymerase to specifically synthesize RNA from a viral RNA template. Picornavirus genomic RNA is released from the capsid into the cytoplasm without replication proteins (save for the viral VPg linked to the 5′ end); therefore, replication of the initial viral genome must follow translation of its coding region to generate the $3D^{pol}$ and other viral replication proteins. The polymerase uses the protein VPg as a primer to initiate RNA synthesis by catalyzing the addition of two uridylate residues to generate VPg-pU-pU. The initiation site of negative-strand RNA synthesis is at the 3′ poly(A) tract of the genomic positive-sense RNA. Viral negative-strand RNA exists in a duplex with its positive-sense RNA template. This double-stranded RNA intermediary is termed the replicative form, or RF. Negative-strand RNA in turn acts as template for the synthesis of positive-strand RNAs. The ratio of positive- to negative-strand RNA in infected cells has been observed to be approximately 50:1, suggesting that a single negative-strand RNA intermediate acts as a template for the production of multiple positive-strand RNAs. Newly synthesized viral positive-sense RNAs will be translated to produce additional viral proteins, used as additional template RNAs for negative-strand RNA synthesis, or packaged into virions for infection of other host cells (**Figure 6**). This section discusses the RNA elements and viral proteins utilized by picornaviruses and their functions in the replicative cycle, the contributions of

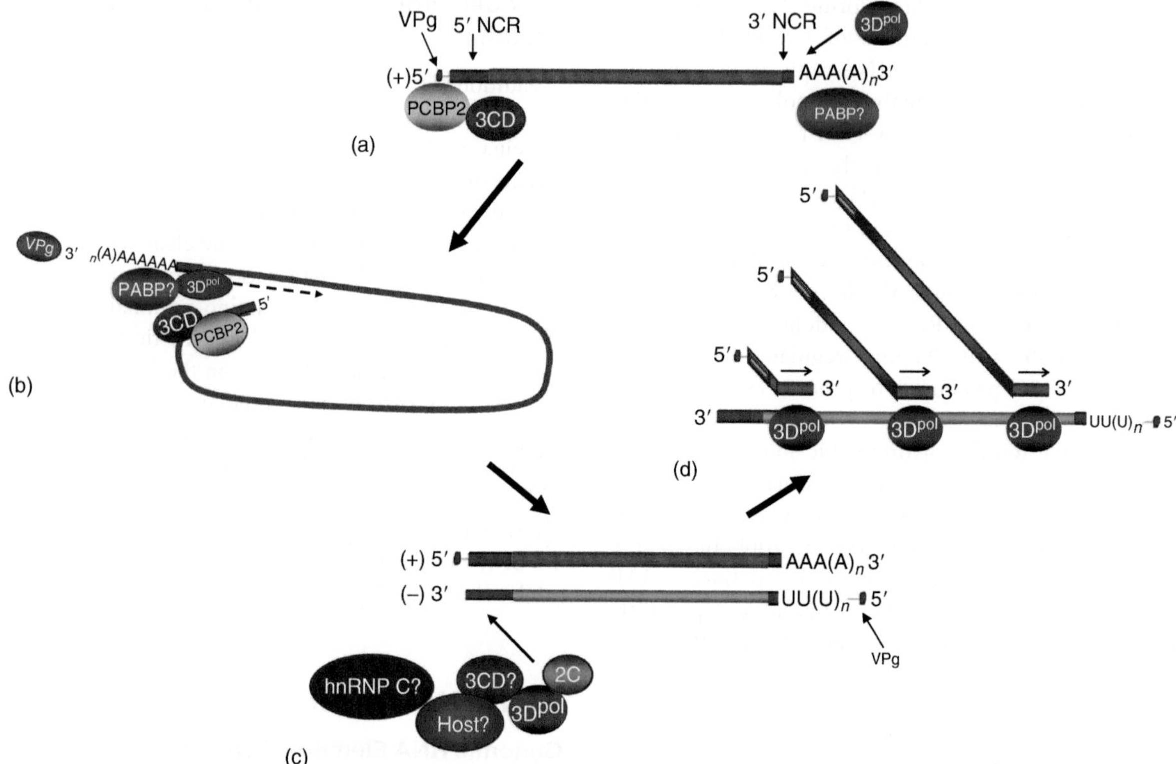

Figure 6 Picornavirus RNA replication. Shown here is a model of the general mechanism of RNA replication of picornaviruses, with the cellular and viral proteins depicted being specific for what is known for poliovirus. (a) The input genomic RNA (positive-sense) associates with viral and cellular factors (including PABP and PCBP2). Viral 3CD and cellular PCBP2 is proposed to interact with the RNA structure at the terminus of the 5′ NCR and contact PABP bound to the poliovirus poly(A) tract. The RNA-dependent RNA polymerase $3D^{pol}$ begins synthesis of a complementary negative-strand RNA. (b) The resulting double-stranded RNA intermediate (termed the replicative form or RF) forms part of the positive-strand RNA replication complex. (c) RNA replication complexes hypothesized to include host proteins (e.g., hnRNP C), viral proteins 3CD, 2C, and $3D^{pol}$. Multiple initiations of $3D^{pol}$ RNA synthesis from the negative-strand RNA template result in the formation of the partially double stranded replicative intermediate, RI. (d) Positive-strand progeny RNAs will then re-enter the replication cycle for translation of additional viral proteins, or to serve as template for the replication of negative-strand RNA, or to be packaged into progeny virion particles.

cellular proteins and structures, and the unanswered questions regarding picornavirus RNA replication.

In addition to the IRES, picornavirus 5′ NCRs contain additional stable RNA secondary structural elements that are critical for replication of viral positive- and negative-strand RNAs. For entero- and rhinovirus genomic RNAs, this structure is termed stem–loop I or cloverleaf, so named for the cruciform secondary structure predicted to form within the first ~100 nt of the RNA. The genomes of other picornaviruses including aphthoviruses and cardioviruses also contain an RNA structure at their 5′ termini, termed stem–loop A, which is analogous in function but not in structural conformation to the cloverleaf. For simplicity, the discussion below will primarily refer to the poliovirus stem–loop I RNA structure at the 5′ terminus. An insertion mutation in stem–loop I resulted in a deficiency in viral RNA synthesis. More importantly, a revertant virus containing a mutation in the genomic region encoding the viral proteinase 3C was isolated and found to have rescued the RNA synthesis defect. Binding of precursor 3CD to stem–loop I was confirmed genetically via mutations of stem–loop I nucleotides designed to destabilize the cruciform. Using electrophoretic mobility shift analysis, the RNA affinity of 3CD for stem–loop I was found to be increased when another protein was bound to the RNA cruciform. This protein was later identified as the cellular protein PCBP2, and it associated with stem–loop I and 3CD to produce a ribonucleoprotein (RNP) structure called the ternary complex. Formation of this RNA–protein complex has been confirmed *in vitro* and in poliovirus-infected cells. Depletion of PCBP2 from HeLa cell S10 extracts reduces levels of poliovirus RNA synthesis, supporting the hypothesis that PCBP2 is required for RNA replication. However, the question remained as to what the 3CD/PCBP2/stem–loop I ternary complex contributed to picornavirus RNA replication.

An intriguing possible answer to the role of the 3CD/PCBP2/stem–loop I complex involves another cellular protein, poly(A) binding protein (PABP). In the cell, PABP stabilizes and protects cellular mRNA transcripts by binding the poly(A) tract at the 3′ end of the mRNA. It also participates in translation through interactions with initiation factors at the 5′ end of the mRNA. *In vitro* binding studies using purified PABP and an RNA probe corresponding to the 3′ NCR and poly(A) tract of poliovirus correlated an ability of PABP to interact with this region with a minimal poly(A) tract. Furthermore, purified PABP could interact with the 3CD/PCBP2/stem–loop I ternary complex *in vitro*. PABP and poliovirus 3CD could be co-immunoprecipitated from extracts of poliovirus-infected HeLa cells, suggesting that PABP and the viral ternary complex may also interact during a poliovirus infection. These interactions would suggest that the 5′ and 3′ ends of positive-strand RNA are in close proximity through a 'protein–protein' bridge formed by viral and cellular factors. End-to-end communication of the positive-strand RNA could facilitate initiation of negative-strand RNA synthesis.

An internal RNA sequence within the picornavirus genome is also important in the picornavirus replicative cycle. This sequence has been termed the *cis*-acting replication element, or *cre*. First discovered in human rhinovirus 14 genomic RNA and later identified for other picornaviruses, the *cre* forms a short (50–100 nt) RNA hairpin structure in the positive-strand RNA of picornaviruses, with a conserved AAACAC sequence in the loop portion of the hairpin. Nucleotide mutations that disrupt the conserved sequence inhibit viral RNA synthesis and are therefore lethal to the virus. The position of the *cre* varies widely among picornaviruses; in enteroviruses, it is located within the 2C coding region, while in cardioviruses the *cre* maps to the VP2 coding region. The *cre* of aphthoviruses is unique among picornaviruses in that it is located outside the coding region in the 5′ NCR. The position of the *cre* varies even among serotypes of the same genus. For example, the human rhinovirus 2 *cre* is found in the 2A proteinase coding region, while human rhinovirus 14 *cre* is found in the VP1 RNA sequence. Studies of lethal mutations introduced in the poliovirus *cre* demonstrated that viral RNA replication could be rescued when a second, wild-type sequence was introduced elsewhere in the genome. The *cre* is considered to be a site at which viral replication proteins bind and uridylylate the VPg protein primer. The poly(A) tract at the end of the 3′ NCR was originally thought to be the template RNA used in VPg uridylylation; however, *in vitro* studies using synthetic VPg indicated that the poliovirus *cre* hairpin as a source of RNA template was much more efficient in stimulating the uridylylation of VPg than was the viral RNA poly(A) tract. Mutational analysis of poliovirus RNA has indicated that the *cre* may only be used in positive-strand RNA synthesis, and that the poly(A) tract is the primary source of template RNA for VPg uridylylation during negative-strand RNA synthesis.

Following the $3D^{pol}$ coding region, the 3′ NCR and the 3′ poly(A) tract are thought to be involved in RNA synthesis by acting as a *cis*-acting signal for negative-strand RNA replication. The poly(A) tract encoded by the viral genome may contribute to stable RNA secondary structures consisting of one to three stem–loop structures in the heteropolymeric 3′ NCR. The poly(A) tract likely stabilizes and protects the viral RNA genome from intracellular degradation enzymes, much as the poly(A) tracts of cellular mRNAs do. The poly(A) tract, in conjunction with bound cellular PABP, $3D^{pol}$, and other viral/cellular proteins may also function in negative-strand RNA synthesis. Human rhinovirus and poliovirus RNA genomes lacking the 3′ NCR (but leaving the genome encoded poly(A) tract intact) are still infectious in tissue culture, indicating the 3′ NCR is dispensable for viral replication. Mutant poliovirus lacking a 3′ NCR displays an RNA replication defect following infection of HeLa cells. This defect was shown to occur at the level of positive-strand RNA synthesis. Interestingly, the RNA replication defect exhibited by the poliovirus 3′ NCR deletion mutant was exacerbated in neuronal cells. Such a cell-specific defect suggested that the presence (or absence) of a cellular factor was important for proper poliovirus 3′ NCR function. The cellular protein nucleolin has been reported as being capable of interaction with the poliovirus 3′ NCR; however, its function in viral RNA replication is unknown. Viral proteins including the 3AB and 3CD protein have been reported to bind to the poliovirus 3′ NCR, and the $3D^{pol}$ of EMCV has been shown to bind to 3′ NCR sequences of the EMCV genome.

Synthesis of Positive- and Negative-Strand Viral RNA

Initiation of viral RNA synthesis requires the uridylylation of VPg by the viral polymerase $3D^{pol}$ and may also involve other viral proteins, including 3CD and the VPg precursor 3AB. Utilizing the 3′ poly(A) tract as a source of template, VPg is uridylylated to form VPg-pU-pU, and the $3D^{pol}$ synthesizes negative-strand RNA complementary to the genomic positive-sense RNA. The double-stranded RNA intermediate (RF) then serves as template for multiple initiation events by the RNA polymerase to synthesize positive-strand RNAs. In the infected cell, negative-strand RNA has only been observed either in a duplex with the positive-strand RNA or with several elongating positive-strand RNAs existing in a partially double-stranded RNA complex termed the replicative intermediate (RI). Similar to negative-strand RNA synthesis, positive-strand RNA synthesis requires the uridylylation of VPg and as noted earlier, *in vitro* studies have indicated that during synthesis of positive-strand RNAs uridylylation of VPg was more effective using a *cre* RNA hairpin as the source of template rather than the poly(A) tract. During positive-strand RNA synthesis, multiple positive-strand RNAs are initiated using a single negative-strand template; therefore, the more efficient *cre* acting as the site of uridylylation may be necessary to add VPg-pU-pU to newly synthesized positive-strand RNAs. RNA secondary structures within the 5′ and 3′ ends of positive- and negative-strand RNAs, respectively, may sufficiently destabilize base-pairing of the RNAs to allow the $3D^{pol}$ access to the negative-strand RNA template. The $3D^{pol}$ of poliovirus has also been reported to have duplex unwinding activity. Interestingly, the poliovirus 2C protein contains NTPase activity and has protein structural similarity to RNA helicases. It has also been shown to bind to the 3′ end of negative-strand RNA, suggesting a role in viral RNA replication. Perhaps $3D^{pol}$ and 2C form a complex capable of simultaneous unwinding and chain elongation of the RNA.

In addition to interacting with picornavirus genomic RNAs to promote virus translation and RNA replication, cellular proteins have also been shown to interact with negative-strand RNA intermediates. Cellular protein hnRNP C can interact with the RNA secondary structures formed by the 3′ end of poliovirus negative-strand RNA as well as with viral proteins 3CD and 2C. Complexes of hnRNP C and poliovirus replication proteins may promote initiation of positive-strand RNA synthesis via such RNA–protein complexes.

Unanswered Questions and Conclusions

Despite decades of research invested into the mechanisms by which members of the family *Picornaviridae* propagate their genome in infected cells, many questions remain unanswered. The discovery of the *cre* RNA structures helped address the mechanism of how VPg is primed for positive-strand RNA synthesis. However, how the negative strand acts as template for positive-strand RNA and what viral and/or cellular factors are necessary for RNA initiation are largely unanswered questions. The question pertaining to the role of cellular factors in this process is further complicated by the tissue tropism primarily determined by virus-specific cell receptor usage and by RNA structure/sequence differences among picornaviruses. Cap-independent RNA translation from the IRES element is known to occur for all picornaviruses, yet distinct differences exist even between IRES classes in the ability of the IRES to direct translation. For example, poliovirus and human rhinovirus utilize a type I IRES and share some similarities in the ability to bind cellular factors to promote translation; yet their tissue tropisms differ. What are the sequence and structural differences in the IRES element responsible for virus species-specific differences in translation, and what are the cellular proteins responsible for these differences?

One long-standing question for picornavirus researchers is how template utilization is controlled for translation versus viral RNA replication. Specifically, translating ribosomes on genomic RNA in the 5′ to 3′ direction would interfere with RNA polymerase synthesizing negative-strand RNA (proceeding along the genomic RNA in a 3′ to 5′ manner). Is there a molecular switch to turn off translation to allow RNA replication? Are viral and/or cellular factors involved in translation modified or relocated to permit RNA replication? Or is it an effect of increased levels of local concentrations of proteins that shifts the template RNA from translation to replication competency. Such mechanistic questions underscore the intricate molecular processes through which these pathogens have evolved to maximize the reproduction of their genomes. Study of these mechanisms of viral gene expression has provided insights into the design of vaccines and therapeutic drugs not only for picornaviruses, but also for other virus families.

Further Reading

Agol VI (2006) Molecular mechanisms of poliovirus variation and evolution. *Current Topics in Microbiology and Immunology* 299: 211–259.

Agol VI, Paul AV, and Wimmer E (1999) Paradoxes of the replication of picornaviral genomes. *Virus Research* 62: 129–147.

Bedard KM and Semler BL (2004) Regulation of picornavirus gene expression. *Microbes and Infection* 6: 702–713.

Belsham GJ (2005) Translation and replication of FMDV RNA. *Current Topics in Microbiology and Immunology* 288: 43–70.

Bubeck D, Filman DJ, Cheng N, Steven AC, Hogle JM, and Belnap DM (2005) The structure of the poliovirus 135S cell entry intermediate at 10-angstrom resolution reveals the location of an externalized polypeptide that binds to membranes. *Journal of Virology* 79: 7745–7755.

Hogle JM (2002) Poliovirus cell entry: Common structural themes in viral cell entry pathways. *Annual Review of Microbiology* 56: 677–702.

Jang SK (2006) Internal initiation: IRES elements of picornaviruses and hepatitis C virus. *Virus Research* 119: 2–15.

Semler BL and Wimmer E (eds.) (2002) *Molecular Biology of Picornaviruses.* Washington, DC: ASM Press.

Stewart SR and Semler BL (1997) RNA determinants of picornavirus cap-independent translation initiation. *Seminars in Virology* 8: 242–255.

Reoviruses: Molecular Biology

K M Coombs, University of Manitoba, Winnipeg, MB, Canada

Glossary

EOP Efficiency of plating; a mathematical description of the relative capacity of temperature-sensitive virus mutants to replicate at a high, nonpermissive temperature compared to replication at a lower, permissive temperature.
Icosahedral A regular three-dimensional geometric shape consisting of 20 equilateral triangles and 12 vertices; the actual architecture of many 'spherical' viruses.
ISVP Intermediate (or infectious) subviral particle; an intermediate form of reovirus that lacks some proteins, compared to the intact virion, but contains additional peptides compared to the innermost core; thought to represent the particle that crosses a cell membrane during viral entry.
PFU Plaque-forming unit; the basic unit used to indicate infectious titer for viruses capable of forming 'plaques' (individual localized regions of dead cells in a cell monolayer, each of which is formed by a single infectious virus).
Reassortant A hybrid virus that contains some genes from one parental virus and other genes from another parental virus; arise after co-infection of cells with two different members of the same virus species.
SCID mice Severe combined immunodeficient mice; so named because such animals lack both humoral-mediated and cell-mediated arms of immunity.
Top component A particle that contains all virus proteins but lacks detectable genomic material; so named because, being less dense than intact virions, it is found above virions in density gradients.

Introduction

Viruses comprising the species *Mammalian orthoreovirus* (MRV) are considered the prototype members of the genus *Orthoreovirus* in the virus family *Reoviridae.* Members of this family have a genome of 9–12 segments of double-stranded RNA (dsRNA) surrounded by 2–3 concentric, nonenveloped protein capsids (**Table 1**). Replication is exclusively cytoplasmic, viral uncoating is incomplete, and the inner capsid serves as the enzymatic complex that produces progeny mRNA. The family *Reoviridae* currently contains 12 genera, and inclusion of an additional genus (*Dinovernavirus*) has been proposed (**Table 2**). In addition to the genus *Orthoreovirus,* the family includes rotaviruses, agents responsible for a significant amount of viral gastroenteritis and numerous deaths annually worldwide, the economically important insect-vectored orbiviruses, and a variety of other viruses that infect animals, fungi, and plants. There have been increasing efforts to better understand other members of this genus, including avian orthoreoviruses and fish reoviruses. This review focuses upon MRV.

Reovirus Structure

The MRV particle comprises 10 dsRNA genome segments encased by two concentric protein capsids built from multiple, nonequivalent copies of eight different proteins (**Figure 1**). Five proteins make up the inner capsid (called 'core') that surrounds the dsRNA genome and transcribes it to produce progeny mRNA. Three additional proteins make up the outer capsid. In addition to the core and virus, an intermediate form, called the intermediate (or infectious) subviral particle (ISVP) is also naturally found. All three particle forms can be isolated *in vivo* or produced easily in the laboratory, and

Table 1 Characteristics of the *Reoviridae*

Structure
About 70–85 nm in diameter
Nonenveloped
Icosahedral
Multiple concentric protein capsids
Innermost capsid serves as transcriptase complex
Outermost capsid serves as gene-delivery system
Genome
Linear double-stranded RNA
9–12 gene segments
Total genome size 18–29 kbp
Segmented genome capable of assortment to produce hybrid reassortant viruses
Most gene segments monocistronic
Replication
Cytoplasmic
Proteolytic processing of intact virion to produce subviral particles
Uncoating is incomplete; innermost core capsid serves to transcribe mRNA

Table 2 Members of the *Reoviridae* family

Genus	*Host Range*	*Prototype Virus*
Turreted[a]		
Orthoreovirus	Vertebrates	Mammalian Reovirus
Mammalian	Mammals	*Reovirus 3*
Avian	Birds	*Avian Reovirus S1113*
Baboon	Baboon	*Baboon Reovirus*
Nelson Bay	Flying fox	*Nelson Bay virus*
Unclassified	Reptiles	
Aquareovirus	Vertebrates	*Golden shiner virus*
Cypovirus	Invertebrates	*Bombyx mori cypovirus 1*
Dinovernavirus[b]	Invertebrates	*Aedes pseudoscutellaris reovirus*
Entomoreovirus	Invertebrates	*Hyposoter exiguae reovirus*
Fijivirus	Invertebrates, plants	*Fiji disease virus*
Mycoreovirus	Fungi	*Mycoreovirus 1*
Oryzavirus	Invertebrates, plants	*Rice ragged stunt virus*
Nonturreted		
Coltivirus	Vertebrates, invertebrates	*Colorado tick fever virus*
Orbivirus	Vertebrates, invertebrates	*Bluetongue virus 10*
Phytoreovirus	Invertebrates, plants	*Wound tumour virus*
Rotavirus	Vertebrates	*Simian rotavirus SA11*
Seadornavirus	Vertebrates, invertebrates	*Banna virus*

[a]Turreted; innermost capsid (core) contains prominent spike projections at icosahedral vertices.
[b]Proposed membership within family.

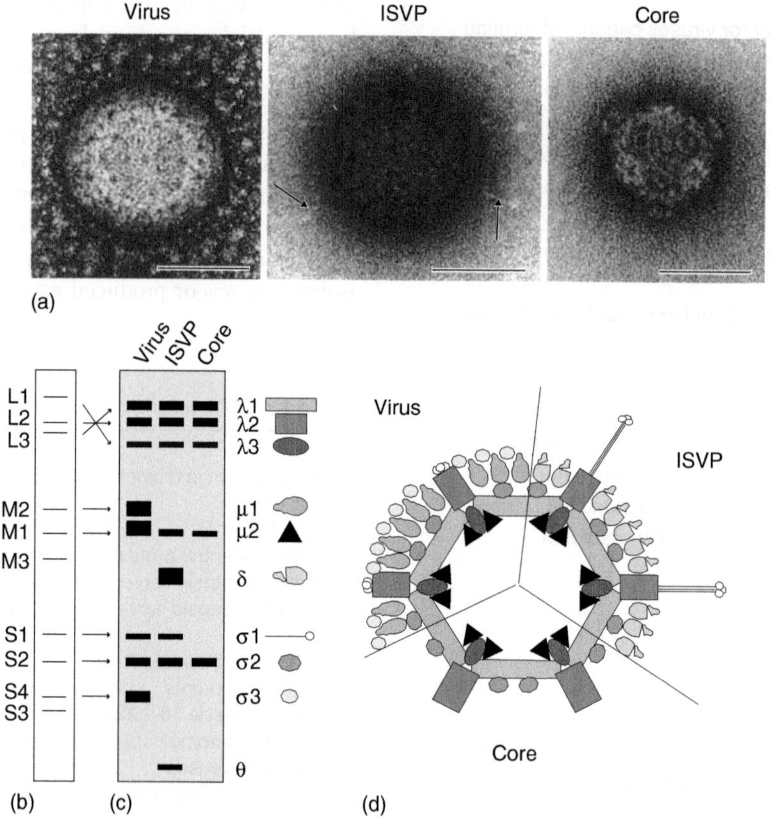

Figure 1 Structure, gene coding, and protein locations within T1L reovirus particle. (a) Electron micrographs of virus, ISVP, and core particles; arrows indicate protruding σ1 molecules on ISVP. (b) Cartoon of T1L dsRNA profile in SDS-PAGE with gene segments L1–S4 labeled on left. (c) Cartoon of protein profiles of virus, ISVP, and core in SDS-PAGE. Each protein is encoded by indicated gene segment in (b) (arrows). (d) Composite cartoon of reovirus virion, ISVP, and core, showing presumptive locations of various structural proteins, and their conversion or removal from each type of particle. Scale = 50 nm (a). Reproduced from Tran AT and Coombs KM (2001) Reoviruses. In: *Encyclopedia of Life Sciences*. New York: Wiley, with permission from John Wiley & Sons Ltd.

the protein compositions of each have been determined. The proteins, and their functions, are discussed more fully below. Three MRV serotypes (designated 1, 2, and 3) have been described. Prototype members of each serotype routinely used are type 1 Lang (T1L), type 2 Jones (T2J), and type 3 Dearing (T3D). In addition, a large number of field isolates that represent distinct clones have been identified. Some recent comparative sequence evidence supports establishment of a fourth serotype, represented by Ndelle virus.

Structure of the Genome

The complete nucleotide sequences of all ten genes of all three prototype MRV serotypes have been determined. The genome consists of three large segments (L1, L2, L3) ranging in size from 3854 to 3916 base pairs (bp), three medium segments (M1, M2, M3) ranging in size from 2203 to 2304 bp, and four small segments (S1, S2, S3, S4) ranging in size from 1196 to 1463 bp (**Figure 2** and **Table 3**), for a total aggregate genomic size of 23 606 bp for T1L, 23 578 bp for T2J, and 23 560 bp for T3D. In addition to the dsRNA genome, purified MRV particles contain large amounts (representing nearly 25% of total RNA) of short single-stranded oligoribonucleotides, which are thought to represent abortive gene transcripts. Every MRV dsRNA gene sequenced to date consists of an open reading frame of variable length (1095 for S4 to 3867 for L2), bounded by a 5′ nontranslated region (NTR) of variable length (ranging from 12 bases in the S1 gene to 32 nucleotides in the S4 gene) and a 3′ NTR ranging from 35 bases in the L1 gene to 83 nucleotides in the M1 gene. Every gene contains a completely conserved consensus GCUA tetranucleotide at the extreme 5′ end of the gene

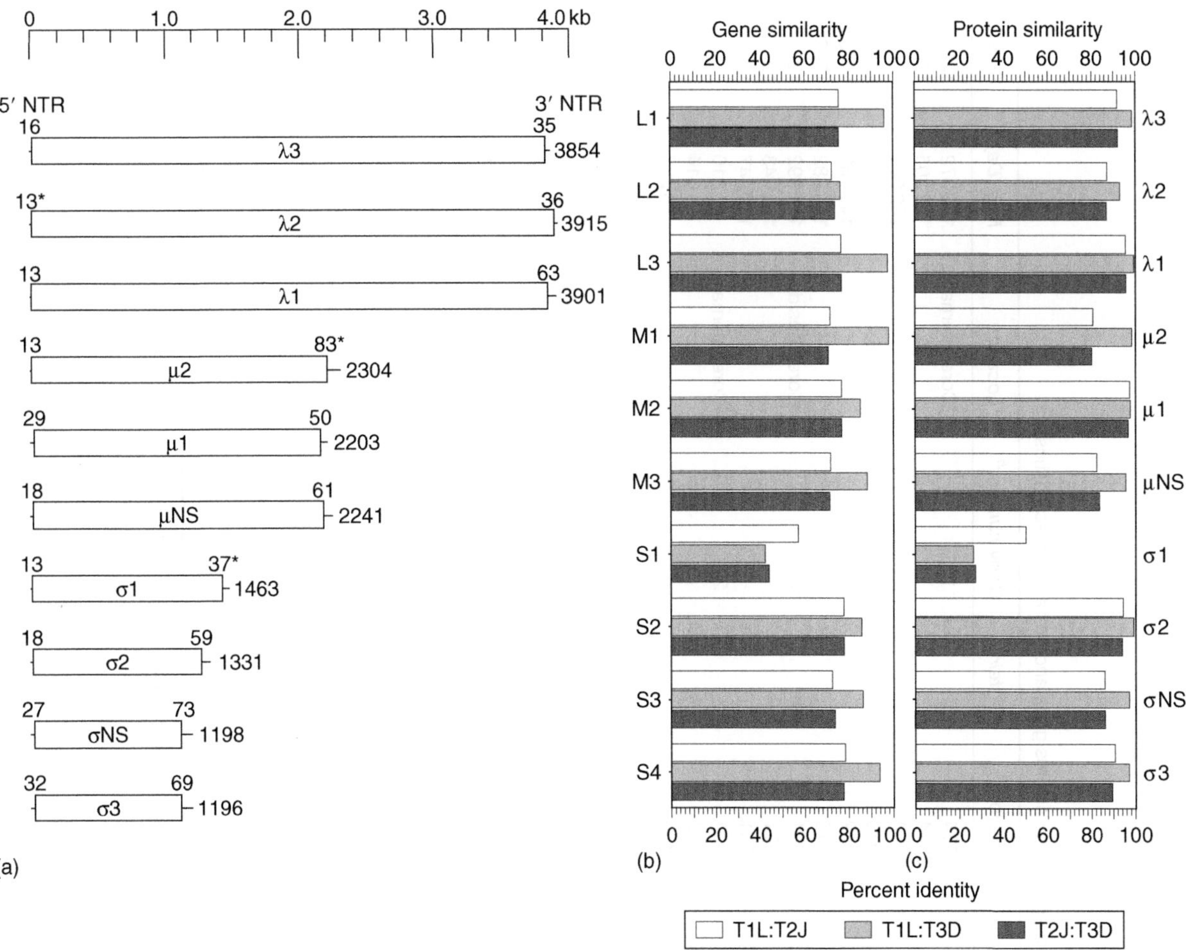

Figure 2 Gene and protein characteristics and similarities. Each of the ten double-stranded RNA genes of mammalian orthoreovirus are shown diagrammatically. (a) Genes, along with kilobase scale bar (top) are indicated. Rectangle corresponds to open reading frame and encoded protein is indicated within box. Total gene length indicated at right. Small numbers above each sequence at extreme left and right correspond to length of nontranslated region (NTR); * indicates variability in NTR length. (b) Pairwise comparisons in nucleotide identity between T1L:T2J, T1L:T3D, and T2J:T3D for each gene; genes indicated between panels A and B. (c) Pairwise comparisons in amino acid identity (indicated at far right) between T1L:T2J, T1L:T3D, and T2J:T3D for each encoded protein.

Table 3 Mammalian orthoreovirus gene and protein characteristics[a]

Gene	Serotype	Size (bp)[a]	Protein	Copy number[b]	Location	MW (Da)[a]	Size (aa)[a]	pI[a]	GenBank #	Functions
L1	T1L	3854	λ3	12	Core internal	142 372	1267	8.10	NC004271	RNA-dependent RNA polymerase
	T2J	3854				142 305	1267	8.17	NC004272	
	T3D	3854				142 287	1267	7.92	NC004282	
L2	T1L	3915	λ2	60	Core spike	143 957	1289	5.08	AF378003	Guanylyltransferase;
	T2J	3912				143 166	1288	4.96	AF378005	methyltransferase
	T3D	3916				144 082	1289	5.01	NC004275	
L3	T1L	3901	λ1	120	Core capsid	141 892	1275	5.92	AF129820	RNA binding; NTPase;
	T2J	3901				142 043	1275	6.08	AF129821	RNA triphosphatase;
	T3D	3901				141 847	1275	5.98	AF129822	helicase
M1	T1L	2304	μ2	24	Core internal	83 310	736	6.75	X59945	RdRp cofactor?
	T2J	2303				84 014	736	7.33	NC004254	NTPase; binds RNA
	T3D	2304				83 230	736	6.72	M27261	
M2	T1L	2203	μ1	600	Outer capsid	76 239	708	4.92	AF490617	Cell penetration;
	T2J	2203				76 152	708	4.90	M19355	transcriptase activation
	T3D	2203				76 339	708	4.91	U24260	
M3	T1L	2241	μNS		Nonstructural	80 174	721	5.92	AF174382	Binds RNA;
	T2J	2240				80 512	721	6.02	AF174383	inclusion formation
	T3D	2241				80 195	721	5.69	AF174384	
S1	T1L	1463	σ1	36	Outer capsid	51 481	470	4.97	M35963	Attachment protein;
	T2J	1440				50 482	462	5.18	M35964	hemmaglutinin;
	T3D	1416				49 072	455	5.07	X01161	type-specific antigen
S2	T1L	1331	σ2	150	Core nodule	47 115	418	8.30	L19774	Binds dsRNA
	T2J	1331				47 050	418	8.31	L19775	
	T3D	1331				47 166	418	8.30	L19776	
S3	T1L	1198	σNS		Nonstructural	41 207	366	6.34	M18389	Binds RNA;
	T2J	1198				41 338	366	6.58	M18390	inclusion formation
	T3D	1198				41 061	366	6.34	X01627	
S4	T1L	1196	σ3	600	Outer capsid	41 159	365	6.69	X61586	Binds ssRNA; role in
	T2J	1196				41 230	365	7.18	X60066	assortment and replication
	T3D	1196				41 164	365	6.84	K02739	

[a]From indicated GenBank nucleotide sequence.
[b]Within purified virion particle.

and a completely conserved consensus UCAUC pentanucleotide (on the plus-sense strand) at the extreme 3′ end of the gene. Although still not known, the current belief is that signals that direct genomic assembly reside in one or both of the NTR. Pairwise comparisons of the T1L, T2J, and T3D NTR show generally high conservation. Most are of gene-specific uniform size (e.g., every L1 gene has a 16-nucleotide (nt) 5′ NTR and 35 nt 3′ NTR and every L3 gene has a 13 nt 5′ NTR and 63 nt 3′ NTR (**Figure 2**, left)) and some NTRs are completely conserved across all three serotypes. For example, the 13 nt L3 5′ NTR is identical in T1L, T2J, and T3D and the 18 nt M3 5′ NTR is 100% conserved across all three serotypes. A few genes show variability in the NTR. For example, the T3D L2 gene 5′ NTR has a 1 nt insertion compared to the T1L and T2J L2 5′ NTR and the T2J M1 3′ NTR has a 1 nt deletion compared to the T1L and T3D M1 3′ NTR.

Pairwise comparisons of gene identity show that, in most cases, T1L and T3D are more closely related to each other than either is to T2J (**Figure 2**, middle). For example, the percent identities of the T1L and T3D L1 genes are ~95% whereas percent identity of the T2J L1 gene compared to either T1L or T3D is <80%. Similar patterns are seen in most other genes, although comparative values range from 78% for T1L:T3D L2 and 76% for T2J:T1L or T3D L2, to ~98% for T1L:T3D M1 and ~70% for T2J: T1L or T3D M1. Only the S1 gene shows a difference, with T1L and T2J being most similar to each other.

The segmented nature of the MRV genome allows genetic mixing if two different reoviruses infect the same cell. Progeny that contain mixtures of parental genome segments (called reassortants) can only arise if the two viruses belong to the same species. Genome segment reassortment may contribute to natural pathogenesis (as seen for rotavirus and influenza virus) and has served as a convenient genetic tool for understanding better reovirus gene/protein structures and functions.

Inner Capsid Structure

The reovirus inner capsid (core) contains the viral genome and is constructed from five proteins, called λ1, λ2, λ3, μ2, and σ2. The core has icosahedral symmetry and a diameter of ~52 nm (excluding the spike 'turrets' which extend outward from each icosahedral fivefold axis (vertex) an additional 5.5 nm) (**Figure 1**). Cores serve as the metabolically active macromolecular 'machines' from which viral mRNA is transcribed, both during replication and *in vitro*. Detailed structural information has been provided from X-ray crystallographic studies of the core, which greatly aid in understanding each protein's structure and/or function.

Full-length λ1 is a 1275-amino acid, 142 kDa protein encoded in the L3 genome segment. Protein λ1 is present in 120 copies within each reovirus core particle (**Table 3**). It is the major structural protein that forms the inner capsid. There is little serotypic variability in this inner capsid protein, implying fairly rigid structural requirements; pairwise protein identity analyses (**Figure 2**, right) show T1L and T3D λ1 share almost 100% identity (nine changes out of 1275 amino acids) and identity between T2J and either T1L or T3D is ~95%. The 120 λ1 proteins are organized as 60 asymmetric dimers in a $T=1$ triangulation lattice to form a thin capsid shell (**Figure 1(d)**). The protein has a zinc-finger motif, a separate region that binds dsRNA, and ATPase activity, which probably is involved in transcriptional events.

Full-length λ2 is a 1289 amino acid (1288 aa for T2J), 144 kDa protein encoded in the L2 genome segment. Sixty copies of this protein are organized as distinctive pentameric turrets at each of the particle's 12 icosahedral vertices (**Figure 1**). This is one of the most variable of core proteins; T1L:T3D identity is ~92% and T2J has ~87% identity to T1L and T3D λ2 (**Figure 2**, right). Protein λ2 binds glucose monophosphate (GMP) and *S*-adenosyl-L-methionine, possesses guanylyltransferase and methyltransferase activities, and serves to attach a type-1 ^{7}mG cap structure to nascent mRNA as the mRNA is extruded from transcribing cores. High-resolution structure determinations, coupled with mutagenesis studies, show λ2 to be a multidomain protein.

Full-length λ3 is a 1267 amino acid, 142 kDa protein encoded in the L1 genome segment. Cores contain 12 copies of this minor protein, which serves as the RNA-dependent RNA polymerase (RdRp). Pairwise protein identity values between any two clones is >90% (**Figure 2**, right). Current structural/functional data indicate that a single copy of this protein is located inside the core shell, slightly offset from center, at each of the icosahedral fivefold axes, located directly below the λ1 shell. Recent X-ray crystallographic structure determinations indicate this protein follows the basic 'right-hand' configuration of most nucleic acid polymerases, but also contains extra domains.

Full-length μ2 is a 736 amino acid, 83 kDa protein encoded in the M1 genome segment. This is the least understood of reovirus structural proteins; there is presently no high-resolution structure determination, and the - precise function(s) remain unknown. The protein may represent an RdRp cofactor and it is present in 20–24 copies, possibly also directly under each of the core vertices. The T1L and T3D μ2 proteins share >98% identity, but the T2J μ2 protein shares only ~80% identity with T1L and T3D μ2. Genetic mapping experiments suggest that μ2 plays a role in determining the severity of cytopathic effect in cultured cells and the level of virus growth in various cells. The protein also is involved in myocarditis, in organ-specific virulence in SCID mice, in *in vitro* transcription of ssRNA, possesses nucleoside triphosphatase activity and binds RNA. It also plays important roles in virus inclusion formation and morphology, a process mediated, in part, by its level of ubiquitination.

Full-length σ2 is a 418 amino acid, 47 kDa protein encoded in the S2 genome segment. One hundred and fifty copies of the σ2 protein decorate the thin λ1 shell (**Figure 1**) and may act as 'clamps' to hold the shell together. Protein σ2 binds RNA and is required for assembly of core capsids. T1L and T3D σ2 share ~99% identity and T2J shows ~94% identity with T1L and T3D σ2.

Outer Capsid Structure

The reovirus outer capsid serves as a 'gene-delivery vehicle'; proteins within this layer are responsible for recognizing host cells and in permitting entry of the viral core into the cellular cytosol. The structures of all outer capsid proteins have been determined by X-ray crystallography. Proteins in the outer capsid are organized in a fenestrated $T = 13(l)$ icosahedral lattice. This lattice is built from only 600 copies of each of two major proteins (μ1, also present as a carboxyl-terminal cleavage product μ1C, and σ3), rather than 780 copies (13 × 60) as seen in nonturreted rotaviruses. This is because vertex positions that would normally contain the 'missing' 180 protein units are occupied by the core λ2 turrets.

Full-length μ1 is a 708 amino acid, N-terminal myristoylated protein encoded in the M2 genome segment. The protein folds into a four-domain structure; three lower predominantly α-helical domains (Domain I, II, and III) plus a fourth jelly roll β barrel head domain (Domain IV). When resolved by standard sodium dodecyl sulfate-polyacrylamide gel electrophoresis (SDS-PAGE), approximately 95% of virion μ1 appears as an ~4 kDa amino-terminal peptide (called μ1N and not resolved in standard SDS-PAGE) and a 72 kDa carboxyl-terminal portion (called μ1C). Much of this μ1N/μ1C cleavage in purified virions, which takes place between Asn_{42} and Pro_{43}, was recently shown to be an artifact of sample preparation prior to SDS-PAGE. The Asn-Pro amino acid pair is found near the N-terminus in all MRV, avian orthoreovirus, and aquareovirus μ1 (and μ1 homolog) sequences determined to date, which suggests this μ1N/μ1C cleavage plays an important role in the replicative cycle. This protein alone accounts for approximately half of the total virion peptide mass, plays important roles in virion stability, and undergoes several specific cleavages during virus entry (discussed more fully below). Unlike the core proteins (discussed above), the T1L, T2J, and T3D μ1 proteins show approximately the same amount of identity to each other (~97%) (**Figure 2**, right).

Protein σ1 is the cell attachment protein and serotype determinant. It is the most variable of reovirus proteins. The T1L and T2J σ1 proteins are more closely related (but share only ~50% identity) whereas the T1L and T3D, and T2J:T3D σ1 proteins, share only ~27% identity. The protein length varies with particular virus type; full-length T1L σ1 is a 470 amino acid, N-terminally acetylated protein encoded by the S1 genome segment, whereas the T2J σ1 is a 462 amino acid protein and the T3D σ1 is a 455 amino acid protein. The σ1 molecules have an overall stalk/knob structure similar to the adenovirus cell attachment fiber. The virion contains 36 copies of σ1, organized as trimers at each of the vertices.

Full-length σ3 is a 365 amino acid, 41 kDa N-terminally acetylated protein encoded in the S4 genome segment. The protein folds into a two-lobed structure, with the small lobe in contact with other major outer capsid protein μ1 and the large lobe exposed on the virion surface. Like the core proteins (discussed above), the T1L and T3D σ3 proteins share greatest identity (~97%), whereas T2J shares ~90% identity with the σ3 proteins of the other strains. The σ3 protein plays important roles in particle stability, in shutting down host macromolecular synthesis, and in downregulation of interferon-induced dsRNA-activated protein kinase (PKR), possibly by virtue of its intracellular distribution and capacity to interact with μ1C. The σ3 protein contains a zinc-finger motif and binds zinc, which appears important both for correct folding and stability, as well as a separate motif that binds dsRNA. Structural analyses indicate that three copies each of μ1 and σ3 form a heterohexameric $(\mu1/\sigma3)_3$ aggregate. The three copies of μ1 are wound around each other to form a 'base', upon which three σ3 monomers sit. Two hundred such heterohexameric aggregates decorate the viral outer capsid, with the μ1 trimer base contacting the core and the σ3 monomers most distally located in the particle. This protein arrangement generates the ~85 nm diameter fenestrated $T = 13(l)$ lattice. Proteolytic processing of σ3, which initiates in a central specific 'hypersensitive region' (HSR) and then proceeds bidirectionally toward the termini plays critical roles in virus entry and uncoating (discussed more fully below).

Reovirus Replication

Reoviruses infect a wide range of cells, both *in vitro* and *in vivo*. The virus usually infects specialized intestinal epithelial cells (M cells) that overlie Peyer's patches *in vivo*. Virus then migrates between and/or through the M cells into mucosal mononuclear cells in the Peyer's patch, and subsequently into a large number of extraintestinal sites, including heart, liver, and central nervous system. Elucidation of steps in MRV replication has been studied primarily in tissue-cultured mouse L929 fibroblast-like cells, although numerous studies have also been performed in a wide, and growing, range of other cells.

MRV replication is primarily cytoplasmic (**Figure 3**). There seems to be no significant nuclear involvement,

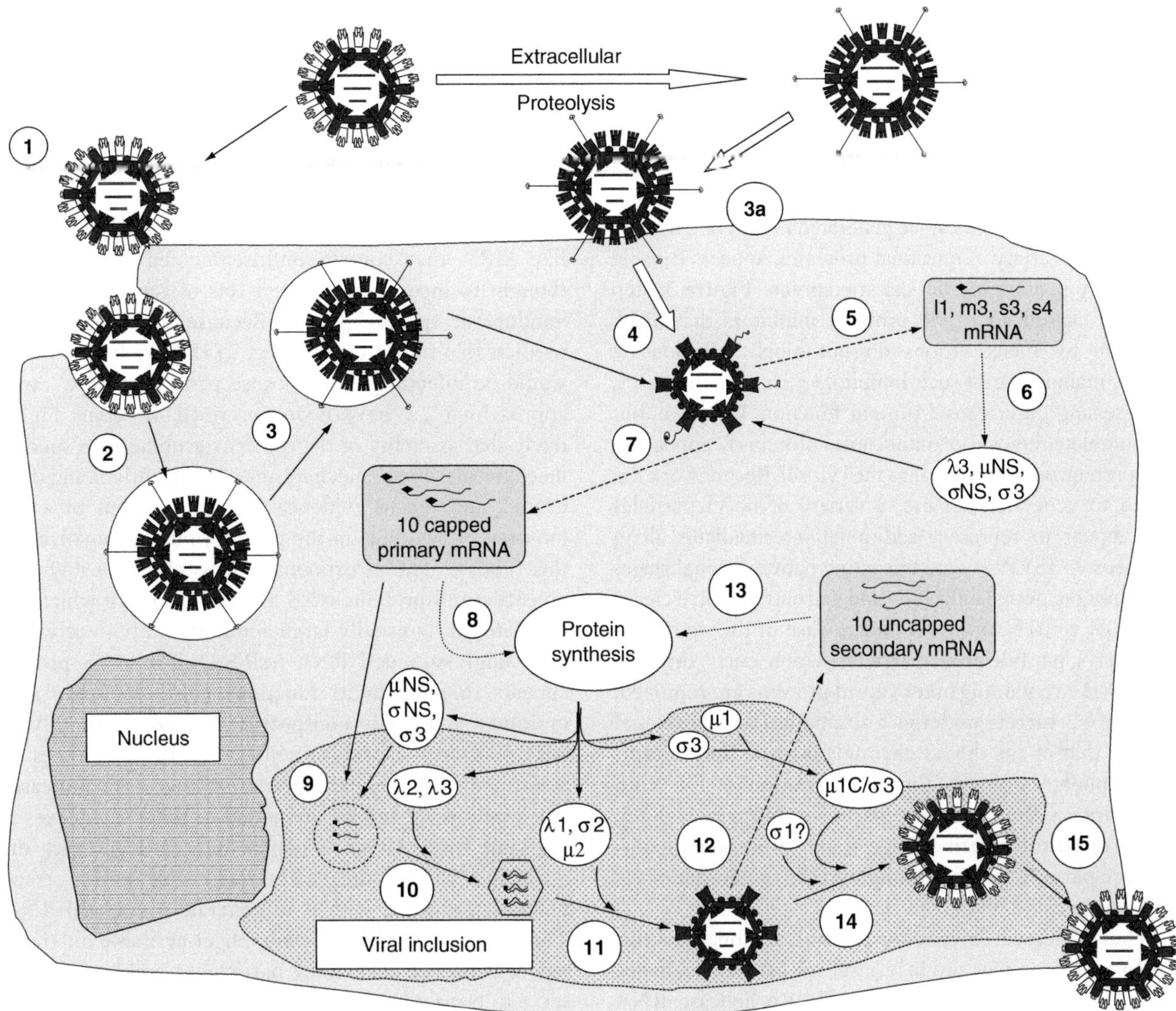

Figure 3 Model of MRV replication cycle. Details are provided in the text. Steps are (1) virus binding (upper left), (2) entry into endosomes where acid-mediated proteolysis occurs to remove outer capsid protein σ3, (3) membrane interaction to allow ISVP to escape endosome, (4) uncoating of ISVP to release transcriptionally active core particle, (5) initial 'pre-early' capped transcription, (6) initial 'pre-early' translation, (7) primary capped transcription, (8) primary translation, (9) assortment of mRNA segments into sets, probably mediated by σ3 and nonstructural proteins, (10) synthesis of negative RNA strands to generate progeny dsRNA (associated with accumulation of core proteins), (11) generation of transcriptase (replicase) complex, (12) secondary uncapped transcription, (13) secondary translation, (14) assembly of outer capsid, which halts transcription, and (15) release (lower right). Core proteins are shown in black and outer capsid proteins are shown in white; ISVP-specific modified proteins are shown in gray. Dashed arrows indicate transcription events leading to production of mRNAs (indicated in gray boxes), dotted arrows indicate translation events leading to production of proteins (indicated in white ovals), and black arrows indicate movement of proteins and viral complexes. An alternate entry mechanism for intermediate subviral particles (ISVPs), that are capable of directly penetrating membranes, is shown with white arrows at the top and in step 3a. Adapted from Tran AT and Coombs KM (2001) Reoviruses. In: *Encyclopedia of Life Sciences*. New York: Wiley, with permission from Wiley.

although nonstructural virus protein σ1s (which is translated from an alternate S1 reading frame) targets the nucleus. The first step in virus replication is binding of virion to susceptible host cells (**Figure 3**, step 1), a process mediated by viral cell attachment protein σ1. The cellular protein(s) with which σ1 interacts is not completely known, but sialic acid appears to be an important component of the receptor(s). Several proteins have been identified as possible receptors, including junction-adhesion molecule. After initial binding, virus enters cells by either of the two mechanisms. Virus may be taken up by receptor-mediated endocytosis (**Figure 3**, step 2), and then converted into the ISVP by both acidification and proteolysis. A variety of agents that perturb either endosomal acidification (e.g., lysosomotropic agents such as ammonium chloride and chloroquine, or specific acidic protease inhibitors such as E-64), can inhibit infection by intact virions. Additional pharmacologic agents that inhibit a variety of other entry mechanisms (including methyl-β-cyclo-dextrin, which perturbs caveolae and lipid rafts, and chlorpromazine, which inhibits clathrin-mediated endocytosis) also reduce virus yields in cell culture. Accumulating evidence suggests that outer capsid protein σ3 must be proteolytically clipped, initially within a central HSR that is susceptible to a wide range of acidic, neutral, and basic proteases. Alternatively, the ISVP, which can be generated *in vitro* or extracellularly by a variety of intestinal proteases, appears capable of directly penetrating the cell membrane (**Figure 3**, step 3a). The acidification and protease inhibitors that block infection by intact virions do not block infection by ISVPs, making use of such inhibitors a convenient means for assessing outer capsid protein function. Cleavage, but not complete removal, of outer capsid protein σ3 appears to be a prerequisite for entry into the cytosol. Recent work has begun to generate, and use, a variety of novel particles that appear to represent additional intermediates along the virus → ISVP → core uncoating pathway. Final stages of membrane permeabilization and cell entry involve cleavage of μ1 to μ1N and μ1C, and release of the myristoylated μ1N peptide from particles. Upon entry into the cytosol, the remaining outer capsid proteins are removed, and the λ2 turrets undergo a dramatic conformational change that opens the λ2 channels further. Both alterations appear necessary for full activation of the viral transcriptase. Resultant core particles (**Figures 1** and **3**, step 4) constitute the final stage of uncoating, and incoming core particles persist throughout the remainder of the replication cycle.

These released cores serve as transcriptionally active RdRp machines that produce viral mRNA. Collectively, the λ2, λ3, and μ2 proteins manifest RdRp, helicase, RNA triphosphatase, methyltransferase, and guanylyltransferase activities to produce the viral mRNA, which is extruded through channels in the modified λ2 spikes. Early work suggested that only four (L1, M3, S3, and S4) of the ten dsRNA genes were initially transcribed to produce mRNA (**Figure 3**, step 5), and that the protein products of these initial transcripts act upon the core by unknown means to promote transcription of all ten genes (**Figure 3**, step 7). However, more recent work questions this 'cascade' interpretation. The ten 'early' transcripts, like parental genomes, contain methylated caps and lack poly(A) tails. The cap structures are provided to nascent mRNAs by the λ2 proteins as the mRNAs are extruded through the λ2 spikes. Viral transcripts are produced in quantities inversely related to transcript length; more 's' transcripts are produced than 'm' transcripts, and more 'm' transcripts are produced than 'l' transcripts. Viral 'early' mRNAs are then translated to produce the full complement of viral proteins (**Figure 3**, step 8). These proteins begin to coalesce and are directed into non-membrane-containing inclusions in the cytoplasm by the σNS, μNS, and/or μ2 proteins (**Figure 3**, step 9). The ten mRNA molecules are also used as templates for progeny minus-strand synthesis (**Figure 3**, step 10), a process most likely mediated by viral proteins. Eventual assembly of a complete progeny virion will require various structural proteins as well as a full complement of the ten dsRNA genes.

The ten different mRNA molecules are believed to be sorted (a process called 'assortment') to ensure that developing viral particles contain correct sets of genes. Viruses, like MRV, that contain multisegmented genomes, are thought to incorporate correct sets of genes by either a 'random' or 'specific' process. Because the random model predicts that only 1 out of every 2755 particles ($10!/10^{10}$) would be infectious, and because particle-to-PFU ratios approaching 2:1 have been described, it seems highly likely that assembly of the reovirus genome uses specific mechanisms. These mechanisms are currently being delineated, and recent evidence indicates regions of a few hundred nucleotides in the gene termini are involved in this process. The assortment process also probably uses nonstructural proteins σNS and μNS, both of which are translated in unusually large amounts and both of which bind single-stranded RNA (ssRNA). The σNS protein appears to perform its functions most efficiently as a multimer and binds nonspecifically most efficiently to ssRNA. Outer capsid protein σ3 also participates in assortment. Immunoprecipitation studies have indicated that proteins λ2 and λ3 are added to these complexes at the same time mRNA is copied into dsRNA. Once the mRNA is copied, the newly transcribed minus strand remains associated with it to generate progeny dsRNA. The plus-sense molecule is no longer available for translation. However, developing particles resemble cores that serve to transcribe additional mRNA, and so presumably also contain other core proteins (**Figure 3**, step 11).

Progeny core-like particles ('transcriptase complexes'), like incoming uncoated cores, are capable of synthesizing viral mRNA (**Figure 3**, step 12). These nascent replicase particles are responsible for the majority of transcription during replication. Transcripts produced during this second wave of transcription are not capped. The switch in translation from cap-dependence to cap-independence remains poorly understood, but is believed to involve intracellular distribution of σ3, its capacity to interact with μ1C, and interactions with several cellular interferon-regulated gene products, including PKR and RNase-L. Assembly of progeny virions requires condensation of the correct numbers of eight different viral proteins with one copy of each of the ten progeny dsRNA genes. *In vitro* 'recoating' studies and studies with assembly-defective, temperature-sensitive reovirus mutants have shed light on pathways by which viral proteins may associate to generate the double-capsid shell that will serve to protect the genome (**Figure 3**, steps 12–15). This is possible because many of the known reovirus temperature-sensitive mutants produce aberrant particles at the nonpermissive temperature (**Table 4**) and

Table 4 Characteristics of selected mammalian reovirus temperature-sensitive mutants

Mutant				*Nonpermissive characteristics*[a]			
Group	*Clone*	*Gene*	*Protein*	*EOP*[b]	*dsRNA*	*Protein*	*Phenotype*
Wild-type T3 Dearing				0.2	+	+	Normal
A	*tsA201*	M2	μ1	0.009	+	+	Normal
	tsA279	M2	μ1	0.0002	+	–	Top component[c]
		and L2	λ2				Spike-less cores
B	*tsB352*	L2	λ2	0.00001	+	+	Cores
C	*tsC447*	S2	σ2	1×10^{-7}	–	–	Empty outer shells
D	*tsD357*	L1	λ3	0.00001	+	+	Top component
E	*tsE320*	S3	σNS	0.002	–	–	No inclusions
G	*tsG453*	S4	σ3	1×10^{-7}	+	+	Cores
H	*tsH11.2*	M1	μ2	1×10^{-7}	–	–	No inclusions
I	*tsI138*	L3	λ1	0.0001	–	–	No inclusions

[a]At nonpermissive temperature of 39.5 °C.
[b]EOP; Titer at 39.5 °C ÷ titer at 33.5 °C.
[c]Top component; genome-deficient double-shelled particles.

the precise genetic lesions (amino acid substitutions fit into available crystal structures) for many of them have been determined. For example, assembly of the core capsid shell requires only major core proteins λ1 and σ2, as indicated by expression studies and analyses of *tsC447*, a mutant defective in σ2. Reversion studies indicated that an $Asn_{383} \rightarrow Asp$ substitution, near the end of a long α-helix that spans residues Thr_{350}–Asn_{386}, is responsible for the inability of *tsC447* to assemble core particles at the nonpermissive temperature. The λ2 proteins then associate with the core shell particles, as implied by generation of 'spike-less' cores by *tsA279*, a double mutant that contains a defective λ2 protein. However, it currently is not known whether λ2 proteins form pentamers before associating with the core shell or whether they associate as monomers and then pentamerize. Addition of σ1 may take place at about this same time because of its intimate association with the λ2 spikes in mature particles. Outer capsid proteins μ1 and σ3 initially associate with each other, as determined both by studies with *tsG453*, a mutant defective in σ3 protein that forms core-like structures (**Table 4**) rather than ISVP-like structures, and by recoating experiments which have, so far, failed to attach only μ1 (in the absence of σ3) to cores. These proteins form heterohexamers ($(\mu1/3)_3$) that then associate with the nascent core particles to complete the double capsid (**Figure 3**, step 14) and turn off transcription. Virus is released when infected cells lyse (**Figure 3**, step 15). The recent development of a 'reverse genetics system' for the reoviruses should prove extremely beneficial to complement some of the above-described transcapsidation and temperature-sensitive mutant analyses.

Further Reading

Coombs KM (2006) Reovirus structure and morphogenesis. *Current Topics in Microbiology and Immunology* 309: 117–167.

Joklik WK (1998) Assembly of the reovirus genome. *Current Topics in Microbiology and Immunology* 233(1): 57–68.

Kobayashi T, Antar AAR, Boehme KW, *et al.* (2007) A plasmid-based reverse genetics system for animal double-stranded RNA viruses. *Cell Host and Microbe* 1: 147–157.

Mertens PPC, Attoui H, Duncan R, and Dermody TS (2005) *Reoviridae*. In: Fauquet CM, Mayo MA, Maniloff J, Desselberger U, and Ball LA (eds.) *Virus Taxonomy. Eighth Report of the International Committee on Taxonomy of Viruses*, pp. 447–454. San Diego, CA: Elsevier Academic Press.

Nibert ML and Schiff LA (2001) Reoviruses and their replication. In: Knipe DM and Howley PM (eds.) *Fields Virology,* 4th edn., pp. 1679–1728. Philadelphia, PA: Lippincott Williams and Wilkins.

Tran AT and Coombs KM (2001) Reoviruses. In *Encyclopedia of Life Sciences*. New York: Wiley, with permission from John Wiley & Sons Ltd.

Tyler KL (2001) Mammalian reoviruses. In: Knipe DM and Howley PM (eds.) *Fields Virology,* 4th edn., pp. 1729–1745. Philadelphia, PA: Lippincott Williams and Wilkins.

Seadornaviruses

H Attoui, Université de la Méditerranée, Marseille, France
P P C Mertens, Pirbright Laboratory, Woking, UK

Glossary

Arthralgia Joint pain.
Cerebral palsy Disability resulting from damage of the brain before, during, or shortly after birth and manifested as muscular and speech disturbances.
Hemiplegia Total or partial paralysis of one side of the body resulting from an affliction of the motor centers in the brain.
Myalgia Muscle pain.
Viremia The presence of virus in the blood.

Introduction

The family *Reoviridae* includes 12 recognized genera and 3 proposed genera (**Table 1**). The genera *Coltivirus*, *Seadornavirus*, and *Orbivirus* include arboviruses that can cause disease in animals (including humans).

In 1991, the International Committee for the Taxonomy of Viruses (ICTV) formally recognized the genus *Coltivirus* (sigla from Colorado tick virus) containing the 12-segmented double-stranded RNA (dsRNA) animal viruses that were previously classified within the genus *Orbivirus*. *Colorado tick fever virus* is the prototype species of the genus *Coltivirus* and includes strains of several tick-borne viruses (including isolates of Colorado tick fever virus (CTFV) from humans and ticks; California hare coltivirus (CTFV-Ca) from a hare in northern California; and Salmon River virus (SRV) from humans in Idaho). The genus *Coltivirus* also contains a second distinct species: *Eyach virus* (EYAV) isolated from ticks in Europe.

The seadornaviruses are 12-segmented mosquito-borne dsRNA viruses that were initially considered as tentative species within the genus *Coltivirus*. However, based on their antigenic properties and nucleotide sequence comparisons to the coltiviruses and other members of the *Reoviridae*, their taxonomic status was reassessed. They are now formally recognized as members of the genus *Seadornavirus* (sigla:

Table 1 Twelve recognized genera and the three proposed genera of family *Reoviridae*

	Type species			*No. of segments*	*Hosts of the member viruses*
Genus					
1. *Idnoreovirus*	*Diadromus pulchellus reovirus*			10	Insects
2. *Cypovirus*	*Cypovirus type 1*			10	Insects
3. *Oryzavirus*	*Rice ragged stunt virus*			10	Plants
4. *Fijivirus*	*Fiji disease virus*			10	Plants
5. *Orthoreovirus*	*Mammalian orthoreovirus*			10	Primates, ruminants, bats, birds, reptiles
6. *Orbivirus*	*Bluetongue virus*			10	Ruminants, equids, rodents, bats, marsupials, birds, humans
7. *Rotavirus*	*Rotavirus A*			11	Mammals, birds
8. *Aquareovirus*	*Aquareovirus A*			11	Aquatic animals
9. *Mycoreovirus*	*Mycoreovirus-1*			11 or 12	Fungi
10. *Phytoreovirus*	*Wound tumor virus*			12	Plants
11. *Coltivirus*	*Colorado tick fever virus*			12	Mammals including man
12. *Seadornavirus*	*Banna virus*	Isolates:	BAV-Ch		
			BAV-In6423		
			BAV-In6969		
			BAV-In7043	12	Humans, porcine, cattle
	Kadipiro virus	Isolates:	KDV-Ja7075		
	Liao ning virus	Isolates:	LNV-NE9712		
			LNV-NE9731		
Proposed genera					
1. Dinovernavirus	Aedes pseudoscutellaris reovirus			9	Insects
2. Mimoreovirus	Micromonas pusilla reovirus			11	Algae
3. Cardoreovirus	Eriocheir sinensis reovirus			12	Crabs

from *South East Asian Dodeca RNA virus*; genus recognized in 2000), which includes the type species *Banna virus* (BAV), isolated from humans; two other species, *Kadipiro virus* (KAD) and *Liao ning virus* (LNV); and several unclassified isolates from mosquitoes. The seadornaviruses have been implicated in a variety of pathological manifestations in humans, including flu-like illness and neurological disorders, as described below.

Historical Overview of Seadornaviruses

The first seadornaviruses were reported in 1980 and 1981, when numerous isolates of 12-segmented dsRNA viruses were made from pooled homogenates of mosquitoes in Indonesia and were initially designated as JKT-6423, JKT-6969, JKT-7043, and JKT-7075. These original viruses were recently renamed as Banna virus (BAV) and Kadipiro virus (KDV). Shortly after the initial identification of BAV in Indonesia, a similar virus was isolated from the serum and cerebrospinal fluid (CSF) (2 isolates) and sera (25 isolates) of human patients with encephalitis in Yunnan province in the south of China and BAV was reported as a causative agent of human viral encephalitis. Seadornaviruses have subsequently been isolated on several occasions from humans, cattle, pigs, and mosquitoes and are vectored by *Anopheles*, *Culex*, and *Aedes* species. BAV is now regarded as endemic in southeast Asia, particularly in Indonesia and China.

Distribution and Epidemiology

BAV was initially isolated from mosquitoes and mammalian species in 1980–81, between latitudes of 39° N and 38° S, in southeast Asia (**Figure 1**). BAV was subsequently isolated from humans in the Yunnan province of southern China (Xishuangbanna prefecture in 1987). A virus identified as BAV was also isolated in 1987 from patients with fever and flu-like manifestation, in Xinjiang province of western China province. In 1992, two 12-segmented dsRNA viruses were isolated from serum of patients with an unknown fever in Mengding county of Yunnan province. Eight further isolates of antigenically Banna-like viruses were made from the sera of 24% of patients with an unidentified fever and viral encephalitis in Xinjiang province of northwest China in 1992. Twelve-segmented dsRNA viruses that are described as antigenically related to BAV have also been isolated from humans in various other provinces in China, including Beijing, Gansu, Hainan, Henan, and Shanshi and consequently BAV is now classified as a BSL3 arboviral agent.

Four viruses initially identified as JKT-6423, JKT-6969, JKT-7043, and JKT-7075, were isolated from *Culex* and *Anopheles* mosquitoes in central Java in Indonesia. The first three isolates were isolates of BAV, while JKT-7075 represents the only known strain of the *Kadipiro virus* species.

Recently, isolates of two distinct serotypes of *Liao ning virus* (LNV) were obtained from *Aedes dorsalis* mosquito in the Liaoning province of north eastern China.

Vectors, Host Range, and Transmission

BAV is the only seadornavirus that has been isolated from humans. Experimentally, seadornaviruses have also been shown to infect and replicate in adult mice and can be detected in the infected mouse blood from 3 to 5 days post infection. LNV replicates readily in several mammalian

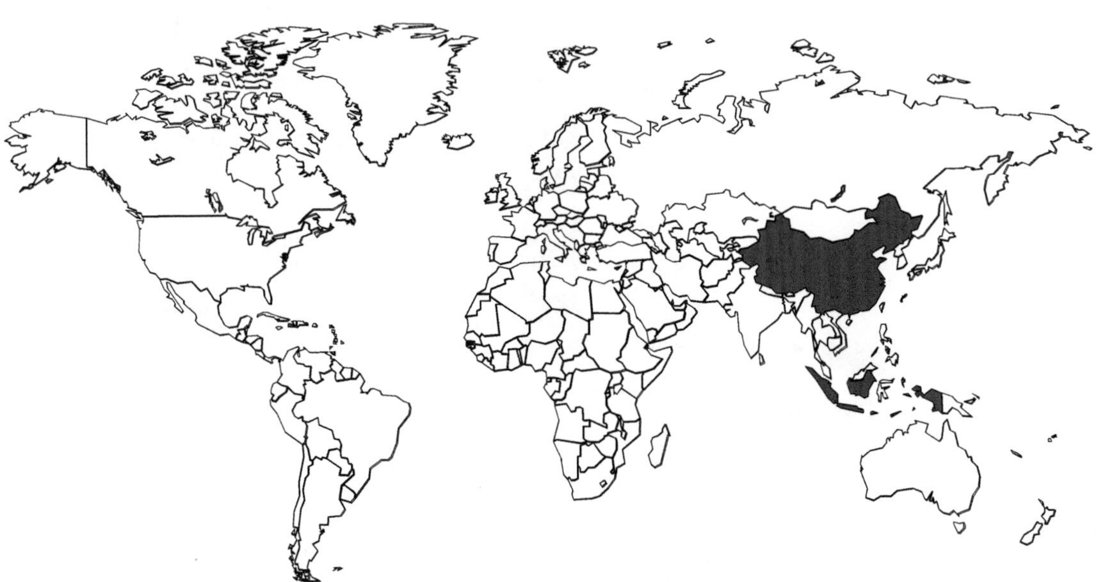

Figure 1 The known world distribution of seadornaviruses (shaded areas): BAV and LNV in China and BAV and KDV in Indonesia.

cell lines with massive lytic effect and kills adult mice. KDV and LNV have only been isolated only from mosquitoes. Indeed seadornaviruses are thought to be transmitted by mosquitoes and have been isolated from *Culex vishnui, Culex fuscocephalus, Anopheles vagus, Anopheles aconitus, Anopheles subpictus*, and *Aedes dorsalis.*

BAV, KDV, and LNV all occur in tropical and subtropical regions, where other mosquito-borne viral diseases, including Japanese encephalitis (JEV) and dengue, are endemic. This suggests that conclusive diagnosis and differentiation of these diseases may be difficult, based on clinical signs alone. Indeed several cases were diagnosed as JE in China (Henan province, Naijing city, and Fujian province in summer–autumn season of 1994), despite the absence of detectable JEV, or JEV-specific antibodies. However, a proportion of these patients (7/98) did show a 4- to16-fold rise in anti-BAV immunoglobulin G (IgG) by enzyme-linked immunosorbent assay (ELISA), indicating a BAV infection. A further 130 sera (from 1141 patients diagnosed with JE or viral encephalitis) from Chinese health institutes were also positive for BAV-specific immunoglobulin M (IgM).

In the summer of 1987, 16 BAV-like isolates were recovered from sera of 53 cattle (30% of the tested animals) and 4 isolates recovered from sera of 13 pigs (30% of the tested animals) collected in a slaughterhouse in the Dai nationality autonomous prefecture of the Xishuangbanna area of Yunnan province, where BAV has been previously isolated from humans.

BAV is regarded as 'a common cause of viral encephalitis and fever in humans during summer–autumn in China. Its high isolation rate and wide geographical distribution make BAV an important public health problem in China. It is likely that seadornaviruses are also endemic in other countries within Southeast Asia.

Virion Properties, Genome, and Replication

Seadornavirus particles, like those of many of the 'unturreted' reoviruses, are approximately spherical in appearance with icosahedral symmetry. The intact virion is 60–70 nm in diameter and has three concentric capsid layers, containing a total of seven structural proteins. The double-layered core-particle contains five proteins (VP1, VP2, VP3, VP8, and VP10) and is approximately 50 nm in diameter, with a central cavity that contains the 12 segments of the viral genome.

VP1 (the polymerase) and VP3 (the capping enzyme) form the transcriptase complex that is present within the core as a dimer of VP1 and a monomer of VP3. This complex is located at each of the 12 fivefold axes of the viral particle. The subcore layer is composed of 120 copies of VP2 and represents the $T=2$ layer. The outer layer of the core is made of 780 copies of VP8 and represents the $T=13$ layer. Based on sequence comparison to other viruses of the *Reoviridae* and in particular the rotaviruses, the seadornavirus VP10 proteins are thought to be anchored at the core surface. The outer capsid layer, which is composed of only two proteins (VP4 and VP9) (**Figure 2**), completely surrounds the core particle.

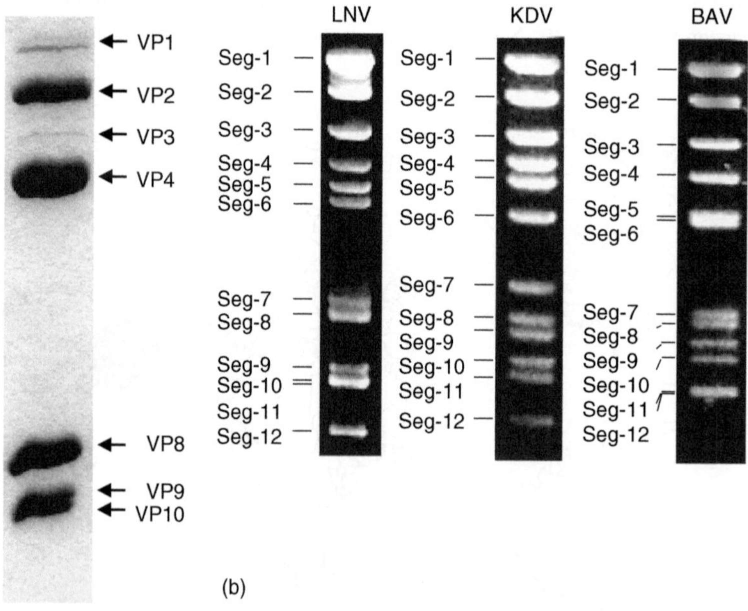

Figure 2 The PAGE profile of the structural proteins of Banna virus and the AGE profile of the genomes of seadornaviruses. (a) The PAGE profile of the structural proteins of Banna virus separated on a 10% polyacrylamide gel: VP4 and VP9 are the outer coat proteins. (b) The AGE profile of the genomes of Liao ning virus (LNV), Kadipiro virus (KDV), and Banna virus (BAV) run in a 1% agarose gel. The distinct electropherotypes of the three species allow to distinguish them.

The BAV genome encodes five nonstructural proteins which are the VP5, VP6, VP7, VP11, and VP12. VP7 contains sequences which are found in various protein kinases. VP12 contains a dsRNA-binding domain and might play a role in circumventing cell defense mechanisms, possibly acting as an anti-PKR component.

Negative staining and electron microscopy have shown that the surface of the virus particles has a well-defined capsomeric structure and icosahedral symmetry. The outer surface also has spikes that are similar to those of the rotaviruses (**Figure 3(a)**). The virus core has a smooth outline that is typical of the nonturreted members of the family *Reoviridae* (**Figure 3(b)**).

Like the orbiviruses, the seadornaviruses are unstable in CsCl and readily lose their outer coat proteins. However, intact virus particles have been purified from infected tissue culture supernatant by ultracentrifugation on colloidal silica gradients (Percoll). Purified virus particles are stable at 4 °C and approximately pH 7.0 for up to 3–4 months. However, increased acidity decreases their infectivity, which is abolished at pH 3.0. The virus is more stable as an unpurified cell culture lysate at 4 °C, which is a convenient way for medium-term storage. For longer periods of storage, viruses are stable at −80 °C, and infectivity can be further conserved by addition of 50% fetal calf serum.

The infectivity of seadornaviruses is decreased considerably by heating to 55 °C. Treatment with sodium dodecyl sulfate (SDS) disrupts the virus particle and consequently abolishes infectivity completely. Infectivity is very significantly reduced by treatment with butanol, although organic solvents such as Freon 113 or Vetrel XF do not affect virion structure or infectivity and have been used successfully during purification of virus particles from cell lysates.

During replication, core-like particles are found within vacuole-like structures in the infected cell cytoplasm (**Figure 3(c)**), and are thought to be involved in virion morphogenesis. These structures appear similar to those found in cells infected with rotaviruses, where progeny particles bud into the endoplasmic reticulum, acquiring the outer coat components and generating mature virus particles.

Seadornaviruses replicate in a number of mosquito cell lines, including C6/36 and AA23 (both from *Aedes albopictus*), A20 (*Aedes aegypti*), and Aw-albus (*Aedes w albus*). Over 40% of the progeny virus particles are liberated in the

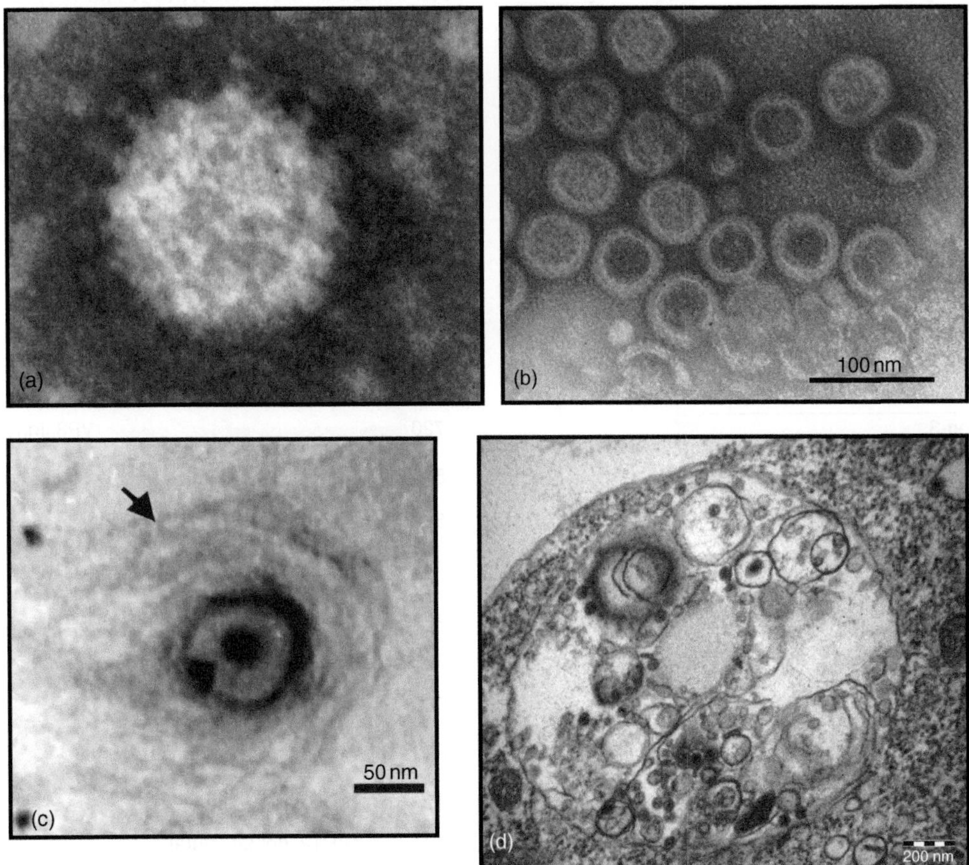

Figure 3 Electron micrographs of Banna virus particles. (a) A whole particle with spike proteins at the surface; (b) double-layered purified cores with smooth outlines; (c) particles surrounded by temporary envelope structure; and (d) thin section of infected C6/36 cells with virus particles seen within vacuole structures.

culture medium (as determined by titration), prior to cell death and massive cytopathic effect (CPE) (fusiform cells), which (with BAV, LNV and KDV) is evident at 48–72 h post infection. Infected cells do not lyse initially and the virus is released by budding, acquiring a temporary membrane envelope in the process (**Figure 3(d)**). However, cell lysis does occur late in infection, as a result of cell death. Intracellular radiolabeling of viral polypeptides has shown that label is incorporated predominantly into viral polypeptides, even in absence of inhibitors of DNA replication such as actinomycin D, demonstrating 'shut off' of host-cell protein synthesis.

Like the other members of the genus, LNV can replicate in several mosquito cell lines, but it is the only seadornavirus that readily replicates in a variety of transformed or primary mammalian cells, including BHK-21, Vero, BGM, Hep-2, and MRC-5, leading to massive cell lysis after 48 h post infection.

The seadornavirus genome consists of 12 segments of dsRNA that are identified as Seg-1 to Seg-12 in the order of decreasing molecular weight and order of migration during agarose gel electrophoresis (AGE). The full-length genome sequence of BAV, KDV, and LNV has been determined. The genome comprises approximately 21 000 bp and the segment length ranges between 3747 and 862 bp. The genomic RNA of BAV and LNV show 6–6 electrophoretic profiles in 1% agarose gel electrophoresis (**Figure 2**). The genome of KDV migrates in 6–5-1 profile. Each genome segment encodes a single protein with an open reading frame spanning almost the whole length of the segment (**Figure 4**).

Antigenic and Phylogenetic Relationships between Seadornaviruses

The earliest data that was generated concerning the relationships between different seadornaviruses were obtained by RNA cross-hybridization analyses, which identified BAV and KDV as members of distinct genogroups. Antigenic relationships between different seadornaviruses were also investigated using polyclonal mouse immune-sera. BAV from southern China and Indonesia, LNV from northeast of China, and KDV from Indonesia, which are classified as distinct species, show no cross-reaction in neutralization tests. Subsequent comparisons of nucleotide and amino acid sequences confirmed that BAV, KDV, and LNV represent three distinct virus species, with AA identities between homologous proteins of 24–42% (**Table 2**). A neighbor-joining tree based on an alignment of the most highly conserved seadornavirus protein, VP1 (the viral polymerase), shows BAV, KDV, and LNV as three distinct phylogenetic groups, confirming their status as distinct species (**Figure 5(a)**).

Antigenic variation between different BAV strains was also investigated using immune-sera, and identified two distinct serotypes, that did not cross-neutralize. These have been named as BAV serotypes A and B. A seroneutralization epitope was identified on the outer coat protein VP9. Sequence analysis of Seg-9 showed only 40% amino acid identity between BAV serotypes A and B, identifying two genotypes: genotype A (represented by isolates BAV-Ch (China) and BAV-In6423 (Indonesia)); and genotype

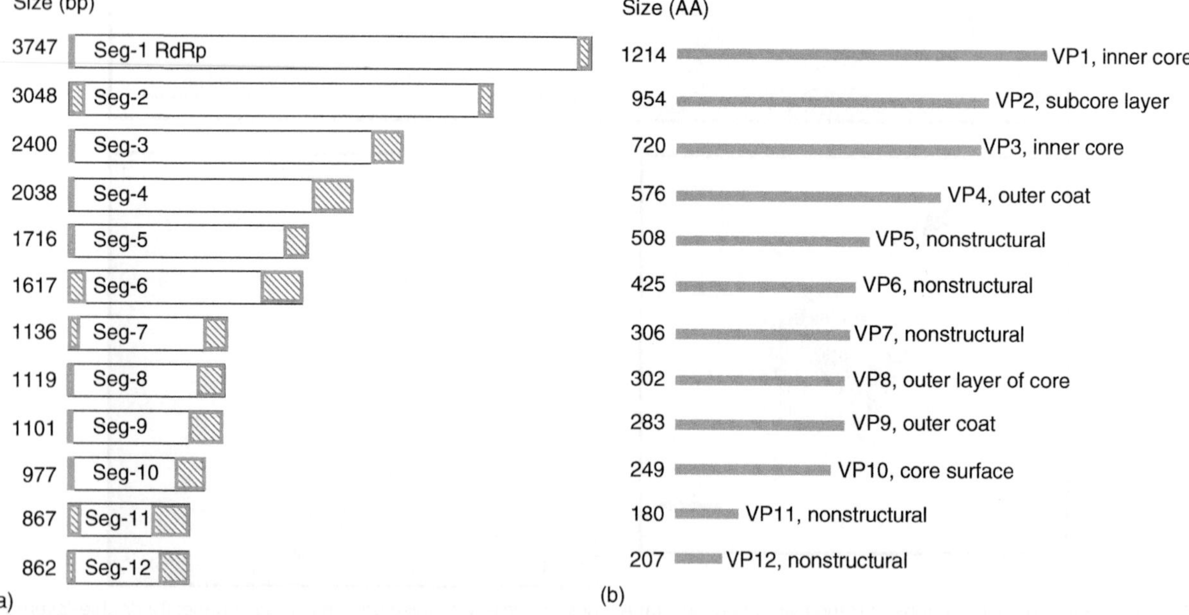

Figure 4 (a) Organization of the genome segments of BAV and (b) their putative encoded proteins. Shaded areas in (a): 5′ and 3′ NCRs. See www.iah.bbsrc.ac.uk/dsRNA_virus_proteins/Bannavirus-Proteins.htm.

Table 2 Correspondence between seadornavirus genes (relative to the type species Banna virus), putative functions, and protein copy numbers per intact virus particle

BAV	*LNV (%AA identity with BAV)*	*KDV (%AA identity with BAV)*	*Function (location)*	*Protein copy number per particle*
Seg-1 (VP1, Pol)	Seg-1 (VP1, Pol) (41)	Seg-1 (VP1, Pol) (42)	RNA-dependent RNA polymerase (core)	24
Seg-2 (VP2, T2)	Seg-2 (VP2, T2) (27)	Seg-2 (VP2, T2) (33)	$T = 2$ protein, nucleotide binding (core)	120
Seg-3 (VP3, Cap)	Seg-3 (VP3, Cap) (37)	Seg-3 (VP3, Cap) (38)	Capping enzyme (core)	12
Seg-4 (VP4)	Seg-4 (VP4) (32)	Seg-4 (VP4) (32)	(Outer coat)	~330
Seg-5 (VP5)	Seg-6 (VP6) (26)	Seg-6 (VP6) (27)	(Nonstructural)	0
Seg-6 (VP6)	Seg-5 (VP5) (26)	Seg-5 (VP5) (26)	NTPase (nonstructural)	0
Seg-7 (VP7)	Seg-7 (VP7) (27)	Seg-7 (VP7) (29)	Protein kinase (nonstructural)	0
Seg-8 (VP8, T13)	Seg-8 (VP8, T13) (23)	Seg-9 (VP9, T13) (21)	(Core)	780
Seg-9 (VP9)	Seg-10 (VP10) (18)	Seg-11 (VP11) (23)	Cell attachment (outer coat)	~310
Seg-10 (VP10)	Seg-9 (VP9) (24)	Seg-10 (VP10) (24)	(Core)	~260
Seg-11 (VP11)	Seg-12 (VP12) (37)	Seg-12 (VP12) (34)	(Nonstructural)	0
Seg-12 (VP12)	Seg-11 (VP12) (35)	Seg-8 (VP8) (26)	dsRNA-binding (nonstructural)	0

BAV, Banna virus, KDV, Kadipiro virus, and LNV, Liao ning virus. The copy number was determined by radiolabeling of the proteins using ^{35}S methionine. The encoded putative proteins are indicated between brackets followed by the nomenclature that is used in the *Eighth Report of the ICTV* and presented on the website www.iah.bbsrc.ac.uk/dsRNA_virus_proteins/protein-comparison.htm.

B (represented by isolates BAV-In6969 and BAV-In7043 (Indonesia)) (which match with virus serotypes A and B).

Seg-7, which codes for nonstructural protein VP7, also shows a significant variation, between the genotypes/serotypes, with only 70% amino-acid-sequence identity between types A and B. The amino-acid-sequence identity (VP1–VP12) between different BAV isolates within a single serotype ranged from 83% to 100%, while sequence identity between different serotypes ranged from 73% to 95% (excluding VP9). This level of genetic divergence is comparable to that observed between serotypes in other insect-transmitted arboviruses of family *Reoviridae*, in particular the orbiviruses. The two identified serotypes of LNV have nucleotide sequence identities ranging between 81% and 90%, while amino acid identity ranging between 80% and 96%. The lowest AA identity (80%) was found in the proteins of the outer capsid VP10 (the homolog of BAV VP9), which defines virus serotype in the seadornaviruses. A neighbor-joining tree, comparing the sequences of cell-attachment outer capsid proteins of BAV (VP9), KDV (VP11), and LNV (VP10) clearly identifies the two serotypes of BAV (A and B) and distinguishes the A and B serotypes of LNV (**Figure 5(b)**).

All 12 segments of BAV, KDV, and LNV have conserved sequences, located at their 3′ and 5′ termini. The motifs $5'\text{-GUAU}^{A}/_{U}{}^{A}/_{U}\text{A}^{A}/_{U}{}^{A}/_{U}{}^{A}/_{U}\text{-}3'$ and $5'\text{-C}^{C}/_{U}\text{GAC-}3'$ (+ve strand) were found in the 5′ noncoding region (NCR) and the 3′ NCR of BAV, respectively, while in KDV, these motifs were $5'\text{-GUAGAA}^{A}/_{U}{}^{A}/_{U}{}^{A}/_{U}{}^{A}/_{U}\text{-}3'$ and $5'\text{-}{}^{C}/_{U}\text{GAC-}3'$ (+ve strand) and in LNV they were $5'\text{-GUUAU}^{A}/_{U}{}^{A}/_{U}{}^{A}/_{U}\text{-}3'$ and $5'\text{-}{}^{C}/_{U}{}^{C}/_{U}\text{GAC-}3'$ (+ve strand), respectively. For the BAV and KDV viruses, the 5′ and 3′ terminal trinucleotides of all segments are inverted complements. LNV has a difference in position 3 of the 5′ ends and therefore only the 5′ and 3′ terminal dinucleotides of all segments are inverted complements.

The genetic relation between BAV, LNV, and KDV is reflected in their morphological characteristics. Electron microscopic analysis showed that the three viruses are morphologically identical.

Within the family *Reoviridae*, the most conserved gene between different genera encodes the RNA-dependent RNA polymerase (RdRp). Values for amino acid identity of lower than 30% can be used to distinguish members of distinct genera. A tree constructed from the alignment of polymerase sequences of representative members of the family *Reoviridae* (**Table 3**) is shown in **Figure 6**. Calculations of amino-acid-identity values based on this alignment show that the coltivirus polymerases exhibit a maximum of 15% identity with those of the seadornaviruses, confirming the status of *Coltivirus* and *Seadornavirus* as a distinct genera. In contrast amino acid identities as high as 28% were detected between the seadornaviruses and the rotaviruses (members of a distinct genus of 11-segmented viruses also within the family *Reoviridae*).

Evolutionary Relationships between BAV and Rotaviruses

A sequence comparison of the structural proteins of BAV (VP1, VP2, VP3, VP4, VP8, VP9, and VP10) to those of other members of the *Reoviridae*, have shown similarities between VP9 and VP10 of BAV and the VP8* and VP5* subunits of the outer coat protein VP4 of rotavirus A, with amino acid identities of 21% and 26%, respectively.

This suggests that an evolutionary jump has occurred between the two genera, involving the expression of two BAV proteins (VP9 and VP10) from two separate genome segments rather than the generation of VP8* and VP5* by the proteolytic cleavage of a single gene product (as seen in the cleavage of rotavirus VP4). VP3 of BAV, which is the guanylyltransferase of the virus also exhibits significant identity (21%, AA identity) with the VP3 (guanylyltransferase) of rotavirus.

Structural Features and Relation to Rotaviruses

Electron micrographs of purified, intact BAV particles show striking similarities to those of the rotaviruses. Numerous protein spikes were observed on the surface of BAV, reminiscent of those of rotaviruses. This morphological resemblance has not previously been reported between rotaviruses and other members of family *Reoviridae*. The evolutionary relationship between the seadornaviruses and rotaviruses was further confirmed when the atomic structure of BAV VP9 was determined by X-ray crystallography. VP9 is a trimeric molecule that is held together by an N-terminal helical bundle (**Figure** 7). The N-terminal tail of the BAV VP9 monomer is reminiscent of the coiled-coils structures of the HIV gp41 protein, and the carboxy terminal of the VP5* of rotavirus (**Figure** 7). In contrast the monomer of BAV VP9 has a head domain made mainly of β-sheets (**Figure** 7), which shows significant structural similarities to rotavirus VP8*. However, VP8* does have a sialic acid-binding domain, which is absent from the head

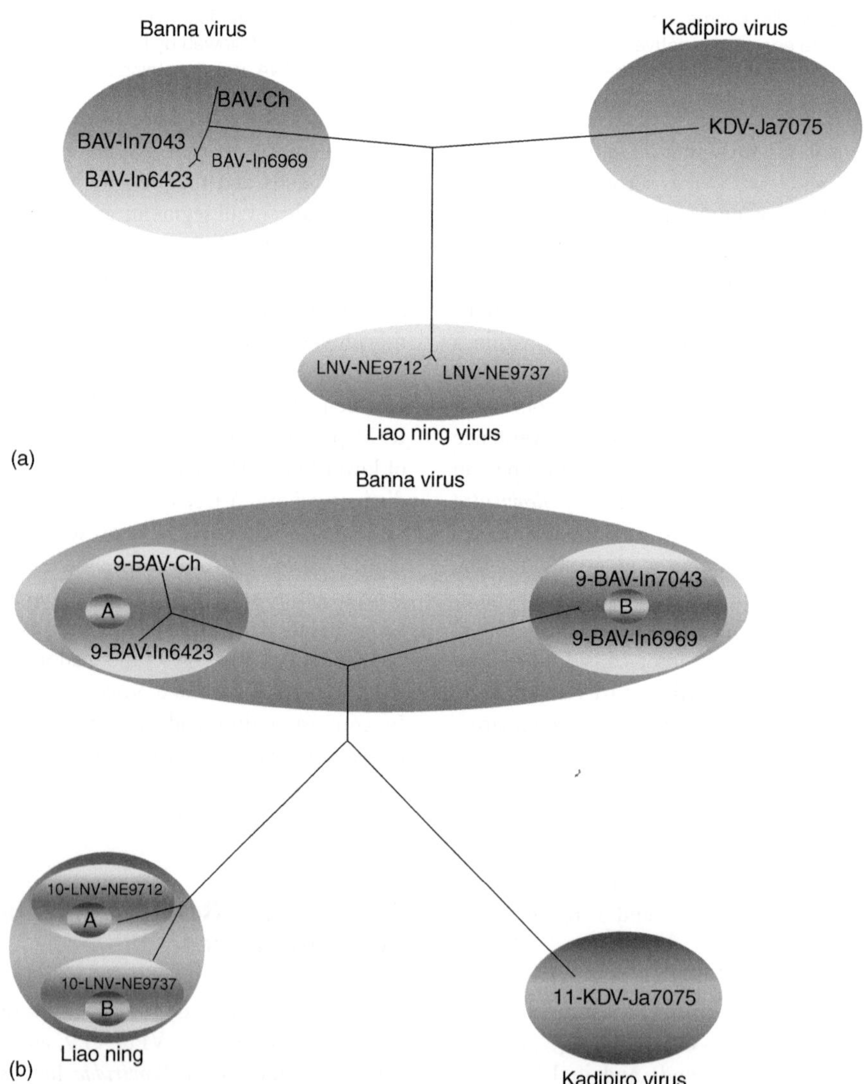

Figure 5 Phylogenetic tree for seadornaviruses based on (a) the polymerase gene and (b) on the cell-attachment (and seroneutralization) protein. The phylogenetic comparison based on the cell-attachment (and seroneutralization) protein distinguishes the the two genotypes (serotypes) of BAV and also those of LNV.

Table 3 Sequences of RdRp used in phylogenetic analysis (**Figure 6**) of various members of family *Reoviridae*

Species	*Isolate*	*Abbreviation*	*Accession number*
Genus *Seadornavirus* (12 segments)			
Banna virus	Ch	BAV-Ch	AF168005
Kadipiro virus	Java-7075	KDV-Ja7075	AF133429
Liao ning viurus	LNV-NE9712	LNV-NE9712	AY701339
Genus *Coltivirus* (12 segments)			
Colorado tick fever virus	Florio	CTFV-Fl	AF134529
Eyach virus	Fr578	EYAV-Fr578	AF282467
Genus *Orthoreovirus* (10 segments)			
Mammalian orthoreovirus	Lang strain	MRV-1	M24734
	Jones strain	MRV-2	M31057
	Dearing strain	MRV-3	M31058
Genus *Orbivirus* (10 segments)			
African horse sickness virus	Serotype 9	AHSV-9	U94887
Bluetongue virus	Serotype 2	BTV-2	L20508
	Serotype 10	BTV-10	X12819
	Serotype 11	BTV-11	L20445
	Serotype 13	BTV-13	L20446
	Serotype 17	BTV-17	L20447
Palyam virus	Chuzan	CHUV	Baa76549
St. Croix river virus	SCRV	SCRV	AF133431
Genus *Rotavirus* (11 segments)			
Rotavirus A	Bovine strain UK	BoRV-A/UK	X55444
	Simian strain SA11	SiRV-A/SA11	AF015955
Rotavirus B	Human/murine strain IDIR	Hu/MuRV-B/IDIR	M97203
Rotavirus C	Porcine Cowden strain	PoRV-C/Co	M74216
Genus *Aquareovirus* (11 segments)			
Golden shiner reovirus	GSRV	GSRV	AF403399
Grass Carp reovirus	GCRV-873	GCRV	AF260511
Chum salmon reovirus	CSRV	CSRV	AF418295
Striped bass reovirus	SBRV	SBRV	AF450318
Genus *Fijivirus* (10 segments)			
Nilaparvata lugens reovirus	Izumo strain	NLRV-Iz	D49693
Genus *Phytoreovirus* (10 segments)			
Rice dwarf virus	Isolate China	RDV-Ch	U73201
	Isolate H	RDV-H	D10222
	Isolate A	RDV-A	D90198
Genus *Oryzavirus* (10 segments)			
Rice ragged stunt virus	Thai strain	RRSV-Th	U66714
Genus *Cypovirus* (10 segments)			
Bombyx mori cytoplasmic polyhedrosis virus 1	Strain I	BmCPV-1	AF323782
Dendrlymus punctatus cytoplasmic polyhedrosis 1	DsCPV-1	DsCPV-1	AAN46860
Lymantria dispar cytoplasmic polyhedrosis 14	LdCPV-14	LdCPV-114	AAK73087
Genus *Mycoreovirus* (11 or 12 segments)			
Rosellinia anti-rot virus	W370	RaRV	AB102674
Cryphonectria parasitica reovirus	9B21	CPRV	AY277888
Genus Mimoreovirus (11 segments)			
Micromonas pusilla reovirus	MPRV	MPRV	DQ126102
Genus Dinovernavirus (9 segments)			
Aedes pseudoscutellaris reovirus	ApRV	ApRV	DQ087277
Genus Cardoreovirus (12 segments)			
Eriocheir sinensis reovirus	Isolate 905	EsRV	AY542965

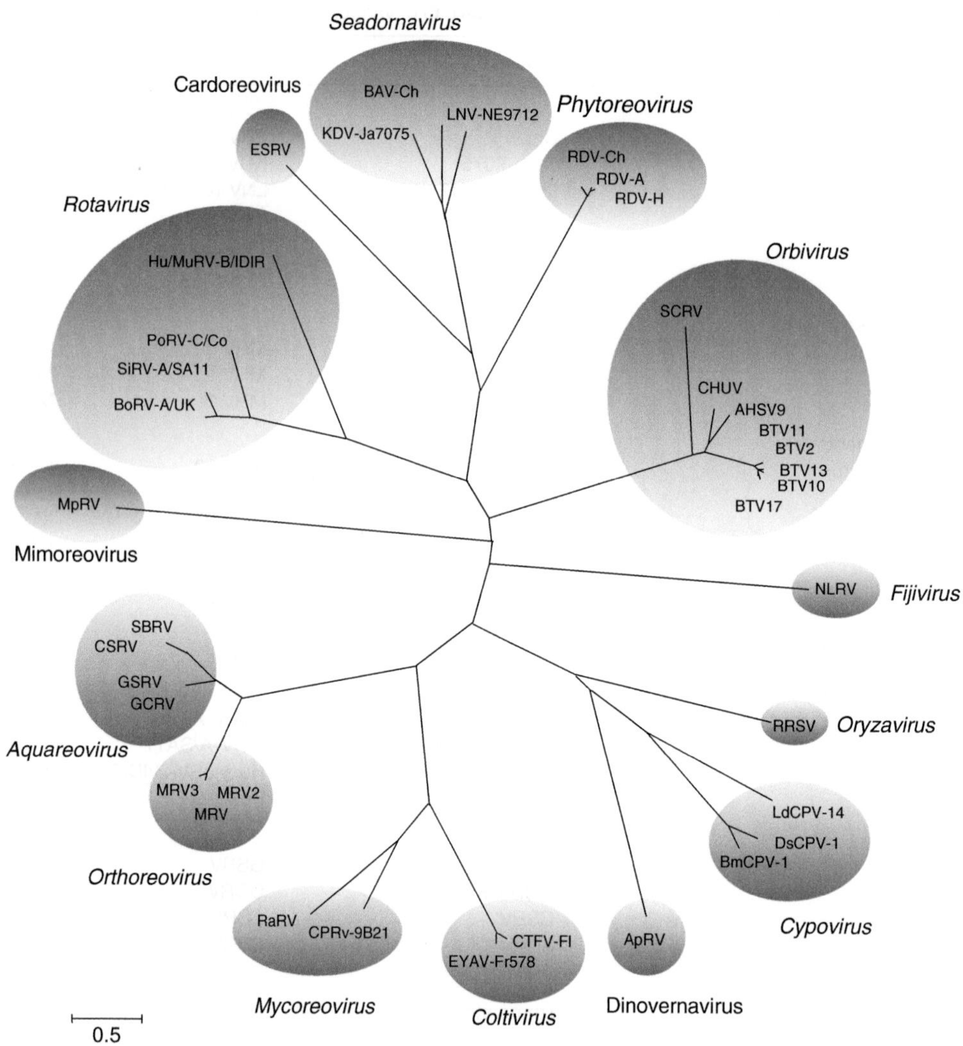

Figure 6 Phylogenetic relations among members of the *Reoviridae* based on the sequences of their putative RNA-dependent RNA polymerases (RdRps). Neighbor-joining phylogenetic tree built with available polymerase sequences (using the Poisson correction or gamma-distribution algorithms) for representative members of 12 recognized and 3 proposed genera of family *Reoviridae*. The abbreviations and accession numbers are those provided in **Table 3**.

domain of the VP9 of BAV. This might explain why rotavirus infectivity can be decreased when cells are pretreated with sialidases although BAV infectivity, to C6/36 mosquito cells, is not altered by a similar treatment.

Functional Studies

Expressed BAV VP9 was used in competition assays with intact virus particles, in an attempt to decrease their infectivity, as demonstrated with the outer capsid proteins of some other reoviruses (e.g., the avian orthoreoviruses). Surprisingly, however, pretreatment of cells with VP9 increased the infectivity of BAV by 10–100 times. VP9 is thought to act as a membrane fusion protein and the N-terminal coiled-coils may be responsible for such an activity. This may explain why soluble trimeric VP9 can increase viral infectivity. VP9 trimers not only bind receptor but also initiate endocytosis, perhaps by receptor oligomerization. The initiation of endocytosis facilitates penetration by virus particles present at or near the cell surface and thereby increases their infectivity.

BAV VP9 also carries the virus seroneutralization epitopes. Anti-VP9 antibodies are highly neutralizing and define the two serotypes A and B of BAV. Animals immunized with VP9 failed to replicate the virus when challenged with the homologous type. Animals immunized with VP9 of the heterologous serotype showed unaltered virus replication and viremia. Similar results were obtained in cell cultures. VP9 could therefore be used as a vaccine subunit for

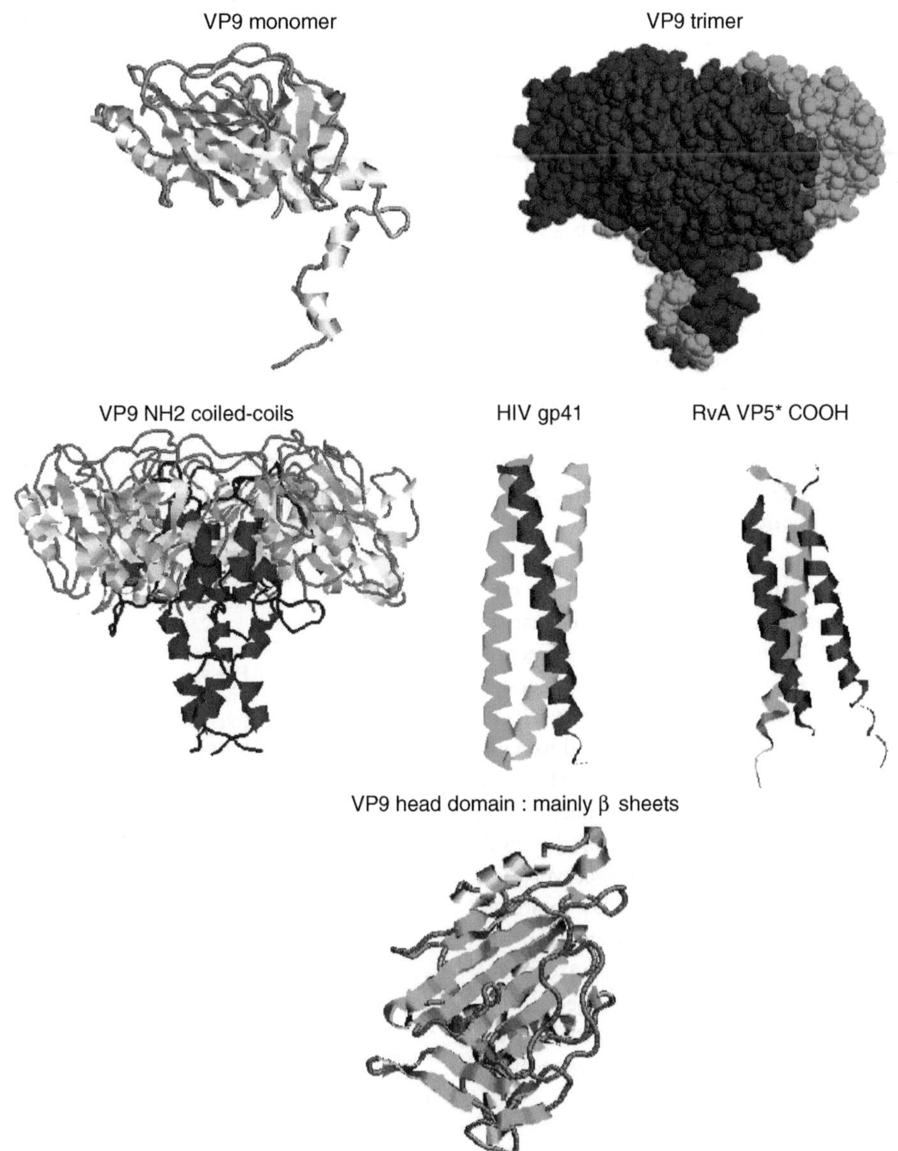

Figure 7 The structure of the VP9 outer capsid protein of Banna virus determined by X-ray crystallography at a resolution of 2.6 angstrom. The NH2 terminal of the BAV VP9 is organized in the form of coiled-coils that are similar to those found in fusion active proteins such as HIV gp41 and VP5* of rotavirus. RvA, Rotavirus A.

immunization. Sequence analysis of BAV VP3 suggests that it is the viral capping enzyme (CaP).

This has been confirmed experimentally by incubating virus particles with α-^{32}P GTP. VP3 was covalently labeled with α-^{32}P derived from GTP. The reaction was found to be dependent on the presence of divalent ions such as magnesium. The recombinant VP3 also covalently bound α-^{32}P GTP and exhibited an inorganic pyrophosphatase activity, possibly as a detoxifying mechanism which removes the inorganic pyrophosphate accumulated during RNA and Cap structure synthesis by the virus-core-associated enzymes. Similar properties were described for the capping enzyme VP4 of Bluetongue virus (genus *Orbivirus*).

Clinical Features and Diagnostic Assay for Seadornaviruses

As mentioned earlier, the only seadornavirus that has been isolated from humans and associated with human disease (to date) is BAV. Humans infected with BAV develop flu-like manifestation, myalgia, arthralgia, fever, and encephalitis. Reports of children born to infected mothers (as revealed by serological assays) show various manifestations including cerebral palsy, hemiplegia, delay of development, and viral encephalitis.

A diagnostic serological assay was developed for different serotype of BAV, based on outer coat protein VP9 (which is responsible for cell-attachment and sero-neutralization).

Molecular diagnostic assays have also been developed for BAV and KDV. These are RT-PCR-based assays, which have been validated using an infected murine model in which viral RNA could be detected in blood as early as 3 days post infection. These RT-PCR assays can also be used to distinguish genotypes A and B of BAV, based on distinct amplicons of different lengths, that are obtained by specific primers at different locations in Seg-9. A quantitative PCR assay has also been developed for BAV based on the sequence of Seg-10. A standard RT-PCR assay was recently developed for LNV based on the sequence of Seg-12, which can be used to detect viral RNA in infected mouse blood.

Severity of Infection in Animal Model and Diversification of Genome Sequence after Replication in Permissive Animals

Banna, Kadipiro, and Liao ning viruses all replicate in mice after intraperitoneal injection. The viral genome can be detected in mouse blood 3 days until 5–7 days post injection. Clearance of the viremia is accompanied by the appearance of antibodies. When animals were injected on a second occasion with the same virus, BAV and KDV failed to replicate, demonstrating the immune status of the mice. However, LNV replicated in the immunized mice causing death with a severe hemorrhagic syndrome. An analysis of the PCR amplicons from the infected mice blood showed very few changes in the sequences of the BAV and KDV genomes. However, there was considerable diversification in the genome of LNV, as demonstrated by analyses of the sequence of Seg-12, leading to formation of a diverse quasispecies. This might explain why LNV can continue to replicate in the immunized mice, as diversification of the sequence might help the virus to 'escape' the immune system.

Treatment and Immunity

There is no specific treatment for BAV infection. Patients infected with BAV have shown a fourfold rise in the anti-BAV antibody titers in paired sera tested by ELISA, showing that an immune response is developed to the virus infection. Mice experimentally infected with BAV developed viremia. The clearance of the virus from the blood circulation occurred concomitantly with the appearance of anti-BAV antibodies. Based on the mouse model, immunization with a recombinant expressed VP9 is likely to provide a significant level of protection against an initial infection with the homologous virus type. Immunization with VP9 could therefore provide a significant level of protection in high-risk areas and it might form the basis for an effective subunit vaccine.

Further Reading

Attoui H, Billoir F, Biagini P, *et al.* (2000) Complete sequence determination and genetic analysis of Banna virus and Kadipiro virus: Proposal for assignment to a new genus (*Seadornavirus*) within the family *Reoviridae*. *Journal of General Virology* 81: 1507–1515.

Attoui H, Charrel R, Billoir F, *et al.* (1998) Comparative sequence analysis of American, European and Asian isolates of viruses in the genus *Coltivirus*. *Journal of General Virology* 79: 2481–2489.

Attoui H, de Lamballerie X, and Mertens PPC (2005) *Coltivirus, Reoviridae*. In: Fauquet CM, Mayo MA, Maniloff J, Desselberger U and Ball LA (eds.) *Virus Taxonomy: Eighth Report of the International Committee on Taxonomy of Viruses*, pp. 497–503. San Diego, CA: Elsevier Academic Press.

Attoui H, de Lamballerie X, and Mertens PPC (2005) *Seadornavirus, Reoviridae*. In: Fauquet CM, Mayo MA, Maniloff J, Desselberger U and Ball LA (eds.) *Virus Taxonomy: Eighth Report of the International Committee on Taxonomy of Viruses*, pp. 504–510. San Diego, CA: Elsevier Academic Press.

Billoir F, Attoui H, Simon S, *et al.* (1999) Molecular diagnosis of group B coltiviruses infections. *Journal of Virological Methods* 81: 39–45.

Brown SE, Gorman M, Tesh RB, and Knudson DL (1993) Coltiviruses isolated from mosquitoes collected in Indonesia. *Virology* 196: 363–367.

Chen BQ and Tao SJ (1996) Arbovirus survey in China in recent ten years. *Chinese Medical Journal (Engl)* 109: 13–15.

Mertens PPC, Attoui H and Bamford DH (eds.) (2002) Identification of comparable proteins of the dsRNA viruses. In: *The RNAs and Proteins of dsRNA Viruses*. www.iah.bbsrc.ac.uk/dsRNA_virus_proteins/protein-comparison.htm (accessed May 2007).

Mertens PPC, Duncan R, Attoui H, and Dermody TS (2005) *Reoviridae*. In: Fauquet CM, Mayo MA, Maniloff J, Desselberger U and Ball LA (eds.) *Virus Taxonomy: Eighth Report of the International Committee on Taxonomy of Viruses*, pp. 447–454. San Diego, CA: Elsevier Academic Press.

Mohd Jaafar F, Attoui H, Bahar MW, *et al.* (2005) The structure and function of the outer coat protein VP9 of Banna virus. *Structure* 13: 17–29.

Mohd Jaafar F, Attoui H, Gallian P, *et al.* (2004) Recombinant VP9-based enzyme-linked immunosorbent assay for detection of immunoglobulin G antibodies to Banna virus (genus *Seadornavirus*). *Journal of Virological Methods* 116: 55–61.

Mohd Jaafar F, Attoui H, Mertens PPC, *et al.* (2005) Structural organisation of a human encephalitic isolate of Banna virus (genus *Seadornavirus*, family *Reoviridae*). *Journal of General Virology* 86: 1141–1146.

Tao SJ and Chen BQ (2005) Studies of coltivirus in China. *Chinese Medical Journal (Engl)* 118: 581–586.

Relevant Website

http://www.iah.bbsrc.ac.uk – The RNAs and Proteins of dsRNA Viruses.

Tetraviruses

J A Speir and J E Johnson, The Scripps Research Institute, La Jolla, CA, USA

Introduction

The *Tetraviridae* are a family of viruses with nonenveloped $T=4$ particles that package single-stranded, positive-sense RNA genomes and infect only a single order of insects, the Lepidoptera (moths and butterflies). This family stands out from the rest of the known viruses based on these properties alone. It is the only RNA virus family with a host range restricted to insects and the only family with nonenveloped icosahedral $T=4$ particles. The unusual symmetry of their capsids has become the basis for their family name (from the Greek *tettares*, four).

The first report of a larvae disease later to be attributed to a tetravirus was from South Africa in 1941. Researchers Tooke and Hubbard described a devastating annual mortality of the emperor pine moth, *Nudaurelia cytherea capensis*, such that the dead larvae formed a carpet beneath the pine trees (*Pinus radiata*). The outbreaks have now reduced the larvae population on the pine plantations to a great extent. Five different viruses were found in the infected larvae that were named nudaurelia α, β, γ, δ, and ε, with β being the most abundant. The nudaurelia β virus (NβV) was extensively studied and became the type member of a new virus group called the nudaurelia β-like viruses (betatetraviruses) that was officially recognized in 1982. These studies showed NβV had a monopartite genome and the first known $T=4$ icosahedral capsid. The most recently characterized betatetravirus is from the USA, named Providence virus, which was discovered by Pringle, Ball, and colleagues in 2003. Importantly, this is the first tetravirus that replicates in cell culture, greatly aiding the study of tetravirus molecular biology (see below).

While the betatetraviruses were the first tetraviruses discovered and form the majority of the family, Hendry and colleagues isolated another virus from *Nudaurelia* larvae in 1985, the nudaurelia ω virus (NωV), which was physically similar to NβV but encapsidated a second genomic RNA. NωV became the type member for a second genus within the *Tetraviridae*, the ω-like genus (omegatetraviruses). A second omegatetravirus was isolated by Hanzlik and colleagues in 1993 from Australia, the helicoverpa armigera stunt virus (HaSV), and a third was discovered in 2005 by Yi and colleagues from the Yunnan Province of China, the dendrolimus punctatus tetravirus (DpTV). Both HaSV and DpTV are closely similar to NωV (see below). Currently, the family consists of three omegatetraviruses, and nine betatetraviruses, as well as eight other possible members (**Table 1**). The tetravirus family has grown little in recent years, with only a few new members identified in the last decade. This is due to their limited host range, the difficult field collections of the insect hosts due to the locations where they thrive, and the difficulty or impossibility of rearing the host insects in the laboratory. Also, several aspects of their life cycle remain poorly characterized due to their inability to grow in cell culture despite extensive efforts covering a wide variety of cell types. In spite of these drawbacks, studies of the tetraviruses have expanded to a great extent. It is possible to raise HaSV from laboratory-grown insects, Providence virus is the first tetravirus found to replicate in cell culture, and virus-like particles (VLPs) can be produced in a baculovirus expression system for NωV, HaSV, and TaV. These systems have allowed for advances in detailed biophysical descriptions of these capsids, including a crystal structure of NωV, and new data on replication, pathology, and the biotechnological potential of this unique virus family.

Genome Organization, Replication, and Capsid Assembly

General Characteristics of Tetravirus Genomic RNAs

The main characteristic separating the betatetraviruses and omegatetraviruses is the RNA encapsidated by the particles in each genus. All tetraviruses analyzed for RNA content share the presence of a large genomic RNA strand of about 5.3–6.5 kbp in length and a smaller genomic or subgenomic RNA strand of about 2.5 kbp. In the monopartite β-like viruses, both the replicase and capsid protein genes are on the large genomic RNA strand (~6.5 kbp). In the bipartite ω-like viruses, the replicase and capsid protein genes are split between the large (~5.3 kbp) and small (~2.5 kbp) genomic RNA strands, respectively.

Although the betatetraviruses are monopartite, the capsids often package two RNA species. A subgenomic mRNA of 2.5 kbp coding for the capsid protein is also encapsidated with the genomic RNA strand in NβV, PrV, TaV, and TnV. Thus, virus particles from both genera package two RNA molecules totaling ~7.8–9.0 kbp. Experimental evidence also suggests that the ω-like viruses package both genomic RNAs in a single particle, as is the case for the nodaviruses, the only other small spherical RNA animal viruses with a bipartite genome. This contrasts with plant viruses having multipartite genomes, which package single genomic RNA strands in different particles.

Table 1 Members of the *Tetraviridae*

Virus	*Acronym*	*Family of Lepidoptera host*	*Geographic location*
Betatetraviruses			
Nudaurelia capensis β virus[a]	NβV	Saturniidae	South Africa
Antheraea eucalypti virus[b]	AeV	Saturniidae	Australia
Dama trima virus	DtV	Limacodidae	Malaysia
Dasychira pudibunda virus[c]	DpV	Lymantriidae	UK
Philosamia cynthia X ricini virus	PxV	Saturniidae	UK
Providence virus	PrV	Noctuidae	USA
Pseudoplusia includens virus	PiV	Noctuidae	USA
Thosea asigna virus[d]	TaV	Limacodidae	Malaysia
Trichoplusia ni virus	TnV	Noctuidae	USA
Omegatetraviruses			
Nudaurelia capensis ω virus[a]	NwV	Saturniidae	South Africa
Dendrolimus punctatus tetravirus	DpTV	Lasiocampidae	China
Helicoverpa armigera stunt virus	HaSV	Noctuidae	Australia
Unassigned possible members[e]			
Acherontia atropas virus	AaV	Sphingidae	Canary Islands
Agraulis vanillae virus	AvV	Nymphalidae	Argentina
Callimorpha quadripuntata virus	CqV	Arctiidae	UK
Eucocytis meeki virus	EmV	Cocytiidae	Papua New Ginea
Euploea corea virus	EcV	Danadidae	Australia/Germany
Hypocritae jacobeae virus	HjV	Arctiidae	UK
Lymantria ninayi virus	LnV	Lymantriidae	Papua New Ginea
Nudaurelia ε virus[f]	NεV	Saturniidae	South Africa

[a]Type virus for genus.
[b]Serological evidence shows identity to NβV.
[c]Host renamed *Calliteara pudibunda.*
[d]Host renamed *Setothosea asigna*.
[e]Viruses showing serological relationship to a known betatetravirus (excluding NεV), but otherwise uncharacterized.
[f]NεV resembles the tetraviruses but is serologically unrelated to any known member.
Reproduced by permission of Elsevier.

Analysis of the HaSV genomic RNAs revealed 5′ cap structures and distinctive tRNA-like structures at the 3′ termini, which are not polyadenylated. A tRNA-like structure also exists at the 3′ terminal of the NβV RNA, and no β-like virus genome has been found to be polyadenylated. However, both the HaSV and NβV RNAs could be polyadenylated, indicating no terminal blockage. Therefore, all tetravirus genomic RNAs are likely to have unblocked, tRNA-like 3′ termini and 5′ caps, although the existence of the cap has not been experimentally tested with any β-like virus.

β-Like Genome Organization and Replication Model

The complete nucleotide sequence of the NβV genome (6625 nt) was elucidated in 1998, and partial cDNA clones of the TaV and PrV genomes have been analyzed more recently (2001 and 2003, respectively). All three show nearly identical organization. The single genomic RNAs are dicistronic and have overlapping open reading frames (ORFs) (**Figure 1**). The 5′ ends encode the replicase and the 3′ ends encode the capsid protein genes, which have extensive overlap with the replicase. The NβV replicase ORF contains 5778 nt encoding 1925 amino acids with a calculated molecular weight of 215 kDa. It contains the three domains characteristic of viruses in the α-like RNA virus superfamily (the N-terminal methyltransferase domain, the nucleotide-binding or helicase domain, and the RNA-dependent RNA polymerase (RdRp) domain), but lacks the motifs for the papainlike protease and proline-rich hinge domains that are implicated in autoprocessing of the replicase proteins of some +RNA viruses of vertebrates. Finally, the genomic RNA appears to be the mRNA for the replicase, since its initiation codon is the first located at the 5′ end of the genomic RNA. Translation studies of other β-like virus genomic RNAs have also produced large polypeptides, supporting this scenario.

The sequence of the putative TaV replicase shows a novel permutation among RNA viruses, and together with birnaviruses and an unassigned possible tetravirus (Euprosterna elaeasa virus) defines a unique, ancient lineage of RdRp's. None of the other sequenced tetravirus replicases have a permutation. The TaV replicase lacks both the methyltransferase and helicase domains and the RdRp has little homology with other RNA virus replicases, which as a group have a universal sequence conservation. In 2002, Gorbalenya and colleagues discovered that the catalytic palm subdomain of the TaV RdRp had permuted the conserved 'A–B–C' motif to 'C–A–B' by moving the C motif, which has two Asp residues

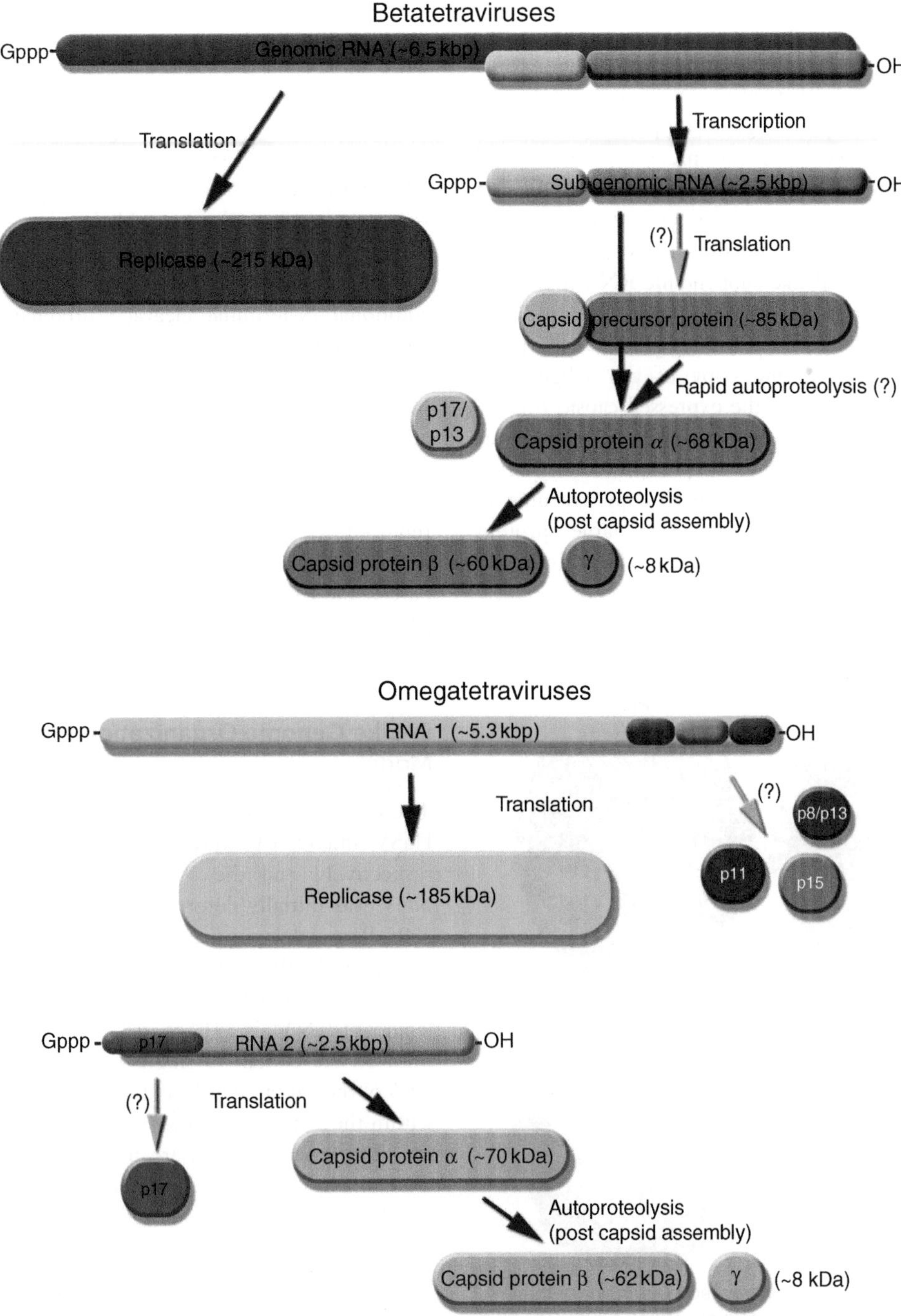

Figure 1 Genome organization and replication strategy of the tetraviruses. Each diagram represents the consensus of available data, but some group members may have slightly different organizations and steps from those shown. The individual events and gene products are discussed in the text. The gray arrows with question marks represent little, if any, translation of those gene products or an unverified event. While autocatalytic processing of the 68–70 kDa capsid protein is nearly identical, significant differences exist between the two groups in their pathways and their potential products. The β-like group utilizes a subgenomic RNA for translation of the capsid precursor protein, which may have two autoproteolytic processing events. The ω-like group has additional ORFs that overlap the end of the replicase gene with unknown functions. Both groups encode 13–17 kDa proteins on the small RNAs in addition to the capsid protein, but it is an additional ORF that would be translated as part of a polyprotein in the β-like group and an overlapping ORF that would be translated separately in the ω-like group.

critical for activity, 110 residues upstream in the sequence. Strikingly, structural modeling demonstrated that this permutation can be accommodated in the conserved RdRp structures by simply rearranging the connectivity of three loops, all of which are positioned away from the active site. Thus, the TaV RdRp is likely to have a conserved fold but with unique connectivity, which might be a major event in the evolution of polymerases.

The 3′ proximal ORF of NβV has 1836 nt and encodes the 66 kDa capsid protein precursor (named α). The capsid protein ORF overlaps the replicase ORF by more than 99% and is in the +1 reading frame relative to the replicase. The capsid protein gene is apparently translated through the production of the previously mentioned 2.5 kbp subgenomic RNA (**Figure 1**), yet replication experiments designed to detect double-stranded RNA (dsRNA) intermediates do not show evidence for a separate complex for the β-like subgenomic RNA indicating that both genomic and subgenomic RNAs are produced from a single full length (–)RNA template. One copy of the genomic (+)RNA, one copy of the subgenomic (+)RNA, and 240 copies of the expressed capsid protein precursor assemble to form an ~400 Å diameter icosahedral $T=4$ virion (**Figure 2**). The capsid protein is labeled a precursor at this stage because each copy executes an autocatalytic cleavage to produce a large protein (β, ~60 kDa) and a small protein (γ, ~6 kDa), but only after the capsid is assembled (**Table 2**). Both the β and γ polypeptides remain part of the virion after cleavage.

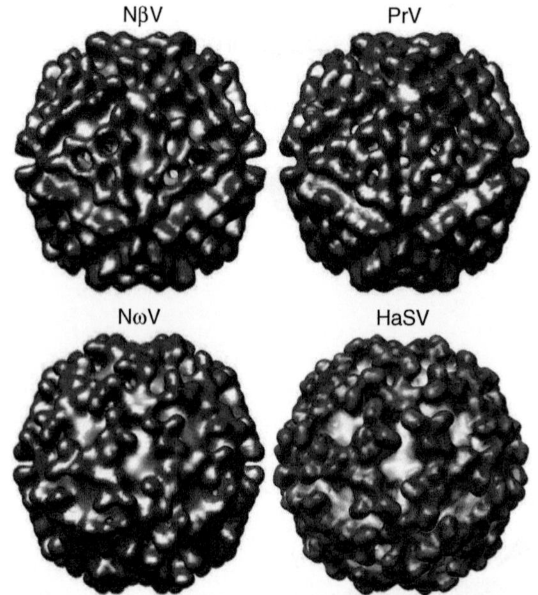

Figure 2 Cryoelectron microscopy reconstructions of the icosahedral tetravirus virions. All four structures are at ~25–30 Å resolution and are viewed down their icosahedral twofold axes. The capsids are ~400 Å in diameter and have $T=4$ quasi-icosahedral symmetry. The capsid morphologies within each group are closely similar (β-like at top, ω-like at bottom), but differ between the groups. Most notably, within the triangular facets, the β-like viruses have a pitted surface where the ω-like surface is filled in. TaV also displays the same pitted surface in negative-stain EM photographs. The recent crystal structure determination of PrV showed this is mainly due to different rotations of the Ig-like domains relative to the β-barrels when compared to the subunits in the NωV crystal structure (see **Figure 3**). This supports the suggested role of the Ig-like domain in determining the restrictive host cell specificity of these viruses by displaying distinctly different surfaces to cellular receptors.

The capsid protein precursors in TaV and PrV may be significantly larger than that of NβV. An additional codon exists upstream from the known N-terminus of the capsid proteins found in the TaV and PrV virions, yet the extra codon is transcribed to their subgenomic RNAs (**Figure 1**). The additional codons would add approximately 17 and 13 kDa to the TaV and PrV capsid protein precursors, respectively. Since the assembled capsids do not have a protein of this size, these precursors possibly undergo two post-assembly, autocatalytic cleavages, with the additional N-terminal cleavage producing polypeptides corresponding to the extra codons (p17, p13). Evidence for this came from expressing the entire capsid protein precursor of TaV in a recombinant baculovirus system as a single polypeptide (p17 + α). TaV VLPs were produced and contained only polypeptides of 58 and 7 kDa, suggesting that the additional 17 kDa had been previously removed by auto-proteolysis. However, it has not yet been shown that the extra codons are translated *in vivo* as part of the capsid protein precursor, and neither p17 nor p13 has been found in assembled particles (authentic or VLPs) of either virus.

ω-Like Genome Organization and Replication Model

The complete sequences of both RNA1 and RNA2 from HaSV and DpTV were determined in 1995 and 2005, respectively, and the complete RNA2 sequence from NωV was initially determined in 1992 and amended in 2005. Partial data for the RNA1 sequence of NωV are also available (**Table 2**). Like the β-like group, the RNAs of the ω-like group have near-identical organization. The length of RNA1 is 5312 nt in HaSV and 5492 in DpTV. Both RNAs contain a single large ORF covering over 90% of their total length, and three smaller ORFs located within the 3′ portion of the large ORF (**Figure 1**). While the three smaller ORFs are in the same reading frame, they are out of frame with the large ORF.

The large ORFs encode the viral replicases with calculated molecular weights of 187 and 180 kDa for HaSV and DpTV, respectively. Like the replicases of the β-like viruses, they also contain the N-terminal methyltransferase domain, the nucleotide-binding or helicase domain, and the RdRp domain, but lack the motifs for the papain-like protease and proline-rich hinge domains. The domains share from 63% to 76% identity between the HaSV and DpTV replicases, and about 36% identity with the same domains from the β-like replicases. The HaSV and NωV replicases are more closely related, sharing close to 90% identity. RNA1 self-replication has been detected before RNA2 replication has been observed, and protein expressed from HaSV RNA1 in a baculovirus system is able to replicate RNA2 transcripts at very low levels.

Table 2 Properties of tetravirus capsid proteins, genomic (g), and subgenomic (sg) RNAs[a]

	Beta (Da)	*Gamma (Da)*			*GenBank accession number(s)*
β-Like virus			*gRNA (bases)*	*sgRNA (bases)*	
NβV	58 448[b]	7975[b]	6625[b]	2656[b]	AF102884
PrV	60 649[b]	7343[b]	6400	2500	AF548354
TaV	58 327[b]	6781[b]	6500	2500	AF282930, AF062037
AeV			Present		
TnV	67 000–68 000		5865[c]		
DtV	62 000–66 000		Present		
PxV	62 400		Present		
DpV	66 000		5555[c]		
PiV	55 000		5865[c]		
ω-Like virus			*gRNA1 (bases)*	*gRNA2 (bases)*	
NωV	62 019[b]	7817[b]	5300	2445[b]	S43937[d]
HaSV	63 378[b]	7309[b]	5312[b]	2478[b]	U18246, L37299
DpTV	62 107[b]	7636[b]	5492[b]	2490[b]	AY594352, AY594353
Unassigned possible members			*gRNA or RNA1*	*sgRNA or RNA2*	
NεV	61 000				

[a]Blank fields represent undetermined or unavailable data.
[b]Data derived from sequence analysis.
[c]Converted from RNA molecular weight using average of 324 Da/nt.
[d]Sequence was amended in 2005 by GenBank accession numbers DQ054382 and DQ054403 to determine the full 5′ terminal sequence of RNA2.
Reproduced by permission of Elsevier.

Table 3 Percentage of sequence identity and similarity between the capsid proteins of six tetraviruses (three ω-like: HaSV, NωV, DpTV; three β-like: PrV, NβV, TaV). The similarity score is in parentheses

	NωV	*HaSV*	*DpTV*	*PrV*	*NβV*	*TaV*
NωV		66 (72)	87 (93)	37 (46)	29 (35)	27 (29)
HaSV	66 (72)		66 (78)	38 (47)	29 (36)	27 (33)
DpTV	87 (93)	66 (78)		37 (55)	25 (39)	26 (41)
PrV	37 (46)	38 (47)	37 (55)		27 (35)	28 (35)
NβV	29 (35)	29 (36)	25 (39)	27 (35)		35 (43)
TaV	27 (29)	27 (33)	26 (41)	28 (35)	35 (43)	

The three small ORFs encode proteins of 11, 15, and 8 kDa in HaSV and 11, 15, and 13 kDa in DpTV. In both genomes, the stop codon for p11 is located immediately to the 5′ side of the p15 ORF, allowing these two ORFs to form a single 26 kDa protein if the stop codon was suppressed. Expression of these proteins would almost certainly require a subgenomic RNA due to the long 5′ leading sequence; however, the existence of this RNA has not been confirmed. The amino acid sequences of all three ORFs show no discernible relationship to other proteins; therefore, no putative functions can be assigned.

The RNA2 lengths in NωV, HaSV, and DpTV are 2445, 2478, and 2490 nt, respectively. Each has a long (~280 nt) 5′ lead sequence followed by two overlapping genes starting at the first two AUG codons. The first ORF encodes a 17 kDa protein (p17) and overlaps the larger second ORF that encodes the 70–71 kDa capsid protein precursor (**Figure 1**). The initiation codon for p17 is in a poor context, suggesting that it will be translated inefficiently by scanning ribosomes. However, *in vitro* translation studies of both NωV and HaSV RNA2's revealed both gene products were produced. The function of p17, if any, remains unclear, but preliminary experimental observations have supported a regulatory role in genomic RNA replication, and/or a function analogous to that of the movement proteins of plant viruses.

The close relationship between the ω-like viruses is seen in their capsid protein precursors (α). The amino acid lengths of the precursors are 644, 647, and 643 for NωV, HaSV, and DpTV, respectively, and they share between 66% and 86% sequence identity (**Table 3**). One copy of the genomic (+)RNA1, one copy of the genomic (+)RNA2, and 240 copies of the expressed capsid protein precursor assemble to form an ~400 Å diameter icosahedral $T=4$ virion (**Figure 2**). As in the β-like viruses, each copy of α undergoes an autocatalytic cleavage to produce a large protein (β, ~62–63 kDa) and a small protein (γ, ~7–8 kDa) after the capsid is assembled (**Table 2**), and both the β and γ polypeptides remain part of the virion after cleavage. Both genomic RNAs are likely to be replicated from separate (–)RNA templates based on larvae infected with NωV having separate dsRNAs corresponding to each of the genomic RNAs. RNAs 1 and 2 represent the complete genomic information required for virus growth based on studies of HaSV assembly and replication in plant and insect cells.

Providence Virus Replication in a Fat Body Cell Line

PrV is currently the only tetravirus shown to replicate in cell culture. The virus was discovered in 2003 as a persistent infection of a midgut cell line (MG8) derived from the corn earworm (*Helicoverpa zea*). In addition to the *H. zea* midgut, PrV was also shown to infect the *H. zea* fat body cell line (FB33) at low doses. A time course study of the PrV infection cycle was conducted using the FB33 cells. Input virus genomic and subgenomic (+)RNA is detected inside the cells at 6 h post infection (h.p.i.). Viral RNA synthesis is first detected at 18 h.p.i. and significantly increases between 18 and 24 h.p.i. At 36 h.p.i., total viral RNA accumulation reaches its maximum level, and (–)RNA also reaches its peak level, which is about 10-fold lower than that of corresponding (+)RNA.

The PrV β capsid protein (~60 kDa) appears at 6 h.p.i. along with the genomic and subgenomic RNAs. The β protein begins increasing in copy number after about 24 h.p.i., and after 36 h the uncleaved capsid protein precursor (α) also accumulates, slightly lagging behind the appearance of the subgenomic RNA that is thought to drive its translation. Although the α protein accumulated in infected cells, it is not detected in purified virions. In contrast, p13 translation was not detected during the time course experiment and it is not found in purified virions. However, proteolytic processing and degradation of this precursor could be extremely rapid as observed with the baculovirus-expressed TaV precursor.

To determine if assembled virions could be seen in infected cells, both MG8 and FB33 cells infected with PrV for 3 days were sectioned and examined by TEM. Both cell lines contained virus particles that were ~400 Å in diameter. The particles were seen only in the cytoplasm and often associated with one or more unidentified cytoplasmic vesicles that appear to be induced by the infection. In a few percent of the FB33 cells with very high virus accumulation, large crystalline particle arrays appeared within the cytoplamic structures.

Capsid Assembly

PrV particles are seen only in the cytoplasm of sectioned fat body cells at all time points measured; thus, it is highly probable they assemble in the cytoplasm and the cytoplasmic structures produced by the infection may aid in the process. In lieu of cell culture systems, assembly of VLPs has been achieved using the capsid proteins from NωV, HaSV, and TaV (the only β-like virus for which successful VLP production has been reported). The VLPs will assemble upon expression of only the capsid protein precursor in heterologous expression systems, namely yeast, plant protoplasts, and baculoviruses. Considerable yields of VLPs can be purified from the baculovirus expression systems and this has enabled a number of biophysical and biochemical studies of tetravirus capsids, particularly those of NωV and HaSV (see below). Like native virions, the VLP capsid protein subunits undergo autocatalytic cleavage from α to β and γ peptides, can specifically package viral RNA, and can bind specifically to a larval midgut cell receptor.

Capsid Structure and Dynamics

High-Resolution Capsid Structure

Both X-ray crystallography and cryoelectron microscopy image reconstruction (cryoEM) have been used to examine the different structural states of tetravirus capsids. The crystal structure of authentic NωV was determined in 1996 to 2.8 Å resolution, and is currently the only tetravirus crystal structure that has been published. Recently, the authors' laboratory has also determined the crystal structure of purified HaSV VLPs, and found it closely similar to that of NωV. The NωV crystal structure showed that the capsid is icosahedral, 420 Å in diameter, and composed of 240 capsid proteins arranged with $T=4$ quasi-symmetry. Thus, the 60 icosahedral asymmetric units each have four subunits, named A–D, in unique chemical environments (**Figure 3(a)**). The 644 amino acid capsid subunit has three domains: an exterior immunoglobulin (Ig)-like fold, a central β-barrel in tangential orientation, and an interior helix bundle (**Figure 3(b)**). The helix bundle is created by the N- and C-termini of the polypeptide before and after the β-barrel fold, and the Ig-like domain is an insert between the E and F β-strands of the barrel. The exterior Ig domain is unique in nonenveloped viruses and is expected to be involved in cell binding and virus tropism, and has the least conserved sequence among tetravirus subunits. The interior helix domain is anticipated to interact with the RNA genome and may have a role in capsid polymorphism (see below). All 240 of the NωV capsid subunits have undergone autoproteolysis between residues N570 and F571 after assembly, leaving the cleaved portion of

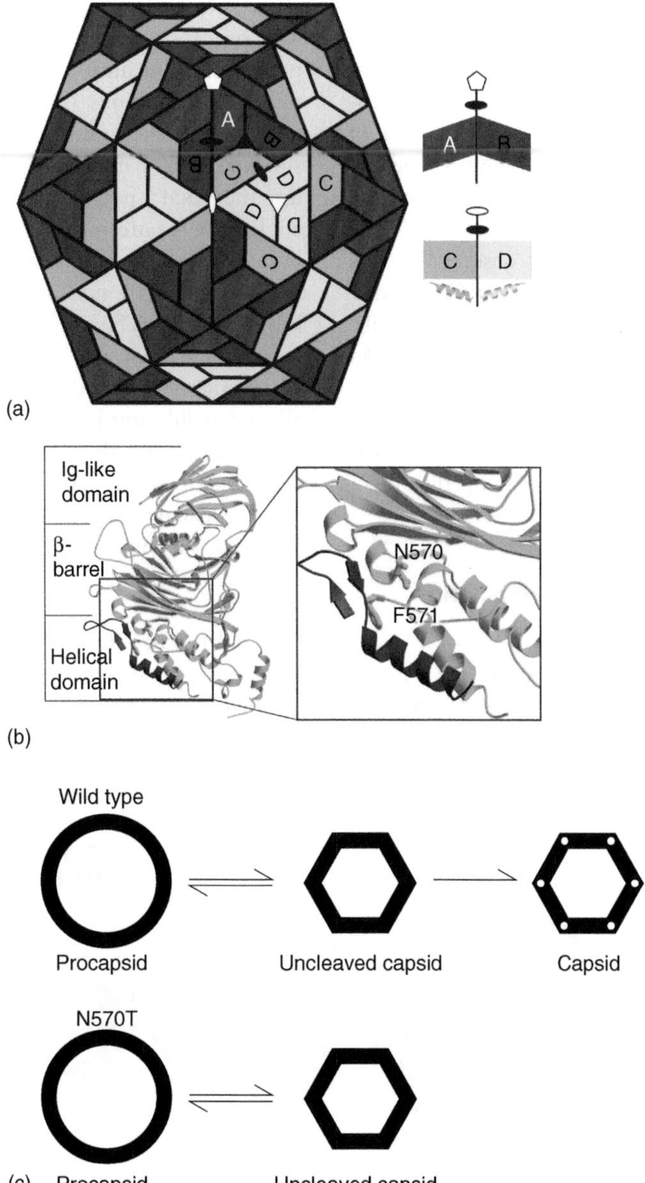

Figure 3 Structure and dynamics of the NωV capsid. (a) The subunit organization in the NωV $T = 4$ capsid crystal structure. There are four subunits in the asymmetric unit (A in blue, B in red, C in green, and D in yellow). Icosahedral symmetry axes are marked with white symbols, quasi-symmetry axes are marked with black symbols. Quasi-twofold dimer interfaces occur with the bent A–B or flat C–D conformations (right). The well-ordered C-terminal helices in the C–D subunits comprise a molecular switch that stabilizes the flat interface. (b) Ribbon diagram of the C subunit illustrates the three domain tertiary structure present in all 240 subunits (exterior of capsid on top, interior on bottom). Maturation cleavage occurs within the helical domain between residues Asn570 and Phe571. The molecular switch, present only in the C or D subunits, is highlighted in magenta. (c) In the wild-type virus, the procapsid converts to uncleaved capsid with the drop of pH, followed by a maturation cleavage, which locks the particle in capsid conformation after ~10% of the subunits are cleaved. In the cleavage-defective mutant (N570T), particle conformational change becomes reversible.

the C-terminus, called the γ peptide, associated with the capsid. The quasi-symmetry is controlled by a molecular switch (**Figures 3(a)** and **3(b)**). In NωV, a segment of the γ peptide (residues 608–641) is only ordered in the C and D subunits, and functions as a wedge between the ABC and DDD morphological units to prevent curvature and create a flat contact. The interface between two ABC units is bent partly due to the lack of ordered γ peptide. Importantly, the subunit structure and autoproteolysis revealed a strong relationship between tetraviruses and the $T = 3$ insect nodaviruses. Indeed, the β-barrel folds and autocatalytic sites superimpose with little variation, and the γ peptide in the nodaviruses is also involved in quasi-equivalent switches.

The crystal structure of PrV, the first high-resolution structure of a β-like virus, was recently determined in the authors' laboratory (data not shown). The PrV subunit retains the three domains and a superimposable auto-proteolysis site (all four subunits show a clear break at the cleavage site), but the Ig domain and the first and last ordered residues of the termini in the helical domain have different positions and structure compared to NωV. The different Ig domain position was first observed in the cryoelectron microscopy reconstruction of NβV as a distinctly different surface structure compared to the ω-like particle surfaces (**Figure 2**). The use of subunit termini in PrV is more like that seen in the nodaviruses. The quasi-symmetry switch in PrV has swapped elements compared to the omegatetraviruses, and represents a new and unique mixture of structural features among the insect viruses.

Capsid Maturation, Auto-Proteolysis, and Large-Scale Transitions

NωV and HaSV coat proteins expressed in the baculovirus system assemble into round and porous intermediate VLPs that are 480–490 Å in diameter, called procapsids, when purified at neutral pH (pH 7.6). The procapsid coat proteins remain in the uncleaved α precursor form until a reduction in pH. When exposed to acidic conditions (pH 5.0), procapsids underwent a large-scale structural rearrangement (maturation) to the smaller, angular mature capsid, where α begins to cleave into the β and γ polypeptides (**Figures 3(c)** and **4**). Small-angle X-ray scattering (SAXS) demonstrated that the pH-induced conformational change occurs in less than 100 ms in NωV, while the autocatalytic cleavage event had a half-life of hours. The relationship of the conformational change and cleavage was also examined using SAXS, which suggested that the conformational change was initially reversible until approximately 10% of the subunits had cleaved, and then the particles were 'locked' in the capsid conformation (**Figure 3(c)**). Finer pH titrations of NωV revealed intermediate capsid states at pH 5.8 and 5.5 (data not shown). Presumably similar intermediates also exist for HaSV.

Both ANS (1-anilino-8-naphthalene sulfonate) and the thiol-reactive fluorophore, maleimide-ANS (MIANS), have been bound to procapsids to further characterize the conformational transition to capsid. Absorption of the covalently attached MIANS (forms covalent thiol-ether linkage

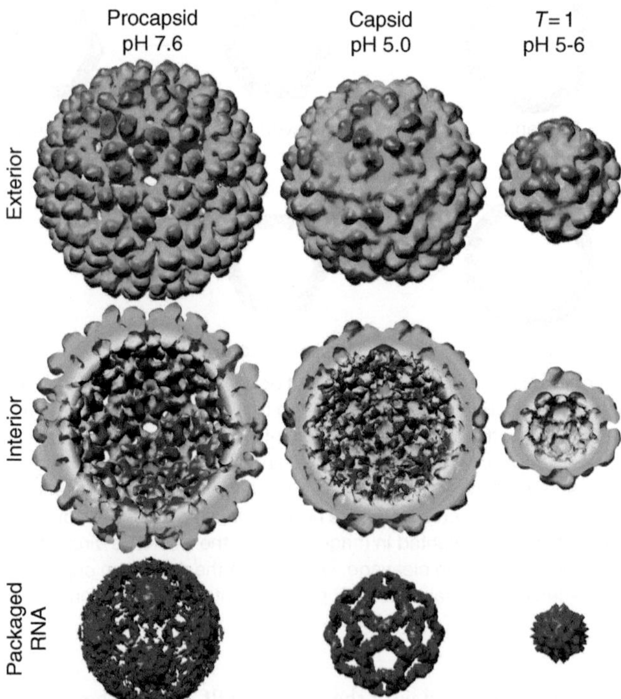

Figure 4 CryoEM reconstructions between 23 and 27 Å resolution of three HaSV polymorphic capsid forms. The polymorphic form and its associated pH are given at the top. Three views of each form are shown with radial coloring from red on the interior to blue on the exterior. The procapsid and capsid diameters are similar to those of NωV (490 and 430 Å, respectively). Note the well-ordered dodecahedral RNA structure in the mature capsid, which has been observed in several insect viruses. The $T = 1$ form has so far been found only in HaSV purifications as an extra peak in the sucrose gradients, and only appears after the acid maturation step (i.e., not found at pH 7.6). The diameter of the $T = 1$ particles is 260Å and the protein shell is composed only of a truncated version of the β subunits (55 kDa).

with cysteines) situated at subunit interfaces can report on environment changes as it is buried during maturation. Unexpectedly, acidification of the labeled particles trapped them at an intermediate stage of maturation and the autocatalytic cleavage was inhibited. The labeled VLPs showed no cleavage after 4 h of incubation at pH 5.0 buffer while wild-type particles were fully cleaved. A cryoEM image reconstruction of the MIANS-labeled VLPs at pH 5.0 revealed the particles are 440 Å in diameter, having only progressed about halfway along the trajectory between procapsid and capsid. This demonstrates that cleavage depends on the conformational change and is not triggered by just lowering the pH, and provides support for the strategy of targeting subunit interfaces with antiviral compounds as a means of disrupting the viral life cycle.

Cleave Site Mutation Allows Reversible Transition and Procapsid Structure Analysis

Mutagenesis studies have shown the asparagine at the scissile bond (N570) in NωV is required for cleavage. Substituting threonine for asparagine (N570T) results in assembly of less stable particles, but they do not cleave and continue to undergo the conformational change reversibly as a function of pH (**Figure 3(c)**). CryoEM structures at 8 Å resolution of cleaved and N570T VLPs showed the role of cleavage in stabilizing the capsid form. There are more extensive inter-subunit contacts between the C-termini and the neighboring N-termini in cleaved particles that results in more ordered C-terminal helices, which accounts for the added stability.

The N570T mutant was also used to study the structure of the uncleaved procapsid state. An 11 Å resolution cryoEM structure of N570T procapsids displays a morphology indistinguishable from procapsids formed by wild-type subunits. The higher resolution showed, at pH 7.6, that portions of the C-terminal regions of the γ polypeptides of the A and B subunits, which are not ordered in the capsid form, are ordered and contribute to a second molecular 'wedge' that prevents hinging between the A and B subunits. It makes the quasi-twofold interactions at the A–B and C–D interfaces closely similar in the procapsid, producing its round morphology and consistent with its role as an assembly intermediate during NωV particle maturation.

Pathobiology

Host Range

Lepidopteran insects are the only known hosts for tetraviruses. No replication in other animals has been detected. The hosts are confined largely to the Noctuidae, Saturniidae, and Limacodidae families within the Lepidoptera order. The range of species that they are able to infect differs between the viruses. HaSV appears unable to infect species outside the Heliothinae subfamily, while TnV and DpV are able to infect insects outside the family of their nominal host.

Transmission and Symptoms

Horizontal transmission via ingestion by larvae has been demonstrated for several tetraviruses with a widely varying range of symptoms. Only slight growth retardation is observed using high doses of TnV, but NβV infection of larvae is lethal, causing cessation of feeding, and moribund, discolored, and flaccid appearance at 7–9 days post infection. A dependence upon the larval life stage has also been seen. HaSV is highly active against neonate larvae in the first three instars of development, needing as little as 5000 particles to cause them to stunt and cease feeding within 24 h and to die within 4 days. In contrast, no detectable symptoms occur in later larval development, signifying a high degree of developmental resistance. It has not been demonstrated whether adults or pupae are capable of being infected by tetraviruses. Vertical transmission of tetraviruses is also believed to occur, but the evidence remains undefinitive.

Histopathology and Tropism

Natural infection by tetraviruses appears to be exclusively in the larval midgut, although PrV has been cultured in fat body cells. In a definitive experiment, Northern blots of RNA extracted from HaSV-infected larvae showed viral RNA only in the midgut tissue, even after virus was injected into the hemocoel.

In a quantitative study of HaSV infection of its host, the African or cotton bollworm, supplying the virus to larvae *per os* during the first three instars of development resulted in an infection restricted to three of the four midgut cell types. The virus initiated infection in closely situated foci that expanded and converged with others over time. In response, the midgut cells increased their rate of sloughing and apoptosis to an extent that incapacitated the midgut due to loss of all but a few cells. Infection of older larvae results in only sparsely situated foci, which fail to expand and eventually disappear presumably due to cell sloughing. These data indicate that cell sloughing is an immune response existing throughout development, but older larvae have an additional mechanism for increased resistance to infection.

Biotechnology

Application as Biopesticides

Small RNA viruses offer attractive properties as biological control agents for pests: high specificity, high efficacy, and potentially self-replicating. A notable drawback is the difficulty in producing large quantities of infectious virions, but the simplicity of the tetraviral genomes makes production of virions in nonhost cells feasible. Agrobacterium-mediated transgenesis has been used to place the genes required for HaSV assembly into tobacco plants, which resulted in stunted, HaSV-infected larvae when they fed on the modified plants. Expanding on this result, Hanzlik and colleagues conducted the first field trial of HaSV as a control agent in 2005 on a sorghum crop in Queensland, Australia. The semipurified HaSV preparation reduced the larval population by 50% at both 3 and 6 days post application (dpa), which was equivalent to a commercial baculovirus preparation. Notably, the amount of HaSV on the sorghum heads increased at 5.5 dpa, probably reflecting dispersal of newly produced virus from cadavers and frass. These results indicate that HaSV could be used effectively as a biopesticide.

Application as Chemical Platforms

A study in 2006 demonstrated that the Ig domains of NωV particles could be modified to possibly alter its tropism and/or to carry specific chemical tags. A $(His)_6$ tag was inserted into the GH loop (between A378 and G379) of a capsid precursor construct that was then expressed in a baculovirus system. The His-tagged p70 self-assembled into VLPs that had similar morphological and RNA encapsidation properties as wild-type VLPs. Two assays using paramagnetic nickel beads confirmed that multiple affinity tags were present on the surface of the modified VLPs and could bind the beads. Thus, the GH loop of NωV appears to be a suitable modification site for further biotechnology applications.

See also: Cryo-Electron Microscopy; Nodaviruses; Structure of Non-Enveloped Virions; Principles of Virion Structure.

Further Reading

Canady MA, Tsuruta H, and Johnson JE (2001) Analysis of rapid, large-scale protein quaternary structural changes: Time resolved X-ray solution scattering of nudaurelia capensis ω virus. *Journal of Molecular Biology* 311: 803–814.
Christian PD, Murray D, Powell R, Hopkinson J, Gibb NN, and Hanzlik TN (2005) Effective control of a field population of *Helicoverpa armigera* by using a small RNA virus Helicoverpa armigera stunt virus (*Tetraviridae, Omegatetravirus*). *Journal of Economic Entomology* 98: 1839–1847.
Gorbalenya AE, Pringle FM, Zeddam J-L, *et al.* (2002) The palm subdomain-based active site is internally permuted in viral RNA-dependent RNA polymerases of an ancient lineage. *Journal of Molecular Biology* 324: 47–62.
Hanzlik TN and Gordon KHJ (1997) The *Tetraviridae*. *Advances in Virus Research* 48: 101–168.
Helgstrand C, Munshi S, Johnson JE, and Liljas L (2004) The refined structure of nudaurelia capensis ω virus reveals control elements for a *T*=4 capsid maturation. *Virology* 318: 192–203.
Johnson JE and Reddy V (1998) Structural studies of nodaviruses and tetraviruses. In: Miller LK and Ball LA (eds.) *The Insect Viruses*, pp. 171–223. New York: Plenum.
Lee KL, Tang J, Taylor D, Bothner B, and Johnson JE (2004) Small compounds targeted to subunit interfaces arrest maturation in a nonenveloped, icosahedral animal virus. *Journal of Virology* 78: 7208–7216.
Maree HJ, Van der Walt E, Tiedt FA, Hanzlik TN, and Appel M (2006) Surface display of an internal His-tag on virus-like particles of nudaurelia capensis omega virus (NωV) produced in a baculovirus expression system. *Journal of Virological Methods* 136: 283–288.
Pringle FM, Johnson KN, Goodmann CL, McIntosh AH, and Ball LA (2003) Providence virus: A new member of the *Tetraviridae* that infects cultured insect cells. *Virology* 306: 359–370.
Taylor DJ, Krishna NK, Canady MA, Schneemann A, and Johnson JE (2002) Large-scale, pH-dependent, quaternary structure changes in an RNA virus capsid are reversible in the absence of subunit autoproteolysis. *Journal of Virology* 76: 9972–9980.

Togaviruses: Molecular Biology

K D Ryman and W B Klimstra, Louisiana State University Health Sciences Center at Shreveport, Shreveport, LA, USA
S C Weaver, University of Texas Medical Branch, Galveston, TX, USA

Glossary

Apoptosis A cascade of cellular responses to a stimulus (e.g., virus infection) resulting in cell death.

Cytopathic effect Destructive changes in cell morphology, structure, or metabolic processes resulting from virus infection.

Full-length cDNA clone A copy of an RNA virus genome that has been reverse-transcribed into DNA and placed into a subcloning vector. With positive-sense RNA viruses, infectious genomic RNA is generally directly synthesized from bacteriophage transcription promoters located upstream of the virus sequences.

Host cell shut-off Interruption or cessation of host transcription, translation, or other processes that occurs within virus-infected cells.
Interferon A pro-inflammatory cytokine that is produced after virus infection and stimulates antiviral activities in cells expressing the IFNAR receptor complex.
Interferon-stimulated gene (ISG) A gene whose transcription is increased after interferon signaling. Some ISGs are directly antiviral and can inhibit virus transcription or translation.
Positive-sense RNA genome The genomic material of the virus resembles a cellular messenger RNA and the open reading frame can be translated on host ribosomes.
Receptor Structures (e.g., protein, sulfated polysaccharide, phospholipid) found upon the surface of host cells that interact with virus attachment proteins and mediate attachment of virus particles to cell surfaces. These structures may also mediate subsequent events in cell entry such as viral protein rearrangements leading to membrane fusion.
Replicon system A replication-competent but propagation-incompetent form of a cloned virus. In many replicon systems, the replicon genome can be packaged into a virus-like particle by *trans*-expression of structural proteins.
RNA-dependent RNA polymerase (RdRp) The primary RNA synthesis enzyme for RNA viruses which uses viral genomic RNA as a template for transcription/replication.

Introduction

The family *Togaviridae* includes the genera *Alphavirus* and *Rubivirus*. The genus *Alphavirus* is comprised of 29 virus species segregated into five antigenic complexes, most of which are transmitted between vertebrate hosts by mosquito vectors and are capable of replicating in a wide range of hosts including mammals, birds, amphibians, reptiles, and arthropods. In contrast, the sole member of the genus *Rubivirus* is Rubella virus, which is limited to human hosts and is primarily transmitted by respiratory, congenital, or perinatal routes. Together, the viruses are a significant worldwide cause of human and livestock disease. Alphavirus disease ranges from mild to severe febrile illness, to arthritis/arthralgia and fatal encephalitis, while Rubella virus is an important cause of congenital abnormalities and febrile illness frequently accompanied by arthralgia/arthritis in the developing world. At the molecular level, the togaviruses are relatively uncomplicated, small, enveloped virions with single-stranded, positive-sense RNA genomes encoding two polyproteins, one translated from the genomic RNA and the other from a subgenomic mRNA transcribed during replication. However, their replication is very tightly regulated and intimately associated with host cell metabolic processes. Currently, an important aspect of togavirus research is designed to exploit their molecular biology for therapeutic and gene therapy applications. Here, we summarize major aspects of togavirus structure, genome organization, replicative cycle, and interactions with host cells.

Virion Structure and Genome Organization

Togavirus virions are small and enveloped, comprising an icosahedral nucleocapsid composed of 240 capsid (C) protein monomers, cloaked in a lipid envelope studded with membrane-anchored glycoprotein components, E1 and E2. Their tightly packed, regular structure has been revealed by high-resolution reconstruction of the particles by cryoelectron microscopy (**Figure 1**). The envelope glycoproteins of alphaviruses are similarly arranged on the outer surface of particles. E1 and E2 form a stable heterodimer, and three heterodimers interact to form 'spikes' distributed on the virion surface in an icosahedral lattice that mirrors the symmetry of the nucleocapsid. The regularity of the spike distribution is determined by the direct hydrophobic interaction of the E2 glycoprotein tail with a pocket on the surface of the C-protein. E2 projects outward from the virion to form the spikes that interact with host-cell attachment receptors, while the E1 glycoprotein, which mediates fusion with host cell membranes, appears to lie parallel to the lipid envelope. The E2 and E1 proteins are glycosylated at one to four positions, depending on the virus strain, during transit through the host-cell secretory apparatus.

The encapsidated alphavirus genome consists of a nonsegmented, single-stranded, positive-sense RNA molecule of approximately 11–12 kb with a 5′-terminal methylguanylate cap and 3′ polyadenylation, resembling cellular mRNAs (**Figure 2**). The genome is divided into two major regions flanked by the 5′ and 3′ nontranslated regions (NTRs) and divided by an internal NTR: (1) the 5′-terminal two-thirds of the genome encode the four nonstructural proteins (nsPs 1–4); and (2) the 3′ one-third of the genome encodes the three structural proteins (C, precursor E2 [PE2], and precursor E1 [6K/E1]). Overall, the capped and polyadenylated 9.5 kb Rubella virus genome is similarly organized (**Figure 2**), but the 5′-proximal open reading frame (ORF) encodes only two nsPs (P150 and P90) and three structural proteins are encoded by the 3′ ORF (C, E1, and E2).

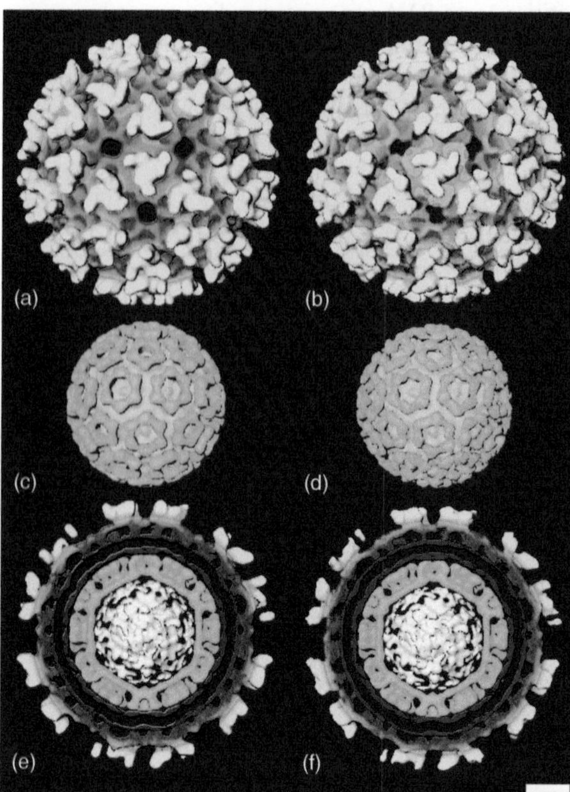

Figure 1 Structure of New World (VEEV) and Old World (SINV) alphaviruses determined by image reconstructions of electron micrographs. Isosurface view along a threefold axis of VEEV (a) and SINV (b) reconstructions showing outer spike trimers (yellow) and envelope skirt region (blue). Isosurface representations of VEEV (c) and SINV (d) nucleocapsids viewed along a threefold-symmetry axis. Cross-sections through VEEV (e) and SINV (f) perpendicular to the threefold axis and in plane with a vertical fivefold axis showing trimers (yellow), skirt region (blue), virus membrane (red), nucleocapsid (green), and RNA genome (white). Scale = 100 Å. Reproduced from Paredes A, Alwell-Warda K, Weaver SC, Chiu W, and Watowich SJ (2007) Venezuelan equine encephalomyelitis virus structure and its divergence from old world alphaviruses. *Journal of Virology* 75: 9532–9537, with permission from American Society for Microbiology.

Infection and Replication Cycle

The togavirus genome encodes very few but, therefore, necessarily multifunctional proteins. The replication of these viruses is very tightly regulated and as little as a single nucleotide change can alter susceptibility to cellular defense mechanisms. The following section outlines molecular aspects of togavirus infection and replication pertinent to the virus–cell interaction. Replication processes of viruses of the family *Togaviridae* have been most extensively studied for the alphaviruses, Sindbis virus (SINV) and Semliki Forest virus (SFV), and extrapolated to the other family members.

Virus Attachment, Entry, and Uncoating

The entry pathways of Rubella virus have undergone only limited study, but there is some evidence that a glycolipid molecule serves as an attachment receptor on some cell types. The nature of the receptor for naturally circulating, 'wild-type' alphaviruses has been examined more extensively, but remains controversial, as is the case with many arthropod-borne viruses. Typically, arthropod-borne viruses are thought to interact either with a single, evolutionarily conserved molecule expressed on most cells, or to utilize different receptors in different hosts and with cells derived from different tissues. The 67 kDa high-affinity laminin receptor (HALR) has been identified as an initial attachment receptor for cell culture-adapted, laboratory strains of SINV and an antigenically related molecule has similarly been identified for Venezuelan equine encephalitis virus (VEEV). In addition, receptor activity for particular cell types has been attributed to several other, as yet uncharacterized, proteins. However, more recent studies have indicated that efficient cell binding by many laboratory strains of alphaviruses is due to interaction with heparan sulfate (HS), a sulfated glycosaminoglycan molecule, and that this phenotype is conferred by positively charged amino acid substitutions in E2 that accompany adaptation to cultured cells. Currently, it is unclear whether or not the 67 kDa HALR or other identified receptor proteins act in concert with HS to bind cell culture-adapted alphavirus strains or if any of these molecules function in the natural replication cycle of wild-type alphaviruses. Moreover, it has become clear that many noncell culture-adapted alphavirus strains bind only weakly to commonly used cell lines, such as those derived from fibroblasts. Recently, DC-SIGN and L-SIGN, C-type lectins, were found to bind carbohydrate modifications on E1 and/or E2 proteins of both HS-binding and non-HS-binding strains of SINV and VEEV, promoting infection. The interaction of the lectins with virion carbohydrates is greatly enhanced by replication of viruses in mosquito cells, most likely due to the retention of high mannose carbohydrate structures, which are processed to complex carbohydrates in mammalian cells. Therefore, C-type lectins may be important attachment receptors for alphaviruses when transmitted by mosquitoes, and receptor utilization could be determined, at least in part, by the cell type in which alphaviruses replicate. Since DC-SIGN and L-SIGN receptors are expressed by a subset of alphavirus-permissive cells, these results further suggest that infection by alphaviruses may be mediated by different receptors on different cell types.

Following receptor interactions, which may initiate uncoating-related conformational changes in E1 and E2, alphavirus and Rubella virus virions enter acidified endosomes via a dynamin-dependent process. The decreasing pH in the endosome results in more dramatic

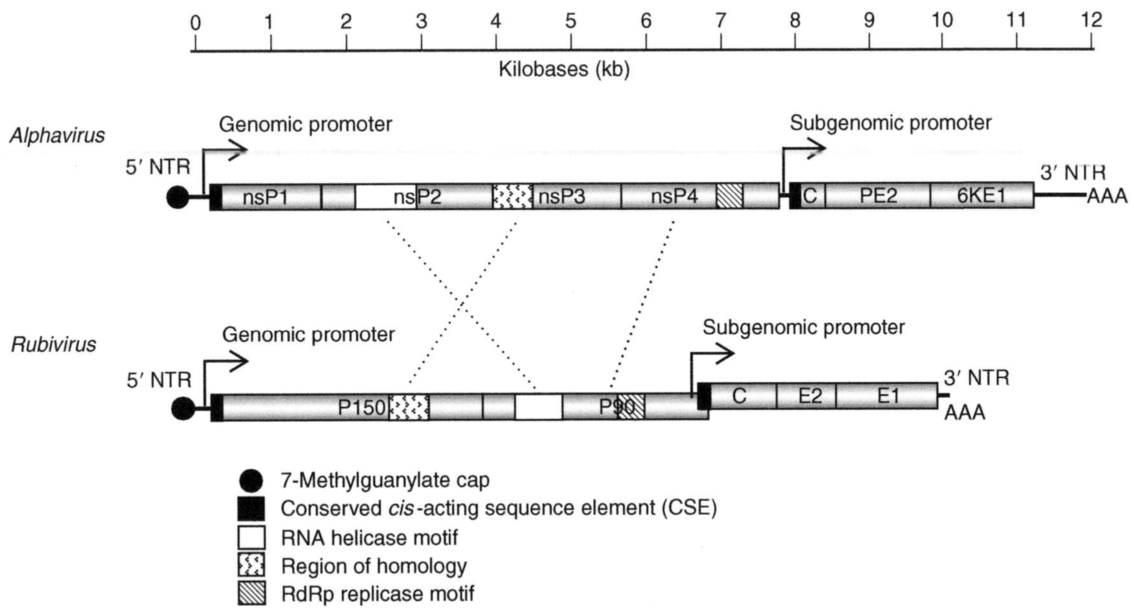

Figure 2 Comparison of *Alphavirus* (SINV) and *Rubivirus* (RUB) genome organization. Regions of nucleotide homology and regions encoding homologous amino acid sequence within the nonstructural protein ORF are shown. ORFs are denoted by boxes and nontranslated regions (NTRs) by lines. Note that for SINV the two ORFs are separated by a NTR region, whereas for RUB the two ORFs overlap in different translational frames. Reproduced from Dominguez G, Wang C-T, and Frey TK (1990) Sequence of the genome RNA of rubella virus: Evidence for genetic rearrangement during togavirus evolution. *Virology* 177: 225–238 with permission from Elsevier.

conformational changes (threshold of ~pH 6.2 with SFV) in the E2/E1 spike leading to formation of E1 homotrimers and exposure of a putative class II fusogenic domain in E1 that is thought to interact with the host cell membrane, promoting fusion of host and virion lipid bilayers. In mammalian cells, SFV membrane fusion is cholesterol-dependent and sphingolipid-dependent, and inhibitors of endosomal acidification block infection. Endosomal fusion leads to a poorly characterized entry of nucleocapsids into the cytoplasm and to their association with ribosomes, most likely directing final uncoating of the genomic RNA and translation initiation.

Genome Translation, Transcription, and Replication

Togavirus genome transcription and replication occur by a strictly positive-sense strategy and are entirely cytoplasmic (**Figure 3**). The alphavirus replicase complex comprises four nsPs encoded in the 5′-terminal ORF, all of which are essential for viral transcription and replication: (1) nsP1 is a guanine-7-methyltransferase and guanylyl-transferase; (2) nsP2 has NTPase, RNA helicase, and RNA triphosphatase activities in its amino-terminal domain, while the carboxy-terminus has highly specific thiol protease activity; (3) nsP3 is a phosphoprotein with unknown function; and (4) nsP4 is the catalytic RNA-dependent RNA polymerase (RdRp). In addition to the final products, some of the processing intermediates have distinct and indispensable functions during the replication process. The 5′-proximal ORF of the Rubella virus genome is also translated as a polyprotein (p200) and cleaved in *cis* by the Rubella virus protease (NS-pro) to form the replication complex comprising P150 and P90. However, the order of its conserved motifs differs from that of the alphavirus nsP polypeptide (**Figure 2**): P150 contains the putative methyltransferase (N-terminal) and protease sequences (C-terminal), while P90 contains the predicted helicase (N-terminal) and RNA polymerase (C-terminal).

Once the nsPs have been translated from the infecting, positive-sense RNA genome, full-length RNAs complementary to the genomic sequence (negative-sense) are synthesized, creating partially double-stranded (ds) RNA replicative intermediates. The RdRp complex then switches to preferential synthesis of positive-sense RNAs, which continues throughout the remainder of the infection replication cycle, while negative-sense RNA transcription ceases entirely. Two positive-sense, capped, and polyadenylated RNAs are synthesized on the negative-sense template: full-length progeny genomes that are packaged into virions and subgenomic (26S) RNAs that are collinear with the 3′ one-third of the genome and encode the structural protein ORF. The 26S RNA accumulates in the cells to 5–20-fold molar excess over genomic RNA and is efficiently translated into a polyprotein that is co-translationally processed to produce the structural proteins.

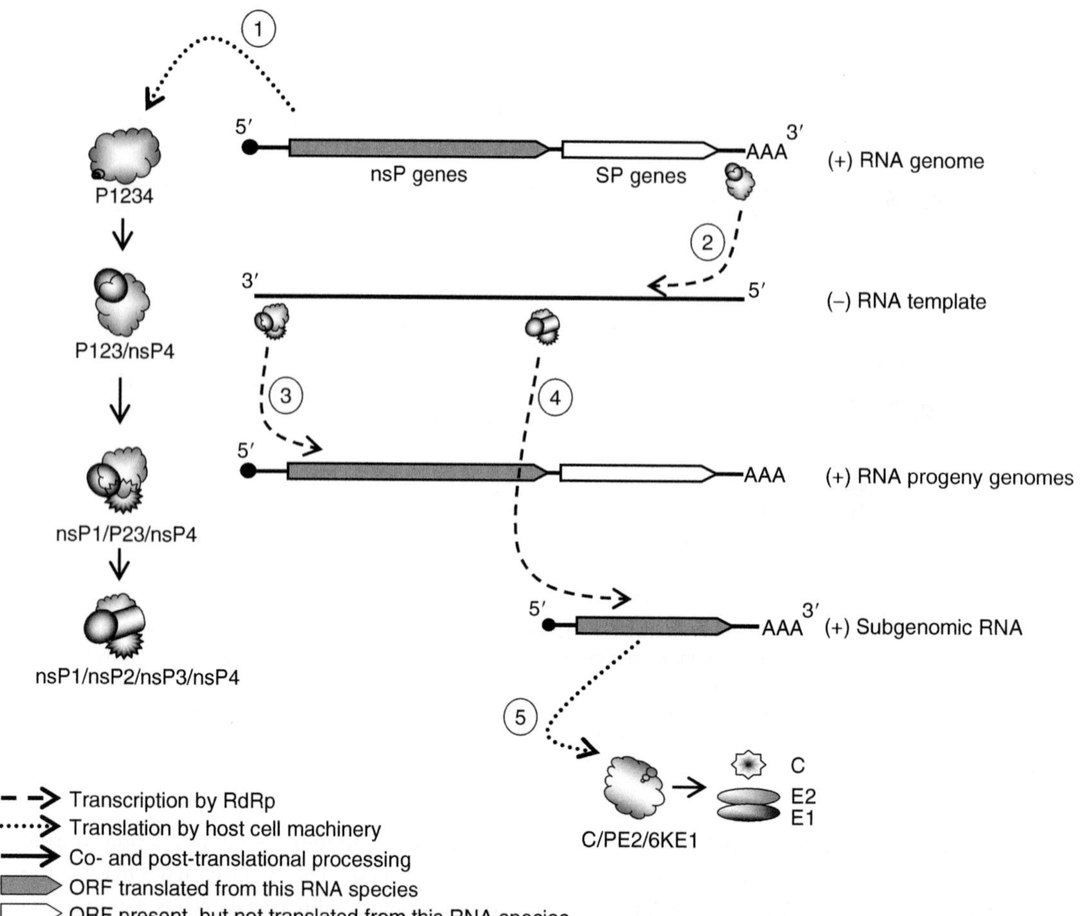

Figure 3 Diagrammatic representation of *Alphavirus* replication in a permissive cell. The replication cycle is depicted as a series of temporally regulated steps: ① Translation and processing of nonstructural polyprotein P1234 or P123; ② Synthesis of complementary, negative-sense RNA by P123/nsP4; ③ Synthesis of progeny genomes by nsP1/P23/nsP4; ④ Synthesis of subgenomic RNAs by nsP1/nsP2/nsP3/nsP4; ⑤Translation and processing of structural polyprotein C/PE2/6KE1. Progeny genomes are then packaged into nucleocapsids and bud at the plasma membrane to release progeny virions.

Alphavirus genome replication and transcription are tightly regulated temporally by sequential processing of the nsPs, which alters replicase composition and RdRp activity. For some alphaviruses (e.g., SFV), the nsPs are translated as P1234, while others (e.g., SINV) translate the nsPs as either P1234 or P123 polyproteins, depending upon read-through of an opal termination codon. The nascent P1234 polyprotein is partially processed by autocatalytic nsP2 protease-mediated cleavage at the 3/4 junction and assembled into the primary P123/nsP4 RdRp complex. This short-lived P123/nsP4 primarily synthesizes genomic, negative-sense RNA replicative intermediates. Membrane-bound P123 is cleaved relatively slowly in *cis* at the 1/2 junction, perhaps to enable formation of protein–protein interactions prior to proteolytic processing, and to form the nsP1–P23–nsP4 replicase that preferentially synthesizes positive-sense RNA. After the aminoterminus of nsP2 has been released, rapid autocatalytic cleavage of P23 produces a mature, stable RdRp complex (nsP1–nsP2–nsP3–nsP4), which transcribes only positive-sense RNA. Evidence suggests that nsP1–P23–nsP4 and the mature replicase may be biased, respectively, toward synthesis of genome-length or subgenomic mRNA, providing further temporal regulation of the replication cycle. Late in infection, when free nsP2 concentrations are high, cleavage at the 2/3 junctions is favored. P1234 polyproteins are rapidly processed into short-lived P12 and P34 precursors, thereby precluding assembly of new replication complexes and terminating negative-strand synthesis. *cis*-Cleavage of the P12 precursor gives rise to nsP1, which targets the plasma membrane, and nsP2. Interestingly, much of nsP2 is transported to the nucleus, suggesting a function beyond the replication and transcription of viral RNAs, as discussed below. P34 is cleaved in *trans* by free nsP2, yielding nsP3, which aggregates in the cytoplasm, and nsP4, which is rapidly degraded by the ubiquitin pathway.

Specificity of togavirus RNA replication by the RdRp replicase is achieved via differential recognition of *cis*-acting conserved sequence elements (CSEs) in the termini of the viral genome and negative-sense genome template. These CSEs are conserved in all alphaviruses, as well as in Rubella virus. A 19-nucleotide (nt) CSE in the 3′ NTR immediately upstream of the poly(A) tail, is the core promoter for synthesis of genome-length, negative-sense RNA replicative intermediates, and probably interacts with the 5′ NTR via translation initiation factors to initiate replication and/or translation. The complement of the 5′ NTR in the minus-strand RNA serves as a promoter for the synthesis of positive-sense genomes. A third *cis*-acting element that is essential for replication, the 51-nt CSE, is also found near the 5′ end of the genomic RNA within the nsP1 gene. The primary sequence and two-stem–loop secondary structure of this CSE are highly conserved among alphaviruses and serve as replication and translation enhancers. Finally, a 24-nt CSE is found upstream of and including the start of the subgenomic RNA, the complement of which in the negative-strand forms the 26S subgenomic promoter and translation enhancer element.

Togavirus RNA replication occurs on cytoplasmic surfaces of endosome-derived vesicles or cytopathic vacuoles (CPVs). Expression of the nsP polyprotein (particularly nsP1) actively modifies intracellular membranes to create these compartments and to mediate membrane association of the replicase via membrane phosphatidylserine and other anionic phospholipids. The surfaces of CPVs have small invaginations or 'spherules', in which the nsPs and nascent RNA are sequestered, creating a microenvironment for RNA replication, synthesis of structural proteins, and assembly of nucleocapsids. The formation of CPVs in togavirus-infected cells may be linked to stress-induced autophagic mechanisms intended for the targeted degradation of cellular proteins and organelles, but exploited by the virus.

Structural Protein Maturation, Genome Packaging, and Egress of Virions

The subgenomic mRNAs of alphaviruses and of Rubella virus are translated as polyproteins; however, processing of the polyprotein differs between the two. With alphaviruses, the capsid protein is cleaved autoproteolytically from the E2 precursor (known as PE2 or P62) in the cytoplasm and the remaining polyprotein is translocated into the ER lumen, where N-linked oligosaccharide addition occurs. The 6K hydrophobic protein, which is primarily located in the ER membrane, is then cleaved from the polyprotein by the host signalase. With Rubella virus, the capsid-E2 and E2–E1 cleavages are all completed by signalase. Furthermore, Rubella virus lacks the 6K protein and mature E2 is produced in the ER. Alphavirus PE2 and E1 proteins form heterodimers in the ER that are anchored by their transmembrane domains and involve intra- and intermolecular disulfide linkages and associations with host chaperone proteins such as Bip and calnexin/calreticulin. The PE2/E1 heterodimers are routed to the cytoplasmic membrane and PE2 is cleaved into mature E2 by a host furin-like protease as a late event, in a structure between the *trans* Golgi network and cell surface. The cleaved E3 fragment is lost from some (e.g., SINV), but not all (e.g., SFV) virus particles. PE2-containing virions appear to bud normally from vertebrate cells, but are often defective in conformational rearrangements associated with cell fusion, supporting the hypothesis that the presence of PE2 stabilizes the glycoprotein heterodimer during low pH exposure in the secretory pathway.

Budding of particles occurs through an interaction between C and the cytoplasmic tail of E2; however, it is unclear at which point in the secretory pathway this interaction occurs and whether the interaction is between C monomers, oligomers, or RNA-containing preformed nucleocapsids. With alphaviruses, RNA packaging is directed by an RNA secondary structure in either the nsP1 or nsP2 genes, depending upon the virus, leading to selective packaging of the genome over the subgenome, which is in molar excess. In the accepted model, final budding occurs at the cytoplasmic membrane at sites enriched in E2/E1 heterodimers and is driven by E2 tail-C interactions that force an extrusion of the host lipid bilayer, envelopment of the particle, and release.

Infectious Clone Technology

Knowledge of the genome organization and replication cycle of the togaviruses have enabled the construction of full genome-length cDNA clones from which infectious RNA molecules can be transcribed *in vitro* and transfected into permissive cells to initiate productive virus replication. This procedure allows easy introduction of specific mutations into the virus genome followed by generations of genetically homogeneous virus populations encoding the mutation(s). Indeed, much of the work described in this article was performed using virus mutants generated from cDNA clones. In addition, 'double-subgenomic promoter viruses' (DP-virus) have been created in which the subgenomic promoter is reiterated immediately upstream of the authentic subgenomic promoter or at the 3′ end of the E1 coding region to drive expression of genes used as reporters of infection or vaccine vectors (e.g., GFP, luciferase, or immunogens from infectious microorganisms) or to test the effect of protein/RNA expression in the context of a togaviral infection (e.g., antiviral or apoptosis-inhibiting proteins, or interfering RNAs). Propagation-defective virus-like particles or 'replicon' systems have also been developed in which the structural protein genes are deleted and replaced with a heterologous protein gene. With alphaviruses, replicon genomes can be packaged to form replicon particles by expressing the structural proteins in *trans*. Typically, the replicon genome RNA is co-electroporated with two 'helper' RNAs, which are replicated and transcribed by the nsPs provided by the replicon genome and separately encode the capsid and PE2/6K/E1 proteins. Use of these bipartite helper systems minimizes the potential for generation of propagation-competent progeny. In contrast, Rubella virus exhibits a dependence of expression on the nsP ORF from the same nucleic acid as the structural proteins, limiting this approach. Replicon particles can infect a single cell and express the heterologous protein at high levels from the subgenomic promoter but, in the absence of structural protein expression, no progeny virions are produced. Therefore, these vectors allow detailed examination of the interaction of the virus with single cells and can also be used as vaccines with little possibility that adverse effects associated with propagating vial vectors will occur in vaccinated hosts. The results of numerous studies have indicated that togavirus DP-viruses and replicons will likely become important vector systems for the delivery of heterologous gene products to cells *in vitro* and *in vivo.*

Togavirus–Host Cell Interactions

The togavirus–cell interaction can be viewed as a conflict between the cell and the virus: the virus must replicate to propagate itself, generally leading to cytopathic effects (CPEs), while the host cell attempts to suppress virus replication in order to avoid CPE or to save the greater organism from death (in which case the individual cell may be sacrificed). In the following section, these interactions are addressed as a continuum from poorly restrained virus replication leading to rapid CPE on one extreme, to host-cell-circumscribed virus replication and prolonged cell survival on the other.

Cell Viability and CPE

Historically, alphaviruses (e.g., prototypic SINV) have been considered to be strongly cytopathic, causing rapid killing of cultured vertebrate cells that are highly permissive of their replication. In contrast, alphavirus replication in mosquito cells is most often associated with persistent infection and limited cell death. Rubella virus is also cytopathic for vertebrate cells at high infection multiplicities, but can develop persistent infection at low multiplicity and does not infect mosquito cells. Only recently has the complex interaction between the virus infection cycle and cell death begun to be unraveled. Infection of many vertebrate cell lines with alphaviruses or Rubella virus results in morphological changes (e.g., chromatin condensation, nuclear fragmentation, and formation of membrane-enclosed apoptotic bodies), as well as molecular changes (e.g., caspase activation and DNA fragmentation), associated with programmed cell death/apoptosis. Expression of SINV structural glycoproteins, particularly the transmembrane domains, can cause apoptosis in rat AT-3 prostatic adenocarcinoma cells. Furthermore, mutation from glutamine to histidine at position 55 of the SINV E2 glycoprotein has been associated with increased apoptosis in cultured cells and neurons in mice. However, although accumulation of viral structural proteins accelerates CPE development, apoptosis also occurs in replicon-infected cells in the absence of structural protein synthesis, linked to replication and nsP2 activities. Complicating this picture further is that interaction of high concentrations of UV-inactivated SINV particles with cell surfaces can initiate cell death pathways in cells of some lines, suggesting that viral replication/gene expression is not always required for killing. Finally, evidence also exists for alphavirus-induced necrotic cell death (i.e., not involving active cellular processes) in subpopulations of infected neurons *in vivo.* Therefore, the stimulus for CPE caused by togaviruses may depend upon the virus genotype, dose, the infected cell type, and/or the infection context. At the molecular level, multiple pathways, including mitochondrial cytochrome *c* release, TNF-α death receptor signaling, sphingomyelinase activation and production of ceramide, and redox-stress and inflammatory-stress pathway activation have been associated with togavirus-induced cell death, further supporting the idea that induction of particular mechanisms may be dependent upon multiple virus and host cell factors.

Host-Protein Synthesis Shut-Off

Togavirus replication in highly permissive vertebrate cells devastates cellular macromolecular synthesis, arresting both cellular transcription and translation by independent mechanisms. Transcriptional downregulation of cellular mRNAs and rRNAs can be directly mediated by the mature SINV nsP2 protein even in the absence of viral replication and is critically involved in the production of CPE. This activity is determined by integrity of the nsP2 carboxy-terminal domain, not by helicase or protease activity and can be greatly reduced by point mutations in this region. Moreover, the degree of nsP2-mediated pro-apoptotic stimulus appears to vary considerably between different togaviruses. Nonstructural protein-mediated inhibition of cellular transcription in highly permissive cells, such as fibroblast cell lines, dramatically suppresses the cell's ability to generate an antiviral stress response and is likely beneficial for togavirus replication and dissemination. However, infection with the noncytopathic nsP2 mutants described above restores the cell's response. In contrast, infection of dendritic cells and macrophages, which may be less permissive to infection than fibroblasts or may express a different ensemble of antiviral response mediators, produces a vigorous stress response after virus exposure and large quantities of new mRNAs are synthesized.

Translation of cellular mRNAs is also dramatically inhibited within hours of togavirus infection in highly permissive cells. This process of host-protein synthesis 'shut-off' occurs independent of transcriptional arrest. The degree to which translational shut-off occurs is inextricably linked with the extent of viral RNA replication, such that incomplete translation inhibition is associated with permissivity and mutations in viral replicase complex proteins, particularly nsP2. As a cellular response, translational inhibition is a stress-induced defense mechanism, able to recognize viral infection, amino acid starvation, iron deficiency, and accumulation of misfolded proteins in the endoplasmic reticulum (ER) via the activation of stress kinases. The activity of four distinct stress kinases converges to phosphorylate eukaryotic translation initiation factor (eIF) 2α, inhibiting GTP–eIF2–tRNAi-Met ternary complex formation and globally suppressing translation initiation. However, the expression of specific stress-inducible cellular proteins continues under conditions of phosphorylated eIF2α and generalized shut-off. If the stress response fails to clear the virus and restore homeostasis, the effects of prolonged shut-off become detrimental to the infected cell, leading to CPE and apoptotic cell death. Phosphorylation of eIF2α during togavirus infection is triggered by the presence of dsRNA primarily through activation of the dsRNA-dependent protein kinase (PKR). In some situations, eIF2α phosphorylation appears to be largely responsible for translation inhibition, since the overexpression of a nonphosphorylatable eIF2α mutant abrogates shut-off despite efficient viral replication. However, the arrest of host protein synthesis has also been shown to occur independently of PKR activity and in the apparent absence of eIF2α phosphorylation, by an unknown mechanism.

In response to this cellular defense mechanism, togaviruses have evolved degrees of tolerance to translational shut-off, enabling them to circumvent this block and redirect the cell's inoperative translational apparatus to the synthesis of viral proteins. Indeed, once shut-off has occurred, virtually the only proteins synthesized by the cell are virus-encoded structural proteins, expressed from the subgenomic mRNA. This implies that host protein synthesis shut-off confers an advantage to the virus, allowing usurpation of cellular translation machinery and potentially dampening the host cell's antiviral stress response. Translation enhancer elements located in the 5′ termini of SINV and SFV 26S (and, very likely, Rubella virus) mRNAs, particularly the highly stable RNA hairpin downstream of the AUG start codon, facilitate the continued expression of alphavirus structural proteins during the translation inhibition imposed by eIF2α phosphorylation. SFV infection has been shown to induce the transient formation of stress granules containing cellular TIA-1/R proteins, which sequester cellular mRNAs, but disassemble in proximity to the viral replicase. The temporal correlation between eIF2α phosphorylation, stress granule assembly and localized disassembly, and the transition from cellular to viral protein synthesis suggest that these may be important processes in generalized shut-off of protein synthesis and avoidance of the shut-off by the 26S mRNA. It is likely that the togaviruses also evade and/or antagonize PKR/phospho-eIF2α-independent translation arrest. Notably, the 26S mRNA also has a low requirement for the translation complex scaffolding protein, eIF4G, although integrity of eIF4G is not known to be affected by togavirus replication.

The Interferon-Mediated Antiviral Response

Antiviral activity of interferon-alpha/beta (IFN-α/β) is a critical determinant of the outcome of togavirus infection both *in vitro* and *in vivo*. In mice, the absence of this response results in greatly increased susceptibility to alphavirus infection and disease. In cell culture, IFN-α/β pretreatment profoundly blocks virus replication and protects cells from CPE induced by all togaviruses tested. However, some alphaviruses (e.g., VEEV) appear to be more resistant to the effects of exogenously added IFN-α/β than others (e.g., SINV). All togaviruses tested also stimulate the production of IFN-α/β from cultured cells to varying degrees. Presumably, IFN-α/β induction results from triggering of cellular pattern recognition receptors (PRRs), such as toll-like receptors (TLRs), PKR and/or cytoplasmic RNA helicases such as RIG-I or MDA-5 by 'pathogen-associated molecular patterns' (PAMPs). In addition to IFN-α/β induction, these receptors stimulate a general inflammatory response in infected cells by activating the NF-κB pathway (which is activated after SINV infection). Historically, cytoplasmic dsRNA, produced as a component of the togavirus replicative cycle has been considered the primary PAMP; however, UV-inactivated preparations of some alphaviruses (e.g., SFV) can elicit inflammatory responses in certain cultured cells suggesting multiple pathways of pathogen detection and response. Secreted IFN-α/β signals through its cognate receptor on infected and uninfected cells to upregulate expression of many IFN-stimulated genes (ISGs), producing antiviral proteins, some of which directly inhibit togavirus replicative processes. In less permissive vertebrate cells, induced IFN-α/β can act upon the infected cell to suppress togavirus replication, prevent overt CPE, and promote a persistent infection.

Several studies have evaluated the individual and combined contribution of the two best-characterized IFN-α/β-inducible antiviral pathways, the PKR pathway (described above), and the coupled 2–5A synthetase/RNase L pathway, to the control of alphavirus replication. The latter pathway is composed of the interferon-inducible 2′–5′ oligoadenylate synthetase (OAS) family of dsRNA-dependent enzymes and dormant, cytosolic RNase L. dsRNA-activated 2′–5′ OAS synthesizes 2′–5′-linked oligoadenylates that specifically bind and activate RNase L which then cleaves diverse RNA substrates, thus inhibiting cellular and viral protein synthesis. As described above, constitutively expressed and IFN-α/β-inducible PKR is activated by dsRNA binding to phosphorylate eIF2α, causing a decline in cap-dependent translation of viral and cellular mRNAs. Infection of mice with targeted disruption of the PKR gene, the RNase L gene or both, revealed that PKR, but not RNase L, is involved in the early control of SINV replication *in vivo* and in primary cell cultures. Surprisingly, IFN-α/β-mediated antiviral responses against SINV are largely intact in the absence of PKR and/or RNase L, confirming the existence of 'alternative' IFN-α/β-induced pathway(s) capable of curtailing togavirus replication. Gene transcription and *de novo* protein synthesis are required for this activity. Although the proteins involved are yet to be characterized, one such pathway involves a novel inhibition of cap-dependent translation of infecting viral genomes in the absence of eIF2α phosphorylation. In addition, the interferon-inducible proteins MxA, ISG15, and the zinc-finger antiviral protein (Zap) have some inhibitory effects upon alphavirus replication. Considering the potent inhibitory effects of IFN-α/β toward numerous togaviruses, the expression of additional mechanisms to evade or antagonize the IFN-α/β response is likely critical for togavirus virulence, although convincing evidence of their existence is yet to be presented.

Conclusions and Perspectives

Research into the molecular biology of togaviruses has two primary goals: (1) to understand the relationship of virus interactions with single cells to replication and disease pathogenesis in natural hosts and, subsequently, to identify targets for therapeutic intervention; and (2) to manipulate virus–cell interactions to capitalize upon the potential for use of these viruses as tools for gene therapy and vaccination. In the examples described above, the dissection of virus attachment, entry, translation, replication, effects upon host cells, and host cell responses to infection all have identified vulnerabilities through which virus infection may be curtailed and many laboratories are currently developing antiviral and/or disease-ameliorating strategies. Furthermore, research into virus stimulation of CPE or stress responses has led to an improved understanding of how to, for example, maximize CPE (important for the development of tumor-destroying vectors) or minimize host antiviral activity (important for expression of immunogens by vaccine vectors). Some important issues that remain to be addressed include the full elucidation of the complex relationship of virus replication to development of CPE, the nature of antiviral stress response mediators capable of blocking virus replication, and the mechanisms through which togaviruses antagonize and/or evade these responses.

Further Reading

Dominguez G, Wang C-T, and Frey TK (1990) Sequence of the genome RNA of rubella virus: Evidence for genetic rearrangement during togavirus evolution. *Virology* 177: 225–238.

Griffin DE (2001) In: Fields BN, Knipe DM, Howley PM, *et al.* (eds.) *Fields Virology,* 4th edn., 917pp. Philadelphia: Lippincott.

Li ML and Stollar V (2004) Alphaviruses and apoptosis. *International Reviews of Immunology* 23: 7.

Paredes A, Alwell-Warda K, Weaver SC, Chiu W, and Watowich SJ (2007) Venezuelan equine encephalomyelitis virus structure and its divergence from old world alphaviruses. *Journal of Virology* 75: 9532–9537.

Schlesinger S and Schlesinger MJ (2001) *Togaviridae*: The viruses and their replication. In: Fields BN, Knipe DM, Howley PM, *et al.* (eds.) *Fields Virology,* 4th edn., 895pp. Philadelphia: Lippincott.

Strauss JH and Strauss EG (1994) The alphaviruses: Gene expression, replication, and evolution. *Microbiology Reviews* 58: 491.

SUBJECT INDEX

Notes

Cross-reference terms in italics are general cross-references, or refer to subentry terms within the main entry (the main entry is not repeated to save space). Readers are also advised to refer to the end of each article for additional cross-references - not all of these cross-references have been included in the index cross-references.

The index is arranged in set-out style with a maximum of three levels of heading. Major discussion of a subject is indicated by bold page numbers. Page numbers suffixed by T and F refer to Tables and Figures respectively. *vs.* indicates a comparison.

This index is in letter-by-letter order, whereby hyphens and spaces within index headings are ignored in the alphabetization. Prefixes and terms in parentheses are excluded from the initial alphabetization.

To save space in the index the following abbreviations have been used

CJD - Creutzfeldt–Jakob disease

CMV - cytomegalovirus

EBV - Epstein–Barr virus

HCMV - human cytomegalovirus

HHV - human herpesvirus

HIV - human immunodeficiency virus

HPV - human papillomaviruses

HSV - herpes simplex virus

HTLV - human T-cell leukemia viruses

KSHV - Kaposi's sarcoma-associated herpesvirus

RdRp - RNA-dependent RNA polymerase

RNP - ribonucleoprotein

SARS - severe acute respiratory syndrome

TMEV - Theiler's murine encephalomyelitis virus

For consistency within the index, the term "bacteriophage" has been used rather than the term "bacterial virus."

A

B

C

D

E

G

I

J

K

L

M

N

O

P

Q

R

S

U

V

W

X

Y

Z

Printed and bound by CPI Group (UK) Ltd, Croydon, CR0 4YY

03/10/2024

01040316-0017